IIT Objective Chemistry

For IIT JEE, AIEEE and Other Engineering Entrance Examinations

Eleventh Edition

Arun Syamal
M.Sc., Ph.D., D.Sc.
Formerly Chairman,
Professor of Chemistry and Director,
National Institute of Technology
(Deemed University), Kurukshetra 136119
Haryana

ATLANTIC
PUBLISHERS & DISTRIBUTORS (P) LTD

Published by
ATLANTIC
PUBLISHERS & DISTRIBUTORS (P) LTD
7/22, Ansari Road, Darya Ganj,
New Delhi-110002
Phones : +91-11-40775252, 23273880, 23275880, 23280451
Fax : +91-11-23285873
Web : www.atlanticbooks.com
E-mail : orders@atlanticbooks.com

Branch Office
5, Nallathambi Street, Wallajah Road,
Chennai-600002
Phones : +91-44-64611085, 32413319
E-mail : chennai@atlanticbooks.com

Second Edition 2008
Third Edition 2010
Fourth Edition 2011
Fifth Edition 2012
Sixth Edition 2013
Seventh Edition 2014
Eighth Edition 2015
Ninth Edition 2016
Tenth Edition 2018
Eleventh Edition 2019

Printed in India at Nice Printing Press, A-33/3A, Site-IV,
Industrial Area, Sahibabad, Ghaziabad, U.P.

Preface

It has been a real pleasure to note the overwhelming response with which the tenth edition of the book *IIT Objective Chemistry* was received. The present edition has been written according to the recently introduced system of Main and Advanced examination in IIT JEE. The objective questions have been designed according to the new system of examination. This eleventh edition has been thoroughly revised, updated and substantially enlarged. Sufficient care has been taken to maintain the projected standard of the IIT JEE examination. The book will serve not only the needs of students appearing for IIT JEE but also for AIEEE and other engineering and medical entrance examinations.

In a book of this size, it is not possible to have comprehensive theoretical discussions and hence only the objective type multiple choice questions and their answers have been provided. In the beginning of every chapter, important relations for tackling numerical problem-oriented questions have been given. The students are advised to solve all the questions themselves in order to test the depth of their knowledge. The book is based on clear understanding of the subject matter and will help the students in confidence building. I have intentionally avoided to give complete solutions to the questions so that the students enthusiastically try to solve them. Attack the problems intelligently and you will be able to solve the questions. You must not leave till you are able to solve them. This approach will lead you to success in IIT JEE. The students are advised to read the theoretical discussions from my following four books: *IIT Chemistry*, *Objective Chemistry*, *CBSE Chemistry XI* and *CBSE Chemistry XII*. The questions in the present book have been selected in such a manner that they cover all concepts and really test the student's understanding of the subject. The number of questions have been kept large enough to offer meaningful practice to the students. Hints to some difficult questions have also been given. The book contains an exhaustive list of up-to-date objective type multiple choice questions asked in the entrance examinations of IIT in the previous years, upto 2018. A further advantage of the book is that it contains very rich Appendices.

Being a teacher of the National Institute of Technology, I have vast experience of dealing with engineering entrance tests and setting papers for such tests. In addition, I have a long experience of teaching Chemistry to aspiring students for IIT JEE and other engineering entrance examinations. This book will be extremely useful to all those students who do not have the facility to receive extra guidance for the preparation to appear in such tests.

I wish to thank Mr. Arun Roy of Arun Roy Classes, Mumbai for making useful suggestions. Thanks are also due to Dr. M. M. Singh, Senior Model School, Kurukshetra University, Dr. Dinesh Kumar, Professor of Chemistry, National Institute of Technology, Kurukshetra, and Mr. R. K. Tomar, Govt. High School, Pehwa, Haryana for checking the answers of questions and for making useful suggestions. As a result of their sincere efforts, the questions and answers in this book are claimed to be error-free. Various suggestions received from students, teachers and coaching institutions are gratefully acknowledged. I will welcome constructive criticisms and suggestions for further improvement of the book.

Arun Syamal

E-mail: arunsyamal@yahoo.co.in

Contents

PART—III
ORGANIC CHEMISTRY

PART—IV
PRACTICE TESTS AND QUESTION PAPERS

APPENDICES

Part—I

Physical Chemistry

1

Atomic Structure

IMPORTANT RELATIONS FOR NUMERICAL PROBLEMS

1. Relation between velocity (c in cm s^{-1}), wavelength (λ in cm), frequency (ν in s^{-1}) and wave number ($\bar{\nu}$): (i) $c = \nu\lambda$ (ii) $\bar{\nu} = 1/\lambda$ (iii) $\bar{\nu} = \nu/c$
2. Relation between energy (E), wavelength (λ) and frequency (ν): $E = h\nu$, $E = hc/\lambda$ where E is expressed in joule, h (Planck constant) = 6.6256×10^{-34} J s
3. Energy of electron in unielectron system: $E = -2\pi^2 me^4Z^2/n^2h^2$ where m = mass of electron, e = electronic charge, n = principal quantum number, Z = atomic number. In S.I. system, $E = -2\pi^2 me^4Z^2/(4\pi\varepsilon_0)^2n^2h^2$ where ε_0 (permittivity constant) = 8.8542×10^{-12} C^2 N^{-1} m^{-2}. $E_n = (-1312Z^2/n^2)$ kJ mol^{-1}. E_1 for H = -21.72×10^{-12} erg = -13.6 eV, E_1 for H-like atom = E_1 for H $\times Z^2$
4. Energy of photon emitted when electron jumps from n_2 to n_1 level: $\Delta E = [2\pi^2 me^4Z^2/h]$ $[(1/n_1^2) - (1/n_2^2)] = 1312Z^2[(1/n_1^2) - (1/n_2^2)]$
5. Frequency and wave number of radiations emitted when electron jumps from n_2 to n_1 level: $\nu = R[(1/n_1^2) - (1/n_2^2)]$ where R (Rydberg constant) = $2\pi me^4/h^3c = 3.289 \times 10^{15}$ s^{-1}. $\bar{\nu} = R'[(1/n_1^2) - (1/n_2^2)]$ where $R' = 2\pi^2 me^2/h^3c = 109677$ cm^{-1}
6. de Broglie equation: $\lambda = h/mv$
7. Heisenberg uncertainty principle: $(\Delta x)(\Delta p) \geq h/4\pi$ where Δx = uncertainty in position, Δp = uncertainty in momentum
8. $r_n = n^2h^2/4\pi^2me^2$, r_1 for H = 0.529 Å, r_1 for H-like atom = r_1 for H/Z
9. Velocity in the nth orbit, $V_n = 2.188 \times 10^6 \times Z/n$ m s^{-1}
10. No. of revolutions in the nth orbit = $V_n/2\pi r_n$
11. Angular momentum of an electron in an orbit = $n(h/2\pi)$
12. Angular momentum of an electron in an orbital = $[l(l + 1)]^{1/2}(h/2\pi)$
13. Spin angular momentum of an electron = $[s(s + 1)]^{1/2}(h/2\pi)$
14. Magnetic moment of an atom = $[n(n + 2)]^{1/2}$ BM where n = No. of unpaired electron
15. Orbital nodes: Radial nodes = $n - l - 1$, angular nodes = l, total nodes = $n - 1$

OBJECTIVE TYPE MULTIPLE CHOICE QUESTIONS

1. The quantum numbers + 1/2 and – 1/2 for electron spin represent
   (A) rotation of the electron in clockwise and anticlockwise direction respectively
   (B) rotation of the electron in anticlockwise and clockwise direction respectively
   (C) magnetic moment of the electron pointing up and down respectively
   (D) two quantum mechanical spin states which have no classical analogue. **(I.I.T. 2001)**
2. The wavelength associated with a golf ball weighing 200 g and moving at a speed of 5 m/h is of the order
   (A) 10^{-10} m (B) 10^{-20} m (C) 10^{-30} m (D) 10^{-40} m **(I.I.T. 2001)**

   Hint: $\lambda = \dfrac{h}{mv} = \dfrac{6.626\times10^{-34}}{200\times10^{-3}\times5/(60\times60)} = \dfrac{6.626\times10^{-34}\times60\times60}{200\times10^{-3}\times5} = 238.5\times10^{-30}$ m
3. The electronic configuration of an element is $1s^22s^22p^63s^23p^63d^54s^1$. This represents its
   (A) excited state (B) anionic form (C) cationic form (D) ground state. **(I.I.T. 2000)**
4. The number of nodal planes in a p_x orbital is
   (A) 1 (B) 2 (C) 3 (D) 0 **(I.I.T. 2000)**

   Hint: Number of nodal planes in p_x orbital is in the yz plane.
5. The electrons identified by the quantum numbers n and l, (i) $n = 4$, $l = 1$ (ii) $n = 4$, $l = 0$ (iii) $n = 3$, $l = 2$ (iv) $n = 3$, $l = 1$ can be placed in order of increasing energy, from the lowest to the highest, as
   (A) iv < ii < iii < i (B) ii < iv < i < iii (C) i < iii < ii < iv (D) iii < i < iv < ii **(I.I.T. 1999)**
6. The energy of an electron in the first orbit of hydrogen atom is – 13.6 eV. The possible energy values of the excited state(s) for electrons in Bohr orbits is/are
   (A) – 3.4 eV (B) – 4.2 eV (C) – 6.8 eV (D) + 6.8 eV **(I.I.T. 1998)**

   Hint: $E_n = -13.6/n^2 = -13.6/2^2 = -13.6/4 = -3.4$ eV
7. For a *d*-electron, the angular momentum is
   (A) $\sqrt{6}(h/2\pi)$ (B) $\sqrt{2}(h/2\pi)$ (C) $h/2\pi$ (D) $2(h/2\pi)$ **(I.I.T. 1997)**
8. The angular momentum of an electron in $2s$ orbital is
   (A) $(1/2)h/2\pi$ (B) 0 (C) $h/2\pi$ (D) $\sqrt{2}(h/2\pi)$ **(I.I.T. 1996)**
9. The first use of quantum theory to explain the structure of atom was made by
   (A) Heisenberg (B) Bohr (C) Planck (D) Einstein **(I.I.T. 1997)**
10. Which of the following statements is correct?
    (A) The electronic configuration of Cr is [Ar] $3d^54s^1$
    (B) The magnetic quantum number may have a negative value
    (C) In Ag atom (At. No. 47), 23 electrons have a spin of one type and 24 of the opposite type
    (D) The oxidation state of nitrogen in HN_3 is – 3 **(I.I.T. 1998)**

11. From the given sets of quantum numbers, the one that is inconsistent with theory is
   (A) $n = 3, l = 2, m_l = -3, m_s = +1/2$
   (B) $n = 4, l = 3, m_l = 3, m_s = +1/2$
   (C) $n = 2, l = 1, m_l = 0, m_s = -1/2$
   (D) $n = 4, l = 3, m_l = 2, m_s = +1/2$ **(I.I.T. 1994)**

12. In which of the following, the Pauli exclusion principle is violated?

	2s	2p		
(A)	↑	↑	↑	↑
(B)	↑↓	↑	↑	↑
(C)	↑↓	↑↓		
(D)	↑↑	↑↓	↑	

(I.I.T. 1993)

13. Two nodal planes are present in
   (A) *p*-orbital (B) *s*-orbital (C) *f*-orbital (D) *d*-orbital. **(I.I.T. 1992)**

14. The wave number of the first line of Balmer series of hydrogen atom is 15200 cm^{-1}. The wave number of the first Balmer line of Li^{2+} ion is
   (A) 136800 cm^{-1} (B) 15200 cm^{-1} (C) 60800 cm^{-1} (D) 76000 cm^{-1} **(I.I.T. 1992)**

15. Which combination of quantum numbers n, l, m_l and m_s for the electron in an atom does not provide a permissible solution of the wave equation?
   (A) 3, 2, – 2, + 1/2 (B) 3, 2, 1, + 1/2 (C) 3, 1, 1, – 1/2 (D) 3, 3, 1, – 1/2 **(I.I.T. 1992)**

16. Which of the following relates to photons both as wave motion and as stream of particles?
   (A) Interference (B) $E = mc^2$ (C) Diffraction (D) $E = h\nu$ **(I.I.T. 1992)**

17. The ground state electronic configuration of nitrogen atom is represented by

(A)	↑↓	↑↓	↑	↑	↑
(B)	↑↓	↑↓	↑	↓	↑
(C)	↑↓	↑↓	↑	↓	↓
(D)	↑↓	↑↓	↓	↓	↓

(I.I.T. 1999)

18. Which of the following does not characterize X-rays?
   (A) The radiation can ionise gases (B) It causes ZnS to fluoresce
   (C) Deflected by electric and magnetic fields
   (D) Have wavelengths shorter than ultraviolet rays. **(I.I.T. 1992)**

19. Atomic nucleus contains
   (A) protons (B) neutrons (C) electrons (D) photons **(I.I.T. 1992)**

20. The quantum number not obtained from the Schrodinger's wave equation of chlorine atom is
   (A) s (B) n (C) m_l (D) l **(I.I.T. 1990)**

21. The correct set of quantum numbers for the unpaired electron of chlorine atom is
   (A) $n = 2, l = 1, m_l = 0$ (B) $n = 2, l = 1, m_l = 1$
   (C) $n = 3, l = 0, m_l = 0$ (D) $n = 3, l = 1, m_l = 1$ **(I.I.T. 1989)**

22. The orbital diagram in which the Aufbau principle is violated is

	2s	2p		
(A)	↑↓	↑↓	↑	
(B)	↑	↑↓	↑	↑
(C)	↑↓	↑	↑	↑
(D)	↑↓	↑↓	↑↓	↑

(I.I.T. 1988)

23. The wavelength of a spectral line for an electronic transition is invariably related to
    (A) the number of electrons undergoing the transition
    (B) the nuclear charge of the atom
    (C) the difference of the energy levels involved in the transition
    (D) the velocity of the electron undergoing the transition. **(I.I.T. 1988)**

24. The triad of nuclei that is isotonic is
    (A) ${}^{14}_{6}C$, ${}^{15}_{7}N$, ${}^{17}_{9}F$ (B) ${}^{12}_{6}C$, ${}^{14}_{7}N$, ${}^{19}_{9}F$ (C) ${}^{14}_{6}C$, ${}^{14}_{7}N$, ${}^{17}_{9}F$ (D) ${}^{14}_{6}C$, ${}^{14}_{7}N$, ${}^{19}_{9}F$ **(I.I.T. 1988)**

25. The ratio of the energy of a photon of 2000 Å wavelength radiation to that of 4000 Å is
    (A) 1/2 (B) 1/4 (C) 2 (D) 4 **(I.I.T. 1986)**

26. Which one of the following sets of quantum numbers is an impossible arrangement?

	n	l	m_l	m_s		n	l	m_l	m_s
(A)	3	2	– 2	+ 1/2	(B)	4	0	0	+ 1/2
(C)	3	2	– 3	+ 1/2	(D)	5	2	0	– 1/2

(I.I.T. 1986)

27. Rutherford's alpha particle scattering experiment led to the conclusion that
    (A) mass and energy are related (B) electrons occupy space around the nucleus
    (C) neutrons are buried deep in the nucleus
    (D) the point of impact with matter can be precisely determined. **(I.I.T. 1986)**

28. Electromagnetic radiation with maximum wavelength is
    (A) ultraviolet (B) radiowave (C) infrared (D) X-rays **(I.I.T. 1985)**

29. The radius of the atomic nucleus is of the order of
    (A) 10^{-10} cm (B) 10^{-13} cm (C) 10^{-15} cm (D) 10^{-8} cm **(I.I.T. 1985)**

30. Bohr model can explain
    (A) the spectrum of hydrogen atom only
    (B) the solar system
    (C) spectrum of an atom or ion containing one electron only
    (D) the spectrum of hydrogen molecule. **(I.I.T. 1985)**

31. The increasing order (lowest first) for the values of *e*/*m* for electron (e), proton (p), neutron (n) and α-particle (α) is
    (A) e, p, n, α (B) n, p, e, α (C) n, p, α, e (D) n, α, p, e **(I.I.T. 1984)**

32. An isotone of ${}^{76}_{32}Ge$ is
    (A) ${}^{77}_{32}Ge$ (B) ${}^{77}_{33}As$ (C) ${}^{77}_{34}Sc$ (D) ${}^{78}_{34}Sc$ **(I.I.T. 1984)**

33. When α-particles are sent through a thin metal foil, most of them go straight through the foil because
(A) α-particles are much heavier than electrons
(B) alpha-particles are positively charged
(C) most part of an atom is empty space
(D) α-particles move with high velocity. **(I.I.T. 1984)**

34. In the Bohr model of the hydrogen atom
(A) the radius of the nth orbit is proportional to n^2
(B) the total energy of the electron in the nth orbit is inversely proportional to n
(C) the magnitude of the potential energy of the electron in any orbit is greater than its kinetic energy
(D) the angular momentum of electron in an orbit is an integral multiple of $h/2\pi$
(I.I.T. 1984)

35. The electron level which allows the hydrogen atom to absorb a photon but not emit a photon is
(A) 1*s* (B) 2*s* (C) 3*s* (D) 2*p* **(I.I.T. 1984)**

36. Many elements have non-integral atomic masses because
(A) they have isotopes (B) their isotopes have different masses
(C) the constituent neutrons, protons and electrons combine to give fractional masses
(D) their isotopes have non-integral masses. **(I.I.T. 1984)**

37. The correct set of four quantum numbers for the valence electron of rubidium (At. No. = 37) is
(A) $n = 5, l = 0, m_l = 0, m_s = +1/2$ (B) $n = 5, l = 1, m_l = 0, m_s = +1/2$
(C) $n = 5, l = 1, m_l = 0, m_s = +1/2$ (D) $n = 6, l = 0, m_l = 0, m_s = +1/2$ **(I.I.T. 1984)**

38. Rutherford's scattering experiment is related to size of the
(A) atom (B) nucleus (C) electron (D) neutron **(I.I.T. 1983)**

39. The principal quantum number of an atom is related to the
(A) spin angular momentum (B) orbital angular momentum
(C) orientation of the orbital in space (D) size of the orbital. **(I.I.T. 1983)**

40. A *p*-orbital can accommodate up to
(A) two electrons with opposite spin (B) two electrons with parallel spin
(C) four electrons (D) six electrons. **(I.I.T. 1983)**

41. Consider the spectral line resulting from the transition $n = 2 \longrightarrow n = 1$ in the atoms and ions given below. The shortest wavelength is produced by
(A) hydrogen atoms (B) deuterium atoms
(C) singly ionised helium (D) doubly ionised lithium. **(I.I.T. 1983)**

42. Rutherford's experiment on scattering of alpha particles showed for the first time that an atom has
(A) electrons (B) nucleus (C) protons (D) neutrons **(I.I.T. 1981)**

43. The number of neutrons in Zn^{2+} ion with mass number 70 is
(A) 34 (B) 36 (C) 38 (D) 40 **(I.I.T. 1979)**

44. If the energy of the first orbit of He^+ ion is – 54.4 eV, then the energy of the second orbit will be
(A) – 5.44 eV (B) – 27.2 eV (C) + 27.2 eV (D) – 13.6 eV
(Roorkee 1996)

45. Which of the following has non-spherical shell of electron?
(A) He (B) Be (C) Al (D) Li **(Roorkee 1992)**

46. Which of the following has non-spherical shell of electron?
(A) He (B) B (C) Be (D) Li **(Roorkee 1993)**

47. The mass number of a nucleus is
(A) always less than its atomic number
(B) always more than its atomic number
(C) sometimes equal to its atomic number
(D) sometimes more than and sometimes equal to its atomic number. **(I.I.T. 1986)**

48. If the speed of electron in the Bohr's first orbit of hydrogen atom be x, then the speed of the electron in the 3rd orbit is
(A) $x/9$ (B) $x/3$ (C) $3x$ (D) $9x$ **(I.I.T. 1990)**

49. An element M has an atomic mass 19 and atomic number 9. Its ion is represented by
(A) M^+ (B) M^{2+} (C) M^- (D) M^{2-} **(I.I.T. 1996)**

50. If wavelength of photon is 2.2×10^{-11} m and $h = 6.6 \times 10^{-34}$ J s, then the momentum of photon is
(A) 3×10^{-23} kg m/s (B) 3.33×10^{22} kg m/s
(C) 1.452×10^{-44} kg m/s (D) 6.89×10^{43} kg m/s **(Roorkee 1990)**

51. Potential energy of the electron of Li^{2+} ion is
(A) $-3e^2/4\pi\varepsilon_0 r$ (B) $3e^2/4\pi\varepsilon_0 r$ (C) $-2e^2/4\pi\varepsilon_0 r^2$ (D) $-e^2/4\pi\varepsilon_0 r$
(Roorkee 1990)

52. An electron has a kinetic energy of 2.8×10^{-23} J. de Broglie wavelength will be ($m_e = 9.1 \times 10^{-31}$ kg)
(A) 9.8×10^{-4} m (B) 9.28×10^{-7} m (C) 9.28×10^{-8} m (D) 9.28×10^{-10} m

53. If the wavelength of photon is 2.2×10^{-11} m, $h = 6.6 \times 10^{-34}$ J s, then the momentum of photon in kg m s^{-1} is
(A) 3.33×10^{22} (B) 1.45×10^{-44} (C) 6.89×10^{43} (D) 3.0×10^{-23}

54. With the increase in quantum number, the energy difference between the adjacent orbits of hydrogen atom
(A) decreases (B) remains constant
(C) increases (D) first increases and then decreases.

55. The ratio of the kinetic energy and potential energy of an electron in a Bohr orbit of a hydrogen-like atom is
(A) 1 (B) – 1 (C) 1/2 (D) – 1/2

56. The z-component of the angular momentum of an electron in an orbital is governed by
(A) magnetic quantum number (B) spin quantum number
(C) azimuthal quantum number (D) principal quantum number.

57. The sign of the lobes in an atomic orbital indicates
(A) sign of the wave function (B) sign of charge
(C) sign of probability distribution (D) presence or absence of electrons in the orbital

58. Identify the incorrect statement.
(A) An orbital is characterized by the quantum numbers n, l, m_l
(B) $(n - 1)d$ sub-shell has higher energy than ns sub-shell
(C) Aufbau principle is not obeyed in the cases where the energy difference between the ns and $(n - 1)d$ sub-shell is large
(D) nf sub-shell has higher energy than $(n - 2)s$ sub-shell.

59. Identify the correct statement.
(A) The special stability of the half-filled sub-shell is attributed to higher exchange energy of stabilisation
(B) The chemical properties of an atom are determined by its nuclear charge
(C) In the β-emission from a nucleus the atomic number of the daughter element decreases by one
(D) The outer electronic configuration in the ground state of Cr is $3d^4 4s^2$

60. Identify the incorrect statement.
(A) The energy of an electron is negative because electron has negative charge
(B) The spectral line obtained when an electron jumps from $n = 6$ to $n = 2$ level in hydrogen-like atom belongs to the Pfund series
(C) The sidewise overlap of two p-orbitals results in the formation of π-molecular orbital
(D) The concept of the wave nature of electron was proposed by de Broglie.

61. Identify the incorrect statement.
(A) The electron density in the xy plane in the $3d_{x^2-y^2}$ orbital is zero
(B) Gamma rays are electromagnetic radiations of wavelengths 10^{-6} m to 10^{-5} m
(C) The overlap of p_y and p_y orbitals gives σ bond
(D) The greater the difference in the energy between two levels, the lower is the wavelength of light emitted, when an electron moves between them.

62. Identify the incorrect statement.
(A) The energy of an electron in $3d$ orbital is more than that in $4s$ orbital
(B) The diameter of an atom is of the order of 10^{-8} cm
(C) The radius of the nucleus is of the order of 10^{-13} cm
(D) $h/2\pi$ is the angular momentum of the electron in the second orbit of He^+ ion.

63. Identify the incorrect statement
(A) The spectrum of hydrogen atom is similar to that of helium atom
(B) An orbital always has spherical symmetry

(C) When the velocity of electron is 1.84×10^3 times that of a proton, the wavelength associated with an electron is equal to the wavelength associated with a proton (mass of proton = 1.6723×10^{-24} g, $m_e = 9.1095 \times 10^{-31}$ kg)

(D) The mass of a particle whose uncertainty in position and velocity are 1×10^{-2} and 5.27×10^{-24} respectively, is 0.10 kg.

64. Identify the incorrect statement.
   (A) The mass of a π-meson is 300 times heavier than that of an electron
   (B) The transition of an electron in hydrogen atom from fourth to first energy state emits a spectral line which falls in Paschen series
   (C) Isotopes have same number of electrons and protons but different number of neutrons
   (D) $(n + 1)d$ sub-shell has lower energy than that of nf sub-shell.

65. In photoelectric effect, the photoelectric current
   (A) does not depend on photon frequency, but only on the intensity of incident beam
   (B) decreases when frequency of incident photon increases
   (C) depends both on intensity and frequency of the incident beam
   (D) increases when frequency of the incident photon increases.

66. The $d_{x^2-y^2}$ orbital possesses the m_l value of
   (A) + 1 (B) – 1 (C) + 2 (D) – 2

67. Spin quantum number (s) has a value of
   (A) ± 1/2 (B) + 1/2 (C) – 1/2 (D) 1/2

68. Magnetic spin quantum number (m_s) has a value of
   (A) 1/2 (B) ± 1/2 (C) + 1/2 (D) – 1/2

69. m_s can be generated from s
   (A) by the internal magnetic field produced due to the rotation of an electron in an orbit
   (B) by the external magnetic field
   (C) by the both internal and external magnetic field (D) none of these.

70. The potential energy of an electron in the hydrogen atom is
   (A) $-Ze^2/r$ (B) Ze^2/r (C) Ze^2/r^2 (D) mv^2/r

71. The condition for a stable electron orbit in hydrogen atom is
   (A) $n\lambda = 2\pi r$ (B) $e^2/r^2 = mv^2/r$ (C) $e/r = mv^2/r$ (D) $e^2/r^2 = (1/2)mv^2/r$

72. The condition for a stable wave in hydrogen atom is
   (A) $n\lambda = 4\pi r$ (B) $n\lambda = 2\pi r$ (C) $e^2/r^2 = mv^2/r$ (D) $n\lambda = (1/2)\pi r$

73. de Broglie waves can be used to explain
   (A) Schrodinger wave equation (B) Bohr's novel postulate
   (C) Rutherford's experiment of α-rays on metal foils (D) none of these.

74. The rule that explains the reason for Cr to have [Ar] $3d^5 4s^1$ configuration instead of [Ar] $3d^4 4s^2$ is
   (A) Hund's rule (B) Pauli exclusion principle
   (C) Aufbau principle (D) Heisenberg uncertainty principle.

75. Which of the following is the correct electronic configuration of the element with At. No. 106?
(A) [Rn] $5f^{14}6d^4 7s^2$ (B) [Rn] $5f^{14}6d^5 7s^1$
(C) [Rn] $5f^{14}6d^6 7s^0$ (D) [Rn] $5f^{14}6d^1 7s^2 7p^3$
76. The electronic configuration of Cu is $3d^{10}4s^1$ and not $3d^9 4s^2$. This is due to the
(A) exchange energy effect (B) inert pair effect
(C) screening effect (D) none of these.
77. The electron distribution in the *p*-orbitals is
(A) symmetrical (B) unsymmetrical
(C) neither symmetrical nor unsymmetrical
(D) none of these.
78. The electron distribution in the *d*-orbitals is
(A) symmetrical (B) unsymmetrical
(C) neither symmetrical nor unsymmetrical
(D) none of these.
79. The angular momentum of an electron in a Bohr orbit is given by
(A) $n(h/2\pi)$ (B) $[l(l + 1)]^{1/2}h/2\pi$
(C) $m(h/2\pi)$ (D) $h(h/2\pi)$
80. The observation of the line spectrum of hydrogen indicates that
(A) the energy of electron is quantised
(B) there is absorption of energy
(C) there occurs emission of energy when hydrogen is heated
(D) the electrons do not lose energy when in stationary states.
81. Different lines of Lyman series of the hydrogen spectrum lie in
(A) UV (B) IR (C) far IR (D) visible
82. The electronic transition in the line spectrum of He^+ that would have the same wavelength as the first Lyman transition in hydrogen atom is
(A) $4 \longrightarrow 2$ (B) $2 \longrightarrow 1$ (C) $4 \longrightarrow 1$ (D) $3 \longrightarrow 2$
83. The number of spherical nodes present in 4*s* orbital is
(A) 0 (B) 1 (C) 2 (D) 3
84. The number of spherical nodes present in 3*p* orbitals is
(A) 1 (B) 2 (C) 3 (D) none of these.
85. The uncertainty in the position of an electron ($m_e = 9.1 \times 10^{-28}$ g) moving with a velocity of 3×10^4 cm s^{-1} accurate up to 0.01% will be ($h = 6.626 \times 10^{-27}$ erg)
(A) 0.175 cm (B) 5.76 cm (C) 7.68 cm (D) 1.92 cm
86. The splitting of spectral line under the influence of magnetic field is called
(A) Zeeman effect (B) photoelectric effect
(C) Stark effect (D) none of these.
87. Boron has two isotopes $^{10}_{5}B$ and $^{11}_{5}B$. If the atomic mass of boron is 10.81, the ratio of $^{10}_{5}B$ to $^{11}_{5}B$ in nature is
(A) 10/11 (B) 15/16 (C) 19/81 (D) 20/53

88. The d_{xy} orbital possesses the m_l value of

(A) – 2 (B) + 2 (C) – 1 (D) + 1

89. An electron of mass m and charge e is accelerated from rest through a potential difference V in vacuum. Its final speed will be

(A) $(eV/2m)^{1/2}$ (B) $(2eV/m)^{1/2}$ (C) $(eV/m)^{1/2}$ (D) $2eV/m$

90. In the photoelectron emission, the energy of the emitted electron is

(A) smaller than the incident photon

(B) same as that of the incident photon

(C) greater than the incident photon

(D) proportional to the intensity of the incident photon.

91. If electron, hydrogen, helium and neon nuclei are all moving with the same velocity, then the wavelength associated with these particles will be in the order

(A) electron > He > H > Ne (B) electron > H > He > Ne

(C) Ne < He < electron < H (D) electron < H < He < Ne

92. Atoms consist of electrons, protons and neutrons. If the mass of neutron were halved and that of electron were doubled, the atomic mass of $^{12}_{6}C$ would be approximately

(A) halved (B) same

(C) reduced by 25% (D) doubled

93. Identify the correct statement.

(A) The spectral line obtained when an electron jumps from $n = 7$ to $n = 3$ belongs to the Paschen series

(B) The number of bonds in N_2 is 1π and 2σ

(C) With the increase in bond order, bond length increases

(D) The probability of finding an electron along x and y axes in the d_{z^2} orbital is zero.

94. The number of nodal planes present in the d_{z^2} orbital is

(A) 0 (B) 1 (C) 2 (D) 3

95. Identify the incorrect statement.

(A) Meson was discovered by Yukawa

(B) Mass defect is related to binding energy

(C) Magnetic quantum number is a measure of orbital angular momentum of an electron

(D) The size of the nucleus is of the order of 10^{-12}–10^{-13} cm.

96. If the uncertainty in position and momentum are equal, the uncertainty in velocity is

(A) $\frac{1}{2m}\sqrt{\frac{h}{\pi}}$ (B) $\sqrt{\frac{h}{\pi}}$ (C) $\sqrt{\frac{h}{2\pi}}$ (D) none of these.

97. An electron with a velocity v has a certain value of de Broglie wavelength. The velocity of the neutron having the same de Broglie wavelength will be

(A) $1840/v$ (B) $1840v$ (C) v (D) $v/1840$

98. The first emission line in the atomic spectrum of hydrogen in the Balmer series appears at
(A) $(5R/36)\ cm^{-1}$ (B) $(3R/4)\ cm^{-1}$ (C) $(9R/400)\ cm^{-1}$ (D) $(7R/144)\ cm^{-1}$

99. If the uncertainty in the position of an electron is zero, the uncertainty in its momentum will be
(A) zero (B) infinite (C) $> h/4\pi$ (D) $< h/4\pi$

100. In the spectrum of the hydrogen atom, which of the following transitions occurs at the shortest wavelength?
(A) $n_2 \longrightarrow n_1$ (B) $n_3 \longrightarrow n_1$ (C) $n_4 \longrightarrow n_1$ (D) $n_4 \longrightarrow n_2$

101. The following quantum numbers are possible for how many orbitals? $n = 3$, $l = 2$, $m_l = +2$
(A) 1 (B) 2 (C) 3 (D) 4

102. The velocity of the de Broglie wave is given by the expression
(A) $v\lambda$ (B) me^2/h (C) h/mv (D) c^2/v

103. For one-electron system, the wave number of any spectral line is proportional to
(A) $n_2 - n_1$ (B) n^2Z^2 (C) $n_1^2 - n_2^2$ (D) $(1/n_1^2) - (1/n_2^2)$

104. If the shortest wavelength of hydrogen atom in Lyman series is x, then the longest wavelength in Balmer series of He^+ ion will be
(A) $5x/9$ (B) $x/4$ (C) $9x/5$ (D) $36x/5$

105. The ionization energy of He^+ ion is 19.6×10^{-18} J per atom. What is the energy of the second stationary state in Li^{2+} ion?
(A) -11.02×10^{-18} J (B) -44.10×10^{-18} J
(C) -5.51×10^{-18} J (D) -19.6×10^{-18} J

Hint: $E = Z^2E_H$. For He^+ ion, $Z = 2$. Hence $E_H = E/Z^2 = -(19.6 \times 10^{-18}/4)$ J
For Li^{2+} ion, $Z = 3$. Hence $E = (3)^2E_H = 9(-4.9 \times 10^{-18}\ J) = -44.1 \times 10^{-18}$ J
Energy of second orbit, $E' = E/n^2 = -44.1 \times 10^{-18}\ J/2^2 = -11.02 \times 10^{-18}$ J

106. The ionisation energy of H atom in terms of Rydberg constant is given by
(A) R_HN_Ahc (B) R_Hhc (C) $2R_Hhc$ (D) R_Hc

107. The expression for velocity of an electron in a Bohr orbit of a hydrogen-like species is given by

(A) $v = \dfrac{2\pi(Ze^2/4\pi\varepsilon_0)}{nh}$ (B) $v = \dfrac{(Ze^2/4\pi\varepsilon_0)}{2\pi nh}$

(C) $v = \dfrac{nh}{2\pi(Ze^2/4\pi\varepsilon_0)}$ (D) $v = \dfrac{2\pi nh}{(Ze^2/4\pi\varepsilon_0)}$

108. The expression for radius of a Bohr orbit in a hydrogen-like species is given by

(A) $r = n^2\left[\dfrac{h^2}{4\pi^2m(Ze^2/4\pi\varepsilon_0)}\right]$ (B) $r = \dfrac{1}{n^2}\left[\dfrac{h^2}{4\pi^2m(Ze^2/4\pi\varepsilon_0)}\right]$

(C) $r = \dfrac{1}{n^2}\left[\dfrac{4\pi^2m(Ze^2/4\pi\varepsilon_0)}{h^2}\right]$ (D) $r = n^2\left[\dfrac{4\pi^2m(Ze^2/4\pi\varepsilon_0)}{h^2}\right]$

109. The expression for energy of an electron in a Bohr orbit of hydrogen-like species is given by

(A) $E = \frac{1}{n^2}\left[\frac{2\pi^2 m(Ze^2/4\pi\varepsilon_0)^2}{h^2}\right]$ (B) $E = \frac{1}{n^2}\left[\frac{h^2}{2\pi^2 m(Ze^2/4\pi\varepsilon_0)^2}\right]$

(C) $E = -\frac{1}{n^2}\left[\frac{2\pi^2 m(Ze^2/4\pi\varepsilon_0)^2}{h^2}\right]$ (D) $E = -\frac{1}{n^2}\left[\frac{2\pi^2 m(Ze^2/4\pi\varepsilon_0)^2}{h^2}\right]$

110. The angle θ made by the angular momentum vector of an electron with z-axis is given by

(A) $\cos\theta = m/\sqrt{l(l+1)}$ (B) $\cos\theta = \sqrt{l(l+1)}/m$

(C) $\cos\theta = \sqrt{l/m}$ (D) $\cos\theta = l/m$

111. The number of radial nodes in the probability distribution curve of many electron system is equal to

(A) $n - 1$ (B) $n + 1$ (C) $n - l - 1$ (D) $n - l - 1$

112. m_l can be generated from l

(A) by the internal magnetic field produced due to the rotation of an electron in an orbit

(B) by the external magnetic field

(C) by the both internal and external magnetic field

(D) none of these.

113. If l is azimuthal quantum number and m_l is magnetic quantum number, then

(A) $l = 2m_l + 1$ (B) $l = (m_l - 1)/2$ (C) $m_l = 2l^2 + 1$ (D) $m_l = l + 2$

114. Which of the following statements is incorrect?

(A) The e/m ratio of anode rays is constant

(B) Radial probability density $\rho_{n,l}(r) = 4\pi r^2 R^2_{n,l}(r)$

(C) The number of nodes in the radial distribution of orbitals is $n - l - 1$

(D) A node is a point in space where the wave function has zero amplitude.

115. For an electron, if the uncertainty in velocity is Δv, the uncertainty in its position is given by

(A) $hm/4\pi\Delta v$ (B) $4\pi/hm\Delta v$ (C) $h/4\pi m\Delta v$ (D) $4\Delta m/h\Delta v$

116. Rutherford's experiment, which established the nuclear model of the atom, used a beam of

(A) β-particles, which impinged on a metal foil and got absorbed

(B) γ-rays, which impinged on a metal foil and got absorbed

(C) helium atoms, which impinged on a metal foil and got scattered

(D) helium nuclei, which impinged on a metal foil and got scattered. **(I.I.T. 2002)**

117. If the nitrogen atom had electronic configuration $1s^7$, it would have energy lower than that of the normal ground state configuration $1s^2 2s^2 2p^3$, because the electrons would be closer to the nucleus. Yet $1s^7$ is not observed because it violates

(A) Heisenberg uncertainty principle (B) Hund's rule

(C) Pauli exclusion principle (D) Bohr postulate of stationary state.

(I.I.T. 2002)

118. If the energy of an electron in the first Bohr orbit of H atom is -13.12×10^5 J/mol, the possible energy value of the excited state for electron in Bohr orbit of H atom in J/mol is

(A) -3.28×10^5 (B) -4.2×10^5 (C) -6.8×10^5 (D) $+6.8 \times 10^5$

119. If the ionisation energy of He^+ ion is 19.6×10^{-18} J atom^{-1}, the energy of first stationary state of Li^{2+} ion in J atom^{-1} is

(A) 44.1×10^{-18} (B) 88.2×10^{-18} (C) 22.05×10^{-18} (D) 48.1×10^{-18}

Hint: E_1 for $He^+ = Z^2 \times E_1$ for H $= 4E_1$. E_1 for $Li^{2+} = Z^2 \times E_1$ for H $= 9E_1$. Hence E_1 for $Li^{2+} = E_1$ for $He^+ \times 9/4 = 19.6 \times 10^{-18} \times 9/4 = 44.1 \times 10^{-18}$ J atom^{-1}.

120. Which of the following species is (are) expected to give line spectrum similar to that of H(g)?

(A) He(g) (B) Li^{2+}(g) (C) He^+(g) (D) D(g)

121. The orbital angular momentum of an electron in 2*s* orbital is

(A) $\sqrt{2}(h/2\pi)$ (B) $h/2\pi$ (C) 0 (D) $(1/2)(h/2\pi)$

122. Which of the following transitions in the H atom produces the fourth line of the second series in the spectrum of H atom?

(A) $n = 5 \longrightarrow n = 2$ (B) $n = 6 \longrightarrow n = 2$

(C) $n = 4 \longrightarrow n = 1$ (D) $n = 5 \longrightarrow n = 1$

Hint: In the Balmer series: first line: $n = 3 \longrightarrow n = 2$; second line: $n = 4 \longrightarrow n = 2$; third line: $n = 5 \longrightarrow n = 2$; fourth line: $n = 6 \longrightarrow n = 2$.

123. The most intense line in the Brackett series in the spectrum of H atom is due to the transition

(A) $n = 5 \longrightarrow n = 4$ (B) $n = 8 \longrightarrow n = 4$

(C) $n = \infty \longrightarrow n = 4$ (D) $n = \infty \longrightarrow n = 1$

Hint: (D) corresponds to Lyman series and others to Brackett series. The intensity of spectral line decreases with increase of *n* (higher). Hence the most intense line in the Brackett series corresponds to $n = 5 \longrightarrow n = 4$.

124. Which of the following orbital is symmetrical about molecular axis (*z*-axis)?

(A) $2p_z$ (B) $2p_x$ (C) $2p_y$ (D) $3d_{x^2-y^2}$

125. The number of orbitals accommodated in the *n*th energy level is given by

(A) $n + 2$ (B) $2n$ (C) $2n^2$ (D) n^2

126. A naturally occurring element E is 78.99% ^{24}E, 10.00% ^{25}E and 11.01% ^{26}E. The atomic mass of E is

(A) 24.9 (B) 25.2 (C) 24.3 (D) 25.3

Hint: At. mass = $(78.99 \times 24)/100 + (10.00 \times 25)/100 + (11.01 \times 26)/100 = 24.3$

127. The nucleus $^{23}_{11}Na$ contains

(A) 11 protons, 12 electrons (B) 11 protons, 12 neutrons

(C) 23 protons, 11 electrons (D) 23 protons, 11 neutrons

128. Which hydrogen-like species will have the same radius as that of Bohr orbit of hydrogen atom?

(A) $n = 2$, Li^{2+} (B) $n = 2$, Be^{3+} (C) $n = 2$, He^+ (D) $n = 3$, Li^{2+} **(I.I.T. 2004)**

Hint: $rZ = n^2 r_{1(H^+)}$. But $r = r_{1(H^+)}$. Hence $n^2 = Z$. Thus $n = 2$, $Z = 4$. Hence $r_{2(Be^{3+})} = r_{1(H^+)}$

129. The number of radial nodes in 3*s* and 2*p* orbitals respectively are

(A) 2 and 0 (B) 1 and 2 (C) 0 and 2 (D) 2 and 1 **(I.I.T. 2005)**

Hint: Number of radial nodes = $n - l - 1$. For 3*s* orbital: No. of radial nodes = 3 – 0 – 1 = 2. For 2*p* orbital: No. of radial nodes = 2 – 1 – 1 = 0.

130. The energy of the second Bohr orbit of H-atom is – 3.41 eV. The energy of the second Bohr orbit of He^+ ion will be

(A) – 6.82 eV (B) – 8.62 eV (C) – 13.46 eV (D) – 13.64 eV

Hint: $\dfrac{E_{He^+}}{E_H} = \dfrac{(Z_{He^+})^2}{(Z_H)^2}$ or $E_{He^+} = E_H\left(\dfrac{Z_{He^+}}{Z_H}\right)^2 = -3.41\left(\dfrac{2}{1}\right)^2 \text{ eV} = -13.64 \text{ eV}$

131. Assuming that Hund's rule is violated, the bond order and magnetic nature of the diatomic molecule B_2 is

(A) 1 and diamagnetic (B) 0 and diamagnetic

(C) 1 and paramagnetic (D) 0 and paramagnetic **(I.I.T. 2010)**

Hint: B_2 (10) = $\sigma_{1s^2}\sigma^*_{1s^2}\sigma_{2s^2}\sigma^*_{2s^2}\pi_{2p_x^2}$

Bond order = $\dfrac{6-4}{2} = 1$

Nature of B_2 molecule is diamagnetic since it does not contain any unpaired electron.

Paragraph for questions 132 to 134

The hydrogen-like species Li^{2+} is in a spherically symmetric state S_1 with one radial node. Upon absorbing light the ion undergoes transition to a state S_2. The state S_2 has one radial node and its energy is equal to the ground state energy of the hydrogen atom.

132. The state S_1 is

(A) 1*s* (B) 2*s* (C) 2*p* (D) 3*s* **(I.I.T. 2010)**

Hint: For S_1 (spherically symmetrical), radial node = 1

Hence $n - 1 = 1$

Or $n = 2$

For S_2, radial node = 1

$$E_{S_2} = \frac{-13.6 \times 2^2}{n^2} = E_H \text{ in ground state} = -13.6$$

$$E_{S_2} = \frac{-13.6 \times 9}{n^2} = -13.6 \text{ Or } n = 3$$

Hence the state S_1 is 2*s* and the state S_2 is 3*p*.

133. Energy of the state S_1 in units of the hydrogen atom ground state energy is

(A) 0.75 (B) 1.50 (C) 2.25 (D) 4.50 **(I.I.T. 2010)**

Hint: $\dfrac{E_{S_1}}{E_{H(ground)}} = \dfrac{-13.6 \times 9}{4 \times (-13.6)} = 2.25$

134. The orbital angular momentum quantum number of the state S_2 is

(A) 0 (B) 1 (C) 2 (D) 3 **(I.I.T. 2010)**

Hint: The azimuthal quantum number for $S_2 = l = 1$

ANSWERS

1. D	2. C	3. D	4. A	5. A	6. A
7. A	8. B	9. B	10. A, B, C	11. A	12. D
13. A	14. A	15. D	16. D	17. A, D	18. C
19. A, B	20. A	21. D	22. B	23. C	24. A
25. C	26. C	27. B	28. B	29. B	30. C
31. D	32. A, B	33. C	34. A, C, D	35. A	36. A, B
37. A	38. B	39. D	40. A	41. D	42. B
43. D	44. D	45. C	46. B	47. C, D	48. C
49. C	50. A	51. A	52. C	53. D	54. A
55. B	56. A	57. A	58. C, D	59. A, B	60. A, B
61. A, B, C	62. D	63. A, B	64. A, B	65. A	66. C
67. D	68. B	69. C	70. A	71. B	72. B
73. B	74. A	75. A	76. D	77. B	78. A
79. B	80. A	81. A	82. A	83. D	84. A
85. A	86. A	87. C	88. A	89. B	90. A
91. B	92. C	93. A	94. A	95. C	96. A
97. D	98. A	99. B	100. C	101. A	102. C
103. D	104. C	105. A	106. B	107. A	108. A
109. D	110. A	111. C	112. C	113. B	114. A
115. C	116. D	117. C	118. A	119. A	120. B, C, D
121. C	122. B	123. A	124. D	125. D	126. C
127. B	128. B	129. A	130. D	131. A	132. B
133. C	134. B				

2

Nuclear Chemistry and Radiochemistry

IMPORTANT RELATIONS FOR NUMERICAL PROBLEMS

1. (i) $\lambda = (2.303/t) \log a/(a - x)$

   (ii) $\lambda = (2.303/t) \log (N_0/N)$

   (iii) $\lambda = (2.303/t) \log$ [initial ratio of $^{14}C/^{12}C$]/[$^{14}C/^{12}C$ at present]

   (iv) $\lambda = (2.303/t) \log [^{238}U + ^{206}Pb]/^{238}U$

   (v) $N_0/N = e^{\lambda t}$ or $N/N_0 = e^{-\lambda t}$

2. $t_{1/2} = 0.693/\lambda$
3. $t_{av} = 1/\lambda = t_{1/2}/0.693 = 1.44 \times t_{1/2}$
4. $t_{1/2} = 0.693/\lambda = 0.693 t_{av}$
5. Binding energy = $931.478 \times \Delta m'$ MeV
6. Binding energy per nucleon = total B.E./No. of nucleons
7. Rate of decay = $\lambda \times$ No. of atoms
8. Amount left after n half-lives = $N_0/2^n$
9. Amount decayed after n half-lives = $N_0(2^n - 1)/2^n$
10. 1 Curie = 3.7×10^{10} dps = 3.7×10^{10} Bq
11. Radioactive equilibrium: $\lambda_A/\lambda_B = N_B/N_A = (t_{1/2})_B/(t_{1/2})_A = (t_{av})_B/(t_{av})_A$
12. $\Delta m = Zm_p + (A - Z)m_n - M$, where Δm = mass defect
13. Packing fraction = (Isotopic mass − Mass number) × 10^4/Mass number
14. Energy released by loss of 1 amu mass = 931.478 MeV
15. Nuclear radius $(r) = R_0 A^{1/3}$ where A = mass number of atom, R_0 = constant = 1.4×10^{-15} m

OBJECTIVE TYPE MULTIPLE CHOICE QUESTIONS

1. The number of neutron accompanying the formation of $^{139}_{54}Xe$ and $^{94}_{38}Sr$ from the absorption of a slow neutron by $^{235}_{92}U$, followed by nuclear fission is
   (A) 0 (B) 2 (C) 1 (D) 3 **(I.I.T. 1999)**

   Hint: $^{235}_{92}U + ^{1}_{0}n \longrightarrow ^{139}_{54}Xe + ^{94}_{38}Sr + x\ ^{1}_{0}n$. Hence 235 + 1 = 139 + 94 + x or x = 336 – 233 = 3.

2. Decrease in atomic number is observed during
   (A) α-emission (B) β-emission
   (C) positron emission (D) electron emission. **(I.I.T. 1998)**

3. Assertion (*A*): Nuclide $^{30}_{13}Al$ is less stable than $^{40}_{20}Ca$.
   Reason (*R*): Nuclides having odd number of protons and neutrons are generally unstable.
   (A) Both *A* and *R* are true and *R* is the correct explanation of *A*
   (B) Both *A* and *R* are true and *R* is not the correct explanation of *A*
   (C) *A* is true but *R* is incorrect (D) *A* is incorrect but *R* is correct. **(I.I.T. 1998)**

4. $^{27}_{13}Al$ is a stable isotope. $^{29}_{13}Al$ is expected to disintegrate by
   (A) α-emission (B) β-emission
   (C) positron emission (D) proton emission. **(I.I.T. 1996)**

5. Nuclear reactions accompanied with the emission of neutrons are
   (A) $^{27}_{13}Al + ^{4}_{2}He \longrightarrow ^{30}_{15}P$ (B) $^{12}_{6}C + ^{1}_{1}H \longrightarrow ^{13}_{7}N$
   (C) $^{30}_{15}P \longrightarrow ^{30}_{14}Si + ^{0}_{1}e$ (D) $^{240}_{96}Cm + ^{4}_{2}He \longrightarrow ^{243}_{97}Bk + ^{0}_{1}e$ **(I.I.T. 1996)**

6. The half-life period of a radioactive element is 140 days. After 560 days, 1 g of the element will reduce to
   (A) (1/2) g (B) (1/4) g (C) (1/8) g (D) (1/16) g **(I.I.T. 1986)**
   Hint: No. of half lives, n = 560/140 = 4. $N = N_0/2^n$ = 1 g/2^4 = (1/16) g

7. The radiations from a naturally occurring radioactive substance as seen after deflection by a magnetic field in one direction are
   (A) definitely α-rays (B) definitely β-rays
   (C) both α and β-rays (D) either α or β-rays. **(I.I.T. 1984)**

8. From the following equations pick up the possible fusion reactions.
   (A) $^{13}_{6}C + ^{1}_{1}H \longrightarrow ^{14}_{6}C + ^{0}_{1}e$ + 4.3 MeV (B) $^{12}_{6}C + ^{1}_{1}H \longrightarrow ^{13}_{7}N$ + 2 MeV
   (C) $^{14}_{7}N + ^{1}_{1}H \longrightarrow ^{15}_{8}O$ + 7.3 MeV
   (D) $^{235}_{92}U + ^{1}_{0}n \longrightarrow ^{140}_{54}Xe + ^{94}_{38}Sr + ^{1}_{0}n + \gamma$ + 200 MeV **(I.I.T. 1984)**

9. The β-rays emitted by a radioisotope are
   (A) electromagnetic radiations (B) the electrons orbiting around the nucleus
   (C) charged particles emitted by the nucleus
   (D) neutral particles. **(I.I.T. 1983)**

10. The equation $4\,{}^{1}_{1}H \longrightarrow {}^{4}_{2}He^{2+} + 2\,{}^{0}_{1}e + 26$ MeV represents
    (A) β-decay (B) γ-decay (C) fusion (D) fission **(I.I.T. 1983)**
11. He atom is two times as heavy as a H_2 molecule. At 298 K the average kinetic energy of He atom is
    (A) twice that of H_2 molecule (B) same as that of H_2 molecule
    (C) four times that of H_2 molecule (D) half that of H_2 molecule. **(I.I.T. 1982)**
12. The uranium (mass No. = 238, At. No. = 92) emits an α-particle. The product has the mass No. and At. No. respectively
    (A) 236, 92 (B) 234, 90 (C) 238, 90 (D) 236, 96 **(I.I.T. 1981)**
13. During negative β-decay
    (A) an atomic electron is ejected
    (B) an electron which is already present within the nucleus is ejected
    (C) a neutron in the nucleus decays emitting an electron
    (D) a part of the binding energy is converted into an electron. **(I.I.T. 1987)**
14. During nuclear fusion reaction
    (A) a heavy nucleus breaks into two fragments itself
    (B) a light nucleus bombarded by thermal neutrons breaks up
    (C) a heavy nucleus bombarded by thermal neutrons breaks up
    (D) two light nuclei combine to give a heavier nucleus and possibly other products. **(I.I.T. 1987)**
15. An element X loses 1α and 2β particles in three successive stages. The resulting element will be
    (A) an isobar of X (B) an isotope of X
    (C) an isotone of X (D) X itself **(Roorkee 1993)**
16. One mole of an α-emitting nuclide ${}^{A}_{Z}X$ ($t_{1/2}$ = 10 hour) was placed in a sealed container. 4.52×10^{23} He atoms will accumulate in the container in
    (A) 4.52 min (B) 9.40 h (C) 10.00 h (D) 20.00 h **(Roorkee 1992)**
    Hint: 1 mole = 6.02×10^{23} atoms = N_0. No. of atoms still present = $6.02 \times 10^{23} - 4.52 \times 10^{23} = 1.5 \times 10^{23}$. $t = \frac{2.303}{\lambda} \log \frac{N_0}{N} = \frac{2.303}{0.693/10} \log \frac{6.02 \times 10^{23}}{1.5 \times 10^{23}} = \frac{2.303 \times 0.06021}{0.0693}$
    = 20 h
17. What will be the half-life period of a nucleus if at the end of 4.2 days $N = 0.798N_0$?
    (A) 10 days (B) 13 days (C) 15 days (D) 20 days
18. Which of the following statements is correct?
    (A) Electron and positron both have same mass
    (B) The smaller the $t_{1/2}$, the lesser is the number of atoms undergoing decay in a given time
    (C) Nuclear reactions are markedly affected by temperature
    (D) Atom bomb and hydrogen bomb are the examples of nuclear fission.

19. In the radioactive decay series, $A \xrightarrow{\alpha} B \xrightarrow{\beta} C \xrightarrow{\beta} D$
    (A) A and D are isobars (B) A and D are isotopes
    (C) A and B are isobars (D) A and C are isotopes.
20. The relation between decay constant and half-life of a radioactive decay reaction is
    (A) $t_{1/2} = \lambda \log_e 2$ (B) $t_{1/2} = (\log_e 2)/\lambda$
    (C) $t_{1/2} = 2.303 \log_{10} 2\lambda$ (D) $t_{1/2} = (2.303/\lambda) \log_e 2$
21. The instability of a nucleus is attributed to
    (A) high proton : neutron ratio (B) high proton : electron ratio
    (C) low proton : neutron ratio (D) low proton : electron ratio
22. If N and N_0 are the number of radioactive particles at time t and $t = 0$ respectively, then
    (A) $\lambda = (2.303/t) \log (N_0/N)$ (B) $\lambda = (t/2.303) \log (N_0/N)$
    (C) $\lambda = (2.303/t) \log (N/N_0)$ (D) $\lambda = (1/t) \log (N_0/N)$
23. Which of the following reactions shows the product incorrectly?
    (A) ${}^{242}_{96}Cm(\alpha, 2n){}^{243}_{97}Bk$ (B) ${}^{10}_{5}B(\alpha, n){}^{13}_{7}N$
    (C) ${}^{14}_{7}N(n, p){}^{14}_{6}C$ (D) ${}^{28}_{14}Si(d, n){}^{29}_{15}P$
24. The end product of (4n + 2) disintegration series is
    (A) ${}^{204}_{82}Pb$ (B) ${}^{206}_{82}Pb$ (C) ${}^{208}_{82}Pb$ (D) ${}^{209}_{82}Pb$
25. The activity of a radioelement (${}^{100}X$) is 6.203 Curie. If its disintegration constant is 3.7×10^4 s^{-1}, the mass of X is
    (A) 10^{-3} g (B) 10^{-6} g (C) 10^{-14} g (D) 10^{-15} g
26. Mesons are responsible for
    (A) attraction between nucleons (B) attraction between nucleus
    (C) disintegration of nucleus (D) repulsion between the protons in nucleus.
27. ${}^{235}_{92}U + {}^{1}_{0}n \longrightarrow {}^{236}_{92}U \longrightarrow$ fission products + neutron + 3.20×10^{-11} J. The energy released when 1 g of ${}^{235}_{92}U$ undergoes fission is
    (A) 6.55×10^6 kJ (B) 8.21×10^7 kJ (C) 12.75×10^8 kJ (D) 18.60×10^9 kJ
28. What is the $t_{1/2}$ of a radioisotope if 87.5% of any given amount of the material disintegrates in 40 min?
    (A) 10 min (B) 13 min 20 s (C) 20 min (D) 160 min
29. ${}^{24}Na$ (15 h, β) is used for
    (A) detection of defect in bone (B) examination of disorder in brain
    (C) estimation of circulation of blood (D) cure of some form of cancer.
30. Identify the correct statements.
    (A) γ-rays have a higher penetrating power than X-rays
    (B) α-particles have a mass of 4 and a charge of + 2
    (C) Emission of β-rays is accompanied by the emission of nutrinos
    (D) All isotopes are radioactive.

31. The protons and neutrons in the nucleus of an atom undergo interconversion through the exchange of
(A) charged mesons (B) neutral mesons
(C) mucons (D) positrons

32. Which of the following is not a mode of self dissociation of radioisotope?
(A) α-emission (B) β-emission (C) Fission (D) Electron capture

33. The radiation having maximum penetrating power is
(A) α-ray (B) β-ray (C) cathode ray (D) γ-ray

34. When $^{235}_{92}U$ is bombarded with slow neutron, the products formed are
(A) Ba, Xe (B) Cd, Kr (C) Pb, Sr (D) Ba, Kr

35. The energy required to break up the nucleus to protons and neutrons is known as
(A) thermal energy (B) binding energy
(C) nuclear energy (D) kinetic energy

36. Which of the following has the same *e*/*m* ratio shown by cathode ray?
(A) α-particle (B) β-particle (C) γ-ray (D) none of these.

37. In which of the following radioactive series $^{226}_{88}Ra$ is found?
(A) $4n$ (B) $4n + 1$ (C) $4n + 2$ (D) $4n + 3$

38. The nuclear isomers differ in their
(A) mass number (B) number of neutrons
(C) number of protons (D) half-life period

39. If a radioisotope has a large value of p/n ratio, it will emit
(A) β-particles (B) α-particles (C) γ-particles (D) neutrons

40. Which of the following is formed when $^{35}_{17}Cl$ undergoes (n, α) reaction?
(A) $^{32}_{15}P$ (B) $^{32}_{16}S$ (C) $^{39}_{19}K$ (D) $^{31}_{15}P$

41. Which of the following has the smallest wavelength?
(A) α-ray (B) β-ray (C) γ-ray (D) all of these.

42. When a radioelement is kept under vacuum, the rate of disintegration per second
(A) decreases considerably (B) increases considerably
(C) increases if the products are gaseous
(D) is not affected.

43. The nucleus of a radioelement possesses
(A) high potential energy (B) high binding energy
(C) low binding energy (D) zero binding energy.

44. The n/p plot of stable isotopes breaks off from the hypothetical 1:1 n/p plot at the atomic number
(A) 10 (B) 40 (C) 30 (D) 20

45. The hydrogen radioisotope tritium emits
(A) neutrons (B) γ-rays (C) β-particles (D) α-particles

46. A rock sample on analysis indicated the presence of equal number of U and Pb. The age of the rock is ($t_{1/2}$ of U = 4.5×10^9 yr)
(A) 2.25×10^9 yr (B) 4.5×10^9 yr (C) 9.0×10^9 yr (D) 13.5×10^9 yr

47. The number of α and β particles emitted in the nuclear reaction $^{228}_{90}Th \longrightarrow {}^{212}_{83}Bi$ is
(A) 4α, 7β (B) 8α, 1(C) 3α, 7β (D) 4α, 1β

48. After the emission of an α-particle from $^{238}_{92}U$, the number of neutrons in the atom will be
(A) 150 (B) 144 (C) 140 (D) 138

49. Ra has atomic weight 226 and a $t_{1/2}$ of 1600 yr. The number of disintegrations produced per second from 1 g is
(A) zero (B) 3.7×10^{10} (C) 9.2×10^{6} (D) 4.8×10^{10}

50. If the total binding energies of $^{2}_{1}H$, $^{4}_{2}He$, $^{56}_{26}Fe$ and $^{235}_{92}U$ are 2.22, 28.3, 492 and 1786 MeV respectively, identify the most stable nucleus out of the following:
(A) $^{235}_{92}U$ (B) $^{56}_{26}Fe$ (C) $^{2}_{1}H$ (D) $^{4}_{2}He$

51. If the half-life period of a radioisotope is $2n$ years, the fraction of a sample will remain undecayed after n years is
(A) 2 (B) 1/2 (C) $1/\sqrt{2}$ (D) $1/\sqrt{3}$

52. 1 gram-atom of an alpha emitting radioelement $^{A}_{Z}Y$ ($t_{1/2}$ = 10 h) was placed in a sealed container. 4.52×10^{13} He atoms will accumulate in the container in
(A) 20 h (B) 10 h (C) 9.40 h (D) 4.82 h

53. The loss of a β-particle is equivalent to
(A) increase of one proton only (B) decrease of one neutron only
(C) both (A) and (B) (D) none of these. **(I.I.T. 1998)**

54. A radioisotope decays by two parallel reactions for which the decay constants are λ_1 and λ_2. The effective decay constant (λ) of the radioisotope is
(A) $\lambda = (1/\lambda_1) + (1/\lambda_2)$ (B) $\lambda = \lambda_1 + \lambda_2$
(C) $\lambda = \lambda_1 - \lambda_2$ (D) $\lambda = \lambda_1/\lambda_2$

55. The most stable nuclei found in the binding energy curve are of
(A) Pb (B) Cu (C) Zn (D) Fe

56. In the nuclear reaction $^{14}_{7}N + {}^{4}_{2}He \longrightarrow {}^{p}_{q}X + {}^{1}_{1}H$, the nuclide X is
(A) nitrogen of mass 16 (B) nitrogen of mass 17
(C) oxygen of mass 16 (D) oxygen of mass 17

57. Identify the correct statement.
(A) Mesons are present in cosmic rays
(B) Mesons are about 200 times as heavy as electron
(C) Mesons are discovered by Yukawa
(D) None of these.

58. The radioactive decay of $^{88}_{35}X$ by a β-emission produces an unstable nucleus which spontaneously emits a neutron. The final product is
(A) $^{88}_{37}X$ (B) $^{89}_{35}Y$ (C) $^{88}_{34}Z$ (D) $^{87}_{36}W$

59. One of the hazards of nuclear explosion is the generation of ^{90}Sr and its subsequent incorporation in bones. ^{90}Sr has a half-life period of 28.1 years. If 1 microgram of it is absorbed by a newly born baby, then its quantity after 20 years will be
(A) 6.1×10^{-7} g (B) 6.1×10^{-6} g (C) 5.1×10^{-7} g (D) 5.1×10^{-6} g

Hint: $t = \frac{2.303}{\lambda}\log(N_0/N)$ or $20 = \frac{2.303 \times 28.1}{0.693}\log\frac{10-6}{N}$ or $N = 6.1 \times 10^{-7}$ g

60. Which of the following is the characteristic of an unstable nucleus?
    (A) High packing fraction (B) Low binding energy
    (C) High n/p ratio (D) None of these.
61. A radioelement has a half-life period of 1500 years. A sealed tube containing 1 g of the sample will contain after 3000 years
    (A) 1 g of the sample (B) 0.5 g of the sample
    (C) 0.25 g of the sample (D) 0.00 g of the sample.
62. Which form of following nuclear radiations most closely resembles X-rays?
    (A) γ-rays (B) α-rays (C) β-rays (D) All of these.
63. Which of the following radioisotopes is used in establishing the reaction mechanism of photosynthesis in plants?
    (A) ${}^{2}_{1}H$ (B) ${}^{14}_{6}C$ (C) ${}^{13}_{6}C$ (D) ${}^{12}_{6}C$
64. Which of the following series is the man-made radioactive disintegration series?
    (A) Actinium series (B) Uranium series
    (C) Neptunium series (D) Thorium series
65. The emission of β-particles occurs from
    (A) inner shell of an atom (B) valence shell of an atom
    (C) nucleus due to the conversion of neutron to proton and electron
    (D) nucleus due to the conversion of proton to neutron and electron.
66. Which of the following radioisotopes is used for the determination of age of dead animals and plants?
    (A) ${}^{2}_{1}H$ (B) ${}^{14}_{6}C$ (C) ${}^{13}_{6}C$ (D) ${}^{12}_{6}C$
67. In a nuclear explosion, the energy released is in the form of
    (A) thermal energy (B) kinetic energy
    (C) potential energy (D) electrical energy.
68. Gamma rays are
    (A) high energy positrons (B) high energy electrons
    (C) low energy electrons (D) high energy electromagnetic waves.
69. Artificial radioactivity was discovered by
    (A) Irene Curie and Joliot (B) Hahn and Strassmann
    (C) Chadwick (D) Becquerel
70. The introduction of a neutron into the nuclear composition of an atom would lead to a change in
    (A) its atomic mass (B) its atomic number
    (C) the chemical nature of the atom (D) the number of the electron.
71. Radioactive disintegration differs from a chemical change in being
    (A) a nuclear process (B) a spontaneous process
    (C) an exothermic change (D) an unimolecular first order reaction.

72. If 8.0 g of a radioelement has a $t_{1/2}$ of 10 h, the $t_{1/2}$ of 2.0 g of the same substance is
(A) 2.5 h (B) 5 h (C) 10 h (D) 40 h

73. If the rate of decay of a radioelement is R_1 at time t_1 and R_2 at time t_2, the mean life of this radioelement is
(A) $R_1 = R_2t_2/t_1$ (B) $R_1 = R_2e^{(t_2 - t_1)T}$
(C) $R_2 = R_1e^{(t_1 - t_2)/T}$ (D) $R_1t_1 = R_2e^{(t_1 - t_2)} \times t_2$

74. The activity of a given mass of a sample is given by
(A) λN (B) $\lambda_1 N_1 = N_0e^{-\lambda t}$ (C) $\dfrac{\lambda \times \text{weight}}{\text{M.W.} \times N_A}$ (D) $\dfrac{\lambda \times \text{weight}}{\text{At. wt.} \times N_A}$

75. Identify the correct expressions.
(A) $t_{1/2} = \ln 2/\lambda$ (B) $T = 1.44t_{1/2}$
(C) $A = A_0e^{-\lambda t}$ (D) $t = t_{1/2}[\log (N_0/N)]/\log 2$

76. Which of the following series are artificial radioactive disintegration series?
(A) Actinium series ($4n + 3$) (B) Uranium series ($4n + 2$)
(C) Neptunium series ($4n + 1$) (D) Thorium series ($4n$)

77. If the relation between time of decay (t) and half-life period ($t_{1/2}$) is $t = 4t_{1/2}$, the relation between t and mean life (T) is
(A) $\ln 2/T^2$ (B) $2T^4 \ln 2$ (C) $4T \ln 2$ (D) $2T \ln 2$

78. Which of the following are β^+-emitters?
(A) ^{34}Cl (B) ^{222}Rn (C) ^{40}K (D) Antineutrino

79. Which of the following are β^--emitters?
(A) ^{3}T (B) ^{60}Co (C) ^{14}C (D) neutron

80. The most stable nuclei are obtained from the combinations
(A) even N and even Z (B) odd N and even Z
(C) odd N and odd Z (D) even N and odd Z

81. In what ratio $^{35}_{17}$Cl and $^{37}_{17}$Cl should be present so as to get $^{35.5}_{17}$Cl?
(A) 1 : 2 (B) 3 : 1 (C) 1 : 3 (D) 1 : 4

82. During the transformation $^a_cX \longrightarrow {}^b_dY$, the number of β-particles emitted is
(A) $2c - d + a - b$ (B) $(a - b)/4$
(C) $d + [(a - b)/2] - c$ (D) $d + [(a + b)/2] + c$

Hint: $^a_cX \longrightarrow {}^b_dY + x\,{}^4_2He + y\,{}^0_{-1}e$. Hence $a = b + 4x$ or $x = (a - b)/4$. $c = d + 2x - y$ or $y = d - c + 2x = d - c + (a - b)/2 = d + [(a - b)/2] - c$

83. A radioelement X emits 3α, 1β, and 1γ particles and forms $^{225}_{76}$Y. The element X is
(A) $^{236}_{80}$X (B) $^{238}_{92}$X (C) $^{236}_{81}$X (D) $^{227}_{81}$X

84. The number of neutrons in the parent nucleus, which gives ^{14}N on β-emission is
(A) 6 (B) 7 (C) 8 (D) 14

85. If the isotopic mass of $^{238}_{92}$U is 235.125 amu, the packing fraction is
(A) 525 (B) 5.31 (C) 52.5 (D) 531

86. If 180 MeV energy is released during the fission of one nucleon of ^{235}U, the energy released by the fission of 0.235 g of ^{235}U is
   (A) 1.73×10^7 kJ (B) 1.73×10^{16} kJ (C) 1.08×10^8 kJ (D) 1.08×10^{20} kJ
87. An atom of Ra combines with two atoms of Cl to give $RaCl_2$. The radioactivity of $RaCl_2$ will be
   (A) as much as that of the same quantity of Ra
   (B) one fourth of the same quantity of Ra
   (C) one half of the same quantity of Ra
   (D) one eighth of the same quantity Ra
88. The radioactive nucleus $^{13}_{7}N$ decays to $^{13}_{6}C$ through the emission of
   (A) neutrons (B) positrons (C) protons (D) electrons
89. If the binding energy per nucleon of ^7Li and ^4He nuclei is 5.60 MeV and 7.06 MeV respectively, the energy of the reaction, $^7Li + p \longrightarrow {}^4_2He$, is
   (A) 17.3 MeV (B) 19.6 MeV (C) 2.4 MeV (D) 8.4 MeV
90. The activity of a radioisotope decreases with
   (A) pressure (B) temperature
   (C) chemical environment (D) none of these.
91. During the emission of a β-particle
   (A) one proton increases (B) one electron increases
   (C) one electron decreases (D) no change occurs.
92. The nucleus resulting from $^{238}_{92}U$ after successive emissions of two α and four β-particles is
   (A) $^{230}_{90}Th$ (B) $^{230}_{92}U$ (C) $^{230}_{94}Pu$ (D) $^{230}_{88}Ra$ **(I.I.T. 1991)**
93. $^{235}_{92}U$ belongs to group 3 of periodic table. If it loses one α-particle, the new element will belong to the group
   (A) 1 (B) 11 (C) 12 (D) 15
94. The energy released in an atom bomb explosion is mainly due to
   (A) release of electrons (B) release of neutrons
   (C) lower mass of products than the initial material
   (D) greater mass of products than the initial material. **(I.I.T. 1980)**
95. The ionisation power of X-rays is
   (A) greater than β and γ-rays (B) greater than γ-rays
   (C) less than β and γ-rays (D) less than only γ-rays.
96. The most important concern for the environmentalists in the future will be
   (A) industrial waste (B) nuclear waste
   (C) household waste (D) hydel waste.
97. Which of the following has a negative charge?
   (A) Positron (B) Antiproton (C) Neutrino (D) Neutron
98. Control rods used in a nuclear reactor are made of
   (A) Cd (B) graphite (C) steel (D) all of these.

99. The half-life period of ^{234}Th is 24 days. What is the amount of Th left after 96 days in a sample that initially had 12 g of this isotope?
(A) 0.75 g (B) 1.5 g (C) 3.0 g (D) 6.0 g

100. What is the binding energy in $^{11}_{5}B$ nucleus, if its mass defect is 0.08181 amu?
(A) 1.2×10^{-4} erg/nucleus (B) 8.2×10^{-4} erg/nucleus
(C) 10.2×10^{-4} erg/nucleus (D) 1.9×10^{-4} erg/nucleus

101. $^{13}_{7}N$ changes to $^{13}_{6}C$ by the emission of
(A) electron (B) neutron (C) positron (D) proton **(Roorkee 1990)**

102. Identify the incorrect statement.
(A) ^{33}P is used in the treatment of leukemia
(B) ^{131}I is used in the treatment of thyroid gland cancer
(C) ^{59}Co cannot be used for the treatment of cancer
(D) Excessive use of radioisotopes is responsible for cancerous growth.

103. If N_0 is the number of nuclei in the beginning, the number of nuclei remaining undecayed at the end of the nth half-life period is
(A) $2^n N_0$ (B) $2^{-n} N_0$ (C) $n^2 N_0$ (D) $n^{-2} N_0$

104. If N_0 and N are the number of radioactive particles at $t = 0$ and at time t, then
(A) $\lambda = \frac{2.303}{t} \log \frac{N_0}{N}$ (B) $\lambda = \frac{t}{2.303} \log \frac{N_0}{N}$
(C) $\lambda = \frac{2.303}{t} \log \frac{N}{N_0}$ (D) $\lambda = \frac{1}{t} \log \frac{N_0}{N}$

105. The mass defect of the nuclear reaction $^{8}_{5}B \longrightarrow {}^{8}_{4}Be + {}^{0}_{+1}e$ is given by
(A) Δ_m = atomic mass of $({}^{8}_{4}Be - {}^{8}_{5}B)$
(B) Δ_m = atomic mass of $({}^{8}_{4}Be - {}^{8}_{5}B)$ + mass of two electrons
(C) Δ_m = atomic mass of $({}^{8}_{4}Be - {}^{8}_{5}B)$ + mass of one electron
(D) Δ_m = atomic mass of $({}^{8}_{4}Be - {}^{8}_{5}B)$ + mass of one positron

106. The triad which is (are) isotonic is (are)
(A) $^{16}_{8}O, {}^{14}_{6}C, {}^{15}_{7}N$ (B) $^{14}_{6}C, {}^{14}_{7}N, {}^{12}_{9}F$ (C) $^{12}_{6}C, {}^{14}_{7}N, {}^{19}_{9}F$ (D) $^{14}_{6}C, {}^{15}_{7}N, {}^{16}_{9}F$

107. In which of the following processes, there occur decrease in atomic number?
(A) Electron capture (B) Positron emission
(C) β^--emission (D) α-emission

108. One atomic mass unit in kilogram is (N = Avogadro's number)
(A) $(12 \times 10^{-3})/N$ (B) $(1 \times 10^{-3})/N$ (C) $12/N$ (D) $1/N$

109. ^{23}Na is the more stable isotope of Na. The process by which $^{24}_{11}Na$ can undergo decay is
(A) β^--emission (B) α-emission (C) β^+-emission (D) K electron capture
(I.I.T. 2003)

110. Tritium has a half-life period of 12.33 yr. What is the age of a bottle of wine whose $^{3}_{1}H$ radiation is about 1/10 that present in a new bottle of wine?
(A) 40 yr (B) 41 yr (C) 42 yr (D) 21 yr

111. An electron and a positron collide and annihilated emitting gamma photons to same energy. The λ of gamma photons is
(A) 2.6×10^{-12} m (B) 2.3×10^{-12} m (C) 2.4×10^{-12} m (D) 2.5×10^{-12} m
Hint: Equate E using the equations: $E = mc^2$, $E = hc/\lambda$

112. A photon of hard gamma radiation knocks a proton out of $^{24}_{12}Mg$ nucleus to form
(A) the isobar of $^{23}_{11}Na$ (B) the nuclide of $^{23}_{11}Na$
(C) the isobar of parent nucleus (D) the isotope of parent nucleus.

Hint: $^{24}_{12}Mg + \gamma \longrightarrow {}^{23}_{11}Na + {}^{1}_{1}H$. Thus the nuclide of $^{23}_{11}Na$ is formed.

113. With what velocity should an α-particle travel towards the nucleus of a copper atom so as to arrive at a distance 10^{-13} m from the nucleus of the Cu atom? ($\varepsilon_0 = 8.85 \times 10^{-12}$ J^{-1} C^2 m^{-1})
(A) 9.87×10^6 m s^{-1} (B) 1.6×10^9 m s^{-1}
(C) 7.89×10^6 m s^{-1} (D) 8.97×10^6 m s^{-1}

Hint: $$V = \frac{Z_1 Z_2 e^2}{4\pi\varepsilon_0 r} = -\frac{29 \times 4(1.6 \times 10^{-19}\ C)2}{4 \times 3.14 \times 8.85 \times 10^{-12}\ J^{-1}\ C^2\ m^{-1} \times 10^{-13}\ m} = 2.67 \times 10^{-13}\ J$$

$$V = \sqrt{\frac{2V}{m}} = \left[\frac{2 \times 2.67 \times 10^{-13}\ J}{4.0 \times 10^{-3}\ kg\ mol^{-1}/6.023 \times 10^{23}\ mol^{-1}}\right]^{1/2} = 8.97 \times 10^6\ m\ s^{-1}$$

114. A position is emitted from $^{23}_{11}Na$. The ratio of the atomic mass and atomic number of the resulting nuclide is
(A) 22/10 (B) 22/11 (C) 23/10 (D) 23/12 **(I.I.T. 2007)**
Hint: When a position is emitted from the nucleus, proton gets converted into neutron. Hence the atomic number decreases by one but the atomic mass remains constant. Thus, atomic mass/atomic number = 23/10.

115. Statement–1: The plot of atomic number (y-axis) versus number of neutrons (x-axis) for stable nuclei shows a curvature towards x-axis from the line of 45° slope as the atomic number is increased.
Statement–2: Proton-proton electrostatic repulsions begin to overcome attractive forces involving protons and neutrons in heavier nuclides.
(A) Statement–1 is true, Statement–2 is true; Statement–2 is a correct explanation for Statement–1
(B) Statement–1 is true, Statement–2 is true; Statement–2 is not a correct explanation for Statement–1
(C) Statement–1 is true, Statement–2 is false
(D) Statement–1 is false, Statement–2 is true. **(I.I.T. 2008)**

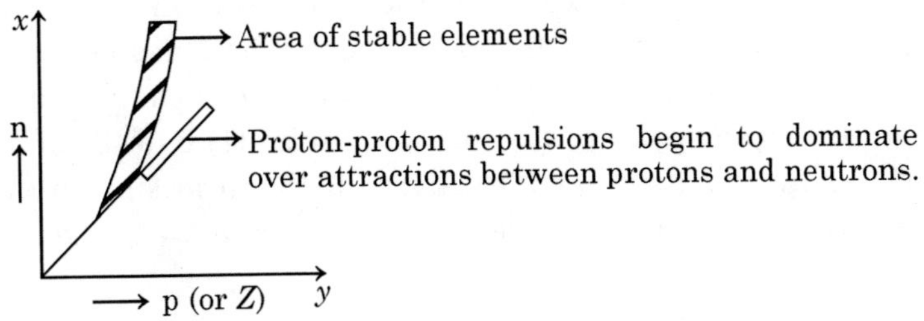

Hint: Proton-proton repulsions begin to dominate over proton-neutron attractions with the increase of value of Z. Hence the Statement–2 is correct. In order to stabilize the heavier nuclides, more number of neutrons are to be added so that the proton-neutron attractions overcome the proton-proton repulsions Hence the Statement–2 is the correct explanation of Statement–1.

116. Given that the abundances of isotopes ^{54}Fe, ^{56}Fe and ^{57}Fe are 5%, 90% and 5% respectively, the atomic mass of Fe is

(A) 55.85 (B) 55.95 (C) 55.75 (D) 56.06 **(I.I.T. 2009)**

Hint: $\overline{A} = \dfrac{\sum A_i X_i}{\sum X_i}$

$\overline{A} = 54 \times 0.05 + 56 \times 0.90 + 57 \times 0.05$ (where $\overline{A}$ is atomic mass of Fe); $\sum x_i = 1$

$\overline{A} = 55.95$

117. Bombardment of aluminium by α-particle leads to its artificial disintegration in two ways, (i) and (ii) as shown. Products X, Y and Z respectively are,

$$\begin{array}{ccc} {}^{27}_{13}Al & \xrightarrow{(ii)} & {}^{30}_{15}P + Y \\ \downarrow (i) & & \downarrow \\ {}^{30}_{14}Si + X & & {}^{30}_{14}Si + Z \end{array}$$

(A) proton, neutron, positron (B) neutron, positron, proton

(C) proton, positron, neutron (D) positron, proton, neutron **(I.I.T. 2011)**

Hint: ${}^{27}_{13}Al + {}^{4}_{2}\alpha \longrightarrow {}^{30}_{14}Si + {}^{1}_{1}p(X)$

${}^{27}_{13}Al + {}^{4}_{2}\alpha \longrightarrow {}^{30}_{15}P + {}^{1}_{0}n(Y)$

${}^{30}_{15}P \longrightarrow {}^{30}_{14}Si + {}^{0}_{+1}\beta(Z)$

Paragraph for Question Nos. 118 to 120

Due to the neutrons present in cosmic rays, the following reaction takes place by the bombardment of nitrogen with neutrons.

${}^{14}_{7}N + {}^{1}_{0}n \longrightarrow {}^{14}_{6}C + {}^{1}_{1}H$; ${}^{14}_{6}C \longrightarrow {}^{14}_{7}N + {}^{0}_{-1}e$

Cosmic rays form ${}^{14}_{6}C$ which gets circulated in the atmosphere as well as in living species. Whenever there is a nuclear explosion, the concentration of ^{14}C increases both in the atmosphere as well as in living species. When a species dies the ^{14}C concentration decreases and hence this decrease can be measured and the time estimated as to when the organism died. For this we require (1) half-life of ^{14}C (5760 years) (2) activity of ^{14}C in living species (3) activity of ^{14}C in fossil. Beyond 30000 years the age of fossil cannot be accurately determined as the activity then would be too low.

118. In radiocarbon dating for finding the age of fossils, the most appropriate statement is

(A) During the life-time ^{14}C assimilated by the human beings is in equilibrium with the ^{14}C that decomposes by β^- emission. At any time in a human life $^{14}C/^{12}C$ is always constant at a particular instant.

(B) ^{14}C dating method is inappropriate for finding the life of a given sample because ^{14}C undergoes β^- emission and hence the ratio of $^{14}C/^{12}C$ in human beings in not constant.

(C) Radiocarbon dating is inappropriate to determine the age because ^{14}C concentration varies from place to place.

(D) None of these. **(I.I.T. 2006)**

Hint: The correct statement is (C). (A) is incorrect because after nuclear explosion $^{14}C/^{12}C$ will not remain constant. (B) is incorrect because ^{14}C will always remain in equilibrium with the surroundings.

119. A fossil whose age can be determined using carbon dating must be

(A) 6 years old (B) 6000 years old (C) 60000 years old

(D) Can be accurately calculated for any age of the sample **(I.I.T. 2006)**

Hint: For effective carbon dating, the fossil must be minimum 1000 years old.

120. Two organisms died on the same day. One died at a place where nuclear explosion had taken place while the other died at a place where no such explosion had occurred. The ratio of ^{14}C during life to that present in the fossil at an instant is C_1 for the former and C_2 for the latter. The age of the former was calculated as t_1 and for the latter as t_2. The correct answer is

(A) There is an increae in t_1 w.r.t. t_2 and $t_1 - t_2 = \frac{1}{\lambda}\ln\frac{C_1}{C_2}$

(B) There is a decreae in t_1 w.r.t. t_2 and $t_1 - t_2 = \frac{1}{\lambda}\ln\frac{C_1}{C_2}$

(C) $\frac{t_1}{t_2} = \frac{C_1}{C_2}$ (D) Cannot be determined **(I.I.T. 2006)**

Hint: $t_1 = \frac{1}{\lambda}\ln C_1$; $t_2 = \frac{1}{\lambda}\ln C_2$. Hence $t_1 - t_2 = \frac{1}{\lambda}\ln\frac{C_1}{C_2}$. Since due to nuclear explosion the initial activity A_0 would be greater, it would take more time to reach the same value of activity A.

ANSWERS

1. D	2. A, C, D	3. A	4. B	5. A, D	6. D
7. D	8. B, C	9. C	10. C	11. B	12. B
13. D	14. D	15. B	16. D	17. B	18. A
19. B	20. B	21. A, C	22. A	23. A	24. B
25. D	26. A	27. B	28. B	29. C	30. A, B, C
31. A	32. C	33. D	34. D	35. B	36. B
37. C	38. D	39. B	40. A	41. B	42. D
43. C	44. D	45. C	46. B	47. D	48. B
49. B	50. B	51. C	52. A	53. B	54. B
55. D	56. D	57. A, B, C	58. D	59. A	60. A, B, C
61. C	62. A	63. B	64. C	65. C	66. B
67. B	68. D	69. A	70. A	71. A	72. C
73. B	74. A	75. B, C, D	76. C	77. C	78. C
79. A, B, C, D	80. A	81. B	82. C	83. D	84. C
85. B	86. A	87. A	88. B	89. A	90. D
91. A	92. B	93. A	94. C	95. A	96. B
97. B	98. A, B	99. A	100. A	101. C	102. A
103. B	104. A	105. B	106. A, D	107. B, D	108. B
109. A	110. B	111. C	112. B	113. D	114. C
115. A	116. B	117. A	118. C	119. B	120. A

3

Chemical Bonding

OBJECTIVE TYPE MULTIPLE CHOICE QUESTIONS

1. The common features among the species CN^-, CO and NO^+ are
   (A) bond order three and isoelectronic
   (B) bond order two and π-acceptors
   (C) bond order three and weak field ligands
   (D) isoelectronic and weak field ligands. **(I.I.T. 2001)**
2. The hybridisation of the central atom in the species NH_3, $[PtCl_4]^{2-}$, PCl_5 and BCl_3 is
   (A) dsp^2, dsp^3, sp^2, sp^3 (B) sp^3, dsp^2, dsp^3, sp^2
   (C) dsp^2, sp^2, sp^3, dsp^3 (D) dsp^2, sp^3, sp^2, dsp^3 **(I.I.T. 2001)**
3. Molecular shapes of SF_4, CF_4 and XeF_4 are
   (A) the same with 2, 0, 1 lone pairs of electrons respectively
   (B) the same with 1, 1, 1 lone pairs of electrons respectively
   (C) different with 0, 1, 2 lone pairs of electrons respectively
   (D) different with 1, 0, 2 lone pairs of electrons respectively. **(I.I.T. 2000)**
4. Assertion (*A*): The electronic structure of O_3 is 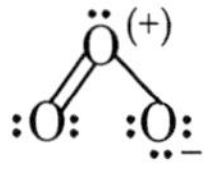
   Reason (*R*): 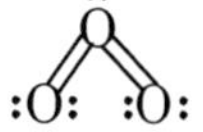 structure is not allowed because the octet around O cannot be expanded.
   (A) Both *A* and *R* are true and *R* is the correct explanation of *A*
   (B) Both *A* and *R* are true and *R* is not the correct explanation of *A*
   (C) *A* is correct but *R* is incorrect (D) *A* is incorrect but *R* is correct. **(I.I.T. 1998)**
5. The hydridization of N in NO_2^+, NO_3^- and NH_4^+ is respectively
   (A) sp, sp^2, sp^3 (B) sp, sp^3, sp^2 (C) sp^2, sp, sp^3 (D) sp^2, sp^3, sp **(I.I.T. 2000)**

6. Among the following the one that is polar and has the central atom with sp^2 hybridisation is
(A) H_2CO_3 (B) SiF_4 (C) BF_3 (D) $HClO_2$ **(I.I.T. 1997)**
7. Among the species NF_3, NO_3^-, BF_3, H_3O^+ and HN_3, identify the isostructural species.
(A) (NF_3, NO_3^-) and (BF_3, H_3O^+) (B) (NF_3, HN_3) and (NO_3^-, BF_3)
(C) (NF_3, H_3O^+) and (NO_3^-, BF_3) (D) (NF_3, H_3O^+) and (HN_3, BF_3) **(I.I.T. 1996)**
8. The number and type of bonds between two carbon atoms in CaC_2 are
(A) 1σ and 1π (B) 1σ and 2π (C) 1σ and 1.5π (D) 1σ **(I.I.T. 1996)**
9. The geometry and the type of hybrid orbitals present around B in BF_3 are
(A) pyramidal, sp^3 (B) tetrahedral, sp^3
(C) trigonal planar, sp^2 (D) linear, sp **(I.I.T. 1998)**
10. Among the following molecules, which one is planar?
(A) BCl_3 (B) SO_2Cl_2 (C) NH_3 (D) NF_3 **(I.I.T. 1994)**
11. Which of the following has zero dipole moment?
(A) ClF (B) PCl_3 (C) SiF_4 (D) $CFCl_3$ **(I.I.T. 1993)**
12. Which one of the following molecules is planar?
(A) NF_3 (B) NCl_3 (C) BF_3 (D) PH_3 **(I.I.T. 1996)**
13. Which of the following have identical bond order?
(A) CN^- (B) O_2^- (C) NO^+ (D) CN^+ **(I.I.T. 1992)**
14. The molecules that will have dipole moment are
(A) 2,2-dimethylpropane (B) *trans*-pent-2-ene
(C) *cis*-hex-3-ene (D) 2,2,3,3-tetramethylbutane **(I.I.T. 1992)**
15. Which of the following compounds has sp^2 hybridisation?
(A) CO_2 (B) SO_2 (C) N_2O (D) CO **(I.I.T. 1997)**
16. Which contains both polar and non-polar bonds?
(A) CH_4 (B) HCN (C) H_2O_2 (D) NH_4Cl **(I.I.T. 1997)**
17. Which of the following species is non-linear?
(A) H_2S (B) NH_3 (C) CO_2 (D) SO_2 **(I.I.T. 1995)**
18. The linear structure is assumed by
(A) $SnCl_4$ (B) NCO^- (C) CS_2 (D) NO_2^+ **(I.I.T. 1991)**
19. The combination of two AOs leads to the formation of
(A) two MOs (B) one MO (C) three MOs (D) four MOs **(I.I.T. 1990)**
20. The enolic form of acetone contains
(A) 9σ bonds, 1π bond and 2 lone pairs
(B) 8σ bonds, 2π bonds and 2 lone pairs
(C) 10σ bonds, 1π bond and 1 lone pair
(D) 9σ bonds, 2π bonds and 1 lone pair. **(I.I.T. 1990)**
21. The geometric form of crystals is the result of orderly arrangement of
(A) molecules only (B) ions only
(C) atoms only (D) molecules, atoms or ions. **(I.I.T. 1989)**

22. The molecule which has zero dipole moment is
(A) CH_2Cl_2 (B) BF_3 (C) NF_3 (D) ClO_2 **(I.I.T. 1989)**
23. Which of the following species is paramagnetic?
(A) O_2^- (B) CN^- (C) NO^+ (D) CO **(I.I.T. 1989)**
24. The molecule which has pyramidal shape is
(A) PCl_3 (B) SO_3 (C) CO_3^{2-} (D) NO_3^- **(I.I.T. 1989)**
25. The species in which the central atom uses sp^2 hybrid orbitals is
(A) PH_3 (B) NH_3 (C) H_3C^+ (D) SbH_3 **(I.I.T. 1988)**
26. The increasing order of bond dissociation energy is
(A) $F_2 < Cl_2 < O_2 < N_2$ (B) $F_2 < Cl_2 < N_2 < O_2$
(C) $Cl_2 < F_2 < O_2 < N_2$ (D) $N_2 < O_2 < F_2 < Cl_2$ **(I.I.T. 1988)**
27. The metallic lustre exhibited by sodium is explained by
(A) diffusion of sodium ions (B) excitation of free protons
(C) oscillation of loose electrons (D) existence of *bcc* lattice. **(I.I.T. 1987)**
28. Dipole moment is shown by
(A) 1,4-dichlorobenzene (B) *cis*-1,2-dichloroethene
(C) *trans*-1,2-dichloroethene (D) *trans*-1,2-dichloropent-2-ene **(I.I.T. 1986)**
29. The bond between two identical non-metal atoms has a pair of electrons
(A) unequally shared between the two
(B) transferred fully from one atom to another
(C) with identical spins (D) equally shared between them. **(I.I.T. 1986)**
30. Which of the following compounds has a linear structure?
(A) CCl_4 (B) SO_2 (C) C_2H_2 (D) C_2H_4 **(I.I.T. 1986)**
31. Which of the following is present in N_2O_3?
(A) Covalent bond (B) Ionic bond
(C) Ionic and covalent bonds (D) Covalent and coordinate bonds. **(I.I.T. 1986)**
32. Carbon dioxide is isostructural with
(A) $HgCl_2$ (B) $SnCl_2$ (C) C_2H_2 (D) NO_2 **(I.I.T. 1986)**
33. The molecule having one unpaired electron is
(A) NO (B) CO (C) CN^- (D) O_2 **(I.I.T. 1985)**
34. Oxygen molecule is paramagnetic because
(A) bonding electrons are more than antibonding electrons
(B) bonding electrons are less than antibonding electrons
(C) bonding electrons are equal to antibonding electrons
(D) it contains unpaired electrons. **(I.I.T. 1984)**
35. Carbon tetrachloride has no dipole moment because of
(A) its planar structure
(B) its tetrahedral structure
(C) similar electron affinities of C and Cl
(D) similar sizes of C and Cl **(I.I.T. 1983)**

36. Which among the following has no hydrogen bond?
(A) Liquid NH_3 (B) Liquid HF (C) Liquid HCl (D) Phenol **(I.I.T. 1983)**
37. The types of bonds present in $CuSO_4 \cdot 5H_2O$ are
(A) electrovalent and covalent (B) covalent and coordinate covalent
(C) electrovalent, covalent and coordinate covalent
(D) electrovalent and coordinate covalent. **(I.I.T. 1983)**
38. Among the following molecules that is linear is
(A) CO_2 (B) NO_2 (C) SO_2 (D) ClO_2 **(I.I.T. 1982, 1988)**
39. The ion that is isoelectronic with CO is
(A) CN^- (B) O_2^+ (C) O_2 (D) N_2^+ **(I.I.T. 1982)**
40. The compound with no dipole moment is
(A) CH_3Cl (B) CCl_4 (C) CH_2Cl_2 (D) $CHCl_3$ **(I.I.T. 1982)**
41. The bond order of individual carbon-carbon bonds in benzene is
(A) one (B) two
(C) between one and two (D) one and two alternatively. **(I.I.T. 1981)**
42. If a molecule MX_3 has zero dipole moment, the σ bonding orbitals used by M (At. No. < 21) are
(A) pure *p* (B) *sp* hybrid (C) sp^2 hybrid (D) sp^3 hybrid **(I.I.T. 1981)**
43. The angle between two covalent bonds is maximum in
(A) H_2O (B) CO_2 (C) NH_3 (D) CH_4 **(I.I.T. 1981)**
44. Element X is strongly electropositive and element Y is strongly electronegative. Both are univalent. The compound formed would be
(A) X^+Y^- (B) X^-Y^+ (C) X—Y (D) $X \rightarrow Y$ **(I.I.T. 1981)**
45. Which of the following is soluble in water?
(A) CS_2 (B) C_2H_5OH (C) CCl_4 (D) $CHCl_3$ **(I.I.T. 1980)**
46. The total number of electrons that take part in forming bonds in N_2 is
(A) 2 (B) 4 (C) 6 (D) 10 **(I.I.T. 1980)**
47. Which of the following compounds is covalent?
(A) H_2 (B) CaS (C) KCl (D) Na_2S **(I.I.T. 1980)**
48. The octet rule is not valid for the molecule
(A) CO_2 (B) SO_2 (C) O_2 (D) $BeCl_2$ **(I.I.T. 1979)**
49. The compound which contains both ionic and covalent bonds is
(A) CH_4 (B) H_2 (C) KCN (D) KCl **(I.I.T. 1979)**
50. The C—H bond distance is the longest in
(A) C_2H_2 (B) C_2H_4 (C) $C_2H_2Br_2$ (D) C_2H_6 **(I.I.T. 1989)**
51. Which of the following has the greatest bond length?
(A) F_2 (B) N_2 (C) Cl_2 (D) O_2 **(I.I.T. 1985)**
52. Pick up the isoelectronic structures from the following: (i) CH_3^+ (ii) H_3O^+ (iii) NH_3 (iv) CH_3^-
(A) (i), (ii) (B) (iii), (iv) (C) (ii), (iii), (iv) (D) (i), (iii) **(I.I.T. 1983)**

53. On hybridisation of one s and one p orbitals, we get
(A) two mutually perpendicular orbitals
(B) two orbitals at 180°
(C) four orbitals directed tetrahedrally
(D) three orbitals in a plane. **(I.I.T. 1984)**

54. The compound in which C uses sp^3 hybrid orbitals for bond formation is
(A) $H\overset{*}{C}OOH$ (B) $(NH_2)_2\overset{*}{C}O$ (C) $(CH_3)_3\overset{*}{C}OH$ (D) $CH_3\overset{*}{C}HO$ **(I.I.T. 1989)**

55. The hybridisation present in OF_2 is
(A) sp (B) sp^2 (C) sp^3 (D) sp^3d **(Roorkee 1996)**

56. sp^3 hybridisation is present in
(A) SO_4^{2-} (B) C_2H_2 (C) BO_3^{3-} (D) NH_4^+ **(Roorkee 1996)**

57. Which type of bond is not present in HNO_2?
(A) Ionic and coordinate (B) Covalent
(C) Coordinate (D) Ionic **(Roorkee 1991)**

58. All bonds in benzene are equal due to
(A) tautomerism (B) isomerism (C) resonance (D) I-effect **(Roorkee 1990)**

59. Which of the following compounds has the largest dipole moment?
(A) CH_3OH (B) CH_3F (C) CH_4 (D) CF_4 **(Roorkee 1997)**

60. NH_3 and BF_3 form adduct readily through
(A) ionic bond between BF_3 and NH_3 (B) coordinate bond between B and N
(C) covalent bond between B and N
(D) H-bonds between F atoms of BF_3 and H-atoms of NH_3 **(Roorkee 1998)**

61. Which of the following has the smallest bond order?
(A) O_2^+ (B) O_2^{2+} (C) O_2^- (D) O_2

62. A compound is called covalent when
(A) its covalent character is 100%
(B) its covalent character is 50% and electrovalent character is 50%
(C) its covalent character is greater than 50% and electrovalent character is less than 50%
(D) its covalent character is greater than 75% and electrovalent character is less than 25%

63. The decreasing order of bond dissociation energy of O_2, O_2^+, O_2^- and O_2^{2-} is
(A) $O_2^- > O_2^{2-} > O_2 > O_2^+$ (B) $O_2^+ > O_2 > O_2^- > O_2^{2-}$
(C) $O_2 > O_2^+ > O_2^{2-} > O_2^-$ (D) $O_2^- > O_2^{2-} > O_2^+ > O_2$

64. In which of the following the bond length is greatest?
(A) O_2 (B) O_2^- (C) O_2^+ (D) O_2^{2-}

65. A σ bond is formed by the overlap of orbitals
(A) s–s (B) p_x–p_x (C) p_y–p_y (D) p_z–p_z

66. A π bond is formed by the overlap of the orbitals
   (A) s–s (B) p_x–p_x (C) p_y–p_y (D) p_z–p_z
67. The molecule which possesses both sp^3 and sp^3d^2 hybridisation is
   (A) $PCl_5(g)$ (B) $PCl_5(s)$ (C) PCl_6 (D) none of these.
68. Identify the incorrect statement.
   (A) Solubility of n-alcohols in water decreases with increase in M.W.
   (B) p-Hydroxybenzaldehyde is more soluble in water than o-hydroxybenzaldehyde
   (C) The valence atomic orbitals on carbon in silver acetylide is sp hybridized
   (D) The presence of polar bonds in a polyatomic molecule suggests that the molecule has non-zero dipole moment.
69. Identify the incorrect statement.
   (A) All molecules with polar bonds have dipole moment.
   (B) ∠H—N—H in NH_3 is greater than the ∠H—As—H in AsH_3
   (C) Linear overlap of two p-orbitals leads to a σ bond
   (D) The order of increasing bond length in H_2, O_2, F_2 and N_2 is: $N_2 < O_2 < F_2 < H_2$
70. Identify the incorrect statement.
   (A) Hydrogen bond is a chemical bond
   (B) There are two π bonds in N_2
   (C) The hybridisation of O in H_2O is sp^3
   (D) Solid NaCl is a bad conductor of electricity.
71. Identify the incorrect statement.
   (A) The bond order of N_2^{2-} is 2.5 (B) The bond order of O_2^- is 2.5
   (C) The bond order of O_2^+ is 2.5
   (D) Among N_2, N_2^+, N_2^{2-} and N_2^-, the highest paramagnetism is shown by N_2^{2-}.
72. Identify the incorrect statement.
   (A) $[Ni(CO)_4]$ is soluble in non-polar solvents and is insoluble in polar solvents
   (B) $[Cu(NH_3)_4]Cl_2$ is soluble in both polar and non-polar solvents
   (C) O_2^{2-} is paramagnetic
   (D) The bond dissociation energy of hydrogen bond is around 4–42 kJ/mol.
73. Identify the incorrect statement.
   (A) The number of electrons in the antibonding molecular orbitals of N_2^{2-} is six
   (B) A π molecular orbital is weaker than a π molecular orbital
   (C) H_2^+ and H_2^- are diamagnetic
   (D) The bond length in H_2^+ is longer than that in H_2.
74. Identify the incorrect statement.
   (A) ClO_2 is paramagnetic (B) Metallic bond is a chemical bond
   (C) The bond dissociation energy of a π bond is less than that of a σ bond
   (D) The bond dissociation energy of a H-bond is less than that of a metallic bond.

75. Identify the incorrect statement.
   (A) The number of sigma bonds in C_2H_4 is five
   (B) Out of CO_2, H_2O and CH_4, the angle between two covalent bonds is minimum in CH_4
   (C) Oxygen can exhibit positive oxidation state only in its fluoride
   (D) sp^3d^3 hybridisation represents pentagonal bipyramidal structure.
76. Identify the incorrect statement.
   (A) The conversion of Na to Na^+ is an endothermic process.
   (B) The experimentally determined N—F bond length in NF_3 is greater than the sum of the single bond covalent radii of N and F
   (C) Maximum polarisation is brought about by a cation of high charge
   (D) Anhydrous $FeCl_3$ cannot exist in the dimeric form.
77. Identify the incorrect statement.
   (A) SF_4 possesses the bond angles of 90°, 102° and 173°
   (B) Solid PCl_5 possesses both sp^3 and sp^3d^2 hybridization
   (C) The O—H bond is weaker than O—D bond
   (D) $MnCl_2$ is linear while $SnCl_2$ is angular.
78. Identify the incorrect statement.
   (A) Four sp^3 hybrid orbitals of N atom are involved in the formation of NH_4^+ ion
   (B) The dipole moment of CH_3OH is higher than that of CH_3SH
   (C) The shape of $[CH_3]^+$ is planar
   (D) In benzene, carbon atom uses all the three *p*-orbitals for hybridisation.
79. Identify the incorrect statement.
   (A) Decrease in bond length means increase in bond strength
   (B) The compound propanediene has both *sp* and sp^2 hybridised C atoms
   (C) $SnCl_2$ is a linear molecule
   (D) sp^2 hybrid orbitals have equal *s* and *p* character.
80. Which of the following is isoelectronic with CN^-?
   (A) O_2^- (B) O_2^{2-} (C) F_2 (D) Si
81. O_2 and N_2 are converted to O_2^+ and N_2^+. Which is wrong?
   (A) In O_2^+, the paramagnetism decreases
   (B) N_2^+ becomes diamagnetic
   (C) In N_2^+, the N—N bond weakens (D) In O_2^+, the O—O bond order increases.
82. The structure of ICl_2^- is
   (A) square planar (B) octahedral
   (C) trigonal pyramidal (D) linear
83. Although in the MO diagram of N_2, there is appreciable σ_s–σ_z interaction, there is no appreciable σ_s–σ_z interaction in the MO diagram of O_2 because
   (A) with the increase in atomic number the difference in energy between σ_s and σ_z orbitals increases
   (B) oxygen is more electronegative than nitrogen

(C) the symmetry of σ_s and σ_z orbitals is different and as a result these orbitals cannot overlap with each other

(D) none of these.

84. In the MO diagram of N_2, the energy of the σ_z^b orbital is higher than that of the degenerate orbitals π_x^b and π_y^b because

(A) in N_2, the difference in energy between the σ_s and σ_z orbitals is small and the repulsive forces between the electrons present in them raise the energy of the σ_z^b orbital above that of π_x^b, π_y^b orbitals

(B) nitrogen has the stable half-filled *p*-orbitals

(C) nitrogen is less electronegative than oxygen

(D) the electron affinity of nitrogen is remarkably less than that expected.

85. Which of the following properties of glucose cannot be attributed to the presence of hydrogen bonds between the molecules

(A) soluble in water (B) solid at room temperature

(C) high latent heat of fusion (D) high enthalpy of combustion.

86. Which of the following is the correct bond order?

(A) $O_2^+ > O_2 > O_2^- > O_2^{2-}$ (B) $O_2^+ > O_2^- > O_2 > O_2^{2-}$

(C) $O_2 > O_2^- > O_2^+ > O_2^{2-}$ (D) $O_2^- > O_2^+ > O_2 > O_2^{2-}$

87. Which of the following has the highest magnetic moment?

(A) NO^+ (B) CN^- (C) Cl_2^+ (D) CO

88. Which of the following has the highest single bond dissociation energy?

(A) F—F (B) C—C (C) N—N (D) O—O

89. The decreasing order of O—O bond length in O_3, O_2 and H_2O_2 is

(A) $O_2 > H_2O_2 > O_3$ (B) $H_2O_2 > O_3 > O_2$

(C) $O_3 > H_2O_2 > O_2$ (D) $O_2 > O_3 > H_2O_2$

90. In the coordination compound [Cu(ethylenediamine)$_2$]Cl_2, the unpaired electron of Cu(II) is

(A) localised (B) retained completely by Cu(II)

(C) transferred to the ligand atom

(D) partially delocalised in the chelate ring system.

91. Which of the following does not possess tetrahedral structure?

(A) SO_2Cl_2 (B) SiO_4^{2-} (C) SO_4^{2-} (D) SF_4

92. The hybridisation of ClF_3 is sp^3d and the Cl atom has two lone pairs of electrons. Which of the following is correct?

(A) Two lone pairs of electrons are situated in the *z*-axis

(B) Two lone pairs of electrons are situated in the trigonal plane

(C) One lone pair of electrons is in the trigonal plane and the other lone pair of electrons is situated in the *z*-axis

(D) None of these.

93. The decreasing order of bond dissociation energy of N_2, N_2^+, N_2^- and N_2^{2-} is

(A) $N_2 > N_2^+ = N_2^- > N_2^{2-}$ (B) $N_2 > N_2^+ > N_2^- > N_2^{2-}$
(C) $N_2 > N_2^- > N_2^+ > N_2^{2-}$ (D) $N_2 > N_2^- > N_2^{2-} > N_2^+$

94. Which of the following exhibits the highest magnetic moment?
(A) CO (B) NO^+ (C) CN^- (D) Cl_2^+

95. Which electron is removed in the formation of N_2^+ from N_2?
(A) a π^b orbital (B) a π^* orbital (C) a σ^b orbital (D) a σ^* orbital

96. The hybridisation of I in ICl_2^+ is
(A) *sp* (B) sp^2 (C) sp^3 (D) dsp^2

97. The hybridisation of N in piperidine 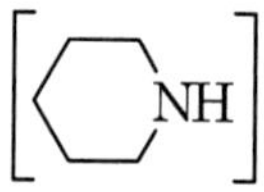 is
(A) *sp* (B) sp^2 (C) sp^3 (D) dsp^2

98. Born–Haber cycle is used to determine
(A) nitrogen cycle (B) lattice energy
(C) electronegativity (D) both (B) and (C)

99. Which of the following possesses the lowest bond angle around the central atom?
(A) I_3^- (B) H_2S (C) O_3 (D) NO_2^-

100. Identify the incorrect order of boiling point.
(A) $CH_4 < GeH_4 < SiH_4$ (B) $Br_2 > Cl_2 > F_2$
(C) $H_2Te > H_2Se > H_2S$ (D) HI > HBr > HCl

101. Identify the correct statements.
1. CH_4 and SeF_4 have identical shape
2. Electron travels at the speed of light
3. $p\pi$–$d\pi$ bonding is present in SO_2
4. $BiCl_5$ and PH_5 do not exist 5. I_3^+ has angular geometry
(A) 1, 3, 5 (B) 2, 3, 5 (C) 1, 2, 4 (D) 3, 4, 5

102. The bond axis of a diatomic molecule is *z*. The overlap of p_x orbital with p_y orbital will lead to the formation of
(A) π molecular orbital (B) σ molecular orbital
(C) δ molecular orbital (D) no bond will be formed.

103. Which of the following is isostructural?
(A) XeF_2, IF_2^- (B) NH_3, BF_3 (C) CO_3^{2-}, SO_3^{2-} (D) PCl_5, ICl_5

104. Which of the following compounds possesses the lowest dipole moment value?
(A) CO (B) NH_3 (C) SO_2 (D) NF_3

105. The N—O bond length is maximum in
(A) NO (B) NO_2^- (C) NO_3^- (D) N_2O_4

106. Which of the following is the correct order of decreasing bond length?
(A) Cl—O > S—O > P—O (B) S—O > Cl—O > P—O
(C) P—O > S—O > Cl—O (D) P—O > Cl—O > S—O

107. Identify the correct order of decreasing bond strength in the bonds formed by 2*s*–2*s*, 2*p*–2*p* and 2*s*–2*p* overlaps.

(A) $2s–2s > 2s–2p > 2p–2p$ (B) $2s–2s > 2p–2p > 2s–2p$
(C) $2p–2p > 2s–2p > 2s–2s$ (D) $2p–2p > 2s–2s > 2s–2p$

108. The phosphorus exhibits sp^3d hybridisation in PCl_5(g). Which *d*-orbital is involved in this hybridisation?
(A) d_{xy} (B) d_{xz} (C) $d_{x^2-y^2}$ (D) d_{z^2}

109. Which of the following has the lowest bond dissociation energy?
(A) Cl—Cl (B) O—O (C) C—Cl (D) C—O

110. Which of the following has the lowest bond dissociation energy?
(A) N—H (B) P—H (C) As—H (D) Sb—H

111. The ∠N—O—N bond angle is maximum in
(A) N_2O_3 (B) NO_2^- (C) NO_2 (D) NO_2^+

112. Amongst BO_3^{3-}, SO_3^{2-}, ClO_3^-, CO_3^{2-}, AsO_3^{3-} and NO_3^-, the non-planar species are
(A) BO_3^{3-}, SO_3^{2-}, NO_3^- (B) BO_3^{3-}, CO_3^{2-}, NO_3^-
(C) SO_3^{2-}, ClO_3^-, AsO_3^{3-} (D) BO_3^{2-}, SO_3^{2-}, CO_3^{2-}

113. Amongst O_3, SO_2, N_2O_5, ClO_2, CO and NO, the pair of molecules which have odd electrons are
(A) ClO_2, NO (B) O_3, SO_2 (C) CO, ClO_2 (D) SO_2, CO

114. In solid ammonia, each NH_3 molecule has six other NH_3 molecules as the nearest neighbours. At the melting point ΔH of sublimation of ammonia is 30.6 kJ mol^{-1}. The estimated ΔH of sublimation in the absence of H-bonding is 14.4 kJ mol^{-1}. The energy of one hydrogen bond in solid ammonia is
(A) 8.1 kJ mol^{-1} (B) 2.7 kJ mol^{-1} (C) 5.4 kJ mol^{-1} (D) 16.2 kJ mol^{-1}
Hint: Energy of six H-bonds = 30.6 – 14.4 = 16.2 kJ mol^{-1}. Energy of one H-bond = 16.2/6 = 2.7 kJ mol^{-1}

115. Specify the coordination geometry around and hybridization of N and B atoms in 1 : 1 complex of BF_3 and NH_3.
(A) N: tetrahedral, sp^3, B: tetrahedral, sp^3
(B) N: pyramidal, sp^3, B: pyramidal, sp^3
(C) N: pyramidal, sp^3, B: planar, sp^2
(D) N: pyramidal, sp^3, B: tetrahedral, sp^3 **(I.I.T. 2002)**

116. Which of the following species has unpaired electron(s)?
(A) N_2 (B) F_2 (C) O_2^- (D) O_2^{2-} **(I.I.T. 2002)**

117. Which of the following has tetrahedral structure and the central atom is sp^3 hybridised?
(A) BF_4^- (B) IF_4^- (C) SiH_4 (D) PF_4^-

118. Which of the following molecules has zero dipole moment?
(A) NaH (B) SO_2 (C) CO (D) C_6H_6

119. Which of the following molecules has a bond angle smaller than tetrahedral bond angle?
(A) NH_4^+ (B) PF_5 (C) SF_6 (D) H_2O

120. Identify the species which have eight electrons in the valence shell.
(A) BF_4^- (B) C_2H_4 (C) NF_3 (D) SiF_4

121. Identify the species which are non-planar.
(A) CH_3^+ (B) NH_4^+ (C) BF_4^- (D) C_2H_4

122. Identify the species which are isostructural with IF_4^-.
(A) $[Ni(CN)_4]^{2-}$ (B) $[Cu(NH_3)_4]^{2+}$ (C) BF_4^- (D) XeF_4

123. Which of the following ions have identical bond order?
(A) CN^+ (B) CN^- (C) NO^+ (D) O_2^-

124.Which of the following species have bent structure and the central atomsp^3 hybridised?
(A) NH_2^- (B) H_2O (C) $SnCl_2$ (D) SO_2

125. Identify the compounds in which the bonds are both ionic and covalent and the central atom is sp^2 hybridised.
(A) $Fe_2(SO_4)_3$ (B) $NaBH_4$ (C) $CaCO_3$ (D) $NaNO_3$

126. Identify the triad in which all the species have same geometry.
(A) NO_3^-, CH_3^+, CO_3^{2-} (B) NH_3, CH_3^-, H_3O^+
(C) NO_2, SO_2, CO_2 (D) CH_3^+, BF_3, NF_3

127. Which of the following are isoelectronic and isostructural?NO_3^-, CO_3^{2-}, ClO_3^-, SO_3
(A) SO_3, NO_3^- (B) NO_3^-, CO_3^{2-} (C) ClO_3^-, CO_3^{2-} (D) CO_3^{2-}, SO_3 **(I.I.T. 2003)**

128. According to MO theory
(A) O_2^+ is paramagnetic and bond order is greater than O_2
(B) O_2^+ is paramagnetic and bond order is less than O_2
(C) O_2^+ is diamagnetic and bond order is more than O_2
(D) O_2^+ is diamagnetic and bond order is less than O_2 **(I.I.T. 2004)**

129. Which of the following contains the maximum number of lone pairs on the central atom?
(A) ClO_3^- (B) XeF_4 (C) SF_4 (D) I_3^- **(I.I.T. 2005)**

Hint:

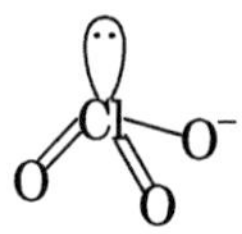
(One lone pair)

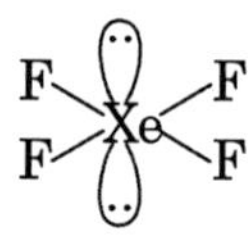
(Two lone pairs)

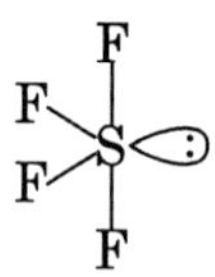
(One lone pair)

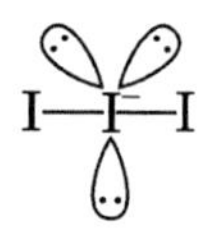
(Three lone pairs)

130. Which of the following diatomic molecules has the $E(\pi_{2p})$ greater than $E(\sigma_{2p})$?
(A) O_2 (B) N_2 (C) C_2 (D) B_2

131. Among the following the paramagnetic compound is
(A) Na_2O_2 (B) O_3 (C) N_2O (D) KO_2 **(I.I.T. 2007)**

Hint: $O_2^{2-} : (\sigma_{1s})^2 (\sigma_{1s}^*)^2 (\sigma_{2s})^2 (\sigma_{2s}^*)^2 (\sigma_{2p_z})^2 (\pi_{2p_x})^2 (\pi_{2p_y})^2 (\pi_{2p_x}^*)^2 (\pi_{2p_y}^*)^2$; No. of unpaired electrons = 0

N═N→O; No. of unpaired electron = 0

$O{=}\overset{+}{O}{-}\overset{-}{O} \leftrightarrow O{=}O{\rightarrow}O$; No. of unpaired electron = 0

$O_2^- : (\sigma_{1s})^2(\sigma_{1s}^*)^2(\sigma_{2s})^2(\sigma_{2s}^*)^2(\sigma_{2p_z})^2(\pi_{2p_x})^2(\pi_{2p_y})^2(\pi_{2p_x}^*)^2(\pi_{2p_y}^*)^1$; No. of unpaired electrons = 1; O_2^- is paramagnetic.

132. The species having bond order different from that in CO is

(A) NO^- (B) NO^+ (C) CN^- (D) N_2 **(I.I.T. 2007)**

Hint: NO^- is 16 electron system and bond order in NO^- is 2. NO^+, CN^- and N_2 are isoelectronic with CO and they all have the same bond order of 3.

133. The bond energy (in kcal mol^{-1}) of a C—C single bond is approximately

(A) 1 (B) 10 (C) 100 (D) 1000 **(I.I.T. 2010)**

Hint: The bond energy of a C—C single bond is approximately 100 kcal mol^{-1}.

134. The species having pyramidal shape is

(A) SO_3 (B) BrF_3 (C) SiO_3^{2-} (D) OSF_2 **(I.I.T. 2010)**

Hint: 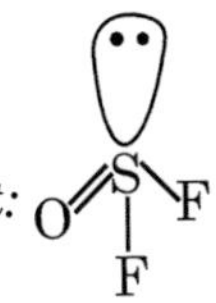

Trignonal pyramid

ANSWERS

1. A	2. B	3. D	4. A	5. A	6. A
7. C	8. B	9. C	10. A	11. C	12. C
13. A, C	14. B, C	15. B	16. C	17. A, B, D	18. B, C
19. A	20. A	21. C	22. B	23. A	24. A
25. C	26. A	27. C	28. B, D	29. D	30. C
31. D	32. A, C	33. A	34. D	35. B	36. C
37. C	38. A	39. A	40. B	41. C	42. C
43. B	44. A	45. B	46. C	47. A	48. D
49. C	50. D	51. C	52. B, C	53. B	54. C
55. C	56. A, D	57. A	58. C	59. B	60. B
61. C	62. C	63. B	64. D	65. A, D	66. B, C
67. B	68. D	69. A, D	70. A	71. B	72. C
73. C	74. B	75. B	76. D	77. C	78. D
79. C, D	80. D	81. B	82. D	83. A	84. A
85. D	86. A	87. C	88. B	89. B	90. D
91. D	92. B	93. A	94. D	95. C	96. C

97. C	98. B	99. B	100. A	101. D	102. D
103. A	104. A	105. C	106. C	107. C	108. D
109. B	110. D	111. D	112. C	113. A	114. B
115. A	116. C	117. A, C, D	118. D	119. B, C, D	120. A, C, D
121. B, C	122. A, B, D	123. B, C	124. A, B	125. C, D	126. A, B
127. B	128. A	129. D	130. A	131. D	132. A
133. C	134. D				

4

Redox Reactions

OBJECTIVE TYPE MULTIPLE CHOICE QUESTIONS

1. Identify the species with an atom in + 6 oxidation state.
   (A) CrO_2Cl_2 (B) $Cr(CN)_6^{3-}$ (C) MnO_4^- (D) NiF_6^{2-} **(I.I.T. 2000)**
2. Number of moles of $KMnO_4$ that is needed to react with one mole of FeC_2O_4 in acidic medium is
   (A) 2/5 (B) 3/5 (C) 4/5 (D) 1 **(I.I.T. 1997)**
3. Number of moles of $KMnO_4$ that is needed to react with one mole of sulphite ion in acidic medium is
   (A) 2/5 (B) 3/5 (C) 4/5 (D) 1 **(I.I.T. 1997)**
4. For the redox reaction, $MnO_4^- + C_2O_4^{2-} + H^+ \longrightarrow Mn^{2+} + CO_2 + H_2O$, the correct coefficients of the reactants for the balanced reaction are

	MnO_4^-	$C_2O_4^{2-}$	H^+		MnO_4^-	$C_2O_4^{2-}$	H^+
(A)	2	5	16	(B)	16	5	2
(C)	5	16	2	(D)	2	16	5

(I.I.T. 1986, 1993)

5. The oxidation states of the most electronegative element in the products of the reaction of BaO_2 with dil. H_2SO_4 are
   (A) 0 and – 1 (B) – 1 and – 2 (C) – 2 and 0 (D) – 2 and + 1 **(I.I.T. 1992)**
6. The number of electrons to balance the equation, $NO^- + 4H^+ + e^- \longrightarrow NO + 2H_2O$ is
   (A) 2 (B) 3 (C) 4 (D) 5 **(I.I.T. 1991)**
7. The equivalent mass of $MnSO_4$ is half of the molecular mass when it is converted to
   (A) Mn_2O_3 (B) MnO_2 (C) MnO_4^- (D) MnO_4^{2-} **(I.I.T. 1988)**
8. The brown ring compound is formulated as $[Fe(H_2O)_5(NO)]SO_4$. The oxidation state of iron is
   (A) 1 (B) 2 (C) 3 (D) 0 **(I.I.T. 1998)**

9. The increasing order of the oxidation number of iodine in the following compounds is
(A) $HI < I_2 < ICl < HIO_4$ (B) $HI < I_2 < HIO_4 < ICl$
(C) $I_2 < HI < ICl < HIO_4$ (D) $ICl < I_2 < HI < HIO_4$ **(I.I.T. 1986)**

10. Zinc-copper couple used as a reducing agent is prepared by
(A) mixing Zn dust with Cu gauze (B) coating Zn with Cu
(C) coating Cu with zinc (D) welding together Zn and Cu **(I.I.T. 1984)**

11. The oxidation number of Mn in $MnSO_4$ is
(A) + 5 (B) + 7 (C) + 4 (D) + 2 **(I.I.T. 1982)**

12. The oxidation number of carbon in CH_2O is
(A) – 2 (B) + 2 (C) 0 (D) + 4 **(I.I.T. 1982)**

13. One mole of N_2H_4 loses ten moles of electrons to form a new compound Y. Assuming that all the nitrogen appears in the new compound, what is the oxidation state of nitrogen in Y? There is no change in the oxidation state of hydrogen.
(A) – 1 (B) – 3 (C) + 3 (D) + 5 **(I.I.T. 1981)**

14. White P reacts with NaOH and the products are PH_3 and NaH_2PO_2. This reaction is an example of
(A) oxidation (B) reduction
(C) disproportionation (D) neutralisation **(I.I.T. 1980)**

15. The compound which could not act both as oxidising as well as reducing agent is
(A) SO_2 (B) MnO_2 (C) Al_2O_3 (D) CrO **(I.I.T. 1991)**

16. The equivalent weight of $Na_2S_2O_3$ in its reaction with I_2 is equal to
(A) M.W. (B) M.W./2 (C) M.W./3 (D) M.W./4

17. In the reaction $Cr_2O_7^{2-} + 3Sn^{2+} + 14H^+ \longrightarrow 2Cr^{3+} + 3Sn^{4+} + 7H_2O$, how many moles of $K_2Cr_2O_7$ are reduced by 1 mole of Sn^{2+}?
(A) 1/3 (B) 1/6 (C) 3 (D) 6

18. $KMnO_4$ acts as an oxidant in neutral, alkaline and acidic media. The final product obtained in the three conditions is respectively
(A) MnO_4^{2-}, Mn^{3+}, Mn^{2+} (B) MnO_2, MnO_2, Mn^{2+}
(C) MnO_2, MnO_2, Mn^{3+} (D) MnO, MnO_2, Mn^{2+}

19. Which of the following statements is incorrect?
(A) $KMnO_4$ is a powerful oxidising agent
(B) $KMnO_4$ is a weak oxidising agent than $K_2Cr_2O_7$
(C) $KMnO_4$ is a stronger oxidising agent than $K_2Cr_2O_7$
(D) $K_2Cr_2O_7$ oxidises secondary alcohol to ketone.

20. The partial ionic equation $Cr_2O_7^{2-} + 14H^+ + 6e^- \longrightarrow 2Cr^{3+} + 7H_2O$ suggests that the equivalent weight of $Cr_2O_7^{2-}$ is equal to its molecular weight divided by
(A) 3 (B) 6 (C) 14 (D) 1

21. In the reaction, $SO_2 + 2H_2S \longrightarrow 3S + 2H_2O$, which choice has value twice that of equivalent weight of oxidising agent?
(A) 64 (B) 32 (C) 48 (D) 16

22. The oxidation state of nitrogen is incorrectly given for
    (A) NH_2OH – 1 (B) Mg_3N_2 – 3
    (C) $(N_2H_5)_2SO_4$ – 2 (D) $[Co(NH_3)_5Cl]Cl_2$ – 2
23. Oxygen has an oxidation state of + 2 in
    (A) H_2O (B) OF_2 (C) H_2O_2 (D) SO_2
24. The oxidation number of Fe in Fe_3O_4 is
    (A) 8/3 (B) 3/2 (C) 5/4 (D) 4/5
25. Identify the incorrect statement.
    (A) In acidic solution, dichromate ions are converted into chromate ions
    (B) $(NH_4)_2Cr_2O_7$ on heating undergoes exothermic decomposition to give Cr_2O_3
    (C) In acidified solution $K_2Cr_2O_7$ liberates iodine from iodides
    (D) $K_2Cr_2O_7$ is used as a titrant for Fe^{2+} ions.
26. In the reaction, $3Cl_2 + 6OH^- \longrightarrow 5Cl^- + ClO_3^- + 3H_2O$, chlorine is
    (A) reduced (B) oxidised
    (C) neither oxidised nor reduced (D) oxidised as well as reduced.
27. The oxidation number of S in $Na_2S_4O_6$ is
    (A) 2 and 4 (three S have + 2, one S has + 4)
    (B) 2.5
    (C) 5 and 0 (two S have + 5, two S have 0)
    (D) 2 and 3 (two S have + 2, two S have + 3)
28. Identify the incorrect statement.
    (A) Mn^{2+} can be oxidised to MnO_4^- by SnO_2
    (B) HCl can reduce $KMnO_4$
    (C) In the thermit process Al is used as a reducing agent
    (D) Nitrogen exhibits the lowest oxidation state of – 3
29. Identify the incorrect statement.
    (A) The oxidation state of P varies from – 3 to + 5
    (B) The oxidation state of S in $S_2O_8^{2-}$ is + 6
    (C) The oxidation No. of O is – 1/2 in superoxides
    (D) HF can reduce $KMnO_4$
30. Which of the following metals exhibit more than one oxidation state?
    (A) Mg (B) Fe (C) Al (D) Na
    Hint: The electrolysis of acidulated water with Mg electrodes is not in conformity of Faraday's law and indicates the presence of both Mg^+ and Mg^{2+} ions.
31. In the reaction, $H_3PO_4 + Ca(OH)_2 \longrightarrow CaHPO_4 + 2H_2O$, the equivalent mass of H_3PO_4 is
    (A) 49.0 (B) 32.7 (C) 196.0 (D) 98.0
32. The oxidation number of Co in $K[Co(CO)_4]$ is
    (A) + 1 (B) – 1 (C) + 3 (D) – 3

33. The oxidation number of oxygen in H_2O_2 is
(A) + 2 (B) – 2 (C) – 1 (D) + 1
34. In the reaction, $2Ag + 2H_2SO_4 \longrightarrow Ag_2SO_4 + 2H_2O + SO_2$, H_2SO_4 acts as
(A) catalyst (B) oxidising agent
(C) reducing agent (D) acid and oxidant.
35. Which of the following involves a redox reaction?
(A) Reaction of H_2SO_4 and NaOH (B) Evaporation of water
(C) Production by lightning of O_3 from O_2 in the atmosphere
(D) Production by lightning of nitrogen oxides from N_2 and O_2 in the atmosphere.
36. Identify the incorrect statement.
(A) The reaction, $2H_2O_2 \longrightarrow 2H_2O + O_2$ is a disproportionation reaction
(B) The reaction, $2S_2O_8^{2-} + 2H_2O \longrightarrow 4SO_4^{2-} + 2H^+ + O_2$ is not a redox reaction
(C) The equivalent mass of a compound can be calculated without considering the reaction it undergoes
(D) The disproportionation reaction, $2Mn^{3+} + 2H_2O \longrightarrow MnO_2 + 4H^+ + Mn^{2+}$ is a redox reaction.
37. Identify the incorrect statement.
(A) The oxidation number can never be a fraction
(B) The equivalent mass of $KMnO_4$ in strongly alkaline medium is M.W./2
(C) The equivalent mass of $Na_2S_2O_8$ would be M.W./2
(D) In the reaction of hydrazine with I_2, nitrogen(*g*) is evolved. The equivalent mass of hydrazine in this reaction is M.W./8.
38. Identify the incorrect statement.
(A) The oxidation state of sulphur in SCN^- is – 1
(B) 1 mole of $K_4[Fe(CN)_6]$ contains a total 5 moles of independent species
(C) Dilute HCl oxidises Fe to Fe^{3+}
(D) Sulphur dioxide is a more powerful oxidising agent in alkaline medium than in acidic medium.
39. Identify the incorrect statement.
(A) The compound $XBa_2Cu_3O_7$ which exhibits superconductivity has Cu in the oxidation state 8/3 (the oxidation state of X = + 3)
(B) The oxidation number of sulphur in Marshall's acid is + 6
(C) The reaction, $2CuSO_4 + 10KCN \longrightarrow 2K_3[Cu(CN)_4] + 2K_2SO_4 + (CN)_2$ is a redox reaction
(D) Disproportionation reaction is a special case of redox reaction.
40. Identify the incorrect statement.
(A) The oxidation number of vanadium in $Rb_4K[HV_{10}O_{28}]$ is + 4
(B) Sodium nitrite can act as both oxidising agent and reducing agent
(C) Mn^{2+} is oxidised to MnO_4^- by SnO_2
(D) The oxidation number of N in N_3H is – 1/3.

41. A compound (X) contains the atoms A, B and C having oxidation number + 2, + 5 and – 2 respectively. The formula of X is
(A) ABC_2 (B) $A_3(BC_4)_2$ (C) $A_3(B_4C)_2$ (D) $B_2(AC_3)_2$

42. In which of the following reactions SO_2 acts as an oxidising agent, while reacting with
(A) acidified C_2H_5OH (B) acidified $KMnO_4$
(C) acidified $K_2Cr_2O_7$ (D) H_2S

43. Identify the incorrect statement.
(A) The oxidation number of iron in $Na_2[Fe(NO)(CN)_5]\cdot 2H_2O$ is + 3
(B) In the reaction, $C + H_2O \longrightarrow CO + H_2$, H_2O acts as an oxidising agent
(C) The oxidation number of sulphur in $H_2S_2O_8$ is + 6
(D) The oxidation number of sulphur in H_2SO_5 is + 6

44. In the balanced equation, $MnO_4^- + Cl^- + H^+ \longrightarrow Mn^{2+} + Cl_2 + H_2O$, the coefficient of Cl^- is
(A) 5 (B) 10 (C) 14 (D) 16

45. In the balanced equation, $MnO_4^- + C_2O_4^{2-} + H^+ \longrightarrow Mn^{2+} + CO_2 + H_2O$, the coefficient of CO_2 is
(A) 4 (B) 5 (C) 6 (D) 10

46. H_2S reacts with halogens and the halogens
(A) form sulphur halides (B) are oxidised
(C) are reduced (D) none of these.

47. Identify the incorrect statement.
(A) In the reaction, $IO_3^- + 2I_2 + 10CN^- + 6H^+ \longrightarrow 5I(CN)_2^- + 3H_2O$, the equivalent mass of IO_3^- is M.W./4
(B) In the reaction, $IO_3^- + 5I^- + 6H^+ \longrightarrow 3I_2 + 3H_2O$, the equivalent mass of IO_3^- is M.W./5
(C) The degree of advancement of a reaction has the unit of mol
(D) The oxidation number of N in Ca_3N_2 is – 2.

48. Identify the incorrect statement.
(A) Oxidation number of O is – 2 in most of its compounds
(B) Oxidation number of O is 1/2 in superoxides
(C) Oxidation number of O is – 1 in peroxides
(D) Oxidation number of O is + 2 in OF_2.

49. The colour of $K_2Cr_2O_7$ changes from orange red to yellow on treatment with aqueous KOH because of
(A) reduction of Cr(VI) to Cr(III) (B) oxidation of KOH to K_2O_2
(C) formation of $Cr(OH)_3$ (D) conversion of $Cr_2O_7^{2-}$ ion to CrO_4^{2-} ion.

50. The strongest reducing agent is
(A) HNO_2 (B) H_2S (C) $SnCl_2$ (D) H_2SO_3

51. The oxidation number of S is + 6 in
(A) sodium thiosulphate (hypo) (B) pyrosulphuric acid (oleum)
(C) peroxodisulphuric acid (Marshall's acid)
(D) peroxomonosulphuric acid (Caro's acid)

52. The strongest oxidising agent is
(A) $K_2Cr_2O_7$ (B) $KMnO_4$ (C) H_2O_2 (D) O_3

53. Iron exhibits an oxidation number of + 1 in
(A) $[FeCl_4]^-$ (B) $[Fe(H_2O)_6]^{3+}$ (C) $Fe_4[Fe[CN)_6]_3$ (D) $[Fe(H_2O)_5NO]SO_4$

54. The strongest reducing agent is
(A) I_2 (B) Mg (C) K (D) Na

55. The equivalent mass of $KBrO_3$ in the reaction, $2BrO_3^- + 12H^+ + 10e^- \longrightarrow Br_2 + 6H_2O$, is
(A) $M/4$ (B) $M/5$ (C) $M/6$ (D) $M/10$

56. The oxidation number of oxygen atom in potassium superoxide is
(A) 0 (B) – 1 (C) – 2 (D) – 1/2

57. The oxidation number of nitrogen atom in N_3H is
(A) – 1 (B) + 2 (C) 1/2 (D) – 1/3

58. The oxidation number of oxygen atom in O_3 is
(A) 0 (B) + 1 and – 1 (C) + 2 and – 2 (D) + 0.5 and – 0.5

59. In the reaction, $xKMnO_4 + yNH_3 \longrightarrow MnO_2 + KNO_3 + KOH + H_2O$
(A) $x = 3, y = 8$ (B) $x = 8, y = 6$ (C) $x = 8, y = 3$ (D) $x = 4, y = 6$

60. Identify the correct order of reducing power of the following reducing agents.
(A) Mg < Na < Zn < Al (B) Na < Mg < Al < Zn
(C) Al < Mg < Zn < Na (D) Zn < Al < Mg < Na

61. In $Ni(CO)_4$, the oxidation state of nickel is
(A) 0 (B) + 2 (C) + 4 (D) + 8

62. The oxidation number of Pt in $[Pt(C_2H_4)Cl_3]^-$ is
(A) + 1 (B) + 2 (C) + 3 (D) + 4

63. The equilibrium $Cr_2O_7^{2-} \rightleftharpoons 2CrO_4^{2-}$ is shifted to right in
(A) basic medium (B) acidic medium
(C) neutral medium (D) it does not exist.

64. The reaction $MnO_4^- + e^- \rightleftharpoons MnO_4^{2-}$ occurs in
(A) basic medium (B) acidic medium
(C) neutral medium (D) both basic and acidic medium.

65. In the reaction, $C + H_2O \longrightarrow CO + H_2$, H_2O acts as
(A) reducing agent (B) oxidising agent
(C) both (A) and (B) (D) none of these.

66. The oxidation number and covalency of sulphur in S_8 are respectively.
(A) 0, 2 (B) 0, 8 (C) 6, 8 (D) 6, 2

67. In the reaction, xCu + yHNO$_3$ $\longrightarrow$ Cu(NO$_3$)$_2$ + NO + NO$_2$ + H$_2$O, the coefficients x and y are

(A) 2, 3 (B) 2, 6 (C) 1, 3 (D) 3, 8

68. In the reaction, $C_2H_6(g) + nO_2 \longrightarrow CO_2(g) + H_2O(l)$, the ratio of the coefficients of CO_2 and H_2O is

(A) 1 : 1 (B) 2 : 3 (C) 3 : 2 (D) 1 : 3

69. Hydrogen behaves as an oxidising agent when it reacts with

(A) N_2 to give NH_3 (B) Br_2 to give HBr

(C) Ca to give CaH_2 (D) S to give H_2S

70. In the reaction, $3OCl^- \longrightarrow 2Cl^- + ClO_3^-$, OCl^- acts as

(A) reducing agent (B) oxidising agent

(C) both (A) and (B) (D) neither oxidising nor reducing agent.

71. The oxidation state of sodium in sodium amalgam is

(A) 0 (B) + 1 (C) – 1 (D) + 2

72. The oxidation state of Mo in $Mo_2Cl_8^{4-}$ is

(A) + 1 (B) + 2 (C) + 4 (D) – 4

73. One mole of $K_2Cr_2O_7$ oxidises the following number of moles of $FeSO_4$ in acidic medium:

(A) 1 (B) 3 (C) 5 (D) 6

74. In the reaction, $xBrO_3^- + yCr^{3+} + zH_2O \longrightarrow Br_2 + HCrO_4^- + H^+$, the coefficients x, y and z are

(A) 6, 8, 22 (B) 6, 10, 22 (C) 6, 10, 11 (D) 6, 10, 20

75. In the reaction $xCr_2O_7^{2-} + yFe^{2+} + zH^+ \longrightarrow Cr^{3+} + Fe^{3+} + H_2O$, the coefficients x, y and z are

(A) 1, 6, 14 (B) 1, 3, 14 (C) 2, 6, 14 (D) 2, 3, 7

76. Consider the reaction: $PbO_2 \longrightarrow PbO$, $\Delta G_{298} < 0$; $SnO_2 \longrightarrow SnO$, $\Delta G_{298} > 0$. The most probable oxidation state of Pb and Sn will be

(A) Pb^{4+}, Sn^{4+} (B) Pb^{4+}, Sn^{2+} (C) Pb^{2+}, Sn^{3+} (D) Pb^{2+}, Sn^{4+}

77. The compound having the lowest oxidation state of iron is

(A) Fe_2O (B) K_2FeO_4 (C) $K_4[Fe(CN)_6]$ (D) $Fe(CO)_5$

78. The nitrate anion can be converted into ammonium ion. The equivalent mass of NO_3^- ion in this reaction is

(A) 20.0 g (B) 15.5 g (C) 6.2 g (D) 7.75 g

79. In the standardisation of $Na_2S_2O_3$ using $K_2Cr_2O_7$ by iodometry, the equivalent mass of $K_2Cr_2O_7$ is

(A) $M/2$ (B) $M/6$ (C) same as M (D) $M/3$ **(I.I.T. 2001)**

Hint: $Cr_2O_7^{2-}$ ion changes to Cr^{3+} ion. Equivalent mass of $K_2Cr_2O_7$

$$= \frac{\text{Mol. mass}}{\text{Decrease in ON} \times \text{No. of Cr atoms}} = \frac{\text{Mol. mass}}{3 \times 6} = \frac{M}{6}$$

80. In which of the following ions chlorine does not undergo disproportionation?
(A) Cl^- (B) ClO_3^- (C) ClO_4^- (D) ClO^-

81. In which of the following species, the metal has the oxidation state of + 3?
(A) $K_3[Fe(CN)_6]$ (B) VO^{3+}
(C) $[Co(NH_3)_4Cl_2]Cl$ (D) $NaAlO_2$

82. In which of the following conversions/reactions, chlorine is oxidised?
(A) $I_2 + Cl_2 \longrightarrow 2ICl$ (B) $OCl^- \longrightarrow Cl^-$
(C) $OCl^- \longrightarrow ClO_2$ (D) $Cl_2 \longrightarrow OCl^-$

83. In which of the following compounds, there is an element with the same oxidation state as that of Cr in Na_2CrO_4?
(A) K_2MnO_4 (B) $MoOCl_3$ (C) $K_4[Fe(CN)_6]$ (D) Cl_2O_7

84. Number of moles of $K_2Cr_2O_7$ reduced by 1 mole of Sn^{2+} is
(A) 1/6 (B) 1/3 (C) 2/3 (D) 1

85. The numerical change in the oxidation state of chlorine in the conversion $Cl^- \longrightarrow ClO_3^-$ is
(A) 6 (B) 5 (C) 4 (D) 1

86. From the two redox reactions: (i) $Mn^{2+} + 4H_2O \longrightarrow MnO_4^- + 8H^+ + 5e^-$ (ii) $S_2O_8^{2-} + 2e^- \longrightarrow 2SO_4^{2-}$, predict how many moles of $S_2O_8^{2-}$ ions are needed to oxidise 1 mole of Mn^{2+} ion.
(A) 2.5 (B) 1.5 (C) 2 (D) 1

87. From the two redox reactions: (i) $Zn \longrightarrow Zn^{2+} + 2e^-$ (ii) $NO_3^- + 4H^+ + 3e^- \longrightarrow NO + 2H_2O$, predict how many moles of NO_3^- ions are reduced by 1 mole of Zn.
(A) 2 (B) 3 (C) 3/2 (D) 2/3

88. Which of the following are formed when I^- and MnO_4^- ions react in strongly basic medium?
(A) MnO_2, O_2, IO^- (B) Mn^{2+}, I_2 (C) MnO_2, IO_3^- (D) Mn, I_2

89. Which of the following conversions requires a reducing agent?
(A) $Al(OH)_3 \longrightarrow AlO_2^-$ (B) $H_2O_2 \longrightarrow O_2$
(C) $CrO_4^{2-} \longrightarrow Cr_2O_7^{2-}$ (D) $ClO_3^- \longrightarrow ClO^-$

90. Which of the following cannot act as a reducing agent?
(A) H_2S (B) SO_2 (C) H_2SO_3 (D) H_2SO_4

91. Identify the compound in which iron has the lowest oxidation state.
(A) $Fe(CO)_5$ (B) $Fe_2(CO)_9$ (C) $Fe_{0.94}O$ (D) $K_4[Fe(CN)_6]$

92. The pair of compounds having metals in their highest oxidation state is
(A) MnO_2, $FeCl_3$ (B) MnO_4^-, CrO_2Cl_2
(C) $[Fe(CN)_6]^{3-}$, $[Co(CN)_3]$ (D) $[NiCl_4]^{2-}$, $[CoCl_4]^{2-}$ **(I.I.T. 2004)**

93. For the reduction of NO_3^- ion in aqueous solution, E^0 is + 0.96 V. Values of E^0 for some metal ions are given below:

$V^{2+}(aq) + 2e^- \longrightarrow V \quad E^0 = -1.19$ V; $Au^{3+}(aq) + 3e^- \longrightarrow Au \quad E^0 = +1.40$ V

$Fe^{3+}(aq) + 3e^- \longrightarrow Fe \quad E^0 = -0.04$ V; $Hg^{2+}(aq) + 2e^- \longrightarrow Hg \quad E^0 = +0.86$ V

The pairs of metals that is(are) oxidised by NO_3^- in aqueous solution is(are)

(A) V and Hg (B) Hg and Fe (C) Fe and Au (D) Fe and V **(I.I.T. 2009)**

Hint: For the reduction to occur, $E^0_{cathode}$ should be greater than E^0_{anode}. Hence, V, Fe and Hg can be oxidised by NO_3^- ion.

94. Reduction of the metal centre in aqueous permanganate ion involves

(A) 3 eletrons in neutral medium (C) 5 electrons in neutral medium

(D) 3 eletrons in alkaline medium (D) 5 electrons in acidic medium **(I.I.T. 2011)**

Hint: In acidic medium, $MnO_4^- + 8H^+ + 5e^- \longrightarrow Mn^{2+} + 4H_2O$

In neutral medium, $MnO_4^- + 2H_2O + 3e^- \longrightarrow MnO_2 + 4OH^-$

Hence the number of electrons involved in acidic and neutral medium is 5 and 3 electrons respectively.

ANSWERS

1. A	2. B	3. A	4. A	5. B	6. B
7. B	8. A	9. A	10. D	11. D	12. C
13. C	14. C	15. C	16. A	17. A	18. B
19. B	20. A	21. B	22. D	23. B	24. A
25. A	26. D	27. B	28. A	29. D	30. A, B
31. A	32. B	33. C	34. D	35. D	36. B, C
37. A, D	38. A, C	39. A	40. A	41. B	42. D
43. A	44. D	45. D	46. A	47. D	48. B
49. D	50. B	51. B, C, D	52. D	53. D	54. C
55. B	56. D	57. D	58. A	59. C	60. D
61. A	62. B	63. A	64. A	65. B	66. A
67. B	68. B	69. C	70. C	71. A	72. B
73. D	74. B	75. A	76. D	77. D	78. D
79. B	80. A, C	81. A, C, D	82. C, D	83. A	84. B
85. A	86. A	87. D	88. C	89. D	90. D
91. A, B	92. B	93. A, B, D	94. A, D		

5

States of Matter I: Gaseous and Liquid States

IMPORTANT RELATIONS FOR NUMERICAL PROBLEMS

1. $PV = nRT = (w/m)RT$
2. $P_1V_1 = P_2V_2$
3. $V_1T_1 = V_2T_2$
4. $P_1V_1/T_1 = P_2V_2/T_2$
5. $PV = (1/3)mNu^2$
6. $(P + an^2/V^2)(V - nb) = nRT$
7. $p = PM/RT$
8. $r_1/r_2 = \sqrt{M_2/M_1} = \sqrt{\rho_2/\rho_1}$
9. K. E. = $(3/2)RT$
10. $\alpha = \sqrt{2RT/M}$
11. $u = \sqrt{3RT/M}$
12. $\upsilon = \sqrt{8RT/\pi M} = 1.124$
13. $\alpha : \upsilon : u = \sqrt{2} : \sqrt{8/\pi} : \sqrt{3}$ $= 1 : 1.128 : 1.224$
14. $\lambda = 1/(\sqrt{2}\pi\sigma^2 N^2)$
15. $Z = (1/\sqrt{2})\pi V^2\sigma^2 N^2$
16. $T_c = 8a/27Rb$
17. $P_c = a/27b^2$
18. $V_c = 3b$
19. $P_cV_c = (3/8)RT_c$
20. $T_b = a/Rb$; $T_i = 2a/Rb$
21. $C_p - C_v = R$; $C_v = (3/2)R$; $C_p = (5/2)R$; $\gamma = C_p/C_v$
22. Absolute density: d = mass/volume; expressed in g L^{-1}
23. Vapour density: M.W. = V.D. × 2
24. Mean free path: $\lambda = \eta\sqrt{3/Pd}$ $= 1/\sqrt{2}\pi\sigma^2 N$
25. Values of R: 8.314×10^7 erg K^{-1} mol^{-1} $= 8.314$ J K^{-1} $mol^{-1} = 1.987$ cal K^{-1} mol^{-1}
26. Compressibility factor, $Z = PV/nRT$
27. Dalton's law of partial pressures: $P = p_1 + p_2 + p_3 + \dots$
28. Partial pressure of any component (say A): $p_A = x_AP$
29. 1 atm = 76 cm Hg = 760 mm Hg = 760 torr = 101325 Pa or N m^{-1}
30. t°C = (5/9) (°F – 32); t°C = $(273 + t)$ K

OBJECTIVE TYPE MULTIPLE CHOICE QUESTIONS

1. The RMS velocity of an ideal gas at constant pressure varies with density (d) as
   (A) d^2 (B) $\sqrt{d}$ (C) $1\sqrt{d}$ (D) d **(I.I.T. 2001)**
2. At 100°C and 1 atm, if the density of water(l) is 1.0 g cm^{-3} and that of water vapour is 0.0006 g cm^{-3}, then the volume occupied by water molecules in 1 L of steam at that temperature is
   (A) 60 cm^3 (B) 0.6 cm^3 (C) 0.06 cm^3 (D) 6 cm^3 **(I.I.T. 2000)**
   Hint: Mass of 1 L of vapour = volume × density = 1000 × 0.0006 = 0.6 g
   Volume of liquid water = mass/density = 0.6/1 = 0.6 cm^3.
3. The RMS velocity of H_2 is $\sqrt{7}$ times the RMS velocity of N_2. If T is the temperature of the gas,
   (A) $T(H_2) = T(N_2)$ (B) $T(H_2) > T(N_2)$
   (C) $T(H_2) < T(N_2)$ (D) $T(H_2) = \sqrt{7}\,T(N_2)$ **(I.I.T. 2000)**

   Hint: Given: $\sqrt{\dfrac{3RT_{H_2}}{2}} = \sqrt{7}\sqrt{\dfrac{3RT_{N_2}}{28}}$ $\left(\text{since } u_{rms} = \sqrt{\dfrac{3RT}{M}}\right)$

   or $\dfrac{3RT_{H_2}}{2} = 7 \times \dfrac{3 \times RT_{N_2}}{28}$

   or $T_{N_2} = 2H_2$ or $T_{N_2} > T_{H_2}$
4. The compressibility of a gas is less than unity at STP. Hence
   (A) $V_m > 22.4$ L (B) $V_m < 22.4$ L (C) $V_m = 22.4$ L (D) $V_m = 44.8$ L **(I.I.T. 2000)**
5. Which of the following statements is/are correct?
   (A) A plot of log K_p versus $1/T$ is linear
   (B) A plot of log [X] versus time is linear for a first order reaction, X $\longrightarrow$ P
   (C) A plot of log P versus $1/T$ is linear at constant volume
   (D) A plot of P versus $1/V$ is linear at constant temperature. **(I.I.T. 1999)**
6. A gas will approach the ideal behaviour at
   (A) low temperature and low pressure
   (B) low temperature and high pressure
   (C) high temperature and low pressure
   (D) high temperature and high pressure. **(I.I.T. 1999)**
7. According to Graham's law, at a given temperature the ratio of the rates of diffusion r_A/r_B of the gases A and B is given by
   (A) $(P_A/P_B)\,(M_A/M_B)^{1/2}$ (B) $(M_A/M_B)\,(P_A/P_B)^{1/2}$
   (C) $(P_A/P_B)\,(M_B/M_A)^{1/2}$ (D) $(M_A/M_B)\,(P_B/P_A)^{1/2}$ **(I.I.T. 1998)**
8. X mL of H_2 gas effuses through a hole in a container in 5 sec. The time taken for the effusion of the same volume of the gas specified below under identical conditions is
   (A) 55 sec: CO_2 (B) 20 sec: O_2 (C) 25 sec: CO (D) 10 sec: He **(I.I.T. 1996)**

9. The ratio between RMS velocity of H_2 at 50 K and that of O_2 at 800 K is
(A) 4 (B) 2 (C) 1 (D) 1/4 **(I.I.T. 1996)**

10. In a closed room of 1000 m^3 a perfume bottle is opened up. The room develops smell. This is due to which property of gases?
(A) Density (B) Viscosity (C) Diffusion (D) None **(I.I.T. 1994)**

11. Equal masses of ethane and H_2 are mixed in an empty container at 25°C. The fraction of the total pressure exerted by H_2 is
(A) 1 : 2 (B) 1 : 1 (C) 1 : 16 (D) 15 : 16 **(I.I.T. 1993)**

Hint: $x_{C_2H_6} = \dfrac{(m/30)}{(m/30)+(m/2)} = \dfrac{1}{16}$; $x_{H_2} = \dfrac{(m/2)}{(m/30)+(m/2)} = \dfrac{15}{16}$

$p_{H_2} = \dfrac{15}{16} P_{Total}$. Hence $p_{H_2} : P_{Total} = 15 : 16$

12. At constant volume for a fixed number of moles of a gas, the pressure of the gas increases with rise in temperature due to
(A) increase in average molecular speed
(B) increase in rate of collisions amongst molecules
(C) increase in molecular attraction
(D) increase in mean free path. **(I.I.T. 1992)**

13. According to the kinetic theory of gases
(A) the pressure exerted by a gas is proportional to the mean velocity of the molecules
(B) the pressure exerted by the gas is proportional to the root mean square velocity of the molecules
(C) the root mean square velocity is inversely proportional to the temperature
(D) the mean translational kinetic energy of the molecules is directly proportional to the absolute temperature. **(I.I.T. 1991)**

14. The rate of diffusion of CH_4 at a given temperature is twice that of a gas. The molecular mass of X is
(A) 64.0 (B) 32.0 (C) 4.0 (D) 8.0 **(I.I.T. 1990)**

15. The values of van der Waals' constant a for the gases O_2, N_2, NH_3 and CH_4 is 1.360, 1.390, 4.170 and 2.253 L^2 atm mol^{-2} respectively. The gas which can most easily be liquefied is
(A) O_2 (B) N_2 (C) NH_3 (D) CH_4 **(I.I.T. 1989)**

16. A bottle of dry ammonia and a bottle of dry HCl connected through a long tube are opened simultaneously at both ends. The white NH_4Cl ring first formed will be
(A) at the centre of the tube (B) near the HCl bottle
(C) throughout the length of the tube (D) near ammonia bottle. **(I.I.T. 1988)**

17. In the van der Waals' equation for a real gas, the term that accounts for intermolecular forces is
(A) $(V - b)$ (B) RT (C) $(P + a/V^2)$ (D) $(RT)^{-1}$ **(I.I.T. 1988)**

18. If a gas expands at constant temperature
   (A) pressure decreases (B) K.E. of the molecules remains the same
   (C) number of molecules of the gas increases
   (D) K.E. of the molecules decreases. **(I.I.T. 1986)**

19. Which of the following is true for the rate of diffusion of a gas?
   (A) It is inversely proportional to the square root of its molecular weight (M.W.)
   (B) It is directly proportional to the square root of its M.W.
   (C) It is directly proportional to its M.W.
   (D) It is directly proportional to its density. **(I.I.T. 1985)**

20. The average velocity of an ideal gas molecule at 27°C is 0.3 m/s. The average velocity at 927°C will be
   (A) 0.6 m/s (B) 0.3 m/s (C) 0.9 m/s (D) 3.0 m/s **(I.I.T. 1986)**

21. A liquid is in equilibrium with its vapour at its boiling point. On the average the molecules in the two phases have equal
   (A) intermolecular forces (B) potential energy
   (C) kinetic energy (D) activation energy. **(I.I.T. 1984)**

22. Equal masses of H_2 and CH_4 are mixed in an empty container at 25°C. The fraction of the total pressure exerted by H_2 is
   (A) 1/2 (B) 8/9 (C) 1/9 (D) 16/17 **(I.I.T. 1984)**

   Hint: $n_{CH_4} = 16/16 = 1$, $n_{H_2} = 16/2 = 8$. Hence $x_{H_2} = 8/9$. $p_{H_2} = x_{H_2} \times P_{Total}$ $= (8/9) \times P_{Total}$

23. When an ideal gas undergoes unrestricted expansion, no cooling occurs because the molecules
   (A) are above the inversion temperature
   (B) exert no attractive force on each other
   (C) do work equal to loss in kinetic energy
   (D) collide without loss of energy. **(I.I.T. 1984)**

24. He atom is two times as heavy as a H_2 molecule. At 298 K the average kinetic energy of He atom is
   (A) twice that of a H_2 molecule (B) same as that of a H_2 molecule
   (C) four times that of a H_2 molecule (D) half that of a H_2 molecule. **(I.I.T. 1982)**

25. The ratio of root mean square velocity to average velocity of a gas molecule at a given temperature is
   (A) 1.086 : 1 (B) 1 : 1.086 (C) 2 : 1.086 (D) 1.086 : 2 **(I.I.T. 1981)**

26. The temperature at which real gases obey the ideal gas laws over a wide range of pressure is called
   (A) critical temperature (B) inversion temperature
   (C) reduced temperature (D) Boyle temperature. **(I.I.T. 1994)**

27. Equal masses of CH_4 and O_2 are mixed in an empty container at 25°C. The fraction of the total pressure exerted by O_2 is
(A) 1/3 (B) 1/2
(C) (1/3) × 273/298 (D) 2/3 **(I.I.T. 1981)**

28. The correct representation of van der Waals' equation is
(A) $PV = (1/3)mnu^2$ (B) $(P + a/V^2)(V - b) = RT$
(C) $(P - a/V^2)(V + b) = RT$ (D) $PV = RT$ **(I.I.T. 1982)**

29. One mole of $N_2O_4(g)$ at 300 K is kept in a closed container under 1 atm pressure. It is heated to 600 K when 20% by mass of $N_2O_4(g)$ decomposes to $NO_2(g)$. The resultant pressure is
(A) 1.2 atm (B) 2.4 atm (C) 2.0 atm (D) 1.0 atm **(I.I.T. 1996)**
Hint: Moles at equil. = Moles of N_2O_4 + moles of NO_2 = (80/92) + (20/46) $P_0V = n_0RT_0$ or $1 \times V = (100/92) \times 0.0821 \times 300$ or $V = 26.77$ L. At equil., $P_2V = n_2RT_2$ or $P_2 \times 26.77$ = [(80/92) + (20/46)] × 0..0821 × 600 or $P_2 \times 26.77 = 1.3 \times 0.0821 \times 600$ or $P_2 = 2.4$ atm.

30. The total K.E. of a monoatomic ideal gas at 27°C is
(A) 300 cal (B) 900 cal (C) 1800 cal (D) none **(I.I.T. 1984)**

31. If a 4 : 1 mixture of N_2 and O_2 is taken and 1% of Ar is added to it, the vapour pressure of mixed gases will
(A) increase (B) decrease
(C) remain constant (D) first increase then decrease.

32. Pressure of a mixture of 4 g of O_2 and 2 g of H_2 confined in a bulb of 1 L at 0°C is
(A) 15.21 atm (B) 25.21 atm (C) 31.20 atm (D) 45.21 atm

33. A container contains CO and another H_2 in equimolar concentration at the same temperature. What is the ratio of average molar K.E. of the two gases?
(A) 1 : 1 (B) $KE_{CO} > KE_{H_2}$ (C) $KE_{H_2} > KE_{CO}$ (D) depends on volume.

34. Identify the incorrect statement regarding gas.
(A) Mass of gas cannot be determined by weighing a container in which it is kept
(B) Confined gas exerts uniform pressure on the wall of the container in all directions
(C) Volume of gas is equal to the volume of container containing the gas
(D) Gases do not have a definite shape and volume.

35. The total pressure of a mixture of 6.4 g of O_2 and 5.6 g N_2 present in a 2 L vessel is 1200 mm. What is the partial pressure of N_2?
(A) 200 atm (B) 600 atm (C) 900 atm (D) 1200 atm

36. Identify the incorrect statement.
(A) Oxygen diffuses slower than nitrogen through an orifice
(B) The average speed of molecules depends only on the absolute temperature
(C) Equal volumes of gases contain equal number of moles at STP
(D) A gas in a closed container will exert much higher pressure due to gravity at the bottom than at the top.

37. Identify the incorrect statement.
   (A) The value of PV for 5.6 L of an ideal gas is $0.25RT$ at STP
   (B) The rate of diffusion is inversely proportional to the square root of density and the square root of molecular mass
   (C) Kinetic energy of gas molecules is zero at 0°C
   (D) The heat capacity of a diatomic gas is lower than that of a monoatomic gas.
38. Identify the incorrect statement.
   (A) 8 g each of O_2 and H_2 at 27°C will have average kinetic energy in the ratio 1 : 16
   (B) The total number of electrons present in 18 mL of water is 9×10^{22}
   (C) The total energy of one mole of a monoatomic ideal gas at 27°C is 900 cal
   (D) The $C_p - C_v$ for an ideal gas is 2.
39. At STP, 2.8 L of H_2S were mixed with 1.6 L of SO_2 and the reaction occurred according to the equation: $2H_2S(g) + SO_2(g) \longrightarrow 2H_2O(l) + 3S(s)$. Which of the following is the volume of the gas remaining after the reaction?
   (A) 0.2 L of $SO_2(g)$
   (B) 0.4 L of $H_2S(g)$
   (C) 1.2 L of $H_2S(g)$
   (D) 1.2 L of $SO_2(g)$
40. Identify the incorrect statement.
   (A) The ratio of average velocity (v) and most probable velocity (α) is 2 : 3
   (B) If a gas is heated at constant pressure from 30°C to 60°C, the gas expands to two times its initial volume
   (C) The molar mass of a gas can be calculated using the equation, $M = (\rho/P)RT$
   (D) An ideal gas cannot be liquefied.
41. Identify the incorrect statement.
   (A) The molecules of an ideal gas move with lower speed at a low pressure in comparison to the gas at a higher pressure
   (B) Liquefaction of gases involves decrease in intermolecular forces
   (C) The viscosity of a gas increases with increase of temperature
   (D) The order of increasing density of O_2 at various temperatures and pressures is: O_2 at 273°C, 1 atm < O_2 at 25°C, 1 atm < O_2 at 0°C, 2 atm
42. Identify the incorrect statement.
   (A) A real gas may be liquefied if the temperature is higher than its critical temperature
   (B) At constant pressure, the density of an ideal gas increases with increase of temperature
   (C) The average velocity of molecules is always less than the root mean square velocity
   (D) The viscosity of a liquid decreases with increase in temperature.
43. Identify the incorrect statement.
   (A) The increasing order of average, root mean square and most probable velocities is: root mean square velocity < average velocity < most probable velocity

(B) The unit of van der Waals' constant a is atm litre^2 mol^{-1}

(C) The absolute temperature of an ideal gas is less than average kinetic energy of gas molecules

(D) The volume occupied by 5 g of acetylene at 50°C and 740 mm pressure is 5.24 L.

44. Identify the incorrect statement.

(A) At a constant temperature when the pressure of a certain mass of gas is doubled, its volume is doubled

(B) Dry air diffuses faster than wet air

(C) The mass of CH_4 in a 9.0 L cylinder at 16 atm and 27°C is 93.5 g

(D) van der Waals' constant b is 4 times the volume occupied by molecules of one mole of a real gas.

45. Identify the incorrect statement.

(A) Out of H_2S, NH_3, CO_2 and SO_2, the largest root mean square velocity at the same temperature is possessed by NH_3

(B) The temperature at which the root mean square velocity of O_2 becomes equal to that of SO_2 at 327°C is 27°C

(C) Gases possessing very low critical temperatures exhibit near ideal behaviour at the room temperature

(D) The average velocity of a gas varies linearly with the increase in temperature.

46. For a gas at given temperature, which of the following is true?

(A) most probable velocity < average velocity < root mean square velocity

(B) root mean square velocity < average velocity > most probable velocity

(C) root mean square velocity < average velocity < most probable velocity

(D) root mean square velocity > average velocity < most probable velocity

47. Identify the correct expression.

(A) RMS velocity = $(8RT/\pi M)^{1/2}$
(B) RMS velocity = $(3kT/M)^{1/2}$
(C) RMS velocity = $(3RT/m)^{1/2}$
(D) RMS velocity = $(3RT/M)^{1/2}$

48. The RMS velocity of an ideal gas in a closed container of fixed volume is increased from 5×10^4 cm s^{-1} to 10×10^4 cm s^{-1}. Which of the following statements correctly explains how the change is accomplished?

(A) By heating the gas, the volume is doubled

(B) By heating the gas, the pressure is doubled

(C) By heating the gas, the temperature is doubled

(D) By heating the gas, the pressure is made four times.

49. The vapour pressure of ice is 3.04 torr at – 5°C and the partial pressure of water dropped to 2 torr. If the temperature remains – 5°C for quite some time, the frost will

(A) sublime
(B) not sublime

(C) not sublime till the temperature rises

(D) not sublime till the pressure of water in air decreases.

50. At STP the decreasing order of RMS velocity of the molecules of H_2, O_2, N_2 and HBr is
(A) $H_2 > N_2 > O_2 >$ HBr (B) $N_2 > O_2 > H_2 >$ HBr
(C) HBr $> H_2 > O_2 > N_2$ (D) HBr $> O_2 > N_2 > H_2$

51. The translational kinetic energy of an ideal gas depends only on its
(A) molar mass (B) temperature (C) force (D) pressure

52. Which of the following is the same for all ideal gases at the same temperature?
(A) number of molecules in 1 g (B) number of molecules in 1 mole
(C) kinetic energy of 1 g (D) kinetic energy of 1 mole

53. Identify the correct expressions on the basis of ideal gas equation.
(A) $PV = MkT/m$ (B) $PV = \rho RT/m$ (C) $PV = NkT$ (D) $PV = NRT/N_A$

54. Air contains 77% nitrogen and 23% oxygen by mass. The percentage of oxygen by volume is
(A) 22.0 (B) 21.8 (C) 20.8 (D) 27.1

55. If m_1 and m_2 are the molecular masses of two ideal gases and u_1 and u_2 are the molecular velocities of these gases, identify the correct expression.
(A) $m_1u_1^2 = m_2u_2^2$ (B) $m_1/u_1 = m_2/u_2$
(C) $m_1/u_1^2 = m_2/u_2^2$ (D) $m_1u_1 = m_2u_2$

56. At low pressure and high temperature, the van der Waals' equation is reduced to
(A) $PV_m = RT$ (B) $(P + a/V_m^2)(V_m - b) = RT$
(C) $P(V_m - b) = RT$ (D) $(P + a/V_m^2)(V_m) = RT$

57. At low pressure, the van der Waals' equation is reduced to
(A) $PV_m = RT$ (B) $Z = PV_m/RT = 1 - a/RT$
(C) $Z = PV_m/RT = 1 + bP/RT$ (D) $Z = PV_m/RT = 1 - aP/RT$

58. The ratio of the root mean square velocity (u), average velocity (v) and most probable velocity (α) of an ideal gas is
(A) $(3)^{1/2} : (8\pi)^{1/2} : 1$ (B) $(8)^{1/2} : (3)^{1/2} : (2)^{1/2}$
(C) $(3)^{1/2} : (2)^{1/2} : 1$ (D) $(3)^{1/2} : (8/\pi)^{1/2} : (2)^{1/2}$

59. However greater is the pressure a gas cannot be liquefied above its
(A) critical temperature (B) room temperature
(C) inversion temperature (D) Boyle temperature

60. The total pressure in a 10 L cylinder which contains 0.4 g of He, 1.6 g of O_2 and 1.4 g of N_2 at 27°C is
(A) 0.429 atm (B) 0.492 atm (C) 249 atm (D) none **(Roorkee 1997)**

61. The ratio of the rates of diffusion of SO_2, O_2 and CH_4 is
(A) $1 : 2 : 2\sqrt{2}$ (B) $\sqrt{2} : 1 : 2$ (C) $1 : 2 : 4$ (D) $1 : \sqrt{2} : 2$

62. The unit of van der Waals' constant a is
(A) atm L^2 mol^{-2} (B) atm L^2 mol^{-1} (C) atm L^2 (D) atm

63. The unit of van der Waals' constant b is
(A) $(cm^3\ mol^{-1})^2$ (B) $cm^3\ mol^{-2}$ (C) $cm^3\ mol^{-1}$ (D) cm^3

64. The root mean square velocity of gaseous molecules is given by: $u = (3P/\rho)^{1/2}$. Which of the following expressions is correct at a given temperature?
(A) $u \propto (P/V_m)^{1/2}$ (B) $u \propto (PV_m)^{1/2}$ (C) $u \propto (1/\rho)^{1/2}$ (D) $u \propto P^{1/2}$

65. Under which of the following conditions a real gas resembles an ideal gas?
(A) *a* and *b* are small (B) *a* and *b* are large
(C) *a* is large, *b* is small (D) *a* is small, *b* is large.

66. Dalton's law of partial pressure does not apply to the mixture of
(A) CO_2 and N_2 (B) N_2 and H_2 (C) N_2 and O_2 (D) NH_3 and HCl

67. At the same pressure and temperature, which of the following gases will have the highest kinetic energy per mole?
(A) H_2 (B) O_2 (C) CH_4 (D) all have the same.

68. The internal energy of one mole of an ideal gas is
(A) $3kT/2$ (B) $RT/2$ (C) $kT/2$ (D) $3RT/2$

69. Five grams each of the following gases at 87°C and 750 mm pressure are taken. Which of them will have the least volume?
(A) HI (B) HBr (C) HCl (D) HF

70. Which of the following gases can most easily be liquefied?
(A) SO_2 (B) CO_2 (C) Cl_2 (D) NH_3

71. The fourth state of matter is
(A) gas (B) liquid (C) solid (D) plasma

72. Identify the false statements.
(A) The root mean square velocity (u) of a gas can be calculated using the relation: $u = (3RT/M)^{1/2}$
(B) The mean kinetic energy of an ideal gas is independent of the pressure of the gas
(C) The relationship between average velocity (v) and root mean square velocity (u) is: $v = 0.9216u$
(D) Avogadro's number $= 6.023 \times 10^{21}$

73. A closed vessel contains equal number of N_2 and O_2 molecules at a pressure of P mm. If N_2 is removed from the system then the pressure will be
(A) P (B) $P/2$ (C) P^2 (D) $2P$

74. Dalton's law of partial pressure is not applicable to which of the following mixture of gases?
(A) H_2 and CO_2 (B) H_2 and Cl_2 (C) CO_2 and Cl_2 (D) H_2 and SO_2

75. Gases begin to conduct electricity at low pressure because
(A) the electrons in atoms can move freely at low pressures
(B) the colliding molecules can acquire higher kinetic energy due to the increased mean free path leading to ionisation of atoms
(C) atoms break up into electrons and protons
(D) at low pressures gases turn into plasma.

76. A closed flask contains water in vapour, liquid and solid states at 0°C. In this situation the average kinetic energy of water molecules will be
(A) same in three states (B) greater in vapour state
(C) greater in solid state (D) greater in liquid than in vapour state.

77. The root mean square velocity of gas molecules at 25 K and 1.5 bar is 100 m s^{-1}. If the temperature is raised to 100 K and the pressure to 6.0 bar, the root mean square velocity becomes
(A) 100 m s^{-1} (B) 200 m s^{-1} (C) 400 m s^{-1} (D) 1600 m s^{-1}

78. The temperature of a gas is raised from 27°C to 927°C. The root mean square velocity is
(A) halved (B) doubled
(C) same as above (D) $(927/27)^{1/2}$ of the earlier value.

79. Dimensions of pressure are the same as that of
(A) energy (B) force
(C) force per unit volume (D) energy per unit volume.

80. The vapour pressure of water at 300 K in a closed container is 0.4 atm. It the volume of the container is doubled, its vapour pressure at 300 K will be
(A) 0.8 atm (B) 0.2 atm (C) 0.4 atm (D) 0.6 atm

81. The average velocity of oxygen at 273 K is equal to that of hydrogen at
(A) higher temperature (B) lower temperature
(C) same temperature (D) none of these.

82. The number of moles of an ideal gas is given by
(A) NN_A (B) N/N_A (C) mnN_A (D) n/V

83. The ratio of the inversion temperature to Boyle temperature of a gas is
(A) 2.0 (B) 3.0 (C) 2.5 (D) 1.5

84. Out of CO_2, N_2, H_2 and CO, the gas which has the least mean free path
(A) has the maximum value of T_b (Boyle temperature)
(B) is easily liquefied
(C) has the maximum value of a (van der Waals' constant)
(D) all of these.

85. The mean free path increases when
(A) H_2 is replaced by He (B) O_2 is replaced by CO_2
(C) He is replaced by O_2 (D) He is replaced by H_2

86. In a spacecraft, the output of carbon dioxide per astronaut is 44 g per hour. A catalytic converter reduces carbon dioxide at the rate of 550 mL per minute into H_2O at STP. The fraction of time the converter has to be operated for keeping the output of CO_2 of one astronaut is
(A) 0.68 (B) 0.34 (C) 0.75 (D) 0.54

Hint: Output of CO_2 per hour = 44 g = 1 mole = 22.4 L at STP
Reduction of CO_2 per hour = 550 × 60 mL at STP = 33.0 L

$$\text{Fraction of time the converter has to be operated} = \frac{CO_2 \text{ output rate}}{CO_2 \text{ reduction rate}} = \frac{22.4}{33.0} = 0.68$$

87. Which of the following volume (V)-temperature (T) plots represents the behaviour of one mole of an ideal gas at one atmospheric pressure?

(A) 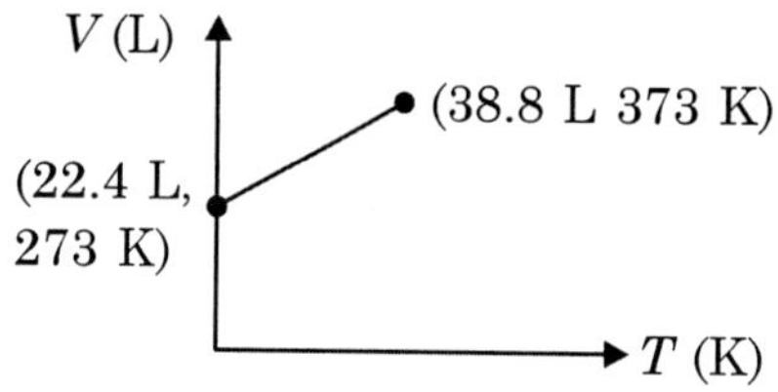

(B) 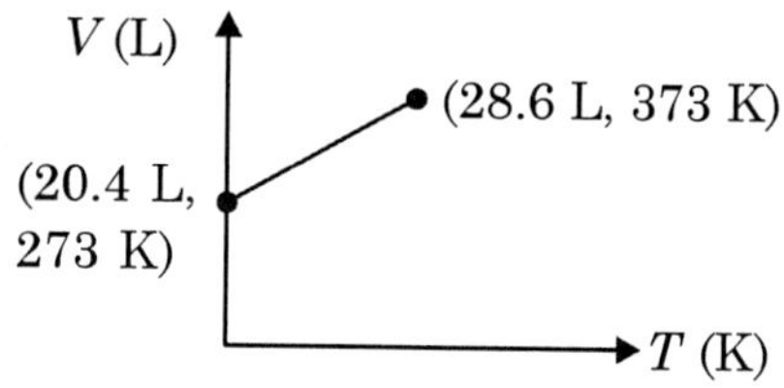

(C) 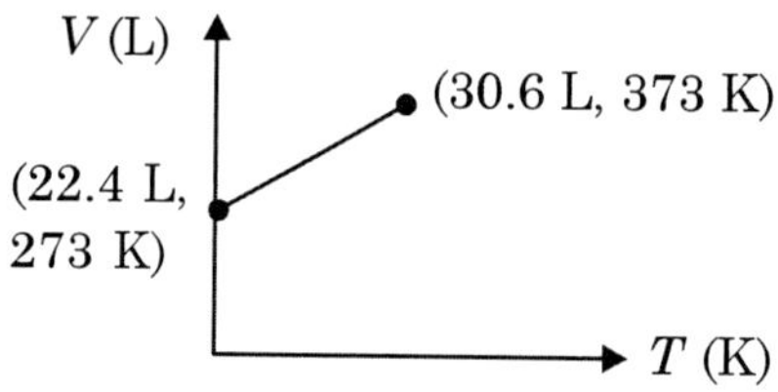

(D) 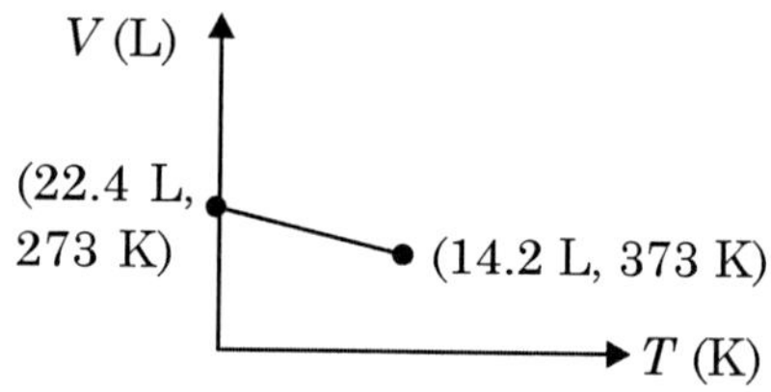

(I.I.T. 2002)

Hint: Volume of 1 mole of an ideal gas at 273 K and 1 atm = 22.4 L. $V = RT/P = 0.082 \times 373/1 = 30.6$

88. Which of the following occupy 3 L at 298 K and 1 atm?
    (A) 20 g of Ar (At. wt. 40) (B) 8 g of CO_2
    (C) 4 g of O_2 (D) 8 g of SO_2

89. Which of the following gases has different root mean square velocity at 298 K than the others?
    (A) CO_2 (B) C_3H_5 (C) N_2O (D) CH_4
    Hint: Molar mass of CO_2, C_3H_5 and N_2O is same (44 g mol^{-1}).

90. The most probable velocity of carbon dioxide molecules is twice that at 323 K at the temperature of
    (A) 1292 K (B) 646 K (C) 623 K (D) 1246 K

91. At what temperature is the average velocity of nitrogen(g) is the same as that of methane(g) at 373 K?
    (A) 653 K (B) 423 K (C) 800 K (D) 203 K

92. At what temperature is the root mean square velocity of $SO_2(g)$ is the same as that of $O_2(g)$ at 150 K?
    (A) 300 K (B) 150 K (C) 750 K (D) 900 K

93. The ratio of root mean square velocity of $O_2(g)$ to that of $H_2(g)$ at 300 K is
    (A) 1/4 (B) 1/8 (C) 1/16 (D) 4

94. The ratio of the root mean square velocity to most probable velocity of a gas at 300 K is
    (A) 1.224 (B) 1.128 (C) 1.114 (D) 1.118

95. Molar volume of an ideal gas
    (A) is independent of P and T (B) depends on T only
    (C) depends on P only (D) depends on T and P

96. Identify the correct relation for an ideal gas

(A) $PV = RT/n$ (B) $V = nRT/P$

(C) PV/T = constant (D) $RT = nPV$

97. Which of the following laws is not an ideal gas law?

(A) Graham's law of effusion (B) Dalton's law of partial pressure

(C) Charles' law (D) Boyle's law

98. The density of a gas A is four times to that of a gas B. If the molecular mass of A is M, what is the molecular mass of B?

(A) $M/4$ (B) $M/8$ (C) $4M$ (D) $2M$

99. What is the number of molecules in 1.12 L of $CO_2(g)$ at STP?

(A) 3×10^{22} (B) 3×10^{23} (C) 2.5×10^{22} (D) 2.5×10^{23}

100. Positive deviation from ideal behaviour takes place because of

(A) molecular interaction between atoms and $PV/nRT > 1$

(B) molecular interaction between atoms and $PV/nRT < 1$

(C) finite size of atoms and $PV/nRT > 1$

(D) finite size of atoms and $PV/nRT < 1$ **(I.I.T. 2003)**

101. For a monoatomic gas, kinetic energy = E. The relation with rms velocity is

(A) $u = \left(\frac{2E}{m}\right)^{1/2}$ (B) $u = \left(\frac{3E}{2m}\right)^{1/2}$ (C) $u = \left(\frac{E}{2m}\right)^{1/2}$ (D) $u = \left(\frac{E}{3m}\right)^{1/2}$ **(I.I.T. 2004)**

Hint: $PV = \frac{1}{3}mc^2$ (where c is rms velocity) or $c = \sqrt{\frac{3PV}{m}}$

For 1 molecule of gas, $PV = kT$. Hence $c = \sqrt{\frac{3kT}{m}}$. But K.E. $= \frac{3}{2}kT$ or 2 K.E. $= 3kT$.

Hence $c = \sqrt{\frac{2\text{ K.E.}}{m}}$

102. If helium and methane are allowed to diffuse out of a container under similar conditions of temperature and pressure, then the ratio of rate of diffusion of helium to methane is

(A) 2.0 (B) 1.0 (C) 0.5 (D) 4.0 **(I.I.T. 2005)**

Hint: $\frac{r_{He}}{r_{CH_4}} = \sqrt{\frac{16}{4}} = \sqrt{4} = 2$

103. One mole of a monoatomic ideal gas expands adiabatically at initial temperature T against a constant external pressure of 1 atm from one litre to two litres. Find out the final temperature. (R = 0.0821 L atm K^{-1} mol^{-1})

(A) T (B) $\frac{T}{2^{[(5/3)-1]}}$

(C) $T - \frac{2}{3 \times 0.0821}$ (D) $T + \frac{2}{3 \times 0.0821}$ **(I.I.T. 2005)**

Hint: Work done against constant pressure $= -P_{ext}(V_2 - V_1)$
Under adiabatic condition, $\Delta q = 0$. Hence $w = \Delta U$

$$\Delta U = nC_{v,m}(T_2 - T_1) = 1 \times \frac{3}{2}R(T_2 - T_1) - P(V_2 - V_1) = \frac{3}{2}R(T_2 - T_1)$$

$$\text{or } T_2 = T_1 - \frac{2}{3R} = T_1 - \frac{2}{3 \times 0.0821} \quad (\text{since } P = 1,\ V_2 - V_1 = 1)$$

104. The pressure exerted by 12 g of an ideal gas at temperature t°C in a vessel of volume V litre is 1 atm. When the temperature is increased by 10 degrees at the same volume, the pressure increases by 10%. The volume V is (M.W. of the gas = 120)
(A) 1.28 L (B) 0.28 L (C) 0.41 L (D) 0.82 L **(I.I.T. 2006)**
Hint: $PV = nRT$

$$(i)\ 1 \times V = \frac{12}{120} \times R(t°C + 273) \qquad (ii)\ 11 \times V = \frac{12}{120} \times R(t°C + 283)$$

By solving Equations (*i*) and (*ii*), we get $t = -173$°C and $V = 0.82$ L.

105. A gas described by van der Waals equation
(A) behaves similar to an ideal gas in the limit of large molar volumes
(B) behaves similar to an ideal gas in the limit of large pressures
(C) is characterized by van der Waals coefficients that are dependent on the identity of the gas but are independent of the temperature
(D) has the pressure that is lower than the pressure exerted by the same gas behaving ideally. **(I.I.T. 2008)**
Hint: A real gas following the van der Waals equation behaves similar to an ideal gas at low pressures (or large molar volumes). The van der Waals coefficients a and b are independent of temperature and characteristic of a gas. The pressure exerted by a real gas is lower than the pressure exerted by the same gas under ideal conditions since some pressure is lost by the real gas due to the presence of intermolecular force of attraction.

106. The term that corrects for the attractive forces present in a real gas in the van der Waals equation is
(A) nb (B) $\frac{an^2}{V^2}$ (C) $-\frac{an^2}{V^2}$ (D) $-nb$ **(I.I.T. 2009)**

Hint: The measure of force of attraction for n moles of real gas $= \frac{an^2}{V^2}$

$$\left(P + \frac{an^2}{V^2}\right)(V - nb) = nRT$$

107. According to kinetic theory of gases
(A) collisions are always elastic
(B) heavier molecules transfer more momentum to the wall of the container
(C) only a small number of molecules have very high velocity
(D) between collisions, the molecules move in straight lines with constant velocities.
Hint: The statements (A), (B), (C) and (D) are correct. **(I.I.T. 2011)**

ANSWERS

1. C	2. B	3. C	4. B	5. A, B, D	6. C
7. C	8. B	9. C	10. C	11. D	12. A, B
13. D	14. A	15. C	16. B	17. C	18. A, B
19. A	20. A	21. C	22. B	23. B	24. B
25. A	26. D	27. A	28. B	29. B	30. B
31. A	32. B	33. A	34. A	35. B	36. D
37. C, D	38. D	39. A	40. A, B	41. A, B	42. A, B
43. A, B	44. A, B	45. D	46. A	47. D	48. D
49. A	50. A	51. B	52. B, D	53. C, D	54. C
55. A	56. A	57. D	58. D	59. A	60. B
61. D	62. A	63. C	64. B	65. A	66. C
67. A	68. D	69. A	70. A	71. D	72. D
73. B	74. B	75. B	76. B	77. B	78. B
79. D	80. C	81. B	82. B	83. A	84. D
85. A	86. A	87. C	88. C, D	89. D	90. A
91. A	92. A	93. A	94. A	95. D	96. B
97. A	98. A	99. A	100. A	101. A	102. A
103. C	104. D	105. A, C, D	106. B	107. A, B, C, D	

6

States of Matter II : Solid State

IMPORTANT RELATIONS FOR NUMERICAL PROBLEMS

1. Relation between neighbour distance (d) and edge length of the unit cell: (i) For simple cube, $d = a$ (ii) For *bcc*, $d = (\sqrt{3}/2)a = 0.866a$ (iii) For *fcc*, $d = (\sqrt{2}/2)a = 0.707a$
2. Relation between atomic radius ($r = d/2$) and a: (i) For simple cube, $r = a/2$ (ii) For *bcc*, $r = (\sqrt{3}/4)a = 0.433a$ (iii) For *fcc*, $r = (\sqrt{2}/4)a = 0.3535a$
3. Density of crystal, $\rho = Mz/a^3N$ where M = molar mass of the substance, z = number of formula units per unit cell, a = edge length of unit cell, N = Avogadro's number
4. $d_{hkl} = a/\sqrt{h^2 + k^2 + l^2}$ where a = edge length of unit cell, h, k, l = Miller indices of the plane
5. Elements of symmetry in cubic crystals: plane of symmetry = 3 + 6 = 9, axis of symmetry = 3 + 4 + 6 = 13, centre of symmetry = 1, total elements of symmetry = 9 + 13 + 1 = 23
6. No. of atoms per unit cell: simple cube – 1, *fcc* – 4, *bcc* – 2
7. Coordination number: simple cube – 6, *fcc* – 12, *bcc* – 8
8. Ionic radii: $r_+ + r_- = a/2$ (for *fcc* structure): $r_+ + r_- = \sqrt{3}a/2$ (for *bcc* structure)
9. In close packing of n spheres, No. of octahedral voids = n, No. of tetrahedral voids = $2n$
10. $n\lambda = 2d \sin\theta$ where n = order of diffraction, λ = wavelength, d = distance between two successive planes in the crystal, θ = angle of reflection

OBJECTIVE TYPE MULTIPLE CHOICE QUESTIONS

1. In a solid AB having the NaCl structure, A atoms occupy the corners of the cubic unit cell. If all the face centred atoms along one of the axes are removed, then the resultant stoichiometry of the solid is
(A) A_3B_4 (B) A_2B (C) A_4B_3 (D) AB_2 **(I.I.T. 2001)**
Hint: The *fcc* atoms along one of the axis being 3, two along the corners and one in the centre of the face, the contribution of A^- ions $= \frac{1}{8} \times 6 \times 5 \times \frac{1}{2} = 3.25 \approx 3$. Hence the stoichiometry of the solid is A_3B_4.

2. Assertion (*A*): In any ionic solid [MX] with Schottky defect, the number of positive and negative ions are same.
Reason (*R*): Equal number of cation and anion vacancies are present.
(A) Both *A* and *R* are correct and *R* is the correct explanation of *A*
(B) Both *A* and *R* are correct and *R* is not the correct explanation of *A*
(C) *A* is true but *R* is false (D) *A* is false but *R* is true. **(I.I.T. 2001)**

3. The coordination number of a metal crystallising in a hexagonal close-packed structure is
(A) 12 (B) 4 (C) 8 (D) 6 **(I.I.T. 1999)**

4. Which of the following statement(s) is (are) correct?
(A) The coordination number of each type of ion in CsCl crystal is 8
(B) A metal that crystallises in *bcc* structure has a coordination number 12
(C) A unit cell of an ionic crystal shares some of its ions with other unit cells
(D) The length of the unit cell in NaCl is 552 pm ($r_{Na^+} = 95$ pm; $r_{Cl^-} = 181$ pm)
(I.I.T. 1998)
Hint: (D) $a = 2(r_{Na^+} + r_{Cl^-}) = 2(95 + 181) = 552$ pm. (A) and (C) are correct.

5. CsBr contains *bcc* structure. The length of its one side is 4.3 Å. The minimum distance between Cs^+ and Br^- ions will be
(A) 0.897 Å (B) 3.72 Å (C) 1.794 Å (D) none **(I.I.T. 1995)**
Hint: The Cs^+ and Br^- ions touch each other along the body diagonal of the cube.
Hence the shortest distance between Cs^+ and $Br^- = \frac{\sqrt{3}}{2} a = \frac{1.732}{2}(4.3 \text{ Å}) = 3.72$ Å

6. FeO has cubic structure and each edge of the unit cell is 5.0 Å. Assuming density of FeO as 4.0 g cm^{-3}, the number of Fe^{2+} and O^{2-} ions present in each unit cell is
(A) $3Fe^{2+}$ and $3O^{2-}$ (B) $4Fe^{2+}$ and $4O^{2-}$
(C) $2Fe^{2+}$ and $4O^{2-}$ (D) $4Fe^{2+}$ and $2O^{2-}$

7. Potassium crystallises with a
(A) *bcc* lattice (B) *fcc* lattice
(C) simple cubic lattice (D) orthorhombic lattice

8. The second order Bragg diffraction of X-rays with $\lambda = 1.00$ Å forms a set of parallel planes in a metal occurs at an angle of 60°. The distance between the scattering planes in the crystal is
   (A) 1.15 Å (B) 2.00 Å (C) 1.00 Å (D) 0.575 Å
9. An example of unit cell with crystallographic dimensions $a \neq b \neq c$, $\alpha = \gamma = 90°$, $\beta \neq 90°$ is
   (A) calcite (B) graphite (C) rhombic S (D) monoclinic S
10. The ratio of close-packed atoms to tetrahedral holes in cubic close-packing is
   (A) 1 : 1 (B) 1 : 2 (C) 2 : 1 (D) 1 : 3
11. Which of the following is a ferroelectric compound?
   (A) Pb_2O_3 (B) $K_4[Fe(CN)_6]$ (C) $BaTiO_3$ (D) none of these.
12. The formula of a spinel is
   (A) Fe_3O_4 (B) $MgAl_2O_4$ (C) Na_3AlF_6 (D) $Na_2SiO_3 \cdot Al_2(SiO_3)_3$
13. A solid has a structure in which W atoms are located at the corners of a cubic lattice. O atoms are located at the centre of edges and Na atoms are at the centre of the cube. The formula of the compound is
   (A) $NaWO_4$ (B) Na_2WO_3 (C) $NaWO_3$ (D) $NaWO_2$
14. Schottky defect in crystals is observed when
   (A) density of the crystal is increased
   (B) unequal number of cations and anions are missing from the lattice
   (C) equal number of cations and anions are missing from the lattice
   (D) an ion leaves its normal site and occupies its interstitial sites.
15. Which substance exhibits antiferromagnetism?
   (A) CdO (B) CrO_2 (C) Mn_2O_3 (D) ZrO_2
16. The edge length of *fcc* cell is 508 pm. If the radius of the cation is 110 pm, the radius of the anion in pm is
   (A) 618 (B) 144 (C) 398 (D) 288
17. Pick up the correct statement.
   (A) The coordination number of Na^+ ion in NaCl is 4
   (B) In ionic compounds having Frenkel defect, the ratio r_+/r_- is high
   (C) The unit cell having crystal parameters $a = b \neq c$, $\alpha = \beta = 90°$, $\gamma = 120°$ is hexagonal
   (D) The ionic crystal of AgBr has Schottky defect.
18. Which of the following is used to calculate the crystal lattice energy of an ionic crystal?
   (A) Carnot cycle (B) Born–Haber cycle
   (C) van't Hoff equation (D) Morse equation
19. In a crystalline state of an ionic compound of a metal M having cation and anion radii r_M and r_X respectively, the coordination number of M will be four if the value of $r_X : r_M$ is
   (A) > 1 to < 1.4 (B) > 1.4 to < 2.4 (C) > 2.4 to < 4.6 (D) > 4.6 to < 5.6

20. In a metal M having *bcc* lattice with the cell edge 400 pm, the atomic radius of M is
(A) 141 pm (B) 173 pm (C) 100 pm (D) 200 pm

21. Identify the correct statement.
(A) Ferro electricity is due to the alignment of dipole in the same direction
(B) In pyroelectric crystals, electricity is produced on heating polar crystals
(C) Piezo electricity is due to net dipole moment (D) All of these.

22. Identify the incorrect statement.
(A) In the wurtzite structure, there is ABAB ... sequence of layers
(B) In the face-centred type of packing there is ABCABC ... sequence of layers
(C) Body centred cubic is not a close-packed arrangement
(D) Cubic packing has a coordination number of 8 whereas body centred cubic packing has a coordination number of 12.

23. The radius ratio limit for a tetrahedral ionic solid is
(A) > 0.732 (B) 0.155–0.225 (C) 0.225–0.414 (D) 0.414–0.732

24. The radius ratio limit for an octahedral ionic solid is
(A) 0.155–0.225 (B) 0.225–0.414 (C) 0.732–0.82 (D) 0.414–0.732

25. The radius ratio limit for a body centred ionic solid is
(A) 0.155–0.225 (B) 0.225–0.414 (C) 0.414–0.732 (D) > 0.732

26. The ABAB ... sequence of layers is present in the type of packing
(A) primitive cube (B) face centred cube
(C) hexagonal close-packed (D) tetragonal

27. The ABCABC ... sequence of layers is present in the type of packing
(A) primitive cube (B) face centred cube
(C) hexagonal close-packed (D) tetragonal

28. A metallic element crystallises into a lattice containing a sequence of layers of ABABAB ... Any packing of spheres leaves out voids in the lattice. The empty space in percentage by volume in this lattice is
(A) 26% (B) 32% (C) 20% (D) 30% **(I.I.T. 1996)**

Hint: In *hcp* unit cell, the volume occupied by the metal atoms $= 6 \times \frac{4}{3} \times \pi \times r^3 = 6 \times 1.33 \times \frac{22}{7} \times r^3 = 25.08r^3$. Base area of *hcp* unit cell $= 6 \times \frac{\sqrt{3}}{4} \times 4r^2$. Height of *hcp* unit cell $= 4r \times \sqrt{2/3}$. Volume of the unit cell = Area × height $= 6 \times \frac{\sqrt{3}}{4} \times 4r^2 \times 4r \times \frac{\sqrt{2}}{3} = 33.94r^3$. Volume of empty space in one unit cell $= 33.94r^3 - 25.14r^3 = 8.80r^3$. Percentage void $= 8.80r^3 \times 100/33.94r^3 = 25.93\%$.

29. Which of the following is the correct expression of Bragg's law?
(A) $n\lambda = 2d \sin \theta$ (B) $2n\lambda = d \sin \theta$ (C) $2\lambda = nd \sin \theta$ (D) $d\lambda = 2n \sin \theta$

30. The density of a crystal does not change due to
(A) F-centres
(B) edge dislocation
(C) Schottky defect
(D) Frenkel defect

31. The density of a crystal is lowered due to
(A) F-centres
(B) edge dislocation
(C) Schottky defect
(D) Frenkel defect

32. The doping of Ga in Ge gives
(A) p-type semiconductor
(B) n-type semiconductor
(C) insulator
(D) conductor comparable to metals.

33. The doping of As in Si gives
(A) p-type semiconductor
(B) n-type semiconductor
(C) insulator
(D) conductor comparable to metals.

34. Boron is chosen as a constituent material along with Si in p-type semiconductor because
(A) B has a lower atomic number than Si
(B) B and Si belong to group 14 of Periodic Table
(C) B has one more electron in the outermost shell than that of Si
(D) B has one less electron in the outermost shell than that of Si.

35. Arsenic is chosen as a constituent material along with Si in n-type semiconductor because
(A) As has a higher atomic number than Si
(B) As and Si belong to group 14 of Periodic Table
(C) As has one less electron in the outermost shell than that of Si
(D) As has one more electron in the outermost shell than that of Si.

36. In semiconductors there occurs
(A) flow of electrons from p-type semiconductor to n-type semiconductor
(B) flow of electrons from n-type semiconductor to p-type semiconductor
(C) equilibrium between electrons and holes
(D) equilibrium between diffusive force and electrostatic force.

37. Suggest a purification method for obtaining highly pure Si used as semiconductor material.
(A) reduction method
(B) oxidation method
(C) electrochemical refining method
(D) zone refining method.

38. Ionic solids with Schottky defects contain in their structure.
(A) cation vacancies
(B) equal number of cation and anion vacancies
(C) cation vacancies and interstitial cations
(D) anion vacancies and interstitial anions.

39. Ni (At. wt = 58.7) crystallises in a face centred cubic lattice. If the edge of the unit cell is 3.52 Å, its density is
    (A) 0.0894 g cm^{-3} (B) 0.00894 g cm^{-3} (C) 0.894 g cm^{-3} (D) 8.94 g cm^{-3}
40. Frenkel defect is caused due to
    (A) the shift of a positive ion from its normal lattice site to an interstitial site
    (B) an ion missing from the normal lattice site creating a vacancy
    (C) an extra negative ion occupying an interstitial position in the lattice
    (D) an extra positive ion occupying an interstitial position in the lattice.
41. The number of atoms in 1000 g of an *fcc* crystal with density = 10 g/cm^3 and unit cell edge = 200 pm is
    (A) 2×10^{25} (B) 1×10^{25} (C) 3×10^{25} (D) 5×10^{25}
42. In a semiconductor device, a crystal of Ge in which 1 atom in 10^9 atoms is an impure atom, weighs 12 g. The number of impure atoms it contains approximately is
    (A) 10^{-9} (B) 10^9 (C) 10^{-14} (D) 10^{14}
43. The conductivity range of semiconductors in ohm^{-1} cm^{-1} is
    (A) 10^6–10^8 (B) 10^{-9}–10^2 (C) 10^8–10^{12} (D) 10^{-13}–10^{-12}
44. The intermetallic compound LiAg crystallises in cubic lattice in which both Li and Ag have coordination number 8. The crystal class is
    (A) primitive cube (B) *fcc*
    (C) *bcc* (D) none of these.
45. An element (At. wt. = 100) having *bcc* structure has unit cell edge 400 pm. The density of the element is
    (A) 2.144 g/cm^3 (B) 7.289 g/cm^3 (C) 5.189 g/cm^3 (D) 10.376 g/cm^3
46. Identify the incorrect statement.
    (A) F-centres contribute to the colour of the crystal
    (B) The electrical conductivity of a semiconductor increases with increase in temperature
    (C) The addition of $CaCl_2$ to AgCl crystal produces Schottky defect
    (D) The void volume in a *bcc* type metallic crystal is 26%
47. Identify the incorrect statement.
    (A) The compounds $M^{II}Fe_2O_4$ having spinel structure are ferrimagnetic solids
    (B) The mixed oxide Fe_3O_4 which is used in recording is a ferromagnetic solid
    (C) In the spinel structure ($M^{II}Fe_2O_4$), the anion lattice is composed of O^{2-} ions
    (D) The coordination number of a metal crystallising in *bcc* lattice is six.
48. Identify the incorrect statement.
    (A) In the NaCl crystal each Na^+ ion is surrounded by $6Cl^-$ ions nearest neighbours and $12Na^+$ ions next nearest neighbours
    (B) The compound $[Cu(CH_3COO)_2{\cdot}H_2O]_2$ is an antiferromagnetic solid
    (C) Superconductors are diamagnetic substances
    (D) The ions in ionic crystals can have the coordination number of 12.

49. Identify the incorrect statement.
    (A) In the fluorite structure, the coordination number of Ca^{2+} ion in CaF_2 is 8
    (B) ZnS can exhibit both zinc blende and wurtzite structure
    (C) The structure of an ionic compound having $r_+/r_- = 0.66$ is *fcc*
    (D) A diamagnetic compound is neither attracted nor repelled by a magnet.
50. Identify the incorrect statement.
    (A) The type of crystal defect caused by adding $SnCl_2$ to NaCl crystal is cation vacancy
    (B) The ABCABC ... type of arrangement is called cubic close-packing
    (C) The ABAB ... type of arrangement is called hexagonal close-packing
    (D) Ionic crystals do not have long range order.
51. Identify the correct expression for NaCl unit cell with lattice parameter a.
    (A) $r_{Na^+} + r_{Cl^-} = \sqrt{3}a/4$ (B) $r_{Na^+} + r_{Cl^-} = a/4$
    (C) $r_{Na^+} + r_{Cl^-} = a/2$ (D) $r_{Na^+} + r_{Cl^-} = 4a$
52. Identify the correct expression for CsCl unit cell with lattice parameter a.
    (A) $r_{Cs^+} + r_{Cl^-} = 3a/2$ (B) $r_{Cs^+} + r_{Cl^-} = \sqrt{3}a/2$
    (C) $r_{Cs^+} + r_{Cl^-} = 2a$ (D) $r_{Cs^+} + r_{Cl^-} = a\sqrt{2}$
53. A normal 2 : 3 spinel structure is possessed by
    (A) $Fe(CO)_5$ (B) Fe_3O_4 (C) $Zn(TiZn)O_4$ (D) $MgAl_2O_4$
54. An inverse 2 : 3 spinel structure is possessed by
    (A) Co_3O_4 (B) Fe_3O_4 (C) $Ni(CO)_4$ (D) Mn_3O_4
55. The number of NaCl molecules in the unit cell of NaCl crystal is
    (A) 2 (B) 4 (C) 6 (D) 8
56. The atomic radius for *bcc* crystal with lattice parameter a is
    (A) a (B) $a/2$ (C) $a\sqrt{2}/4$ (D) $a\sqrt{3}/4$
57. Which of the following crystals contains only one Bravais lattice?
    (A) Monoclinic (B) Rhombohedral
    (C) Triclinic (D) Hexagonal
58. Which of the following crystals contains the maximum number of Bravais lattices?
    (A) Tetragonal (B) Orthorhombic (C) Triclinic (D) Cubic
59. Potassium fluoride crystallises in *fcc* lattice. The edge length of its unit cell is 537.6 pm. The distance between K^+ and F^- is
    (A) 268.0 pm (B) 538.8 pm (C) 1080.4 pm (D) 280.0 pm
60. The atomic radius for *fcc* unit cell with lattice parameter a is
    (A) $a/2$ (B) $a\sqrt{2}$ (C) $a/2\sqrt{2}$ (D) $\sqrt{3}a/4$
61. The atomic radius for the primitive cubic unit cell with lattice parameter a is
    (A) $a/2$ (B) $a/2\sqrt{2}$ (C) $4a/\sqrt{3}$ (D) $\sqrt{3}a/4$
62. The fraction of volume occupied by atoms in *fcc* unit cell is
    (A) 0.74 (B) 0.68 (C) 0.48 (D) 0.32

63. The fraction of volume occupied by atoms in *bcc* unit cell is
(A) 0.74 (B) 0.68 (C) 0.48 (D) 0.32

64. The fraction of volume occupied by atoms in the primitive cubic unit cell is
(A) 0.74 (B) 0.68 (C) 0.48 (D) 0.32

65. In the closest packing of atoms A (radius: r_A), the radius (r_B) of atom B which can be fitted into octahedral void is given by
(A) $r_B = 0.155r_A$ (B) $r_B = 0.225r_A$ (C) $r_B = 0.414r_A$ (D) $r_B = 0.732r_A$

66. In the closest packing of atoms A (radius: r_A), the radius (r_B) of atom B which can be fitted into tetrahedral void is given by
(A) $r_B = 0.155r_A$ (B) $r_B = 0.225r_A$ (C) $r_B = 0.414r_A$ (D) $r_B = 0.732r_A$

67. Graphite is
(A) paramagnetic (B) diamagnetic (C) ferromagnetic (D) antiferromagnetic

68. Graphite conducts electricity due to the presence of
(A) free electrons (B) mobile electrons
(C) lone pair of electrons (D) $p\pi$–$p\pi$ bonded electrons.

69. The 8 : 8 type of packing is present in
(A) MgF_2 (B) KCl (C) NaCl (D) CsCl

70. Schottky defect is found in
(A) NaCl (B) KCl (C) CsCl (D) $BaCl_2$

71. Frenkel defect is found in
(A) AgCl (B) ZnS (C) Agl (D) $MgCl_2$

72. Schottky defect is found in solids
(A) which are ionic (B) which have high coordination number
(C) which have almost same size of cation and anion
(D) none of these.

73. TiO_2 exhibits 6 : 3 coordination and possesses rutile structure. Which of the following exhibits a rutile-like structure?
(A) CaF_2 (B) CsCl (C) SnO_2 (D) MnO_2

74. 8 : 4 coordination is present in fluorite structure. Which of the following exhibits a fluorite-like structure?
(A) $BaCl_2$ (B) $SnCl_2$ (C) PbF_2 (D) ThO_2

75. 4 : 8 coordination is present in antifluorite structure. Which of the following exhibits an antifluorite structure?
(A) Na_2S (B) K_2S (C) Li_2O (D) HgF_2

76. Zinc blende structure exhibits 4 : 4 coordination. Which of the following exhibits a zinc blende-like structure?
(A) BeS (B) AgI (C) CuBr (D) NH_4F

77. Which of the following possesses a wurtzite structure?
(A) NH_4F (B) CdS (C) AlN (D) K_2S

78. An element occurring in the *bcc* structure has 12.08×10^{23} unit cells. The total number of atoms of the elements in these cells will be
(A) 6.04×10^{23} (B) 12.08×10^{23} (C) 24.16×10^{23} (D) 36.18×10^{23}

79. The pink colour of non-stoichiometric LiCl crystal is due to
(A) Li^+ in the lattice (B) Cl^- in the lattice
(C) electrons in the lattice (D) none of these.

80. The empty space present in the hexagonal close-packed structure is about
(A) 26% (B) 32% (C) 36% (D) 38%

81. To get *n*-type semiconductor, the impurity to be added to silicon should have the following number of valence electrons
(A) 2 (B) 3 (C) 1 (D) 5

82. The *hcp* and *ccp* arrangements are described respectively as
(A) ABABAB ..., ABCABCABC ... (B) ABBABA ..., ABCABCABC ...
(C) ABABAB ..., ACBABCABC ... (D) ABCACBABC ..., ACBACBABA ...

83. Ferromagnetism is in
(A) $\uparrow\downarrow\uparrow\downarrow$ (B) $\uparrow\uparrow\uparrow\uparrow\uparrow$ (C) $\uparrow\uparrow\uparrow\downarrow\downarrow$ (D) none of these.

84. Ferrimagnetism is in
(A) $\uparrow\downarrow\uparrow\downarrow$ (B) $\uparrow\uparrow\uparrow\uparrow\uparrow$ (C) $\uparrow\uparrow\uparrow\downarrow\downarrow$ (D) none of these.

85. A substance A_xB_y crystallises in a *fcc* lattice in which atoms A occupy each corner of the cube and atoms B occupy the centres of each face of the cube. Identify the composition of A_xB_y.
(A) AB_3 (B) A_4B_3
(C) A_3B (D) Composition cannot be specified.

(I.I.T. 2002)

Hint: No. of corner atoms (A) = $8 \times 1/8 = 1$, No. of face centred atoms (B) = $6 \times 1/2 = 3$. Hence the composition is AB_3.

86. Frenkel defect is not found in
(A) KCl (B) AgBr (C) AgI (D) ZnS

87. Which of the following oxides is ferromagnetic?
(A) CrO_2 (B) Cr_2O_3 (C) Fe_2O_3 (D) Fe_3O_4

88. Which of the following magnetism occurs due to the spontaneous alignment of electrons in the same direction in an atom or ion?
(A) Ferromagnetism (B) Antiferromagnetism
(C) Diamagnetism (D) Paramagnetism

89. Which of the following oxides is diamagnetic and amphoteric?
(A) V_2O_5 (B) VO_2 (C) TiO (D) FeO

90. Schottky defect is not found in
(A) AgBr (B) NaCl (C) KCl (D) CsCl

91. Doping of $CdCl_2$ in AgCl crystals gives
(A) interstitial defect (B) Frenkel defect
(C) Schottky defect (D) *p*-type semiconductor

92. Which of the following metals do not crystallise in hexagonal close packed structure?
   (A) Cu (B) Cd (C) Zn (D) Mg

93. Which of the following *fcc* structures contains cations in alternate tetrahedral voids?
   (A) NaCl (B) ZnS (C) Na_2O (D) CaF_2 **(I.I.T. 2005)**

   Hint: In ZnS, the anions (S^{2-}) are placed in *fcc* manner and the cations (Zn^{2+}) are placed in the alternate tetrahedral voids.

94. NaCl crystals are doped with 10^{-3} mol per cent of $SrCl_2$. The concentration of cation vacancies is
   (A) 5.05×10^{18} mol^{-1} (B) 2.06×10^{18} mol^{-1}
   (C) 3.01×10^{18} mol^{-1} (D) 6.02×10^{18} mol^{-1}

   Hint: Each Sr^{2+} ion replaces two Na^+ ions, but occupies only one Na^+ lattice point. This leads to a cation vacancy of one. No. of moles of cation vacancies in 100 moles of NaCl $= 10^{-3}$. No. of moles of cation vacancies in 1 mole of NaCl $= 10^{-3}/100 = 10^{-5}$. Total cation vacancies $= 10^{-5} \times N_A = 10^{-5} \times 6.02 \times 10^{23}$ mol^{-1} $= 6.02 \times 10^{18}$ mol^{-1}.

94. NaCl crystals are doped with 10^{-3} mol per cent of $SrCl_2$. The concentration of cation vacancies is

95. Statement–1: Band gap in germanium is small.

   because

   Statement–2: The energy spread of each germanium atomic energy level is infinitesimally small.
   (A) Statement–1 is true, Statement–2 is true; Statement–2 is a correct explanation for Statement–1
   (B) Statement–1 is true, Statment–2 is true; Statement–2 is not a correct explanation for Statement–1
   (C) Statement–1 is true, Statment–2 is false
   (D) Statement–1 is false, Statment–2 is true. **(I.I.T. 2007)**

96. The correct statement(s) regarding defects in solids is (are)
   (A) Frenkel defect is usually favoured by a very small difference in sizes of cation and anion.
   (B) Frankel defect is a dislocation effect.
   (C) Trapping of an electron in the lattice leads to the formation of F-centre.
   (D) Schottky defects have no effect on the physical properties of solids. **(I.I.T. 2009)**

   Hint: Frenkel defect is a dislocation defect, observed when the size of cation and anion differs significantly. F-center is formed when an anion is lost from the lattice and vacancy is filled by trapping an electron. Schottky defect changes the density of a crytalline solid.

97. The packing efficiency of the two-dimensional square unit cell shown below is

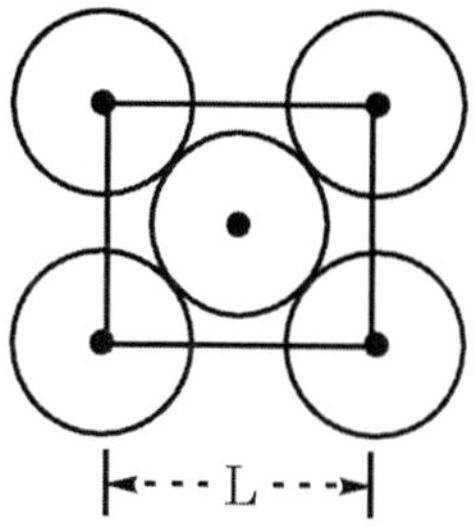

(A) 39.27% (B) 68.02% (C) 74.05% (D) 78.54% **(I.I.T. 2010)**

Hint:

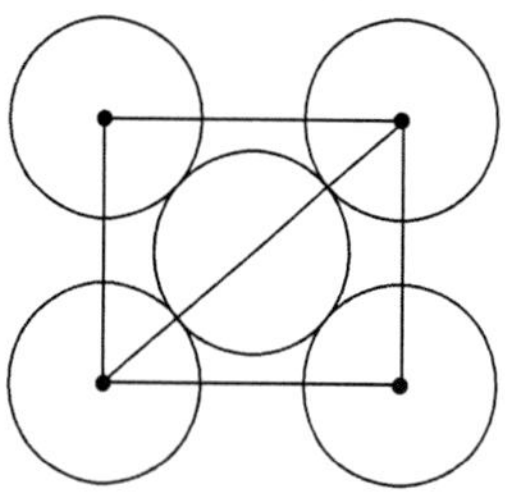

$a = (2\sqrt{2}r)$

$$P.F = \frac{2\times\pi r^2}{\left(2\sqrt{2}r\right)^2} = \frac{2\pi r^2}{8r^2} = \frac{\pi}{4} = \frac{3.14159}{4} = 0.7854$$. Hence the packing efficiency is 78.54%.

ANSWERS

1. A	2. A	3. A	4. A, C, D	5. B	6. B
7. A	8. A	9. D	10. B	11. C	12. B
13. C	14. C	15. C	16. B	17. C	18. B
19. C	20. B	21. D	22. D	23. C	24. D
25. D	26. C	27. B	28. A	29. A	30. D
31. C	32. A	33. B	34. D	35. D	36. B, C, D
37. D	38. B	39. D	40. A	41. D	42. D
43. B	44. C	45. C	46. D	47. B, D	48. D
49. D	50. D	51. C	52. B	53. C	54. B
55. B	56. D	57. B, C, D	58. B	59. A	60. C
61. A	62. A	63. B	64. C	65. C	66. B
67. B	68. B, D	69. D	70. A, B, C	71. A, B, C	72. A, B, C
73. C, D	74. A, B, C	75. A, B, C	76. A, B, C	77. A, B, C	78. C
79. C	80. A	81. D	82. A	83. B	84. C
85. A	86. A	87. A	88. A	89. A	90. A
91. D	92. A	93. B	94. D	95. C	96. B, C
97. D					

7

Solutions

IMPORTANT RELATIONS FOR NUMERICAL PROBLEMS

1. Mole fraction, $x_A = x_A/(n_A + n_B)$, $x_B = n_B/(n_A + n_B)$ where n_A and n_B are the number of moles of A and B respectively.
2. $p_{total} = p_A + p_B = x_A p_A^o + x_B p_B^o$
3. $(p° - p)/p° = wM/Wm$
4. $\pi V = nRT$; $m = wRT/\pi V$; for isotonic solutions, $\pi_1 = \pi_2$
5. $\Delta T_b = K_b m$; $m = K_b \times w \times 1000/W \times \Delta T_b$
6. $\Delta T_f = K_f m$; $m = K_f \times w \times 1000/W \times \Delta T_f$
7. $K_f = 0.002(T_o)^2/l_f$ where T_o = freezing point of the solvent, l_f = latent heat of fusion (in cal g^{-1}) of the solvent
8. Degree of association, $\alpha = (i - 1)/(n - 1)$ where i = van't Hoff factor, n = number of molecules which combine to give one associated molecule
9. Degree of dissociation, $\beta = (1 - i)/[1 - (1/n)]$
10. Variation of vapour pressure with temperature: $dP/dT = q/T(V_B - V_A)$ where P = vapour pressure, T = temperature, V_A and V_B are the molar volumes of the pure substance in the phases A and B respectively, q = heat change
11. Normality of the mixture of two solutions of the same solute, $N = (N_1V_1 + N_2V_2)/(V_1 + V_2)$
12. Volume of water $(V_2 - V_1)$ to be added to get a solution of normality N_2 from V_1 mL of solution of normality N_1: $V_2 - V_1 = [(N_1 - N_2)/N_2]V_1$
13. Normality of the acidic (basic) mixture = $(N_aV_a + N_bV_b)/(V_a + V_b)$
14. Normality = $\dfrac{W \times 1000}{V \times \text{Eq.W.}}$
15. Normality = molarity × M.W./Eq.W.
16. Eq. mass of acid = M.W./basicity; Eq. mass of base = M.W./acidity

OBJECTIVE TYPE MULTIPLE CHOICE QUESTIONS

1. In the depression of freezing point experiment, it is found that
   (A) only solute molecules solidify at freezing point
   (B) only solvent molecules solidify at freezing point
   (C) vapour pressure of the solution is more than that of pure solvent
   (D) vapour pressure of the solution is less than that of pure solvent. **(I.I.T. 1999)**
2. A 0.2 molal aqueous solution of a weak acid (HX) is 20% ionised. The freezing point of the solution is
   (A) – 0.45°C (B) 0.90°C (C) – 0.31°C (D) – 0.53°C **(I.I.T. 1995)**

   Hint: $HX \rightleftharpoons H^+ + X^-$
   $(1 - 0.2)0.2m = 0.16m$; $0.2 \times 0.2m = 0.04m$; $0.2 \times 0.2m = 0.04m$

   Total molality of solution = $0.16m + 0.04m + 0.04m = 0.24m$
   $\Delta T_f = - K_f m = -$ (1.86 K kg mol^{-1}) (0.24 mol kg^{-1}) = – 0.4464 K = – 0.45 K = – 0.45°C.
3. How many moles of Fe^{2+} ions are formed when excess of iron is treated with 50 mL of 4.0*M* HCl under inert atmosphere? Assume no change in volume.
   (A) 0.1 (B) 0.2 (C) 0.4 (D) 0.8 **(I.I.T. 1993)**
4. The vapour pressure of a solution of 5 g of non-electrolyte in 100 g of water at a particular temperature is 2985 N/m^2. The vapour pressure of water is 3000 N/m^2. The molecular mass of the solute is
   (A) 60 (B) 120 (C) 180 (D) 380 **(I.I.T. 1993)**
5. Increasing temperature of an aqueous solution will cause
   (A) decrease in molality (B) decrease in molarity
   (C) decrease in mole fraction (D) decrease in % w/w **(I.I.T. 1993)**
6. The freezing point of equimolal aqueous solution will be highest for
   (A) $C_6H_5NH_3Cl$ (B) $Ca(NO_3)_2$ (C) $La(NO_3)_3$ (D) $C_6H_{12}O_6$ **(I.I.T. 1990)**
7. In which mode of expression, the concentration of the solution remains independent of temperature?
   (A) molarity (B) normality (C) formality (D) molality **(I.I.T. 1988)**
8. If two compounds have the same empirical formula but different molecular formulae, they must have
   (A) different percentage composition (B) different molecular composition
   (C) same viscosity (D) same vapour density. **(I.I.T. 1987)**
9. When mercury(II) iodide is added to an aqueous solution of potassium iodide, the
   (A) freezing point is lowered (B) freezing point is raised
   (C) freezing point does not change (D) boiling point does not change. **(I.I.T. 1987)**
10. At the boiling point, a liquid is in equilibrium with vapour. The molecules in the two phases have equal
    (A) kinetic energy (B) potential energy
    (C) intermolecular forces (D) total energy. **(I.I.T. 1984)**

11. Which of the following $0.1M$ aqueous solutions will have the lowest freezing point?
(A) K_2SO_4 (B) NaCl (C) glucose (D) urea **(I.I.T. 1984)**
12. An azeotropic solution of two liquids has boiling point lower than either of them when it
(A) shows negative deviation from the Raoult's law
(B) shows no deviation from the Raoult's law
(C) shows a positive deviation from the Raoult's law
(D) is saturated. **(I.I.T. 1981)**
13. If 0.2 mole of Na_3PO_4 is mixed with 0.5 mole of $BaCl_2$, the maximum number of moles of $Ba_3(PO_4)_2$ that can be formed is
(A) 0.1 (B) 0.2 (C) 0.5 (D) 0.7 **(I.I.T. 1981)**
14. When a non-volatile solute is dissolved in a solvent, the relative lowering of vapour pressure is equal to
(A) concentration of solute in g/L (B) concentration of solute in g/100 mL
(C) mole fraction of solute (D) mole fraction of solvent. **(I.I.T. 1983)**
15. For a dilute solution, Raoult's law states that
(A) the lowering of vapour pressure is equal to the mole fraction of the solute
(B) the relative lowering of vapour pressure is equal to the mole fraction of the solute
(C) the relative lowering of vapour pressure is proportional to the amount of solute in solution
(D) the vapour pressure of the solution is equal to the mole fraction of the solvent. **(I.I.T. 1985)**
16. $0.01M$ solution each of urea, NaCl and Na_2SO_4 are taken. The ratio of depression of freezing point is
(A) 1 : 1 : 1 (B) 1 : 2 : 1 (C) 1 : 2 : 3 (D) 2 : 2 : 3 **(Roorkee 1990)**
17. The osmotic pressure of a solution is 0.0821 atm at 300 K. Find the concentration in mol/L.
(A) 0.033 (B) 0.066 (C) 0.33×10^{-2} (D) 3 **(Roorkee 1990)**
18. The osmotic pressure of sucrose solution containing 1.75 g in 150 mL of solution at 17°C in atm is
(A) 0.81 (B) 0.081 (C) 1.81 (D) 0.18 **(Roorkee 1985)**
19. A X molal solution of a compound in benzene has mole fraction of solute = 0.2. The value of X is
(A) 14 (B) 3.2 (C) 1.4 (D) 2
20. The vapour pressure of water at 300 K in a closed container is 0.4 atm. If the volume of the container is doubled, its vapour pressure at 300 K will be
(A) 0.08 atm (B) 0.2 atm (C) 0.4 atm (D) 0.6 atm
21. A 5.13% solution of cane sugar ($C_{12}H_{22}O_{11}$) is isotonic with a 0.9% solution of an unknown solute. The molar mass of the solute in g mol^{-1} is
(A) 120 (B) 180 (C) 342 (D) 60

22. The depression of freezing point of 1*M* urea, 1*M* glucose and 1*M* NaCl is in the ratio
   (A) 1 : 2 : 3 (B) 3 : 2 : 2 (C) 1 : 1 : 2 (D) none of these.
23. The vapour pressure of benzene at a certain temperature is 640 mm. A non-volatile and non-electrolytic solid weighing 2.175 g is added to 39.08 g of benzene. If the vapour pressure of the solution is 600 mm, what is the molecular mass of the substance?
   (A) 49.50 (B) 59.60 (C) 69.46 (D) 79.80
24. Which is the characteristic of a solution showing positive deviation from Raoult's law?
   (A) A—B interaction is stronger than A—A and B—B interactions
   (B) A—B interaction is weaker than A—A and B—B interactions
   (C) $\Delta H_{mixing} = 0$ (D) $\Delta V_{mixing} < 0$
25. A non-volatile solute (A) is dissolved in a volatile solvent B. The vapour pressure of the resultant solution is p_s and of the pure solvent is p_B^o. If x_B is the mole fraction of the solvent in solution, which of the following is correct?
   (A) $p_s = x_A p_B^o$ (B) $p_B^o = x_B p_s$ (C) $p_s = x_B p_B^o$ (D) $p_s^o = x_A p_s$
26. When blood cells are placed in a solution of KCl whose osmotic pressure is greater than that of the cell fluid, there occurs
   (A) shrinking in size of the cell (B) increase in size of the cell
   (C) no change in size
   (D) first shrinking and then regaining of original size.
27. Identify the incorrect statement.
   (A) When an azeotropic mixture is distilled its volume remains unchanged
   (B) The isopiestic solutions are solutions which have the same vapour pressure
   (C) When a liquid and its vapours are at equilibrium and the pressure is suddenly decreased, cooling occurs
   (D) Equimolar aqueous solutions of urea and NaCl are isotonic.
28. An aqueous solution of urea containing 18 g urea in 1500 mL water has a density equal to 1.052. If the molecular mass of urea is 60, then the molality of the solution is
   (A) 1.200 (B) 0.100 (C) 0.200 (D) 0.192
29. Which of the following statements is correct?
   (A) The relative lowering of vapour pressure does not depend upon the solvent at a given concentration of the solute
   (B) Lowering of vapour pressure depends upon the nature of the solute
   (C) Lowering of vapour pressure does not depend upon the solvent at a given concentration of the solute
   (D) Lowering of vapour pressure takes place only in ideal solutions.
30. The osmotic pressure of equimolar solutions of $BaCl_2$, NaCl and sucrose will be in the order
   (A) $BaCl_2$ > NaCl > sucrose (B) NaCl > $BaCl_2$ > sucrose
   (C) sucrose > $BaCl_2$ > NaCl (D) sucrose > NaCl > $BaCl_2$

31. What is the freezing point of a solution containing 8.1 g HBr in 100 mL water assuming the acid to be 90% ionised (K_f for water = 1.86 K kg/mol)?
(A) – 0.35°C (B) – 3.53°C (C) 0°C (D) 0.85°C

32. Which of the following salts would have the same value of the van't Hoff factor (i) as that of $K_3[Fe(CN)_6]$?
(A) NaCl (B) Na_2SO_4 (C) $Al_2(SO_4)_3$ (D) $Al(NO_3)_3$

33. The vapour pressure of pure benzene is 640 mm at 929 K. A solution of a solute in benzene shows a vapour pressure of 630 mm at the same temperature. The mole fraction of the solute is
(A) 0.25 (B) 0.20 (C) 0.18 (D) 0.016

34. 5 mL of N HCl, 20 mL of $N/2$ H_2SO_4 and 30 mL of $N/3$ HNO_3 are mixed together and the volume made to 1 L. The normality of the resulting solution is
(A) $N/40$ (B) $N/20$ (C) $N/10$ (D) $N/5$

35. The density of an NH_4OH solution is 0.6 g/mL. It contains 34% by weight of NH_4OH. Calculate the normality of the solution.
(A) 10N (B) 5.8N (C) 4.8N (D) 0.5N

36. If the mole fraction of the component A in vapour phase is x_1 and the mole fraction of component A in liquid mixture is x_2 (p_A^o = vapour pressure of pure A, p_B^o = vapour pressure of pure B), then the total vapour pressure of liquid mixture is
(A) $\dfrac{p_A^o x_2}{x_1}$ (B) $\dfrac{p_B^o x_2}{x_1}$ (C) $\dfrac{p_B^o x_1}{x_2}$ (D) $\dfrac{p_A^o x_1}{x_2}$

37. For an ideal solution containing a non-volatile solute, identify the correct expression.
(A) $\Delta T_b = K_b^m$ where $K_b = M_1 R T_b^2 \Delta H_{m,vap}$
(B) $\Delta T_b = K_b^m$ where $K_b = M_1 T_b^2 / R\Delta H_{m,vap}$
(C) $\Delta T_b = K_b^m$ where $K_b = RT_b^2 / M_1 \Delta H_{m,vap}$
(D) $\Delta T_b = K_b^m$ where $K_b = M_1 R T_b^2 / \Delta H_{m,vap}$
(M_1 = molar mass of solvent)

38. For a weak electrolyte A_xB_y, the van't Hoff factor (i) and degree of dissociation are related by
(A) $\alpha = (x + y + 1)/(i - 1)$ (B) $\alpha = (x + y - 1)/(i - 1)$
(C) $\alpha = (i - 1)/(x + y + 1)$ (D) $\alpha = (i - 1)/(x + y - 1)$

39. For a dilute solution containing a non-volatile solute, the molar mass (m) of the solute obtained from the elevation of boiling point is given by
(A) $m = (K_b \times w \times 1000)/(W \times \Delta T_b)$ (B) $m = (\Delta T_b \times W)/(K_b \times w \times 1000)$
(C) $m = (K_b \times W \times 1000)/(W \times \Delta T_b)$ (D) $m = (\Delta T_b \times w)/(K_b \times W \times 1000)$

40. For a dilute solution containing a non-volatile solute, the molar mass (m) of the solute obtained from the osmotic pressure measurement is given by
(A) $m = wRT/\pi V$ (B) $m = \pi V/wRT$ (C) $m = w\pi V/RT$ (D) $m = \pi VR/wT$

41. Identify the incorrect statement.
(A) Equimolar solution of NaCl and urea are not isotonic

(B) In an ordinary chemical reaction, the number of moles of the reactants always equals the number of moles of the products

(C) For an ideal solution, ΔV_{mixing} and ΔH_{mixing} are each equal to zero

(D) Lattice energy opposes the tendency of any solute to dissolve.

42. Identify the incorrect statement,

(A) If $\Delta V_{mixing} < 0$ and $\Delta H_{mixing} > 0$, the non-ideal solutions exhibit negative deviation

(B) Water forms azeotropic mixture with HCOOH

(C) Non-electrolytes can form solid solutions

(D) Mole fraction is independent of temperature.

43. Identify the incorrect statement.

(A) The molarity of a solution is independent of temperature

(B) The pair of solutions $0.1M$ $Ca(NO_3)_2$ and $0.1M$ K_2SO_4 is isotonic at the same temperature

(C) Red blood cells neither shrink nor swell when placed in $0.16M$ NaCl solution

(D) A liquid which is permanently supercooled is called a glass.

44. Identify the incorrect statement.

(A) A solution of acetone and chloroform is an example of ideal solution

(B) If $\Delta V_{mixing} < 0$ and $\Delta H_{mixing} < 0$, the non-ideal solutions exhibit negative deviation

(C) In case of association, the van't Hoff factor is < 1

(D) The mole fraction of water in 20% aqueous solution of H_2O_2 solution is 0.88.

45. Identify the incorrect statement.

(A) The ideal solutions always exhibit lower vapour pressure than those of non-ideal solutions

(B) The osmotic pressure of sucrose solution containing 1.75 g in 150 mL of solution at 290 K is 0.81 atm

(C) If a solution is just heated, the molarity of the solution decreases

(D) In case of dissociation the van't Hoff factor is > 1.

46. Identify the incorrect statement.

(A) A solution containing benzene and ethanol is an example of ideal solution

(B) The limiting value of the van't Hoff factor for $K_4[Fe(CN)_6]$ is 11

(C) The value of K_f for water is larger than that of K_b for water

(D) Water has maximum density at 4°C.

47. Identify the incorrect statement.

(A) The unit of K_f is K kg^{-1}

(B) The molarity of water at 4°C is 55.6 mol L^{-1}

(C) If the van't Hoff factor (i) of a weak electrolyte (AB) in a solution is 1.1, the degree of dissociation (α) is 0.1

(D) The increasing order of boiling point of $0.001M$ NaCl, $0.01M$ NaCl, $0.001M$ urea and $0.001M$ $MgCl_2$ is $0.001M$ urea < $0.001M$ NaCl < $0.001M$ $MgCl_2$ < $0.01M$ NaCl.

48. Identify the incorrect statement.
   (A) Lowering of vapour pressure is a colligative property
   (B) For an ideal binary liquid solution, the plot of P_{total} *vs.* x_A (mole fraction of A in liquid phase) is nonlinear
   (C) The value of the van't Hoff factor (i) of a benzene solution of benzoic acid, which dimerises in this solvent, is 0.5
   (D) The unit of K_b is K kg mol^{-1}.
49. Increasing the temperature of an aqueous solution will cause
   (A) molality to increase (B) molarity to increase
   (C) mole fraction to decrease (D) % by weight to increase.
50. The azeotropic mixture of HCl and H_2O has
   (A) 20.2% HCl (B) 22.2% HCl (C) 36.0% HCl (D) 48.0% HCl.
51. An organic compound Y undergoes tetramerisation in an organic solvent. The van't Hoff factor of Y is
   (A) 0.25 (B) 0.125 (C) 2.0 (D) 4.0
52. Which of the following 0.1M solutions exhibits the highest osmotic pressure?
   (A) Na_2SO_4 (B) $BaCl_2$ (C) $C_6H_{12}O_6$ (D) $Al_2(SO_4)_3$
53. Which of the following solutions has the highest boiling point?
   (A) 0.1M KCl (B) 0.1M $BaCl_2$ (C) 0.1M $FeCl_3$ (D) 0.1M $Fe_2(SO_4)_3$
54. A binary liquid (AB) exhibits positive deviation from the Raoult's law when
   (A) $\Delta V_{mix} > 0$ (B) $\Delta H_{mix} > 0$
   (C) intermolecular forces: A—A > A—B and B—B > A—B
   (D) $p_A > p^o_A x_A$ and $p_A > p^o_B x_B$
55. 0.1 mole of glucose is dissolved in 10 moles of water. The vapour pressure of water is
   (A) increased by 1% (B) increased by 10%
   (C) decreased by 1% (D) decreased by 10%
56. When 20 g of naphthoic acid ($C_{11}H_8O_2$) is dissolved in 50 g of benzene (K_f = 1.72 K kg mol^{-1}), a freezing point depression of 2 K is observed. The van't Hoff factor (i) is
   (A) 0.5 (B) 1 (C) 2 (D) 3 **(I.I.T. 2007)**
   Hint: $\Delta T_f = K_f \times$ molality $\times\, i$ or $2 = 1.72 \times (20/172) \times (100/50) \times i$ or $i = 0.5$
57. The freezing point of a 0.05m solution of a non-electrolyte in water (K_f = 1.86 K kg mol^{-1}) is
   (A) – 1.86°C (B) – 0.93°C (C) – 0.093°C (D) 0.03°C
58. The molal freezing point constant (K_f) for water is 1.86°C. The freezing point of 0.1m NaCl solution in water is
   (A) – 1.86°C (B) – 0.186°C (C) – 0.372°C (D) 0.372°C
59. A binary solution of n-heptane and ethanol is an example of
   (A) an ideal solution
   (B) a solution exhibiting positive deviation at lower concentrations of C_2H_5OH and negative deviation at higher concentration of C_2H_5OH

(C) non-ideal solution with positive deviation from the Raoult's law
(D) non-ideal solution with negative deviation from the Raoult's law.

60. If a solute dimerises and trimerises in an organic solvent, the van't Hoff factor (i) is respectively
(A) 0.50, 0.33 (B) 0.50, 1.50 (C) 1.50, 0.33 (D) 0.25, 0.67

61. Due to osmosis the volume of the solution
(A) decreases only (B) increases only
(C) increases or decreases (D) remains constant.

62. The van't Hoff factor (i) of NaCl assuming 100% dissociation is
(A) 1 (B) 0.5 (C) 2 (D) 3

63. Blood is isotonic with
(A) 30% NaCl (B) 50% NaCl
(C) conc. NaCl soln. (D) 0.16M NaCl

64. An azeotropic mixture of two liquids has boiling point lower than either of two liquids. It
(A) is saturated
(B) exhibits positive deviation from the Raoult's law
(C) exhibits negative deviation from the Raoult's law
(D) exhibits no deviation from the Raoult's law.

65. At boiling point, a liquid is in equilibrium with its vapours. The molecules in the two phases on the average have equal
(A) intermolecular forces (B) potential energy
(C) kinetic energy (D) total energy.

66. The vapour pressure of a dilute solution of a solute is influenced by
(A) melting point of the solute (B) mole fraction of the solute
(C) degree of dissociation of the solute
(D) temperature of the solution.

67. Equal volume of 0.5M glucose and 0.5M urea is mixed. The mixture will have
(A) same osmotic pressure (B) higher osmotic pressure
(C) lower osmotic pressure (D) none of these.

68. Identify the correct statement.
(A) An ideal solution obeys Raoult's law
(B) a deliquescent compound absorbs water in more amount than a hygroscopic compound
(C) a mixture of water and glycerol is known as antifreeze mixture
(D) all of these.

69. The relation between molality (m) and mole fraction of solute (x_A) in a solution is (where M_B is the molar mass of solvent)
(A) $x_A = (1 - mM_B)/mM_B$ (B) $x_B = (1 + mM_B)/mM_B$
(C) $x_A = mM_B/(1 + mM_B)$ (D) $x_B = mM_B/(1 - mM_B)$

70. The relation between molality (m), molality (M) and density (ρ) of a solution is (where M_1 is the molar mass of solute)
(A) $m = (\rho - MM_1)/M_1$ (B) $m = (\rho + MM_1)/M_1$
(C) $m = M/(\rho - MM_1)$ (D) $m = M/(\rho + MM_1)$

71. For an ideal binary liquid solution with $p_A^o > p_B^o$, the relation between mole fraction of A in liquid phase (x_A) and mole fraction of A in vapour phase (y_A) is given by
(A) $x_A < y_A$ (B) $x_A > y_A$
(C) $x_A = y_A$ (D) There is no relation between x_A and y_A.

72. The freezing point of $0.1m$ solution of glucose is – 1.86°C. If an equal volume of $0.3m$ glucose solution is added, the freezing point of the mixture will be
(A) – 2.79°C (B) – 3.72°C (C) – 5.58°C (D) – 7.44°C

73. When mango is placed in a dilute aqueous solution of HCl, it
(A) shrinks (B) swells (C) bursts (D) nothing happens.

74. In a pair of immiscible liquids a solute dissolves in both and the equilibrium is reached. The concentration of the solute in the lower layer is
(A) same as the upper layer (B) lower than the upper layer
(C) higher than the upper layer (D) in fixed ratio with that in the upper layer.

75. Which is not a colligative property?
(A) Elevation of boiling point (B) Lowering of vapour pressure
(C) Freezing point (D) Depression of freezing point.

76. The molarity of a solution which contains 49 g H_3PO_4 in 2.0 L of a solution is
(A) $0.25M$ (B) $0.50M$ (C) $0.75M$ (D) $1.0M$

77. What would happen if a thin slice of sugar beet is placed in a concentrated solution of NaCl in water?
(A) Sugar beet will dissolve in solution
(B) Sugar beet will lose water from solution
(C) Sugar beet will lose water from its cells
(D) Sugar beet will neither lose nor absorb water.

78. The van't Hoff factor (i) of more than unity indicates that the solute in solution has
(A) associated (B) dissociated
(C) both (A) and (B) (D) cannot say anything.

79. All form ideal solution except
(A) C_2H_5I and C_2H_5OH (B) C_6H_5Cl and C_6H_5Br
(C) C_2H_5Br and C_2H_5I (D) C_6H_6 and $C_6H_5CH_3$

80. A pressure cooker reduces cooking time because
(A) the boiling point of water inside the cooker is increased
(B) the heat is more evenly distributed
(C) a longer flame is used
(D) the high pressure tenderises the food.

81. The vapour pressure of pure A is 10.0 torr and at the same temperature when 1 g of B is dissolved in 20 g of A, its vapour pressure is reduced to 9.0 torr. If the molecular mass of A is 200, then the molecular mass of B is

(A) 75 (B) 90 (C) 100 (D) 120

82. The van't Hoff factor of $0.1M$ $Ba(NO_3)_2$ solution is 2.74. The degree of dissociation is

(A) 91.3% (B) 87% (C) 100% (D) 74% **(I.I.T. 1999)**

Hint: $\alpha = \frac{i-1}{n-1} = \frac{2.74-1}{3-1} = 0.87$ or 87%.

83. The fundamental cause of all collegative properties is

(A) lower enthalpy of solution relative to pure solvent
(B) lower entropy of solution relative to pure solvent
(C) higher enthalpy of solution relative to pure solvent
(D) higher entropy of solution relative to pure solvent.

84. The degree of dissociation α of a weak electrolyte is given by (N = No. of ions given by 1 mole of electrolyte)

(A) $\frac{i-1}{N-1}$ (B) $\frac{N+1}{i-1}$ (C) $\frac{N-1}{i-1}$ (D) $\frac{i-1}{N+1}$

85. When the temperature is increased, surface tension of water

(A) increases (B) decreases
(C) remains constant (D) shows irregular behaviour. **(I.I.T. 2002)**

86. Which of the following forms a minimum boiling point azeotrope?

(A) $CH_3COCH_3 + CHCl_3$ (B) $H_2O + HCl$
(C) $H_2O + H_2SO_4$ (D) H_2O + ethanol

87. Which of the following exhibits a negative deviation from the Raoult's law?

(A) $CS_2 + CCl_4$ (B) $CH_3OH + C_2H_5OH$
(C) $H_2O + HNO_3$ (D) heptane + hexane

88. Which of the following forms a maximum boiling point azeotrope?

(A) $H_2O + C_2H_5OH$ (B) $C_6H_6 + CHCl_3$
(C) $C_6H_6 + C_6H_5CH_3$ (D) cyclohexane + cyclohexanol

89. Which of the following does not exhibit a positive deviation from the Raoult's law?

(A) $H_2O + C_2H_5OH$ (B) $C_2H_5OH + CHCl_3$
(C) $CH_3OH + CCl_4$ (D) $HCl + (C_2H_5)_2O$

90. Which of the following has ΔV_{mix} negative?

(A) $C_6H_6 + C_6H_5CH_3$ (B) $C_6H_6 + CHCl_3$
(C) $H_2O + C_2H_5OH$ (D) cyclohexane + cyclohexanol

91. Which of the following has $\Delta H_{mix} = 0$?

(A) $C_6H_6 + CH_3COCH_3$ (B) $CH_3OH + C_2H_5OH$
(C) $H_2O + CH_3COOH$ (D) $CHCl_3 + CH_3COCH_3$

92. During depression of freezing point in a solution, the following are in equilibrium.

(A) liquid solvent, solid solvent (B) liquid solvent, solid solute
(C) liquid solute, solid solute (D) liquid solute, solid solvent **(I.I.T. 2003)**

93. 0.004M Na_2SO_4 is isotonic with 0.01M glucose. Degree of dissociation of Na_2SO_4 is
(A) 75% (B) 50% (C) 25% (D) 85% **(I.I.T. 2004)**
Hint: $\pi_{Na_2SO_4} = \pi_{glucose} = 0.01 \times RT$ or $0.01RT = i \times 0.004RT$
or $i = 2.5$; $\underset{1-\alpha}{Na_2SO_4} \rightleftharpoons \underset{2\alpha}{2Na^+} + \underset{\alpha}{SO_4^{2-}}$; $i = 1 + 2\alpha = 2.5$ or $\alpha = 0.75$ or $\alpha = 75\%$.

94. The elevation in boiling point, when 13.44 g of freshly prepared $CuCl_2$ is added to one kilogram of water, is
(A) 0.05 (B) 0.1 (C) 0.16 (D) 0.21 **(I.I.T. 2005)**

Hint: $\Delta T_b = i \times K_b \times m = 3 \times 0.52 \times \left(\dfrac{13.44}{134.4} \times \dfrac{1000}{1000}\right) = 0.16$

95. A 0.1 molal solution of tartaric acid freezes at – 0.205°C. The ionisation constant K_{a_1} for tartaric acid is (K_f = 1.86 K kg mol^{-1})
(A) 1.52×10^{-3} (B) 1.11×10^{-4} (C) 1.45×10^{-3} (D) 1.11×10^{-3}
Hint: $TH \rightleftharpoons T^- + H^+$; van't Hoff factor, $i = [1 + (y - 1)x] = 1 + x$ where x = degree of ionisation, y = No. of ions from one mole of solute = 2 (T^- and H^+). $\Delta T = 0 - (0 - 0.205°)$ = 0.205°. $\Delta T_f = K_f mi$ or $0.205 = 1.86 \times 0.1 \times (1 + x)$ or $x = 0.1$.

$$K_a = \frac{cx^2}{(1-x)} = \frac{0.1 \times (0.1)^2}{1 - 0.1} = 1.11 \times 10^{-3}$$

96. Properties such as boiling point, freezing point and vapour pressure of a pure solvent change when solute molecules are added to get homogeneous solution. These are called colligative properties. Applications of colligative properties are very useful in day-to-day life. One of its examples is the use of ethylene glycol and water mixture as anti-freezing liquid in the radiator of automobiles. A solution M is prepared by mixing ethanol and water. The mole fraction of ethanol in the mixture is 0.9.
Given: Freezing point depression constant of water (K_f^{water}) = 1.86 K kg mol^{-1}; Freezing point depression constant of ethanol ($K_f^{ethanol}$) = 2.0 K kg mol^{-1}; Boiling point elevation constant of water (K_b^{water}) = 0.52 K kg mol^{-1}; Boiling point elevation constant of ethanol ($K_b^{ethanol}$) = 1.2 K kg mol^{-1}; Standard freezing point of water = 273 K; Standard freezing point of ethanol = 155.7 K; Standard boiling point of water = 373 K; Standard boiling point of ethanol = 351.5 K; Vapour pressure of pure water = 32.8 mmHg; Vapour pressure of pure ethanol = 40 mmHg; Molecular weight of water = 18 g mol^{-1}; Molecular weight of ethanol = 46 g mol^{-1}.
In answering the following questions, consider the solutions to be ideal dilute solutions and solutes to be non-volatile and non-dissociative.
a. The freezing point of the solution M is
(A) 268.7 K (B) 268.5 K (C) 234.2 K (D) 150.9 K **(I.I.T. 2008)**
Hint: Since the mole fraction of ethanol in the mixture is 0.9, ethanol is the solvent and water is the solute. Assuming water to be a non-volatile solute, the freezing

point of the solution should be less than that of the pure solvent (ethanol). Thus, the freezing point of the solution is < 155.7 K.

b. The vapour pressure of the solution M is

(A) 39.3 mmHg (B) 36.0 mmHg (C) 29.5 mmHg (D) 28.8 mmHg **(I.I.T. 2008)**

Hint: It is given that the solute is to be treated as non-volatile and non-dissociative. Thus, vapour pressure of solution = $P^0_{\text{solvent}}\, x_{\text{solvent}} = 40 \times 0.9 = 36.0$ mmHg.

c. Water is added to the solution M such that the mole fraction of water in the solution becomes 0.9. The boiling point of this solution is

(A) 380.4 K (B) 376.2 K (C) 375.5 K (D) 354.7 K **(I.I.T. 2008)**

Hint: Since the mole fraction of water in solution is 0.9, water is the solvent and ethanol is solute and has to be considered as non-volatile and non-dissociative. Mole fraction of ethanol in solution = 0.1. Molality of ethanol in solution = 0.1/(0.9 × 18) = 6.17. $\Delta T_b = K_b\, m = 0.52 \times 6.17 = 3.2$. Hence the boiling point of solution = 373 + 3.2 = 376.2 K.

97. The Henry's law constant for the solubility of N_2 gas in water at 298 K is 1.0×10^5 atm. The mole fraction of N_2 in air is 0.8. The number of moles of N_2 from air dissolved in 10 moles of water at 298 K and 5 atom presure is

(A) 4.0×10^{-4} (B) 4.0×10^{-5} (C) 5.0×10^{-4} (D) 4.0×10^{-6} **(I.I.T. 2009)**

Hint: According to Henry's law

$P_{N_2} = K_{N_2} X_{N_2}$ where K is the Henry's constant (in atm) and X_{N_2} is mole fraction of N_2.

$$P_{N_2} = X_{N_2} P = 0.8 \times 5 \text{ atm} = 4.0 \text{ atm}$$

Hence 4 atm = 1.0×10^5 atm × X_{N_2}

$$4 \times 10^{-5} = \frac{n_{N_2}}{n_{N_2} + n_{H_2O}} = \frac{n_{N_2}}{n_{N_2} + 10}$$

$n_{N_2} = 4 \times 10^{-4}$ moles

98. The freezing point (in °C) of a solution containing 0.1 g of $K_3[Fe(CN)_6]$ (Mol. Wt. 329) in 100 g of water (K_f = 1.86 K kg mol^{-1}) is

(A) -2.3×10^{-2} (B) -5.7×10^{-2} (C) -5.7×10^{-3} (D) -1.2×10^{-2}

(I.I.T. 2011)

Hint: $K_3[Fe(CN)_6] \longrightarrow 3K^+ + [Fe(CN)_6]^{3-}$

$i = 4$

$$\Delta T_f = K_f \times i \times \frac{m}{M} \times \frac{1000}{W} = 1.86 \times 4 \times \frac{0.1}{329} \times \frac{1000}{100} = 2.3 \times 10^{-2}$$

$T_f = -2.3 \times 10^{-2}$

ANSWERS

1. B, D	2. A	3. A	4. C	5. B	6. D
7. D	8. B	9. B	10. A	11. A	12. C
13. A	14. C	15. B	16. C	17. C	18. A
19. B	20. C	21. D	22. C	23. C	24. B
25. C	26. A	27. D	28. D	29. C	30. A
31. B	32. C	33. D	34. A	35. B	36. A
37. D	38. D	39. A	40. A	41. B	42. A
43. A	44. A	45. A	46. A, B	47. A	48. A, B
49. B	50. A	51. A	52. D	53. D	54. A, B, C, D
55. C	56. A	57. C	58. C	59. C	60. A
61. C	62. C	63. D	64. B	65. C	66. B, C, D
67. A	68. D	69. C	70. C	71. A	72. B
73. A	74. A	75. C	76. A	77. C	78. B
79. A	80. A	81. C	82. B	83. C	84. A
85. B	86. D	87. C	88. B	89. D	90. B
91. B	92. A	93. A	94. C	95. D	96a. D
96b. B	96c. B	97. A	98. A		

8

Chemical Equilibrium

IMPORTANT RELATIONS FOR NUMERICAL PROBLEMS

1. For a reversible reaction, $aA + bB \rightleftharpoons cC + dD$, the equilibrium constant, $K = [C]^c[D]^d/[A]^a[B]^b$
2. $K_p = p_C^c p_D^d / p_A^a p_B^b$
3. $K_p = K_c(RT)^{\Delta n}$ where Δn = moles of products – moles of reactants
4. $\Delta G^0 = -2.303RT \log K$; at equilibrium $K = 1$, $\Delta G^0 = 0$
5. $\log (K_2/K_1) = (\Delta H/2.303R)[(T_2 - T_1)/(T_1T_2)]$
6. Relation between degree of dissociation (α) of a gaseous substance, vapour density and density of the gaseous mixture at equilibrium: $\alpha = (D - d)/d(n - 1)$ where n = change in the number of molecules due to the reaction

OBJECTIVE TYPE MULTIPLE CHOICE QUESTIONS

1. At constant temperature, K_p for the equilibrium $N_2O_4 \rightleftharpoons 2NO_2$ is expressed by $K_p = (4x^2P)/(1 - x^2)$, where P = pressure, x = extent of decomposition. Which of the following statements is true?
   (A) K_p increases with increase of P (B) K_p increases with increase of x
   (C) K_p remains constant with change in P and x
   (D) K_p increases with decrease of x **(I.I.T. 2001)**
2. For the reaction $N_2(g) + 3H_2(g) \rightleftharpoons 2NH_3(g)$ at 500°C, K_p is 1.44×10^{-5} when partial pressure is measured in atm. The value of K_c, with concentration in mol L^{-1}, is
   (A) $1.44 \times 10^{-5}/(0.082 \times 773)^{-2}$ (B) $1.44 \times 10^{-5}/(8.314 \times 773)^{-2}$
   (C) $1.44 \times 10^{-5}/(0.082 \times 500)^{-2}$ (D) $1.44 \times 10^{-5}(0.082 \times 773)^{2}$ **(I.I.T. 2000)**
   Hint: $\Delta n = 2 - 4 = -2$. $K_p = K_c \times (RT)^{\Delta n}$ or $1.44 \times 10^{-5} = K_c \times (0.0821 \times 773)^{-2}$ or $K_c = [1.44 \times 10^{-5}/(0.0821 \times 773)^{-2}]$ $(\text{mol/L})^{-2}$
3. For the reaction $3X(g) + Y(g) \rightleftharpoons X_3Y(g)$, the amount of X_3Y at equilibrium is affected by
   (A) temperature and pressure (B) temperature, pressure and catalyst
   (C) temperature only (D) pressure only. **(I.I.T. 1999)**
4. For the reaction $CO(g) + H_2O(g) \rightleftharpoons CO_2(g) + H_2(g)$ at a given temperature the equilibrium amount of $CO_2(g)$ can be increased by
   (A) adding a suitable catalyst (B) adding an inert gas
   (C) decreasing the volume of the container
   (D) increasing the amount of CO(g) **(I.I.T. 1998)**
5. One mole of $N_2O_4(g)$ at 300 K is kept in a closed container at 1 atm. It is heated to 600 K when 20% by mass of $N_2O_4(g)$ decomposes to $NO_2(g)$. The resultant pressure is
   (A) 1.2 atm (B) 2.0 atm (C) 2.4 atm (D) 1.0 atm **(I.I.T. 1996)**

Hint:	$N_2O_4(g) \rightleftharpoons$	$2NO_2(g)$
Initial mole:	100/92	0
Moles at equil.:	80/92	20/46

   At $t = 0$: $P_0V = n_0RT_0$ or $1 \times V = (100/92) \times 0.0821 \times 300$ or $V = 26.77$ L
   At equil.: $PV = nRT$ or $P \times 26.77 = [(80/92) + (20/46)] \times 0.0821 \times 600$ or $P \times 26.77 = 1.3 \times 0.0821 \times 600$ or $P = 2.4$ atm.
6. An equilibrium mixture for the reaction, $2H_2S(g) \rightleftharpoons 2H_2(g) + S_2(g)$ had 1 mole of H_2S, 0.2 mole of H_2 and 0.8 mole of S_2 in a 2 L flask. The value of K_c in mol/L is
   (A) 0.004 (B) 0.080 (C) 0.160 (D) 0.016 **(I.I.T. 1992)**
7. The following equilibrium is established when hydrogen chloride is dissolved in acetic acid: $HCl + CH_3COOH \rightleftharpoons 2Cl^- + CH_3COOH_2^+$. The set that characterises the conjugate acid-base pair is
   (A) (HCl, CH_3COOH) and $(CH_3COOH_2^+, Cl^-)$
   (B) $(HCl, CH_3COOH_2^+)$ and (CH_3COOH, Cl^-)
   (C) $(CH_3COOH_2^+, HCl)$ and (Cl^-, CH_3COOH)
   (D) (HCl, Cl^-) and $(CH_3COOH_2^+, CH_3COOH)$ **(I.I.T. 1992)**

8. The reaction which proceeds in the forward direction is
   (A) $Fe_2O_3 + 6HCl = 2FeCl_3 + 3H_2O$ (B) $NH_3 + H_2O + NaCl = NH_4Cl + NaOH$
   (C) $SnCl_4 + Hg_2Cl_2 = SnCl_2 + 2HgCl_2$ (D) $2CuI + I_2 + 4K^+ = 2Cu^{2+} + 4KI$ **(I.I.T. 1991)**
9. For which of the following reactions, $K_p = K_c$?
   (A) $H_2(g) + Cl_2(g) \rightleftharpoons 2HCl(g)$ (B) $2NOCl(g) \rightleftharpoons 2NO(g) + Cl_2(g)$
   (C) $N_2(g) + 3H_2(g) \rightleftharpoons 2NH_3(g)$ (D) $N_2O_4(g) \rightleftharpoons 2NO_2(g)$ **(I.I.T. 1991)**
10. For the reaction, $PCl_5(g) \rightleftharpoons PCl_3(g) + Cl_2(g)$ the forward reaction at constant temperature is favoured by
    (A) introducing an inert gas at constant volume
    (B) introducing chlorine gas at constant volume
    (C) introducing an inert gas at constant pressure
    (D) increasing the volume of container (E) introducing PCl_5 at constant volume.
    (I.I.T. 1991)
11. The equilibrium, $SO_2Cl_2(g) \rightleftharpoons SO_2(g) + Cl_2(g)$, is attained at 25°C in a closed container and an inert gas He is introduced. Which of the following statements are incorrect?
    (A) $[SO_2]$, $[Cl_2]$ and $[SO_2Cl_2]$ change (B) More chlorine is formed
    (C) Concentration of SO_2 is reduced (D) More SO_2Cl_2 is formed. **(I.I.T. 1989)**
12. When $NaNO_3$ is heated in a closed vessel, oxygen is liberated and $NaNO_2$ is left behind. At equilibrium
    (A) addition of $NaNO_2$ favours reverse reaction
    (B) addition of $NaNO_3$ favours forward reaction
    (C) increasing temperature favours forward reaction
    (D) increasing pressure favours forward reaction. **(I.I.T. 1986)**
13. An example of reversible reaction is
    (A) $Pb(NO_3)_2(aq) + 2NaI(aq) = PbI_2(s) + 2NaOH(aq)$
    (B) $2Na(s) + 2H_2O(l) = 2NaOH(aq) + H_2(g)$
    (C) $KNO_3(aq) + NaCl(aq) = KCl(aq) + NaNO_3(aq)$
    (D) $AgNO_3(aq) + HCl(aq) = AgCl(s) + HNO_3(aq)$ **(I.I.T. 1985)**
14. The ammonia is placed in a vessel at a temperature where its dissociation constant is appreciable. In this reaction
    (A) α does not change with pressure (P)
    (B) K_p does not change significantly with P
    (C) Concentration of ammonia does not change with pressure
    (D) Concentration of hydrogen is less than that of nitrogen. **(I.I.T. 1984)**
15. For the gas phase reaction, $C_2H_4 + H_2 \rightleftharpoons C_2H_6$, $\Delta H = -136.8$ kJ/mol, carried out in a vessel, the equilibrium concentration of C_2H_4 can be increased by
    (A) increasing T (B) decreasing P
    (C) removing some H_2 (D) adding some C_2H_6 **(I.I.T. 1984)**
16. For the equilibrium, $H_2(g) + I_2(g) \rightleftharpoons 2HI(g)$, K_p changes with
    (A) temperature (B) amount of H_2 and I_2
    (C) total pressure (D) catalyst **(I.I.T. 1981)**

17. The oxidation of SO_2 by O_2 to SO_3 is an exothermic reaction. The yield of SO_3 will be maximum if
    (A) temperature is increased and pressure is kept constant
    (B) temperature is reduced and pressure is increased
    (C) both temperature and pressure are increased
    (D) both temperature and pressure are decreased. **(I.I.T. 1981)**
18. Chemical equilibrium is dynamic in nature because
    (A) the equilibrium is maintained quickly
    (B) the concentrations of reactants and products become same at equilibrium
    (C) the concentrations of reactants and products are constant but different
    (D) both forward and backward reactions occur at all times with same speed. **(I.I.T. 1977)**
19. In a reactions the rate of reaction is proportional to its active mass. This statement is termed as
    (A) Law of mass action (B) Le Chatelier's principle
    (C) Faraday's law of electrolysis (D) Law of constant proportion. **(I.I.T. 1979)**
20. In the reaction, $A_2(g) + 4B_2(g) \rightleftharpoons 2AB_4(g)$, $\Delta H < 0$, the formation of $AB_4(g)$ will be favoured at
    (A) low T and high P (B) high T and low P
    (C) low T and low P (D) high T and high P **(I.I.T. 1990)**
21. Consider the equilibrium, $NaNO_3(s) \xrightleftharpoons{\text{heat}} NaNO_2(s) + O_2(g)$. Which of the following is true?
    (A) Increasing pressure favours reverse the reaction
    (B) Decreasing temperature favours the forward reaction
    (C) Addition of $NaNO_3$ favours the forward reaction
    (D) Addition of $NaNO_2$ favours the reverse reaction. **(Roorkee 1995)**
22. For the reaction, $C(s) + CO_2(g) \rightleftharpoons 2CO(g)$, the partial pressures of CO_2 and CO are 2.0 and 4.0 atm respectively. The K_p for the reaction is
    (A) 0.5 (B) 4.0 (C) 8.0 (D) 32.0 **(Roorkee 1990)**
23. K for the synthesis of HI is 50. K for the dissociation of HI is
    (A) 50 (B) 0.2 (C) 0.02 (D) 5 **(Roorkee 1990)**
24. The equilibrium constant (K_p) for the reaction $PCl_5(g) \rightleftharpoons PCl_3(g) + Cl_2(g)$ is 16. If the volume of the container is reduced to one-half of its original volume, the value of K_p for the reaction at the same temperature will be
    (A) 32 (B) 64 (C) 16 (D) 4
25. In the reaction $A + 2B \rightleftharpoons 2C$, if 2.0 mole of B and 2.0 mole of C are placed in a 2 L flask and equilibrium concentration of C is 0.5 mol/L, the equilibrium constant for the reaction is
    (A) 0.073 (B) 0.147 (C) 0.05 (D) 0.026

26. Identify the incorrect relation for equilibrium constant K.

(A) $\log K = \log A - \dfrac{\Delta H}{2.303RT}$ (B) $E^0_{cell} = \dfrac{2.303RT \log K}{nF}$

(C) $\Delta G^0 = -2.303RT \log K$ (D) $K = \dfrac{\Sigma[\text{product}]}{\Sigma[\text{reactant}]}$

27. The equilibrium mixture [$2NO_2$ (brown) $\rightleftharpoons$ N_2O_4 (colourless)] is slightly brown due to the presence of NO_2. What happens to the colour if pressure is increased?
(A) No change in colour intensity
(B) Colour intensity first increases and then decreases
(C) Colour intensity increases (D) Colour intensity decreases.

28. The equilibrium constant for the reaction $SO_2(g) + (1/2)O_2(g) \rightleftharpoons SO_3(g)$ is 4×10^{-3} atm$^{-1/2}$. The equilibrium constant for the reaction $2SO_3(g) \rightleftharpoons 2SO_2(g) + O_2$ will be
(A) 6.25×10^4 atm (B) 0.25×10^4 atm
(C) 4×10^3 atm (D) 250 atm

29. The equilibrium constant for the reaction $H_2(g) + I_2(g) \rightleftharpoons 2HI(g)$ is 64 at a certain temperature. The equilibrium concentrations of H_2 and HI are 2 mol/L and 16 mol/L respectively. What is the equilibrium concentration of I_2?
(A) 16 mol/L (B) 4 mol/L (C) 8 mol/L (D) 2 mol/L

30. The equilibrium constant for the reaction $A(g) + B(g) \rightleftharpoons C(g) + D(g)$ is 100 at 25°C. Consider the following statements in this regard. If the initial concentration of all the four species were 1.0*M*, then the equilibrium concentration of
1. A would be 0.182 mol 2. C would be 0.818 mol 3. D would be 1.818 mol.
Of these statements
(A) 1, 2 are correct (B) 1, 3 are correct
(C) 2, 3 are correct (D) 1, 2, 3 are correct

31. Identify the incorrect statement.
(A) For the reaction, $N_2O_4 \rightleftharpoons 2NO_2$, $\Delta H = 58.0$ kJ mol^{-1}, the density of the equilibrium mixture increases with the increase of temperature
(B) In the dissociation reaction, $HgO(s) \rightleftharpoons Hg(g) + (1/2)O_2(g)$, the dissociation pressure is 0.51 atm at 693 K. K_p for the reaction is 0.14 atm$^{3/2}$
(C) The solubility of sodium hydroxide in water increases with the increase in temperature
(D) If the equilibrium constant for the forward reaction, $A_2 + B_2 \rightleftharpoons 2AB$ is K, then for the backward reaction $AB \rightleftharpoons (1/2)A_2 + (1/2)B_2$, the equilibrium constant is $1/K$.

32. Identify the incorrect statement.
(A) A ten-fold increase in pressure on the reaction, $N_2(g) + 3H_2(g) \rightleftharpoons 2NH_3(g)$ at equilibrium results in increase in K_p
(B) For the reaction $N_2(g) + 3H_2(g) \rightleftharpoons 2NH_3(g)$ under certain conditions of temperature and partial pressure of the reactants, the rate of formation of NH_3 is 0.001 kg h^{-1}. The rate of conversion of H_2 under the same condition is 0.0015 kg h^{-1}

(C) The unit of the equilibrium constant for the reaction, $N_2(g) + 3H_2(g) \rightleftharpoons 2NH_3(g)$ is atm^2

(D) The solubility of a gas increases with increase in pressure.

33. Identify the incorrect statement.

(A) Pressure has no effect in the reaction in which $\Delta n = 0$

(B) When $\Delta n > 0$, $K_p = K_c$

(C) A negative value of ΔG indicates that the reaction is spontaneous in the backward direction

(D) The melting point of ice is lowered when pressure is increased.

34. Identify the incorrect statement.

(A) A catalyst has no effect on the equilibrium constant of a reaction

(B) The position of equilibrium in a reaction favours the formation of a weak Bronsted acid and a weak Bronsted base

(C) For the reaction, $2SO_2(g) + O_2(g) \rightleftharpoons 2SO_3(g)$, the value of K_c increases when the volume of the reaction vessel is increased

(D) For the reaction, $H_2(g) + I_2(g) \rightleftharpoons 2HI(g)$, the value of the equilibrium constant is 32.67 when $[H_2]$ = 8.0 mol/L, $[I_2]$ = 3.0 mol/L and [HI] = 28.0 mol/L

35. The energy of activation for the forward reaction in the equation $H_2(g) + F_2(g) \rightleftharpoons 2HF(g)$, $\Delta H = -$ 536 kJ/mol H_2, is 208 kJ/mol H_2. The energy of activation per mole of HF for the backward reaction is

(A) 264 kJ (B) 328 kJ (C) 372 kJ (D) 750 kJ

36. The equilibrium constant for the reaction *cis*-$C_2H_2Cl_2 \rightleftharpoons$ *trans*-$C_2H_2Cl_2$ is 0.6 at 500 K. The equilibrium constant for the reaction *trans*-$C_2H_2Cl_2 \rightleftharpoons$ *cis*-$C_2H_2Cl_2$ at 500 K is

(A) 0.6 (B) 1.78 (C) 1.67 (D) 1.06

37. HI was heated in a sealed tube at 440°C till the equilibrium was reached. HI had undergone 22% decomposition. The equilibrium constant for the dissociation is

(A) 0.0199 (B) 1.09 (C) 0.0282 (D) 0.0796

38. When 3 moles of C_2H_5OH are mixed with 3 moles of CH_3COOH, 2 moles of ester are formed at equilibrium, $CH_3COOH(l) + C_2H_5OH(l) \rightleftharpoons CH_3COOC_2H_5(l) + H_2O(l)$. The equilibrium constant for this reaction is

(A) 1 (B) 2/9 (C) 4.9 (D) 4

39. 1.1 mole of A is mixed with 2.2 mole of B and the mixture is kept in a 1 L flask till the equilibrium, $A + 2B \rightleftharpoons 2C + D$ is reached. At equilibrium 0.2 mole of C is formed. The equilibrium constant for the reaction is

(A) 0.01 (B) 0.001 (C) 0.002 (D) 0.003

40. 28 g of N_2 and 6 g of H_2 were kept at 400°C in a 1 L vessel. The equilibrium mixture contained 27.54 g of NH_3. The approximate value of K_c for this reaction is

(A) 200 $mol^{-2}\ L^2$ (B) 75 $mol^{-2}\ L^2$ (C) 150 $mol^{-2}\ L^2$ (D) 30 $mol^{-2}\ L^2$

41. Assuming that the degree of dissociation of PCl_5, at a certain temperature and at atmospheric pressure is 0.2, the pressure at which it would be 50% dissociated is

(A) 1.0 atm (B) 0.5 atm (C) 0.46 atm (D) 0.125 atm

42. A reaction $A + 2B \rightleftharpoons 2C + D$ was studied using an initial concentration of B which was 1.5 times to that of A. But the equilibrium concentrations of A and B were found to be equal. The value of K_c for the equilibrium is
(A) 10 (B) 5 (C) 7 (D) 4

43. In the reaction $A(g) \rightleftharpoons 2B(g) + C(g)$, the equilibrium is established at constant temperature. If the initial pressure of A were 200 mm and the extent of dissociation were 20%, the equilibrium pressure is
(A) 310 mm (B) 220 mm (C) 250 mm (D) 280 mm

44. A mixture of 0.3 mol of H_2 and 0.3 mol of I_2 is allowed to react in a 10 L evacuated flask at 500°C. K for the reaction $H_2 + I_2 \rightleftharpoons 2HI$ is found to be 64. The amount of unreacted I_2 at equilibrium is
(A) 0.06 mol (B) 0.02 mol (C) 1.5 mol (D) 0.25 mol

45. 1 mol of SO_3 was heated in 1 L closed vessel at a certain temperature. The equilibrium $2SO_3 \rightleftharpoons 2SO_2 + O_2$ was established. At equilibrium 0.6 mol of SO_2 was formed. The equilibrium constant for the reaction is
(A) 0.675 (B) 0.65 (C) 0.35 (D) 0.25

46. 64 g of HI are present in a 2 L vessel. The active mass of HI is
(A) 0.5 (B) 0.25 (C) 1.0 (D) none of these.

47. The equilibrium constant for the reaction, $N_2(g) + O_2(g) \rightleftharpoons 2NO(g)$ is 4×10^{-4} at 2000 K. The equilibrium is reached ten times faster in the presence of a catalyst. The equilibrium constant at 2000 K in the presence of a catalyst is
(A) 40×10^{-4} (B) 4×10^{-3} (C) 4×10^{-4} (D) 0.4×10^{-4}

48. According to Le Chatelier's principle adding heat to a solid and liquid in equilibrium will cause the
(A) amount of liquid to decrease (B) amount of solid to decrease
(C) temperature to fall (D) temperature to rise.

49. In a reversible reaction, two substances are at equilibrium. If the concentration of each one is doubled, the equilibrium constant will be
(A) doubled (B) constant
(C) reduced to half (D) reduced to one fourth.

50. The decomposition of N_2O_4 to NO_2 is carried out at 280°C in $CHCl_3$. When equilibrium is reached, 0.2 mol of N_2O_4 and 2×10^{-3} mol of NO_2 are present in a 2 L solution. The equilibrium constant for the reaction $N_2O_4 \rightleftharpoons 2NO_2$ is
(A) 1×10^{-5} (B) 2×10^{-2} (C) 1×10^{-3} (D) 2×10^{-5}

51. If the equilibrium constant for the reaction, $Fe^{2+}(aq) + SCN^-(aq) \rightleftharpoons Fe(SCN)^+(aq)$ is 130 at 300 K, the equilibrium constant for the reaction, $2Fe^{2+}(aq) + 2SCN^-(aq) \rightleftharpoons 2Fe(SCN)^+(aq)$, is
(A) 65 (B) 130 (C) 160 (D) 16900

52. 4.5 moles each of H_2 and I_2 are heated in a sealed 10 L vessel. At equilibrium 3 moles of HI were formed. The equilibrium constant for the reaction, $H_2(g) + I_2(g) \rightleftharpoons 2HI(g)$, is
(A) 0.33 (B) 1 (C) 5 (D) 10

53. For the reaction, $A + B \rightleftharpoons C + D$, the initial concentrations of A and B are equal, but the equilibrium concentration of C is twice that of the equilibrium concentration of A. The equilibrium constant is
(A) 1/4 (B) 1/9 (C) 4 (D) 9

54. If K_1 and K_2 are the respective equilibrium constant for the two reactions: (i) $XeF_6(g) + H_2O(g) \rightleftharpoons XeOF_4(g) + 2HF$ (ii) $XeO_4(g) + XeF_6(g) \rightleftharpoons XeOF_4(g) + XeO_3F_2(g)$, the equilibrium constant for the reaction, $XeO_4(g) + 2HF(g) \rightleftharpoons XeO_3F_2(g) + H_2O(g)$, is
(A) K_1K_2 (B) K_1/K_2 (C) $K_1/(K_2)^2$ (D) K_2/K_1

Hint: $K_1 = \dfrac{[XeOF_4][HF]^2}{[XeF_6][H_2O]}$, $K_2 = \dfrac{[XeOF_4][XeO_2F_2]}{[XeO_4][XeF_6]}$

$$K_2/K_1 = \frac{[XeO_3F_2][H_2O]}{[XeO_4][HF]^2} = K \text{ (for third reaction)}$$

55. If K_p for the reaction, $AB(g) \rightleftharpoons A(g) + B(g)$, is equal to four times the total pressure, the number of moles of A formed is
(A) $\sqrt{5}$ (B) $2\sqrt{5}$ (C) 2/5 (D) $3\sqrt{5}$

56. At 298°C, the pressure necessary to obtain 80% dissociation of PCl_5 is equal to
(A) $K_p/17.7$ (B) $K_p/1.77$ (C) $K_p/0.177$ (D) $K_p/(0.177)^2$

57. Identify the correct expression for the reaction, $C(s) + H_2O(g) \rightleftharpoons CO(g) + H_2(g)$.
(A) $p_{H_2} \propto (p_{H_2O})^{1/2}$ (B) $p_{H_2} \propto p_{H_2O}$ (C) $p_{H_2} \propto p^2_{H_2O}/p_{CO}$ (D) $p_{H_2} \propto p^2_{H_2O}$

58. K_p for the reaction, $H_2(g) + I_2(g) \rightleftharpoons 2HI(g)$, is dependent on
(A) temperature (B) pressure
(C) amount of H_2 and I_2 (D) catalyst

59. K_p for the reaction, $HgO(s) \rightleftharpoons Hg(g) + (1/2)O_2(g)$, is given by
(A) $K_p = P/3^{2/3}$ (B) $K_p = P^{3/2}/3^{2/3}$ (C) $K_p = 2P^{1/2}/3^{1/2}$ (D) $K_p = 2P^{3/2}/3^{3/2}$

60. If α is the degree of dissociation, K_p for the reaction, $A(g) \rightleftharpoons B(g) + C(g)$, is given by
(A) $K_p = \alpha^3 P$ (B) $K_p = \alpha^2(K_p + P)$
(C) $K_p = \alpha(K_p + P)/P$ (D) $K_p = \alpha(K_p + P + 1)$

61. If α for the reaction, $AB(g) \rightleftharpoons A(g) + B(g)$ is 33%, K_p for the reaction is given by
(A) $K_p = P$ (B) $K_p = P/2$ (C) $K_p = P/3$ (D) $K_p = P/8$

62. What will happen when pressure is applied to the equilibrium, $Ice(s) \rightleftharpoons H_2O(l)$?
(A) More $H_2O(l)$ will be formed (B) $H_2O(l)$ will evaporate
(C) More ice(*s*) will be formed (D) Equilibrium will not be reached.

63. If the equilibrium constant for the reaction, $2NO_2(g) + (1/2)O_2(g) \rightleftharpoons N_2O_5(g)$ is K_p, the equilibrium constant (K'_p) for the reaction, $2N_2O_5(g) \rightleftharpoons 4NO_2(g) + O_2(g)$ is given by
(A) $K'_p = 1/(K_p)^{1/2}$ (B) $K'_p = 1/K_p^2$ (C) $K'_p = 2/K_p$ (D) $K'_p = 1/K_p^2$

64. In what manner will increase of pressure affect the equilibrium, $C(s) + H_2O(g) \rightleftharpoons CO(g) + H_2(g)$?
(A) Shift in the forward direction (B) Shift in the reverse direction
(C) Increase the yield of H_2 (D) No effect. **(I.I.T. 1994)**

65. The reaction, $SO_2 + Cl_2 \rightleftharpoons SO_2Cl_2$ is exothermic and reversible. A mixture of $SO_2(g)$, $Cl_2(g)$ and $SO_2Cl_2(l)$ is at equilibrium in a closed container. Now a certain quantity of extra SO_2 is introduced into the container, the volume remaining the same. Which of the following is/are true?
(A) Temperature will decrease (B) Temperature will increase
(C) Temperature will not change
(D) Pressure inside the container will not change.

66. For the reaction, $PCl_3(g) + Cl_2(g) \rightleftharpoons PCl_5(g)$, the value of K_c at 250°C is 26. The value of K_p at 250°C will be
(A) 0.46 (B) 0.61 (C) 0.57 (D) 0.83

67. In a vessel containing SO_3, SO_2 and O_2 at equilibrium, some He gas is introduced so that the total pressure increases while temperature and volume remain constant. According to Le Chatelier's principle, the dissociation of SO_3
(A) decreases (B) increases
(C) remains unaltered (D) changes unpredictably.

68. If the equilibrium constant for the reaction, $N_2 + 3H_2 \rightleftharpoons 2NH_3$, is K, the equilibrium constant for the reaction, $(1/2)N_2 + (3/2)H_2 \rightleftharpoons NH_3$, will be
(A) $2K$ (B) $K/2$ (C) K^2 (D) $K_{1/2}$

69. If helium is added to the equilibrium, $PCl_3(g) + Cl_2(g) \rightleftharpoons PCl_5(g)$, in a closed vessel, then
(A) the concentration of PCl_3 will increase
(B) the concentration of PCl_5 will increase
(C) the concentration of Cl_2 will increase
(D) the equilibrium concentrations will remain unaffected.

70. The equilibrium constant (K_c, K_p) for the reaction, $CaCO_3(s) \rightleftharpoons CaO(s) + CO_2(g)$ is given by
(A) $K_p = p_{CO_2}$ (B) $K_c = [CO_2]$
(C) $K_c = [CaCO_3]/[CaO]\,[CO_2]$ (D) $K_c = [CaO]\,[CO_2]/[CaCO_3]$

71. If helium gas is added to the equilibrium, $2HI \rightleftharpoons H_2 + I_2$, in a closed vessel, the degree of dissociation will be
(A) increased (B) doubled (C) suppressed (D) unaltered.

72. If the equilibrium constants for the reactions: (i) $SO_2 + (1/2)O_2 \rightleftharpoons SO_3$, (ii) $2SO_2 + O_2 \rightleftharpoons 2SO_3$, are K_1 and K_2 respectively, then K_1 and K_2 are related by the expression
(A) $K_1 = K_2$ (B) $K_1 = (K_2)^{1/2}$ (C) $K_2 = (K_1)^{1/2}$ (D) $K_2 = K_1^2$

73. The value of K_c for the reaction, $2NO(g) \rightleftharpoons N_2(g) + O_2(g)$ is 1.8×10^{-2}. The value of K_p for the reaction, $(1/2)N_2(g) + (1/2)O_2(g) \rightleftharpoons NO(g)$ is
(A) 2 (B) 6 (C) 7.5 (D) 8.5

74. For the reaction, $2NO_2(g) \rightleftharpoons 2NO(g) + O_2(g)$, $K_c = 1.8 \times 10^{-2}$ at 185°C. At 185°C, the value of K_c for the reaction, $NO(g) + (1/2)O_2(g) \rightleftharpoons NO_2(g)$ is
(A) 0.9×10^6 (B) 7.5×10^2 (C) 1.95×10^{-3} (D) 1.95×10^3

75. At 25°C the value of K_c for the reaction $1/2N_2 + O_2 \rightleftharpoons NO_2$ is 2×10^4. What is the value of K_c for the reaction $2NO_2 \rightleftharpoons N_2 + 2O_2$?
(A) $\sqrt{2} \times 10^2$ (B) 2×10^4 (C) $1/(4 \times 10^8)$ (D) 1×10^4
Hint: $K_c = 1/K^2 = 1/(2 \times 10^4)^2 = 1/(4 \times 10^8)$
76. For the reaction, $2A(g) + B(g) \rightleftharpoons 3C(g) + D(g)$, two moles each of A and B were taken in a flask. The following must always be true when the system attained equilibrium.
(A) [A] = [B] (B) [A] < [B] (C) [B] = [C] (D) [A] > [B]
77. In which of the following cases, does the reaction go farthest to completion?
(A) $K = 10^2$ (B) $K = 10^{-2}$ (C) $K = 10$ (D) $K = 1$
78. If K_p for a reaction, $A(g) + 2B(g) \rightleftharpoons 3C(g) + D(g)$, is 0.05 atm at 1000 K, its K_c in terms of R will be
(A) $2000R$ (B) $5 \times 10^{-5}R$ (C) $0.02R$ (D) $(5 \times 10^{-5})/R$
79. A system attains equilibrium when ΔG^0 is
(A) zero (B) negative and large
(C) negative and small (D) positive and large.
80. In which of the following gaseous equilibria, K_p is less than K_c?
(A) $H_2 + I_2 \rightleftharpoons 2HI$ (B) $CO + H_2O \rightleftharpoons CO_2 + H_2$
(C) $N_2 + 3H_2 \rightleftharpoons 2NH_3$ (D) $N_2 + O_2 \rightleftharpoons 2NO$
81. A 500 mL vessel contains 1.5M each of A, B, C and D at equilibrium. If 0.5M each of C and D are taken out, the value of K_c for the reaction $A + B \rightleftharpoons C + D$ will be
(A) 1.0 (B) 1/9 (C) 4/9 (D) 8/9
Hint: $K_c = 1.5 \times 1.5/(1.5 \times 1.5) = 1$
82. Consider the equilibrium $N_2O_4(g) \rightleftharpoons 2NO_2(g)$ in a closed container. At a fixed temperature, the volume of the reaction container is halved. For this change, which of the following statements holds true regarding the equilibrium constant (K_p) and degree of dissociation (α)?
(A) neither K_p nor α changes (B) both K_p and α change
(C) K_p changes, but α does not change
(D) K_p does not change, but α changes. **(I.I.T. 2002)**
83. When the concentration of NH_3 is doubled for the system $NH_4Cl(s) \rightleftharpoons NH_3(g) + HCl(g)$, the equilibrium constant
(A) remains same (B) is halved
(C) decreases, but by less than a factor of 2
(D) increases, but by less than a factor of 2.
84. Which of the following reactions will be favoured, if the pressure is reduced?
(A) $H_2(g) + I_2(g) \rightleftharpoons 2HI(g)$ (B) $N_2(g) + O_2(g) \rightleftharpoons 2NO_2(g)$
(C) $N_2(g) + 3H_2(g) \rightleftharpoons 2NH_3(g)$ (D) $PCl_5(g) \rightleftharpoons PCl_3(g) + Cl_2(g)$
85. The K_p value for the reaction, $H_2(g) + I_2(g) \rightleftharpoons 2HI(g)$ is 65. The value of K_p for the reaction, $2HI(g) \rightleftharpoons H_2(g) + I_2(g)$ is
(A) 13 (B) 1/13 (C) 65 (D) 1/65

86. In which of the following reactions, K_p does not change with change of pressure?
(A) $N_2(g) + 2O_2(g) \rightleftharpoons 2NO_2(g)$ (B) $N_2(g) + O_2(g) \rightleftharpoons 2NO(g)$
(C) $2NH_3(g) \rightleftharpoons N_2(g) + 3H_2(g)$ (D) $PCl_3(g) + Cl_2(g) \rightleftharpoons PCl_5(g)$

87. If α is the degree of dissociation of N_2O_5, the total number of moles at equilibrium for the reaction, $N_2O_5(g) \rightleftharpoons 2NO_2(g) + 1/2\ O_2(g)$ will be
(A) $1 - \alpha$ (B) $1 + 3\alpha/2$ (C) $1 - 3\alpha/2$ (D) $1 + \alpha$

88. 5 moles of SO_2 and 5 moles of O_2 are allowed to react to form SO_3 in a closed vessel. At the equilibrium stage, 60% SO_2 is used up. The total number of moles of SO_2, O_2 and SO_3 in the vessel is
(A) 3.9 (B) 10.5 (C) 8.5 (D) 10.0
Hint: Moles at equil.: SO_2: $5 - (60 \times 5/100) = 2$; $O_2 = 5 - 1.5 = 3.5$; $SO_3 = 3$
Total number of moles = 2 + 3.5 + 3 = 8.5

89. A quantity of PCl_5 was heated in a 10 dm³ vessel at 250°C: $PCl_5(g) \rightleftharpoons PCl_3(g) + Cl_2(g)$. At equilibrium the vessel contains 0.1 mole of PCl_5 and 0.2 mole of Cl_2. K_p for the reaction is
(A) 0.04 (B) 0.025 (C) 0.02 (D) 0.05
Hint: At equil.: $[PCl_5] = 0.01/10$; $[PCl_3] = 0.2/10$; $[Cl_2] = 0.2/10$ mol dm^{-3}. $K_p = 0.02 \times 0.02/0.01 = 0.04$.

90. 9.2 g of $N_2O_4(g)$ is taken in a closed 1 L vessel and heated till the equilibrium: $N_2O_4(g) \rightleftharpoons 2NO_2(g)$ is reached. At equilibrium, 50% $N_2O_4(g)$ is dissociated. What is the equilibrium constant (in mol L^{-1})?
(A) 0.1 (B) 0.2 (C) 0.3 (D) 0.4
Hint: At equil.: $[N_2O_4] = 0.05M$, $[NO_2] = 0.1M$. $K_p = (0.1)^2/0.05 = 0.2$

91. 3 mol of a reactant A and one mole of a reactant B are mixed in a vessel of 1 L. If 1.5 mol of C is formed at equilibrium, the value of K_c for the equilibrium $A + B \rightleftharpoons 2C$ is
(A) 0.50 (B) 0.25 (C) 0.12 (D) 4.00
Hint: Moles at equil.: A = 2.25, B = 0.25, C = 1.5. $K_c = (1.5)^2/(2.25 \times 0.25) = 4.0$

92. The reaction $I_2 + I^- \rightleftharpoons I_3^-$ is set up in aqueous medium. Starting with 1 mole of I_2 and 0.5 mole of I^- in 1 L flask, the reaction was allowed to reach equilibrium. After equilibrium is reached, excess of $AgNO_3$ gave 0.25 mole of a yellow precipitate. The equilibrium constant is
(A) 2.00 (B) 3.00 (C) 1.33 (D) 2.66

93. At some temperature and under a pressure of 4 atm, PCl_5 is 10% dissociated. Temperature remaining the same, what is the pressure at which PCl_5 will be 20% dissociated?
(A) 1.79 atm (B) 1.97 atm (C) 0.79 atm (D) 0.97 atm
Hint:

PCl_5	$\rightleftharpoons PCl_3$	$+ Cl_2$
1	0	0
$1 - \alpha$	α	α

$$K_p = \frac{\alpha^2}{1-\alpha}\left[\frac{P}{1+\alpha}\right] = \frac{\alpha^2 P}{1-\alpha^2}$$. For two pressures, $\frac{\alpha_1^2 P_1}{1-\alpha_1^2} = \frac{\alpha_2^2 P_2}{1-\alpha_2^2}$ or $P_2 = \frac{1-\alpha_2^2}{1-\alpha_1^2} \times \frac{\alpha_1^2}{\alpha_2^2}$

$\times P_1 = \frac{1-(0.2)^2}{1-(0.1)^2} \times \frac{(0.1)^2}{(0.2)^2} \times 4 = 0.97$ atm

ANSWERS

1. C	2. A	3. A	4. D	5. C	6. D
7. D	8. A	9. A	10. C, D, E	11. A, B, C, D	12. C
13. C	14. B	15. A, B, C, D	16. A	17. B	18. D
19. A	20. A	21. A	22. C	23. C	24. C
25. C	26. D	27. D	28. A	29. D	30. B
31. A	32. A, C	33. B, C	34. C	35. C	36. C
37. A	38. D	39. B	40. B	41. D	42. D
43. D	44. A	45. A	46. B	47. C	48. B
49. B	50. A	51. D	52. B	53. C	54. D
55. B	56. C	57. A	58. A	59. D	60. B
61. D	62. A	63. B	64. B	65. B	66. A
67. C	68. D	69. D	70. A, B	71. D	72. B, D
73. C	74. B	75. C	76. B	77. A	78. D
79. A	80. C	81. A	82. D	83. A	84. D
85. D	86. B	87. B	88. C	89. A	90. B
91. D	92. C	93. D			

9

Ionic Equilibrium

IMPORTANT RELATIONS FOR NUMERICAL PROBLEMS

1. Degree of dissociation, α = No. of dissociated molecules/total no. of molecules
2. $K_a = c\alpha^2/(1 - \alpha)$
3. $K_b = c\alpha^2/(1 - \alpha)$
4. $\alpha = (K/c)^{1/2}$; $\alpha = (K \times V)^{1/2}$
5. Strength of acid HA_1/strength of acid $HA_2 = (K_{a_1}/K_{a_2})^{1/2}$
6. Strength of base $(BOH)_1$/strength of base $(BOH)_2 = (K_{b_1}/K_{b_2})^{1/2}$
7. $p\text{H} = -\log [\text{H}^+]$, $[\text{H}^+] - 10^{-p\text{H}}$ = antilog $[-p\text{H}]$
8. $p\text{H} + p\text{OH} = 14 = pK_w$
9. Henderson equation: $p\text{H} = pK_a + \log$ [salt]/[acid]; $p\text{OH} = pK_b + \log$ [salt]/[base]; $p\text{H} = 14 - pK_b + \log$ [salt]/[base]
10. Buffer capacity = No. of moles of acid or base added to 1 L of buffer solution/change in $p\text{H} = n/\Delta p\text{H}$.
11. Hydrolysis of salt of a weak acid and a strong base: $K_h = K_w/K_a$; $h = (K_w/K_a c)^{1/2}$; $p\text{H} = (1/2)pK_w + (1/2)pK_a + (1/2)\log c = 7 + (1/2)pK_a + (1/2)\log c$
12. Hydrolysis of a salt of weak acid and a weak base: $K_h = K_w/K_b$; $h = (K_w/K_b c)^{1/2}$; $p\text{H} = (1/2)pK_w - (1/2)pK_b - (1/2)\log c = 7 - (1/2)pK_b - (1/2)\log c$
13. Hydrolysis of a salt of weak acid and a weak base: $K_h = K_w/K_a \times K_b$; $h = (K_w/K_a \times K_b)^{1/2}$; $p\text{H} = (1/2)pK_w + (1/2)pK_a - (1/2)pK_b$
14. For AB type salt, $K_{sp} = s^2$; for AB_2 type salt, $K_{sp} = 4s^3$ (mol/L)^3; for AB_3 type salt, $K_{sp} = 27s^4$ (mol/L)^4; for A_3B_2 type salt, $K_{sp} = 108s^5$ (mol/L)^5 where s = solubility.

OBJECTIVE TYPE MULTIPLE CHOICE QUESTIONS

1. For the sparingly soluble salt A_pB_q, the relationship of its solubility product (L_s) with its solubility (S) is
   (A) $L_s = S^{p+q}.p^p.q^q$ (B) $L_s = S^{p+q}.p^q.q^p$
   (C) $L_s = S^{pq}.p^p.q^q$ (D) $L_s = S^{pq}.(pq)^{p+q}$ **(I.I.T. 2001)**
   Hint: $A_pB_q \rightleftharpoons pA^+ + qB^-$; $K_{sp} = (A^+)^p \times (B^-)^q = (ps)^p \times (qs)^q = (s^{p+q}).p^p.p^q$
2. The correct order of acidic strength is
   (A) $Cl_2O_7 > SO_2 > P_4O_{10}$ (B) $Na_2O > MgO > Al_2O_3$
   (C) $K_2O > CaO > MgO$ (D) $CO_2 > N_2O_5 > SO_3$ **(I.I.T. 2000)**
3. A buffer solution can be prepared from a mixture of
   (A) CH_3COONa and CH_3COOH in water
   (B) CH_3COONa and HCl in water
   (C) NH_4OH and NH_4Cl in water (D) NH_4OH and NaOH in water. **(I.I.T. 1999)**
4. Which of the following statements is/are correct?
   (A) The pH of 1.0×10^{-8} M solution of HCl is 8
   (B) The conjugate base of $H_2PO_4^-$ is HPO_4^{2-}
   (C) Autoprotolysis constant of water increases with temperature
   (D) When a solution of weak monoprotic acid is titrated against a strong base, at half-neutralisation point $p\text{H} = (1/2)pK_a$. **(I.I.T. 1998)**
5. The solubility of A_3X_2 is y mol dm^{-3}. Its solubility product is
   (A) $108y^5$ (B) $36y^5$ (C) $64y^4$ (D) $6y^4$ **(I.I.T. 1997)**
6. The pH of 0.1M solution of the following salts increases in the order:
   (A) NaCl < NH_4Cl < NaCN < HCl (B) HCl < NH_4Cl < NaCl < NaCN
   (C) NaCN < NH_4Cl < NaCl < HCl (D) HCl < NaCl < NaCN < NH_4Cl **(I.I.T. 1999)**
7. The pH of an aqueous solution of 1.0M ammonium formate is (use pK_a of CH_3COOH = 3.8 and pK_b of ammonia = 4.8)
   (A) 6.5 (B) 13 (C) 3.25 (D) 9.75 **(I.I.T. 1995)**
8. The pK_a of HCN is 9.30. The pH of a solution prepared by mixing 2.5 moles of HCN and 0.25 mole of KCN in water and making up the total volume to 500 mL is
   (A) 9.30 (B) 7.30 (C) 8.30 (D) 10.30 **(I.I.T. 1993)**
   Hint: $p\text{H} = pK_a + \log[\text{salt}]/[\text{acid}] = 9.30 + \log (0.25)/(2.5) = 9.30 + \log 0.1 = 9.30 - 1 = 8.30$.
9. What is the decreasing order of strength of the bases OH^-, NH_2^-, $H-C\equiv C^-$ and $CH_3CH_2^-$?
   (A) $CH_3-CH_2^- > NH_2^- > H-C\equiv C^- > OH^-$
   (B) $H-C\equiv C^- > CH_3-CH_2^- > NH_2^- > OH^-$
   (C) $OH^- > NH_2^- > H-C\equiv C^- > CH_3-CH_2^-$
   (D) $NH_2^- > H-C\equiv C^- > OH^- > CH_3-CH_2^-$ **(I.I.T. 1993)**

10. The pH of pure water at 25°C is close to
(A) 2.0 (B) 10.0 (C) 7.0 (D) 12.0 **(I.I.T. 1992)**

11. The pK_a of acetyl salicylic acid (aspirin) is 3.5. The pH of gastric juice in human stomach is about 2–3 and the pH in the small intestine is about 8. Aspirin will be
(A) unionised in the small intestine and in the stomach
(B) completely ionised in the small intestine and in the stomach
(C) ionised in the stomach and almost unionised in the small intestine
(D) ionised in the small intestine and almost unionised in the stomach. **(I.I.T. 1988)**

12. When equal volumes of the following solutions are mixed, precipitation of AgCl (K_{sp} = 1.8×10^{-10}) will occur only with
(A) 10^{-4} M (Ag^+) and 10^{-4} M (Cl^-) (B) 10^{-5} M (Ag^+) and 10^{-5} M (Cl^-)
(C) 10^{-5} M (Ag^+) and 10^{-6} M (Cl^-) (D) 10^{-10} M (Ag^+) and 10^{-10} M (Cl^-) **(I.I.T. 1988)**

13. The conjugate acid of NH_2^- is
(A) NH_3 (B) NH_2OH (C) NH_4^+ (D) N_2H_4 **(I.I.T. 1985)**

14. A certain weak acid has a dissociation constant 1×10^{-4}. The equilibrium constant for its reaction with a strong base is
(A) 1×10^{-4} (B) 1×10^{-10} (C) 1×10^{10} (D) 1×10^{14} **(I.I.T. 1984)**

15. The compound whose 0.1M solution is basic is
(A) CH_3COONH_4 (B) CH_3COONa (C) $(NH_4)_2SO_4$ (D) NH_4Cl **(I.I.T. 1986)**

16. A certain buffer solution contains equal concentration of X^- and HX. The K_b for X^- is 10^{-10}. The pH of the buffer solution is
(A) 4 (B) 7 (C) 10 (D) 14 **(I.I.T. 1984)**

17. A precipitate of CaF_2(K_{sp} = 1.7×10^{-8}) is obtained when equal volumes of the following are mixed.
(A) 10^{-4} M Ca^{2+} and 10^{-4} M F^- (B) 10^{-2} M Ca^{2+} and 10^{-3} M F^-
(C) 10^{-5} M Ca^{2+} and 10^{-3} M F^- (D) 10^{-3} M Ca^{2+} and 10^{-5} M F^- **(I.I.T. 1982)**

18. At 90°C pure water has $[H_3O^+] = 10^{-6}$ mol L^{-1}. The value of K_w at 90°C will be
(A) 10^{-6} (B) 10^{-8} (C) 10^{-12} (D) 10^{-14} **(I.I.T. 1981)**

19. In the given anions the strongest Bronsted base is
(A) ClO^- (B) ClO_2^- (C) ClO_3^- (D) ClO_4^- **(I.I.T. 1981)**

20. An acidic buffer solution can be prepared by mixing the solutions of
(A) CH_3COONH_4 and CH_3COOH (B) NH_4Cl and NH_4OH
(C) H_2SO_4 and Na_2SO_4 (D) NaCl and NaOH **(I.I.T. 1981)**

21. The pH of 10^{-8} molar solution of HCl in water is
(A) between 7 and 8 (B) between 6 and 7
(C) – 8 (D) 8 **(I.I.T. 1981)**
Hint: $[H^+] = 10^{-8}$ (HCl) + 10^{-7} (H_2O) = 10^{-8} (1 + 10) = 11×10^{-8}. pH = – log 11×10^{-8} = 8 – 1.04 = 6.96

22. In which of the following the value of pH will be 12?
(A) 0.01M KOH (B) 1M NaOH (C) 1M $Ca(OH)_2$ (D) 1M KOH
(Roorkee 1995)

23. When CH_3COONa is added to an aqueous CH_3COOH solution
(A) pH increases (B) pH remains unchanged
(C) pH decreases (D) pH becomes zero. **(Roorkee 1994)**
24. Solubility product of A_2B is 4×10^{-9} $(mol/L)^3$. Its solubility is
(A) $2 \times 10^{-5}\,M$ (B) $10^{-4}\,M$ (C) $(4)^{1/3} \times 10^{-3}\,M$ (D) $10^{-3}\,M$ **(Roorkee 1990)**
25. Which of the following salts undergoes anionic hydrolysis?
(A) $CuSO_4$ (B) NH_4Cl (C) Na_2CO_3 (D) $FeCl_3$ **(I.I.T. 1983)**
26. Which of the following mixture of solutions has pH = 1.0?
(A) 100 mL (M/10)HCl + 100 mL (M/10)NaOH
(B) 55 mL (M/10)HCl + 45 mL (M/10)NaOH
(C) 10 mL (M/10)HCl + 90 mL (M/10)NaOH
(D) 75 mL (M/5)HCl + 25 mL (M/5)NaOH **(I.I.T. 1992)**
27. Fear or excitement generally causes one to breathe rapidly and it results in the decrease of CO_2 concentration in blood. In what way will it change the pH of blood?
(A) pH will increase (B) pH will decrease
(C) pH will adjust to 7 (D) No change **(Roorkee 1993)**
28. At 25°C the dissociation constants of CH_3COOH and NH_4OH in aqueous solution are almost the same (10^{-5}). The pH of a solution of CH_3COOH is 4.0 at 25°C. The pH of NH_4OH solution of same concentration at the same temperature is
(A) 10.0 (B) 10.5 (C) 3.0 (D) 4.0 **(Roorkee 1990)**

Hint: pH of CH_3COOH solution = 4. Hence $[H_3O^+] = \sqrt{c \times K_a} = 10^{-4}$ or $c \times 10^{-5} = 10^{-8}$ or $c = 10^{-3}M$ $[OH^-] = \sqrt{10^{-3} \times 10^{-5}} = 10^{-4}$. Hence pOH = 4 and pH = 14 – 4 = 10.

29. What is the correct representation of the solubility product of SnS_2?
(A) $[Sn^{2+}]\,[S^{2-}]^2$ (B) $[Sn^{4+}]\,[2S^{2-}]^2$ (C) $[Sn^{4+}]\,[S^{2-}]^2$ (D) $[Sn^{2+}]\,[2S^{2-}]$
30. Which of the following can act as both an acid and a base?
(A) NH_3 (B) NH_4^+ (C) HCO_3^- (D) HSO_4^-
31. The concentrations of H^+ and OH^- of a 0.1M aqueous solution of 2% ionised weak acid are (K_w of water = 1×10^{-14})
(A) $3 \times 10^{-2}\,M$, $4 \times 10^{-13}\,M$ (B) $2 \times 10^{-3}\,M$, $5 \times 10^{-12}\,M$
(C) $1 \times 10^{-3}\,M$, $3 \times 10^{-11}\,M$ (D) $0.02 \times 10^{-3}\,M$, $5 \times 10^{-11}\,M$
32. A monoprotic acid in a 0.1M solution ionises to 0.001%. Its ionisation constant is
(A) 1.0×10^{-11} (B) 1.0×10^{-8} (C) 1.0×10^{-6} (D) 1.0×10^{-3}
33. A buffer solution is prepared by mixing 10 mL of 1.0M CH_3COOH and 20 mL of 0.5M CH_3COONa and then diluted to 100 mL. If the pK_a of CH_3COOH is 4.76, the pH of the buffer solution is
(A) 4.21 (B) 4.34 (C) 4.76 (D) 5.21
34. The pH of 0.01M HS^- is given by
(A) $pH = 7 + \dfrac{(pK_a - pK_b)}{2}$ (B) $pH = 7 - \dfrac{pK_b}{2} + \dfrac{\log c}{2}$
(C) $pH = \dfrac{pK_1 + pK_2}{2}$ (D) $pH = 7 + \dfrac{pK_a}{2} + \dfrac{\log c}{2}$

35. The degree of dissociation of a weak electrolyte in aqueous solution
   (A) is directly proportional to the initial concentration of electrolyte
   (B) depends on the equilibrium concentration of electrolyte
   (C) does not depend on initial concentration of electrolyte
   (D) is inversely proportional to the initial concentration of electrolyte.
36. The *p*H of an aqueous solution of CH_3COONa is
   (A) 3 (B) 7 (C) 9 (D) 14
37. Identify the incorrect statement.
   (A) AgCl is more soluble in concentrated NaCl solution than in pure water
   (B) A weak base becomes weaker in the presence of its salt
   (C) CH_3COOH is a weaker acid than H_2SO_4 because it gives only one H^+ per molecule
   (D) The addition of NH_4Cl to a solution of NH_4OH does not affect the *p*H of the solution of NH_4OH.
38. Identify the incorrect statement.
   (A) The addition of HCl to a solution of benzoic acid suppresses the ionisation of benzoic acid
   (B) A solution of $CuSO_4$ is neutral
   (C) The *p*H of a 0.50*M* aqueous NaCN solution is 11.5 (pK_b of $CN^- = 4.70$)
   (D) If the ionic product of water is 2.5×10^{-14} at 40°C, the $[H^+]$ of water at 40°C is 1.6×10^{-7} mol/L.
39. Identify the incorrect statement.
   (A) The *p*H of 1.0*M* solution of a weak monoprotic acid which is 1% ionised is 3
   (B) The *p*H of an aqueous solution of HCl is > 7
   (C) If the dissociation constant of acetic acid is 1.8×10^{-5}, the degree of dissociation of 0.5*M* solution is 0.6%.
   (D) The ion SH^- is a stronger base than S^{2-}.
40. Identify the incorrect statement.
   (A) $AlCl_3$ is a Lewis acid because it can donate electrons
   (B) The solubility of $Zn(OH)_2$ is greater in an acidic medium than in neutral water
   (C) The base strength follows the order $I^- > Br^- > Cl^- > F^-$
   (D) If the dissociation constant of NH_4OH is 1.8×10^{-5}, the hydrolysis constant of NH_4^+ ions at 25°C is 0.34 mol.
41. How much sodium acetate should be added to a 0.1*M* solution of CH_3COOH to give a solution of *p*H 5.5 (pK_a of $CH_3COOH = 4.5$)?
   (A) 10.0*M* (B) 1.0*M* (C) 0.2*M* (D) 0.1*M*
42. In a mixture of weak acid and its salt, the ratio of concentration of acid to salt is increased ten-fold. The *p*H of the solution
   (A) increases ten-fold (B) increases by one
   (C) decreases by 1/10 (D) decreases by one.

43. Decimolar aqueous solutions of NaOH, NH_4OH, NaCl and NH_4Cl arranged in the ascending order of pH values follows the sequence
(A) NH_4OH, NH_4Cl, NaOH, NaCl (B) NaOH, NH_4Cl, NaCl, NH_4OH
(C) NH_4Cl, NaCl, NH_4OH, NaOH (D) NaOH, NH_4OH, NaCl, NH_4Cl

44. The pH of 100 mL of an acid is 1. The pH of the same acid of 200 mL with different concentration is 2. The pH of the mixture is
(A) 3 (B) 2 (C) 2.5 (D) 3.4

45. The strongest base among the following is
(A) S^{2-} (B) O^{2-} (C) OH^- (D) NH_2^-

46. H_2S is passed in molar solution of each of the following compounds. Which of the following is in the decreasing order of sulphide ion concentration?
(A) CH_3COOH, NaOH, HCl, NH_4OH (B) NaOH, NH_4OH, CH_3COOH, HCl
(C) NH_4OH, CH_3COOH, NaOH, HCl (D) HCl, NH_4OH, CH_3COOH, NaOH

47. Which of the following will occur if a 0.1M solution of weak acid is diluted to 0.01M at constant temperature?
(A) percentage ionisation will increase
(B) K_a will increase
(C) pH will decrease (D) $[H^+]$ will decrease to 0.01M

48. The dissociation constant of CH_3COOH is 1.6×10^{-5}. The degree of dissociation (% α) of 0.01M CH_3COOH in the presence of 0.1M HCl is equal to
(A) 0.4 (B) 0.04 (C) 0.016 (D) 0.16

49. The solubility product of AgCl is 4.0×10^{-10} at 298 K. The solubility of AgCl in 0.04M $CaCl_2$ will be
(A) 2.2×10^{-4} mol/L (B) 5.0×10^{-9} mol/L
(C) 1.0×10^{-5} mol/L (D) 2.0×10^{-4} mol/L

50. The solubility product of AgCl in water is 2.17×10^{-8}. Its solubility in 0.01M NaCl solution is
(A) 21.70×10^{-6} g/L (B) 0.217×10^{-6} g/L
(C) 4.34×10^{-6} g/L (D) 2.17×10^{-6} g/L **(Roorkee 1995)**

51. The pH of milk lies in the range
(A) 2.6–4.4 (B) 6.6–6.9 (C) 7.0–7.5 (D) 7.3–7.4

52. The pH at which $Mg(OH)_2$ begins to precipitate from a solution containing 0.1M Mg^{2+} ion is (K_{sp} of $Mg(OH)_2 = 1 \times 10^{-11}$)
(A) 6 (B) 7 (C) 8 (D) 9

53. On addition of 0.002 mole of an acid to 250 mL of a buffer solution, the pH of the solution decreases by 0.02 units. The buffer capacity of the buffer solution is
(A) 0.04 (B) 0.4 (C) 4 (D) 2

54. At what molar concentration, the pH of HNO_2 would be 2? ($K_a = 4.5 \times 10^{-4}$)
(A) 0.222M (B) 22.2M (C) 222M (D) 0.022M

55. The pH of a solution formed by mixing equal volumes of 0.1M propanoic acid and 0.1M sodium propanoate is
(A) 1.62 (B) 3.85 (C) 4.89 (D) 5.58

56. The degree of dissociation of 0.01M CH_3COOH in the presence of 0.1M HCl is (K_a of $CH_3COOH = 1.6 \times 10^{-5}$)
(A) 0.016 (B) 0.16 (C) 0.8 (D) 0.08

57. An aqueous solution is 0.01M with respect to NH_4OH and 0.1M with respect to NH_4Cl. pK_b of aqueous ammonia is 5. The pH of the solution is
(A) 6.5 (B) 7.5 (C) 8 (D) 8.5

58. Which of the following expressions is correct for calculating the pH of sodium acetate solution?
(A) $pH = (1/2)pK_w - (1/2)pK_a(CH_3COOH) + (1/2) \log [salt]$
(B) $pH = (1/2)pK_a - (1/2)pK_a(CH_3COOH) - (1/2) \log [salt]$
(C) $pH = (1/2)pK_w + (1/2)pK_a(CH_3COOH) + (1/2) \log [salt]$
(D) $pH = (1/2)pK_w + (1/2)pK_a(CH_3COOH) - (1/2) \log [salt]$

59. Which of the following expressions is correct for calculating the pH of NH_4Cl solution?
(A) $pH = (1/2)pK_w - (1/2)pK_b(NH_4OH) - (1/2) \log [salt]$
(B) $pH = (1/2)pK_w + (1/2)pK_b(NH_4OH) + (1/2) \log [salt]$
(C) $pH = (1/2)pK_w + (1/2)pK_b(NH_4OH) - (1/2) \log [salt]$
(D) $pH = (1/2)pK_w - (1/2)pK_b(NH_4OH) + (1/2) \log [salt]$

60. Which of the following is correct for an acidic buffer whose $pH = pK_a$?
(A) [salt] < [acid] (B) [salt] > [acid]
(C) [salt] = [acid] (D) [salt] + [acid] is minimum.

61. The maximum pH of a solution which is 0.1M in Mg^{2+} from which $Mg(OH)_2$ will not be precipitated is ($K_{sp}(Mg(OH)_2) = 1.2 \times 10^{-11}$)
(A) 4.95 (B) 5.95 (C) 7.65 (D) 9.04

62. What will happen to the pH value, if ammonium chloride is added to 0.1M CH_3COOH solution?
(A) No change in pH value (B) Decrease in pH value
(C) Increase in pH value (D) Change in pH cannot be predicted.

63. What will happen to the pH value, if sodium acetate is added to 0.1M CH_3COOH solution?
(A) No change in pH value (B) Decrease in pH value
(C) Increase in pH value (D) Change in pH cannot be predicted.

64. What will happen to the pH value, if a buffer solution is diluted with water?
(A) Decrease in pH value (B) Increase in pH value
(C) pH remains constant (D) Change in pH cannot be predicted.

65. What will happen to the pH value, if some HCl is added to a buffer solution containing NH_4OH and NH_4Cl?
(A) Decrease a little (B) Increase a little
(C) pH remains constant (D) Change in pH cannot be predicted.

66. The pink colour of the indicator phenolphthalein in an alkaline solution is due to
    (A) OH^- of the alkali
    (B) anionic form of phenolphthalein
    (C) acidic form of phenolphthalein
    (D) non-conjugated structure of phenolphthalein.
67. The number of H^+ ions present in 1 mL of a solution whose pH = 13 is
    (A) 1×10^{13} (B) 6.023×10^{7} (C) 6.023×10^{8} (D) 6.023×10^{9}
68. Identify the incorrect statement.
    (A) $[H^+] = [OH^-] = 10^{-7} M$ for a neutral solution at all temperatures
    (B) $[H^+] = [OH^-] = (K_w)^{1/2}$ for a neutral solution
    (C) $[H^+] < (K_w)^{1/2}$ and $[OH^-] > (K_w)^{1/2}$ an alkaline solution
    (D) $[H^+] > (K_w)^{1/2}$ and $[OH^-] < (K_w)^{1/2}$ for an acidic solution.
69. The H^+ ion concentration of acetic acid is decreased by the addition of
    (A) $Al_2(SO_4)_3$ (B) NH_4Cl (C) $AgNO_3$ (D) NaCN
70. What will happen if sodium hydroxide is added to a solution of acetic acid?
    (A) $[CH_3COO^-]$ increases
    (B) $[CH_3COOH]$ increases
    (C) pH value of solution decreases
    (D) $[H^+]$ increases.
71. What will be the pH of a solution prepared by mixing 2 mL of a strong acid (pH = 3.0) and 3 mL of a strong base (pH = 10.0)?
    (A) 7.0 (B) 5.5 (C) 3.5 (D) 4.5
72. Which of the following will have nearly equal H^+ ion concentration?
    (A) 100 mL of 0.1M HCl mixed with 50 mL of water
    (B) 50 mL of 0.1M HCl mixed with 50 mL of water
    (C) 50 mL of 0.1M H_2SO_4 mixed with 100 mL of water
    (D) 50 mL of 0.1M HCl mixed with 100 mL of water.
73. The K_{sp} for HgS, Ag_2S and PbS is 10^{-31}, 10^{-45} and 10^{-50} respectively. The solubility of these salts is in the order
    (A) HgS > Ag_2S > PbS
    (B) HgS < PbS < Ag_2S
    (C) PbS > Ag_2S > HgS
    (D) Ag_2S > HgS > PbS
74. If equal volumes of two solutions of pH 3 and 4 are mixed, the pH value of the resulting solution will be
    (A) 2.96 (B) 3.26 (C) 3.5 (D) 3.75
75. The conjugate acid of HPO_3^{2-} ion is
    (A) HPO_4^{2-} (B) $H_2PO_4^-$ (C) $H_2PO_3^-$ (D) PO_3^{3-}
76. Which of the following is the strongest acid?
    (A) H_2SO_3 (B) H_2SO_4 (C) $HClO_3$ (D) $HClO_4$
77. The solubility of AgCl in 0.01M NaCl will
    (A) decrease
    (B) increase
    (C) remain same
    (D) AgCl undergoes complete dissociation in 0.01M NaCl

78. The pH of a solution produced when an aqueous solution of pH = 6 is mixed with an equal volume of an aqueous solution of pH = 3 is

(A) 4.5 (B) 4.0 (C) 3.3 (D) 4.3

79. The pK_a values of two acids A and B are 4 and 5 respectively. The strengths of these two acids are related as

(A) the strengths of two acids cannot be compared

(B) acid B is ten times stronger than acid A

(C) strength of acid A : strength of acid B = 4 : 5

(D) acid A is 10 times stronger than acid B.

80. The stepwise ionisation constants of polybasic acid H_3PO_4 are given by: $H_3PO_4 + H_2O \rightleftharpoons H_2PO_4^- + H_3O^+$, K_{a_1}; $H_2PO_4^- \rightleftharpoons HPO_4^{2-} + H_3O^+$, K_{a_2}; $HPO_4^{2-} \rightleftharpoons PO_4^{3-} + H_2O$, K_{a_3}. Which of the following relations is correct?

(A) $K_{a_1} < K_{a_2} > K_{a_3}$ (B) $K_{a_1} < K_{a_2} < K_{a_3}$

(C) $K_{a_1} > K_{a_2} > K_{a_3}$ (D) $K_{a_1} > K_{a_2} < K_{a_3}$

81. The per cent ionisation of a weak acid is calculated using the formula

(A) $\dfrac{100}{1+10(pK_a - p\text{H})}$ (B) $100\sqrt{\dfrac{K_a}{C}}$

(C) both (A) and (B) (D) none of these.

82. The pH of blood is maintained by a buffering system which is

(A) HCO_3^-/H_2CO_3 (B) NaCl/HCl

(C) NH_4Cl/NH_4OH (D) citric acid/sodium citrate.

83. Identify the correct order of solubility of Na_2S, CuS and ZnS in aqueous medium.

(A) CuS > ZnS > Na_2S (B) ZnS > Na_2S > CuS

(C) Na_2S > CuS > ZnS (D) Na_2S > ZnS > CuS **(I.I.T. 2002)**

84. Identify the correct order of acid strengths of CO_2, CuO, CaO and H_2O.

(A) CaO < CuO < H_2O < CO_2 (B) H_2O < CuO < CaO < CO_2

(C) CaO < H_2O < CuO < CO_2 (D) H_2O < CO_2 < CaO < CuO **(I.I.T. 2002)**

85. The conjugate base of $N_2H_5^+$ is

(A) N_2H_6 (B) $N_2H_4^-$ (C) N_2H_4 (D) $N_2H_4^+$

86. The conjugate acid of HPO_4^{2-} is

(A) H_3PO_3 (B) $H_2PO_4^{2-}$ (C) $H_2PO_4^-$ (D) PO_4^{3-}

87. H_2S is a conjugate acid of

(A) HSO_4^- (B) HS^- (C) SO_3^{2-} (D) S^{2-}

88. Which of the following when added to water causes an increase in pH?

(A) K_2SO_4 (B) KCl (C) NH_4Cl (D) CH_3COOK

89. A buffer solution can be prepared by mixing the solutions of

(A) Na_2HPO_4, NaH_2PO_4 (B) NaCl, CH_3COOH

(C) NH_4Cl, NH_3 (D) $NaNO_3$, HNO_3

90. Which of the following are conjugate acid-base pairs?

(A) H_3O^+, OH^- (B) HCl, Cl^- (C) NH_4^+, NH_2^- (D) HCO_3^-, CO_3^{2-}

91. An acidic buffer of *p*H close to 3.2 can be prepared by using the pairs
    (A) CH_3COOH, CH_3COO^- (B) HCO_3^-, CO_3^{2-}
    (C) HCOOH, $HCOO^-$ (D) NH_4^+, NH_3
92. Which of the solutions has the highest *p*H?
    (A) 0.1*M* Na_2CO_3 (B) 0.1*M* KNO_2 (C) 0.1*M* KNO_3 (D) 0.1*M* $HClO_4$
93. Which of the following has the lowest *p*H?
    (A) 0.1*M* HCOOH (B) 0.1*M* CH_3COOH
    (C) 0.1*M* HCl (D) 0.1*M* $HClO_4$
94. A buffer solution can be prepared by mixing the solutions of
    (A) CH_3COOH, KCl (B) HCl, KOH
    (C) H_2SO_4, NaOH (D) CH_3COOH, NH_4OH
95. Which of the following is an ampholyte?
    (A) H_2O (B) $HCOO^-$ (C) C_6H_5COOH (D) NH_4^+
96. The strongest acid out of the following is
    (A) $HClO_4$ (B) H_2SO_4 (C) H_3PO_4 (D) HNO_3
97. Which of the following species cannot exist in aqueous solution?
    (A) NH_3 (B) NH_2^- (C) NO_2^- (D) NH_4^+
98. In aqueous solution, which of the following is a stronger acid than HCOOH?
    (A) H_3PO_4 (B) H_2SO_3 (C) CH_3COOH (D) NH_4^+
99. Identify the Lewis acid and Lewis base in the reaction: $BeF_2 + 2F^- \longrightarrow BeF_4^{2-}$
    (A) BeF_2, F^- (B) BeF_4^{2-}, F^- (C) BeF_2, BeF_4^{2-} (D) F^-, F^-
100. In HS^-, I^-, RNH_2 and NH_3, the order of proton accepting tendency is
    (A) $HS^- < RNH_2 < NH_3 < I^-$ (B) $I^- < HS^- < RNH_2 < NH_3$
    (C) $I^- < HS^- < NH_3 < RNH_2$ (D) $I^- < NH_3 < RNH_2 < HS^-$
101. The *p*H range of methyl red indicator is
    (A) 4.2–6.3 (B) 8.3–10.0 (C) 8.0–9.6 (D) 6.8–8.4
102. The *p*H of a 0.1*M* aqueous solution of a weak acid HA is 3. What is its degree of dissociation?
    (A) 1% (B) 10% (C) 50% (D) 25%
    Hint: $[H^+] = c\alpha = 10^{-3}$ or $0.1 \times \alpha = 10^{-3}$ or $\alpha = 10^{-2} = 1\%$.
103. Hydrogen ion concentration of an aqueous solution is 1×10^{-4} *M*. The solution is diluted with equal volume of water. The hydroxyl ion concentration of the resultant solution in mol dm^{-3} is
    (A) 0.5×10^{-10} (B) 2×10^{-10} (C) 1×10^{-6} (D) 1×10^{-8}
    Hint: After dilution, $[H^+] = 10^{-4}/2$. Hence $[OH^-] = 10^{-14}/[H^+] = 10^{-14} \times 2/10^{-4} = 2 \times 10^{-10}$ mol dm^{-3}.
104. If pK_b for fluoride ion at 25°C is 10.83, the ionisation constant of HF at 25°C is
    (A) 1.74×10^{-5} (B) 3.52×10^{-3} (C) 6.76×10^{-4} (D) 5.38×10^{-2} **(I.I.T. 1997)**
    Hint: $pK_a + pK_b = pK_w$ or $pK_a + 10.83 = 14$ or $pK_a = 3.17$ or $\log K_a = -3.17$ or $K_a = 6.76 \times 10^{-4}$

105. HX is a weak acid ($K_a = 10^{-5}$). It forms a salt on reacting with sodium hydroxide. The degree of hydrolysis of NaX is

(A) 0.01% (B) 0.0001% (C) 0.1% (D) 0.5% **(I.I.T. 2004)**

Hint: $K_h = K_w/K_a = 10^{-14}/10^{-5} = 10^{-9}$; $K_h = c\alpha^2 = 10^{-9}$ or $\alpha^2 = 10^{-9}/0.1 = 10^{-8}$ or $\alpha = 10^{-4}$. % degree of dissociation $= 10^{-4} \times 100 = 0.01\%$.

106. CH_3NH_2 (0.1 mole, $K_b = 5 \times 10^{-4}$) is added to 0.08 mole of HCl and the solution is diluted to one litre. The resulting hydrogen ion concentration is

(A) 1.6×10^{-11} (B) 8×10^{-11} (C) 5×10^{-5} (D) 8×10^{-2} **(I.I.T. 2005)**

Hint:	CH_3NH_2 +	$HCl \longrightarrow$	$CH_3NH_3^+ + Cl^-$
Initial mole	0.1	0.08	–
Moles at equil.	0.02	–	0.08

$$[OH^-] = K_b \frac{[CH_3NH_2]}{[CH_3NH_3^+]} = \frac{5 \times 10^{-4} \times 0.02}{0.08} = \frac{5}{4} \times 10^{-4}$$

$$[H^+] = \frac{K_w}{[OH^-]} = \frac{10^{-14} \times 4}{5 \times 10^{-4}} = 8 \times 10^{-11}$$

107. 2.5 mL of (2/5) *M* weak monoacidic base ($K_b = 1 \times 10^{-12}$ at 25°C) is titrated with (2/15) *M* HCl in water at 25°C. The concentration of H^+ at equivalence point is ($K_w = 1 \times 10^{-14}$ at 25°C)

(A) $3.7 \times 10^{-13} M$ (B) $3.2 \times 10^{-7} M$ (C) $3.2 \times 10^{-2} M$ (D) $2.7 \times 10^{-2} M$ **(I.I.T. 2008)**

Hint: $BOH + HCl \longrightarrow BCl + H_2O$

$$2.5 \times 10^{-3} \times \frac{2}{5} = x \times 10^{-3} \times \frac{2}{15}$$

or $x = 7.5$ mL

$$[BCl] = \frac{2.5 \times \frac{2}{5}}{10} = 0.1M$$

$$\underset{0.1(1-h)}{B^+} + H_2O \rightleftharpoons \underset{0.1h}{BOH} + \underset{0.1h}{H^+}$$

$$K_h = \frac{K_w}{K_b} = \frac{[BOH][H^+]}{[B^+]} = \frac{0.1h^2}{(1-h)}$$

$$\frac{1 \times 10^{-14}}{1 \times 10^{-12}} = 1 \times 10^{-2} = \frac{0.1h^2}{(1-h)}$$

or $h = 0.27$ (We cannot ignore the value of h with respect to 1 because it is more than 0.1)

Hence $[H^+] = 0.1 \times 0.27 = 2.7 \times 10^{-2} M$

108. Aqueous solutions of HNO_3, KOH, CH_3COOH and CH_3COONa of identical concentrations are provided. The pair(s) of solutions which form a buffer upon mixing is (are)

(A) HNO_3 and CH_3COOH (B) KOH and CH_3COONa
(C) HNO_3 and CH_3COONa (D) CH_3COOH and CH_3COONa **(I.I.T. 2010)**

Hint: In the option (C), if HNO_3 is present in limiting amount, then the mixture will behave as a buffer. In the option (D), the mixture contains a weak acid (CH_3COOH) and its salt CH_3COONa and this mixture behaves as a buffer.

109. The equilibrium, $2Cu^{I} \rightleftharpoons Cu^{0} + Cu^{II}$ in aqueous medium at 25°C shifts towards the left in the presence of

(A) NO_3^- (B) Cl^- (C) SCN^- (D) CN^- **(I.I.T. 2011)**

Hint: Cu^{2+} ions will react with CN^- and SCN^- forming $[Cu(CN)_4]^{3-}$ and $[Cu(SCN)_4]^{3-}$ leading the reaction in the backward direction.

$Cu^{2+} + 2CN^- \longrightarrow Cu(CN)_2$

$2Cu(CN)_2 \longrightarrow 2CuCN + (CN)_2$

$CuCN + 3CN^- \longrightarrow [Cu(CN)_4]^{3-}$

$Cu^{2+} + 4SCN^- \longrightarrow [Cu(SCN)_4]^{3-}$

Cu^{2+} also combines with $CuCl_2$ which reacts with Cu to produce CuCl pushing the reaction in the backward direction.

$CuCl_2 + Cu \longrightarrow 2CuCl \downarrow$

ANSWERS

1. A	2. A	3. A, B, C	4. B, C	5. A	6. B
7. A	8. C	9. A	10. C	11. D	12. A
13. A	14. C	15. B	16. A	17. B	18. C
19. A	20. A	21. B	22. A	23. A	24. D
25. C	26. D	27. D	28. A	29. B	30. A, C, D
31. B	32. A	33. C	34. C	35. D	36. C
37. C, D	38. B	39. B, D	40. A, C	41. B	42. D
43. C	44. D	45. B	46. B	47. A	48. C
49. B	50. D	51. B	52. D	53. B	54. A
55. C	56. A	57. C	58. C	59. A	60. C
61. D	62. B	63. C	64. C	65. A	66. B
67. B	68. A	69. D	70. A	71. C	72. A, C
73. D	74. B	75. C	76. D	77. A	78. C
79. D	80. C	81. C	82. A	83. D	84. A
85. C	86. C	87. B	88. D	89. A, C	90. B, D
91. C	92. A	93. D	94. D	95. A	96. A
97. A	98. B	99. A	100. C	101. A	102. A
103. B	104. C	105. A	106. B	107. D	108. C, D
109. B, C, D					

10

Chemical Kinetics

IMPORTANT RELATIONS USED FOR NUMERICAL PROBLEMS

1. Instantaneous rate expression for a reaction, $a\text{A} + b\text{B} \longrightarrow c\text{C} + d\text{D}$:

$$\text{rate} = -\frac{1}{a}\frac{d[\text{A}]}{dt} = -\frac{1}{b}\frac{d[\text{B}]}{dt} = \frac{1}{c}\frac{d[\text{C}]}{dt} = \frac{1}{d}\frac{d[\text{D}]}{dt}$$

2. Rate law: rate = $k[\text{A}]^x/[\text{B}]^y$ (where k = rate constant)
3. For a first order reaction, $k = (2.303/t)\log a/(a - x)$
4. For a second order reaction, $k = (1/t)[2.303/(a - b)]\log [b(a - x)]/[a(b - x)]$
5. For a zero order reaction, $k = [a - (a - x)]/t$
6. $k = Ae^{-E_a}/RT$
7. $\log (k_2/k_1) = (E_a/2.303R)(T_2 - T_1)/T_1T_2$; $E_a = [2.303RT_1T_2/(T_2 - T_1)]\log (k_2/k_1)$
8. For zero order reaction: $t_{1/2} = a/2k$ (unit of k = mol L^{-1} time^{-1}); for first order reaction: $t_{1/2} = 0.693/k$ (unit of k = time^{-1}); for second order reaction: $t_{1/2} = 1/ka$ (unit of k = L mol^{-1} time^{-1}); general formula for $t_{1/2}$; $t_{1/2} \propto 1/k(a^{n-1})$; general formula for unit of k: $\text{L}^{n-1}\ \text{mol}^{1-n}\ \text{time}^{-1}$.
9. General expression for integrated rate equation:

$$-\frac{d[\text{A}]}{dt} = k[\text{A}]^n \text{ (where } n \neq 1)$$

$$t_{1/2} = \frac{2^{n-1} - 1}{k(n-1)a^{n-1}}$$

OBJECTIVE TYPE MULTIPLE CHOICE QUESTIONS

1. If l is the intensity of absorbed light and C is the concentration of AB for the photochemical process AB + $h\upsilon \longrightarrow AB^*$, the rate of formation of AB^* is directly proportional to
   (A) C (B) l (C) l^2 (D) C × l **(I.I.T. 2001)**

2. The rate constant for the reaction, $2N_2O_5 \longrightarrow 4NO_2 + O_2$ is 3.0×10^{-5} s^{-1}. If the rate is 2.4×10^{-5} mol L^{-1} s^{-1}, the concentration in mol L^{-1} is
   (A) 0.8 (B) 0.04 (C) 1.2 (D) 1.4 **(I.I.T. 2000)**

   Hint: Rate of reaction $= -\dfrac{d[N_2O_5]}{dt} = k[N_2O_5]$ or 2.40×10^{-5} mol L^{-1} s^{-1} = 3.0×10^{-5} $s^{-1} \times [N_2O_5]$ or $[N_2O_5] = 2.4 \times 10^{-5}/3.0 \times 10^{-5} = 0.8$ mol L^{-1}.

3. When A and B are mixed to give C and D, the reaction quotient Q at the initial stages of the reaction
   (A) increases with time (B) decreases with time
   (C) is independent of time (D) is zero. **(I.I.T. 2000)**

4. For a first order reaction
   (A) the degree of dissociation = $1 - e^{-kt}$
   (B) the pre-exponential factor for the Arrhenius equation has the dimension of time, t^{-1}
   (C) the time taken for the completion of 75% reaction is thrice the $t_{1/2}$ of the reaction
   (D) a plot of reciprocal concentration of the reactant *vs.* time gives a straight line. **(I.I.T. 1998)**

5. The rate constant, activation energy and Arrhenius parameter of a reaction at 25°C are 3.0×10^{-4} s^{-1}, 104.4 kJ/mol and 6.0×10^{14} s^{-1} respectively. The value of the rate constant at T $\longrightarrow \infty$ is
   (A) 2.0×10^{18} s^{-1} (B) 6.0×10^{14} s^{-1} (C) 3.6×10^{30} s^{-1} (D) ∞ **(I.I.T. 1996)**

   Hint: $\log\dfrac{k_2}{k_1} = \dfrac{E_a}{2.303R}\left[\dfrac{1}{T_1} - \dfrac{1}{T_2}\right]$ or $\log\dfrac{k_2}{3\times10^{-4}} = \dfrac{104.4\times10^3}{2.303\times8.314}\left[\dfrac{1}{298} - \dfrac{1}{T}\right]$

   As $T \longrightarrow \infty$, $1/T \longrightarrow 0$. Hence $\log\dfrac{k_2}{3\times10^{-4}} = \dfrac{104.4\times10^3}{2.303\times8.314\times298} = 18.297$ or 1.98×10^{18} or $\dfrac{R_2}{3\times10^{-4}} = 1.98\times10^{18}$ or $k_2 = (1.98 \times 10^{18}) \times (3 \times 10^{-4}) = 6 \times 10^{14}$ s^{-1}.

6. The rate law for the reaction RCl + NaOH(*aq*) $\longrightarrow$ ROH + NaCl is given by, rate = k[RCl]. The rate of reaction will be
   (A) doubled on doubling the concentration of NaOH
   (B) halved on reducing the concentration of alkyl halide to one half
   (C) increased on increasing the temperature of the reaction
   (D) unaffected by increasing the temperature of the reaction. **(I.I.T. 1988)**

7. The rate constant of a first order reaction is dependent on
(A) time (B) temperature
(C) concentration of reactions (D) concentration of products **(I.I.T. 1983)**

8. The rate constant of a reaction is dependent on
(A) time (B) initial concentration of the reactants.
(C) extent of reaction (D) temperature **(I.I.T. 1981)**

9. The following rate data are obtained for the reaction A + B $\longrightarrow$ products.

[A]	[B]	Initial rate (mol L^{-1} s^{-1}]	[A]	[B]	Intial rate (mol L^{-1} s^{-1})
0.1	0.1	4.0×10^4	0.5	0.1	1.0×10^{-2}
0.2	0.2	1.6×10^{-3}	0.5	0.5	1.0×10^{-2}

The order of reaction with respect to A and B is respectively
(A) 2, 0 (B) 0, 2 (C) 2, 1 (D) 1, 0 **(Roorkee 1995)**

10. The rate of a gaseous reaction is given by the expression k[A] [B]. If the volume of the reaction vessel is reduced to 1/4th of the initial volume, the reaction rate relating to original rate will be
(A) 1/10 (B) 1/8 (C) 8 (D) 16 **(Roorkee 1992)**

11. The half-life period of a first order reaction is 10 min. If the initial amount is 0.08 mol L^{-1} and the concentration at some instant is 0.01 mol L^{-1}, then t =
(A) 10 min (B) 30 min (C) 20 min (D) 40 min **(Roorkee 1990)**

12. If the $t_{1/2}$ for a reaction is 4 min, the time after which the reaction is 99.9% completed is
(A) 16 min (B) 8 min (C) 32 min (D) 40 min

13. For a reversible reaction, if the concentration of reactants is doubled, how will the rate constant and to what extent it is affected at STP? The reaction depends only on concentration of one of the reactants
(A) halves (B) doubles
(C) remains same (D) nothing can be inferred.

14. The half-life period of a zero order reaction is
(A) proportional to initial concentration
(B) independent of concentration of reactants
(C) proportional to square of concentration
(D) inversely proportional to initial concentration.

15. When the temperature is raised, the rate of reaction increases because of
(A) lowering of activation energy (B) increase in number of collisions
(C) decrease in number of active molecules
(D) decrease in number of collisions.

16. In a reaction A $\longrightarrow$ B, the rate of reaction increases two times on increasing the concentration of reactants four times. The order of reaction is
(A) 0 (B) 1/2 (C) 4 (D) 2

17. Identify the incorrect statement.
    (A) The rate of chemical change is directly proportional to concentration
    (B) The hydrolysis of ethyl acetate in acidic medium is a second order reaction
    (C) For a first order reaction the rate of reaction doubles as the concentration of the reactant(s) doubles
    (D) For a second order reaction, the rate of reaction doubles as the concentration of the reactant(s) doubles.
18. The activation energy of a reaction is 9 kcal/mol. The increase in the rate constant when the temperature is raised from 295 to 300 K is
    (A) 10% (B) 29% (C) 50% (D) 100%
19. A substance A decomposes in solution following first order kinetics. Flask 1 contains 1 L of a 1*M* solution of A and flask II contains 0.1 L of a 0.6*M* solution. After 8 h the concentration of A in flask I has become 0.25*M*. What will be the time taken for the concentration of A in flask II to become 0.3*M*?
    (A) 0.4 h (B) 2.4 h (C) 4.0 h
    (D) Cannot be calculated since the rate constant is not given.
20. For the reaction, $2A + 2B \longrightarrow$ products, the initial rates in mol L^{-1} s^{-1} were as follows: rate = 0.46 when [A] = 1*M*; [B] = 0.2*M*; rate = 1.84 when [A] = [B] = 0.2*M*; rate = 0.92 when [A] = 0.2*M* and [B] = 0.1*M*. The specific rate constant is
    (A) 23 s^{-1} (B) 0.46 L mol^{-1} s^{-1}
    (C) 2.3×10^2 L^2 mol^{-2} s^{-1} (D) 1.25×10^2 L^2 mol^{-2} s^{-1}
21. In a reversible reaction, $2NO_2 \underset{k_2}{\overset{k_1}{\rightleftharpoons}} N_2O_4$, the rate of disappearance of NO_2 is equal to
    (A) $(2k_1/k_2)[NO_2]^2$ (B) $2k_1[NO_2]^2 - 2k_2[N_2O_4]$
    (C) $2k_1[NO_2]^2 - k_2[N_2O_4]$ (D) $(2k_1 - k_2)[NO_2]$
    Hint: rate $= -(1/2)d[NO_2]/dt = -k_1[NO_2]^2 - k_2[N_2O_4]$. Rate of disappearance of NO_2 $= -d[NO_2]/dt = 2k_1[NO_2]^2 - 2k_2[N_2O_4]$
22. For the first order decomposition of N_2O_5: (i) $N_2O_5(g) \longrightarrow 2NO_2(g) + O_2(g)$, rate $= k[N_2O_5]$ (ii) $2N_2O_5(g) \longrightarrow 4NO_2(g) + O_2(g)$, rate $= k_1[N_2O_5]$, which of the following is correct?
    (A) $k = k'$ (B) $k' > k$ (C) $k' > 2k$ (D) $k = 2k'$
23. A plot of log *k* *vs.* 1/*T* of a reaction gives a straight line with
    (A) a non-zero intercept and negative slope
    (B) a zero intercept and negative slope
    (C) a non-zero intercept and positive slope
    (D) a zero intercept and positive slope.
24. For a first order reaction, a plot of log [A] *vs.* time gives a striaght line with
    (A) a non-zero intercept and negative slope
    (B) a zero intercept and negative slope

(C) a non-zero intercept and positive slope

(D) a zero intercept and positive slope.

25. For a second order reaction $2A \longrightarrow$ products, the plot of 1/[A] *vs.* time gives a straight line with

(A) a non-zero intercept and negative slope

(B) a zero intercept and negative slope

(C) a nonzero intercept and positive slope

(D) a zero intercept and positive slope.

26. The diazonium salt decomposes as: $C_6H_5N_2^+Cl^- \longrightarrow C_6H_5Cl + N_2$. At 0°C, the evolution of N_2 becomes two times faster when the initial concentration of the salt is doubled. Therefore, it is

(A) a first order reaction (B) a second order reaction

(C) a zero order reaction

(D) independent of the initial concentration of the salt.

27. In a first-order reaction, 75% of the reactant disappeared in 1.386 h. Calculate the rate constant of the reaction.

(A) $2.8 \times 10^{-4}\ s^{-1}$ (B) $3.6 \times 10^{-3}\ s^{-1}$ (C) $1.8 \times 10^{-3}\ s^{-1}$ (D) $7.2 \times 10^{-3}\ s^{-1}$

28. The temperature coefficient of a reaction is

(A) the specific rate constant at 25°C

(B) the ratio of rate constants at 35°C and 25°C

(C) the ratio of rate constants at two temperatures differing by 1°C

(D) the rate of reaction at 25°C.

29. In a photochemical reaction

(A) ΔG is positive (B) free energy change does not occur

(C) ΔG is comparable to that in a thermal reaction

(D) none of these.

30. In photochemical reactions

(A) temperature has no effect

(B) temperature has the same effect as in thermal reaction

(C) temperature has opposite effect as in thermal reaction

(D) temperature may have a little effect in certain cases.

31. A photosensitiser is a substance which

(A) stops the photochemical reaction (B) initiates a photochemical reaction

(C) decreases the speed of a photochemical reaction

(D) increases the speed of a photochemical reaction.

32. The rate constant (k) on one reaction is double to that of the rate constant (k'') of another reaction. The relationship between the corresponding activation energies (E_a', E_a'') is given by

(A) $E_a' > E_a''$ (B) $E_a' < E_a''$ (C) $E_a' = E_a''$ (D) $E_a' = E_a''$

33. Which of the rate laws has an overall order of 0.5 for reaction involving substance x, y and z?

(A) rate = $k\,(C_x)\,(C_y)\,(C_z)$
(B) rate = $k\,(C_x)^{0.5}\,(C_y)^{0.5}\,(C_z)^{0.5}$
(C) rate = $k\,(C_x)^{1.5}\,(C_y)^{-1}\,(C_z)^{0}$
(D) rate = $k\,(C_x)\,(C_z)^{2}/(C_y)^{2}$

34. Identify the incorrect statement.

(A) The slope of the linear plot of log [A] *vs.* time for a first order reaction is $k/2.303$
(B) The hydrolysis of an ester in the presence of H^+ is first order with respect to ester and first order with respect to H_2O
(C) A reaction with zero activation energy is independent of temperature
(D) $t_{1/2}$ of a zero order reaction is directly proportional to the initial concentration of the reacting species.

35. Identify the incorrect statement.

(A) The hydrolysis of ethyl ethanoate in alkaline medium is a first order reaction
(B) The rate constant of a zero order gaseous reaction is dependent on the gaseous pressure
(C) For a first order reaction, the rate of reaction doubles as the concentration of the reactant(s) doubles
(D) The rate of an exothermic reaction increases with increasing temperature.

36. Identify the incorrect statement.

(A) In the inversion of sucrose, $C_{12}H_{22}O_{11} \longrightarrow C_6H_{12}O_6$ (glucose) + $C_6H_{12}O_6$ (fructose), the order of reaction is 2
(B) The decomposition of HI on the gold surface is a first order reaction
(C) The photochemical reaction, $H_2(g) + Cl_2(g) \xrightarrow{h\nu} 2HCl(g)$ is a zero order reaction
(D) Chlorophyll acts as a photosensitiser in the reaction of CO_2 and H_2O in the presence of sunlight to form glucose.

37. Energy of activation for the reaction (i) $N_2O_5(g) \longrightarrow 2NO_2(g) + (1/2)O_2(g)$ and (ii) $2N_2O_5(g) \longrightarrow 4NO_2(g) + O_2(g)$ is E'_a and E''_a respectively. Which of the following is correct?

(A) $E''_a = 2E'_a$ (B) $E''_a < E'_a$ (C) $E''_a > E'_a$ (D) $E'_a = E''_a$

38. For which of the following reactions, the units of rate constant and rate of reaction are identical?

(A) zero order reaction
(B) first order reaction
(C) second order reaction
(D) fractional order reaction.

39. For a first order reaction, the concentration of A after n half-lives is

(A) $[A]_0(1/2)^n$ (B) $[A]_0(1/2)^{n-1}$ (C) $[A]_0/(1/2)^n$ (D) none of these.

40. Identify the incorrect statement.

(A) A first order reaction reaches the completion stage after lapse of some time
(B) An endothermic reaction may have zero activation energy
(C) A fractional order reaction cannot occur in one elementary step
(D) The average rate of appearance of a product decreases with the progress of a reaction.

41. For the reaction, $2N_2O_5 \longrightarrow 4NO_2 + O_2$, $-\dfrac{\Delta[N_2O_5]}{\Delta t} = k_1[N_2O_5]$, $\dfrac{\Delta[NO_2]}{\Delta t} = k_2[N_2O_5]$ and $\dfrac{\Delta[O_2]}{\Delta t} = k_3[N_2O_5]$.
The relation between k_1, k_2 and k_3 is given by
(A) $2k_1 = 4k_2 = k_3$ (B) $k_1 = k_2 = k_3$ (C) $2k_1 = k_2 = 4k_3$ (D) none of these.

42. A rate of reaction is given by $k = 1.5 \times 10^{15} \exp(-25000/RT)\ \text{sec}^{-1}$. It means that
(A) $t_{1/2}$ will be smaller at room temperature
(B) the plot of log k *vs.* $1/T$ will give a straight line having the slope of -25000
(C) $t_{1/2}$ will be smaller at high temperature
(D) the plot of log k *vs.* T will give a straight line.

43. The half-life period for a zero order reaction is given by
(A) $t_{1/2} = a/2k$ (B) $t_{1/2} = 2k/a$ (C) $t_{1/2} = a^2/2k$ (D) $t_{1/2} = k/2a^2$

44. The rate constant for a zero order reaction is given by
(A) $k = C_0/C_t$ (B) $k = C_0/2t$
(C) $k = (C_0 - C_t)/t$ (D) $k = \ln[(C_0 - C_t)/2t]$

45. The order of reaction for the saponification of ethyl acetate is
(A) zero (B) first (C) second (D) third

46. When the initial concentration of a reaction is increased to four times, the half-life period of the reaction increases four-fold. The order of reaction is
(A) zero (B) one (C) two (D) four

47. The reaction A $\longrightarrow$ B follows the second order kinetics. What will happen to the rate of formation of B, if the concentration of A is doubled?
(A) will increase by 2 times (B) will increase by 4 times
(C) will increase by 1/2 time (D) will increase by 1/4 time

48. If the concentration of Y is kept constant and that of X is tripled in the reaction, $2X + Y \longrightarrow M + N$, the rate of reaction will
(A) decrease 9 times (B) increase 9 times
(C) decrease 3 times (D) increase 3 times

49. The order of a reaction with respect to OH^- is -1. The species OH^- acts as
(A) a promoter (B) an inhibitor (C) a catalyst (D) none of these.

50. For a first order reaction, the slope of the plot of log $[A]_t$ versus t is
(A) $-k/2.303$ (B) $-k$ (C) k (D) $k/2.303$

51. For a zero order reaction, the slope of the plot of $[A]_t$ versus t is
(A) $-k/2.303$ (B) $-k$ (C) k (D) $k/2.303$

52. The slope of plot of $1/[A]_t$ versus t for a second order reaction is
(A) $-k/2.303$ (B) $-k$ (C) k (D) $k/2.303$

53. The plot of log k versus $1/T$ is linear with a slope of
(A) $-E_a/2.303R$ (B) $E_a/2.303R$ (C) $-E_a/R$ (D) E_a/R

54. The half-life period of a second order reaction is given by
(A) $t_{1/2} = 1/k[A]_0$ (B) $t_{1/2} = [A]_0/k$ (C) $t_{1/2} = k/[A]_0$ (D) $t_{1/2} = 0.693/k$

55. For a first order reaction, the ratio of times to complete 99.9% and 50% of the reaction is
(A) 7 (B) 8 (C) 9 (D) 10

56. For a reversible reaction, the energy of activation of the forward reaction is 50 kcal. The energy of activation of the reverse reaction will be
(A) 50 kcal (B) > 50 kcal (C) < 50 kcal
(D) either greater than or less than 50 kcal

57. When the concentration of A is doubled in a reaction A $\longrightarrow$ B, the rate of reaction increases by a factor of 8. The reaction is
(A) first order (B) second order (C) third order (D) fourth order

58. The decomposition of NH_3 on the surface of tungsten is a
(A) zero order reaction (B) first order reaction
(C) second order reaction (D) fractional order reaction.

59. A reaction $X_2(g) \longrightarrow Z(g) + (1/2)Y(g)$ exhibits an increase in pressure from 150 mm to 170 mm in 10 min. The rate of disappearance of X_2 per minute is
(A) 2 mm (B) 4 mm (C) 6 mm (D) 8 mm

60. The half-life period of a first order reaction is 1 h. What is the time taken for 87.5% completion of the reaction?
(A) 1 h (B) 2 h (C) 3 h (D) 4 h

61. For a reaction, A + B $\longrightarrow$ products, the rate of reaction was doubled when the concentration of A was doubled. When the concentration of A and B was doubled, the rate was again doubled. The order of reaction with respect to A and B is
(A) 1, 0 (B) 0, 1 (C) 1, 1 (D) 2, 0

62. The rate of spontaneous reactions is very slow. This is due to the fact that
(A) the activation energy of the reactions is large
(B) the equilibrium constant of the reactions is less than 1
(C) the reactions are exothermic (D) the reactions are endothermic.

63. In the presence of a catalyst, the heat absorbed or evolved during a reaction
(A) decreases (B) increases
(C) remains unaffected (D) decreases tremendously.

64. For the reaction, $H_2(g) + I_2(g) \rightleftharpoons 2HI(g)$, k is the rate constant of forward reaction and k_{-1} is the rate constant of backward reaction. The equilibrium constant of the reaction is given by
(A) $K = k_1/k_{-1}$ (B) $K = k_1k_{-1}$ (C) $K = 1/k_1k_{-1}$ (D) $K = k_{-1}/k$

65. 75% of the first order reaction was completed in 32 min. When was 50% of the reaction completed?
(A) 4 min (B) 8 min (C) 16 min (D) 24 min

66. In a reaction, $2A + B \longrightarrow A_2B$, the reactant A will disappear at
(A) half the rate at that B will decrease
(B) the same rate at that B will decrease
(C) the same rate at that A_2B will form
(D) twice the rate at that B will decrease.

67. The half-life period for the reaction, $N_2O_5 \rightleftharpoons 2NO_2 + (1/2)O_2$ is 24 h at 30°C. Starting with 10 g of N_2O_5 how many grams of N_2O_5 will remain after a period of 96 h?
(A) 0.5 g (B) 0.625 g (C) 1.25 g (D) 1.77 g

68. For a reaction, $2NO(g) + Cl_2(g) \longrightarrow 2NOCl(g)$, when the concentration of Cl_2 is doubled, the rate of reaction becomes two times of the original. When the concentration of NO is doubled, the rate becomes four times. The order of reaction is
(A) 1 (B) 2 (C) 3 (D) 0

69. The activation energy of a reaction is zero. The rate constant of the reaction
(A) increases with increase of temperature
(B) decreases with increase of temperature
(C) decreases with decrease of temperature
(D) is nearly independent of temperature.

70. The reaction, $2H_2O_2 \longrightarrow 2H_2O + O_2$, is a
(A) zero order reaction (B) first order reaction
(C) second order reaction (D) third order reaction.

71. For the reaction, $2N_2O_5 \longrightarrow 4NO_2 + O_2$, the rate and rate constant are 1.02×10^{-4} mol L^{-1} s^{-1} and 3.4×10^{-5} s^{-1} respectively. The concentration of N_2O_5 at that time will be
(A) $1.732M$ (B) $3M$ (C) 1.02×10^{-4} M (D) 3.4×10^{-5} M

72. When a biochemical reaction is carried out in the laboratory from outside of human body in the absence of an enzyme, then the rate of reaction is 10^{-6} times. The activation energy in the reaction in the presence of the enzyme is
(A) $6/RT$ (B) P is required
(C) different from E_a obtained in the laboratory
(D) none of these.

73. For a first order reaction with rate constant k, which expression gives the half-life period? (Initial concentration = a)
(A) $\ln 2/k$ (B) $1/ka$ (C) $0.693/ka$ (D) $3/2ka^2$

74. Which of the following is the Arrhenius equation?
(A) $k = Ae^{-E_a/RT}$ (B) $d\ln k/dt = E_a/RT^2$
(C) $d\ln k/dT = -E_a/RT^2$ (D) $d\ln k/dT = -E_a/RT$

75. The photochemical reaction between hydrogen and chlorine on the surface of water is a reaction of
(A) zero order (B) first order (C) second order (D) third order

76. For the reaction, $N_2 + 3H_2 \longrightarrow 2NH_3$, if $\Delta[NH_3]/\Delta t = 2 \times 10^{-4}$ mol L^{-1} s^{-1}, the value of $-\Delta[H_2]/\Delta t$ will be
(A) 1×10^{-4} mol L^{-1} s^{-1} (B) 3×10^{-4} mol L^{-1} s^{-1}
(C) 4×10^{-4} mol L^{-1} s^{-1} (D) 5×10^{-4} mol L^{-1} s^{-1}

77. Which of the following statement(s) is (are) correct?
(A) A plot of log K_p versus $1/T$ is linear
(B) A plot of log [X] versus time is linear for a first order reaction $X \longrightarrow P$
(C) A plot of log P versus $1/T$ is linear at constant volume
(D) A plot of log P versus $1/V$ is linear at constant temperature. **(I.I.T. 1999)**

78. The rate determining step for the reaction $2Fe^{2+} + H_2O_2 \longrightarrow 2Fe^{3+} + 2OH^-$, $\dfrac{d[Fe^{3+}]}{dt} = k[Fe^{2+}][H_2O_2]$, is
(A) $Fe^{2+} \longrightarrow Fe^{3+} + e^-$ (B) $2Fe^{2+} + H_2O_2 \longrightarrow 2Fe^{3+} + 2OH^-$
(C) $Fe^{2+} + OH \longrightarrow Fe^{3+} + OH^-$ (D) $Fe^{2+} + H_2O_2 \longrightarrow Fe^{3+} + OH^- + OH$

79. Consider the reaction, $N_2(g) + 3H_2(g) \rightarrow 2NH_3(g)$. The rate of this reaction can be expressed in terms of time derivative of concentration of $N_2(g)$, $H_2(g)$ or $NH_3(g)$. Identify the correct relationship amongst the rate expressions.
(A) Rate $= -d[N_2]/dt = -1/3d[H_2]/dt = 1/2d[NH_3]/dt$
(B) Rate $= -d[N_2]/dt = -3d[H_2]/dt = 2d[NH_3]/dt$
(C) Rate $= d[N_2]/dt = 1/3d[H_2]/dt = 1/2d[NH_3]/dt$
(D) Rate $= -d[N_2]/dt = -d[H_2]/dt = d[NH_3]/dt$ **(I.I.T. 2002)**

80. 1 dm^3 of $2M$ CH_3COOH is mixed with 1 dm^3 of $3M$ ethanol to form ester. The decrease in the initial rate, if each solution is diluted with an equal volume of water would be
(A) 2 times (B) 4 times (C) 0.25 time (D) 0.50 time
Hint: $r = k[CH_3COOH][C_2H_5OH]$. $r_1 = k(a)(b)$, $r_2 = k(a/2)(b/2)$. Hence $r_2/r_1 = 1/4$ or $r_2 = 0.25r_1$.

81. The decomposition of a substance follows first order kinetics. If its concentration is reduced to 1/8th of its initial value in 24 min, the rate constant of the decomposition process is
(A) 1/24 min^{-1} (B) 0.692/24 min^{-1}
(C) (2.303/24) log(1/8) min^{-1} (D) (2.303/24) log (8/1) min^{-1}
Hint: k = (2.303/24 min) log [1/(1/8)] = (2.303/24) log (8/1) min^{-1}

82. For a reaction, $A + B \rightleftharpoons C + D$, doubling the concentration of both the reactants increases the reaction rate by 8 times and doubling initial concentration of only B simply doubles the reaction rate. The rate law is given by
(A) $r = k[A][B]$ (B) $r = k[A]^2[B]$ (C) $r = k[A][B]^2$ (D) $r = k[A]^{1/2}[B]^{1/2}$
Hint: $r = [A]^x[B]^x$. Hence $(2a)^x(2b)^y/(a)^x(b)^y = 8/1$ or $2^{x+y} = 2^3$ or $x + y = 3$. $(a)^x(2b)^y/(a)^x(b)^y = 2/1$ or $2^y = 2$ or $y = 1$. Hence $x = 2$. Thus $r = [A]^2[B]$

83. For a zero order reaction, a plot of concentration *vs.* time is linear with
   (A) a –ve slope and a zero intercept (B) a +ve slope and a zero intercept
   (C) a –ve slope and a non-zero intercept
   (D) a +ve slope and a non-zero intercept
84. At 400 K, a gaseous reaction, $A(g) \longrightarrow B(g) + C(g)$, is found to be first order. Starting with pure A, the total pressure at the end of 10 min was 176 mm Hg and after long time when A was completely dissociated, the total pressure was 270 mm Hg. Calculate the pressure at the end of 10 min.
   (A) 96 mm Hg (B) 45 mm Hg (C) 47 mm Hg (D) 100 mm Hg
   Hint: After 10 min, $[A] = P - p$, $[B] = 2p$, $[C] = p$. After long time: $[A] = 0$, $[B] = 2p$, $[C] = P$. Total pressure after 10 min $= P - p + 2p + p = p + 2p = 176$ mm Hg. Total pressure after long time $= 3P = 270$ mm Hg. Hence $P = 90$ mm Hg. Then $p = (176 - P)/2 = (176 - 90)2 = 43$ mm Hg. Pressure of A after 10 min $= P - p = 90 - 43 = 47$ mm Hg.
85. The reaction, $2O_3 \longrightarrow 3O_2$ proceeds in the following two steps:
   (i) $O_3 \longrightarrow O_2 + [O]$ (fast) (ii) $[O] + O_3 \longrightarrow 2O_2$ (slow). The rate law for the overall reaction is
   (A) $r = k[O_3]\,[O_2]$ (B) $r = k[O_3]^2$
   (C) $r = k[O_3]^2\,[O_2]^{-1}$ (D) $r = k[O_3][O_2]^2$
   Hint: rate $= k[O][O_3]$. $K = [O_2][O]/[O_3]$ or $[O] = K[O_3]/[O_2]$. Hence rate $= k \times K[O_3] \times [O_3]/[O_2] = k'[O_3]^2[O_2]^{-1}$
86. In a first order reaction the concentration of reactant decreases from 800 mol dm^{-3} to 50 mol dm^{-3} in 2×10^4 s . The rate constant of reaction in s^{-1} is
   (A) 2×10^4 (B) 3.45×10^{-5} (C) 1.386×10^{-4} (D) 2×10^{-4} **(I.I.T. 2003)**
   Hint: $k = \dfrac{2.303}{t}\log\dfrac{[A]_0}{[A]_t} = \dfrac{2.303}{2\times10^4}\log\dfrac{800}{50} = 1.1515\times10^{-4}\log 16 = 1.1515\times10^{-4}\times 1.2041$
   $= 1.386 \times 10^{-4}$ s^{-1}
87. (A) follows the first order reaction, (A) $\longrightarrow$ product. The concentration of A changes from $0.1M$ to $0.025M$ in 40 min. Find the rate of reaction of A when $[A] = 0.01M$.
   (A) $3.47 \times 10^{-4}\,M$ min^{-1} (B) $3.47 \times 10^{-5}\,M$ min^{-1}
   (C) $1.73 \times 10^{-4}\,M$ min^{-1} (D) $1.73 \times 10^{-5}\,M$ min^{-1}
   Hint: [A] changes from $0.1M$ to $0.025M$ in 40 min. Hence $2t_{1/2} = 40$ min or $t_{1/2} = 20$ min; $r = k[A] = (0.693/20) \times 0.01 = 3.47 \times 10^{-4}\,M$ min^{-1}
88. Which of the following statements is incorrect about order of reaction?
   (A) Order of reaction is determined experimentally
   (B) It is the sum of power of concentration terms in the rate law expression
   (C) It does not necessarily depend on stoichiometric coefficients
   (D) Order of reaction cannot have fractional value. **(I.I.T. 2005)**
   Hint: Order of reaction is determined experimentally and it may be fractional.

89. Consider a reaction $aG + bH \longrightarrow$ Products. When concentration of both the reactants G and H is doubled, the rate increases by eight times. However, when concentration of G is doubled keeping the concentration of H fixed, the rate is doubled. The overall order of the reaction is

(A) 0 (B) 1 (C) 2 (D) 3 **(I.I.T. 2007)**

Hint: $aG + bH \longrightarrow$ Products; rate $\propto [G]^a[H]^b$ where $a = 1$ and $b = 2$. Hence overall order of reaction $= a + b = 1 + 2 = 3$.

90. Under the same reaction conditions, initial concentration of 1.386 mol dm^{-1} of a substance becomes half in 40 seconds and 20 seconds through first order and zero order kinetics, respectively. Ratio (k_1/k_0) of the rate constants for first order (k_1) and zero order (k_0) of the reactions is

(A) 0.5 mol^{-1} dm^3 (B) 1.0 mol dm^{-3} (C) 1.5 mol dm^{-3} (D) 2.0 mol^{-1} dm^3 **(I.I.T. 2008)**

Hint: For Ist order reaction, $k_1 = \dfrac{0.693}{t_{1/2(\text{first})}} = \dfrac{0.693}{40}\ \text{sec}^{-1}$

For zero order reaction, $k_0 = \dfrac{[A_0]}{2t_{1/2(\text{zero})}} = \dfrac{1.386}{2 \times 20}\ \text{mol dm}^{-3}\ \text{sec}^{-1}$

$$\frac{k_1}{k_0} = \frac{0.693 \times 2 \times 20}{40 \times 1.386} = \frac{1}{2} = 5\ \text{mol}^{-1}\ \text{dm}^3$$

91. Plots showing the variation of the rate constant (k) with temperature (T) are given below. The plot that follows Arrhenius equation is

(A)

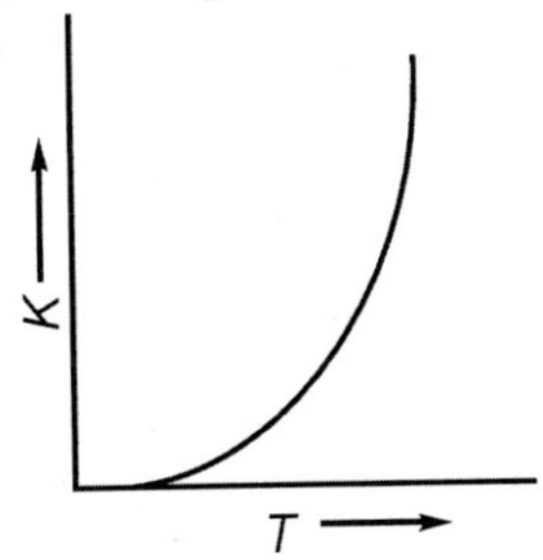

(B)

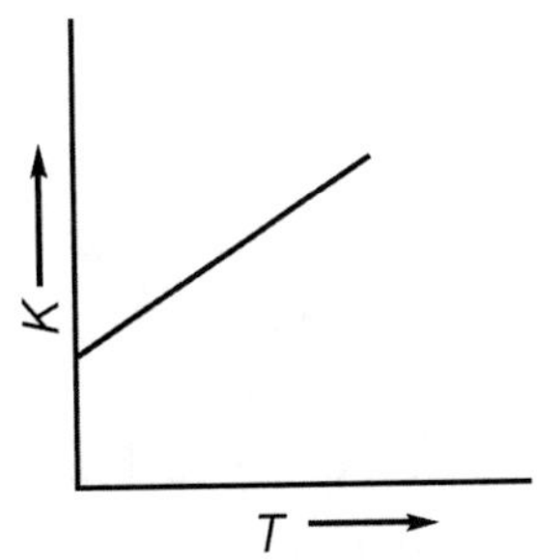

(C)

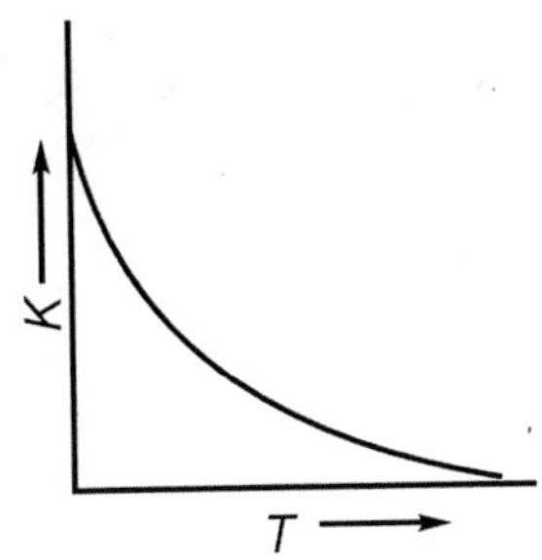

(D)

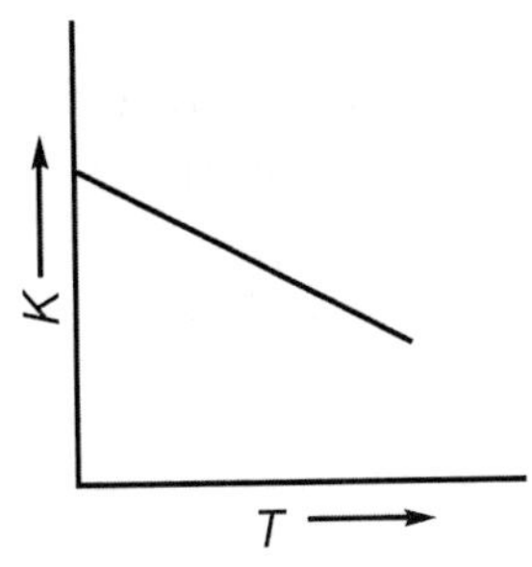

(I.I.T. 2010)

Hint: $k = Ae^{-E_a/RT}$

As $T \uparrow$; $k \uparrow$ exponentially

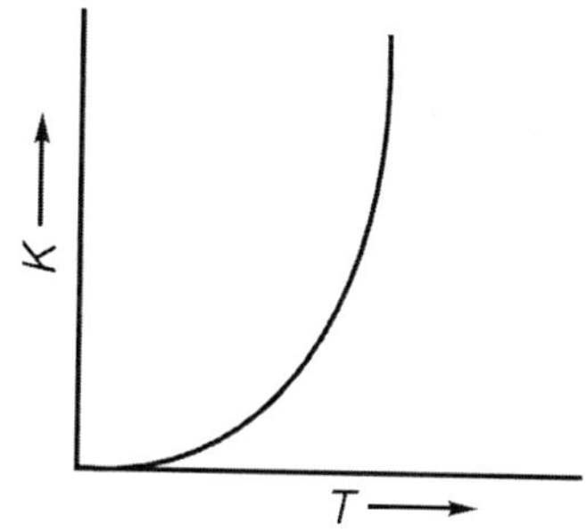

92. For a first order reaction A $\longrightarrow$ P, the temperature (T) dependent rate constant (k) was found to follow the equation $\log k = -(2000)\frac{1}{T} + 6.0$. The pre-exponential factor A and the activation energy E_a, respectively, are

(A) 1.0×10^6 s^{-1} and 9.2 kJ mol^{-1} (B) 6.0 s^{-1} and 16.6 kJ mol^{-1}
(C) 1.0×10^6 s^{-1} and 16.6 kJ mol^{-1} (D) 1.0×10^6 s^{-1} and 38.3 kJ mol^{-1} **(I.I.T. 2010)**

Hint: $\log k = \log A - \frac{E_a}{2.303\, RT}$; $\log k = 6.0 - (2000)\frac{1}{T}$; $\log A = 6.0$; $A = 10^6\,\text{s}^{-1}$

$\frac{E_a}{2.303R} = 2000$ or $E_a = 2000 \times 2.303 \times 8.314 = 38.3$ kJ mol^{-1}.

93. For the first order reaction, $2N_2O_5(g) \longrightarrow 4NO_2(g) + O_2(g)$

(A) the concentration of the reactant decreases exponentially with time
(B) the half-life of the reaction decreases with increasing temperature
(C) the half-life of the reaction depends on the initial concentration of the reactant
(D) the reaction proceeds to 99.6% completion in eight half-life duration.

(I.I.T. 2011)

Hint: For first order reaction, $[A] = [A]_0 e^{-kT}$

Hence the concentration of $[NO_2]$ decreases exponentially.

Also, $t_{1/2} = \frac{0.693}{K}$, which is independent of concentration and $t_{1/2}$ decreases with the increase of temperature.

$$t_{99.6} = \frac{2.303}{K}\log\left(\frac{100}{0.4}\right)$$

$$t_{99.6} = \frac{2.303}{K}(2.4) = 8 \times \frac{0.693}{K} = 8t_{1/2}$$

ANSWERS

1. B	2. A	3. A	4. A, B	5. B	6. B, C
7. B	8. D	9. A	10. D	11. B	12. D
13. C	14. A	15. B	16. B	17. B, D	18. B
19. C	20. C	21. C	22. D	23. A	24. A
25. C	26. A	27. A	28. B	29. A	30. A
31. B	32. B	33. C	34. A, B	35. A, B	36. A, B
37. D	38. A	39. A	40. A, B	41. C	42. C
43. A	44. C	45. C	46. A	47. B	48. B
49. B	50. A	51. B	52. C	53. A	54. A
55. D	56. A	57. C	58. A	59. B	60. C
61. A	62. A	63. C	64. A	65. C	66. D
67. B	68. C	69. D	70. C	71. A	72. C
73. A	74. A, B	75. A	76. B	77. A, B, D	78. D
79. A	80. C	81. D	82. B	83. C	84. C
85. C	86. C	87. A	88. D	89. D	90. A
91. A	92. D	93. A, B, D			

11

Catalysis

OBJECTIVE TYPE MULTIPLE CHOICE QUESTIONS

1. A catalyst
   (A) increases the average kinetic energy of the reacting molecules
   (B) increases the activation energy
   (C) alters the reaction mechanism
   (D) increases the frequency of collisions of the reacting species. **(I.I.T. 1984)**
2. A catalyst is a substance which
   (A) increases the equilibrium concentration of the product
   (B) changes the equilibrium constant of the reaction
   (C) shortens the time to reach the equilibrium
   (D) supplies the energy to the reaction. **(I.I.T. 1983)**
3. The order of a reaction with respect to OH^- ion is – 1. The OH^- ion
   (A) acts as a promoter (B) acts as a catalyst
   (C) acts as an inhibitor (D) none of these.
4. The catalyst used in the lead chamber process in the oxidation of SO_2 to SO_3 is
   (A) Pt (B) V_2O_5 (C) FeO (D) NO
5. The process which is catalysed by one of the products is called
   (A) negative catalysis (B) acid-base catalysis
   (C) autocatalysis (D) none of these.
6. The catalyst used in the inversion of sucrose is
   (A) Pt (B) Fe (C) zymase (D) invertase
7. Which of the statements is wrong?
   (A) The catalyst increases the rate of forward reaction to a greater extent than that of backward reaction
   (B) Increase in temperature by nearly 10° increases the rate of reaction by 2–3 times

(C) The reactions of higher molecularity are rare in nature
(D) Osmotic pressure measurements are preferred for the determination of molecular mass of macromolecules.

8. The catalyst used in the manufacture of H_2SO_4 by the contact process is
(A) Al powder (B) anhy. Al_2O_3
(C) platinised asbestos (D) Ni powder

9. The reactivity of a catalyst depends on
(A) solubility (B) mass (C) particle size (D) none of these.

10. When a small amount of As_2O_3 is present, the platinised asbestos does not help the formation of SO_3 from SO_2 and O_2. As_2O_3 acts as a/an
(A) poison (B) autocatalyst
(C) positive catalyst (D) negative catalyst.

11. Identify the incorrect statement.
(A) Catalyst is deactivated by heating it to high temperature
(B) Catalyst makes a reaction more exothermic
(C) The catalyst used in the manufacture of H_2 by Bosch process is Fe_2O_3 and Cr_2O_3
(D) The decolourisation of $KMnO_4$ by oxalic acid is catalysed by Mn^{2+} ion.

12. Identify the incorrect statement.
(A) The catalyst influences the forward and reverse reactions to the same extent
(B) The catalyst alters the value of equilibrium constant of the reaction
(C) A catalyst helps in attaining the equilibrium more rapidly
(D) A catalyst decreases the enthalpy change of a reaction.

13. Identify the incorrect statement.
(A) A catalyst never enters into a chemical reaction
(B) A catalyst increases the rate of forward reaction and decreases the rate of backward reaction
(C) The order of an enzyme catalysed reaction with respect to the enzyme is first order
(D) The concentration of a catalyst remains constant throughout the progress of the reaction.

14. Identify the correct statement.
(A) A catalyst controls the speed of reaction
(B) A catalyst cannot start a chemical reaction
(C) A catalyst can retard the rate of a chemical reaction
(D) A catalyst accelerates the rate of a chemical reaction.

15. A catalyst increases the rate of reaction by
(A) decreasing enthalpy (B) decreasing internal energy
(C) decreasing activation energy (D) increasing activation energy.

16. In the manufacture of nitric acid by the Ostwald process, the catalyst used is
(A) Pt, Rh (B) Rh (C) Mo (D) Fe

17. In which of the following processes shape selective catalyst catalyses the reaction?
   (A) Synthesis of NH_3 (B) Polymerisation of ethylene
   (C) Conversion of alcohol to petrol (D) Synthesis of CH_3OH from CO and H_2
18. Which of the following reactions is an example of homogeneous catalysis?
   (A) $2SO_2(g) + O_2(g) \xrightarrow{V_2O_5} 2SO_3(g)$ (B) $C_2H_4(g) + H_2(g) \xrightarrow{Ni} C_2H_6(g)$
   (C) $2CH_3OH(g) + O_2(g) \xrightarrow{Au} 2HCHO(g) + 2H_2O(g)$
   (D) $CH_3COOH(l) + C_2H_5OH(l) \xrightarrow{\text{conc. } H_2SO_4} CH_3COOC_2H_5(l) + H_2O(l)$
19. The catalytic activity of enzymes is maximum at
   (A) 300 K (B) 310 K (C) 320 K (D) 330 K
20. In the contact process of manufacture of H_2SO_4, V_2O_5 is used in place of Pt as catalyst because
   (A) V_2O_5 is not easily poisoned (B) V_2O_5 is cheaper than Pt
   (C) both (A) and (B) (D) none of these.
21. Identify the correct statement.
   (A) A catalyst accelerates the reaction by bringing down the energy of activation
   (B) A catalyst makes the equilibrium constant of the reaction more favourable for the forward reaction
   (C) A catalyst makes the reaction more feasible by making the ΔG^0 more negative
   (D) A catalyst does not take part in the reaction mechanism.
22. Which of the following process does not use a catalyst?
   (A) Haber process (B) Contact process
   (C) Thermit process (D) Ostwald process.
23. A chemical reaction is catalysed by X. Hence X
   (A) does not affect the equilibrium constant of the reaction
   (B) reduces the enthalpy of reaction
   (C) decreases the rate constant of the reaction
   (D) increases the activation energy of the reaction.
24. Identify the incorrect statement.
   (A) In the reaction, $2CH_3OH(g) + O_2(g) \longrightarrow 2HCHO(g) + 2H_2O(l)$, the catalyst used is Ag
   (B) The catalyst used in the manufacture of chlorine by Deacon process is $CuCl_2$
   (C) The catalyst used in the cracking of n-decane to n-pentane is Pt
   (D) The catalyst used in the manufacture of CH_3OH from water gas is $ZnO–Cr_2O_3$.
25. Identify the incorrect statement.
   (A) In the Bergius process of manufacture of synthetic petrol, the catalyst used is anhy. $AlCl_3$
   (B) In the Fischer–Tropsch process of manufacture of synthetic petrol the catalyst is 100 parts CO, 8 parts MgO, 5 parts ThO_2 and 200 parts keiselguhr
   (C) In the manufacture of HCHO from CO and H_2 the catalyst used is Cu
   (D) A catalyst lowers the activation energy of a reaction.

26. The activation energy of a reaction is lowered by
    (A) decreasing the potential energy (B) increasing the temperature
    (C) decreasing the temperature (D) adding a catalyst.
27. The use of a catalyst in a reaction cannot change
    (A) equilibrium constant of a reaction
    (B) activation energy of a reaction
    (C) rate of reaction (D) enthalpy of a reaction.
28. The decomposition of H_2O_2 is slowed by the addition of a small amount of acetanilide. Acetanilide act as a/an
    (A) inhibitor (B) promoter (C) stopper (D) detainer.
29. An example of autocatalytic reaction is
    (A) decomposition of nitroglycerine (B) break down of $^{14}_{6}C$
    (C) thermal decomposition of $KClO_3$ and MnO_2 mixture
    (D) hydrogenation of vegetable oil using Ni catalyst.
30. Which of the following statements is true about enzymes?
    (A) Enzymes lack in nucleophilic group
    (B) Enzymes are highly specific both in binding chiral substance and in catalysing their reactions
    (C) Enzymes catalyse chemical reactions by lowering the activation energy
    (D) Pepsin is a proteolytic enzyme.
31. Organic catalysts differ from inorganic catalysts in the fact that
    (A) they are consumed (B) they are proteinaceous in nature
    (C) they are not much active (D) they have no difference.
32. Identify the reaction with the correct catalyst.
    (A) $CO + H_2 \xrightarrow{Cu} HCHO$ (B) $2SO_2 + O_2 \xrightarrow{Fe_2O_3} 2SO_3$
    (C) $2CH_3CHO + O_2 \xrightarrow{Cu} 2CH_3COOH$
    (D) $N_2 + 3H_2 \xrightarrow{V_2O_5} 2NH_3$
33. The oxidation of $CHCl_3$ is inhibited by
    (A) HCHO (B) CCl_4 (C) C_2H_5OH (D) CH_3COOH
34. The catalyst used in the preparation of S from H_2S present in natural gas is
    (A) V_2O_5 (B) CdS (C) RuO_2 (D) Rh_2O_3
35. Wilkinson's catalyst used in the hydrogenation of alkenes is
    (A) $RhCl_3$ (B) $(Ph_3P)_3RhCl$
    (C) $(Ph_3PO)_3RhCl$ (D) Rh_2O_3
36. Which catalyst(s) is (are) used in polymerisation of alkenes?
    (A) $(C_2H_5)_3Al + TiCl_3$ (B) Fe, Co
    (C) $(Ph_3P)_3RhCl$ (D) $HgSO_4$
37. Zeolites have shape selectivity depending upon
    (A) atomic structure (B) molecular structure
    (C) pore structure (D) none of these.

38. Which of the following small sized elements can replace Si and Al in the framework of zeolites?
    (A) P (B) B (C) Mg (D) all of these.

ANSWERS

1. C	2. A	3. C	4. D	5. B	6. D
7. A	8. C	9. C	10. A	11. A, B	12. B, D
13. A, B	14. D	15. C	16. A	17. C	18. B
19. B	20. C	21. A	22. C	23. A	24. A
25. A	26. D	27. A, D	28. A	29. A	30. B, C, D
31. B	32. A	33. C	34. B, C	35. B	36. A
37. C	38. D				

12

Colloids and Surface Phenomenon

OBJECTIVE TYPE MULTIPLE CHOICE QUESTIONS

1. Peptisation is a process of
   (A) precipitation of colloids (B) purification of colloids
   (C) dispersing precipitate into colloidal sol
   (D) movement of colloidal particles in the electric field. **(Roorkee 1992)**
2. Emulsifying agent in milk is
   (A) fat (B) lactic acid (C) protein casein (D) enzyme.
3. Which of the following statements is false?
   (A) The particles of lyophilic sols are highly hydrated
   (B) Lyophobic sols are reversible
   (C) Much larger quantities of electrolytes are needed to cause precipitation of lyophilic sols in comparison to lyophobic sols
   (D) Lyophilic sols have higher viscosity than that of the medium.
4. Adsorption is multilayer in case of
   (A) chemisorption (B) physical adsorption
   (C) both (A) and (B) (D) none of these.
5. Milk, blood and blood clot are the colloidal state parlance, respectively
   (A) aerosol, sol, gel (B) emulsion, sol, gel
   (C) emulsion, gel, sol (D) gel, sol, emulsion.
6. Which are associated colloids?
   (A) Substances which behave like colloids only at higher concentration
   (B) Substances which behave as colloids only at lower temperature
   (C) Substances which consist of aggregates of colloids and molecules with diameter less than 1 cm
   (D) None of these.

7. When a few typical solutes like protein particles, blood capsules are separated by a particular selective membrane, the process is called
   (A) osmosis (B) diffusion (C) transportation (D) dialysis.
8. How many layers are adsorbed in chemical adsorption?
   (A) 0 (B) 1 (C) 2 (D) many
9. Milk can be preserved by adding a few drops of aqueous solution of
   (A) acetaldehyde (B) formic acid (C) formaldehyde (D) acetic acid
10. As_2O_3 sol is
   (A) neutral (B) positively charged
   (C) negatively charged (D) none of these.
11. Which of the following is less than zero during adsorption?
   (A) ΔH (B) ΔG (C) ΔS (D) All of these.
12. For the adsorption of a gas on a solid, the plot of log (x/m) *vs.* log P is linear with the slope equal to
   (A) $1/n$ (B) n (C) k (D) log k
13. Identify the incorrect statement.
   (A) The addition of gelatin to gold sol leads to the destabilisation of the gold sol
   (B) Chemisorption is irreversible
   (C) Adsorption is endothermic
   (D) The process of removing an adsorbate from the surface of an adsorbent is known as desorption.
14. Identify the incorrect statement.
   (A) The difference between lyophobic and lyophilic colloids is in their behaviour towards dispersion medium
   (B) The protective action of colloids is expressed in terms of gold number
   (C) The size of colloidal particles lies between 10^{-5}–10^{-7} cm
   (D) Soap forms a colloidal solution in water and removes the greasy matter by coagulation.
15. Identify the incorrect statement.
   (A) Colloidal silver is obtained due to the reduction of $AgNO_3$ by glucose
   (B) The expression for Freundlich isotherm is $\log (x/m) = \log k + (1/n) \log P$
   (C) The mixing of the colloidal solution of As_2S_3 and Fe_2O_3 leads to the peptisation of the sols
   (D) The sols of metals are hydrophilic.
16. In which of the following solvents starch behaves as a lyophilic and lyophobic colloid respectively?
   (A) water, petrol (B) petrol, water
   (C) benzene, cold water (D) hot water, benzene
17. Colloidal sulphur is prepared by reaction of the following
   (A) H_2SO_4 and $H_2S_2O_8$ (B) I_2 and $Na_2S_2O_3$
   (C) $FeCl_2$ and Na_2S (D) H_2S and $FeCl_3$

18. The addition of $SOCl_2$ to dry sodium salicylate (equal amount) drop by drop with stirring produces a
    (A) gel of salicylic acid (B) gel of S
    (C) sol of S (D) gel of NaCl
19. Which of the following is a heteropolar sol?
    (A) Rubber in toluene (B) Protein in water
    (C) Starch in water (D) Sulphur in water.
20. Which of the following happens when a lyophobic sol is subjected to repeated hydrolysis?
    (A) The concentration increases (B) The concentration decreases
    (C) The sol becomes unstable (D) The sol becomes more stable.
21. A sol has a very low osmotic pressure because of
    (A) its charge (B) its dissociation
    (C) large size of particles (D) its association.
22. To which of the following pumice stone belongs?
    (A) emulsion (B) foam (C) aerosol (D) gel
23. The silver iodide sol is
    (A) positively charged (B) negatively charged
    (C) neutral (D) positive or negative.
24. When alcohol is added to a saturated aqueous solution of $Ca(CH_3COO)_2$, first a solution is formed and then a gelatinous substance is formed. The gelatinous substance is a/an
    (A) foam (B) gel (C) solid sol (D) aerosol.
25. The Tyndall effect is more pronounced in
    (A) lyophilic sol (B) lyophobic sol
    (C) both (A) and (B) (D) starch solution.
26. Identify the incorrect statement.
    (A) 0.1*M* $AlCl_3$ solution is more effective than 0.1*M* NaCl solution in coagulating an As_2S_3 sol
    (B) The rate of adsorption always increases with an increase in pressure
    (C) The rate of adsorption decreases with an increase in temperature
    (D) Soap is an emulsifier.
27. Identify the incorrect statement.
    (A) The addition of $Fe(OH)_3$ sol to As_2S_3 sol results in the precipitation of both of them
    (B) Colloidal $Fe(OH)_3$ absorbs a large volume of $H_2(g)$
    (C) Iron(III) hydroxide sol is a negatively charged sol
    (D) A dispersion of $Fe(OH)_3$ in water is an example of hydrophobic sol
28. Identify the incorrect statement.
    (A) 0.1*M* $AlCl_3$ solution is less effective than 0.1*M* Na_3PO_4 solution in coagulating a Fe_2O_3 sol

(B) The gold number of haemoglobin is 0.03
(C) The viscosity of a lyophilic sol is same as the viscosity of the dispersion medium
(D) Peptisation is the reverse of coagulation.

29. Identify the incorrect statement.
(A) Lyophilic sols are irreversible
(B) The osmotic pressure of a true solution is lower than that of a colloidal solution
(C) The charge of As_2S_3 sol is negative
(D) Butter is an emulsion of water in fat.

30. Identify the incorrect statement.
(A) The colour of a colloidal solution depends on the size of colloidal particles
(B) The viscosity of a lyophobic sol is less than that of the dispersion medium
(C) The colloids which behave as electrolytes in low concentration and as colloidal solutions in high concentration are called associated colloids
(D) When a liquid is dispersed in solid, the colloid is called gel.

31. Which of the following is an example of an associated colloid?
(A) Soap and water (B) Protein and water
(C) Rubber and benzene (D) As_2O_3 and $Fe(OH)_3$

32. In physical adsorption, the forces associated are
(A) ionic (B) covalent (C) van der Waals (D) H-bonding

33. At the critical micelle concentration (cmc), the surface molecules
(A) decompose (B) dissociate
(C) associate (D) become completely soluble.

34. On addition of one mL solution of 10% NaCl to 10 mL gold sol in the presence of 0.025 g of starch, the coagulation is just prevented. Starch has the gold number
(A) 0.025 (B) 0.25 (C) 2.5 (D) 25

35. Which of the following is a hydrophilic colloidal sol?
(A) Barium sulphate solution (B) Arsenic(III) sulphide
(C) Starch sol (D) Silver iodide sol.

36. Which is not correct regarding the adsorption of a gas on the surface of a solid?
(A) On increasing temperature adsorption increases continuously
(B) Enthalpy change and entropy change are negative
(C) Adsorption is more for some specific substances
(D) It is a reversible reaction.

37. When a solution of sodium chloride is added to $Fe(OH)_3$ sol
(A) $[Fe(OH)_3]Na^+$ is formed (B) $[Fe(OH)_3]Cl^-$ is formed
(C) $[Fe(OH)_3]Fe^{3+}$ is formed (D) $Fe(OH)_3$ is coagulated

38. At isoelectric point
(A) the pH of a colloidal solution becomes 7
(B) a colloidal solution becomes electrically neutral
(C) a colloidal solution coagulates
(D) the colloidal particles migrate towards oppositely charged electrodes.

39. Adsorption is accompanied by
    (A) $T\Delta S$ value for the process to be positive
    (B) decrease in free energy of the system
    (C) increase in entropy of the system
    (D) decrease in enthalpy of the system
40. Adsorption is
    (A) an endothermic process (B) exothermic process
    (C) may be endothermic or exothermic
    (D) neither endothermic nor exothermic.
41. In the reaction, $N_2(g) + 3H_2(g) \xrightarrow{Fe(s)} 2NH_3(g)$
    (A) the molecules split into more reactive atomic species due to adsorption
    (B) the concentrations of reactants increase on the surface of Fe due to adsorption
    (C) adsorption on surface of Fe provides the activation energy for the reaction
    (D) Fe increases the rate of forward reaction more than the rate of backward reaction
42. Rate of physisorption increases with
    (A) decrease in temperature (B) increase in temperature
    (C) decrease in pressure (D) decrease in surface area. **(I.I.T. 2003)**
43. Spontaneous adsorption of a gas in solid surface is an exothermic process because
    (A) ΔH increases for system (B) ΔS increases for gas
    (C) ΔS decreases for gas (D) ΔG increases for gas. **(I.I.T. 2004)**
    Hint: For a spontaneous adsorption ΔG is negative. Since the degree of randomness of gas molecules decreases, ΔS is negative. Hence $T\Delta S$ is negative. Thus ΔH (= $\Delta G + T\Delta S$) is negative.
44. Which of the following is correct for lyophilic sols?
    (A) They are irreversible (B) They are formed by inorganic substances
    (C) They are readily coagulated by addition of electrolytes
    (D) They are self stabilised. **(I.I.T. 2005)**
    Hint: Lyophilic sols are solvent loving in nature. Due to this property, lyophilic sols are self stabilised.
45. Statement–1: Micelles are formed by surfactant molecules above the critical micellar concentration (CMC).
    because
    Statement–2: The conductivity of a solution having surfactant molecules decreases sharply at the CMC.
    (A) Statement–1 is true, Statement–2 is true; Statement–2 is a correct explanation for Statement–1
    (B) Statement–1 is true, Statement–2 is true; Statement–2 is not a correct explanation for Statement–1
    (C) Statement–1 is true, Statement–2 is false
    (D) Statement–1 is false, Statement–2 is true. **(I.I.T. 2007)**

Hint: The formation of micelles occurs only above a particular temperature called Kraft temperature (T_k) and above a particular concentration called critical micellar concentration. Since each micelles contains at least 100 molecules, the conductivity of the solution decreases sharply at the critical micellar concentration.

46. Among the electrolytes Na_2SO_4, $CaCl_2$, $Al_2(SO_4)_3$ and NH_4Cl, the most effective coagulating agent for Sb_2S_3 sol is

(A) Na_2SO_4 (B) $CaCl_2$ (C) $Al_2(SO_4)_3$ (D) NH_4Cl **(I.I.T. 2009)**

Hint: Since Sb_2S_3 is a negative sol, $Al_2(SO_4)_3$ will be the most effective coagulant due to higher charge density on Al^{3+} in accordance with Hardy–Schulze rule. Order of effectiveness of cations: $Al^{3+} > Ca^{2+} > Na^+ > NH_4^+$.

47. Among the electrolytes Na_2SO_4, $CaCl_2$, $Al_2(SO_4)_3$ and NH_4Cl, the most effective coagulating agent for Sb_2S_3 sol is

(A) Na_2SO_4 (B) $CaCl_2$ (C) $Al_2(SO_4)_3$ (D) NH_4Cl **(I.I.T. 2009)**

Hint: Since Sb_2S_3 is a negative sol, $Al_2(SO_4)_3$ will be the most effective coagulant due to higher charge density on Al^{3+} in accordance with Hardy–Schulze rule. Order of effectiveness of cations: $Al^{3+} > Ca^{++} > Na^+ > NH_4^+$.

48. The correct statement(s) pertaining to the adsorption of a gas on a solid surface is (are)

(A) Adsorption is always exothermic

(B) Physisorption may transform into chemisorption at high temperature

(C) Physisorption increases with increasing temperature but chemisorption decreases with increasing temperature

(D) Chemisorption is more exothermic than physisorption, however it is very slow due to higher energy of activation. **(I.I.T. 2011)**

Hint: The statements (A), (B) and (D) are correct.

ANSWERS

1. C	2. C	3. B	4. B	5. B	6. A
7. D	8. B	9. C	10. C	11. D	12. A
13. A, C	14. D	15. C, D	16. D	17. D	18. D
19. B	20. C	21. D	22. B	23. D	24. B
25. B	26. B	27. B, C	28. C	29. A, B	30. B
31. A	32. C	33. C	34. D	35. C	36. A
37. D	38. C, D	39. B, D	40. B	41. A, B, C	42. A
43. C	44. D	45. B	46. C	47. C	48. A, B, D

13

Chemical Thermodynamics and Thermochemistry

IMPORTANT RELATIONS FOR NUMERICAL PROBLEMS

1. Efficiency of heat engine, $\eta = w/Q_2 = 1 - (T_1/T_2)$
2. $\Delta E = q_v$
3. $\Delta H = q_p$
4. $w = -P\Delta V$: isothermal conditions: $w = -2.303nRT \log (V_2/V_1) = -2.303nRT \log (P_1/P_2)$
5. $H = E + PV$; $\Delta H = \Delta E + P\Delta V$
6. $\Delta H = \Delta E + \Delta n(g)RT$ where $\Delta n(g) = n_P - n_R$ = change in number of gaseous molecules
7. $\Delta H^0 = \Sigma H^0_{\text{products}} - \Sigma H^0_{\text{reactants}}$
8. First law of thermodynamics: $\Delta E = q + w$
9. $q_p = q_v + \Delta n(g)RT$
10. $\Delta H_{vap}/T_b \approx 88$ J mol^{-1} K
11. Kirchhoff's equation: $\Delta H_2 - \Delta H_1 = \Delta C_p(T_2 - T_1)$; $\Delta E_2 - \Delta E_1 = \Delta C_v(T_2 - T_1)$
12. $\Delta S = q_{rev}/T$; $\Delta S = 2.303nR \log (V_2/V_1)$
13. Entropy change for isothermal reversible expansion of an ideal gas, $\Delta S = R \ln (V_2/V_1)$ $= R \ln (P_1/P_2)$
14. Entropy change for an isobaric process, $\Delta S = C_v \ln (T_1/T_2)$
15. $\Delta S^0 = \Sigma S^0_{\text{products}} - \Sigma S^0_{\text{reactants}}$
16. $\Delta S_{fusion} = \Delta H_{fusion}/T_m$ where T_m = melting point
17. $\Delta S_{vap} = \Delta H_{vap}/T_b$ where T_b = boiling point
18. Entropy change during phase transition, $\Delta S = S_2 - S_1 = q_{rev}/T = \Delta H/T$
19. $\Delta G = \Delta H - T\Delta S$
20. $\Delta G^0 = \Sigma G^0_{\text{products}} - \Sigma G^0_{\text{reactants}}$

21. $\Delta G^0 = 2.303RT \log K_c$; $K_p = e^{-\Delta G^0/RT}$
22. $\Delta G^0 = -nFE^0$
23. $\Delta G^0 = w$ (useful)
24. 1 cal = 4.18 J = 4.184×10^7 erg
25. Clausius–Clapeyron equation: $\log (P_2/P_1) = (\Delta H_v/2.303R)[(T_2 - T_1)/T_1T_2]$
26. $C_p - C_v = R$; $C_p/C_v = \gamma$
27. Heat of neutralisation: $H^+ + OH^- \rightleftharpoons H_2O$, $\Delta H = -13.7$ kcal = 57.3 kJ

OBJECTIVE TYPE MULTIPLE CHOICE QUESTIONS

1. In thermodynamics, a process is called reversible when
   (A) the surroundings and system change into each other
   (B) there is no boundary between system and surroundings
   (C) the surroundings are always in equilibrium with the system
   (D) the system changes into the surroundings spontaneously. **(I.I.T. 2001)**
2. Which of the following statements is false?
   (A) Work is a state function (B) Temperature is a state function
   (C) Change in the state is completely defined when the initial and final states are specified
   (D) Work appears at the boundary of the system. **(I.I.T. 2001)**
3. ΔH_f^0 for $CO_2(g)$, $CO(g)$ and $H_2O(g)$ is – 393.5, – 110.5 and – 241.8 kJ/mol respectively. ΔH^0 (in kJ) for the reaction, $CO_2(g) + H_2(g) \longrightarrow CO(g) + H_2O(g)$ is
   (A) + 524.1 (B) + 41.2 (C) – 262.5 (D) 41.2 **(I.I.T. 2000)**
   Hint: $\Delta H^0 = [\Delta H_f^0(CO) + \Delta H_f^0(H_2O)] - [\Delta H_f^0(CO_2) + \Delta H_f^0(H_2)] = [-110.5 + (-241.8) - (-393.5 + 0)]$ kJ $= -352.3 + 393.5 = +41.2$ kJ
4. Which of the following is (are) endothermic reactions?
   (A) Conversion of graphite to diamond
   (B) Dehydrogenation of ethane to ethylene
   (C) Decomposition of water (D) Combustion of methane. **(I.I.T. 1999)**
5. The standard molar enthalpy of formation of CO_2 is equal to
   (A) zero
   (B) the standard molar enthalpy of combustion of carbon (graphite)
   (C) the standard molar enthalpy of combustion of C(gaseous)
   (D) the sum of the standard molar enthalpies of CO and CO_2 **(I.I.T. 1997)**
6. Molar heat capacity of water in equilibrium with ice at constant pressure is
   (A) infinity (B) 40.45 kJ K^{-1} mol^{-1}
   (C) 75.48 J K^{-1} mol^{-1} (D) zero **(I.I.T. 1997)**
7. Identify the intensive quantities from the following:
   (A) enthalpy (B) temperature
   (C) refractive index (D) volume **(I.I.T. 1993)**

8. Which of the following has the smallest heat of hydrogenation per mole?
   (A) *cis*-but-2-ene (B) *trans*-but-2-ene
   (C) buta-1,3-diene (D) but-1-ene. **(I.I.T. 1993)**
9. Which of the following equations correctly represents the standard heat of formation (ΔH_f^0) of methane?
   (A) C (diamond) + $2H_2(g)$ = $CH_4(g)$ (B) C (graphite) + $2H_2(g)$ = $CH_4(l)$
   (C) C (graphite) + $2H_2(g)$ = $CH_4(g)$ (D) C (graphite) + 4H = $CH_4(g)$ **(I.I.T. 1992)**
10. For an endothermic reaction where ΔH represents the enthalpy of the reaction in kJ/mol, the minimum value for the energy of activation will be
   (A) less than ΔH (B) more than ΔH
   (C) equal to ΔH (D) zero **(I.I.T. 1992)**
11. The difference between heats of reaction at constant pressure and constant volume for the reaction, $2C_6H_6(l) + 15O_2(g) \longrightarrow 12CO_2(g) + 6H_2O(l)$ at 25°C in kJ is
   (A) – 7.43 (B) + 3.72 (C) – 3.72 (D) + 7.43 **(I.I.T. 1991)**
12. Which of the following reactions is an endothermic reaction?
   (A) $2H_2 + O_2 \longrightarrow 2H_2O$ (B) $2NaOH + H_2SO_4 \longrightarrow Na_2SO_4 + 2H_2O$
   (C) $C_2H_5OH + 3O_2 \longrightarrow 2CO_2 + 3H_2O$ (D) $N_2 + O_2 \longrightarrow 2NO$ **(I.I.T. 1989)**
13. The enthalpy of combustion of cyclohexane, cyclohexene and H_2 is – 3920, – 3800 and – 241 kJ/mol respectively. The heat of hydrogenation (in kJ/mol) of cyclohexene is
   (A) – 121 (B) + 121 (C) – 242 (D) + 242 **(I.I.T. 1989)**
14. The products of combustion of an aliphatic thiol (RSH) at 298 K are
   (A) $CO_2(g)$, $H_2O(g)$ and $SO_2(g)$ (B) $CO_2(g)$, $H_2O(l)$ and $SO_2(g)$
   (C) $CO_2(l)$, $H_2O(l)$ and $SO_2(g)$ (D) $CO_2(g)$, $H_2O(l)$ and $SO_2(l)$ **(I.I.T. 1992)**
15. The standard molar heat of formation of ethane, CO_2 and $H_2O(l)$ is respectively – 21.1, – 94.1 and – 68.3 kcal. The standard molar heat of combustion (in kcal) of ethane will be
   (A) – 372 (B) 162 (C) – 240 (D) 183.5 **(I.I.T. 1986)**
16. The heat released when NH_4OH and HCl neutralise is
   (A) 13.7 kcal (B) > 13.7 kcal (C) < 13.7 kcal (D) none **(Roorkee 1990)**
17. The relation, $\Delta H^0_{sub} = \Delta H^0_{fus} + \Delta^0_{vap}$ is valid at all
   (A) pressures (B) temperatures
   (C) pressures and temperatures (D) temperatures and 1 atm
18. In which of the following conditions, the process is non-feasible?
   (A) $\Delta H > 0$, $\Delta S > 0$ (B) $\Delta H < 0$, $\Delta S > 0$
   (C) $\Delta H > 0$, $\Delta S < 0$ (D) $\Delta H < 0$, $\Delta S < 0$
19. In what proportion 1*M* NaOH and 0.5*M* H_2SO_4 are mixed so as to release maximum amount of energy and to form 100 mL solution?
   (A) 33 and 67 (B) 67 and 33 (C) 40 and 60 (D) 50 and 50
20. For an isothermal expansion of one mole of an ideal gas from 1 dm^3 to 4 dm^3 at 27°C, the value of q is
   (A) 57 cal (B) 831.9 cal (C) 13 cal (D) 0

21. The enthalpy of combustion of glucose is – 2840 kJ/mol. The enthalpy of formation of 60 g of glucose is
(A) 2840 kJ (B) 946.6 kJ (C) – 946.6 kJ (D) 1420 kJ
22. The heat of reaction at constant pressure is given by
(A) $H_R - H_P$ (B) $E_P - E_R$ (C) $E_R - E_P$ (D) $H_P - H_R$
23. Internal energy does not include
(A) energy of gravitational pull (B) vibrational energy
(C) rotational energy (D) nuclear energy.
24. If ΔG is zero for a reaction, then
(A) $\Delta H = 0$ (B) $\Delta S = 0$ (C) $K = 0$ (D) $K = 1$
25. In an isochoric process, the increase in thermal energy is equal to
(A) work done (B) heat absorbed
(C) heat evolved (D) the sum of the heat absorbed and work done.
26. In the reaction: $S + (3/2)O_2 \longrightarrow SO_2 + 2x$ kcal; $SO_2 + (1/2)O_2 \longrightarrow SO_3 + y$ kcal, the heat of formation of SO_2 is
(A) $2x + y$ (B) $2x - y$ (C) $x + y$ (D) $x - y$
27. Which of the following is not a spontaneous process?
(A) Reaction between H_2 and O_2 to form water
(B) Flow of current from lower potential to higher potential
(C) Water flowing down a hill
(D) Dissolution of $CuSO_4 \cdot 5H_2O$ in water.
28. The process of evaporation of a liquid is accompanied by
(A) no change in free energy (B) increase in enthalpy
(C) increase in entropy (D) decrease in free energy.
29. Which of the following is true for the first law of thermodynamics?
(A) $\Delta E = q + PV$ (B) $\Delta E = q - w$ (C) $\Delta E = q + w$ (D) $\Delta E = q - PV$
30. For the reaction, $CO(g) + (1/2)O_2(g) \longrightarrow CO_2(g)$, which statement is correct at constant T and P?
(A) $\Delta H = \Delta E$ (B) $\Delta H < \Delta E$
(C) $\Delta H > \Delta E$ (D) ΔH is independent of the physical state of the reactants.
31. Find ΔG for the reaction: $2ZnS + 3O_2 \longrightarrow 2ZnO + 2SO_2$. Given: $2Zn + O_2 \longrightarrow 2ZnO$, $\Delta G_1 = -616$ J; $2Zn + S_2 \longrightarrow 2ZnS$, $\Delta G_2 = -293$ J; $S_2 + 2O_2 \longrightarrow 2SO_2$, $\Delta G_3 = -408$ J
(A) – 731 J (B) + 731 J (C) – 1317 J (D) – 501 J
32. The enthalpy change of a reaction does not depend on
(A) initial and final enthalpy change (B) nature of the reactant and product
(C) different intermediate reactions (D) state of the reactant and product.
33. For an endothermic reaction, ΔS is positive. The reaction is
(A) feasible when $T\Delta S > \Delta H$ (B) feasible when $\Delta H > T\Delta S$
(C) feasible at all temperatures (D) not feasible.

34. Identify the incorrect statement.
    (A) First law of thermodynamics is adequate in predicting the direction of the process
    (B) The enthalpy of fusion of KCl is greater than that of naphthalene
    (C) The standard enthalpy of the diamond is zero at 0°C and 1 atm
    (D) The heat evolved when the requisite amounts of ions in the gaseous state combine to give 1 mole of crystal is known as lattice energy.
35. Identify the incorrect statement.
    (A) The exothermic reactions always become faster as we approach absolute zero
    (B) The heat of neuralisation of a strong acid and a strong base is equal to the heat of formation of water
    (C) A system is said to be isolated if it can neither exchange matter nor energy with the surroundings
    (D) If the heats of combustion of $C_2H_4(g)$, $C_2H_6(g)$ and $H_2(g)$ are – 1409.5, – 1558 and – 285.6 kJ respectively, the heat of hydrogenation of ethylene is – 137.1 kJ.
36. When ideal liquids are mixed, the ΔH_{system}
    (A) decreases (B) increases
    (C) remains unaffected (D) may increase or decrease.
37. The energy change of reaction, $C_2H_6(g) \longrightarrow 2C(g) + 6H(g)$ is X kJ. The bond energy of C—H bond is
    (A) X kJ (B) $X/3$ kJ (C) $X/6$ kJ (D) indeterminate.
38. For the reaction $2Cl(g) \longrightarrow 2Cl_2(g)$, the signs of ΔH and ΔS are
    (A) +, – (B) +, + (C) –, – (D) –, +
39. For the reaction, $A \longrightarrow B$, ΔH and ΔS are positive. The most favourable condition for the reaction is
    (A) high concentration (B) very low temperature
    (C) high temperature (D) low temperature.
40. Given: $C + 2S \longrightarrow CS_2$, $\Delta H_f^0 = + 117.0$ kJ/mol; $C + O_2 \longrightarrow CO_2$, $\Delta H_f^0 = - 393$ kJ/mol; $S + O_2 \longrightarrow SO_2$, $\Delta H_f^0 = - 297$ kJ/mol. The heat of combustion of $CS_2 + 3O_2 \longrightarrow 2SO_2 + CO_2$ is
    (A) + 807 kJ/mol (B) – 807 kJ/mol (C) + 1104 kJ/mol (D) – 1104 kJ/mol
41. The heat of hydrogenation of benzene is 50 kcal/mol. The resonance energy of benzene is 36 kcal/mol. The heat of hydrogenation of cyclohexene is
    (A) 5 kcal/mol (B) 14 kcal/mol (C) 17 kcal/mol (D) 29 kcal/mol
42. Given that the heats of combustion of ethylene, hydrogen and ethane are – 337.0, – 68.4 and – 373.0 kcal respectively at 25°C. The heat of reaction of $C_2H_4(g) + H_2(g) \longrightarrow C_2H_6(g)$ at 25°C will be
    (A) – 32.4 kcal (B) – 36.3 kcal (C) – 37.8 kcal (D) – 39.2 kcal
43. When Zn dust is added to sufficiently large volume of aqueous solution of $CuSO_4$, 3.175 g of Cu is formed and 20 J of heat is evolved. The ΔH of the reaction, $Zn(s) + CuSO_4(aq) \longrightarrow ZnSO_4(aq) + Cu(s)$ is (At. wt.: Zn = 65.3, Cu = 63.5)
    (A) 20 J (B) 65.3 J (C) 200 J (D) – 400 J

44. If $S + O_2 \longrightarrow SO_2$, $\Delta H = -298.2$ kJ; $SO_2 + (1/2)O_2 \longrightarrow SO_3$, $\Delta H = -98.7$ kJ; $SO_3 + H_2O \longrightarrow H_2SO_4$, $\Delta H = -130.2$ kJ; $H_2 + (1/2)O_2 \longrightarrow H_2O$, $\Delta H = -287.3$ kJ, then the enthalpy of formation of H_2SO_4 at 298 K is

    (A) + 320.5 kJ (B) – 433.5 kJ (C) – 650.3 kJ (D) – 814.4 kJ

45. The heats of combustion of C, H_2 and CH_4 are – 94, – 68, – 150 kcal respectively. The heat of formation of CH_4 is

    (A) – 80 kcal (B) + 80 kcal (C) + 150 kcal (D) – 150 kcal

46. Given: $C(s) + O_2(g) \longrightarrow CO_2(g)$, $\Delta H = -94.0$ kcal; $H_2(g) + (1/2)O_2(g) \longrightarrow H_2O(l)$, $\Delta H = -68.0$ kcal; $CH_3COOH(l) + 2O_2(g) \longrightarrow 2CO_2(g) + 2H_2O(l)$, $\Delta H = -210.0$ kcal. The heat of formation of CH_3COOH is

    (A) – 114.0 kcal (B) + 114.0 kcal (C) – 116.0 kcal (D) + 116.0 kcal

47. Given: $2C(s) + 2O_2(g) \longrightarrow 2CO_2(g)$, $\Delta H = -787$ kJ; $H_2(g) + (1/2)O_2(g) \longrightarrow H_2O(g)$, $\Delta H = -286$ kJ; $C_2H_2(g) + (5/2)O_2(g) \longrightarrow 2CO_2(g) + H_2O(l)$, $\Delta H = -1301$ kJ. The heat of formation of C_2H_2 is

    (A) + 228 kJ (B) – 228 kJ (C) + 1802 kJ (D) – 1802 kJ

48. Given: $C(s) + O_2(g) \longrightarrow CO_2(g)$, $\Delta H = -395$ kJ; $S(s) + O_2(g) \longrightarrow SO_2(g)$, $\Delta H = -295$ kJ; $CS_2(l) + 3O_2(g) \longrightarrow CO_2(g) + 2SO_2(g)$, $\Delta H = -1110$ kJ. The heat of formation of CS_2 is

    (A) 31.25 kJ (B) 62.5 kJ (C) 125 kJ (D) 250 kJ

49. Identify the incorrect statement.

    (A) For the reaction, $N_2(g) + 2O_2(g) \longrightarrow 2NO_2(g)$, $\Delta H = 66$ kJ. The heat of formation of $NO_2(g)$ is 60 kJ

    (B) At constant temperature and pressure, if $\Delta H < 0$, the process is spontaneous

    (C) One would expect that the heat produced in the reaction, $H_2(g) + (1/2)O_2(g) \longrightarrow H_2O(l)$ and $2C_2H_2 + 5O_2 \longrightarrow 4CO_2 + 2H_2O$ would have one and the same numerical value

    (D) At constant temperature and pressure, if $\Delta G < 0$, the process is spontaneous.

50. Identify the incorrect statement.

    (A) The entropy of glass at absolute zero is zero

    (B) The energy of a substance at absolute zero is known as zero point energy

    (C) Entropy is an extensive property

    (D) Enthalpy is an intensive property.

51. For a reversible process, the total change in entropy is equal to

    (A) zero (B) negative

    (C) ΔS (system) – ΔS (surroundings) (D) ΔS (system) + ΔS (surroundings)

52. Identify the correct statement.

    (A) When ΔG is positive, the process is known as endergonic

    (B) When ΔG is negative, the process is known as exergonic

    (C) When ΔG is less than zero, the process is spontaneous

    (D) When ΔG is zero, the system is in equilibrium.

53. Which of the following occurs when a solid melts?

    (A) decrease in internal energy (B) no change in enthalpy

    (C) decrease in free energy (D) increase in enthalpy.

54. Identify the elements which are thermodynamically stable.
(A) C (graphite) (B) C (diamond) (C) P_4 (red) (D) P_4 (white)

55. For an ideal gas the process is isothermal when
(A) $E_1 = E_2$ (B) $H_1 \neq H_2$ (C) $w = 0$ (D) $P_1V_1 = P_2V_2$

56. For the reactions: (i) C (diamond) + $2H_2(g) \longrightarrow CH_4(g)$, ΔH_1 (ii) C (graphite) + $4H(g) \longrightarrow CH_4$, ΔH_2, which of the following is correct?
(A) $\Delta H_1 < \Delta H_2$ (B) $\Delta H_1 > \Delta H_2$
(C) $\Delta H_1 = \Delta H_2$ (D) $\Delta H_1 = \Delta H_2 + \Delta H_{vap}(C) + \Delta H_{diss}(H_2)$

57. Which of the following is correct for the reaction, $2Ag(s) + (1/2)O_2(g) \xrightarrow{298\text{ K}} 2Ag_2O(s)$?
(A) $\Delta H < \Delta E$ (B) $\Delta H > \Delta E$
(C) $\Delta H = \Delta E$ (D) ΔH and ΔE are not related.

58. Entropy change of the system and surroundings in equilibrium
(A) decreases (B) increases (C) (A) or (B) (D) is constant.

59. At a given temperature, the energy of activation of two reactions is same if
(A) the specific rate constant for the two reactions is the same
(B) the temperature coefficient for the specific rate constant for the two reactions is the same
(C) ΔH of the two reactions is the same but not zero
(D) ΔH of the two reactions is zero. **(I.I.T. 1993)**

60. The enthalpy of vaporisation of liquid water using the data: (i) $H_2(g) + (1/2)O_2(g) \longrightarrow H_2O(l)$, $\Delta H = -285.77$ kJ/mol, (ii) $H_2(g) + (1/2)O_2(g) \longrightarrow H_2O(g)$, $\Delta H = -241.84$ kJ/mol, in kJ/mol is
(A) + 43.93 (B) – 43.93 (C) + 527.61 (D) – 527.61 **(I.I.T. 1993)**

61. Identify the state function among the following.
(A) q (B) $q - w$ (C) q/w (D) $q + w$ **(I.I.T. 1993)**

62. ΔH^0 of $CO_2(g)$, $CO(g)$, $N_2O(g)$ and $NO_2(g)$ in kJ/mol is – 393, – 110, 81 and 34 respectively. Calculate ΔH for the reaction: $2NO_2(g) + 3CO(g) \longrightarrow N_2O(g) + 3CO_2(g)$.
(A) + 836 kJ (B) – 836 kJ (C) + 1460 kJ (D) – 1460 kJ

63. In a spontaneous process, the system suffers
(A) increase in internal energy (B) decrease in free energy
(C) no change in energy (D) decrease in entropy.

64. Given that $C + O_2 \longrightarrow CO_2$, $\Delta H = -x$ kJ; $2CO + O_2 \longrightarrow 2CO_2$, $\Delta H = -y$ kJ. The enthalpy of formation of CO is
(A) $y - 2x$ (B) $(y - 2x)/2$ (C) $2x - y$ (D) $(2x - y)/2$

65. The bond energy of an O—H bond is 109 kcal/mol. When a mole of water is formed, one of the following occurs.
(A) 218 kcal is released (B) 109 kcal is released
(C) 218 kcal is absorbed (D) 109 kcal is absorbed

66. Given: C(*s*, graphite) $\longrightarrow$ C(*g*), $\Delta H = 716.6$ kJ; C(*s*, diamond) $\longrightarrow$ C(*g*), $\Delta H = 714.8$ kJ. The ΔH for the reaction, C(*s*, graphite) $\longrightarrow$ C(*s*, diamond), is
   (A) + 1.9 kJ (B) – 1.9 kJ (C) zero (D) 714.8 kJ
67. In a flask colourless N_2O_4 is in equilibrium with brown coloured NO_2. At equilibrium, when the flask is heated at 100°C, the brown colour deepens and on cooling it becomes less coloured. The change in enthalpy (ΔH) for this system is
   (A) negative (B) positive (C) zero (D) undefined.
68. Which of the following is true for the reaction $H_2O(l) \longrightarrow H_2O(g)$ at 100°C and 1 atm pressure?
   (A) $\Delta H = \Delta E$ (B) $\Delta E = 0$ (C) $\Delta H = 0$ (D) $\Delta H = T\Delta S$
69. In the combustion of 2.0 g of methane 25 kcal heat is liberated. Heat of combustion of methane would be
   (A) 100 kcal (B) 200 kcal (C) 300 kcal (D) 400 kcal
70. Heat of formation of H_2O(*g*, 25°C) is – 243 kJ. ΔE for the reaction, $H_2(g) + (1/2)O_2(g) \longrightarrow H_2O(g)$ at 25°C is
   (A) + 241.8 kJ (B) – 241.8 kJ (C) – 243.0 kJ (D) + 243.0 kJ
71. The first ionisation energy for Li is 5.4 eV and electron affinity of Cl is 3.61 eV. The ΔH for the reaction, $Li(g) + Cl(g) \longrightarrow Li^+(g) + Cl^-(g)$, if the resulting ions do not combine with each other is (1 eV = 1.6×10^{-19} J)
   (A) 70 kJ/mol (B) 100 kJ/mol (C) 172.5 kJ/mol (D) 270 kJ/mol
   Hint: $\Delta H = (5.4 - 3.61) \times (1.6 \times 10^{-19}) \times (6.023 \times 10^{23}) \times 10^{-3} = 172.5$ kJ/mol
72. $\Delta G = \Delta G^0$ when the reaction quotient (Q) is
   (A) 0 (B) < 1 (C) > 1 (D) 1
73. Which of the following is correct for entropy?
   (A) $\lim_{T \to 0} S = 0$ (B) $\lim_{T \to 0} S = \infty$ (C) $\lim_{T \to \infty} S = 0$ (D) $S_{liq} > S_{vap}$
74. In an electrochemical cell $\Delta G = \Delta H$, when
   (A) $\left(\dfrac{dE_{cell}}{dT}\right)_P = 0$ (B) $T \longrightarrow 0$ (C) $T \longrightarrow \infty$ (D) both (A) and (B)
75. Inversion temperature is given by
   (A) $\dfrac{a}{Rb}$ (B) $\dfrac{2a}{Rb}$ (C) $\dfrac{Rb}{a}$ (D) $\dfrac{Rb}{2a}$
76. Which of the following is correct for an ideal gas?
   (A) $\left(\dfrac{\partial E}{\partial T}\right)_V = 0$ (B) $\left(\dfrac{\partial E}{\partial P}\right)_T = 0$ (C) $\left(\dfrac{\partial E}{\partial T}\right)_P = 0$ (D) All of these. **(I.I.T. 2000)**
77. One mole of an ideal gas undergoes a change of state (2.0 atm, 3.0 L, 95 K) $\longrightarrow$ (4.0 atm, 5.0 L, 245 K) with a change in internal energy, $\Delta U = 30.0$ L atm. The change in enthalpy (ΔH) of the process in L atm is
   (A) 40.0 (B) 42.3 (C) 44.0
   (D) not defined, because pressure is not constant. **(I.I.T. 2002)**
   Hint: To calculate ΔH, pressure should remain constant.

78. ΔH for solid to liquid transitions for proteins A and B are 2.73 kcal/mol and 3.0 kcal/mol. The two melting points are 0°C and 30°C respectively. The entropy changes ΔS_A and ΔS_B at two transition temperatures are related as
(A) $\Delta S_A = \Delta S_B$ (B) $\Delta S_A < \Delta S_B$ (C) $\Delta S_B < \Delta S_A$ (D) $\Delta S_B = 300\Delta S_A/273$
Hint: $\Delta S_A = 273/2.73 = 0.01$; $\Delta S_B = 3.0/303 = 0.0009$. Hence $\Delta S_B < \Delta S_A$.
79. For the melting of a solid at 25°C, the fusion process requires energy equivalent to 2906 J to be added to system considering the process to be reversible at fusion point. The entropy change of the process is
(A) 9.75 J K^{-1} (B) 11.272 J K^{-1} (C) 2.33 J K^{-1} (D) insufficient data
Hint: $\Delta S = 2906/298 = 9.75$ J K^{-1}
80. In an adiabatic expansion of an ideal gas
(A) $w = -\Delta E$ (B) $w = \Delta E$ (C) $\Delta E = 0$ (D) $w = 0$
81. Which of the following does not result in an increase in entropy?
(A) Vaporisation of camphor (B) Conversion of ice to water
(C) Crystallisation of sucrose from solution
(D) Rusting of iron.
82. The final temperature in an adiabatic expansion is
(A) less than the initial temperature (B) half of the initial temperature
(C) same as the initial temperature (D) greater than the initial temperature.
83. For an adiabatic expansion of an ideal gas
(A) $TV^{\gamma-1}$ = constant (B) PV^{γ} = constant
(C) $TP^{1-\gamma}$ = constant (D) none of these.
84. Identify the intensive properties.
(A) Enthalpy (B) Density (C) Pressure (D) Temperature
85. Which of the following is not a state property?
(A) Enthalpy (B) Heat (C) Volume (D) Internal energy
86. Which of the following process is favoured by the tendency for minimum energy?
(A) Melting of ice
(B) Condensation of water vapours to liquid water
(C) Dissolution of sugar in water
(D) Dissociation of Cl_2 molecule to chlorine atoms.
87. The standard free energy change ΔG^0 is related to equilibrium constant K_p as
(A) $K_p = -RT \ln \Delta G^0$ (B) $K_p = (e/RT)^{\Delta G^0}$
(C) $K_p = -\Delta G^0/RT$ (D) $K_p = e^{-\Delta G^0/RT}$
88. The heat of combustion of graphite and CO are – 393.5 kJ mol^{-1} and – 283.0 kJ mol^{-1} respectively. Heat of formation of CO in kJ mol^{-1} is
(A) + 172.5 (B) – 110.5 (C) – 10760 (D) – 676.5
Hint: ΔH_f (CO) = ΔH_{comb} (C) – ΔH_{comb} (CO) = – 393.5 – (– 283.0) = – 110.5 kJ mol^{-1}

89. When 1 mole of a gas is heated at constant volume, temperature is raised from 298 K to 308 K. Heat supplied to gas is 500 J. Identify the correct statement.
    (A) $q = w = 500$ J; $\Delta E = 0$ (B) $q = \Delta E = 500$ J; $w = 0$
    (C) $q = w = 500$ J; $\Delta E = 500$ (D) $\Delta E = 0$, $q = w = -500$ J
90. Enthalpy change of the reaction, $CH_4 + 1/2O_2 \longrightarrow CH_3OH$ is negative. If enthalpy of combustion of methane and CH_3OH is x and y respectively, which of the following is correct?
    (A) $x > y$ (B) $x < y$ (C) $x = y$ (D) $x \geq y$
91. Which of the reactions defines ΔH_f^0?
    (A) C (diamond) + $O_2(g) \longrightarrow CO_2(g)$ (B) $1/2H_2(g) + 1/2F_2(g) \longrightarrow HF(g)$
    (C) $N_2(g) + 3H_2(g) \longrightarrow 2NH_3(g)$ (D) $CO(g) + 1/2O_2(g) \longrightarrow CO_2(g)$
92. $\Delta H_{vap} = 30$ kJ mol^{-1} and $\Delta S_{vap} = 75$ J mol^{-1} K^{-1}. Find the temperature of vapour at 1 atmosphere.
    (A) 400 K (B) 350 K (C) 298 K (D) 250 K **(I.I.T. 2004)**
    Hint: $\Delta G = \Delta H - T\Delta S$; At equilibrium $\Delta G = 0$; $T = \Delta H/\Delta S = 30 \times 10^3/75 = 400$ K
93. 2 mol of an ideal gas expanded isothermally and reversibly from 1 L to 10 L at 300 K. What is the enthalpy change?
    (A) 4.98 kJ (B) 11.47 kJ (C) – 11.47 kJ (D) 0 kJ **(I.I.T. 2004)**
    Hint: $H = E + PV$; $\Delta H = \Delta E + \Delta(PV)$ or $\Delta H = \Delta E + nR\Delta T = 0 + nR \times 0$ (since $\Delta T = 0$, $\Delta E = 0$) or $\Delta H = 0$.
94. The value of $\log_{10} K$ for a reaction $A \rightleftharpoons B$ is (Given: $\Delta H^0_{298\,K} = -54.0$ kJ mol^{-1}, $\Delta S^0_{298\,K} = 10$ J K^{-1} mol^{-1} and $R = 8.314$ J K^{-1} mol^{-1}; $2.303 \times 8.314 \times 298 = 5705$)
    (A) 5 (B) 10 (C) 95 (D) 100 **(I.I.T. 2007)**
    Hint: $\Delta G^0 = \Delta H^0 - T\Delta S^0 = -54.07 \times 1000 - 298 \times 10 = -57050$ J mol^{-1}
    $\Delta G^0 = -2.303\, RT \log K$ or $-57050 = -5705 \log K$ or $\log K = 10$
95. For the process $H_2O(l)$ (1 bar, 373 K) $\longrightarrow H_2O(g)$ (1 bar, 373 K), the correct set of thermodynamic parameters is
    (A) $\Delta G = 0$, $\Delta S = +$ve (B) $\Delta G = 0$, $\Delta S = -$ve
    (C) $\Delta G = +$ve, $\Delta S = 0$ (D) $\Delta G = -$ve, $\Delta S = +$ve **(I.I.T. 2007)**
    Hint: $H_2O(l)$ (1 bar, 373 K) $\rightleftharpoons$ $H_2O(g)$ (1 bar, 373 K)
    At 373 K, $H_2O(l)$ is in equilibrium with $H_2O(g)$. Hence $\Delta G = 0$. Since liquid molecules are converted into gaseous molecules, $\Delta S = +$ve.
96. Statement–1: For every chemical reaction at equilibrium, standard Gibbs energy of reaction is zero.
    Statement–2: At constant temperature and pressure, chemical reactions are spontaneous in the direction of decreasing Gibbs energy.
    (A) Statement–1 is true, Statement–2 is true; Statement–2 is a correct explanation for Statement–1
    (B) Statement–1 is true, Statement–2 is true; Statement–2 is not a correct explanation for Statement–1

(C) Statement–1 is true, Statement–2 is false

(D) Statement–1 is false, Statement–2 is true. **(I.I.T. 2008)**

Hint: For a chemical reaction at equilibrium, Gibbs free energy change of reaction (ΔG) is zero but the standard Gibbs free energy change of reaction (ΔG^0) may not be zero. Hence the Statement–1 is false but the Statement–2 is true.

97. Among the following, the state function is(are)

(A) internal energy (B) irreversible expansion work

(C) reversible expansion work (D) molar enthalpy. **(I.I.T. 2009)**

98. The species which by definition has ZERO standard molar enthalpy of formation at 298 K is

(A) $Br_2(g)$ (B) $Cl_2(g)$ (C) $H_2O(g)$ (D) $CH_4(g)$ **(I.I.T. 2010)**

Hint: Chlorine is a gas at 298 K and bromine is a liquid at 298 K.

99. Among the following, the intensive property is (properties are)

(A) molar conductivity (B) electromotive force

(C) resistance (D) heat capacity **(I.I.T. 2010)**

Hint: Resistance and heat capacity are mass dependent properties and hence are extensive property.

ANSWERS

1. C	2. A	3. B	4. A, B, C	5. B	6. A
7. B, D	8. C	9. C	10. B	11. A	12. D
13. A	14. B	15. A	16. C	17. D	18. C
19. D	20. B	21. B	22. D	23. A	24. D
25. B	26. B	27. B	28. B, C, D	29. C	30. B
31. A	32. C	33. A	34. A, C	35. A, B	36. D
37. D	38. C	39. C	40. D	41. D	42. A
43. D	44. D	45. A	46. A	47. A	48. C
49. A, C	50. A, D	51. A	52. A, B, C, D	53. C, D	54. A, C
55. A, D	56. B	57. A	58. D	59. A	60. A
61. D	62. B	63. B	64. B	65. A	66. A
67. B	68. D	69. B	70. B	71. C	72. D
73. A	74. D	75. B	76. B	77. D	78. C
79. A	80. A	81. C	82. A	83. A, B, C	84. B, C, D
85. B	86. B	87. D	88. B	89. B	90. A
91. B	92. A	93. D	94. B	95. A	96. D
97. A, D	98. B	99. C, D			

14

Electrochemistry

IMPORTANT RELATIONS FOR NUMERICAL PROBLEMS

1. Conductance = $1/R$
2. Potential difference = $I \times R$
3. κ (specific conductivity) = $1/\rho$ (resistivity)
4. Cell constant = l/A
5. κ = conductance × cell constant
6. $\Lambda_{eq} = 1000\ \kappa/C_{eq}$ where C_{eq} is equivalent L^{-1}
7. $\Lambda_m = 1000\ \kappa/C$ where C is molar concentration (mol/L)
8. $\Lambda_0 = \lambda_0^+ + \lambda_0^-$
9. $\Lambda_m^\infty = \upsilon^+\lambda_+^\infty + \upsilon^-\lambda_-^\infty$ where υ is the number of ions and λ^∞ is the conductivity at infinite dilution.
10. Degree of dissociation, $\alpha = \Lambda_m^C/\Lambda_m^0$
11. $E_{cell} = E_{cath} - E_{an}$
12. Nernst equation: (i) For a reaction, $M^{n+}(aq) + ne^- \longrightarrow M(s)$, $E_{M^{n+}/M} = E^0_{M^{n+}/M} + (0.0591/n)$ $\log [M^{n+}]/[M]$ (ii) For a cell reaction, $aA + bB \longrightarrow cC + dD$, $E_{cell} = E^0_{cell} + (0.0591/n) \log$ $[A]^a[B]^b/[C]^c[D]^d$
13. $E^0_{cell} = (0.0591/n) \log K_c$ where K_c is the equilibrium constant.
14. Faraday's laws of electrolysis: (i) $W = Z \times C \times t$ (ii) $W_1/W_2 = E_1/E_2$, $Z_1/Z_2 = E_1/E_2$
15. Charge = nF/N_A
16. $\Delta G^0 = -nFE^0$
17. Unit of conductance (C): ohm^{-1} or S; unit of specific conductance (κ): $ohm^{-1}\ cm^{-1}$ or $S\ cm^{-1}$; unit of equivalent conductance (Λ_{eq}) : $S\ cm^2\ equiv^{-1}$; unit of molar conductance (Λ_m) : $S\ cm^2\ mol^{-1}$

OBJECTIVE TYPE MULTIPLE CHOICE QUESTIONS

1. Saturated solution of KNO_3 is used to make salt-bridge because
(A) velocity of K^+ is greater than that of NO_3^-
(B) velocity of NO_3^- is greater than that of K^+
(C) velocities of K^+ and NO_3^- are nearly the same
(D) KNO_3 is highly soluble in water. **(I.I.T. 2001)**

2. The order of equivalent conductance at infinite dilution of LiCl, NaCl and KCl is
(A) LiCl > NaCl > KCl (B) KCl > NaCl > LiCl
(C) NaCl > KCl > LiCl (D) LiCl > KCl > NaCl **(I.I.T. 2001)**

3. For the cell M | M^+ || X^- | X, E^0 (M^+/M) = 0.44 V and E^0 (X/X^-) = 0.33 V. One can deduce
(A) M + X ⟶ M^+ + X^- is spontaneous reaction
(B) M^+ + X^- ⟶ M + X is spontaneous reaction
(C) E_{cell} = – 0.77 V (D) E_{cell} = 0.77 V **(I.I.T. 2000)**
Hint: (*i*) X^- ⟶ X + e^-, E^0 = – 0.334 V (*ii*) M^+ + e^- ⟶ M, E^0 = 0.44 V. By adding Equations. (*i*) and (*ii*), we get M^+ + X^- ⟶ M + X, E^0_{cell} = 0.106 V. Since E^0_{cell} is positive, the reaction is spontaneous.

4. A gas X at 1 atm is bubbled through a solution containing 1*M* Y^- and 1*M* Z^- at 25°C. If the reduction potential of Z > Y > X, then
(A) Y will oxidise X and not Z (B) Y will oxidise Z and not X
(C) Y will oxidise both X and Z (D) Y will reduce both X and Z **(I.I.T. 1999)**

5. The standard reduction potential values of three metallic cations of X, Y and Z are 0.52, – 3.03 and – 1.18 V respectively. The order of reducing power of the corresponding metals is
(A) Y > Z > X (B) X > Y > Z (C) Z > Y > X (D) Z > X > Y **(I.I.T. 1998)**

6. The standard oxidation potentials for the half-cell reactions are: Zn = Zn^{2+} + $2e^-$, E^0 = + 0.76 V; Fe = Fe^{2+} + $2e^-$, E^0 = + 0.41 V. The emf of the cell reaction Fe^{2+} + Zn ⟶ Zn^{2+} + Fe is
(A) – 0.35 V (B) + 0.35 V (C) + 1.17 V (D) – 1.17 V **(I.I.T. 1988)**

7. A solution of sodium sulphate in water is electrolysed using inert electrodes. The products at the cathode and anode are respectively
(A) H_2, O_2 (B) O_2, H_2 (C) O_2, Na (D) O_2, SO_2
(I.I.T. 1987, 1996)

8. When a lead storage battery is discharged
(A) SO_2 is evolved (B) lead sulphate is consumed
(C) sulphuric acid is consumed (D) lead is formed. **(I.I.T. 1987)**

9. The electric charge for electrode deposition of one gram equivalent of a substance is
(A) one ampere per second (B) charge on one mole of electrons
(C) one ampere for one hour (D) 96500 coulombs per second. **(I.I.T. 1984)**

10. A solution containing one mole per litre of each $Cu(NO_3)_2$, $AgNO_3$, $Hg_2(NO_3)_2$, $Mg(NO_3)_2$ is electrolysed by using inert electrodes. The values of standard electrode potentials

in volts are: Ag^+/Ag = + 0.80, Hg_2^{2+}/Hg = + 0.79, Cu^{2+}/Cu = + 0.34, Mg^{2+}/Mg = – 2.37. With increasing voltage, the sequence of deposition of metals on the cathode will be

(A) Ag, Hg, Cu, Mg (B) Mg, Cu, Hg, Ag
(C) Ag, Hg, Cu (D) Cu, Hg, Ag **(I.I.T. 1984)**

11. Faraday's laws of electrolysis are related to
(A) atomic member of the cation (B) atomic number of the anion
(C) equivalent mass of the electrolyte
(D) speed of the cation. **(I.I.T. 1983)**

12. The standard electrode potentials in volts are: Zn^{2+}/Zn = – 0.762, Cr^{3+}/Cr = – 0.740, $2H^+/H_2$ = 0.00, Fe^{3+}/Fe^{2+} = + 0.770. Which is the strongest reducing agent?
(A) Zn(*s*) (B) Cr(*s*) (C) $Fe^{2+}(aq)$ (D) $H_2(g)$ **(I.I.T. 1981)**

13. The electrode potential (V) of hydrogen electrode at *p*H = 10 is
(A) 0 (B) – 0.0591 (C) – 0.591 (D) + 0.591 **(Roorkee 1996)**

14. When a copper wire is placed in a solution of $AgNO_3$, the solution acquires blue colour. This is due to the formation of
(A) Cu^{2+} ions (B) soluble complex of copper with $AgNO_3$
(C) Cu^- ions by the reduction of Cu (D) Cu^+ ions **(Roorkee 1989)**

15. The reaction, $(1/2)H_2(g)$ AgCl(*s*) $\longrightarrow$ $H^+(aq)$ + $Cl^-(aq)$ + Ag(*s*) occurs in the galvanic cell
(A) Ag | AgCl(*s*) | KCl(*soln*) | $AgNO_3$(*soln*) | Ag
(B) Pt | $H_2(g)$ | HCl(*soln*) | $AgNO_3$(*soln*) | Ag
(C) Pt | $H_2(g)$ | HCl(*soln*) | AgCl(*s*) | Ag
(D) Pt | $H_2(g)$ | KCl(*soln*) | AgCl(*s*) | Ag **(I.I.T. 1985)**

16. For a Ag–Zn button cell, the net reaction is represented as Zn(*s*) + $Ag_2O(s)$ $\longrightarrow$ ZnO(*s*) + 2Ag(*s*), ΔG_f^0 (Ag_2O) = – 11.21 kJ mol^{-1}, ΔG_f^0 (ZnO) = – 318.30 kJ mol^{-1}. E^0_{cell} of the button cell is
(A) 1.59 V (B) – 1.59 V (C) 11.21 V (D) 3.18 V

17. The reaction taking place in a fuel cell is
(A) $2H_2(g) + 4OH^-(aq) \longrightarrow 4H_2O(l) + 4e^-$
(B) $O_2(g) + 2H_2O(l) + 4e^- \longrightarrow 4OH^-(aq)$
(C) both (A) and (B) (D) none of these.

18. The standard electrode potentials are: $E^0_{Fe^{3+}/Fe^{2+}}$ = 0.77 V, $E^0_{Sn^{4+}/Sn^{2+}}$ = 0.15 V. The reaction that is possible is
(A) $FeCl_2 + SnCl_4 \longrightarrow FeCl_3 + SnCl_2$ (B) $FeCl_3 + SnCl_2 \longrightarrow FeCl_2 + SnCl_4$
(C) $FeCl_2 + SnCl_2 \longrightarrow FeCl_3 + SnCl_4$ (D) $FeCl_3 + SnCl_4 \longrightarrow FeCl_2 + SnCl_2$

19. The reaction occurring at anode during electrolysis of an aqueous solution containing Na_2SO_4 and $CuSO_4$ using Pt electrode is
(A) $2H_2O \longrightarrow O_2 + 4H^+ + 4e^-$ (B) $2Na^+ + e^- \longrightarrow Na$
(C) $Cu \rightarrow Cu^{2+} + 2e^-$ (D) $2SO_4^{2-} + 2H_2O \longrightarrow 2H_2SO_4 + O_2 + 4e^-$

20. The emf of a Daniell cell, Zn(*s*) | $Zn^{2+}(aq)$ || $Cu^{2+}(aq)$ | Cu(*s*) can be increased by
(A) using dil. H_2SO_4 instead of $CuSO_4$ solution

(B) increasing the concentration of $ZnSO_4$ solution
(C) increasing the concentration of $CuSO_4$ solution
(D) increasing the area of the electrodes.

21. Which of the following is true for an electrochemical cell?
(A) Oxidation occurs at Cu electrode (B) Reduction occurs at H_2 electrode
(C) H_2 is anode and Cu is cathode (D) H_2 is cathode and Cu is anode.

22. A cell reaction is spontaneous if the cell potential is
(A) 0 (B) negative (C) positive (D) infinite

23. The electrolysis of K_2SO_4 yields
(A) KSO_4 (B) $K_2S_2O_4$ (C) K_2SO_8 (D) $K_2S_2O_8$

24. Identify the correct statement regarding fuel cells.
(A) They are more efficient (B) They are free from pollution
(C) They run till the reactants are active
(D) All of these.

25. The reaction taking place at the graphite anode in a dry cell is
(A) $Mn(s) \longrightarrow Mn^{2+} + 2e^-$ (B) $Zn^{2+} + 2e^- \longrightarrow Zn(s)$
(C) $Zn(s) \longrightarrow Zn^{2+} + 2e^-$ (D) $Mn^{2+} + 2e^- \longrightarrow Mn(s)$

26. What is the quantity of electricity in coulombs required to deposit all the silver from 250 mL of 1*M* $AgNO_3$ in aqueous solution?
(A) 48250 (B) 12062.5 (C) 24125 (D) 96500

27. Identify the incorrect statement.
(A) The dependence of electrode potential for the electrode M^{n+}/M with concentration under STP conditions is given by the experssion: $E = E^0 + (0.0591/n) \log [M^{n+}]$
(B) The fuel cells convert 75% of chemical energy to electrical energy
(C) The reduction potential of an electrode can be increased by increasing the concentration of ions
(D) Electrons move through salt bridge.

28. Identify the incorrect statement.
(A) The electrolysis of $KF(aq)$ gives F_2 at the anode
(B) Copper metal reduces Fe^{2+} ion in acidic medium
(C) The electrolysis of molten NaOH liberates $H_2(g)$ at the cathode
(D) The electrical conductivity of a solution of CH_3COOH will be increased if a solution of NaOH is added.

29. Identify the incorrect statement.
(A) Dilute H_2SO_4 on electrolysis liberates O_2 at the anode
(B) Cell voltage is an intensive property
(C) The electrolysis of aqueous NaCl produces O_2 and H_2 at the anode and cathode respectively
(D) The electrolysis of molten NaCl leads to the formation of H_2 and O_2

30. Identify the incorrect statement.
    (A) The discharge potential of H^+ ions is more than that of Na^+ ions
    (B) When a Zn rod is dipped into a $CuSO_4$ solution, Cu is deposited because the standard reduction potential of zinc is greater than that of copper.
    (C) In an electrolytic cell, the electrons enter through cathode and leave through anode
    (D) Ag is a noble metal but in presence of I^- or S^{2-} ions, it can behave as an active metal.
31. Identify the incorrect statement.
    (A) The value of zero for the electrode potential of a standard hydrogen electrode holds good at all temperatures
    (B) On decreasing concentration, the molar conductivity decreases
    (C) The electrolysis of KF(*aq*) does not give F_2 at the anode because F^- ions have lower oxidation potential than water
    (D) The chemical formula of rust is $Fe_2O_3 \cdot xH_2O$
32. The equivalent conductance at infinite dilution of a weak acid, HF
    (A) can be determined by the measurement of very dilute solution of HF
    (B) can be determined from the measurements on dilute solution of NaF, NaCl and HCl
    (C) can be determined by the extrapolation of measurements on dilute solutions of HCl, HBr and HI
    (D) is an undefined quantity.
33. The E^0 values of OCl^-/Cl^- and $Cl^-/(1/2)Cl_2$ are 0.94 V and – 1.36 V respectively. The E^0 value for $OCl^-/(1/2)Cl_2$ is
    (A) 1.04 V (B) 0.52 V (C) – 2.20 V (D) – 0.42 V
34. The conductivity of a saturated solution of $BaSO_4$ is 3.06×10^{-6} ohm^{-1} cm^{-1} and its equivalent conductance is 1.53 ohm^{-1} cm^2 eq^{-1}. The K_{sp} of $BaSO_4$ is
    (A) 4×10^{-6} (B) 2.5×10^{-13} (C) 2.5×10^{-9} (D) 4×10^{-12}
35. A current of 2 ampere is passed for 5 hours through a molten salt depositing 22.2 g of metal (At. wt. = 177). The oxidation state of the metal is
    (A) + 4 (B) + 3 (C) + 2 (D) + 1
36. Equivalent conductance (Λ) of NaCl, HCl and CH_3COOH is 126.45, 426.16 and 91 ohm^{-1} cm^2 eq^{-1} respectively. Λ of CH_3COOH at infinite dilution in ohm^{-1} cm^2 eq^{-1} would be
    (A) 678.90 (B) 390.71 (C) 253.62 (D) 101.38
37. For the cell reaction, Cu^{2+} [$C_1(aq)$] + Zn(*s*) $\longrightarrow$ Zn^{2+} [$C_2(aq)$] + Cu(*s*) of an electrochemical cell, the change in free energy (ΔG) at a given temperature is a function of
    (A) ln (C_1) (B) ln (C_2) (C) ln (C_2/C_1) (D) ln ($C_1 + C_2$)
38. In which of the following the free energy of a chemical reaction is directly converted to electricity?
    (A) Lead storage battery (B) Fuel cell
    (C) Concentration cell (D) Lechlanche cell

39. E^0 for the cell, Zn | $Zn^{2+}(aq)$ || $Cu^{2+}(aq)$ | Cu is 1.10 V at 25°C. The equilibrium constant for the reaction, Zn + $Cu^{2+}(aq) \rightleftharpoons Zn^{2+}(aq)$ + Cu is of the order
(A) 10^{-28} (B) 10^{18} (C) 10^{17} (D) 10^{37}
40. An electrochemical cell is set up as follows: Pt (H_2, 1 atm) | 0.1*M* HCl | 10.1*M* CH_3COOH | (H_2, 1 atm) Pt. The emf of the cell will not be zero because
(A) the emf of a cell depends on molarities of acids used
(B) acids used in two compartments are different
(C) the *p*H of 0.1*M* HCl and 0.1*M* CH_3COOH is different
(D) the temperature is constant.
41. In a galvanic cell
(A) chemical energy is converted into heat
(B) electrical energy is converted into heat
(C) chemical energy is converted into electrical energy
(D) electrical energy is converted into chemical energy.
42. In an electrolytic cell
(A) electrical energy is converted into mechanical energy
(B) chemical energy is converted into mechanical energy
(C) electrical energy is converted into chemical energy
(D) chemical energy is converted into electrical energy.
43. Calculate the mass of Cu that will be deposited at the cathode in the electrolysis of a 0.2*M* solutions of $CuSO_4$ when a quantity of electricity equal to that required to liberate 2.24 L of H_2 from a 0.2*M* aqueous H_2SO_4, is passed
(A) 12.70 g (B) 6.35 g (C) 3.18 g (D) 1.59 g
44. When a rod of metal A is dipped in an aqueous solution of metal B (concentration of B^{2+} ion being 1*M*) at 25°C (E^0 of $A^{2+}/A = -0.76$ V, $B^{2+}/B = 0.34$ V)
(A) water will decompose into H_2 and O_2
(B) no reaction will occur
(C) B will deposit on A (D) B will gradually dissolve.
45. The anode in an electrolytic cell and galvanic cell is
(A) positively charged in electrolytic cell and negatively charged in galvanic cell
(B) negatively charged in electrolytic cell and positively charged in galvanic cell
(C) positively charged in both cells (D) negatively charged in both cells.
46. Three Faraday of electricity are passed through molten Al_2O_3, aqueous solution of $CuSO_4$ and molten NaCl taken in different electrolytic cells. The amount of Al, Cu and Na deposited at the cathodes will be in the ratio of
(A) 1.5 mol : 2 mol : 3 mol (B) 1 mol : 1.5 mol : 3 mol
(C) 3 mol : 2 mol : 1 mol (D) 1 mol : 2 mol : 3 mol
47. A depolariser used in dry cell batteries is
(A) Na_3PO_4 (B) KOH (C) MnO_2 (D) NH_4Cl

48. The standard reduction potentials for Fe^{2+}/Fe and Sn^{2+}/Sn electrodes are – 0.44 V and 0.14 V respectively. For the cell reaction: $Fe^{2+} + Sn \longrightarrow Fe + Sn^{2+}$, the standard emf is
(A) + 0.30 V (B) – 0.58 V (C) + 0.58 V (D) – 0.30 V **(I.I.T. 1990)**
Hint: $E^0_{cell} = E^0_{Fe^{2+}/Fe} - E^0_{Sn^{2+}/Sn} = -0.44\text{ V} - (-0.14\text{ V}) = -0.30\text{ V}$

49. Cu^+ ion is not stable in aqueous solution because of disproportionation reaction. E^0 value for the disproportionation of Cu^+ is ($E^0_{Cu^{2+}/Cu^+} = +0.15$ V, $E^0_{Cu^+/Cu} = +0.34$ V)
(A) – 0.19 V (B) – 0.38 V (C) + 0.19 V (D) 0.38 V **(I.I.T. 1995)**
Hint: (*i*) $Cu^{2+} + 2e^- \longrightarrow Cu$, $E^0_1 = 0.34$ V, $\Delta G^0_1 = -2FE^0_1$ (*ii*) $Cu^{2+} + e^- \longrightarrow Cu^+$, $E^0_2 = 0.15$ V, $\Delta G^0_2 = -1FE^0_2$ (*iii*) $Cu^+ + e^- \longrightarrow Cu$, $E^0_3 = ?$, $\Delta G^0_3 = -1FE^0_3$
$\Delta G^0_3 = \Delta G^0_1 - \Delta G^0_2$ or $-FE^0_3 = -2FE^0_1 - (-FE^0_2)$ or $E^0_3 = 2E^0_1 - E^0_2 = 2 \times 0.34$ V – 0.15 = 0.53 V. Hence $E^0_{cell} = E^0_{Cu^+/Cu} - E^0_{Cu^{2+}, Cu^+/Pt} = 0.53\text{ V} - 0.15\text{ V} = 0.38$ V.

50. A standard hydrogen electrode has zero electrode potential because
(A) hydrogen is easier to oxidise
(B) this electrode potential is assumed to be zero
(C) hydrogen atom has only one electron
(D) hydrogen is the lightest element. **(I.I.T. 1997)**

51. The standard reduction potentials of Cu^{2+}/Cu and Cu^{2+}/Cu^+ are 0.337 V and 0.153 V respectively. The standard electrode potential of Cu^+/Cu half-cell is
(A) 0.184 V (B) 0.627 V (C) 0.521 V (D) 0.491 V **(I.I.T. 1997)**
Hint: Apply $\Delta G^0 = -nFE^0$; $E^0_{Cu^+/Cu} = 2E^0_{Cu^{2+}/Cu} - E^0_{Cu^{2+}/Cu^+} = 2 \times 0.337 - 0.153 = 0.521$ V

52. The dependence of electrode potential for the electrode M^{n+}/M with concentration under STP conditions is given by
(A) $E = E^0 + (0.0591/n) \log [M^{n+}]$ (B) $E = E^0 + (0.0591/2n) \log [M^{n+}]$
(C) $E = E^0 - (0.0591/n) \log [M^{n+}]$ (D) $E = E^0 + (0.591/n) \log [M^{n+}]$ **(I.I.T. 1993)**

53. Identify the incorrect statement.
(A) The electrical conductivity of a solution of CH_3COOH will be decreased if a solution of NaOH is added
(B) During electrolysis 48250 coulombs of electricity will deposit 1/2 equivalent of Ag from Ag^+ ions
(C) Dilute H_2SO_4 on electrolysis liberates O_2 at the cathode
(D) The electrolysis of molten NaH liberates H_2 at the anode.

54. Identify the incorrect statement.
(A) Zinc is able to displace silver from $AgNO_3$ solution because its standard reduction potential is more negative than that of Ag
(B) The best electrical conductor is Ag
(C) The more negative the standard reduction potential, the greater is its ability to displace hydrogen from acid
(D) In the electrolysis of molten NaCl, the Cl^- ion is reduced at the anode.

55. The charge in coulomb on 1 gm ion of N^{3-} is
(A) 2.89×10^5 (B) 1.6×10^{-19} (C) 4.8×10^{-19} (D) 6.02×10^{19}

56. If $0.01M$ solution of an electrolyte has a resistance of 40 ohm in a cell having a cell constant of 0.4 cm^{-1}, its molar conductance in $ohm^{-1}\ cm^2\ mol^{-1}$ is
(A) 10 (B) 10^2 (C) 10^3 (D) 10^4

57. On passing a current of 1.0 amp for 16 min and 5 sec through 1 L solution of $CuCl_2$, all copper was deposited at the cathode. The strength of $CuCl_2$ solution was
(A) $0.2N$ (B) $0.01N$ (C) $0.01M$ (D) $0.02M$

58. The hydrogen electrode is dipped in a solution of pH 3 at 25°C. The potential of the cell would be (the value of $2.303RT/F$ is 0.059 V at 25°C)
(A) 0.177 V (B) – 0.177 V (C) 0.059 V (D) 0.087 V

59. When electricity is passed through a solution of $AlCl_3$, 13.5 g of Al are deposited. The number of Faraday involved is
(A) 0.5 (B) 1.0 (C) 1.5 (D) 3

60. The resistance of a $0.1N$ CH_3COOH solution is 250 ohm, when measured in a cell of cell constant 1.15 cm^{-1}. The equivalent conductance of $0.1N$ CH_3COOH in $ohm^{-1}\ cm^2\ eq^{-1}$ is
(A) 2.3 (B) 4.6 (C) 9.2 (D) 46

61. Electrolytic conduction differs from metallic conduction from the fact that in the former
(A) the resistance increases with increasing temperature
(B) the resistance decreases with increasing temperature
(C) the resistance remains constant with increasing temperature
(D) the resistance is independent of the length of the conductor.

62. Identify the false statement.
(A) Rusting of iron is enhanced by wet air
(B) Rusting of iron is prevented if the article is connected with a wire to Mg metal
(C) Rusting of iron is prevented on coating with Zn
(D) Rusting of iron is prevented in the presence of dissolved salts.

63. The standard reduction potential of hydrogen half-cell will be negative if
(A) $[H^+] = 1M$ and $p_{H_2} = 2$ atm (B) $[H^+] = 1M$ and $p_{H_2} = 1$ atm
(C) $[H^+] = 2M$ and $p_{H_2} = 1$ atm (D) $[H^+] = 2M$ and $p_{H_2} = 2$ atm

64. Which of the following statements is correct?
(A) Oxidation number of oxygen in KO_2 is + 1
(B) Sn^{2+} oxidises Fe^{2+}
(C) The specific conductance of an electrolyte solution decreases with increase in dilution
(D) $Zn/ZnSO_4$ is a reference electrode.

65. A cell constructed by coupling copper electrode and a standard magnesium electrode has emf of 2.7 V. If the standard reduction potential of copper electrode is + 0.34 V, that of magnesium electrode is
   (A) + 3.04 V (B) – 3.04 V (C) + 2.36 V (D) – 2.36 V
66. On passing 3 amp of electricity for 50 min, 1.8 g metal is deposited. The equivalent mass of metal is
   (A) 20.53 (B) 25.8 (C) 19.3 (D) 30.7
67. In an electroplating experiment m g of silver is deposited when 4 amp of current flow for 2 min. The amount (in g) of silver deposited by 6 amp of current flowing for 40 sec will be
   (A) 4m (B) m/2 (C) m/4 (D) 2m
68. Which of the following statements is not applicable to electrolytic conductors?
   (A) New products show up at the electrodes
   (B) Ions are responsible for carrying current
   (C) They show a positive temperature coefficient for conductance
   (D) A single stream of electrons flows from cathode to anode.
69. The standard emf for the cell reaction, $Zn + Cu^{2+} \longrightarrow Zn^{2+} + Cu$ is 1.1 V at 25°C. The emf of the cell reaction when 0.1*M* Cu^{2+} and 0.1*M* Zn^{2+} solution is used, at 25°C is
   (A) + 1.10 V (B) + 0.10 V (C) – 1.10 V (D) – 0.11 V
70. A depolariser used in dry cell batteries is
   (A) NH_4Cl (B) MnO_2 (C) KOH (D) Na_3PO_4
71. The ionic conductances of Ba^{2+} and Cl^- are respectively 127 ohm^{-1} and 76 ohm^{-1} at infinite dilution. The equivalent conductance (in ohm^{-1} cm^2 eq^{-1}) of $BaCl_2$ at infinite dilution will be
   (A) 203 (B) 279 (C) 101.5 (D) 139.5
72. The specific conductance of a 0.1*N* KCl solution at 23°C is 0.012 ohm^{-1} cm^{-1}. The resistance of the cell containing the solution at the same temperature was 35 ohm. The cell constant is
   (A) 0.142 cm^{-1} (B) 0.616 cm^{-1} (C) 0.918 cm^{-1} (D) 1.12 cm^{-1}
73. Given: $E^0_{Fe^{2+}/Fe} = -0.44$ V and $E^0_{Fe^{3+}/Fe} = +0.77$ V. If Fe^{2+}, Fe^{3+} and Fe blocks are kept together, then
   (A) Fe^{3+} increases (B) Fe^{3+} decreases
   (C) Fe^{2+}/Fe^{3+} remains constant (D) Fe^{2+} decreases.
74. If the solution, in a hydrogen electrode, whose *p*H = 0 is neutralised to *p*H = 7, the standard electrode potential
   (A) decreases by 0.41 V (B) increases by 0.41 V
   (C) decreases by 0.059 V (D) increases by 0.059 V
75. Assuming the species in their standard states, which of the following ions is the strongest oxidising agent?
   (A) Co^{3+} (B) Fe^{2+} (C) Cu^{2+} (D) Cl^-

76. The emf of the cell, Ag(s) | 0.1*M* KCl satd. with AgCl || 0.1*M* $AgNO_3$ | Ag is 0.455 V at 298 K. The solubility product of AgCl at 298 K is

(A) 1.95×10^{-10} (B) 1.95×10^{10} (C) 1.95×10^{-13} (D) 1.95×10^{-16}

77. Standard electrode potential data are useful for understanding the suitability of an oxidant in a redox titration. Some half-cell reactions and their standard potentials are given below:

$MnO_4^-(aq) + 8H^+(aq) + 5e^- \longrightarrow Mn^{2+}(aq) + 4H_2O(l)$ $E^0 = 1.51$ V

$Cr_2O_7^{2-}(aq) + 14H^+(aq) + 6e^- \longrightarrow 2Cr^{3+}(aq) + 7H_2O(l)$ $E^0 = 1.38$ V

$Fe^{3+}(aq) + e^- \longrightarrow Fe^{2+}(aq)$ $E^0 = 0.77$ V; $Cl_2(g) + 2e^- \longrightarrow 2Cl^-(aq)$ $E^0 = 1.40$ V

Identify the only incorrect statement regarding the quantitative estimation of aqueous $Fe(NO_3)_2$.

(A) MnO_4^- can be used in aqueous HCl

(B) $Cr_2O_7^{2-}$ can be used in aqueous HCl

(C) MnO_4^- can be used in aqueous H_2SO_4

(D) $Cr_2O_7^{2-}$ can be used in aqueous H_2SO_4 **(I.I.T. 2002)**

Hint: $MnO_4^- + 16H^+ + 10Cl^- \longrightarrow Mn^{2+} + 8H_2O + 5Cl_2$

For Pt, Cl_2 (1 atm) | Cl^- || MnO_4^-, Mn^{2+}, H^+ | Pt, $E^0_{cell} = 1.51 - 1.40 = 0.11$ V. Since E^0_{cell} is positive, ΔG^0 will be negative. Hence the reaction is feasible.

78. In the cell reaction, $Cu(s) + 2Ag^+(aq)^- \longrightarrow Cu^{2+}(aq) + 2Ag(s)$, $E^0_{cell} = 0.46$ V. By doubling the concentration of Cu^{2+}, E^0_{call} is

(A) doubled (B) halved

(C) unchanged (D) decreases by small fraction.

79. Aluminium displaces hydrogen from acids, but copper does not. A galvanic cell prepared by combining Cu/Cu^{2+} and Al/Al^{3+} has an emf of 2.0 V at 298 K. If the potential of copper electrode is + 0.34 V, that of aluminium electrode is

(A) – 2.3 V (B) + 2.34 V (C) – 1.66 V (D) + 1.66 V

Hint: $E_{cell} = E_{Cu^{2+}/Cu} - E_{Al^{3+}/Al}$ or $2.0 \text{ V} = 0.34 \text{ V} - E_{Al^{3+}/Al}$ or $E_{Al^{3+}/Al} = -1.66$ V

80. The half-cell reaction with the standard reduction potentials are: (i) $Pb^{2+} + 2e^- \longrightarrow Pb$ ($E^0 = -0.13$ V) (ii) $Ag^+ + e^- \longrightarrow Ag$ ($E^0 = +0.80$ V). Which of the following reactions will occur?

(A) $Pb^{2+} + 2Ag \longrightarrow 2Ag^+ + Pb$ (B) $Pb^{2+} + H_2 \longrightarrow Pb + 2H^+$

(C) $2Ag^+ + Pb \longrightarrow Pb^{2+} + 2Ag$ (D) $2Ag + Pb \longrightarrow 2Ag^+ + Pb^{2-}$

Hint: E^0_{cell} should be positive for spontaneous cell reaction. $Ag^+ + e^- \longrightarrow Ag$ ($E^0 = +0.80$ V), $Pb \longrightarrow Pb^{2+} + 2e^-$ ($E^0 = +0.13$ V). Net reaction: $2Ag^+ + Pb \longrightarrow 2Ag + Pb^{2+}$

81. In the electrolysis of NaCl when Pt electrode is taken, H_2 is liberated at cathode while with Hg cathode, it forms Na/Hg. Which of the following is correct?

(A) Hg is more inert than Pt

(B) More voltage is required to reduce H^+ ion at Hg than at Pt

(C) Na is dissolved in Hg while it does not dissolve in Pt

(D) Concentration of H^+ ions is larger when Pt electrode is taken

82. In the electrolytic cell, flow of electrons is from
(A) cathode to anode in solution
(B) cathode to anode through external supply
(C) cathode to anode through internal supply
(D) anode to cathode through internal supply. **(I.I.T. 2003)**

83. The emf of the cell Zn/Zn^{2+} ($a = 0.1M$) || Fe^{2+} ($a = 0.01M$) | Fe is 0.2905 V. Equilibrium constant for the cell reaction is
(A) $10^{0.32/0.591}$ (B) $10^{0.32/0.0295}$ (C) $10^{0.26/0.0295}$ (D) $10^{0.32/0.295}$ **(I.I.T. 2004)**
Hint: $E = E^0 - (0.0591/n) \log K$ or $0.2905 = E^0 - (0.0591/2) \log (0.1/0.01)$ or $E^0 = 0.2905 + 0.0295 = 0.32$. Hence $0.32 = (0.0591/2) \log K$ or $K = 10^{0.32/0.0295}$

84. The half-cell reactions for rusting of iron are: $2H^+ + \frac{1}{2} O_2 + 2e^- \longrightarrow 2H_2O$; $E^0 = + 1.23$ V, $Fe^{2+} + 2e^- \longrightarrow Fe(s)$; $E^0 = - 0.44$ V. ΔG^0 (in kJ) for the reaction is
(A) – 76 (B) – 322 (C) – 122 (D) – 176 **(I.I.T. 2005)**
Hint: $2H^+ + \frac{1}{2} O_2 + 2e^- \longrightarrow H_2O$; $E^0 = + 1.23$ V
$Fe^{2+} + 2e^- \longrightarrow Fe(s)$; $E^0 = - 0.44$ V

$Fe(s) + 2H^+ + \frac{1}{2} O_2 \longrightarrow Fe^{2+} + H_2O$; $E^0_{cell} = 1.67$ V

$\Delta G^0 = - nFE^0_{cell} = - 2 \times 96500 \times 1.67 = - 322$ kJ

85. Which of the following is correct for the reaction occurring in a galvanic cell?
(A) If $E^0_{cell} < 0$, $K_{eq} > 1$ (B) If $E^0_{cell} < 0$, $K_{eq} > 1$
(C) If $E^0_{cell} < 0$, $\Delta G < 0$ (D) If $E^0_{cell} > 0$, $\Delta G < 0$

86. A 4.0 molar aqueous solution of NaCl is prepared and 500 mL of this solution is electrolysed. This leads to the evolution of chlorine gas at one of the electrodes. The total number of moles of chlorine gas evolved is
(A) 0.5 (B) 1.0 (C) 2.0 (D) 3.0 **(I.I.T. 2007)**
Hint: $NaCl \longrightarrow Na^+ + Cl^-$
At anode: $2Cl^- \longrightarrow Cl_2$
In 500 ml of solution, mole of Cl^- ion = 2. Hence moles of Cl_2 evolved = 1.

87. A 4.0 molar aqueous solution of NaCl is prepared and 500 mL of this solution is electrolysed. This leads to the evolution of chlorine gas at one of the electrodes. If the cathode is a Hg electrode, the maximum weight (g) of amalgam formed from this solution is (atomic mass: Na = 23, Hg = 200)
(A) 200 (B) 225 (C) 400 (D) 446 **(I.I.T. 2007)**
Hint: Na/Hg formed = 2 moles at the cathode. Maximum weight of amalgam formed = 23 × 2 + 200 × 2 = 446 g

88. A 4.0 molar aqueous solution of NaCl is prepared and 500 mL of this solution is electrolysed. This leads to the evolution of chlorine gas at one of the electrodes. The total charge (C) required for complete electrolysis is (1 F = 96500 C)
(A) 24125 (B) 48250 (C) 96500 (D) 193000 **(I.I.T. 2007)**
Hint: 2 moles of electrons (= 2 Faraday) are required. 2 Faraday = 2 × 96500 C = 193000 C

89. While Fe^{3+} is stable, Mn^{3+} is not stable in acid solution because ($Mn^{3+} + e^- \longrightarrow Mn^{2+}$, $E^0 = 1.5$ V; $O_2 + 4H^+ + 4e^- \longrightarrow 2H_2O$, $E^0 = 1.23$ V; $Fe^{3+} + e^- \longrightarrow Fe^{2+}$, $E^0 = 0.77$ V)
(A) O_2 oxidises Mn^{2+} to Mn^{3+} (B) O_2 oxidises both Mn^{2+} and Fe^{2+} to Fe^{3+}
(C) Fe^{3+} oxidises H_2O to O_2 (D) Mn^{3+} oxidises H_2O to O_2 **(I.I.T. 2007)**
Hint: The reaction of Mn^{3+} with H_2O is spontaneous due to the above E^0 values.

90. Among the following, identify the correct statement ($I_2 + 2e^- \longrightarrow 2I^-$, $E^0 = 0.54$ V; $Cl_2 + 2e^- \longrightarrow 2Cl^-$, $E^0 = 1.36$ V; $Mn^{3+} + e^- \longrightarrow Mn^{2+}$, $E^0 = 1.50$ V; $Fe^{3+} + e^- \longrightarrow Fe^{2+}$, $E^0 = 0.77$ V)
(A) Chloride ion is oxidised by O_2 (B) Fe^{2+} is oxidised by iodine
(C) Iodide ion is oxidised by chlorine (D) Mn^{2+} is oxidised by chlorine. **(I.I.T. 2007)**
Hint: $E^0_{I_2/I^-}$ is less than $E^0_{Cl_2/Cl^-}$

91. For the electrolytic cell, M(s) | M^+(*aq*; 0.05 molar) || M^+(*aq*; 1 molar) | M(s), the magnitude of the cell potential is 70 mV.
For the above cell
(A) $E_{cell} < 0; \Delta G > 0$ (B) $E_{cell} > 0; \Delta G < 0$ (C) $E_{cell} < 0; \Delta G^0 > 0$ (D) $E_{cell} > 0; \Delta G^0 > 0$
(I.I.T. 2010)

Hint: $M(s) + M^+_{(aq)1M} \longrightarrow M^+_{(aq).05M} + M(s)$

According to Nernst equation, $E_{cell} = 0 - \dfrac{2.303RT}{F}\log\dfrac{M^+_{.05M}}{M^+_{1M}}$

$$= 0 - \frac{2.303RT}{F}\log(5\times10^{-2}) = +\text{ve}$$

Hence $|E_{cell}| = E_{cell} = 0.70$ V and $\Delta G < 0$ for the feasibility of the reaction.

92. For the electrolytic cell, M(s) | M^+ (*aq*; 0.05 molar) || M^+ (*aq*; 1 molar) | M(s), the magnitude of the cell potential is 70 mV.
If the 0.05 molar solution of M^+ is replaced by 0.0025 molar M^+ solution, then the magnitude of the cell potential would be
(A) 35 mV (B) 70 mV (C) 140 mV (D) 700 mV **(I.I.T. 2010)**

Hint: From the above equation $\dfrac{2.303RT}{F} = 0.0538$. Hence $E_{cell} = E^0_{cell} - \dfrac{0.0538}{1}\log$

$$0.0025 = 0 - \frac{0.0538}{1}\log 0.0025 \approx 0.13988 \text{ V} \approx 140 \text{ mV}$$

93. $AgNO_3(aq)$ was added to an aqueous KCl solution gradually and the conductivity of the solution was measured. The plot of conductance (Λ) versus the volume of $AgNO_3$ is

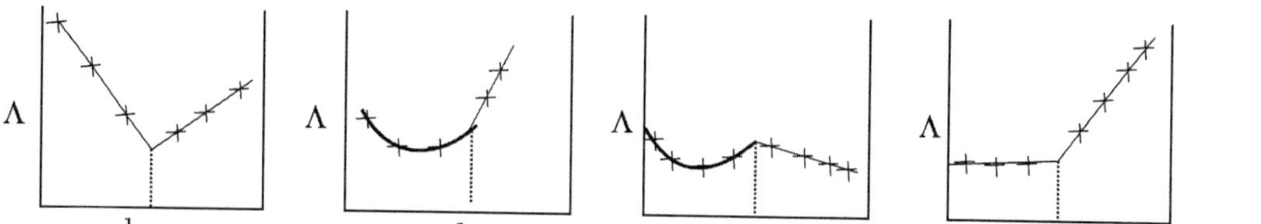

volume (P) | volume (Q) | volume (R) | volume (S)

(A) (P) (B) (Q) (C) (R) (D) (S) **(I.I.T. 2011)**

Hint: $AgNO_3 + KCl(aq) \longrightarrow AgCl(s) + KNO_3(aq)$

Initially there is aqueous KCl solution. As solution of $AgNO_3$ is added, AgCl(*s*) is formed. Hence the conductivity of solution is almost compensated (or slightly increased) by the formation of KNO_3. After the end point conductivity increases more repidly because of addition of excess $AgNO_3$ solution.

94. Consider the following cell reaction: $2Fe(s) + O_2(g) + 4H^+(aq) \longrightarrow 2Fe^{2+}(aq) + 2H_2O(l)$
$E^0 = 1.67$ V

At $[Fe^{2+}] = 10^{-3}$ *M*, $P(O_2) = 0.1$ atm and *p*H = 3, the cell potential at 25°C is

(A) 1.47 V (B) 1.77 V (C) 1.87 V (D) 1.57 V **(I.I.T. 2011)**

$2Fe(s) + O_2(g) + 4H^+(aq) \longrightarrow 2Fe^{+2}(aq) + 2H_2O(l)$

$N = 4$ (no. of moles of electron involved)

$$E_{cell} = E^0_{cell} - \frac{0.0591}{n}\log Q = 1.67 - \frac{0.0591}{4}\log\frac{(10^{-3})^2}{0.1\times(10^{-3})^4} \quad \{\because [H^+] = 10^{-pH}\}$$

$$= 1.67 - 0.10 = 1.57 \text{ V}$$

ANSWERS

1. C	2. B	3. B	4. A	5. A	6. B
7. A	8. C	9. B	10. C	11. C	12. A
13. C	14. A	15. C	16. A	17. C	18. B
19. A	20. C	21. C	22. C	23. D	24. D
25. C	26. C	27. A, D	28. A, B	29. D	30. A, B
31. A, B	32. B	33. D	34. A	35. B	36. B
37. C	38. B	39. D	40. C	41. C	42. C
43. B	44. C	45. A	46. B	47. C	48. D
49. D	50. B	51. C	52. A	53. A, C	54. D
55. A	56. C	57. B	58. A	59. C	60. D
61. B	62. D	63. A	64. C	65. D	66. C
67. B	68. D	69. A	70. B	71. D	72. B
73. B	74. A	75. A	76. A	77. A	78. D
79. C	80. C	81. B	82. C	83. B	84. B
85. D	86. B	87. D	88. D	89. D	90. C
91. B	92. C	93. D	94. D		

Part—II
Inorganic Chemistry

15

Periodic Properties of Elements

OBJECTIVE TYPE MULTIPLE CHOICE QUESTIONS

1. The set representing the correct order of first ionisation energy is
(A) $K > Na > Li$ (B) $Be > Mg > Ca$ (C) $B > C > N$ (D) $Ge > Si > C$ **(I.I.T. 2001)**

2. The correct order of radii is
(A) $N < Be < B$ (B) $F^- < O^{2-} < N^{3-}$
(C) $Na < Li < K$ (D) $Fe^{3+} < Fe^{2+} < Fe^{4+}$ **(I.I.T. 2000)**

3. Assertion (*A*): F—F bond in F_2 is strong.
Reason (*R*): F atom is small in size.
(A) Both *A* and *R* are correct and *R* is the correct explanation of *A*
(B) Both *A* and *R* are correct and *R* is not the correct explanation of *A*
(C) *A* is correct but *R* is false (D) *A* is incorrect but *R* is correct. **(I.I.T. 1998)**

4. Assertion (*A*): F atom has a less negative electron affinity than Cl atom.
Reason (*R*): Additional electrons are repelled more effectively by $3p$ electrons in Cl atom than by $2p$ electrons in F atom.
(A) Both *A* and *R* are correct and *R* is the correct explanation of *A*
(B) Both *A* and *R* are correct and *R* is not the correct explanation of *A*
(C) *A* is correct but *R* is false (D) *A* is incorrect but *R* is correct. **(I.I.T. 1998)**

5. The incorrect statement among the following is
(A) first IE of Al < first IE of Mg
(B) second IE of Mg > second IE of Na
(C) first IE of Na < first IE of Mg
(D) third IE of Mg > third IE of Al **(I.I.T. 1997)**

6. The decreasing order of the second IE of K, Ca and Ba is
(A) $K > Ca > Ba$ (B) $Ca > Ba > K$ (C) $Ba > K > Ca$ (D) $K > Ba > Ca$ **(I.I.T. 1993)**

7. The size of the following is in the order
(A) $Mg^{2+} < Na^+ < Al < F^-$ (B) $F^- < Al < Na^+ < Mg^{2+}$
(C) $Al < Mg^{2+} < F^- < Na^+$ (D) $Na^+ < Al < F^- < Mg^{2+}$ **(I.I.T. 1990)**

8. The statement that is not correct for periodic classification of elements is
   (A) The properties of elements are a periodic function of their atomic numbers
   (B) Non-metallic elements are less in number than metallic elements
   (C) The first ionisation energies of elements along a period do not vary in a regular manner with increase in atomic number
   (D) For transition elements, the *d*-sub-shells are filled with electrons monotonically in atomic number. **(I.I.T. 1992)**
9. The elements which exist in liquid state at room temperature are
   (A) Na (B) Hg (C) Br (D) Ca **(I.I.T. 1991)**
10. Amongst the following elements (whose electronic configurations are given below), the one having the highest IE is
   (A) [Ar] $3d^{10}4s^24p^2$ (B) [Ne] $3s^23p^3$ (C) [Ne] $3s^23p^2$ (D) [Ne] $3s^23p^1$ **(I.I.T. 1990)**
11. The first IE of nitrogen and oxygen atoms are related as follows:
   (A) The IE of oxygen is less than that of nitrogen
   (B) The two IE values are comparable
   (C) The IE of nitrogen is greater than that of oxygen
   (D) The difference between the two ionisation energies is too large. **(I.I.T. 1989)**
12. Which of the following statements are true for the long form of periodic table?
   (A) It reflects the sequence of filling electrons in order of sub-energy levels *s*, *p*, *d* and *f*
   (B) It helps to predict the stable valency states of the elements
   (C) It reflects trends in physical and chemical properties of the elements
   (D) It helps to predict the relative ionicity of the bond between any two elements. **(I.I.T. 1988)**
13. The first ionisation energies of Na, Mg, Al and Si are in the order
   (A) Na < Mg > Al < Si (B) Na > Mg > Al > Si
   (C) Na > Mg > Al < Si (D) Na < Al < Mg < Si **(I.I.T. 1988)**
14. Which of the following is the smallest size?
   (A) N^{3-} (B) O^{2-} (C) F^- (D) Na^+ **(I.I.T. 1989)**
15. Atomic radii of fluorine and neon in angstrom unit are respectively
   (A) 0.72, 1.60 (B) 1.60, 1.60 (C) 0.72, 0.72 (D) none. **(I.I.T. 1987)**
16. The electronegativity of the following elements increases in the order
   (A) C, N, Si, P (B) N, Si, C, P (C) Si, P, C, N (D) P, Si, N, C **(I.I.T. 1987)**
17. Which of the following values in electron volt per atom represent the first IE of oxygen and nitrogen atoms respectively?
   (A) 14.6, 13.6 (B) 13.6, 14.6 (C) 13.6, 13.6 (D) 14.6, 14.6 **(I.I.T. 1987)**
18. Ionic size decreases in the order
   (A) $N^{3-} > O^{2-} > F^- > Na^+ > Mg^{2+}$ (B) $N^{3-} > O^{2-} > F^- > Mg^{2+} > Na^+$
   (C) $N^{3-} > F^- > O^{2-} > Na^+ > Mg^{2+}$ (D) $O^{2-} > N^{3-} > F^- > Na^+ > Mg^{2+}$ **(I.I.T. 1991)**

19. Ionic size increases in the order
(A) $Ca^{2+} < Cl^- < S^{2-} < Ar$ (B) $Ar < Ca^{2+} < Cl^- < S^{2-}$
(C) $Cl^- < Ca^{2+} < Ar < S^{2-}$ (D) $S^{2-} < Cl^- < Ca^{2+} < Ar$ **(I.I.T. 1986)**

20. The element with the highest IE is
(A) B (B) C (C) N (D) O **(I.I.T. 1992)**

21. The correct order of second ionisation energy of carbon, nitrogen, oxygen and fluorine is
(A) C > N > O > F (B) O > N > F > C
(C) O > F > N > C (D) F > O > N > C **(I.I.T. 1981)**

22. Which of the following has the lowest electron affinity?
(A) Be (B) Cl (C) N (D) B **(Roorkee 1995)**

23. The decreasing order of second ionisation energy of K, Ca and Ba is
(A) K > Ca > Ba (B) Ca > Ba > K (C) Ba > K > Ca (D) K > Ba > Ca **(I.I.T. 1993)**

24. Which of the following sets of ions represents a collection of isoelectronic species?
(A) K^+, Cl^-, Ca^{2+}, Sc^{3+} (B) Ba^{2+}, Sr^{2+}, K^+, Ca^{2+}
(C) N^{3-}, O^{2-}, F^-, S^{2-} (D) Li^+, Na^+, Mg^{2+}, Ca^{2+}
Hint: The isoelectronic species are K^+, Cl^-, Ca^{2+} and Sc^{3+} each having 18 electrons.

25. The ascending order of electron affinity is
(A) O < Se < S (B) Se < O < S (C) S < O < Se (D) S < Se < O
(Roorkee 1990)

26. The set containing isoelectronic species is
(A) C_2^{2-}, NO^+, CN^-, O_2^{2+} (B) CO, NO, O_2, CN^-
(C) CO_2, NO_2, O_2, N_2O_5 (D) CO, CO_2, NO, NO_2

27. The radii of F, F^-, O and O^{2-} are in the order
(A) $O^{2-} > F^- > F > O$ (B) $F^- > O^{2-} > F > O$
(C) $O^{2-} > O > F^- > F$ (D) $O^{2-} > F^- > O > F$

28. The electron affinity of the halogens are: F = 332, Cl = 349, Br = 324, I = 295 kJ/mol. The higher value for Cl compared to F is due to
(A) smaller electronegativity of F (B) more vacant *p* sub-shell in Cl
(C) weaker electron-electron repulsion in Cl
(D) higher atomic radius of F

29. The ionic size decreases as
(A) $K^+ > S^{2-} > Sc^{3+} > V^{5+} > Mn^{7+}$ (B) $S^{2-} > K^+ > Sc^{3+} > V^{5+} > Mn^{7+}$
(C) $Mn^{7+} > V^{5+} > Sc^{3+} > K^+ > S^{2-}$ (D) $Mn^{7+} > V^{5+} > Sc^{3+} > S^{2-} > K^+$

30. The ionic size decreases as
(A) $Al^{3+} < O^{2-} < F^- < Cl^-$ (B) $F^- < O^{2-} < Cl^- < Al^{3+}$
(C) $Al^{3+} < F^- < O^{2-} < Cl$ (D) $Al^{3+} < F^- < Cl^- < O^{2-}$

31. Which of the following has the highest electronegativity?
(A) Al (B) S (C) Si (D) P

32. Which of the following is most electropositive?
    (A) Si (B) Al (C) Mg (D) Na
33. Which of the following is a semi-metal?
    (A) Mg (B) B (C) C (D) As
34. In the periodic table, which of the following trends is not correct in a period?
    (A) metallic to non-metallic (B) oxidising to reducing
    (C) reducing to oxidising (D) base to acid forming.
35. In the periodic table, which of the following trends is correct in a period?
    (A) electron affinity decreases (B) chemical reactivity increases
    (C) electronegative character decreases
    (D) electropositive character increases.
36. The electronegativity of an element
    (A) is always constant (B) depends on the valency of the element
    (C) is independent of the valency of the element
    (D) is related to nuclear binding energy.
37. Which of the following statements is correct?
    (A) Electronegativity of an atom is its tendency to attract an electron in a molecule towards itself.
    (B) Electronegativity increases on moving from top to bottom in any particular group.
    (C) Electronegativity decreases on moving from left to right in a period.
    (D) Electronegativity of an atom is its tendency to attract the shared electrons in a molecule towards itself.
38. In the sixth period, the orbitals are filled as
    (A) 6*s*, 5*f*, 6*d*, 6*p* (B) 6*s*, 4*f*, 5*d*, 6*p* (C) 5*s*, 5*p*, 5*d*, 6*p* (D) 6*s*, 6*p*, 6*d*, 6*f*
39. As the atomic number increases, the ionic radii of *f*-block elements in a period decreases more in comparison to the decrease of ionic radii of *d*-block elements in a period from left to right. This is due to
    (A) similar shape of *d*-orbitals and similar shape of *f*-orbitals
    (B) similar shape of *d*-orbitals and different shapes of *f*-orbitals
    (C) inefficient screening of outermost electrons by *d*-orbitals and inefficient screening of outermost electrons by *f*-orbitals
    (D) efficient screening of outermost electrons by *d*-orbitals and inefficient screening of outermost electrons by *f*-orbitals.
40. In the descending order of a vertical group in the periodic table, which of the following would be true?
    1. Electronegativity decreases 2. Gram atomic volume increases
    3. All the atoms have the same number of valence electrons
    4. Metallic character decreases and the basic nature of their oxides decreases
    (A) 2, 3 and 4 (B) 1, 2 and 4 (C) 1, 3 and 4 (D) 1, 2 and 3

41. Which of the following is a representative element?
(A) Ar (B) Al (C) La (D) Cr

42. The electronic configuration of the element with atomic number 106 is
(A) [Rn] $5f^{14}6d^5 7s^1$ (B) [Rn] $5f^{14}6d^4 7s^2$
(C) [Rn] $5f^{14}6d^6 7s^0$ (D) [Rn] $5f^{14}6d^1 7s^2 7p^3$

43. Al^{3+} has a lower ionic radius than Mg^{2+} because
(A) Al has a lower IE than Mg (B) Al has a higher IE than Mg
(C) Al^{3+} has a higher nuclear charge than Mg^{2+}
(D) their electronegativities are different.

44. The correct order of electronegativity is
(A) F > O > N > C (B) F > N > O > C
(C) F < N < O < C (D) F > N > O < C

45. Which of the following has the highest first IE?
(A) Si (B) Al (C) K (D) P

46. In the periodic table, the ionisation energy of elements
(A) (i) increases from left to right across a period and (ii) decreases on descending a group
(B) (i) decreases from left to right across a period and (ii) increases on descending a group
(C) decreases (i) from left to right across a period and (ii) decreases on descending a group
(D) increases (i) from left to right across a period and (ii) on descending a group.

47. The size of the species I, I^+ and I^- decreases in the order
(A) $I^+ > I^- > I$ (B) $I^- > I > I^+$ (C) $I^- > I^+ > I$ (D) $I > I^+ > I^-$

48. Which of the following has the lowest IE?
(A) O_2 (B) O_2^+ (C) O_2^- (D) O

49. Identify the false statement.
(A) The first IE of C is greater than that of B, whereas the reverse is true for the second IE.
(B) The first IE of Mg is smaller than that of Na and greater than that of Al.
(C) The element with atomic number 26 occurs in the 4th period and Group VIII.
(D) The electron affinity of C, N, O and F decreases in the order: F > O > C > N.

50. Identify the false statement.
(A) On the Pauling electronegativity scale the element next to F is oxygen.
(B) On Mulliken scale, the average of IE and electron affinity is known as electronegativity.
(C) The energy released when an electron is added to a neutral gaseous atom is called EA of the atom.
(D) Beryllium and magnesium impart colour in Bunsen flame.

51. Identify the false statement.
    (A) In Group 1 the IE decreases down the group. Hence Li is a strong reducing agent.
    (B) The softness of Group 1 metals increases down the group with increasing atomic number.
    (C) In aqueous solution chlorine is stronger oxidising agent than fluorine.
    (D) The increasing IE of N, O and F is in the order: $O < N < F$.
52. The ionisation energy of nitrogen is more than that of oxygen because of
    (A) the greater attraction of electrons by the nucleus
    (B) the extra stability of the half-filled *p*-orbitals
    (C) the smaller size of nitrogen (D) more penetration effect.
53. The process requiring absorption of energy is
    (A) $F \longrightarrow F^-$ (B) $Cl \longrightarrow Cl^-$
    (C) $O \longrightarrow O^{2-}$ (D) $H \longrightarrow H^-$ **(Roorkee 1990)**
54. Which of the following are isoelectronic species?
    (I) $\overset{+}{C}H_3$ (II) NH_2^- (III) NH_4^+ (IV) NH_3
    (A) I, II, III (B) II, III, IV (C) I, II, IV (D) II and I
55. Highest ionisation energy is shown by
    (A) halogens (B) alkali metals (C) nobal gases (D) transition metals
56. With reference to ionisation energy which of the following sets is correct?
    (A) $Li > K < Cs$ (B) $B > Li > K$ (C) $Cs > Li > B$ (D) $Cs < Li < K$
57. Which of the following pairs of atoms/ions has identical electronic configuration?
    (A) Li and He (B) Cl^- and Ar (C) Na and K (D) F and Ne
58. Which of the following has the highest IP?
    (A) Li^+ (B) Mg^+ (C) Al^+ (D) Ne
59. The decreasing order of the size of the following cations is
    (A) $Li^+ > H^+ > H^-$ (B) $H^+ > H^- > Li^+$ (C) $H^- > Li^+ > H^+$ (D) $H^- > H^+ > Li^+$
60. The order of electron affinity of B, C, N and O is
    (A) $O > C > N > B$ (B) $B > N > C > O$
    (C) $O > C > B > N$ (D) $O > B > C > N$
61. The decreasing order of the size of Mg^{2+}, Al^{3+} and O^{2-} is
    (A) $Mg^{2+} = Al^{3+} > O^{2-}$ (B) $Al^{3+} > Mg^{2+} > O^{2-}$
    (C) $O^{2-} > Mg^{2+} > Al^{3+}$ (D) $Mg^{2+} > Al^{3+} > O^{2-}$
62. The decreasing order of electron affinity of N, O, S is
    (A) $S > N > O$ (B) $N > O > S$ (C) $O > S > N$ (D) $S > O > N$
63. The increasing order of ionic radius of K^+, Li^+, Mg^{2+} and Al^{3+} is
    (A) $Al^{3+} < Li^+ < Mg^{2+} < K^+$ (B) $Li^+ < K^+ < Mg^{2+} < Al^{3+}$
    (C) $K^+ < Mg^{2+} < Li^+ < Al^{3+}$ (D) $K^+ < Li^+ < Mg^{2+} < Al^{3+}$

64. The atomic radius of elements of the following series should be nearly the same.
(A) Na, K, Rb, Cs (B) Li, Be, B, C
(C) Fe, Co, Ni, Cu (D) F, Cl, Br, I

65. Which of the following ionic radii would be the maximum?
(A) C^{4-} (B) N^{3-} (C) O^{2-} (D) Mg^{2+}

66. On going from right to left in a period in the periodic table, the electronegativity of the elements
(A) increases (B) decreases
(C) remains unchanged (D) decreases first then increases.

67. Which has the smallest size?
(A) Na^+ (B) Mg^{2+} (C) Al^{3+} (D) P^{5+}

68. Gradual addition of electronic shells in the noble gases causes a decrease in their
(A) ionisation energy (B) atomic radius
(C) boiling point (D) density.

69. Which of the following configurations represents atoms of elements having the highest second ionisation energy ?
(A) $1s^22s^2p^4$ (B) $1s^22s^2p^6$ (C) $1s^22s^2p^63s^1$ (D) $1s^22s^2p^63s^2$

70. Which of the following is not arranged in the correct sequence?
(A) d^5, d^3, d^1, d^4 — increasing magnetic moment
(B) MO, M_2O_3, MO_2, M_2O_5 — decreasing basic strength
(C) Sc, V, Cr, Mn — increasing number of oxidation states
(D) Co^{3+}, Fe^{3+}, Cr^{3+}, Sc^{3+} — increasing stability

71. The correct order of first ionisation energy among the elements Be, B, C, N and O is
(A) $B < Be < C < O < N$ (B) $B < Be < C < N < O$
(C) $Be < B < C < N < O$ (D) $Be < B < C < O < N$

72. The maximum number of elements occur in which period of the periodic table?
(A) 4th (B) 5th (C) 6th (D) 7th

73. For which of the following, the hydration energy of M^{n+} is larger?
(A) Na^+ (B) Al^{3+} (C) Be^{2+} (D) Cr^{3+}
Hint: Charge : radius: Na^+, 1.05; Be^{2+}, 6.4; Al^{3+}, 5.7; Cr^{3+}, 4.3. Higher the charge : radius ratio, larger is the hydration energy.

74. Among the following groupings, which represents the collection of isoelectronic species?
(A) NO^+, C_2^{2-}, CO_2^-, CO (B) N_2, C_2^{2-}, CO, NO
(C) CO, NO^+, CN^-, C_2^{2-}, (D) NO, CN^-, N_2, O_2

75. The values of electronegativity of atoms A and B are 1.20 and 4.0 respectively. The percentage ionic character of A—B bond is
(A) 50.0% (B) 72.24% (C) 55.3% (D) 43.0%
Hint: % ionic character = $16(4 - 1.2) + 3.5(4 - 1.2)^2 = 72.24$

76. The element with atomic number 118 (Uno) (group 0) has been discovered recently. Which of the following is not expected for this element?
(A) It is a radioactive element (B) It is more reactive than Xe
(C) It is a solid at room temperature
(D) Its ionisation energy is the lowest in group 0

77. Astatine is the last element in group 17. It is expected to be
(A) limited to oxidation number of – 1 in its compounds
(B) a solid at room temperature
(C) more electronegative than iodine
(D) composed of diatomic molecules.

78. Identify the least stable ion amongst the following.
(A) Li^- (B) Be^- (C) B^- (D) C^- **(I.I.T. 2002)**

79. In which of the following triads the melting point of the elements does not increase with increase in atomic number?
(A) Si, P, S (B) He, Li, Be (C) Na, Mg, Al (D) Be, B, C

80. In which of the following triads, the atomic radius of the elements is almost same?
(A) Cr, Mn, Fe (B) Mg, Ca, Sr (C) B, C, N (D) Ga, Ge, Si

81. Which of the following statements is incorrect?
(A) Mg has higher IE_1 than that of Al.
(B) Na has higher IE_1 than that of Al.
(C) K has higher IE_1 than that of Cs.
(D) Li has higher IE_1 than that of Na.

82. In which of the following triads, the elements exhibit both positive and negative oxidation states?
(A) H, O, F (B) H, Cl, Br (C) He, Li, Be (D) Na, Mg, Al

83. Identify the block in the periodic table in which the elements have similar electronegativities.
(A) *s*-block (B) *p*-block (C) *d*-block (D) *f*-block

84. Out of the four elements belonging to four different groups, which element can form + 3 oxidation state with electronic configuration of a noble gas?
(A) In (B) Sr (C) Sb (D) Y

85. The weakest base among the following is
(A) $Mg(OH)_2$ (B) $Zn(OH)_2$ (C) NaOH (D) RbOH

86. Identify the electronic configuration of an element which has the highest electron affinity.
(A) $1s^22s^22p^4$ (B) $1s^22s^22p^5$ (C) $1s^22s^22p^63s^1$ (D) $1s^22s^22p^63s^2$

87. The ionisation energy of an element A(*g*) is *I* and the electron affinity of $A^+(g)$ is *E*. *I* and *E* are related as
(A) $IE = 1$ (B) $IE = -1$ (C) $I = E$ (D) $I = -E$

88. Which of the following sets exhibits inert pair effect?
(A) He, Ne, Ar (B) Ge, Sn, Pb (C) B, Al, Ga (D) Fe, Co, Ni

ANSWERS

1. B	2. B	3. D	4. C	5. B	6. A
7. A	8. C	9. B, C, D	10. B	11. A, B	12. A, C, D
13. D	14. D	15. A	16. C	17. B	18. A
19. A	20. C	21. C	22. A	23. A	24. A
25. A	26. A	27. D	28. C	29. B	30. C
31. B	32. D	33. D	34. B	35. B	36. B
37. D	38. B	39. D	40. D	41. B	42. B
43. C	44. A	45. D	46. A	47. B	48. C
49. B	50. D	51. C	52. B	53. C	54. B
55. C	56. B	57. B	58. A	59. C	60. C
61. C	62. D	63. A	64. C	65. A	66. B
67. D	68. A	69. C	70. A	71. A	72. C
73. C	74. C	75. B	76. C	77. B	78. B
79. A	80. A, D	81. A	82. B	83. C	84. D
85. B	86. B	87. D	88. B		

16

Hydrogen and Its Compounds

OBJECTIVE TYPE MULTIPLE CHOICE QUESTIONS

1. Which of the following pairs will not form hydrogen gas?
   (A) Cu + dil. HCl (B) Fe + H_2SO_4 (C) Mg + steam (D) Na + alcohol **(I.I.T. 1994)**
2. The maximum possible number of hydrogen bonds, a water molecule can form is
   (A) 1 (B) 2 (C) 3 (D) 4 **(I.I.T. 1992)**
3. The sum of protons, electrons and neutrons in the heaviest isotope of hydrogen is
   (A) 6 (B) 5 (C) 4 (D) 3 **(I.I.T. 1986)**
4. Hydrogen will not reduce
   (A) heated CuO (B) heated Fe_2O_3 (C) heated SnO_2 (D) heated Al_2O_3 **(I.I.T. 1985)**
5. The adsorption of hydrogen by palladium is commonly known as
   (A) occlusion (B) chemisorption
   (C) physical adsorption (D) Langmuir's adsorption **(I.I.T. 1983)**
6. Heavy water is
   (A) $H_2{}^{18}O$ (B) D_2O
   (C) water at 4°C (D) water obtained by repeated distillation.
   (I.I.T. 1983)
7. The mass of hydrogen atom is
   (A) 1.66×10^{-27} kg (B) 1.066×10^{-27} kg
   (C) 1.606×10^{-27} kg (D) 1.66×10^{-28} kg
8. The colour of hydrogen is
   (A) black (B) yellow (C) orange (D) red
   (E) none of these.
   (I.I.T. 1981)
9. Which of the following metals when treated with cold dilute HNO_3 generates H_2?
   (A) Mn (B) Al (C) Fe (D) Cu **(I.I.T. 1980)**

10. When the same amount of Zn is treated separately with excess of H_2SO_4 and excess of NaOH, the ratio of volumes of H_2 evolved is
    (A) 1 : 1 (B) 1 : 2 (C) 2 : 1 (D) 9 : 4 **(I.I.T. 1980)**
11. The adsorption of H_2 by platinum is known as
    (A) reduction (B) hydrogenation
    (C) dehydrogenation (D) occlusion. **(Roorkee 1989)**
12. The metal which displaces hydrogen from a boiling NaOH solution is
    (A) As (B) Zn (C) Mg (D) Fe **(Roorkee 1989)**
13. Hydrogen does not combine with
    (A) Sb (B) Na (C) Bi (D) He **(Roorkee 1989)**
14. Hydrogen behaves as an oxidising agent in its reaction with
    (A) chlorine (B) sodium (C) nitrogen (D) sulphur
15. Which of the following will not displace hydrogen?
    (A) Sn (B) Ba (C) Pb (D) Hg
16. The property of hydrogen which distinguishes it from alkali metals is its
    (A) electropositive character (B) non-metallic character
    (C) reducing character (D) affinity for non-metal.
17. In ortho hydrogen
    (A) electrons of the two atoms spin in opposite directions
    (B) electrons of both the atoms spin in same direction
    (C) nuclei of the two atoms spin in opposite directions
    (D) nuclei of both the atoms spin in same direction.
18. The hydride ion is a stronger base than the hydroxide ion. Which of the reactions will occur if NaH is dissolved in water
    (A) $H^-(aq) + H_2O(l) \longrightarrow OH^- + H_2$ (B) $H^- + H_2O \longrightarrow$ no-reaction
    (C) $H^+(aq) + H_2O(l) \longrightarrow H_3O^+$ (D) none of these.
19. The metal which gives H_2 on treatment with acid as well as NaOH is
    (A) Cu (B) Zn (C) Fe (D) none of these.
20. Which of the following is correct for hydrogen?
    (A) It has a very high IE (B) It has oxidation number + 1 and – 1
    (C) It is always collected at anode (D) It has the same IE as alkali metals.
21. Hydrogen at room temperature contains
    (A) 25% para + 75% ortho (B) 25% ortho + 75% para
    (C) 50% ortho + 50% para (D) 66% ortho + 34% para
22. Identify the incorrect statement.
    (A) Mass of H-atom is 1.66×10^{-27} kg.
    (B) Cu dissolves in H_2SO_4 and liberates H_2.
    (C) H_2 is liberated by the action of Al with conc. NaOH.
    (D) CO combines with H_2 when heated to 150-450°C in the presence of Cu + ZnO to give methanol.

23. Identify the incorrect statement.
   (A) Hydrogen exists as H^+ and H^- in different chemical compounds.
   (B) Hydrogen atom is the only species which has no neutron in the nucleus.
   (C) Hydrogen ion consists of one proton and one electron.
   (D) Heavy water is unstable because hydrogen is substituted by its isotope deuterium.

24. Identify the correct statement.
   (A) Protium does not contain a neutron.
   (B) Water is an ionic hydride.
   (C) Tritium is an isomer of hydrogen.
   (D) Para hydrogen has lower energy than ortho hydrogen.

25. When water gas and steam are passed over a catalyst at 500°C, CO is oxidised to CO_2. The catalyst is
   (A) Cr (B) Cr_2O_3
   (C) Fe_2O_3 (D) a mixture of Cr_2O_3 and Fe_2O_3

26. The electrolysis of acidified water between Ni anode and Fe cathode gives H_2 at the cathode and O_2 at the anode. The optimum concentration of acid in water is
   (A) 1% (B) 5% (C) 12% (D) 15%

27. An acidified solution of $KMnO_4$ undergoes reduction with
   (A) H_2 (B) Zn–Cu couple and H_2O
   (C) Na + absolute alcohol (D) Zn/H_2SO_4

28. Hydrogen directly combines with
   (A) Au (B) Ni (C) Cu (D) Ca **(Roorkee 1990)**

29. Which of the following evolves hydrogen on reacting with cold dilute HNO_3?
   (A) Mg (B) Al (C) Fe (D) Cu

30. Identify the incorrect statement.
   (A) Ortho and para hydrogen do not resemble in their chemical properties
   (B) At absolute zero only ortho hydrogen exists
   (C) Pure ortho hydrogen has not been isolated
   (D) The reaction of atomic hydrogen with $[Cu(CH_3COO)_2 \cdot H_2O]_2$ leads to the formation of metallic Cu.

31. The first ionisation energy of hydrogen is
   (A) 485 kJ (B) 520 kJ (C) 1312 kJ (D) 1680 kJ

32. On treatment of hard water with zeolite, sodium ions of zeolite are exchanged with
   (A) H^+ ions (B) OH^- ions (C) Mg^{2+} ions (D) Ca^{2+} ions

33. The relative abundance of deuterium in hydrogen is
   (A) 0.016% (B) 0.1% (C) 0.4% (D) 1%

34. Which of the following, heated oxides are reduced by hydrogen?
   (A) SnO_2 (B) Al_2O_3 (C) Fe_2O_3 (D) CuO

35. In which of the following, hydrogen is obtained as a by-product?
   (A) Lane process (B) Bosch process
   (C) Manufacture of NaOH (D) Electrolysis of H_2O

36. Which of the following is obtained when hydrogen is passed through an electric arc?
    (A) para hydrogen (B) ortho hydrogen
    (C) nascent hydrogen (D) atomic hydrogen.
37. Which of the following particles are emitted by tritium?
    (A) neutrons (B) γ-rays (C) β-particles (D) α-particles
38. Which one of the following pairs of substances on reaction will not evolve H_2 gas?
    (A) Fe and $H_2SO_4(aq)$ (B) Fe and steam
    (C) Cu and HCl(*aq*) (D) Na and C_2H_5OH
39. Boiling point of heavy water is
    (A) 110°C (B) 101.4°C (C) 100°C (D) 99°C
40. Which of the following groups of ions makes the water hard?
    (A) Na^+, HCO_3^- (B) Mg^{2+}, Cl^- (C) K^+, SO_4^{2-} (D) NH_4^+, Cl^-
41. Interstitial hydride is formed by
    (A) Ca (B) K (C) Be (D) Mg
42. Identify the correct statement.
    (A) Hydrogen exhibits an oxidation number of – 1.
    (B) Hydrogen is electropositive.
    (C) Hydrogen will not be liberated at the anode.
    (D) Hydrogen has the same ionisation energy as sodium.
43. Identify the incorrect statement.
    (A) The reactivity of hydrogen towards halogens is in the order: $Cl_2 > Br_2 > I_2 > F_2$.
    (B) The oxides of hydrogen and deuterium are neutral.
    (C) Molecular hydrogen reduces ZnO and CuO to Zn and Cu respectively.
    (D) Atomic hydrogen reduces $KMnO_4$ to metallic manganese.
44. Atomic hydrogen can reduce $KMnO_4$ to
    (A) Mn(VI) (B) Mn(IV) (C) Mn(II) (D) Mn(0)
45. Atomic hydrogen can reduce $K_2Cr_2O_7$ to
    (A) Cr(VI) (B) Cr(III) (C) Cr(II) (D) Cr(0)
46. The life period of atomic hydrogen is
    (A) 1/3 sec (B) 3 sec (C) 5 sec (D) 10 sec
47. The conjugate acid of H^- is
    (A) OH^- (B) H_3O^+ (C) H_2O (D) H_2
48. Hydrogen cannot be obtained by
    (A) Zn + dil. H_2SO_4 (B) Zn + dil. HNO_3
    (C) Mg + H_2O (D) Mg + dil. H_2SO_4
49. In which of the following reactions, hydrogen is acting as an oxidising agent?
    (A) with Li to give LiH (B) with I_2 to give HI
    (C) with N_2 to give NH_3 (D) with S to give H_2S

50. Which of the following metals adsorb hydrogen?
    (A) Pd (B) Pt (C) Na (D) Zn
51. Which of the following statements is not correct?
    (A) Atomic hydrogen is a strong reducing agent
    (B) Dissociation of hydrogen is an endothermic process
    (C) Atomic hydrogen exists for 10 seconds
    (D) Mg on reaction with dil. HNO_3 liberates H_2.
52. Which of the following hydrides does not decompose to the constituent elements on strong heating?
    (A) LiH (B) NaH (C) CaH_2 (D) MgH_2
53. Which of the following hydrides conducts electricity?
    (A) GeH_4 (B) SiH_4 (C) B_2H_6 (D) NaH
54. Which of the following metals does not form metallic hydride?
    (A) Cu (B) Ti (C) Cr (D) Ca
55. Which of the following metals form non-stoichiometric hydride?
    (A) Ca (B) Cu (C) V (D) La
56. The conversion of atomic hydrogen to molecular hydrogen is a/an
    (A) photochemical change (B) nuclear change
    (C) exothermic change (D) endothermic change.
57. Pure hydrogen is prepared by
    (A) reaction of saline hydrides with water
    (B) mixing water gas with steam in the presence of Fe_2O_3 at 723 K
    (C) passing steam over red hot coke (D) electrolysis of acidulated water
58. In which of the following hydrides, hydrogen exists in negative oxidation state?
    (A) HCl (B) NaH (C) CaH_2 (D) HBr
59. H^+, D^+ and T^+ differ from each other except in
    (A) number of neutrons (B) number of electrons
    (C) ionic mass (D) all are correct.
60. Hydrogen peroxide ionises as: $H_2O_2 \rightleftharpoons H^+ + HO_2^-$. If pH of H_2O_2 at 25°C is 5.91, the ionic product (auto-protolysis constant) of H_2O_2 is
    (A) 1.51×10^{-12} (B) 1.15×10^{-12} (C) 1×10^{-14} (D) 1.1×10^{-12}
    Hint: $[H^+] = 10^{-5.91} = 1.23 \times 10^{-6}\ M$; $[HO_2^-] = 1.23 \times 10^{-6}\ M$; $K = [H^+][HO_2^-] = (1.23 \times 10^{-6})^2 = 1.51 \times 10^{-12}$

ANSWERS

1. A	2. D	3. C	4. D	5. A	6. B
7. A	8. E	9. A	10. A	11. D	12. B
13. D	14. B	15. D	16. B	17. D	18. A
19. B	20. B	21. A	22. B	23. C, D	24. A

25. D	26. D	27. B, C, D	28. D	29. A	30. A, B
31. C	32. C, D	33. A	34. A, C, D	35. C	36. D
37. C	38. C	39. B	40. B	41. C, D	42. A, B
43. A, D	44. C	45. B	46. A	47. D	48. B
49. A	50. C, D	51. C	52. A	53. D	54. D
55. C, D	56. C	57. D	58. B, C	59. B	60. A

17

Group 1 Elements: Alkali Metals

OBJECTIVE TYPE MULTIPLE CHOICE QUESTIONS

1. Assertion (*A*): LiCl is predominantly a covalent compound.
   Reason (*R*): Electronegativity difference between Li and Cl is too small.
   (A) Both *A* and *R* are true and *R* is the correct explanation of *A*
   (B) Both *A* and *R* are true and *R* is not the correct explanation of *A*
   (C) *A* is true but *R* is false
   (D) *A* is not true but *R* is correct. **(I.I.T. 1998)**
2. Which of the following statements is correct for $CsBr_3$?
   (A) It is a covalent compound
   (B) It contains Cs^+, Br^- and lattice Br_2 molecule
   (C) It contains Cs^{3+} and Br^-
   (D) It contains Cs^+ and Br_3^- **(I.I.T. 1996)**
3. KF combines with HF to form KHF_2. The compound contains the species
   (A) K^+, F^- and H^+
   (B) K^+, F^- and HF
   (C) K^+ and $[HF_2]^-$
   (D) $[KHF]^+$ and F^- **(I.I.T. 1996)**
4. The following compounds have been arranged in order of their increasing thermal stabilities. Identify the order. K_2CO_3(I), $MgCO_3$(II), $CaCO_3$(III), $BeCO_3$(IV)
   (A) I < II < III < IV
   (B) IV < II < III < I
   (C) IV < II < I < III
   (D) II < IV < III < I **(I.I.T. 1996)**
5. Among KO_2, AlO_2^-, BaO_2 and NO_2^+, the unpaired electron is present in
   (A) KO_2 and AlO_2^-
   (B) NO_2^+ and BaO_2
   (C) KO_2
   (D) BaO_2 **(I.I.T. 1997)**
6. Sodium thiosulphate is prepared by
   (A) reducing Na_2SO_4 solution with H_2S
   (B) neutralising $H_2S_2O_3$ solution with NaOH
   (C) boiling Na_2SO_3 solution with S in alkaline medium
   (D) boiling Na_2SO_3 solution in acidic medium. **(I.I.T. 1996)**

7. Sodium thiosulphate is used in photography because of its
   (A) reducing property (B) oxidising property
   (C) complex forming property (D) reaction with light. **(I.I.T. 1981)**
8. The increasing order of basic character in the oxides MgO, SrO, K_2O, NiO and Cs_2O is
   (A) NiO < MgO < SrO < K_2O < Cs_2O (B) NiO < MgO < SrO < Cs_2O < K_2O
   (C) NiO < MgO < K_2O < Cs_2O < SrO (D) NiO < MgO < K_2O < Cs_2O < SrO
   (I.I.T. 1991)
9. An aqueous solution of sodium sulphate is electrolysed using inert electrodes. The products at the cathode and anode are respectively
   (A) H_2, O_2 (B) O_2, H_2 (C) O_2, Na (D) O_2, SO_2 **(I.I.T. 1987)**
10. Metallic lustre exhibited by sodium is explained by
    (A) diffusion of sodium ions (B) oscillation of loose electrons
    (C) existence of body centred cubic lattice
    (D) excitation of free protons. **(I.I.T. 1987)**
11. The pair of compounds which cannot exist together in solution is
    (A) $NaHCO_3$, NaOH (B) $NaHCO_3$, H_2O
    (C) $NaHCO_3$, Na_2CO_3 (D) $NaHCO_3$, NaCl **(I.I.T. 1986)**
12. Solid sodium chloride is
    (A) ionic (B) covalent (C) non-conductor (D) brittle. **(I.I.T. 1986)**
13. Which of the following elements has the lowest first ionisation energy?
    (A) Mg (B) Al (C) Si (D) Na **(I.I.T. 1985)**
14. The oxide that gives H_2O_2 on treatment with dilute acid is
    (A) PbO_2 (B) Na_2O_2 (C) MnO_2 (D) TiO_2 **(I.I.T. 1985)**
15. Molten sodium chloride conducts electricity due to the presence of
    (A) free electron (B) free ions
    (C) free molecules (D) atoms of sodium and chlorine. **(I.I.T. 1981)**
16. The solution of sodium metal in liquid ammonia is strongly reducing due to the presence of
    (A) sodium atoms (B) solvated electrons
    (C) sodium amide (D) NaOH **(I.I.T. 1981)**
17. Which of the following is purely ionic?
    (A) Sodium chloride (B) Beryllium chloride
    (C) Lithium chloride (D) CCl_4 **(I.I.T. 1980)**
18. Sodium reacts with water more vigorously than Li because
    (A) Na is more electropositive (B) Na is more electronegative
    (C) atomic mass of Na is higher than that of Li
    (D) sodium is a metal. **(I.I.T. 1981)**
19. Molecular formula of Glauber's salt is
    (A) $MgSO_4 \cdot 7H_2O$ (B) $ZnSO_4 \cdot 7H_2O$ (C) $FeSO_4 \cdot 7H_2O$ (D) $Na_2SO_4 \cdot 10H_2O$
    (I.I.T. 1985)

20. The most ionic compound out of the following is
(A) KCl (B) $CaCl_2$ (C) $MgCl_2$ (D) NaCl **(Roorkee 1996)**

21. The alkali metal that reacts with nitrogen directly to form nitride is
(A) Li (B) Na (C) Rb (D) K **(Roorkee 1992)**

22. The bond length in LiF is
(A) less than that of NaF (B) equal to that of KF
(C) more than that of KF (D) more than that of NaF

23. The atomic radii of alkali metals (M) lie in the order Li < Na < K < Rb, but the ionic radii of M^+ ions in aqueous solution lie in the reverse order: $Li^+ > Na^+ > K^+ > Rb^+$. The reason is
(A) increasing weakness of metallic bond
(B) gradual increase in ionisation energy
(C) decreasing degree of hydration (D) increasing electrophilic character.

24. Which alkali metal ion in aqueous solution is the best conductor of electricity?
(A) Li^+ (B) Cs^+ (C) Na^+ (D) K^+

25. Na_2SO_4 is soluble in water but $BaSO_4$ is sparingly soluble because
(A) the hydration energy of Na_2SO_4 is more than its lattice energy
(B) the lattice energy of $BaSO_4$ is less than its hydration energy
(C) the hydration energy of Na_2SO_4 is less than its lattice energy
(D) the lattice energy of Na_2SO_4 is more than that of $BaSO_4$.

26. In the reactions: $NaCl + CO_2 + NH_3 + H_2O \longrightarrow X + NH_4Cl$; $2X \xrightarrow{\text{heat}} Y + CO_2 + H_2O$, the products X and Y are respectively
(A) NaOH, $NaHCO_3$ (B) $NaHCO_3$, Na_2CO_3
(C) Na_2CO_3, NaOH (D) $NaNH_2$, NaOH

27. Sodium hydroxide reacts slowest with
(A) I_2 (B) Br_2 (C) Cl_2 (D) F_2

28. The thermal stability of alkali metal chlorides is in the order
(A) LiCl > KCl > NaCl > CsCl (B) CsCl > KCl > NaCl > LiCl
(C) NaCl > KCl > LiCl > CsCl (D) NaCl > KCl > CsCl > LiCl

29. The raw materials used in the manufacture of Na_2CO_3 by Le Blanc process are
(A) NaCl, H_2SO_4, limestone and coke
(B) NaCl, HCl, limestone and coke
(C) $NaNO_3$, H_2SO_4, limestone and coke
(D) HCl, limestone, ammonia and coke.

30. When chlorine is passed through hot and conc. KOH solution, the compound formed is
(A) KClO (B) $KClO_2$ (C) $KClO_3$ (D) $KClO_4$

31. Which of the following compounds does not show colouration with phenolphthalein?
(A) Na_2CO_3 (B) $NaHCO_3$ (C) K_2CO_3 (D) NH_4OH

32. Which of the following salts is weakly basic in aqueous solution?
(A) $NaHCO_3$ (B) NaCl (C) NH_4HCO_3 (D) $NaHSO_4$

33. Potassium reacts with water more vigorously than lithium because it
(A) has a higher nuclear charge (B) has a larger size and greater surface area
(C) is more electronegative (D) is more electropositive

34. In the potassium atom, the energy of the orbitals is in the order
(A) $4s < 3d$ (B) $4s > 3d$ (C) $4s > 4p$ (D) $4s < 2p$ **(Roorkee 1996)**

35. The violet flame shown by potassium in Bunsen flame is due to jumping of the electron from
(A) $1s$ to $4p$ (B) $1s$ to $5p$ (C) $4p$ to $4s$ (D) $5p$ to $4s$

36. The yellow flame shown by sodium in Bunsen flame is due to jumping of the electron from
(A) $3p$ to $3s$ (B) $3p$ to $1s$ (C) $2p$ to $1s$ (D) $1s$ to $3p$

37. When sodium metal is dissolved in liquid ammonia, a blue colour is formed. The blue colour is due to the electronic transition
(A) $1s \longrightarrow 2p$(N) (B) $1s \longrightarrow 3p$(N) (C) $1s \longrightarrow 3s$(N) (D) $3s \longrightarrow 3p$(N)

38. The alkali metals react with liquid ammonia in the presence of catalyst to form metal amides and hydrogen. The catalyst is
(A) Pt (B) Pd (C) Ni (D) Fe_2O_3

39. Which of the following metals forms amide when ammonia is passed over it at 300°C?
(A) Mg (B) Al (C) Pb (D) K

40. The ionisation energies of the alkali metals are in the order: Cs < Rb < K < Na < Li. But Li is the strongest reducing agent. This is due to
(A) small size and low heat of hydration
(B) small size and high heat of hydration
(C) small size and high ionisation energy
(D) none of these.

41. Which of the following alkali metals does not form alums?
(A) Rb (B) K (C) Na (D) Li

42. Which of the following metals is used in the photoelectric cell?
(A) Si (B) Mg (C) Li (D) Cs

43. The electrolysis of molten NaH gives the products at the anode and cathode respectively
(A) H_2 and O_2 (B) H_2 and Na (C) Na and H_2 (D) O_2 and H_2

44. Which of the following compounds decomposes to M_2O on heating?
(A) NaOH (B) KOH (C) RbOH (D) LiOH

45. When sodium is heated with excess of O_2, which of the following is formed?
(A) NaO (B) NaO_2 (C) Na_2O (D) Na_2O_2

46. Which of the following fluorides has the highest lattice energy?
(A) NaF (B) KF (C) CsF (D) LiF

47. LiF is less soluble in water than KF because
(A) LiF is more covalent than KF (B) LiF has higher lattice energy than KF

(C) Li^+ ions are not extensively hydrated as K^+

(D) LiF has higher enthalpy of hydration than KF.

48. When NaCl is electrolysed in Nelson cell, hydrogen is liberated at the cathode and not sodium because

(A) H^+ is smaller than Na^+ (B) H^+ does not react with H_2O

(C) H^+ has lower electrode potential (D) H^+ has greater ionic mobility than Na^+.

49. In the disproportionation reaction of chlorine with NaOH solution, the main reaction products are

(A) NaCl, NaClO (B) NaCl, $NaClO_2$ (C) NaCl, $NaClO_3$ (D) NaCl, $NaClO_4$

50. Identify the false statement.

(A) Glauber's salt is $Na_2SO_4{\cdot}10H_2O$ (B) Washing soda is $Na_2CO_3{\cdot}H_2O$

(C) Sodium forms sodium superoxide on heating in large excess of oxygen.

(D) Sodium ethoxide is prepared by reacting ethanol with aqueous NaOH.

51. Identify the false statement.

(A) Solid NaOH is deliquescent (B) $^{23}_{11}Na$ and $^{23}_{12}Mg$ are isobars.

(C) $NaHCO_3$ is used in foam type fire extinguisher.

(D) The phenomenon of crumbling to powder of $Na_2CO_3{\cdot}10H_2O$ crystals on exposure to air is known as deliquescence.

52. Identify the false statement.

(A) Li is a stronger reducing agent than the other Group 1 metals.

(B) Sodium metal when burnt in excess of oxygen gives only Na_2O.

(C) The order of covalent character in the lithium halides is: LiCl > LiBr > LiI.

(D) The Group 1 metals are obtained by the electrolysis of their molten salts and not by the electrolysis of the aqueous solutions of their salts.

53. Identify the false statement.

(A) LiI is more covalent than LiF.

(B) The softness of Group 1 metals decreases down the group.

(C) Metallic sodium is used for drying diethyl ether but not ethanol.

(D) Both Na_2CO_3 and K_2CO_3 are obtained by Solvay process.

54. Identify the false statement.

(A) LiI is more soluble in ethanol than KI.

(B) The transition metals are less electropositive than Group 1 metals.

(C) The thermal stability of alkali metal hydrides increases with the increasing size of cations.

(D) $LiNO_3$ on reaction with D_2O forms ND_3 and LiOD.

55. K_2CO_3 is not obtained by a method similar to Solvay method because

(A) KCl is not available in pure form

(B) $KHCO_3$ is not stable in aqueous solution

(C) $KHCO_3$ is less soluble in water than $NaHCO_3$

(D) $KHCO_3$ is more soluble in water than $NaHCO_3$.

56. When a solution of $AgNO_3$ is added to a concentrated solution of $Na_2S_2O_3$, which of the following is formed?
(A) Black precipitate of Ag (B) Black precipitate of Ag_2S
(C) A soluble complex $Na_3[Ag(S_2O_3)_2]$ (D) White precipitate of $Ag_2S_2O_3$
57. Microcosmic salt is
(A) $Na(NH_4)HPO_4 \cdot 4H_2O$ (B) $Na(NH_4)HPO_4 \cdot H_2O$
(C) $Na(NH_3)HPO_4 \cdot 4H_2O$ (D) $K(NH_4)HPO_3 \cdot 2H_2O$
58. Table salt absorbs water because
(A) it is mixed with KI (iodised salt) which is hygroscopic
(B) it is contaminated with KCl which is hygroscopic
(C) NaCl is hygroscopic in nature
(D) it is contaminated with $MgCl_2$ which is hygroscopic.
59. Potassium metal is commercially prepared by the reduction of molten KCl with sodium at 850°C. This method is based upon the principle
(A) sodium and potassium form an alloy at this temperature
(B) sodium prefers to bind to chloride ions in preference to potassium ions
(C) potassium being more volatile distils off thus shifting the reaction in the forward direction
(D) sodium is more reactive than potassium at this temperature.
60. Bromine reacts with hot and conc. NaOH solution to produce
(A) $NaOBr + NaBr + H_2O$ (B) $NaBr + NaBrO_3 + H_2O$
(C) $NaBr + H_2O + O_2$ (D) $NaBrO_3 + H_2O + O_2$
61. Which is paramagnetic?
(A) KO_2 (B) Na_2O (C) Na_2O_2 (D) H_2O
62. The raw materials used in the manufacture of Na_2CO_3 by Solvay process are
(A) sodium chloride, limestone and ammonia
(B) sodium chloride, limestone and carbon dioxide
(C) sodium chloride and carbon dioxide
(D) ammonia and carbon dioxide.
63. Which of the following is used as an air purifier in spacecraft?
(A) Quicklime (B) Slacked lime
(C) Potassium superoxide (D) Anhydrous $CaCl_2$
64. In certain matters lithium differs from other alkali metals. The main reason for this is
(A) small size of Li atom and Li^+ ion (B) extremely high electronegativity of Li
(C) greater hardness of Li (D) hydration of Li^+ ion.
65. Lithium and sodium are placed in dry air. Which of the following compounds will be formed?
(A) Li_2O, Li_3N, Na_2O (B) Li_2O, Li_3N, Na_2O, NH_3
(C) Li_2O, Na_2O (D) Li_2O, Na_2O, NaOH

66. There occurs loss of mass when a mixture of Li_2CO_3 and $Na_2CO_3 \cdot 10H_2O$ is heated strongly. The loss of mass is due to decomposition of
   (A) $Na_2CO_3.10H_2O$ (B) Li_2CO_3
   (C) both (A) and (B) (D) none of these.
67. Which of the following statements is incorrect?
   (A) Sodium benzoate is used as a food preservative.
   (B) Sodium hydrogen carbonate is used as an antacid.
   (C) NaOH is a stronger base than KOH.
   (D) Sodium on exposure to air gets tarnished due to the formation of a layer of carbonate and oxide.
68. Which of the following statements is incorrect?
   (A) Sodium burns in dry air to form Na_2O and Na_2O_2
   (B) Na_2CO_3 decomposes readily on heating.
   (C) $NaNO_3$ decomposes to $NaNO_2$ on heating.
   (D) Electrolysis of molten NaH liberates H_2 at the anode.
69. Which of the following nitrates give NO_2 on heating?
   (A) $NaNO_3$ (B) $Pb(NO_3)_2$ (C) KNO_3 (D) $LiNO_3$
70. When potassium is heated with excess of O_2, which of the following is formed?
   (A) KO (B) K_2O (C) KO_2 (D) K_2O_2
71. Which of the following metals does not react with nitrogen?
   (A) Ca (B) Mg (C) Li (D) K
72. When sodium is heated in a current of dry NH_3, which of the following is formed?
   (A) Sodium azide (B) Sodium amide
   (C) Sodium nitrate (D) Sodium nitrite
73. Sodalime is made from
   (A) CaO + NaOH(*aq*) (B) $CaCO_3$ + NaOH(*aq*)
   (C) CaO + Na + H_2O (D) CaO + $NaHCO_3$
74. During electrolysis of aqueous NaCl, which of the following reactions occurs at cathode?
   (A) Water is reduced. (B) Water is oxidised.
   (C) Na^+ is reduced. (D) Cl^- is oxidised.
75. Which of the following does not co-exist in solid state?
   (A) $NaHCO_3$ (B) $NaHSO_3$ (C) $LiHCO_3$ (D) $CaCO_3$
76. Which of the following is formed when sulphur is reacted with conc. NaOH?
   (A) Na_2S_2 (B) $Na_2S_2O_3$ (C) Na_2SO_4 (D) Na_2SO_3
77. Sodium oxide cannot be obtained by heating which of the following?
   (A) NaOCl (B) $NaHCO_3$ (C) $NaNO_3$ (D) all of these.
78. Which of the following statements is not correct for lithium compounds?
   (A) LiCl is non-conducting in molten state.
   (B) Li_3N is a stable compound. (C) LiCl is predominantly covalent.
   (D) Li_2CO_3 decomposes into Li_2O while other Group 1 carbonates are thermally stable.

79. Statement–1: Alkali metals dissolve in liquid ammonia to give blue solution.
because
Statement–2: Alkali metals in liquid ammonia give solvated species of the type $[M(NH_3)_n]^+$ (M = alkali metals).
(A) Statement–1 is true, Statement–2 is true; Statement–2 is a correct explanation for Statement–1
(B) Statement–1 is true, Statement–2 is true; Statement–2 is not a correct explanation for Statement–1
(C) Statement–1 is true, Statement–2 is false
(D) Statement–1 is false, Statement–2 is true. **(I.I.T. 2007)**
Hint: The blue colour is due to solvated electrons.

80. The compound(s) formed upon combination of sodium metal in excess air is(are)
(A) Na_2O_2 (B) Na_2O (C) NaO_2 (D) NaOH **(I.I.T. 2009)**
Hint: Combustion of sodium metal in excess of air yields Na_2O and Na_2O_2.

$$4Na + O_2 \xrightarrow{\Delta} 2Na_2O$$

$$2Na + O_2 \xrightarrow{\Delta} Na_2O_2$$

ANSWERS

1. C	2. D	3. C	4. B	5. C	6. D
7. C	8. A	9. A	10. B	11. A	12. A, C, D
13. D	14. B	15. B	16. B	17. A	18. A
19. D	20. A	21. A	22. A	23. C	24. B
25. A	26. B	27. A	28. B	29. A	30. C
31. B	32. A	33. D	34. A	35. D	36. A
37. A	38. D	39. D	40. B	41. D	42. D
43. B	44. D	45. C, D	46. D	47. B	48. C
49. A	50. B, C, D	51. D	52. B, C	53. D	54. C
55. D	56. B	57. A	58. D	59. B	60. B
61. A	62. A	63. C	64. A	65. A	66. C
67. C	68. B	69. B, D	70. C	71. D	72. B
73. A	74. A	75. C	76. B	77. D	78. A
79. B	80. A, B				

18

Group 2 Elements: Alkaline Earth Metals

OBJECTIVE TYPE MULTIPLE CHOICE QUESTIONS

1. Assertion (*A*): The first ionisation energy of Be is greater than that of B.
   Reason (*R*): 2*p* orbital has lower energy than 2*s*.
   (A) Both *A* and *R* are true and *R* is the correct explanation of *A*
   (B) Both *A* and *R* are true and *R* is not the correct explanation of *A*
   (C) *A* is true and *R* is false (D) Both *A* and *R* are false. **(I.I.T. 2000)**
   Hint: Boron contains one extra electron in 2*p* orbital in comparison to beryllium. The removal of 2*p* electron requires less energy in comparison to the removal of an electron from more stable completely filled 2*s* orbital.
2. Property of the alkaline earth metals that increases with their atomic number is
   (A) ionisation energy (B) solubility of their hydroxides
   (C) solubility of their sulphates (D) electronegativity. **(I.I.T. 1997)**
3. When zeolite, which is hydrated sodium aluminium silicate, is treated with hard water, the sodium ions are exchanged with
   (A) H^+ ions (B) Ca^{2+} ions (C) SO_4^{2-} ions (D) Mg^{2+} ions **(I.I.T. 1999)**
4. The compound insoluble in acetic acid is
   (A) $Ca(OH)_2$ (B) $CaCO_3$ (C) CaC_2O_4 (D) CaO **(I.I.T. 1986)**
5. Which of the following has the smallest ionic radius?
   (A) Mg^{2+} (B) O_2^- (C) Na^+ (D) F^- **(I.I.T. 1985)**
6. Amongst the following hydroxides, the one which has the lowest value of K_{sp} at ~ 25°C is
   (A) $Mg(OH)_2$ (B) $Ca(OH)_2$ (C) $Ba(OH)_2$ (D) $Be(OH)_2$ **(I.I.T. 1990)**
7. The compound which is not a Lewis acid is
   (A) BF_3 (B) $AlCl_3$ (C) $BeCl_2 \cdot 2H_2O$ (D) $SnCl_4$ **(I.I.T. 1985)**

8. The hydration energy of Mg^{2+} is greater than that of
(A) Al^{3+} (B) Na^+ (C) Be^{2+} (D) Mg^{3+} **(I.I.T. 1984)**
9. Calcium is obtained by the
(A) electrolysis of molten $CaCl_2$ (B) electrolysis of an aqueous solution of $CaCl_2$
(C) reduction of $CaCl_2$ with carbon (D) roasting of limestone. **(I.I.T. 1980)**
10. The temporary hardness of water due to calcium bicarbonate is removed by adding
(A) $CaCO_3$ (B) $Ca(OH)_2$ (C) $CaCl_2$ (D) HCl **(I.I.T. 1979)**
11. The chemical formula of gypsum is
(A) $CaSO_3$ (B) $CaSO_4{\cdot}0.5H_2O$ (C) $2CaSO_4{\cdot}H_2O$ (D) $CaSO_4{\cdot}2H_2O$ **(I.I.T. 1978)**
12. A fire work gives out crimson coloured light. It consists of a salt of
(A) Ca (B) Na (C) Sr (D) Ba **(Roorkee 1989)**
13. Which of the following statements is correct?
(A) Gypsum contains a higher percentage of Ca than plaster of Paris.
(B) Plaster of Paris can be obtained by hydration of gypsum.
(C) Gypsum contains lower percentage of Ca than plaster of Paris.
(D) Gypsum is obtained by heating plaster of Paris.
14. A metal is burnt in air and the ash on moistening smells of NH_3. The metal is
(A) Na (B) Fe (C) Al (D) Mg
15. The flame test is not shown by
(A) NaCl (B) $MgCl_2$ (C) $CaCl_2$ (D) KCl **(Roorkee 1996)**
16. Identify the incorrect statement about Be.
(A) Its halides are covalent (B) Its carbonate decomposes on heating
(C) It is easily attacked by water (D) If forms an ionic carbide.
17. The order of increasing ionic character is
(A) $BeCl_2 < MgCl_2 < CaCl_2 < BaCl_2$ (B) $BeCl_2 < MgCl_2 < BaCl_2 < CaCl_2$
(C) $BeCl_2 < BaCl_2 < MgCl_2 < CaCl_2$ (D) $BaCl_2 < CaCl_2 < MgCl_2 < BeCl_2$
18. Portland cement consists of
(A) $CaO{\cdot}SiO_2{\cdot}SO_3{\cdot}Fe_2O_3$ (B) $CaO{\cdot}SO_3{\cdot}MgO{\cdot}Na_2O$
(C) $SiO_2{\cdot}MgO{\cdot}Na_2O{\cdot}SO_3$ (D) $CaO{\cdot}SiO_2{\cdot}Al_2O_3{\cdot}Fe_2O_3{\cdot}MgO{\cdot}SO_3{\cdot}Na_2O$
19. Which of the following is obtained when nitrogen reacts with calcium carbide?
(A) Calcium carbonate (B) Calcium cyanamide
(C) Calcium acetate (D) Calcium cyanate
20. A metal salt solution forms a yellow precipitate with K_2CrO_4 in CH_3COOH, a white precipitate with dil. H_2SO_4, but gives no precipitate with NaCl or NaI. The white precipitate obtained when Na_2CO_3 is added to the metal salt solution will consist of
(A) $PbCO_3$ (B) $SrCO_3$ (C) $BaCO_3$ (D) basic lead carbonate
21. Which of the following statements is false regarding saline hydrides?
(A) They are used as reducing agents
(B) They are covalent in nature (C) In the molten state they conduct electricity
(D) They dissolve in water giving H_2

22. Which of the following has a greater affinity for oxygen?
    (A) B (B) Na (C) K (D) Mg
23. Identify the false statements.
    (A) Ca is less reactive than Mg.
    (B) Ca and Mg can be obtained by chemical reduction of their compounds.
    (C) Mg can be burnt in the atmosphere of SO_2 and CO_2.
    (D) Alkaline earth metals are denser and harder than alkali metals.
24. Identify the false statement.
    (A) M.P. and B.P. of Ca are lesser than those of Mg.
    (B) Ca is heavier than Mg.
    (C) Magnesium is used as a reducing agent and deoxidiser.
    (D) Metallic bond in alkaline earth metals is stronger than that in alkali metals.
25. Which of the following statements is not correct?
    (A) IE of Be is larger than that of Mg.
    (B) Be shows both covalency and electrovalency.
    (C) Heating of $MgCl_2 \cdot 6H_2O$ gives MgO.
    (D) Anhydrous $MgCl_2$ is obtained by heating the hydrated salt with HCl gas.
26. Which of the following statements is not correct?
    (A) $Ca(OH)_2$ is more soluble in water than $Mg(OH)_2$.
    (B) $Be(OH)_2$ is a stronger base than $Ba(OH)_2$.
    (C) The order of ionic character of the following chlorides is:
    $BeCl_2 < MgCl_2 < CaCl_2 < BaCl_2 < SrCl_2$
    (D) The order of basic character of the following oxides is:
    $NiO < MgO < SrO < K_2O < Cs_2O$
27. Which of the following statements is not correct?
    (A) In the electrolysis of CaH_2, hydrogen is liberated at the cathode.
    (B) $MgCl_2 \cdot 6H_2O$ on heating gives anhydrous $MgCl_2$.
    (C) NaCl is added during electrolysis of anhydrous $MgCl_2$.
    (D) MgO is used for the lining of furnace for steel making.
28. Which of the following statements is not correct?
    (A) Alkaline earth metal salts are diamagnetic.
    (B) An aqueous solution of $BeCl_2$ is acidic.
    (C) Plaster of Paris is made by heating gypsum to 100°C and its formula is $CaSO_4 \cdot H_2O$
    (D) Anhydrous $MgCl_2$ is obtained by heating the hydrated salt with NH_4Cl.
29. Mg reacts with dil. HNO_3 to give
    (A) N_2O (B) NO (C) NO_2 (D) NH_4^+
30. Which of the following compounds will give anhydrous compounds on heating?
    (A) $SnCl_2 \cdot 2H_2O$ (B) $AlCl_3 \cdot H_2O$ (C) $CaCl_2 \cdot 2H_2O$ (D) $MgCl_2 \cdot 2H_2O$
31. Which of the following is present in highest percentage in portland cement?
    (A) Al_2O_3 (B) SiO_2 (C) CaO (D) MgO

32. Sorel's cement has the composition
(A) $MgCO_3 \cdot 5MgO \cdot nH_2O$ (B) $MgCO_3 \cdot 5H_2O$
(C) $MgCl_2 \cdot 5H_2O$ (D) $MgCl_2 \cdot 5MgO \cdot nH_2O$

33. Epsom salt has the composition
(A) $MgSO_4 \cdot 7H_2O$ (B) $Mg(OH)_2$ (C) $2CaSO_4 \cdot H_2O$ (D) $BaSO_4$

34. Dow metal is an alloy of
(A) Al, Mg, Zn (B) Cu, Zn (C) Mg, Zn (D) Al, Zn

35. Magnalium is
(A) an alloy of Mg and Al (B) C_2H_5MgI
(C) Mg_3N_2 (D) $Mg(HCO_3)_2$

36. Superphosphate of lime has the composition
(A) $Ca_3(PO_4)_2$ (B) $Ca(H_2PO_4)_2$ (C) $CaHPO_4$ (D) $H_3PO_4 + Ca_3(PO_4)_2$

37. Electrolysis of $KCl \cdot MgCl_2 \cdot 6H_2O$ gives
(A) Mg only (B) K only (C) K and Mg (D) Mg and Cl_2

38. The highly effective method of obtaining Be is
(A) reduction of $BeCl_2$ with Mg (B) reduction of BeO with carbon
(C) electrolysis of fused $BeCl_2$ (D) dissociation of Be_2C

39. Identify the false statement.
(A) $Ba(OH)_2$ is more soluble in water than $Mg(OH)_2$.
(B) $BaCO_3$ melts at a higher temperature than $CaCO_3$.
(C) $Be(OH)_2$ is more basic than $Ba(OH)_2$.
(D) Sr decomposes water readily than Be.

40. Identify the false statement.
(A) The bicarbonates of Ca and Mg are stable crystalline compounds.
(B) $BaCO_3$ is thermally more stable than $MgCO_3$.
(C) The carbonates of Ca and Mg decompose below their M.P.
(D) Milk of magnesia is an aqueous solution of $Mg(OH)_2$.

41. Portland cement contains the following amount of CaO.
(A) 70–80% (B) 50–60% (C) 20–35% (D) 5–15%

42. BaO_2 reacts with ozone to give
(A) Ba_2O_2 (B) BaO (C) $Ba(OH)_2$ (D) Ba

43. Halides of alkaline earth metals form hydrates such as $MgCl_2 \cdot 6H_2O$, $CaCl_2 \cdot 2H_2O$, $BaCl_2 \cdot 2H_2O$ and $SrCl_2 \cdot 2H_2O$. This shows that the halides of Group 2 elements
(A) are hygroscopic in nature (B) act as dehydrating agent
(C) can absorb moisture from air (D) all of these. **(Roorkee 2000)**

44. Which of the following is incorrect?
(A) Mg burns in air releasing dazzling light rich in UV rays.
(B) $CaCl_2 \cdot 2H_2O$ when mixed with ice gives freezing mixture.
(C) Be can form complex due to its very small size.
(D) Mg cannot form complex. **(Roorkee 1998)**

45. Magnesium burns in air to give
(A) MgO (B) $MgCO_3$ (C) Mg_3N_2 (D) $MgO + Mg_3N_2$
46. Which of the following will liberate H_2 by its reaction with HCl?
(A) Cu (B) P (C) Hg (D) Mg
47. An important ore of magnesium is
(A) malachite (B) cassiterite (C) carnallite (D) galena. **(I.I.T. 1993)**
48. Mortar is a mixture of slaked lime and sand in the ratio of
(A) 3 : 1 (B) 1 : 3 (C) 2 : 3 (D) 1 : 4
49. Mortar is a mixture of
(A) slaked lime, sand and water (B) slaked lime, plaster of Paris and water
(C) magnesium chloride, tar and lime (D) lime, portland cement and water.
50. The increasing order of solubility is
(A) $CaCO_3 < KHCO_3 < NaHCO_3$ (B) $NaHCO_3 < KHCO_3 < CaCO_3$
(C) $KHCO_3 < NaHCO_3 < CaCO_3$ (D) $CaCO_3 < NaHCO_3 < KHCO_3$
51. $MgCl_2 \cdot 6H_2O$ when heated gives
(A) magnesium oxychloride (B) magnesium dichloride
(C) magnesium oxide (D) magnesium chloride.
52. The decreasing order of lattice energy of the carbonates of Ca, Mg, Sr and Ba is
(A) $SrCO_3 > MgCO_3 > CaCO_3 > BaCO_3$
(B) $BaCO_3 > SrCO_3 > CaCO_3 > MgCO_3$
(C) $CaCO_3 > SrCO_3 > BaCO_3 > MgCO_3$
(D) $MgCO_3 > CaCO_3 > SrCO_3 > BaCO_3$
53. The decreasing order of extent of hydration of the ions Be^{2+}, Mg^{2+}, Ca^{2+}, Sr^{2+} and Ba^{2+} is
(A) $Sr^{2+} > Ba^{2+} > Ca^{2+} > Be^{2+} > Mg^{2+}$ (B) $Ba^{2+} > Sr^{2+} > Ca^{2+} > Mg^{2+} > Be^{2+}$
(C) $Be^{2+} > Mg^{2+} > Ca^{2+} > Sr^{2+} > Ba^{2+}$ (D) $Ca^{2+} > Ba^{2+} > Sr^{2+} > Be^{2+} > Mg^{2+}$
54. For two ionic solids CaO and KI, identify the wrong statement.
(A) The lattice energy of CaO is much higher than that of KI.
(B) KI is more soluble in water. (C) KI has higher melting point.
(D) CaO has higher melting point.
55. Which of the following is amphoteric?
(A) BeO (B) Ag_2O (C) SnO_2 (D) CO_2
56. Which of the following oxides possesses highest melting point?
(A) MgO (B) BeO (C) CaO (D) SrO
57. Which of the following ions exhibits lowest ionic mobility?
(A) Be^{2+} (B) Mg^{2+} (C) Ca^{2+} (D) Sr^{2+}
58. Which of the following metals exhibits lowest melting point?
(A) Be (B) Mg (C) Ca (D) Sr
59. The charge to radius ratio of Be^{2+} to Al^{3+} is respectively
(A) 4.6, 4.5 (B) 6.4, 6.0 (C) 5.4, 5.1 (D) 6.2, 6.1

60. Bleaching powder is obtained by the reaction of chlorine and
(A) dilute solution of $Ca(OH)_2$ (B) concentrated solution of $Ca(OH)_2$
(C) dry slaked lime (D) dry CaO

61. Which of the following ions is most stable?
(A) $[Sr(H_2O)_4]^{2+}$ (B) $[Be(H_2O)_4]^{2+}$ (C) $[Mg(H_2O)_4]^{2+}$ (D) $[Ca(H_2O)_4]^{2+}$

62. Identify the correct statement.
(A) BeH_2 contains three-centre-two-electron bond.
(B) CaH_2, BaH_2 and SrH_2 are ionic.
(C) Beryllium and magnesium hydrides are covalent and polymeric.
(D) All of these.

63. Which of the following statements is correct?
(A) CaF_2 is soluble in water. (B) Ca is more reactive than Mg.
(C) Out of Ca and Mg, only Mg reacts with N_2 to form nitride.
(D) Magnesium can form + 1 oxidation state.
Hint: Electrolysis of acidulated water with Mg electrodes is not in conformity with Faraday's law and indicates the presence of both + 2 and + 1 ions.

64. Which of the following statements is incorrect?
(A) Atomic radius of Na is greater than that of Mg.
(B) Stability of carbonates of Group 2 elements increases down the group.
(C) Solubility of hydroxides of Group 2 elements increases down the group.
(D) Melting and boiling points of Mg are higher than those of Ca.

65. Which of the following statements is incorrect?
(A) Reactivity of Group 2 elements with water increases down the group.
(B) Magnalium is an alloy of Mg.
(C) Magnesium is extracted from sea water by Dow process.
(D) Magnesium is extracted by self-reduction method.

66. Which of the following chlorides of Group 2 elements will form precipitate with a solution of Na_2SO_4?
(A) $BaCl_2$ (B) $CaCl_2$ (C) $MgCl_2$ (D) $BeCl_2$

67. Magnesium is obtained by
(A) electrolysis of fused $MgCl_2$ (B) reduction of $MgCl_2$ with carbon
(C) electrolysis of aqueous solution of $MgCl_2$
(D) roasting of $MgCO_3$

68. Electrolysis of fused $CaCl_2$ and NaCl yields
(A) sodium at cathode (B) calcium at cathode
(C) both Na and Ca at cathode (D) Na at cathode and Cl_2 at anode

69. Carbon cannot be used as a reducing agent in the metallurgy of Mg because
(A) carbon reacts with Mg to form carbides
(B) Mg does not react with carbon
(C) Mg is less electropositive than carbon
(D) carbon is not a strong reducing agent.

70. Magnesium does not readily react with water, although $E^0_{Mg^{2+}/Mg}$ is favourable, because
(A) Mg has higher IE than hydrogen
(B) Mg has low enthalpy of hydration
(C) Mg forms MgO which is reduced to Mg by hydrogen
(D) Mg forms a protective layer of MgO.

71. A sodium salt of an unknown anion when heated with $MgCl_2$ gives a white precipitate only on boiling. The anion is
(A) SO_4^{2-} (B) HCO_3^- (C) CO_3^{2-} (D) NO_3^- **(I.I.T. 2004)**

Hint: $MgCl_2 + 2NaHCO_3 \longrightarrow Mg(HCO_3)_2 + 2NaCl$; $Mg(HCO_3)_2 \xrightarrow{heat} MgCO_3(s) + H_2O + CO_2$

72. Ammoniacal solution of $MgSO_4$ in the presence of NH_4Cl is heated with Na_2HPO_4. A white crystalline precipitate is formed of
(A) $Mg(NH_4)PO_4$ (B) $Mg_3(PO_4)_2$ (C) $MgCl_2 \cdot MgSO_4$ (D) $MgSO_4 \cdot MgPO_4$ **(I.I.T. 2006)**

Hint: $Mg^{2+} + NH_3 + HPO_4^{2-} \longrightarrow Mg(NH_4)PO_4 \downarrow$

73. The reagent(s) used for softening the temporary hardness of water is(are)
(A) $Ca_3(PO_4)_2$ (B) $Ca(OH)_2$ (C) Na_2CO_3 (D) NaOCl **(I.I.T. 2010)**

Hint: $Ca(HCO_3)_2 + Ca(OH)_2 \rightarrow 2CaCO_3 \downarrow + 2H_2O$
(Clarke's method)

$NaOCl + H_2O \rightleftharpoons HOCl + NaOH$

$OH^- + HCO_3^- \rightarrow CO_3^{2-} + H_2O$

$Ca(HCO_3)_2 + Na_2CO_3 \rightarrow CaCO_3 \downarrow + 2NaHCO_3$

ANSWERS

1. C	2. B	3. B, D	4. C	5. A	6. D
7. C	8. B	9. A	10. B	11. D	12. C
13. C	14. D	15. B	16. C	17. A	18. D
19. B	20. C	21. B	22. D	23. A, B	24. A, B
25. B	26. B	27. A	28. C	29. D	30. C
31. C	32. D	33. A	34. A	35. A	36. B
37. D	38. B	39. C	40. A	41. B	42. B
43. D	44. D	45. D	46. D	47. C	48. B
49. A	50. D	51. C	52. D	53. C	54. C
55. A	56. A	57. A	58. B	59. B	60. C
61. B	62. D	63. B, D	64. A, D	65. D	66. A
67. A	68. B	69. A	70. D	71. B	72. A
73. B, C, D					

19

Group 13 Elements: Boron Family

OBJECTIVE TYPE MULTIPLE CHOICE QUESTIONS

1. The process of converting hydrated alumina into anhydrous alumina is called
   (A) smelting (B) dressing (C) calcination (D) roasting **(Roorkee 2000)**
2. In the commercial electrochemical process for Al extraction, the electrolyte used is
   (A) a molten mixture of $AlO(OH) + Al(OH)_3$
   (B) an aqueous solution of $Al_2(SO_4)_3$
   (C) a molten mixture of $Al_2O_3 + Na_3AlF_6$
   (D) $Al(OH)_3$ in NaOH solution. **(I.I.T. 1999)**
3. Assertion (*A*): $Al(OH)_3$ is amphoteric in nature.
   Reason (*R*): Al—O and O—H bonds can be broken with equal ease in $Al(OH)_3$.
   (A) Both *A* and *R* are true and *R* is the correct explanation of *A*
   (B) Both *A* and *R* are true and *R* is not the correct explanation of *A*
   (C) *A* is correct but *R* is false (D) *A* is false but *R* is correct. **(I.I.T. 1998)**
4. Which of the following is most acidic in water?
   (A) $AlCl_3$ (B) $NiCl_2$ (C) $FeCl_3$ (D) $BeCl_2$ **(I.I.T. 1995)**
5. Which of the following minerals does not contain Al?
   (A) Cryolite (B) Mica (C) Fluorspar (D) Feldspar **(I.I.T. 1992)**
6. The reagent NH_4Cl and aqueous NH_3 will precipitate
   (A) Ca^{2+} (B) Bi^{3+} (C) Zn^{2+} (D) Al^{3+} **(I.I.T. 1991)**
7. In the electrolysis of alumina, cryolite is added to
   (A) lower the melting point of alumina
   (B) increase the electrical conductivity
   (C) remove impurities from alumina
   (D) minimise anodic effect. **(I.I.T. 1986)**

8. Hydrogen gas will not reduce
   (A) heated CuO (B) heated Fe_2O_3 (C) heated SnO_2 (D) heated Al_2O_3
   (I.I.T. 1985)
9. Which of the following statements about anhydrous aluminium chloride is correct?
   (A) It exists as $AlCl_3$ molecule (B) It is not easily hydrolysed
   (C) It sublimes at 100°C under vacuum
   (D) It is a strong Lewis base. **(I.I.T. 1981)**
10. The function of fluorspar in the electrolytic reduction of alumina dissolved in fused cryolite is
    (A) as a catalyst
    (B) to lower the temperature of melt and to make the fused mixture very conducting
    (C) to decrease the rate of oxidation of carbon at the anode
    (D) none of these. **(I.I.T. 1993)**
11. Electrolytic reduction of alumina to aluminium by Hall–Heroult process is carried out
    (A) in the presence of NaCl (B) in the presence of fluorite
    (C) in the presence of cryolite which forms a melt with lower melting temperature
    (D) in the presence of cryolite which forms a melt with high melting temperature.
    (I.I.T. 2000)
12. On the addition of mineral acid to an aqueous solution of borax, the compound formed is
    (A) orthoboric acid (B) metaboric acid
    (C) pyroboric acid (D) boron hydride **(Roorkee 1992)**
13. Borax is prepared by treating colemanite with
    (A) Na_2CO_3 (B) $NaNO_3$ (C) $NaHCO_3$ (D) NaCl **(Roorkee 1990)**
14. Boric acid $[B(OH)_3]$ and BF_3 have the same number of electrons. The former is a solid and the latter is a gas. The reason is
    (A) BF_3 is a Lewis acid while $B(OH)_3$ is not
    (B) They have different geometrical isomers
    (C) In BF_3, F^- is smaller in size than OH^- in $B(OH)_3$
    (D) No molecular association is possible in BF_3 while it is possible in $B(OH)_3$.
15. Which of the following is the order of decreasing basic nature of oxides?
    (A) Al_2O_3, MgO, Cl_2O_7, SO_3 (B) Cl_2O_7, SO_3, Al_2O_3, MgO
    (C) SO_2, Cl_2O_7, MgO, Al_2O_3 (D) MgO, Al_2O_3, SO_3, Cl_2O_7
16. When alumina is heated with carbon in the atmosphere of N_2, the products formed are
    (A) Al, CO_2 (B) AlN, CO (C) Al, CO, N_2 (D) Al, CO
17. Al has a great affinity for oxygen and its oxidation is an exothermic process. This fact is made use of in
    (A) preparing thin film of Al (B) preparing duralumin alloy
    (C) thermite welding (D) making utensils.

18. Which of the following is formed when borax is strongly heated?
    (A) $Na_2B_4O_7$ (B) $NaBO_2$ (C) B_2O_3 (D) B_2O_3 and $NaBO_2$
19. Which of the following statements is not true about potash alum?
    (A) Its empirical formula is $KAl(SO_4)_2 \cdot 12H_2O$
    (B) It is used in dyeing industry
    (C) Its aqueous solution is acidic in nature
    (D) On heating it melts in its water of crystallisation.
20. Concentrated aqueous solution of NaOH can separate a mixture of
    (A) Al^{3+} and Fe^{3+} (B) Al^{3+} and Zn^{2+} (C) Zn^{2+} and Pb^{2+} (D) Al^{3+} and Sn^{2+}
21. Aluminium phosphide reacts with water to give
    (A) PH_3 (B) H_3PO_4 (C) $AlPO_4$ (D) phosgene
22. Orthoboric acid on strong heating to red hot gives
    (A) borax (B) B_2O_3
    (C) tetraboric acid (D) metaboric acid
23. Chemical reduction is not suitable for converting
    (A) haematite into Fe (B) ZnO into Zn
    (C) cuprite into Cu (D) bauxite into Al
24. Which of the following statements is correct?
    (A) Anhydrous $AlCl_3$ fumes in air because of hydration.
    (B) Thermite is a mixture of 1 part of powdered Al and 3 parts of appropriate metal oxide (Fe_2O_3, Cr_2O_3 or Mn_3O_4).
    (C) The ionisation energy of B is less than that of Au.
    (D) $p\pi$-$d\pi$ back bonding occurs in the halides of Al.
25. Which of the following statements is incorrect?
    (A) The first ionisation energies of Al and B are lower than that of Be.
    (B) K_2HgI_4 made alkaline with NaOH is used to test ammonia.
    (C) Anhydrous $CuSO_4$ is used for testing water.
    (D) Al is a weaker reducing agent than Zn.
26. Which of the following statements is incorrect?
    (A) Al powder mixed with linseed oil is used as a paint.
    (B) Out of Cu and Al, aluminium is a better conductor of electricity.
    (C) In the extraction of Ca and Mg, aluminium is used as a reducing agent.
    (D) Duralumin covered with a thin coating of Al is known as alcad.
27. Which of the following statements is incorrect?
    (A) Conc. HNO_3 attacks Al liberating NO_2.
    (B) Al dissolves in HCl and not in conc. NaOH.
    (C) The corrosion of Al is prevented by the formation of a coating of its oxide on its surface.
    (D) Dil. H_2SO_4 does not attack Al.

28. Zinc displaces all metals except

    (A) Hg (B) Al (C) Cu (D) Ag

29. Identify the correct statement regarding Al_2Cl_6.

    (A) All bond angles are equal. (B) All bond lengths are equal.

    (C) All bond angles and bond lengths are different.

    (D) The terminal Al—Cl bond length and terminal ∠Cl—Al—Cl are different from those involved in bridged atoms.

30. Identify the incorrect statement.

    (A) At a very high temperature, aluminium chloride exists as a monomer.

    (B) Aluminium chloride exists as a monomer in benzene.

    (C) Fused aluminium chloride conducts electricity.

    (D) Anhydrous aluminium chloride reacts with $NH_3(g)$ to form $AlCl_3 \cdot 6NH_3$.

31. Which of the following is used in artificial gems?

    (A) PbO_2 (B) B_2O_3 (C) SnO_2 (D) Al_2O_3

32. In the purification of Al by Hoope's process, the impurities of Si and Cu are added to molten alumina in order to

    (A) lower the M.P. of melt (B) make the melt heavier

    (C) smooth deposit of Al (D) make the melt conducting.

33. Identify the false statement about boric acid.

    (A) It is a weak tribasic acid.

    (B) It is prepared by acidifying an aqueous solution of borax.

    (C) It has a layer structure in which planar BO_3 units are joined by H-bonds.

    (D) It does not act as proton donor but acts as a Lewis acid by accepting hydroxyl ion.

34. Which of the following statements is correct?

    (A) Both $AlCl_3$ and BCl_3 are not Lewis acids.

    (B) $AlCl_3$ and BCl_3 are both equally strong Lewis acids.

    (C) $AlCl_3$ and BCl_3 are both Lewis acids and $AlCl_3$ is stronger than BCl_3.

    (D) $AlCl_3$ and BCl_3 are both Lewis acids and BCl_3 is stronger than $AlCl_3$.

35. Anhydrous $AlCl_3$ fumes in moist air because

    (A) it is very volatile (B) it is covalent

    (C) HCl is formed in moist air (D) it is highly hygroscopic.

36. Identify the incorrect statement regarding the structure of diborane.

    (A) There are two bridging H-atoms and four terminal H-atoms in diborane.

    (B) Each boron atom forms four bonds in diborane.

    (C) Hydrogen atoms are not in the same plane in diborane.

    (D) All B—H bonds in diborane are similar.

37. The reactivity of borazole is greater than that of benzene because

    (A) borazole is a nonpolar compound (B) borazole is a polar compound

    (C) borazole is an electron deficient compound

    (D) of localised electrons in it.

38. The hybridisation of boron in diborane is
(A) *sp* (B) sp^2 (C) sp^3 (D) sp^3d^2

39. During Serpeck's process silica is eliminated as
(A) Si (vapour) (B) SiO_2 (vapour) (C) $SiO_2(s)$ (D) Si(*s*)

40. Of the following, the metals that cannot be obtained by electrolysis of the aqueous solution of their salts are
(A) Ag (B) Mg (C) Al (D) Cr **(I.I.T. 1990)**

41. In diborane, the banana bond is formed between
(A) 2 electrons, 3 atoms (B) 2 electrons, 1 atom
(C) 2 electrons, 2 atoms (D) 1 electron, 2 atoms.

42. Which of the following statements is not correct for diborane?
(A) It is an electron deficient molecule.
(B) It has a bridged structure. (C) It is paramagnetic.
(D) It has three-centre bond.

43. Purification of aluminium by electrolytic refining is known as
(A) Baeyer's process (B) Hall's process
(C) Hoope's process (D) Serpeck's process.

44. Which of the following compounds are formed when boron trichloride is treated with water?
(A) $H_3BO_3 + HCl$ (B) $B_2H_6 + HCl$ (C) $B_2O_3 + HCl$ (D) none of these.

45. Which of the following statements is incorrect?
(A) The hydroxide of Al is more acidic than that of B.
(B) The hydroxide of B is basic while that of Al is amphoteric.
(C) The hydroxide of B is acidic while that of Al is amphoteric.
(D) The hydroxides of B and Al are amphoteric.

46. Thermit process is used to extract metals
(A) when their oxides cannot be reduced by carbon
(B) when their carbonates do not yield oxides by thermal decomposition
(C) when their sulphides cannot be converted into oxides by roasting
(D) when their melting points are very high.

47. Aluminium is more reactive than iron. But aluminium is less easily corroded than iron because
(A) aluminium is a noble metal
(B) oxygen forms a protective oxide layer on Al
(C) iron undergoes reaction with water easily
(D) iron forms mono and divalent ions.

48. In the extraction of aluminium, the electrolyte is
(A) fused cryolite with feldspar (B) fused cryolite with fluorspar
(C) pure alumina with bauxite and molten cryolite
(D) pure alumina in molten cryolite. **(I.I.T. 1999)**

49. Which of the following metals is used in high temperature thermometry?
   (A) Hg (B) Tl (C) Na (D) Ga
50. Al and Ga have the same covalent radius because of
   (A) greater shielding power of *s*-electrons of Ga atoms
   (B) poor shielding power of *s*-electrons of Ga atoms
   (C) poor shielding power of *d*-electrons of Ga atoms
   (D) greater shielding power of *d*-electrons of Ga atoms.
51. Identify the false statement.
   (A) Boric acid behaves as a strong acid in the presence of catechol.
   (B) Boron forms a stable ion in $[BF_6]^{3-}$.
   (C) $Tl(OH)_3$ is amphoteric.
   (D) The first ionisation energies of B and Al are lower than that of Be.
52. Identify the false statement.
   (A) B and Al combine with nitrogen to form nitrides.
   (B) Al—Cl bonds in Al_2Cl_6 are equivalent.
   (C) BF_3 is a stronger acid than BBr_3.
   (D) In the thermit process, Al is used as a reducing agent.
53. Identify the false statement.
   (A) All boron compounds are covalent.
   (B) Boron is a non-metal.
   (C) Boron has a much lower atomic radius than the other Group 13 elements.
   (D) Boron has a maximum covalency of 6.
54. Borazole $B_3N_3H_6$ is related to benzene as
   (A) isoelectronic with benzene (B) isostructural with benzene
   (C) isoelectronic and isostructural with benzene
   (D) none of these.
55. In the purification of bauxite by Hall's process
   (A) bauxite is fused with Na_2CO_3
   (B) bauxite is heated with NaOH solution at 150°C
   (C) bauxite is heated with lime
   (D) bauxite is heated at 1800°C in a current of N_2.
56. The ionic character of the given chlorides is in the order
   (A) $SiCl_4 < AlCl_3 < CaCl_2 < KCl$ (B) $SiCl_4 > AlCl_3 < KCl < CaCl_2$
   (C) $AlCl_3 < SiCl_4 < KCl < CaCl_2$ (D) $AlCl_3 < SiCl_4 < CaCl_2 < KCl$
57. Treatment of an aqueous solution of borax with a mineral acid produces
   (A) boron hydride (B) pyroboric acid
   (C) orthoboric acid (D) metaboric acid.
58. Boric acid undergoes polymerisation due to
   (A) its monobasic nature (B) its geometry
   (C) its acidic nature (D) the presence of hydrogen bonding.

59. The B—H bonds in B_2H_6 are
(A) of two types (B) equivalent (C) of three types (D) of four types.

60. Decreasing order of basic strength of $Ca(OH)_2$, $La(OH)_3$ and $Al(OH)_3$ is
(A) $Ca(OH)_2 > Al(OH)_3 > La(OH)_3$ (B) $Ca(OH)_2 > La(OH)_3 > Al(OH)_3$
(C) $La(OH)_3 > Ca(OH)_2 > Al(OH)_3$ (D) $Al(OH)_3 > La(OH)_3 > Ca(OH)_2$

61. In the Baeyer's process for purification of bauxite, the ore is roasted first. It is to
(A) oxidise impurities like carbon and sulphur
(B) oxidise ion(II) oxide to iron(III) oxide
(C) remove moisture (D) all of these.

62. Identify the incorrect statement.
(A) All B—H bonds in B_2H_6 are similar.
(B) H-atoms are not in same plane in B_2H_6.
(C) Each B atom forms four bonds in B_2H_6.
(D) There are two bridging H atoms in B_2H_6.

63. Which of the following is inorganic benzene?
(A) $B_3N_3H_6$ (B) $B_3N_3H_3$ (C) $B_6N_6H_3$ (D) $B_6N_6H_6$

64. Which of the following ores does not contain Al?
(A) mica (B) fluorspar (C) feldspar (D) cryolite

65. In the purification of bauxite, it is treated with hot aqueous NaOH solution when it dissolves to form
(A) $NaAlO_2$ (B) $Na_3[Al(OH)_6]$ (C) $Al(OH)_3$ (D) AlO(OH)

66. Which of the following is not formed at the anode during electrolysis of fused alumina containing cryolite?
(A) CO (B) CO_2 (C) O_2 (D) F_2

67. Which of the following is formed when $AlCl_3{\cdot}6H_2O$ is heated?
(A) $AlCl_3{\cdot}H_2O$ (B) $AlCl_3{\cdot}4H_2O$ (C) $AlCl_3$ (D) $HCl + Al_2O_3$

68. Which of the following occurs when aluminium is reacted with dil. NaOH?
(A) H_2 is evolved (B) O_2 is evolved (C) H_2O is formed (D) $Al(OH)_3$ is formed.

69. Which of the following is formed when a hot solution of sodium meta aluminate is cooled?
(A) AlO(OH) (B) $Al(OH)_3$ (C) Al_2O_3 (D) $Na_3[Al(OH)_6]$

70. H_3BO_3 is
(A) monobasic and weak Lewis acid (B) monobasic and weak Bronsted acid
(C) monobasic and strong Lewis acid
(D) tribasic and weak Bronsted acid **(I.I.T. 2003)**

71. The stability of monovalent and trivalent cations of Group 13 elements is in the order
(A) $Ga^+ < In^+ < Tl^+$ (B) $Ga^+ > In^+ > Tl^+$
(C) $Ga^{3+} < In^{3+} < Tl^{3+}$ (D) $Ga^{3+} > In^{3+} > Tl^{3+}$

72. In the reaction, $B(OH)_3 + H_2O \longrightarrow [B(OH)_4]^- + H^+$
   (A) $B(OH)_3$ is a Lewis base
   (B) $B(OH)_3$ is a Lewis acid
   (C) $B(OH)_3$ is amphoteric
   (D) none of these.
73. The molecule(s) which possess three-centre-two electron bonds and three-centre-four electron bonds are
   (A) BF_6 and XeF_6 (B) B_2H_6
   (C) B_2H_6 and SiF_4 (D) B_2H_6 and XeF_6
74. The products formed when boron is fused with NaOH are
   (A) B_2O_3 and H_2
   (B) Na_3BO_3 and H_2
   (C) $NaBO_2$ and H_2
   (D) $Na_2B_4O_7$ and H_2
75. On heating Na_3BO_3 with conc. H_2SO_4 and methanol, a green-edged flame is formed. The green colour is due to
   (A) H_3BO_3 (B) B_2O_3 (C) $(CH_3O)_3B$ (D) $(CH_3)_3B$
76. In aqueous solution, Ga^+ ion changes to
   (A) Ga (B) Ga^{3+} (C) Ga and Ga^{3+} (D) Ga^{2+} and Ga^{3+}
77. $B(OH)_3 + NaOH \rightleftharpoons Na[B(OH)_4]$
   How can this reaction be made to proceed in forward direction?
   (A) Addition of *cis*-1,2-diol
   (B) Addition of borax
   (C) Addition of *trans*-1,2-diol
   (D) Addition of Na_2HPO_4 **(I.I.T. 2006)**

   Hint: $Na[B(OH)_4]$ is removed from the equilibrium due to the formation of chelate compound with *cis*-1,2-diol and the reaction moves in the forward direction.

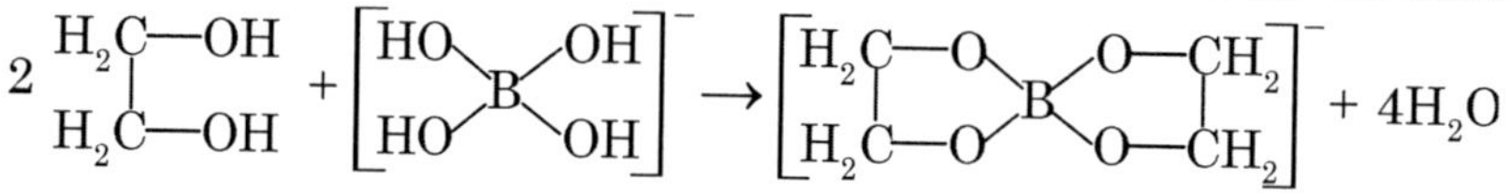

78. Statement–1: Boron always forms covalent bond.
   because
   Statement–2: The small size of B^{3+} favours the formation of covalent bond.
   (A) Statement–1 is true, Statement–2 is true; Statement–2 is a correct explanation for Statement–1
   (B) Statement–1 is true, Statement–2 is true; Statement–2 is not a correct explanation for Statement–1
   (C) Statement–1 is true, Statement–2 is false
   (D) Statement–1 is false, Statement–2 is true. **(I.I.T. 2007)**

   Hint: According to the Fajan's rule small cations having high charge density always have tendency to form covalent bond.
79. Statement–1: In water, orthoboric acid behaves as a weak monobasic acid.
   because
   Statement–2: In water, orthoboric acid acts as a proton donor.
   (A) Statement–1 is true, Statement–2 is true; Statement–2 is a correct explanation for Statement–1
   (B) Statement–1 is true, Statement–2 is true; Statement–2 is not a correct explanation for Statement–1

(C) Statement–1 is true, Statement–2 is false

(D) Statement–1 is false, Statement–2 is true. **(I.I.T. 2007)**

Hint: Orthoboric acid (H_3BO_3) is a weak Lewis acid ($H_3BO_3 + H_2O \rightleftharpoons B(OH)_4^- + H^+$). It does not donate proton but accepts OH^- ion from water.

80. In the reaction, $2X + B_2H_6 \rightarrow [BH_2(X)_2]^+[BF_4]^-$, the amine(s) X is(are)

(A) NH_3 (B) CH_3NH_2 (C) $(CH_3)_2NH$ (D) $(CH_3)_3N$ **(I.I.T. 2009)**

Hint: Due to bulkiness of $(CH_3)_3N$, it does not react.

ANSWERS

1. C	2. C	3. A	4. C	5. C	6. B, D
7. A, B	8. D	9. C	10. B	11. C	12. A
13. A	14. D	15. D	16. B	17. C	18. D
19. D	20. A	21. A	22. D	23. D	24. A, D
25. D	26. C	27. A, B	28. B	29. D	30. B, C
31. D	32. B	33. A	34. D	35. C	36. D
37. B	38. C	39. A	40. B, C	41. A	42. C
43. C	44. A	45. C	46. A	47. B	48. D
49. D	50. C	51. B, C	52. B, C	53. D	54. B
55. A	56. A	57. C	58. D	59. A	60. B
61. B	62. A	63. A	64. B	65. A	66. D
67. D	68. A	69. B	70. A	71. A, D	72. B
73. B	74. B	75. C	76. C	77. A	78. A
79. C	80. A, B, C				

20

Group 14 Elements: Carbon Family

OBJECTIVE TYPE MULTIPLE CHOICE QUESTIONS

1. Assertion (*A*): Between $SiCl_4$ and CCl_4, only $SiCl_4$ reacts with water.
Reason (*R*): $SiCl_4$ is ionic and CCl_4 is covalent.
(A) Both *A* and *R* are correct and *R* is the correct explanation of *A*
(B) Both *A* and *R* are correct and *R* is not the correct explanation of *A*
(C) *A* is true but *R* is false (D) *A* is false but *R* is true. **(I.I.T. 2001)**

2. The order of increasing C—O bond length in CO, CO_3^{2-} and CO_2 is
(A) $CO_3^{2-} < CO_2 < CO$ (B) $CO < CO_2 < CO_3^{2-}$
(C) $CO_2 < CO_3^{2-} < CO$ (D) $CO < CO_3^{2-} < CO_2$ **(I.I.T. 1999)**

3. Which of the following halides is least stable and has doubtful existence?
(A) GeI_4 (B) SnI_4 (C) PbI_4 (D) CCI_4 **(I.I.T. 1996)**

4. The number and type of bonds between C atoms in CaC_2 are
(A) 1σ, 1π (B) 1σ, 1.5π (C) 1σ (D) 1σ, 2π **(I.I.T. 1996)**

5. Which of the following oxides is neutral?
(A) CO (B) SnO_2 (C) ZnO (D) SiO_2 **(I.I.T. 1996)**

6. The element exhibiting most stable + 2 oxidation state among the following is
(A) Pb (B) Sn (C) Fe (D) Ag **(I.I.T. 1995)**

7. One recently discovered allotrope of carbon (e.g., C_{60}) is commonly known as
(A) fullerene (B) graphite (C) diamond (D) C-60 **(I.I.T. 1994)**

8. The hydrolysis of trialkylchlorosilane, R_3SiCl yields
(A) $R_3Si(OH)$ (B) $R_2Si(OH)_2$ (C) $RSi(OH)_3$ (D) $Si(OH)_4$ **(I.I.T. 1994)**

9. The material used in solar cells contains
(A) Cs (B) Si (C) Sn (D) Ti **(I.I.T. 1993)**

10. Which is wrongly matched?
(A) Wollastonite — $Ca_3(Si_3O_9)$ (B) Tremolite — $CaMg_3(SiO_4)_2$
(C) Chrysotile — $Al_2(OH)_4Si_2O_5$ (D) Albite — $NaAlSi_3O_8$ **(I.I.T. 1993)**

11. The increasing order of extent of hydrolysis in the chlorides CCl_4, $MgCl_2$, $AlCl_3$, PCl_5 and $SiCl_4$ is
(A) $CCl_4 < MgCl_2 < AlCl_3 < PCl_5 < SiCl_4$
(B) $CCl_4 < MgCl_2 < AlCl_3 < SiCl_4 < PCl_5$
(C) $CCl_4 < MgCl_2 < PCl_5 < AlCl_3 < SiCl_4$
(D) $MgCl_2 < CCl_4 < AlCl_3 < PCl_5 < SiCl_4$ **(I.I.T. 1991)**

12. The compounds used as refrigerant are
(A) NH_3 (B) CCl_4 (C) CF_4 (D) CF_2Cl_2 **(I.I.T. 1990)**

13. The increasing order of acidic character in the oxides CO_2, N_2O_5, SiO_2 and SO_3 is
(A) $SiO_2 < CO_2 < N_2O_5 < SO_3$ (B) $CO_2 < SiO_2 < N_2O_5 < SO_3$
(C) $SiO_2 < CO_2 < SO_3 < N_2O_5$ (D) $CO_2 < SiO_2 < SO_3 < N_2O_5$ **(I.I.T. 1988)**

14. Which one of the following pairs of ions cannot be separated by H_2S in dilute hydrochloric acid?
(A) Bi^{3+}, Sn^{4+} (B) Al^{3+}, Hg^{2+} (C) Zn^{2+}, Cu^{2+} (D) Ni^{2+}, Cu^{2+} **(I.I.T. 1986)**

15. The ion that cannot be precipitated both by HCl and H_2S is
(A) Pb^{2+} (B) Cu^{+} (C) Ag^{+} (D) Sn^{2+} **(I.I.T. 1982)**

16. Moderate electrical conductivity is shown by
(A) carborundum (B) graphite (C) diamond (D) silica **(I.I.T. 1982)**

17. If two compounds have different crystal structures and analogous formulae, they are called
(A) isomorphs (B) isotopes (C) isomers (D) isobars **(I.I.T. 1981)**

18. Lead pencil contains
(A) Pb (B) FeS (C) graphite (D) PbS **(I.I.T. 1980)**

19. Which of the following oxides is/are amphoteric?
(A) Na_2O (B) CaO (C) Al_2O_3 (D) SnO_2 **(I.I.T. 1993)**

20. The increasing order of oxidising power of the following compounds is
(A) $GeCl_4 < SnCl_4 < PbCl_4$ (B) $GeCl_4 < PbCl_4 < SnCl_4$
(C) $SnCl_4 < PbCl_4 < GeCl_4$ (D) $PbCl_4 < SnCl_4 < GeCl_4$

21. The increasing order of reducing power in the following compounds is
(A) $PbCl_2 < SnCl_2 < GeCl_2$ (B) $SnCl_2 < PbCl_2 < GeCl_2$
(C) $GeCl_2 < PbCl_2 < SnCl_2$ (D) $PbCl_2 < GeCl_2 < SnCl_2$

22. $PbBr_4$ does not exist because
(A) IE of Pb^{4+} is large (B) Pb^{4+} is a reducing agent
(C) Pb^{4+} is an oxidising agent
(D) the electronegativity difference between Pb and Br is less.

23. $PbCl_2$ is solid and $PbCl_4$ is liquid because
(A) $PbCl_2$ is ionic and $PbCl_4$ is covalent
(B) $PbCl_2$ is angular and $PbCl_4$ is tetrahedral
(C) $PbCl_2$ is polymeric and $PbCl_4$ is monomeric
(D) $PbCl_2$ exhibits resonance and $PbCl_4$ does not exhibit resonance.

24. $Pb(NO_3)_2$ when heated at 400°C gives
    (A) $PbO + NO_2 + O_2$
    (B) $PbO + NO + O_2$
    (C) $Pb_3O_4 + NO + O_2$
    (D) $Pb_3O_4 + NO_2 + O_2$
25. Identify the false statement.
    (A) Lead forms $PbCl_2$, a powerful oxidant.
    (B) Lead forms PbI_4, a powerful oxidant.
    (C) Lead reacts with dil. NaOH to form $Pb(OH)_2$.
    (D) Lead reacts with conc. HCl to form $PbCl_2$.
26. Identify the incorrect statement.
    (A) Both CCl_4 and $SiCl_4$ act as Lewis acid.
    (B) The order of reducing power of the following Group 14 halides is: $PbCl_2 < SnCl_2 < GeCl_2$.
    (C) Graphite is a better lubricant on the moon than on the earth.
    (D) H_2O_2 restores the white colour of a painting made with white lead which has turned black on long exposure to atmosphere.
27. $SnCl_4$ is liquid and $SnCl_2$ is solid because
    (A) $SnCl_4$ is covalent and $SnCl_2$ is ionic
    (B) $SnCl_4$ is tetrahedral and $SnCl_2$ is angular
    (C) $SnCl_4$ is monomeric and $SnCl_2$ is polymeric
    (D) $SnCl_4$ does not exhibit resonance and $SnCl_2$ exhibits resonance.
28. Which of the following is not correct for fluorocarbons?
    (A) Neither carbon nor fluorine has *d*-orbitals.
    (B) Neither carbon nor fluorine can expand its octet.
    (C) Fluorine atoms cannot effectively shield the carbon atoms.
    (D) In C—F bonds, carbon may be supposed to have undergone oxidation so that fluorocarbons are not susceptible to air.
29. Graphite
    (A) is paramagnetic
    (B) has unpaired electrons
    (C) has mobile electrons
    (D) is diamagnetic.
30. Identify the incorrect statement.
    (A) The pigment chrome red is $PbO{\cdot}PbCrO_4$.
    (B) The composition of silica gel is SiO_2.
    (C) The catalyst used in the conversion of CO and H_2 to CH_3OH is $ZnO–CrO_3$.
    (D) Oxalic acid on heating with conc. H_2SO_4 gives SO_2.
31. Identify the incorrect statement.
    (A) The IE_I of C is greater than that of B but the reverse is true for IE_{II}
    (B) Crystalline Si is prepared from $SiCl_4$.
    (C) Cassiterite is an ore of lead.
    (D) Hydrolysis of $(CH_3)_2SiCl_2$ gives silicone polymers.

32. Identify the incorrect statement.
    (A) $SnCl_2$ dissolves in HNO_3.
    (B) $SiCl_4$ is readily hydrolysed but CCl_4 is stable.
    (C) CO_2 is a gas while SiO_2 is high melting solid.
    (D) Argon is more abundant than CO_2 in the atmosphere.
33. Identify the incorrect statement.
    (A) Silicones are attacked by ozone.
    (B) Silicon hydrides are known as silicones.
    (C) Liquid silicones act as lubricants.
    (D) Silicones have good thermal and oxidative stability.
34. Identify the correct statement regarding Group 14 elements as their atomic number increases.
    (A) Their metallic character decreases.
    (B) Their ionisation potential increases.
    (C) The stability of + 2 oxidation state increases.
    (D) Their atomic radius decreases.
35. Which of the following melts in boiling water?
    (A) Monel metal (B) Bell metal (C) Wood's metal (D) Gun metal
36. In the manufacture of glass, the addition of MnO_2 gives
    (A) red colour (B) violet colour (C) pink colour (D) yellow colour
37. $PbCl_4$ exists but $PbBr_4$ and PbI_4 do not because of
    (A) Br^- and I^- are bigger in size (B) chlorine is a gas
    (C) more electropositive nature of bromine and iodine
    (D) inability of bromine and iodine to oxidise Pb^{2+} to Pb^{4+}
38. Identify the correct statement.
    (A) Only $SnBr_4$ is coloured (B) Only $SnCl_4$ is coloured
    (C) Both $SnCl_4$ and $SnBr_4$ are coloured
    (D) $SnCl_4$, $SnBr_4$ and SnI_4 are all coloured.
39. Water passing through lead pipes is not suitable for drinking because
    (A) lead reacts with dissolved O_2 to form PbO which is poisonous
    (B) lead reacts with water to form PbO which is poisonous
    (C) lead reacts with dissolved carbonates to form $PbCO_3$ which is poisonous
    (D) lead reacts with water and dissolved O_2 to form soluble $Pb(OH)_2$ which is poisonous.
40. Carbogen is
    (A) pure form of carbon (B) mixture of O_2 and CO_2
    (C) mixture of CO and CO_2 (D) $COCl_2$
41. Identify the false statement.
    (A) $Ni(CO)_4$ is diamagnetic (B) BI_3 is a stronger Lewis acid than BF_3
    (C) CCl_4 is hydrolysed whereas BCl_3 is inert
    (D) graphite conducts electricity whereas diamond does not.

42. The material used in warfares for making smoke screens for military purposes is
(A) CCl_4 (B) $SiCl_4$ (C) PCl_5 (D) PH_3
43. An oxide of lead which is used in lead storage batteries, in safety matches and is a powerful oxidising agent, is
(A) Pb_3O_4 (B) PbO_2 (C) PbO (D) $2PbO{\cdot}PbO_2$
44. Which metal does not form nitrate when heated with conc. HNO_3?
(A) Sn (B) Pb (C) Mg (D) Zn
45. PbO_2 is
(A) basic (B) acidic (C) neutral (D) amphoteric.
46. Identify the false statement.
(A) GeO_2 dissolves in acids (B) $Ge(OH)_2$ is amphoteric
(C) $GeCl_2$ is more stable than $GeCl_4$ (D) In HCl, $GeCl_4$ forms $[GeCl_6]^{2-}$ ion.
47. Which of the following statements is incorrect?
(A) CO reduces iron(III) to iron(0) (B) Zn dissolves in hot NaOH solution
(C) HgI_2 dissolves in KI solution (D) $SnCl_4$ is made by dissolving Sn in conc. HCl
48. Which of the following is the strongest base?
(A) $Si(OH)_4$ (B) $Be(OH)_2$ (C) $Mg(OH)_2$ (D) $Al(OH)_3$
49. White lead is
(A) $PbCO_3$ (B) $PbCO_3{\cdot}PbO$ (C) $PbSO_4{\cdot}PbO$ (D) $Pb(OH)_2{\cdot}2PbCO_3$
50. When monosilane comes in contact with air, it burns with a luminous flame and produces vortex rings. These rings are of
(A) Si (B) SiO (C) SiO_2 (D) H_2SiO_3
51. Which of the following are covalent?
(A) $SnCl_2$ (B) $SiCl_4$ (C) $SnCl_4$ (D) KCl
52. When tin is treated with conc. HNO_3
(A) tin becomes passive (B) $Sn(NO_3)_4$ is formed
(C) $Sn(NO_3)_2$ is formed (D) metastannic acid is formed.
53. Ultraviolet rays are checked by
(A) Crooke's glass (B) pyrex glass (C) flint glass (D) soda glass
54. Which of the following bonds is the weakest?
(A) Si—Si (B) Si—O (C) Si—H (D) Si—Cl
55. Which of the following Group 14 elements has the lowest melting point?
(A) Pb (B) Sn (C) Ge (D) Si
56. Which of the products is formed when $SiCl_4$ vapour is passed over Mg?
(A) $MgSiCl_6$ (B) $MgCl_2$ + Si (C) $Mg_2Si + Cl_2$ (D) $MgCl_2 + SiCl_4$
57. Lead sulphate is soluble in
(A) conc. HNO_3 (B) conc. HCl
(C) CH_3COONH_4 solution (D) water.

58. Germanium(II) compounds are powerful reducing agents whereas lead(IV) compounds are strong oxidising agents. It is due to
    (A) Pb is more electronegative than Ge.
    (B) the ionisation energy of Pb is less than that of Ge.
    (C) the ionic radii of Pb^{2+} and Pb^{4+} are less than those of Ge^{2+} and Ge^{4+}.
    (D) more pronounced inert pair effect in Pb than in Ge.

59. Sugar of lead is
    (A) $PbCO_3{\cdot}Pb(OH)_2$ (B) $PbCO_3$
    (C) $2PbSO_4{\cdot}PbO$ (D) $Pb(CH_3COO)_2$

60. Identify the correct statement(s) regarding CO.
    (A) It combines with water to form carbonic acid.
    (B) It reacts with haemoglobin in red blood cells.
    (C) It is a powerful oxidising agent.
    (D) It is used to prepare aerated drinks.

61. Graphite conducts electricity due to the presence of
    (A) unpaired electrons (B) sigma electrons
    (C) mobile electrons (D) $p\pi$-$p\pi$ electrons.

62. Which of the following does not exist?
    (A) $[SiCl_6]^{2-}$ (B) $[GeCl_6]^{2-}$ (C) $[CCl_6]^{2-}$ (D) $[SnCl_6]^{2-}$

63. On heating with conc. H_2SO_4, sucrose gives
    (A) CO and CO_2 (B) CO and SO_2
    (C) CO, CO_2 and SO_2 (D) none of these.

64. Which of the following statements is false?
    (A) Zinc dissolves in hot NaOH solution.
    (B) CO reduces iron(III) to iron(0).
    (C) $SnCl_2$ is prepared by dissolving Sn in conc. HCl.
    (D) HgI_2 dissolves in excess KI solution.

65. When sodium oxalate is treated with dil. H_2SO_4, the gas/gases produced is/are
    (A) CO (B) CO_2 (C) CO + CO_2 (D) CO_2 + SO_2

66. Butter of tin is
    (A) $SnBr_4{\cdot}5H_2O$ (B) $SnBr_2{\cdot}5H_2O$ (C) $SnCl_4{\cdot}5H_2O$ (D) $SnCl_2{\cdot}5H_2O$

67. Tin plague is
    (A) generation of a metallic sound when a tin plate is bent
    (B) conversion of Sn(II) salt to Sn(IV) salt
    (C) conversion of white tin into grey tin
    (D) electroplating of metals by tin.

68. Which of the following is formed when tin burns in air?
    (A) SnO_2 (B) SnO (C) Sn_3O_4 (D) White tin

69. Which of the following Group 14 oxides oxidises HNO_3 to oxygen?
 (A) SnO (B) SnO_2 (C) PbO (D) PbO_2

70. Which of the following oxides is formed when lead is heated at a temperature below 773 K?
 (A) Pb_2O_3 (B) Pb_3O_4 (C) PbO_2 (D) PbO

71. Which of the following is formed when tin(IV) oxide dissolves in dil. NaOH?
 (A) $Na[Sn(OH)_3]$ (B) $Na_2[Sn(OH)_6]$ (C) $Sn(OH)_4$ (D) $Na[Sn(H_2O)_3(OH)_3]$

72. Which of the following is correct regarding plumbo-solvency?
 (A) It is enhanced in the presence of NH_4^+ and NO_3^- ions.
 (B) It is retarded due to the formation of protective layer of $Pb(OH)_2 \cdot PbCO_3$.
 (C) It is enhanced in the presence of SO_4^{2-} and PO_4^{3-} ions.
 (D) All of these.

73. Which of the following metal oxides can be reduced by carbon?
 (A) CuO (B) MgO (C) SnO_2 (D) Al_2O_3

74. Buckministerfullerene is
 (A) diamagnetic (B) paramagnetic
 (C) has unpaired electrons
 (D) has lower electrical conductivity than graphite.

75. $(Me)_2SiCl_2$ on hydrolysis will produce
 (A) $(Me)_2Si(OH)_2$ (B) $(Me)_2Si{=}O$
 (C) $\lbrack O-(Me)_2Si-O \rbrack_n$ (D) $Me_2SiCl(OH)$ **(I.I.T. 2003)**

76. Which silicates are formed from $[SiO_4]^{4-}$ tetrahedral units by sharing three oxygen atoms?
 (A) Sheet silicates (B) Pyro silicates
 (C) Linear chain silicates (D) 3-Dimensional silicates. **(I.I.T. 2005)**

 Hint:

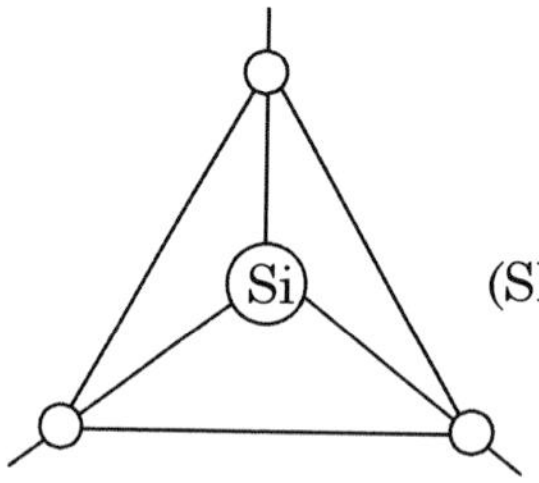

(Sheet structure)

77. The species present in solution when CO_2 is dissolved in water are
 (A) CO_2, H_2CO_3 (B) CO_3^{2-}, HCO_3^- (C) H_2CO_3, CO_3^{2-} (D) CO_2, H_2CO_3, HCO_3^-, CO_3^{2-}
 (I.I.T. 2006)

 Hint: $CO_2 + H_2O \rightleftharpoons H_2CO_3 \rightleftharpoons H^+ + HCO_3^-$; $HCO_3^- \rightleftharpoons H^+ + CO_3^{2-}$

78. Statement–1: Pb^{4+} compounds are stronger oxidizing agents than Sn^{4+} compounds.
 Statement–2: The higher oxidation states for the group 14 elements are more stable for the heavier members of the group due to inert pair effect.

(A) Statement–1 is true, Statement–2 is true; Statement–2 is a correct explanation for Statement–1
(B) Statement–1 is true, Statement–2 is true; Statement–2 is not a correct explanation for Statement–1
(C) Statement–1 is true, Statement–2 is false
(D) Statement–1 is false, Statement–2 is true. **(I.I.T. 2008)**

Hint: Statement–1 is true and Statement–2 is false. Since Pb^{4+} ion is less stable than Sn^{4+} ion, Pb^{4+} ion is a stronger oxidising agent than Sn^{4+} ion. Due to the inert pair effect, the stability of higher oxidation state of group 14 elements decreases on moving down the group.

ANSWERS

1. C	2. B	3. C	4. D	5. A	6. A
7. A	8. A	9. B	10. C	11. A	12. A, D
13. A	14. A	15. B	16. B	17. A	18. C
19. C, D	20. A	21. A	22. C	23. A	24. A
25. B, C	26. A	27. A	28. C	29. C, D	30. D
31. B	32. A, D	33. A, B	34. C	35. C	36. C
37. D	38. B	39. D	40. B	41. C	42. B
43. B	44. A	45. A	46. A	47. D	48. C
49. D	50. C	51. B, C	52. D	53. A	54. A
55. A	56. B	57. A	58. D	59. D	60. B
61. C, D	62. C	63. D	64. C	65. C	66. C
67. A	68. A	69. D	70. B	71. B	72. D
73. C	74. A, D	75. C	76. A	77. D	78. C

21

Group 15 Elements: Nitrogen Family

OBJECTIVE TYPE MULTIPLE CHOICE QUESTIONS

1. The number of P—O—P bonds in cyclic metaphosphoric acid is
   (A) zero (B) two (C) three (D) four **(I.I.T. 2000)**
2. Which of the following is the most suitable drying agent for ammonia gas?
   (A) Quicklime (B) Anhy. $CaCl_2$ (C) Slaked lime (D) P_4O_{10} **(I.I.T. 1978, 2000)**
3. Ionic radii of
   (A) $P^{3+} > P^{5+}$ (B) $Ti^{4+} < Mn^{7+}$ (C) $^{35}Cl^- < {}^{37}Cl^-$ (D) $K^+ > Cl^-$ **(I.I.T. 1999)**
   Hint: Larger the effective nuclear charge, smaller is the ionic radius.
4. Which of the following compounds is formed when ammonia is reacted with OCl^-?
   (A) NH_2NH_2 (B) NH_4Cl (C) HNO_2 (D) NO **(I.I.T. 1999)**
5. One mole of calcium phosphide on reaction with excess water gives
   (A) 1 mole of PH_3 (B) 2 moles of H_3PO_4
   (C) 1 mole of P_2O_5 (D) 2 moles of PH_3 **(I.I.T. 1999)**
6. In compounds of the type ECl_3 (E = B, P, As, Bi), the angles Cl—E—Cl for different E are in the order
   (A) B > P = As = Bi (B) B < P = As = Bi
   (C) B < P < As < Bi (D) B > P > As > Bi **(I.I.T. 1999)**
7. Ground state electronic configuration of nitrogen atom can be represented by
   (A) [↑↓] [↑↓] [↑ | ↑ | ↑] (B) [↑↓] [↑↓] [↑ | ↓ | ↑]
   (C) [↑↓] [↑↓] [↑ | ↓ | ↓] (D) [↑↓] [↑↓] [↓ | ↓ | ↓] **(I.I.T. 1999)**

8. Which of the following is formed when ammonium dichromate is strongly heated?
   (A) N_2 (B) NO (C) N_2O (D) NO_2 **(I.I.T. 1999)**
9. Sodium nitrate decomposes above ~ 800°C to give
   (A) N_2 (B) O_2 (C) NO_2 (D) Na_2O **(I.I.T. 1998)**
10. White phosphorus (P_4) has
    (A) six P—P single bonds (B) four P—P double bonds
    (C) four lone pairs of electrons (D) P—P—P angle of 60° **(I.I.T. 1998)**
11. Assertion (*A*): HNO_3 is a stronger acid than HNO_2.
    Reason (*R*): In HNO_3 there are two nitrogen-to-oxygen bonds while in HNO_2 there is only one.
    (A) Both *A* and *R* are true and *R* is the correct explanation of *A*
    (B) Both *A* and *R* are true but *R* is not the correct explanation of *A*
    (C) *A* is true but *R* is incorrect (D) *A* is incorrect but *R* is true. **(I.I.T. 1998)**
12. Among the species NF_3, BF_3, NO_3^-, H_3O^+ and HN_3, identify the isostructural pairs.
    (A) (NF_3, NO_3^-) and (BF_3, H_3O^+) (B) (NF_3, H_3O^+) and (NO_3^-, BF_3)
    (C) (NF_3, HN_3) and (NO_3^-, BF_3) (D) (NF_3, H_3O^+) and (NH_3, BF_3) **(I.I.T. 1996)**
13. In P_4O_{10} the number of oxygen atoms bonded to each P atom is
    (A) 2 (B) 2.5 (C) 3 (D) 4 **(I.I.T. 1995)**
14. Assertion (*A*): Although PF_3, PCl_3 and PBr_3 are known, the pentahalides of nitrogen have not been observed.
    Reason (*R*): Phosphorus has lower electronegativity than nitrogen.
    (A) Both *A* and *R* are true and *R* is the correct explanation of *A*
    (B) Both *A* and *R* are true and *R* is not the correct explanation of *A*
    (C) *A* is true but *R* is false (D) *A* is false but *R* is true. **(I.I.T. 1994)**
15. The oxidation state of phosphorus in $Ba(H_2PO_2)_2$ is
    (A) + 3 (B) + 2 (C) + 1 (D) – 1 **(I.I.T. 1990)**
16. Which of the following is the strongest base?
    (A) AsH_3 (B) NH_3 (C) PH_3 (D) SbH_3 **(I.I.T. 1989, 1990)**
17. Nitrogen(I) oxide is produced by
    (A) thermal decomposition of NH_4NO_3
    (B) disproportionation of N_2O_4 (C) thermal decomposition of NH_4NO_2
    (D) interaction of NH_2OH and HNO_2. **(I.I.T. 1989)**
18. Concentrated HNO_3 reacts with I_2 to give
    (A) HI (B) HOI
    (C) $HOIO_2$ (D) $HOIO_3$ **(I.I.T. 1989, Roorkee 1995)**
19. Lightning bolts in the atmosphere cause the formation of
    (A) NO (B) NH_3
    (C) NH_4OH (D) NH_2OH **(I.I.T. 1987, 1988)**
20. Which of the following oxides of nitrogen is a coloured gas?
    (A) N_2O (B) NO (C) N_2O_4 (D) NO_2 **(I.I.T. 1987)**

21. Among the trihalides of nitrogen, which one is least basic?
(A) NF_3 (B) NCl_3 (C) NBr_3 (D) NI_3 **(I.I.T. 1987)**

22. The bonds present in N_2O_5 are
(A) only ionic (B) covalent and coordinate
(C) only covalent (D) covalent and ionic. **(I.I.T. 1986)**

23. Which of the following is most acidic?
(A) ZnO (B) Na_2O_2 (C) P_4O_{10} (D) MgO **(I.I.T. 1985)**

24. Nitrogen dioxide cannot be obtained by heating
(A) KNO_3 (B) $Pb(NO_3)_2$ (C) $Cu(NO_3)_2$ (D) $AgNO_3$ **(I.I.T. 1985)**

25. Ammonia solution dissolves fairly
(A) $Cu(OH)_2$ (B) Hg_2Cl_2 (C) $PbCl_2$ (D) AgI **(I.I.T. 1983)**

26. The element with the highest first ionisation energy is
(A) B (B) C (C) N (D) O **(I.I.T. 1982)**

27. Which of the following is coloured?
(A) NO (B) N_2O (C) SO_3 (D) CO
(E) none of these. **(I.I.T. 1980)**

28. Ammonia gas can be dried by
(A) conc. H_2SO_4 (B) P_4O_{10} (C) anhy. $CaCl_2$ (D) anhy. $CuSO_4$
(E) none of these. **(I.I.T. 1980)**

29. The reddish brown gas formed when nitric oxide is oxidised by air is
(A) N_2O_3 (B) N_2O_4 (C) NO_2 (D) N_2O_5 **(I.I.T. 1979, 2000)**

30. Which are the incorrect statements?
(A) NO is heavier than O_2. (B) The formula of heavy water is H_2O.
(C) Nitrogen diffuses faster than oxygen through an orifice.
(D) NH_3 can be used as a refrigerant. **(I.I.T. 1978)**

31. Mark the element which gives M^{3-} ion.
(A) P (B) N (C) As (D) Sn **(I.I.T. 1980)**

32. The ammonium compound which on heating does not give NH_3 is
(A) $(NH_4)_2CO_3$ (B) $(NH_4)_2SO_4$ (C) NH_4NO_3 (D) $(NH_4Cl$ **(Roorkee 1993)**

33. When ammonia is passed over heated CuO, it is oxidised to
(A) N_2 (B) NO_2 (C) HNO_3 (D) N_2O **(Roorkee 1992)**

34. When $AgNO_3$ is heated strongly, the products formed are
(A) NO and NO_2 (B) NO_2 and O_2 (C) NO_2 and N_2O (D) NO and O_2
(Roorkee 1990)

35. Which of the following compounds will liberate NO_2 on heating?
(A) $NaNO_3$ (B) KNO_3 (C) $Cu(NO_3)_2$ (D) $Pb(NO_3)_2$
(Roorkee 1996)

36. The increasing order of forming + 5 oxidation state in Group 15 elements is
(A) Bi < Sb < As < P < N (B) Sb < Bi < As < P < N
(C) N < P < As < Sb < Bi (D) N < P < As < Bi < Sb **(Roorkee 1987)**

37. Phosphine is formed from
    (A) calcium superphosphate (B) calcium hypophosphite
    (C) potassium phosphide (D) calcium phosphide **(Roorkee 1995)**
38. The acid which forms two series of salts is
    (A) H_3PO_4 (B) H_3PO_3 (C) H_3BO_3 (D) H_3PO_2
39. Which of the following acids possesses oxidising, reducing and complex forming properties?
    (A) HNO_3 (B) HNO_2 (C) H_2SO_4 (D) HCl
40. Which of the following oxides of nitrogen is paramagnetic?
    (A) NO_2 (B) NO (C) N_2O (D) N_2O_4
41. The cyclic structure is possessed by
    (A) $H_6P_4O_{13}$ (B) $H_5P_5O_{15}$ (C) $H_5P_3O_{10}$ (D) $H_7P_5O_{16}$
42. Which oxide of nitrogen is formed when phosphorus is heated with conc. HNO_3?
    (A) NO (B) NO_2 (C) N_2O_3 (D) N_2O_5
43. Heating of P_4O_{10} with conc. HNO_3 gives
    (A) NO (B) N_2O (C) NO_2 (D) N_2O_5
44. Phosphorous acid contains
    (A) 1 hydrogen atom attached to P (B) 2 hydrogen atoms attached to P
    (C) 3 hydrogen atoms attached to P (D) no hydrogen atom attached to P.
45. Orthophosphorous acid is
    (A) neutral (B) monobasic (C) dibasic (D) tribasic.
46. Identify the false statement.
    (A) Phosphorus tetraoxide produces phosphoric acid when treated with water.
    (B) When phosphorus pentaoxide is treated with conc. HNO_3, the product is NO_2.
    (C) Phosphorus pentaoxide is an anhydride of metaphosphoric acid.
    (D) Hypophosphorous acid has two hydrogen atoms attached to phosphorus.
47. Identify the incorrect statement.
    (A) Pyrophosphoric acid forms four series of salts ($NaH_3P_2O_7$, $Na_2P_2O_7$, $Na_3HP_2O_7$, $Na_4P_2O_7$).
    (B) Phosphorus tetraoxide produces H_3PO_4 when treated with water.
    (C) Orthophosphorous acid is formed when P_4O_6 is dissolved in cold water.
    (D) Heating of pyrophosphoric acid gives metaphosphoric acid.
48. Identify the false statement.
    (A) Pyrophosphoroic acid is a tribasic acid.
    (B) Hypophosphoric acid is a tetrabasic acid.
    (C) Number of hydrogen atoms attached to each P atom in hypophosphoric acid is zero.
    (D) Number of hydrogen atoms attached to P atom in orthophosphoric acid is zero.

49. Which of the oxides of nitrogen is a white solid?
    (A) NO (B) N_2O (C) N_2O_3 (D) N_2O_5
50. Identify the incorrect statement.
    (A) Dimerised nitrogen dioxide is paramagnetic.
    (B) N_2O_5 is planar.
    (C) Covalent nitrates are less stable than ionic nitrates
    (D) NO_2 is an angular molecule with $\angle O—N—O = 132°$.
51. Hypophosphorous acid has
    (A) 1 hydrogen atom attached to phosphorus
    (B) 2 hydrogen atoms attached to phosphorus
    (C) 3 hydrogen atoms attached to phosphorus
    (D) no hydrogen atom attached to phosphorus.
52. White P reacts with NaOH to form PH_3 and NaH_2PO_4. The reaction is an example of
    (A) oxidation (B) reduction
    (C) neutralisation (D) oxidation-reduction. **(I.I.T. 1980)**
53. Which of the following is formed when a mixture of H_2 and NO is passed over heated Pt?
    (A) N_3H (B) NH_3 (C) NH_2OH (D) NH_2NH_2
54. Identify the incorrect statement.
    (A) Red P is more volatile than white P because the former has tetrahedral structure.
    (B) Orthophosphorous acid is a dibasic acid.
    (C) $\angle H—As—H$ in AsH_3 is smaller than $\angle H—N—H$ in NH_3.
    (D) In P_4O_6 the number of oxygen atoms bonded to each P atom is three.
55. Identify the incorrect statement.
    (A) NH_4NO_2 on heating at 250-300°C produces NO_2
    (B) The conjugate acid of HPO_4^{2-} is $H_2PO_4^-$
    (C) In $Ba(H_2PO_2)_2$ the oxidation number of P is + 1
    (D) Orthophosphorous acid disproportionates on heating, giving PH_3 and orthophosphoric acid.
56. Pollution due to NO can be controlled by
    (A) reduction to NH_3 (B) oxidation to HNO_3
    (C) reduction to N_2 (D) none of these.
57. Identify the incorrect statement.
    (A) PCl_3 is polar while BCl_3 is non-polar.
    (B) $(CF_3)_3N$ is a weaker base than $(CH_3)_3N$.
    (C) Both NCl_5 and PCl_5 are stable compounds.
    (D) H_3AsO_3 is a stronger acid than H_3AsO_4.
58. The order of N—O bond length in NO, NO_3^-, NO_3^- and N_2O_4 is
    (A) $NO_3^- > NO_2^- > N_2O_4 > NO$ (B) $NO > N_2O_4 > NO_2^- > NO_3^-$
    (C) $NO > NO_3^- > N_2O_4 > NO_2^-$ (D) $N_2O_4 > NO_2^- > NO_3^- > NO$

Hint: N—O bond length (Å) in NO, N_2O_4, NO_2^- and NO_3^- is 1.15, 1.17, 1.21 and 1.22 respectively.

59. Identify the incorrect statement.
    (A) HNO_2 reacts with H_2SO_4 to give NO_2 and SO_2.
    (B) Both PCl_5 and PI_5 are stable compounds.
    (C) An oxide of nitrogen which is isoelectronic and isostructural with CO is NO.
    (D) The reaction of Cu with conc. HNO_3 gives NO_2.

60. Identify the incorrect statement.
    (A) NO_2 dissolves in water to form a mixture of HNO_2 and HNO_3.
    (B) Heating a mixture of KNO_2 and NH_4Cl gives NO.
    (C) NO_2 dissolves in water to form HNO_3.
    (D) NO_2 dissolves in water to form HNO_2 and O_2.

61. Which of the following oxides is most acidic?
    (A) N_2O_5 (B) P_2O_5 (C) As_2O_5 (D) Sb_2O_5

62. Among the following ions, $p\pi$-$d\pi$ overlap is present in
    (A) NO_3^- (B) PO_4^{3-} (C) CO_3^{2-} (D) NO_2^-

63. Which set of compounds does not exist?
    (A) NCl_5, N_4, PO_2 (B) NCl_5, PCl_5, PO_2
    (C) NCl_5, N_4, NO_2 (D) all of these.

64. Identify the correct statement regarding $P_4O_6$and P_4O_{10}.
    (A) Both P_4O_6 and P_4O_{10}form oxoacids H_3PO_3 and H_3PO_4 respectively.
    (B) In P_4O_6 each P is joined to three O atoms and in P_4O_{10} each P is joined to five O atoms.
    (C) In P_4O_6 each P is joined to four O atoms and in P_4O_{10} each P is joined to four O atoms.
    (D) In P_4O_6 each P is joined to four O atoms and in P_4O_{10} each P is joined to six O atoms.

65. Which of the following sets of oxides of nitrogen is paramagnetic?
    (A) NO, NO_2 (B) NO, NO_2, N_2O_5
    (C) N_2O_2, N_2O_4, N_2O_5 (D) all of these.

66. Which of the following can be used to prepare N_2O?
    (A) Cold dilute HNO_3 reacts with Cu and Zn
    (B) $SnCl_2$ reacts with conc. HNO_3 in the presence of conc. HCl
    (C) Moist iron reacts with NO (D) all of these.

67. Which of the following oxoacids of phosphorus contains maximum number of P—H bonds?
    (A) H_3PO_4 (B) $H_4P_2O_7$ (C) H_3PO_3 (D) H_3PO_2

68. Which of the following statements is incorrect for pyrophosphoric acid ($H_4P_2O_7$)?
    (A) It is a tribasic acid (B) It contains P in +5 oxidation state
    (C) It contains one P—O—P bond (D) It does not contain a P—H bond

69. Polyphosphates are used as water softening agents because they
    (A) form soluble complexes with anionic species
    (B) form soluble complexes with cationic species
    (C) precipitate cationic species (D) precipitate anionic species. **(I.I.T. 2002)**
70. Which of the following statements is incorrect?
    (A) H_3PO_4 on gentle heating forms pyrophosphoric acid whose basicity is 4.
    (B) In P_4O_{10}, $p\pi$-$d\pi$ back bonding is present.
    (C) P_4O_{10} hydrolyses in water to form phosphorous acid.
    (D) P_4O_6 is less acidic than P_4O_{10}.
71. Which of the following is not isostructural with PO_4^{3-} ion?
    (A) SO_4^{2-} (B) ClO_4^- (C) CrO_4^{2-} (D) SF_4
72. Identify the acid in which phosphorus exists in lowest oxidation state.
    (A) Hypophophorous acid (B) Pyrophosphoric acid
    (C) Phosphorous acid (D) Hypophosphoric acid
73. Identify the acid which acts as a reducing agent.
    (A) Hypophosphorous acid (B) Hypophosphoric acid
    (C) Orthophosphoric acid (D) Pyrophosphoric acid
74. Which of the following acids of phosphorous contains P—P bond?
    (A) Hypophosphorous acid (B) Hypophosphoric acid
    (C) Pyrophosphoric acid (D) Perdiphosphoric acid
75. For H_3PO_3 and H_3PO_4, the correct choice is
    (A) H_3PO_3 is dibasic and reducing (B) H_3PO_3 is dibasic and non-reducing
    (C) H_3PO_4 is tribasic and reducing (D) H_3PO_3 is tribasic and non-reducing.
    (I.I.T. 2003)
76. One mole of which of the following acids neutralises the least number of moles of KOH?
    (A) H_3PO_2 (B) H_3PO_3 (C) $H_4P_2O_7$ (D) H_3PO_4
77. Ammonium dichromate on heating gives a gas which is also given by
    (A) heating NH_4NO_2 (B) heating NH_4NO_3
    (C) $Mg_3N_2 + H_2O$ (D) Na (comp.) + H_2O_2 **(I.I.T. 2004)**

    Hint: $(NH_4)_2Cr_2O_7 \xrightarrow{\text{heat}} Cr_2O_3 + 4H_2O + N_2$; $NH_4NO_2 \xrightarrow{\text{heat}} 2H_2O + N_2$
78. Which of the following allotropes of phosphorus is thermodynamically most stable?
    (A) Red (B) White (C) Black (D) Yellow **(I.I.T. 2005)**

    Hint: Black phosphorus is most stable due to the presence of layered structure.
79. A pale blue liquid which is obtained by reacting equimolar mixture of two gases at – 30°C is
    (A) N_2O (B) N_2O_3 (C) N_2O_4 (D) N_2O_5 **(I.I.T. 2005)**

    Hint: $NO + NO_2 \xrightarrow{-30°C} N_2O_3$ (Pale blue)

80. The percentage of p-character in the orbitals forming P—P bonds in P_4 is

(A) 25 (B) 33 (C) 50 (D) 75 **(I.I.T. 2007)**

Hint: The hybridisation of P in P_4 is sp^3. Hence the percentage of p-character is 75%.

81. A solution of colourless salt H on boiling with excess NaOH produces a non-flammable gas. The gas evolution ceases after some time. Upon addition of Zn dust to the same solution, the gas evolution restarts. The colourless salt(s) H is(are)

(A) NH_4NO_3 (B) NH_4NO_2 (C) NH_4Cl (D) $(NH_4)_2SO_4$ **(I.I.T. 2008)**

Hint: Since the colourless salt produces NH_3 gas (non-inflammable) on boiling with excess of NaOH, it is an ammonium salt. On addition of zinc dust (a reducing agent) to this solution, only the nitrites and nitrates will liberate NH_3 gas again.

$$4Zn + NO_3^- + 7OH^- \longrightarrow 4ZnO_2^{2-} + 2H_2O + NH_3$$

$$3Zn + NO_2^- + 5OH^- \longrightarrow 3ZnO_2^{2-} + H_2O + NH_3$$

82. Phosphine is a flammable gas and is prepared from white phosphorus. White phosphorus on reaction with NaOH gives PH_3 as one of the products. This is a

(A) dimerization reaction (B) disproportionation reaction

(C) condensation reaction (D) precipitation reaction. **(I.I.T. 2008)**

Hint: $\overset{0}{P}_4 + 3NaOH + 3H_2O \longrightarrow 3NaH_2\overset{+1}{P}O_2 + \overset{-1}{P}H_3$

This is a disproportionation reaction.

83. The reaction of P_4 with X leads selectively to P_4O_6. The X is

(A) dry O_2 (B) a mixture of O_2 and N_2

(C) moist O_2 (D) O_2 in the presence of aquesus NaOH.

(I.I.T. 2009)

Hint: $P_4 + 3O_2 \xrightarrow{N_2} P_4O_6$ (exclusively)

(N_2 is used to retard the further oxidation)

84. The nitrogen oxide(s) that contain(s) N—N bond(s) is(are)

(A) N_2O (B) N_2O_3 (C) N_2O_4 (D) N_2O_5 **(I.I.T. 2009)**

Hint: N_2O, N_2O_3 and N_2O_4 contain N—N bonds. Their structures are as follows:

N_2O N≡N→O N_2O_3 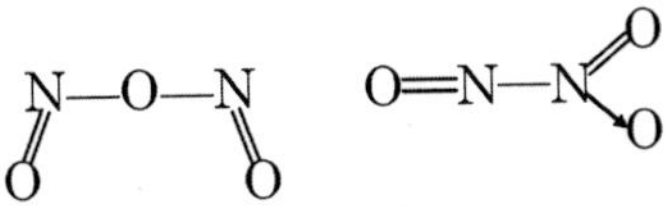

N_2O_4 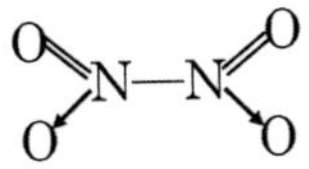 N_2O_5 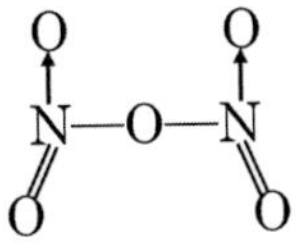

85. Extra pure N_2 can be obtained by heating

(A) NH_3 with CuO (B) NH_4NO_3

(C) $(NH_4)_2Cr_2O_7$ (D) $Ba(N_3)_2$ **(I.I.T. 2011)**

Hint: $Ba(N_3)_2 \xrightarrow{\Delta} Ba + 3N_2$

86. Among the following, the correct statement is

(A) Phosphates have no biological significance in humans.

(B) Between nitrates and phosphates, phosphates are less abundant in earth's crust.

(C) Between nitrates and phosphates, nitrates are less abundant in earth's crust.

(D) Oxidation of nitrates is possible in soil. **(I.I.T. 2008)**

Hint: Nitrates are less abundant than phosphates in earth's crust as they are soluble in water. Moreover, they can also be reduced by numerous microbes present in the earth's crust.

87. Among the following, the correct statement is

(A) Between NH_3 and PH_3, NH_3 is a better electron donor because the lone pair of electrons occupies spherical '*s*' orbital and is less directional.

(B) Between NH_3 and PH_3, PH_3 is a better electron donor because the lone pair of electrons occupies spherical sp^3 orbital and is more directional.

(C) Between NH_3 and PH_3, NH_3 is a better electron donor because the lone pair of electrons occupies spherical sp^3 orbital and is more directional.

(D) Between NH_3 and PH_3, PH_3 is a better electron donor because the lone pair of electrons occupies spherical '*s*' orbital and is less directional. **(I.I.T. 2008)**

Hint: NH_3 is a better electron donor than PH_3 because the lone pair of electrons on N-atom occupies sp^3 orbital and is more directional due to smaller size of N-atom.

ANSWERS

1. C	2. A	3. A	4. A, B	5. D	6. D
7. A, D	8. A	9. B	10. A, C, D	11. C	12. B
13. D	14. B	15. C	16. B	17. A	18. C
19. A	20. D	21. A	22. B	23. C	24. A
25. A	26. C	27. E	28. E	29. C	30. A, B
31. A, B	32. C	33. A	34. B	35. D	36. A
37. C, D	38. B	39. B	40. A	41. B	42. B
43. D	44. A	45. C	46. A, B	47. A, B	48. A
49. D	50. A	51. B	52. D	53. B	54. A
55. A	56. B	57. C, D	58. A	59. B, C	60. B, D
61. A	62. B	63. A	64. A	65. A	66. D
67. D	68. A	69. B	70. C	71. D	72. A
73. A	74. B	75. A	76. A	77. A	78. C
79. B	80. D	81. A, B	82. B	83. B	84. C
85. D	86. C	87. C			

22

Group 16 Elements: Oxygen Family

OBJECTIVE TYPE MULTIPLE CHOICE QUESTIONS

1. The number of S—S bonds in sulphur trioxide trimer (S_3O_9) is
   (A) 3 (B) 2 (C) 1 (D) 0 **(I.I.T. 2001)**
2. The geometry of H_2S and its dipole moment are
   (A) angular and non-zero (B) angular and zero
   (C) linear and non-zero (D) linear and zero.
   (I.I.T. 1999, Roorkee 2000)
3. The oxidation number of sulphur in S_8, S_2F_2 and H_2S is respectively
   (A) 0, + 1, – 2 (B) + 2, + 1, – 2
   (C) 0, + 1, + 2 (D) – 2, + 1, + 2 **(I.I.T. 1999)**
4. Hydrolysis of one mole of peroxodisulphuric acid produces
   (A) two moles of sulphuric acid (B) two moles of peroxomonosulphuric acid
   (C) one mole of sulphuric acid, one mole of peroxomonosulphuric acid and one mole of H_2O_2
   (D) one mole each of sulphuric acid and peroxomonosulphuric acid. **(I.I.T. 1996)**
5. The critical temperature of water is higher than that of O_2 because the H_2O molecule has
   (A) dipole moment (B) V-shape
   (C) two covalent bonds (D) fewer electrons than O_2. **(I.I.T. 1997)**
6. Which compound acts as an oxidising as well as reducing agent?
   (A) SO_2 (B) MnO_2 (C) Al_2O_3 (D) CrO_3 **(I.I.T. 1991)**
7. There is no S—S bond in
   (A) $S_2O_4^{2-}$ (B) $S_2O_5^{2-}$ (C) $S_2O_3^{2-}$ (D) $S_2O_7^{2-}$
   (I.I.T. 1991, Roorkee 2000)

8. The oxidation states of the most electronegative element in the products of the reaction of BaO_2 with dil. H_2SO_4 are
   (A) – 1, – 2 (B) – 2, 0 (C) – 2, + 1 (D) 0, + 1 **(I.I.T. 1991)**
9. The compound which gives off oxygen on moderate heating is
   (A) CuO (B) HgO (C) ZnO (D) Al_2O_3 **(I.I.T. 1986)**
10. The volume strength of 1.5*N* H_2O_2 is
    (A) 4.8 (B) 8.4 (C) 3.0 (D) 8.0 **(I.I.T. 1991)**
11. The hybridisation of sulphur in SO_2 is
    (A) *sp* (B) sp^3 (C) sp^2 (D) dsp^2 **(I.I.T. 1986)**
12. A gas which is not collected over water is
    (A) N_2 (B) O_2 (C) SO_2 (D) PH_3 **(I.I.T. 1985)**
13. HCl is added to the following oxides. Which one would give H_2O_2?
    (A) MnO_2 (B) BaO (C) PbO_2 (D) None. **(I.I.T. 1980)**
14. A black sulphide is formed by the action of H_2S on
    (A) NaCl (B) $CaCl_2$ (C) $ZnCl_2$ (D) $CuCl_2$ **(I.I.T. 1978)**
15. From which of the following is SO_2 formed?
    (A) Reaction of dil. H_2SO_4 with O_2 (B) Hydrolysis of dil. H_2SO_4
    (C) Reaction of conc. H_2SO_4 on $Na_2S_2O_3$
    (D) All of these. **(Roorkee 1995)**
16. When sulphur is boiled with Na_2SO_3 solution, the compound formed is
    (A) Na_2S (B) Na_2SO_4 (C) $Na_2S_2O_8$ (D) $Na_2S_2O_3$ **(Roorkee 1992)**
17. The species which do not contain peroxide ions are
    (A) PbO_2 (B) H_2O_2 (C) SrO_2 (D) BaO_2 **(I.I.T. 1992)**
18. Oxygen is more electronegative than sulphur. Yet H_2S is acidic while H_2O is neutral. This is because
    (A) H_2S is gas while H_2O is a liquid (B) H—S bond is weaker than H—O bond
    (C) H_2O is highly associated compound
    (D) molecular mass of H_2S is more than that of H_2O.
19. Sulphur has the highest electronegativity in
    (A) S_2Cl_2 (B) SCl_2 (C) SCl_4 (D) same.
20. The final acid obtained during the manufacture of H_2SO_4 is
    (A) conc. H_2SO_4 (B) dil. H_2SO_4 (C) $H_2S_2O_7$ (D) $H_2S_2O_6$
21. Identify the incorrect statement about an acidic solution of H_2O_2.
    (A) It reduces $MnO_4^- \longrightarrow Mn^{2+}$ (B) It oxidises $I^- \longrightarrow I_2$
    (C) It oxidises $Fe^{2+} \longrightarrow Fe^{3+}$ (D) It reduces $H_3AsO_4 \longrightarrow H_3AsO_3$
22. Which of the following is a true peroxide?
    (A) Na_2O_2 (B) MnO_2 (C) BaO_2 (D) SO_2 **(Roorkee 1995)**
23. The highest number of oxo group is possessed by
    (A) H_2SO_4 (B) H_2SO_3 (C) H_3PO_4 (D) H_3PO_3

24. Sulphuric acid reacts with PCl_5 to give
    (A) thionyl chloride
    (B) sulphur tetrachloride
    (C) sulphur monochloride
    (D) sulphuryl chloride.
25. H_2O_2 oxidises
    (A) $KMnO_4$ (B) PbS (C) MnO_2 (D) H_2S **(Roorkee 1995)**
26. The temperature at which the heavy water has maximum density is
    (A) 1.16°C (B) 4.0°C (C) 4.60°C (D) 11.60°C
27. Which of the following catalysts is used for the manufacture of H_2SO_4 by lead chamber process?
    (A) V_2O_5
    (B) Mo
    (C) A mixture of NO and NO_2
    (D) A mixture of NO and N_2O
28. Identify the correct statement.
    (A) SO_2 is the acid anhydride of H_2SO_4.
    (B) The $\angle$O—S—O in SO_2 is 119.5°.
    (C) Sulphur exhibits a maximum covalency of four.
    (D) Plastic, rhombic and monoclinic forms of sulphur are all allotropic forms of sulphur.
29. Identify the incorrect statement.
    (A) SO_2 acts as both oxidising and reducing agent.
    (B) H_2O_2 acts as both oxidising and reducing agent.
    (C) H_2S is a weaker reducing agent than H_2O.
    (D) The bond angle in H_2O is smaller than that in H_2S.
30. Identify the incorrect statement.
    (A) H_2O_2 is a better oxidising agent than water.
    (B) In a sample of 10 volume H_2O_2, gram of H_2O_2 per litre is 30.35.
    (C) The percentage of H_2O_2 in a sample labelled 10 volume is 3.035.
    (D) The formation of $O^{2-}(g)$ from O(*g*) is exothermic whereas that of $F^-(g)$ from F(*g*) is endothermic.
31. Identify the incorrect statement.
    (A) The O—O bond dissociation energy in O_2 is larger than that of S—S bond in S_8.
    (B) Sulphur acts as a cross-linking agent in vulcanisation of rubber.
    (C) Metallic oxides react with water to form bases and non-metallic oxides react with water to form acids.
    (D) The precipitation of the second group sulphides in qualitative analysis is carried out with H_2S in the presence of dil. HNO_3.
32. Silver bromide dissolves in a solution of $Na_2S_2O_3$ to give
    (A) $Ag(S_2O_3)^-$ (B) $Ag(S_2O_3)_2^{3-}$ (C) $Ag(S_2O_3)_3^{5-}$ (D) $Ag(S_2O_3)_4^{7-}$
33. Which is formed when H_2S is passed through dil. HNO_3?
    (A) plastic sulphur
    (B) amorphous sulphur
    (C) monoclinic sulphur
    (D) rhombic sulphur

34. Identify the false statement.
   (A) Monoclinic sulphur is stable at room temperature.
   (B) Rhombic sulphur is stable at room temperature.
   (C) Both monoclinic and rhombic sulphur are soluble in CS_2.
   (D) Rhombic sulphur is converted into monoclinic form but the reverse is not possible.
35. Sulphur reacts with conc. HNO_3 to give
   (A) $H_2SO_4 + H_2O + NO_2$ (B) $H_2SO_4 + H_2O + NO$
   (C) $SO_2 + H_2O + NO$ (D) $SO_2 + H_2O + NO_2$
36. The highest bond dissociation energy is possessed by
   (A) S—S bond (B) O—O bond (C) Se—Se bond (D) Te—Te bond
37. The catalyst used in the manufacture of H_2SO_4 by the contact process is
   (A) MnO_2 (B) Fe_2O_3 (C) Cr_2O_3 (D) V_2O_5
38. Which of the following contains the S—S group?
   (A) $H_2S_2O_6$ (B) $H_2S_2O_8$ (C) $H_2S_2O_7$ (D) $H_2S_2O_5$
39. Which of the following when heated with dil. HCl gives a yellow precipitate and a pungent gas?
   (A) Na_2S (B) Na_2SO_3 (C) $Na_2S_2O_3$ (D) Na_2SO_4
40. Which of the following makes lead pipes corrosion resistant?
   (A) S (B) Se (C) Zn (D) Te
41. Identify the incorrect statement.
   (A) Conc. H_2SO_4 is used for drying H_2 gas.
   (B) Thermal stability of Group 16 hydrides decreases with the increasing size of cations.
   (C) H_2SO_4 acts as a dehydrating agent in its reaction with sugar.
   (D) O_2^{2-} is paramagnetic and O_2^- is diamagnetic.
42. A dilute solution of H_2O_2 cannot be concentrated by
   (A) fractional distillation. (B) vacuum distillation
   (C) treatment with P_4O_{10} (D) careful and slow evaporation at 60°C.
43. An aqueous solution of H_2O_2 is
   (A) weakly acidic (B) strongly acidic
   (C) basic (D) neutral.
44. The reaction, $2H_2O_2 \longrightarrow 2H_2O + O_2$ shows that
   (A) it is an oxidising agent (B) H_2O_2 is decomposed
   (C) it is a reducing agent (D) none of these.
45. When SO_3 is heated with D_2O, we get
   (A) monodeuterium sulphuric acid (B) deuterosulphuric acid
   (C) D_2 and H_2SO_4 (D) none of these.
46. In the contact process of manufacture of H_2SO_4, the impurity of arsenic is removed by
   (A) Fe_2O_3 (B) $Fe(OH)_3$ (C) $Cr(OH)_3$ (D) $Al(OH)_3$

47. With respect to O_2 and O_3, which statement is false?
   (A) O_2 has 2 bonds and O_3 has 3 bonds.
   (B) O_2 is colourless and O_3 is coloured.
   (C) Oxygen valency is 2 in both.
   (D) Both have similar reactivity in water.
48. The products of the chemical reaction between $Na_2S_2O_3$, Cl_2 and H_2O are
   (A) $S + NaClO_3 + H_2O$ (B) $S + Na_2SO_3 + HCl$
   (C) $S + Na_2SO_4 + HCl$ (D) $S + H_2S + HCl$
49. When H_2S is passed over acidified $KMnO_4$, one of the products formed is
   (A) $KHSO_3$ (B) K_2SO_3 (C) S (D) MnO_2
50. Amongst H_2O, H_2S, H_2Se and H_2Te, the one with the highest boiling point is
   (A) H_2O because of H-bonding (B) H_2Te because of higher M.W.
   (C) H_2S because of H-bonding (D) H_2Se because of lower M.W. **(I.I.T. 2000)**
51. Which of the following is obtained from SO_2?
   (A) Sulphamic acid (B) Sodium thiosulphate
   (C) Sodium sulphite (D) Sodium bisulphite
52. Which of the following exhibits highest bond strength?
   (A) O—O bond (B) S—S bond (C) Se—Se bond (D) Te—Te bond
53. CCl_2F_2 destroys ozone by liberating
   (A) F_2 (B) Cl_2 (C) NO_2 (D) all of these.
54. The compound formed in the depletion of ozone layer by chlorofluorocarbons is
   (A) OF_2 (B) OCl^- (C) OCl_2 (D) O_2F_2
55. Which of the following gases is formed when KO_2 reacts with CO_2?
   (A) H_2 (B) N_2 (C) CO (D) O_2
56. The lowest boiling point is possessed by
   (A) H_2O (B) H_2S (C) H_2Se (D) H_2Te
57. Which of the following has the highest boiling point?
   (A) S (B) Se (C) Te (D) Po
58. The process which is used for the production of SO_2 is
   (A) Mg with dil. H_2SO_4 (B) Al with dil. H_2SO_4
   (C) Cu with dil. H_2SO_4 (D) C with dil. H_2SO_4
59. When H_2S reacts with limited O_2, the compounds formed are
   (A) $S + H_2O$ (B) $S + H_2SO_4$ (C) $SO_3 + H_2$ (D) $SO_2 + H_2O$
60. Which of the following contains the dithio (S—S) group?
   (A) $H_2S_2O_6$ (B) $H_2S_2O_7$ (C) $H_2S_2O_5$ (D) $H_2S_2O_8$
61. Which of the following contains the sulpho (S→S) group?
   (A) H_2SO_4 (B) $H_2S_2O_4$ (C) $H_2S_2O_6$ (D) $H_2S_2O_5$
62. The SO_3 molecule contains
   (A) 2 covalent and 1 ionic bonds (B) 2 covalent and 2 ionic bonds

(C) 1 covalent and 2 coordinate bonds
(D) 2 covalent and 2 coordinate bonds.

63. In ozone there are
(A) 2σ and 1π bonds (B) 1σ and 2π bonds
(C) 2σ and 1π bonds and one lone pair of electrons
(D) 2σ and 2π bonds.

64. The structure of H_2O_2 is
(A) planar (B) non-planar (C) spherical (D) linear.

65. Point out in which of the following properties, oxygen differs from the rest of the members of Group 16.
(A) oxidation states (+ 4, + 6) (B) formation of hydrides
(C) high value of ionisation energy (D) polymorphism.

66. All the elements of the oxygen family are
(A) non-metals (B) semi-metals (C) radioactive (D) polymorphic.

67. The decomposition of H_2O_2 is retarded by
(A) formic acid (B) acetamide (C) acetanilide (D) glycerol.

68. The bond angle $\angle O—O—H$ in H_2O_2 is
(A) 94.8° (B) 97° (C) 109°28′ (D) 105°

69. The bond angle $\angle O—O—O$ in ozone is
(A) 95° (B) 109°28′ (C) 116.8° (D) 120°

70. Identify the incorrect statements.
(A) Hydrogen peroxide is more stable in ether than in water.
(B) Hydrogen peroxide acts as both oxidising and reducing agent.
(C) Hydrogen peroxide is a pale blue syrupy liquid.
(D) Two hydroxyl groups lie in the same plane in hydrogen peroxide.

71. A dilute solution of H_2O_2 is concentrated by
(A) heating the dilute solution under reduced pressure.
(B) drying the dilute solution over anhydrous $CuSO_4$.
(C) drying the dilute solution over conc. H_2SO_4.
(D) drying the dilute solution over anhydrous $CaCl_2$.

72. Which of the following has peroxo linkage?
(A) $H_2S_4O_6$ (B) H_2SO_5 (C) $H_2S_2O_8$ (D) $H_2S_2O_3$

73. Hydrogen peroxide decomposes to produce O_2 according to the reaction $2H_2O_2 \rightarrow 2H_2O + O_2$. The number of moles of pure hydrogen peroxide required to produce 100 L of O_2 at STP is
(A) 18.85 (B) 8.93 (C) 13.39 (D) 4.46

74. Identify the bond angles present in SF_4.
(A) 120° (B) 90°, 120° (C) 109°28′ (D) 90°, 120°, 180°

75. Which of the following statements is incorrect?
(A) Hybridisation of S in SO_3 is sp^2. (B) SO_3 is pyramidal in shape.

(C) $S_2O_3^{2-}$ ion is tetrahedral in shape.
(D) SO_3 exists in three polymorphic forms.

76. Which of the following statements is incorrect?
(A) Cl_2 reacts with aqueous solution of $Na_2S_2O_3$ to form $NaHSO_4$ and HCl.
(B) $AgNO_3$ reacts with a solution of $Na_2S_2O_3$ to form a black precipitate.
(C) The reducing action of H_2SO_3 in acid solution is due to SO_2.
(D) SO_3 acts as a strong oxidising agent when hot.

77. Sulphur trioxide oxidises phosphorus to
(A) P_4O_{10} (B) P_2O_3 (C) H_3PO_4 (D) PH_3

78. Which of the following compounds does not exist?
(A) SO_2Cl_2 (B) SCl_4 (C) SCl_6 (D) S_2Cl_2

79. Which of the following statements is correct?
(A) Cl_2 reacts with aqueous $Na_2S_2O_3$ to give SO_3 and NaCl.
(B) Cl_2 reacts with aqueous $Na_2S_2O_3$ to give SO_2 and NaCl.
(C) Cl_2 reacts with aqueous $Na_2S_2O_3$ to give $Na_2S_4O_6$ and NaCl.
(D) Cl_2 reacts with aqueous $Na_2S_2O_3$ to form HCl and $NaHSO_4$

80. Which of the following on heating gives SO_3 gas?
(A) $Fe_2(SO_4)_3$ (B) Na_2SO_3 (C) $CaSO_4$ (D) Na_2SO_4

81. [X] + $H_2SO_4 \longrightarrow$ [Y] (a colourless gas with irritating smell); [Y] + $K_2Cr_2O_7$ + $H_2SO_4 \longrightarrow$ green solution. [X] and [Y] are
(A) CO_3^{2-}, CO_2 (B) S^{2-}, H_2S (C) SO_3^{2-}, SO_2 (D) Cl^-, HCl **(I.I.T. 2003)**

82. Which of the following has —O—O— linkage?
(A) $H_2S_2O_6$ (B) $H_2S_2O_8$ (C) $H_2S_2O_3$ (D) $H_2S_4O_6$ **(I.I.T. 2004)**

83. Which gas is evolved when PbO_2 is treated with conc. HNO_3?
(A) NO_2 (B) O_2 (C) N_2 (D) N_2O **(I.I.T. 2005)**
Hint: $PbO_2 + 2HNO_3 \longrightarrow Pb(NO_3)_2 + H_2O + \frac{1}{2}O_2$

84. Which of the following species have non-distorted octahedral structure?
1. SF_6 2. PF_6^- 3. SiF_6^{2-} 4. SeF_6 5. XeF_6
(A) 1, 2, 4 (B) 1, 2, 3 (C) 1, 3, 4 (D) 2, 3, 4

85. Aqueous solution of $Na_2S_2O_3$ on reaction with Cl_2 gives
(A) $Na_2S_4O_6$ (B) $NaHSO_4$ (C) NaCl (D) NaOH **(I.I.T. 2008)**
Hint: Chlorine is a stronger oxidising agent than iodine. Hence $Na_2S_2O_3$ on oxidation forms $NaHSO_4$ rather than sodium tetrathionate.
$Na_2S_2O_3 + 4Cl_2 + 5H_2O \longrightarrow 2NaHSO_4 + 8HCl$

ANSWERS

1. D	2. A	3. A	4. D	5. A	6. A
7. D	8. A	9. B	10. B	11. C	12. C

13. D	14. D	15. C	16. D	17. A, C	18. B
19. C	20. C	21. D	22. A, C	23. A	24. D
25. B, D	26. D	27. C	28. B, D	29. C, D	30. D
31. D	32. C	33. B	34. A, D	35. A	36. D
37. D	38. A, D	39. C	40. D	41. A, D	42. C
43. D	44. B	45. B	46. B	47. D	48. D
49. C	50. A	51. B, C, D	52. D	53. B	54. B
55. D	56. B	57. C	58. C, D	59. A	60. A, C
61. A	62. D	63. C	64. B	65. A, C	66. D
67. C, D	68. B	69. C	70. D	71. A	72. B, C
73. B	74. D	75. B	76. B	77. A	78. C
79. D	80. A	81. C	82. B	83. B	84. A
85. B					

23

Group 17 Elements: Halogen Family

OBJECTIVE TYPE MULTIPLE CHOICE QUESTIONS

1. The reaction, $3ClO^-(aq) \longrightarrow ClO_3^-(aq) + 2Cl^-(aq)$, is an example of
   (A) oxidation reaction (B) reduction reaction
   (C) disproprotionation reaction (D) decomposition reaction. **(I.I.T. 2001)**
2. Which of the following statements is/are correct when a mixture of NaCl and $K_2Cr_2O_7$ is warmed with conc. H_2SO_4?
   (A) A deep red vapour is evolved
   (B) The vapour when passed into NaOH solution gives a yellow solution of Na_2CrO_4
   (C) Cl_2 gas is evolved
   (D) Chromyl chloride is formed. **(I.I.T. 1998)**
3. Which of the following species is not a pseudohalide?
   (A) CNO^- (B) $RCOO^-$ (C) OCN^- (D) NNN^- **(I.I.T. 1997)**
4. The following acids have been arranged in the order of decreasing acid strength. Identify the correct order. HOCI (I), HOBr (II), HOI (III)
   (A) I > II > III (B) II > I > III (C) III > II > I (D) I > III > II **(I.I.T. 1996)**
5. H_2SO_4 cannot be used to obtain HBr from KBr because
   (A) H_2SO_4 oxidises HBr (B) H_2SO_4 reduces HBr
   (C) HBr disproportionates (D) KBr reacts more slowly. **(I.I.T. 1995)**
6. Which of the following pairs when mixed produces Cl_2 gas at room temperature?
   (A) NaCl + conc. H_2SO_4 (B) NaCl + MnO_2
   (C) Conc. HCl + $KMnO_4$ (D) NaCl + conc. HNO_3 **(I.I.T. 1995)**
7. The order of electron affinity of F, Cl and Br is
   (A) Br < F < Cl (B) F < Br < Cl (C) F > Cl > Br (D) Br < Cl < F **(I.I.T. 1993)**
8. The halogen most easily reduced is
   (A) F_2 (B) Cl_2 (C) Br_2 (D) I_2 **(I.I.T. 1992)**

9. The element(s) which exist(s) in the liquid state is/are
   (A) bromine (B) mercury (C) gallium (D) sodium. **(I.I.T. 1991)**
10. The correct increasing order of ionic size of N^{3-}, Na^+, F^-, O^{2-}, and Mg^{2+} is
    (A) $Mg^{2+} < Na^+ < F^- < O^{2-} < N^{3-}$ (B) $Na^+ < Mg^{2+} < F^- < O^{2-} < N^{3-}$
    (C) $N^{3-} < O^{2-} < F^- < Na^+ < Mg^{2+}$ (D) $F^- < N^{3-} < O^{2-} < Mg^{2+} < Na^+$ **(I.I.T. 1991)**
11. Which of the following is the strongest acid?
    (A) $ClO_3(OH)$ (B) $ClO_2(OH)$ (C) $SO_2(OH)_2$ (D) $SO(OH)_2$ **(I.I.T. 1989)**
12. The correct increasing order of bond dissociation energy for N_2, O_2, F_2 and Cl_2 is
    (A) $N_2 < O_2 < F_2 < Cl_2$ (B) $F_2 < Cl_2 < O_2 < N_2$
    (C) $F_2 < Cl_2 < N_2 < O_2$ (D) $N_2 < Cl_2 < F_2 < O_2$ **(I.I.T. 1988)**
13. Which of the following has the highest boiling point?
    (A) N_2 (B) O_2 (C) F_2 (D) Cl_2 **(I.I.T. 1988)**
14. Which of the following is thermally most stable?
    (A) HOCl (B) $HOClO_2$ (C) $HOClO_3$ (D) HOClO **(I.I.T. 1988)**
15. The increasing order of thermal stability is
    (A) $HOCl < HOClO < HOClO_2 < HOClO_3$
    (B) $HOClO_3 < HOClO_2 < HOClO < HOCl$
    (C) $HOClO < HOCl < HOClO_2 < HOClO_3$
    (D) $HOClO_2 < HOClO_3 < HOClO < HOCl$ **(I.I.T. 1988)**
16. Bromine can be liberated from KBr solution by the action of
    (A) Iodine solution (B) Chlorine water
    (C) NaCl (D) KCl **(I.I.T. 1987)**
17. In which of the following iodine has the highest oxidation number?
    (A) I_2 (B) HI (C) HIO_4 (D) ICl **(I.I.T. 1986)**
18. Which of the following bonds is the strongest?
    (A) H—Cl (B) H—Br (C) H—F (D) H—I **(I.I.T. 1986)**
19. The increasing order of acid strength is
    (A) $HClO < HClO_2 < HClO_3 < HClO_4$
    (B) $HClO_2 < HClO < HClO_3 < HClO_4$
    (C) $HClO_4 < HClO_3 < HClO < HClO_2$
    (D) $HClO_3 < HClO_4 < HClO_2 < HClO$ **(I.I.T. 1986, 2001)**
20. Hydrogen bonding is the strongest in
    (A) O—H···S (B) S—H···O (C) F—H···F (D) F—H···O **(I.I.T. 1986)**
21. Chlorine acts as a bleaching agent in the presence of
    (A) pure oxygen (B) moisture (C) sunlight (D) dry air. **(I.I.T. 1983)**
22. HBr and HI reduce sulphuric acid: HCl reduces $KMnO_4$. HF can reduce
    (A) H_2SO_4 (B) $KMnO_4$ (C) $K_2Cr_2O_7$ (D) none. **(I.I.T. 1981)**
23. Among the following, the weakest acid is
    (A) HF (B) HCl (C) HBr (D) HI **(I.I.T. 1980)**

24. A solution of KBr is treated with each of the following. Which one will liberate bromine?
(A) Cl_2 (B) HI (C) I_2 (D) SO_2 **(I.I.T. 1980)**
25. Which of the following is most acidic?
(A) $HOClO_3$ (B) $HOClO_2$ (C) HOClO (D) HOCl **(I.I.T. 1986)**
26. Which of the following is thermally most stable?
(A) HF (B) HCl (C) HBr (D) HI **(I.I.T. 1980)**
27. Which of the following has the largest size?
(A) Br^- (B) I^+ (C) I (D) I^- **(Roorkee 1999)**
28. Which of the following is correct about the reaction: $3NaClO \xrightarrow{\text{heat}} NaClO_3 + 2NaCl$?
(A) It is a disproportionation reaction
(B) Oxidation number of Cl decreases and increases
(C) The reaction is used in the manufacture of halates
(D) All of these. **(Roorkee 1999)**
29. A greenish-yellow gas reacts with an alkali metal hydroxide to form a halate, which can be used in fire works and safety matches. The gas and halate are respectively
(A) Br_2, $KBrO_3$ (B) Cl_2, $KClO_3$ (C) I_2, $NaIO_3$ (D) Cl_2, $NaClO_3$
(Roorkee 2000)

30. Fluorine is the best oxidising agent because
(A) it has the highest electron affinity
(B) it has the lowest electron affinity
(C) it has the highest E^0_{red} (D) it has the highest E^0_{ox}
31. When iodine is passed through aqueous solution of KCl, KF and KBr
(A) Cl_2 and Br_2 are evolved (B) Cl_2, Br_2 and F_2 are evolved
(C) Cl_2 is evolved (D) none of these.
32. Which of the following properties increases on going down the group in Group 17?
(A) Electronegativity (B) Oxidising power
(C) Volatile nature (D) Ionic radius.
33. Which of the following is true about ClO_2?
(A) It dimerises in liquid phase (B) It is an angular molecule
(C) It is paramagnetic (D) It is very reactive.
34. On heating $KClO_3$, we get
(A) ClO_2 (B) ClO_3 (C) Cl_2O_7 (D) Cl_2O
35. Which of the following property is incorrect with respect to the property indicated?
(A) Oxidising power: $F_2 > Cl_2 > Br_2$ (B) Bond energy: $F_2 > Cl_2 > Br_2$
(C) En: $F_2 > Cl_2 > Br_2$ (D) EA: $Cl_2 > F_2 > Br_2$
36. Among Cl^-, Br^- and I^-, the correct order for being oxidised to dihalogen is
(A) $Cl^- > Br^- > I^-$ (B) $I^- > Br^- > Cl^-$ (C) $I^- > Cl^- > Br^-$ (D) $Br^- > I^- > Cl^-$

37. The order of decreasing polarisability among halogens is
(A) Cl > F > Br > I
(B) F > Cl > Br > I
(C) I > Cl > Br > F
(D) I > Br > Cl > F

38. Which of the following is most volatile?
(A) HCl
(B) HF
(C) HBr
(D) HI

39. Which of the following oxides of chlorine is paramagnetic?
(A) Cl_2O
(B) ClO_2
(C) Cl_2O_6
(D) Cl_2O_7

40. Which of the following is the strongest acid?
(A) HI
(B) HCl
(C) HBr
(D) HF **(Roorkee 1995)**

41. Preparation of F_2 by direct electrolysis of HF(*l*) is not practical because
(A) very low temperature is required
(B) very high temperature is required
(C) electrolytic products are different from F_2
(D) HF is a non-electrolyte.

42. The purpose of addition of KF to HF in the preparation of F_2 by electrolysis is
(A) to decrease the oxidation potential of HF
(B) to increase the concentration of F_2
(C) to increase the conductance of HF
(D) none of these.

43. Identify the incorrect statement.
(A) Bromine can be prepared from carnallite.
(B) Chlorine can bleach a wet piece of cloth.
(C) Bromine is liberated when iodine is passed through an acidic KBr solution.
(D) Iodine can be removed by hypo solution.

44. Identify the incorrect statement.
(A) IUPAC name of ClO_3^- is trioxochlorate(V).
(B) O^{2-} has a higher ionic radius than F^-.
(C) Chlorine atom is smaller than sulphur atom.
(D) IF_5 possesses zero dipole moment.

45. Identify the incorrect statement.
(A) Chlorine displaces iodine from KI.
(B) Tincture of iodine contains KI, I_2 and rectified spirit.
(C) The reaction between H_2 and F_2 is very fast.
(D) Fluorine forms a number of oxoacids.

46. Identify the incorrect statement.
(A) HF is a weak acid while HCl, HBr and HI are strong acids.
(B) Cl_2 bleaches flowers permanently while SO_2 bleaches flowers temporarily.
(C) When blue litmus paper is dipped into a solution of HOCl, it first turns red and is then decolourised.
(D) In the preparation of HI from alkali iodides, H_2SO_4 is preferred to H_3PO_4.

47. Identify the incorrect statement.
   (A) The increase in solubility of iodine in aqueous solution of KI is due to the formation of KI_3
   (B) When iodine reacts with hot NaOH solution, the products are NaI and $NaIO_3$
   (C) Chlorine is the most reactive halogen in the halogenation of alkanes under sunlight
   (D) AgF is sparingly soluble in water.
48. Identify the incorrect statement.
   (A) Bromine turns starch paper blue (B) I^- is a better nucleophile than Br^-
   (C) Dilute HCl oxidises Fe to Fe^{2+} (D) HCl gas is dried by passing it through P_2O_5.
49. Which of the following is formed when chlorine reacts with cold dilute sodium hydroxide?
   (A) NaCl, $NaClO_3$ (B) $NaClO_3$ (C) NaCl, NaClO (D) NaClO, $NaClO_3$
50. Which of the following is formed when iodine reacts with hot and concentrated solution of NaOH?
   (A) I^-, IO_3^- (B) I^-, IO^- (C) I^-, IO_4^- (D) $I^-, I_2O_9^{4-}$
51. Which of the following is formed when iodine reacts with cold dilute solution of NaOH?
   (A) I^-, IO^- (B) I^-, IO_3^- (C) I^-, IO_4^- (D) $I^-, I_2O_9^{4-}$
52. Identify the false statement regarding Group 17 elements.
   (A) Electron affinity decreases with the increase of atomic number.
   (B) Ionisation energy decreases with the increase of atomic number.
   (C) Enthalpy of fusion increases with the increase of atomic number.
   (D) Electronegativity decreases with the increase of atomic number.
53. Which of the following reactions will not occur?
   (A) $I_2 + Cl^- \longrightarrow$ (B) $Br_2 + I^- \longrightarrow$ (C) $Cl_2 + Br^- \longrightarrow$ (D) $F_2 + Cl^- \longrightarrow$
54. Identify the false statement.
   (A) In acidic medium, chlorites are stable and in alkaline medium they are unstable.
   (B) Hypochlorous ion disproportionates in aqueous medium.
   (C) Chlorous acid is a strong acid.
   (D) H_3PO_3 is reducing in character.
55. The strongest Bronsted base is
   (A) ClO_4^- (B) ClO_3^- (C) ClO_2^- (D) ClO^-
56. The hybridisation of Cl in ClO_2^- is
   (A) *sp* (B) sp^2 (C) sp^3 (D) none. **(I.I.T. 1992)**
57. Chlorine is prepared by the
   (A) action of HCl on MnO_2 (B) electrolysis of aqueous NaCl
   (C) evaporation of sea water which contains metal chlorides
   (D) action of conc. H_2SO_4 on NaCl in the presence of MnO_2

58. The increasing order of bond dissociation energy is
(A) $I_2 < F_2 < Br_2 < Cl_2$ (B) $F_2 < Cl_2 < Br_2 < I_2$
(C) $I_2 < F_2 < Cl_2 < Br_2$ (D) $Cl_2 < I_2 < F_2 < Br_2$

59. Which of the following halogens does not form oxoacids?
(A) Fluroine (B) Chlorine (C) Bromine (D) Iodine

60. Identify the correct statement for the compounds HF, HBr, H_2Te and H_3P.
(A) HF (strong acid) (B) H_2Te (strong alkali)
(C) HBr (strong acid) (D) H_3P (strong alkali)

61. Iodine oxidises red phosphorus in aqueous solution to
(A) H_2PO_3 (B) H_3PO_2 (C) HPO_3 (D) H_3PO_4

62. The hybridisation of I in IF_5 is
(A) sp^3d^2 (B) sp^3d^3 (C) sp^3d (D) none of these.

63. Identify the incorrect statement.
(A) MgF_2 is covalent but $MgCl_2$ is ionic.
(B) The bond dissociation energy of Cl—Cl bond is higher than that of F—F bond.
(C) Aluminium chloride is ionic in aqueous solution.
(D) HBr is a stronger acid than HI due to the presence of hydrogen bonding in HBr.

64. The colour of the halogens is due to the electronic transition in the molecular orbitals
(A) $\pi^- \longrightarrow \pi^*$ (B) $\pi^{*-} \longrightarrow \sigma^*_{2p_z}$ (C) $\pi^- \longrightarrow \sigma^*_{2p_z}$ (D) $\sigma \longrightarrow \sigma^*$

65. A gas X_2 is passed through water to form a saturated solution. The aqueous solution on treatment with $AgNO_3$ gives a white precipitate. The saturated aqueous solution dissolves magnesium ribbon with evolution of a colourless gas Y. Identify X and Y.
(A) X = CO_2, Y = Cl_2 (B) X = Cl_2, Y = CO_2
(C) X = Cl_2, Y = H_2 (D) X = H_2, Y = Cl_2 **(I.I.T. 2002)**

Hint: $\underset{X}{Cl_2} + H_2O \longrightarrow HCl + HOCl$

$HCl + AgNO_3 \longrightarrow \underset{\text{White}}{AgCl\downarrow} + HNO_3$

$2HCl + Mg \longrightarrow MgCl_2\downarrow + H_2$

Hence X = Cl_2, Y = H_2

66. Which of the following statements is not correct?
(A) The order of oxidising character of ClO_4^-, BrO_4^- and IO_4^- is:
$ClO_4^- < IO_4^- < BrO_4^-$
(B) The whole of Cl_2 dissolved in water exists as HOCl and HCl
(C) ClO^- ion disproportionates in aqueous solution.
(D) HOCl is known only in aqueous solution.

67. Which of the following are formed when Cl_2 reacts with water?
(A) HCl + $HClO_3$ (B) HCl + $HClO_2$ (C) HCl + HOCl (D) HCl + O_2

68. Which of the following are formed when bleaching powder slowly decomposes?
(A) $CaCO_3 + Cl_2$ (B) $CaCl_2 + Ca(OCl)_2$
(C) $CaO + Cl_2$ (D) $CaCl_2 + Ca(ClO_3)_2$

69. Which of the following is formed when I_2 reacts with cold dil. NaOH?
(A) $Na + HIO_3$ (B) $NaI + HOI$ (C) $NaI + NaIO_3$ (D) $NaI + NaOI$

70. When I^- ion is oxidised by MnO_4^- ion in alkaline medium, I^- ion converts into
(A) IO_3^- (B) I_2 (C) IO_4^- (D) IO^- **(I.I.T. 2004)**
Hint: When I^- ion is oxidised by MnO_4^- ion in alkaline solution, I^- ion is converted into IO_3^- ion.
$KI + 2KMnO_4 + H_2O \longrightarrow 2MnO_2 + KIO_3 + 2KOH$

71. Identify the correct statements.
1. On heating, $Rb(ICl)_2$ will decompose giving RbCl and ICl
2. CsI_3 will ionise to give $Cs^+I_3^-$
3. KBrICl will ionise to give KCl and IBr
4. In the reaction $I^- + I_2 \longrightarrow I_3^-$, the Lewis acid is I_2
(A) 1, 3, 4 (B) 1, 2, 3 (C) 2, 3, 4 (D) 1, 2, 3, 4

72. Which of the following is an interhalogen compound?
(A) FCl_5 (B) FI_7 (C) BrF_5 (D) ICl_3

ANSWERS

1. C	2. A, B, D	3. B	4. A	5. A	6. C
7. A	8. A	9. A, B, C	10. A	11. A	12. B
13. D	14. C	15. A	16. B	17. C	18. C
19. A	20. C	21. B	22. D	23. A	24. A
25. A	26. A	27. D	28. D	29. B	30. C
31. D	32. D	33. C	34. A	35. B	36. B
37. D	38. A	39. B	40. A	41. D	42. C
43. C	44. D	45. D	46. D	47. D	48. A
49. C	50. A	51. A	52. A	53. A	54. A, C
55. D	56. C	57. A, B, D	58. A	59. A	60. C
61. A	62. A	63. D	64. B	65. C	66. B
67. C	68. D	69. D	70. A	71. A	72. C, D

24

Group 18 Elements: Noble Gases

OBJECTIVE TYPE MULTIPLE CHOICE QUESTIONS

1. The density of neon will be highest at
   (A) 0°C, 2 atm (B) 273°C, 1 atm (C) 273°C, 2 atm (D) STP **(I.I.T. 1990)**
2. The shape of XeF_4 is
   (A) square planar (B) tetrahedral
   (C) trigonal pyramidal (D) linear. **(Roorkee 1998)**
3. The hybridisation of Xe in XeF_4 is
   (A) sp^2 (B) sp^3 (C) sp^3d (D) sp^3d^2
4. The hybridisation of Xe in XeF_6 is
   (A) sp^3 (B) sp^3d (C) sp^3d^2 (D) sp^3d^3
5. The hybridisation of Xe in XeF_2 is
   (A) sp (B) sp^2 (C) sp^3 (D) sp^3d
6. The hybridisation of Xe in XeO_6^{4-} is
   (A) sp^3 (B) sp^3d (C) sp^3d^2 (D) sp^3d^3
7. The hybridisation of Xe in $XeOF_4$ is
   (A) sp^3 (B) sp^3d (C) sp^3d^2 (D) sp^3d^3
8. The ease of liquefaction of the noble gases is in the order
   (A) Ne < He < Xe < Kr < Ar (B) He < Ne < Ar < Kr < Xe
   (C) Xe < Kr < Ar < Ne < He (D) Ar < He < Xe < Kr < Ne
9. XeF_4 on hydrolysis produces
   (A) XeF_4 (B) $XeOF_2$ (C) $XeOF_4$ (D) XeO_3
10. XeF_6 on hydrolysis produces
    (A) Xe (B) XeO (C) XeO_2 (D) XeO_3

11. Assertion (*A*): Xenon forms fluorides.
    Reason (*R*): It is because 5*d* orbitals are available for valence shell expansion.
    (A) Both *A* and *R* are correct and *R* is the correct explanation of *A*
    (B) Both *A* and *R* are correct and *R* is not the correct explanation of *A*
    (C) *A* is true but *R* is false (D) Both *A* and *R* are false.
12. The forces of attraction in liquid He are
    (A) covalent (B) ionic (C) metallic (D) van der Waals
13. Which of the following is/are paramagnetic?
    (A) Ne^{2+} (B) Be^- (C) Cl^- (D) As^+
14. Argon is used in place of nitrogen in incandescent filament bulbs because
    (A) argon is more inert than nitrogen
    (B) nitrogen forms tungsten nitride
    (C) the thermal conductivity of Ar is lower than that of nitrogen
    (D) none of these.
15. The discovery of the first noble gas compound $Xe^+[PtF_6]^-$ was based on
    (A) high electronegativity of fluorine (B) high reactivity of fluorine
    (C) almost same IE of Xe and O_2 (D) high bond energy of Xe—F bond.
16. The reaction of XeF_6 with SiO_2 gives
    (A) XeF_4 and SiF_4 (B) $XeOF_2$ and SiF_4
    (C) $XeOF_4$ and SiF_4 (D) none of these.
17. Helium-oxygen mixture is used by deep sea divers in preference to N_2–O_2 mixture because
    (A) He is much less soluble in blood than N_2
    (B) N_2 is much less soluble in blood than He
    (C) nitrogen is highly soluble in water
    (D) due to high pressure deep under the sea N_2 and O_2 react to give poisonous NO.
18. The noble gas which is the least polarisable is
    (A) He (B) Ne (C) Xe (D) Kr
19. Identify the incorrect statement.
    (A) All noble gases form compounds with F_2.
    (B) XeF_2 has linear structure.
    (C) The most abundant of the noble gases in the atmosphere is Ne.
    (D) The atomicity of the noble gases is one because they have completely filled valence shells.
20. Helium is found in radioactive minerals because
    (A) it is radioactive in nature (B) it reacts with radioactive elements
    (C) it is formed by disintegration of radioactive elements present in minerals like monazite etc. and remains enclosed
    (D) none of these.

21. Identify the incorrect statement.
 (A) Kr is obtained during radioactive disintegration.
 (B) He is used to produce very low temperature.
 (C) The half-life period of Rn is 3.8 days.
 (D) Argon is used in electric bulbs.
22. Sodium perxenate oxidises Mn^{2+} to MnO_4^- in the presence of catalyst.
 (A) Ni (B) Pd (C) H^+ (D) no catalyst is necessary.
23. The structure of $XeOF_4$ is
 (A) tetrahedral (B) octahedral (C) square planar (D) square pyramidal.
24. The structure of XeF_2 is
 (A) trigonal pyramidal (B) trigonal planar
 (C) square planar (D) linear.
25. The structure of XeF_6 is
 (A) trigonal pyramidal (B) tetrahedral
 (C) distorted octahedral (D) square pyramidal.
26. Which of the following oxides is explosive?
 (A) MgO (B) Na_2O (C) XeO_3 (D) Al_2O_3
27. In the experiments on superconductivity, which of the following gases is used in its liquid form?
 (A) Ne (B) He (C) N_2 (D) H_2
28. Identify the incorrect statement.
 (A) The most abundant noble gas found in the atmosphere is He.
 (B) Rn is obtained by the decay of Ra.
 (C) The reaction of XeF_4 and SbF_5 gives $[XeF_3]^+[SbF_6]^-$.
 (D) The reaction of XeF_6 and KF gives $K[XeF_7]$.
29. Identify the incorrect statement.
 (A) Out of H_2, He, Ne, Xe and CH_4, the highest boiling point is possessed by He.
 (B) In an electric discharge tube containing He, the species present are He^{2+}, HeH^+, HeH^{2+}.
 (C) XeF_6 on reaction with dil. NaOH gives $Na_4[XeO_6]$.
 (D) The spectrum of He is similar to that of H.
30. Identify the incorrect statement.
 (A) The percentage of argon in air is about 2%.
 (B) Xenon can form compounds with F_2 and Cl_2.
 (C) The most polarisable noble gas is Xe.
 (D) In the welding of Al, the noble gas which is used to protect the metal surface is argon.
31. Which of the following fluorides does not exist?
 (A) HeF_2 (B) CF_4 (C) SF_6 (D) XeF_4

32. Which cannot be formed?
    (A) He^{2+} (B) He^+ (C) He_2 (D) He
33. Which of the following is square planar?
    (A) SiF_4 (B) SF_4 (C) XeF_4 (D) NF_3
34. Identify the correct statement(s).
    (A) XeF_2 is a linear molecule. (B) XeF_4 has a square planar structure.
    (C) XeF_2 has a square planar structure.
    (D) XeF_4 is a linear molecule.
35. Identify the incorrect statement.
    (A) Xe is obtained during radioactive disintegration.
    (B) Xe forms $XeOF_2$
    (C) Ne is used in electric bulbs. (D) Ar is used to produce very low temperature.
36. Which of the following is planar?
    (A) XeO_4 (B) XeO_3F (C) XeO_2F_2 (D) XeF_4
37. Which of the following is most abundant in air?
    (A) CO_2 (B) Ne (C) Ar (D) He
38. Which of the following exhibits the weakest intermolecular forces?
    (A) NH_3 (B) HCl (C) H_2O (D) He
39. Which of the following is used by deep sea divers for respiration?
    (A) Mixture of He, O_2 (B) Mixture of Ne, O_2
    (C) Mixture of Ar, O_2 (D) Mixture of N_2, O_2
40. Identify the correct statement(s).
    (A) XeF_6 on hydrolysis with dil. NaOH gives Na_4XeO_6.
    (B) Na_4XeO_6 oxidises Mn^{2+} ion to MnO_4^- ion.
    (C) XeF_6 reacts with SiO_2 to give SiF_4 and $XeOF_4$.
    (D) Krypton is obtained during radioactive disintegration of $^{238}_{92}U$.
41. Which of the following liquids is most volatile?
    (A) Liquid Ne (B) Liquid N_2 (C) Liquid ether (D) Liquid He
42. Which of the following is used from the preparation of Rn?
    (A) Passing electric discharge through air
    (B) Fractional distillation of liquid air
    (C) Radioactive decay of Ra (D) Fractionation of natural gas.
43. The spectrum of He is similar to that of
    (A) Li^+ (B) H (C) He^+ (D) Na
44. Number of lone pair(s) in $XeOF_4$ is
    (A) 0 (B) 1 (C) 2 (D) 3 **(I.I.T. 2004)**

    Hint: Number of lone pair on Xe in $XeOF_4$ is 1. The structure of $XeOF_4$ is as follows:

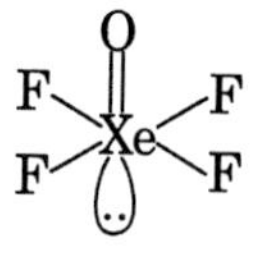

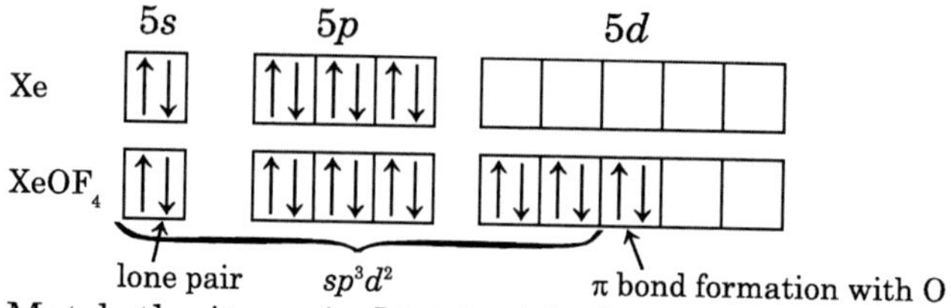

45. Match the items in List I with the items in List II using the codes given below the list.

List I	*List II*
M. XeF_4	1. Distorted octahedral
N. XeF_6	2. Tetrahedral
O. XeO_3	3. Square planar
P. XeO_4	4. Trigonal pyramidal

	M	N	O	P
(A)	1	2	3	4
(B)	1	3	2	4
(C)	3	1	4	2
(D)	2	4	1	3

46. Identify the incorrect statements.
   1. The hybridisation of Xe in XeO_6^{4-} is sp^3d^2.
   2. The structure of XeF_2 is square planar.
   3. Sodium perxenate (Na_4XeO_4) oxidises Mn(II) to Mn(VI).
   4. The hybridisation of Xe in XeO_4 is sp^3.
   5. The structure of XeO_4 is square pyramidal.

   (A) 1, 2, 3 (B) 3, 4, 5 (C) 1, 4, 5 (D) 2, 3, 5

47. Identify the incorrect statement.
   (A) The hybridisation of Xe in XeO_3 is sp^2.
   (B) XeF_6 on hydrolysis with dil. NaOH gives Na_4XeO_6.
   (C) The colour of $O_2[PtF_6]$ is yellow and that of $Xe[PtF_6]$ is orange yellow.
   (D) The catalyst used in the preparation of XeF_2 by heating Xe and F_2 at 400°C is Ne.

48. Identify the correct statement.
   (A) He is more soluble in blood than nitrogen.
   (B) Kr is obtained during radioactive disintegration.
   (C) The reaction of XeF_6 and SiO_2 gives SiF_4 and $XeOF_4$.
   (D) In an electric discharge tube containing He, the species present are He^{2+}, HeH^+, HeH^{2+}.

49. Argon is used in arc welding because of its
   (A) low reactivity with metal (B) ability to lower the melting point of metal
   (C) flammability (D) high calorific value. **(I.I.T. 2007)**

Hint: Argon is used to provide inert atmosphere in high temperature metallurgy (arc welding of metals and alloys).

50. The structure of XeO_3 is
(A) linear (B) planar (C) pyramidal (D) T-shaped. **(I.I.T. 2007)**

Hint: The hybridisation of Xe in XeO_3 is sp^3 and XeO_3 has a pyramidal structure:

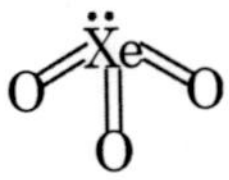

51. XeF_4 and XeF_6 are expected to be
(A) oxidising (B) reducing (C) unreactive (D) strongly basic. **(I.I.T. 2007)**

Hint: $6XeF_4 + 12H_2O \longrightarrow 4Xe + 2XeO_3 + 24HF + 3O_2$; $XeF_6 + 3H_2O \longrightarrow XeO_3 + 6HF$

ANSWERS

1. C	2. A	3. D	4. D	5. D	6. C
7. C	8. B	9. D	10. D	11. A	12. D
13. A, B, D	14. B, C	15. C	16. C	17. A	18. A
19. A, C	20. C	21. A	22. D	23. D	24. D
25. C	26. C	27. B	28. A	29. A, D	30. A, B
31. A	32. C	33. C	34. A, B	35. A, C, D	36. D
37. C	38. D	39. A	40. A, B, C	41. D	42. C
43. A	44. B	45. C	46. A	47. A	48. C, D
49. A	50. C	51. A			

25

Transition Elements

OBJECTIVE TYPE MULTIPLE CHOICE QUESTIONS

1. The complex ion which has no *d*-electrons in the central metal atom is
   (A) $[Co(NH_3)_6]^{3+}$ (B) $[Fe(CN)_6]^{3-}$ (C) $[Cr(H_2O)_6]^{3+}$ (D) $[MnO_4]^-$ **(I.I.T. 2001)**
2. Amongst the following, identify the species with an atom in + 6 oxidation state.
   (A) MnO_4^- (B) $Cr(CN)_6^{3-}$ (C) CrO_2Cl_2 (D) NiF_6^{2-} **(I.I.T. 2000)**
3. In the dichromate dianion
   (A) 4 Cr—O bonds are equivalent (B) all Cr—O bonds are non-equivalent
   (C) 6 Cr—O bonds are equivalent (D) all Cr—O bonds are equivalent.
   (I.I.T. 1999)
4. Which of the following is coloured?
   (A) Ag_2SO_4 (B) CuF_2 (C) MgF_2 (D) CuCl **(I.I.T. 1997)**
5. Ammonium dichromate is used in some fire works. The green coloured powder blowing in the air is
   (A) CrO_3 (B) Cr_2O_3 (C) $CrO(O_2)$ (D) Cr **(I.I.T. 1997)**
6. An aqueous solution of $FeSO_4$, $Al_2(SO_4)_3$ and chrome alum is heated with excess of Na_2O_2 and filtered. The materials obtained are
   (A) a colourless filtrate and a green residue
   (B) a yellow filtrate and a green residue
   (C) a yellow filtrate and a brown residue
   (D) a green filtrate and a brown residue **(I.I.T. 1996)**
7. Which of the following has the maximum number of unpaired electrons?
   (A) Mg^{2+} (B) Ti^{3+} (C) V^{3+} (D) Fe^{2+} **(I.I.T. 1996)**
8. Which of the following oxides is ionic?
   (A) P_4O_{10} (B) CrO_3 (C) MnO (D) Mn_2O_4 **(I.I.T. 1995)**

9. The number of d-electrons of Cr in $[Cr(H_2O)_6]^{3+}$ ion is
   (A) 2 (B) 3 (C) 4 (D) 5 **(I.I.T. 1993)**
10. Among the following ions which one has the highest paramagnetism?
   (A) $[Cr(H_2O)_6]^{3+}$ (B) $[Fe(H_2O)_6]^{2+}$ (C) $[Cu(H_2O)_6]^{2+}$ (D) $[Zn(H_2O)_6]^{2+}$ **(I.I.T. 1993)**
11. Copper(II) sulphate solution reacts with KCN to give
   (A) $Cu(CN)_2$ (B) $K_2[Cu(CN)_4]$ (C) $K_3[Cu(CN)_4]$ (D) CuCN **(I.I.T. 1996)**
12. Identify the complexes which are coloured.
   (A) $[Ti(NO_3)_4]$ (B) $[Cu(NCCH_3)_4]^+[BF_4]^-$
   (C) $[Cr(NH_3)_6]Cl_3$ (D) $K_3[VF_6]$ **(I.I.T. 1994)**
13. Among $Ni(CO)_4$, $[No(CN)_4]^{2-}$ and $[NiCl_4]^{2-}$
   (A) $[NiCl_4]^{2-}$ and $Ni(CO)_4$ are diamagnetic and $[Ni(CN)_4]^{2-}$ is paramagnetic
   (B) $[NiCl_4]^{2-}$ and $[Ni(CN)_4]^{2-}$ are diamagnetic and $Ni(CO)_4$ is paramagnetic
   (C) $Ni(CO)_4$ and $[Ni(CN)_4]^{2-}$ are diamagnetic and $[NiCl_4]^{2-}$ is paramagnetic
   (D) $Ni(CO)_4$ is diamagnetic and $[NiCl_4]^{2-}$ and $[Ni(CN)_4]^{2-}$ are paramagnetic. **(I.I.T. 1991)**
14. Which of the following is not expected to be paramagnetic?
   (A) $[Cu(NH_3)_4]Cl_2$ (B) $[Ag(NH_3)_2]Cl$ (C) NO (D) NO_2 **(I.I.T. 1991)**
15. The oxidation state of Mo in its oxo complex species $[Mo_2O_4(C_2O_4)_2(H_2O)_2]^{2-}$ is
   (A) 2 (B) 3 (C) 4 (D) 5 **(I.I.T. 1991)**
16. The ground state electronic configuration of chromium is
   (A) [Ar] $3d^54s^1$ (B) [Ar] $3d^44s^2$ (C) [Ar] $3d^64s^0$ (D) [Ar] $4d^55s^1$ **(I.I.T. 1989, 1994)**
17. Aqueous solution of the following salts will be coloured in the case of
   (A) $Zn(NO_3)_2$ (B) $LiNO_3$ (C) $Co(NO_3)_2$ (D) $CrCl_3$ **(I.I.T. 1990)**
18. Among the following, the lowest degree of paramagnetism per mole of compound at 298 K will be shown by
   (A) $MnSO_4{\cdot}4H_2O$ (B) $CuSO_4{\cdot}5H_2O$ (C) $FeSO_4{\cdot}6H_2O$ (D) $NiSO_4{\cdot}6H_2O$ **(I.I.T. 1988)**
19. Potassium manganate (K_2MnO_4) is formed when
   (A) chlorine is passed into aqueous $KMnO_4$ solution
   (B) $KMnO_4$ reacts with conc. H_2SO_4 (C) MnO_2 is fused with KOH in air
   (D) formaldehyde reacts with $KMnO_4$ in the presence of strong alkali. **(I.I.T. 1988)**
20. How many unpaired electrons are there in Ni(II)?
   (A) 0 (B) 4 (C) 2 (D) 8 **(I.I.T. 1981)**
21. Which of the following dissolves in hot conc. NaOH solution?
   (A) Fe (B) Zn (C) Cu (D) Ag **(I.I.T. 1980)**
22. Transition elements have the general electronic configuration
   (A) $(n-1)s^2p^6d^{0-9}ns^{1-2}$ (B) $(n-1)s^2p^6d^{1-9}ns^{1-2}$
   (C) $(n-1)s^2p^6d^{0-10}ns^{1-2}$ (D) $(n-1)s^2p^6d^{1-10}ns^{1-2}$

23. The general electronic configuration of the lanthanoids is
(A) $(n-2)s^2p^6d^{10}f^{0-14}(n-1)s^2p^6d^{0-1}ns^{0-2}$
(B) $(n-2)s^2p^6d^{10}f^{1-14}(n-1)s^2p^6d^{0-1}ns^{0-2}$
(C) $(n-2)s^2p^6d^{10}f^{0-14}(n-1)s^2p^6d^{0-1}ns^2$
(D) $(n-2)s^2p^6d^{10}f^{1-13}(n-1)s^2p^6d^{0-1}ns^2$

24. An example of double salt is
(A) bleaching powder (B) $K_4[Fe(CN)_4]$
(C) potash alum (D) hypo. **(Roorkee 1990)**

25. Which of the following does not represent the correct sequence of property indicated?
(A) $Sc^{3+} > Cr^{3+} > Fe^{3+} > Mn^{3+}$: stability of + 3 ions
(B) Sc < Ti < Cr < Mn : number of oxidation states
(C) $Mn^{2+} < Ni^{2+} < Co^{2+} < Fe^{2+}$: magnetic moment
(D) FeO > CoO > NiO > CuO : basic character

26. For Zr^{4+} the ionic radius is 0.86 Å which is almost the same as that for Hf^{4+}. This is due to
(A) diagonal relationship (B) lanthanoid contraction
(C) difference in coordination number of Zr(IV) and Hf(IV)
(D) Hf(IV) forming compounds having lower degree of ionic character.

27. Which of the following oxides of chromium is amphoteric in nature?
(A) CrO (B) CrO_3 (C) Cr_2O_3 (D) CrO_5

28. Which of the following is used as the starting material for the preparation of $K_2Cr_2O_7$?
(A) $K_2SO_4 \cdot Cr_2(SO_4)_3 \cdot 24H_2O$ (chrome alum)
(B) $PbCrO_4 \cdot PbO$ (chrome red)
(C) $PbCrO_4$ (chrome yellow) (D) $FeCr_2O_4$ (chromite)

29. Identify the incorrect statement.
(A) The colour of $(NH_4)_2SO_4 \cdot Fe_2(SO_4)_3 \cdot 24H_2O$ is violet.
(B) $NiSO_4 \cdot 7H_2O$ when dissolved in water gives two ions.
(C) $FeSO_4 \cdot 7H_2O$ is isostructural with $ZnSO_4 \cdot 7H_2O$.
(D) Both $K_3[Fe(CN)_6]$ and $K_4[Fe(CN)_6]$ are diamagnetic.

30. Which of the following statements is incorrect regarding lanthanoids?
(A) Lanthanoids are separated from one another by the ion exchange methods.
(B) The ionic radii of the trivalent lanthanoids increase steadily with increase in the At. No.
(C) Most characteristic oxidation state of lanthanoid elements is + 3.
(D) All lanthanoids are highly dense elements.

31. Transition elements are less reactive because of
(A) high heat of vapourisation
(B) low heat hydration of ions
(C) high ionisation energy
(D) all of these.

32. The tendency of 3d metal ions to form stable complexes is due to
    (A) strong electronegative nature (B) very low ionisation energies
    (C) variable valency
    (D) high charge/radius ratio and vacant d-orbitals.
33. Formation of interstitial compounds makes the transition metal
    (A) more ductile (B) more hard (C) more soft (D) more metallic.
34. In the reaction $Na_2[Fe(CN)_5(NO)] + S^{2-} \longrightarrow Na_4[Fe(CN)_5(NOS)]$ (purple), the oxidation state of iron
    (A) changes from + 2 to 0 (B) changes from + 2 to + 3
    (C) changes from + 3 to + 2 (D) does not change.
35. The decrease of ionic radii of lanthanoid elements in a row is due to
    (A) different shapes of f-orbitals
    (B) the attraction of ns^2 electrons by the nucleus
    (C) efficient shielding by f-orbitals for the attraction of ns^2 electrons by the nucleus
    (D) inefficient shielding by f-orbitals for the attraction of ns^2 electrons by the nucleus.
36. The transition elements exhibit variable valency due to
    (A) different energy of the ns^2 and $(n-1)d^n$ electrons ($n = 1, 2, 3, ...$)
    (B) same energy of the ns^2 and $(n-1)d^n$ electrons
    (C) different energy of the ns and np electrons
    (D) different energy of the $(n-1)d^n$ electrons.
37. Ni(II) has an ionic radius (0.69 Å) smaller than that of Cu(II) (0.72 Å) in contrast to the expectation of ionic radius of Ni(II) being greater than that of Cu(II). This is due to
    (A) the nuclear charge is lower in Cu(II) than that in Ni(II)
    (B) the nuclear charge is higher in Cu(II) than that in Ni(II)
    (C) the repulsion of electrons in the d_{xy}, d_{xz}, d_{yz} orbitals of Cu(II) with the ligand electrons
    (D) the repulsion of electrons in the $d_{x^2-y^2}$ orbital of Cu(II) with the ligand electrons.
38. The magnetic moment (μ) of transition elements is given by the relation $\mu = [n(n+2)]^{1/2}$ BM when n is the number of unpaired electrons. Which of the following statements is true?
    (A) V^{3+} has $\mu = 2.83$ BM (B) Cr^{3+} has $\mu = 4.90$ BM
    (C) Co^{2+} has $\mu = 1.73$ BM (D) Ni^{2+} has $\mu = 1.73$ BM
39. Above Curie temperature iron is
    (A) diamagnetic (B) ferromagnetic
    (C) paramagnetic (D) superconductor.
40. Identify the incorrect statement.
    (A) Number of unpaired electrons in Ni^{2+} is 2.
    (B) Chromium exhibits an oxidation state of – 1.

(C) The general electronic configuration of the lanthanoid elements is $(n-2)s^2p^6d^{10}f^{1-13}(n-1)s^2p^6d^{0-1}ns^2$.

(D) The general electronic configuration of the transition elements is $(n-1)s^2p^6d^{1-10}ns^{1-2}$.

41. Identify the incorrect statement.
 (A) Dilute HCl oxidises Fe to Fe(III).
 (B) Fe^{3+} is more stable than Fe^{2+}.
 (C) A solution of $FeCl_3$ in water gives a brown precipitate on standing.
 (D) A small quantity of H_2SO_4 is added while making a standard solution of Mohr salt since H_2SO_4 prevents the oxidation of Fe(II) to Fe(III).

42. Identify the incorrect statement.
 (A) Transition elements exhibit colour as they absorb energy for the d–p transition.
 (B) The separation of the lanthanoid ions is difficult since the ionic radii of the lanthanoid ions are very close to each other.
 (C) Zr and Hf occur in nature together since their ionic radii (M^{4+}) are almost the same.
 (D) Fluorine helps the transition metal ions to achieve the highest coordination number.

43. Identify the incorrect statement.
 (A) f-orbitals offer inefficient screening in comparison to the d-orbitals.
 (B) Cast iron is an interstitial compound.
 (C) Magnetic moment expected for four unpaired electrons is 4.93 BM.
 (D) Tetrahedral transition metal compounds are intensely coloured than octahedral transition metal compounds.

44. Identify the incorrect statement.
 (A) The hybridisation of Co in $[Co(H_2O)_6]Cl_2$ is d^2sp^3.
 (B) Most transition elements do not displace hydrogen from dilute acids.
 (C) Ionisation energies of the first transition series elements vary much with increasing atomic number.
 (D) Transition metal ions exhibit low enthalpy of hydration.

45. Identify the incorrect statement.
 (A) The hybridisation of Ni in $[Ni(NH_3)_6]Cl_2$ is d^2sp^3.
 (B) Number of d-electrons in Fe^{3+} is equal to the p-electrons in Cl.
 (C) Transition metal compounds are always paramagnetic.
 (D) Ce exhibits the oxidation states of + 2, + 3 and + 4.

46. Identify the false statement.
 (A) The hybridisation of Co in $[Co(H_2O)_6]Cl_3$ is sp^3d^2.
 (B) Both $K_2[PtCl_6]$ and $K_2[NiCl_6]$ are stable coordination compounds.
 (C) The paramagnetic character in $3d$ transition series ions (M^{2+}) increases up to Mn and then decreases.
 (D) Zn is not considered a transition metal.

47. Which of the following coordination compounds is diamagnetic?
(A) $K_2[Ni(CN)_4]$ (B) $[Ni(NH_3)_6]Cl_2$ (C) $Cs_2[NiCl_4]$ (D) $K_3[CoF_6]$
48. Which of the following is formed when $KMnO_4$ undergoes reduction in strongly basic medium?
(A) Mn_2O_7 (B) MnO_2 (C) MnO_4^{2-} (D) Mn^{2+}
49. Amongst the following, the compound that is both paramagnetic and coloured is
(A) $K_2Cr_2O_7$ (B) $(NH_4)_2[FeCl_6]$ (C) $VOSO_4$ (D) $K_3[Cu(CN)_4]$ **(I.I.T. 1997)**
50. Which of the following is not correct for $[Co(NH_3)_6]Cl_3$?
(A) It gives white precipitate with $AgNO_3$
(B) It is diamagnetic
(C) It is paramagnetic (D) It is colourless.
51. Which of the following hydroxides is most basic?
(A) $Lu(OH)_3$ (B) $La(OH)_3$ (C) $Er(OH)_3$ (D) $Sm(OH)_3$
52. Which of the following metals is present in Zeigler–Natta catalyst?
(A) Zr (B) Zn (C) Ti, Al (D) V
53. Which of the following is paramagnetic?
(A) $La(NO_3)_3$ (B) $Lu(NO_3)_3$ (C) $Sc(NO_3)_3$ (D) None of these.
54. Identify the oxides which are basic in nature.
(A) TiO_2 (B) V_2O_5 (C) NiO (D) FeO
55. The magnetic moment (μ) of transition metal compounds is calculated using the formula (where n = number of unpaired electrons)
(A) $\mu = [n(n + 1)]^{1/2}$ BM (B) $\mu = [n(n + 2)]^{1/2}$ BM
(C) $\mu = [n(n - 1)]^{1/2}$ BM (D) $\mu = [n(n - 2)]^{1/2}$ BM
56. Which of the following possesses sp^3d^2 hybridisation?
(A) $K_3[CoF_6]$ (B) $[Co(NH_3)_6]Cl_3$ (C) $K_3[Fe(CN)_6]$ (D) $K_4[Fe(CN)]_6$
57. Transition metal compounds exhibit
(A) paramagnetism (B) diamagnetism
(C) ferromagnetism (D) all of these.
58. Which of the following elements shows maximum number of different oxidation states in its compounds?
(A) La (B) Am (C) Eu (D) Gd
59. The lanthanoid contraction is responsible for the fact that
(A) Zr and Y have about the same radius
(B) Zr and Nb have similar oxidation states
(C) Zr and Hf have about the same radius
(D) Zr and Zn have the same oxidation states.
60. Acidified solution of chromic acid on treatment with H_2O_2 yields
(A) $CrO_3 + H_2O + O_2$ (B) $Cr_2O_3 + H_2O + O_2$
(C) $H_2Cr_2O_7 + H_2O + O_2$ (D) $CrO_5 + H_2O$

61. The dimeric complex $[Cu(CH_3COO)_2{\cdot}H_2O]_2$ exhibits
   (A) paramagnetism (B) ferromagnetism
   (C) ferrimagnetism (D) antiferromagnetism.
62. Wilkinson's catalyst is a transition metal complex in which the metal is
   (A) Co (B) Rh (C) Ru (D) Fe
63. The magnetic moments of the complexes $K_3[FeF_6]$ and $K_3[Fe(CN)_6]$ are respectively
   (A) 5.92 BM, 1.73 BM (B) 1.73 BM, 1.73 BM
   (C) 4.90 BM, 1.73 BM (D) 3.97 BM, 1.73 BM
64. Which of the following compounds exhibits the lowest ionic radius of Cr?
   (A) K_2CrO_4 (B) CrO_2 (C) $CrCl_3$ (D) CrF_3
65. Identify the correct statement(s).
   (A) $La(OH)_3$ is less basic than $Lu(OH)_3$.
   (B) In lanthanoid series, the ionic radius of Ln^{3+} ions decreases.
   (C) La is an element of transition series rather than lanthanoid series.
   (D) Atomic radius of Zr and Hf is same due to the lanthanoid contraction.
66. The first ionisation energy of the elements of the first transition series (Ti to Cu)
   (A) increases as the atomic number increases
   (B) decreases as the atomic number increases
   (C) does not show any change as the addition of electrons takes place in the inner $(n-1)d$ orbitals
   (D) increases from Ti to Mn and then decreases from Mn to Cu.
67. Identify the incorrect statement.
   (A) The magnetic moment of $K_3[FeF_6]$ is 1.73 B.M.
   (B) MnO is an antiferromagnetic substance.
   (C) Fe_3O_4 is a ferrimagnetic substance.
   (D) The general electronic configuration of transition elements is $(n-1)s^2p^6d^{1-9}ns^{1-2}$.
68. Identify the incorrect statement.
   (A) $[Ni(CO)_4]$ is a paramagnetic substance.
   (B) The magnetic moment of $K_3[CoF_6]$ is 2.89 B.M.
   (C) $Cu(CH_3COO)_2{\cdot}H_2O$ is an antiferromagnetic substance.
   (D) The general electronic configuration of actinoids is $(n-2)s^2p^6d^{10}f^{1-14}(n-1)s^2p^6d^{0-1}ns^2$.
69. The compound $Fe(CO)_5$ is
   (A) paramagnetic (B) diamagnetic (C) ferromagnetic (D) antiferromagnetic.
   Hint: The hybridisation of Fe in $Fe(CO)_5$ is dsp^3.
70. The magnetic moment of $K_3[FeF_6]$ is given by
   (A) $\sqrt{3}$ B.M. (B) $\sqrt{35}$ B.M. (C) $\sqrt{24}$ B.M. (D) $\sqrt{8}$ B.M.
71. Which of the following cannot act as a ligand in the formation of coordination compounds?
   (A) NH_4^+ (B) $CH_3CH_2NH_2$ (C) $C_6H_5NH_2$ (D) $^+H_3NNH_2$

72. Which of the following statements is incorrect?
   (A) The colour of metallic Cu and Au is due to $d \longrightarrow s$ transition.
   (B) The colour of $KMnO_4$ is due to *d–d* transition.
   (C) The colour of $KMnO_4$ is due to charge transfer transition.
   (D) Tetrahedral complexes are more intensely coloured than octahedral complexes.
73. Which of the following statements is incorrect?
   (A) The order of stability constants of metal complexes involving same ligand is: $Mn^{2+} < Fe^{2+} < Co^{2+} < Ni^{2+} < Cu^{2+} < Zn^{2+}$.
   (B) Copper is coloured but Ag is colourless because of latter's inability to undergo $d \longrightarrow s$ transition since the atomic radius of Ag is greater than that of Cu.
   (C) Cl^- ion can act as a bridging ligand.
   (D) $Ac(OH)_3$ is more basic than $La(OH)_3$.
74. Which of the following statements is correct?
   (A) Lanthanoid ions (M^{3+}) form alums.
   (B) Lanthanoids form oxometal species.
   (C) V_2O_5 is amphoteric.
   (D) Lanthanoids can act as efficient catalysts similar to transition metals.
75. Which of the following statements is incorrect?
   (A) The melting point order of halides of lanthanoids is: $LnCl_3 < LnBr_3 < LnI_3$.
   (B) In lanthanoid compounds, the ligands bond in the order of electronegativity: $Cl^- < NO_3^- < H_2O < OH^- < F^-$.
   (C) The ligand exchange reaction of the type: $[M(H_2O)_6]^{3+} + 6D_2O \longrightarrow [M(D_2O)_6]^{3+} + 6H_2O$ is more rapid in lanthanoid ions than in transition metal ions.
   (D) The ligand field effect in lanthanoid ions is of the same order as that in transition metal ions.
76. Which of the following statements is incorrect?
   (A) Bond strengths in transition metal compounds are determined by orbital interaction in the order: $F^- < OH^- < H_2O < NH_3 < CN^-$
   (B) Lanthanoid hydroxides are highly soluble in water.
   (C) Transition metal contraction is of lesser degree than lanthanoid contraction.
   (D) The common oxidation state of both transition metals and actinoids is + 3.
77. Which of the following statements is incorrect?
   (A) The orbital contributions to the magnetic moment in transition metal compounds is insignificant.
   (B) The spin and orbital contributions to the magnetic moment in transition metal compounds are given by the equation: $\mu = \sqrt{4S(S + 1) + L(L + 1)}$ B.M. where $S = \Sigma m_s$ and $L = \Sigma m_l$.
   (C) Both $[Au(CN)_4]^-$ and $[Co(CN)_6]^{3-}$ form white crystals when combined with colourless cations.
   (D) $[AuBr_4]^-$ is more stable than $[AuBr_2]^-$ in aqueous solution.

78. When MnO_2 is fused with KOH, a coloured compound is formed. The product and its colour are

(A) K_2MnO_4, purple green (B) $KMnO_4$, purple
(C) Mn_2O_3, brown (D) Mn_3O_4, black. **(I.I.T. 2003)**

79. Spin-only magnetic moment of the compound $Hg[Co(NCS)_4]$ is

(A) $\sqrt{3}$ (B) $\sqrt{15}$ (C) $\sqrt{24}$ (D) $\sqrt{8}$ **(I.I.T. 2004)**

Hint: Hybridisation: sp^3; $n = 3$; $\mu_s = \sqrt{n(n+2)} = \sqrt{3 \times 5} = \sqrt{15}$

80. Which of the following pairs is expected to exhibit same colour in solution?

(A) $VOCl_2$; $FeCl_2$ (B) $CuCl_2$; $VOCl_2$ (C) $MnCl_2$; $FeCl_2$ (D) $FeCl_2$; $CuCl_2$ **(I.I.T. 2005)**

Hint: Both Cu^{2+} and V^{4+} ions have one unpaired electron.

81. Which of the following will not be oxidised by ozone?

(A) KI (B) $FeSO_4$ (C) $KMnO_4$ (D) K_2MnO_4 **(I.I.T. 2005)**

Hint: Since manganese can exhibit the maximum possible oxidation state of + 7, $KMnO_4$ containing Mn(VII) cannot be oxidised by any oxidising agent.

82. Identify the incorrect statement.

(A) Fe_3O_4 behaves as a ferromagnetic substance above 850 K
(B) CrO_3 is an acid anhydride of chromic acid
(C) The magnetic moment of $[VO(H_2O)_5]SO_4$ is 1.73 B.M.
(D) $K_3[CoF_6]$ is diamagnetic.

83. Lanthanoid contraction is caused due to

(A) the appreciable shielding of outer electrons by 5*d* orbitals from the attraction of nuclear charge
(B) the same effective nuclear charge from Ce to Lu
(C) the imperfect shielding of outer electrons from the nuclear charge
(D) the appreciable shielding of outer electrons by 4*f* orbitals from the nuclear charge.

84. The spin-only magnetic moment value (in Bohr Megneton units) of $Cr(CO)_6$ is

(A) 0 (B) 2.84 (C) 4.90 (D) 5.92

Hint: $Cr(CO)_6$

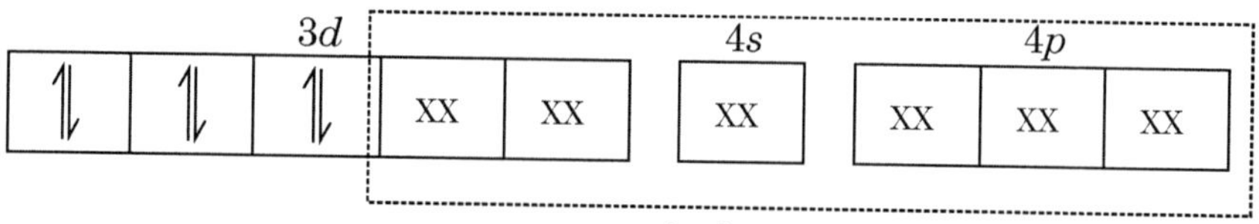

d^2sp^3 hybridisation

CO is a strong field ligand and it forces the unpaired electrons of Cr to pair up. As the coordination compound does not have any unpaired electron, its magnetic moment is zero.

ANSWERS

1. D	2. D	3. C	4. B	5. B	6. C
7. D	8. C	9. B	10. B	11. C	12. C, D
13. C	14. B	15. D	16. A	17. C, D	18. B

19. C, D	20. C	21. B	22. B	23. D	24. C
25. C	26. B	27. C	28. D	29. D	30. B
31. C	32. D	33. B	34. D	35. D	36. A
37. D	38. A	39. C	40. B, D	41. A, D	42. A
43. C	44. A, C	45. A, B, C	46. A, B	47. A	48. C
49. B, C	50. C, D	51. B	52. C	53. D	54. C
55. B	56. A	57. D	58. B	59. C	60. D
61. D	62. B	63. A	64. A	65. B, C, D	66. A
67. A	68. A, D	69. B	70. B	71. A	72. B
73. A	74. C	75. D	76. B	77. A	78. A
79. B	80. B	81. C	82. A, D	83. C	84. A

26

Coordination Compounds and Organometallic Compounds

OBJECTIVE TYPE MULTIPLE CHOICE QUESTIONS

1. The geometry of $Ni(CO)_4$ and $Ni(Ph_3P)_2Cl_2$ is respectively
   (A) both square planar (B) both tetrahedral
   (C) tetrahedral and square planar (D) square planar and tetrahedral. **(I.I.T. 1999)**
2. Which of the following is an organometallic compound?
   (A) Lithium methoxide (B) Lithium acetate
   (C) Lithium dimethylamide (D) Methyl lithium **(I.I.T. 1997)**
3. The IUPAC name of $[Co(NH_3)_6]Cl_3$ is
   (A) hexaamminecobalt(III) chloride (B) trichlorohexaamminecobalt(III)
   (C) cobalt(III) hexaammine chloride (D) hexaamminecobalt(III) trichloride. **(I.I.T. 1994)**
4. The type of magnetism exhibited by $[Mn(H_2O)_6]^{2+}$ ion is
   (A) diamagnetism (B) paramagnetism
   (C) ferromagnetism (D) antiferromagnetism. **(I.I.T. 1994)**
5. Which of the following complexes is likely to show optical activity?
   (A) $[Cr(H_2O)_6]^{3+}$ (B) *cis*-$[Co(NH_3)_2en_2]^{3+}$
   (C) *trans*-$[Co(NH_3)_4Cl_2]^+$ (D) *trans*-$[Co(NH_3)_2en_2]^{3+}$
6. The hybridisation of Ni in $[Ni(CO)_4]$ is
   (A) sp (B) sp^3 (C) dsp^2 (D) sp^3d
7. Which of the following is a π-acid ligand?
   (A) BCl_3 (B) CN^- (C) PH_3 (D) Cyclopentadiene
8. Which of the following is a π-acid ligand?
   (A) CO (B) NH_3 (C) F^- (D) $H_2NCH_2CH_2NH_2$

9. Which of the following compounds exhibits geometrical and optical isomerism?
(A) $[Pt(NH_3)Cl_3]$ (B) $[Pt(NH_3)_2Cl_2]$ (C) $[Pten_2Cl_2]$ (D) $[Pten_3]^{2+}$
10. The coordination number of a metal ion represents
(A) the number of *d*-orbitals present
(B) the number of sigma bonds formed by the metal ion
(C) the number of π bonds formed by the metal ion (D) none of these.
11. $[Pt(NH_3)_4Cl_2]Br_2$ and $[Pt(NH_3)_4Br_2]Cl_2$ are related to each other as
(A) linkage isomers (B) optical isomers
(C) ionisation isomers (D) coordination isomers.
12. The IUPAC name for the compound $K_4[Fe(CN)_6]$ is
(A) potassium ferricyanide (B) potassium hexacyanoferrate(II)
(C) potassium hexacyanoferrate(III) (D) prussian blue.
13. The IUPAC name of $Na_3[Co(NO_2)_6]$ is
(A) sodium hexanitrocobaltate(III) (B) sodium hexanitritocobaltate(III)
(C) sodium hexanitrocobalt(III) (D) sodium cobaltinitrite.
14. The IUPAC name of $[Co(NH_3)_3(H_2O)_2Cl]Cl_2$ is
(A) diaquachloronitroamminecobalt(III) chloride
(B) triamminediaquachlorocobalt(III) chloride
(C) chlorotriamminediaquacobalt(II) chloride
(D) triamminediaquachlorocobalt(II) chloride.
15. The IUPAC name $[Pt(NH_3)_4NO_2Cl]SO_4$ is
(A) chloronitrotetraammineplatinum(IV) sulphate
(B) tetraamminechloronitroplatinum(IV) sulphate
(C) nitrochlorotetraammineplatinum(IV) sulphate
(D) platinum(IV) tetraamminenitrochloro sulphate.
16. The IUPAC name of $K_3[Al(C_2O_4)_3]$ is
(A) potassium aluminium(III) oxalate
(B) potassium alumino oxalate
(C) potassium tris(oxalato)aluminate(III)
(D) potassium tris(oxalato)aluminate(VI).
17. The IUPAC name of $[Ni(CO)_4]$ is
(A) tetracarbonylnickel(0) (B) tetracarbonylnickelate(0)
(C) tetracarbonylnickel(II) (D) tetracarbonylnickelate(II).
18. Which of the following is not correctly matched?
(A) Effective atomic number of Pt in $[PtCl_6]^{2-}$ is 28
(B) Absorption peak for $[Cr(NH_3)_6]^{3+}$ is 21680 cm^{-1}
(C) Example of weak ligand field for d^5 configuration is $[MnF_6]^{4-}$
(D) Crystal field stabilisation energy of d^2 in weak ligand field = $-0.8\Delta_0$.
19. The formula of sodium nitroprusside is
(A) $Na_2[Fe(CN)_5NO]$ (B) $Na_4[Fe[CN)_5NOS]$
(C) $NaFe[Fe(CN)_6]$ (D) $Na_2[Fe(CN)_5NO_2]$

20. Which of the following is not true for metal-ligand complex?
   (A) Greater the ionisation energy of the central metal atom, the stronger is the bond
   (B) Highly charged ligand forms stronger bonds
   (C) Larger the ligand, more stable is the metal complex
   (D) Larger the permanent dipole moment of the ligand, the more stable is the bond.
21. Which of the following ions forms most stable coordination compound?
   (A) Cu^{2+} (B) Fe^{2+} (C) Ni^{2+} (D) Mn^{2+}
   Hint: The stability order is: $Mn^{2+} < Fe^{2+} < Co^{2+} < Ni^{2+} < Cu^{2+} > Zn^{2+}$
22. Identify the false statement.
   (A) The stability of the metal halide complexes is in the order: $I^- < Br^- < Cl^- < F^-$
   (B) The stability of $[Cu(NH_3)_4]SO_4$ is larger than that of $[Cuen_2]SO_4$
   (C) The stability of $[Co(NH_3)_6]Cl_3$ is larger than that of $[Co(NH_3)_6]Cl_2$
   (D) The stability of $[Cu(NH_3)_4]SO_4$ is larger than that of $[Cd(NH_3)_4]SO_4$
23. In haemoglobin, iron is in the oxidation state
   (A) + 1 (B) + 2 (C) + 3 (D) + 4
24. The hybridisation of Fe in $[Fe(H_2O)_5NO]SO_4$ is
   (A) dsp^2 (B) sp^3d (C) sp^3d^2 (D) d^2sp^3
25. Which of the following complexes will exhibit optical activity?
   (A) $K_3[Co(CN)_5NC]$ (B) $[Co(H_2O)_4Cl_2]Cl$
   (C) $[Co(H_2O)_2(NH_3)_2Cl_2]Cl$ (D) $[Co(NH_3)_6]Cl_3$
26. The complex $[Cu(NH_3)_4]SO_4$ is obtained by
   (A) adding $(NH_4)_2SO_4$ to a mixture of Cu turnings and H_2SO_4
   (B) adding ammonia to Cu turnings followed by acidification with H_2SO_4
   (C) adding conc. NH_4OH solution to a solution of $CuSO_4$
   (D) mixing equimolar amounts of $CuSO_4$ and $(NH_4)_2SO_4$
27. The possible number of isomers of $[MCl_2Br_2]SO_4$ is
   (A) 1 (B) 2 (C) 4 (D) 5
28. The number of geometrical isomers of $[Co(NO_2)_3(NH_3)_3]$ is
   (A) 0 (B) 2 (C) 3 (D) 4
29. Which of the following is paramagnetic?
   (A) $[Ni(CN)_4]^{2-}$ (B) $[NiCl_4]^{2-}$ (C) $[Ni(CO)_4]$ (D) $[Co(NH_3)_6]^{3+}$
30. Which of the following is a chelating ligand?
   (A) H_2O (B) NH_3 (C) Pyridine (D) $C_2O_4^{2-}$
31. The complexes $[Cr(NH_3)_6][Cr(NCS)_6]$ and $[Cr(NH_3)_4(SCN)_2][Cr(NH_3)_2(SCN)_4]$ are the examples of
   (A) ligand isomers (B) linkage isomers
   (C) ionisation isomers (D) coordination isomers.
32. Which of the following is an example of bimetallic catalyst system?
   (A) Grignard reagent (B) Wilkinson's catalyst
   (C) Zeigler–Natta catalyst (D) None of these.

33. Vitamin B_{12} is a coordination compound of
(A) Co(0) (B) Co(II) (C) Co(III) (D) Cr(III)
34. The number of geometrical isomers of MA_4B_2 (where A ≠ B = monodenate ligand) type complex is
(A) 1 (B) 2 (C) 3 (D) 4
35. A *trans* isomer
(A) can exhibit optical isomerism (B) cannot exhibit optical isomerism
(C) exhibits higher dipole moment than that of the *cis* isomer
(D) none of these.
36. How many isomers do you expect from the compound $[CoCl_2(NH_3)_4]Cl$?
(A) 1 (B) 2 (C) 3 (D) 4
37. Geometrical isomerism is not shown by
(A) square planar complexes (B) tetrahedral complexes
(C) octahedral complexes (D) trigonal bipyramidal complexes.
38. The hybridisation of Co in $Na_4[Co(NO_2)_6]$ is
(A) sp^3 (B) dsp^3 (C) d^2sp^3 (D) sp^3d^2
39. Which of the following complexes possesses dsp^2 hybridisation?
(A) $[Ni(CO)_4]$ (B) $[Ni(CN)_4]^{2-}$ (C) $[NiCl_4]^{2-}$ (D) $[Ni(PF_3)_4]$
40. The coordination compound $Cs_2[CuCl_4]$ is
(A) square planar and paramagnetic (B) tetrahedral and paramagnetic
(C) square planar and diamagnetic (D) tetrahedral and diamagnetic.
41. The coordination compounds $K_4[Fe(CN)_6]$ and $K_3[Fe(CN)_6]$
(A) have different structures but same magnetic property
(B) have same structure but different magnetic property
(C) are both paramagnetic and octahedral
(D) are both diamagnetic and octahedral.
42. The total number of geometrical and optical isomers of $[CoCl_2(ethylenediamine)_2]Cl$ is
(A) 1 (B) 2 (C) 3 (D) 4
43. Ethylenediaminetetraacetic acid coordinates to Ca^{2+} through
(A) two nitrogen atoms and four oxygen atoms
(B) four nitrogen atoms and two oxygen atoms
(C) four oxygen atoms (D) four nitrogen atoms.
44. Dimethylglyoxime is coordinated to Ni^{2+} through
(A) two oxygen atoms (B) two nitrogen atoms
(C) one oxygen atom and one nitrogen atom
(D) two oxygen atoms and one nitrogen atom.
45. Identify the incorrect statement.
(A) $[Pt(NH_3)_5Br]NO_2$ and $[Pt(NH_3)_5NO_2]Br$ are ionisation isomers.
(B) It is not possible to have low-spin complexes in tetrahedral structure.

(C) In naming a coordination complex, the positive ligands are listed last, after the negative and neutral ligands have been listed.
(D) The ligand H_2O is named as aquo.

46. Identify the incorrect statement.
(A) Number of unpaired electrons in $K_3[CoF_6]$ is 4.
(B) Ferrocene is a π bonded organometallic compound.
(C) Nature of ligand affects the stability of a complex ion.
(D) $[Ti(H_2O)_6]^{3+}$ and $[Fe(H_2O)_6]^{3+}$ are colourless.

47. Identify the incorrect statement.
(A) $\Delta_{tetrahedral}$ is 4/9th $\Delta_{octahedral}$
(B) Number of unpaired electrons in $K_3[CoF_6]$ is 0
(C) The hybridisation of Cr in $[Cr(H_2O)_6]SO_4$ is sp^3d^2
(D) In a tetrahedral complex the d_{xy}, d_{xz}, d_{yz} orbitals set has lower energy than the $d_{x^2-y^2}$, d_{z^2} orbitals set.

48. Identify the incorrect statement.
(A) $[Co(trimethylenediamine)_2Cl_2]Cl$ and $[Co(propylenediamine)_2Cl_2]Cl$ are ligand isomers.
(B) Zeigler–Natta catalyst is known as bimetallic catalyst.
(C) *trans*-$[PtCl_2(NH_3)_2]$ is used in chemotherapy for the treatment of cancer.
(D) The order of stability of metal complexes of the ions is: $Mn^{2+} < Fe^{2+} < Co^{2+} < Ni^{2+} < Cu^{2+} < Zn^{2+}$.

49. Which of the following is Wilkinson's catalyst?
(A) $(Ph_3P)_3RhCl$ (B) $(C_6H_6)_3RhCl$ (C) $(H_5C_2)_3RhCl$ (D) $(H_3C)_3RhCl$

50. Wilkinson's catalyst is used for
(A) hydrogenation of alkenes (B) hydrogenation of alkynes
(C) hydrogenation of alkanes (D) hydrogenation of unsaturated oils.

51. Zeigler–Natta catalyst (trialkylaluminium + $TiCl_4$) is used for
(A) polymerisation of alkenes (B) preparation of C_2H_5OH
(C) hydrogenation of alkenes (D) antiknock agent.

52. Which of the following ligands accepts electrons from the metal atom in a complex?
(A) NH_3 (B) Pyridine (C) H_2O (D) CO

53. Which of the following ligands can act as bridging ligand?
(A) NH_3 (B) H_2O (C) Cl^- (D) CO

54. A bridging ligand may contain
(A) one lone pair of electrons (B) two lone pairs of electrons
(C) three lone pairs of electrons (D) four lone pairs of electrons.

55. In the silver mirroring of glass, formaldehyde is used to reduce the coordination complex
(A) $[Ag(NH_3)_2]^+$ (B) $Na[Ag(CN)_2]$ (C) $Na[Ag(SCN)_2]$ (D) $Na_3[Ag(S_2O_3)_2]$

56. In the silver plating of metals, the electrolyte consists of the coordination compound
(A) $Na[Ag(CN)_2]$ (B) $[Ag(NH_3)_2]^+$ (C) $Na[Ag(SCN)_2]$ (D) $Na_3[Ag(S_2O_3)_2]$

57. The formula of cisplatin used as medicine for the treatment of cancer is
(A) *cis*-$[PtCl_2(NH_3)_2]$ (B) *trans*-$[PtCl_2(NH_3)_2]$
(C) *cis*-$[Pt(NH_3)_2Cl_4]$ (D) *cis*-$[PtCl_2(NH_3)_4]$

58. The tetrahedral splitting (Δ_t) is
(A) 9/4th of the octahedral splitting (Δ_0)
(B) 4/9th of the octahedral splitting (Δ_0)
(C) 1/2 of the octahedral splitting (Δ_0)
(D) same as the octahedral splitting (Δ_0)

59. The lobes of the d_{xy}, d_{xz} and d_{yz} orbitals in octahedral complexes point
(A) in the midway between the axes
(B) not in the midway between the axes
(C) towards the axes (D) none of these.

60. The lobes of the $d_{x^2-y^2}$ and d_{z^2} orbitals in octahedral complexes point
(A) towards the axes (B) in the midway between the axes
(C) not in the midway between the axes
(D) none of these.

61. Which of the following complex ions is most stable?
(A) $[Ni(H_2O)_6]^{2+}$ (B) $[Ni(NH_3)_6]^{2+}$
(C) $[Ni(ethylenediamine)_3]^{2+}$ (D) $[Ni(ethylenediaminetetraacetic\ acid)]^{2-}$

62. Which of the following will give maximum number of isomers?
(A) $[Co(NH_3)_4Cl_2]$ (B) $[Ni(en)(NH_3)_4]^{2+}$
(C) $[Ni(C_2O_4)(en)_2]$ (D) $[Cr(SCN)_2(NH_3)_4]^+$

63. Which of the following is diamagnetic?
(A) $K_3[TiF_6]$ (B) $K_3[FeF_6]$ (C) $K_3[CoF_6]$ (D) chlorophyll

64. Of the following complex ions, the one with the largest value of crystal field splitting is
(A) $[Fe(CN)_6]^{3-}$ (B) $[Ru(CN)_6]^{3-}$ (C) $[Fe(H_2O)_6]^{2+}$ (D) $[Fe(NH_3)_6]^{3+}$

65. The shape of $Fe(CO)_5$ molecule is
(A) trigonal bipyramidal (B) square pyramidal
(C) tetrahedral (D) octahedral.

66. The *d*-orbitals in an octahedral complex contain
(A) a five-fold degenerate set
(B) a four-fold degenerate set and an orbital with different energy from the four-fold degenerate set
(C) a two-fold degenerate set and a three-fold degenerate set
(D) none of these.

67. Which of the following is paramagnetic?
(A) $K_4[Fe(CN)_6]$ (B) $[Ni(CO)_4]$ (C) $K_2[Ni(CN)_4]$ (D) $K_3[CoF_6]$

68. Which of the following is a π-complex?
   (A) $[Ni(CO)_4]$ (B) Ferrocene (C) C_2H_5MgI (D) $K_2[Ni(CN)_4]$
69. Identify the incorrect statement.
   (A) $Fe(CO)_5$ has trigonal bipyradimal structure
   (B) $[Co(H_2NCH_2CH_2CH_2NH_2)_2Cl_2]Cl$ and $[Co(H_2NCH_2\underset{\displaystyle CH_3}{\underset{|}{C}}HNH_2)_2Cl_2]Cl$ are ligand isomers
   (C) The hybridisation of Ni in $Cs_2[NiCl_4]$ is sp^3
   (D) The compound $Fe(CO)_5$ possesses appreciable dipole moment.
70. The most common coordination number of Cr(II) and Pt(IV) is respectively
   (A) 6, 6 (B) 4 (square planar), 6
   (C) 4 (tetrahedral), 6 (D) 4 (square planar), 4 (square planar)
71. Vitamin B_{12} is a
   (A) coordination compound (B) organometallic compound
   (C) diamagnetic compound (D) paramagnetic compound.
72. Identify the incorrect statement(s).
   (A) Chlorophyll is a Co(II) complex (B) Chlorophyll is a Mg(II) complex
   (C) Chorophyll is an organometallic compound
   (D) Chlorophyll is a diamagnetic compound.
73. In the bridged complex ion $[(en)_2Co\begin{smallmatrix}\diagup NH \diagdown\\ \diagdown OH \diagup\end{smallmatrix}Co(en)_2]^{n+}$ (when en = ethylenediamine), the oxidation state of Co is + 3. The value of n is
   (A) 1 (B) 2 (C) 3 (D) 4
74. Vitamins B_{12}, a complex of cobalt, is
   (A) diamagnetic (B) paramagnetic
   (C) ferromagnetic (D) antiferromagnetic.
75. The complex ion which has no d-electrons in the central metal atom is
   (A) MnO_4^- (B) $[Co(NH_3)_6]^{3+}$ (C) $[Fe(CN)_6]^{3-}$ (D) $[Cr(H_2O)_6]^{3+}$ **(I.I.T. 2001)**
76. The stability constant (K) for some coordination entity is: $[Cu(H_2O)_6]^{2+}$, 9.5×10^8; $[Cu(NH_3)_4]^{2+}$, 4.5×10^{11}; $[Cu(CN)_4]^{2-}$, 2.0; $[Cu(en)_2]^{2+}$, 3.0×10^{15}. The strongest ligand among the following is
   (A) H_2O (B) NH_3 (C) CN^- (D) en
77. Haemoglobin is
   (A) diamagnetic (B) paramgnetic
   (C) iron(II) complex (D) iron(III) complex.
78. Oxyhaemoglobin is
   (A) diamagnetic (B) paramagnetic (C) ferromagnetic (D) antiferromagnetic.
79. The coordination number of iron in haemoglobin and oxyhaemoglobin is repectively
   (A) 6, 6 (B) 6, 7 (C) 5, 6 (D) 6, 8

80. Which of the following statements is incorrect?
(A) Haemoglobin reduces oxygen rapidly.
(B) Myoglobin reduces oxygen rapidly and does not transport O_2.
(C) Heme ring in myoglobin is substituted and is unsubstituted in haemoglobin.
(D) Myoglobin is an iron(II) complex.

81. Haemoglobin reacts with CO to form a coordination compound which is
(A) non-toxic (B) more stable than oxyhaemoglobin
(C) less stable than oxyhaemoglobin (D) seven coordinated.

82. *cis-trans* isomerism is not shown by
(A) square planar complexes (B) tetrahedral complexes
(C) octahedral complexes (D) trigonal bipyramidal complexes.

83. A square planar MA_2B_2 type complex forms
(A) two geometrical isomers (B) three geometrical isomers
(C) three optical isomers (D) four geometrical isomers.

84. An octahedral complex of the type MA_5B exhibits the number of geometrical isomers as
(A) zero (B) one (C) two (D) three.

85. Mixture X = 0.02 mol of $[Co(NH_3)_5SO_4]Br$ and 0.2 mol of $[Co(NH_3)_5Br]SO_4$ was prepared in 2 L of solution. 1 L of mixture X + excess of $AgNO_3 \longrightarrow$ Y, 1 L of mixture X + excess $BaCl_2 \longrightarrow$ Z. No. of moles of Y and Z are
(A) 0.02, 0.01 (B) 0.01, 0.01 (C) 0.01, 0.02 (D) 0.2, 0.2 **(I.I.T. 2003)**
Hint: Ag^+ (excess) + Br^- (0.01 mol) $\longrightarrow$ AgBr (0.01 mol); Ba^{2+} (excess) + SO_4^{2-} (0.01 mol) $\longrightarrow$ $BaSO_4$ (0.01 mol).

86. The compound with tetrahedral geometry is
(A) $[Ni(CN)_4]^{2-}$ (B) $[Pd(CN)_4]^{2-}$ (C) $[PdCl_4]^{2-}$ (D) $[NiCl_4]^{2-}$ **(I.I.T. 2004)**

87. Which type of isomerism is shown by $[Co(NH_3)_4Br_2]Cl$?
(A) Geometrical and ionisation (B) Optical and ionisation
(C) Geometrical and optical (D) Geometrical only. **(I.I.T. 2005)**
Hint: $[Co(NH_3)_4Br_2]Cl$, can exhibit both geometrical isomerism (*cis*-$[Co(NH_3)_4Br_2]Cl$, *trans*-$[Co(NH_3)_4Br_2]Cl$) and ionisation isomerism ($[Co(NH_3)_4Br_2]Cl$, $[Co(NH_3)_4BrCl]Br$).

88. Identify the correct statement.
(A) Cyclopentadiene is a π acid ligand.
(B) $Pb(C_2H_5)_4$ is a π bonded organometallic compound.
(C) Methyl lithium is a π bonded organometallic compound.
(D) The effective atomic number of Pt in $K_2[PtCl_6]$ is 86.

89. Identify the correct statement.
(A) The effective atomic number of Co in $[Co(NH_3)_6]Cl_3$ is 35.
(B) Platinum(IV) exhibits coordination number 4.
(C) The most common coordination number of Pt(II) is 4 and the next common is 6.
(D) The most common coordination number of Zn(II) is 6.

90. Identify the correct statement.
   (A) The complex $K_2[PtCl_6]$ is known but the corresponding Ni(IV) complex is not known.
   (B) In both the complexes $[Co(NH_3)_6]Cl_3$ and $K_3[CoF_6]$ the number of unpaired electrons is 4.
   (C) In a neutral or cationic complex, the ligands are named first starting with the negative ions.
   (D) The IUPAC name of $[Pt(NH_3)_2Cl_2]$ is dichlorodiammineplatinum(II).

91. If the bond length of CO bond in carbon monoxide is 1.128 Å, then what is the value of CO bond length in $Fe(CO)_5$?
   (A) 1.13 Å (B) 1.128 Å (C) 1.10 Å (D) 1.115 Å **(I.I.T. 2006)**
   Hint: Due to the back donation of electrons from M to CO, the M—C bond becomes stronger and the C≡O bond becomes weaker leading to a shorter M—C bond and longer C—O bond as compared to M—C single bond and C≡O triple bond respectively.

92. $Ag^+ + NH_3 \rightleftharpoons [Ag(NH_3)]^+$; $k_1 = 6.8 \times 10^3$; $[Ag(NH_3)]^+ + NH_3 \rightleftharpoons [Ag(NH_3)_2]^+$; $k_2 = 1.6 \times 10^3$. The overall formation constant of $[Ag(NH_3)_2]^+$ is
   (A) 1.08×10^7 (B) 1.08×10^3 (C) 6.8×10^6 (D) 6.8×10^7 **(I.I.T. 2006)**
   Hint: $\beta_k = k_1 \times k_2 = 6.8 \times 10^3 \times 1.6 \times 10^3 = 1.08 \times 10^7$

93. The coordination number of Ni^{2+} is 4 in A and B.
   $NiCl_2$ + KCN (excess) ⟶ A (cyano complex)
   $NiCl_2$ + conc. HCl (excess) ⟶ B (chloro complex)
   The IUPAC names of A and B are respectively
   (A) potassium tetracyanonickelate(II), potassium tetrachloronickelate(II)
   (B) potassium tetracyanonickel(II), potassium tetrachloronickel(II)
   (C) potassium tetracyanonickel, potassium tetrachloronickel
   (D) potassium tetracyanonickelate(II), potassium tetrachloronickel(II) **(I.I.T. 2006)**
   Hint: A = $K_2[Ni(CN)_4]$, B = $K_2[NiCl_4]$

94. The coordination number of Ni^{2+} is 4 in A and B.
   $NiCl_2$ + KCN (excess) ⟶ A (cyano complex)
   $NiCl_2$ + conc. HCl (excess) ⟶ B (chloro complex)
   Predict the magnetic nature of A and B.
   (A) Both are diamagnetic (B) Both are paramagnetic
   (C) A is paramagnetic and B is diamagnetic
   (D) A is diamagnetic and B is paramagnetic. **(I.I.T. 2006)**
   Hint: CN^- ion being a strong field ligand, it forms low-spin square planar (dsp^2 hybridrisation) $K_2[Ni(CN)_4]$ which is diamagnetic. Cl^- ion being a weak field ligand, it forms high-spin tetrahedral (sp^3 hybridisation) $K_2[NiCl_4]$ which is paramagnetic with two unpaired electrons.

95. The coordination number of Ni^{2+} is 4 in A and B.
   $NiCl_2$ + KCN (excess) ⟶ A (cyano complex)

$NiCl_2$ + conc. HCl (excess) ⟶ B (chloro comples)

The hybridisation of A and B are respectively

(A) dsp^2, sp^3 (B) sp^3, sp^3 (C) dsp^2, dsp^2 (D) sp^3, dsp^2 **(I.I.T. 2006)**

Hint: $K_2[Ni(CN)_4]$: 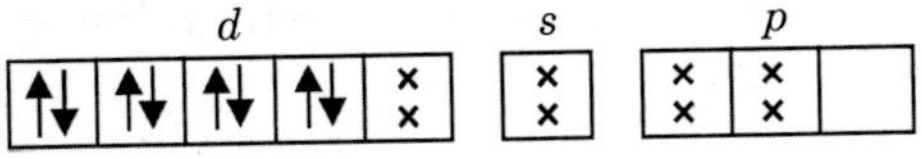 (dsp^2 hybridisation)

$K_2[NiCl_4]$: 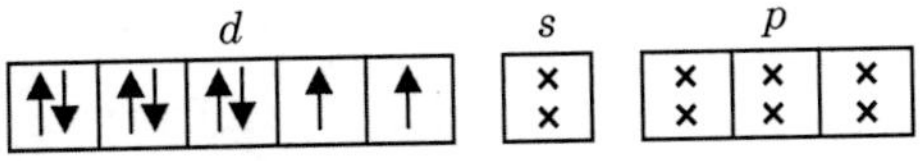 (sp^3 hybridisation)

96. In $Fe(CO)_5$, the Fe—C bond possesses

(A) both σ and π characters (B) ionic character

(C) σ character only (D) π character only

Hint: In metal carbonyls, the ligand CO first forms a dative σ bond with the metal and then the metal (with a filled orbital) forms a dative π bond with CO having an empty *p*-orbital.

97. Among the following metal carbonyls, the C—O bond order is lowest in

(A) $[Mn(CO)_6]^+$ (B) $[Fe(CO)_5]$ (C) $[Cr(CO)_6]$ (D) $[V(CO)_6]^-$ **(I.I.T. 2007)**

Hint: (A) Mn^+: $3d^54s^1$. In presence of strong field ligand CO, the effective configuration of Mn^+ is $3d^64s^0$. Three lone pairs of electrons are involved in back bonding with vacant orbital of C in CO.

(B) Fe^0: $3d^64s^2$. In presence of strong field ligand CO, the effective configuration of Fe^0 is $3d^8$. Four lone pairs of electrons are involved in back bonding with vacant orbital of C in CO.

(C) Cr^0: $3d^54s^1$. In presence of strong field ligand CO, the effective configuration of Cr^0 is $3d^6$. Three lone pairs of electrons are involved in back bonding with vacant orbital of C are in CO.

(D) V^-: $3d^44s^2$. In presence of strong field ligand CO, the effective configuration of V^- is $3d^6$. Three lone pairs of electrons are involved in back bonding with vacant orbital of C in CO.

Maximum back bonding is present in $Fe(CO)_5$ leading to lowest CO bond order.

98. Ammonia forms large number of complexes with transition metal ions. Hybridization easily explains the ease of sigma donation capability of NH_3 and PH_3. Among the following, the correct statement is

(A) Between NH_3 and PH_3, NH_3 is a better electron donor because the lone pair of electrons occupies spherical *s* orbital and is less directional

(B) Between NH_3 and PH_3, PH_3 is a better electron donor because the lone pair of electrons occupies sp^3 orbital and is more directional

(C) Between NH_3 and PH_3, NH_3 is a better electron donor because the lone pair of electrons occupies sp^3 orbital and is more directional

(D) Between NH_3 and PH_3, PH_3 is a better electron donor because the lone pair of electrons occupies spherical *s* orbital and is less directional. **(I.I.T. 2008)**

Hint: NH_3 is a better electron donor than PH_3 because the lone pair of electrons on the nitrogen atom occupies an sp^3 hybridised orbital and is more directional due to the smaller size of *N* atom.

99. The compound(s) that exhibit(s) geometrical isomerism is(are)
(A) $[Pt(en)Cl_2]$ (B) $[Pt(en)_2]Cl_2$ (C) $[Pt(en)_2Cl_2]Cl_2$ (D) $[Pt(NH_3)_4Cl_2]$ **(I.I.T. 2009)**

Hint:
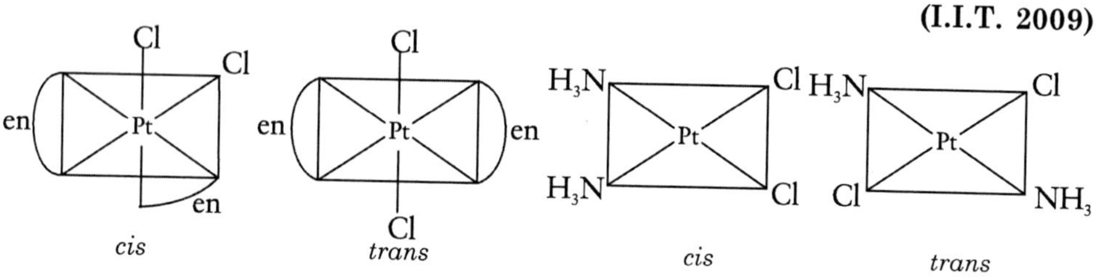

100. The spin-only magnetic moment value (in Bohr Magneton units) of $Cr(CO)_6$ is
(A) 0 (B) 2.84 (C) 4.90 (D) 5.92 **(I.I.T. 2009)**

Hint: CO being a strong field ligand forces the unpaired electron of Cr to pair up in d^2sp^3 hybridisation. Since $Cr(CO)_6$ does not have any unpaired electron, its magnetic moment is 0 B.M.

101. The ionization isomer of $[Cr(H_2O)_4Cl(NO_2)]Cl$ is
(A) $[Cr(H_2O)_4(O_2N)]Cl_2$ (B) $[Cr(H_2O)_4Cl_2](NO_2)$
(C) $[Cr(H_2O)_4Cl(ONO)]Cl$ (D) $[Cr(H_2O)_4Cl_2(NO_2)]H_2O$ **(I.I.T. 2010)**

Hint: Cl^- ion is replaced by NO_2^- ion in the coordination sphere.

102. The complex showing a spin-only magnetic moment of 2.82 B.M. is
(A) $Ni(CO)_4$ (B) $[NiCl_4]^{2-}$ (C) $Ni(PPh_3)_4$ (D) $[Ni(CN)_4]^{2-}$ **(I.I.T. 2010)**

Hint: $[NiCl_4]^{2-}$, oxidation state of Ni = + 2

Ni (28) = $3d^84s^2$

$Ni^{2+} = 3d^8$

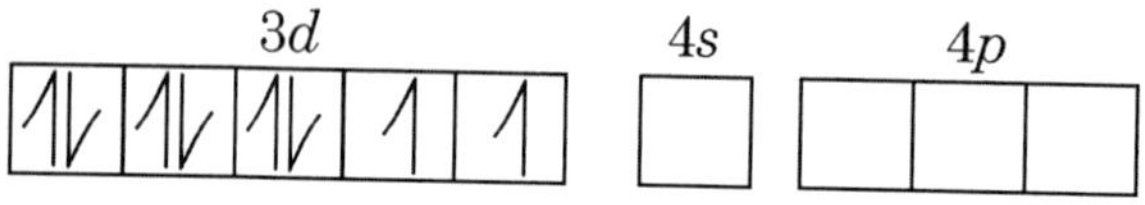

Cl^- being a weak ligand, outer orbitals are involved in the hybridisation.

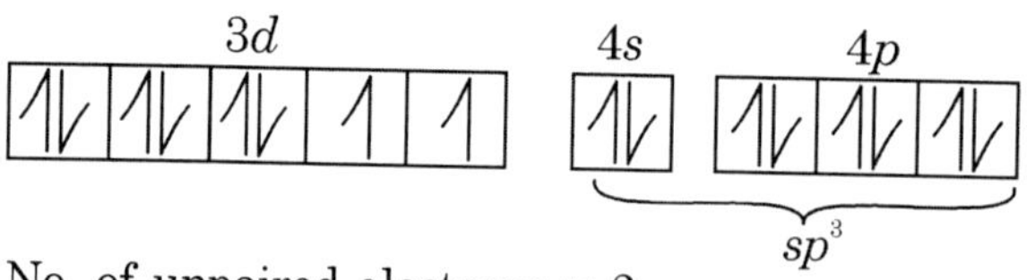

No. of unpaired electrons = 2
Magnetic moment μ = 2.82 B. M.

103. Geometrical shapes of the complexes formed by the reaction of Ni^{2+} with Cl^-, CN^- and H_2O, respectively, are

(A) octahedral, tetrahedral and square planar

(B) tetrahedral, square planar and octahedral

(C) square planar, tetrahedral and octahedral

(D) octahedral, square planar and tetrahedral **(I.I.T. 2011)**

Hint: $[NiCl_4]^{2-} \longrightarrow$ Tetrahedral

$[Ni(CN)_4]^{2-} \longrightarrow$ Square planar

$[Ni(H_2O)_6]^{2+} \longrightarrow$ Octahedral

104. Among the following complexes (K – P)

$K_3[Fe(CN)_6]$ (K), $[Co(NH_3)_6]Cl_3$ (L), $Na_3[Co(oxalate)_3]$ (M), $[Ni(H_2O)_6]Cl_2$ (N), $K_2[Pt(CN)_4]$ (O) and $[Zn(H_2O)_6](NO_3)_2$ (P), which are diamagnetic?

(A) K, L, M, N (B) K, M, O, P (C) L, M, O, P (D) L, M, N, O

(I.I.T. 2011)

Hint: Following compounds are diamagnetic: L: $[Co(NH_3)_6]Cl_3$; M: $Na_3[Co(ox)_3]$; O: $K_2[Pt(CN)_4]$; P: $[Zn(H_2O)_6](NO_3)_2$.

ANSWERS

1. B	2. D	3. A	4. B	5. B	6. B
7. B	8. A	9. C	10. B	11. C	12. B
13. A	14. B	15. B	16. C	17. A	18. A
19. A	20. C	21. A	22. B	23. B	24. C
25. B, C	26. D	27. D	28. B	29. B	30. D
31. D	32. D	33. C	34. B	35. B	36. C
37. B	38. C	39. B	40. A	41. B	42. C
43. A	44. B	45. C, D	46. D	47. B, D	48. C, D
49. A	50. A	51. A	52. D	53. B, C, D	54. B
55. A	56. A	57. A	58. B	59. A	60. A
61. D	62. D	63. D	64. B	65. A	66. C
67. D	68. B	69. D	70. A	71. A, B, C	72. A, C
73. C	74. A	75. A	76. C	77. B, C	78. A
79. C	80. A	81. B	82. B	83. A	84. A
85. B	86. D	87. A	88. D	89. C	90. A
91. A	92. A	93. A	94. D	95. A	96. A
97. B	98. C	99. C, D	100. A	101. B	102. B
103. B	104. C				

27

General Principles of Metallurgy

OBJECTIVE TYPE MULTIPLE CHOICE QUESTIONS

1. Among the following statements, the incorrect one is
   (A) calamine and siderite are carbonates.
   (B) malachite and azurite are ores of Cu.
   (C) zinc blende and pyrites are sulphides.
   (D) argentite and cuprite are oxides. **(I.I.T. 1997)**
2. In the aluminothermit process, aluminium acts as
   (A) an oxidising agent (B) a solder
   (C) a reducing agent (D) a flux. **(I.I.T. 1983)**
3. Which of the following metals are used for reduction in metallurgical processes?
   (A) Al (B) Fe (C) Zn (D) Ni **(Roorkee 1980)**
4. In the metallurgical process, the flux used for removing acidic impurities is
   (A) silica (B) Na_2CO_3 (C) limestone (D) NaCl **(Roorkee 1991)**
5. Which of the following is not an ore?
   (A) Bauxite (B) Malachite (C) Zinc blende (D) Pig iron **(I.I.T. 1982)**
6. The metallurgical process in which a metal is obtained in a fused state is called
   (A) smelting (B) roasting (C) calcination (D) froth floatation. **(I.I.T. 1978)**
7. Which of the following statements is correct regarding the slag obtained during the extraction of a metal like copper or iron?
   (A) The slag is lighter and has higher melting point than the metal.
   (B) The slag is lighter and has lower melting point than the metal.
   (C) The slag is heavier and has higher melting point than the metal.
   (D) The slag is heavier and has lower melting point than the metal.
8. Identify the process when an ore is heated in the absence of oxygen.
   (A) Calcination (B) Smelting (C) Roasting (D) Carbonisation

9. The reagent used to facilitate froth floatation process is called
(A) depressant (B) activator (C) flux (D) both (A) and (B)

10. The process of obtaining molten metal from oxide ore by heating with carbon and flux is known as
(A) smelting (B) roasting (C) calcination (D) melting.

11. When the impurities are removed from an ore, the metal becomes liable to
(A) reduction (B) oxidation (C) corrosion (D) disproportionation.

12. A compound which is used in the metallurgy of Al and which is also used to remove the impurities is
(A) basic flux (B) acidic flux (C) flux (D) slag.

13. Other than Cr(III) and Mn(IV), which metal ions can be reduced by Al in Goldschmidt thermit process?
(A) Ba(II) (B) Fe(III) (C) Ni(II) (D) Pb(II)

14. Other than aluminium which metals can be used as reducing agent in Goldschmidt thermit process?
(A) Na (B) Ca (C) Ba (D) Mg

15. Thermite is a mixture of
(A) oxides of Cr and Mn (B) oxides of Cr and Al
(C) Al powder and oxide of Cr (D) Al powder and oxide of Mn.

16. An ore of tin containing $FeCrO_4$ is concentrated by
(A) froth floatation (B) magnetic separation
(C) gravity separation (D) electrostatic method.

17. Ethyl xanthate and potassium ethyl xanthate are used as collectors in the froth floatation process because
(A) they get attached to the ore particles through H-bonding
(B) they get attached to the grains of ore containing polar groups
(C) they facilitate frothing (D) they attract the ore particles.

18. In the metallurgy of iron, when limestone is added to the blast furnace, the calcium ions end up in
(A) $CaCO_3$ (B) gangue (C) metallic Ca (D) slag. **(I.I.T. 1982)**

19. The principal reducing agent in the metallurgy of tin is
(A) Al (B) CO_2 (C) CO (D) C

20. Mg is extracted by
(A) leaching with a solution of NaCN followed by reduction
(B) electrolytic method
(C) self-reduction method (D) carbon reduction method.

21. Pb is extracted by
(A) leaching with a solution of NaCN followed by reduction
(B) electrolytic method
(C) self-reduction method (D) carbon reduction method.

22. In which of the following processes, the metal is purified by the formation of a volatile coordination compound?
    (A) Thermit process (B) Van Arkel process
    (C) Mond process (D) Carter and Brown process.
23. Desilverisation of Pb is carried out by
    (A) cyanide process (B) parting process
    (C) Parke process (D) MacArthur–Forest process.
24. Which of the following ores contains both calcium and magnesium?
    (A) Gypsum (B) Talc (C) Epsomite (D) Dolomite
25. Which of the following is the composition of cuprite?
    (A) Cu_2O (B) Cu_2S (C) $CuFeS_2$ (D) $Cu(OH)_2{\cdot}CuCO_3$
26. Which of the following ores does not contain iron?
    (A) Siderite (B) Copper pyrites
    (C) Magnetite (D) Magnesite
27. Which of the following is preserved under water?
    (A) Na (B) Fe (C) Zn (D) P
28. Which of the following metals is obtained by electrolytic reduction method?
    (A) Ag (B) Cu (C) Al (D) Fe
29. Cassiterite is concentrated by
    (A) liquation (B) froth floatation
    (C) levigation (D) electromagnetic separation.
30. Zone refining is used for the
    (A) purification of an ore (B) purification of metal
    (C) concentration of an ore (D) reduction of metal oxide.
31. Heating pyrites in air to remove sulphur is known as
    (A) calcination (B) smelting (C) fluxing (D) roasting.
32. Froth floatation process is used to increase the concentration of the mineral in
    (A) chalcopyrites (B) bauxite (C) haematite (D) calamine. **(I.I.T. 1989)**
33. Van Arkel method of purification of metals involves converting the metal to a
    (A) volatile stable compound (B) volatile unstable compound
    (C) non-volatile stable compound (D) none of these.
34. Which of the following metals cannot be obtained by electrolysis?
    (A) Al (B) Ca (C) Mg (D) Cr
35. Hausmannite is the ore of
    (A) Zn (B) Ni (C) Mn (D) Hg
36. If the impurity in a metal has a greater affinity for oxygen and is more easily oxidised than the metal, the purification of metal is carried out by
    (A) zone refining (B) cupellation (C) electrolysis (D) poling.
37. Which of the following is not an ore of aluminium?
    (A) Bauxite (B) Corundum (C) Diaspore (D) Anhydrite

38. The highest temperature is achieved in which type of furnace?
(A) Reverberatory (B) Muffle (C) Blast (D) Electric.
39. Reverberatory furnace is employed in the metallurgical process mainly for
(A) conversion of chloride to sulphate (B) conversion of chloride to metal
(C) smelting of sulphide ores (D) getting magnetic materials.
40. Roasting is done generally in case of
(A) silicate ore (B) sulphide ore (C) oxide ore (D) carbonate ore.
41. The main function of roasting is
(A) oxidation of ore (B) to remove the volatile matter
(C) both (A) and (B) (D) to make slag.
42. Semiconductor grade silicon is purified by
(A) electrolytic refining (B) zone refining
(C) Van Arkel method (D) none of these.
43. The process of extraction of a metal in the molten state is called
(A) roasting (B) calcination
(C) pyrometallurgy (D) hydrometallurgy.
44. Discharge potential relates to the potential at which
(A) the metal is discharged at the anode
(B) the metal is discharged at the cathode
(C) the metal is reduced (D) any ion is discharged.
45. Corundum is
(A) SrO_2 (B) $CaCl_2$ (C) Cu_2Cl_2 (D) Al_2O_3
46. In a metallurgical process, the impure metal is heated under vacuum in an electric arc furnace leading to the volatilisation of impurities. The process is known as
(A) Van Arkel process (B) Van Bolten process
(C) Mond process (D) Pyroscopy.
47. In which of the following processes, a metal is extracted by heating its volatile iodide?
(A) Zone refining (B) Mond process
(C) Van Arkel process (D) Carter and Brown process.
48. Which of the following alloys possesses almost zero coefficient of expansion?
(A) Duralumin (B) Gun metal (C) Invar (D) Solder
49. Which of the following alloys does not contain Zn?
(A) Gun metal (B) Type metal (C) Muntz metal (D) German silver
50. Which of the following alloys does not contain Ni?
(A) Monel metal (B) Muntz metal (C) Nichrome (D) German silver
51. The components of duralumin are
(A) Al and Mg (B) Al, Mg and Ni (C) Al and Ni (D) Al, Cu, Mg and Mn
52. Which of the following alloys is silvery-white having low coefficient of friction?
(A) Babbitt metal (B) Wood's metal (C) Duralumin (D) Phosphor bronze

53. Which of the following alloys melts in boiling water?
    (A) Wood's metal (B) Gun metal (C) Bell metal (D) Monel metal
54. Which of the following is the major component of Babbitt metal?
    (A) Zn (B) Sn (C) Sb (D) Cu
55. In the froth floatation process, the ore particles float because
    (A) their surface is not easily wetted by water
    (B) they are light
    (C) they are insoluble (D) they bear electrostatic charge.
56. Chromium is obtained from Cr_2O_3 by reduction with
    (A) carbon (B) Al (C) CO (D) $CO + H_2$
57. Copper is obtained by
    (A) extraction of copper pyrites with aqueous NaCN followed by reduction
    (B) electrolysis of molten anhydrous $CuCl_2$
    (C) carbon reduction method (D) self reduction method.
58. Which of the following reactions occurs in the zone of heat absorption of the blast furnace in the metallurgy of iron?
    (A) $C + CO_2 \longrightarrow 2CO$ (B) $C + O_2 \longrightarrow CO_2$
    (C) $2C + O_2 \longrightarrow 2CO$ (D) $C + 2O_2 + 2CO \longrightarrow 3CO_2$
59. Liquation is used in the purification of
    (A) K (B) Cu (C) Pb (D) Sn
60. Which of the following concentration processes is used when the gangue is light?
    (A) Froth floatation (B) Magnetic separation
    (C) Gravity separation (D) None of these.
61. In the roasting process, the ore is heated
    (A) below its melting point (B) for removing moisture and volatile matter
    (C) in the presence of flux in order to remove gangue
    (D) none of these.
62. In which of the following, smelting is carried out?
    (A) Bessemer converter (B) Open hearth process
    (C) Blast furnace (D) None of these.
63. In which of the following, roasting is carried out?
    (A) Bessemer converter (B) Reverberatory furnace
    (C) Muffle furnace (D) Electric oven
64. The component having low melting point is separated in an impure metal by
    (A) froth floatation (B) washing with water
    (C) magnetic separation (D) liquation.
65. Bearing metal contains
    (A) Pb, Mg, Li, K (B) Zn, Pb, Ca, Li
    (C) Pb, Ca, Li, K (D) Pb, Ca, Li, Na

66. Which of the following processes is used in the extractive metallurgy of magnesium?
(A) Fused salt electrolysis (B) Thermite reduction
(C) Aqueous solution electrolysis (D) Self reduction. **(I.I.T. 2002)**

67. Magnesium(II) oxide is used in furnace lining because
(A) it is an electrical insulator (B) it has high melting point
(C) it is a good heat conductor (D) all of these.

68. Which of the following processes is used for the manufacture of magnesium from sea water?
(A) Solvay process (B) Dow process
(C) Deacon process (D) Pidgeon process.

69. In the metallurgy of iron, limestone acts as
(A) slag (B) oxidiser (C) deoxidiser (D) flux.

70. In the metallurgy of iron in the blast furnace, the highest temperature is in the
(A) fusion zone (B) combustion zone
(C) slag zone (D) reduction zone.

71. Which of the elements present in iron is responsible for cold brittleness and low tensile strength?
(A) P (B) C (C) S (D) Si

72. Presence of impurities in the pig iron makes it
(A) brittle (B) hard
(C) unsuitable for welding (D) all of these.

73. Which of the following processes involves smelting process?
(A) $ZnCO_3 \longrightarrow ZnO + CO_2$ (B) $2PbS + 3O_2 \longrightarrow 2PbO + 2SO_2$
(C) $Fe_2O_3 + 3C \longrightarrow 2Fe + 3CO$ (D) $Al_2O_3.2H_2O \longrightarrow Al_2O_3 + 2H_2O$

74. Purification of aluminium by electrorefining is known as
(A) Serpeck process (B) Baeyer process
(C) Hoope process (D) Hall process.

75. The methods chiefly used for the extraction of lead and tin from their ores are respectively
(A) self reduction and carbon reduction
(B) self reduction and electrolytic reduction
(C) electrolytic reduction and self reduction
(D) carbon reduction and self reduction. **(I.I.T. 2004)**

76. Which of the following ores contains both copper and iron?
(A) Cuprite (B) Chalcocite (C) Chalcopyrite (D) Malachite **(I.I.T. 2005)**
Hint: Chalcopyrite ($CuFeS_2$)

77. Out of ZnS, HgS, Ag_2S and Cu_2S, roasting will convert ores into metal in case of
(A) Ag_2S, HgS (B) Ag_2S, CuS
(C) ZnS, Cu_2S (D) ZnS, HgS

78. In the reactions, $Ag_2S + NaCN \longrightarrow X \xrightarrow{Zn} Y$, Y is a metal. X and Y are
   (A) $Na_3[Ag(CN)_4]$, Ag (B) $Na_2[Ag(CN)_4]$, Ag
   (C) $Na[Ag(CN)_2]$, Ag (D) $Na_2[Zn(CN)_4]$, Zn

79. Match List I (ores) with List II (metals) and select the correct answer using the codes given below:

List I	*List II*
(a) Bauxite	1. Al
(b) Haematite	2. Pb
(c) Galena	3. Zn
(d) Calamine	4. Fe

	a	b	c	d		a	b	c	d
(A)	1	4	3	2	(B)	1	4	2	3
(C)	4	1	2	3	(D)	4	1	3	2

80. Match List I (alloys) with List II (constituents) and select the correct answer using the codes given below.

List I	*List II*
(a) German silver	1. Cu, Sn
(b) Brass	2. Cu, Zn, Ni
(c) Monel metal	3. Cu, Ni
(d) Bronze	4. Cu, Zn

	a	b	c	d		a	b	c	d
(A)	4	2	3	1	(B)	4	2	1	3
(C)	2	4	1	3	(D)	2	4	3	1

81. Match List I (method) with List II and select the correct answer using the codes given below.

List I	*List II*
(a) Poling	1. Refining of Ag
(b) Cupellation	2. Purification of Cu
(c) Solvay process	3. Manufacture of Na_2CO_3
(d) Van Arkel method	4. Purification of Ti
	5. Manufacture of NaOH

	a	b	c	d		a	b	c	d
(A)	2	1	3	4	(B)	5	1	3	4
(C)	4	3	2	5	(D)	2	3	5	4

82. Chromite ($FeCr_2O_4$) is an ore of chromium. The preparation of some chromium compounds and their reactions are as follows:
   (i) $FeCr_2O_4 + NaOH + O_2 \longrightarrow Fe_2O_3 + A$
   (ii) $A + B \longrightarrow Na_2Cr_2O_7$
   (iii) $Na_2Cr_2O_7 + C \xrightarrow{heat} Cr_2O_3$

(iv) $Cr_2O_3 + D \xrightarrow{heat} Cr$

C and D are

(A) C, C (B) Al, C (C) C, Al (D) Al, Al

83. Extraction of zinc from zinc blende is achieved by

(A) electrolytic reduction (B) roasting followed by reduction with carbon

(C) roasting followed by reduction with another metal

(D) roasting followed by self-reduction. **(I.I.T. 2007)**

Hint: $2ZnS + 3O_2 \rightarrow 2ZnO + 2SO_2$; $ZnO + C \xrightarrow{1400°C} Zn + CO$, $ZnO + CO \xrightarrow{1400°C} Zn + CO_2$

84. There are some deposits of nitrates and phosphates in earth's crust. Nitrates are more soluble in water. Nitrates are difficult to reduce under the laboratory conditions but microbes do it easily. Among the following, the correct statement is

(A) Phosphates have no biological significance in humans

(B) Between nitrates and phosphates, phosphates are less abundant in earth's crust

(C) Between nitrates and phosphates, nitrates are less abundant in earth's crust

(D) Oxidation of nitrates is possible in soil. **(I.I.T. 2008)**

Hint: Nitrates are less abundant than phosphates in the earth's crust because they are soluble in water. Moreover, they can also be reduced by the numerous microbes present in the earth's crust.

85. The extraction of copper from chalcopyrite involves partial roasting, removal of iron and self-reduction.

Partial roasting of chalcopyrite produces

(A) Cu_2S and FeO (B) Cu_2O and FeO (C) CuS and Fe_2O_3 (D) Cu_2O and Fe_2O_3 **(I.I.T. 2010)**

Hint: $2CuFeS_2 + O_2 \rightarrow Cu_2S + 2FeS + SO_2 \uparrow$

$2Cu_2S + 3O_2 \rightarrow 2Cu_2O + 2SO_2 \uparrow$

$2FeS + 3O_2 \rightarrow 2FeO + 2SO_2 \uparrow$

86. The extraction of copper from chalcopyrite involves partial roasting, removal of iron and self-reduction.

Iron is removed from chalcopyrite as

(A) FeO (B) FeS (C) Fe_2O_3 (D) $FeSiO_3$ **(I.I.T. 2010)**

Hint: $FeO + SiO_2 \rightarrow FeSiO_3$ (slag)

87. The extraction of copper from chalcopyrite involves partial roasting, removal of iron and self-reduction.

In self-reduction, the reducing species is

(A) S (B) O^{2-} (C) S^{2-} (D) SO_2 **(I.I.T. 2010)**

Hint: $Cu_2S + 2Cu_2O \rightarrow \underset{\text{(Blister copper)}}{6Cu} + SO_2 \uparrow$

$S^{2-} \rightarrow S^{4+}$ is oxidation, hence S^{2-} is the reducing agent.

88. Extraction of metal from the cassiterite involves

(A) carbon reduction of an oxide ore (B) self-reduction of a sulphide ore

(C) removal of copper impurity (D) removal of iron impurity

(I.I.T. 2011)

Hint: $SnO_2 + 2C \longrightarrow 2CO + Sn$

The ore cassiterite contains the impurity of Fe, Mn, W and traces of Cu.

89. Oxidation states of the metal in the minerals haematite and magnetite, respectively, are

(A) II, III in haematite and III in magnetite (B) II, III in haematite and II in magnetite

(C) II in haematite and II, III in magnetite (D) III in haematite and II, III in magnetite

(I.I.T. 2011)

Hint: Haematite: Fe_2O_3: $2x + 3 \times (-2) = 0$

$x = 3$

Magnetite: Fe_3O_4 [an equimolar mixture of FeO and Fe_2O_3]

FeO: $x - 2 = 0 \Rightarrow x = 2$

Fe_2O_3: $x = 3$

ANSWERS

1. D	2. C	3. A, B	4. C	5. D	6. A
7. B	8. A	9. D	10. A	11. A	12. A
13. A, B, C, D	14. A, B, D	15. C, D	16. B	17. B	18. D
19. C	20. B	21. C	22. C	23. C	24. D
25. A	26. D	27. D	28. C	29. D	30. B
31. D	32. A	33. A	34. D	35. C	36. B
37. D	38. D	39. B	40. B	41. C	42. B
43. C	44. B	45. D	46. B	47. C	48. C
49. B	50. B	51. D	52. A	53. A	54. B
55. A	56. B	57. D	58. A	59. C, D	60. C
61. A, B	62. C	63. B	64. D	65. D	66. A
67. D	68. B	69. D	70. A	71. A	72. D
73. C	74. C	75. A	76. C	77. A	78. B
79. B	80. D	81. A	82. C	83. B	84. C
85. B	86. D	87. C	88. A, C, D	89. D	

28

Iron and Steel

OBJECTIVE TYPE MULTIPLE CHOICE QUESTIONS

1. The chemical reactions in the production of steel from haematite ore involves
   (A) oxidation (B) reduction
   (C) reduction followed by oxidation (D) oxidation followed by reduction.
   (I.I.T. 2000)

2. Addition of high proportions of Mn makes steel useful in making rails of railroads, because Mn
   (A) helps the formation of oxides of iron
   (B) can show highest oxidation state of + 7
   (C) can remove oxygen and sulphur
   (D) gives hardness to steel. **(I.I.T. 1998)**

3. Which of the following statements is/are correct with respect to Fe^{2+} and Fe^{3+} ions?
   (A) Fe^{3+} gives brown colour with $K_3[Fe(CN)_6]$
   (B) Fe^{2+} gives blue precipitate with $K_3[Fe(CN)_6]$
   (C) Fe^{2+} gives brown colour with NH_4SCN
   (D) Fe^{3+} gives red colour with KSCN. **(I.I.T. 1998)**

4. In nitroprusside ion, the iron and NO exist as Fe(II) and NO^+ rather than Fe(III) and NO. These forms can be differentiated by
   (A) thermally decomposing the compound
   (B) measuring the concentration of CN^-
   (C) measuring the solid state magnetic moment
   (D) estimating the concentration of iron. **(I.I.T. 1998)**

5. Identify the reducing agent for the reduction of FeO in the blast furnace.
   (A) C (B) CO (C) $CaCO_3$ (D) SiO_2 **(I.I.T. 1987)**

6. In the metallurgy of iron, the Ca^{2+} of the limestone added ends up in
(A) $CaCO_3$ (B) slag (C) gangue (D) Ca **(I.I.T. 1982)**
7. Iron is rendered passive by treatment with concentrated
(A) H_2SO_4 (B) H_3PO_4 (C) HNO_3 (D) HCl **(I.I.T. 1982)**
8. A solution of $FeCl_3$ in water on standing gives a brown precipitate whose composition is
(A) $Fe(OH)_3$ (B) $Fe_2O_3{\cdot}xH_2O$ (C) $FeCl_3{\cdot}xH_2O$ (D) Fe_2O_3 **(I.I.T. 1980)**
9. Purest form of iron is
(A) white cast iron (B) grey cast iron
(C) wrought iron (D) steel.
10. Steel resistant to acid is
(A) high carbon steel (B) molybdenum steel
(C) stainless steel (D) nickel steel.
11. Spiegeleisen is an alloy of
(A) Fe, Ni, Cr (B) Fe, Ni, W (C) Fe, Cr, C (D) Fe, Mn, C
12. Which of the following alloys is used to make vessels for transporting conc. H_2SO_4?
(A) Ferromanganese (B) Ferrosilicon
(C) Silica bronze (D) Tinitron.
13. Steel differs from pig iron and wrought iron in that it contains
(A) no carbon at all (B) less carbon than either
(C) an amount of carbon intermediate between two
(D) more carbon than either.
14. Nitriding is the process of surface hardening steel by treating it in an atmosphere of
(A) O_3 (B) N_2 (C) NH_3 (D) H_2S
15. Which of the following is a mixed oxide?
(A) Fe_2O_3 (B) Fe_3O_4 (C) Cr_2O_3 (D) Al_2O_3
16. In the manufacture of cast iron in the blast furnace the flux used is
(A) CaO (B) $CaCO_3$ (C) $CaSiO_3$ (D) $CaSO_4$
17. Best quality steel is obtained from
(A) Bessemer process (B) Open hearth process
(C) Deacon process (D) L. D. process.
18. Which of the following contains the least percentage of carbon?
(A) wrought iron (B) mild steel (C) stainless steel (D) pig iron
19. The percentage of carbon in steel is
(A) 0.05–0.1% (B) 0.10–1.5% (C) 1.5–25% (D) 2.5–3.5%
20. In which of the following, the percentage of carbon is the highest?
(A) mild steel (B) cast iron (C) wrought iron (D) stainless steel.
21. The increase in carbon content in steel leads to
(A) increase in hardness (B) decrease in ductility
(C) increase in tensile strength (D) none of these.

22. The process of heating steel to red hot and then cooling slowly is known as
(A) annealing (B) nitriding (C) tempering (D) case hardening.
23. The process of heating wrought iron with powdered coke and then quenching in oil is known as
(A) tempering (B) annealing
(C) case hardening (D) galvanisation.
24. Which form of iron cannot be tempered and welded?
(A) mild steel (B) cast iron (C) wrought iron (D) stainless steel.
25. The form of iron which is the softest is
(A) wrought iron (B) mild steel (C) pig iron (D) nichrome steel.
26. The form of iron which has the highest melting point is
(A) wrought iron (B) pig iron (C) mild steel (D) nichrome steel.
27. The addition of vanadium to steel increases
(A) hardness (B) tensile strength
(C) fatigue resistance (D) all of these.
28. Which of the following is achieved by the addition of tungsten to steel?
(A) Corrosion resistance (B) Cutting hardness
(C) Paramagnetic property (D) None of these.
29. Which of the following pairs forms alloy?
(A) Fe and Hg (B) Hg and Pt (C) Fe and C (D) Zn and Pb
30. Which of the following is the hardest?
(A) High carbon steel (B) Low carbon steel
(C) Medium carbon steel (D) Mild steel.
31. Pig iron reacts with C or CO and forms a carbide called cementite whose compositionis
(A) FeC (B) Fe_2C (C) Fe_3C (D) Fe_3C_2
32. Siderite is
(A) oxide ore of iron (B) silicate ore of iron
(C) sulphide ore of iron (D) carbonate ore of iron.
33. Thoms slag is
(A) $FeSiO_3$ (B) $MnSiO_3$ (C) $CaSiO_3$ (D) $Ca_3(PO_4)_2 \cdot CaO$
34. Which of the following processes is carried out for the removal of impurities from cast iron in the Bessemer converter?
(A) Liquation (B) Zone refining
(C) Van Bolten process (D) Oxidative refining.
35. Melting point of pure iron is
(A) more than that of pig iron (B) less than that of pig iron
(C) 1435°C (D) 1535°C
36. The charge introduced in the blast furnace consists of
(A) calcined ore + SiO_2 + C (B) calcined ore + $CaCO_3$
(C) calcined ore + $CaCO_3$ + C (D) calcined ore + $CaCO_3$ + $Mg(NO_3)_2$

37. Which of the following is produced when molten cast iron is rapidly cooled?
    (A) Grey cast iron (B) White cast iron
    (C) Wrought iron (D) Quenched steel.
38. The increasing order of carbon content in cast iron, wrought iron and mild steel is
    (A) mild steel < wrought iron < cast iron
    (B) mild steel < cast iron < wrought iron
    (C) mild steel < pig iron < wrought iron
    (D) wrought iron < mild steel < cast iron.
39. In the metallurgy of iron, the material obtained from the bottom of the blast furnace is
    (A) wrought iron (B) cast iron (C) pig iron (D) slag.
40. Iron is extracted by
    (A) leaching with NaCN solution followed by reduction
    (B) electrolytic method
    (C) self-reduction method (D) carbon reduction method.
41. In the ring test for the nitrate ion, the brown compound formed is
    (A) $[Fe(H_2O)_4(NO)_2]SO_4$ (B) $[Fe(H_2O)_5(NO)]SO_4$
    (C) $[Fe(H_2O)_3(NO)_3]SO_4$ (D) $[Fe(H_2O)_2(NO)_4]SO_4$
42. Iron(III) oxide is
    (A) acidic oxide (B) basic oxide (C) neutral oxide (D) amphoteric oxide.
43. Which of the following is an interstitial compound?
    (A) Solder (B) Brass (C) Gun metal (D) Steel
44. Blow holes of steel are removed by adding
    (A) sand (B) C (C) Ni (D) spiegeleisen.
45. Identify the false statement.
    (A) On strong heating, $FeSO_4$ decomposes to Fe_2O_3, SO_2 and SO_3
    (B) On standing the colour of green vitriol ($FeSO_4 \cdot 7H_2O$) remains unchanged
    (C) $FeSO_4 \cdot 7H_2O$ is a primary standard
    (D) $FeSO_4 \cdot 7H_2O$ is a reducing agent.
46. The underlining of blast furnace is made of
    (A) graphite (B) silica rocks
    (C) fire clay bricks (D) basic bricks.
47. Bessemer converter is used in the manufacture of
    (A) pig iron (B) wrought iron (C) cast iron (D) steel.
48. If a steel article is heated to redness and then suddenly cooled by plunging into oil, it becomes
    (A) hard and brittle (B) hard and ductile
    (C) soft and ductile (D) soft and brittle.
49. The ratio of calcined ore : carbon : flux introduced in blast furnace is
    (A) 2 : 2 : 1 (B) 6 : 4 : 1 (C) 8 : 4 : 1 (D) 8 : 2 : 2

50. In the metallurgy of iron, the function of cup and cone feeder in the blast furnace is
    (A) to prevent the loss of heat from the furnace
    (B) to prevent the escape of gases
    (C) to maintain the required pressure in the furnace (D) none of these.
51. The mixed oxide Fe_3O_4 which is used in magnetic tapes is
    (A) ferromagnetic (B) paramagnetic (C) ferrimagnetic (D) antiferromagnetic
52. When a pure iron rod is dipped in fuming nitric acid, the products formed are
    (A) $Fe(NO_3)_3 + NO$ (B) $Fe_2O_3 + NO_2$
    (C) $Fe(NO_3)_3 + NO_2$ (D) $Fe(NO_3)_2$
53. When SO_2 gas is passed through a $FeCl_3$ solution
    (A) $FeCl_3$ is reduced to Fe and SO_2 is oxidised
    (B) $FeCl_3$ is reduced to $FeCl_2$ and SO_2 is oxidised
    (C) $FeSO_4$ is formed and SO_2 is oxidised (D) none of these.
54. When ozone is passed through $K_4[Fe(CN)_6]$, the product formed is
    (A) $K_3[Fe(CN)_6] + KOH + O_2$ (B) ferri-ferrocyanide
    (C) $Fe_2O_3 + K_2O$ (D) none of these.
55. Anhydrous iron(III) chloride is prepared by
    (A) heating hydrated iron(III) chloride at a high temperature in a stream of air
    (B) heating metallic iron in a stream of dry Cl_2 gas
    (C) reaction of metallic iron with HCl(*aq*)
    (D) reaction of metallic iron with HCl(*g*). **(I.I.T. 2002)**

    Hint: Anhydrous iron(III) chloride is prepared by heating metallic iron in a stream of dry chlorine gas.

    $$2Fe + 3Cl_2 \xrightarrow{\text{heat}} 2FeCl_3$$
56. The yellow colour of aqueous solution of iron(III) salts is due to
    (A) $[Fe(H_2O)_5OH]^{2+}$ (B) $[Fe(H_2O)_6]^{2+}$
    (C) FeO(OH) (D) $Fe(OH)_3$
57. Which of the following is formed when steam is passed over red hot iron?
    (A) Fe_3O_4, H_2O (B) Fe_3O_4, H_2 (C) FeO, H_2 (D) Fe_2O_3, H_2
58. Which of the following compounds is formed when $Fe(OH)_2$ dissolves in a strong solution of NaOH?
    (A) $Fe_2O_3{\cdot}xH_2O$ (B) $Fe(OH)_3$ (C) $Na_3[Fe(OH)_6]$ (D) $Na_4[Fe(OH)_6]$
59. Which of the following is formed when iron(III) sulphate is strongly heated?
    (A) S_8 (B) SO_3 (C) SO_2 (D) SO_2, SO_3
60. Iron reacts with dil. HNO_3 to form
    (A) NH_4NO_3 (B) N_2O (C) NO_2 (D) NO
61. Interstitial compounds are formed by
    (A) Fe, Ni (B) Fe, Co (C) Co, Ni (D) Fe, Co, Ni
62. The oxidation state of Fe in $[Fe(CN)_5(NO^+)]^{2-}$ is + 2. This can be decided by
    (A) colour (B) hybridisation
    (C) colligative property (D) magnetic measurement.

63. Chromite is an ore containing Fe and Cr. The preparation of some compounds from chromite and their reactions are as follows:

(i) $FeCr_2O_4 + NaOH + O_2 \xrightarrow{heat} Fe_2O_3 + A$

(ii) $A + B \longrightarrow Na_2Cr_2O_7$

(iii) $Na_2Cr_2O_7 + C \xrightarrow{heat} Cr_2O_3$

(iv) $Cr_2O_3 + D \xrightarrow{heat} C_6$

A and B are

(A) $Na_4[Fe(OH)_6]$, H_2SO_4
(B) Na_4CrO_5, H_2SO_4
(C) $Na_2Cr_2O_7$, HCl
(D) Na_2CrO_4, H_2SO_4

ANSWERS

1. D	2. C, D	3. B, D	4. C	5. B	6. B
7. C	8. B	9. C	10. B	11. D	12. B
13. C	14. C	15. B	16. B	17. B	18. A
19. B	20. B	21. A, C	22. A	23. C	24. B
25. C	26. A	27. D	28. B	29. C	30. A
31. C	32. D	33. D	34. D	35. A, D	36. C
37. B	38. D	39. C	40. D	41. B	42. D
43. D	44. D	45. B, C	46. C	47. D	48. A
49. C	50. B	51. C	52. A	53. B	54. A
55. B	56. A	57. B	58. D	59. B	60. A
61. D	62. D	63. D			

29

Coinage Metals (Copper, Silver and Gold)

OBJECTIVE TYPE MULTIPLE CHOICE QUESTIONS

1. The chemical composition of slag formed during the smelting process in the extraction of copper is
   (A) $Cu_2O + FeS$ (B) $Cu_2S + FeO$ (C) $FeSiO_3$ (D) $CuFeS_2$ **(I.I.T. 2001)**
2. Which of the following alloys contain(s) Cu and Zn?
   (A) Bronze (B) Brass (C) Gun metal (D) Type metal **(I.I.T. 1993)**
3. Of the following, the metal that can be obtained by electrolysis of aqueous solution is
   (A) Ag (B) Cu (C) Mg (D) Al **(I.I.T. 1990)**
4. Zinc and not copper is used for the recovery of metallic silver from the complex $Na[Ag(CN)_2]$. This is due to the fact that
   (A) Zn is a stronger reducing agent than Cu
   (B) Zn is a weaker reducing agent than Cu
   (C) Cu cannot displace Ag (D) none of these. **(I.I.T. 1985)**
5. One of the constituents of German silver is
   (A) Ag (B) Cu (C) Mg (D) Al **(I.I.T. 1980)**
6. The function of AgBr in photography is
   (A) due to formation of coordination compound
   (B) oxidising agent
   (C) reducing agent (D) none of these. **(Roorkee 1996)**
7. In copper (atomic number = 29)
   (A) 13 electrons have spin in one direction and 16 electrons in other direction
   (B) 14 electrons have spin in one direction and 15 electrons in other direction
   (C) 12 electrons have spin in one direction and 17 electrons in other direction
   (D) none of these.

8. Which of the following species is present in the cyanide bath used for the purification of gold?
   (A) $Au(CN)_2^-$ (B) $Au(CN)_4^-$ (C) $Au_3(CN)_2^-$ (D) $Au(CN)^+$
9. Flux used in the smelting of copper ore is
   (A) coke (B) limestone (C) SiO_2 (D) MgO
10. When $CuCl_2$ is exposed to moist air, it turns green due to the formation of
    (A) $CuCl{\cdot}Cu(OH)_2$ (B) $CuCl_2{\cdot}Cu(OH)_2$
    (C) $Cu(OH)_2$ (D) $CuCl_2$
11. Which of the following does not react with AgCl?
    (A) $Na_2S_2O_3$ (B) $NaNO_3$ (C) NH_4OH (D) Na_2CO_3
12. The anode mud obtained during electrolytic refining of silver contains
    (A) Zn, Cu, Au (B) Cu, Zn (C) Zn, Au (D) Au
13. The anode mud obtained during electrolytic refining of copper contains
    (A) Zn, Ag (B) Zn, Au (C) Zn (D) Ag, Au
14. The black image on an exposed and developed photographic film is composed of
    (A) $Ag(S_2O_3)_2^{3-}$ (B) Ag_2O (C) AgBr (D) Ag
15. When silver nitrate is added to a concentrated solution of sodium thiosulphate we get
    (A) a soluble complex $Na_3[Ag(S_2O_3)_2]$ (B) a black precipitate of Ag
    (C) a black precipitate of Ag_2S (D) a white precipitate of $Ag_2S_2O_3$
16. When silver nitrate is added to a dilute solution of sodium thiosulphate we get
    (A) a soluble complex $Na_3[Ag(S_2O_3)_2]$ (B) a black precipitate of Ag
    (C) a black precipitate of Ag_2S (D) a white precipitate of $Ag_2S_2O_3$
17. The composition of copper matte is
    (A) a mixture of molten Cu and FeS (B) a mixture of molten Cu_2S and FeS
    (C) a mixture of Cu_2S and molten slag
    (D) molten copper.
18. In the electrorefining of copper, copper is deposited on the cathode because
    (A) oxidation potential of $Zn^{2+} > Cu^{2+}$
    (B) reduction potential of $Zn^{2+} > Cu^{2+}$
    (C) Zinc does not react with cathode (D) none of these.
19. Which of the following gives copper from $CuSO_4$?
    (A) Heating with NaOH (B) Reducing with a solution of glucose
    (C) Addition of Fe filings to it (D) Heating with Na_2CO_3
20. The extraction of copper from poor copper pyrites is done by
    (A) smelting process (B) cementation
    (C) carbon reduction process (D) hydrometallurgical process.
21. Which of the following is the main difficulty in the extraction of copper from copper pyrites?
    (A) Cu_2O cannot be reduced by C (B) Copper is contaminated with iron
    (C) Unless the whole of sulphur is removed Cu_2O cannot be obtained
    (D) None of these.

22. Which of the following methods is used for the extraction of copper from non-sulphide ores?
(A) Electrolytic method (B) Thermal decomposition
(C) Reduction by hydrogen (D) Leaching with H_2SO_4
23. Identify the incorrect statement.
(A) Silver is produced by hydrometallurgical process
(B) Steel is produced by bessemerisation
(C) Production of copper does not involve bessemerisation
(D) Aluminium is produced by Serpeck process.
24. Roasting of copper ore is carried out in
(A) electric arc furnace (B) refractory furnace
(C) blast furnace (D) reverberatory furnace
25. The ratio of Cu : Ni : Zn in German silver is
(A) 1 : 2 : 1 (B) 1 : 1 : 2 (C) 2 : 1 : 1 (D) none of these.
26. Parke's process of extraction of silver from argentiferous lead is based on
(A) Ag is less volatile than Zn (B) Ag is more volatile than Zn
(C) Ag is more soluble in molten Zn than in molten Pb
(D) Zn–Ag alloy solidifies earlier than molten Pb.
27. A current of air is passed in the cyanide process of extraction of Ag. The current of air
(A) catalyses the forward reaction (B) prevents the formation of colloidal Ag
(C) oxidises Ag_2S to Ag_2SO_4 (D) prevents the backward reaction.
28. Identify the correct statement.
(A) Ag is insoluble in molten Pb. (B) Ag is insoluble in molten Zn.
(C) Ag is less soluble in molten Zn than molten Pb.
(D) Ag is more soluble in molten Zn than molten Pb.
29. The treatment of Zn dust to a solution of $Na[Ag(CN)_2]$ leads to the precipitation of Ag because
(A) Zn is more electropositive than Ag
(B) Zn is less electropositive than Ag
(C) Zn forms a complex with cyanide
(D) both Zn^{2+} and Ag^+ have d^{10} configuration.
30. Which of the following principles is involved in Pattinson's process of extraction of Ag?
(A) Elevation of boiling point (B) Depression of freezing point
(C) Gay-Lussac's law (D) Le Chatelier's principle.
31. Which of the following reacts with conc. H_2SO_4?
(A) Pb (B) Pt (C) Ag (D) Au
32. Silver containing lead as impurity is removed by
(A) distillation (B) levigation (C) poling (D) cupellation.

33. Which of the following is correct?
 (A) German silver: Cu + Au + C (B) Solder: Pb + Al
 (C) Gun metal: Cu + Sn + Zn (D) Duralumin: Al + Cu + Mg + Ag
34. AgCl on fusion with Na_2CO_3 gives
 (A) Ag_2O (B) Ag (C) Ag_2CO_3 (D) silver carbide.
35. CuCl, CuCN and CuSCN are the only cuprous compounds stable in water due to
 (A) diamagnetic nature (B) high hydration energy of Cu^+
 (C) insolubility in water (D) tendency for disproportionation.
36. AgO formed by the oxidation of Ag_2O is
 (A) diamagnetic (B) paramagnetic (C) ferromagnetic (D) antiferromagnetic.
37. Which of the following is used to remove lead from silver?
 (A) Leaching with dil. NaCN solution
 (B) Leaching with dil. NaCN solution in the presence of air
 (C) Electrolytic purification using $AgNO_3$ (D) Parke's process.
38. On heating $AgNO_3$ strongly we get
 (A) silver nitride (B) Ag_2O (C) $AgNO_2$ (D) Ag
39. Which of the following methods is used to obtain silver from argentite (Ag_2S)?
 (A) Fused mixture of Ag_2S and KCl is electrolysed
 (B) Ag_2S is roasted to Ag_2O followed by reduction with C
 (C) Ag_2S is reduced with CO
 (D) Treating argentite with NaCN solution followed by metal displacement with Zn.
40. Which of the following is formed when copper is treated with dil. HNO_3?
 (A) $Cu(NO_3)_2 + NO + H_2O$ (B) $CuNO_3 + H_2$
 (C) $CuNO_2 + H_2$ (D) $Cu(NO_3)_2 + NO_2 + H_2O$
41. Although both copper and silver have d^{10} configuration, copper is brownish yellow but silver is colourless. This is due to
 (A) smaller atomic radius of Cu in comparison to that of Ag making the $(n-1)d^{10}ns^1 \longrightarrow (n-1)d^9ns^2$ transition possible in Cu
 (B) presence of d–p transition in Cu and absence of d–p transition in Ag
 (C) poor shielding effect of f-orbitals in Ag and larger shielding effect of d-orbitals in Cu
 (D) presence of lanthanoid contraction in Ag and absence of lanthanoid contraction in Cu.
42. A metal which is not affected by conc. HNO_3, H_2SO_4 or alkalis forms a compound X. The compound X can be used to give a complex which finds application for toning in photography. The metal is
 (A) Au (B) Ag (C) Hg (D) Cu
43. Which of the following does not react with AgCl?
 (A) $Na_2S_2O_3$ (B) $NaNO_3$ (C) NH_4OH (D) Na_2CO_3

44. In the reaction, $4M + 8CN^- + 2H_2O + O_2 \longrightarrow 4[M(CN)_2]^- + 4OH^-$, the metal M is
(A) Cu (B) Ag (C) Au (D) Zn

45. AgF_2 is
(A) diamagnetic (B) paramagnetic (C) ferromagnetic (D) antiferromagnetic.

46. Which of the following is used in silvering of mirror?
(A) Ammoniacal $AgNO_3$ + HCHO + $FeCl_3$
(B) Ammoniacal $AgNO_3$ + $FeCl_3$
(C) Ammoniacal $AgNO_3$ + HCHO + Pb_3O_4 (D) Ammoniacal $AgNO_3$.

47. Which of the following silver(I) salts is soluble in water?
(A) Ag_2S (B) Ag_2CO_3 (C) AgCl (D) AgF

48. Which of the following silver(I) salts is least soluble in water?
(A) AgF (B) AgCl (C) Ag_2S (D) AgI

49. Which of the following silver(I) salts is used for the preparation of marking ink?
(A) $AgNO_3$ (B) AgBr (C) AgI (D) AgF

50. Molten silver absorbs a large volume of O_2 which is evolved on cooling leading to scattering of silver particles. The phenomenon is called
(A) frosting of silver (B) hairing of silver
(C) spitting of silver (D) silvering of mirror.

51. Cu^{2+} ion forms a reddish-brown precipitate with potassium ferrocyanide. The reddish-brown compound is
(A) $Cu_3[Fe(CN)_6]_2$ (B) $Cu_3[Fe(CN)_6]$ (C) $Cu_2[Fe(CN)_6]$ (D) $Cu_4[Fe(CN)_6]$

52. When copper is exposed to moist air, it is slowly oxidised to form a green coating whose composition is
(A) $CuCO_3$ (B) $CuSO_4{\cdot}5H_2O$ (C) $Cu(OH)_2$ (D) $CuCO_3{\cdot}Cu(OH)_2$

53. Copper reacts with conc. HNO_3 to evolve
(A) N_2O_3 (B) NO_2 (C) NO (D) N_2O

54. In the process of extraction of gold, roasted ore + CN^- + $H_2O \xrightarrow{O_2}$ [X] + OH^-, [X] + Zn $\longrightarrow$ [Y] + Au, identify the complexes [X] and [Y].
(A) $[X] = [Au(CN)_2]^-$, $Y = [Zn(CN)_4]^{2-}$ (B) $X = [Au(CN)_4]^{2-}$, $Y = [Zn(CN)_4]^{2-}$
(C) $X = [Au(CN)_2]^-$, $Y = [Zn(CN)_4]^{4-}$ (D) $X = [Au(CN)_4]^-$, $Y = [Zn(CN)_4]^{2-}$ **(I.I.T. 2003)**

Hint: $2Au + 4CN^- + H_2O + \frac{1}{2}O_2 \longrightarrow 2[Au(CN)_2]^- + 2OH^-$
$2[Au(CN)_2]^- + Zn \longrightarrow [Zn(CN)_4]^{2-} + 2Au$
Hence $X = [Au(CN)_2]^-$ and $Y = [Zn(CN)_4]^{2-}$

55. Identify the correct statement for the reaction of KCN with an aqueous solution of copper(II) sulphate pentahydrate.
(A) KCN acts as a reducing agent
(B) KCN acts as a coordinating agent
(C) $K_3[Cu(CN)_4]$ is formed by reduction of copper(II)
(D) all of these

56. CuCl, CuCN and CuSCN are the only compounds of copper(I) which are stable in aqueous medium due to
    (A) insolubility in water (B) high hydration energy of Cu^+ ion
    (C) diamagnetic nature (D) tendency to disproportionation.
57. For what purpose moist Ag_2O is used in organic reactions?
    (A) Replacement of CH_3 group by OH group
    (B) Replacement of halogen group by OH group
    (C) Replacement of NH_2 group by OH group
    (D) Replacement of NO_2 group by OH group.
58. Native silver metal forms a water soluble complex with a dilute aqueous solution of NaCN in the presence of
    (A) nitrogen (B) oxygen (C) carbon dioxide (D) argon **(I.I.T. 2008)**

    Hint: Silver is oxidised to Ag^+ by oxygen in the presence of cyanide ion. The reaction of Ag^+ with CN^- gives a complex ion, $[Ag(CN)_2]^-$.

Paragraph for Question Nos. 59-61

When a metal rod M is dipped into an aqueous colourless concentrated solution of compound N, the solution turns light blue. Addition of aqueous NaCl to the blue solution gives a white precipitate O. Addition of aqueous NH_3 dissolves O and gives an intense blue solution.

59. The metal rod M is
    (A) Fe (B) Cu (C) Ni (D) Co **(I.I.T. 2011)**

    Hint: $\underset{M}{Cu} + \underset{N}{2AgNO_3} \longrightarrow \underset{Blue}{Cu(NO_3)_2} + 2Ag$

    While Cu is partially oxidized to $Cu(NO_3)_2$, the remaining $AgNO_3$ reacts with NaCl.
60. The compound N is
    (A) $AgNO_3$ (B) $Zn(NO_3)_2$ (C) $Al(NO_3)_3$ (D) $Pb(NO_3)_2$ **(I.I.T. 2011)**

    Hint: The compound N is silver nitrate.
61. The final solution contains
    (A) $[Pb(NH_3)_4]^{2+}$ and $[CoCl_4]^{2-}$ (B) $[Al(NO_3)_4]^{3+}$ and $[Cu(NH_3)_4]^{2+}$
    (C) $[Ag(NH_3)_2]^+$ and $[Cu(NH_3)_4]^{2+}$ (D) $[Ag(NH_3)_2]^+$ and $[Ni(NH_3)_6]^{2+}$ **(I.I.T. 2011)**

    Hint: $\underset{N}{AgNO_3} + NaCl \longrightarrow \underset{O}{AgCl} \downarrow + NaNO_3$

    $AgCl + 2NH_3 \longrightarrow [Ag(NH_3)_2]^+Cl^-$

    $Cu(NO_3)_2 + 4NH_4OH \longrightarrow [Cu(NH_3)_4]^{2+}$

ANSWERS

1. C	2. B, C	3. C, D	4. A	5. B	6. A
7. B	8. A	9. C	10. B	11. B	12. D
13. D	14. D	15. A	16. C	17. B	18. B
19. C	20. D	21. C	22. D	23. C	24. D
25. C	26. A, C, D	27. C, D	28. D	29. A	30. B
31. C	32. D	33. C	34. B	35. D	36. B
37. D	38. D	39. D	40. A	41. A	42. A
43. A	44. B, C	45. D	46. C	47. D	48. C
49. A	50. C	51. C	52. D	53. B	54. A
55. D	56. D	57. B	58. B	59. B	60. A
61. C					

30

Zinc, Cadmium and Mercury

OBJECTIVE TYPE MULTIPLE CHOICE QUESTIONS

1. Assertion (*A*): Zn^{2+} is diamagnetic.
   Reason (*R*): The electrons are lost from 4*s* orbital to form Zn^{2+}
   (A) Both *A* and *R* are true and *R* is the correct explanation of *A*
   (B) Both *A* and *R* are true and *R* is not the correct explanation of *A*
   (C) *A* is correct but *R* is false (D) *A* is false but *R* is correct. **(I.I.T. 1998)**
2. Which of the following is not the characteristics of zinc?
   (A) It is a volatile metal (B) It dissolves in alkali forming Na_2ZnO_2
   (C) Zinc dust is used as a reducing agent
   (D) It is brittle at very high temperature.
3. A white solid halide of Hg forms a black mixture with NH_4OH. The halide is
   (A) $HgCl_2$ (B) Hg_2Cl_2 (C) Hg_2I_2 (D) HgI_2
4. What is the product obtained in the reaction of $HgCl_2$ and $Hg(CN)_2$?
   (A) Hg(CN)Cl (B) $Hg[Hg(CN)_2Cl_2]$
   (C) $HgCl_2{\cdot}Hg(CN)_2$ (D) $(CN)_2$
5. The heating of HgO with ammonia solution gives a pale yellow precipitate whose formula is
   (A) 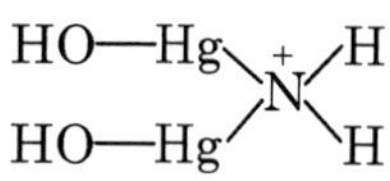
   (B) 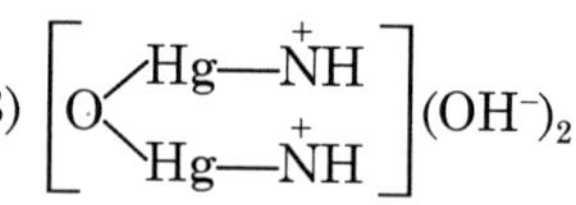
   (C) $Hg_2OH{\cdot}NH_3{\cdot}H_2O$ (D) $Hg_2O{\cdot}NH_2OH$
6. Philosopher's wool when heated with BaO at 1100°C gives
   (A) Ba + ZnO (B) $BaCdO_2$ (C) BaO_2 + Zn (D) $BaZnO_2$
7. Identify the false statement.
   (A) Pure Zn reacts with water forming H_2 and ZnO.

(B) Zn–Cu couple decomposes water forming H_2.
(C) Zn gets tarnished in moist air. (D) Zn is malleable and ductile.

8. Identify the incorrect statement.
(A) Zn^{2+} is a transition metal ion. (B) ZnO is an amphoteric oxide.
(C) $[ZnCl_2(H_2O)_2]$ has a square planar structure.
(D) SnO_2 is a basic oxide.

9. Zinc spelter is
(A) 100% Zn (B) ZnO (C) an ore of Zn (D) Zn + Zn dust + ZnO

10. Which of the following acids is used in leaching ZnO before electrorefining of zinc?
(A) dil. HCl (B) dil. H_2SO_4 (C) dil. HNO_3 (D) conc. HNO_3

11. Zinc and silver are separated from one another by
(A) treatment with dil. HNO_3 (B) distillation
(C) heating with conc. NaOH (D) fractional crystallisation.

12. Heating of ZnO with $Co(NO_3)_2$ forms a compound whose name is
(A) butter of zinc (B) verdigris (C) cobalt zincate (D) Rinman's green

13. H_2S gas is passed through an aqueous solution of $CuSO_4$ and $ZnSO_4$ in the presence of dilute NH_4OH. Which of the following is precipitated?
(A) CuS (B) ZnS (C) ZnS and CuS (D) No precipitate.

14. Pure zinc is not soluble in acids because
(A) zinc has a low overvoltage (B) zinc has a high overvoltage
(C) zinc has a positive standard electrode potential
(D) zinc has a negative standard electrode potential.

15. In the extraction of Hg from cinnabar, the ore is roasted with
(A) 2% coke + 2% CaO (B) 3% coke + 3% CaO
(C) 2% coke + 5% CaO (D) 5% coke + 5% CaO

16. Zinc reacts with cold and dilute HNO_3 to give
(A) NO (B) NO_2 (C) NH_4NO_3 (D) $Zn(NO_3)_2$

17. The reducing agent used in the extraction of mercury from cinnabar is
(A) $SnCl_2$ (B) H_2 (C) CO + H_2 (D) none of these.

18. Mercury is used in the
(A) extraction of Zn (B) extraction of Ag
(C) extraction of Al (D) manufacture of NaOH.

19. Which of the following methods is used to prepare calomel?
(A) Sublimation of Hg and $HgCl_2$ (B) Addition of excess of NaCl to $Hg_2(NO_3)_2$
(C) Addition of HCl to Hg (D) Addition of Cl_2 to Hg

20. The trace metal present in insulin is
(A) Co (B) Mn (C) Zn (D) Fe

21. Identify the incorrect statement.
(A) Silver exhibits the oxidation states of + 1, + 2, + 3.
(B) Zinc blende mainly contains $ZnCl_2$.

(C) Copper pyrites also contains Fe_2S_3.
(D) Silver glance mainly contains Ag_2S.

22. Identify the incorrect statement regarding corrosive sublimate.
(A) It is prepared by heating Hg and Cl_2.
(B) It oxidises $SnCl_2$.
(C) It reduces $SnCl_4$. (D) It sublimes readily.

23. Which of the following does not form an amalgam?
(A) Co (B) Zn (C) Ag (D) Fe

24. A metal gives two chlorides X and Y. X gives black precipitate and Y gives white precipitate with NH_4OH. With KI, Y gives a red precipitate which is soluble in excess of KI. X and Y are respectively
(A) $ZnCl_2$, $HgCl_2$ (B) $HgCl_2$, Hg_2Cl_2 (C) $HgCl_2$, $ZnCl_2$ (D) Hg_2Cl_2, $HgCl_2$

25. Mercury on heating with aqua regia gives
(A) Hg_2Cl_2 (B) $HgCl_2$ (C) $Hg(NO_2)_2$ (D) $Hg(NO_3)_2$

26. Which of the following does not respond to chromyl chloride test?
(A) $ZnCl_2$ (B) $CuCl_2$ (C) $HgCl_2$ (D) $C_6H_5NH_3^+Cl^-$

27. The pair of compounds which sublimes on heating is
(A) $HgCl_2$, Ag_2C (B) $HgCl_2$, NH_4Cl (C) ZnO, CuO (D) NH_4OH, NH_4Cl

28. In the metallurgical extraction of Zn from ZnO, the reducing agent is
(A) NO (B) CO_2 (C) H_2 (D) CO

29. Identify the incorrect statement.
(A) Mercury can evolve H_2 from H_2S.
(B) Mercury has high specific heat.
(C) Mercury is less reactive than hydrogen.
(D) Zinc exhibits coordination number of 4, 5 and 6.

30. Identify the incorrect statement.
(A) Zinc forms square planar complexes.
(B) Zinc is less reactive than cadmium.
(C) Poisoning by Hg vapour is prevented by sprinkling sulphur powder.
(D) Al/Hg is more reactive than aluminium.

31. In the metallurgy of which of the following liquation is carried out?
(A) Cu (B) Zn (C) Na (D) Al

32. The metal which is purified by distillation is
(A) Fe (B) Cu (C) Ni (D) Zn

33. Which of the following is used for the purification of zinc spelter?
(A) Fractional distillation (B) Liquation
(C) Electrolysis (D) None of these.

34. Zn–Cu couple reacts with water to give
(A) $ZnCl_2 + CuCl_2$ (B) $H_2 + O_2$ (C) H_2 (D) none of these.

35. Which of the following gases is evolved when ZnS is treated with dil. HCl?
(A) H_2S (B) H_2 (C) SO_2 (D) Cl_2

36. Mercury(II) chloride on reaction with KI gives a product P which is soluble in excess KI solution to form another compound Q. P and Q are
(A) Hg_2I_2, K_2HgI_4 (B) HgI_2, K_2HgI_4 (C) HgI_2, K_2HgI_3 (D) HgI_2, HgI_3^-
37. When H_2S is passed through an aqueous solution of $CuSO_4$ and $ZnSO_4$ in the presence of dil. NH_4OH, the precipitate formed is
(A) CuS (B) ZnS (C) CuS + ZnS (D) no precipitate is formed.
38. The compound used as a flux in soldering is
(A) ZnO (B) $ZnCO_3$ (C) $ZnSO_4$ (D) $ZnCl_2$
39. The reaction of mercuric chloride with ammonium hydroxide gives
(A) $Hg(NH_3)_2Cl_2$ (B) $Hg(NH_2)_4Cl$ (C) $Hg(NH_2)Cl$ (D) $NH_2—Hg—Hg—Cl$
40. In moist air, zinc forms a protective layer on its surface. The layer consists of
(A) ZnO (B) $Zn(OH)_2$ (C) $ZnSO_4$ (D) $ZnCO_3·Zn(OH)_2$
41. Metal oxide(s) which decomposes on heating is/are
(A) HgO (B) Ag_2O (C) ZnO (D) Al_2O_3
42. Which of the following is white vitriol?
(A) $ZnSO_4·2H_2O$ (B) $ZnSO_4·7H_2O$ (C) $CuSO_4·5H_2O$ (D) $FeSO_4·7H_2O$
43. Zinc oxide is
(A) acidic oxide (B) basic oxide (C) neutral oxide (D) amphoteric oxide.
44. Zinc reacts with conc. H_2SO_4 to evolve
(A) $SO_2(g)$ (B) $O_2(g)$ (C) $SO_3(g)$ (D) $H_2(g)$
45. Zn^{2+} ion reacts with dil. NaOH to form a white precipitate which dissolves in excess NaOH to form
(A) $Na_2[Zn(OH)_4]$ (B) $[Zn(H_2O)_4]^{2+}$ (C) $Zn(OH)_2$ (D) Na_2ZnO_2
46. When zinc blende is roasted in air, the species formed are
(A) Zn, SO_2 (B) ZnS, SO_2 (C) ZnO, SO_2 (D) ZnO, ZnS, SO_2
47. Which of the following mercury compounds are poisonous?
(A) $HgCl_2$ (B) Hg_2Cl_2 (C) $HgSO_4$ (D) All of these.
48. Which of the following elements do not show the properties characteristics of *d*-block elements?
(A) Fe, Co, Ni (B) Sc, Ti, V (C) Zn, Cd, Hg (D) Cu, Ag, Au
49. An aqueous solution containing 1 mol. HgI_2 and 2 mol. NaI is orange in colour. On addition of excess NaI the solution becomes colourless. The orange colour reappears on addition of NaOCl and is due to
(A) Na_2HgI_4 (B) HgI_2 (C) NaOCl (D) NaI_3
Hint: $HgI_2 + 2NaI \rightleftharpoons Na_2HgI_4$ (coloured due to presence of residual HgI_2)
HgI_2 + excess NaI $\longrightarrow Na_2HgI_4$ (colourless)
$3Na_2HgI_4 + 2NaOCl + 2H_2O \longrightarrow 3HgI_2$ (orange) $+ 2NaCl + 4NaOH + 2NaI_3$
50. The reaction of zinc with cold and very dilute HNO_3 gives
(A) $Zn(NO_3)_2 + NO$ (B) $Zn(NO_3)_2 + NH_4NO_3$
(C) $Zn(NO_3)_2 + N_2O$ (D) $Zn(NO_3)_2 + NO_2$

ANSWERS

1. B	2. D	3. B	4. D	5. A	6. D
7. A	8. A, C, D	9. D	10. B	11. B, C	12. D
13. C	14. B	15. C	16. C, D	17. D	18. B, D
19. A, B	20. C	21. B	22. C	23. D	24. D
25. B	26. C	27. B	28. D	29. A	30. A, B
31. B	32. D	33. C	34. C	35. A	36. B
37. C	38. D	39. C	40. D	41. A, B	42. B
43. D	44. A	45. A	46. C	47. D	48. C
49. B	50. B				

31

Mole Concept and Chemical Arithmetic

IMPORTANT RELATIONS FOR NUMERICAL PROBLEMS

1. Mass of one atom of element = gram atomic mass/N where = N = Avogadro number
2. Mass of X atoms of element = gram atomic mass × X/N
3. Mass of one molecule of substance = gram molecular mass/N
4. Mass of X molecules of substance = gram molecular mass × X/N
5. Number of atoms in x gram of an element = xN/gram atomic mass
6. Number of atoms in x moles of an element = $n \times N$
7. Number of moles of element in y atoms = y/N
8. Number of molecules in x gram of a substance = xN/gram molecular mass
9. Number of molecules in n moles of a substance = $n \times N$
10. Number of moles in y molecules = y/N
11. For the elements with atomicity > 1, the formulae are modified as:

    Number of atoms in n moles = $n \times N \times$ atomicity

    Number of moles in y atoms = $y \times$ atomicity/N
12. Weight % of an element = At. wt. × No. of atoms per molecule × 100/M.W. of the compound
13. Equivalent weight of an element = At. wt./valency
14. Equivalent weight of compound = M.W./total charge on cation or anion
15. Equivalent weight of acid = M.W./basicity
16. Equivalent weight of base = M.W./acidity
17. Equivalent weight of acid salt = M.W./No. of replaceable H atom

18. Normality = equivalent of solute/volume of solvent in litre
19. Molarity = moles of solute/volume of solvent in litre
20. Molality = moles of solute/volume of solvent in kg
21. Formality = weight of ionic solute/(formula weight of solute × volume of solute in litre)
22. Mole fraction of solute = moles of solute/(moles of solute + moles of solvent)
23. Mole fraction of solvent = moles of solvent/(moles of solute + moles of solvent)
24. % by weight = weight of solvent × 100/weight of solution
25. % by volume = weight of solvent × 100/volume of solution
26. % by strength = volume of solvent × 100/volume of solution
27. M.W. = vapour density × 2 (for gases only)
28. No. of equivalents = normality × volume in L = wt./Eq. wt.
29. No. of milliequivalents = normality × volume in mL = wt. × 1000/Eq. wt.
30. No. of moles = molarity × volume in L = wt./M.W.
31. No. of millimoles = molarity × volume in mL = wt. × 1000/M.W.
32. Eq. wt. of an oxidising agent = Formula weight/No. of electrons gained per molecule
33. Eq. wt. of a reducing agent = Formula weight/No. of electrons lost per molecule
34. Eq. wt. of metal = wt. of metal × 1.008/wt. of hydrogen displaced
    Eq. wt. of metal = wt. of metal × 8.0/wt. of oxygen
    Eq. wt. metal = wt. of metal × 35.5/wt. of chlorine
35. 1 mole of atoms = 6.023×10^{23} atoms = gram-atomic mass of element
36. 1 gram-atom = gram-atomic weight = weight of Avogadro number of atoms
37. Number of gram-atoms = weight/atomic weight
38. 1 mole of molecules = 6.023×10^{23} molecules = gram-molecular mass of substance
39. 1 gram-mole = gram-molecular weight = weight of Avogadro number of molecules
40. Number of gram moles of substance = weight/molecular weight
41. 1 u = (1/12)th of mass of ^{12}C atom = 1.66×10^{-24} g
42. Molecular formula = n × empirical formula
43. Dulong–Petits law: At. wt. × specific heat × 6.4 (for metals only)
44. Strength (S) = N × Eq. wt.

OBJECTIVE TYPE MULTIPLE CHOICE QUESTIONS

1. The law of multiple proportions was proposed by
   (A) Lavoisier (B) Dalton (C) Gay-Lussac (D) Proust. **(I.I.T. 1992)**
2. 1.12 mL of a gas was produced at STP by the action of 4.12 mg of alcohol (ROH) with CH_3MgI. The molecular mass of ROH is
   (A) 16.0 (B) 41.2 (C) 82.4 (D) 156.0
   (Roorkee 1992, I.I.T. 1993)
3. The sulphate of a metal M contains 9.87% of M. The sulphate is isomorphous with $ZnSO_4 \cdot 7H_2O$. The atomic mass of M is
   (A) 24.3 (B) 11.3 (C) 36.3 (D) 40.3 **(I.I.T. 1991)**
4. The mass of 1×10^{22} molecules of $CuSO_4 \cdot 5H_2O$ is
   (A) 41.44 g (B) 4.144 g (C) 414.4 g (D) none. **(I.I.T. 1991)**
5. In which mode of expression, the concentration of a solution remains independent of temperature?
   (A) Molarity (B) Normality (C) Formality (D) Molalilty. **(I.I.T. 1998)**
6. A molal solution is one that contains one mole of a solute in
   (A) 1000 g of the solvent (B) 1 L of the solvent
   (C) 1 L of the solution (D) 22.4 L of the solution. **(I.I.T. 1986)**
7. If 0.5 mole of $BaCl_2$ is mixed with 0.2 mole of Na_3PO_4, the maximum number of moles of $Ba_3(PO_4)_2$ that can be formed is
   (A) 0.7 (B) 0.5 (C) 0.3 (D) 0.1 **(I.I.T. 1981)**
8. M is the M.W. of $KMnO_4$. The equivalent mass of $KMnO_4$ when it is converted into K_2MnO_4 is
   (A) M (B) M/3 (C) M/5 (D) M/7 **(I.I.T. 1980)**
9. 2.76 g of silver carbonate on being strongly heated yield a residue weighing
   (A) 2.16 g (B) 2.48 g (C) 2.32 g (D) 2.64 g **(I.I.T. 1979)**
10. A gaseous mixture contains oxygen and nitrogen in the ratio of 1 : 4 by mass. The ratio of their number of molecules is
    (A) 1 : 4 (B) 1 : 8 (C) 7 : 32 (D) 3 : 16 **(I.I.T. 1979)**
11. The largest number of molecules is in
    (A) 36 g H_2O (B) 54 g N_2O_5 (C) 46 g C_2H_5OH (D) 28 g CO **(I.I.T. 1979)**
12. 27 g aluminium will react completely with
    (A) 8 g O_2 (B) 16 g O_2 (C) 24 g O_2 (D) 40 g O_2 **(I.I.T. 1978)**
13. Which of the following has the highest mass?
    (A) 20 g of phosphorus (B) 5 moles of water
    (C) 2 equivalents of Na_2CO_3 (D) 12×10^{24} atoms of hydrogen **(I.I.T. 1978)**
14. 10 g of $CaCO_3$ contain
    (A) 10 moles of $CaCO_3$ (B) 1 g atom of Ca
    (C) 6×10^{22} atoms of Ca (D) 0.1 g equiv. of Ca **(I.I.T. 1978)**

15. How much lime (CaO) can be obtained by heating 200 kg of limestone which is 95% pure?
    (A) 106.4 kg (B) 53.2 kg (C) 10.64 kg (D) 5.32 kg **(I.I.T. 1975)**
16. How much AgCl will be formed by adding 200 mL of 5*M* HCl to a solution containing 1.7 g $AgNO_3$?
    (A) 1.435 g (B) 1.7175 g (C) 0.1435 g (D) 2.87 g **(Roorkee 1975)**
17. The number of molecules in one litre of a gas at STP is known as
    (A) Loschmidt number (B) Berzelius number
    (C) Avogadro number (D) Gay-Lussac number.
18. Two gaseous samples were analysed and both contained 6% H and 94% S. These figures illustrate
    (A) law of definite proportions (B) law of multiple proportions
    (C) law of conservation of mass (D) Boyle's law.
19. Different proportions of oxygen in the various oxides of nitrogen prove the law of
    (A) multiple proportions (B) conservation of matter
    (C) equivalent proportions (D) constant proportions.
20. X and Y are two allotropes of an element. 1 g of X will differ from 1 g of Y in
    (A) atomic arrangement (B) oxidation number
    (C) total number of atoms (D) chemical composition.
21. Two samples of lead oxide were separately reduced by heating in a current of H_2. The mass of Pb from one oxide was half the mass of Pb obtained from the other oxide. This illustrates the law of
    (A) multiple proportions (B) reciprocal proportions
    (C) constant proportions (D) equivalent proportions.
22. Consider the statements: 1. In chemical reactions, the number of moles of reactants always equals the number of moles of products. 2. When a liquid and its vapours are in equilibrium, if the pressure is suddenly decreased, cooling occurs. 3. Isopiestic solutions are solutions which have the same vapour pressure.
    (A) 1, 2 (B) 1, 3 (C) 1, 2, 3 (D) 2, 3
23. Which is heaviest?
    (A) 25 g of Hg (B) 2 moles of H_2O
    (C) 2 moles of CO_2 (D) 4 gram-atoms of O_2
24. The number of molecules in 8.96 L of a gas at 0°C and 1 atm pressure is
    (A) 6.02×10^{23} (B) 12.04×10^{23} (C) 18.06×10^{23} (D) 2.408×10^{23}
25. The ratio of the rate of diffusion of a gaseous element to that of He is 4. What is the atomic mass of the element?
    (A) 0.25 (B) 2 (C) 4 (D) 7
26. Assume that the nucleus of F atom is a sphere of radius 5×10^{-13} cm. The density of the matter in the nucleus is
    (A) 5.75×10^{13} g/mL (B) 6.02×10^{23} g/mL
    (C) 12.04×10^{23} g/mL (D) 12.04×10^{13} g/mL

27. Assume the density of water to be 1 g/mL. The volume occupied by one molecule of water is
(A) 3.72×10^{-20} mL (B) 6.0×10^{23} mL
(C) 6.0×10^{-23} mL (D) 3.0×10^{-24} mL
28. If 1.12 L of CH_4 contains 3×10^{22} molecules at STP, the number of molecules in 5.56 L of O_2 at STP is
(A) 1.5×10^{23} (B) 6.0×10^{22} (C) 3.0×10^{22} (D) 0.15×10^{22}
29. The mass of 350 mL of a diatomic gas at 2 atm and 0°C is 1 g. The mass of 1 atom of the gas is
(A) 3.0×10^{-23} g (B) 2.66×10^{-23} g (C) 4.3×10^{-23} g (D) 0.15×10^{-23} g
30. What is the volume of CO_2 liberated at STP when 2.12 g of Na_2CO_3 is treated with excess dil. HCl?
(A) 0.448 L (B) 2.28 L (C) 22.4 L (D) 44.8 L
31. 12 g of an alkaline earth metal gives 14.8 g of its nitride. Atomic mass of the metal is
(A) 20 (B) 40 (C) 12 (D) 14.8
32. The number of moles of a solute in its solution is 20 and total number of moles is 80. The mole fraction of the solute is
(A) 0.5 (B) 0.75 (C) 0.25 (D) 1.0
33. The number of grams of oxygen in 32.2 g of $Na_2SO_4{\cdot}10H_2O$ is
(A) 22.4 (B) 20.8 (C) 2.08 (D) 2.24
34. For the reaction, $BaCO_3(s) \longrightarrow BaO(s) + CO_2(g)$, if 98.7 g of $BaCO_3$ is decomposed at STP, the reaction on completion gives the amount of CO_2 (At, wt. of Ba = 137.4)
(A) 22.4 L (B) 11.2 L (C) 44.8 L (D) 0.56 L
35. Molarity of a solution that has 49 g H_3PO_4 in 2 L solution is
(A) 1.0 (B) 0.25 (C) 0.50 (D) 0.75
36. The mass of all the atoms in a compound $C_{70}H_{22}$ is
(A) 1.43×10^{-21} g (B) 6.023×10^{23} g (C) 5.025×10^{23} g (D) 842 g
37. In a sample of pure compound Na = 0.0887 mole, O = 0.132 mole, C = 2.65×10^{22} atoms. The empirical formula of the compound is
(A) NaCO (B) $Na_3C_5O_2$ (C) Na_2CO_3 (D) $Na_{0.0887}C_{2.65 \times 10^{22}}O_{0.132}$
38. Which has the highest mass?
(A) 10 L of Hg (B) 1 m^3 of water (C) An adult man (D) All same.
39. If N is the Avogadro number, then the number of valence electrons in 4.2 g of N^{3-} is
(A) $3.2N$ (B) $2.4N$ (C) $4.2N$ (D) $1.6N$
40. What is the molarity of H_2SO_4 solution that has a density of 1.8 g/cc at 35°C and contains 98% by mass?
(A) $4.18M$ (B) $8.14M$ (C) $18M$ (D) $18.4M$

41. How many litres of O_2 at STP are needed to complete the combustion of 39 g benzene(l)? The combustion reaction is given by $2C_6H_6(l) + 15O_2(g) \longrightarrow 12CO_2(g) + 6H_2O(g)$.
(A) 11.2 L (B) 22.4 L (C) 24 L (D) 84 L

42. Which is heavy in mass?
(A) 2 g equiv. of Na_2CO_3 (B) 1 mole of H_2
(C) 64 g of O_2 (D) 20 g of phosphorus

43. The mole fraction of water in 36 g water and 34 g ammonia solution is
(A) 0.005 (B) 0.05 (C) 0.02 (D) 0.5

44. 50 cc of a hydrocarbon (X) required 45 cc of O_2 for complete combustion and 30 cc of CO_2 is formed. The formula of X is
(A) C_2H_4 (B) C_2H_6 (C) C_3H_8 (D) C_4H_{10}

45. One molecule of H_2O has the mass
(A) 1.5×10^{-26} kg (B) 2.5×10^{-26} kg (C) 3.0×10^{-25} kg (D) 3.0×10^{-26} kg

46. 0.175 g of a diacid base requires 28 mL of (N/8) H_2SO_4 for complete neutralisation. The molecular mass of the base is
(A) 50 (B) 100 (C) 175 (D) 200

47. The total number of atoms of all the elements present in 1 mole of $(NH_4)_2Cr_2O_7$ is
(A) 14 (B) 19 (C) 6×10^{23} (D) 114×10^{23}

48. Which of the following contains the largest number of ions?
(A) 50 g NH_4Cl (M.W. = 53.5) (B) 50 g $CaCl_2$ (M.W. = 111)
(C) 50 g K_2SO_4 (M.W. = 174) (D) 50 g NaCl (M.W. = 58.5)

49. 100 mL of 0.1N $Na_2S_2O_3$ decolourised iodine by the addition of x gram $CuSO_4{\cdot}5H_2O$ to excess of KI. The value of x is (M.W. of $CuSO_4{\cdot}5H_2O$ = 250)
(A) 1.25 g (B) 2.5 g (C) 4.0 g (D) 5.0 g

50. The number of atoms is greatest in
(A) 4 g H_2 (B) 48 g Mg (C) 127 g I_2 (D) 71 g Cl_2

51. 3.36 g of metal(II) displace from an acid 1120 mL of H_2 at STP. The equivalent mass of metal is
(A) 3.36 (B) 16.8 (C) 33.6 (D) 67.2

52. A metal oxide has the formula M_2O_3. It can be reduced by H_2 to give free metal and water. 0.1596 g of M_2O_3 required 6 mg of H_2 for complete combustion. The atomic mass of the metal is
(A) 27.9 (B) 55.8 (C) 79.8 (D) 159.8

53. In a vessel 2 g O_2, 2 g H_2 and 2 g N_2 are present. Which has the largest number of atoms?
(A) H_2 (B) O_2 (C) N_2 (D) all same.

54. An element M forms MCl_3. In that 20% of metal is present. The atomic mass of the metal is
(A) 13 (B) 20 (C) 26.6 (D) 40

55. The number of molecules in 4.25 g NH_3 is
(A) 0.5×10^{23} (B) 1.5×10^{23} (C) 2.5×10^{23} (D) 3.5×10^{23}

56. Which of the following contains the least number of molecules?
(A) 4 g of CO_2 (B) 4 g of N_2O (C) 3 g of PH_3 (D) 10 g of $COCl_2$

57. Which of the following pairs of gases contains same number of molecules?
(A) 11 g CO_2, 7 g N_2 (B) 22 g CO_2, 72 g N_2
(C) 44 g CO_2, 7 g N_2 (D) 11 g CO_2, 28 g N_2

58. Which of the following has the smallest number of molecules?
(A) 22.4×10^3 mL of $CO_2(g)$ (B) 22 g of $CO_2(g)$
(C) 11.2 L of $CO_2(g)$ at NTP (D) 0.1 mole of $CO_2(g)$

59. 0.16 g of a dibasic acid required 25 mL of decinormal NaOH solution for complete neutralisation. The molecular mass of the acid is
(A) 32 (B) 64 (C) 128 (D) 256

60. Molarity of pure water is
(A) 18 (B) 50 (C) 55.6 (D) 100

61. Haemoglobin contains 0.33% iron by mass. If the molecular mass of haemoglobin is 68000, how many iron atoms are present in each molecule of haemoglobin?
(A) 1 (B) 2 (C) 3 (D) 4

62. How many moles of electron weigh one kilogram?
(A) 6.023×10^{23} (B) $(1/9.108) \times 10^{31}$
(C) $(6.023/9.108) \times 10^{54}$ (D) $(1/9.108 \times 6.023) \times 10^8$ **(I.I.T. 2002)**
Hint: 1 mole of electron weighs $9.108 \times 10^{-31} \times 6.023 \times 10^{23}$ kg. No. of moles of electron in 1 kg = $1/9.108 \times 10^{-31} \times 6.023 \times 10^{23} = (1/9.108 \times 6.023) \times 10^8$.

63. Which has maximum number of atoms?
(A) 24 g of C (12) (B) 56 g of Fe (56)
(C) 27 g of Al (D) 108 g of Ag (108) **(I.I.T. 2003)**
Hint: 24 g carbon amounts to 2 mole, i.e. $2 \times 6.023 \times 10^{23}$ atoms while all others amount to 1 mole.

64. The law of equivalent is expressed as (where m_1 and m are mass of metal A and B, and E_1 and E_2 are equivalent mass of A and B)
(A) $m_1 + m_2 = E_1 + E_2$ (B) $m_1E_2 = m_2E_1$
(C) $E_1E_2 = m_1m_2$ (D) $m_1/m_2 = E_2/E_1$

65. A plant virus consists of uniform cylindrical particles of 150 Å in diameter and 5000 Å long. The specific volume of the virus is 0.75 cm^3/g. If the virus is considered to be a single particle, its molecular mass is
(A) 7.09×10^7 g mol^{-1} (B) 7.90×10^7 g mol^{-1}
(C) 9.07×10^7 g mol^{-1} (D) 9.70×10^7 g mol^{-1}
Hint: Volume of virus = $\pi r^2 h = 3.1416 \times (75 \times 10^{-8})^2 \times (5000 \times 10^{-8}) = 8.836 \times 10^{-17}$. Mass of single virus = volume/specific volume = $8.836 \times 10^{-17}/0.75 = 1.178 \times 10^{-16}$. Molecular mass = $1.178 \times 10^{-16} \times 6.023 \times 10^{23} = 7.09 \times 10^7$ g mol^{-1}.

ANSWERS

1. B	2. C	3. A	4. B	5. D	6. A
7. D	8. A	9. A	10. C	11. A	12. C
13. C	14. C	15. A	16. A	17. A	18. A
19. A	20. A	21. A	22. D	23. C	24. D
25. A	26. A	27. A	28. A	29. B	30. A
31. B	32. C	33. A	34. B	35. B	36. A
37. C	38. B	39. B	40. C	41. D	42. A
43. D	44. A	45. D	46. B	47. D	48. A
49. B	50. A	51. C	52. B	53. A	54. C
55. B	56. C	57. A	58. D	59. C	60. C
61. D	62. D	63. A	64. B	65. A	

32

Analytical Chemistry

IMPORTANT RELATIONS FOR NUMERICAL PROBLEMS

1. Strength of solution (g/litre) = normality × Eq. wt.
2. Normality = weight × 1000/Eq. wt. × volume of solvent in litre
3. No. of moles = weight in gram/M.W. = (volume in litre/22.4 L) at NTP (for gases only)
4. No. of millimoles = weight in gram × 1000/M.W. = molarity × volume in mL
5. No. of equivalents = weight in gram × x × No. of moles/Eq. wt. (where x = M.W./Eq. wt. = valency)
6. No. of milliequivalents = weight in gram × 1000/Eq. wt. = normality × volume in mL
7. Normality = x × No. of millimoles = x × molarity = (strength in g/L)/Eq. wt.
8. Normality formula, $N_1V_1 = N_2V_2$
9. Normality of the mixture of two solutions of the same solute, $N = (N_1V_1 + N_2V_2)/(V_1 + V_2)$
10. Volume of water $(V_2 - V_1)$ to be added to get a solution of normality (N_2) from V_1 mL of solution of normality (N_1): $V_2 - V_1 = [(N_1 - N_2)/N_2]V_1$
11. Normality of the acidic (basic) mixture = $(N_aV_a + N_bV_b)/(V_a + V_b)$
12. Weight of solute (W) = (normality × Eq. wt. × volume in mL)/1000
13. Weight of solute (W) = (molarity × M.W. × volume in mL)/1000

OBJECTIVE TYPE MULTIPLE CHOICE QUESTIONS

1. An aqueous solution of 6.3 g oxalic acid dihydrate is made up to 250 mL. The volume of 0.1*N* NaOH required to completely neutralise 10 mL of this solution is
(A) 40 mL (B) 20 mL (C) 10 mL (D) 4 mL **(I.I.T. 2001)**
2. In the standardisation of $Na_2S_2O_3$ using $K_2Cr_2O_7$ by iodometry, the equivalent mass of $K_2Cr_2O_7$ is
(A) M.W./2 (B) M.W./6 (C) M.W./3 (D) M.W. **(I.I.T. 2001)**
3. The normality of 0.3*M* phosphorous acid (H_3PO_3) is
(A) 0.6 (B) 0.3 (C) 0.1 (D) 0.9 **(I.I.T. 1999)**
4. Assertion (*A*): Sulphate is estimated by $BaSO_4$ and not as $MgSO_4$.
Reason (*R*): Ionic radius of Mg^{2+} is greater than that of Ba^{2+}
(A) Both *A* and *R* are true and *R* is the correct explanation of *A*
(B) Both *A* and *R* are true and *R* is not the correct explanation of *A*
(C) *A* is true but *R* is incorrect (D) *A* is incorrect but *R* is correct. **(I.I.T. 1998)**
5. The only cations present in a slightly acidic solution are Fe^{3+}, Zn^{2+} and Cu^{2+}. The reagent that when added in excess to this solution would identify and separate Fe^{3+} in one step is
(A) 2*M* HCl (B) 6*M* NH_3 (C) 6*M* NaOH (D) $H_2S(g)$ **(I.I.T. 1997)**
6. Which compound is formed when excess KCN is added to aqueous solution of $CuSO_4$?
(A) $Cu(CN)_2$ (B) $K_2[Cu(CN)_4]$ (C) $K[Cu(CN)_2]$ (D) $K_3[Cu(CN)_4]$ **(I.I.T. 1996)**
7. Which compound does not dissolve in hot, dilute HNO_3?
(A) HgS (B) PbS (C) CuS (D) CdS **(I.I.T. 1996)**
8. An aqueous solution contains Hg^{+}, Pb^{2+} and Cd^{2+}. The addition of 6*N* HCl precipitates
(A) Hg_2Cl_2 only (B) $PbCl_2$ only
(C) $Hg_2Cl_2 + PbCl_2$ (D) $PbCl_2 + HgCl_2$ **(I.I.T. 1995)**
9. A substance on treatment with dil. H_2SO_4 liberates a colourless gas which gives (i) turbidity with baryta water and (ii) turns acidic dichromate solution green. These indicate the presence of
(A) CO_3^{2-} (B) S^{2-} (C) SO_3^{2-} (D) NO_2^- **(I.I.T. 1992)**
10. The reagent NH_4Cl and NH_4OH will precipitate
(A) Ca^{2+} (B) Al^{3+} (C) Mg^{2+} (D) Zn^{2+} **(I.I.T. 1991)**
11. Which one among the following pairs cannot be separated by H_2S in dil. HCl?
(A) Bi^{3+}, Sn^{4+} (B) Al^{3+}, Hg^{2+} (C) Zn^{2+}, Cu^{2+} (D) Ni^{2+}, Cu^{2+} **(I.I.T. 1986)**
12. The best indicator for the detection of end point in the titration of weak acid and strong base is
(A) methyl red (*p*H 4–6) (B) bromothymol blue (*p*H 6–7.5)
(C) methyl orange (*p*H 3–4) (D) phenolphthalein (*p*H 8.2–9.6) **(I.I.T. 1985)**
13. The indicator used in titrating weak acid (e.g., oxalic acid) with a strong base (e.g., NaOH) is
(A) methyl orange (B) methyl red (C) fluorescin (D) phenolphthalein.
(I.I.T. 1985)

14. What product is formed by mixing a solution of $K_4[Fe(CN)_6]$ with a solution of $FeCl_3$?
(A) Ferro-ferricyanide (B) Ferric-ferrocyanide
(C) Ferric-ferricyanide (D) none of these. **(Roorkee 1989)**

15. To an acid solution of an anion, a few drops of $KMnO_4$ solution are added. Which of the following will not decolourise the $KMnO_4$ solution?
(A) CO_3^{2-} (B) NO_2^- (C) S^{2-} (D) Cl^-

16. Which of the following sulphides is completely precipitated when an acidic solution is made dilute?
(A) PbS (B) CdS (C) HgS (D) CuS

17. Borax bead test is dependent on the formation of a bead of
(A) boron metal (B) boron oxide (C) boric acid (D) metal metaborate.

18. The *p*H indicators are
(A) salts of weak acid and weak base
(B) salts of strong acid and strong base
(C) either strong acids or strong bases
(D) either weak acids or weak bases.

19. Number of moles of $KMnO_4$ required to oxidise 1 mole of FeC_2O_4 in acidic medium is
(A) 0.6 (B) 1.67 (C) 0.2 (D) 0.4

20. H_2S is passed through molar solutions of each of the following: (1) HCl (2) NH_4OH (3) CH_3COOH (4) NaOH. The decreasing order of $[S^{2-}]$ in these solutions is
(A) 2 > 4 > 3 > 1 (B) 2 > 4 > 1 > 3 (C) 4 > 2 > 3 > 1 (D) 4 > 2 > 1 > 3

21. A 0.5 g sample containing MnO_2 is treated with HCl, liberating Cl_2 which is passed into a solution of KI. 30 mL of 0.1 *M* $Na_2S_2O_3$ are required to titrate the liberated I_2. The percentage of MnO_2 in the sample is
(A) 26.1 (B) 21.6 (C) 23.1 (D) 21.3

22. Which of the following metals will give blue ash when its salt is heated with solid Na_2CO_3 and $Co(NO_3)_2$ on charcoal?
(A) Zn (B) Mg (C) Al (D) Cu

23. An aqueous solution of a colourless metal sulphate gives white precipitate with NH_4OH. This is soluble in excess of NH_4OH. On passing H_2S through this solution a white precipitate is formed. The metal in the salt is
(A) Al (B) Ba (C) Zn (D) Ca

24. In the precipitation of the iron group in qualitative analysis, NH_4Cl is added before adding NH_4OH to
(A) increase the concentration of Cl^- ions
(B) increase the concentration of NH_4^+ ions
(C) decrease the concentration of OH^- ions
(D) to prevent the interference by PO_4^{3-} ions.

25. Which of the following precipitates is obtained on adding HCl to a solution of Sb_2S_3 in yellow ammonium sulphide?
(A) Sb_2S_3 (B) Sb_2S_5 (C) SbS_2 (D) SbS

26. Of the following sulphides, which is soluble in dilute acid, but insoluble in alkali?
(A) CdS (B) PbS (C) Sb_2S_3 (D) FeS

27. Which of the following can be used in place of NH_4Cl for the identification of group III radicals?
(A) $(NH_4)_2SO_4$ (B) NH_4NO_3 (C) $(NH_4)_2CO_3$ (D) NaCl

28. In group IV, $(NH_4)_2CO_3$ is added to precipitate out the carbonates. We do not add Na_2CO_3 because
(A) $CaCO_3$ is soluble in Na_2CO_3 (B) $MgCO_3$ will be precipitated
(C) Na_2CO_3 increases the solubility of group V elements
(D) none of these.

29. Which of the following is obtained when HCl is added to a solution of SnS in yellow ammonium sulphide?
(A) SnS (B) SnS_2 (C) Sn_2S_3 (D) $(NH_4)_2SnS_2$

30. The yellow ammonium sulphide solution is a suitable reagent for the separation of
(A) PbS and Bi_2S_3 (B) HgS and PbS
(C) CdS and As_2S_3 (D) Bi_2S_3 and CuS

31. Which gives violet colour in the borax bead test?
(A) Co (B) Fe (C) Ni (D) Mn

32. Which gives blue colour in the borax bead test?
(A) Co (B) Zn (C) Cr (D) Fe

33. Which of the following metal salts does not give a borax bead test?
(A) Cr(III) salt (B) Fe(II) salt (C) Na salt (D) Co(II) salt

34. Conc. HNO_3 is added before proceeding to test for group III elements. This is to
(A) convert Fe(II) to Fe(III) (B) oxidise any remaining H_2S
(C) increase the ionisation of NH_4OH
(D) form nitrates which give granular precipitates.

35. Which of the following anions is not easily removed from aqueous solution by precipitation?
(A) SO_4^{2-} (B) Cl^- (C) NO_3^- (D) CO_3^{2-} **(I.I.T. 1995)**

36. The oxidation of oxalic acid by acidic $KMnO_4$ is an example of autocatalysis. It is due to the presence of
(A) MnO_2 (B) MnO_4^- (C) MnO_4^{2-} (D) Mn^{2+}

37. Which of the following statements is correct about the equivalent mass of $KMnO_4$?
(A) It is one third of M.W. in neutral medium.
(B) It is one fifth of M.W. in alkaline medium.
(C) It is equal to M.W. in acidic medium.
(D) It is one third of M.W. in acidic medium.

38. Which of the following cannot be separated by H_2S and dil. HCl?
(A) $Cu^{2+} + Ni^{2+}$ (B) $Cu^{2+} + Zn^{2+}$ (C) $Cu^{2+} + Cd^{2+}$ (D) $Hg^{2+} + Al^{3+}$

39. In neutral medium, the equivalent mass of $KMnO_4$ is
(A) M/2 (B) M/3 (C) M/5 (D) M/6 **(Roorkee 1996)**

40. A 100 mL solution of 0.1*N* HCl was titrated with 0.2*N* NaOH solution. The titration was discontinued after adding 30 mL of NaOH solution. The remaining titration was completed by adding 0.25*N* KOH solution. The volume of KOH required for completing the titration is
(A) 70 mL (B) 32 mL (C) 35 mL (D) 16 mL

41. What mass of hydrated oxalic acid should be added for complete neutralisation of 100 mL of 0.2*N* NaOH solution?
(A) 0.90 g (B) 0.45 g (C) 1.08 g (D) 1.26 g

42. The maximum amount of $BaSO_4$ precipitated on mixing 0.05*M* $BaCl_2$ with 1*M* H_2SO_4 corresponds to
(A) 0.05*M* (B) 1.0*M* (C) 1.5*M* (D) 2.0*M*

43. HCl solutions of A and B have concentrations 0.5*N* and 0.1*N* respectively. The volume of solutions A and B required to make 2 L of 0.2*N* HCl is
(A) 0.5 L of A, 1.5 L of B (B) 1.5 L of A, 0.5 L of B
(C) 1.0 L of A, 1.0 L of B (D) 0.75 L of A, 1.25 L of B

44. 1.25 g of a solid dibasic acid is completely neutralised by 25 mL of 0.25*M* $Ba(OH)_2$ solution. Molecular mass of the acid is
(A) 100 (B) 150 (C) 120 (D) 200

45. The pink colour of phenolphthalein in alkaline medium is
(A) due to OH^- ions (B) due to neutral form
(C) due to negative form (D) due to positive form.

46. Identify the incorrect statement.
(A) A metallic oxide contains 60% metal. The equivalent mass of the metal is 12.
(B) The normality of solution mixture containing 40 mL *N*/20 HCl and 60 mL *N*/10 H_2SO_4 is 0.08*N*.
(C) The mass of $K_2Cr_2O_7$ required to prepare 500 mL of 0.1*N* solution in acid medium is 4.90 g.
(D) The equivalent mass of a divalent metal is 32.7. The molecular mass of the metallic chloride is 136.4.

47. Identify the incorrect statement.
(A) AgCl is more soluble in concentrated NaCl solution than in water.
(B) 4.0 g of a metal on heating with O_2 gave 4.5 g of metal oxide. The equivalent mass of M is 64.
(C) The number of moles of $BaCO_3$ which contains 1.5 moles of oxygen atom is 1.5.
(D) On heating 0.5 g of a metal oxide in hydrogen, 0.4 g of M is obtained. The equivalent mass of M is 32.

48. Identify the incorrect statement.
(A) From the solution containing Cu^{2+} and Zn^{2+} ions, Cu^{2+} can be selectively precipitated using sodium sulphide.

(B) The mass of oxalic acid dihydrate required to prepare 500 mL of 0.2N solution is 6.3 g.

(C) The volume of 0.1M $K_2Cr_4O_7$ solution required to oxidise 35 mL of 0.5M $FeSO_4$ solution in acid medium is 29.17 mL.

(D) Phenolphthalein can be used as an indicator in iodometry.

49. The colour of methyl orange and methyl red in acid solution is respectively
(A) pink, red (B) yellow, yellow (C) yellow, red (D) red, yellow.

50. The colour of methyl orange and methyl red in alkaline solution is respectively
(A) pink, red (B) yellow, yellow (C) yellow, red (D) red, yellow.

51. An aqueous solution of a substance gives a white precipitate on treatment with dil. HCl, which dissolves on heating. When H_2S is passed through the hot acidic solution, a black precipitate is obtained. The substance is
(A) Pb^{2+} salt (B) Hg_2^{2+} salt (C) Cu^{2+} salt (D) Ag^+ salt. **(I.I.T. 2002)**

52. Identify the correct order of solubility of Na_2S, CuS and ZnS in aqueous medium.
(A) CuS > ZnS > Na_2S (B) ZnS > Na_2S > CuS
(C) Na_2S > CuS > ZnS (D) Na_2S > ZnS > CuS **(I.I.T. 2002)**

Hint: Sodium sulphide is the strongest electrolyte among Na_2S, ZnS and CuS, and sodium being an alkali metal, Na_2S is highly soluble in water. The solubility product of ZnS and CuS is in the order: CuS < ZnS. Hence CuS is less soluble than ZnS. Thus the order of solubility is: Na_2S > ZnS > CuS.

53. A solution which is 10^{-3} M each in Mn^{2+}, Fe^{2+}, Zn^{2+} and Hg^{2+} is treated with 10^{-16} M sulphide ion. If K_{sp} of MnS, FeS, ZnS and HgS is 10^{-15}, 10^{-23}, 10^{-20} and 10^{-54} respectively, which one will precipitate first?
(A) FeS (B) MnS (C) HgS (D) ZnS **(I.I.T. 2003)**

Hint: The precipitation of a salt occurs when the ionic product exceeds the solubility product. Hence HgS having the lowest K_{sp} among the given salts will precipitate first.

54. 0.1 M and 2.5 L NaOH solution is mixed with another 0.5 M and 3 L NaOH solution. Molarity of the resultant solution is
(A) 0.80M (B) 1.0M (C) 0.73M (D) 0.50M

55. As_2S_3 dissolves in a solution of yellow ammonium sulphide to give
(A) $(NH_4)_2AsO_2$ (B) $(NH_4)_3AsS_3$ (C) $(NH_4)_3AsS_4$ (D) $(NH_4)_5AsS_5$

56. When bismuth chloride is added to an excess of water, the white precipitate formed is
(A) BiOCl (B) Bi_2OCl_3 (C) Bi_2O_3 (D) $Bi(OH)_2$

57. When $NaNO_2$ is added to a cobalt salt solution in ethanoic acid, the yellow precipitate formed is
(A) $Na_4[Co(NO_2)_6]$ (B) $Na_3[Co(NO_2)_6]$
(C) $NaCo[Co(NO_2)_6]$ (D) $Na[Co(NO_2)_4(H_2O)_2]$

58. Which of the following is/are primary standard?

(A) $KMnO_4$ (B) KOH
(C) $Na_2CO_3 \cdot 10H_2O$ (D) $Na_2B_4O_7 \cdot H_2O$

59. When dil. NaOH is added to a solution of $KMnO_4$, the green colour formed is due to
(A) MnO_2 (B) Mn_2O_7 (C) Mn^{2+} (D) MnO_4^{2-}

60. A salt produced a green bead in borax bead test. The salt is a
(A) iron salt (B) nickel salt
(C) copper salt (D) manganese salt.

61. A metal sulphide is soluble in yellow ammonium sulphide solution. Identify the metal.
(A) Sb (B) Cu (C) Hg (D) Pb

62. Which of the following metal ions form yellow precipitate with $K_2Cr_2O_7$?
(A) Fe^{2+} (B) Cu^{2+} (C) Pb^{2+} (D) Ba^{2+}

63. Which of the following ions will not be precipitated on passing H_2S gas through a conc. HCl solution?
(A) As^{3+} (B) Sb^{3+} (C) Cu^{2+} (D) Cd^{2+}

64. When dil. HCl is added to a solution of an inorganic salt, a white precipitate is formed. The metal ion present in the salt is
(A) Cd^{2+} (B) Zn^{2+} (C) Hg^{2+} (D) Hg^{+}

65. In which of the following acid-base titrations, there occurs two *p*H breaks?
(A) NaOH–HCl (B) Na_2CO_3–HCl (C) $NaHCO_3$–HCl (D) NaOH–CH_3COOH

66. A metal nitrate gives black precipitate with KI and on adding excess of KI it gives orange colour. It is
(A) Hg^{2+} (B) Bi^{3+} (C) Sn^{2+} (D) Pb^{2+} **(I.I.T. 2005)**
Hint: $Bi^{3+} + KI \longrightarrow BiI_3$; $BiI_3 + KI \longrightarrow K[BiI_4]$ (orange)

67. When excess of NH_4SCN is added to an aqueous solution of iron(III) alum a red colour is formed due to
(A) $[Fe(NCS)]^{2+}$ (B) $[Fe(NCS)_6]^{3-}$ (C) $[Fe(SCN)_3]^{-}$ (D) $[Fe(NCS)_3]$
Hint: $Fe^{3+} + 6SCN^-$ (excess) $\longrightarrow [Fe(NCS)_6]^{3-}$. In $[Fe(NCS)_6]^{3-}$, thiocyanate ion is N-bonded as evidenced by X-ray structure analysis and is not S-bonded as erroneously given in many books. The compound formed is not $[Fe(SCN)]^{2+}$ as erroneously given in many books.

68. In the reactions, $K_2Cr_2O_7 + KCl + \text{conc. } H_2SO_4 \xrightarrow{\Delta} A \xrightarrow{NaOH} B$. *A* is reddish-brown gas soluble in sodium hydroxide solution forming *B*. *A* and *B* are
(A) CrO_2Cl_2, Na_2CrO_4 (B) CrO_2Cl_2, Na_2CrO_3
(C) CrO_2Cl, Na_2CrO_4 (D) CrO_2Cl_2, K_2CrO_4

69. Which of the following cations give precipitate on reaction with KI?
(A) Ag^+, Ca^{2+}, Sr^{2+} (B) Na^+, Ca^{2+}, Mg^{2+}
(C) Cu^+, Ni^{2+}, Zn^{2+} (D) Hg_2^{2+}, Ag^+, Pb^{2+}

70. Which of the following is formed when CO_2 gas is passed through an aqueous solution of sodium chromate?

(A) $Cr(OH)_3$ is precipitated (B) Yellow solution of $Cr_2(CO_3)_2$ is formed
(C) Orange solution of $Na_2Cr_2O_7$ is formed
(D) No reaction.

71. When excess of NH_4SCN is added to a solution of cobalt(II) chloride, a blue colour is formed due to
(A) $NH_4[Co(NCS)_4]$ (B) $(NH_4)_2[Co(NCS)_4]$
(C) $(NH_4)_2[Co(SCN)_6]$ (D) $(NH_4)_2[Co(SCN)_4]$
Hint: Co^{2+} + excess $NH_4SCN \longrightarrow (NH_4)_2[Co(NCS)_4]$ (blue). In $(NH_4)_2[Co(NCS)_4]$, thiocyanate ion is N-bonded as evidenced by X-ray structure analysis and is not S-bonded as erroneously given in many books.

72. Which of the following cations (Zn^{2+}, Cd^{2+}, Cu^{2+}, Fe^{2+}, Fe^{3+}) can be detected using a solution of $K_4[Fe(CN)_6]$?
(A) Zn^{2+}, Cd^{2+}, Cu^{2+}, Fe^{3+} (B) Zn^{2+}, Cd^{2+}, Cu^{2+}, Fe^{2+}
(C) Zn^{2+}, Cu^{2+}, Fe^{3+} (D) Fe^{2+}, Fe^{3+}

73. Prussian blue and Turnbull blue are respectively
I. $Fe^{III}[Fe^{II}(CN)_6]^-$ II. $Fe^{II}[Fe^{III}(CN)_6]^-$
III. $Fe^{III}[Fe^{III}(CN)_6]$ IV. $Fe^{II}[Fe^{II}(CN)_6]^{2-}$
(A) III, IV (B) IV, III (C) I, (I + IV) (D) II, I

74. An acid type indicator HIn differs in colour from its conjugate base, In^-. The human eye is sensitive to colour difference only when the ratio $[In^-]/[HIn]$ is greater than 10 or smaller than 0.1. The minimum change in pH of the solution in order to observe a complete colour change will be ($K_a = 1.0 \times 10^{-5}$)
(A) 1 (B) 2 (C) 1.5 (D) 3
Hint: When the colour of In^- would be just visible, $[H^+] = K_a/10 = 1 \times 10^{-6}$ *M*. Hence pH = 6. When the colour of HIn would be just visible, $[H^+] = K_a \times 10 = 1 \times 10^{-4}$ *M*. Hence pH = 4. Thus the minimum change in pH = 6 – 4 = 2.

75. Which of the following is stable to H_2S?
(A) $K_3[Cu(CN)_4]$ (B) $(NH_4)_2[Sn(C_2O_4)_3]$
(C) both (A) and (B) (D) $K_2[Cd(CN)_4]$

76. The blue solution of $CuSO_4$ is decolourised on addition of excess KCN. The product is
(A) $[Cu(CN)_4]^{2-}$ (B) CuCN
(C) $Cu(OH)_2$
(D) Cu^{2+} is reduced by CN^- to Cu^+ which forms complex $[Cu(CN)_4]^{3-}$. **(I.I.T. 2006)**
Hint: $CuSO_4 + 4KCN \longrightarrow 2CuCN + 2K_2SO_4 + (CN)_2$
$CuCN + 4KCN \longrightarrow [Cu(CN)_4]^{3-} + 4K^+ + CN^-$

77. A solution when diluted with H_2O and boiled, it gives a white precipitate. On addition of excess NH_4Cl/NH_4OH, the volume of precipitate decreases leaving behind a white gelatinous precipitate. Identify the precipitate which dissolves in NH_4OH/NH_4Cl.
(A) $Zn(OH)_2$ (B) $Al(OH)_3$ (C) $Mg(OH)_2$ (D) $Ca(OH)_2$ **(I.I.T. 2006)**

Hint: White precipitate of $Zn(OH)_2$ dissolves in NH_4OH/NH_4Cl to form $[Zn(NH_3)_4](OH)_2$ ($Zn(OH)_2 + 4NH_4OH \longrightarrow [Zn(NH_3)_4](OH)_2 + 4H_2O$).

78. Consider a titration of potassium dichromate solution with acidified Mohr's salt solution using diphenylamine as indicator. The number of moles of Mohr's salt required per mole of dichromate is

(A) 3 (B) 4 (C) 5 (D) 6 **(I.I.T. 2007)**

Hint: $CrO_7^{2-} + Fe^{2+} \longrightarrow Fe^{3+} + Cr^{3+}$; n factor of $CrO_7^{2-} = 6$ and n factor of $Fe^{2+} = 1$. Hence to reduce one mole of dichromate six moles of Fe^{2+} are required.

79. A solution of metal ion when treated with KI gives a red precipitate which dissolves in excess KI to give a colourless solution. Moreover, the solution of metal ion on treatment with a solution of cobalt(II) thiocyanate gives rise to a deep blue crystalline precipitate. The metal ion is

(A) Pb^{2+} (B) Hg^{2+} (C) Cu^{2+} (D) Co^{2+} **(I.I.T. 2007)**

Hint: $Hg^{2+} + 2KI \longrightarrow HgI_2$ ((Red precipitate)

$$\underset{\text{(excess)}}{HgI_2 + 2KI} \longrightarrow \underset{\text{Soluble}}{K_2HgI_4}$$

$$HgI_4^{2+} + Co(NCS)_2 \longrightarrow \underset{\text{Crystalline blue precipitate}}{Hg[Co(NCS)_4]}$$

80. The correct structure of ethylenediaminetetraacetic acid (EDTA) is

(A) 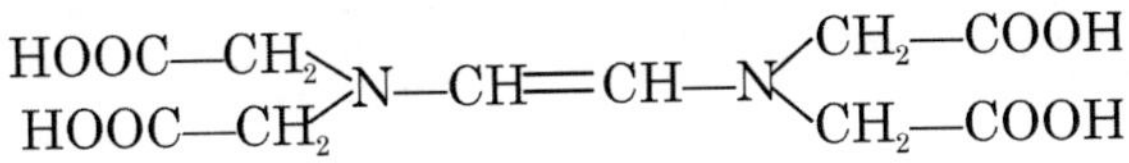

(B) 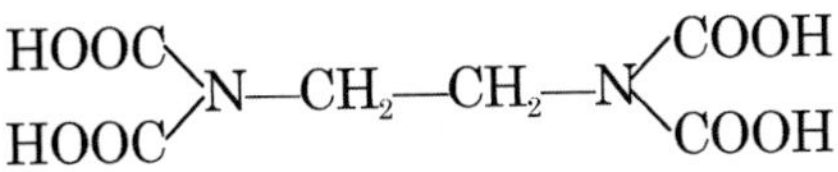

(C) 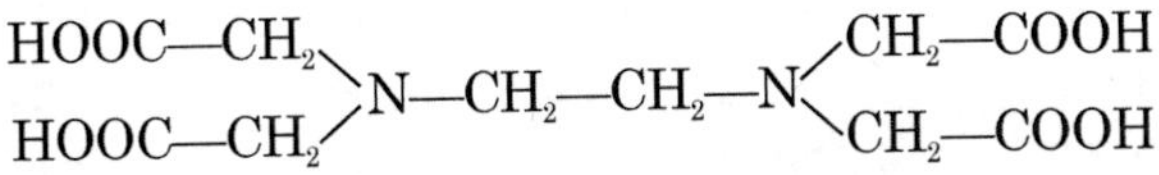

(D) 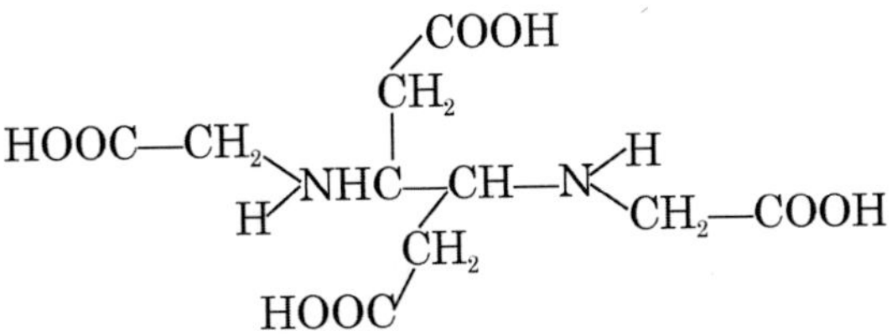 **(I.I.T. 2010)**

Hint: The correct structure of ethylenedianminetetraacetic acid (EDTA) is (C).

81. Dissolving 120 g of urea (mol. wt. 60) in 1000 g of water gave a solution of density 1.15 g/mL. The molarity of the solution is

(A) 1.78M (B) 2.00M (C) 2.05M (D) 2.22M **(I.I.T. 2011)**

Hint: Total mass of solution = 1000 + 120 = 1120 g

$$\text{Total volume of solution in L} = \frac{1120}{1.15} \times 10^3$$

$$M = \frac{W}{M} \times \frac{1}{V(\text{in L})} = \frac{120}{60} \times \frac{1.15 \times 10^3}{1120} = 2.05M$$

82. Passing H_2S gas into a mixture of Mn^{2+}, Ni^{2+}, Cu^{2+} and Hg^{2+} ions in an acidified aqueous solution precipitates

(A) CuS and HgS (B) MnS and CuS (C) MnS and NiS (D) NiS and HgS

(I.I.T. 2011)

Hint: H_2S in presence of aqueous acidified solution precipitates as sulphide of Cu and Hg other than Pb^{2+}, Bi^{3+}, Cd^{2+}, As^{3+}, Sb^{3+} and Sn^{2+}.

ANSWERS

1. A	2. B	3. A	4. B	5. B	6. D
7. A	8. C	9. C	10. B	11. A	12. D
13. D	14. B	15. A	16. B	17. D	18. D
19. A	20. C	21. A	22. C	23. C	24. C
25. B	26. C	27. B	28. B	29. B	30. C
31. D	32. A	33. C	34. A	35. C	36. D
37. A	38. C	39. B	40. D	41. D	42. A
43. A	44. D	45. C	46. C	47. D	48. A, D
49. A	50. B	51. A	52. D	53. C	54. C
55. C	56. A	57. B	58. C, D	59. D	60. C
61. A	62. C, D	63. D	64. D	65. B	66. B
67. B	68. A	69. D	70. C	71. B	72. A
73. C	74. B	75. C	76. D	77. A	78. D
79. B	80. C	81. C	82. A		

Part—III
Organic Chemistry

33

Purification and Characterisation of Organic Compounds

OBJECTIVE TYPE MULTIPLE CHOICE QUESTIONS

1. In the qualitative test for sulphur, sodium sulphide gives a violet colour on reaction with sodium nitroprusside. The violet colour is due to
   (A) $Na_4[Fe(CN)_5(NOS)]$ (B) $Na_3[Fe(CN)_5(NS)]$
   (C) $Fe(SCN)_3$ (D) $Na_3[Fe(CN)_6]$ **(I.I.T. 1995)**
2. Carbon and hydrogen are estimated by
   (A) Liebig method (B) Carius method
   (C) Duma's method (D) none of these. **(I.I.T. 1993)**
3. If two compounds have the same empirical formula but different molecular formulae, they must have
   (A) different percentage composition (B) different molecular masses
   (C) same viscosity (D) same vapour density. **(I.I.T. 1987)**
4. A solid organic compound on heating is converted into vapour directly, which on cooling solidifies. Which of the following is the correct method of purifying it?
   (A) Sublimation (B) Steam distillation
   (C) Distillation (D) Distillation at reduced pressure.
5. Two organic compounds *A* and *B* both containing C and H only, on quantitative analysis gave same percentages of C and H. *A* decolourises bromine water while *B* does not. *A* and *B* are
   (A) C_2H_2, C_6H_6 (B) C_2H_6, C_6H_6 (C) C_2H_4, C_6H_4 (D) C_2H_2, C_2H_6
6. The decomposition of organic compounds in the presence of oxygen and without the development of odoriferous substances, is called
   (A) N_2-fixation (B) decay (C) nitrification (D) denitrification.

7. If 0.2 g of an organic compound containing C, H and O on combustion yielded 0.147 g CO_2 and 0.12 g water, what will be the oxygen content in the substance?
(A) 73.29% (B) 78.45% (C) 83.23% (D) 89.50%
8. A mixture contains solid compounds A, B, C and D. On heating only C changes from solid to vapour. C can be separated from the rest in the mixture by
(A) fractional distillation (B) sublimation
(C) crystallisation (D) distillation.
9. Empirical formula of a compound is CH_2O. Its molecular mass is 180. Its molecular formula is
(A) $C_3H_6O_3$ (B) $C_4H_{10}O_4$ (C) $C_5H_{10}O_5$ (D) $C_6H_{12}O_6$
10. Beilstein test is used for the detection of
(A) N (B) CO_2 (C) Na (D) Cl
11. Assertion (*A*): Duma's method is of more general application to nitrogen containing organic compounds than Kjeldahl's method.
Reason (*R*): Kjeldahl's method does not give satisfactory results for compounds in which nitrogen is directly linked to oxygen.
(A) Both *A* and *R* are true and *R* is the correct explanation of *A*
(B) Both *A* and *R* are true and *R* is not the correct explanation of *A*
(C) *A* is true but *R* is false (D) Both *A* and *R* are false.
12. Which of the following is a technique most suitable for purification of cyclohexanone from a mixture of benzoic acid, isoamyl alcohol, cyclohexane and cyclohexanone?
(A) IR spectroscopy (B) Crystallisation
(C) Gas chromatography (D) Sublimation.
13. Which of the following is not used for the purification of solid impurities?
(A) Distillation (B) Sublimation
(C) Crystallisation (D) None of these.
14. Lassaigne's extract is boiled with dil. HNO_3 before testing for halogens because
(A) AgCN is soluble in HNO_3 (B) Silver halides are insoluble in HNO_3
(C) Na_2S and NaCN are decomposed by HNO_3
(D) Ag_2S is soluble in HNO_3
15. Complete combustion of a sample of hydrocarbon gives 0.66 g CO_2 and 0.36 g H_2O. The empirical formula of the hydrocarbon is
(A) CH_2 (B) C_3H_4 (C) C_3H_8 (D) C_6H_8
16. Distillation under reduced pressure is used to purify liquids which
(A) are explosive (B) are highly volatile
(C) have high boiling point (D) decompose below their boiling point.
17. Which of the following compounds is formed when an organic compound containing N and S is fused with sodium?
(A) $NaCNO + Na_2S$ (B) $Na_2SO_3 + NaCN$
(C) $NaCN + Na_2S$ (D) NaSCN

18. Which of the following compounds cannot be estimated for nitrogen by Kjeldahl's method?
    (A) Nitrobenzene (B) Benzenediazonium chloride
    (C) Pyridine (D) All of these.
19. In the gravimetric estimation of phosphorus in an organic compound, the compound weighed is
    (A) $Mg_3(PO_4)_2$ (B) $Mg(NH_4)PO_4$ (C) $Mg_2P_2O_7$ (D) $MgCl_2$
20. Which one of the following gases is evolved in Duma's method?
    (A) NH_3 (B) N_2 (C) NO (D) CO_2
21. Halogen is tested with a copper wire. The test is known as
    (A) Liebig's method (B) Beilstein test
    (C) Fusion test (D) Flame test.
22. The catalyst used in the Kjeldahl's method is
    (A) H_2SO_4 (B) K_2SO_4 (C) $CuSO_4{\cdot}5H_2O$ (D) none of these.
23. Which of the following compounds gives blood red colour in Lassaigne's test for N?
    (A) NH_2CH_2COOH (B) $CH_3CH_2SO_3H$
    (C) $NH_2CH_2SO_3H$ (D) Ethylenediaminetetraacetic acid
24. Which of the following is used for the detection of Br^- and I^- in a mixture?
    (A) CS_2 and chlorine water (B) CS_2 and bromine water
    (C) CCl_4 and chlorine water (D) CCl_4 and bromine water.
25. In the qualitative test for Cl^-, the white precipitate of AgCl gets solubilised with NH_4OH. The soluble silver compound has the formula
    (A) $[Ag(NH_3)_4]Cl$ (B) $[Ag(NH_3)]Cl$ (C) $[Ag(NH_3)_2]Cl$ (D) $[Ag(NH_3)_3]Cl$
26. When ammonium molybdate and conc. HNO_3 are added to the fusion extract of an organic compound containing phosphorus, a yellow precipitate is formed. The formula of the yellow compound is
    (A) $(NH_4)_3PO_4{\cdot}6MoO_3$ (B) $(NH_4)_3PO_4{\cdot}12MoO_3$
    (C) $(NH_4)_3PO_4{\cdot}10MoO_3$ (D) $(NH_4)_3PO_4{\cdot}24MoO_3$
27. The reagent used for the qualitative analysis in an organic compound is
    (A) $Na_3[Fe(CN)_6]$ (B) $Na_4[Fe(CN)_6]$
    (C) $Pb(CH_3COO)_2$ (D) $Na_2[Fe(CN)_5NO]$
28. For which of the following compounds, the Lassaigne's test for the detection of nitrogen will fail?
    (A) H_2NCONH_2 (B) $C_6H_5N{=}NC_6H_5$
    (C) $H_2NNH_2{\cdot}2HCl$ (D) $H_2NCONHNH_2{\cdot}HCl$ **(I.I.T. 1994)**
29. The substance which is used as an adsorbent in the column chromatography is
    (A) Al_2O_3 (B) Na_2O (C) Na_2SO_4 (D) alum
30. Absolute alcohol can be obtained
    (A) by removing water using P_4O_{10} (B) by distilling over plenty of quicklime
    (C) by removing water using conc. H_2SO_4
    (D) by distilling with appropriate amount of benzene.

31. 0.2 g of an organic compound on complete combustion produces 0.18 g of water. The percentage of hydrogen in it is
(A) 5 (B) 10 (C) 15 (D) 20

32. If 0.32 g of an organic compound containing sulphur produces 0.233 g of $BaSO_4$, then the percentage of sulphur in it is
(A) 10 (B) 15 (C) 20 (D) 25

33. An organic compound containing C, H and N gave the analysis: C = 40.0%, H = 13.33%, N = 46.67%. The empirical formula of the compound is
(A) C_2H_7N (B) $C_2H_7N_2$ (C) CH_4N (D) CH_5N

34. Which of the following compounds is formed in Kjeldahl's method?
(A) N_2 (B) NH_3 (C) $(NH_4)_2SO_4$ (D) NO_2

35. The compound that does not give a blue colour in Lassaigne's test is
(A) aniline (B) glycine (C) hydrazine (D) urea

36. Anthracene is purified by
(A) filtration (B) crystallisation (C) distillation (D) sublimation.

37. Distillation is used to separate liquids which differ in their boiling point by
(A) 5°C (B) 10°C (C) 30–50°C (D) 15°C

38. The best method to separate a mixture of *ortho* and *para* nitrophenol (1 : 1) is
(A) steam distillation (B) crystallisation
(C) vaporisation (D) colour spectrum.

39. The principle involved in chromatographic separation of organic compounds is based on
(A) differential absorption (B) differential adsorption
(C) different solubility (D) different density.

40. The blue or green colour formed in Lassaigne's test for nitrogen is due to
(A) $Na_4[Fe(CN)_6]$ (B) $Fe_4[Fe(CN)_6]_3$ (C) $K_4[Fe(CN)_6]$ (D) NaCN

41. Impure glycerine is purified by
(A) vacuum distillation (B) steam distillation
(C) simple distillation (D) extraction with a solvent.

42. The desiccant used for absorbing water during Liebig's method for the estimation of C and H is
(A) anhy. $CaCl_2$ (B) $MgSO_4 \cdot 7H_2O$ (C) anhy. Na_2SO_4 (D) $Mg(ClO_4)_2$

43. Sodium extract of an organic compound gives blood red colour with $FeCl_3$. It contains
(A) N (B) S and Cl (C) N and S (D) S

44. Sodium fusion extract, obtained from aniline, on treatment with iron(II) sulphate and H_2SO_4 in the presence of air gives a Prussian blue precipitate. The blue colour is due to the formation of
(A) $Fe_4[Fe(CN)_6]_3$ (B) $Fe_3[Fe(CN)_6]_2$ (C) $Fe_4[Fe(CN)_6]_2$ (D) $Fe_3[Fe(CN)_6]_3$

(I.I.T. 2007)

Hint: $Na + C + N \longrightarrow NaCN$; $2Na + 2H_2O \longrightarrow 2NaOH + H_2$; $FeSO_4 + 2NaOH \longrightarrow Fe(OH)_2 + Na_2SO_4$; $Fe(OH)_2 + 6NaCN \longrightarrow Na_4[Fe(CN)_6] + 2NaOH$; $3Na_4[Fe(CN)_6] + 2Fe_2(SO_4)_3 \longrightarrow Fe_4[Fe(CN)_6]_3 + 6Na_2SO_4$

ANSWERS

1. A	2. A	3. B	4. A	5. A	6. D
7. A	8. A	9. D	10. D	11. B	12. C
13. A	14. C	15. C	16. D	17. C, D	18. D
19. C	20. B	21. B	22. C	23. C	24. A, C
25. C	26. B	27. C, D	28. B, C	29. A	30. D
31. B	32. A	33. C	34. C	35. C	36. D
37. C	38. A	39. B	40. B	41. A	42. A, D
43. C	44. A				

34

Classification, Nomenclature, Structure and Isomerism

OBJECTIVE TYPE MULTIPLE CHOICE QUESTIONS

1. The IUPAC name of $CH_2{=}CH{-}C(CH_3)_3$ is

(A) 3,3-dimethylbut-1-ene (B) 2,2-dimethylbut-3-ene

(C) 3,3,3-trimethylprop-1-ene (D) 1,1,1-trimethylprop-2-ene **(I.I.T. 1984)**

2. The IUPAC name of 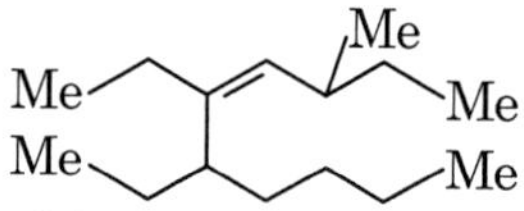 (Me = methyl) is

(A) 5,6-diethyl-3-methyldec-4-ene (B) 5,6-diethyl-8-methyldec-7-ene

(C) 3-methyl-5,6-diethyldec-5-ene (D) 5,6-diethyl-8-methyldec-6-ene **(I.I.T. 1990)**

3. The IUPAC name of 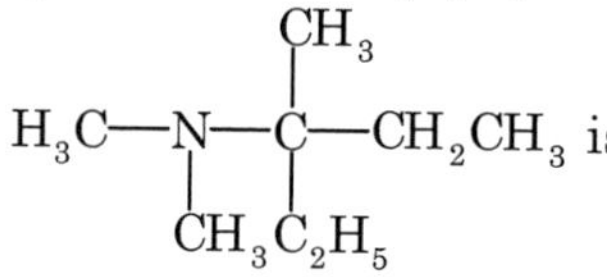 is

(A) 3,*N*,*N*-trimethyl-3-pentanamine (B) 3-dimethylamino-3-methylpentane

(C) 3-dimethylamino-3-ethylbutane (D) 3-(*N*,*N*-dimethylamino)-3-ethylbutane **(I.I.T. 1991)**

4. The IUPAC name of $CH_2{=}CH{-}CH(CH_3)_2$ is

(A) 2-vinylpropane (B) 3-methylbut-1-ene

(C) 1-isopropylethylene (D) 1,1-dimethylprop-2-ene **(I.I.T. 1987)**

5. The IUPAC name of 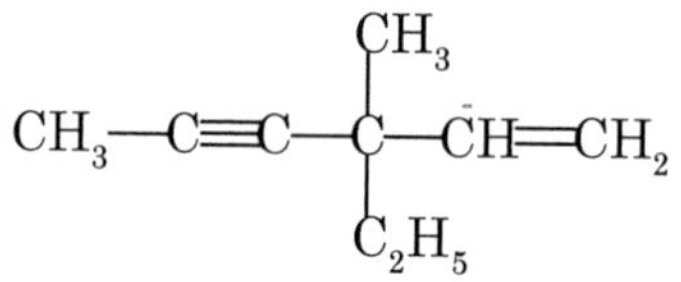 is

(A) 3-ethyl-3-methylhex-1-en-4-yne (B) 3-methylene-4-methylhepta-5-yne

(C) 5-methylene-5-ethyl-4-methylhepta-2-yne
(D) 4-ethyl-4-methylhex-5-en-2-yne

6. Only two isomeric monochloro derivatives are possible for
(A) *n*-butane (B) benzene
(C) 2,4-dimethylpentane (D) 2-methylpropane **(I.I.T. 1986)**

7. Which of the following show *cis-trans* isomerism?
(A) but-2-ene (B) 1-phenylpropene
(C) 2-methylbut-2-ene (D) propene **(I.I.T. 1998)**

8. Keto-enol tautomerism is observed in
(A) $C_6H_5COCH_3$ (B) $C_6H_5COC_6H_5$
(C) C_6H_5CHO (D) $C_6H_5COCH_2COCH_3$ **(I.I.T. 1988)**

9. Which of the following has zero dipole moment?
(A) *cis*-1,2-dichloroethylene (B) 1,1-dichloroethylene
(C) *trans*-1,2-dichloroethylene (D) None of these. **(I.I.T. 1987)**

10. Which of the following have asymmetric carbon atoms?

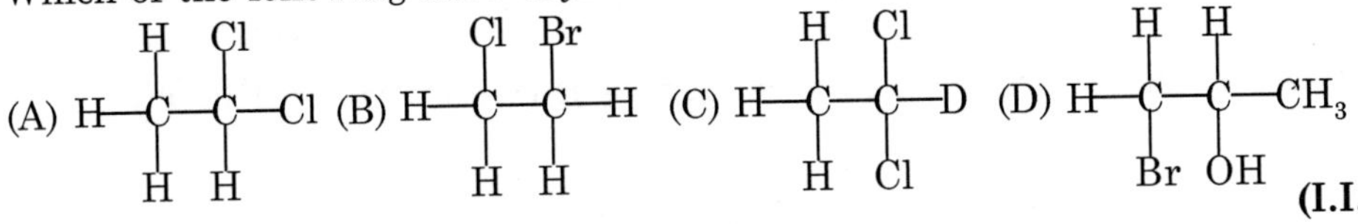

(I.I.T. 1989)

11. How many optical isomers are there for butane-2,3-diol?
(A) 1 (B) 4 (C) 2 (D) 3 **(I.I.T. 1997)**

12. Number of enantiomeric pairs that can be produced during monochlorination of 2-methylbutane is
(A) 2 (B) 4 (C) 1 (D) 3 **(I.I.T. 1997)**

13. The correct statement about a, b and c is

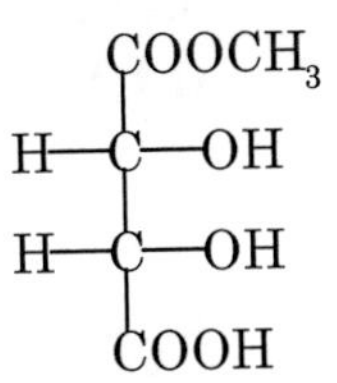

a

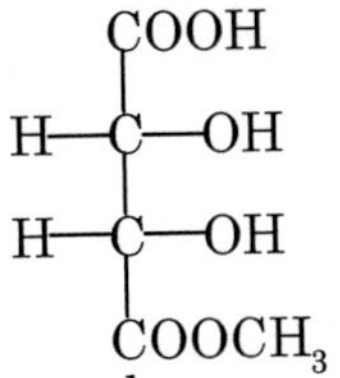

b

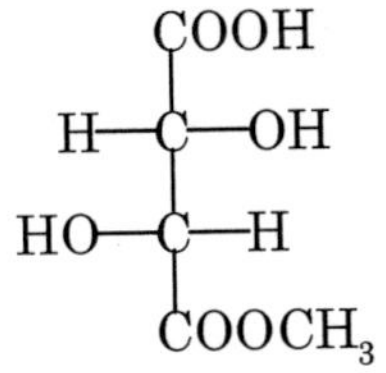

c

(A) a and c are enantiomers (B) a and b are identical
(C) a and b are diastereoisomers (D) a and b are enantiomers. **(I.I.T. 1997)**

14. The following compound can exhibit

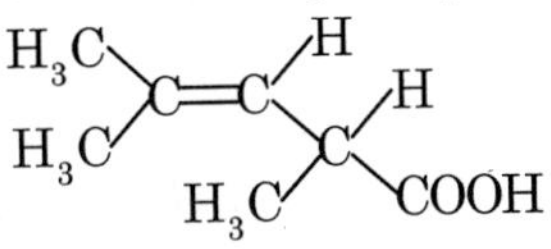

(A) geometrical and optical isomerism
(B) geometrical isomerism (C) optical isomerism
(D) tautomerism **(I.I.T. 1995)**

15. Tautomerism is exhibited by

(A) 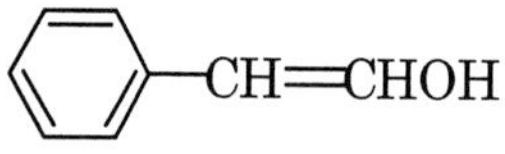

(B) 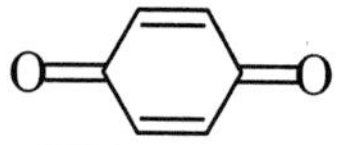

(C) 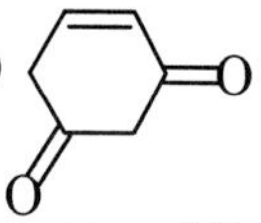

(D) 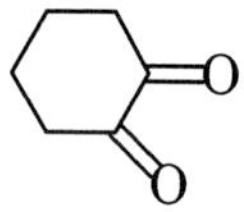

(I.I.T. 1998)

16. Assertion (*A*): Addition of Br_2 to but-1-ene gives two optical isomers.
Reason (*R*): The product contains one asymmetric carbon.
(A) *A* and *R* are true and *R* is the correct explanation of *A*
(B) *A* and *R* are true and *R* is not the correct explanation of *A*
(C) *A* is true but *R* is incorrect (D) *A* is incorrect but *R* is correct. **(I.I.T. 1998)**

17. A solution of (+)-2-chloro-2-phenylethane in toluene racemises slowly in the presence of $SbCl_5$ due to the formation of
(A) free radical (B) carbocation (C) carbanion (D) carbene **(I.I.T. 1999)**

18. In the compound $CH_2{=}CH{-}CH_2{-}C{\equiv}CH$, the $C_2{-}C_3$ bond is of the type
(A) sp-sp^2 (B) sp-sp^3 (C) sp^2-sp^3 (D) sp^3-sp^3 **(I.I.T. 1999)**

19. The optically active tartaric acid is named as D-(+)-tartaric acid because it has a positive
(A) optical rotation only when substituted by deuterium
(B) optical rotation and is derived from D-glucose
(C) optical rotation and is derived from D-(+)-glyceraldehyde
(D) *p*H in organic solvent. **(I.I.T. 1999)**

20. Which of the following will exhibit geometrical isomerism?
(A) 1,1-Diphenylprop-1-ene (B) 2-Phenylbut-1-ene
(C) 3-Phenylbut-1-ene (D) 1-Phenylbut-2-ene **(I.I.T. 2000)**

21. The number of isomers for the compound with molecular formula $C_2BrClFI$ is
(A) 3 (B) 4 (C) 5 (D) 6 **(I.I.T. 2001)**

Hint: 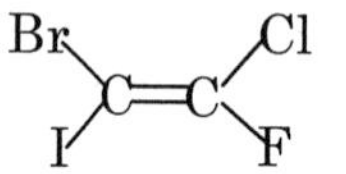 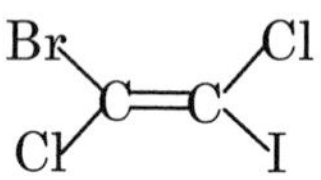 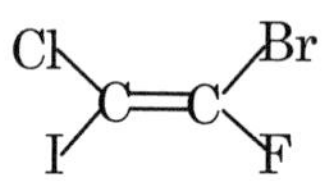

(Z & E) (Z & E) (Z & E)

22. Arrange the following compounds in the order of increasing dipole moment: toluene (I), *m*-dichlorobenzene (II), *o*-dichlorobenzene (III), *p*-dichlorobenzene (IV).
(A) I < IV < II < III (B) IV < I < II < III
(C) IV < I < III < II (D) IV < II < I < III **(I.I.T. 1996)**

23. An aromatic molecule will
(A) have $4n\pi$ electrons (B) have $(4n + 2)\pi$ electrons
(C) be cyclic (D) be planar. **(I.I.T. 1999)**

24. Optical activity is expected for
(A) methyl-2-methylpropanoate (B) 2-methylpropanoic acid
(C) methyl-2-chloropropanoate (D) propanoic acid **(I.I.T. 1992)**

25. The C—C bond length of the following molecules is in the order
(A) $C_2H_6 > C_2H_2 > C_6H_6 > C_2H_4$ (B) $C_2H_2 < C_2H_4 < C_6H_6 < C_2H_6$
(C) $C_2H_6 > C_2H_4 > C_6H_6 > C_2H_2$ (D) $C_2H_4 > C_2H_6 > C_2H_2 > C_6H_6$ **(I.I.T. 1991)**

26. The number of π electrons in cyclopentadienyl anion $[(C_4H_4)^{2-}]$ is
(A) 4 (B) 2 (C) 6 (D) 8 **(I.I.T. 1991)**

27. The hybridisation of carbon atoms in C—C single bond of H—C≡C—CH=CH$_2$ is
(A) sp^2-sp^3 (B) sp^3-sp^3 (C) sp-sp^2 (D) sp^3-sp **(I.I.T. 1991)**

28. The compound in which carbon uses only sp^3 hybrid orbitals for bond formation is
(A) $(NH_2)_2CO$ (B) HCOOH (C) $(CH_3)_3COH$ (D) CH_3CHO **(I.I.T. 1989)**

29. Which of the following pairs shows isomerism?
(A) CH_2Cl_2 and $CHCl_3$ (B) C_2H_4 and C_3H_6
(C) C_6H_6 and C_3H_8 (D) CH_3OCH_3 and C_2H_5OH **(I.I.T. 1990)**

30. The isomers which can be interconverted through rotation around a single bond are
(A) conformers (B) enantiomers
(C) diastereoisomers (D) position isomers. **(I.I.T. 1992)**

31. The compound with an isopropyl group is
(A) 2,2-dimethylpentane (B) 2,2,3-trimethylpentane
(C) 2,2,2,3-tetramethylpentane (D) 2-methylpentane **(I.I.T. 1989)**

32. The number of σ and π bonds in but-1-en-3-yne is
(A) 8σ, 2π (B) 5σ, 5π (C) 7σ, 3π (D) 6σ, 4π **(I.I.T. 1989)**

33. The number of structural and configurational isomers of a bromo compound C_5H_9Br, formed by the addition of HBr to pent-2-yne respectively is
(A) 2 and 1 (B) 2 and 4 (C) 1 and 2 (D) 4 and 2 **(I.I.T. 1988)**

34. The ∠Cl—C—Cl angle in 1,1,2,2-tetrachloroethane and tetrachloromethane respectively is about
(A) 120° and 109.5° (B) 109.5° and 90°
(C) 109.5° and 120° (D) 90° and 109.5° **(I.I.T. 1988)**

35. The bond between carbon atom (1) and carbon atom (2) in compound $N \equiv \overset{1}{C} - \overset{2}{C}H = CH_2$ involves the hybridisation as
(A) sp^3, sp (B) sp^2, sp^2 (C) sp, sp^2 (D) sp, sp **(I.I.T. 1987)**

36. Which of the following has least hindered rotation about carbon-carbon bond?
(A) Ethane (B) Acetylene
(C) Hexachloroethane (D) Ethylene **(I.I.T. 1987)**

37. Which of the following shows *cis-trans* isomerism?
(A) But-2-ene (B) Butan-2-ol (C) But-1-yne (D) Butanol

38. The compound buta-1,2-diene has
(A) only sp^2 hybridised carbon atoms
(B) sp, sp^2 and sp^3 hybridised carbon atoms
(C) both sp and sp^2 hybridised carbon atoms
(D) only sp hybridised carbon atoms. **(I.I.T. 1983)**

39. The maximum number of isomers for an alkene with molecular formula C_4H_8 is
(A) 2 (B) 5 (C) 4 (D) 3 **(I.I.T. 1982)**

40. The compound $C_4H_{10}O$ can show
(A) position isomerism (B) metamerism
(C) functional isomerism (D) all types. **(I.I.T. 1981)**

41. The molecule in which the distance between the two adjacent C atoms is largest is
(A) ethane (B) ethyne (C) ethene (D) benzene **(I.I.T. 1981)**

42. The compound having both *sp* and *sp*^{2} hybridised carbon atoms is
(A) propyne (B) propene (C) propadiene (D) none of these. **(I.I.T. 1981)**

43. The most strained cycloalkane is
(A) cyclopropane (B) cyclopentane (C) cyclobutane (D) cyclohexane **(I.I.T. 1981)**

44. The maximum number of stereoisomers possible for 2-hydroxy-2-methylbutanoic acid is
(A) 2 (B) 1 (C) 3 (D) 4 **(Roorkee 1992)**

45. Which has zero dipole moment?
(A) *cis*-but-2-ene (B) *trans*-but-2-ene
(C) 2-methylprop-1-ene (D) but-1-ene **(I.I.T. 1994)**

46. The number of isomers of C_6H_{14} is
(A) 4 (B) 5 (C) 6 (D) 7 **(I.I.T. 1987, 2007)**

Hint: 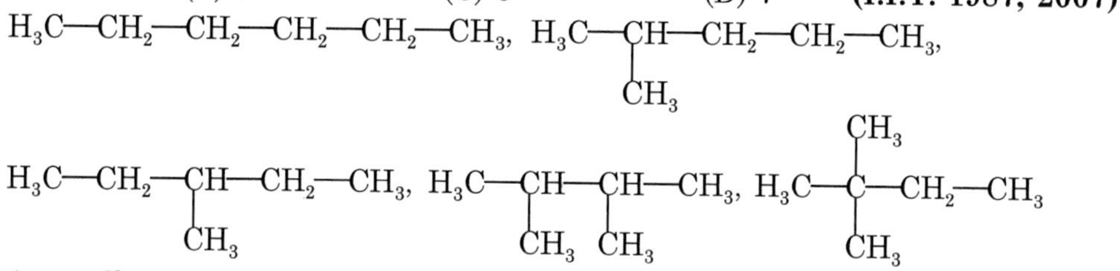

47. There is an alkene which reacts with one molecule of H_2 and gives a compound which contains only one asymmetric carbon atom. The possible structure of the alkene is
(A) $H_3C—CH{=}CH—CH_3$ (B) $H_3C—CH{=}CH_2$
(C) 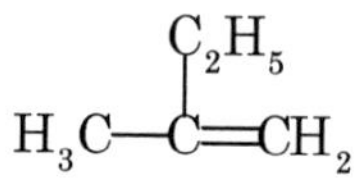 (D) 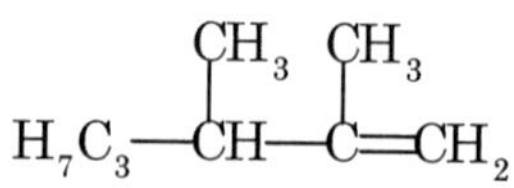 **(Roorkee 1992)**

48. In the pyrolysis of alkane the C—C bond breaks but not the C—H bond. This is due to
(A) the C—C bond energy is less than C—H bond energy
(B) the energy of activation of the C—H bond is more
(C) the C—C bond is the attacking site
(D) none of these. **(Roorkee 1996)**

49. Which of the following exhibits metamerism?
(A) $C_2H_5SC_2H_5$ (B) $CH_3OC_2H_5$ (C) CH_3OCH_3 (D) $CH_3COC_2H_5$

50. Racemic tartaric acid is optically inactive because of
    (A) external compensation (B) internal compensation
    (C) steric hindrance (D) loss of asymmetric centre.
51. Mesotartaric acid possesses
    (A) a plane of symmetry (B) a centre of symmetry
    (C) an angle of symmetry (D) an axis of symmetry.
52. The C—H bond distance is longest in
    (A) $C_2H_2Br_2$ (B) C_2H_4 (C) C_2H_6 (D) C_2H_2 **(I.I.T. 1988)**
53. Two enantiomers of an optically active compound are separated by
    (A) the use of suitable optically active reagent
    (B) fractional distillation
    (C) chromatography (D) fractional crystallisation.
54. The most stable conformation of n-butane is
    (A) skew boat (B) eclipsed (C) gauche (D) staggered-anti
55. Which of the following bonds is the strongest?
    (A) >C—C< (B) >C=C< (C) >C—C= (D) —C≡C—
56. Aromatic character of benzene is proved by
    (A) resonance theory (B) molecular orbital theory
    (C) aromatic sextate theory (D) all of these.
57. Which of the following statements is correct for 2-chlorobutane and 3-chlorobutane?
    (A) The number of Cl atoms present in 2-chlorobutane and 3-chlorobutane is different
    (B) 2-Chlorobutane and 3-chlorobutane are structural isomers
    (C) 2-Chlorobutane is less reactive than 3-chlorobutane
    (D) Both 2-chlorobutane and 3-chlorobutane are the same compound since 3-chlorobutane is named incorrectly.
58. An optically active compound is
    (A) 1-bromobutane (B) β-bromobutyric acid
    (C) 2-bromo-2-methylpropane (D) 1-bromo-2-methylpropane **(Roorkee 1993)**
59. An organic molecule necessarily shows optical activity if it
    (A) contains asymmetric carbon atoms
    (B) is not superimposable on its mirror image
    (C) is superimposable on its mirror image
    (D) is non-polar. **(I.I.T. 1993)**
60. How many structural isomers (including stereoisomers) could be obtained by replacing one hydrogen atom of propene with chlorine?
    (A) 2 (B) 4 (C) 3 (D) 5
61. How many isomers are possible for C_4H_8O?
    (A) 3 (B) 4 (C) 5 (D) 9
62. Which of the following can exist in optically active form?
    (A) But-1-ene (B) Butan-2-ol (C) Pentan-3-ol (D) Heptan-4-ol

63. Which of the following is optically active?
(A) 2-methylbutan-1-ol (B) 3-methylpentane
(C) *n*-pentane (D) butane

64. Which of the following orders is correct regarding the electronegativity of carbon atom?
(A) $sp < sp^2 > sp^3$ (B) $sp < sp^3 < sp^2$ (C) $sp > sp^2 > sp^3$ (D) $sp > sp^2 < sp^3$

65. Which of the following orders is correct regarding the size of hybrid orbitals of carbon?
(A) $sp < sp^2 < sp^3$ (B) $sp < sp^2 > sp^3$ (C) $sp > sp^2 < sp^3$ (D) $sp > sp^2 > sp^3$

66. Which of the following hydrocarbons has the lowest dipole moment?
(A) 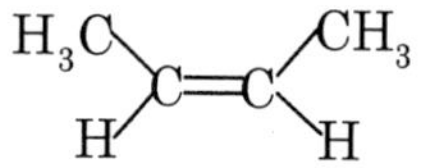

(B) $CH_3C{\equiv}CCH_3$
(C) $CH_3CH_2C{\equiv}CH$ (D) $CH_2{=}CH{-}C{\equiv}CH$ **(I.I.T. 2002)**

67. The nodal plane in the π-bond of ethene is located in
(A) the molecular plane
(B) a plane parallel to the molecular plane
(C) a plane perpendicular to the molecular plane which bisects the carbon-carbon σ-bond at right angle
(D) a plane perpendicular to the molecular plane which contains the carbon-carbon σ-bond. **(I.I.T. 2002)**

68. Which of the following exhibits stereoisomerism?
(A) 2-Methylbut-1-ene (B) 3-Methylbut-1-ene
(C) 3-Methylbutanoic acid (D) 2-Methylbutanoic acid **(I.I.T. 2002)**

69. Which of the following alkanes has the maximum C—H bond dissociation energy?
(A) $(CH_3)_3C{-}H$ (B) $(CH_3)_2CH{-}H$ (C) $CH_3CH_2{-}H$ (D) $H_3C{-}H$

70. Which of the following conformations of cyclohexane has the minimum strain energy?
(A) Boat (B) Twist-boat (C) Half-chair (D) Chair

71. Potential energy difference between the staggered and eclipsed conformations of ethane is – 12.55 kJ mol^{-1}. Ignoring the entropy effect, the ratio of staggered to eclipsed conformation at 298 K is
(A) 255 (B) 195 (C) 158 (D) 145
Hint: $\log K = -\Delta G^0/2.303RT = 12.55 \times 10^3$ J mol^{-1}/(2.303 × 8.314 J K^{-1} mol^{-1} × 298 K) = 2.1993 or K = 158.2

72. The number of isomers of dimethylcyclopropane is
(A) 2 (B) 3 (C) 4 (D) 5

Hint:

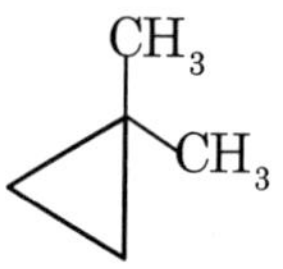

1,1-Dimethylcyclopropane

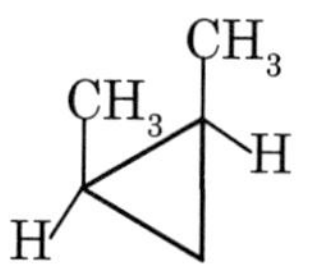

cis-1,2-Dimethylcyclopropane

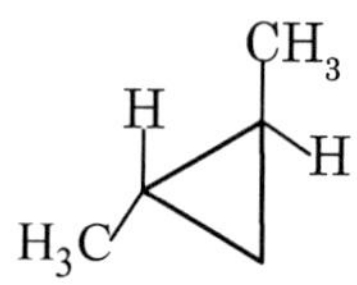

trans-1,2-Dimethylcyclopropane

73. Potential energy difference between the *anti* and *gauche* conformations of butane is – 3.8 kJ mol^{-1}. Ignoring the entropy effect, the ratio of *anti* to *gauche* conformations at 298 K is

(A) 4.6 (B) 2.6 (C) 5.5 (D) 6.2

Hint: $\log K = -\Delta G^0/2.303RT = 3.8 \times 10^3$ J mol^{-1}/2.303 × 8.314 J K^{-1} mol^{-1} × 298 K = 0.666 or K = 4.63

74. Identify the incorrect statement.

(A) Cycloheptane and cyclooctane are more stable than cyclohexane

(B) Ring strain in cyclobutane is less than that in cyclopropane

(C) Number of trichloro derivatives of cyclopropane is four

(D) The C—H bond length in cyclopropane is shorter than that in propane.

Hint: The statement (A) is incorrect and all others are correct:

(A) Both cycloheptane and cyclooctane are slightly less stable than cyclohexane.

(B) The statement is correct.

(C) No. of trichloro derivatives of cyclopropane is four: 1,1,2-; 1,1,3-; 1,2,3-; 1,2,4-trichloropentane.

(D) The ∠H—C—H in cyclopropane is 114° indicating that C atoms in cyclopropane is sp^2 hybridised. The C—H bond length is in the order: sp^2 (more *s*-character, cyclopropane) < sp^3 (less *s*-character, propane).

75. Identify the incorrect statement.

(A) Eclipsed conformation of ethane has lower potential energy in comparison to the staggered conformation

(B) The ∠H—C—H bond angle in methyl radical is 120°

(C) The carbon atom in methyl radical is sp^2 hybridised

(D) The *sp* hybrid orbital is more electronegative than sp^2 hybrid orbital.

76. Which of the following rotations is allowed in Fischer projection structure?

(A) A 180° rotation in the plane of paper

(B) A 90° rotation in the plane of paper

(C) A 180° vertical rotation outside the plane of paper

(D) A 180° horizontal rotation outside the plane of paper.

77. Two enantiomers differ in their

(A) solubilities in a solvent (B) optical activities

(C) chemical properties towards achiral reagents

(D) boiling point and melting point.

78. The number of stereoisomers of 1,2,4-trimethylcyclohexane is

(A) 2 (B) 3 (C) 4 (D) 5

Hint:

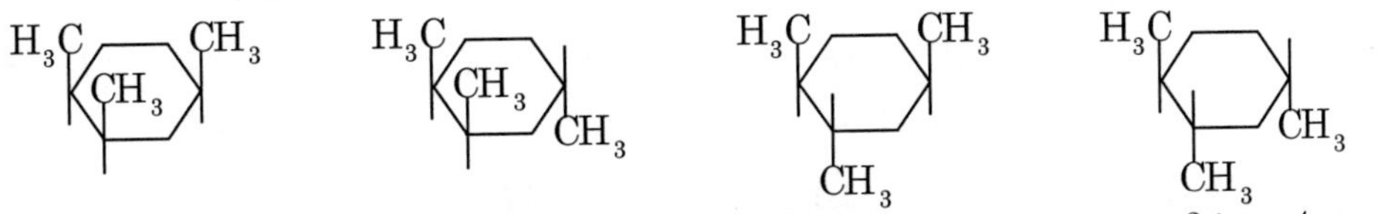

cis-cis-1,2,4-trimethyl-cyclohexane; *cis-trans*-4-trimethyl-cyclohexane; *trans*-2-*cis*-4-trimethyl-cyclohexane; *trans*-2-*trans*-4-trimethyl-cyclohexane

79. The number of stereoisomers of 2,3,4-trihydroxypentane is

    (A) 6 (B) 8 (C) 4 (D) 2

    Hint: No. of stereoisomers = $2^n = 2^3 = 8$. Out of eight stereoisomers, two stereoisomers have identical configurations with two other stereoisomers. Hence the total number of stereoisomers = 8 – 2 = 6.

80. The number of stereoisomers of butane-2,3-diol is

    (A) 8 (B) 4 (C) 3 (D) 1

81. Identify the incorrect statement regarding a *meso* compound.

    (A) A *meso* compound is formed on mixing equal amounts of two enantiomers
    (B) A *meso* compound is achiral
    (C) A *meso* compound has at least one pair of similar stereoisomers
    (D) A *meso* compound has either a plane of symmetry or point of symmetry.

82. In a compound having four chiral centres, the number of stereoisomers is

    (A) 4 (B) 8 (C) 16 (D) 2

83. Identify the operation on the Fischer projection formula 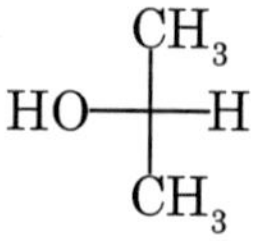 which does not lead to a change in absolute configuration.

    (A) Exchanging a horizontal group with a vertical group
    (B) Exchanging the groups across the vertical bond and also across the horizontal bond
    (C) Exchanging the groups across the vertical bond
    (D) Exchanging the groups across the horizontal bond.

84. Which of the following structures is/are non-aromatic?

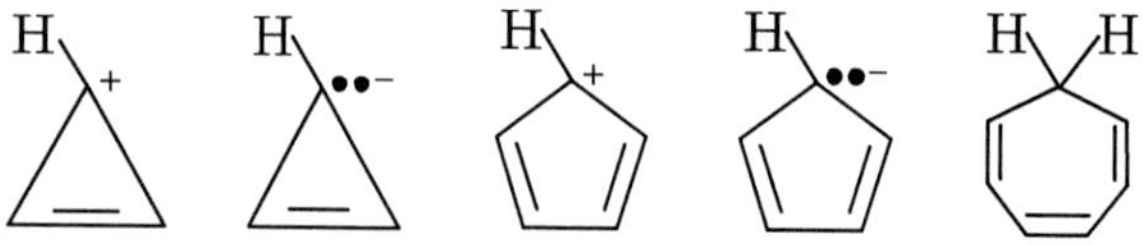

I II III IV V

    (A) II, V (B) III (C) I (D) IV

85. Identify the compound(s) in which the C═C bond length is 134 pm.

    (A) But-2-ene (B) But-1-ene
    (C) 1,2,3-Hexatriene (D) Cyclohexatriene

86. The number of stereoisomers obtained by the free radical monochlorination of 3-chloro-2-methylbutane is

    (A) 3 (B) 4 (C) 6 (D) 5

Hint:

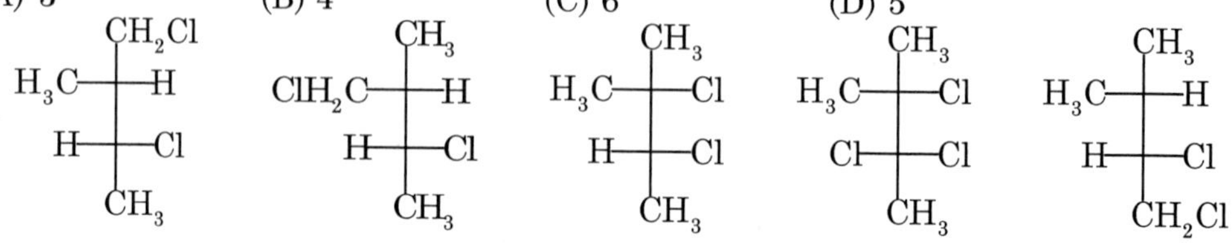

Optically inactive

87. Identify the incorrect statement.

(A) The optically active 2-iodobutane is eventually converted into racemic mixture in the presence of I^- ion

(B) *trans*-1,2-dimethylcyclopropane exists as a pair of isomers

(C) *cis*-1,2-dimethylcyclopropane is a meso compound

(D) The Fischer structure of the Newman structure 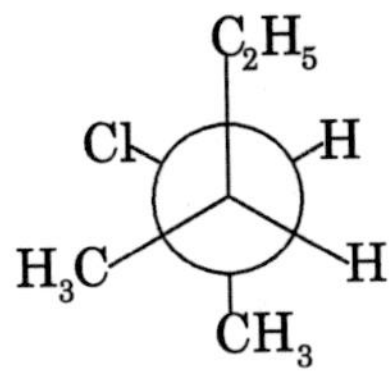 is 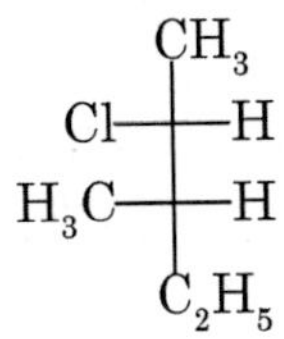

Hint: The statement (D) is incorrect and all others are correct.

(A) The optically active RI reacts with I^- ion with inversion at C and is racemised.

(B) The statement is correct.

(C) The statement is correct.

(D) 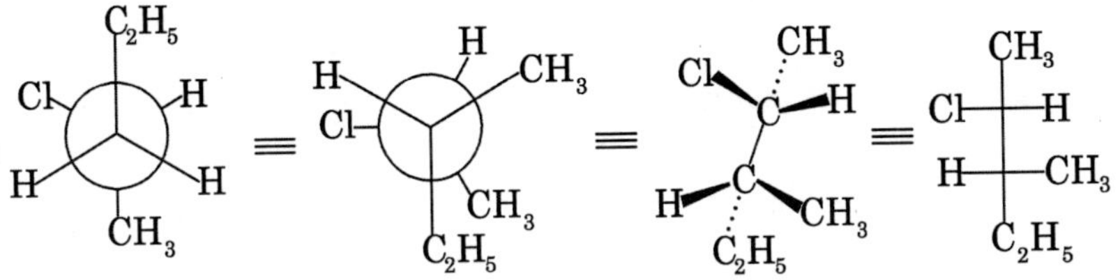

88. Identify the incorrect statement.

(A) Optically active 2-iodobutane on treatment with NaI in propanone gives a product which exhibits optical activity

(B) The specific rotation of a pure enantiomer is + 10°. The percentage composition of this isomer in a mixture having rotation of – 7° is 15%

(C) The specific rotation of a pure enantiomer is + 10°. The isomer is obtained from a reaction with 80% racemisation and 20% inversion. Its observed angle of rotation will be – 2.0°

(D) The specific rotation of a pure enantiomer is + 10°. The isomer is obtained from a reaction with 20% racemisation and 80% retention. Its observed angle of rotation will be + 8.0°.

Hint: The statement (A) is incorrect and all others are correct.

(A) Since there occurs breaking and reforming of C—I bond, two enantiomers with opposite signs of rotation will be formed. The resulting racemic mixture does not exhibit optical activity.

(B) If x is the fraction of the given isomer in the mixture, then

$x(+10°) + (1 - x)(-10°) = -7°$

or $x = 3/20$

Hence the mixture contains (3/20) × 100 = 15% of this isomer.

(C) The isolated mixture contains 40% of this isomer (from 80% racemic mixture) and a total of 60% of its inversion. Hence the observed bond angle will be

$$\alpha = (0.40)\ (+10°) + (0.60)\ (-10°) = 4.6° - 6.0° = -2.0°$$

(D) The isolated mixture contains 90% of this isomer (10% from the racemic mixture and 80% due to retention) and 10% of its inversion. Hence

$$\alpha = (0.90)\ (+10°) + (0.10)\ (-10°) = 9.0° - 1.0° = +8.0°$$

89. The correct order of priority sequence of groups OH, COOH, CHO and CH_2OH in absolute configuration of an enantiomer is

(A) $CH_2OH < CHO < COOH < OH$ (B) $CH_2OH < CHO < OH < COOH$
(C) $COOH < CH_2OH < CHO < OH$ (D) $CH_2OH < COOH < OH < CHO$

90. The correct order of priority sequence of groups F, NO_2, NH_2 and CH_3 in absolute configuration of an enantiomer is

(A) $CH_3 < NH_2 < NO_2 < F$ (B) $CH_3 < NO_2 < NH_2 < F$
(C) $NH_2 < NO_2 < CH_3 < F$ (D) $CH_3 < NH_2 < F < NO_2$

91. The number of optically active compounds in the isomers of $C_3H_5Cl_3$ is

(A) 4 (B) 3 (C) 2 (D) 1

92. The number of isomers of the compound $C_3H_5Cl_3$ is

(A) 5 (B) 4 (C) 3 (D) 1

93. How many isomers are possible for $C_4H_{10}O$?

(A) 10 (B) 8 (C) 6 (D) 4

94. The number of optically active compound in the isomers of a hydroxy acid $C_4H_8O_3$ is

(A) 5 (B) 4 (C) 3 (D) 2

95. The compound 2,3-dibromobutane has

(A) two pairs of isomers (B) four stereoisomers
(C) one pair of enantiomers and two meso compounds
(D) one pair of enantiomers and one meso compound.

96. Two enantiomers of an organic compound are separated by

(A) chromatographic techniques (B) use of suitable optically active reagents
(C) fractional crystallisation (D) fractional distillation.

97. Identify the incorrect statement.

(A) The chair form of cyclohexane is rigid and resists distortion
(B) In the chair form of cyclohexane all the hydrogen atoms are in the eclipsed orientations
(C) The chair form of cyclohexane is more stable than the twist boat form
(D) The energy difference between the chair and half-chair forms of cyclohexane is 46 kJ mol^{-1}.

98. Staggered conformation of ethane is

(A) impossible to twist (B) equally stable as eclipsed conformation
(C) less stable than the eclipsed conformation
(D) more stable than the eclipsed conformation.

99. Identify the incorrect statement.
   (A) The relative energies of various conformations of*n*-butane are in the order: partially eclipsed < fully eclipsed < *gauche* < *anti*
   (B) *Gauche* conformation of butane is 3.35 kJ mol^{-1} less stable than the *anti* conformation
   (C) Partially eclipsed conformation of butane is 14.2 kJ mol^{-1} stable than *anti* conformation
   (D) Fully eclipsed conformation of butane is 18.4 kJ mol^{-1} less stable than *anti* conformation.

100. Which of the following represents the given mode of hybridisation sp^2-sp^2-sp^3-sp from left to right?
   (A) $H_2C=CH-CH-C\equiv N$ (B) $HC\equiv C-C\equiv CH$
   (C) $H_2C=C=C=CH_2$ (D) $H_2C=CH-CH=CH_2$

101. 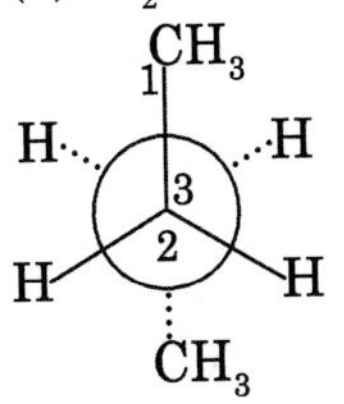

C_2 is rotated anticlockwise 120° about C_2-C_3 bond. The resulting conformer is
   (A) partially eclipsed (B) eclipsed
   (C) *gauche* (D) staggered **(I.I.T. 2004)**

Hint: When C_2 is rotated anticlockwise 120° about C_2-C_3 bond, the resulting conformer is a gauche conformer.

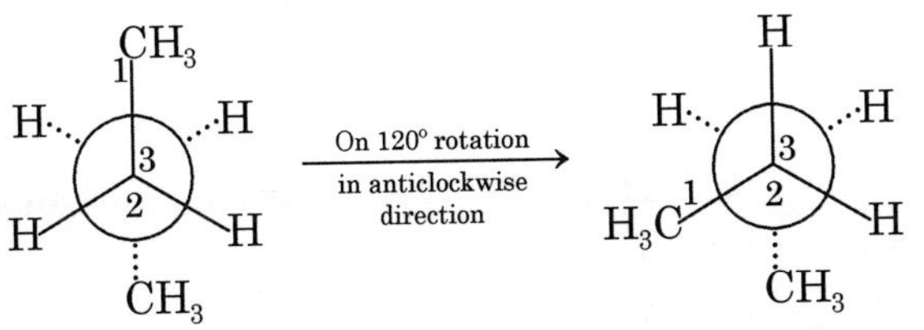

Gauche conformer

102. How many chiral compounds are possible on monochlorination of 2-methylbutane?
   (A) 2 (B) 4 (C) 6 (D) 8 **(I.I.T. 2004)**

Hint: $Cl-CH_2-\overset{*}{C}H(CH_3)-CH_2-CH_3$, $(CH_3)_2CH-\overset{*}{C}H(CH_3)-CH_3$,
$(CH_3)_2\overset{*}{C}(Cl)-CH_2-CH_3$, $(CH_3)_2\overset{*}{C}H-CH_2-CH_2-Cl$

103. For 1-methoxybuta-1,3-diene, which of the following resonating structures is the least stable?
   (A) $H_2\bar{C}-\overset{+}{C}H-CH=CH-O-CH_3$ (B) $H_2\bar{C}-CH=CH-CH=\overset{+}{O}-CH_3$
   (C) $H_2C=CH-\overset{+}{C}H-\bar{C}H=O-CH_3$ (D) $H_2C=CH-\bar{C}H-CH=\overset{+}{O}-CH_3$

Hint: The point of difference is nature of carbanion. 2° carbanions are less stable than 1° carbanions.

104. Which of the following operations of switching ligands or rotating Fischer structures changes its absolute configuration?
   (A) Exchanging ligands across the horizontal bond as well as the vertical bond
   (B) Rotating the Fischer projection by 180° in the plane of the paper
   (C) An odd number of switches (D) An even number of switches.
105. Identify the incorrect statement.
   (A) Diastereoisomers are observed in a molecule containing one chiral carbon atom
   (B) Mesotartaric acid has a plane of symmetry
   (C) Benzene has about 78% contribution from Kekule structures and 22% contribution from Dewar structures
   (D) In the compound D-(+)-glyceraldehyde, D stands for the absolute configuration.

   Hint: The statement (A) is incorrect and all others are correct.

   (A) Diastereoisomers are observed in a molecule containing more than one chiral carbon atoms.
106. Identify the incorrect statement.
   (A) Resonance energy is the difference between the enthalpy of formation of the molecule and that of resonating structure having the largest energy
   (B) Metamerism is due to the difference in arrangement of groups attached to a functional group
   (C) The size of sp^3 hybrid orbital is greater than that of sp hybrid orbital
   (D) An sp hybrid orbital is more electronegative than an sp^2 hybrid orbital.

   Hint: The statement (A) is incorrect and all other statements are correct.

   (A) Resonance energy is the difference between the enthalpy of formation of the molecule and that of resonating structure having the lowest energy.
107. Identify the incorrect statement.
   (A) Out of $(CH_3)_3CH$, $(CH_3)_2CHCH_3$, $CH_2{=}CHCH_3$ and $C_6H_5CH_3$, the maximum ease of abstraction of H atom is exhibited by $(CH_3)_3CH$
   (B) With increase in temperature the relative rates of substitution of 1°, 2° and 3° hydrogen atom by chlorine atom become equal, i.e. 1 : 1 : 1
   (C) In the vapour phase reaction at 300°C, the relative rates of substitution of 1°, 2° and 3° hydrogen atom by chlorine atom are 1.00 : 3.25 : 4.43.
   (D) Chlorine has lesser tendency than bromine to replace 3° hydrogen atom in preference to 1° or 2° hydrogen atom

   Hint: The statement (A) is incorrect and all others are correct.

   (A) The maximum ease of abstraction of H atom is shown by $C_6H_5CH_3$. The order of ease of abstraction of H atom is: CH_4, vinylic < 1° < 2° < 3° < allylic, benzylic.
108. Identify the incorrect statement.
   (A) 1.79×10^{-3} g of an organic compound having molar mass 90 g mol^{-1} produces 1.34 mL of methane gas when reacted with methyl magnesium bromide. The number of active hydrogen atoms in the organic compound is two

(B) When CH_3MgBr is added to 4.11×10^{-3} g of an alcohol, 1.56 mL of methane gas is liberated at STP. The molar mass of the alcohol is 59.0 g mol^{-1}

(C) The relative rates of abstraction of 1°, 2° and 3° hydrogen atoms are 1.0 : 3.8 : 5.0

(D) The number of monochloroderivatives of 2-methylpentane is five.

Hint: The statement (A) is incorrect and all others are correct.

(A) The number of active H atoms in the compound is three.

$$\text{Amount of } CH_4(g) \text{ evolved} = \frac{1.34 \text{ mL}}{22400 \text{ mL mol}^{-1}} = 5.98 \times 10^{-5} \text{ mol}$$

$$\text{Amount of organic compound} = \frac{1.79 \times 10^{-3} \text{ g}}{90 \text{ g mol}^{-1}} = 1.99 \times 10^{-5} \text{ mol}$$

$$\text{Number of active H atoms} = \frac{5.98 \times 10^{-5} \text{ mol}}{1.99 \times 10^{-5} \text{ mol}} = 3$$

(B) $$\text{Amount of } CH_4(g) \text{ evolved} = \frac{1.56 \text{ mL}}{22400 \text{ mL mol}^{-1}} = 6.96 \times 10^{-5} \text{ mol}$$

Amount of alcohol = 6.96×10^{-5} mol

$$\text{Molar mass of alcohol} = \frac{4.11 \times 10^{-3} \text{ g}}{6.96 \times 10^{-5} \text{ mol}} = 59.0 \text{ g mol}^{-1}$$

(C) The relative rates of abstraction of 1°, 2° and 3° H atoms are: 1° < 2° < 3°.

(D) Number of monochloroderivatives of 2-methylpentane = 5

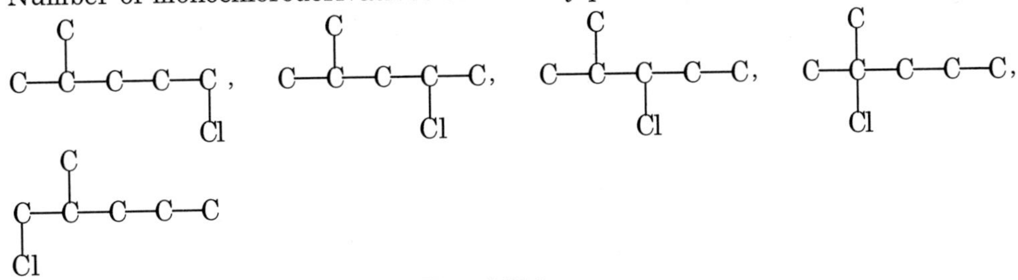

109. The type of conformation shown by X and Y is

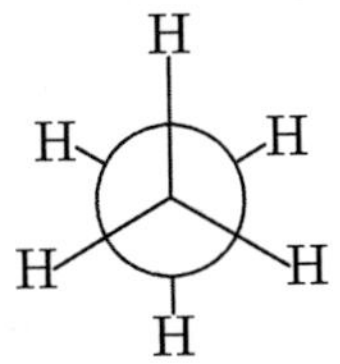

X

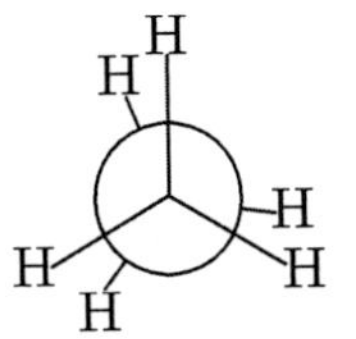

Y

(A) both are staggered
(B) both are eclipsed
(C) Y is eclipsed, X is staggered
(D) X is eclipsed, Y is staggered.

110. What type of isomers are X and Y?

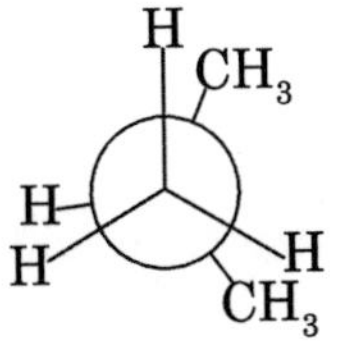

X

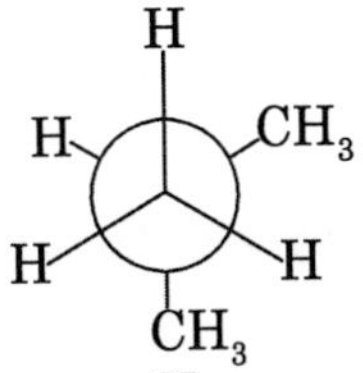

Y

(A) geometrical (B) optical (C) position (D) conformation

111. The number of stereoisomers obtained by bromination of *trans*-but-2-ene is

(A) 1 (B) 2 (C) 3 (D) 4 **(I.I.T. 2007)**

Hint: 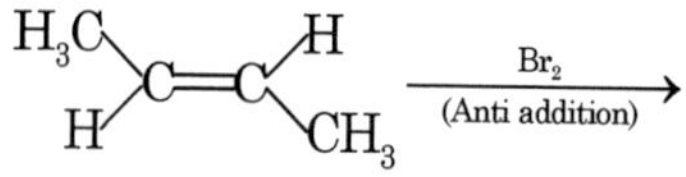

Meso product

112. Statement–1: Molecules that are not superimposable on their mirror images are chiral.
because

Statement–2: All chiral molecules have chiral centres.

(A) Statement–1 is true, Statement–2 is true; Statement–2 is a correct explanation for Statement–1

(B) Statement–1 is true, Statement–2 is true; Statement–2 is not a correct explanation for Statement–1

(C) Statement–1 is true, Statement–2 is false

(D) Statement–1 is false, Statement–2 is true. **(I.I.T. 2007)**

113. Hyperconjugation involves overlap of the following orbitals

(A) $\sigma-\sigma$ (B) $\sigma-p$ (C) $p-p$ (D) $\pi-\pi$ **(I.I.T. 2008)**

Hint: Hyperconjugation involves the overlap of σC—H bonding electrons with the *p*-orbital of unsaturated system, carbocation or carbon free radical.

114. The correct statement(s) about the compound given below is(are)

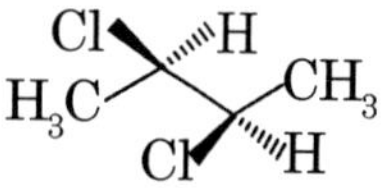

(A) The compound is optically active

(B) The compound possesses centre of symmetry

(C) The compound possesses plane of symmetry

(D) The compound possesses axis of symmetry. **(I.I.T. 2008)**

Hint: The given compound is represented in saw-horse projection as

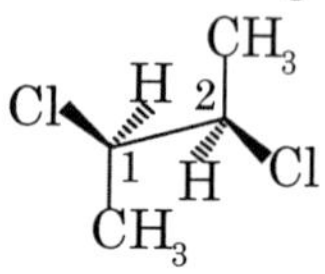

It does not have a plane of symmetry or a centre of symmetry. Hence it is optically active. The compound on rotation about C_1–C_2, keeping C_2 fixed and C_1 rotated by 180°, gives the following compound:

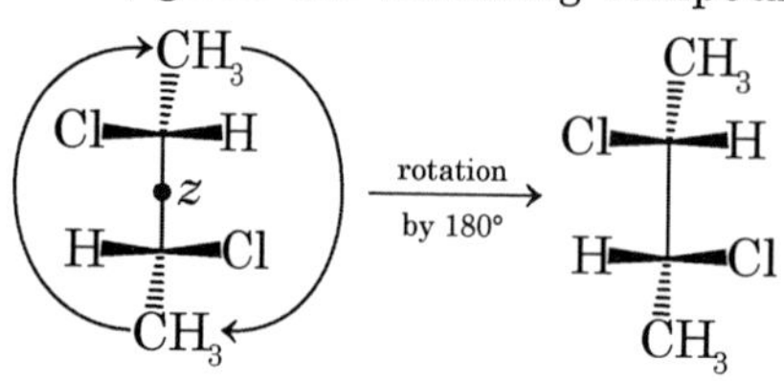

The compound possesses 180° axis of symmetry (two fold axis of symmetry) passing

through the point z perpendicular to the plane of paper.

115.In the Newman projection for 2,2-dimethylbutane,

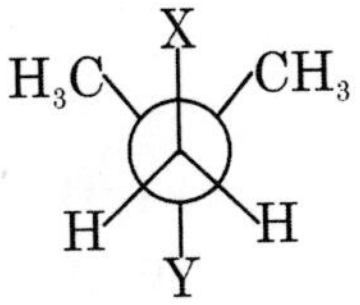

X and Y can respectively be

(A) H and H (B) H and C_2H_5 (C) C_2H_5 and H (D) CH_3 and CH_3

(I.I.T. 2010)

Hint: 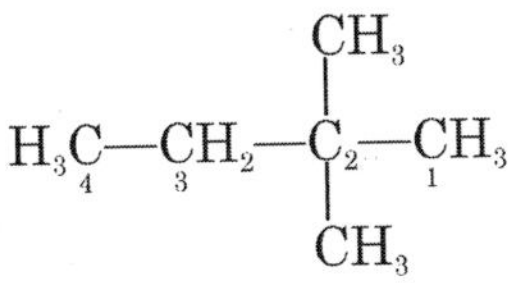

On $C_2—C_3$ bond axis, X = CH_3, Y = CH_3 and on $C_1—C_2$ bond axis, X = H, Y = C_2H_5.

116.The IUPAC name of the following compound is

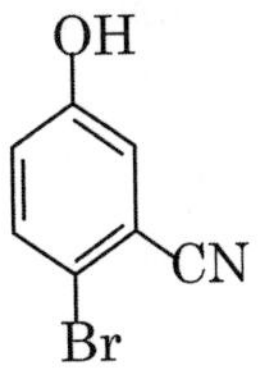

(A) 4-Bromo-3-cyanophenol

(B) 2-Bromo-5-hydroxybenzonitrile

(C) 2-Cyano-4-hydroxybromobenzene

(D) 6-Bromo-3-hydroxybenzonitrile **(I.I.T. 2009)**

Hint: The priority of CN is the highest.

117. Amongst the given options, the compound(s) in which all the atoms are in one plane in all the possible conformations (if any), is (are)

(A) 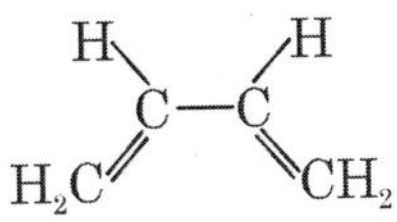

(B) 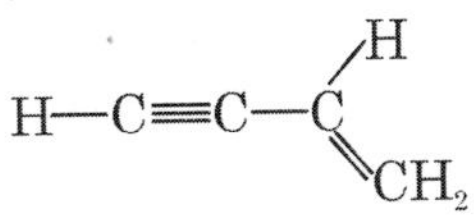

(C) $H_2C═C═O$ (D) $H_2C═C═CH_2$ **(I.I.T. 2011)**

Hint: Along C—C single bond conformations are possible in butadience in which all the atoms may not lie in the same plane. In $H_2C═C═CH_2$, the hydrogen atoms are not in the same plane.

ANSWERS

1. A	2. A	3. A	4. B	5. A	6. A, D
7. A, B	8. D	9. C	10. C	11. C	12. A
13. D	14. C	15. A, C, D	16. A	17. B	18. C
19. C	20. D	21. D	22. B	23. B, C	24. C
25. B	26. C	27. C	28. C	29. D	30. A
31. D	32. C	33. B	34. A	35. C	36. A
37. A	38. B	39. C	40. D	41. A	42. C
43. A	44. A	45. B	46. B	47. D	48. A
49. A	50. A	51. A	52. C	53. A	54. D
55. D	56. D	57. D	58. B	59. B	60. B
61. D	62. B	63. A	64. C	65. A	66. B
67. A	68. D	69. D	70. D	71. C	72. B
73. A	74. A	75. A	76. A	77. B	78. C
79. A	80. C	81. A	82. C	83. B	84. A, B
85. A, B	86. D	87. D	88. A	89. A	90. A
91. D	92. A	93. B	94. C	95. D	96. B
97. B	98. D	99. A	100. A	101. C	102. B
103. C	104. C	105. A	106. A	107. A	108. A
109. C	110. D	111. A	112. C	113. B	114. A, D
115. B, D	116. C	117. B, C			

35

Hydrocarbon—I: Sources and Synthesis

OBJECTIVE TYPE MULTIPLE CHOICE QUESTIONS

1. When isobutyl magnesium bromide in dry ether is treated with absolute alcohol, the products formed are
   (A) $(CH_3)_3CH + CH_3CH_2OMgBr$ (B) $(CH_3)_2CHCH_2OH + CH_3CH_2MgBr$
   (C) $(CH_3)_3CH + CH_2{=}CH_2 \cdot MgBr(OH)$ (D) $(CH_3)_2CHCH_2CH_2CH_3 + MgBr(OH)$
   (I.I.T. 1991)

2. 1-Chlorobutane on reaction with alcoholic KOH gives
   (A) but-1-ene (B) 1-butanol (C) but-2-ene (D) 2-butanol **(I.I.T. 1991)**

3. A mixture of ethyl iodide and *n*-propyl iodide is subjected to Wurtz reaction. The hydrocarbon which will not be formed is
   (A) *n*-butane (B) *n*-propane (C) *n*-pentane (D) *n*-hexane **(I.I.T. 1990)**

4. The dehydrohalogenation of neopentyl bromide with alcoholic KOH gives mainly
   (A) 2-methylbut-1-ene (B) 1,1-dimethylcyclopropane
   (C) 2,2-dimethylbut-1-ene (D) but-2-ene **(I.I.T. 1990)**

5. *n*-Propyl bromide on treatment with ethanolic KOH produces
   (A) propane (B) propene (C) propyne (D) propanol **(I.I.T. 1987)**

6. Identity Y in the reactions, $CH_3CH_2CH_2OH \xrightarrow{PCl_5} X \xrightarrow{alc.\ KOH} Y$
   (A) propene (B) propane (C) 1-propanol (D) propyne **(I.I.T. 1981)**

7. Marsh gas consists of mainly
   (A) C_2H_2 (B) CH_4 (C) H_2S (D) CO **(I.I.T. 1980)**

8. Ethyl alcohol is heated with conc. H_2SO_4. The product formed is
   (A) $CH_3COOC_2H_5$ (B) C_2H_6 (C) C_2H_4 (D) C_2H_2 **(I.I.T. 1980)**

9. The octane number has zero value for
   (A) isooctane (B) *n*-octane (C) *n*-heptane (D) TEL **(Roorkee 1989)**

10. Gasoline is the name of
    (A) crude oil (B) the gaseous constitutents of petroleum
    (C) the mixture of uncondensed gases produced in the distillation of crude oil
    (D) the mixture of residue and gas oil obtained in the distillation of crude oil.
    (Roorkee 1989)
11. In the fractional distillation of crude petroleum
    (A) petrol condenses at the bottom of the column
    (B) the gases condense at the top of the column
    (C) high boiling constituents condense at the bottom of the column
    (D) high boiling constituents condense at the top of the column. **(Roorkee 1989)**
12. In gasoline of high octane number, there is
    (A) more quantity of branched chain hydrocarbons
    (B) more quantity of straight chain hydrocarbons
    (C) more quantity of *n*-octane (D) more quantity of isooctane.
13. Consider the following statements about petroleum: (1) It is non-renewable energy source. (2) It consists mainly glycerol esters of long chain fatty acids. (3) Its composition differs from location to location. (4) It does not contain either alicyclic or aromatic compounds. Which statements are correct?
    (A) 1, 2 (B) 1, 3 (C) 1, 3, 4 (D) 2, 3
14. Ether solution of which compound when treated with Na gives 2,3-dimethylbutane?
    (A) isopropyl bromide (B) ethyl bromide
    (C) *n*-propyl bromide (D) 1-bromobutane
15. 2-Bromopentene is heated with C_2H_5OK in ethanol. The major product formed is
    (A) pent-1-ene (B) *cis*-pent-2-ene
    (C) *trans*-pent-2-ene (D) 2-ethoxypentene
16. The reagent used for the conversion, $CH_3CH_2COOH \longrightarrow CH_3CH_2OH$ is
    (A) $LiAlH_4$ (B) sodalime
    (C) Zn/Hg and conc. HCl (D) red P and conc. HCl
17. Identify Y in the reactions, $CH_3CH{=}CH_2 \xrightarrow{HBr} X \xrightarrow{\text{(i) Mg/ether (ii) } H_2O\text{, heat}} Y$
    (A) CH_3CH_3 (B) $CH_3CH_2CH_2CH_3$
    (C) $CH_3CH_2CH_3$ (D) $CH_3CH(Br)CH_3$
18. Which is formed when chloroform is heated with Ag powder?
    (A) Ethyne (B) Ethene (C) Ethane (D) Methanoic acid
19. The reaction product from the reaction of methyl magnesium bromide and ethanol is
    (A) ethane (B) methane (C) butane (D) propane
20. The order of appearance of the following during the refining of crude oil is
    (A) gasoline, kerosene oil, diesel (B) diesel, gasoline, kerosene oil
    (C) gasoline, diesel, kerosene oil (D) kerosene oil, gasoline, diesel.
21. Cetane number of diesel will increase with the addition of
    (A) *n*-hexadecane (B) *n*-pentane
    (C) *n*-decane (D) α-methylnaphthalene

22. In the process of cracking
    (A) organic compounds decompose into their constituent elements
    (B) hydrocarbons decompose into carbon and hydrogen
    (C) high molecular mass organic compounds decompose to give low molecular mass organic compounds
    (D) hydrocarbons yield alkyl radicals and hydrogen. **(Roorkee 1989)**
23. The inorganic origin of petroleum is indicated by the fact that
    (A) its constituents can be separated by fractional distillation
    (B) C and H can combine by absorption of solar energy to give hydrocarbons
    (C) petroleum contains traces of chlorophyll
    (D) oil fields are located with the help of seismograph. **(Roorkee 1990)**
24. Petrol used as an automobile fuel is a mixture of
    (A) hydrocarbons (B) alcohols
    (C) carbohydrates (D) hydrocarbons and alcohols **(Roorkee 1989)**
25. Which of the following is not true of natural gas?
    (A) It is a mixture of CO and H_2 (B) It is a mixture of gaseous hydrocarbons
    (C) It is a fuel (D) It is used in the manufacture of fertilizers.
26. Other than tetraethyllead which of the following is used as antiknock compound?
    (A) tetramethyllead (B) $Te(C_2H_5)_2$
    (C) Mn(cyclopentadiene)$(CO)_3$ (D) none of these.
27. The antiknocking property of non-leaded petrol is achieved by adding to petrol
    (A) $Te(C_2H_5)_2$ (B) Mn(cyclopentadiene)$(CO)_3$
    (C) benzene (D) none of these.
28. Which of the following has the highest octane number?
    (A) *n*-heptane (B) 2,3-dimethylbutane
    (C) 2,6-dimethylpentane (D) 2,3,3-trimethylpentane
29. Which is the correct order of decrease of knocking?
    (A) straight chain paraffins > branched chain paraffins > olefins > cycloparaffins > aromatic hydrocarbons
    (B) branched chain paraffins > straight chain paraffins > olefins > cycloparaffins > aromatic hydrocarbons
    (C) olefins > cycloparaffins > straight chain paraffins > branched chain paraffins > aromatic hydrocarbons
    (D) cycloparaffins > aromatic hydrocarbons > olefins > straight chain paraffins > branched chain paraffins.
30. Which of the following is used as catalyst in the catalytic cracking of hydrocarbons?
    (A) CO (B) V_2O_5 (C) H_3PO_4 (D) ZrO_2 and clay
31. Which of the following is used as a petrol additive for the removal of Pb deposited in the cylinder?
    (A) C_2H_5Br (B) $CH_3(CH_2)_4Br$ (C) $CH_2{=}CHBr$ (D) $BrCH_2CH_2Br$

32. Which of the following is/are used for increasing the cetane number of diesel?
(A) n-Pentane
(B) Eithyl nitrite
(C) Isoamyl nitrite
(D) Acetone peroxide

33. Catalyst used in the Bergius process for the manufacture of synthetic petrol is
(A) ZrO_2 and clay (B) V_2O_5 (C) zinc oleate (D) nickel oleate

34. Catalyst used in Fischer–Tropsch process for the manufacture of synthetic petrol is
(A) Co, ThO_2, MgO, keiselguhr
(B) Fe_2O_3
(C) V_2O_5
(D) tin oleate

35. Which of the following will convert 1, 2-dichloropropane to propyne?
(A) Zn dust (B) Cu, 100°C (C) Ag powder (D) None of these.

36. Identify X in the reactions $X \xrightarrow{Na} Y \xrightarrow{CaO/NaOH} CH_4$
(A) CH_3CH_2OH (B) CH_3CHO (C) CH_3COOH (D) $CH_3CH_2CH_3$

37. In the acid catalysed dehydration of alcohols to olefins the species involved is
(A) a free radical intermediate
(B) a carbocation intermediate
(C) a carbanion intermediate
(D) the expulsion of OH^- ion.

38. Indicate the formula of the organic product when C_2H_5MgBr is treated with D_2O.
(A) C_2H_5D (B) C_2D_6 (C) C_2H_5OD (D) $C_2H_5—C_2H_5$

39. When 2-bromopentene is heated with C_2H_5OK in ethanol, the major product is
(A) *cis*-pent-2-ene
(B) *trans*-pent-2-ene
(C) pent-1-ene
(D) 2-ethoxypentene

40. When an aqueous solution of sodium ethanoate and sodium propanoate is electrolysed, we get
(A) butane (B) propane (C) ethane (D) all of these.

41. Which of the following has the least octane number?
(A) 2,4-Dimethylbutane
(B) Cetane
(C) Triptane
(D) 2,2,4-Trimethylpentane

42. The major product of the reaction, $H_3C—CHBr—CH_2—CH_3 \xrightarrow{alc.\ KOH}$ is
(A) but-1-ene (B) but-2-ene (C) but-1-yne (D) none of these.

43. The formation of alkane by the action of Zn on haloalkane is known as
(A) Kolbe reaction
(B) Cannizzaro reaction
(C) Wurtz reaction
(D) Frankland reaction.

44. Ethyne is prepared industrially by passing electric discharge through graphite electrodes in the presence of
(A) N_2 (B) H_2 (C) air (D) CO_2

45. To prepare n-hexane using Na in dry ether as one reactant, the other reactant will be
(A) n-propyl bromide
(B) ethyl bromide and n-butyl bromide
(C) methyl bromide and n-pentyl bromide
(D) ethyl chloride and butyl chloride

46. In the reaction, $CH_3—CHX—CH_2—CH_3 \xrightarrow{H_2SO_4,\ 475\ K} CH_3—CH{=}CH—CH_3 +$ $CH_2{=}CH—CH_2—CH_3$
   (A) the product ratio is dependent on X
   (B) both are formed in equal amounts
   (C) $CH_2{=}CH—CH_2—CH_3$ predominates
   (D) $CH_3—CH{=}CH—CH_3$ predominates.
47. Which of the following is expected to give readily a hydocarbon in good yield?
   (A) $RCOOAg + Br_2 \longrightarrow$ (B) $CH_3CH_3 + Cl_2 \xrightarrow{hv}$
   (C) $(CH_3)_3CCl + C_2H_5OH \longrightarrow$ (D) $RCOOK \xrightarrow{electrolysis}$
48. When a mixture of 1-chloropropane and 2-chloropropane is treated with alcoholic KOH, we get
   (A) isopropylene (B) prop-2-ene (C) prop-1-ene (D) all of these.
49. Petrol for aviation purpose must contain
   (A) aromatic hydrocarbons (B) olefinic hydrocarbons
   (C) straight chain hydrocarbons (D) highly branched chain hydrocarbons.
50. The ease of dehydrohalogenation of alkyl halides is in the order
   (A) $CH_3CH_2Br > CH_3CH_2I > CH_3CH_2Cl$
   (B) $CH_3CH_2I > CH_3CH_2Cl > CH_3CH_2Br$
   (C) $CH_3CH_2Cl > CH_3CH_2Br > CH_3CH_2I$
   (D) $CH_3CH_2I > CH_3CH_2Br > CH_3CH_2Cl$
51. When 3,3-dimethylbutan-2-ol is heated with H_2SO_4, the major product is
   (A) 2,3-dimethylbutene (B) 3,3-dimethylbutene
   (C) 2,3-dimethylbutane (D) *cis* and *trans* isomers of (C)
52. The compound obtained on heating *n*-heptane with $CrO_3–Al_2O_3$ at 600°C is
   (A) benzene (B) toluene (C) cyclohexane (D) cyclohexene
53. The reaction of C_2H_5MgI with water gives
   (A) ethanal (B) ethanol (C) methane (D) ethane
54. Electrolysis of an aqueous solution of sodium propanoate gives
   (A) propane (B) butane (C) methane (D) ethane
55. The major product in the dehydrohalogenation of 2-bromo-2,3-dimethylbutane is
   (A) 2,3-dimethylbut-2-ene (B) 3,3-dimethylbutene
   (C) 4-methylpent-2-ene (D) 2,3-dimethylbutene
56. Dehydrohalogenation of 2-bromobutane with alcoholic KOH produces
   (A) but-1-ene as the major product (B) but-2-ene as the major product
   (C) but-1-ene only (D) but-2-ene only.
57. Dehydration of 2-methylbutanol with conc. H_2SO_4 will give
   (A) pent-2-ene (B) pentene
   (C) 2-methylbut-2-ene as the major product
   (D) 2-methylbut-1-ene as the major product.

58. Ethyne is formed during the electrolysis of
    (A) potassium carbonate (B) potassium acetate
    (C) potassium fumarate (D) potassium succinate
59. Propyne is produced during the hydrolysis of
    (A) SiC (B) Mg_2C_3 (C) Al_4C_3 (D) CaC_2
60. Which catalyst is used in the polymerisation of ethyne to benzene?
    (A) HCl (B) $NbCl_5$ (C) $HgSO_4$ (D) $AlCl_3$
61. Heating a mixture of sodium benzoate and sodalime gives
    (A) calcium benzoate (B) benzene
    (C) methane (D) sodium benzoate
62. Heating a mixture of sodium acetate and sodalime gives
    (A) CH_3CH_3 (B) CH_4 (C) $Ca(CH_3COO)_2$ (D) CH_3COOH
63. Identify the incorrect statement.
    (A) When ethanol is treated with HI, ethane is formed
    (B) Methane is a major constituent of natural gas
    (C) Ammonia acts as a base in the reaction: $CH_3MgBr + NH_3 \longrightarrow CH_4 + Mg(NH_2)Br$
    (D) Electrolysis of a concentrated solution of sodium ethanoate gives ethane.
64. Identify the incorrect statement.
    (A) In the reaction, $CH_3MgBr + CH_3OH \longrightarrow CH_4 + Mg(OCH_3)Br$, methanol is a stronger acid than CH_4
    (B) Wurtz reaction converts haloalkane into alkane when it is reacted with Na in dry ether
    (C) Percentage of methane in coal gas is 45
    (D) Marsh gas is methane.
65. Which of the following alkanes cannot be prepared by Wurtz reaction?
    (A) $CH_3CH_2CH_2CH_3$ (B) $(CH_3)_2CHCH_3$
    (C) $CH_3CH(CH_3)CH_3$ (D) CH_3CH_3
66. Which of the following reactions will give methane?
    (A) Electrolysis of potassium ethanoate
    (B) Aluminium carbide reacts with water
    (C) Reduction of iodomethane (D) Sodium ethanoate is heated with sodaline
67. Dehydrohalogenation is carried out by using
    (A) Zinc dust (B) conc. H_2SO_4 (C) alcoholic KOH (D) aqueous KOH
68. In the Kolbe's electrolysis, the hydrocarbon is obtained
    (A) at anode (B) at cathode
    (C) first at cathode and then at anode
    (D) first at anode and then at cathode.

69. Which of the following is/are formed when bromoethane is fused with sodium in the presence of dry ether?
    (A) $CH_3\overset{\displaystyle CH_3}{\overset{|}{C}}HCH_3$ (B) $CH_3CH_2CH_2CH_3$
    (C) $CH_2{=}CH_2$ (D) CH_3CH_3
70. Which of the following is/are obtained by the dehydration of butan-1-ol?
    (A) $CH_2{=}CH{-}CH{=}CH_2$ (B) $HC{\equiv}CCH_2CH_3$
    (C) $CH_3CH{=}CHCH_3$ (D) $CH_2{=}CHCH_2CH_3$
71. Which of the following is not obtained in the reaction of chloroethane with sodium in the presence of dry ether?
    (A) C_3H_8 (B) C_2H_4 (C) C_2H_6 (D) C_4H_{10}
72. Kolbe's electrolysis occurs through
    (A) electrophilic substitution mechanism
    (B) free radical mechanism
    (C) electrophilic addition mechanism (D) nucleophilic substitution mechanism.
73. Which of the following cannot be obtained by Kolbe's electrolytic method?
    (A) Methane (B) Ethane (C) Butane (D) Ethene
74. Which of the following does not react with Grignard's reagent to form alkane?
    (A) Ethyne (B) Ethanal (C) Ethanoic acid (D) Ethanol
75. Which of the following compounds on electrolysis of its concentrated solution gives ethane?
    (A) Ethyl ethanoate (B) Potassium ethanoate
    (C) Acetamide (D) Ethanoic acid
76. In the reaction, $CH_3COONa + NaOH(CaO) \longrightarrow CH_4$, the carboxylate group is eliminated as
    (A) $CaCO_3$ (B) Na_2CO_3 (C) CO (D) CO_2
77. What happens to the *p*H of electrolyte during Kolbe's electrolytic method of preparation of alkane?
    (A) Remains constant throughout the reaction
    (B) Decreases progressively as the reaction proceeds
    (C) Increases progressively as the reaction proceeds
    (D) May decrease if the concentration of the electrolyte is not very high.
78. In the Friedel–Crafts alkylation, the function of anhydrous $AlCl_3$ is to
    (A) produce nucleophile (B) produce electrophile
    (C) absorb HCl (D) absorb water.
79. In the Friedel–Crafts alkylation of benzene with bromomethane, the electrophile is
    (A) $AlCl_3^-$ (B) $\overset{+}{C}H_3CO$ (C) $C_6H_5^+$ (D) $\overset{+}{C}H_3$
80. Bromomethane on heating with zinc in a close tube gives
    (A) methanol (B) ethyne (C) ethane (D) methane

81. In the Friedel–Crafts alkylation, besides $AlCl_3$, the other reactants are
(A) $C_6H_6 + NH_3$ (B) $C_6H_6 + CH_3Cl$ (C) $C_6H_6 + CH_4$ (D) $C_6H_6 + CH_3COCl$

82. In the formation of cyclohexane from hex-1-ene, the entropy of cyclohexane is
(A) equal to that of hex-1-ene (B) less than that of hex-1-ene
(C) more than that of hex-1-ene
(D) less or more than that of hex-1-ene depending upon the condition of reaction.

83. In the reaction, $CH_3CH_2CCl_2CH_3 \xrightarrow{X} CH_3C{\equiv}CCH_3$, X is
(A) Na (B) HCl/H_2O (C) Zn (D) alcoholic KOH

84. Alkanes and alkenes both are obtained by
(A) Williamson synthesis (B) Kolbe's electrolytic method
(C) Frankland reaction (D) Wurtz reaction.

85. Heating of *o*-$O_2NC_6H_4I$ with Cu yields
(A) benzene (B) iodobenzene
(C) *o*-nitrobenzene (D) 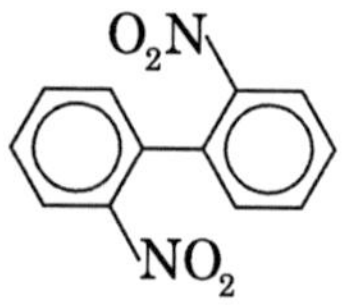

86. Aluminium carbide reacts with water to form
(A) ethane (B) methane (C) ethene (D) ethyne

87. The conversion of haloalkane into alkane by Wurtz reaction occurs in the presence of
(A) Zn in dry ether (B) Zn in ethanol
(C) Na in dry ether (D) Na in ethanol

88. Which of the following is formed as the major product when $CH_3CH_2CHBrCH_3$ is reacted with alcoholic KOH?
(A) But-1-yne (B) Butane (C) But-2-ene (D) But-1-ene

89. Identify the product Y in the sequence of reactions, $CH_3CH_2CH_2OH \xrightarrow{PCl_5} X \xrightarrow{\text{alc. KOH}} Y$
(A) Propanol (B) Propane (C) Propene (D) Propyne

90. The conversion of 2,3-dibromobutane to but-2-ene with Zn is
(A) redox reaction (B) α-elimination
(C) β-elimination (D) both α-elimination and redox reaction.

91. Identify the product formed when benzene vapour is passed through a red hot tube at 600–800°C.
(A) Cyclohexane (B) Cyclobutadiene
(C) $C_6H_5{-}C_6H_5$ (D) C_2H_4 and C_3H_6

92. $(CH_3)_2CHBr \xrightarrow[\text{(iii) } (CH_3)_2CHCH_2Br]{\text{(i) Li (ii) Cu}} X$
This is Corey–House method of synthesis of X. In this reaction, X is
(A) $(CH_3)_2CHCH_2CH_2CH_3$ (B) $(CH_3)_2CHCH_2CH(CH_3)_2$
(C) $(CH_3)_2CHCH_2CH_2CH_2CH_3$ (D) none of these.

93. In the reaction, 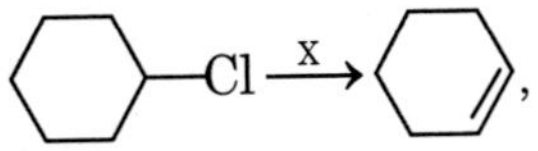, X is

(A) alcoholic KOH (B) NH_3
(C) Et_3N (D) all of these.

94. In the reaction, 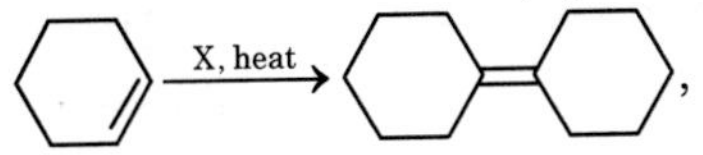, X is

(A) *t*-BuOK (B) Et_3N (C) alcoholic KOH (D) conc. H_2SO_4

95. Identify the correct statement.
(A) The treatment of but-2-yne with one mole of H_2/Pt gives *cis*-but-2-ene
(B) The treatment of 1,1-dibromopentane with $NaNH_2$ followed by acid produces pent-2-yne
(C) The treatment of 1,1-dibromopentane with solid KOH at 200°C gives pent-1-yne
(D) The treatment of but-2-yne with Na/C_2H_5OH gives *cis*-but-2-ene.

Hint: The statement (A) is correct and all others are incorrect.

(A) $H_3C-C\equiv C-CH_3 \xrightarrow{H_2/Pt}$ 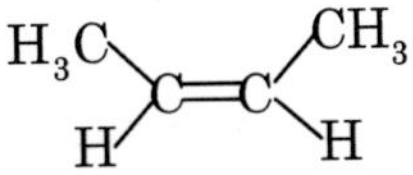

(B) Pent-1-yne is formed. The strong base $NaNH_2$ removes the terminal proton from 1,1-dibromopentane and the dehydrohalogenation product cannot rearrange.
$CH_3CH_2CH_2CH_2CHBr_2 \xrightarrow[-2HBr]{NaNH_2} CH_3CH_2CH_2C\equiv C^- \xrightarrow{H_3O^+} CH_3CH_2CH_2C\equiv CH$

(C) The initially formed pent-1-yne is converted into pent-2-yne because the more highly substituted alkyne is more stable.
$CH_3CH_2CH_2CH_2CHBr_2 \xrightarrow[-2HBr]{KOH(s),\ 200°C} CH_3CH_2CH_2C\equiv CHO \xrightarrow{\text{rearrangement}}$
$CH_3CH_2C\equiv CCH_3$

(D) $H_3C-C\equiv C-CH_3 \xrightarrow[{[H]}]{Na/C_2H_5OH}$ 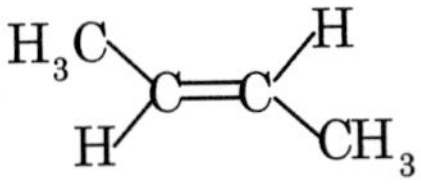

96. 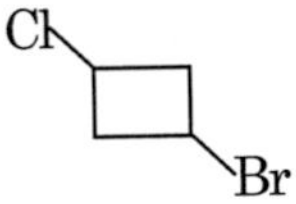 $\xrightarrow{Mg/ether} X \xrightarrow{D_2O} Y \xrightarrow{Na/ether} Z$. Z is

(A) 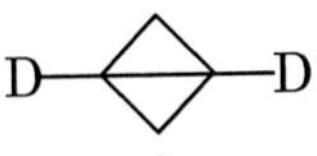 (B) 
(C) 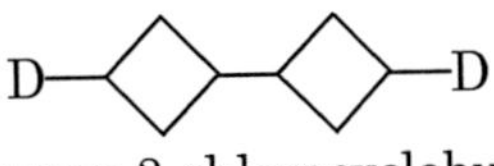 (D) 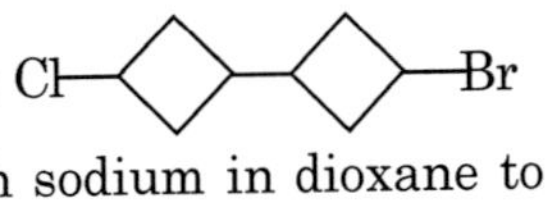

97. 1-Bromo-3-chlorocyclobutane reacts with sodium in dioxane to give

(A)  (B) 
(C)  (D) 

Hint: 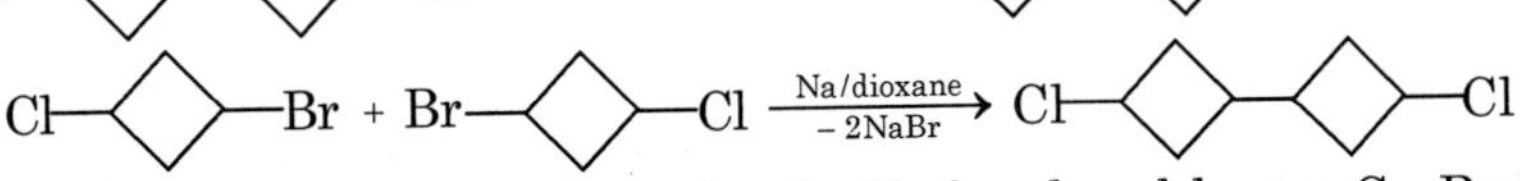

The C—Cl bond is stronger than the C—Br bond and hence C—Br bond is broken.

98. The reagent(s) for the conversion, $Br—CH_2—CH_2—Br \xrightarrow{?} H—C\equiv C—H$ is/are

(A) alcoholic KOH
(B) alcoholic KOH followed by $NaNH_2$
(C) aqueous KOH followed by $NaNH_2$
(D) Zn/CH_3OH **(I.I.T. 2007)**

Hint: $Br—CH_2—CH_2—Br \xrightarrow{\text{alc. KOH}} H_2C=CH—Br \xrightarrow{NaNH_2} HC\equiv CH$

Since $CH_2=CH—Br$ has partial C—Br double bond character, it requires a more stronger base to remove HBr.

99. In the following carbocation, H/CH_3 that is most likely to migrate to the positively charged carbon is

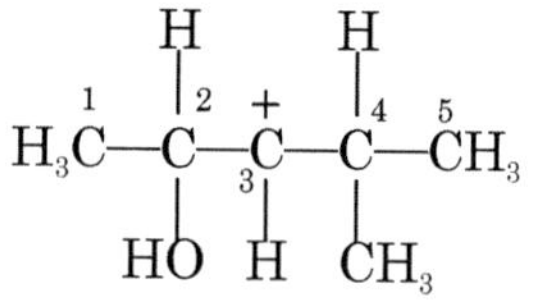

(A) CH_3 at C–4
(B) H at C–4
(C) CH_3 at C–2
(D) H at C–2 **(I.I.T. 2009)**

Hint: The migrating tendency of hydride is greater than that of alkyl group due to its smaller size. The migration of hydride from second carbon gives more stable carbocation (stabilized by + R effect of OH group and + I effect of methyl group).

100. The synthesis of 3-octyne is achieved by adding a bromoalkane into a mixture of sodium amide and alkyne. The bromoalkane and alkyne respectively are

(A) $BrCH_2CH_2CH_2CH_2CH_3$ and $CH_3CH_2C\equiv CH$
(B) $BrCH_2CH_2CH_3$ and $CH_3CH_2CH_2C\equiv CH$
(C) $BrCH_2CH_2CH_2CH_2CH_3$ and $CH_3C\equiv CH$
(D) $BrCH_2CH_2CH_2CH_3$ and $CH_3CH_2C\equiv CH$ **(I.I.T. 2010)**

Hint: $CH_3—CH_2—C\equiv CH \xrightarrow{NaNH_2} CH_3—CH_2—C\equiv C^-Na^+$

$CH_3—CH_2—CH_2—CH_2\boxed{Br + Na^+}{}^-C\equiv C—CH_2—CH_3 \longrightarrow$

$CH_3—CH_2—CH_2—CH_2—C\equiv C—CH_2—CH_3$

3-octyne

ANSWERS

1. A	2. A	3. B	4. A	5. B	6. A
7. B	8. C	9. C	10. B	11. C	12. A
13. B	14. A	15. D	16. D	17. C	18. A
19. B	20. A	21. A	22. C	23. B	24. A
25. A	26. B, C	27. C	28. D	29. A	30. D
31. D	32. B, C, D	33. D	34. A	35. D	36. C
37. B	38. A	39. D	40. D	41. A	42. B

43. D	44. B	45. A	46. A	47. D	48. C
49. D	50. D	51. B	52. B	53. D	54. B
55. A	56. B	57. C	58. D	59. B	60. D
61. B	62. B	63. C	64. C	65. B, C	66. B, C, D
67. C	68. A	69. B, C, D	70. D	71. A	72. B
73. A	74. B	75. B	76. B	77. C	78. B
79. D	80. C	81. B	82. B	83. D	84. B
85. D	86. B	87. C	88. C	89. B	90. A
91. C	92. B	93. D	94. D	95. A	96. C
97. D	98 B	99. D	100. D		

36

Hydrocarbons—II: Properties and Reactions

OBJECTIVE TYPE MULTIPLE CHOICE QUESTIONS

1. The reaction of propene with HOCl proceeds via the addition of
   (A) H^+ in the first step (B) Cl^+ in the first step
   (C) Cl^+ and OH^- in a single step (D) OH^- in the first step. **(I.I.T. 2001)**
2. In the presence of peroxide, HCl and HI do not give anti-Markownikoff addition to alkenes because
   (A) both are highly ionic
   (B) one of the steps is endothermic in both the cases
   (C) all the steps are exothermic in both the cases
   (D) one is oxidising and the other is reducing. **(I.I.T. 2001)**
3. Hydrogenation of the compound 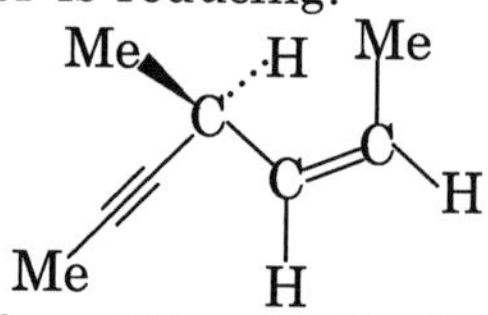 in presence of Pd catalyst gives
   (A) an optically active compound (B) an optically inactive compound
   (C) a diastereoisomeric mixture (D) a racemic mixture. **(I.I.T. 2001)**
4. An S_{N^2} reaction at an asymmetric carbon of a compound always gives
   (A) an enantiomer of the substrate (B) a mixture of diastereoisomers
   (C) a single stereoisomer (D) a product with opposite optical rotation. **(I.I.T. 2001)**
5. Assertion (*A*): Addition of bromine to *trans*-but-2-ene yields *meso*-2,3-bromobutane.
   Reason (*R*): Bromine addition to an alkene is an electrophilic addition.
   (A) Both *A* and *R* are correct and *R* is the correct explanation of *A*

(B) Both *A* and *R* are correct and *R* is not the correct explanation of *A*
(C) *A* is true but *R* is false (D) *A* is false but *R* is true. **(I.I.T. 2001)**

6. Assertion (*A*): But-1-ene on reaction with HBr in the presence of peroxide produces 1-bromobutane.
Reason (*R*): It involves the formation of a primary radical.
(A) Both *A* and *R* are correct and *R* is the correct explanation of *A*
(B) Both *A* and *R* are correct and *R* is not the correct explanation of *A*
(C) *A* is correct but *R* is incorrect (D) *A* is incorrect but *R* is correct. **(I.I.T. 2000)**

7. Which of the following alkenes will react fastest with H_2 under catalytic hydrogenation condition?
(A) 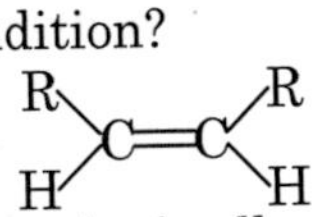 (B) 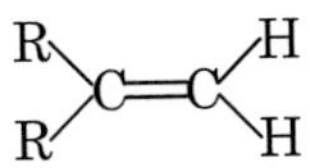 (C) 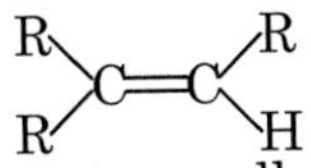 (D) 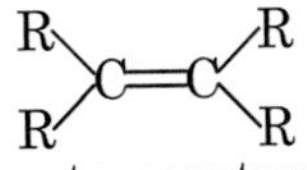 **(I.I.T. 2000)**
Hint: A *cis*-alkene is more reactive than a *trans*-alkene due to greater strain at the double bond and least steric hindrance. The smaller the number of R substituents, lesser is the steric hindrance and faster is the rate of hydrogenation.

8. Which of the following has the highest nucleophilicity?
(A) CH_3^- (B) NH_2^- (C) OH^- (D) F^- **(I.I.T. 2000)**
Hint: The increasing order of basic strength (nucleophilicity) is:
$F^- < OH^- < NH_2^- < CH_3^-$
Hence CH_3^- has the highest nucleophilicity.

9. Toluene when treated with Br_2/Fe, gives *p*-bromotoluene as the major product because the CH_3 group
(A) is *para* directing (B) is *meta* directing
(C) activates the ring (D) deactivates the ring. **(I.I.T. 1999)**

10. The products obtained via oxomercuration ($HgSO_4 + H_2SO_4$) of but-1-yne would be
(A) $CH_3CH_2CH_2CHO$ (B) $CH_3CH_2COCH_3$
(C) $CH_3CH_2CHO + HCHO$ (D) $CH_3CH_2COOH + HCOOH$ **(I.I.T. 1999)**

11. Which of the following compounds is most acidic?
(A) CH_3Cl (B) CH_3OH (C) $H_2C{=}CH_2$ (D) $HC{\equiv}CH$ **(I.I.T. 1998)**

12. Benzyl chloride ($C_6H_5CH_2Cl$) can be prepared from toluene by chlorination with
(A) SO_2Cl_2 (B) $SOCl_2$ (C) Cl_2 (D) NaOCl **(I.I.T. 1998)**

13. The reaction of 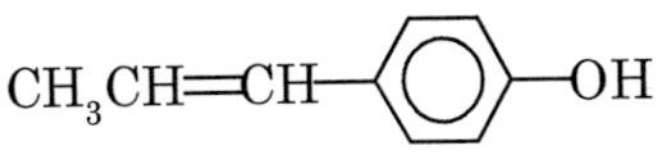 with HBr gives

(A) 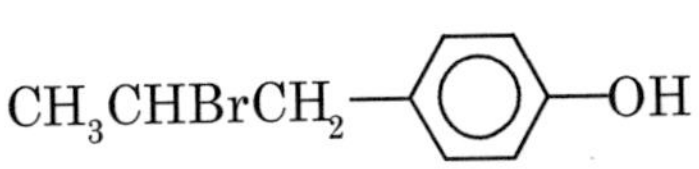 (B) 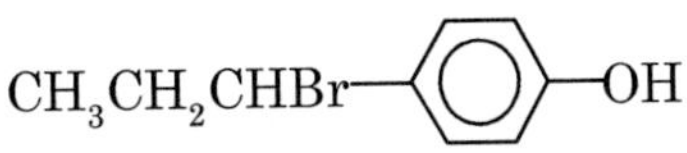

(C) 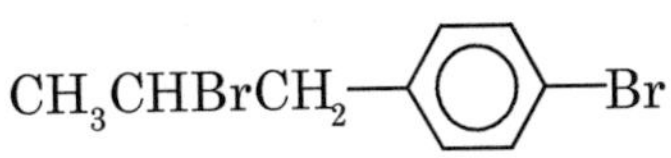 (D) 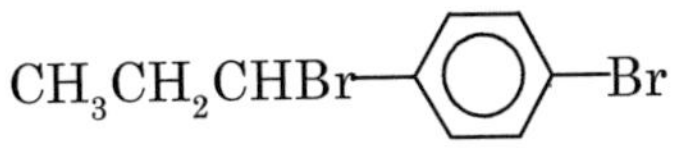 **(I.I.T. 1998)**

14. When cyclohexane is poured in water, it floats because
(A) cyclohexane is less denser than water

(B) cyclohexane is in chair form
(C) cyclohexane is in crown form
(D) cyclohexane is in boat form. **(I.I.T. 1997)**

15. The intermediate during the addition of HCl to propene in presence of peroxide is
(A) $CH_3\dot{C}HCH_2Cl$ (B) $CH_3\overset{+}{C}HCH_3$ (C) $CH_3\dot{C}H_2CH_2$ (D) $CH_3CH_2\overset{+}{C}H_2$ **(I.I.T. 1997)**

16. In the groups —OAc (I), —OMe (II), —OSO_2Me (III) and —OSO_2CF_3 (IV), the order of leaving group ability is
(A) I > II > III > IV (B) IV > III > I > II
(C) III > II > I > IV (D) II > III > IV > I **(I.I.T. 1997)**

17. What is the chief product when *n*-butane is treated with Br_2 in the presence of light at 130°C?
(A) $CH_3CH_2CH_2CH_2Br$ (B) $CH_3CH_2CHBrCH_3$
(C) $(CH_3)_2CHCH_2Br$ (D) $(CH_3)_3CBr$ **(I.I.T. 1995)**

18. The prominent product formed when 3-methylpent-2-ene reacts with HOCl is
(A) $CH_3CH_2\underset{CH_3}{\underset{|}{C}}Cl\underset{OH}{\underset{|}{C}}HCH_3$ (B) $(CH_3)_3C\underset{OH}{\underset{|}{C}}HCH_3$
(C) $CH_3CH_2\underset{CH_3}{\underset{|}{C}}Cl\underset{Cl}{\underset{|}{C}}HCH_3$ (D) $CH_3CH_2\underset{CH_3}{\underset{|}{C}}(OH)\underset{Cl}{\underset{|}{C}}HCH_3$ **(I.I.T. 1995)**

19. Propyne and propene can be distinguished by
(A) $AgNO_3$ in NH_4OH (B) dil. $KMnO_4$
(C) conc. H_2SO_4 (D) Br_2 in CCl_4 **(I.I.T. 2000)**

20. The major product of nitration of benzoic acid is
(A) 3-nitrobenzoic acid (B) 4-nitrobenzoic acid
(C) 2-nitrobenzoic acid (D) 2,4-dinitrobenzoic acid **(I.I.T. 1993)**

21. In the addition of HBr to propene in the absence of peroxide, the first step involves the addition of
(A) H^+ (B) Br^- (C) $\cdot H$ (D) $\cdot Br$ **(I.I.T. 1993)**

22. Which is the decreasing order of strength of the bases OH^-, NH_2^-, $H—C{\equiv}C^-$ and $CH_3—CH_2^-$?
(A) $CH_3—CH_2^- > NH_2^- > H—C{\equiv}C^- > OH^-$
(B) $H—C{\equiv}C^- > CH_3—CH_2^- > NH_2^- > OH^-$
(C) $OH^- > NH_2^- > H—C{\equiv}C^- > CH_3—CH_2^-$
(D) $NH_2^- > H—C{\equiv}C^- > OH^- > CH_3—CH_2^-$ **(I.I.T. 1993)**

23. Which is the decreasing order of stability?
(I) $CH_3\overset{+}{C}HCH_3$ (II) $CH_3\overset{+}{C}HOCH_3$ (III) $CH_3\overset{+}{C}HCOCH_3$
(A) I < II < III (B) I > II > III (C) III > II > I (D) II > III > I **(I.I.T. 1993)**

24. The homolytic fission of C—C bond in ethane gives an intermediate in which C is
(A) *sp* hybridised (B) sp^2 hybridised
(C) sp^3 hybridised (D) sp^2d hybridised. **(I.I.T. 1992)**

25. Which behaves both as nucleophile and electrophile?
(A) CH_3NH_2 (B) CH_3Cl (C) CH_3CN (D) CH_3OH **(I.I.T. 1991)**

26. The compound which gives most stable carbocation on dehydration is
(A) $(CH_3)_2CH—CH_2OH$ (B) $(CH_3)_3COH$
(C) $CH_3—CH_2—CH_2—CH_2OH$ (D) $CH_3—CH(OH)—CH_2—CH_3$ **(I.I.T. 1989)**

27. The increasing order of reactivity towards sulphonation with fuming H_2SO_4 for benzene, toluene, methoxybenzene and chlorobenzene is
(A) chlorobenzene < benzene < toluene < methoxybenzene
(B) benzene < chlorobenzene < toluene < methoxybenzene
(C) methoxybenzene < chlorobenzene < benzene < toluene
(D) chlorobenzene < methoxybenzene < benzene < toluene **(I.I.T. 1988)**

28. The increasing order of boiling point in *n*-butane, *n*-butanol, *n*-butyl chloride and isobutane is
(A) isobutane < *n*-butane < *n*-butyl chloride < *n*-butanol
(B) *n*-butane < isobutane < *n*-butyl chloride < *n*-butanol
(C) *n*-butyl chloride < isobutane < *n*-butane < *n*-butanol
(D) *n*-butyl chloride < *n*-butane < isobutane < *n*-butanol **(I.I.T. 1988)**

29. The highest boiling point is expected for
(A) isooctane (B) *n*-octane
(C) 2,2,3,3-tetramethylpropane (D) 2-methylbutane **(I.I.T. 1986)**

30. The reaction of toluene with Cl_2 in the presence of $FeCl_3$ gives predominantly
(A) benzoyl chloride (B) *m*-chlorotoluene
(C) *o*- and *p*-chlorotoluene (D) benzyl chloride **(I.I.T. 1986)**

31. The reaction conditions leading to the best yield of C_2H_5Cl are
(A) C_2H_6 (excess) + $Cl_2 \xrightarrow{\text{UV light}}$ (B) $C_2H_6 + Cl_2 \xrightarrow{\text{dark, room temp.}}$
(C) $C_2H_6 + Cl_2$ (excess) $\xrightarrow{\text{UV light}}$ (D) $C_2H_6 + Cl_2 \xrightarrow{\text{UV light}}$ **(I.I.T. 1986)**

32. Acidic hydrogen is present in
(A) ethyne (B) ethene (C) benzene (D) ethane **(I.I.T. 1985)**

33. Anti-Markownikoff addition of HBr is not observed in
(A) propene (B) but-1-ene (C) but-2-ene (D) pent-2-ene **(I.I.T. 1985)**

34. Which of the following compounds will not react with KCN?
(A) Ethyl chloride (B) Acetyl chloride
(C) Chlorobenzene (D) Benzaldehyde **(I.I.T. 1984)**

35. In the nitration of benzene the active species involved is
(A) nitrate ion (B) nitrite ion (C) nitronium ion (D) nitrogen peroxide
(I.I.T. 1983)

36. Which of the following compounds does not dissolve in conc. H_2SO_4 on warming?
(A) Ethene (B) Benzene (C) Hexane (D) Aniline **(I.I.T. 1983)**

37. The compound with the highest boiling point is
(A) *n*-hexane (B) *n*-pentane (C) 2,2-dimethylpropane
(D) 2-methylbutane **(I.I.T. 1982)**

38. The compound that is most reactive towards electrophilic nitration is
(A) toluene (B) benzene (C) benzoic acid (D) nitrobenzene **(I.I.T. 1985)**

39. The halogen which is most reactive in halogenation of alkanes in sunlight is
(A) Cl_2 (B) Br_2 (C) I_2 (D) none of these. **(I.I.T. 1981)**

40. Among the given cations, the most stable carbonium ion is
(A) *sec*-butyl (B) *t*-butyl (C) *n*-butyl (D) none of these. **(I.I.T. 1981)**

41. Which of the following decolourises alkaline $KMnO_4$ solution?
(A) C_3H_8 (B) C_2H_4 (C) CH_4 (D) CCl_4 **(I.I.T. 1980)**

42. Dilute $KMnO_4$ at room temperature reacts with R—CH═CH—R to give
(A) R—CHO (B) R—COOH
(C) RCH(OH)CH(OH)R (D) $CO_2 + H_2O$ **(Roorkee 1992)**

43. Aqueous H_2SO_4 reacts with 2-methylbut-2-ene to give predominantly
(A) isobutyl hydrogen sulphate (B) 2-methylbutan-2-ol
(C) *sec*-butyl hydrogen sulphate (D) 2-methylbutan-1-ol **(Roorkee 1992)**

44. The product when but-1-ene is reacted with HBr in the presence of peroxide is
(A) 1-bromobutane (B) 2-bromobutane
(C) 1,1-dibromobutane (D) 1,2-dibromobutane **(Roorkee 1990)**

45. When acetylene is passed through dil. H_2SO_4 in the presence of $HgSO_4$, the product is
(A) acetaldehyde (B) acetone (C) acetic acid (D) C_2H_5OH **(Roorkee 1992)**

46. Arrange (I) $C_6H_5CH_3$, (II) C_6H_6, (III) $C_6H_5OCH_3$ and (IV) $C_6H_5CF_3$ in the order of electrophilic substitution reaction
(A) I > II > III > IV (B) IV > I > II > III
(C) III > IV > II > I (D) II > IV > III > I

47. Isomerisation in alkane is brought about by the reagent
(A) conc. H_2SO_4 (B) $AlCl_3$ + HCl (C) Fe_2O_3 (D) Al_2O_3

48. Alkanes undergo iodination in the presence of
(A) HIO_3 (B) HI (C) PI_3 (D) I_2 + P

49. The photobromination of 2-methylbutane gives predominantly
(A) 2-bromo-2-methylbutane (B) 2-bromo-3-methylbutane
(C) 1-bromo-1-methylbutane (D) 1-bromo-2-methylbutane

50. The chlorination of alkanes involves
(A) $\cdot CH_4$ free radicals (B) $\cdot Cl$ free radicals
(C) Cl^+ ion (D) Cl^- ion

51. Which of the following has the smallest heat of hydrogenation per mole?
(A) But-1-ene (B) *trans*-but-2-ene
(C) *cis*-but-2-ene (D) Buta-1,3-diene **(I.I.T. 1993)**

52. The molecules that will have dipole moment are
    (A) 2,2-dimethylpropane (B) *trans*-pent-2-ene
    (C) 2,2,3,3-tetramethylbutane (D) *cis*-hexene **(I.I.T. 1992)**
53. Which of the following will have zero dipole moment?
    (A) 1,1-Dichloroethene (B) *cis*-1,2-dichloroethene
    (C) *trans*-1,2-dichloroethene (D) none of these. **(I.I.T. 1987)**
54. Baeyer's reagent is
    (A) alkaline $KMnO_4$ solution (B) acidified $KMnO_4$ solution
    (C) neutral $KMnO_4$ solution (D) aqueous Br_2 solution. **(I.I.T. 1984)**
55. The peroxide effect involves
    (A) homolytic fission of double bond of alkene
    (B) heterolytic fission of double bond of alkene
    (C) free radical mechanism (D) ionic mechanism.
56. But-1-yne and but-2-yne are distinguished by
    (A) Cl_2 gas (B) alkaline $KMnO_4$
    (C) bromine water (D) Tollen reagent.
57. In order to get *trans* $H_2CCH{=}CHCH_3$ from $H_3CC{\equiv}CCH_3$, the reagents used are
    (A) Na in liquid NH_3 (B) H_2/Pd–C, $BaSO_4$
    (C) H_2/Pd–C, $Pd(OAc)_2$ (D) $LiAlH_4$ in H_2O
58. Propan-1-ol is obtained from propene by treating with
    (A) $B_2H_6 + H_2O_2/OH^-$ (B) $LiAlH_4 + H_2O^+$
    (C) $HCl + H_2O/H^+$ (D) B_2H_6 + NaOH
59. A mixture of ethyne and HCN in the ratio 2 : 1 is heated in a red hot tube. The product is
    (A) pyrrole (B) benzene (C) pyridine (D) thiophene
60. Which of the following is an electrophilic substitution reaction?
    (A) Kolbe's electrolytic synthesis (B) Dehydrohalogenation
    (C) Wurtz reaction (D) Friedel–Crafts reaction.
61. Chlorination of toluene in the presence of light and heat followed by treatment with aq. NaOH gives
    (A) *o*-cresol (B) *p*-cresol
    (C) 2,4-dihydroxytoluene (D) benzyl alcohol
62. Consider the reaction: 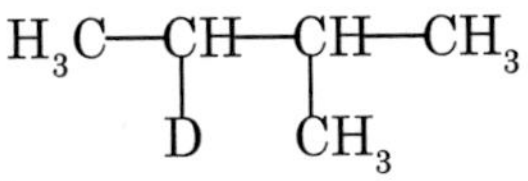 + Br· → X + HBr. Identify the structure of the major product X.

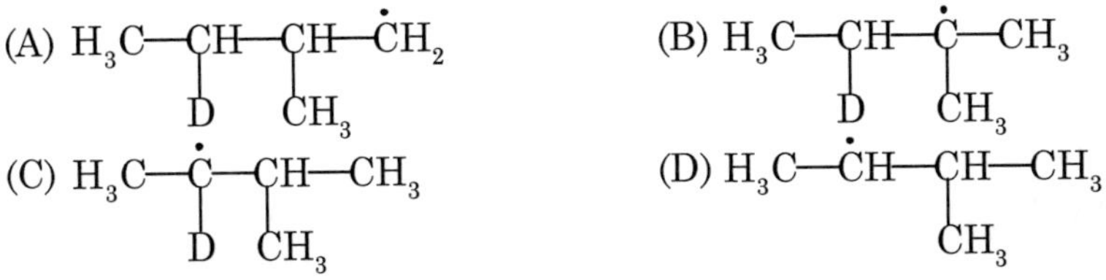

(I.I.T. 2002)

Hint: The tertiary free radical 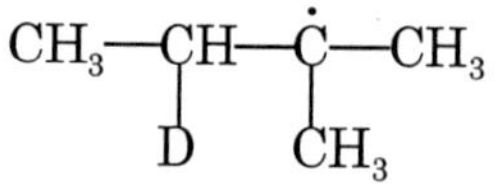 is more stable among all other free radicals because the order of stability of free radicals is: primary < secondary < tertiary.

63. Identify the correct order of reactivity in electrophilic substitution reactions of the following compounds.

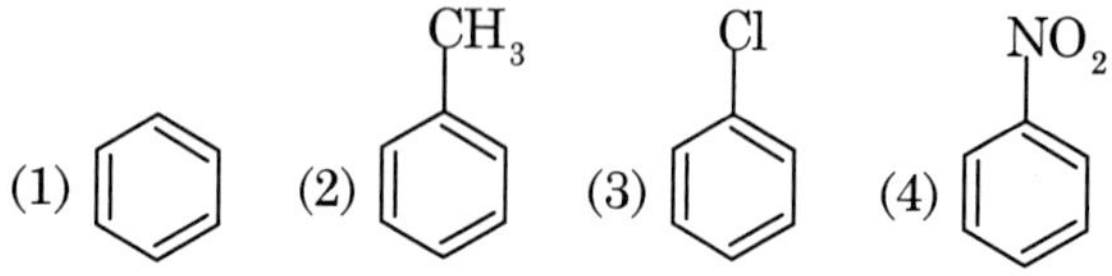

(A) 1 > 2 > 3 > 4 (B) 4 > 3 > 2 > 1 (C) 2 > 1 > 3 > 4 (D) 2 > 3 > 1 > 4

(I.I.T. 2002)

Hint: The $—NO_2$ group is an electron-withdrawing group and hence it deactivates the benzene ring: The —Cl group also deactivates the benzene ring but its deactivation of the benzene ring is less than that of the $—NO_2$ group. The $—CH_3$ group is an electron-releasing group and hence it activates the benzene ring. Thus on the basis of activation and deactivation of the benzene ring, the order of reactivity in electrophilic substitution reaction is: 2 > 1 > 3 > 4.

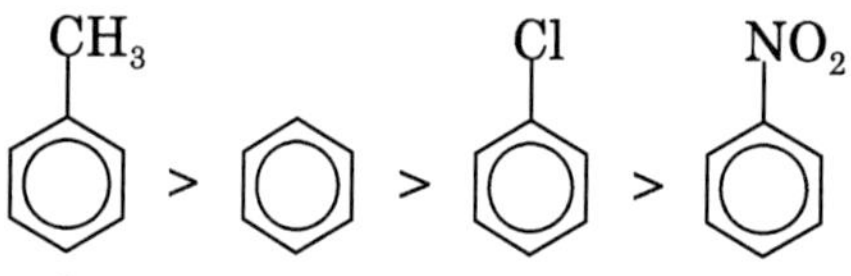

2 1 3 4

64. Identify the reagent from the following list which can easily distinguish between but-1-yne and but-2-yne.

(A) bromine, CCl_4 (B) H_2, Lindlar catalyst

(C) dil. H_2SO_4, $HgSO_4$ (D) ammoniacal Cu_2Cl_2 solution. **(I.I.T. 2002)**

Hint: Acidic hydrogen is present in but-1-ene and is absent in but-2-ene. Hence but-1-yne gives a red precipitate with ammoniacal Cu_2Cl_2 solution while but-2-yne does not form such a precipitate.

$$\underset{\text{But-1-yne}}{2CH_3—CH_2≡CH} + Cu_2Cl_2 + 2NH_4OH \rightarrow \underset{\text{Red}}{2CH_3—CH_2—C≡CCu\downarrow} + 2NH_4Cl + 2H_2O$$

$$CH_3—C≡C—CH_3 + Cu_2Cl_2 + 2NH_4OH \rightarrow \text{No reaction}$$

65. Chlorination of *n*-butane gives

(A) 2-chlorobutane as the major product and 1-chlorobutane as the minor product

(B) 1-chlorobutane as the major product and 2-chlorobutane as the minor product

(C) 2-chlorobutane as the major product

(D) 1-chlorobutane as the major product.

66. Rate of bromination of an alkane in comparison to rate of chlorination of the alkane is

(A) faster (B) slower (C) equal
(D) equal or different depending on the temperature.

67. Which of the following has the minimum boiling point?
(A) 2,2-Dimethylpropane (B) 2-Methylbutane
(C) *n*-Pentane (D) *n*-Butane

68. The reaction, $CH_3 + X_2 \rightarrow CH_3X + X$ has the maximum negative value of ΔH^0 when X_2 is
(A) I_2 (B) Br_2 (C) Cl_2 (D) F_2

69. Identify the incorrect statement.
(A) Octane number of a mixture containing 85% isooctane and 15% heptane is 15
(B) Octane number of isooctane is 100
(C) Octane number of *n*-heptane is zero
(D) Hydrogen atoms in cyclopropane is more acidic than propane

70. Identify the incorrect statement.
(A) Free radical chlorination of methane occurs faster than that of CD_4
(B) Bromination of alkanes occurs with great difficulty than chlorination
(C) Chlorination of $CH_3-\overset{\overset{\large Br}{|}}{C}H-CH_3$ gives $CH_3-\underset{\underset{\large Cl}{|}}{\overset{\overset{\large Br}{|}}{C}}-CH_3$
(D) The order of reactivity of hydrogen atom in alkane towards replacement by a halogen atom is: 1° < 2° < 3°.

71. Identify the incorrect statement.
(A) Reactivity of H atom in alkane towards replacement by a halogen atom follows the order: primary < secondary < tertiary
(B) Vinyl free radical is more stable than secondary free radical
(C) The more stable the free radical, the more easily it is formed
(D) The order of reactivity of halogen towards alkanes is: $Br_2 < Cl_2 < F_2$.

72. Percentage of 1-chloropropane obtained in the chlorination of propane is
(A) 44 (B) 56 (C) 36 (D) 64
Hint: 1-chloropropane/2-chloropropane = No. of 1° H × Reactivity of 1° H/No. of 2° H × reactivity of 2° H = 6 × 1/(2 × 3.8) = 6.0/7.6. Percentage of 1-chloropropane = [6.0/(6.0 + 7.6)] × 100 = 44.

73. Percentage of 1-chloro-2-methylpropane formed in the chlorination of 2-methylpropane is
(A) 64 (B) 56 (C) 44 (D) 36
Hint: 1° isomer/3° isomer = No. of 1° H × Reactivity of 1° H/No. of 3° H × Reactivity of 3° H = 9 × 1/(1 × 5.0) = 9/5. Percentage of 1° isomer = [9/(9 + 5)] × 100 = 64.3.

74. If the relative reactivities of tertiary, secondary and primary H atoms are 1600 : 82 : 1, the percentage of 1-bromopropane formed in the bromination of propane is
(A) 44 (B) 3.5 (C) 4 (D) 14

Hint: 1° isomer/2° isomer = No. of 1° H × Reactivity of 1° H/No. of 2° H × Reactivity of 2° H = 6 × 1/(2 × 82) = 6/164. Percentage of 1° isomer = [6/(6 + 164)] × 100 = 3.5.

75. Identify the false statement.
    (A) Alkanes burn in air with luminous flame
    (B) The more stable the free radical, the lesser the enthalpy of formation of the free radical from the corresponding alkane
    (C) The relative stability of free radicals is: CH_3 < primary < secondary < tertiary
    (D) Oxygen acts as an inhibitor in the photochemical reaction between methane and chlorine.
76. Identify the false statement.
    (A) Lower alkanes are soluble in water but higher ones are insoluble
    (B) Iodine does not react with methane
    (C) Chlorination of methane can be stopped at monochlorination stage if methane is taken in large excess in comparison to chlorine
    (D) Alkanes are less denser than water.
77. Identify the incorrect statement.
    (A) Methane does not react with chlorine in the dark
    (B) Fluorination of methane requires the use of light or heat
    (C) The relative rate of abstraction of primary, secondary and tertiary hydrogen atoms is minimum for primary H atoms
    (D) Free radical monochlorination of 2-bromo-2-methylpropane gives 1-bromo-2-chloro-2-methylpropane.
78. Identify the incorrect statement.
    (A) The reaction of methane and chlorine proceeds via ionic mechanism
    (B) The energy of activation of $\dot{C}l + CH_4 \longrightarrow \dot{C}H_3 + HCl$ is smaller than that of $\dot{B}r + CH_4 \longrightarrow \dot{C}H_3 + HBr$
    (C) The relative rate of abstraction of primary, secondary and tertiary H atoms is maximum for tertiary H atoms
    (D) Out of allyl, vinyl, primary, secondary, tertiary and benzyl free radicals, the least stable one is vinyl free radical.
79. Free radical monochlorination of butane gives
    (A) *rac*-2-chlorobutane (B) *meso*-2-chlorobutane
    (C) *S*-2-chlorobutane (D) *R*-2-chlorobutane
80. In the reaction, *R*-$CH_3CHIC_2H_5$ (α = – 15.90°) + $^{137}I \longrightarrow CH_3CHICH_3$, the percentage of molecules which undergo inversion is
    (A) 20% (B) 2% (C) 4% (D) 10%

    Hint: Fraction of molecules which have undergone inversion = x (x) (15.9°) + (1 – x) (– 15.9°) = – 15.26° or x = 0.02 = 2%.
81. Identify the correct statement
    (A) A $-I$ group destabilises a carbanion
    (B) A $-I$ group stabilises a carbocation

(C) A $+I$ group stabilises a carbanion

(D) A $+I$ group stabilises a carbocation.

82. Identify the correct order regarding the $-I$ effect of substituents.

(A) $—F < —OCH_3 > —NH_2$ (B) $—F > —OCH_3 > —N(CH_3)_2$

(C) $—F > —OCH_3 > —N(CH_3)_2$ (D) $—F < —OCH_3 < —N(CH_3)_2$

83. Identify the most stable carbocation.

(A) $(CH_3)_3\overset{+}{C}$ (B) $(CH_3)_2\overset{+}{C}H$ (C) $CH_3\overset{+}{C}H_2$ (D) $\overset{+}{C}H_3$

84. Identify the $+I$ group among the following:

(A) $—CH_3$ (B) $—COOH$ (C) $—OCH_3$ (D) $—OH$

85. Identify the species which is not a nucleophile.

(A) NH_3 (B) OH^- (C) CN^- (D) BF_3

86. Which of the following alkenes gives adipic acid on oxidative ozonolysis?

(A) Hex-3-ene (B) 1,2-Dimethylcyclobutene

(C) 1-Methylcyclopentene (D) Cyclohexene

87. Which of the following compounds is formed on addition of bromine to *trans*-but-2-ene?

(A) *meso*-2,3-dibromobutane (B) (–)-2,3-dimethylbutane

(C) (+)-2,3-dibromobutane (D) *rac*-2,3-dibromobutane

88. Which of the following alkenes gives propanoic acid and CO_2 on treatment with hot acidified $KMnO_4$?

(A) $(CH_3)_2C{=}CH_2$ (B) $CH_3CH{=}CHCH_3$

(C) $CH_3CH_2CH{=}CH_2$ (D) $CH_3CH{=}CH_2$

89. Which of the following alkenes gives two molecules of $OHCCH_2CHO$ on reductive ozonolysis?

(A) 1-Methyl-1,3-cyclopentadiene (B) 1,4-Cyclohexadiene

(C) 1,3-Cyclohexadiene (D) 2,4-Hexadiene

90. Propene and propyne are distinguished by

(A) ammoniacal $AgNO_3$ (B) dilute $KMnO_4$

(C) bromine in CCl_4 (D) conc. H_2SO_4

91. Which of the following is formed when propyne is passed through a hot iron tube at 400°C?

(A) Mesitylene (B) Xylene (C) Toluene (D) Benzene

92. Water is added to ethyne in the presence of

(A) H_2SO_4 and $HgSO_4$ (B) neutral medium

(C) alkaline medium (D) acidic medium

93. Which of the following order regarding the nitration is true?

(A) $k_{C_6H_6} = k_{C_6D_6} = k_{C_6T_6}$ (B) $k_{C_6T_6} > k_{C_6D_6} < k_{C_6H_6}$

(C) $k_{C_6H_6} < k_{C_6D_6} < k_{C_6T_6}$ (D) $k_{C_6T_6} < k_{C_6D_6} < k_{C_6H_6}$

94. The smallest aromatic species is

(A) benzene (B) cyclopentadiene

(C) cyclobutadiene (D) cyclopropenyl cation.

95. Toluene reacts with chloromethane in the presence of anhydrous $AlCl_3$ to give mainly *m*-xylene. This is because
    (A) CH_3 group in *m*-directing (B) CH_3 group has + *I* effect
    (C) of hyperconjugation effect of the CH_3 group
    (D) *m*-xylene is thermodynamically more stable.

96. In the reaction, $CH{\equiv}CH \xrightarrow{X} Cl_2CHCHCl_2$, X is
    (A) NaCl (B) chlorine water
    (C) Cl_2/CCl_4 (D) NaOCl

97. Which of the following is formed when $CH_3CH{=}CH_2$ is reacted with B_2H_6 followed by oxidation with alkaline H_2O_2?
    (A) CH_3CH_2CHO (B) $CH_3CH_2CH_2OH$
    (C) CH_3COCH_3 (D) $CH_3CH(OH)CH_2OH$

98. In the reaction, $C_6H_5CH_2CH_3 \xrightarrow{h\nu, Cl_2\ 273\,K}$ P, the product P is
    (A) $C_6H_5CHClCH_2Cl$ (B) $C_6H_5CCl_2CH_3$
    (C) $C_6H_5CH_2CH_2Cl$ (D) $C_6H_5CHClCH_3$

99. The product of acid hydrolysis of P and Q can be distinguished by

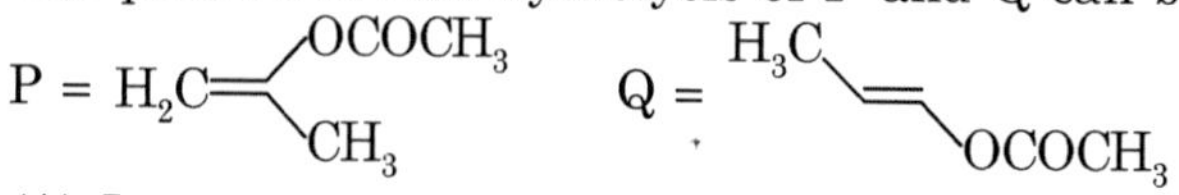

    (A) Lucas reagent (B) 2,4–DNP
    (C) Fehling's solution (D) $NaHSO_3$ **(I.I.T. 2003)**

Hint:

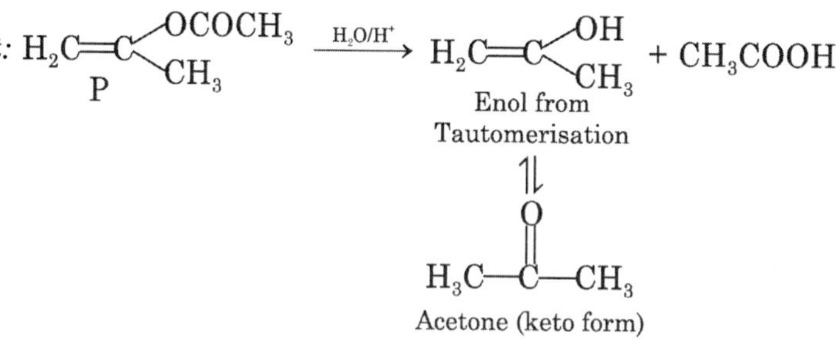

Acetone does not reduce Fehling solution.

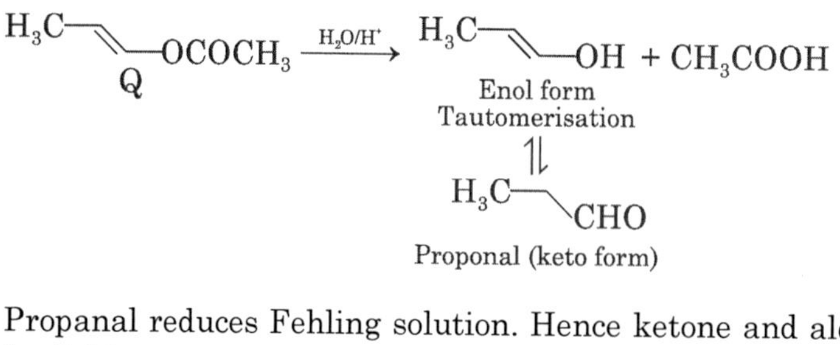

Propanal reduces Fehling solution. Hence ketone and aldehyde can be distinguished by Fehling solution.

100. In the reaction, $Ph—C{\equiv}C—CH_3 \xrightarrow{Hg^{2+}/H^+}$ A, A is

(A) 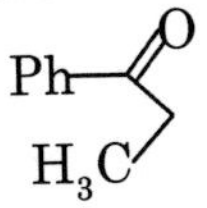 (B) 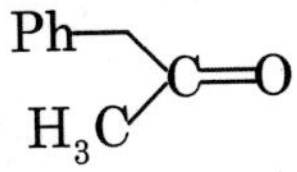 (C) 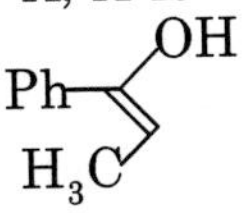 (D) 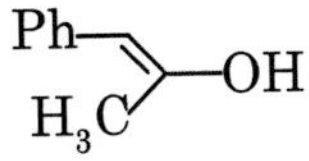

(I.I.T. 2003)

Hint:

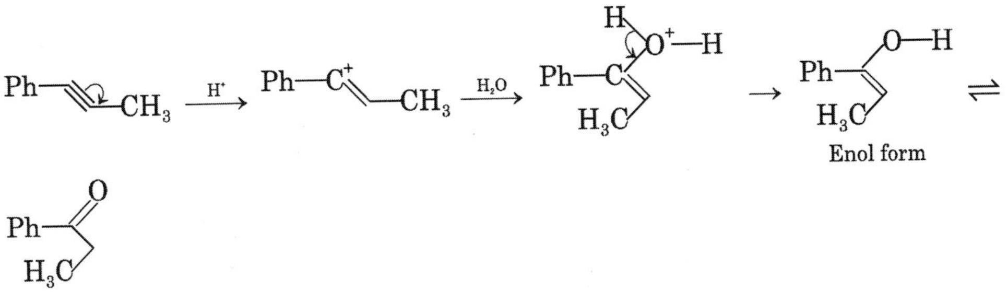

101. Hex-2-yne gives *trans*-hex-2-ene on treatment with

(A) Li/NH_3 (B) $Pd/BaSO_4$ (C) $LiAlH_4$ (D) Pt/H_2 **(I.I.T. 2004)**

Hint: Li/NH_3 brings about *trans* addition of H_2.

102. 1-Bromo-3-chlorocyclobutane when treated with two equivalents of Na in the presence of ether, which of the following will be formed?

(A) 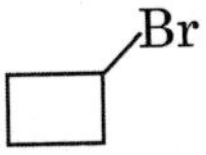 (B) 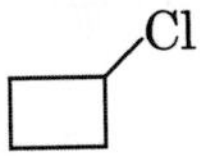 (C) 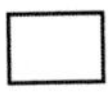 (D) 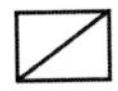 **(I.I.T. 2005)**

Hint: This is an intramolecular Wurtz reaction.

103. The addition of bromine dissolved in methanol to cyclohexene gives
(A) only dibromocyclohexane
(B) a mixture of *cis*- and *trans*-dibromocyclohexane
(C) a mixture of *trans*-dibromocyclohexane and *trans*-1-bromo-2-methoxycyclohexane
(D) A mixture of *cis*-1-bromo-2-methoxycyclohexane and *trans*-1-bromo-2-methoxycyclohexane.

104. The addition of bromine to cyclohexane in the presence of NaCl gives
(A) *cis*-1,2-dibromocyclohexane and *cis*-1-bromo-2-chlorocyclohexane
(B) *trans*-1,2-dibromocyclohexane and *trans*-1-bromo-2-chlorocyclohexane
(C) only *cis*-1,2-dibromocyclohexane (D) only *trans*-1,2-dibromocyclohexane

105. Identify the correct statement.
(A) Dehydrobromination of *meso*-2,3-dibromobutane gives *cis*-2-bromobut-2-ene
(B) Dehalogenation of *meso*-2,3-dibromobutane gives *cis*-but-2-ene
(C) Hydroxylation of *trans*-2,3-dimethylbutene with Baeyer's reagent gives *meso*-2,3-dihydroxybutane
(D) Hydroxylation of *cis*-2,3-dimethylbutene with Baeyer's reagent gives racemic 2,3-dihydroxybutene.

Hint: The statement (A) is correct and all others are incorrect.

(A) 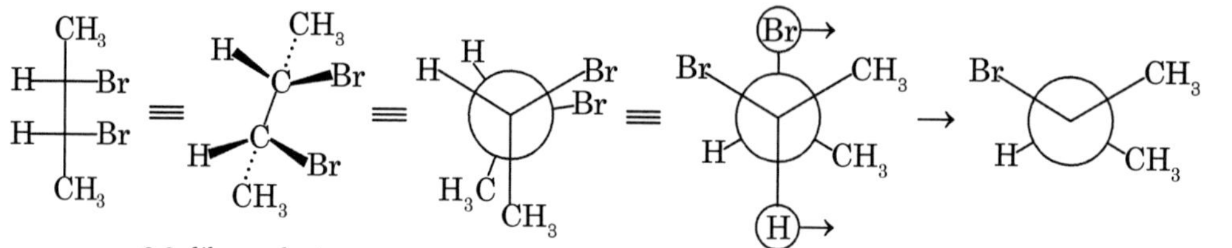

meso-2,3-dibromobutane *cis*-2-bromobut-2-ene

(B) *trans*-but-2-ene is formed

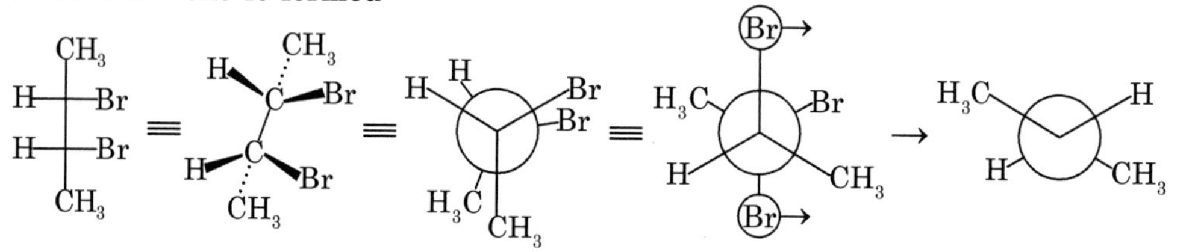

meso-2,3-dibromobutane *trans*-but-2-ene

(C) Racemic mixture is formed. The addition of two hydroxyl groups involves *syn*-addition on the same side of the double bond. Thus the hydroxylation of *trans*-2,3-dimethylbutene produces a racemic mixture.

(D) *meso* compound is formed. The addition of two hydroxyl groups involves *syn*-addition on the same side of the double bond. Thus the hydroxylation of *cis*-2,3-dimethylbutene gives racemic 2,3-dihydroxybutane.

106. Identify the correct statement.

(A) Addition of H_2 to *trans*-2,3-dibromobutene gives racemic 2,3-dibromobutane

(B) Addition of H_2 to *cis*-2,3-dibromobutene gives racemic 2,3-dibromobutene

(C) Addition of bromine to *trans*-but-2-ene gives racemic 2,3-dibromobutane

(D) Addition of bromine to *cis*-but-2-ene gives *meso*-2,3-dibromobutane

Hint: The statement (A) is correct and all others are incorrect.

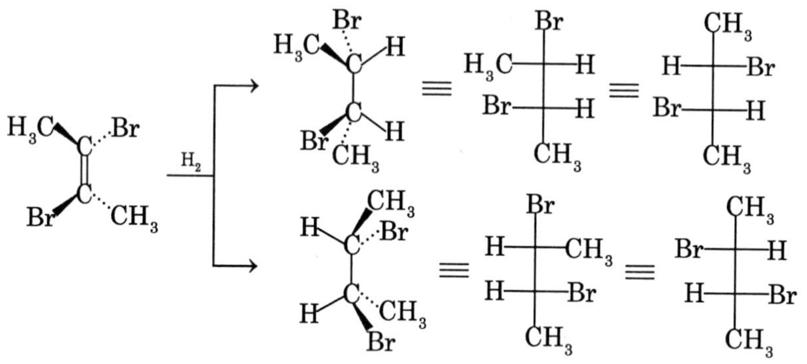

(racemic mixture)

(B) *meso*-2,3-dibromobutane is formed. The addition of two H atoms involves *syn*-addition on the same side of the double bond.

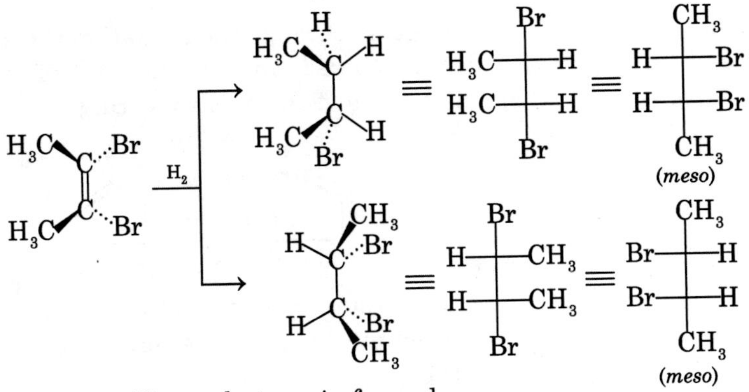

(C) *meso*-2,3-dibromobutane is formed.

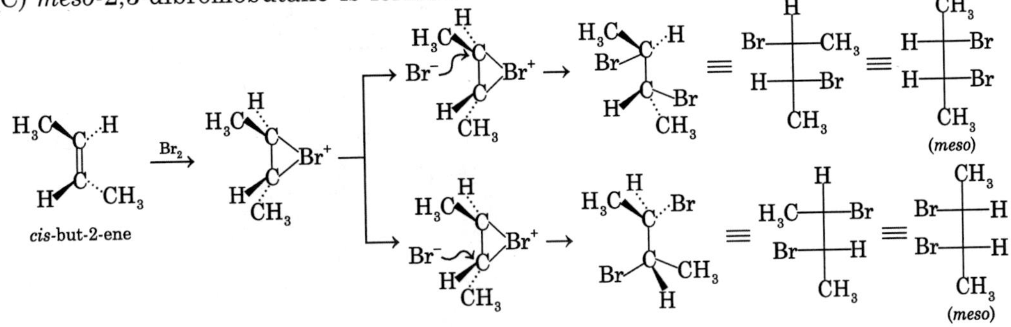

(D) Racemic 2,3-dibromobutane is formed.

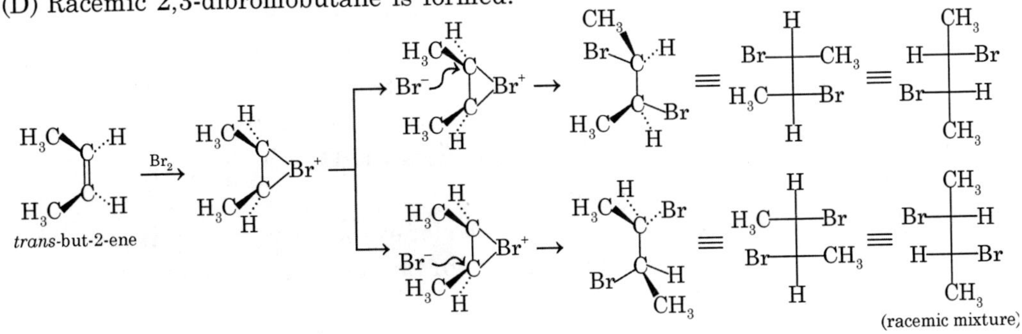

107. Identify the incorrect statement.

(A) Acid-catalysed dehydration of 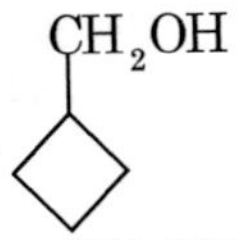 gives 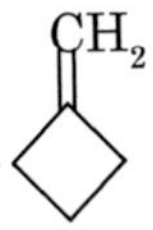

(B) Acid-catalysed dehydration of 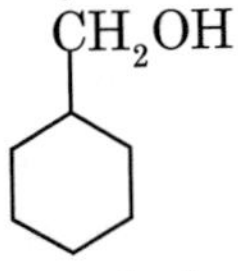 gives 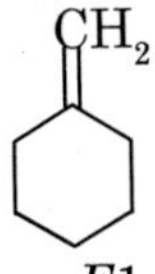

(C) Acid-catalysed dehydration of an alcohol follows *E*1 mechanism

(D) Addition of HBr to $CH_3OCH{=}CHCH_3$ gives $CH_3OCH_2CH_2Br$

Hint: The statement (C) is correct and all others are incorrect.

(A) Cyclopentene is formed. The reaction occurs through the formation of a secondary carbocation and formation of a five-membered ring (by the shift of side chain CH_2) which is more stable than the original four-membered ring.

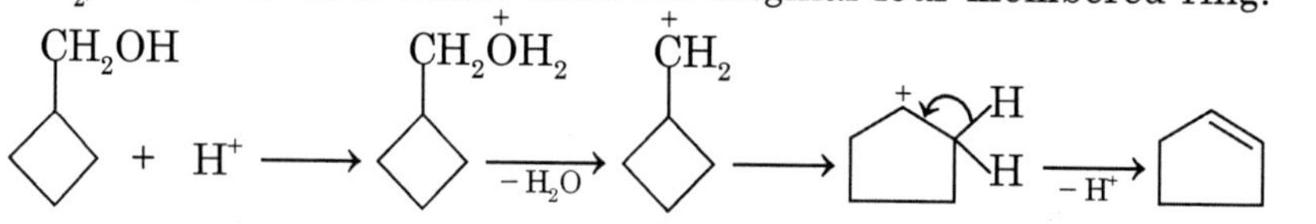

2° carbocation cyclopentene

(B) 1-Methylcyclohexene is formed. There occurs H shift and as a result the less stable primary carbocation is converted into more stable tertiary carbocation.

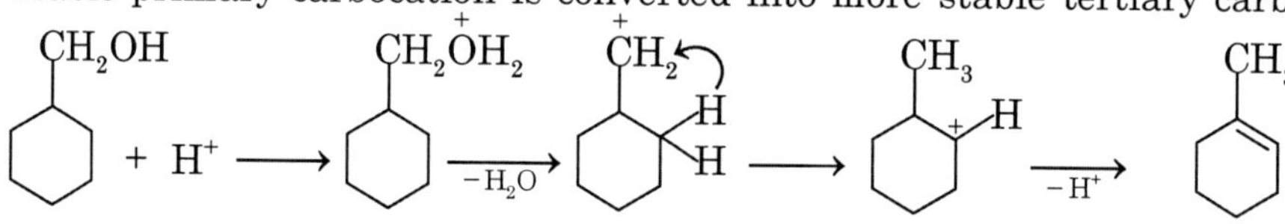

1° carbocation 3° carbocation 1-Methylcyclohexene

(C) The acid catalysed dehydration of an alcohol occurs through $E1$ mechanism.

$$H_3C-\underset{\displaystyle OH}{\underset{|}{C}H}-CH_3 + H^+ \rightleftharpoons H_3C-\underset{\displaystyle \overset{+}{O}H_2}{\underset{|}{C}H}-CH_3 \xrightarrow[-H_2O]{E_1}$$

Propan-2-ol

$$H_3C-\overset{+}{C}H-\underset{\displaystyle H}{\underset{|}{C}H_2} \xrightarrow{HSO_4^-} CH_3-CH{=}CH_2 + H_2SO_4$$

Propene

(D) $CH_3OCHBrCH_2CH_3$ is formed by Markownikoff addition. The electron-withdrawing-I-effect of $-OCH_3$ group is superseded by the extended π-delocalisation of an unshared pair of electrons from O atoms to the adjacent C^+ and as a result the carbocation $CH_3O\overset{+}{C}HCH_2CH_3$ is stabilised. The nucleophilic attack of Br^- on the carbocation produces $CH_3OCHBrCH_2CH_3$.

108. Identify the incorrect statement.

(A) Addition of bromine to chlorocyclohexene gives a mixture of *trans*-dibromocyclohexane and *trans*-1-bromo-2-chlorocyclohexane and *trans*-dichlorocyclohexane

(B) Addition of bromine to cyclohexene gives *trans*-1,2-dibromocyclohexane

(C) Addition of HBr to $BrCH{=}CH_2$ produces Br_2CHCH_3

(D) Addition of HBr to $F_3C-CH{=}CH_2$ produces $F_3C-CH_2-CH_2-Br$

Hint: The statement (A) is incorrect and all others are correct.

(A) Bromination of chlorocyclohexene produces a mixture of*trans*-dibromocyclohexane and *trans*-1-bromo-2-chlorocyclohexane. Due to the non-availability of Cl^+ ion, *trans*-dichlorocyclohexane is not formed.

(B) Addition of bromine to an alkene occurs through anti–addition and *trans*-1,2-dibromocyclohexane is formed.

(C) Addition of HBr to $BrCH{=}CH_2$ occurs according to Markownikoff rule and Br_2CHCH_3 is formed.

(D) The electron-withdrawing fluorine attached to the carbon atom makes the carbocation $F_3CCH_2\overset{+}{C}H_2$ more stable than $F_3C\overset{+}{C}HCH_3$ and as a result the addition occurs according to anti–Markownikoff rule.

109. Identify the incorrect statement.

(A) The order of reactivity of ethylbromide *n*-propyl bromide and isobutyl bromide towards *E*2 dehydrohalogenation is: isobutyl < *n*-propyl < ethyl

(B) The major product in the dehydrohalogenation of 2-chloro-2,3-dimethylbutane is 2,3-dimethylbut-2-ene

(C) The reaction of propene with *N*-bromosuccinimide in aqueous dimethyl sulphide gives 1-bromopropan-2-ol as the major product

(D) Propan-2-ol can be produced by treating propene with 80% H_2SO_4 followed by heating with water.

Hint: The statement (A) is incorrect and all others are correct.

(A) The order of reactivity is: ethyl < *n*-propyl < isobutyl. The reactivity of haloalkanes towards *E*2 dehydrohalogenation depends on the stability of the alkene formed.

$$CH_3CH_2Br \xrightarrow{-HBr} CH_2{=}CH_2$$

Ethyl bromide — unsubstituted

$$CH_2CH_2CH_2Br \xrightarrow{-HBr} CH_3{-}CH{=}CH_2$$

n-Propyl bromide — monosubstituted

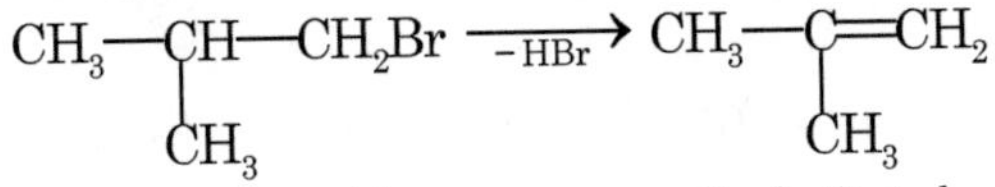

Isopropyl bromide — disubstituted

Since according to the Saytzeff's rule, the stability of alkenes increases with increase in the number of substituents on the doubly bonded carbon atoms, the order of reactivity is: ethyl < *n*-propyl < isobutyl.

(B) The major product formed is 2,3-dimethylbut-2-ene. This is because according to the Saytzeff's rule the stability of alkanes increases with increase in the number of substituents on the doubly bonded carbon atoms.

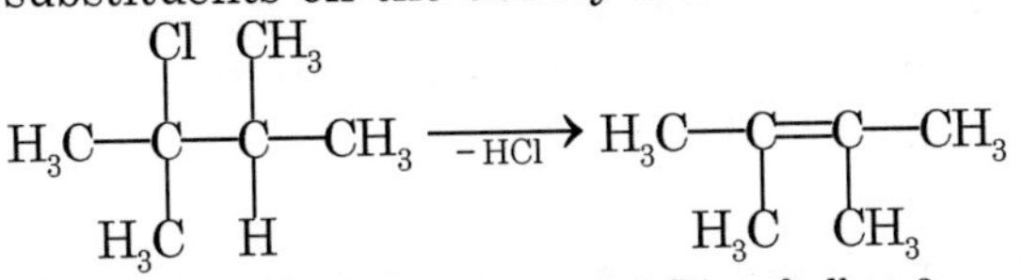

2-Chloro-2,3-dimethylbutane — 2,3-Dimethylbut-2-ene

(C) The major product formed is 1-bromoporpan-2-ol.

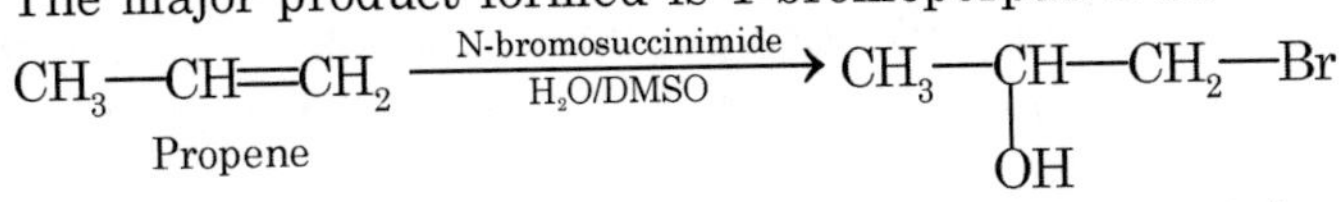

Propene

1-Bromopropan-2-ol

(D) Propan-2-ol is prepared by treating propene with 80% H_2SO_4 followed by heating with water. The addition of H_2SO_4 to double bond occurs according to the Markownikoff's rule.

110. Identify the correct statement.

(A) Higher energy is evolved in the hydrogenation of *cis*-but-2-ene in comparison to that of *trans*-but-2-ene

(B) Addition of aqueous bromine to ethene in the presence of NaCl gives only 1,2-dibromoethene

(C) During addition of bromine to alkenes a carbocation is formed as an intermediate

(D) 1,2 Elimination reaction occurring through $E1$ mechanism does not involve the rearrangement of carbon skeleton.

Hint: The statement (A) is correct and all others are incorrect.

(A) Since *cis*-but-2-ene is less stable than *trans*-but-2-ene, the former lies at higher level in comparison to *trans*-but-2-ene in the potential energy *vs*. reaction coordinate diagram. Hence the difference $H_{\text{butane}} - H_{cis}$ is greater than the difference $H_{\text{butane}} - H_{trans}$.

(B) The addition of aqueous bromine to ethene in the presence of NaCl gives 1,2-dibromoethene, 1-bromo-2-chloroethene and bromoethanol.

(C) A bromonium ion is formed as the intermediate instead of a carbocation.

(D) 1,2 Elimination reaction occurring through $E1$ mechanism involves the formation of a carbocation which rearranges to form the most stable carbocation.

111. Identify the incorrect statement.

(A) Addition of HBr to buta-1,4-diene at 40°C gives 3-bromobut-1-ene as the major product

(B) Addition of HBr to buta-1,4-diene at – 80°C gives 3-bromobut-1-ene

(C) Addition of HBr to buta-1,4-diene gives both 1-bromobut-2-ene and 3-bromobut-1-ene which are in equilibrium with each other and their relative amount is dependent upon temperature

(D) Addition of HBr to an alkene occurs through an ionic mechanism.

Hint: The statement (A) is incorrect and all others are correct.

(A) The major product is 1-bromobut-2-ene since at higher temperature the 1,4-adduct predominates over the 1,2–adduct.

(B) The major product is 3-bromobut-1-ene since at lower temperature, the 1,2-adduct predominates over the 1,4-adduct.

(C) The statement is correct.

(D) Addition of HBr to an alkene occurs through an ionic mechanism.

112. Identify the incorrect statement.

(A) Out of but-1-ene, *cis*-but-2-ene and *trans*-but-2-ene, the alkene having the lowest enthalpy of combustion is but-1-ene and that having the highest enthalpy of combustion is *trans*-but-2-ene

(B) Out of 3-bromopentane and 2-bromo-2-methylbutane, the alkane which is more reactive towards dehydrohalogenation is bromo-2-methylbutane

(C) Low concentration of halogen instead of high temperature favours substitution reaction in propene instead of addition reaction across the double bond

(D) Out of 2-bromo-2-methylbutane and 3-bromo-2-methylbutane, the former is more reactive towards dehydrohalogenation.

Hint: The statement (A) is incorrect and all others are correct.

(A) The order of enthalpy of combustion is:
trans-but-2-ene < *cis*-but-2-ene < but-1-ene. The *trans* isomer is more stable (lower energy) than the *cis* isomer (higher energy). The lower energy content indicates that more heat would be required to break the compound and hence the enthalpy of combustion would be minimum.

(B) The statement is correct.

(C) The statement is correct.

(D) 2-Bromo-2-methylbutane is more reactive than 3-bromo-2-methylbutane towards dehydrohalogenation.

113. Identify the incorrect statement.

(A) Out of the dienes $CH_2{=}CH{-}CH{=}CH{-}CH_3$ and $CH_2{=}C{=}CH{-}CH_2{-}CH_3$, the more stable diene is $CH_2{=}C{=}CH{-}CH_2{-}CH_3$

(B) Propene reacts faster with acids than ethene

(C) Isobutyl alcohol cannot be obtained by reacting 2-methylprop-1-ene with 80% H_2SO_4 followed by heating with water

(D) 1,2-Dichloroethene is less reactive towards addition of H_2SO_4 in comparison to vinyl chloride.

Hint: The statement (A) is incorrect and all others are correct.

(A) A conjugated diene is more stable than a cumulated diene. Hence $CH_2{=}CH{-}CH{=}CH{-}CH_3$ is more stable.

(B) Propene forms a secondary carbocation while ethene forms a primary carbocation. Since a secondary carbocation is more reactive than a primary carbocation, propene reacts faster with acids than ethene.

(C) *tert*-butyl alcohol is formed.

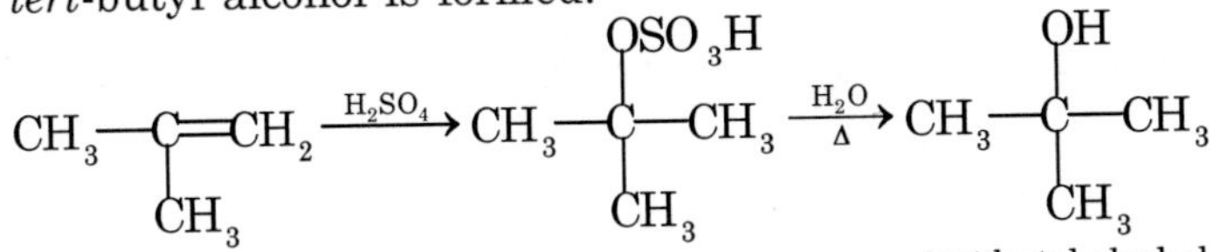

Isobutylene *tert*-butyl alcohol

(D) 1,2-Dichloroethene is more reactive than vinyl chloride towards addition of H_2SO_4. Two chlorine atoms resent in 1,2-dichloroethene destabilises the intermediate carbocation more than the monochloro substituted ethene (vinyl chloride).

114. Identify the correct statement.

(A) Addition of HBr to penta-1,3-diene occurs at a faster rate than that to penta-1,4-diene

(B) The reaction of hept-3-yne with $KMnO_4$ under alkaline condition at high temperature produces $CH_3CH_2COCOCH_2CH_2CH_3$

(C) The reaction of hept-3-yne with $KMnO_4$ under neutral condition at room temperature produces CH_3CH_2COOH and $CH_3CH_2CH_2COOH$

(D) The treatment of pent-1-en-4-yne with one mole of HBr gives 2-bromopenta-1,4-diene.

Hint: The statement (A) is correct and all others are incorrect.

(A) Addition of HBr to an alkene occurs through the formation of a carbocation and an allylic carbocation is more stable due to resonance. The formation of allylic carbocation requires less ΔH in comparison to secondary carbocation. Hence the allylic carbocation is formed at a faster rate making the addition of HBr faster.

$$\underset{\text{Buta-1,4-diene}}{CH_2{=}CH{-}CH_2{-}CH{=}CH_2} \xrightarrow{H^+} \underset{\text{Secondary carbocation}}{CH_3{-}\overset{+}{C}H{-}CH_2{-}CH{=}CH_2}$$

$$CH_2{=}CH{-}CH{=}CH{-}CH_3 \xrightarrow{H^+} \underset{\text{Allylic carbocation}}{CH_3{-}\overset{+}{C}H{-}CH{=}CH{-}CH_3}$$

$$CH_3{-}\overset{+}{C}H{-}CH{=}CH{-}CH_3 \leftrightarrow CH_3{-}CH{=}CH{-}\overset{+}{C}H{-}CH_3 \text{ or}$$

$$\underset{\text{Resonance hybrid}}{CH_3{-}\overbrace{CH{\overset{....}{=}}CH{\overset{....}{=}}CH}^{+}{-}CH_3}$$

(B) Under alkaline condition at high temperature, the tripple bond in hept-3-yne is cleaved to produce two carboxylic acids.

$$CH_3CH_2C{\equiv}C(CH_2)_2CH_3 \xrightarrow[\text{[O]}]{KMnO_4/OH^-} CH_3CH_2COOH + CH_3(CH_2)_2COOH$$

(C) Under neutral condition at room temperature, a diketone is formed.

$$CH_3CH_2C{\equiv}C(CH_2)_2CH_3 \longrightarrow [CH_3CH_2\overset{OH}{\overset{|}{C}}{=}\overset{OH}{\overset{|}{C}}(CH_2)_2CH_3] \longrightarrow$$

$$[CH_3CH_2\overset{OH}{\overset{|}{C}H}{-}\overset{O}{\overset{\|}{C}}(CH_2)_2CH_3] + [CH_3CH_2\overset{O}{\overset{\|}{C}}{-}\overset{OH}{\overset{|}{C}}(CH_2)_2CH_3] \xrightarrow{[O]}$$

$$[CH_3CH_2\overset{OH}{\overset{|}{C}}{-}\overset{O}{\overset{\|}{C}}(CH_2)_2CH_3$$

(D) $H_2C{=}CHCH_2C{\equiv}CH + HBr \longrightarrow H_2C{=}CHCH_2\overset{Br}{\overset{|}{C}}{=}CH_2$

115. Identify the incorrect statement.

(A) The reaction of 4-chlorotoluene with aqueous NaOH gives exclusively 4-hydroxytoluene

(B) Addition of one mole of H_2/Pt to $C_6H_5CH{=}CH{-}CH{=}CHCH_3$ produces $C_6H_5CH{=}CHCH_2CH_2CH_3$

(C) The treatment of benzene with 1-chloropropane in the presence of anhydrous $AlCl_3$ gives $PhCH_2CH_2CH_3$ and $PhCH(CH_3)_2$

(D) Although $-N(CH_3)_2$ is an *ortho* and *para*-director, it influences the formation of *meta* nitrated product when HNO_3/H_2SO_4 is used as the electrophilic reagent.

116. Identify the incorrect statement.

(A) The C—H bond in benzene is stronger than C—D bond in deuterated benzene

(B) In spite of electron-attracting inductive effect of the phenyl group, the electrophylic substitution of biphenyl occurs at *ortho* and *para*-positions

(C) The phenyl anion is more stable than cyclohexyl anion

(D) 4-Chlorotoluene reacts with aqueous NaOH at 340°C to form 4-hydroxytoluene. No such replacement occurs for 2,6–dimethylchlorobenzene.

Hint: The statement (A) is incorrect and all others are correct.

(A) C—D bond is stronger than C—H bond because D is a heavier isotope than H.

(B) In the attack at the *ortho*- and *para*-positions, the ring stabilises the + charge on the attached C atom of the σ complex by delocalisation and the + charge is delocalised to both the rings.

(C) The electron pair in $C_6H_{11}^-$ is present in the sp^3 hybridised orbital while in $C_6H_5^-$ the electron pair is present in the sp^2 hybridised orbital. Thus the $C_6H_5^-$ carbanion has *s*-character and as a result it is more stable.

(D) Since there is no *ortho* H atom in 2,6-dimethylchlorobenzene, benzyne cannot form.

117. An organic compound containing 85.0% C and 14.4% H and having molecular mass 56.1 g mol^{-1} decolourises Br_2/CCl_4 and reacts with alkaline $KMnO_4$ to form only propanoic acid. The organic compound is

(A) but-1-yne (B) but-1-ene (C) but-2-ene (D) hex-3-ene

Hint: The compound is but-1-ene.

n_H = 14.4 and n_C = 85.6/12 = 7.13. Hence $n_C : n_H$ = 7.13 : 14.4 = 1 : 2.013. Thus its empirical formula is CH_2 and its empirical formula mass is 14. Hence

$$n = \frac{MW}{EFW} = \frac{56.1}{14} = 4$$

Molecular formula = $(CH_2)_n = (CH_2)_4 = C_4H_8$ and it is an alkene (C_nH_{2n}). Since the compound decolourises Br_2/CCl_4, it is an unsaturated compound. Since it reacts with $KMnO_4/OH^-$ to form propanoic acid (loss of one C atom), the double bond is present at the terminal position. Thus, it is but-1-ene and not but-2-ene.

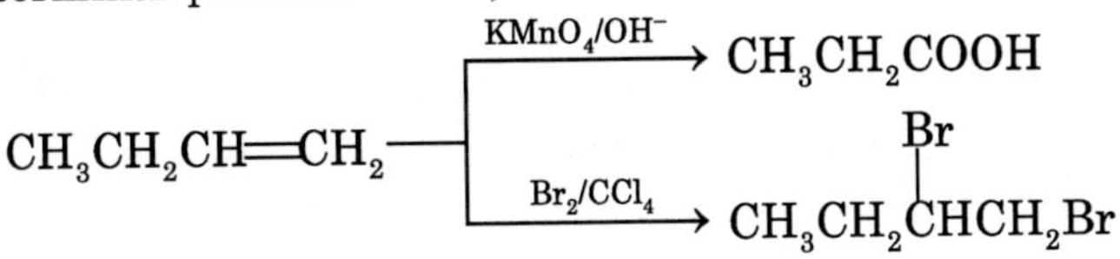

118. In the reactions, 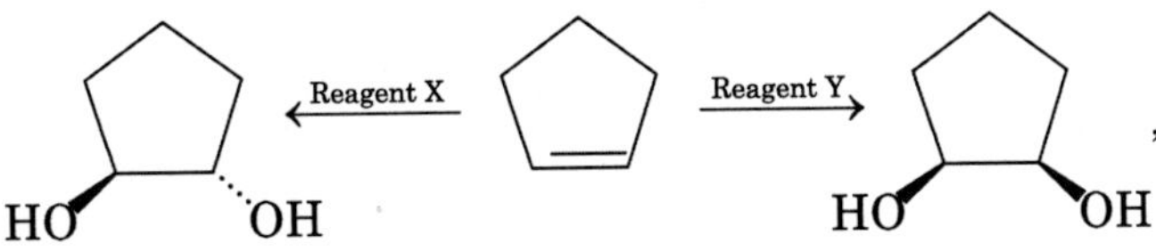

, X and Y are respectively

(A) $C_6H_5CO_3H$, HCO_3H (B) cold alkaline $KMnO_4$, $C_6H_5CO_3H$

(C) cold alkaline $KMnO_4$, HCO_3H (D) cold alkaline $KMnO_4$, OsO_4/H_2O_2

119. In the reaction, 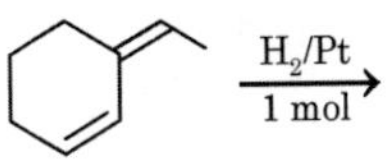

X, X is

(A) 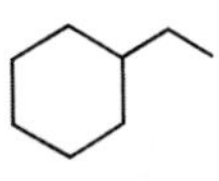 (B) 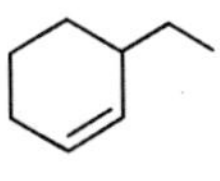 (C) 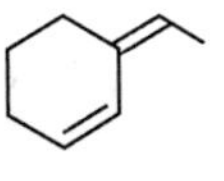 (D) 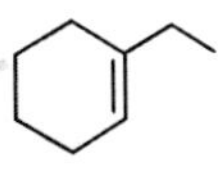

120. In the reaction, 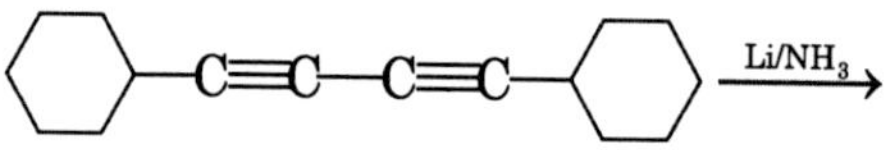 X, X is

(A) 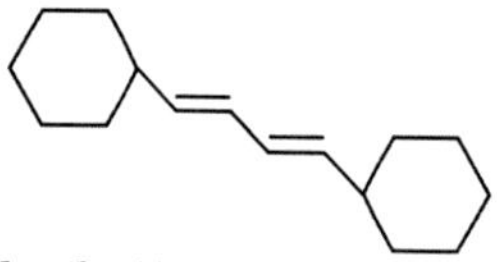

(B) 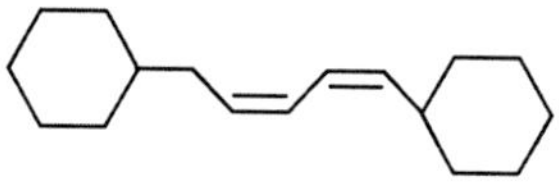

(C) both (A) and (B)

(D) none of these.

121. In the reaction, 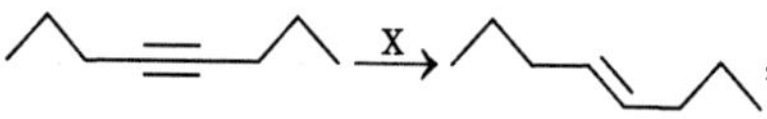, X can be

I. Li/NH_3, II. $LiAlH_4$, III. $H_2/Pd/CaCO_3$, IV. H_2/Ni

(A) I or II (B) II or III (C) I or III (D) I, II, III or IV

122. In the reaction, 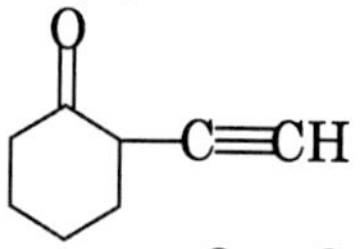 $\xrightarrow[\text{(ii) } H_2O_2/OH^-]{\text{(i) } BH_3/THF}$ X, X is

(A) 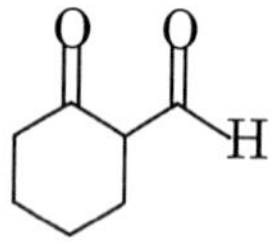 (B) 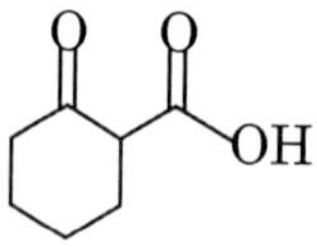 (C) 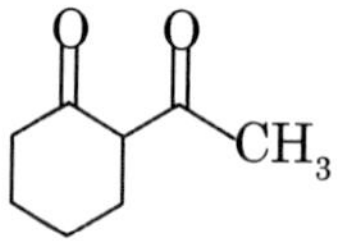 (D) 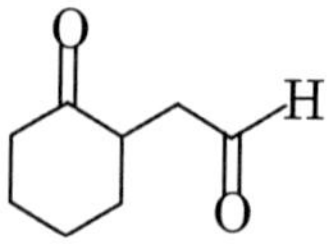

123. In the reaction, 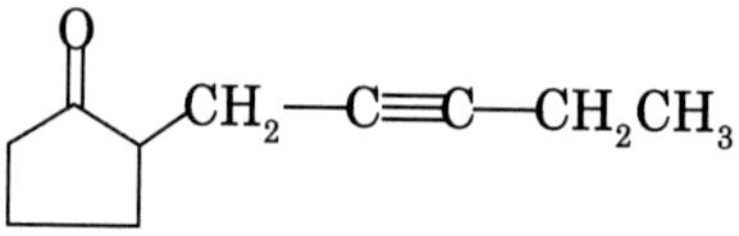 $\xrightarrow{H_2/Ni_2B}$ X, X is

(A) 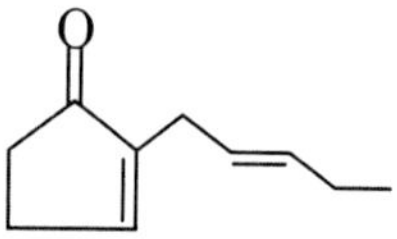

(B) 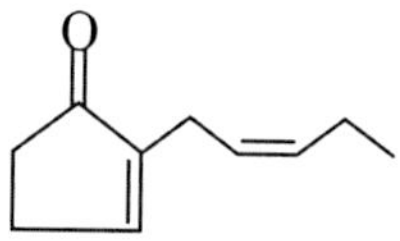

(C) 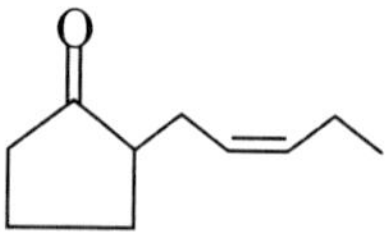

(D) 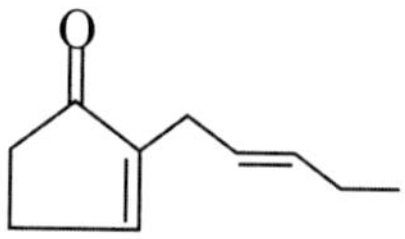

124. $CH_3—CH{=}CH_2 + NOCl \rightarrow P$

Identify the adduct.

(A) $CH_3—\underset{\displaystyle Cl}{\underset{|}{C}H}—\underset{\displaystyle NO}{\underset{|}{C}H_2}$

(B) $CH_3—\underset{\displaystyle NO}{\underset{|}{C}H}—\underset{\displaystyle Cl}{\underset{|}{C}H_2}$

(C) $CH_3—CH_2—\overset{\displaystyle NO}{\overset{|}{\underset{\displaystyle Cl}{\underset{|}{C}H}}}$

(D) $\underset{\displaystyle NO}{\underset{|}{C}H_2}—CH_2—\underset{\displaystyle Cl}{\underset{|}{C}H_2}$ **(I.I.T. 2006)**

Hint: The reaction occurs through Markownikoff addition.

$CH_3—CH{=}CH_2 + NO^+ \rightarrow CH_3—\overset{+}{C}H—\underset{\displaystyle NO}{\underset{|}{C}H_2} \xrightarrow{Cl^-} CH_3—\underset{\displaystyle Cl}{\underset{|}{C}H}—\underset{\displaystyle NO}{\underset{|}{C}H_2}$

The 2° carbocation ($CH_3—\overset{+}{C}H═CH_2—NO$) formed is more stable than 1° carbocation ($CH_3—CH_2═\overset{+}{C}H_2$). The chloride ion attacks the more stable 2° carbocation to give the major product (A).

125. Cyclohexene on ozonolysis followed by reaction with zinc dust and water gives compound E. Compound E on further treatment with aqueous KOH yields compound F. Compound F is

(A) 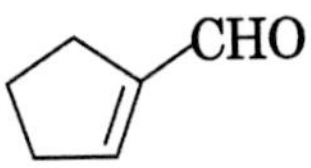 (B) 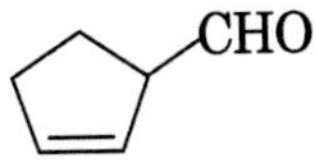 (C) 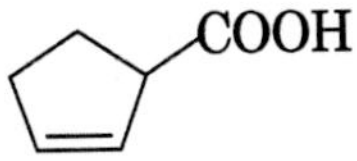 (D) 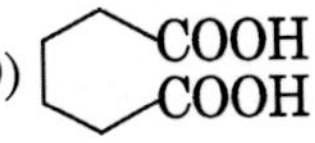 **(I.I.T. 2007)**

Hint: 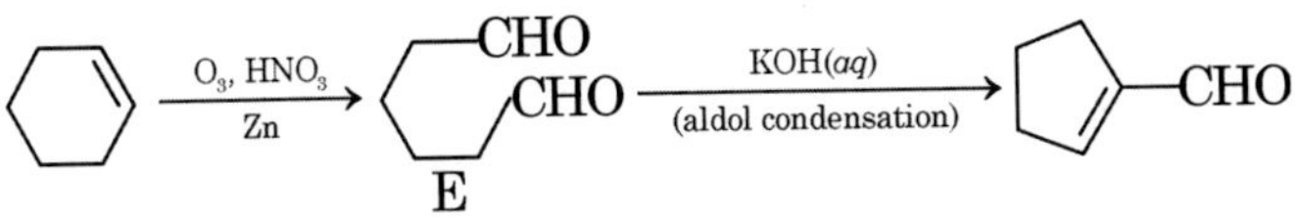

126. The correct statement(s) about the compound $H_3C(HO)HC—CH—CH—CH(OH)CH_3$ (X) is(are)

(A) The total number of stereoisomers possible for X is 6

(B) The total number of diastureoisomers possible for X is 3

(C) If the stereochemistry about the double bond is *trans*, the number of enantiomers possible for X is 4

(D) If the stereochemistry about the double bond in X is *cis*, the number of enantiomers possible for X is 2. **(I.T.I. 2009)**

Hint: $CH_3—\overset{*}{C}H(OH)—CH═CH—\overset{*}{C}H(OH)—CH_3$

(I)	*d*	*cis*	*d*
(II)	*d*	*trans*	*d*
(III)	*l*	*cis*	*l*
(IV)	*l*	*trans*	*l*
(V)	*d*	*cis*	*l*
(VI)	*d*	*trans*	*l*

I and II are enantiomers, II and IV are enatiomers, V is meso due to the presence of plane of symmetry while VI is meso due to the presence of center of symmetry. A total of six stereoisomers are possible while number of diastereomers are four (I or III, II or IV, V and VI). If double bond in X is *trans*, number of enantiomers is two (II and IV). If double bond in X is *cis*, number of enantiomers is two (I and III).

Paragraph for Question Nos. 127-128

An acyclic hydrocarbon P, having molecular formula C_6H_{10}, gave acetone as the only organic product through the following sequence of reactions, in which Q is an intermediate organic compound.

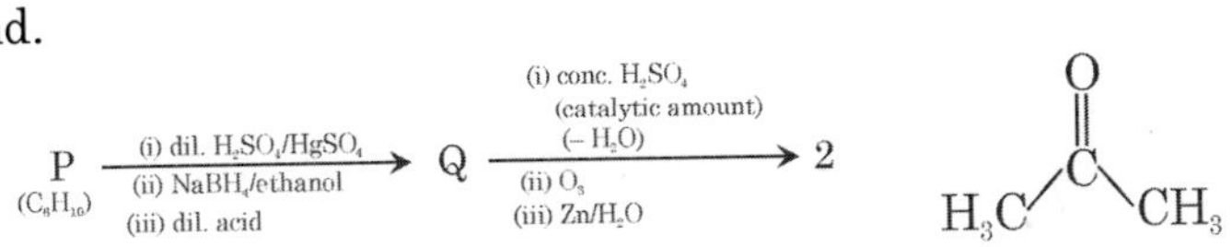

127. The structure of compound P is

(A) $CH_3CH_2CH_2CH_2—C≡C—H$ (B) $H_3CH_2C—C≡C—CH_2CH_3$

(C) 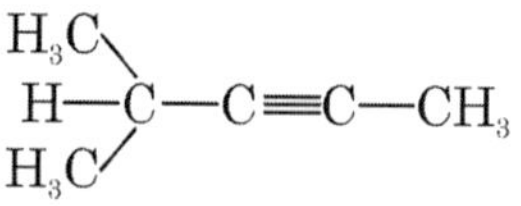

(D) 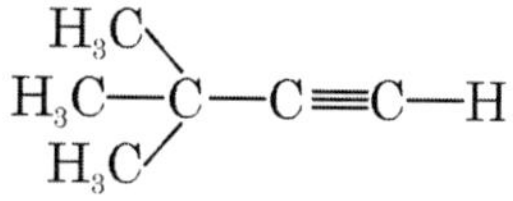

(I.I.T. 2011)

128. The structure of the compound Q is

(A) 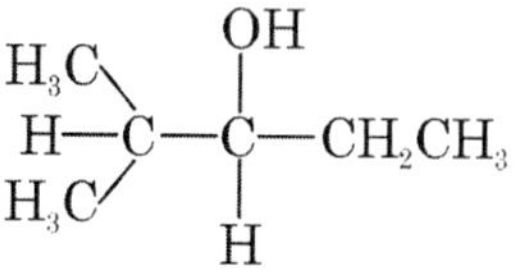

(B) 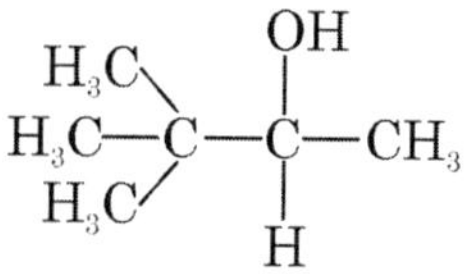

(C) 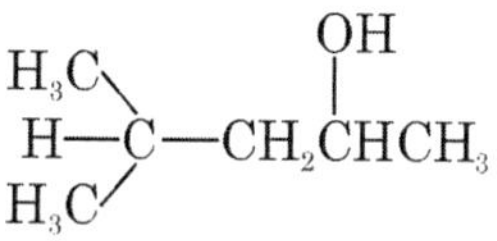

(D) $CH_3CH_2CH_2CH(OH)CH_2CH_3$

(I.I.T. 2011)

Hint: For the Question Nos. 127-128

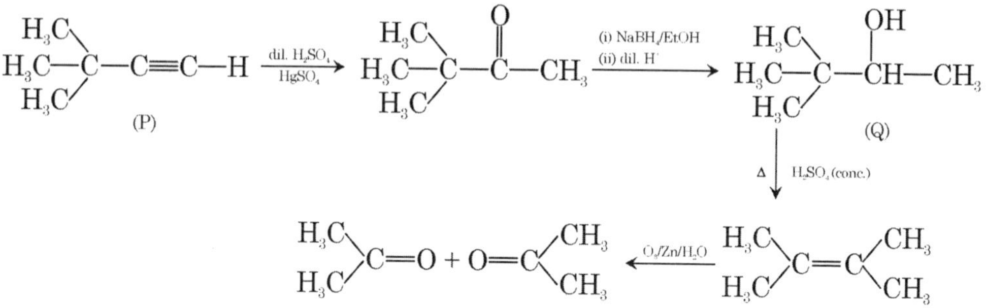

ANSWERS

1. B	2. B	3. B	4. C	5. B	6. C
7. A	8. A	9. A, C	10. B	11. B	12. A, C
13. B	14. A	15. B	16. B	17. B	18. D
19. A	20. A	21. A	22. A	23. B	24. B
25. C	26. B	27. A	28. A	29. B	30. C
31. A	32. A	33. C	34. C	35. C	36. C
37. A	38. A	39. A	40. B	41. B	42. C
43. B	44. A	45. A	46. A	47. B	48. A
49. A	50. B	51. D	52. B, D	53. C	54. A
55. C	56. D	57. A	58. A	59. C	60. D
61. D	62. B	63. C	64. D	65. A	66. A
67. A	68. D	69. A	70. C	71. B	72. A

73. A	74. B	75. A	76. A	77. B	78. A
79. A	80. B	81. D	82. C	83. A	84. A
85. D	86. D	87. A	88. C	89. B	90. A
91. A	92. A	93. A	94. D	95. D	96. C
97. B	98. B, D	99. C	100. A	101. A	102. D
103. C	104. B	105. A	106. A	107. C	108. A
109. A	110. A	111. A	112. A	113. A	114. A
115. A	116. A	117. B	118. C	119. D	120. A
121. A	122. C	123. B	124. A	125. A	126. A, D
127. D	128. B				

37

Alkyl and Aryl Halides

OBJECTIVE TYPE MULTIPLE CHOICE QUESTIONS

1. Benzoyl chloride is prepared from benzoic acid by
(A) Cl_2, $h\nu$ (B) Cl_2, H_2O (C) SO_2Cl_2 (D) $SOCl_2$ **(I.I.T. 2000)**
2. The ether $C_6H_5OCH_2C_6H_5$ when treated with HI produces
(A) $C_6H_5CH_2I$ (B) $C_6H_5CH_2OH$ (C) C_6H_5OH (D) C_6H_5I **(I.I.T. 1999)**
3. Which of the following is the order of reactivity sequence of the alkyl halide in S_{N^2} reaction?
(A) $CH_3I < CH_3Br < CH_3Cl < CH_3F$ (B) $CH_3Cl < CH_3Br < CH_3I < CH_3F$
(C) $CH_3F < CH_3Cl < CH_3Br < CH_3I$ (D) $CH_3F < CH_3I < CH_3Cl < CH_3Br$
(I.I.T. 2000)
4. During debromination of *meso*-dibromobutane, the major compound formed is
(A) *n*-butane (B) *trans*-but-2-ene
(C) *cis*-but-2-ene (D) but-1-ene **(I.I.T. 1997)**
5. $(CH_3)_3CMgCl$ on reaction with D_2O produces
(A) $(CH_3)_3CD$ (B) $(CH_3)_3OD$ (C) $(CD_3)_3CD$ (D) $(CD_3)OD$ **(I.I.T. 1997)**
6. In the reaction of *p*-chlorotoluene with KNH_2 in liquid NH_3, the major product is
(A) *o*-toluidine (B) *m*-toluidine
(C) *p*-toluidine (D) *p*-chloroaniline **(I.I.T. 1997)**
7. The products of reaction of alcoholic silver nitrite with C_2H_5Br are
(A) ethane (B) ethene (C) nitroethane (D) ethyl nitrite **(I.I.T. 1991)**
8. Aryl halides are less reactive towards nucleophilic substitution as compared to alkyl halides due to
(A) formation of less stable carbocation
(B) sp^2 hybridised C attached to halogen
(C) larger carbon halogen bond
(D) resonance stabilisation. **(I.I.T. 1990)**

9. How many chiral carbon atoms are present in 2,3,4-trichloropentane?
   (A) 1 (B) 2 (C) 3 (D) 4 **(I.I.T. 1990)**
10. Which of the following have asymmetric carbon atom?
   (A) $CH_2Cl—CH_2Br$ (B) $CH_3—CHCl_2$
   (C) $CH_2BrCH(OH)CH_3$ (D) CH_3CHDCl **(I.I.T. 1989)**
11. The reaction of CH_3MgI with acetone followed by hydrolysis gives
   (A) *sec*-butanol (B) *tert*-butanol (C) *n*-butanol (D) none of these. **(I.I.T. 1987)**
12. Only two isomeric monochloro derivatives are possible for
   (A) *n*-butane (B) 2,4-dimethylpentane
   (C) benzene (D) 2-methylpropane **(I.I.T. 1986)**
13. Carbylamine test is performed in alcoholic KOH by heating a mixture of
   (A) trihalogenated methane and a primary amine
   (B) an alkyl cyanide and a primary amine
   (C) an alkyl halide and a primary amine
   (D) chloroform and Ag powder. **(I.I.T. 1984)**
14. Which of the following compounds is formed when chloral reacts with aqueous NaOH?
   (A) $CHCl_3$ (B) CH_2Cl_2 (C) $Cl_3CCH(OH)_2$ (D) $COCl_2$ **(I.I.T. 1984)**
15. Diethyl ether on heating with concentrated HI gives two moles of
   (A) ethyl iodide (B) methyl iodide (C) iodoform (D) ethanol **(I.I.T. 1983)**
16. Which is used as a refrigerant?
   (A) CF_4 (B) $CHCl_3$ (C) CCl_4 (D) CCl_2F_2 **(Roorkee 1989)**
17. Which of the following will react with water?
   (A) $CHCl_3$ (B) CCl_3CHO (C) CCl_4 (D) $ClCH_2CH_2Cl$ **(I.I.T. 1998)**
18. Benzyl chloride ($C_6H_5CH_2Cl$) can be prepared from toluene by chlorination with
   (A) SO_2Cl_2 (B) $SOCl_2$ (C) Cl_2 (D) NaOCl **(I.I.T. 1998)**
19. Chloroform on reduction with Zn and HCl forms
   (A) methylene chloride (B) chloropicrin
   (C) formic acid (D) phosgene
20. Benzene reacts with Cl_2 to form benzene hexachloride in the presence of
   (A) $AlCl_3$ (B) sunlight (C) Zn (D) Ni
21. Chloretone is obtained by reaction of acetone with
   (A) $CHCl_3$ (B) C_6H_5Cl (C) $ClCH_2CH_2Cl$ (D) Cl_2
22. The reaction $CH_3(CH_2)_5CH(CH_3)Br \xrightarrow{OH^-} CH_3(CH_2)_5CH(CH_3)OH$ is described as
   (A) S_{E^2} (B) S_{N^1} (C) S_{N^0} (D) S_{N^2}
23. In the reaction, $C_2H_5MgBr + H_2\overset{O}{C—C}H_2 \longrightarrow$ A, A is
   (A) C_2H_5CHO (B) $C_2H_5CH_2CH_2OH$
   (C) $C_2H_5CH_2OH$ (D) $C_2H_5CH_2CHO$
24. The reactivity of halogen atom is minimum in
   (A) propyl iodide (B) propyl chloride
   (C) isopropyl chloride (D) isopropyl bromide

25. The reactivity of methyl chloride (X), propyl chloride (Y) and chlorobenzene (Z) is in the order
   (A) X > Y > Z (B) X > Z > Y (C) Y > X > Z (D) Z > Y > X
26. A compound *A* ($C_3H_6Cl_2$) on reaction with alkali can give *B* (C_3H_6O) or *C* (C_3H_4). *B* on oxidation gave a compound of the formula $C_3H_6O_2$. *C* on reaction with dil. H_2SO_4 containing Hg^{2+} ion gave *D* (C_3H_6O), which with Br_2 and alkali gave the sodium salt of $C_2H_4O_2$. *A* is
   (A) $CH_3CH_2CHCl_2$ (B) $CH_3CCl_2CH_3$
   (C) $CH_2ClCH_2CH_2Cl$ (D) $CH_3CHClCH_2Cl$
27. To produce chlorobenzene, which of the following is used first?
   (A) Chlorobenzene (B) Nitrobenzene
   (C) Benzaldehyde (D) C_6H_6 **(Roorkee 1995)**
28. Identify Z in the reactions, $CH_3CH_2CH_2Br \xrightarrow{\text{aq.NaOH}} X \xrightarrow{Al_2O_3,\text{ heat}} Y \xrightarrow{Cl_2/H_2O} Z$
   (A) $CH_3CHClCH_2Cl$ (B) $CH_3CHClCH_2OH$
   (C) $CH_3CHOHCH_2Cl$ (D) mixture of $CH_3CHClCH_2Cl$ and $CH_3CHOHCH_2Cl$
29. An unknown alkyl halide *A* reacts with alcoholic KOH to produce a hydrocarbon *B* (C_4H_8). Ozonolysis of *B* affords one mole of propionaldehyde and one mole of formaldehyde. *A* is
   (A) $CH_3(CH_2)_3Br$ (B) $CH_3CH_2CHBrCH_3$
   (C) $Br(CH_2)_4Br$ (D) $CH_3CHBrCHBrCH_3$
30. When HCl gas is treated with propene in the presence of benzoyl peroxide, the product is
   (A) 2-chloropropane (B) allyl chloride
   (C) *n*-propyl chloride (D) no reaction.
31. Which of the following gives chloroethane on reaction with ethyl alcohol?
   (A) Cl_2 (B) Cl_2/P (C) conc. HCl (D) PCl_5
32. The green house effect is because of the
   (A) presence of CO_2 in the atmosphere
   (B) presence of O_3 and CH_4 in the atmosphere
   (C) presence of gases which in general are strong infrared absorbers in the atmosphere
   (D) N_2O and chloroflurocarbons in the atmosphere.
33. The products formed when chloroethane is treated with AgCN are
   (A) C_2H_5CN + AgCl (B) C_2H_5NC + AgCl
   (C) C_2H_4 + AgCl (D) C_2H_5ONO + AgCl
34. Iodoethane reacts with dry Ag_2O to give
   (A) ethane (B) diethyl ether (C) ethanol (D) ethanal
35. When chlorine is passed through propene at 500°C, the product is
   (A) allyl chloride (B) vinyl chloride (C) $CH_3CHClCH_2Cl$
   (D) PVC

36. The S_{N^1} mechanism for the hydrolysis of an alkyl halide to an alcohol involves the formation of
(A) carbocation (B) carbanion (C) free radical
(D) pentavalent carbon in the transition state.
37. Total number of acyclic isomers including stereoisomers of C_4H_7Cl is
(A) 9 (B) 10 (C) 12 (D) 8
38. Chloroethane reacts with Y to form NaCl and Z. One mole of Z reacts with two moles of HI to form C_2H_5I and H_2O. Y is
(A) CH_3COOH (B) $C_2H_5OC_2H_5$ (C) CH_3CHO (D) C_2H_5ONa
39. Chloroethane reacts with Mg in dry ether to form X. When X is hydrolysed a carbon compound Y and Z are formed. Y is
(A) C_2H_2 (B) C_2H_4 (C) C_2H_6 (D) C_6H_6
40. Isopropyl chloride undergoes hydrolysis by the mechanism
(A) S_{N^2} (B) S_{N^1} and S_{N^2} (C) S_{N^1} (D) neither S_{N^1} nor S_{N^2}
41. When an alkyl halide reacts with an alkoxide, the product is
(A) hydrocarbon (B) ether (C) alcohol (D) ester
42. 1,3-Dibromopropane reacts with zinc to form
(A) propene (B) propane (C) cyclopropane (D) hexane
43. Chlorobenzene on heating with aqueous NH_3 under pressure in the presence of Cu_2Cl_2 gives
(A) benzamide (B) nitrobenzene (C) aniline (D) chloroaminobenzene
44. The treatment of ammonia with excess of chloroethane gives
(A) ethane (B) methylamine
(C) diethylamine (D) tetraethyl ammonium chloride
45. Bromoethane reacts with sodium–lead alloy to give
(A) tetraethyl bromide (B) tetraethyllead
(C) both (A) and (B) (D) none of these.
46. 1-Chlorobutane on reaction with alcoholic potash gives
(A) but-1-ene (B) butan-1-ol (C) but-2-ene (D) butan-2-ol
47. 2-Bromopentane is heated with potassium ethoxide in ethanol. The product is
(A) 2-ethoxypentane (B) pent-1-ene
(C) *cis*-pent-2-ene (D) *trans*-pent-2-ene
48. Chlorobenzene reacts with Mg in dry ether to give *A* which further reacts with ethanol to give
(A) ethylbenzene (B) benzene (C) phenol (D) phenyl methyl ether
49. Replacement of Cl in chlorobenzene to give phenol requires drastic conditions but Cl of 2,4-dinitrochlorobenzene is readily replaced because
(A) NO_2 makes the ring electron rich at *o*- and *p*-positions
(B) NO_2 withdraws electrons from *meta* position
(C) NO_2 donates electrons at *meta* position
(D) NO_2 withdraws electrons from *ortho* and *para* positions.

50. Tetrabromoethane on heating with zinc gives
(A) CH_3CH_3 (B) $HC\equiv CH$ (C) C_2H_4 (D) C_2H_5Br

51. An organic compound A (C_4H_9Cl) on reaction with Na/diethyl ether gives a hydrocarbon which on monochlorination gives only one chloro derivative. A is
(A) *tert*-butyl chloride (B) *sec*-butyl chloride
(C) isobutyl chloride (D) *n*-butyl chloride

52. The reaction of haloalkane with NH_3 gas gives
(A) primary amine (B) primary, secondary and tertiary amine
(C) primary and secondary amine
(D) primary, secondary and tertiary amine, and quarternary ammonium salt.

53. Identify the set of reagent/reaction conditions X and Y in the following set of transformations:

$$CH_3—CH_2—CH_2Br \xrightarrow{X} \text{Product} \xrightarrow{Y} CH_3—CHBr—CH_3$$

(A) X = dilute aqueous NaOH, 20°C; Y = HBr/acetic acid, 20°C
(B) X = concentrated alcoholic NaOH, 80°C; Y = HBr/acetic acid, 20°C
(C) X = dilute aqueous NaOH, 20°C; Y = $Br_2/CHCl_3$, 0°C
(D) X = concentrated alcoholic NaOH, 80°C; Y = $Br_2/CHCl_3$, 0°C **(I.I.T. 2002)**

Hint:

$$CH_3—CH_2—CH_2Br \xrightarrow[80^\circ]{\text{alcoholic NaOH}} CH_3—CH=CH_2 \xrightarrow[20^\circ]{HBr/CH_3COOH} CH_3—\underset{Br}{\underset{|}{C}H}—CH_3$$

Hence X = alcoholic NaOH, 80°C and Y = HBr/CH_3COOH, 20°C.

54. The most effective reagent for the replacement of —OH group of an alcohol by a —Cl group is
(A) $SOCl_2$ (B) SOCl (C) conc. HCl (D) HCl gas

55. Total number of isomers of dibromobutane is
(A) 8 (B) 6 (C) 5 (D) 4

56. Chlorination of benzene occurs through
(A) addition-elimination mechanism
(B) elimination-addition mechanism
(C) electrophilic substitution mechanism
(D) nucleophilic substitution mechanism

57. The treatment of CH_3MgBr with C_2H_5CHO gives
(A) $C_2H_5CH(OH)C_2H_5$ (B) $CH_3CH(OH)C_2H_5$
(C) $CH_3CH(OH)CH_3$ (D) CH_3OH

58. Identify an endothermic reaction out of the following reactions.
(A) Iodination of benzene (B) Bromination of benzene
(C) Fluorination of benzene (D) None of these.

59. Identify the product when toluene is heated with excess of chlorine in the presence of light.

(A) Benzotrichloride (B) Benzyl chloride
(C) 4-Chlorotoluene (D) 2-Chlorotoluene

60. Identify the product when iodomethane is decomposed by light.
(A) CH_3CHI_2 (B) CH_3CH_2I (C) CH_4 (D) CH_3CH_3

61. Which of the following orders regarding ease of formation of Grignard reagent is correct?
(A) $CH_3CH_2Br < CH_3CH_2CH_2Br < CH_3Br$
(B) $CH_3CH_2CH_2Br < CH_3CH_2Br < CH_3Br$
(C) $CH_3CH_2Br < (CH_3)_2CHBr < CH_3Br$
(D) $CH_3Br < (CH_3)_2CHBr < (CH_3)_3CBr$

62. Identify the correct order regarding ease of formation of Grignard reagent for a given haloalkane.
(A) RBr < RCl < RI (B) RI < RBr < RCl
(C) RCl > RBr < RI (D) RCl < RBr < RI

63. Which of the following is formed when bromine reacts with silver benzoate?
(A) 1,4-dibromobenzene (B) bromobenzene
(C) benzyl bromide (D) benzoyl bromide

64. Iodination of benzene occurs in the presence of iodine and
(A) HgO (B) HIO_3 (C) HNO_3 (D) any of these.

65. Alkaline hydrolysis of chloroform yields
(A) $CHCl_2OH$ (B) CH_3OH (C) $HCOO^- + CO$ (D) $HCOO^-$

66. Which of the following is formed when a haloalkane (R′X) is heated with an ethanolic solution of silver salt of a carboxylic acid (RCOOH)?
(A) RCOOR′ (B) R—R (C) R′—H (D) R—R′

67. The chlorination of side chain of toluene occurs in the presence of
(A) Cl_2/air (B) Cl_2/heat and light
(C) $Cl_2/FeCl_3$ (D) halogen carrier.

68. Which of the following is formed when toluene reacts with excess of chlorine in the presence of heat and light and the product is heated with a solution of $Ca(OH)_2$?
(A) Benzoic acid (B) *o*-Cresol (C) *p*-Cresol (D) Benzyl alcohol

69. Which of the following is formed when iodobenzene is heated with copper powder in a sealed tube?
(A) 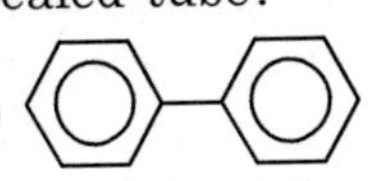 (B) 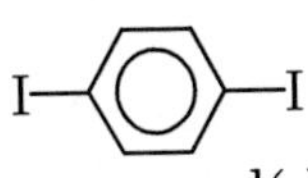 (C) 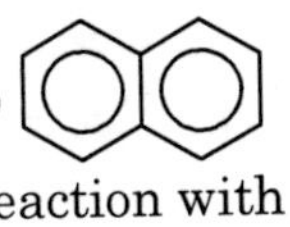 (D) 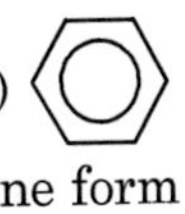

70. Which of the following compound(s) on reaction with bromine form bromo compounds?
(A) CH_3CH_2OH (B) CH_3COOH (C) CH_3COOAg (D) $CH_2{=}CH_2$?

71. Identify the compound(s) which on reaction with zinc dust and methanol gives alkane.
(A) $CH_3CCl_2CH_3$ (B) $CH_3CHBrCH_2Br$
(C) $BrCH_2CH_2Br$ (D) $CH_3CH_2CHCl_2$

72. The reaction of haloalkanes with RCOOAg produces

(A) esters (B) ethers (C) aldehydes (D) ketones

73. An organic compound, on treatment with Br_2 in CCl_4 gives bromoderivative of an alkene. The compound is

(A) $CH_3CH{=}CH_2$ (B) $H_2C{=}CH_2$ (C) $HC{\equiv}CH$ (D) $CH_3CH{=}CHCH_3$

74. The compound $CH_3CH_2CHClCH_3$ obtained by chlorination of *n*-butane, is

(A) *meso*-form (B) racemic form (C) *d*-form (D) *l*-form

75. Identify the compound(s) which on reaction with Cl_2 and alkali form chloroform.

(A) CH_3COCH_3 (B) CH_3CHO (C) CH_4 (D) CH_3CH_2OH

76. When $CH_3CH_2CHCl_2$ is treated with $NaNH_2$, the product formed is

(A) $CH_3{-}C{\equiv}CH$ (B) $CH_3{-}CH{=}CH_2$

(C) $CH_3CH_2CH(NH_2)_2$ (D) None of these.

77. In the reaction, $CH_3{-}\underset{\text{Cl}}{\underset{|}{C}H}{-}CH_3 \xrightarrow{\text{Mg/ether}} A \xrightarrow{D_2O} B$, B is

(A) $CH_3CH_2CH_2D$ (B) $CH_3CH_2CH_2Mg(OD)$

(C) CH_3CHDCH_3 (D) No reaction from A to B.

78. Among the following, the molecule with the highest dipole moment is

(A) CH_3Cl (B) CH_2Cl_2 (C) $CHCl_3$ (D) CCl_4 **(I.I.T. 2003)**

79. 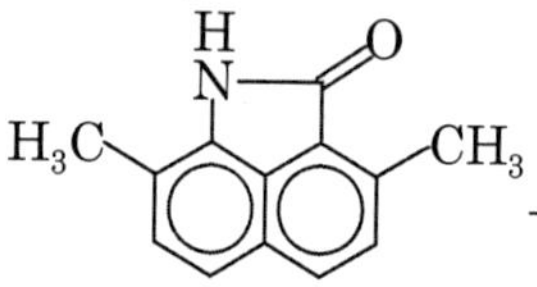 $\xrightarrow{Fe/Br_2}$. The product of monobromination of this compound is

(A) 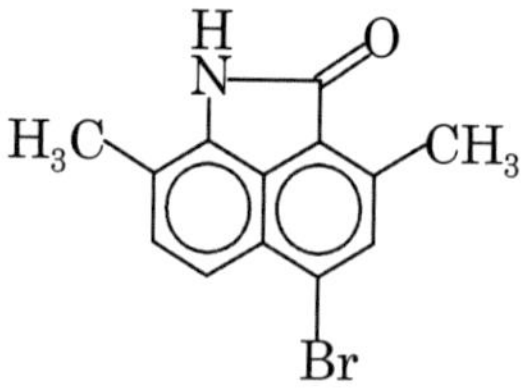

(B) 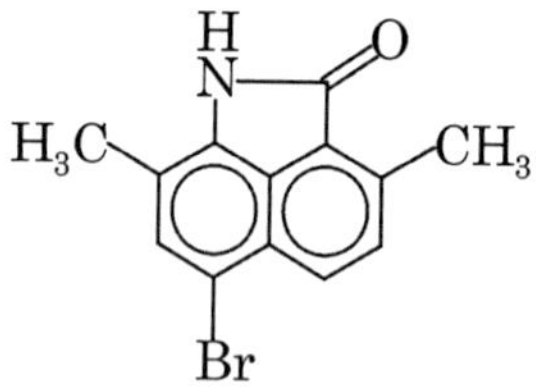

(C) 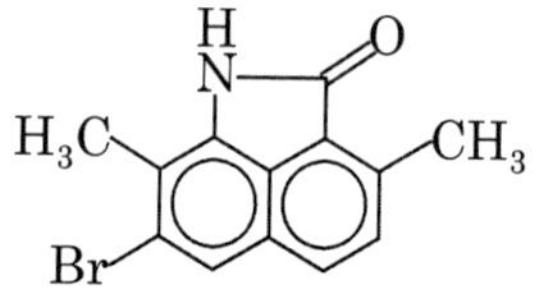

(D) 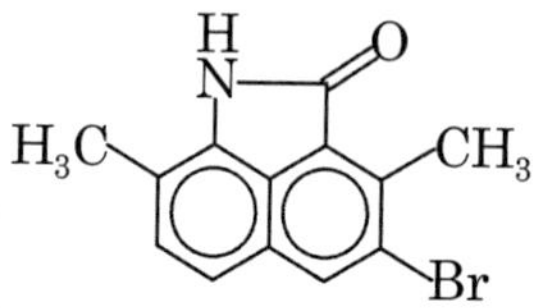

Hint: The ring with maximum electron density will be substituted by an electrophile. The ring attached to —NH— will have rich electron density due to resonance. As the *ortho* position is blocked, the electrophile attacks the *para* position.

80. When phenyl magnesium bromide reacts with tertiary butanol, which of the following is formed?

(A) Tertiary butyl methyl ether (B) Benzene

(C) Tertiary butyl benzene (D) Phenol

Hint: $PhMgBr + Me_3COH \rightarrow Ph{-}H + Me_3COMgBr$

81. 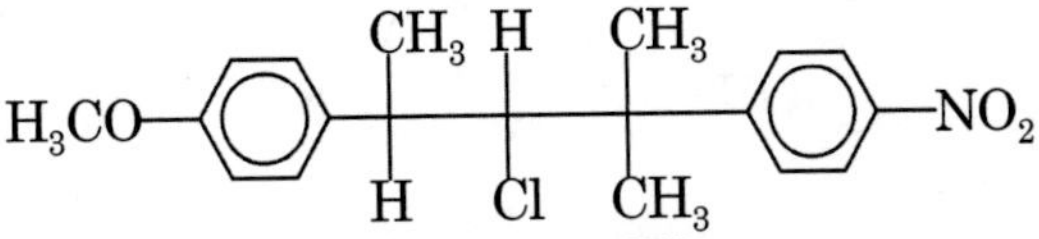 on hydrolysis in aqueous acetone gives

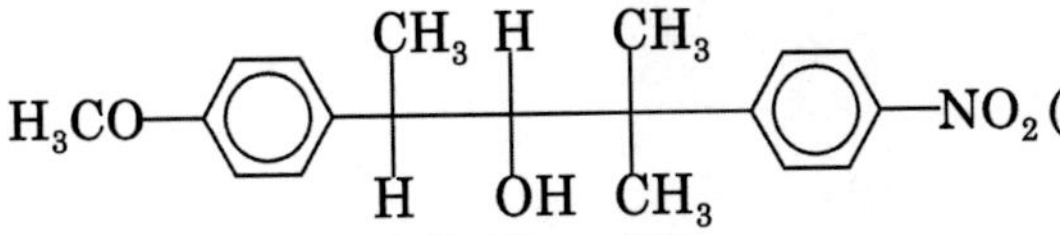 (K)

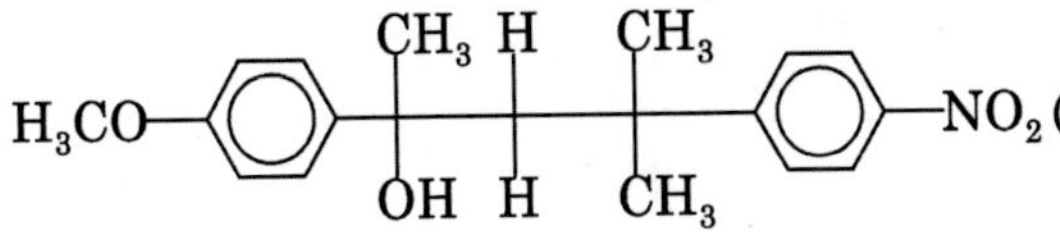 (L)

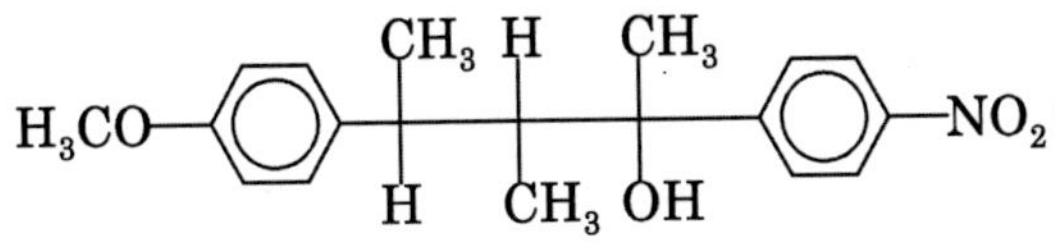 (M)

It mainly gives

(A) K and L (B) only K (C) L and M (D) only M **(I.I.T. 2005)**

Hint: Both S_{N^1} and S_{N^2} reactions are possible due to aqueous acetone solution.

82. The increasing order of hydrolysis of the following halogen derivatives is

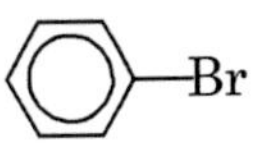 I

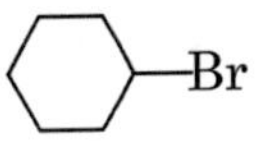 II

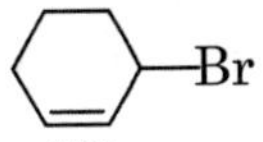 III

$(CH_3)_3C—Br$ IV

(A) I < II < IV < III (B) IV < III < II < I

(C) I < IV < II < III (D) I < II < III < IV

83. In the reaction, 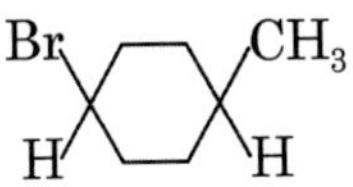 $+ OH^- \xrightarrow{S_{N^2}} X$, X is

(A) 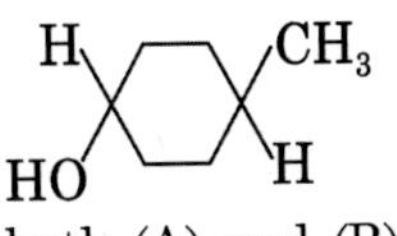 (B) 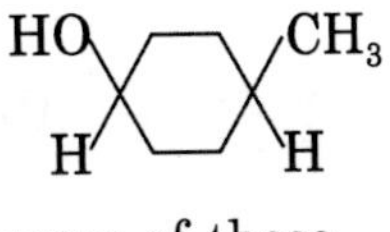

(C) both (A) and (B) (D) none of these.

84. In the reaction, 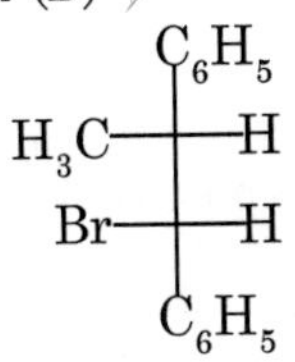 $\xrightarrow[E2]{KOH/C_2H_5OH} X$, X is

(A) 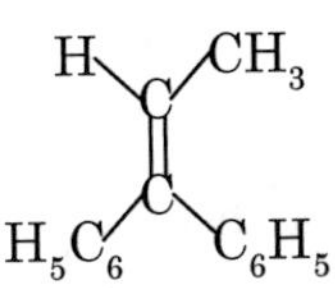 (B) 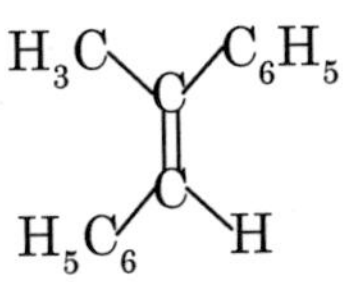 (C) 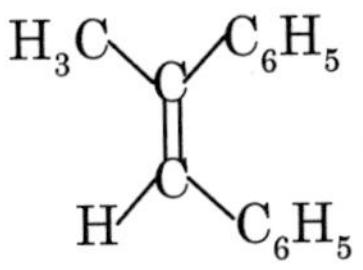 (D) none of these.

85. In the reaction, 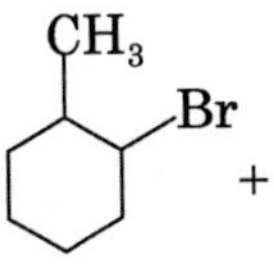 $+ CH_3ONa \xrightarrow{CH_3OH} X$, X is

(A) 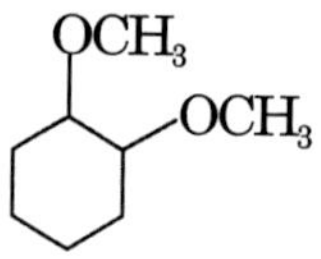 (B) 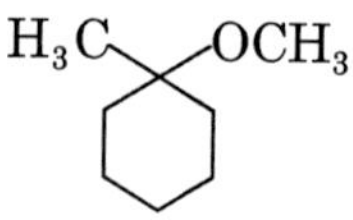 (C) 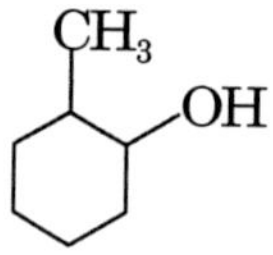 (D) 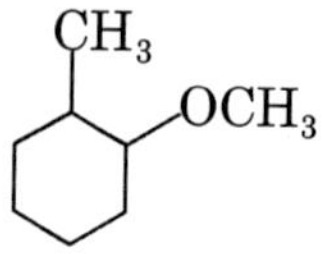

86. In the reaction, 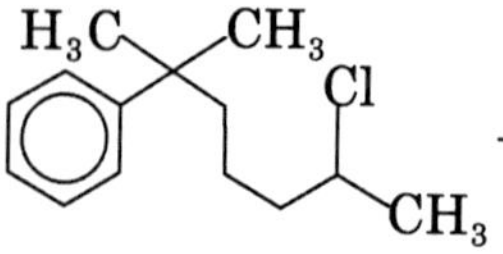 $\xrightarrow[CH_3NO_2,\ 0°C]{FeCl_3}$ X, X is

(A) 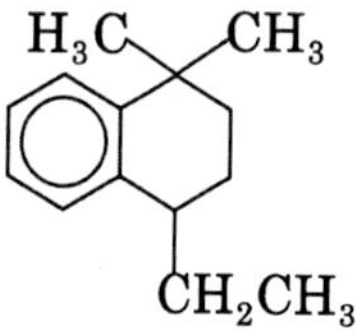

(B) 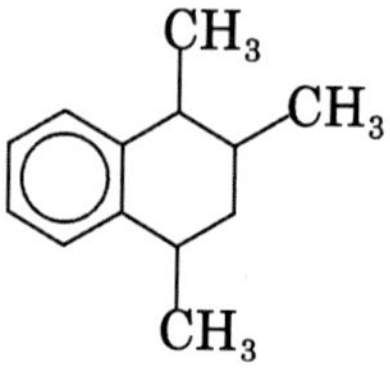

(C) 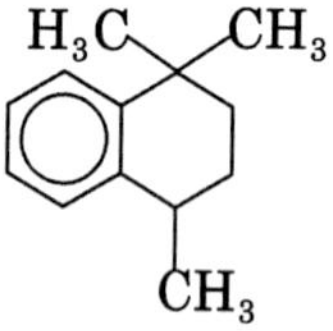

(D) none of these.

87. In the sequence of reactions, 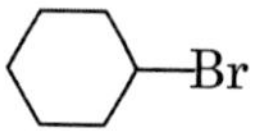 $\xrightarrow{NH_3}$ X $\xrightarrow{O_3/H_2O}$ Y $\xrightarrow[\Delta]{BaO}$ Z, Z is

(A) $HOOC(CH_2)_4COOH$

(B) 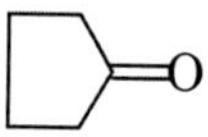

(C) 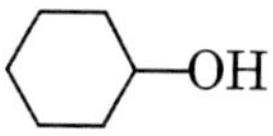

(D) 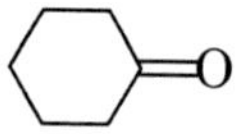

88. In the reaction, 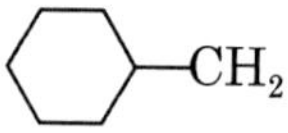 $\xrightarrow{Br_2/H_2O}$ X, X is

(A) 

(B) 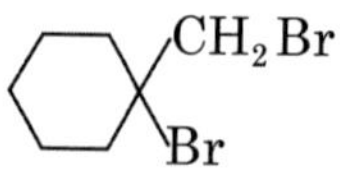

(C) 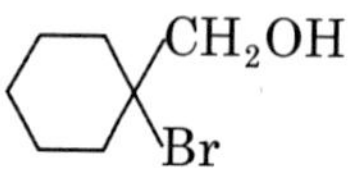

(D) 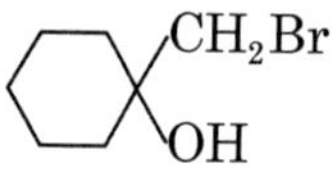

89. Which of the following is the best method for the preparation of 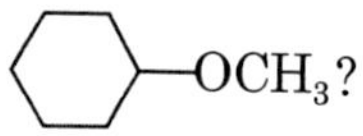?

I. 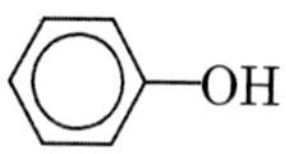 + CH_2N_2 ⟶ II. 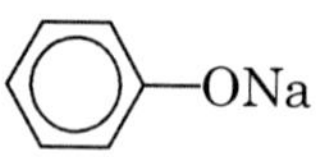 + CH_3Br ⟶

III. 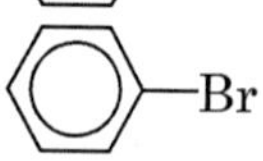 + CH_3ONa ⟶

(A) III (B) II (C) I (D) All of these.

90. In the reaction, 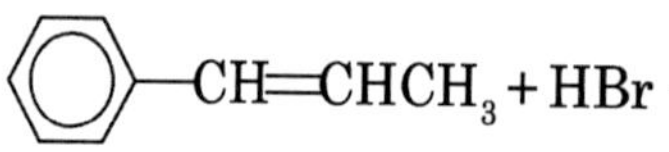 + HBr ⟶, the major product is

(A) 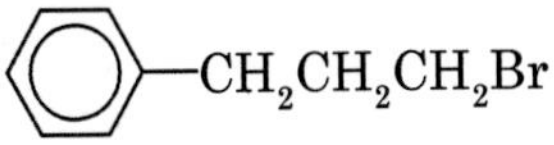

(B) 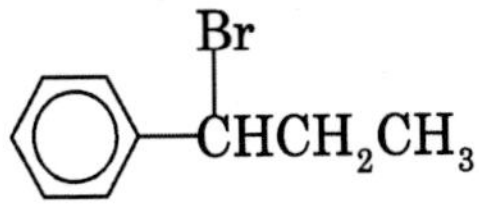

(C) 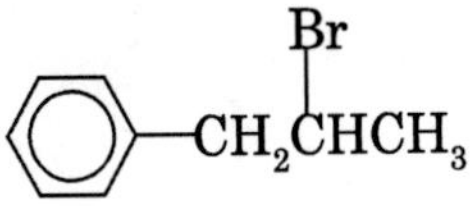

(D) no reaction.

91. 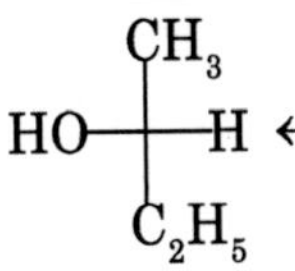 $\xleftarrow[\text{II}]{OH^-}$ 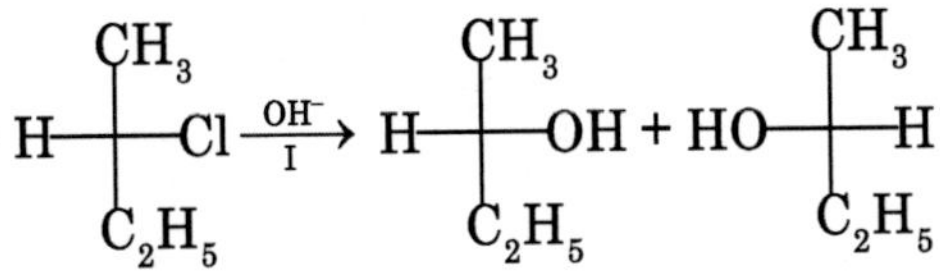

The steps I and II are respectively

(A) S_{N^2}, S_{N^1} (B) S_{N^1}, S_{N^2} (C) S_{N^2}, S_{N^2} (D) S_{N^1}, S_{N^1}

92. In the reaction, $CH_3CH(Br)CH(CH_3)CH_3 + C_2H_5O^- \xrightarrow{S_{N^1}}$ X, the major product X is

(A) $CH_3CH_2C(CH_3)(OC_2H_5)CH_3$ (B) $CH_3CH(CH_3)CH_2CH_2OC_2H_5$

(C) $CH_3CH(OC_2H_5)CH(CH_3)CH_3$ (D) none of these.

93. In the reaction, $CH_3CH_2CH(Br)CH_2CH{=}CH_2 \xrightarrow{KOH/C_2H_5OH}$ X, the major product X is

(A) $CH_3CH_2CH_2CH_2C{\equiv}CH$ (B) $CH_2{=}CHCH_2CH_2CH{=}CH_2$

(C) $CH_3CH{=}CHCH_2CH{=}CH_2$ (D) $CH_3CH_2CH{=}CHCH{=}CH_2$

94. In the reaction, $H_3C{-}C(CH_3)_2{-}Cl + C_2H_5ONa \rightarrow$ X, the major product X is

(A) $CH_3CH(CH_3)CH_2OC_2H_5$ (B) $CH_3{-}C(CH_3){=}CH_2$

(C) $CH_3C(CH_3)_2OC_2H_5$ (D) none of these.

95. Which of the following haloarenes undergoes replacement of —Cl by —OH by merely warming the haloarene with aqueous sodium hydroxide?

(A) 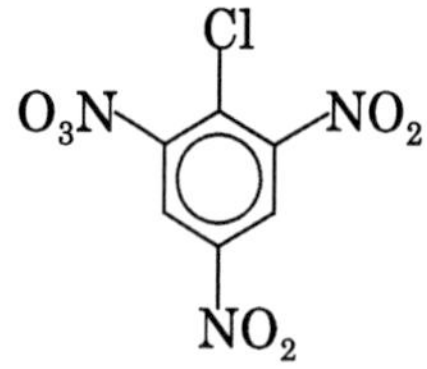

(B) 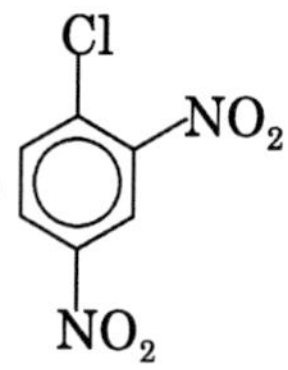

(C) 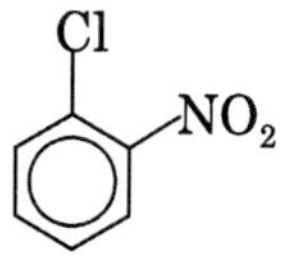

(D) 

96. 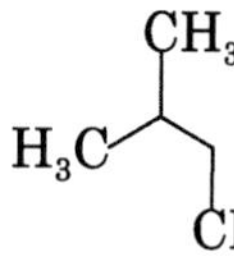 $\xrightarrow{Cl_2,\ h\nu}$ N isomers of $C_5H_{11}Cl$ $\xrightarrow{\text{fractional distillation}}$ M distilled products.

The values of N and M are:

(A) 6,6 (B) 6,4 (C) 4,4 (D) 3,3 **(I.I.T. 2006)**

Hint: 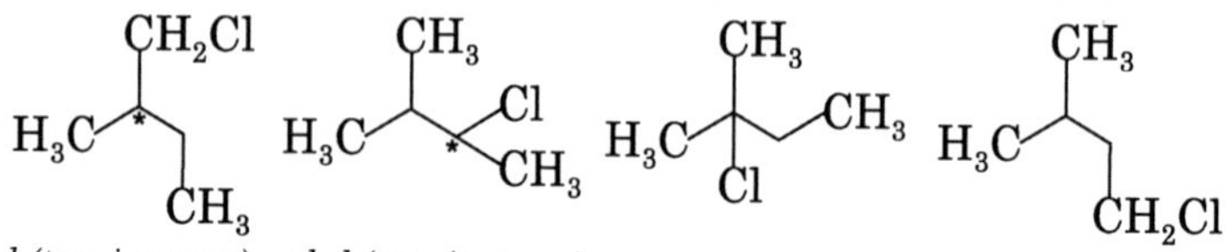

d, l (two isomers) *d, l* (two isomers)

M: two *d, l* compounds (four isomers) cannot be separated by fractional distillation.

97. The major product of the following reaction is

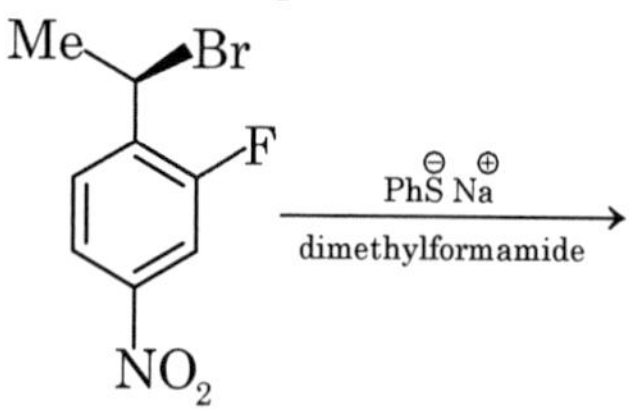

(A) 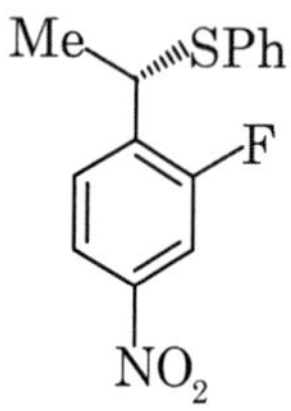 (B) 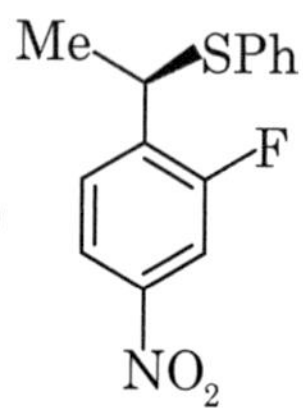 (C) 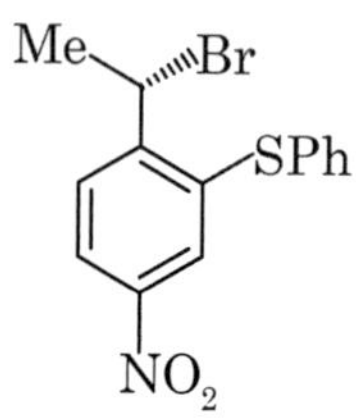 (D) 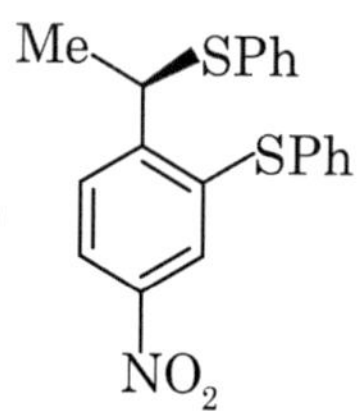 **(I.I.T. 2008)**

Hint: The given organic compound is a haloalkane as well as haloarene. Since PhS^- is a weak base, it cannot cause nucleophilic substitution by benzyne mechanism. Since the NO_2 group is present at the *meta* position with respect to the fluorine atom, S_NAr mechanism is also ruled out. Hence, the given compound reacts with PhS^- by S_{N^2} mechanism.

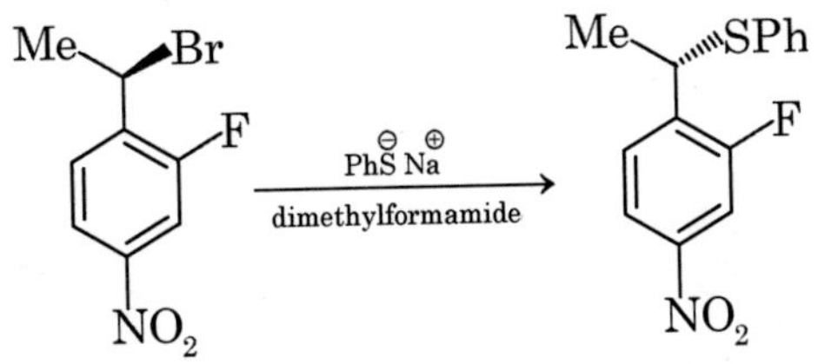

98. Statement–1: Bromobenzene upon reaction with Br_2/Fe gives 1,4-dibromobenzene as the major product.

Statement–2: In bromobenzene, the inductive effect of the bromo group is more dominant than the mesomeric effect in directing the incoming electrophile.

(A) Statement–1 is true, Statement–2 is true; Statement–2 is a correct explanation for Statement–1

(B) Statement–1 is true, Statement–2 is true; Statement–2 is not a correct explanation for Statement–1

(C) Statement–1 is true, Statement–2 is false

(D) Statement–1 is false, Statement–2 is true. **(I.I.T. 2008)**

Hint: The Statement–2 is correct because the inductive effect is a permanent effect and the inductive effect of halogen atom always dominates its mesomeric effect. But the Statement–2 cannot explain the formation of 1,4-dibromobenzene as the major product. The reason for the formation of 1,4-dibromobenzene is that Br is an *ortho* and *para* directing group due to the resonance stabilization of arenium ion. The prominent attack of Br^+ occurs at the *para* position due to the steric effect.

ANSWERS

1. D	2. A, C	3. C	4. B	5. A	6. B
7. C, D	8. B, D	9. B	10. C, D	11. B	12. A, D
13. A	14. A	15. A	16. D	17. B, D	18. A, C
19. A	20. B	21. A	22. D	23. B	24. C
25. A	26. A	27. D	28. C	29. A	30. A
31. B, C, D	32. A, B, D	33. B	34. B	35. A	36. A
37. A	38. D	39. C	40. B	41. B	42. C
43. C	44. D	45. B	46. A	47. D	48. B
49. D	50. B	51. A	52. D	53. B	54. A
55. B	56. C	57. B	58. A	59. A	60. D
61. B	62. D	63. B	64. D	65. C	66. A
67. B	68. D	69. A	70. C, D	71. A, B, D	72. A
73. C	74. B	75. A, B, D	76. A	77. C	78. A
79. B	80. B	81. A	82. A	83. A	84. C
85. B	86. A	87. B	88. D	89. B	90. B
91. B	92. A	93. D	94. B	95. A	96. B
97. A	98. B				

38

Compounds with Functional Groups Containing Oxygen—I: Alcohols, Phenols and Ethers

OBJECTIVE TYPE MULTIPLE CHOICE QUESTIONS

1. Propan-1-ol and propan-2-ol can be distinguished by
   (A) oxidation with alkaline $KMnO_4$ followed by reaction with Fehling solution
   (B) oxidation with acidic $K_2Cr_2O_7$ followed by reaction with Fehling solution
   (C) oxidation by heating with Cu followed by reaction with Fehling solution
   (D) oxidation with conc. H_2SO_4 followed by reaction with Fehling solution.

   (I.I.T. 2001)

   Hint: Heating of an alcohol with copper causes dehydration.
   1-Propanol on dehydration gives propanol while 2-propanol on dehydration gives propanone. Only propanol reacts with Fehling solution.

$$\underset{\text{1-Propanol}}{CH_3—CH_2—CH_2—OH} \xrightarrow{\text{Cu, 300°C}} CH_3—CH_2—CHO \xrightarrow[\text{Fehling solution}]{2CuO} \underset{\text{Red}}{Cu_2O\downarrow}$$

$$\underset{\text{2-Propanol}}{CH_3—\underset{\substack{|\\OH}}{CH}—CH_3} \xrightarrow{\text{Cu, 300°C}} \underset{\text{Propanone}}{CH_3—CO—CH_3} \xrightarrow[\text{Fehling solution}]{2CuO} \text{No reaction}$$

2. Benzenediazonium chloride on reaction with phenol in weakly basic medium gives
   (A) diphenyl ether (B) *p*-hydroxyazobenzene
   (C) chlorobenzene (D) benzene **(I.I.T. 1998)**
3. The order of reactivity of the following alcohols towards HCl is

(I) $FCH_2\underset{\substack{|\\ CH_3}}{C}HOH$ (II) $FCH_2CH_2\underset{\substack{|\\ CH_3}}{C}HOH$ (III) $(CH_3)_2CHOH$ (IV) $C_6H_5CH_2OH$

(A) I > II > III > IV (B) I > III > II > IV

(C) IV > III > II > I (D) IV > III > I > II **(I.I.T. 1997)**

4. The order of acidity of (I) phenol, (II) m-nitrophenol, (III) *p*-nitrophenol and (IV) *p*-methylphenol is

(A) III > II > I > IV (B) II > III > I > IV

(C) I > III > II > IV (D) IV > I > II > III **(I.I.T. 1996)**

5. The products formed in the reaction, $C_6H_5OCH_3 + HI \xrightarrow{\text{heat}}$ are

(A) $C_6H_5OH + CH_3I$ (B) $C_6H_5I + CH_3OH$

(C) $C_6H_5CH_3 + HOI$ (D) $C_6H_5I + CH_3OI$

6. Which of the following species is most stable?

(A) $p\text{-}O_2N—C_6H_4—\overset{+}{C}H_2$ (B) $p\text{-}Cl—C_6H_4—\overset{+}{C}H_2$

(C) $C_6H_5\overset{+}{C}H_2$ (D) $p\text{-}CH_3O—C_6H_4—\overset{+}{C}H_2$ **(I.I.T. 1995)**

7. When phenol is treated with $CHCl_3$ and NaOH followed by acidification, salicylaldehyde is obtained. Which of the following species is involved in the above reaction as intermediate?

(A) 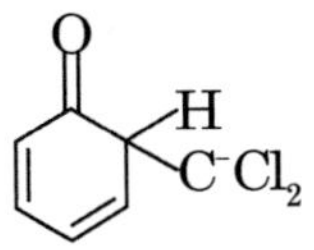 (B) 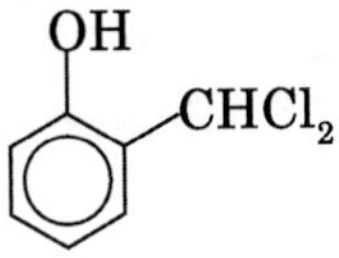 (C) 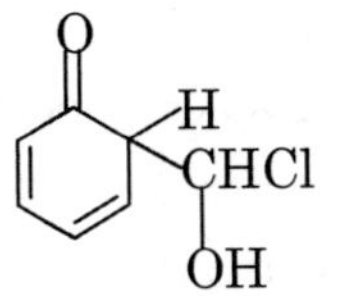 (D) 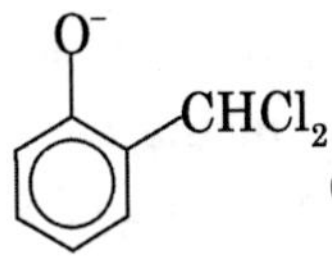 **(I.I.T. 1995)**

8. An organic compound, C_3H_6O does not give a precipitate with 2,4-dinitrophenylhydrazine and does not react with metallic sodium. It could be

(A) CH_3CH_2CHO (B) $CH_2{=}CHCH_2OH$

(C) CH_3COCH_3 (D) $CH_2{=}CHOCH_3$ **(I.I.T. 1993)**

9. The products of combustion of an aliphatic thiol at 298 K are

(A) $CO_2(g)$, $H_2O(g)$ and $SO_2(g)$ (B) $CO_2(g)$, $H_2O(l)$ and $SO_2(g)$

(C) $CO_2(l)$, $H_2O(l)$ and $SO_2(g)$ (D) $CO_2(g)$, $H_2O(l)$ and $SO_2(l)$ **(I.I.T. 1992)**

10. The acids which do not contain —COOH group are

(A) ethanoic acid (B) picric acid

(C) lactic acid (D) *p*-toluenesulphonic acid **(I.I.T. 1991)**

11. In the reaction of phenol with $CHCl_3$ and aq. NaOH at 70°C, the electrophile attacking the ring is

(A) $CHCl_3$ (B) $\cdot CHCl_2$ (C) $:CCl_2$ (D) $COCl_2$ **(I.I.T. 1991)**

12. The compound which gives most stable carbocation during dehydration is

(A) $(CH_3)_2CHCH_2OH$ (B) $(CH_3)_3COH$

(C) $CH_3CH_2CH_2CH_2OH$ (D) $CH_3CH(OH)CH_2CH_3$ **(I.I.T. 1989)**

13. In ethyl alcohol, the bond that undergoes hydrolytic cleavage most readily is

(A) C—C (B) C—O (C) C—H (D) O—H **(I.I.T. 1988)**

14. Hydrogen bonding is maximum in

(A) triethylamine (B) diethyl ether (C) ethyl chloride (D) ethanol **(I.I.T. 1987)**

15. Phenol reacts with bromine in CS_2 at low temperature to give
(A) *m*-bromophenol (B) *o*- and *p*-bromophenol
(C) *p*-bromophenol (D) 2,4,6-tribromophenol **(I.I.T. 1988)**

16. Statement (*S*): Solubility of *n*-alcohols in water decreases with increase in molecular mass.
Explanation (*E*): The relative proportion of the hydrocarbon part in alcohols increases with increasing molecular mass which permits enhanced hydrogen bonding water.
(A) Both *S* and *E* are true and *E* is the correct explanation of *S*
(B) Both *S* and *E* are true but *E* is not the correct explanation of *S*
(C) *S* is true but *E* is false (D) *S* is false but *E* is true. **(I.I.T. 1988)**

17. Phenol is less acidic than
(A) acetic acid (B) *p*-methoxyphenol
(C) *p*-nitrophenol (D) ethanol **(I.I.T. 1986)**

18. HBr reacts fastest with
(A) 2-methylpropan-2-ol (B) propan-1-ol
(C) propan-2-ol (D) 2-methylpropan-1-ol **(I.I.T. 1986)**

19. An industrial method of preparation of methanol is
(A) catalytic hydrogenation of CO in the presence of $ZnO–Cr_2O_3$
(B) by reducing CH_4 with steam at 900°C with Ni catalyst
(C) by reduction of HCHO with $LiAlH_4$
(D) by reaction of HCHO with dil. NaOH **(I.I.T. 1984)**

20. When phenol is treated with excess bromine water, it gives
(A) *m*-bromophenol (B) *o*- and *p*-bromophenol
(C) 2,4-dibromophenol (D) 2,4,6-tribromophenol **(I.I.T. 1984)**

21. Diethyl ether on heating with concentrated HI gives two moles of
(A) ethanol (B) iodoform
(C) ethyl iodide (D) methyl iodide **(I.I.T. 1983)**

22. Ethyl alcohol is heated with conc. H_2SO_4 at 170°C. The product formed is
(A) $CH_3COOC_2H_5$ (B) C_2H_6 (C) C_2H_4 (D) C_2H_2 **(I.I.T. 1981)**

23. The compound which reacts fastest with Lucas reagent at room temperature is
(A) 2-methylpropan-1-ol (B) 2-methylpropan-2-ol
(C) propan-1-ol (D) propan-2-ol **(I.I.T. 1981)**

24. Which one/ones of the following reactions will yield propan-2-ol?
I. $CH_2{=}CH{-}CH_3 + H_2O \xrightarrow{H^+}$ II. $CH_3CHO \xrightarrow{\text{(i)}CH_3MgI\ \text{(ii)}H_2O}$
III. $HCHO \xrightarrow{\text{(i)}CH_3MgI\ \text{(ii)}H_2O}$ IV. $CH_2{=}CH{-}CH_3 \xrightarrow{\text{neutral } KMnO_4}$
(A) I and II (B) II and III (C) I and III (D) II and IV **(I.I.T. 1994)**

25. Which of the following is basic?
(A) CH_3CH_2OH (B) $HOCH_2CH_2OH$

(C) CH_3COOH (D) H_2O_2 **(I.I.T. 1980)**

26. Phenol on reduction with H_2 in the presence of Ni catalyst gives
(A) benzene (B) toluene (C) cyclohexane (D) cyclohexanol **(Roorkee 1993)**

27. Which of the following undergoes easy dehydration?
(A) 3-Methylbutan-2-ol (B) Ethanol
(C) 2-Methylpropan-2-ol (D) 2-Methylbutan-2-ol **(Roorkee 1995)**

28. Which of the following undergoes iodoform test?
(A) CH_3CH_2OH (B) CH_3COCH_3
(C) $CH_3CH_2COCH_2CH_3$ (D) $(CH_3)_2CHCOCH_2CH_3$ **(Roorkee 1995)**

29. After the reaction of $C_6H_5SO_3H$ with NaOH, it is hydrolysed by acid. The product formed is
(A) phenol (B) benzoic acid (C) benzene (D) disodium benzaldehyde

30. Intermediate involved in the Reimer–Tiemann reaction is a
(A) carbocation (B) carbanion (C) free radical (D) carbene

31. Dehydration of alcohol is an example of
(A) addition reaction (B) elimination reaction
(C) substitution reaction (D) redox reaction.

32. The reaction of propan-2-ol with hot Cu at 570 K produces
(A) propanone (B) propan-1-ol (C) propene (D) propanal

33. The order of reactivity of HX (X = Cl, Br, I) with ROH is
(A) HI < HBr > HCl (B) HI > HBr < HCl
(C) HI < HBr < HCl (D) HI > HBr > HCl

34. The appearance of cloudiness in the Lucas test for alcohols is due to the formation of
(A) alkyl chlorides (B) acid chlorides
(C) aldehydes (D) ketones

35. The alcohol which does not react with Lucas reagent is
(A) isobutyl alcohol (B) *n*-butanol
(C) *tert*-butyl alcohol (D) *sec*-butyl alcohol

36. Which of the following is formed when butan-1-ol is dehydrated?
(A) 2-methylpropene (B) but-1-ene as the major product
(C) but-2-ene as the major product (D) equal amounts of but-1-ene and but-2-ene

37. The compound obtained by the reaction of ethene with diborane followed by hydrolysis with alkaline H_2O_2 is
(A) ethanol (B) propanol (C) ethanal (D) triethyl boride

38. An organic compound *A* on treatment with acidified $K_2Cr_2O_7$ gives a compound *B* which reacts with I_2 and Na_2CO_3 to form triiodomethane. *B* is
(A) CH_3OH (B) CH_3COCH_3 (C) CH_3CHO (D) $CH_3CH(OH)CH_3$

39. Which set of the following reagents would you select to convert acetophenone to $C_6H_5C(OH)(CH_3)CH(CH_3)_2$?

(A) $(CH_3)_2CHMgBr$ and acid hydrolysis
(B) $CH_3CH_2CH_2MgBr$ and acid hydrolysis
(C) $CH_3CH(OH)CH_3$, Zn (D) $CH_3CH(Br)CH_3$, $AlCl_3$

40. An alcohol on oxidation gives CH_3COOH and CH_3CH_2COOH. The alcohol is
(A) $CH_3CH(OH)CH_2CH_3$ (B) $CH_3CH_2CH_2OH$
(C) $(CH_3)_2C(OH)CH_2CH_3$ (D) $CH_3(CH_2)_3CH_2OH$

41. In the reactions, $(CH_3)_2CHOH \xrightarrow{\text{mild oxid.}} X \xrightarrow{\text{(i) } CH_3MgBr \text{ (ii) } H_2O} Y$, Y is
(A) isobutylene (B) isobutyl alcohol
(C) *tert*-butyl alcohol (D) *sec*-butyl alcohol

42. In the reactions, $CH_2{=}CH_2 \xrightarrow{O_2,\ Ag} X \xrightarrow{473\text{ K, steam}} Y$, Y is
(A) ethanal (B) ethanol
(C) ethylene glycol (D) epoxyethane

43. A compound is soluble in conc. H_2SO_4. It does not decolourise bromine in CCl_4 but is oxidised by chromic anhydride in aqueous H_2SO_4 in two seconds, turning orange solution to blue green, then opaque. The compound is
(A) an alkene (B) an ether
(C) a primary alcohol (D) a secondary alcohol

44. Which of the following will not be soluble in sodium carbonate solution?
(A) 2,4,6-trinitrophenol (B) benzoic acid
(C) benzensulphonic acid (D) *o*-nitrophenol

45. In the reactions, $C_2H_5OH \xrightarrow{PBr_3} X \xrightarrow{\text{alc. KOH}} Y \xrightarrow{\text{(i) } H_2SO_4\text{, (ii) } H_2O\text{, heat}} Z$, Z is
(A) $CH_2{=}CH_2$ (B) $C_2H_5OC_2H_5$ (C) C_2H_5OH (D) none of these.

46. The compound $\begin{matrix} CH_2OOCCH_3 \\ | \\ CH_2OOCCH_3 \end{matrix}$ is obtained by the reaction of
(A) ethanol and ethanal (B) acetone and glycol
(C) glycerol and $(CH_3CO)_2O$ (D) glycol and CH_3COCl

47. The ether $C_6H_5OCH_2C_6H_5$ when treated with HI produces
I. $C_6H_5CH_2I$ II. $C_6H_5CH_2OH$ III. C_6H_5I IV. C_6H_5OH
(A) I and II (B) I and III (C) I, II and III (D) all of these. **(I.I.T. 1999)**

48. Which of the following reagents will convert propanoic acid to propon-1-ol?
(A) $KMnO_4$ (B) $LiAlH_4$ (C) MnO_2 (D) Cr_2O_3

49. Which one of the following will most readily be dehydrated in acidic conditions?
(A) $CH_3COCH_2CH(OH)CH_3$ (B) $CH_3CH(OH)CH_2CH_2CH_3$
(C) $CH_3COCH_2CH_2CH(OH)CH_3$ (D) $CH_3COCH(OH)CH_2CH_3$ **(I.I.T. 2000)**

50. To prepare 3-ethylpentan-3-ol, the reagents needed are
(A) $CH_3CH_2MgBr + CH_3COCH_2CH_3$ (B) $CH_3MgBr + CH_3CH_2CH_2COCH_2CH_3$
(C) $CH_3CH_2MgBr + CH_3CH_2COCH_2CH_3$
(D) $CH_3CH_2CH_2MgBr + CH_3COCH_2CH_3$

51. Which of the following reactions does not yield an alkyl halide?
    (A) Diethyl ether + Cl_2 (B) Diethyl ether + HI
    (C) Diethyl ether + PCl_5 (D) Diethyl ether $\xrightarrow{\text{reduction}}$ X $\xrightarrow{SO_2Cl_2}$
52. An aromatic ether is not cleaved by HI even at 525 K. The ether is
    (A) $C_6H_5OCH_3$ (B) $C_6H_5OC_6H_5$ (C) $C_6H_5OC_3H_7$ (D) tetrahydrofuran
53. Which of the following are the starting materials for the Grignard's synthesis of *tert*-butanol?
    (A) $CH_3MgBr + CH_3COCH_3$ (B) $CH_3CH_2MgBr + CH_3COCH_3$
    (C) $CH_3MgBr + CH_3CHOHCH_3$ (D) $CH_3MgBr + (CH_3)_3CH$
54. *p*-Phenolsulphonic acid is formed as the major product when phenol is treated with
    (A) dil. H_2SO_4 at 373 K (B) conc. H_2SO_4 at 373 K
    (C) dil. H_2SO_4 at 293 K (D) conc. H_2SO_4 at 293 K
55. An organic compound, C_7H_8O dissolves in NaOH and gives a characteristic colour when treated with neutral $FeCl_3$ solution. On reaction with Br_2, it gives a tribromo derivative ($C_7H_5OBr_3$). The compound is
    (A) *o*-cresol (B) *m*-cresol (C) *p*-cresol (D) benzyl alcohol
56. The reagents used to prepare salicylaldehyde from phenol is
    (A) (CaO + NaOH), heat (B) ($CHCl_3$ + NaOH), 340 K
    (C) (CCl_4 + NaOH), 340 K (D) (CO_2 + NaOH), 410 K
57. Which of the following does not produce acrolein when heated with glycerol?
    (A) $KHSO_4$ (B) Conc. H_2SO_4 (C) P_4O_{10} (D) Anhy. $ZnCl_2$
58. Glycerol on warming with a small amount of HI is converted to
    (A) $CH_3CH{=}CH_2$ (B) CH_3CHICH_3 (C) $CH_2{=}CHCH_2I$ (D) ICH_2CHICH_2I
59. Oxalic acid is converted into formic acid in one step by heating it
    (A) with $KHSO_4$ (B) with glycerol at 503 K
    (C) with glycerol at 383 K (D) at 473 K
60. Diethyl ether is prepared by passing vapours of ethanol over a catalyst under high pressure and temperature. The catalyst is
    (A) SiO_2 (B) CuO (C) Ag_2O (D) Al_2O_3
61. Which compound is obtained on passing ethanol vapours on heated Al_2O_3?
    (A) Diethyl ether (B) Acetone (C) Acetaldehyde (D) Ethane
62. The reaction of ethylene glycol with PI_3 gives
    (A) ICH_2CH_2I (B) $CH_2{=}CH_2$ (C) $CH_2{=}CHI$ (D) $ICH{=}CHI$
63. Identify the incorrect statement.
    (A) (+)-3-Methylhexan-3-ol on reacting with conc. HBr forms (–)-3-bromo-3-methylhexane (major) and (±)-3-bromo-3-methylhexane (minor)
    (B) (+)-2-Hexanol on reacting with conc. HBr forms (–)-2-bromohexane
    (C) The reaction of HBr with 2-methyl propan-2-ol follows S_{N^2} mechanism
    (D) The reaction of HBr with butan-1-ol follows S_{N^2} mechanism.

    Hint: The statement (C) is incorrect and all others are correct.
    (A) The reaction occurs through S_{N^1} mechanism and formation of a carbocation. The

carbocation being planar, the attack of Br^- is equally probable from the front side and back side. The H_2O shields $\overset{+}{C}$ from the frontal attack. There is no steric hindrance from the back side and as a result excess of (–)-isomer is formed.

(B) The reaction occurs through S_{N^2} mechanism and inversion of configuration. Br^- ion attacks from back side of the C, away from the departing H_2O.

(C) The reaction occurs through S_{N^1} mechanism.

$$(CH)_3COH + HBr \xrightleftharpoons{\text{fast}} (CH_3)_3C\overset{+}{O}H_2 + Br^-$$

$$(CH_3)_3C\overset{+}{O}H_2 \xrightarrow[-H_2O]{\text{slow}} (CH_3)_3\overset{+}{C} \xrightarrow[\text{fast}]{Br^-} (CH_3)_3CBr$$

(D) $C_4H_9OH + HBr \xrightleftharpoons{\text{fast}} C_4H_9\overset{+}{O}H_2 + Br^-$
Butan-1-ol

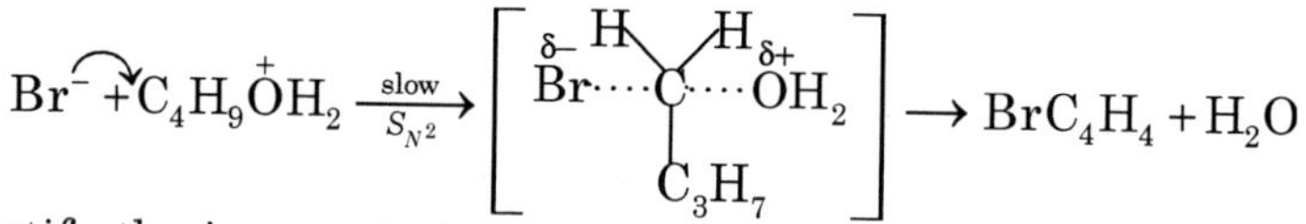

64. Identify the incorrect statement.
(A) The $\angle C—O—H$ in methanol is smaller than that of $\angle C—S—H$
(B) ROH is a stronger base than RSH
(C) The reaction of $(C_6H_5)_2C{=}CHCH_3$ with BH_3 in tetrahydrofuran followed by H_2O_2/OH^- gives $(C_6H_5)_2CHCH(OH)CH_3$
(D) The oxidation of $C_6H_5CH_2OH$ with MnO_2 produces C_6H_5CHO

Hint: The statement (A) is incorrect and all others are correct.

(A) The bond angles are as follows: $\angle C—O—H = 109°$, $\angle C—S—H = 96°$. The hybridisation of O is sp^3 while S uses *p*-orbitals for bonding with H and C atoms.

(B) Because of smaller size (more charge density) of O atom in comparison to S atom (less charge density), O atom is more capable of attracting H^+ ion.

(C) The hydration occurs according to anti-Markownikoff addition.

$$(C_6H_5)_2C{=}CHCH_3 \xrightarrow[\text{(ii) } H_2O_2/OH^-]{\text{(i) } BH_3/THF} (C_6H_5)_2CHCH(OH)CH_3$$

(D) $C_6H_5CH_2OH \xrightarrow[\text{[O]}]{MnO_2} C_6H_5CHO$

65. HBr reacts slowest with
(A) 4-$ClC_6H_5CH_2OH$
(B) 4-$CH_3OC_6H_4CH_2OH$
(C) 4-$O_2NC_6H_4CH_2OH$
(D) $C_6H_5CH_2OH$

66. Iodoform test is not given by
(A) $CH_3CH_2CH(OH)CH_3$
(B) 1-methylcyclohexan-1-ol
(C) $C_6H_5CH(OH)CH_2CH_3$
(D) $(CH_3)_2CH(OH)CH_3$

67. The reaction of 3,3-dimethylbutan-2-ol with conc. HCl gives
(A) 3,3-dimethyl-1-chlorobutane
(B) 2,3-dimethyl-2-chlorobutane
(C) 3,3-dimethyl-2-chlorobutane
(D) a mixture of 2,3-dimethyl-2-chlorobutane and 3,3-dimethyl-2-chlorobutane

68. Propanone can be converted into 2,3-dimethylbutane-2,3-diol by
(A) Wolff–Kishner reduction
(B) Mg/Hg/H_2O
(C) Zn/Hg/Cl
(D) none of these.

69. Ethoxyethane reacts with conc. H_2SO_4 to give
(A) oxonium ion (B) oxoanion
(C) alkyl free radical (D) zwitterion

70. Which of the following compounds will not form ether as the major product on reaction with sodium ethoxide?
(A) 1-Chloropropane (B) 2-Chloro-2-methylpropane
(C) 1-Chloroethane (D) 2-Chloropropane
Hint: In Williamson's synthesis a tertiary haloalkane does not form ether but undergoes *E*2 elimination to form an alkene.

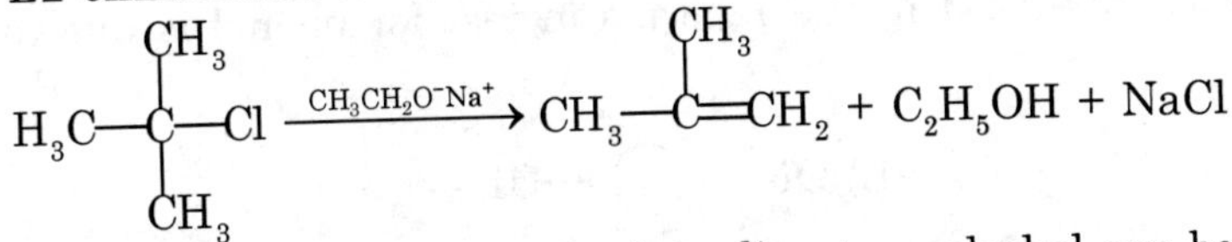

71. The conversion of propanone into diacetone alcohol can be effected by
(A) heat (B) $Ba(OH)_2$ (C) conc. H_2SO_4 (D) dry HCl gas

72. The reaction of ethanol with conc. H_2SO_4 at 130°C gives
(A) $CH_2{=}CH_2$ (B) $CH_3CH_2OCH_2CH_3$
(C) $CH_3CH_2OSO_3H$ (D) $CH_3CH_2OH_2^+HSO_4^-$

73. The reaction of 2,2-dimethylpropan-1-ol with conc. HCl gives
(A) 2-methylbut-2-ene (B) 2-chloro-2-methylbutane
(C) 1-chloro-2,2-dimethylpropane
(D) a mixture of 2-methylbut-2-ene and 1-chloro-2,2-dimethylbutane.

74. A secondary alcohol undergoes nucleophilic substitution reaction by
(A) S_{N^2} (B) S_{N^1}
(C) both S_{N^1} and S_{N^2} (D) neither S_{N^1} nor S_{N^2}

75. Which of the following is formed when $C_6H_5CH_2CH(OH)CH(CH_3)_2$ reacts with conc. H_2SO_4?

(A) 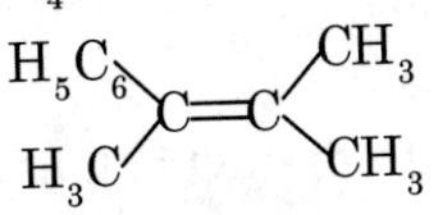
(B) 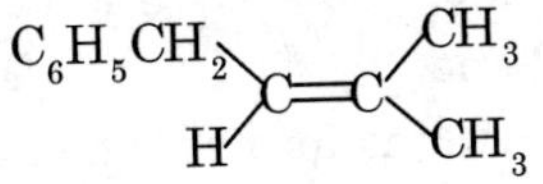
(C) 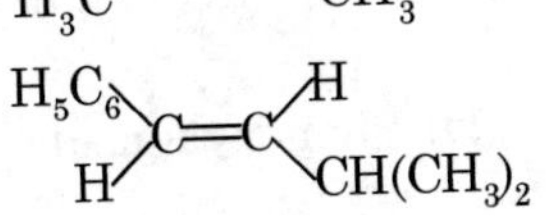
(D) 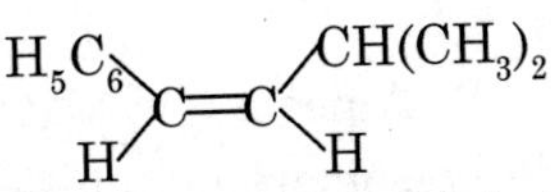

76. Which of the following test(s)/reagent(s) is used to distinguish between 2-methylpropan-1-ol and 2-methylpropan-2-ol?
(A) Oxidation with Cu at 573 K (B) Victor–Meyer test
(C) HCl/anhy. $ZnCl_2$ (D) I_2/NaOH

77. Which of the following reaction(s) is suitable to prepare methoxybenzene?
(A) $C_6H_5OMgBr + CH_3Br$ (B) $C_6H_5Cl + NaOCH_3$
(C) $C_6H_5OH + CH_2N_2$ (D) $C_6H_5ONa + CH_3Br$

78. Which of the following is not suitable for detecting phenolic group?
(A) Br_2/H_2O (B) dil. NaOH (C) I_2/NaOH (D) neutral $FeCl_3$

79. Which of the following is formed when phenol is exposed to air?
    (A) *o*-Benzoquinone (B) *p*-Benzoquinone
    (C) Phenoquinone (D) *o*- and *p*-Benzoquinones
80. Which of the following is formed when phenol reacts with cold dil. HNO_3?
    (A) 4-Nitrophenol (B) 2-Nitrophenol
    (C) 3-Nitrophenol (D) 2-Nitrophenol + 4-nitrophenol
81. Ethoxyethane does not react with
    (A) HI (B) HCl (C) CH_3COOH (D) H_2SO_4
82. The blue or green colour produced in the Libermann test for phenol is due to the formation of

    (A) 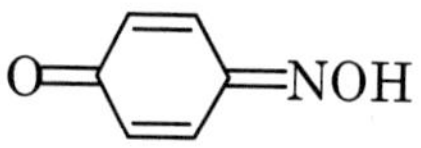 (B) 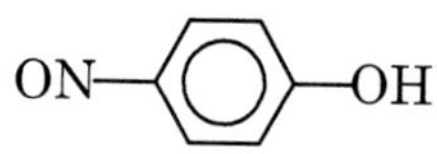

    (C) 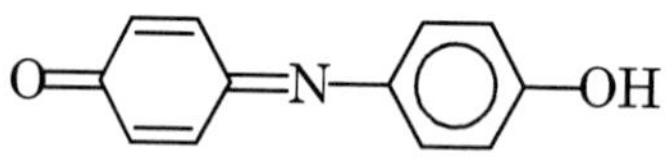 (D) 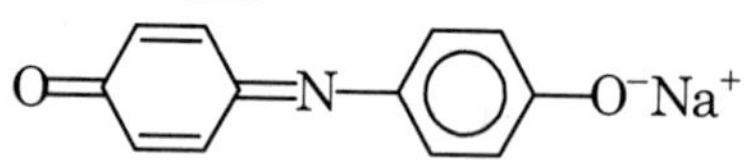

83. The reaction, $2CH_3CH_2OH \xrightarrow{H^+,\ 413\ K} CH_3CH_2OCH_2CH_3$ occurs through the formation of
    (A) $CH_3\overset{+}{C}H_2$ (B) $CH_3CH_2\overset{+}{O}H_2$ (C) $CH_3\dot{C}H_2$ (D) $CH_3CH_2{-}\overset{+}{\ddot{O}}(H){-}CH_2CH_3$
84. Which of the following reagents is suitable for the conversion of $CH_2{=}CH{-}CHO$ into $CH_2{=}CHCH_2OH$?
    (A) $LiAlH_4$ (B) H_2/Ni (C) Zn/HCl (D) Na/C_2H_5OH
85. Which of the following is formed when glycerol is heated with oxalic acid at 503 K?
    (A) Glyceric acid (B) Acrolein (C) Allyl alcohol (D) Methanoic acid
86. Which of the following is formed when tetrahydrofuran is treated with HBr at 373 K?
    (A) 4-Bromobut-1-ene (B) 4-Bromobutan-1-ol
    (C) 1-Bromobut-2-ene (D) 1,4-Dibromobutane
87. Which of the following is formed when phenol is treated with diethyl sulphate in the presence of NaOH?
    (A) Ethoxyethane (B) Anisole (C) Phenetole (D) Phenoxybenzene
88. The number of stereoisomers of pent-3-en-2-ol is
    (A) 2 (B) 3 (C) 4 (D) 5
    Hint: One chiral C atom; two geometrical and two optical isomers.
89. Which of the following is formed when ethane-1,2-diol is heated with a mixture of conc. HNO_3 and conc. H_2SO_4?
    (A) $CH_2ONO_2{-}CH_2OH$ (B) $CH_2ONO_2{-}CH_2ONO_2$ (C) $COOH{-}COOH$ (D) $CO + H_2$
90. In the reaction, $(CH_3)_3CCH{=}CH_2 \xrightarrow{H_2O,\ Hg(OAc)_2} X$, the compound X is

(A) $(CH_3)_2C(OH)CHCH_2CH_3$ (B) $(CH_3)_2CHCH(OH)CH_2CH_3$

(C) $(CH_3)_3CCHOHCH_3$ (D) $(CH_3)_3CCH_2CH_2OH$

91. In the reactions Y 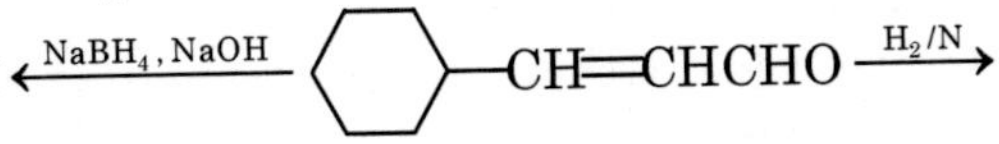 X, X and Y are respectively

(A) 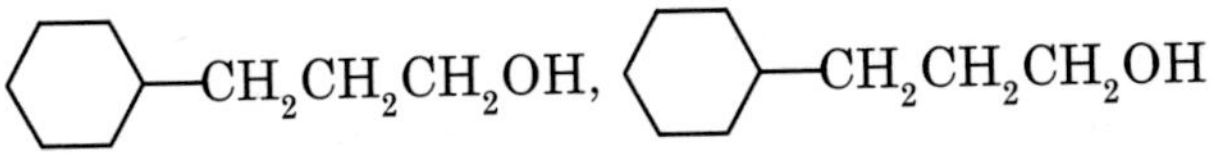

(B) 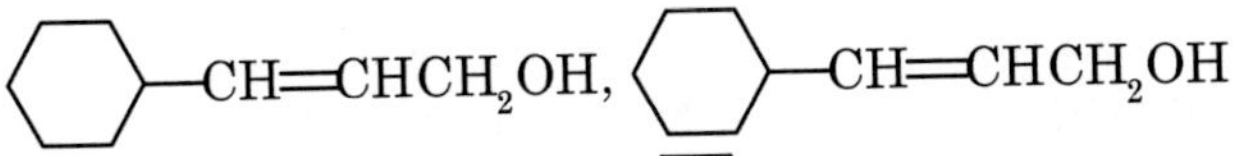

(C) 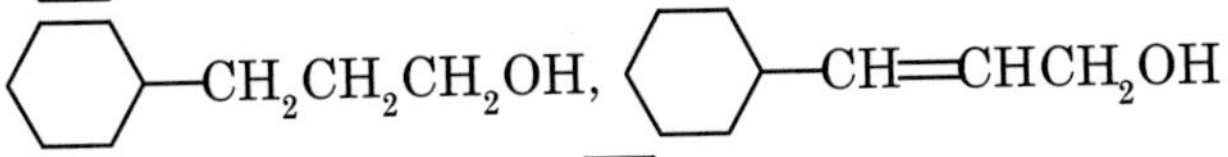

(D) 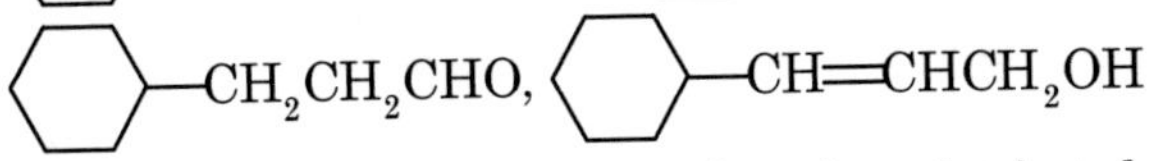

92. The best method of preparation for phenyl *t*-butyl ether is

(A) 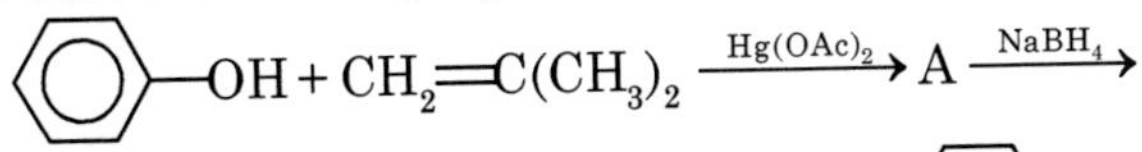

(B) 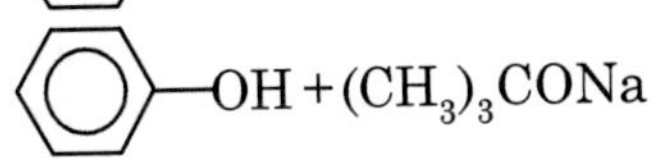 (C) 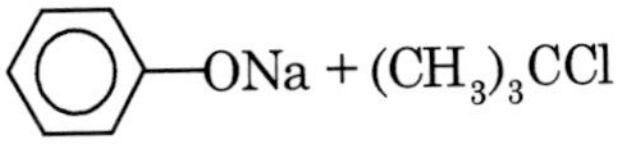

(D) none of these.

93. In the reaction,  X, X is

(A) $CH_3OCH_2CH(OH)CH_2Cl$ (B) $CH_3OCH_2CH(OH)CH_2Cl$

(C)  (D) 

94. Identify the correct statement.

(A) The reaction of $C_6H_5OCH_2CH_3$ with one equivalent of HI produces PhI and CH_3CH_2OH

(B) The reaction of $CH_3CH_2OCH_2CH_2CH_3$ with one equivalent of HI gives CH_3CH_2I and $CH_3CH_2CH_2OH$

(C) Bromination of phenol in non-aqueous medium gives 2,4,6-tribromophenol which in aqueous medium gives monobromophenol

(D) Sodium carbonate reacts with phenol because it is more acidic than phenol.

Hint: The statement (D) is correct and all others are incorrect.

(A) The products formed are C_6H_5OH and CH_3CH_2I. Aryl ethers do not cleave on the aryl side but cleave on the alkyl side.

(B) Unsymmetrical ethers produce a mixture of two sets of products: (i) CH_3CH_2OH, $CH_3CH_2CH_2I$, (ii) CH_3CH_2I, $CH_3CH_2CH_2OH$.

(C) In non-aqueous medium, the ionisation of phenol is not possible and the monobromination of phenol occurs. In aqueous medium, the ionisation of phenol leads to the formation of $C_6H_5O^-$ ion. The $C_6H_5O^-$ ion being more reactive than C_6H_5OH, it reacts with bromine to give 2,4,6-tribromophenol.

(D) $C_6H_5OH + CO_3^{2-} \rightarrow C_6H_5O^- + HCO_3^-$

$pK_a = 10$ (stronger) $pK_b = 3.7$ (stronger) $pK_b = 4$ (weaker) $pK_a = 10.3$ (weaker)

Since both the products are weaker than the corresponding reactants, the reaction proceeds in the forward direction.

95. Which of the following statements is correct?

(A) Hydration of cyclobutylethene in dil. H_2SO_4 gives 1-cyclobutylethanol

(B) Reaction of $4\text{-}O_2NC_6H_4CH_2COOH$ with $LiAlH_4$ gives $4\text{-}O_2NC_6H_4CH_2CH_2OH$

(C) Oxidation of $C_6H_5CH(OH)CH_2CH_2OH$ with MnO_2 yields $C_6H_5COCH_2CHO$

(D) Oxidation of $CH_3CH{=}CHCH_2OH$ with MnO_2 gives $C_6H_5COCH_2CH_2OH$

Hint: The statement (D) is correct and all others are incorrect.

(A) *cis*- and *trans*-methylcyclopentanol are formed. Addition of H^+ to the ethylenic C atom produces a secondary carbocation which rearranges by ring expansion to produce substituted cyclopropanol.

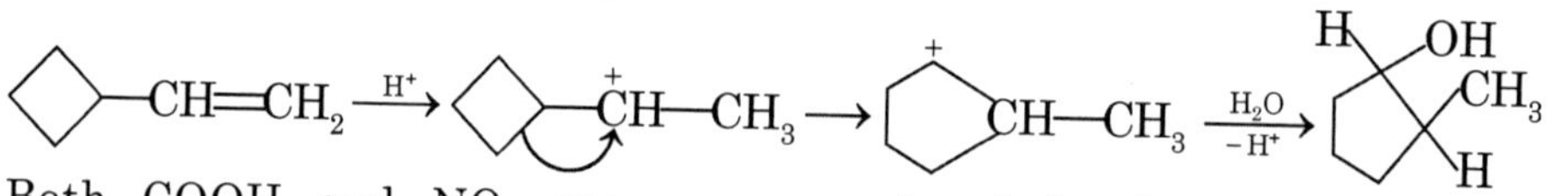

(B) Both COOH and NO_2 groups are reduced by $LiAlH_4$. The product is $4\text{-}H_2NC_6H_4CH_2CH_2OH$.

(C) MnO_2 oxidises only the allylic OH group and the product is $CH_3CH{=}CHCHO$

96. Which of the following statements is correct?

(A) Hydration of 3-phenylbut-1-ene in dil. H_2SO_4 yields 3-phenylbutan-2-ol

(B) S_{N^2} substitution of ROH not only brings inversion of geometry but also the rearrangement of carbon skeleton.

(C) S_{N^1} substitution of ROH leads to rearrangement of carbon skeleton

(D) Reaction of pentan-3-ol with HBr to give a mixture of 2- and 3-bromopentane occurs *via* S_{N^1} mechanism.

Hint: The statement (C) is correct and all others are incorrect.

(A) The hydration of 3-phenylbut-1-ene gives 2-phenylbutan-2-ol.

(B) There occurs inversion and not rearrangement.

(C) The carbocation formed undergoes rearrangement in going from a less stable carbocation to a more stable carbocation.

(D) The reaction occurs via both S_{N^1} and S_{N^2} mechanisms.

97. Which of the following statements is incorrect?

(A) Reaction of a primary or secondary alcohol with PBr_3 to give bromoalkane occurs with inversion of configuration

(B) $(C_6H_5)_2CHCH_2OH$ reacts with HBr to form $C_6H_5CHBrCH_2C_6H_5$

(C) Rate of dehydration of $(CH_3)_2C(OH)CH(CH_3)_2$ with H_2SO_4 is slower than that with $(CH_3)_2C(OH)CH_2CH_2CH_3$

(D) Rate of dehydration of $CH_3CH_2CH(OH)CH(CH_3)_2$ with H_2SO_4 is faster than that with $CH_3CH_2CH(OH)CH_2CH_2CH_3$

Hint: The statement (C) is incorrect and all others are correct.

(A) $3ROH + PBr_3 \xrightarrow[-3HBr]{} R{-}O{-}P(OR)_2 \xrightarrow{HBr} \overset{-}{Br} + R{-}\overset{+}{\underset{H}{O}}{-}P(OR)_2 \rightarrow$

$RBr + HOP(OR)_2$

(B) The reaction occurs through the formation of oxonium ion followed by its rearrangement.

$C_6H_5\underset{C_6H_5}{CH}CH_2OH \xrightarrow{H^+} C_6H_5\underset{C_6H_5}{CH}CH_2\overset{+}{O}H_2 \xrightarrow[-H_2O]{} \left[C_6H_5CH{-}CH_2 \cdots \overset{\delta+}{O}H_2 \; \overset{\delta+}{C_6H_5}\right] \rightarrow$

$C_6H_5\overset{+}{C}H{-}CH_2C_6H_5 \xrightarrow{Br^-} C_6H_5CHBrCH_2C_6C_5$

(C) Both the alcohols are tertiary alochol. The alcohol $(CH_3)_2C(OH)CH(CH_3)_2$ forms more substituted alkene (Saytzeff rule) and hence reacts faster.

(D) Both the alcohols are secondary alcohol. The alcohol $CH_3CH_2CH(OH)CH(CH_3)_2$ forms more substituted alkene (Saytzeff rule) and hence reacts faster.

98. In the reaction, $C_6H_5OH + C_2H_5I \xrightarrow{C_2H_5O^-,\ anhy.\ C_2H_5OH} P$, the product P is

(A) $C_6H_5OC_2H_5$ (B) $C_2H_5OC_2H_5$ (C) $C_6H_5OC_6H_5$ (D) C_6H_5I **(I.I.T. 2003)**

Hint: $C_2H_5O^-$ abstracts a proton from phenol to form $C_6H_5O^-$. But $C_2H_5O^-$ is a better nucleophile than $C_6H_5O^-$ since in $C_2H_5O^-$ the charge is localised on oxygen atom while in $C_6H_5O^-$ the charge is delocalised over the whole ion. Hence $C_2H_5O^-$ will make nucleophilic attack and not $C_6H_5O^-$.

99. 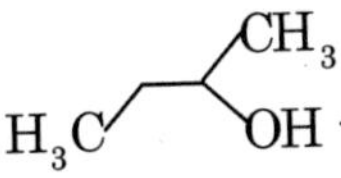 $\xrightarrow[-H_2O]{H^+}$ [F] $\xrightarrow{Br_2/CCl_4}$ C_4H_8Br (5 such products are possible). How many structures of F are possible?

(A) 2 (B) 5 (C) 3 (D) 6 **(I.I.T. 2003)**

Hint :

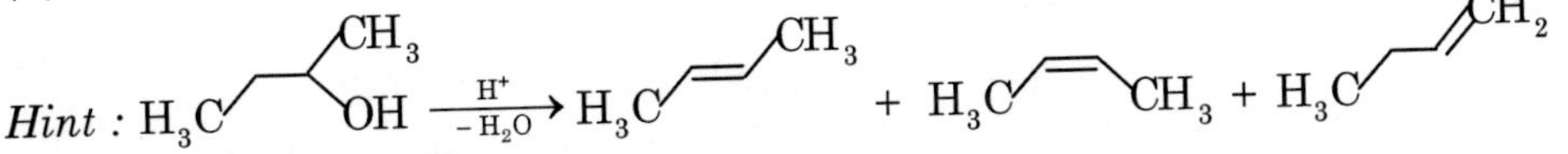

100.

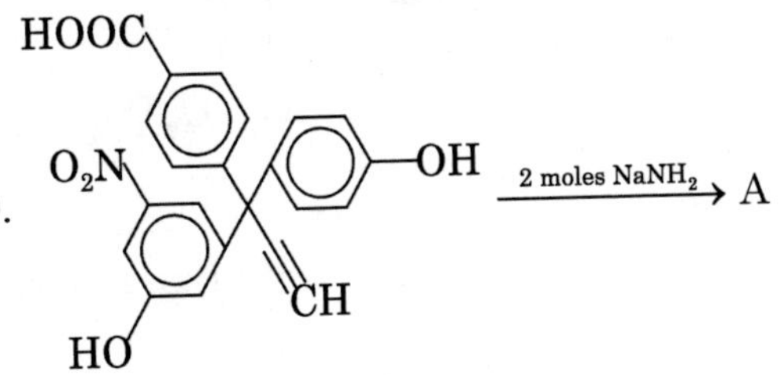

The product A will be

(A)

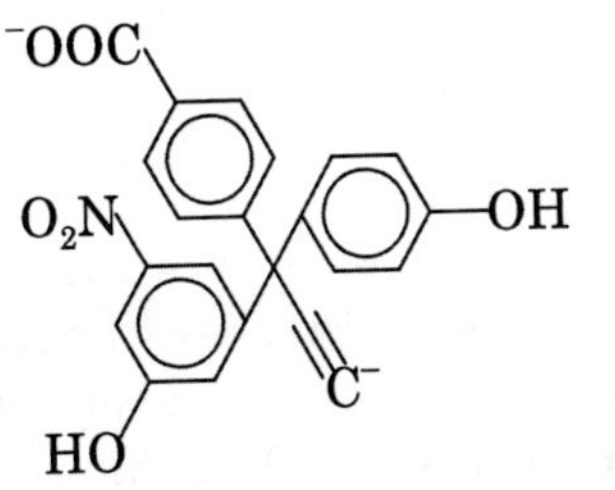

(B)

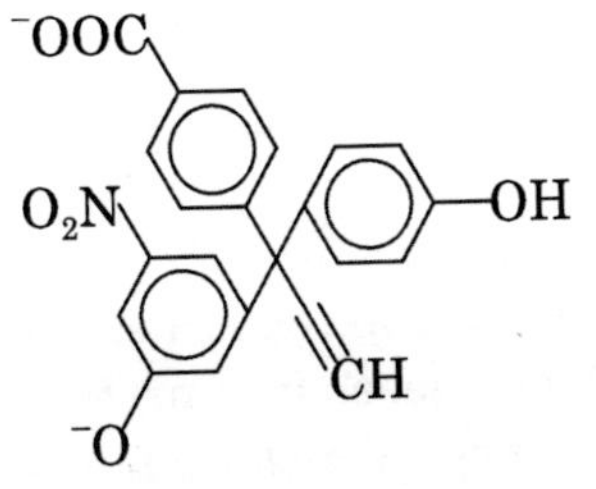

(C) 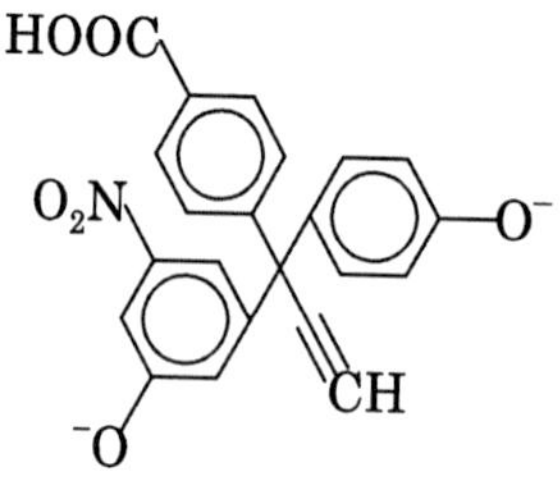 (D) 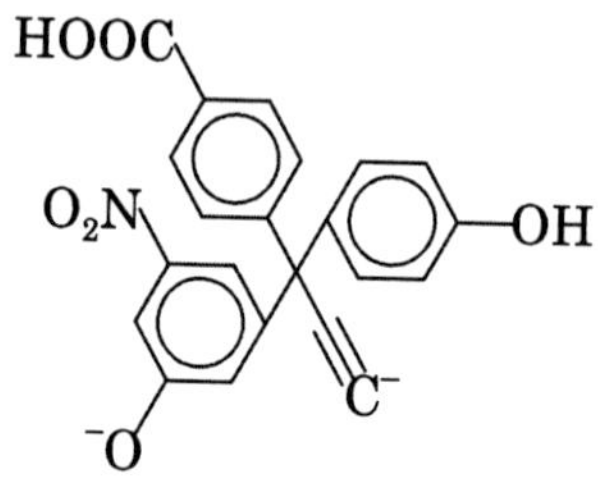 **(I.I.T. 2003)**

Hint: Two moles of NH_2^- will abstract two moles of most acidic H out of four moles of H present per mole of the compound. The acidic strength is in the order: alkynic H < —OH (phenolic) < —OH (phenolic deactivated by NO_2 group) < —COOH .

101. 2-Phenylpropene on acid hydration gives

(A) 2-phenylpropan-2-ol (B) 2-phenylpropan-1-ol

(C) 3-phenylpropan-1-ol (D) 1-phenylpropan-1-ol **(I.I.T. 2004)**

Hint: 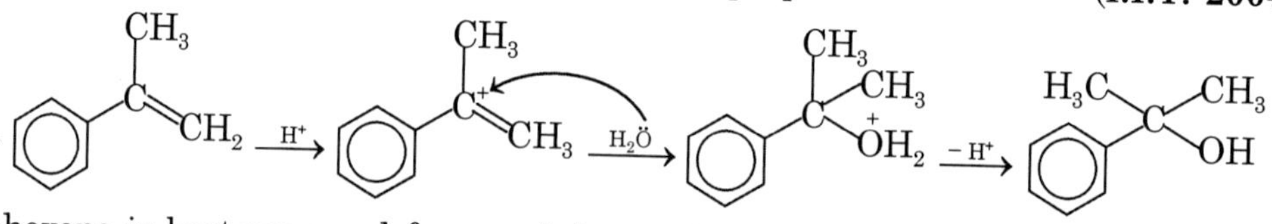

102. Cyclohexene is best prepared from cyclohexanol by which of the following:

(A) conc. H_3PO_4 (B) conc. HBr (C) conc. HCl (D) conc. $HCl/ZnCl_2$

(I.I.T. 2005)

Hint: 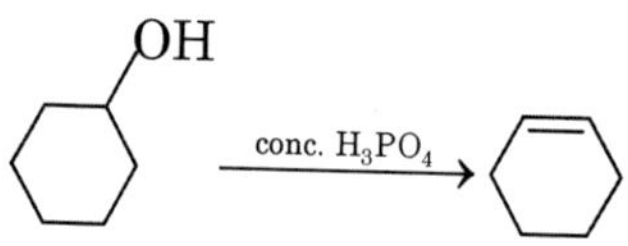

H_3PO_4 acts as a dehydrating agent.

103. Identify the incorrect statement.

(A) Dehydration of cyclobutylmethanol produces cyclobutylethene

(B) Thiols are less soluble in water than the corresponding alcohols

(C) RS^- in a protic solvent, acts as a stronger nucleophile than RO^-

(D) The increasing order of reactivity of benzyl alcohol with HBr is: p-$O_2NC_6H_4CH_2OH$ < p-$ClC_6H_4CH_2OH$ < $C_6H_5CH_2OH$ < p-$CH_3OC_6H_4CH_2OH$

Hint: The statement (A) is incorrect and all others are correct.

(A) The expansion of the four-membered ring to the five-membered ring relieves the strain in the ring.

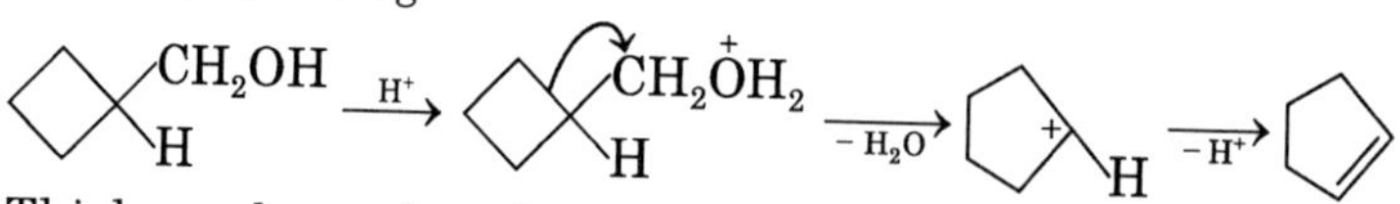

(B) Thiols are less polar than alcohols and hence they are less soluble in water than alcohols.

(C) Due to the bigger size of S atom, the negative charge on sulphur atom is more easily polarised making it a stronger nucleophilic agent. Further, the nucleophilicity of RO^- decreases in a protic solvent due to more stronger H-bonding arising out of large negative charge density on the smaller sized O atom.

(D) The electron-attracting NO_2 group destabilises the benzilic carbocation while the CH_3O group stabilises it. The chloro group destabilises the benzylic carbocation due to the – I-effect.

104. The increasing order of dehydration of the following alcohols is

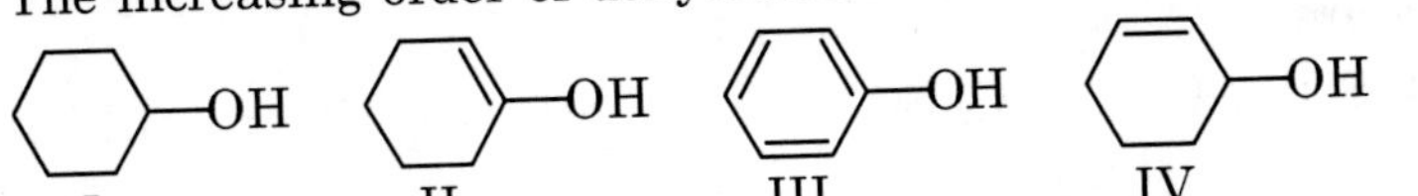

I II III IV

(A) II < III < IV < I (B) III < II < I < IV

(C) IV < III < II < I (D) I < II < III < IV

105. In the reaction, 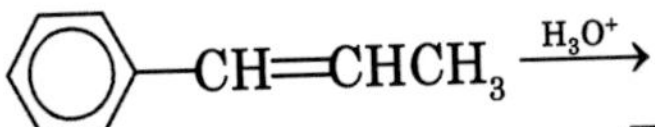 $\xrightarrow{H_3O^+}$ X (major product), X is

(A) 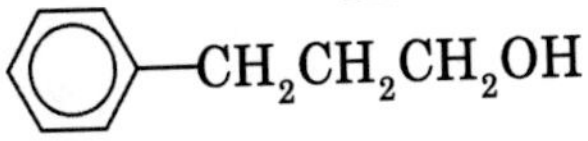 (B) 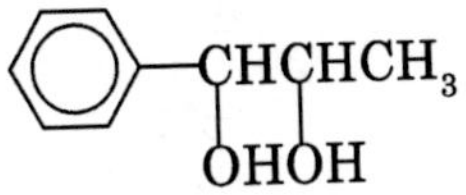

(C) 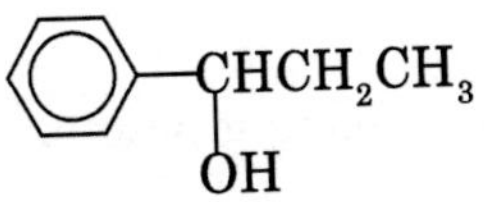 (D) 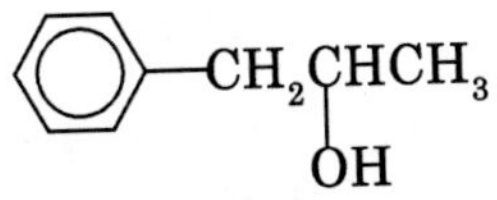

106. In the reaction, $CH_2{=}CH\overset{O}{\overset{\|}{C}}CH_2CH_2OH \xrightarrow{MnO_2}$ X, X is

(A) $CH_2{=}CH\overset{O}{\overset{\|}{C}}CH_2\overset{O}{\overset{\|}{C}}{-}OH$ (B) $CH_2{=}CH\overset{O}{\overset{\|}{C}}CH_2CHO$

(C) $CH_2{=}CH\underset{OH}{\underset{|}{C}}HCH_2CHO$ (D) $CH_2{=}CH\overset{O}{\overset{\|}{C}}H_2CH_2OH$

107. In the reaction, 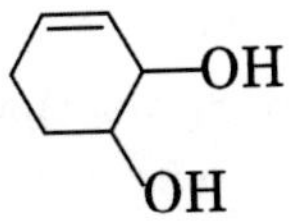 $\xrightarrow{MnO_2}$ X, X is

(A) 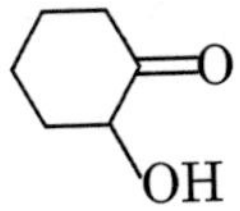 (B) 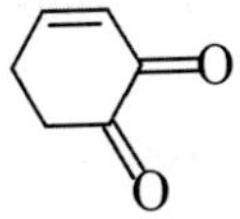 (C) 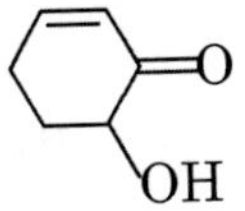 (D) 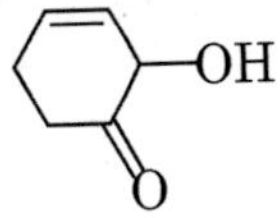

108. In the reaction, $CH_3\overset{CH_3}{\underset{CH_3}{C}}{-}\overset{OH}{C}HCH_3 \xrightarrow{HBr}$ X (major product), X is

(A) $(CH_3)_2\overset{Br}{\overset{|}{C}}CH(CH_3)_2$ (B) $(CH_3)_3C\overset{Br}{\overset{|}{C}}HCH_3$

(C) both (A) and (B) (D) none of these.

109. In the reaction, $(CH_3)_3CCH{=}CH_2 \xrightarrow{R} (CH_3)_3C\underset{OH}{\underset{|}{C}}HCH_3$, R is

(A) $Hg(OAc)_2/NaBH_4$, NaOH (B) BH_3, THF/H_2O_2, OH^-
(C) H_3O^+ (D) none of these.

110. When benzenesulphonic acid and *p*-nitrophenol are treated with $NaHCO_3$, the gases released respectively are

(A) SO_2, NO_2 (B) SO_2, NO (C) SO_2, CO_2 (D) CO_2, CO_2 **(I.I.T. 2006)**

Hint:

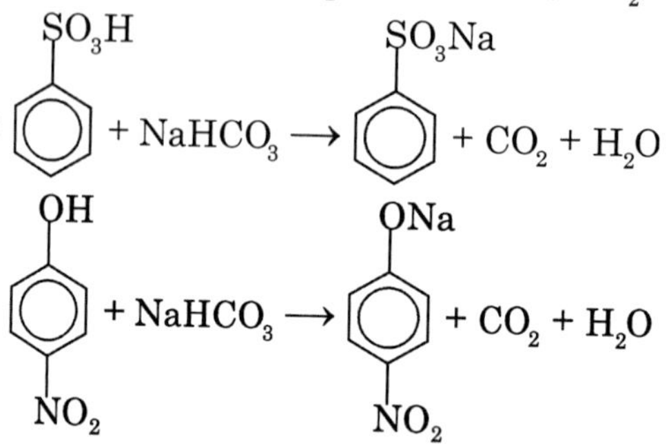

111. Arrange the following compounds according to decreasing boiling points.

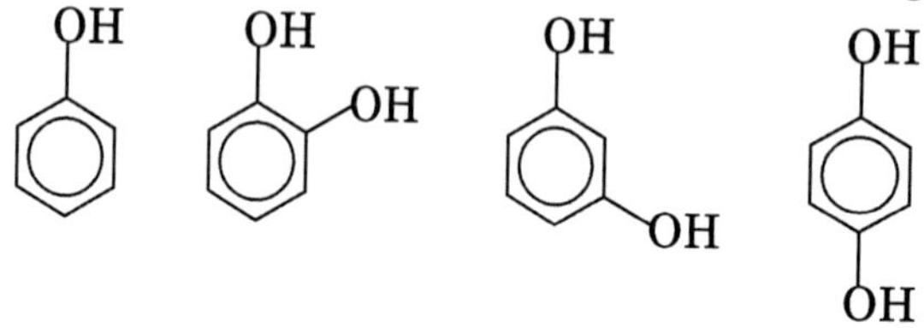

I II III IV

(A) IV > III > II > I (B) III > IV > II > I
(C) I > II > III > IV (D) II > III > I > IV **(I.I.T. 2006)**

Hint: Increase in the number of hydroxyl groups increases the boiling point. Among the isomeric dihydroxybenzenes, as the distance between the hydroxyl groups increases, the repulsion between them decreases and hence the boiling point increases.

112. In the reaction 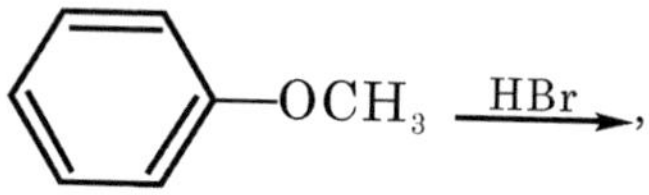, the products are

(A) 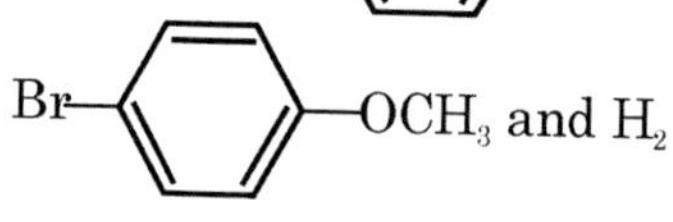

(B) 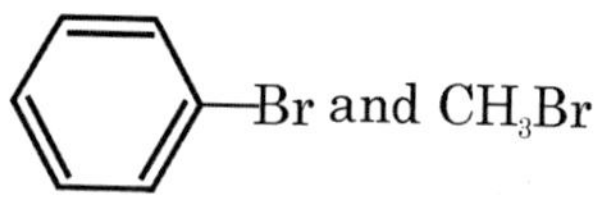

(C) 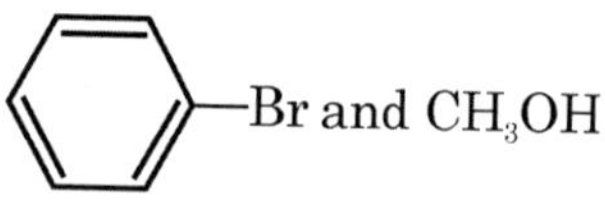

(D) 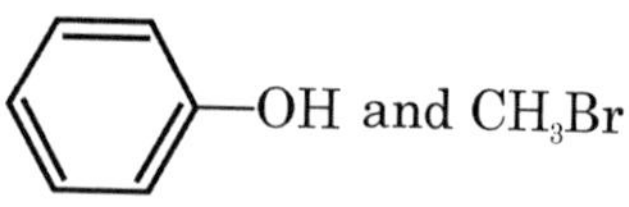

(I.I.T. 2010)

Hint: 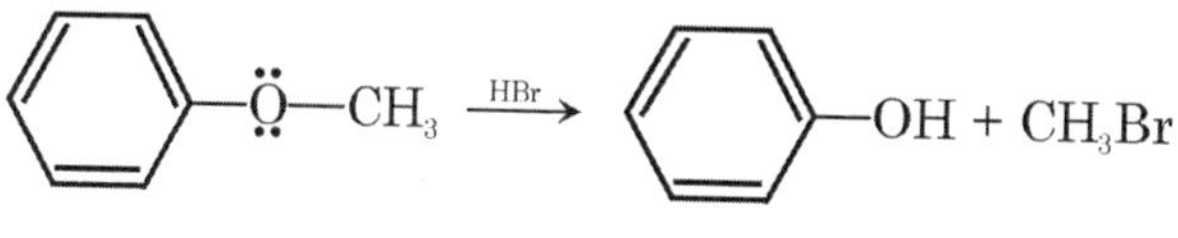

113. In the reaction 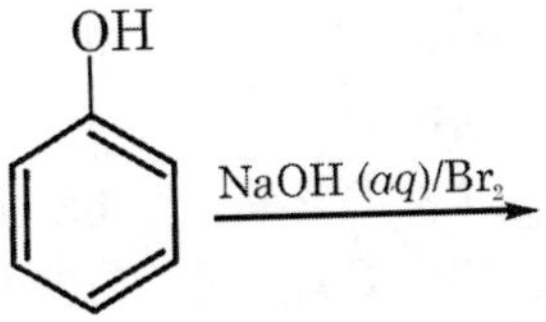 the intermediate(s) is(are)

(A) 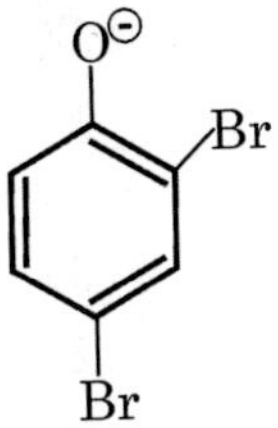

(B) 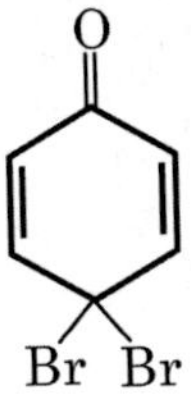

(C) 

(D) 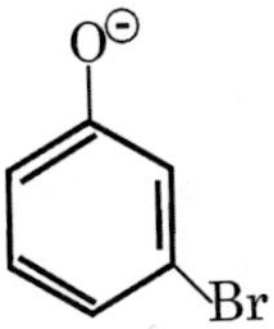

(I.I.T. 2010)

Hint: The phenoxide ion is para* and ortho directing (*preferably),

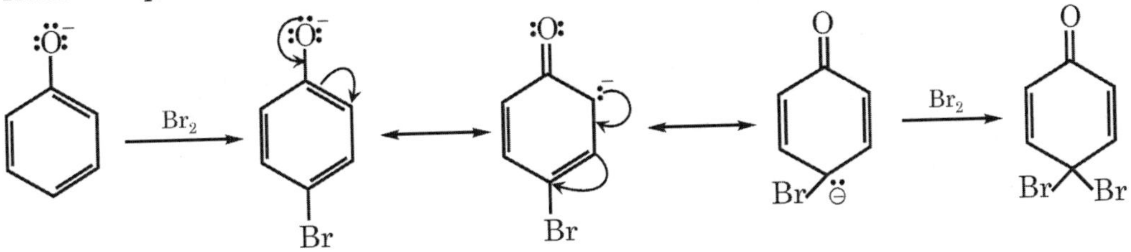

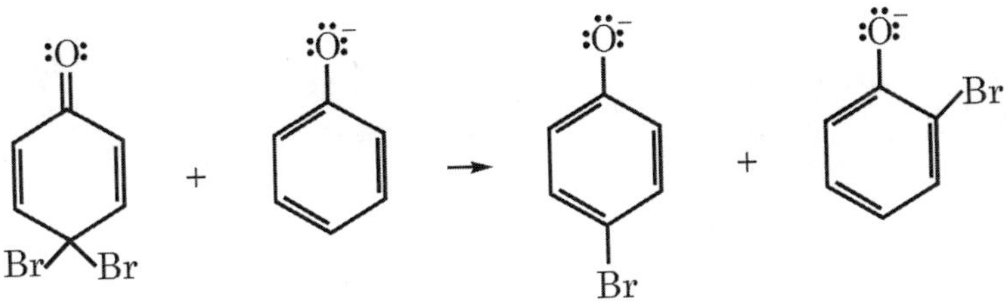

ANSWERS

1. C	2. B	3. C	4. A	5. A	6. D
7. A, D	8. D	9. B	10. B, D	11. C	12. B
13. D	14. D	15. B	16. C	17. A, C	18. A
19. A	20. D	21. C	22. C	23. B	24. A
25. A	26. D	27. C, D	28. A	29. A	30. D
31. B	32. A	33. D	34. A	35. B	36. C

37. A	38. D	39. A	40. A	41. C	42. C
43. C	44. D	45. C	46. D	47. A	48. B
49. A	50. C	51. A	52. B	53. A	54. B
55. B	56. B	57. C	58. C	59. C	60. D
61. A	62. B	63. C	64. A	65. C	66. C
67. B	68. B	69. A	70. B	71. B	72. B
73. B	74. C	75. C	76. A, B, C	77. C, D	78. C
79. D	80. D	81. C	82. D	83. A	84. A
85. C	86. D	87. C	88. C	89. B	90. C
91. C	92. A	93. C	94. D	95. D	96. C
97. C	98. A, B	99. C	100. B	101. A	102. A
103. A	104. B	105. C	106. D	107. C	108. A
109. A	110. D	111. A	112. D	113. A, B, C	

39

Compounds with Functional Groups Containing Oxygen—II: Aldehydes and Ketones

OBJECTIVE TYPE MULTIPLE CHOICE QUESTIONS

1. Which of the following has the most acidic hydrogen?
   (A) hexane-2,4-dione (B) hexane-2,3-dione
   (C) hexane-2,5-dione (D) hexan-3-one **(I.I.T. 2000)**
2. The reagent for the following transformation is

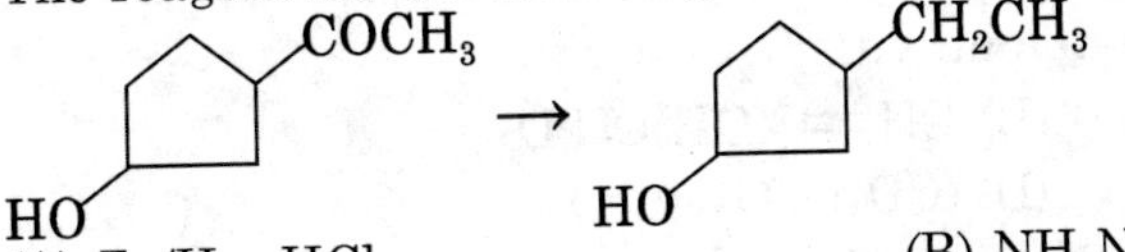

   (A) Zn/Hg, HCl (B) NH_2NH_2, OH^-
   (C) $NaBH_4$ (D) H_2/Ni **(I.I.T. 2000)**
3. A mixture of benzaldehyde and formaldehyde on heating with aqueous NaOH solution gives
   (A) benzyl alcohol + sodium formate (B) sodium benzoate + methanol
   (C) benzyl alcohol + methanol (D) sodium benzoate + sodium formate
   (I.I.T. 2001)
4. Assertion (*A*): Dimethyl sulphide is commonly used for the reduction of an ozonide of an alkene to get carbonyl compounds.
   Reason (*R*): It reduces the ozonide giving water soluble dimethyl sulphoxide and excess of it evaporates.
   (A) Both *A* and *R* are correct and *R* is the correct explanation of *A*
   (B) Both *A* and *R* are correct and *R* is not the correct explanation of *A*
   (C) *A* is true but *R* is false (D) *A* is false but *R* is true. **(I.I.T. 2001)**

5. Which of the following will undergo aldol condensation?
   (A) CH_3CHO (B) CD_3CHO (C) CH_3CH_2CHO (D) C_6H_5CHO **(I.I.T. 1998)**
6. A new C—C bond formation is possible in
   (A) Cannizzaro reaction (B) Friedel–Crafts reaction
   (C) Clemmensen reduction (D) Reimer–Tiemann reaction. **(I.I.T. 1998)**
7. Which of the following will react with water?
   (A) $CHCl_3$ (B) Cl_3CCHO (C) $ClCH_2CH_2Cl$ (D) CCl_4 **(I.I.T. 1998)**
8. In the Cannizzaro reaction, the intermediate that will be the best hydride donor is

   (A) 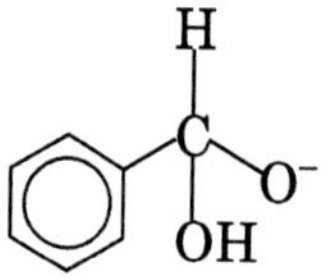

   (B) 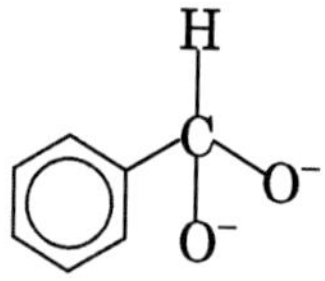

   (C) 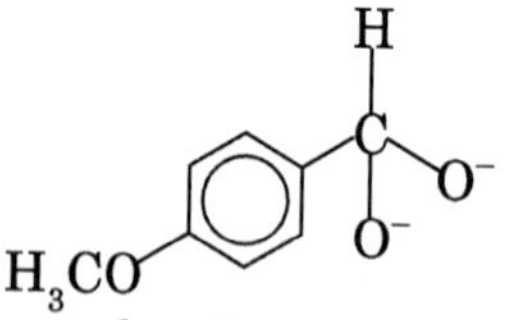

   (D) 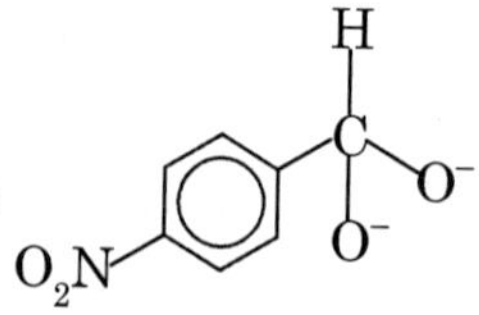

   (I.I.T. 1997)
9. Among the given compounds the most susceptible to nucleophilic attack of carbonyl group is
   (A) CH_3COCl (B) CH_3CHO (C) CH_3COOCH_3 (D) $CH_3COOCOCH_3$
   (I.I.T. 1997)
10. Among the following compounds which will react with acetone to give a product containing $>C=N$?
    (A) $C_6H_5NH_2$ (B) $C_6H_5NHNH_2$ (C) $C_6H_5NHC_6H_5$ (D) $(CH_3)_3N$ **(I.I.T. 1998)**
11. The enol form of acetone after treatment with D_2O gives
    (A) $CH_2=C(OD)CH_3$ (B) $CH_2=C(OH)CH_2D$
    (C) $CD_2=C(OD)CD_3$ (D) $(CD_3)_2CO$
    (I.I.T. 1999)
12. In the Cannizzaro reaction, $2C_6H_5CHO + OH^- \longrightarrow C_6H_5OH + C_6H_5CO_2^-$, the slowest step is
    (A) attack of OH^- at the carbonyl group
    (B) transfer of hydride to the carbonyl group
    (C) abstraction of H^+ from carboxylic acid
    (D) deprotonation of $C_6H_5CH_2OH$
    (I.I.T. 1996)
13. Which is/are the products(s) formed when an equimolar mixture of benzaldehyde and formaldehyde is heated with conc. NaOH?
    (A) $C_6H_5CH_2OH + HCOONa$ (B) $C_6H_5COONa + CH_3OH$
    (C) $C_6H_5CH_2COONa$ (D) $C_6H_5COOH + CH_3ONa$ **(I.I.T. 1995)**
14. Which of the following will give yellow precipitate with I_2/NaOH?
    (A) CH_3CONH_2 (B) $ICH_2COCH_2CH_3$
    (C) $CH_3COOCOCH_3$ (D) $CH_3CH(OH)CH_2CH_3$ **(I.I.T. 1997)**

15. The structure of enol form of $H_3CCOCH_2COCH_3$ with intramolecular H-bonding is

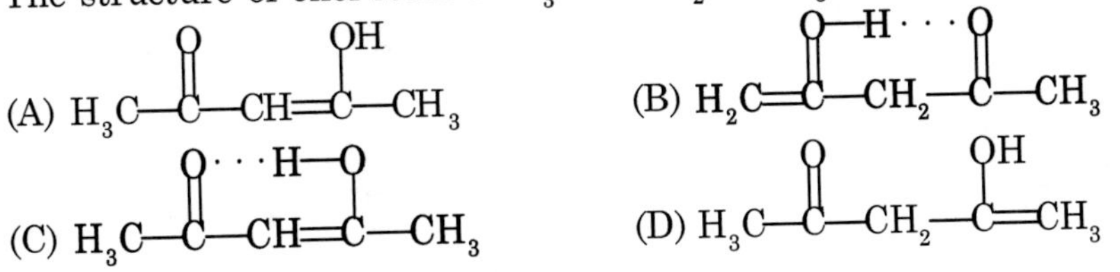

(I.I.T. 1993)

16. 'The IUPAC name of 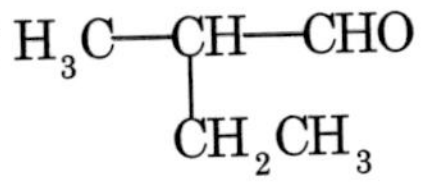 is

(A) butan-2-aldehyde (B) 2-methylbutanal
(C) 2-ethylpropanal (D) 3-methylbutyraldehyde **(I.I.T. 1993)**

17. The reaction of R—C(=O)—NH_2 with Br_2 and KOH gives R—NH_2 as the main product. The intermediate involved in this reaction is
(A) R—C(=O)—NHBr (B) R—N=C=O
(C) R—NH—Br (D) R—C(=O)—$N(Br)_2$ **(I.I.T. 1992)**

18. Under Wolff–Kishner reduction conditions, the conversions which may be brought about are
(A) cyclohexanone into cyclohexane (B) benzaldehyde into benzyl alcohol
(C) cyclohexanone into cyclohexanol (D) benzophenone into diphenyl methane
(I.I.T. 1992)

19. Hydrogenation of benzoyl chloride in the presence of Pd and $BaSO_4$ gives
(A) benzyl alcohol (B) benzaldehyde (C) benzoic acid (D) phenol **(I.I.T. 1992)**

20. Identify Z in the reactions, $CH_3CH{=}CHCHO \xrightarrow{NaBH_4} X \xrightarrow{HCl,\ ZnCl_2} Y \xrightarrow{H^+,\ HCN} Z$
(A) $CH_3CH{=}CHCH_2COOH$ (B) $CH_3CH{=}CHCH_2OH$
(C) $CH_3CH{=}CHCH_2Cl$ (D) $CH_3CH{=}CHCH_2CHO$ **(I.I.T. 1991)**

21. *m*-Chlorobenzaldehyde on reaction with conc. KOH at room temperature gives
(A) potassium *m*-chlorobenzoate and *m*-hydroxybenzaldehyde
(B) *m*-hydroxybenzaldehyde and *m*-chlorobenzyl alcohol
(C) *m*-chlorobenzyl alcohol and hydroxybenzyl alcohol
(D) potassium *m*-chlorobenzoate and *m*-chlorobenzyl alcohol. **(I.I.T. 1991)**

22. The formation of cyanohydrin from a ketone is an example of
(A) electrophilic addition (B) nucleophilic addition
(C) nucleophilic substitution (D) electrophilic substitution. **(I.I.T. 1990)**

23. Which of the following are the examples of aldol condensation?
(A) $CH_3CHO \xrightarrow{\text{dil. NaOH}} CH_3CH(OH)CH_2CHO$
(B) $2CH_3COCH_3 \xrightarrow{\text{dil. NaOH}} (CH_3)_2C(OH)CH_2COCH_3$
(C) $HCHO \xrightarrow{\text{dil. NaOH}} CH_3OH$
(D) $C_6H_5CHO + HCHO \xrightarrow{\text{dil. NaOH}} C_6H_5CH_2OH$ **(I.I.T. 1989)**

24. Keto-enol tautomerism is observed in
(A) C_6H_5CHO (B) $C_6H_5COCH_3$ (C) $C_6H_5COC_6H_5$ (D) $C_6H_5COCH_2COCH_3$
(I.I.T. 1988)

25. The polarisation of electrons in acrolein may be written as
(A) $\overset{\delta-}{C}H_2=CH-\overset{\delta+}{C}H=O$ (B) $\overset{\delta-}{C}H_2=CH-CH=\overset{\delta+}{O}$
(C) $\overset{\delta-}{C}H_2=\overset{\delta+}{C}H-CH=O$ (D) $\overset{\delta+}{C}H_2=CH-CH=\overset{\delta-}{O}$
(I.I.T. 1988)

26. Identify Y in the reactions, $2CH_3COCH_3 \xrightarrow{\text{base catalyst}} X \xrightarrow{\text{acid catalyst}} Y$
(A) $(H_3C)_2C=CHCOCH_3$ (B) $(H_3C)_2C(OH)CH_2COCH_3$
(C) $H_3CCH(OH)CH(CH_3)COCH_3$ (D) none of these.
(I.I.T. 1988)

27. Identify X in the reaction, $X \xrightarrow{\text{NaOH}} C_6H_5CH=CHCHO$
(A) C_6H_5CHO (B) CH_3CHO
(C) C_6H_5CHO and CH_3CHO (D) none of these.
(I.I.T. 1986)

28. The compounds that will not give iodoform test on treatment with alkali and iodine is
(A) acetone (B) ethanol
(C) isopropyl alcohol (D) diethyl ketone
(I.I.T. 1985)

29. Which of the following compounds reacts with ethanolic KCN?
(A) Ethyl chloride (B) Benzaldehyde
(C) Benzoyl chloride (D) Chlorobenzene
(I.I.T. 1984)

30. Which of the following compounds will give a yellow precipitate with iodine and alkali?
(A) 2-Hydroxypropanone (B) Benzophenone
(C) Methylamine (D) Acetamide
(I.I.T. 1984)

31. The base catalysed aldol condensation will occur with
(A) C_6H_5CHO (B) CH_3CH_2CHO
(C) $(CH_3)_2CHCHO$ (D) $(CH_3)_3CCHO$
(I.I.T. 1984)

32. The Cannizzaro reaction is not given by
(A) $(CH_3)_3CCHO$ (B) acetaldehyde (C) benzaldehyde (D) formaldehyde
(I.I.T. 1983)

33. Acetaldehyde when heated with Fehling solution gives a precipitate of
(A) Cu (B) CuO (C) Cu_2O (D) Cu + CuO + Cu_2O
(I.I.T. 1983)

34. A compound that gives the iodoform test is
(A) pentan-1-ol (B) pentan-2-one (C) pentan-3-one (D) pentanal **(I.I.T. 1982)**

35. The compound which is not isomeric with diethyl ether is
(A) *n*-propyl methyl ether (B) butan-1-ol
(C) 2-methylpropan-2-ol (D) butanone
(I.I.T. 1981)

36. The reagent which reacts with both acetaldehyde and acetone is
(A) Fehling solution (B) Grignard reagent
(C) Schiff reagent (D) Tollen reagent.
(I.I.T. 1981)

37. For the reduction of ketones to hydrocarbons, the appropriate reagent is
(A) Zn/Hg, HCl (B) H_2SO_4 (C) red P (D) HI **(Roorkee 1995)**

38. Which of the following reactions cannot be used for the reduction of $RCOR \rightarrow RCH_2R$?
(A) Clemmensen reaction (B) Wolff–Kishner reaction
(C) Wurtz reaction (D) HI and red P at 200°C. **(Roorkee 1993)**

39. Which of the following aldehydes/ketone is most reactive towards nucleophilic addition reactions?
(A) HCHO (B) CH_3CHO (C) C_2H_5CHO (D) CH_3COCH_3
(Roorkee 1992)

40. The treatment of ketone with Grignard reagent followed by hydrolysis gives
(A) *sec*-butanol (B) *pri*-alcohol (C) *tert*-alcohol (D) aldehyde
(Roorkee 1990)

41. Z in the reactions, $C_6H_5COOH \xrightarrow{X} C_6H_5COCl \xrightarrow{NH_3} Y \xrightarrow{Z} C_6H_5CN$ is
(A) PCl_5 (B) P_4O_{10} (C) H^+, HCN (D) KCN **(Roorkee 1988)**

42. Z in the reactions, $C_6H_5OH \xrightarrow{Et_2SO_4,\ NaOH} X \xrightarrow{HCN/HCl,\ AlCl_3} Y \xrightarrow{C_6H_5NHNH_2} Z$, is

(A) 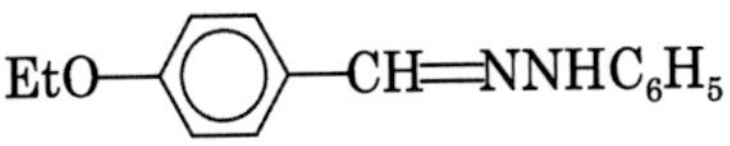 (B) 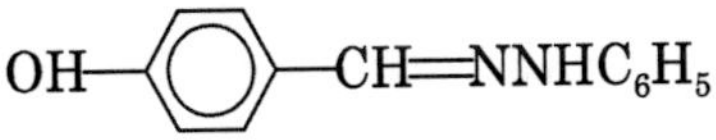

(C) 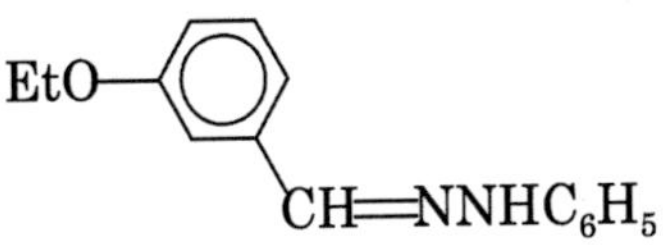 (D) 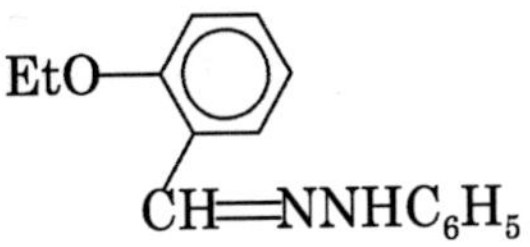 **(Roorkee 1988)**

43. In the reaction, $CH_3COCH_3 \xrightarrow{SeO_2} P + Se + H_2O$, P is
(A) CH_3COOCH_3 (B) CH_3COCH_2OH (C) $CH_3COCOOH$ (D) CH_3COCHO **(I.I.T. 1995)**

44. The reaction, $CH_3CHO + H_2NOH \rightarrow CH_3CH{=}NOH$ occurs at
(A) $pH = 12$ (B) $pH = 4.5$ (C) $pH = 1$ (D) any value of pH
(I.I.T. 1997)

45. The reaction of HCOOH with conc. H_2SO_4 gives
(A) $CO + H_2O$ (B) HCHO (C) CH_3COOH (D) $CO_2 + H_2O$

46. Which of the following will give acetophenone on heating?
(A) Calcium acetate (B) Calcium acetate + calcium benzoate
(C) Calcium formate + calcium benzoate
(D) Calcium formate + calcium acetate

47. The reaction $H_2C{=}CH{-}CHO + HCN \rightarrow NCCH_2CH_2CHO$ involves
(A) free radical addition across double bond
(B) electrophilic addition across double bond
(C) nucleophilic addition across double bond
(D) no effect of the carbonyl group on the C=C bond.

48. Acetaldehyde differs from benzaldehyde in that
(A) benzaldehyde does not react with Fehling solution

(B) benzaldehyde does not react with Tollen reagent
(C) benzaldehyde does not react with sodium bisulphite
(D) benzaldehyde does not react with HCN

49. Acetone on reaction with Cl_2 gives
(A) monochloroacetone (B) dichloroacetone
(C) trichloroacetone (D) hexachloroacetone

50. An unknown alkyl halide (*A*) reacts with alc. KOH to produce a hydrocarbon (C_4H_8). Ozonolysis of the hydrocarbon affords one mole of propionaldehyde and one mole of formaldehyde. Suggest which organic compound among the following is *A*.
(A) $CH_3(CH_2)_3Br$ (B) $CH_3CH(Br)CH(Br)CH_3$
(C) $CH_3CH_2CH(Br)CH_3$ (D) $Br(CH_2)_4Br$

51. Treatment of propanal with dil. NaOH solution gives
(A) $CH_3CH_2COOCH_2CH_2CH_3$ (B) $CH_3CH_2CH(OH)CH(CH_3)CHO$
(C) $CH_3CH_2CH(OH)CH_2CH_2CHO$ (D) $CH_3CH_2COCH_2CH_2CHO$

52. Which of the following will form a stable hydrate?
(A) Trichloroacetaldehyde (B) Propionaldehyde
(C) Acetaldehyde (D) Formaldehyde

53. $(CH_3)_2C{=}CHCOCH_3$ can be oxidised to $(CH_3)_2C{=}CHCOOH$ by
(A) NaOI (B) chromic acid (C) Cu at 300°C (D) $KMnO_4$

54. Benzaldehyde is prepared by the hydrolysis of
(A) benzal chloride (B) benzyl chloride
(C) benzonitrile (D) benzotrichloride

55. Which of the following aldehydes gives Cannizzaro reaction when heated with strong alkali?
(A) Ethanal (B) Propanal (C) Benzaldehyde (D) All of these.

56. In the presence of aluminium ethoxide, aldehydes get converted into esters. The reaction is known as
(A) Tischenko reaction (B) Aldol condensation
(C) Beckmann rearrangement (D) Schimdt reaction.

57. Which of the following statements about benzaldehyde is true?
(A) It undergoes aldol condensation (B) It undergoes Cannizzaro reaction
(C) It reduces Tollen reagent
(D) It does not form an addition compound with $NaHSO_3$

58. In which of the following reactions, both oxidised and reduced forms of the same compound are obtained?
(A) Cannizzaro reaction (B) Aldol condensation
(C) Riemer–Tiemann reaction (D) Kolbe's synthesis.

59. Which of the following are the starting materials for the Grignard's synthesis of *tert*-butanol?
(A) $CH_3MgBr + CH_3COCH_3$ (B) $CH_3CH_2MgBr + CH_3COOH_3$
(C) $CH_3MgBr + CH_3CH(OH)CH_3$ (D) $CH_3MgBr + (CH_3)_3COH$

60. Assertion (*A*): Lower aldehydes and ketones are soluble in water but the solubility decreases as the molecular mass increases.
    Reason (*R*): Distinction between aldehydes and ketones can be made by Tollen's test.
    (A) Both *A* and *R* are true and *R* is the correct explanation of *A*
    (B) Both *A* and *R* are true and *R* is not the correct explanation of *A*
    (C) *A* is true but *R* is false (D) Both *A* and *R* are false.

61. Which of the following is incorrect?
    (A) $FeCl_3$ is used in the detection of phenol
    (B) $NaHSO_3$ is used in the detection of carbonyl compound
    (C) Tollen reagent is used in the detection of unsaturation
    (D) Fehling solution is used in the detection of glucose.

62. Compound *A* (C_3H_8O) is treated with $K_2Cr_2O_7/H^+$ to form a product *B* (C_3H_6O). *B* forms a shining mirror on warming with ammoniacal silver nitrate. *B* when treated with an aqueous solution of $H_2NCONHNH_2 \cdot HCl$ and sodium acetate gives *C*. Identify the structure of *C*.
    (A) $CH_3CH_2CH{=}NNHCONH_2$ (B) $(CH_3)_2C{=}NNHCONH_2$
    (C) $(CH_3)_2C{=}NCONHNH_2$ (D) $CH_3CH_2CH{=}NCONHNH_2$ **(I.I.T. 2002)**

    Hint: Since compound B forms a shining mirror on warming with ammoniacal silver nitrate, B is an aldehyde. Since the compound A on oxidation gives an aldehyde B, the compound A is a primary alcohol. Hence B (C_3H_6O) is CH_3CH_2CHO and A (C_3H_8O) is $CH_3CH_2CH_2OH$. The compound B reacts with an aqueous solution of $H_2NCONHNH_2$ and sodium acctate to form C. The compound C is $CH_3CH_2CH{=}NNHCONH_2$. The involved reactions are as follows:

$$\underset{\text{A}}{CH_3CH_2CH_2OH} \xrightarrow{K_2Cr_2O_7/H^+} \underset{\text{B}}{CH_3CH_2CHO} \xrightarrow[\text{AgNO}_3]{\text{Ammoniacal}} Ag\downarrow + CH_3CH_2COOH$$

$$CH_3CH_2CHO \xrightarrow{H_2NCONHNH_2} \underset{\text{C}}{CH_3CH_2CH{=}NNHCONH_2}$$

63. Which of the following carbonyl compounds will have maximum enol content in ethanol?
    (A) $CH_3COCH(CH_3)COCH_3$ (B) $CH_3COCH_2COCH_3$
    (C) $CH_3COCH_2COOC_2H_5$ (D) $CH_3COCH_2COOCH_3$

64. In the reaction, 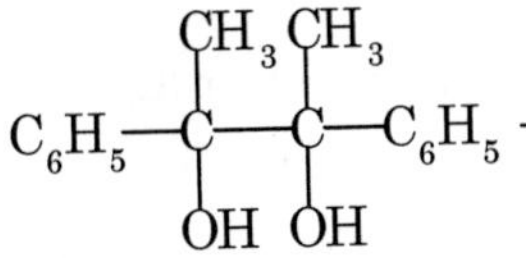 $\xrightarrow{\text{conc. } H_2SO_4}$ P, the product P is

    (A) 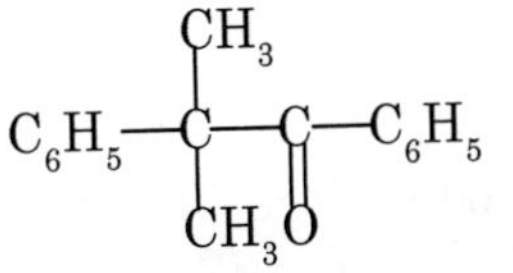

    (B) 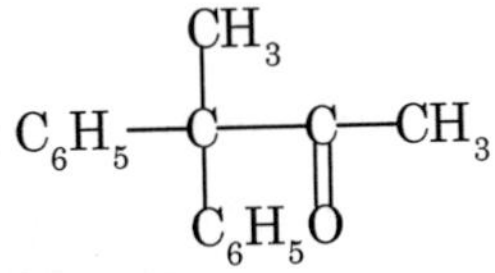

(C) 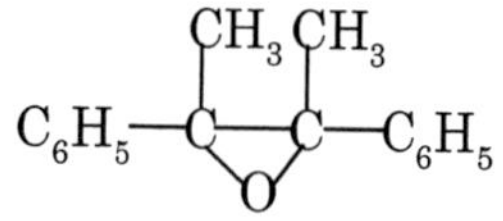   (D) 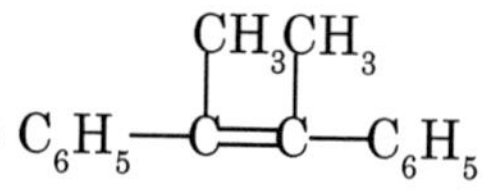

65. Which of the following reagent(s) will distinguish ethanal and benzaldehyde?
 (A) Phenylhydrazine (B) $NaHSO_3$
 (C) Fehling's solution (D) Tollen's reagent

66. In the reaction, 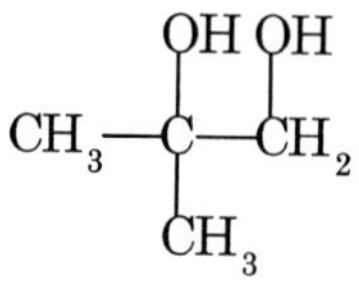 $\xrightarrow{\text{conc. } H_2SO_4}$ P, the product P is
 (A) 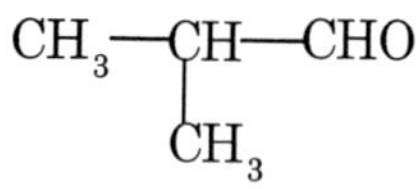
 (B) 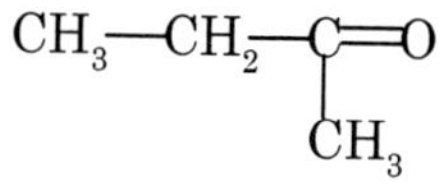
 (C) 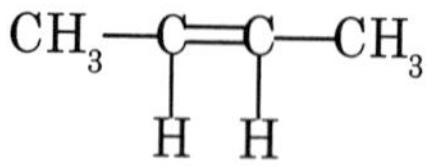
 (D) 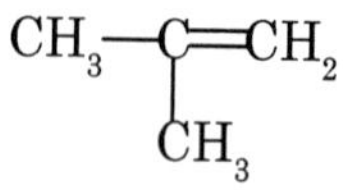

67. Which of the following will undergo iodoform reaction?
 (A) Ethanamide (B) Methyl ethanoate
 (C) Acetophenone (D) 2-Hydroxypropane

68. Which of the following statements is not correct?
 (A) The reaction of 1,2-cyclohexanedione with HIO_4 gives $HOOC(CH_2)_4COOH$
 (B) The reaction of CH_3COCHO with HIO_4 gives CH_3COOH and HCOOH
 (C) The reaction of $C_6H_5CH{=}CHCOCH_3$ with hydrazine in alkaline medium gives $C_6H_5CH{=}CHCH_2CH_3$
 (D) The reaction of $C_6H_5CH{=}CHCOCH_3$ with Li in NH_3 in ether gives $C_6H_5CH{=}CHCH(OH)CH_3$.

Hint: The statement (D) is incorrect and all others are correct.

(A) 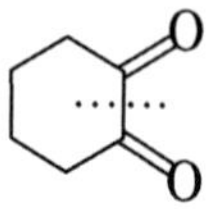 $\xrightarrow{HIO_4}$ $HOOC(CH_2)_4COOH$
Adipic acid

(B) 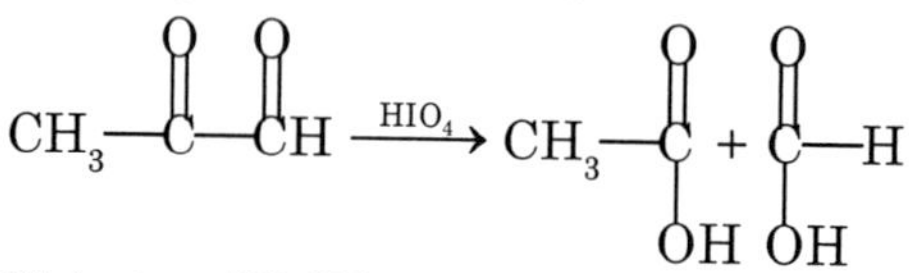

(C) This is a Wolff–Kishner reduction and the C═O group is reduced.
 $C_6H_5CH{=}CHCOCH_3 \xrightarrow{N_2H_4,\ OH^-} C_6H_5CH{=}CHCH_2CH_3$

(D) Instead of C═O group, the C═C of α,β-unsaturated carbonyl compound is reduced.
 $C_6H_5CH{=}CHCOCH_3 \xrightarrow{Li,\ NH_3} C_6H_5CH_2CH_2COCH_3$

69. Which of the following statements is not correct?
 (A) In the reaction, $CH_3CH(CH_3)COCl \xrightarrow{LiAlH_4}$ P, the product P is $CH_3CH(CH_3)CHO$
 (B) In the reaction, $C_6H_5CH_2CH_2COCl + H_2 \xrightarrow[\text{S or quinoline}]{Pd/BaSO_4}$ P, the product P is $C_6H_5CH_2CH_2CHO$

(C) The value of enthalpy of combustion of butanal is higher than that of butan-2-one

(D) The enol content of $CH_3COCH_2COCH_3$ is less than that of $C_6H_5COCH_2COC_6H_5$.

Hint: The statement (A) is incorrect and all others are correct.

(A) $CH_3CH(CH_3)COCl \xrightarrow{LiAlH_4} CH_3CH(CH_3)CH_2OH$

(B) $C_6H_5CH_2CH_2COCl + H_2 \xrightarrow[\text{S or quinoline}]{Pd/BaSO_4} C_6H_5CH_2CH_2CHO$

(C) Due to the + I-effect of alkyl groups, they stabilise the C=O group by releasing electrons towards the sp^2 hybridised atom. The enthalpy of the aldehyde is higher because it has only one alkyl group while the ketone has two alkyl groups.

(D) Due to the electron delocalisation to the benzene ring, the enolisation of $C_6H_5COCH_2COC_6H_5$ is enhanced. Due to the electron-releasing + I-effect of CH_3 groups, the enolisation of $CH_3COCH_2COCH_3$ is less.

70. Which one of the following will most readily be dehydrated in acidic condition?

(A) 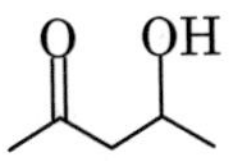 (B) 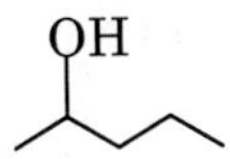 (C) 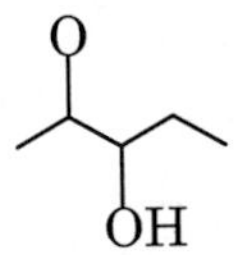 (D) 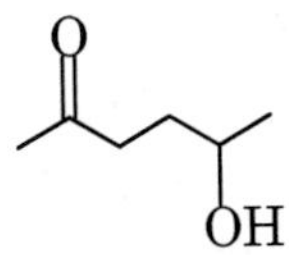 **(I.I.T. 2000)**

Hint: 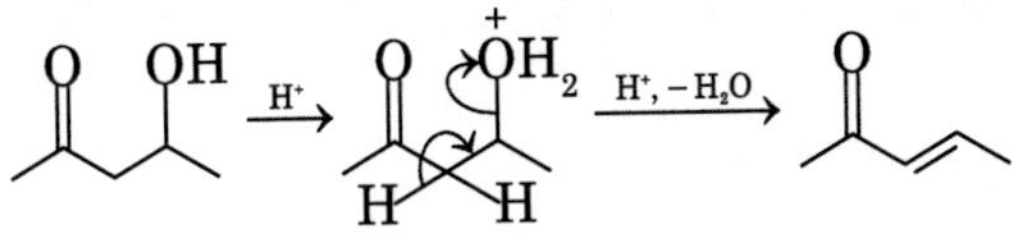

71. Which of the following has the most acidic hydrogen?

(A) Hexan-3-one (B) Hexane-2,4-dione

(C) Hexan-2,5-dione (D) Hexane-2,3-dione **(I.I.T. 2000)**

72. Which of the following reducing agent is used for the reduction of $CH_3COCH_2CH_2COOC_2H_5$ to $CH_3CH(OH)CH_2CH_2COOC_2H_5$?

(A) H_2/Ni (B) $Pd/BaSO_4$, S (C) $LiAlH_4$ (D) $NaBH_4$

73. Which oxidising agent is used for the conversion of CH_3CHO to OHC—CHO?

(A) alk. $KMnO_4$ (B) $Br_2/NaOH$ (C) SeO_2 (D) CrO_3

74. Which of the following reacts with Fehling solution?

(A) $C_6H_5CH_2CHO$ (B) CH_3COCH_3 (C) CH_3CHO (D) C_6H_5CHO

75. The reduction of $>C{=}O$ to $>CH_2$ is carried out with

(A) catalytic reduction (B) Na/C_2H_5OH

(C) Wolff–Kishner reduction (D) $LiAlH_4$

76. Which of the following reagents converts C_6H_5COCHO to $C_6H_5CH(OH)COOH$?

(A) Aq. NaOH (B) Acidic Na_2SO_3

(C) Na_2CrO_4/H_2SO_4 (D) $NaNO_2/HCl$ **(Roorkee 1998)**

Hint: It is an intramolecular Cannizzaro reaction.

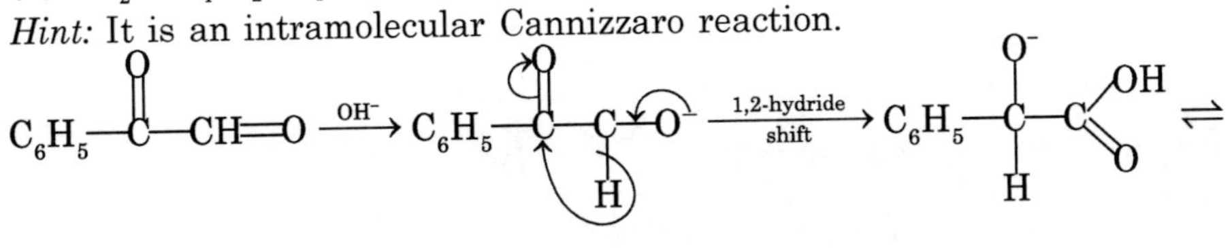

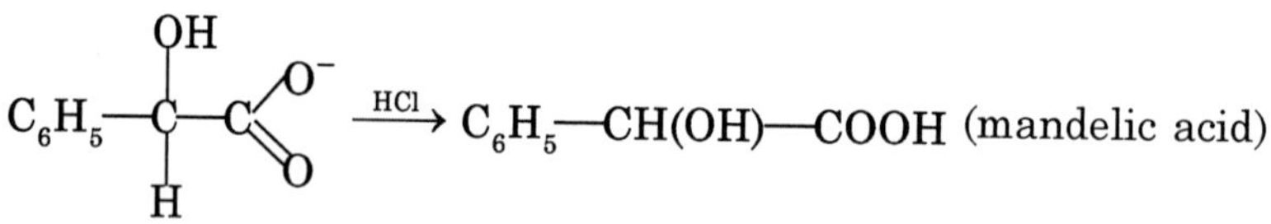

77. Which of the following is used to oxidise $(CH_3)_2C{=}CHCOCH_3$ to $(CH_3)_2C{=}CHCOOH$?
(A) H_2CrO_4 (B) NaOI (C) Cu at 573 K (D) $KMnO_4$

78. Which of the following is formed when ethanoyl chloride is treated with dimethylcadmium(II)?
(A) Acetophenone (B) Ethanoic acid (C) Ethanol (D) Propanone

79. Which of the following is formed when vapours of methanol is passed over red hot Cu at 573 K?
(A) Propanone (B) Methane (C) Methanol (D) Methanoic acid

80. Which of the following statements is correct?
(A) Pent-3-en-2-one is thermodynamically more stable than pent-4-en-2-one
(B) $LiAlH_4$ is a weaker reducing agent than lithium-*t*-butoxyaluminium hydride
(C) Propanone reacts with one equivalent of Br_2 in H_3O^+ to give $CH_3CBr(OH)CH_3$
(D) The enol content of pentane-2,4-dione in water is greater than that of pentane-2,4-dione in hexane.

81. Which of the following statements is correct?
(A) In the Beckmann rearrangement reaction, 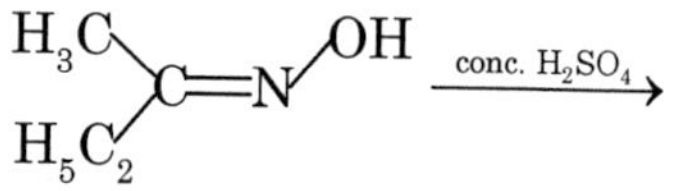 P, the product P is 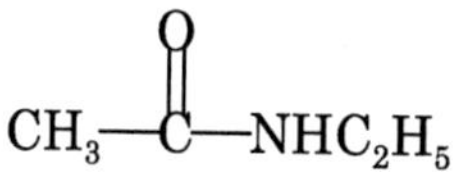
(B) Cannizzaro reaction is shown by an aldehyde containing alpha hydrogen atom
(C) The —CHO group in benzaldehyde is *o*- and *p*-directing
(D) The formation of oximes from aldehyde requires slightly alkaline medium.

82. The reduction of propanone to propane with Zn(Hg) and conc. HCl is an example of
(A) Rosenmund reduction (B) Cannizzaro reaction
(C) Clemmensen reduction (D) Wolf–Kishner reduction.

83. In the reaction, 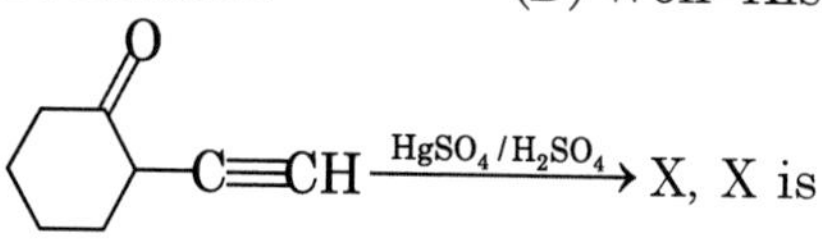 X, X is

(A) 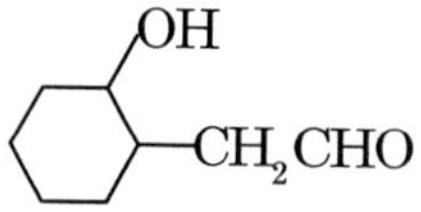

(B) 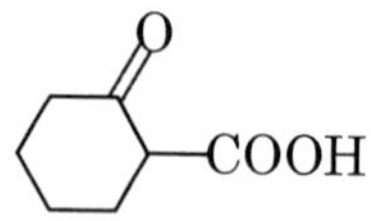

(C) 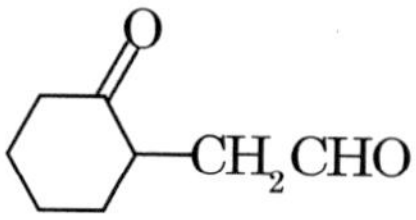

(D) 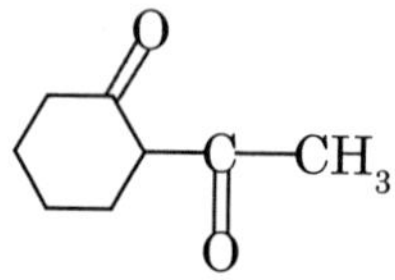

84. Which of the following reactions will give CH_2ClCHO?

(A) $HC\equiv CH \xrightarrow[KMnO_4]{Cl_2}$ (B) $HC\equiv CH \xrightarrow{Cl_2}$

(C) $HC\equiv CH \xrightarrow[peroxide]{HOCl}$ (D) $HC\equiv CH \xrightarrow{HOCl}$

85. Which of the following carbonyl compounds is most reactive in nucleophilic addition reactions?

(A) $CH_3COC_2H_5$ (B) CH_3COCH_3 (C) CH_3CHO (D) HCHO

86. Which of the following is paraldehyde?

(A) Hexamer of ethanol (B) Hexamer of methanal

(C) Trimer of ethanal (D) Dimer of methanal.

87. When propanone is saturated with HCl gas, the final product formed is

(A) benzene (B) mesityl oxide (C) phorone (D) diacetone alcohol

88. Which of the following is formed when ethanol is reacted with $Al(OC_2H_5)_3$?

(A) $CH_3COOC_2H_5$ (B) $CH_3CH{=}CHCHO$

(C) $CH_3CH(OH)CH_2CHO$ (D) None of these.

89. Which of the following will reduce both Fehling solution and Tollen reagent?

(A) $(CH_3)_3CCOCH_3$ (B) $CH_3COCH_2CH_3$

(C) C_6H_5CHO (D) CH_3CH_2CHO

90. Which of the following reagents is used for the reduction of $CH_3CH{=}CHCHO$ to $CH_3CH{=}CHCH_2OH$?

(A) Zn(Hg)/HCl (B) H_2/Ni (C) $NaBH_4$ (D) $[(CH_3)_2CHO]_3Al$

91. Which of the following reagents is used for the conversion of $CH_3CH_2C\equiv N$ to CH_3CH_2CHO?

(A) $NaBH_4/ether/H_3O^+$ (B) $SnCl_2/HCl/H_3O^+$

(C) $H_2/Pd–BaSO_4$, S (D) $LiAlH_4/H_3O^+$

92. Formylation of benzene is effected by

(A) $HCONH_2$ + CO (B) $HCOOCH_3$ + HCl

(C) CO + HCl (D) HCHO + HCl

93. Which of the following is formed when propanone is oxidised with SeO_2?

(A) CH_3COOH (B) CH_3COCHO

(C) CH_3COOCH_3 (D) $CH_3COCOOH$

94. Acid-catalysed aldol condensation occurs by

(A) the formation of stable tertiary carbanion

(B) the transfer of hydride ion

(C) the formation of carbanion

(D) the reaction between conjugate acid and enol form of carbonyl compound.

Hint: $(CH_3)_2C{=}O \xrightarrow{H^+} (CH_3)_2\overset{+}{C}{-}OH$

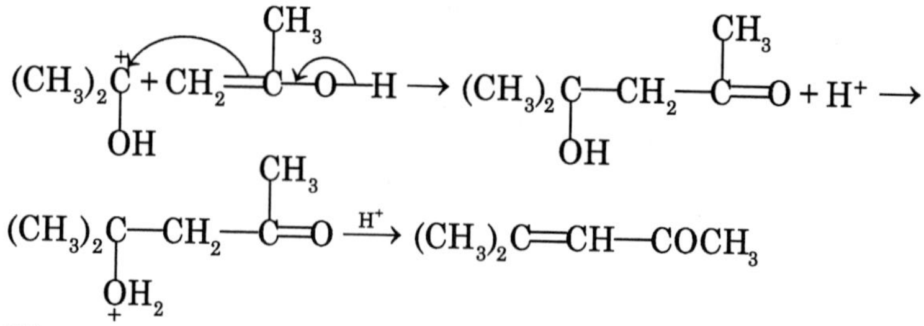

95. The reagent used to distinguish acetophenone and benzophenone is
(A) I_2/NaOH
(B) Benedict's reagent
(C) $LiAlH_4$
(D) 2,4-dinitrophenylhydrazine

96. Which of the carbonyl compounds does not form addition product with sodium hydrogen sulphate?
(A) Propanone (B) Ethanal (C) Acetophenone (D) Benzophenone

97. In the reaction, 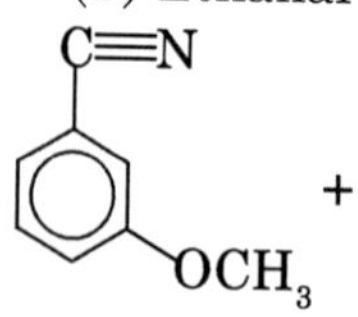 + $CH_3MgBr \xrightarrow{H_3O^+}$ P, the product P is

(A) 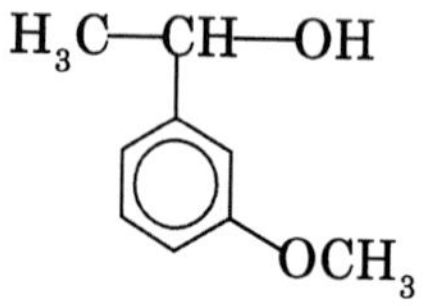
(B) 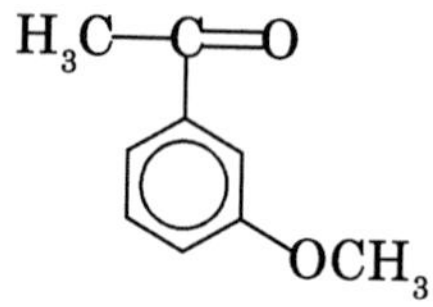
(C) 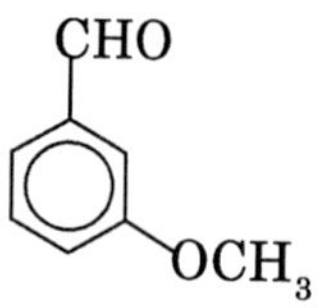
(D) 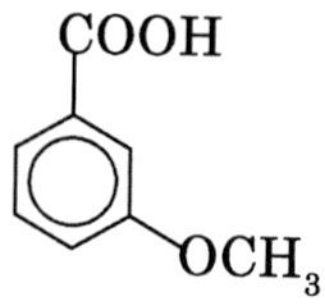

98. Methylethyl ketone is reduced to butane by
(A) HI/red P at 423 K
(B) Mg/Hg/H_2O
(C) Wolff–Kishner reduction
(D) Meerwein–Ponndorf reduction.

99. Which of the following is formed when acetophenone reacts with Br_2/ether at 273 K in the presence of anhy. $AlCl_3$?
(A) Phenacylbromide
(B) 4-Bromoacetophenone
(C) 3-Bromoacetophenone
(D) 2-Bromoacetophenone

100. Tautomerism is exhibited by

(A) 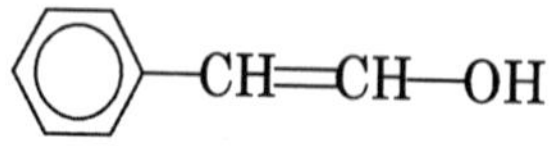
(B) 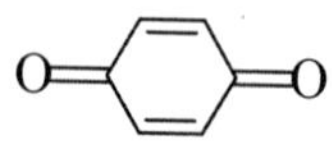
(C) 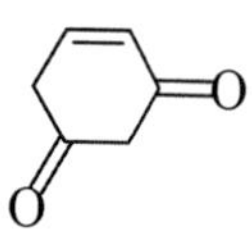
(D) 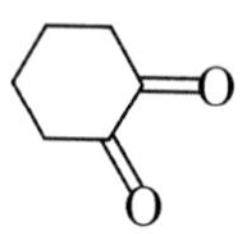

(I.I.T. 1998)

101. In the reaction, 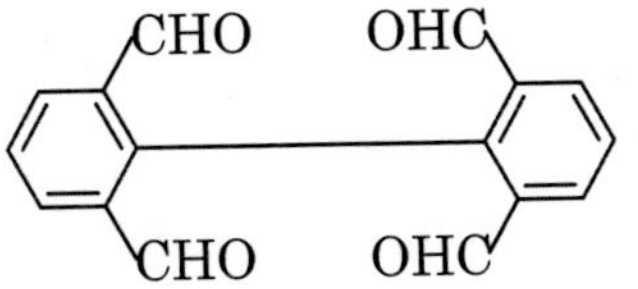 $\xrightarrow[\text{(ii) } H^+/H_2O]{\text{(i) NaOH/100°C}}$ P, the major product P is

(A) 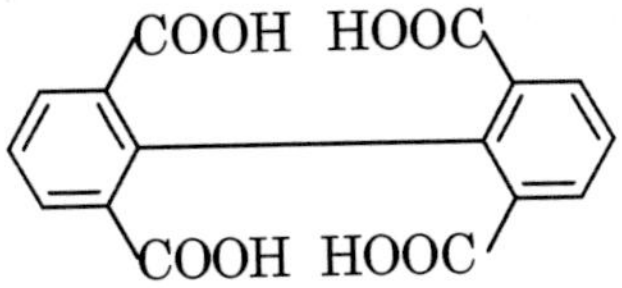

(B) 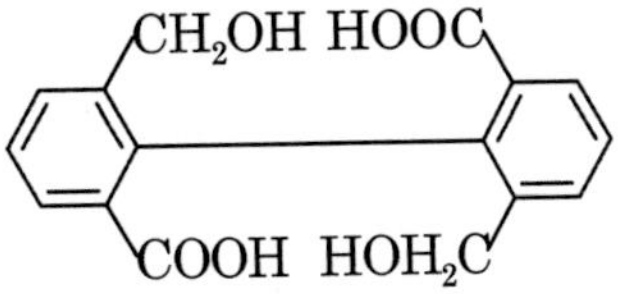

(C) 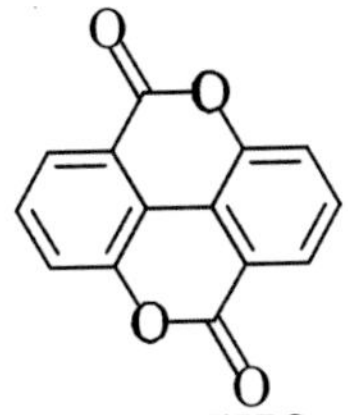

(D) 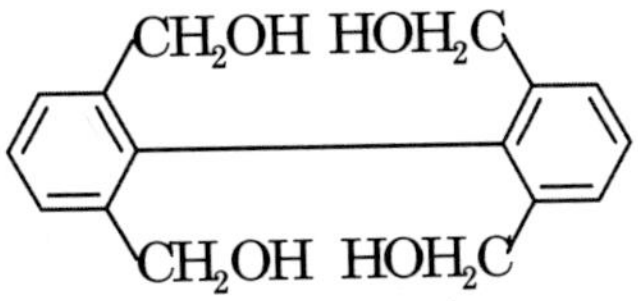

(I.I.T. 2003)

Hint: 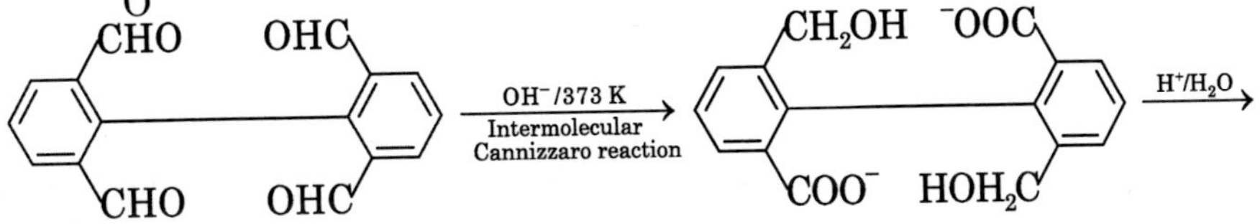

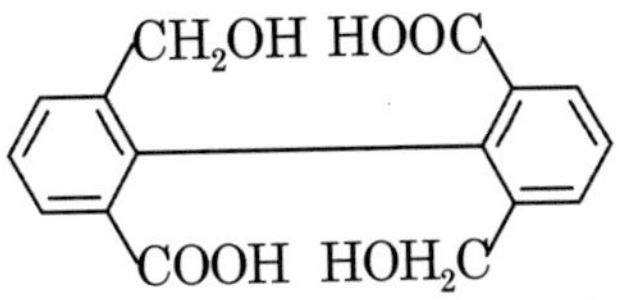

102. The order of reactivity of phenyl magnesium bromide with the following compounds is

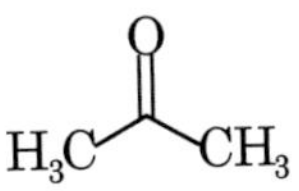
I

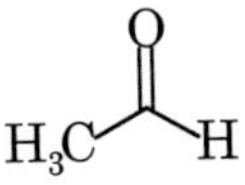
II

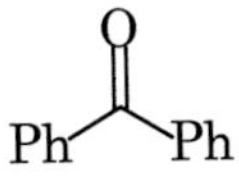
III

(A) II > III > I
(B) I > III > II
(C) II > I > III
(D) All react with same rate. **(I.I.T. 2004)**

Hint: Nucleophile attacks the most electrophilic site first. Among aldehyde and ketone, the aldehydes are more electrophilic as in the ketones the$\delta+$ charge on carbonyl carbon is decreased by the + I effect of both the alkyl groups. Moreover, in the tetrahedral intermediate aldehyde the steric repulsion is less than that in ketones. This facilitates the formation of intermediate in the aldehyde. Hence the order of reactivity is: II > I > III.

103. In the reaction, 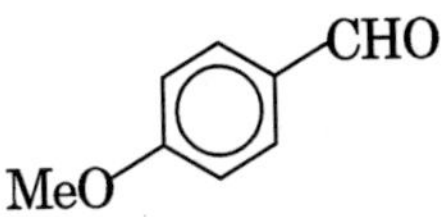 + X $\xrightarrow{CH_3COONa}$ 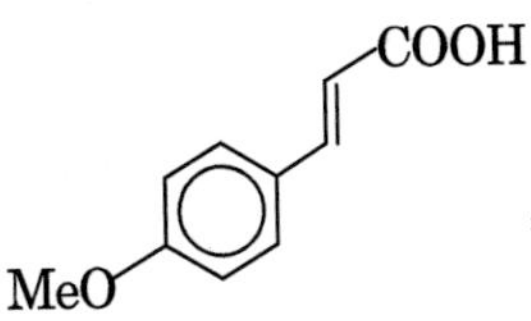, what is X?

(A) CH_3COOH (B) $BrCH_2COOH$ (C) $(CH_3CO)_2O$ (D) OHC—COOH

(I.I.T. 2005)

Hint: 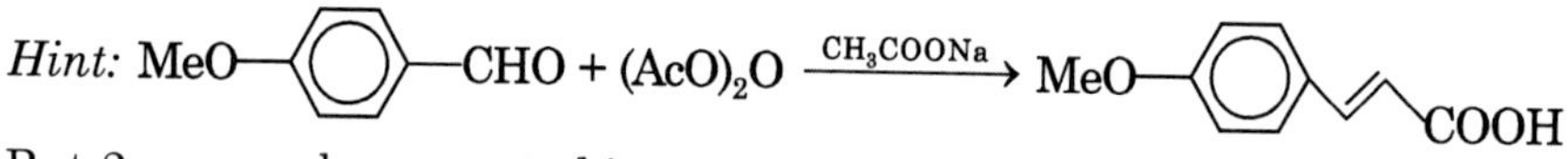

104. But-2-one can be converted into propanoic acid by which of the following:

(A) NaOH, NaI/H^+ (B) Fehling solution

(C) NaOH, I_2/H^+ (D) Tollen reagent **(I.I.T. 2005)**

Hint: 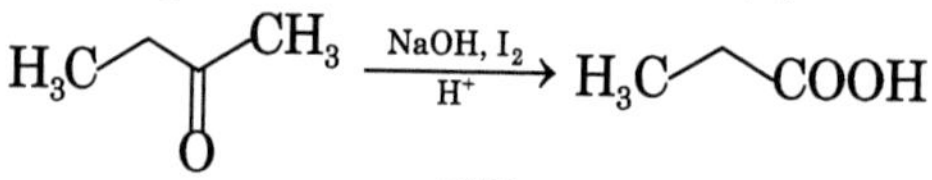

105. In the reaction, 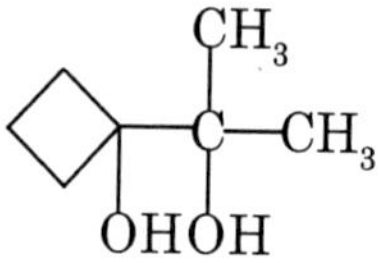 $\xrightarrow{\text{conc. } H_2SO_4}$ X, X is

(A) 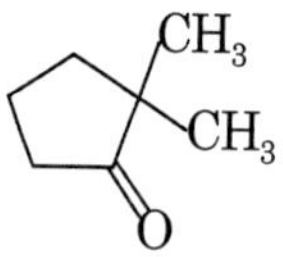

(B) 

(C) 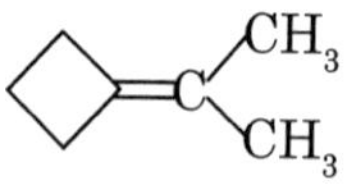

(D) 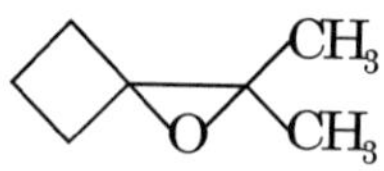

106. In the reaction, $2D{-}\underset{\underset{D}{|}}{C}{=}O + OH^- \xrightarrow{\text{Cannizzaro}}$ A and B (D is deuterium, B is alcohol), A and B are

(A) $H{-}\overset{\overset{O}{\|}}{C}{-}C^-$, $D{-}\overset{\overset{D}{|}}{\underset{\underset{D}{|}}{C}}{-}OH$

(B) $D{-}\overset{\overset{O}{\|}}{C}{-}O^-$, $D{-}\overset{\overset{D}{|}}{\underset{\underset{H}{|}}{C}}{-}OH$

(C) $D{-}\overset{\overset{O}{\|}}{C}{-}O^-$, $D{-}\overset{\overset{D}{|}}{\underset{\underset{D}{|}}{C}}{-}OH$

(D) none of these.

107. In the reaction, $CH_3\underset{\underset{O}{\|}}{C}CH_2CH_2\underset{\underset{O}{\|}}{C}OCH_2CH_3 \xrightarrow[\text{(ii) } H_3O^+]{\text{(i) } CH_3MgBr \text{ (1 mol)}}$ X, X is

(A) $CH_3\overset{\overset{CH_3}{|}}{\underset{\underset{CH_3}{|}}{C}}CH_2CH_2\overset{\overset{CH_3}{|}}{\underset{\underset{OH}{|}}{C}}CH_3$

(B) 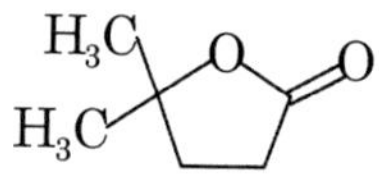

(C) $CH_3\underset{\underset{O}{\|}}{C}CH_2CH_2\underset{\underset{O}{\|}}{C}CH_3$

(D) $CH_3\overset{\overset{CH_3}{|}}{\underset{\underset{OH}{|}}{C}}CH_2CH_2\underset{\underset{O}{\|}}{C}OCH_2CH_3$

108. In the reaction, $CH_2{=}CH{-}\underset{\underset{O}{\|}}{C}{-}H + CH_3MgBr \xrightarrow{H_3O^+}$ X (1 : 4 addition), X is

(A) $CH_3CH_2CH_2CHO$ (B) $CH_3{-}CH{=}CHCH_2OH$

(C) $CH_2{=}CH{-}\underset{OH}{\overset{CH_3}{C}}{-}H$ (D) none of these.

109. Identify the correct statement.
   (A) Benzyl alcohol is a weaker acid than 4-nitrobenzyl alcohol
   (B) The reaction, $ROH + OH^- \rightleftharpoons RO^- + H_2O$ lies far towards the left side
   (C) The reaction, $RSH + OH^- \rightleftharpoons RS^- + H_2O$ lies far towards the left side
   (D) ROH is a weaker base than RSH.

   Hint: The statement (A) is correct and all others are incorrect.
   (A) The nitro group being an electron-withdrawing group, stabilises the resultant anion and as a result 4-nitrobenzyl alcohol is a stronger acid than benzyl alcohol.
   (B) The differences in acidities and basicities are only slight and as a result all the species are present at equilibrium.
   (C) The OH^- ion is a much stronger base than RS^- and RSH is a much stronger acid than HOH. Thus the differences in acidities and basicities are significant. An acid-base equilibrium lies towards the weaker acid and weaker base. Hence the equilibrium lies far towards the forward direction.
   (D) ROH is a stronger base than RSH. Due to the small size (more charge density) and greater electronegativity in comparison to S atom (less charge density and lower electronegativity), the O atom in ROH has greater capability of attracting H^+ ion.

110.  $+ Cl{-}CH_2{-}CH_2{-}CH_3 \xrightarrow{AlCl_3} P \xrightarrow[(ii)\,H_3O^+]{(i)\,O_2/\Delta} Q + \text{Phenol}$

The major products P and Q are

(A) 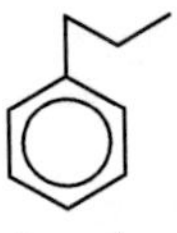 and CH_3CH_2CHO (B) 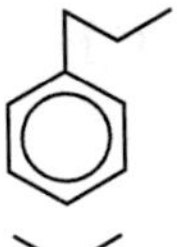 and CH_3COCH_3

(C) 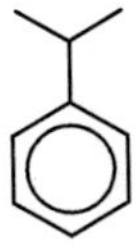 and CH_3COCH_3 (D) 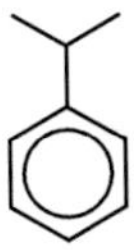 and CH_3CH_2CHO **(I.I.T. 2006)**

Hint: 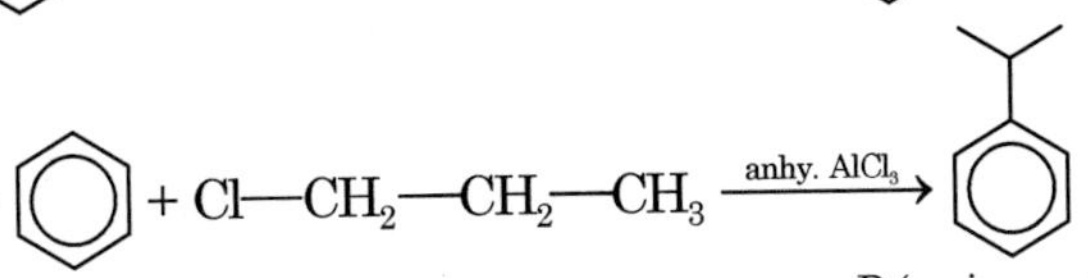

P (major product)

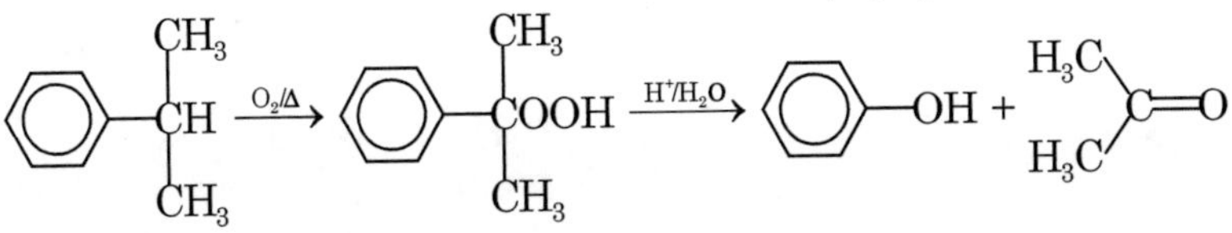

111. Which of the following compounds on reaction with aqueous alkali followed by acidification gives the following lactone?

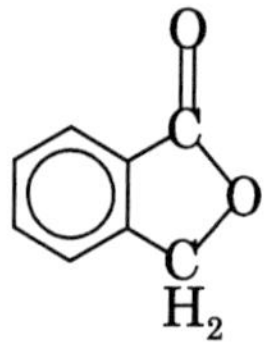

(A) 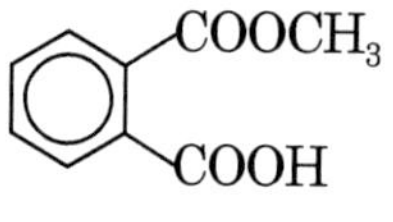

(B) 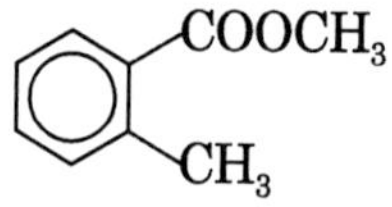

(C) 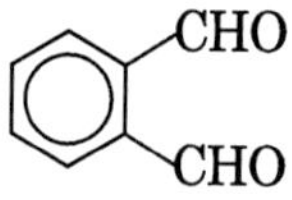

(D) 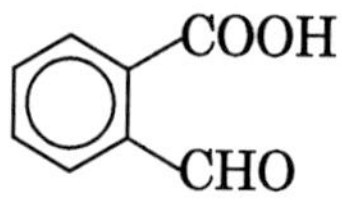

(I.I.T. 2006)

Hint: 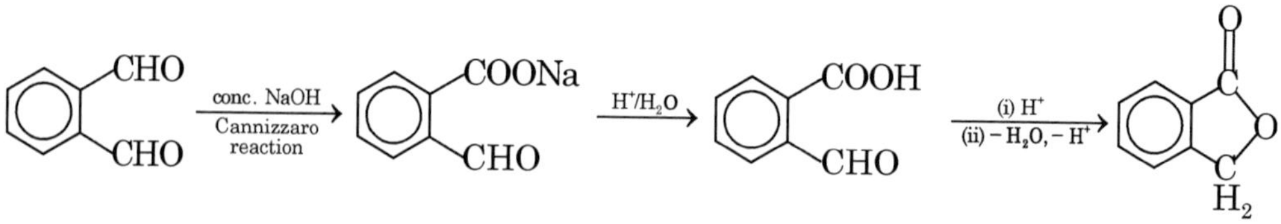

112. The lowest molecular mass ketone and its next homologue are reacted with NH_2OH to form oximes. Which of the following is/are correct?

(A) Two oximes are formed (B) Two oximes are optically active

(C) One oxime is optically active (D) Three oximes are formed. **(I.I.T. 2006)**

Hint: 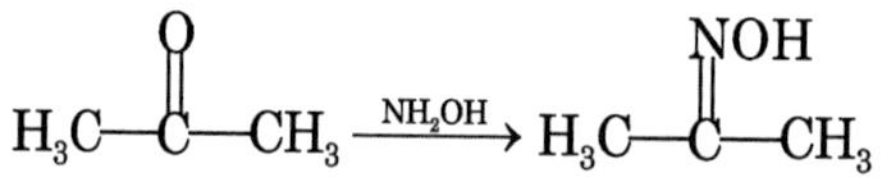

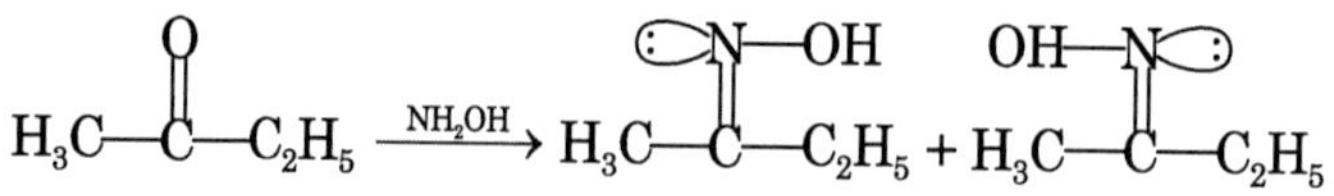

113. In the reaction, 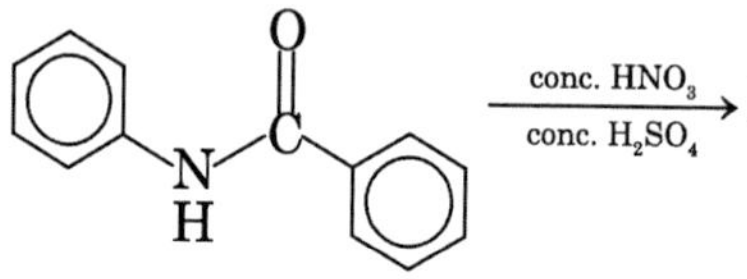 X, the structure of the major product X is

(A) 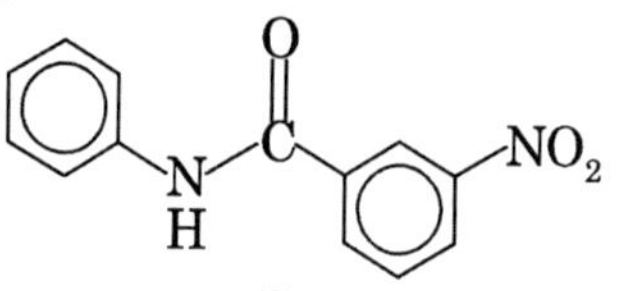

(B) 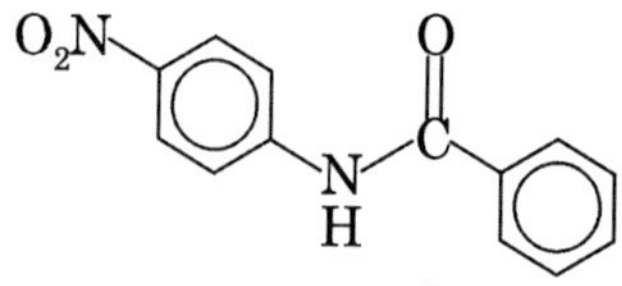

(C) 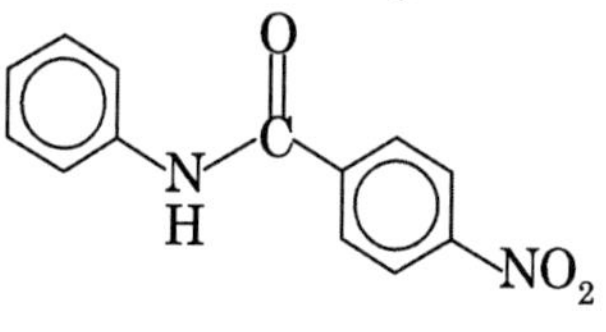

(D) 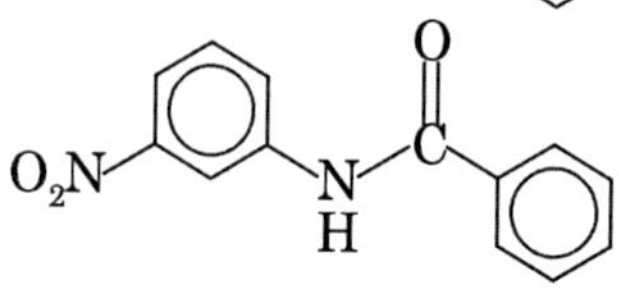

(I.I.T. 2007)

Hint: 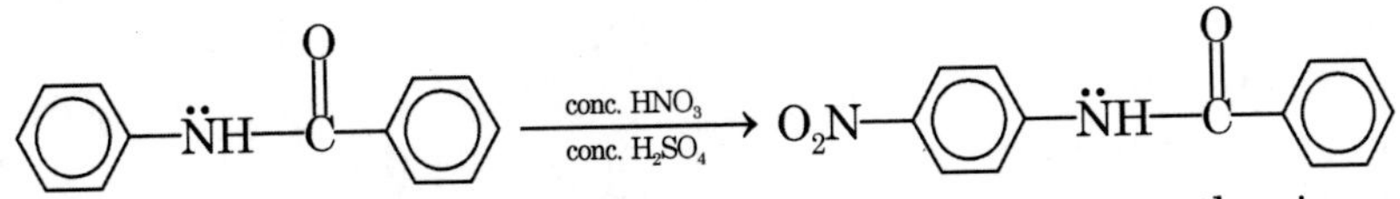

Due to the presence of lone pair of electrons on nitrogen atom, the ring gets activated and the intermediate carbocation is stabilised at *o*- and *p*-positions.

114. Statement–1: Glucose gives a reddish-brown precipitate with Fehling's solution.
because
Statement–2: Reaction of glucose with Fehling's solution gives CuO and gluconic acid.
(A) Statement–1 is true, Statement–2 is true; Statement–2 is a correct explanation for Statement–1
(B) Statement–1 is true, Statement–2 is true; Statement–2 is not a correct explanation for Statement–1
(C) Statement–1 is true, Statement–2 is false
(D) Statement–1 is false, Statement–2 is true. **(I.I.T. 2007)**

Hint: $C_6H_{12}O_6$ + Fehling solution $\longrightarrow (C_6H_{11}O_7)^- + Cu_2O\downarrow$ (Red precipitate)

115. Which one of the following reagents is used in the reaction,

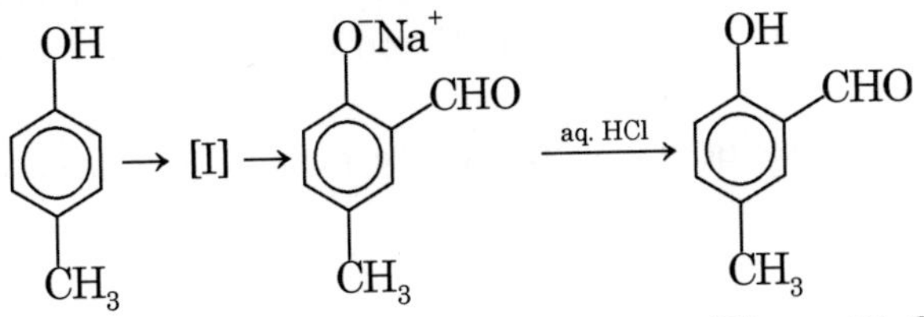

(A) aq. NaOH + CH_3Cl (B) aq. NaOH + CH_2Cl_2
(C) aq. NaOH + $CHCl_3$ (D) aq. NaOH + CCl_4 **(I.I.T. 2007)**

116. The electrophile in the reaction, 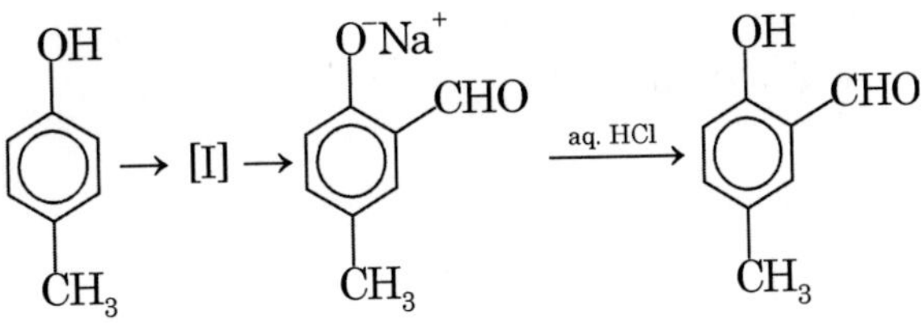 is

(A) :CHCl (B) $\overset{+}{C}HCl_2$ (C) :CCl_2 (D) $\cdot CCl_3$ **(I.I.T. 2007)**

Hint: $OH^-_{aq} + CHCl_3 \rightleftharpoons \bar{C}Cl_3 + H_2O$; $\bar{C}Cl_3 \longrightarrow Cl^- + :CCl_2$ (Dichloro carbene intermediate)

117. The structure of the intermediate (I) in the reaction is,

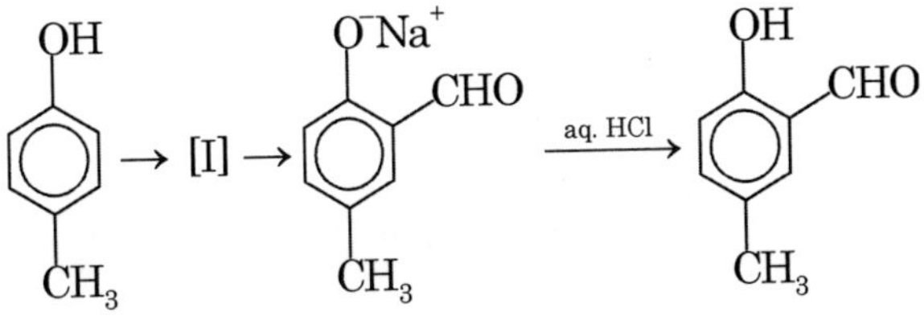 is

(A) 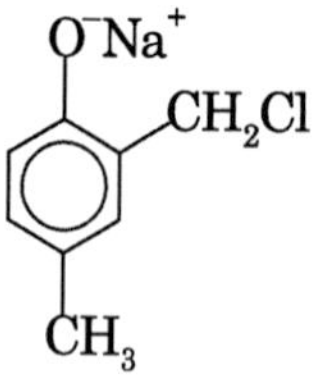 (B) 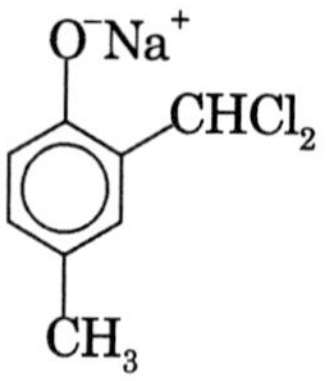 (C) 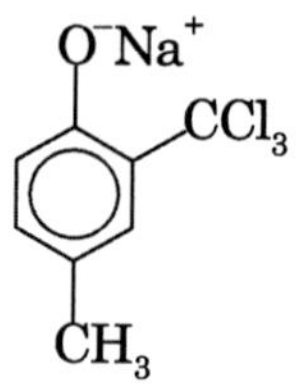 (D) 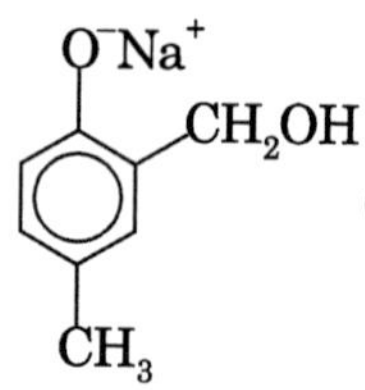 **(I.I.T. 2007)**

Hint: 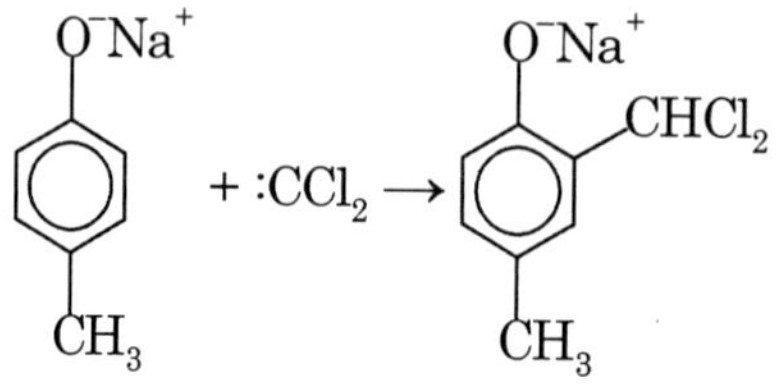

118. The correct statement(s) concerning the structures E, F and G is(are)

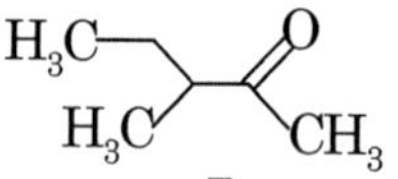
E

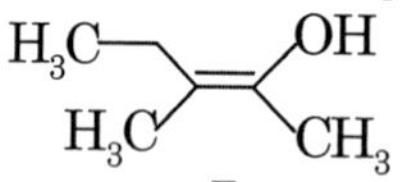
F

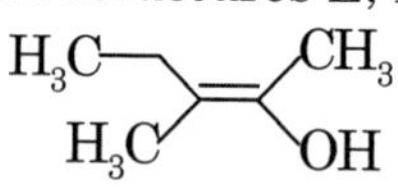
G

(A) E, F and G are resonance structures (B) E, F and G are tautomers

(C) F and G are geometrical isomers (D) F and G are diastereomers. **(I.I.T. 2008)**

Hint: E is a keto form and F and G are its enol forms. F and G are also geometrical isomers and diastereomers.

119. In the following reaction sequence, products I, J and L are formed. K represents a reagent.

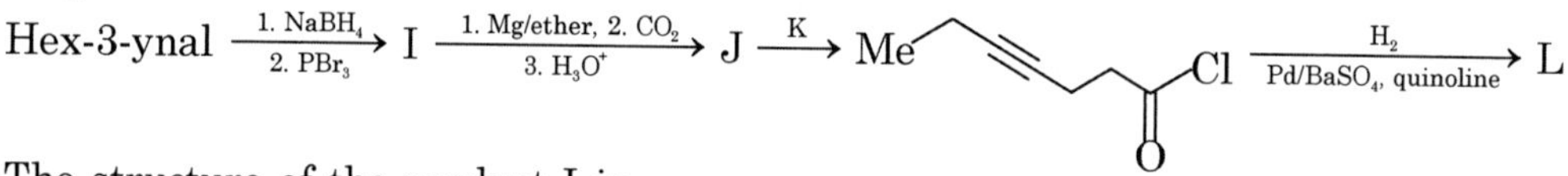

The structure of the product I is

(A) 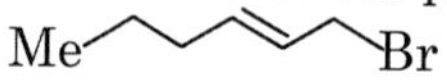 (B) 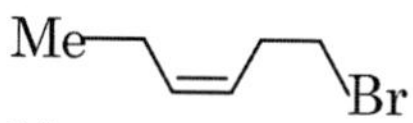

(C) 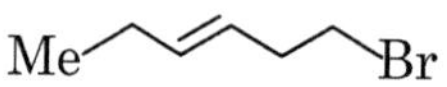 (D)  **(I.I.T. 2008)**

Hint: 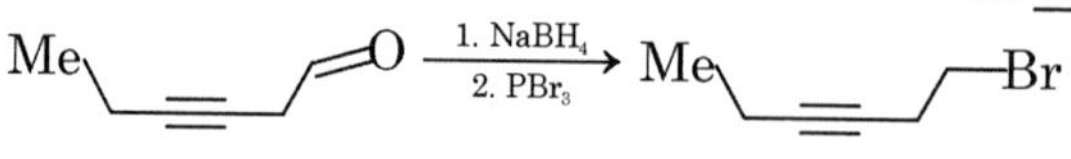

I

Paragraph for Question Nos. 120 to 122

A carbonyl compound P, which gives positive iodoform test, undergoes reaction with MeMgBr followed by dehydration to give an olefin Q. Ozonolysis of Q leads to a dicarbonyl compound R, which undergoes intramolecular aldol reaction to give predominantly S.

$$P \xrightarrow[\substack{2.\ H^+,\ H_2O \\ 3.\ H_2SO_4,\ \Delta}]{1.\ MeMgBr} Q \xrightarrow[2.\ Zn,\ H_2O]{1.\ O_3} R \xrightarrow[2.\ \Delta]{1.\ OH^-} S$$

120. The structure of the carbonyl compound P is

(A) 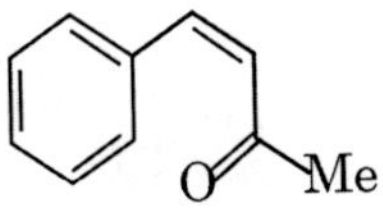

(B) 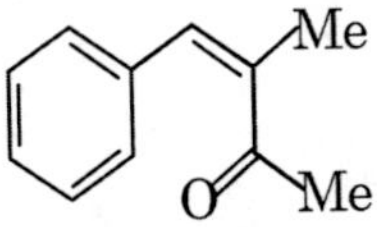

(C) 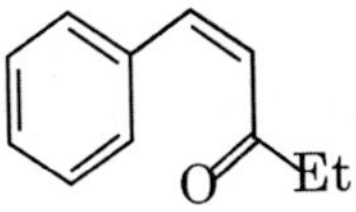

(D) 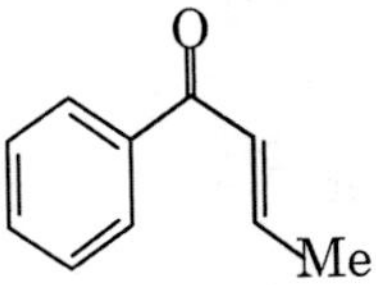

(I.I.T. 2009)

121. The structures of the products Q and R, respectively, are

(A) 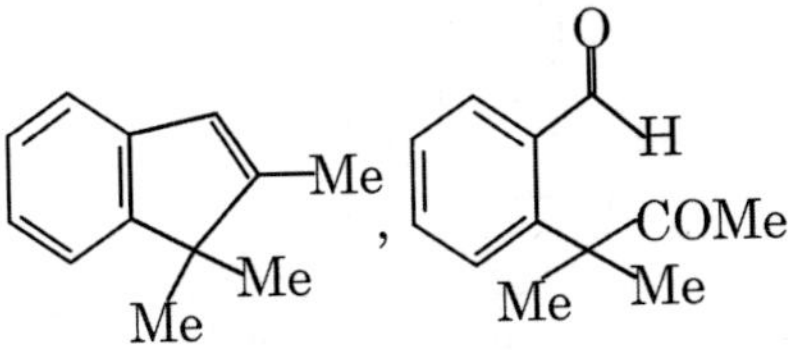

(B) 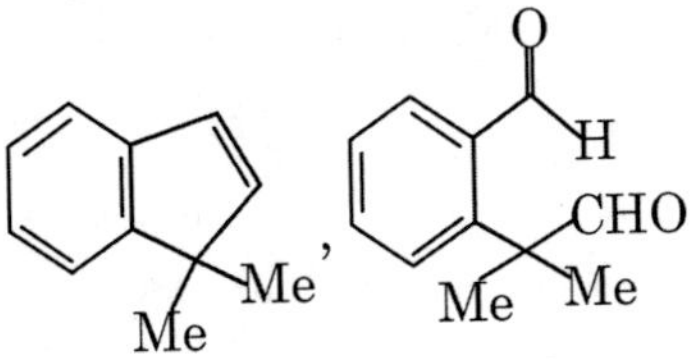

(C) 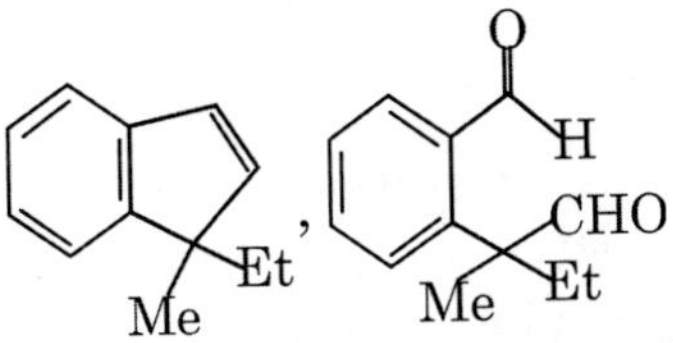

(D) 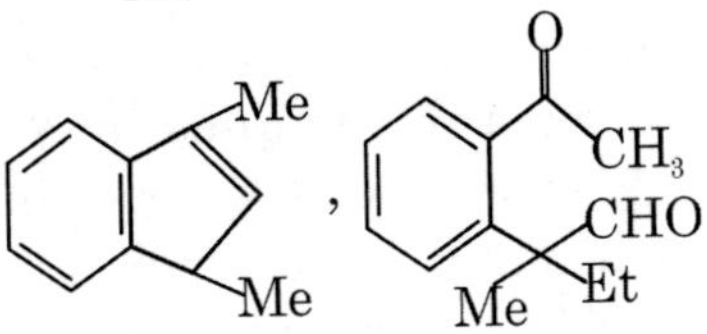

(I.I.T. 2009)

122. The structure of the product S is

(A) 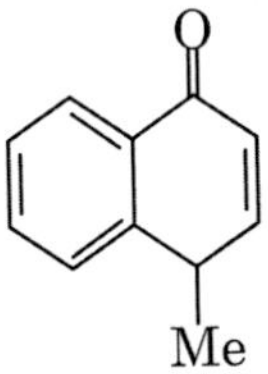

(B) 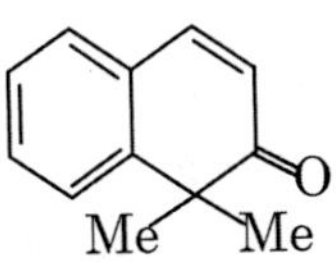

(C) 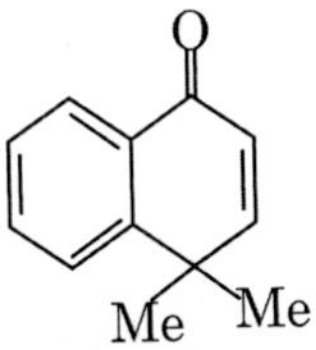

(D) 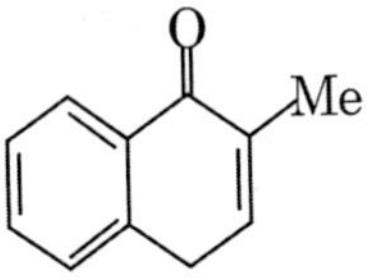

(I.I.T. 2009)

Hints for the Question Nos. 120 to 122

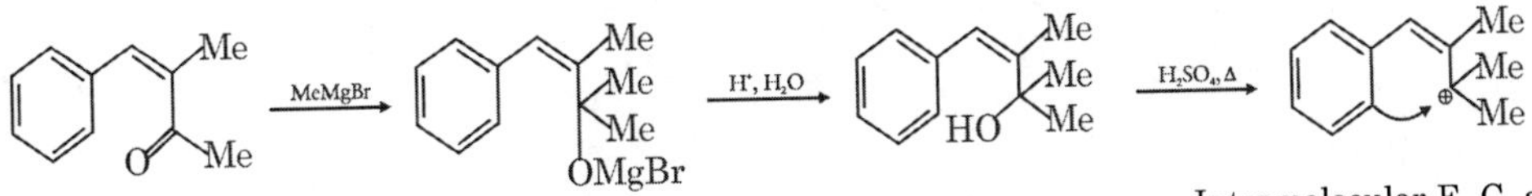

(P)

Intramolecular F. C. alkylation

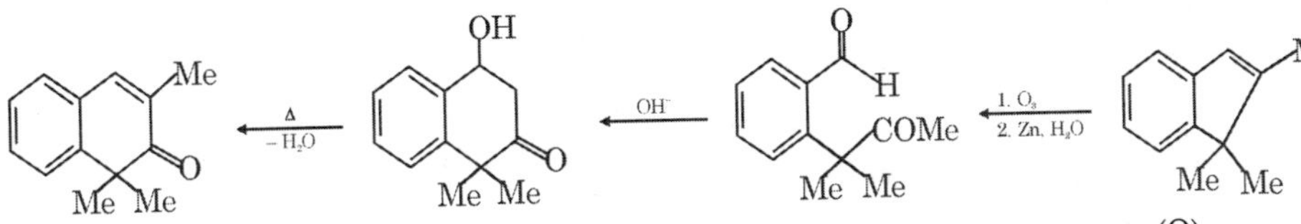

(S) (R) (Q)

123. In the following carbocation, H/CH_3 that is most likely to migrate to the positively charged carbon is

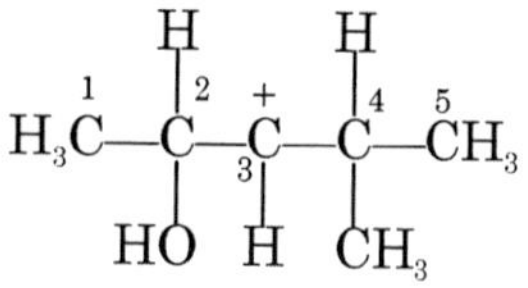

(A) CH_3 at C–4 (B) H at C–4 (C) CH_3 at C–3 (D) H at C–2 **(I.I.T. 2009)**

Hint: 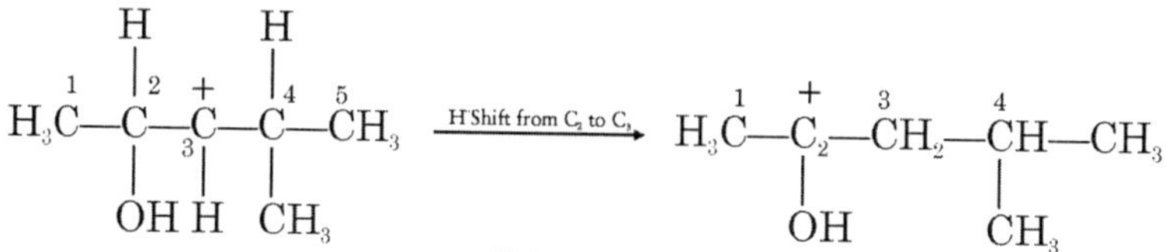

(driving force is conjugation from oxygen)

Migrating tendency of hydride is greater than that of alkyl group. Migration of hydride from second carbon gives more stable carbocation (stabilized by + R effect of OH group and + I effect of methyl group).

Paragraph for questions 124 to 126

Two aliphatic aldehydes P and Q react in the presence of aqueous K_2CO_3 to give compound R, which upon treatment with HCN provides compound S. On acidification and heating, S gives the product shown as:

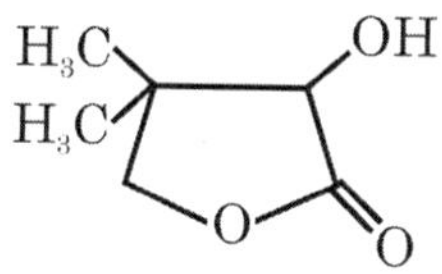

124.The compounds P and Q respectively are

(A) 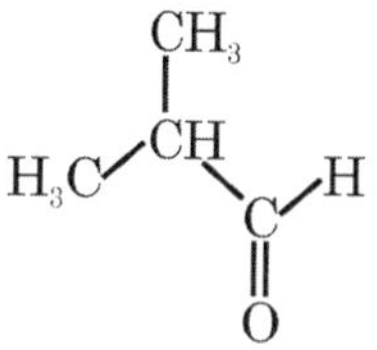 and 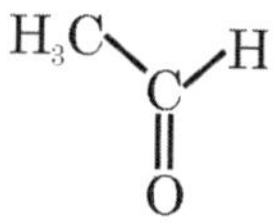

(B) 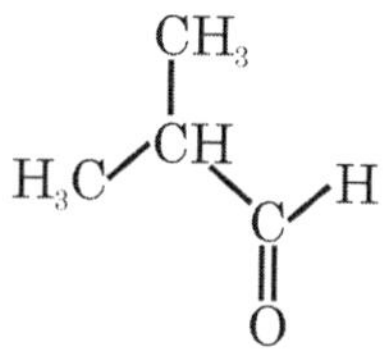 and 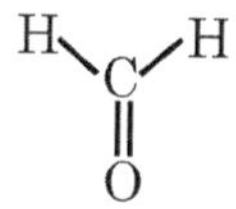

(C) 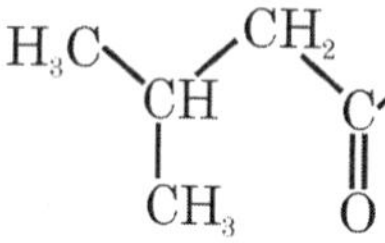 and 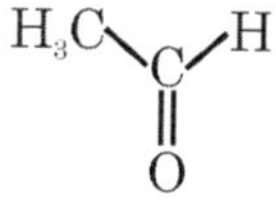

(D) 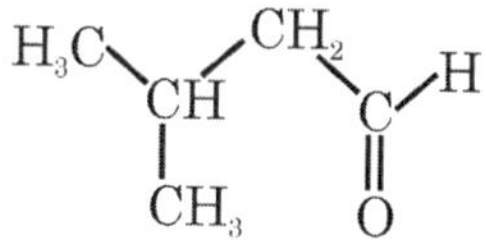 and 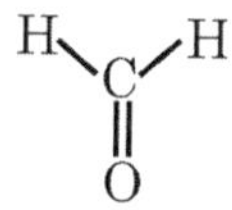

(I.I.T. 2010)

125. The compound R is

(A)

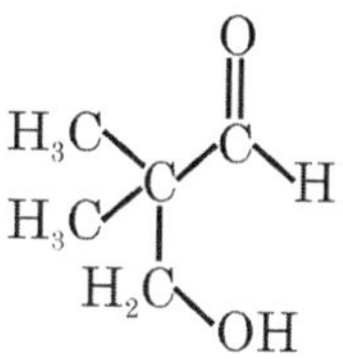

(B)

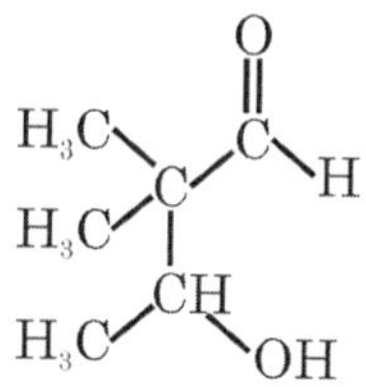

(C) 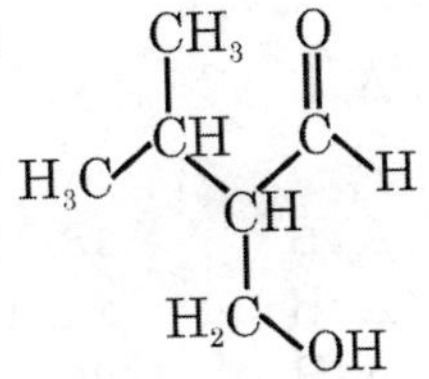

(D) 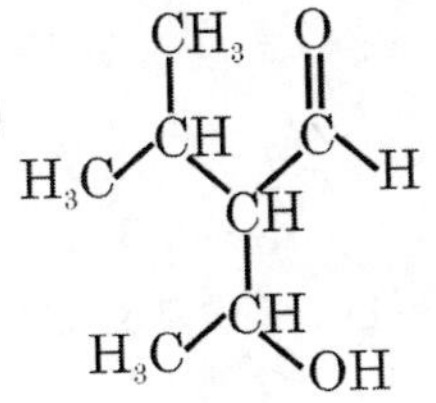

(I.I.T. 2010)

126. The compound S is

(A) 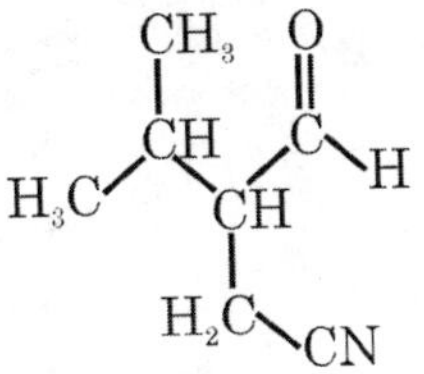

(B) 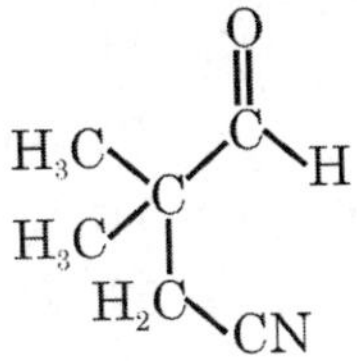

(C) 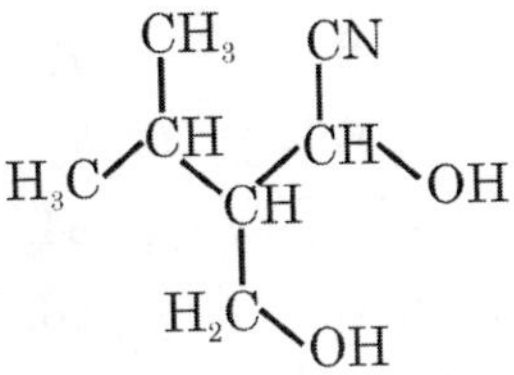

(D) 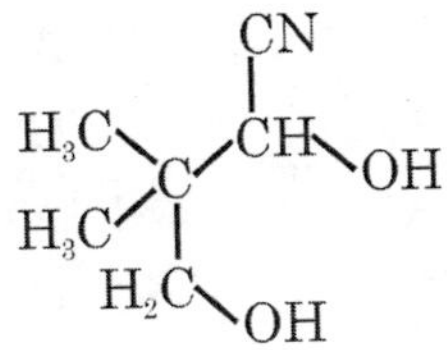

(I.I.T. 2010)

Hints to Question 124 to 126

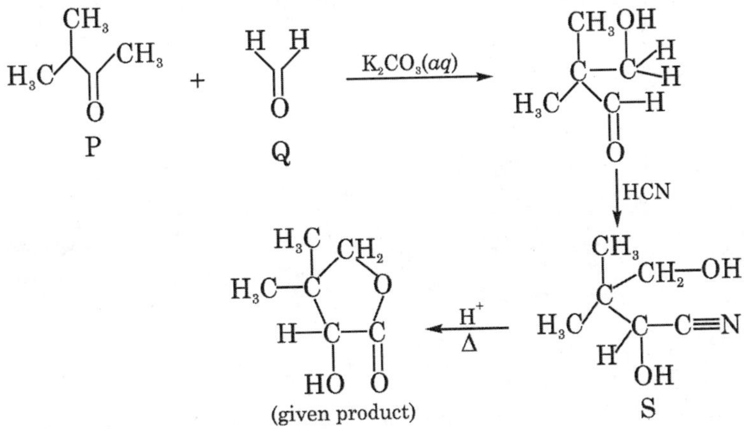

127. The major product of the following reaction is

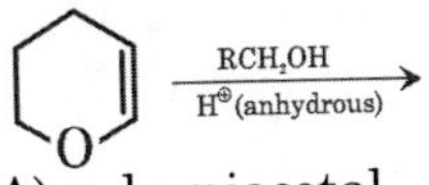

(A) a hemiacetal (B) an acetal (C) an ether (D) an ester

Hint: 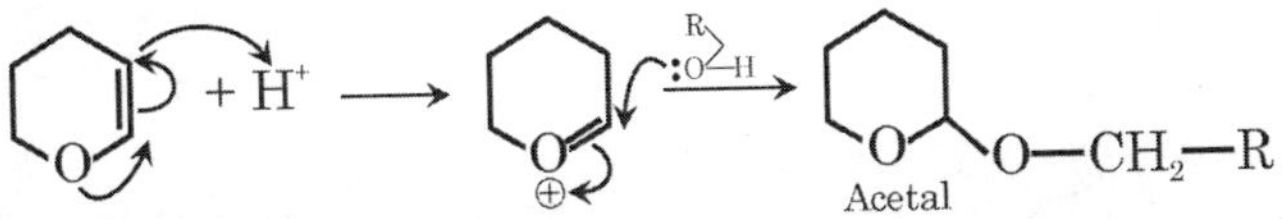

ANSWERS

1. A	2. B	3. A	4. A	5. A, B, C	6. B, D
7. B, C	8. C	9. A	10. A, B	11. D	12. B
13. A	14. B, D	15. C	16. B	17. A, B	18. A, D
19. B	20. A	21. D	22. B	23. A, B	24. B, D
25. D	26. A	27. C	28. D	29. A, B	30. A
31. B, C	32. B	33. C	34. B	35. D	36. B
37. A	38. C	39. A	40. C	41. B	42. A
43. D	44. B	45. A	46. B	47. C	48. A
49. A	50. A	51. B	52. A	53. A	54. A
55. C	56. A	57. B, C	58. A	59. A	60. B
61. C	62. A	63. B	64. B	65. C	66. A
67. C, D	68. D	69. A	70. A	71. B	72. D
73. C	74. A, C	75. C	76. A	77. B	78. D
79. C	80. A	81. A	82. C	83. C	84. C
85. D	86. C	87. C	88. A	89. D	90. C
91. B	92. C	93. B	94. D	95. A	96. D
97. B	98. A, C	99. A	100. A, C, D	101. B	102. C
103. C	104. C	105. A	106. C	107. B	108. A
109.A	110. C	111. C	112. D	113. B	114. C
115.C	116. C	117. B	118. B, C, D	119. D	120. B
121. A	122. B	123. D	124. B	125. A	126. D
127. B					

40

Compounds with Functional Groups Containing Oxygen—III: Carboxylic Acids and their Derivatives

OBJECTIVE TYPE MULTIPLE CHOICE QUESTIONS

1. The IUPAC name of succinic acid is
   (A) 1,4-butanedioic acid (B) 1,4-dibutanoic acid
   (C) 1,4-butanedicarboxylic acid (D) butane-1,4-dicarboxylic acid **(I.I.T. 1994)**
2. The IUPAC name of $CH_3CH{=}CHCOOH$ is
   (A) but-2-en-1-oic acid (B) but-3-en-1-oic acid
   (C) but-2,3-en-1-oic acid (D) butene-2-carboxylic acid **(I.I.T. 1986)**
3. Which of the following pairs forms strongest hydrogen bonding?
   (A) SiH_4 and SiF_4 (B) CH_3COCH_3 and $CHCl_3$
   (C) HCOOH and CH_3COOH (D) H_2O and H_2O_2 **(I.I.T. 1981)**
4. Benzoic acid and phenol are distinguished by their reaction with
   (A) aq. NaOH (B) aq. $NaHCO_3$ (C) neutral $FeCl_3$ (D) aq. NH_3 **(I.I.T. 1992)**
5. Chlorination of toluene in the presence of light and heat followed by treatment with aq. NaOH gives
   (A) *o*-cresol (B) *p*-cresol
   (C) 2,4-dihydroxytoluene (D) benzoic acid **(I.I.T. 1990)**
6. Benzoic acid reacts with conc. HNO_3 and conc. H_2SO_4 to give
   (A) *o*-nitrobenzoic acid (B) *p*-nitrobenzoic acid
   (C) *m*-nitrobenzoic acid (D) *o*-, *p*-dinitrobenzoic acid **(I.I.T. 1993)**
7. The increasing order of acid strength in the acids $ClCH_2COOH$, CH_3CH_2COOH, $ClCH_2CH_2COOH$, $(CH_3)_2CHCOOH$ and CH_3COOH is
   (A) $(CH_3)_2CHCOOH < CH_3CH_2COOH < ClCH_2CH_2COOH < CH_3COOH < ClCH_2COOH$

(B) $CH_3CH_2COOH < (CH_3)_2CHCOOH < ClCH_2CH_2COOH < CH_3COOH < ClCH_2COOH$

(C) $(CH_3)_2CHCOOH < CH_3CH_2COOH < ClCH_2CH_2COOH < ClCH_2COOH < CH_3COOH$

(D) $(CH_3)_2CHCOOH < ClCH_2CH_2COOH < CH_3CH_2COOH < CH_3COOH < ClCH_2COOH$

(I.I.T. 1991)

8. Which of the following carboxylic acids undergoes decarboxylation easily?

(A) $C_6H_5COCH_2COOH$ (B) $C_6H_5COCOOH$

(C) $C_6H_5CH(OH)COOH$ (D) $C_6H_5CH(NH_2)COOH$ **(I.I.T. 1985)**

9. When CH_3CH_2COOH is treated with aq. $NaHCO_3$, CO_2 is evolved. The C of CO_2 comes from

(A) —CH_3 group (B) —COOH group

(C) —CH_2 group (D) $NaHCO_3$ **(I.I.T. 1999)**

10. *Assertion (A):* Acetic acid does not undergo haloform reaction.

Reason (R): Acetic acid has no α-hydrogen atom.

(A) Both *A* and *R* are correct and *R* is the correct explanation of *A*

(B) Both *A* and *R* are correct and *R* is not the correct explanation of *A*

(C) *A* is correct but *R* is incorrect (D) *A* is incorrect but *R* is correct. **(I.I.T. 1998)**

11. Identify Z in the reactions, $C_6H_6 \xrightarrow{H_2SO_4} W \xrightarrow[(ii)\,H^+]{(i)\,NaOH} X \xrightarrow{(i)\,CO_2,\,KOH\,(ii)\,H^+} Y \xrightarrow{CH_3COCl} Z$

(A) 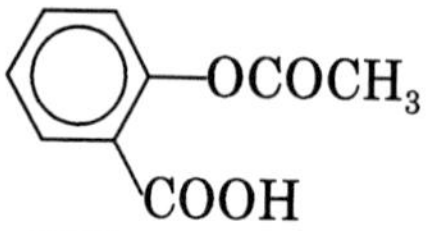

(B) 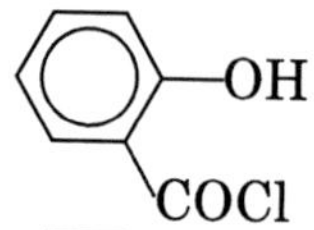

(C) 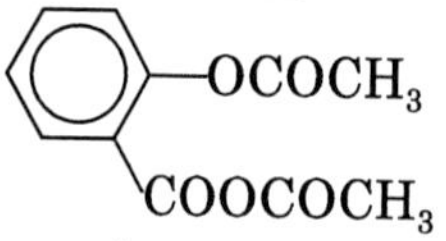

(D) 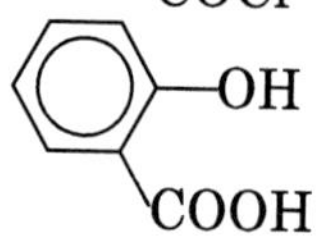

(Roorkee 1987)

12. The weaker acid among the following is

(A) CH_3COOH (B) $ClCH_2COOH$ (C) Cl_3CCOOH (D) $(CH_3)_2CHCOOH$

(Roorkee 1992)

13. When an alcohol reacts with any acid, it produces

(A) acid with higher C (B) secondary alcohol

(C) ester and water (D) alkane **(Roorkee 1995)**

14. An inorganic compound producing an organic compound on heating is

(A) ammonium cyanate (B) sodalime

(C) sodamide (D) potassium cyanide **(Roorkee 1992)**

15. A colourless liquid at room temperature reacts with sodalime to form sodium salt of carboxylic acid and NH_3 gas. The liquid is

(A) propanoic acid (B) methyl ethanoate

(C) propanamide (D) formamide **(Roorkee 1998)**

16. Reaction of ethyl formate with excess of CH_3MgI followed by hydrolysis gives

(A) ethanol (B) *n*-propyl alcohol

(C) propanal (D) isopropyl alcohol **(I.I.T. 1992)**

17. Which of the following converts acetic acid to acetyl chloride?
(A) NaCl (B) PCl_5 (C) Cl_2/P (D) HCl

18. RCH_2CH_2OH can be converted to RCH_2CH_2COOH. The sequence of reagents is
(A) PBr_3, KCN, H^+ (B) PBr_3, KCN, H_2
(C) HCN, PBr_3, H^+ (D) KCN, H^+

19. In the Hofmann bromamide reaction, the reagents are
(A) $Br_2/NaOH$ (B) Br_2/HNO_2 (C) CH_3Br (D) HNO_3

20. In the reaction, $CH_3CH_2COOH \xrightarrow{Cl_2/\text{red P}} A \xrightarrow{\text{alc. KOH}} B$, B is
(A) CH_3CH_2COCl (B) $CH_3CHClCOOH$
(C) $CH_2{=}CHCOOH$ (D) CH_3CH_2OH

21. When urea is heated slowly, then NH_3 and one other substance is formed. The substance is
(A) $NH_2CONHNO_2$ (B) carbonic acid
(C) cyanic acid (D) biuret

22. In the reaction $C_6H_5COOCH_3 \xrightarrow{LiAlH_4,\ H^+/H_2O} X$, X is
(A) $C_6H_5COOH + CH_3OH$ (B) $C_6H_5CH_2OH + CH_3OH$
(C) $C_6H_5CHO + CH_3COOH$ (D) all of these. **(I.I.T. 1995)**

23. Ethyl acetate reacts with CH_3MgBr to form
(A) secondary alcohol (B) tertiary alcohol
(C) primary alcohol and acid (D) acid

24. Formic acid is stronger than acetic acid because of
(A) stronger + I-effect in formic acid (B) stronger – R-effect in formic acid
(C) stronger + I-effect in acetic acid (D) weaker + I-effect in acetic acid.

25. Which of the following is more sensitive to nucleophilic attack at carbonyl group?
(A) $CH_3COOC_2H_5$ (B) CH_3CHO (C) CH_3COCl (D) $CH_3COOCOCH_3$

26. Acetylation is not given by
(A) C_6H_5COOH (B) $C_6H_5CH_3$ (C) C_6H_5Cl (D) C_6H_6

27. Benzoic acid is less acidic than salicylic acid because of
(A) hydrogen bond (B) inductive effect
(C) resonance (D) none of these.

28. In a reaction involving ring substitution of C_6H_5X, the major product is *meta* isomer. X is
(A) $—NH_2$ (B) —COOH (C) $—CH_3$ (D) —Cl

29. The reaction $2CH_3COOC_2H_5 \xrightarrow{C_2H_5O^-} CH_3COCH_2COOC_2H_5$ is called
(A) Perkin reaction (B) Tischenko reaction
(C) Claisen condensation (D) Schotten–Baumann reaction.

30. Which of the following will convert ethanoic acid to propanone?
(A) $KMnO_4$ (B) $LiAlH_4$ (C) Cr_2O_3 (D) MnO

31. Silver benzoate reacts with Br_2 to form
    (A) C_6H_6 (B) C_6H_5COOBr (C) C_6H_5Br (D) m-$C_6H_4(Br)(COOAg)$
32. An organic liquid of the composition $C_4H_8O_2$ yields a sodium salt of $C_3H_6O_2$ and methanol on boiling with NaOH solution. The given organic liquid is
    (A) $HCOOC_3H_7$ (B) $CH_3COOC_2H_5$
    (C) $CH_3CH_2CH_2COOH$ (D) $CH_3CH_2COOCH_3$
33. The molecular weight of benzoic acid in benzene as determined by depression in freezing point method corresponds to
    (A) ionisation of benzoic acid (B) dimerisation of benzoic acid
    (C) trimerisation of benzoic acid (D) solvation of benzoic acid **(I.I.T. 1996)**
34. Formic acid and acetic acid are distinguished by reaction with
    (A) sodium ethoxide (B) sodium
    (C) $HgCl_2$ (D) 2,4-dinitrophenylhydrazine
35. When a carboxylic acid is warmed with hydrazoic acid in presence of conc. H_2SO_4, it produces
    (A) alkyl cyanide (B) primary amine
    (C) secondary amine (D) acid amide
36. An acid amide is converted into an acid and N_2 by
    (A) NaOH (B) P_4O_{10} (C) HNO_2 (D) HCl
37. Which of the following esters is obtained by the esterification of propan-2-ol with ethanoic acid?
    (A) $(CH_3)_2CHCOOCH_2CH_3$ (B) $CH_3COOCH(CH_3)_2$
    (C) $(CH_3)_2CHCOOCH_3$ (D) $CH_3COOCH_2CH_3$
38. Compounds X and Y give effervescence with Na_2CO_3 solution. X gives a white precipitate with ammoniacal $AgNO_3$ while Y gives a sweet smelling compound on heating with alcohol. X and Y are
    (A) HCOOH and CH_3COOH (B) CH_3COCH_3 and HCOOH
    (C) CH_3CHO and CH_3COOH (D) CH_3COOH and CH_3COCH_3
39. Which of the following statements is wrong?
    (A) Formic acid is stronger than acetic acid
    (B) Benzaldehyde does not reduce Fehling solution
    (C) *o*-Bromobenzoic acid is weaker than *o*-chlorobenzoic acid
    (D) Lactic acid does not answer silver mirror test.
40. Heating a mixture of sodium benzoate and sodalime gives
    (A) calcium benzoate (B) benzene
    (C) sodium benzoate (D) methane
41. In the esterification reaction of alcohols
    (A) H^+ is replaced by Na (B) OH^- is replaced by C_6H_5OH
    (C) OH^- is replaced by CH_3COO^- (D) OH^- is replaced by chlorine

42. Hydrolysis of acetamide gives
 (A) acetic acid (B) formic acid (C) methylamine (D) acetaldehyde
43. Diazomethane reacts with carboxylic acids to give
 (A) esters (B) amines (C) alcohols (D) imines
44. Acrylic acid reacts with HBr to give
 (A) $Br—CH_2—CHBr—COOH$ (B) $Br—CH_2—CH_2—COOH$
 (C) $CH_3—CHBr—COOH$ (D) $CH_2=CHCOBr$
45. On strong heating ammonium acetate gives
 (A) acetamide (B) urea (C) formamide (D) methyl cyanide
46. What is the main reason for the fact that carboxylic acids undergo ionisation?
 (A) Hydrogen bonding (B) Resonance stabilisation of COO^- ion
 (C) Absence of α-hydrogen atom (D) High reactivity of α-hydrogen atom.
47. The reaction involving condensation of acid anhydride with an aromatic anhydride catalysed by a carboxylate ion is an example of
 (A) Claisen reaction (B) Perkin reaction
 (C) aldol condensation (D) none of these.
48. On heating a mixture of potassium cyanate and ammonium chloride, it gives
 (A) urea (B) methanamide (C) ethanamide (D) ethanamine
49. Hell–Volhard–Zelinsky reaction is used to synthesise
 (A) α-haloacids (B) ketones (C) aldehydes (D) acid halides
50. Which of the following is the strongest acid?
 (A) CH_3OH (B) CH_3CH_2OH (C) $C_6H_5SO_3H$ (D) C_6H_5COOH
51. Identify the correct order of boiling points of the following compounds:
 (1) $CH_3CH_2CH_2CH_2OH$ (2) $CH_3CH_2CH_2CHO$ (3) $CH_3CH_2CH_2COOH$
 (A) 1 > 2 > 3 (B) 3 > 1 > 2 (C) 1 > 3 > 2 (D) 3 > 2 > 1 **(I.I.T. 2002)**
52. Which of the following reducing agents is used to reduce CH_3COOH to CH_3CH_2OH?
 (A) $NaBH_4$ (B) $LiAlH_4$ (C) Na/C_2H_5OH (D) BF_3/THF followed by H_2O
53. Which of the following does not react with nitrous acid?
 (A) $HCONH_2$ (B) $CH_3CON(CH_3)—C_2H_5$
 (C) $CH_3CONHCH_3$ (D) CH_3CONH_2
54. Which of the following carboxylic acids cannot be reduced to the primary alcohol by $LiAlH_4$?
 (A) Cyclohexanecarboxylic acid (B) $4\text{-}O_2NC_6H_4COOH$
 (C) $4\text{-}CH_3C_6H_4COOH$ (D) *Z*-$CH_3CH=CHCH_2COOH$
55. Identify the correct statement.
 (A) A tertiary amide liberates N_2 on reaction with HNO_2
 (B) A secondary amide liebrates N_2 on reaction with HNO_2
 (C) A primary amide liberates N_2 on reaction with HNO_2

(D) None of primary, secondary and tertiary amides liberates N_2 on reaction with HNO_2

56. Identify the correct order of reactivity of acid derivatives toward a nucleophile.
(A) ester < acid chloride < acid anhydride
(B) acid anhydride < ester < acid chloride
(C) acid chloride < acid anhydride < ester
(D) ester < acid anhydride < acid chloride

57. Identify the correct order of the basic strength of a leaving group in a reaction of an acid derivative with a nucleophile.
(A) $CH_3COO^- < Cl^- < CH_3O^-$ (B) $Cl^- < CH_3COO^- < CH_3O^-$
(C) $CH_3COO^- < CH_3O^- < Cl^-$ (D) $CH_3O^- < CH_3COO^- < Cl^-$

58. Which of the following statements is not correct?
(A) Amides are more acidic than ammonia
(B) Peroxy acids (RCOOOH) are more acidic than carboxylic acids (RCOOH)
(C) The C═O bond length in CH_3COOH is longer than that in CH_3CHO
(D) 2-Hydroxybenzoic acid is a stronger acid than 2-methoxybenzoic acid.

Hint: The statement (B) is incorrect and all others are correct.

(A) The amides are more acidic than ammonia because of the stabilisation of amidates due to resonance.

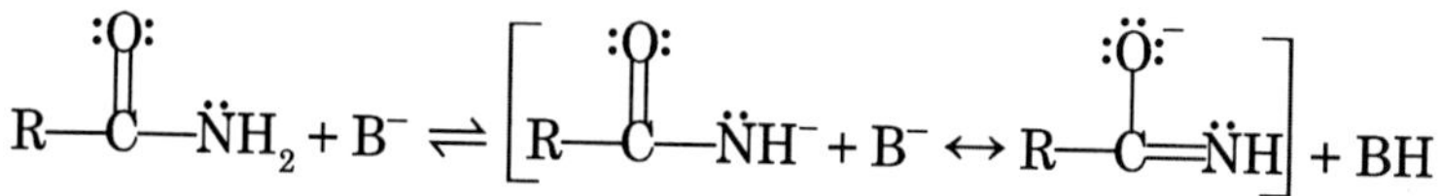

(B) The conjugate base of a carboxylic acid (RCOOH) is stabilised due to resonance. No such stabilisation occurs in the conjugate base of a peroxy acid (RCOOOH).

(C) Due to resonance in CH_3COOH, the C═O bond acquires some single-bond character. As a result the C═O bond is longer than that in CH_3CHO.

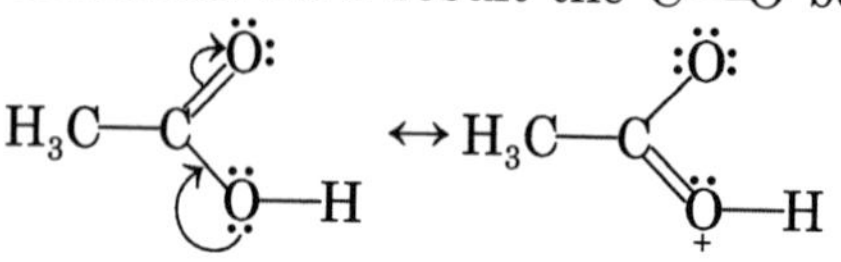

(D) Due to hydrogen bonding and resonance salicylate ion is stabilised and 2-hydroxybenzoic acid is a stronger acid than 2-methoxybenzoic acid.

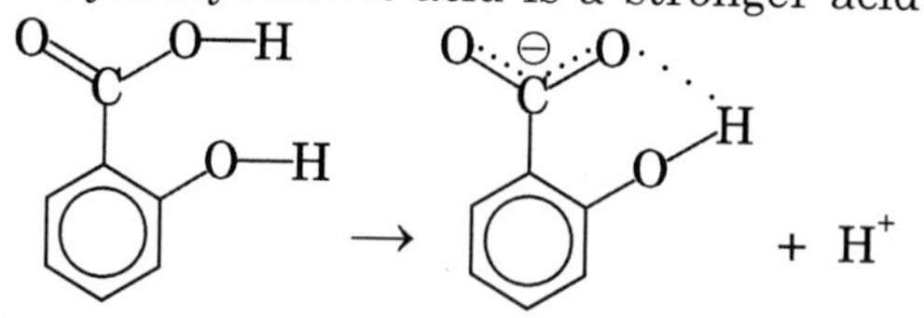

59. Which of the following statements is not correct?
(A) CH_3COCl can be converted into CH_3CONH_2 but the reverse is not possible
(B) 4-Chlorobenzoic acid is a stronger acid than 3-chlorobenzoic acid
(C) Reactivity of the C═O in CH_3COOH towards a nucleophile is less than that in CH_3CHO

(D) 3-Methylbenzoic acid is more acidic than 4-methylbenzoic acid.

Hint: The statement (B) is incorrect and all others are correct.

(A) A more reactive acid derivative is converted into a less reactive one by reaction with a nucleophile. The order of reactivity is: $CH_3CONH_2 < CH_3COCl$.

(B) In 4-chlorobenzoic acid, there is a negative charge on the ring carbon atom bonded to COO^-. But in 3-chlorobenzoic acid the negative charge is on the ring carbon atom *ortho* to the carbon atom bonded to COO^-. Further, the electron-withdrawing – I-effect of chlorine is more pronounced in 3-chlorobenzoic acid due to the proximity to COO^-. Hence 4-chlorobenzoic acid is less acidic than 3-chlorobenzoic acid.

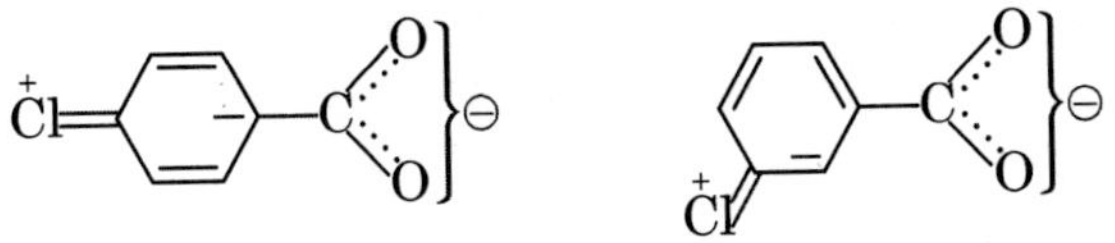

(C) In CH_3COOH, there is positive charge on the C=O carbon atom making this carbon atom to be less electrophilic and less reactive towards nucleophiles.

(D) Methyl group is an electron-donating (+ I-effect) and acid-weakening group. In 4-methylbenzoic acid, there is negative charge on the ring carbon atom bonded to COO^- while in 3-methylbenzoic acid the negative charge is on the ring carbon atom *ortho* to the carbon atom bonded to COO^-. As a result 4-chlorobenzoic acid is less acidic than 3-chlorobenzoic acid.

60. Methanoic acid reacts with conc. H_2SO_4 to yield
   (A) HCHO (B) $CO + H_2O$ (C) $CO_2 + H_2O$ (D) CH_3COOH

61. In the reaction sequence, $CH_3CH_2Cl \xrightarrow{NaCN} X \xrightarrow{H_2/Ni} Y \xrightarrow{(CH_3CO)_2O} Z$, is
   (A) $CH_3CH_2CH_2NHCOCH_3$ (B) $CH_3CH_2CH_2NH_2$
   (C) $CH_3CH_2CH_2CONHCH_3$ (D) $CH_3CH_2CH_2CONHCOCH_3$

62. Which of the following products is formed when urea is heated rapidly?
   (A) 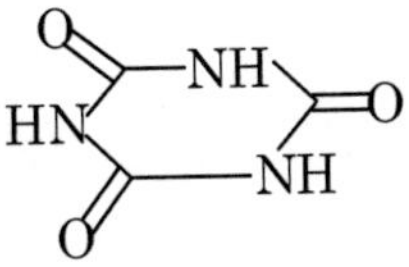 (cyanouric acid) (B) H—N=C=O (cyanic acid)
   (C) $H_2N\overset{O}{\overset{\|}{C}}NH\overset{O}{\overset{\|}{C}}NH_2$ (biuret) (D) all of these.

63. Identify the incorrect statement.
   (A) Urea is a diacidic base
   (B) A cyclic anhydride cannot be prepared by heating adipic acid
   (C) Benzamide is less acidic than phthalimide
   (D) The rate of esterification of Et_3CCOOH is faster than that of $(i\text{-}Pr)_2CHCOOH$.

Hint: The statement (A) is incorrect and all others are correct.

(A) Urea is a monoacidic base and its protonated form exists as a resonance hybrid of the following resonance structures.

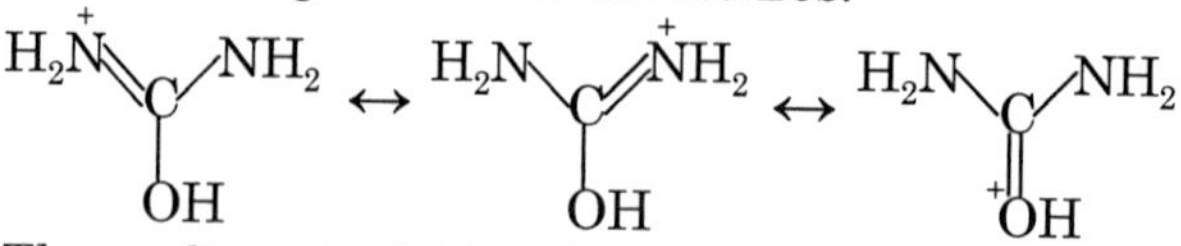

(B) The cyclic anhydride of adipic acid is not formed because there will be a large amount of strain in the seven-membered ring. Heating of adipic acid gives cyclopentanone.

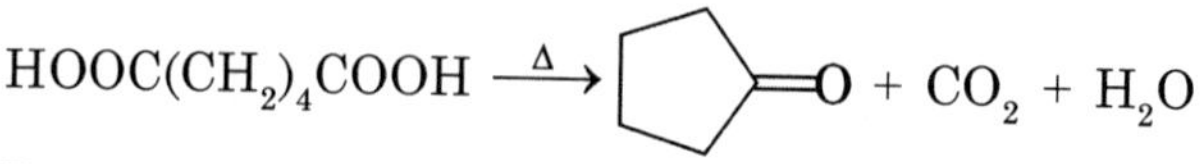

(C) Benzamide has only one carbonyl group to stabilise the imide while phthalimide has two carbonyl groups. Thus there is more stabilisation in phthalimide.

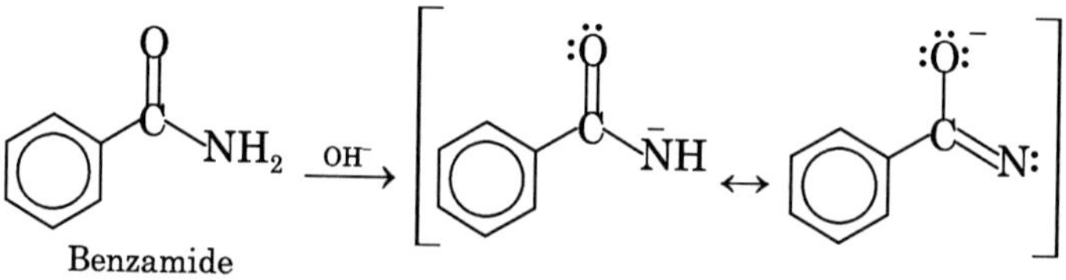

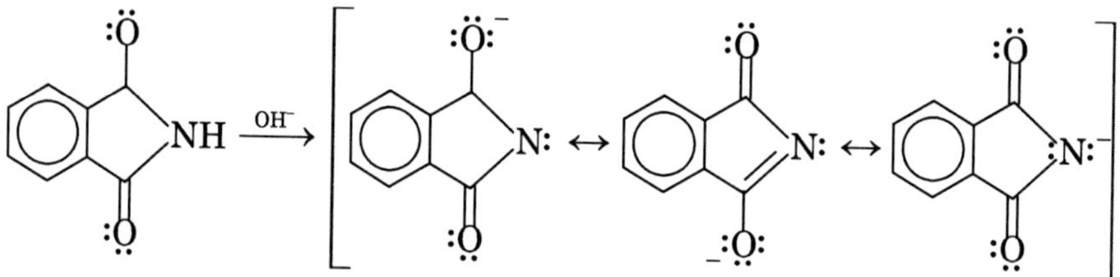

(D) With increase of the size of the substituent on the α-carbon atom, the tetrahedrally bound intermediate gets more crowded. The greater the crowding, the larger is the ΔH of the transition state and the slower is the reaction.

64. Trans-esterification involves a reaction between
    (A) acid anhydride and alcohol (B) ester and alcohol
    (C) acyl chloride and alcohol (D) two molecules of ester.

65. Which of the following gases is evolved when an acid amide reacts with HNO_2?
    (A) NH_3 (B) N_2 (C) N_2O (D) O_2

66. Ethyl ethanoate reacts with excess of CH_3MgBr to form
    (A) propan-1-ol (B) propan-2-ol
    (C) 2-methylpropan-1-ol (D) 2-methylpropan-2-ol

67. Which of the following is formed when ethyl ethanoate reacts with sodium ethoxide?
    (A) Acetophenone (B) Ethane nitrile (C) Ethylacetoacetate (D) None of these.

68. The reaction of two moles of ethyl ethanoate in the presence of $NaOC_2H_5$ to form ethylacetoacetate is an example of
    (A) haloform reaction (B) aldol condensation
    (C) MPV reaction (D) Claisen condensation.

69. Hofmann bromamide reaction is an example of
    (A) molecular rearrangement (B) elimination reaction
    (C) substitution reaction (D) addition reaction.
70. The treatment of an ester with $LiAlH_4$ followed by acid hydrolysis produces
    (A) two aldehydes
    (B) one molecule of alcohol and one molecule of carboxylic acid
    (C) two alcohols (D) two acids.
71. Which of the following esters cannot undergo Claisen self-condensation?
    (A) $CH_3CH_2CH_2CH_2COOC_2H_5$ (B) $C_6H_5COOC_2H_5$
    (C) $C_6H_{11}CH_2COOC_2H_5$ (D) $C_6H_5CH_2COOC_2H_5$
72. Which of the following compounds does not undergo Hofmann bromamide reaction?
    (A) $(CH_3)_2CHCONH_2$ (B) CH_3CONH_2
    (C) $CH_3NH\underset{\underset{O}{\|}}{C}—CH_3$ (D) $C_6H_5CONH_2$
73. Identify the pair of reagents in which each member of the pair reacts separately with ethanoic acid.
    (A) NH_3, Cl_2 (B) PCl_3, H_2SO_4 (C) PCl_3, NH_3 (D) NaOH, H_2SO_4
74. Identify the intermediate which is formed in Hofmann bromamide reaction.
    (A) $CH_3—N{=}C{=}O$ (B) $[CH_3CONBr]^-K^+$
    (C) $CH_3CONHBr$ (D) All of these.
75. Which of the following compounds will give acetylation?
    (A) CH_3COCH_3 (B) $CH_3CH_2CH_2OH$
    (C) CH_3COOH (D) $CH_3CH_2NH_2$
76. In the reaction, $CH_3CH_2CONH_2 \xrightarrow{P_4O_{10},\ heat} P$, the product P is
    (A) $CH_3CH_2CH_2NH_2$ (B) $CH_3CH_2C{\equiv}N$
    (C) $CH_3CH_2NH_2$ (D) $(CH_3CO)_2O$
77. In the reaction, $CH_3COOC_2H_5 \xrightarrow{(i)\ C_2H_5ONa\ (ii)\ H^+/H_2O} P$, the product P is
    (A) $CH_3COCH_2CH_2COOH$ (B) $CH_3COCH{=}CHCOOH$
    (C) $CH_3CH{=}CHCOOH$ (D) CH_3COCH_2COOH
78. In the reaction, $C_6H_5CONH_2 \xrightarrow{Na/C_2H_5OH} P$, the product P is
    (A) $C_6H_5CH_3$ (B) $C_6H_5CH_2NH_2$
    (C) $C_6H_5COOC_2H_5$ (D) $C_6H_5CH_2OH$
79. In the reaction, $C_6H_5CHO + CH_3COOC_2H_5 \xrightarrow{(i)\ C_2H_5OH\ (ii)\ H^+/H_2O} P$, the product P is
    (A) $C_6H_5CH{=}CHCOOH$ (B) $C_6H_5CH{=}CHCOOC_2H_5$
    (C) $C_6H_5CH{=}CH_2$ (D) $C_6H_5COOC_2H_5$
80. Identify the product formed when ketene is passed into glacial ethanoic acid.
    (A) $CH_3COCOOH$ (B) CH_3COCH_3 (C) $(CH_3CO)_2O$ (D) CH_3COOCH_3

81. Which of the following compounds undergoes Hofmann bromamide reaction?

(A) $CH_3CONHOH$ (B) $C_6H_5CONH_2$

(C) $CH_3C(=O)-\overset{-}{O}\overset{+}{N}H_3$ (D) $CH_3CONHCH_3$

82. Identify the reagent which does not reduce ester into primary alcohol.

(A) H_2/Ni (B) $NaBH_4$ (C) $LiAlH_4$ (D) Na/C_2H_5OH

83. Which of the following is formed when ethanoyl chloride reacts with hot alcoholic potash?

(A) $(CH_3CO)_2O$ (B) CH_3COOK (C) $CH_2{=}C{=}O$ (D) CH_3COOH

84. Which of the following is formed when ethanoyl chloride reacts with one molecule of CH_3MgBr?

(A) Propanone (B) Butan-2-ol (C) Butanone (D) 2-Methylpropan-2-ol

85. In the reaction, $CH_3COCl + C_2H_5OC_2H_5 \xrightarrow{\text{anhy. } ZnCl_2} P$, the product P is

(A) ethyl methanoate (B) ethanoic acid

(C) ethanoic anhydride (D) ethyl ethanoate

86. In the reaction, $CH_2{=}CHCH_2Cl \xrightarrow{\text{CuCN, NaOH}} P$, the product P is

(A) $CH_3CH{=}CHCOOH$ (B) $CH_2{=}CHCH_2COOH$

(C) $CH_3CH{=}CHCN$ (D) $CH_2{=}CHCH_2CN$

87. Identify the acid which yields a cyclic ketone on heating.

(A) $HOOC(CH_2)_4COOH$ (B) $HOOC(CH_2)_3COOH$

(C) $CH_2COOH-CH_2COOH$ (D) $HOOCCH_2COOH$

88. The rate law for the acid-catalysed hydrolysis of an ester is given by

(A) rate = $k[\text{ester}][H^+]^0$ (B) rate = $k[\text{ester}][H^+]$

(C) rate = $k[\text{ester}][H^+]^2$ (D) rate = $k[\text{ester}]$

89. Identify the correct statement.

(A) ΔG^0 for the reaction $CH_3CONH_2 \xrightarrow{H^+} CH_3CONH_3^+$ is less than the reaction $CH_3NH_2 \xrightarrow{H^+} CH_3NH_3^+$

(B) An aqueous solution of CH_3CONH_2 is alkaline

(C) pK_{a_1} of a dicarboxylic acid is smaller than the pK_a of a monoalkylcarboxylic acid

(D) *cis*-4-*t*-butylcyclohexanecarboxylic acid is a stronger acid than *trans*-4-*t*-butylcyclohexane carboxylic acid.

Hint: The statement (C) is correct and all other statements are incorrect.

(A) ΔG^0 for the reaction $CH_3CONH_2 \xrightarrow{H^+} CH_3CONH_3^+$ is more than the reaction $CH_3NH_2 \xrightarrow{H^+} CH_3NH_3$

(B) An aqueous solution of CH_3CONH_2 is alkaline due to slow hydrolysis: $CH_3CONH_2 + H_2O \longrightarrow CH_3COOH + NH_3$

(C) The first ionisation constant (K_{a_1}) of a dicarboxylic acid is larger than that of a monoalkylcarboxylic acid due to the electron-withdrawing COOH group of dicarboxylic acid. The K_a value of a monoalkylcarboxylic acid is smaller also due to the +I-effect of the alkyl group. Hence the pK_{a_1} of a dicarboxylic acid is smaller than the pK_a of a monoalkylcarboxylic acid.

(D) *trans*-4-*t*-butylcyclohexanecarboxylic acid is a stronger acid than the *cis* isomer. In the *trans* isomer, the —COO^- is less shielded and hence is more stabilised by the solvent molecules.

90. Which of the following is formed when as ester $CH_3COOC_2H_5$ undergoes catalytic hydrogenation in the presence of copper(II) oxide and copper(II) chromite?
(A) $CH_3OH + C_2H_5CH_2OH$ (B) $CH_3COOH + C_2H_5OH$
(C) $CH_3OH + C_2H_5COOH$ (D) $CH_3CH_2OH + C_2H_5OH$

91. Which of the following carboxylic acids does not undergo reduction to primary alcohol by BH_3/THF?
(A) *Z*-$CH_3CH{=}CHCH_2COOH$ (B) 3-$O_2NC_6H_4COOH$
(C) 4-$CH_3C_6H_4COOH$ (D) $CH_3CH(CH_3)CH_2COOH$

92. The carboxylic acid which undergoes dicarboxylation easily is
(A) $C_6H_5\underset{\underset{NH_2}{|}}{C}HCOOH$ (B) $C_6H_5\underset{\underset{OH}{|}}{C}HCOOH$
(C) $C_6H_5\underset{\underset{O}{\|}}{C}COOH$ (D) $C_6H_5\underset{\underset{O}{\|}}{C}CH_2COOH$

93. The reaction, $CH_3CH_2CH_2COOH \xrightarrow{\text{red P/Br}_2} CH_3CH_2CH(Br)COOH$ is an example of
(A) Reimer–Tiemann reaction (B) Hell–Volhard–Zelinsky reaction
(C) Perkin reaction (D) Cannizzaro reaction.

94. The reaction, $CH_3COOAg + Br_2 \xrightarrow{CCl_4,\ \text{heat}} CH_3Br + AgBr + CO_2$ is an example of
(A) Kolbe reaction (B) Friedel–Crafts reaction
(C) Hunsdiecker reaction (D) Wurtz reaction.

95. The reaction, $CH_3CH{=}CH_2 + H_2O + CO \xrightarrow{H_3PO_4,\ 675\ K} (CH_3)_3CHCOOH$ is an example of
(A) Kolbe reaction (B) Koch reaction (C) Carbonylation reaction (D) None of these.

96. Which of the following reagents is used for the conversion of $CH_3CH{=}CHCHO$ to $CH_3CH{=}CHCOOH$?
(A) SeO_2 (B) Ammoniacal $AgNO_3$
(C) $K_2Cr_2O_7/H^+$ (D) $KMnO_4/OH^-$

97. Identify the incorrect statement.
(A) Benzoic acid reacts with conc. H_2SO_4 to form 3-carboxybenzenesulphonic acid
(B) Benzoic acid reacts with $Br_2/FeBr_3$ to form 3-bromobenzoic acid
(C) Heating of $HOOC(CH_2)_4COOH$ gives a cyclic anhydride
(D) Heating of $HOOC(CH_2)_3COOH$ gives a cyclic anhydride.

98. Which of the following statements is incorrect?
(A) Methanoic acid on heating gives methanoic anhydride
(B) Ethanoic anhydride is prepared by the reaction of CH_3COOH with ketene
(C) Maleic acid is a stronger acid than fumaric acid

(D) 4-$O_2NC_6H_4COOH$ is a stronger acid than 3-$O_2NC_6H_4COOH$

Hint:

(A) Methanoic anhydride cannot be prepared by heating methanoic acid. The dehydration occurs intramolecularly rather than intermolecularly.

(B) $CH_3COOH \xrightarrow{AlPO_4,\ 700°C} H_2O + \underset{\text{Ketene}}{CH_3{=}C{=}O} \xrightarrow{CH_3COOH} \underset{\text{Acetic anhydride}}{(CH_3CO)_2O}$

(C) Maleic acid is a stronger acid than fumaric acid.

This is because of the stabilisation of the mono anion of maleic acid due to hydrogen bonding between the *cis* COOH and COO^-. In the *trans* isomer (fumarate anion), no such H-bonding and stabilisation are possible.

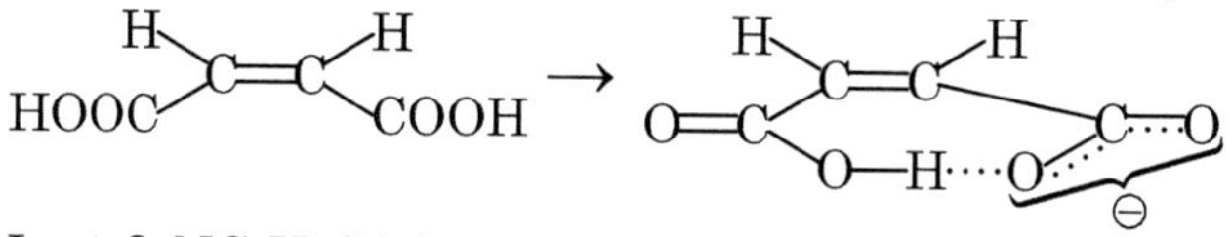

(D) In 4-$O_2NC_6H_5COOH$, the resonance interaction of the NO_2 group with the ring induces positive charge on the ring carbon atom bonded to COO^-. This induces a strong electron-withdrawing – I-effect. As a result the anion is stabilised and its power to withdraw electron from COO^- is high. Hence 4-$O_2NC_6H_5COOH$ is more acidic. In 3-$O_2NC_6H_5COOH$, the positive charge is induced at the carbon atom *ortho* to the ring carbon atom bonded to COO^-. Hence its power to withdraw electron from COO^- is less. Thus 3-$O_2NC_6H_5COOH$ is less acidic.

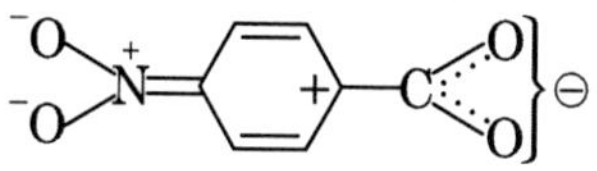

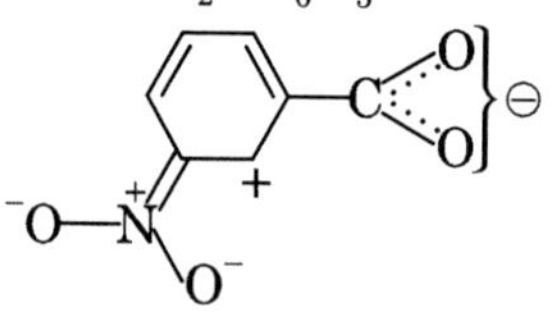

99. A linear polymeric anhydride is formed by heating

(A) $HOOC(CH_2)_6COOH$ (B) $HOOC(CH_2)_5COOH$

(C) $HOOC(CH_2)_4COOH$ (D) $HOOC(CH_2)_3COOH$

100. Which of the following is formed when propanoic acid is reacted with Cl_2 in the presence of red phosphorus?

(A) CH_2ClCH_2COOH (B) $CH_3CHClCOOH$

(C) $CH_3CH_2CH_2Cl$ (D) CH_3CH_2COCl

101. Identify the correct order of acid strength.

(A) $CH_3OH < HOH < HC{\equiv}CH < CH_3COOH$

(B) $CH_3OH < HC{\equiv}CH < HOH < CH_3COOH$

(C) $HC{\equiv}CH < CH_3OH < HOH < CH_3COOH$

(D) $HC{\equiv}CH < HOH < CH_3OH < CH_3COOH$

102. Which of the following acids undergoes decarboxylation to form ketone?

(A) COOH–COOH (B) CH_3COOH

(C) $CH_3COCH_2CH_2COOH$ (D) $CH_3CH_2COCH_2COOH$

Hint: 3-Ketoacids undergo decarboxylation readily due to the formation of a six membered cyclic transition state due to intramolecular H-bonding.

103. In the reaction, 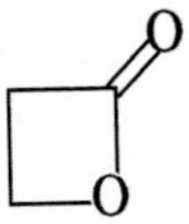 $\xrightarrow{CH_3OH}$ P, the product P is

(A) 

(B) 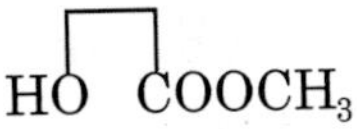

(C) 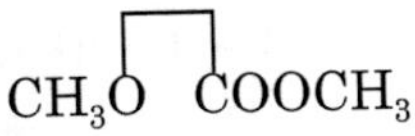

(D) 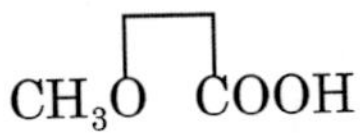

Hint:

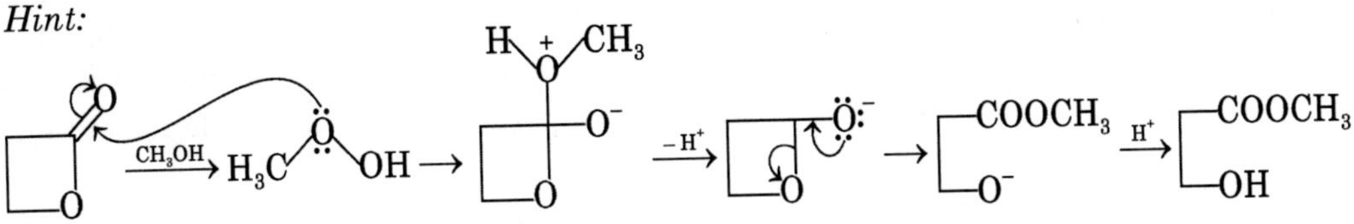

104. In the reaction, $CH_3COOAg + I_2 \xrightarrow{CS_2}$ P, the product P is

(A) CH_3COOCH_3 (B) CH_3COI (C) CHI_3 (D) CH_3I

Hint: $CH_3COOAg + I_2 \xrightarrow{CS_2} CH_3I + AgI + CO_2$; $CH_3I + CH_3COOAg \longrightarrow CH_3COOCH_3 + AgI$

105. Identify the correct order of esterification of the following three acids:

I. 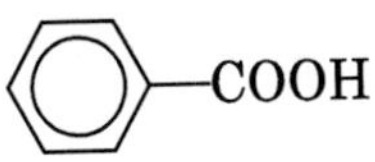 II. 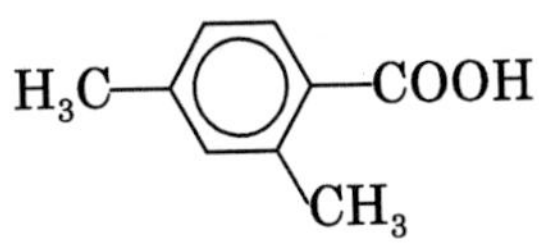 III. 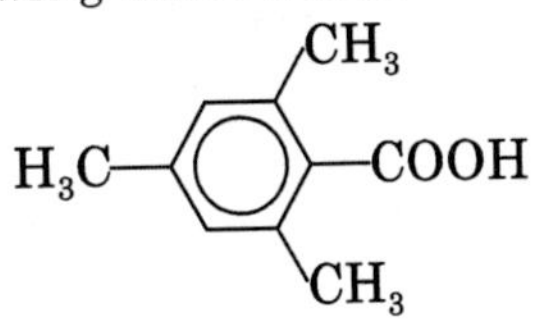

(A) I < III < II (B) II < I < III (C) III < II < I (D) II < III < I

Hint: The higher the number of electron-releasing methyl group at the α-carbon atom, the lesser is the susceptibility of the acid to undergo nucleophilic attack.

106. Which of the following products is formed when 4-chlorobutanoic acid reacts with KOH?

(A) 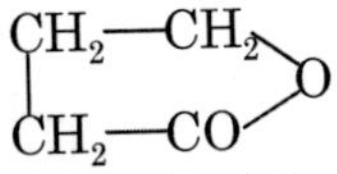

(B) $CH_2{=}CH{-}CH_2COOH$

(C) $HOCH_2CH_2COOH$

(D) $CH_3CH{=}CHCOOH$

Hint: In 4-chlorobutanoic acid, the Cl and COOH groups are apart and hence each of them reacts immediately with KOH.

107. Which of the following is obtained when ethane is heated with CO and steam at 673 K under pressure in the presence of H_3PO_4?

(A) CH_3CH_2CHO (B) CH_3COOH (C) CH_3COCH_3 (D) CH_3CH_2COOH

Hint: $CH_2{=}CH_2 + CO + H_2O \xrightarrow{H_3PO_4,\ 673\ K} CH_3CH_2COOH$

108. Which of the following acids has the smallest dissociation constant?

(A) $CH_3CHFCOOH$

(B) FCH_2CH_2COOH

(C) $BrCH_2CH_2COOH$

(D) $CH_3CHBrCOOH$

(I.I.T. 2002)

Hint: (a) Br has lesser – I-effect than F. (b) Br atom is further away from —COOH group. Acidic strength is proportional to dissociation constant (K_a) of acid. In the given acids, the order of increasing acidic strength is as follows:

$BrCH_2CH_2COOH < FCH_2CH_2COOH < CH_3CHBrCOOH < CH_3CHFCOOH$

Hence the acid $BrCH_2CH_2COOH$ has the smallest dissociation constant.

109. Identify the correct order of boiling point of $CH_3CH_2CH_2OH$ (1), $CH_3CH_2CH_2CHO$ (2) and $CH_3CH_2CH_2COOH$ (3).

(A) 3 > 1 > 2 (B) 1 > 2 > 3 (C) 1 > 3 > 2 (D) 3 > 2 > 1 **(I.I.T. 2002)**

Hint: The compound 3 forms intermolecular hydrogen bonding and exists as a dimer. The compound 1 forms intermolecular hydrogen bonds. The compound 2 is not involved in hydrogen bonding. Hence the decreasing order of molecular mass of these compounds is: 3 > 2 > 1. So they follow the same order in their boiling point: 3 > 2 > 1.

110. Ethyl ester $\xrightarrow{\text{excess } CH_3MgBr}$ P. The product P will be

(A) 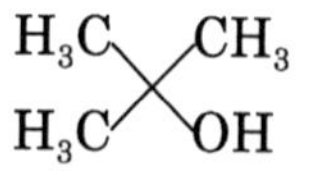 (B) 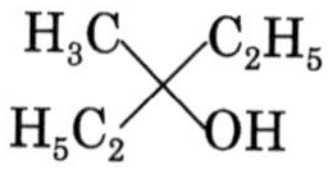

(C) 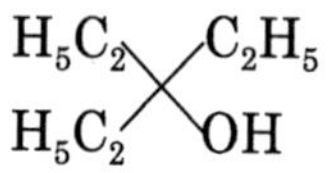 (D) 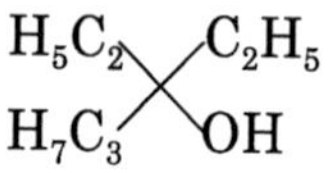 **(I.I.T. 2003)**

Hint: 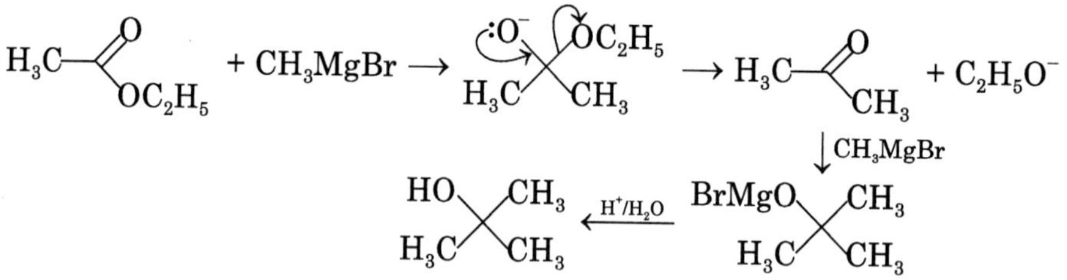

111. An enantiomerically pure acid is treated with racemic mixture of an alcohol having one chiral carbon. The ester formed will be

(A) optically active mixture (B) pure enantiomer

(C) meso compound (D) racemic mixture. **(I.I.T. 2003)**

Hint: The optically active pure acid reacts with *d* and *l* forms of alcohol present in the racemic mixture to form two types of isomeric esters. In each type, the configuration of the chiral centre of acid remains the same. Hence the mixture will be optically active.

112. In 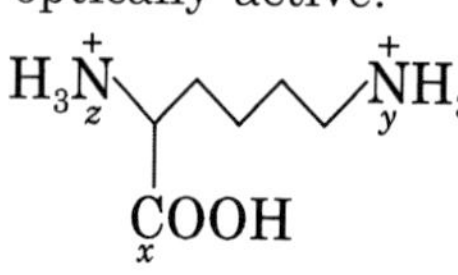, arrange in the order of increasing acidic strength.

(A) $x > z > y$ (B) $z < x > y$ (C) $x > y > z$ (D) $z > x > y$ **(I.I.T. 2004)**

Hint: The pK_a value of carboxylic group is less than the pK_a value of $—\overset{+}{N}H_3$ (*y*) in amino acid and $—\overset{+}{N}H_3$ (*z*) will have comparatively less pK_a value than $—\overset{+}{N}H_3$ (*y*) due to the – I-effect of the carboxylic group.

113. Benzamide on treatment with $POCl_3$ gives

(A) aniline (B) benzonitrile

(C) chlorobenzene (D) benzylamine **(I.I.T. 2004)**

Hint: $POCl_3$ is a dehydrating agent. Hence dehydration of benzamide leads to the formation of benzonitrile.

114. Which of the following is obtained when 4-methylbenzenesulphonic acid is reacted with excess of sodium acetate solution?

(A) 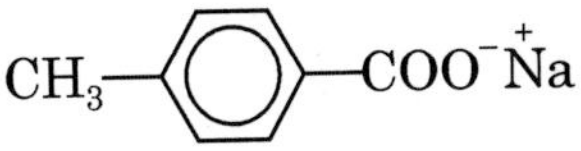

(B) 

(C) 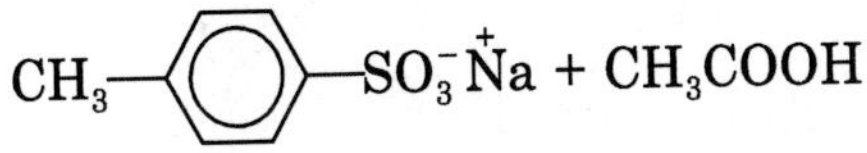

(D) 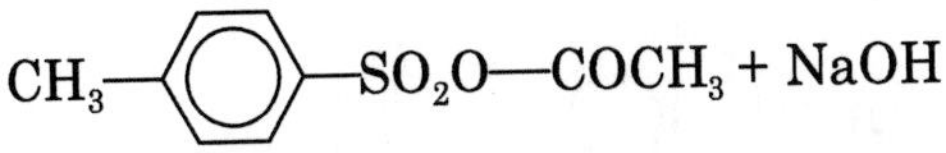 **(I.I.T. 2005)**

Hint: 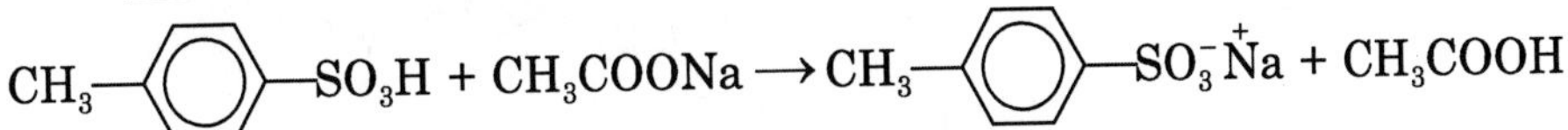

115. In which of the following compounds the acidic hydrolysis is maximum?

(A) 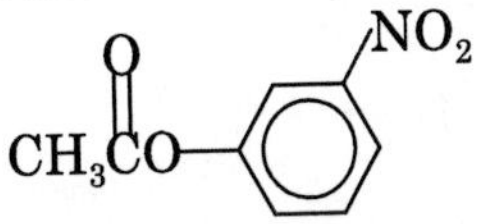

(B) 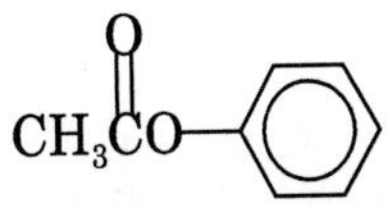

(C) 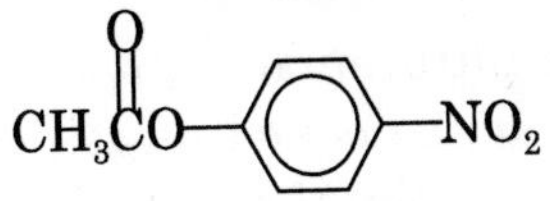

(D) 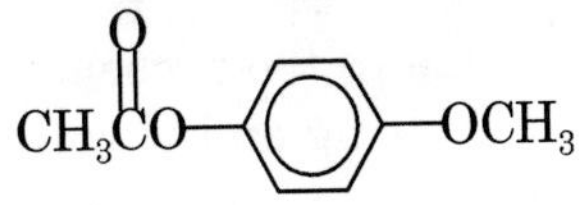

116. In which of the following compounds the alkaline hydrolysis is maximum?

(A) 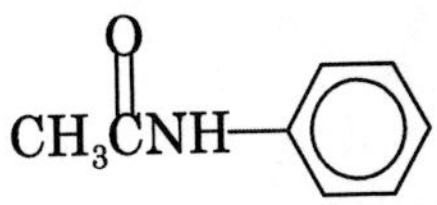

(B) 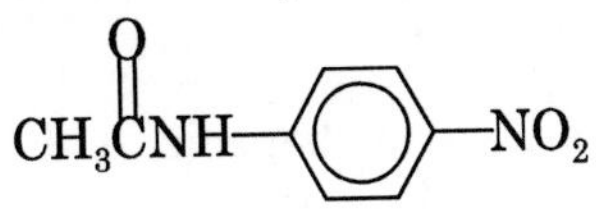

(C) 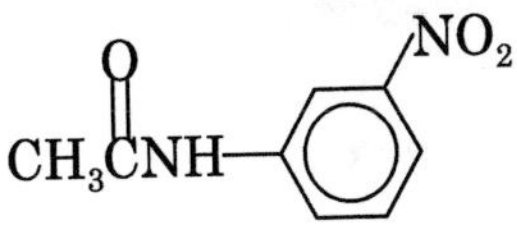

(D) 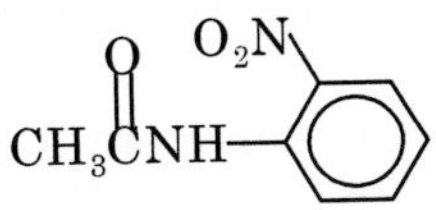

117. In the reaction 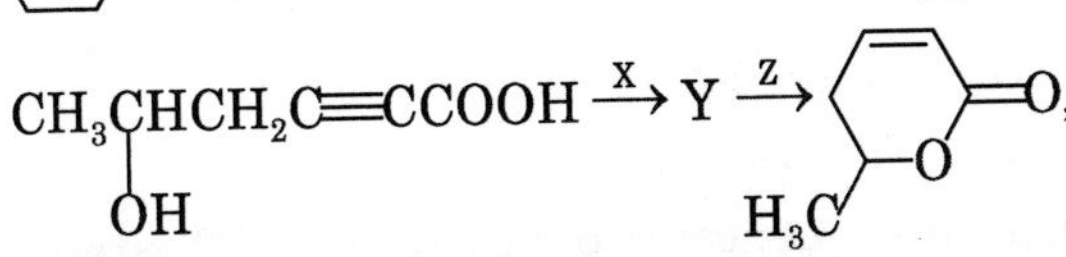, X and Z are

(A) H_2O/Na, H_3O^+
(B) H_2/Lindlar catalyst, H_3O^+
(C) Li/NH_3, H_3O^+
(D) Ni/H_2, H_3O^+

118. In the reactions,

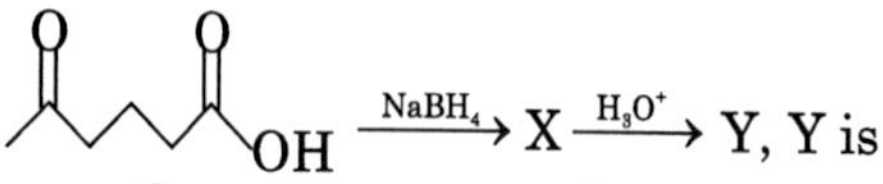

(A) 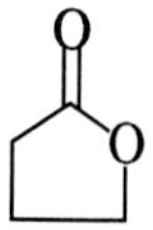 (B) 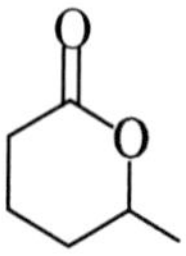 (C) 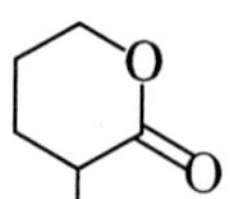 (D) 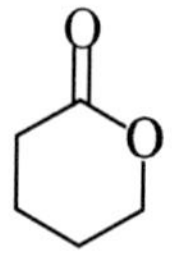

119. In the reaction, 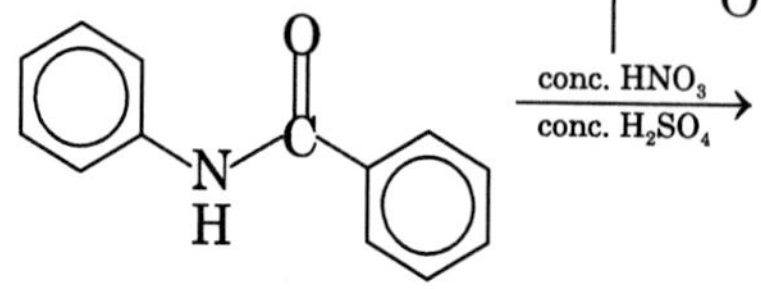 X, the structure of the major product X is

(A) 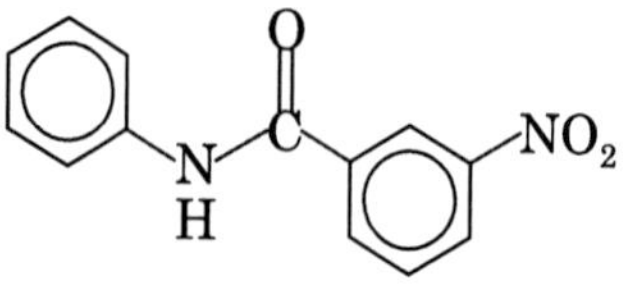

(B) 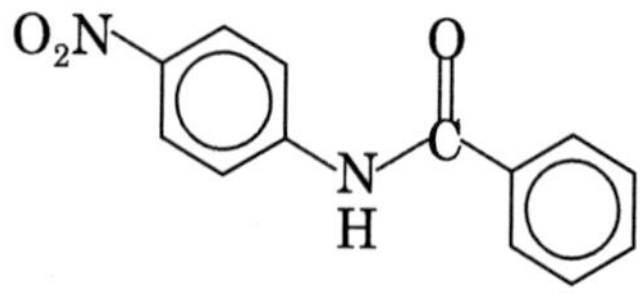

(C) 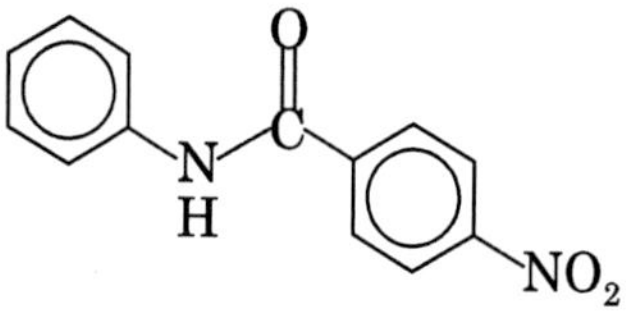

(D) 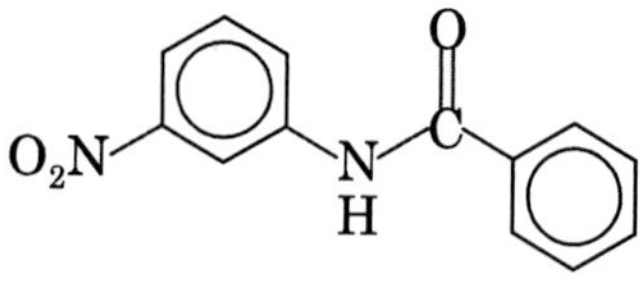

(I.I.T. 2007)

Hint: 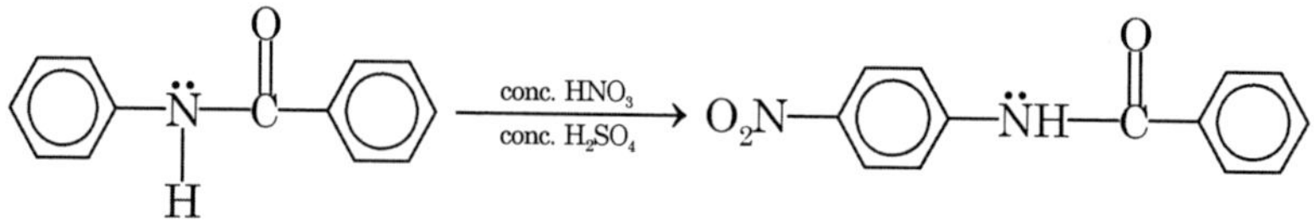

Due to the presence of a lone pair of electrons on the nitrogen atom, it will activate the ring and it will stabilise the intermediate cation at the *o*- and *p*-positions.

120. Statement–1: *p*-Hydroxybenzoic acid has a lower boiling point than *o*-hydroxybenzoic acid.

because

Statement–2: *o*-Hydroxybenzoic acid has intramolecular hydrogen bonding.

(A) Statement–1 is true, Statement–2 is true; Statement–2 is a correct explanation for Statement–1
(B) Statement–1 is true, Statement–2 is true; Statement–2 is not a correct explanation for Statement–1
(C) Statement–1 is true, Statement–2 is false
(D) Statement–1 is false, Statement–2 is true. **(I.I.T. 2007)**

121. In the following reaction sequence, products I, J and L are formed. K represents a reagent.

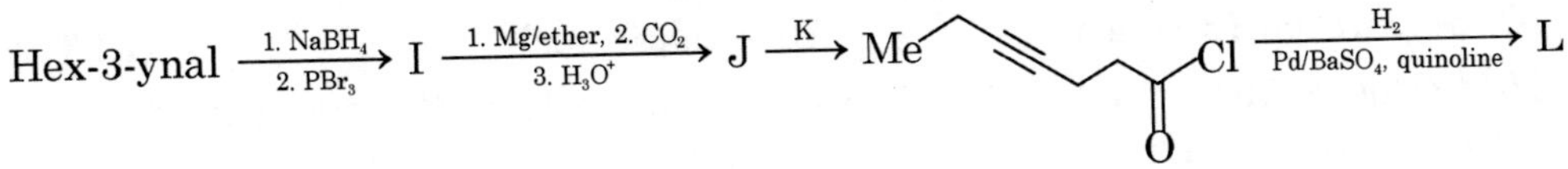

The structures of compounds J and K, respectively, are

(A) 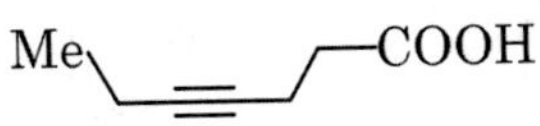 and $SOCl_2$ (B) 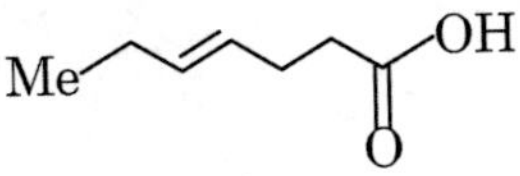 and SO_2Cl_2

(C) 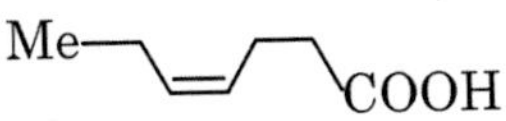 and $SOCl_2$ (D) 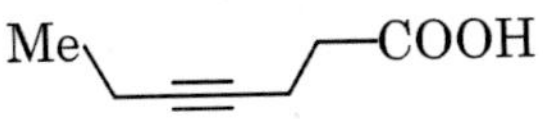 and CH_3SO_2Cl

(I.I.T. 2008)

Hint: 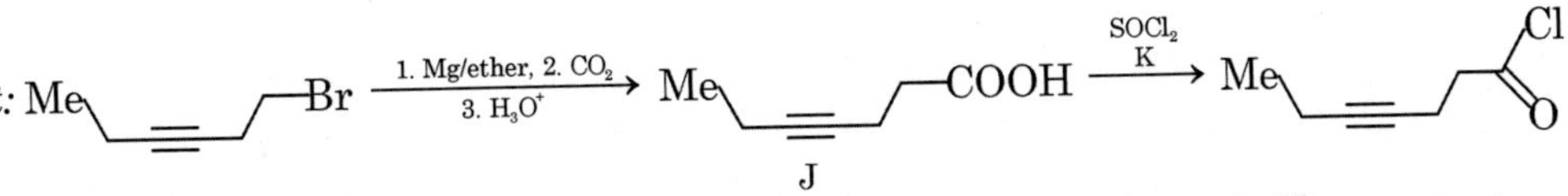

122. In the following reaction sequence, products I, J and L are formed. K represents a reagent.

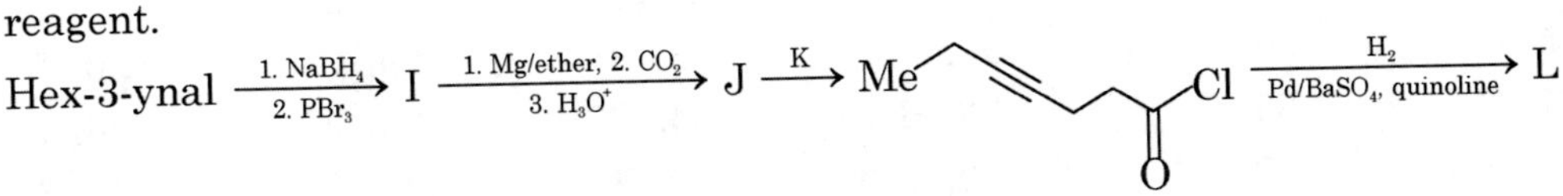

The structure of product L is

(A) 

(B) 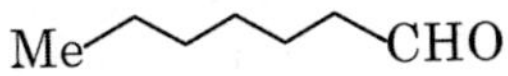

(C) 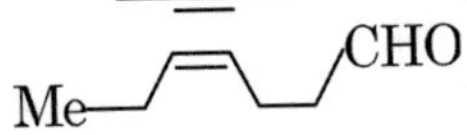

(D) 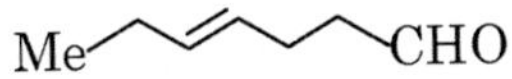

(I.I.T. 2008)

Hint: 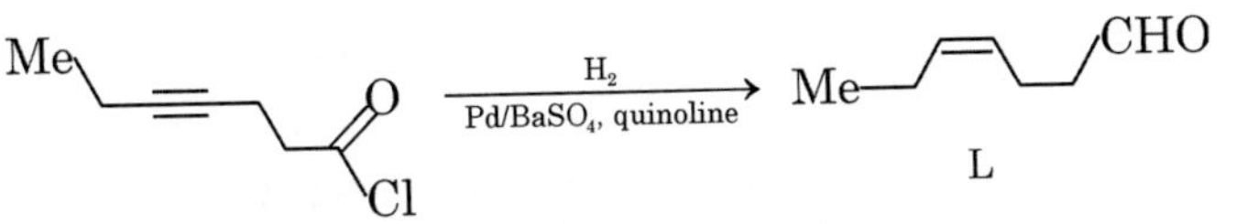

123. The correct acidity order of the following is

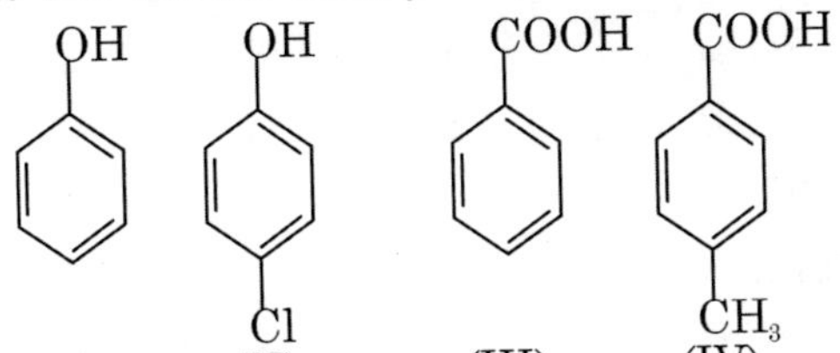

(I) (II) (III) (IV)

(A) (III) > (IV) > (II) > (I)

(B) (IV) > (III) > (I) > (II)

(C) (III) > (II) > (I) > (IV)

(D) (II) > (III) > (IV) > (I)

(I.I.T. 2009)

Hint: Carboxylic acid is a stronger acid than phenol. The presence of electron donating methyl group decreases acidic strength while presence of electron withdrawing halogen increases acidic strength. Decreasing order of acid strength: III > IV > II.

124. Among the following compounds, the most acidic is

(A) *p*-nitrophenol

(B) *p*-hydroxybenzoic acid

(C) *o*-hydroxybenzoic acid (D) *p*-toluic acid. **(I.I.T. 2011)**

Hint: Due to the ortho effect *o*-hydroxybenzoic acid is the strongest acid and the correct order of decreasing K_a is

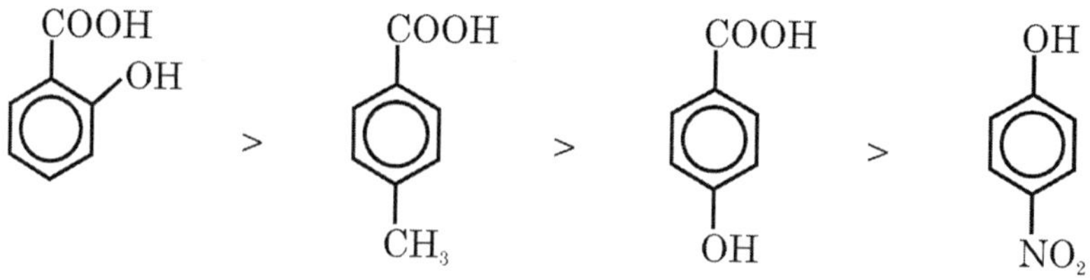

ANSWERS

1. A	2. A	3. C	4. B, C	5. D	6. C
7. A	8. A	9. D	10. C	11. A	12. D
13. C	14. A	15. D	16. D	17. B	18. A
19. A	20. C	21. D	22. B	23. B	24. C
25. C	26. A	27. D	28. B	29. C	30. D
31. C	32. D	33. B	34. C	35. B	36. C
37. B	38. A	39. C	40. B	41. C	42. A
43. A	44. B	45. A	46. B	47. D	48. A
49. A	50. C	51. B	52. B, D	53. B	54. B, C
55. C	56. D	57. B	58. B	59. B	60. B
61. A	62. A	63. A	64. B	65. B	66. D
67. B	68. D	69. A	70. C	71. B	72. C
73. A, C	74. D	75. B, D	76. B	77. D	78. B
79. A	80. C	81. B	82. B	83. C	84. A
85. D	86. A	87. A	88. B	89. C	90. D
91. A	92. D	93. B	94. C	95. B	96. B
97. C	98. A	99. A	100. B	101. C	102. D
103.B	104. A	105. C	106. A	107. D	108. C
109.A	110. A	111. A	112. A	113. B	114. C
115.D	116. B	117. B	118. B	119. B	120. D
121. A	122. C	123. A	124. C		

41

Compounds with Functional Groups Containing Nitrogen: Cyanides, Isocyanides, Nitro Compounds and Amines

OBJECTIVE TYPE MULTIPLE CHOICE QUESTIONS

1. The compound that will react most readily with NaOH to form methanol is
   (A) $(CH_3)_4N^+I^-$ (B) CH_3OCH_3 (C) $(CH_3)_3S^+I^-$ (D) $(CH_3)_3CCl$ **(I.I.T. 2001)**

2. The order of basicities of the compounds: 1. $CH_3-C(=NH)NH_2$ 2. $CH_3CH_2NH_2$ 3. $(CH_3)_2NH$ 4. CH_3CONH_2 is
   (A) 2 > 1 > 3 > 4 (B) 1 > 3 > 2 > 4 (C) 3 > 1 > 2 > 4 (D) 1 > 2 > 3 > 4
   (I.I.T. 2001)

3. Assertion (*A*): In strongly acidic solutions, aniline becomes more reactive towards electrophilic reagents.
   Reason (*R*): The amino group being completely protonated in strongly acidic solution, the lone pair of electrons on the nitrogen is no longer available for resonance.
   (A) Both *A* and *R* are correct and *R* is the correct explanation of *A*
   (B) Both *A* and *R* are correct and *R* is not the correct explanation of *A*
   (C) *A* is true but *R* is false (D) *A* is false but *R* is true. **(I.I.T. 2001)**

4. Among the following the strongest base is
   (A) $C_6H_5CH_2NH_2$ (B) *p*-$O_2NC_6H_4NH_2$
   (C) *m*-$O_2NC_6H_4NH_2$ (D) $C_6H_5NH_2$ **(I.I.T. 2000)**

5. A positive carbylamine test is given by
   (A) *N,N*-dimethylaniline (B) 2,4-dimethylaniline
   (C) *N*-methyl-*o*-methylaniline (D) *p*-methylbenzylamine **(I.I.T. 1999)**

6. The most unlikely representation of structure of *p*-nitrophenoxide ion is

(A) 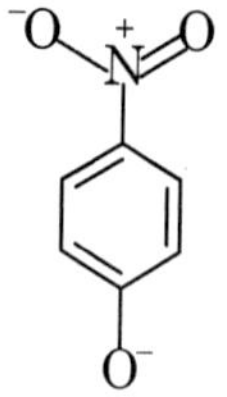 (B) 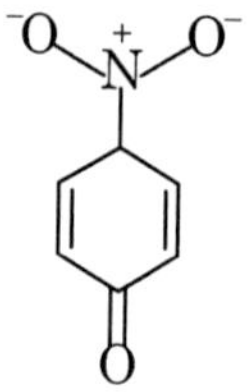 (C) 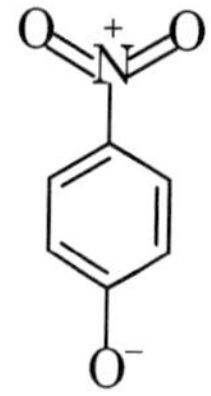 (D) 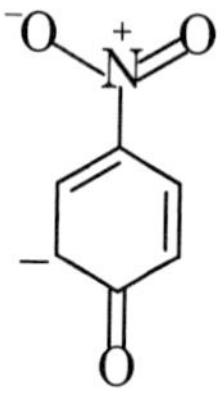 **(I.I.T. 1999)**

7. *p*-Chlorobenzene and aniline hydrochloride can be distinguished by
   (A) Sandmeyer reaction (B) carbylamine test
   (C) $NaHCO_3$ (D) $AgNO_3$ **(I.I.T. 1998)**

8. Identify the false statement.
   (A) The rate of nitration of benzene is almost the same as that of hexadeuterobenzene
   (B) The rate of nitration of toluene is greater than that of benzene
   (C) The rate of nitration of benzene is greater than that of hexadeuterobenzene
   (D) Nitration is an electrophilic substitution. **(I.I.T. 1997)**

9. In the compounds 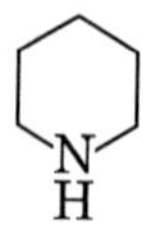 I,  II, 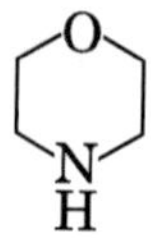 III, and  IV, the order of basicity is
   (A) IV > I > III > II (B) III > I > IV > II
   (C) II > I > III > IV (D) I > III > II > IV **(I.I.T. 1997)**

10. Nitrobenzene can be prepared from benzene by using a mixture of conc. NHO_3 and conc. H_2SO_4 in the nitrating mixture. HNO_3 acts as a
    (A) base (B) catalyst
    (C) reducing agent (D) acid. **(I.I.T. 1997)**

11. Assertion (*A*): Benzonitrile is prepared by the reaction of chlorobenzene with KCN.
    Reason (*R*): CN^- ion is a strong nucleophile.
    (A) Both *A* and *R* are true and *R* is the correct explanation of *A*
    (B) Both *A* and *R* are true and *R* is not the correct explanation of *A*
    (C) *A* is true but *R* is false (D) *A* is false and *R* is correct. **(I.I.T. 1998)**

12. In the following compounds the order of acidity is

 (I),  (II), 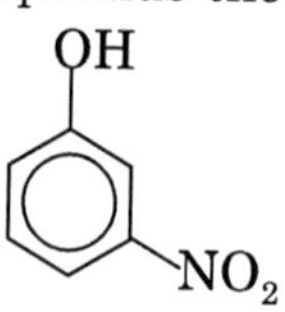 (III),  (IV)

    (A) III > IV > I > II (B) I > IV > III > II
    (C) II > I > III > IV (D) IV > III > I > II **(I.I.T. 1996)**

13. Allyl cyanide $CH_2{=}CH{-}C{\equiv}N$ has bonds
    (A) 9σ, 4π (B) 9σ, 3π (C) 8σ, 5π (D) 8σ, 3π **(I.I.T. 1995)**

14. Examine the following two structures for the anilinium ion and choose the correct statement.

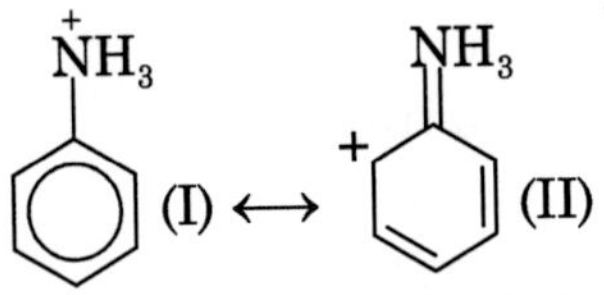

(A) It is not a canonical structure because the carbonium ions are less stable than the ammonium ions
(B) It is not an acceptable canonical structure because it is non-aromatic
(C) It is not an acceptable canonical structure because the nitrogen has ten valence electrons
(D) It is an acceptable canonical structure. **(I.I.T. 1993)**

15. Towards electrophilic substitution, the most reactive is
(A) aniline (B) aniline hydrochloride
(C) *N*-acetylaniline (D) nitrobenzene **(I.I.T. 1992)**

16. When nitrobenzene is treated with Br_2 in presence of $FeBr_3$, the major product formed is *m*-bromonitrobenzene. Statements which are related to obtain the *m*-isomer are:
(A) Easier loss of H^+ occurs to regain aromaticity from the *meta* position than from *ortho* and *para* positions
(B) The intermediate carbonium ion formed after initial attack of Br^+ at the *meta* position is least destabilised
(C) Loss of aromaticity occurs when Br^+ attacks at the *ortho* and *para* positions and not at *meta* position
(D) The electron density on *meta* carbon is more than that on *ortho* and *para* positions. **(I.I.T. 1992)**

17. Reaction of R—CO—NH_2 with a mixture of Br_2 and KOH gives R—NH_2 as the main product. The intermediates involved in this reaction are
(A) R—CO—NHBr (B) R—CO—NBr_2
(C) R—N=C=O (D) RNHBr **(I.I.T. 1992)**

18. Butanenitrile may be prepared by heating
(A) propyl alcohol with KCN (B) butyl alcohol with KCN
(C) butyl chloride with KCN (D) propyl chloride with KCN **(I.I.T. 1992)**

19. Identify Z in the reactions, $C_6H_5COOH \xrightarrow{PCl_3} X \xrightarrow{NH_3} Y \xrightarrow{P_4O_{10}} C_6H_5CN \xrightarrow{Ni/H_2} Z$
(A) $C_6H_5CH_2NH_2$ (B) $C_6H_5CONH_2$
(C) $C_6H_5CH_2CH_2NH_2$ (D) C_6H_5COCl **(I.I.T. 1991)**

20. The most basic compound amongst the following is
(A) benzylamine (B) acetanilide (C) *m*-nitroaniline (D) aniline **(I.I.T. 1990)**

21. The basicity of the amines decreases in the order
(A) dimethylamine > methylamine > *N*-methylaniline > aniline
(B) aniline > dimethylamine > methylamine > *N*-methylaniline

(C) *N*-methylaniline > aniline > dimethylamine > methylamine
(D) methylamine > dimethylamine > *N*-methylaniline > aniline **(I.I.T. 1988)**

22. The basicity of the following amines decreases in the order
(A) *N,N*-dimethyl-*p*-toluidine > *p*-toluidine > aniline > *p*-nitroaniline
(B) *p*-nitroaniline > *N,N*-dimethyl-*p*-toluidine > *p*-toluidine > aniline
(C) *p*-toluidine > *N,N*-dimethyl-*p*-toluidine > aniline > *p*-nitroaniline
(D) *N,N*-dimethyl-*p*-toluidine > aniline > *p*-toluidine > *p*-nitroaniline **(I.I.T. 1986)**

23. The increasing use of hydrolysis in the compounds $CH_3COOC_2H_5$, CH_3COCl, $(CH_3CO)_2O$ and CH_3CONH_2 is
(A) $CH_3CONH_2 < CH_3COOC_2H_5 < (CH_3CO)_2O < CH_3COCl$
(B) $CH_3COOC_2H_5 < (CH_3CO)_2O < CH_3COCl < CH_3CONH_2$
(C) $CH_3CONH_2 < (CH_3CO)_2O < CH_3COCl < CH_3COOC_2H_5$
(D) $(CH_3CO)_2O < CH_3CONH_2 < CH_3COOC_2H_5 < CH_3COCl$ **(I.I.T. 1986)**

24. In the nitration of benzene with conc. HNO_3 and conc. H_2SO_4, the active species involved is
(A) nitrate ion (B) nitrite ion (C) nitronium ion (D) nitric acid **(I.I.T. 1985)**

25. The compound which is most reactive towards electrophilic nitration is
(A) toluene (B) nitrobenzene (C) benzoic acid (D) benzene **(I.I.T. 1985)**

26. Chlorobenzene can be prepared by treating aniline with
(A) hydrochloric acid (B) nitrous acid followed by heating with CuCl
(C) chlorine in presence of anhydrous NH_4Cl
(D) copper(I) chloride **(I.I.T. 1984)**

27. Acetamide is treated with the following reagents. Which of these would give methylamine?
(A) PCl_5 (B) $NaOH + Br_2$
(C) Sodalime (D) Hot conc. H_2SO_4 **(I.I.T. 1983)**

28. Aniline on reaction with acetyl chloride gives
(A) benzene (B) acetamide (C) acetanilide (D) phenol **(Roorkee 1993)**

29. Which is most readily nitrated?
(A) nitrobenzene (B) aniline (C) benzene (D) phenol **(Roorkee 1992)**

30. When nitrobenzene is treated with Zn dust and NH_4Cl, we get
(A) aniline (B) benzene
(C) *N*-phenylhydroxylamine (D) azoxybenzene

31. Bromination of aniline gives
(A) 2,6-dibromoaniline (B) 2,4,6-tribromoaniline
(C) 4-bromoaniline (D) 2-bromoaniline

32. Hofmann's method for the separation of amines uses the reagent
(A) diethyl oxalate (B) benzeneisocyanide
(C) *p*-toluenesulphonic acid (D) benzenesulphonyl chloride

33. When an aqueous solution of benzenediazonium chloride is warmed, the product formed is
    (A) phenol (B) amide (C) aniline (D) benzene
34. Which of the following statements is/are not correct?
    (A) Aromatic amines are stronger bases than ammonia
    (B) Tertiary amines exhibit intermolecular H-bonding
    (C) Aliphatic amines are stronger bases than ammonia
    (D) Secondary amines exhibit intermolecular H-bonding.
35. Which of the following amines will dissolve in a solution of NaOH after its reaction with Hinsberg reagent?
    (A) CH_3NH_2 (B) $C_6H_5NHC_6H_5$ (C) $(CH_3)_2NC_6H_5$ (D) $(C_2H_5)_2NH$
36. Which of the following amines will dissolve in a solution of NaOH after its reaction with Hinsberg reagent?
    (A) $C_6H_5NH_2$ (B) *p*-$ClC_6H_4NH_2$ (C) CH_3NH_2 (D) C_6H_5NHR
37. Hydrolysis of $CH_3CH_2NO_2$ with 85% H_2SO_4 gives
    (A) C_2H_6 (B) CH_3COOH (C) C_2H_5OH (D) $CH_3CH{=}NOH$
38. Which of the following does not react with $NaNO_2$ and HCl?
    (A) $C_6H_5NH_2$ (B) $(CH_3)_3CNO_2$ (C) $(CH_3)_2CHNO_2$ (D) C_6H_5OH
39. The reaction, $C_6H_5NO_2$ + KOH (solid) $\xrightarrow{\text{heat}}$ *o*-$O_2NC_6H_4OH$ + *p*-$O_2NC_6H_4OH$ is
    (A) electrophilic substitution (B) electrophilic addition
    (C) free radical substitution (D) nucleophilic substitution.
40. Which of the following compounds will show tautomerism?
    (A) $(CH_3)_2NH$ (B) RCH_2NO_2 (C) R_3CNO_2 (D) $(CH_3)_3CNO$
41. The reaction of nitrobenzene with zinc and alkali gives
    (A) hydrazobenzene (B) aniline
    (C) phenylhydroxylamine (D) nitrosobenzene
42. Electrolytic reduction of nitrobenzene in weakly acidic medium produces
    (A) *p*-aminophenol (B) phenylhydroxylamine
    (C) aniline (D) azoxybenzene
43. Electrolytic reduction of nitrobenzene in weakly acidic medium gives
    (A) *p*-aminophenol (B) phenylhyroxylamine
    (C) aniline (D) azoxybenzene
44. Oxidation of aniline with MnO_2 and H_2SO_4 gives
    (A) *p*-benzoquinone (B) phenol
    (C) nitrobenzene (D) phenylhydroxylamine
45. Aniline when treated with conc. NHO_3 + H_2SO_4 gives
    (A) nitrobenzene (B) *m*-nitroaniline
    (C) *p*-benzoquinone (D) *o*-nitroaniline + *p*-nitroaniline
46. Aniline reacts with phosgene to give
    (A) C_6H_5NCO (B) C_6H_5OH (C) C_6H_5COCl (D) $C_6H_5NHCOCl$

47. When $(CH_3)_2NCOCH_3$ is refluxed with acid, the products formed are
   (A) $(CH_3)_2NCOOH + CH_4$ (B) $(CH_3)_2NH + CH_3COOH$
   (C) $2CH_3NH_2 + CH_3COOH$ (D) $2CH_3OH + CH_3CONH_2$
48. Beckmann rearrangement is effected by
   (A) PCl_5 (B) H_2SO_4
   (C) polyphosphoric acid (D) any of these.
49. A compound *A* (C_7H_7NO) on treatment with Br_2 and KOH gives an amine *B* which gives carbylamine test. *B* gives an azo dye on diazotisation and coupling with phenol. *A* is
   (A) $C_6H_5CONH_2$ (B) $C_6H_5NO_2$
   (C) *o*-, *m*- and *p*-$C_6H_4(NH_2)CHO$ (D) $C_6H_5CONHCOCH_3$
50. Benzene on treatment with a mixture of conc. HNO_3 and conc. H_2SO_4 at 100°C gives
   (A) nitrobenzene (B) *o*-dinitrobenzene
   (C) *m*-dinitrobenzene (D) *p*-dinitrobenzene
51. When primary amine is heated with CS_2 in presence of excess of $HgCl_2$, it gives isothiocyanate. The reaction is called
   (A) Perkin reaction (B) Hoffmann elimination reaction
   (C) Hoffmann mustard oil reaction (D) Hoffmann bromamide reaction.
52. Which of the following is produced by reaction of RCN in sodium and alcohol?
   (A) RCH_2NH_2 (B) $RCONH_2$ (C) $R(CH_2)_3NH_2$ (D) $RCOONH_4$
53. Which of the following statements is correct?
   (A) Boiling points of haloalkanes are more than the corresponding alkanes
   (B) $C_6H_5NH_2$ is weaker base than ammonia
   (C) In water the solubility of $CH_3OH > C_2H_5OH > C_6H_5OH$
   (D) All the statements are correct.
54. Which of the following will give primary amine on hydrolysis?
   (A) Oxime (B) Alkyl cyanide
   (C) Alkyl isocyanide (D) Nitroparaffin
55. Propanone oxime on catalytic hydrogenation gives
   (A) 1-propanamine (B) 2-propanamine
   (C) ethyl methyl amine (D) ethanamine + CH_4
56. In the reduction of nitrobenzene, which of the following is the intermediate?
   (A) $C_6H_5NHNHC_6H_5$ (B) $C_6H_5N{=}NC_6H_5$
   (C) $C_6H_5N{=}O$ (D) $C_6H_5N{=}\overset{\overset{O}{\uparrow}}{N}C_6H_5$
57. In the reactions, $CH_3CN \xrightarrow{SnCl_2/HCl} X \xrightarrow{H_2O,\ boil} Y$, Y is
   (A) ethanamine (B) acetaldehyde (C) acetone (D) dimethylamine
58. In the reactions, $CH_3CN \xrightarrow{Na/C_2H_5OH} X \xrightarrow{HNO_2} Y \xrightarrow{\text{Tollen reagent}} Z$, Z is
   (A) CH_3COOH (B) CH_3CONH_2 (C) CH_3CHO (D) CH_3CH_2NHOH

59. Which of the following reactions will not give primary amine?

(A) $CH_3CONH_2 \xrightarrow{KOH, Br_2}$ (B) $CH_3CN \xrightarrow{LiAlH_4}$

(C) $CH_3CONH_2 \xrightarrow{LiAlH_4}$ (D) $CH_3NC \xrightarrow{LiAlH_4}$

60. In the diazotisation of aniline with $NaNO_2$ and HCl, excess of HCl is used primarily to

(A) supress the concentration of free aniline

(B) supress the hydrolysis to phenol

(C) ensure stoichiometric amount of NHO_2

(D) neutralise the base liberated.

61. Nitrobenzene is converted into azoxybenzene by reduction with

(A) $Zn/NH_4/HCl$ (B) Sn/HCl

(C) $Zn/NaOH/CH_3OH$ (D) $Na_3AsO_3/NaOH/H_2O$

62. Which of the following reagents is used for the conversion $C_6H_5CH{=}CHNO_2 \rightarrow C_6H_5CH_2CH_2NO_2$?

(A) $LiAlH_4$ (B) Pd/H_2 (C) $NaBH_4$ (D) $[(C_6H_5)_3P]_3RhCl/H_2$

63. Ethylamine on heating with CS_2 in the presence of $HgCl_2$ forms

(A) C_2H_5NCS (B) $(C_2H_5)_2S$ (C) $(C_2H_5)_2CS$ (D) $(C_2H_5)_2(CS)_2$

64. Which of the following gives a ketone when boiled with HCl?

(A) $(CH_3)_2CHNO_2$ (B) $CH_3CH_2NO_2$ (C) $(CH_3)_3CNO_2$ (D) $(CH_3)_3CCH_2NO_2$

65. Which of the following is formed when dimethylamine reacts with HNO_2?

(A) $(CH_3)_2NNO$ (B) $(CH_3)_2NH^+NO_2^-$

(C) CH_3OH (D) $CH_3OH + N_2$

66. Identify the compound which does not form a bipolar ion.

(A) $H_2NCH_2CH_2CH_2OH$ (B) H_2NCH_2COOH

(C) 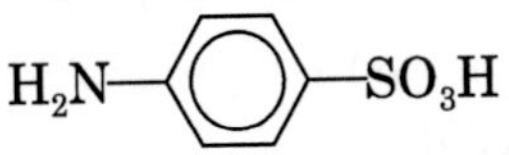

(D) 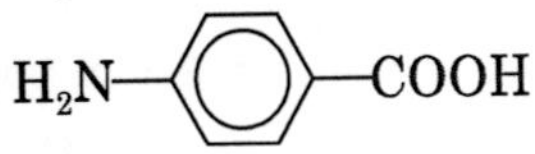

67. Which of the following statements is incorrect?

(A) 4-Nitroaniline is a weaker base than 4-cyanoaniline

(B) Aniline is less basic than pyridine

(C) 2-Substituted anilines are less basic than 4-subsituted anilines regardless of the electronic effect of the subsitutent

(D) The ability of coordination compound formation of $(CH_3)_3N$ with $B(CH_3)_3$ is greater than that of $(CH_3)_2NH$.

68. Which of the following statements is incorrect?

(A) Amides are weaker bases than amines

(B) The ability of coordination compound formation of CH_3NH_2 with $B(CH_3)_3$ is lower than that of $(CH_3)_2NH$

(C) $(CH_3)_3N$ is a stronger base than CH_3NH_2 in aqueous solution

(D) $(CH_3)_3N$ is weaker base than $(CH_3)_2NH$ in aqueous solution.

69. Which of the following statements is incorrect?
   (A) 4-Aninabenzoic acid is not a bipolar ion
   (B) Bromination of aniline gives 2-bromoaniline and 4-bromoaniline
   (C) Aliphatic amines do not form stable diazonium salts with HNO_2
   (D) An electron-releasing substituent increases the basicity of aniline.
70. Which of the following statements is incorrect?
   (A) The increasing order of basicity of CH_3CN, $CH_3CH{=}NCH_3$ and CH_3NH_2 is: $CH_3CN < CH_3CH{=}NCH_3 < CH_3NH_2$
   (B) $(CH_3)_2NH$ is a weaker base than CH_3NH_2 in aqueous solution
   (C) Increasing order of basicity of three isomers of nitroaniline is: 2-nitroaniline < 4-nitroalinine < 3-nitroaniline
   (D) Increasing order of basicity of three isomers of methoxyaniline is: 3-$CH_3OC_6H_4NH_2$ < 2-$CH_3OC_6H_4NH_2$ < 4-$CH_3OC_6H_4NH_2$
71. Which of the following compounds forms a secondary amine on reduction?
   (A) CH_3NC (B) CH_3CN (C) CH_3NO_2 (D) $C_6H_5NO_2$
72. Which of the following is not formed in the reaction, $CH_3NO_2 + Cl_2 + NaOH \longrightarrow ?$
   (A) CH_3 (B) $ClCH_2NO_2$ (C) Cl_2CHNO_2 (D) Cl_3CNO_2
73. Which of the following statements is incorrect?
   (A) Nitrobenzene is converted into azoxybenzene by glucose/NaOH
   (B) Aniline is a weaker base than ethylamine
   (C) Anilline is a stronger base than 4-methoxyaniline
   (D) The presence of $—CH_3$ or $—OCH_3$ group increases the basic strengths of amines.
74. Which of the following reactions is not suitable for the preparation of benzylamine?
   (A) $C_6H_5NC \xrightarrow{LiAlH_4}$ (B) $C_6H_5CN \xrightarrow{LiAlH_4}$
   (C) $C_6H_5CONH_2 \xrightarrow{LiAlH_4}$ (D) $C_6H_5CH_2CONH_2 + Br_2 + KOH \longrightarrow$
75. Which of the following are formed in the hydrolysis of methyl isocyanide?
   (A) $CH_3NH_2 + (CH_3)_2CHCOOH$ (B) $CH_3NH_2 + HCOOH$
   (C) $CH_3NH_2 + CH_3CH_2COOH$ (D) $CH_3NH_2 + CH_3COOH$
76. In the reaction, $CH_3CONHCH_3 \xrightarrow{POCl_3,\ pyridine} P + H_2O$, the product P is
   (A) CH_3NC (B) CH_3CN
   (C) $CH_3CH{=}NOH$ (D) $CH_3—N{=}C{=}O$
77. Which of the following reagents is used to convert iodoethane to ethyl isocyanide?
   (A) CuCN (B) AgCN (C) KCN (D) HCN
78. Which of the following poisonous substances caused Bhopal tragedy in 1984?
   (A) $CH_3—O—N \rightleftarrows C$ (B) $CH_3—CH{=}N{=}S$
   (C) $CH_3—N{=}C{=}S$ (D) $CH_3—N{=}C{=}O$
79. Which of the following methods is not suitable for the preparation of benzonitrile?
   (A) Benzaldoxine is heated with ethanoic anhydride
   (B) Sodium benzenesulphonate is fused with KCN

(C) Benzenediazonium chloride is heated with KCN/C_2H_5OH
(D) Bromobenzene is heated with KCN/C_2H_5OH

80. In the reaction, $CH_3-C(CH_3)_2-CH_2NH_2 \xrightarrow{HNO_2}$ X (major product), X is

(A) $CH_3-CH(CH_3)-CH=CH_2$
(B) $CH_3-C(CH_3)_2-CH_2NH_3^+$
(C) $CH_3-C(CH_3)(OH)-CH_2CH_3$
(D) $CH_3-C(CH_3)_2-CH_2OH$

81. Which of the following statements is incorrect?
(A) C_6H_5CN is prepared by reacting C_6H_5Br with aqueous alcoholic NaCN
(B) Alkyl cyanides are as poisonous as KCN
(C) CH_3CN is soluble in water but CH_3NC is not
(D) Catalytic reduction of methyl isocyanide gives dimethylamine

82. Which of the following is formed when a secondary nitroalkane reacts with HNO_2?
(A) Ketone
(B) Pseudonitrole
(C) Carboxylic acid
(D) Nitrolic acid

83. Tollen's reagent is reduced by
(A) C_6H_5NHOH (B) $C_6H_5NH_2$ (C) C_6H_5NO (D) $C_6H_5NO_2$

84. Which of the following amines exhibits the highest pK_b value?
(A) 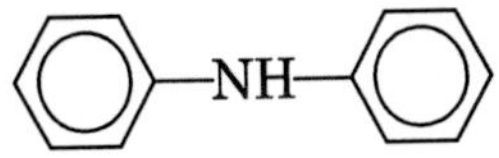
(B) $(CH_3)_2NH$
(C) $(CH_3CH_2)_2NH$
(D) 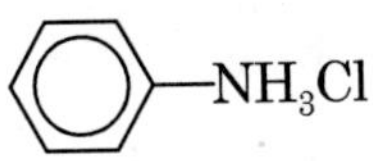

85. Which of the following amines is the strongest base?
(A) 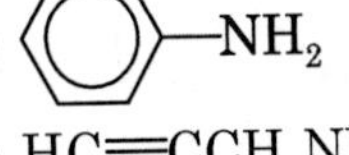
(B) $CH_3CH_2CH_2NH_2$
(C) $HC\equiv CCH_2NH_2$
(D) $CH_2=CHCH_2NH_2$

86. Which of the following amines is the strongest base?
(A) $(H_2N)_2\overset{+}{C}-OH$
(B) $(H_2N)_2C=O$
(C) $(H_2N)_2C=NH_2^+$
(D) $(H_2N)_2C=NH$

87. Which of the following reactions can be used to prepare benzenesulphonic acid?

(A) $C_6H_6 \xrightarrow{SO_2Cl_2,\ pyridine}$ (B) $C_6H_6 \xrightarrow{ClSO_3H}$

(C) $C_6H_6 \xrightarrow{fuming\ H_2SO_4,\ 473-523\ K}$ (D) $C_6H_6 \xrightarrow{conc.\ H_2SO_4,\ 353\ K}$

88. Which of the following is Sandmeyer reaction?

(A) $C_6H_5NO_2 + 6[H] \xrightarrow{Sn/HCl} C_6H_5NH_2 + 2H_2O$

(B) $C_6H_5OH \xrightarrow{Zn\ dust,\ heat} C_6H_6 + ZnO$

(C) $2C_6H_5Cl \xrightarrow{CuCl_2,\ HCl} C_6H_6 + Cl_2$

(D) $C_6H_5N_2Cl \xrightarrow{CuCl,\ heat} C_6H_5Cl + N_2$

89. Which of the following is formed when 4-clorotoluene is reacted with KNH_2 is liquids ammonia?

(A) 4-Chloroaniline (B) 4-Aminotoluene

(C) 3-Aminotoluene (D) 2-Aminotoluene

90. Which of the following Schieman's reaction?

(A) $C_6H_5N_2Cl \xrightarrow{CuCN,\ H_3O^+} C_6H_5COOH$

(B) $C_6H_5N_2Cl \xrightarrow{HBF_4,\ heat} C_6H_5F$

(C) $2C_6H_5N_2HSO_4 \xrightarrow{Cu,\ C_2H_5OH} C_6H_5—C_6H_5$

(D) $C_6H_5N_2Cl \xrightarrow{CuI,\ KI} C_6H_5I$

91. Which of the following amines will not dissolve in a solution of NaOH after its reaction with Hinsberg reagent?

(A) *N,N*-Dimethylaniline (B) *N*-Methylethanamine

(C) Propan-2-amine (D) Diethylamine

92. Which of the following is Gattermann's reaction?

(A) $C_6H_5N_2Cl \xrightarrow{Cu\ powder,\ HCl}$

(B) $C_6H_5N_2HSO_4 \xrightarrow{KI} C_6H_5I$

(C) $2C_6H_5N_2HSO_4 \xrightarrow{heat} C_6H_5—C_6H_5$

(D) $C_6H_5N_2HSO_4 \xrightarrow{CuBr} C_6H_5Br$

93. Which of the following amines does not liberate N_2 gas on reaction with HNO_2?

(A) Methanamine (B) Aniline (C) Benzylamine (D) Ethanamine

94. Activation of benzene ring by amino group in aniline is reduced by treating aniline with

(A) ethanoyl chloride (B) ethanoic acid

(C) ethanol (D) dil. HCl

95. The compound which on reaction with aqueous HNO_2 at low temperature produces an oily nitrosamine is

(A) methylamine (B) ethylamine (C) diethylamine (D) trimethylamine

96. Benzenediazonium chloride on reaction with phenol in weakly basic medium gives
(A) diphenyl ether (B) *p*-hydroxyazobenene
(C) chlorobenzene (D) benzene **(I.I.T. 1998)**

97. Which of the following gives red colour on reaction with dilute solution of sodium hydroxide?
(A) $(CH_3)_3C—NO$ (B) $(CH_3)_3C—NO_2$
(C) 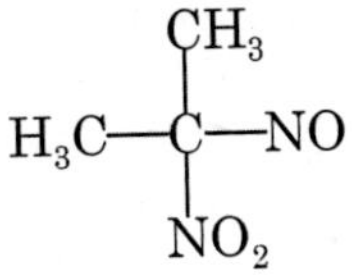 (D) 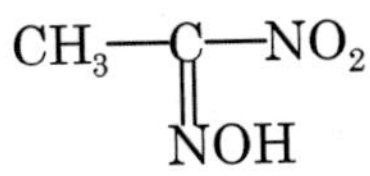

98. In the sequence of reactions, 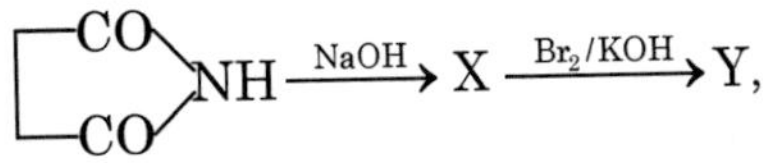, Y is
(A) γ-aminobutyric acid (B) ethylenediamine
(C) α-alanine (D) β-alanine

Hint: 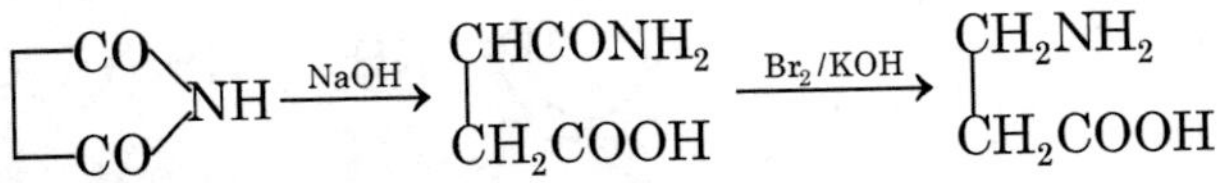

99. The base with lowest K_b value is
(A) $H_2NCH_2CH_2OH$ (B) NH_3
(C) $(C_2H_5)_3N$ (D) H_2NCH_2CN
Hint: Due to the presence of strong electron-withdrawing (– I-effect) of —CN group, D is the weakest base.

100. In the sequence of reactions, 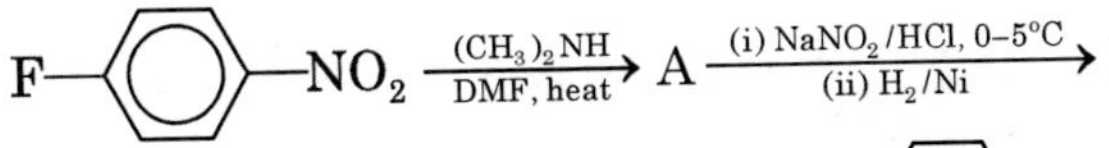 B, B is
(A) 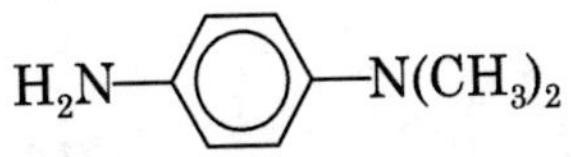 (B) 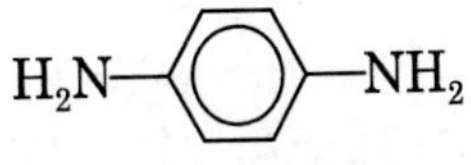
(C) 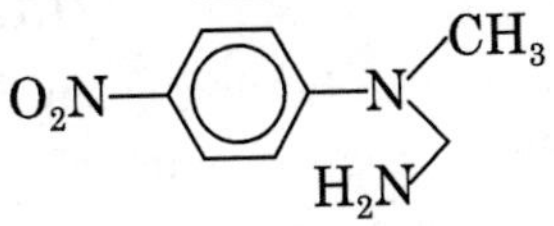 (D) 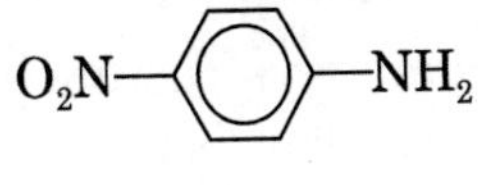

Hint: 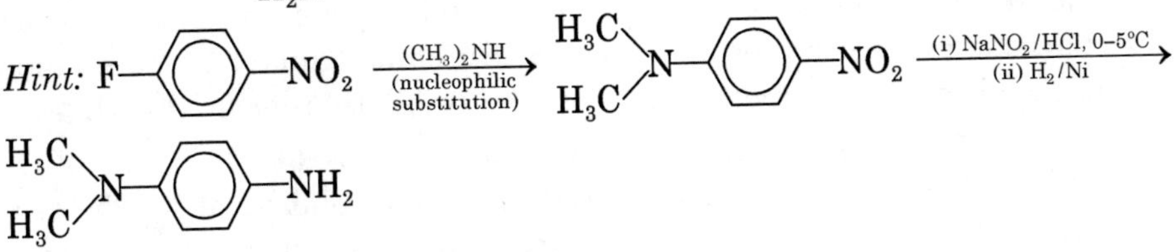

101. $RCH_2CH_2NH_2$ does not undergo *E*2 elimination because
(A) NH_2^- is a very good leaving group
(B) NH_2^- is a very poor leaving group
(C) of formation of unstable alkene (D) of highly basic nature.

102. What are the constituent amines formed when the mixture of 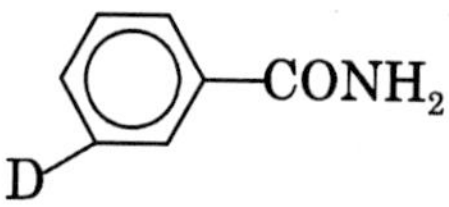 and 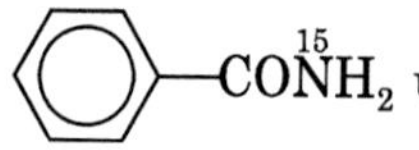 undergoes Hoffmann bromamide degradation?

(A) 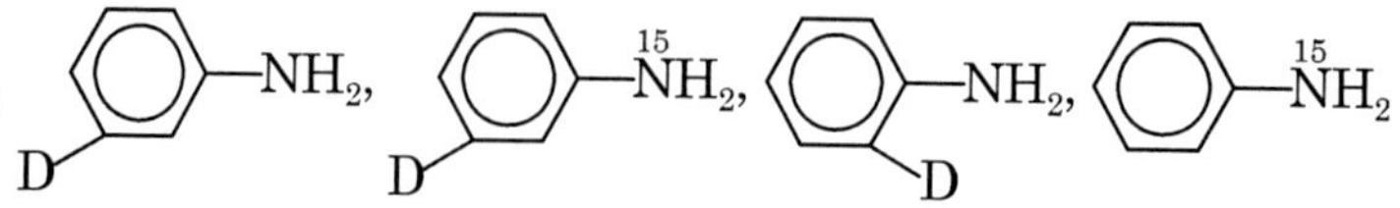

(B) 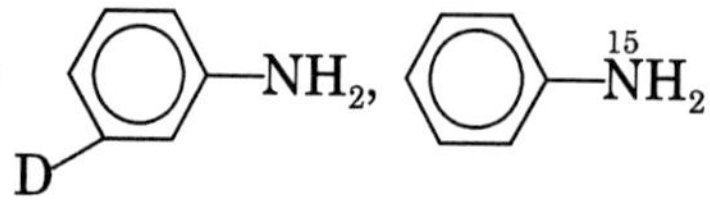

(C) 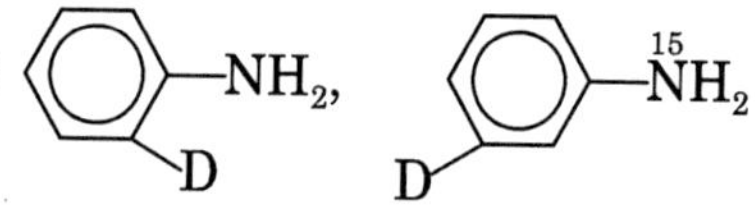

(D) 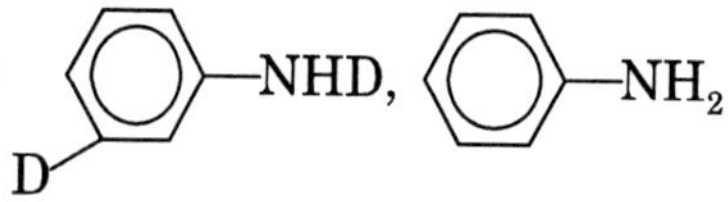

(I.I.T. 2006)

Hint: 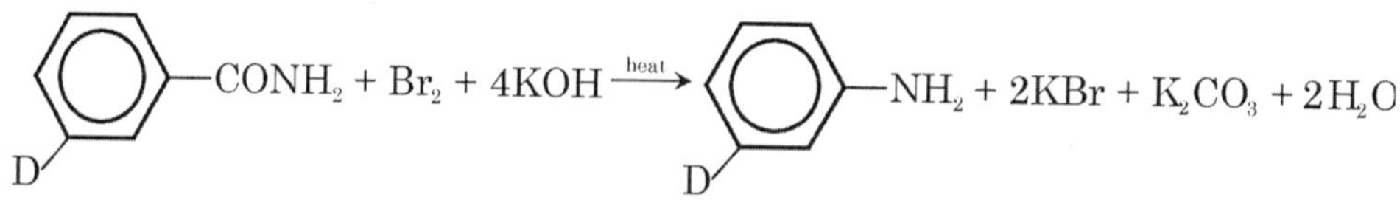

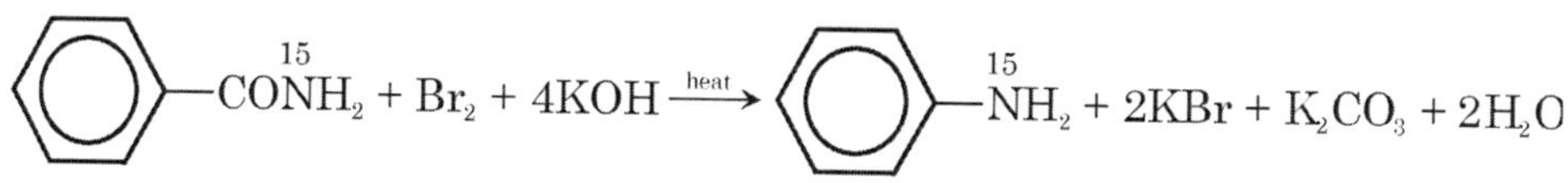

103. The correct stability order of the following resonance structures is

$H_2C{=}\overset{+}{N}{=}\overset{-}{N}$ (I) $\quad H_2\overset{+}{C}{-}N{=}\overset{-}{N}$ (II) $\quad H_2\overset{-}{C}{-}\overset{+}{N}{\equiv}N$ (III) $\quad H_2\overset{-}{C}{-}N{=}\overset{+}{N}$ (IV)

(A) I > II > IV > III (B) I > III > II > IV

(C) II > I > III > IV (D) III > I > IV > II **(I.I.T. 2009)**

Hint: The structure having complete octet of all the atoms and having greater number of cavalent bonds are more stable than the others. The structure with the negative charge on more electronegative atom is more stable than the one in which the negative charge is present on less electronegative atom. Thus, I is more stable. On the basis of stability of resonance structures, the stability order is : I > III > II > IV.

Paragraph for Question Nos. 104 to 106

p-Amino-*N*,*N*-dimethylaniline is added to a strongly acidic solution of X. The resulting solution is treated with a few drops of aqueous solution of Y to yield blue coloration due to the formation of methylene blue. Treatement of the aqueous solution of Y with the reagent potassium hexacyanoferrate(II) leads to the formation of an intense blue precipitate. The precipitate dissolves on excess addition of the reagant. Similarly, treatment of the solution of Y with the solution of potassium hexacyanoferrate(III) leads to a brown coloration due to the formation of Z.

104. The compound X is

(A) $NaNO_3$ (B) NaCl (C) Na_2SO_4 (D) Na_2S **(I.I.T. 2009)**

Hint: $Na_2S + 2H^+ \longrightarrow H_2S + 2Na^+$

$FeCl_3 + H_2S \longrightarrow FeCl_2 + 2HCl + S$

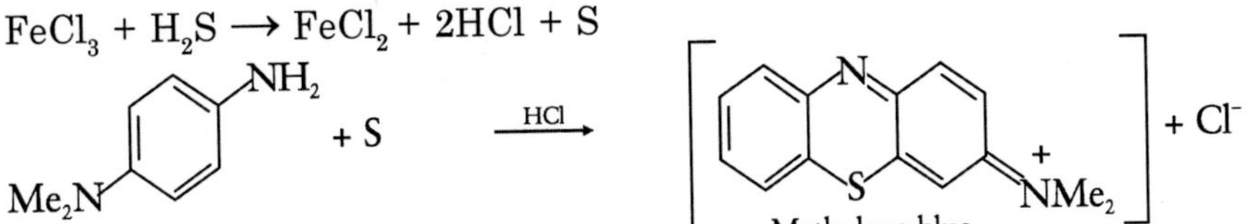

Methylene blue

105. The compound Y is

(A) $MgCl_2$ (B) $FeCl_2$ (C) $FeCl_3$ (D) $ZnCl_2$ **(I.I.T. 2009)**

Hint: The compound Y is $FeCl_3$ because when it is treated with $K_4[Fe(CN)_6]$ it gives intense blue precipitate of $Fe_4[Fe(CN)_6]_3$, which dissolves in excess $K_4[Fe(CN)_6]$.

$$4Fe^{3+} + 3\left[Fe(CN)_6\right]^{4-} \longrightarrow Fe_4\left[Fe(CN)_6\right]_3$$

106. The compound Z is

(A) $Mg_2[Fe(CN)_6]$ (B) $Fe[Fe(CN)_6]$ (C) $Fe_4[Fe(CN)_6]_3$ (D) $K_2Zn_3[Fe(CN)_6]_2$

(I.I.T. 2009)

Hint: The compound Z is $Fe[Fe(CN)_6]$ (brown colour) and it is formed due to reaction of $K_3[Fe(CN)_6]$ and $FeCl_3$ (Y).

$$Fe^{3+} + \left[Fe(CN)_6\right]^{3-} \longrightarrow Fe\left[Fe(CN)_6\right]$$

107. The compounds P, Q and S were separately subjected to nitration using HNO_3/H_2SO_4 mixture. The major product formed in each case respectively, is

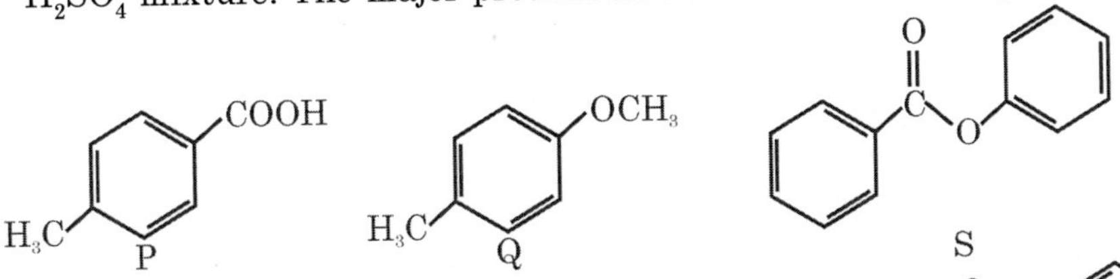

(A)

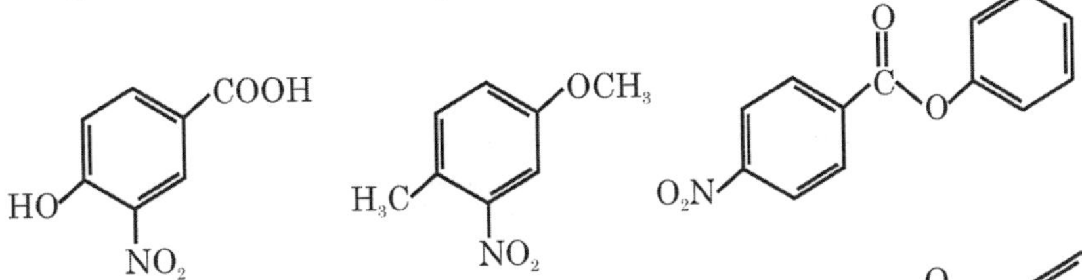

(B)

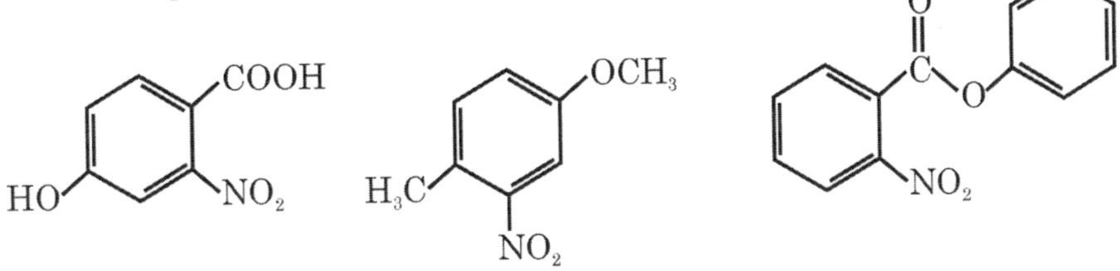

(C) 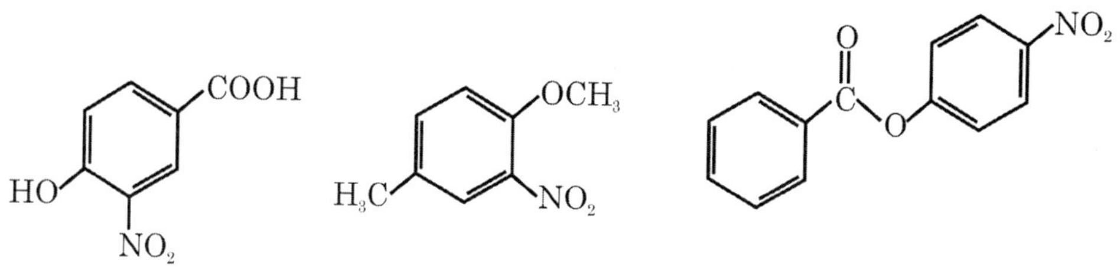

(D) 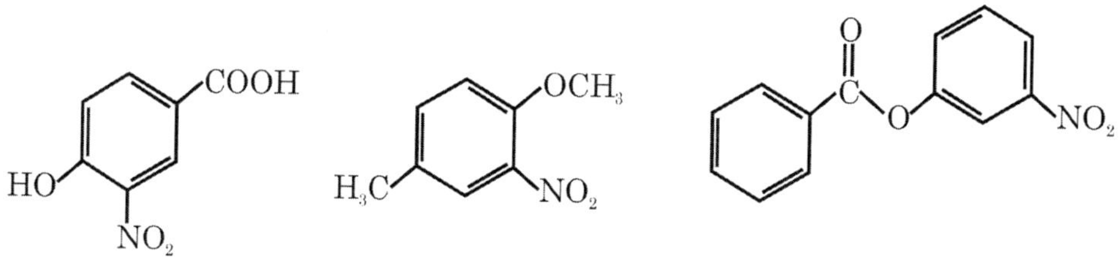

(I.I.T. 2010)

Hint:

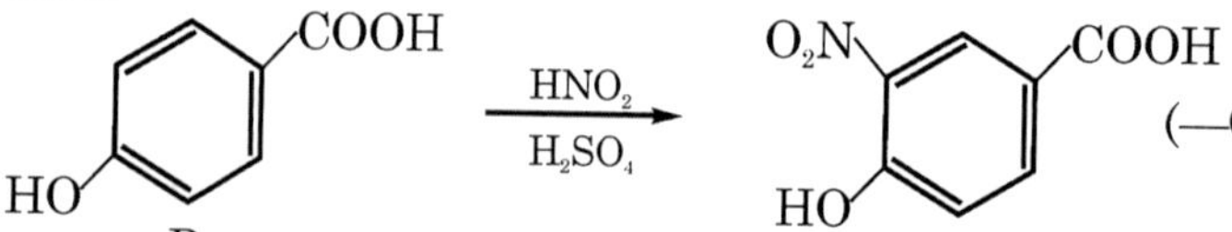

P

(—OH group is more activating)

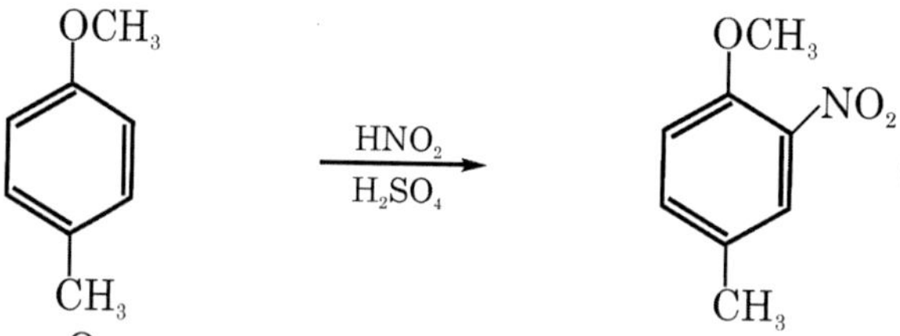

Q

(—OCH_3 group is more activating)

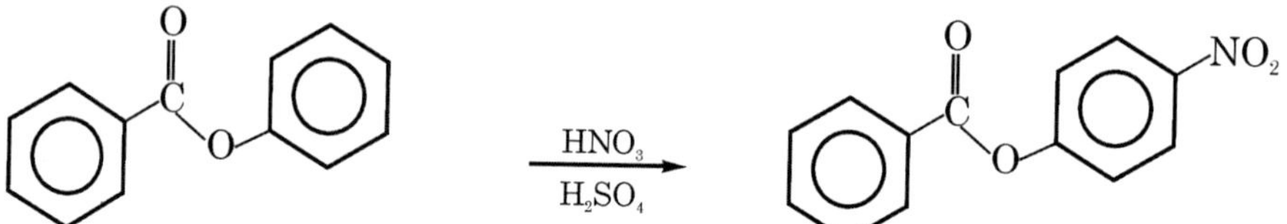

108. In the reaction 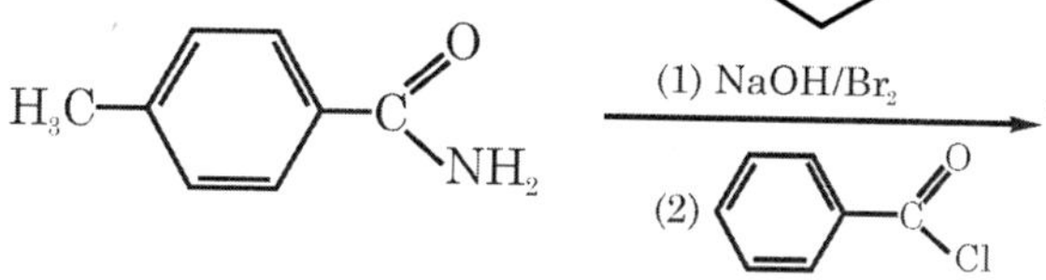 T, the structure of the product T is

(A) 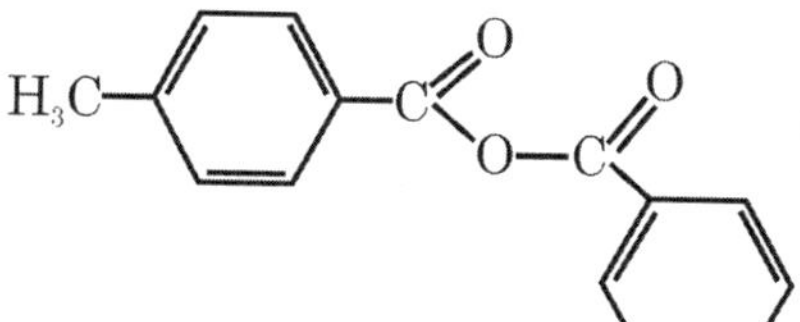

(B) 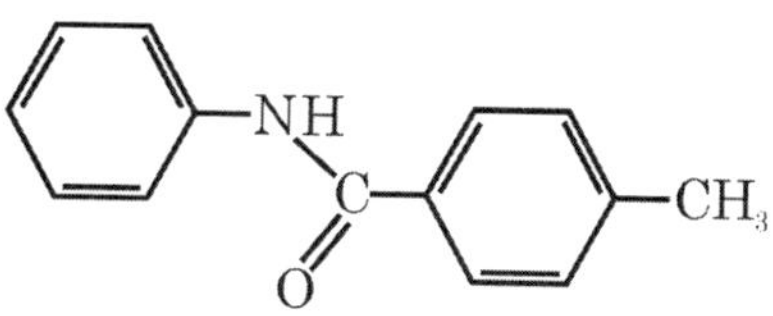

(C) 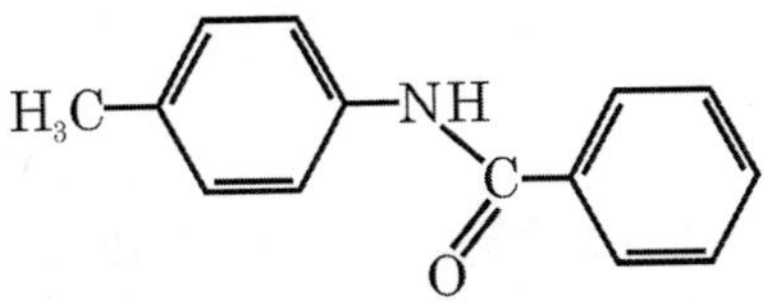

(D) 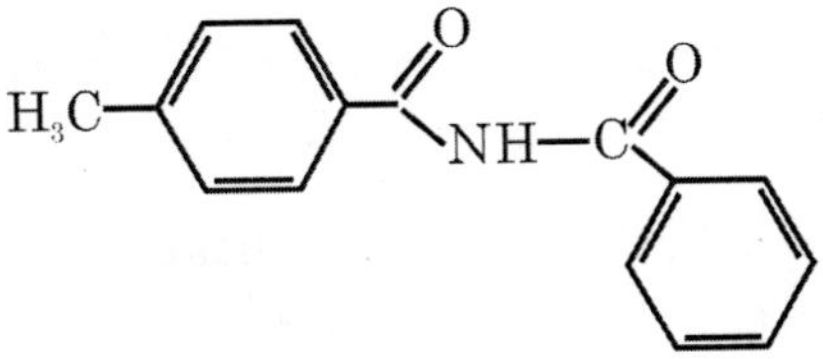

(I.I.T. 2010)

Hint: 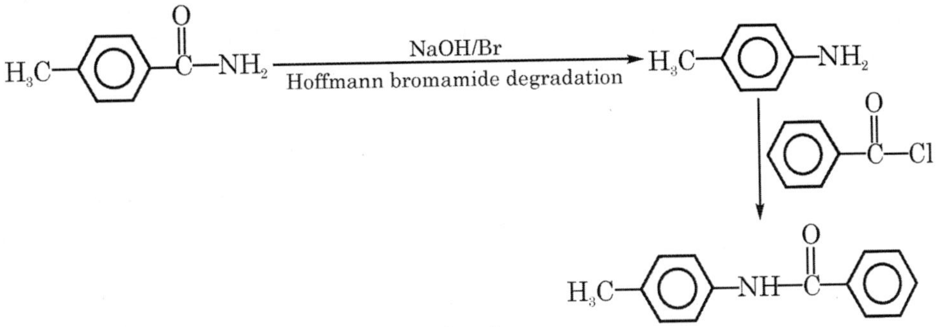

109. The major product of the following reaction is

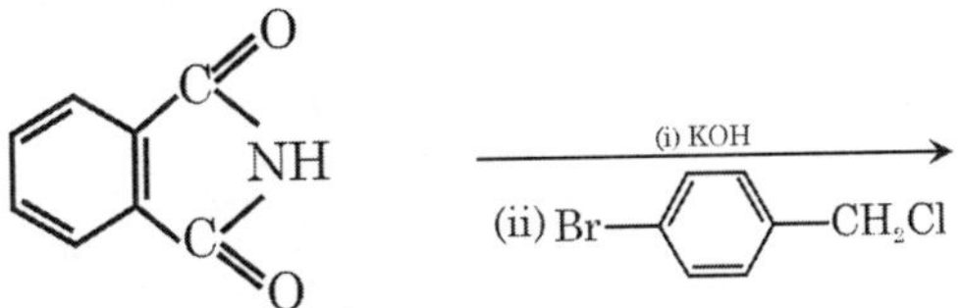

(A) 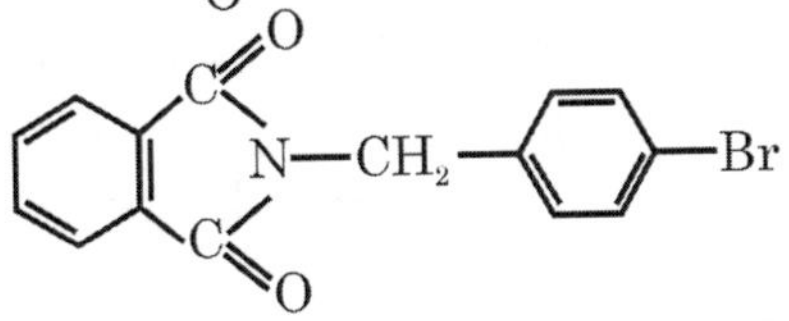

(B) 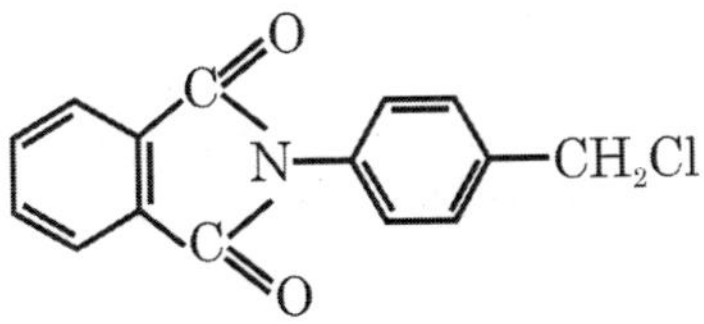

(C) 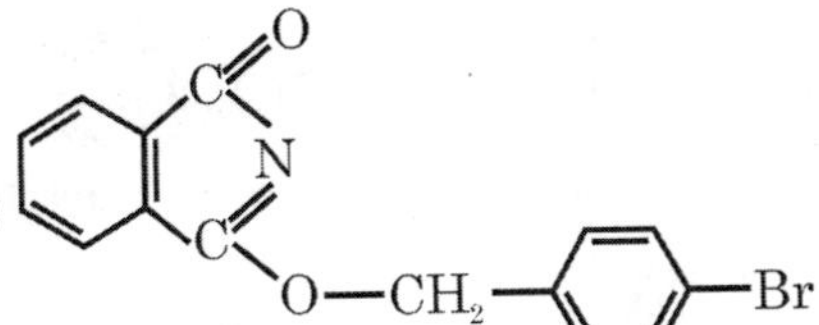

(D) 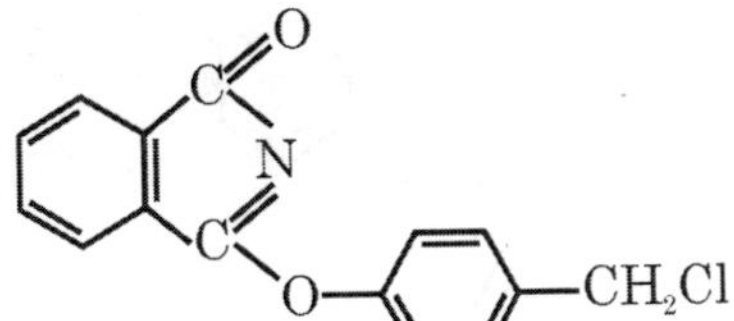

(I.I.T. 2011)

Hint: 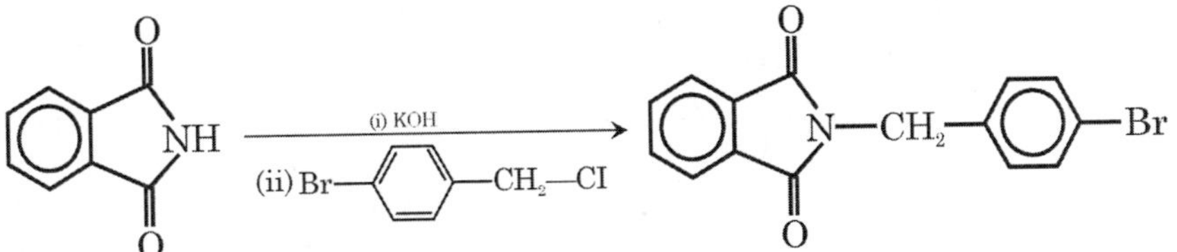

The partial double bond character along C—Br bond prevents the attack of nucleophile at phenylic position.

110. The correct functional group X and the reagent/reaction conditions Y in the following scheme are

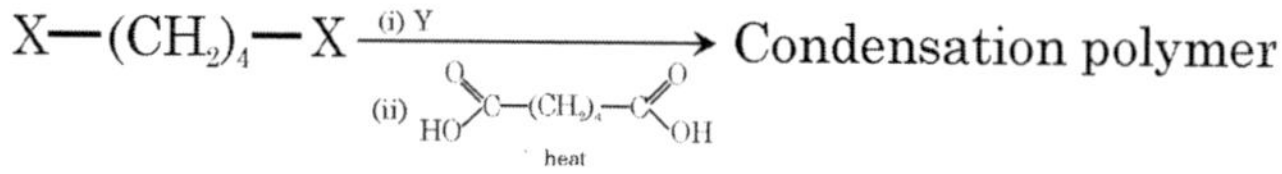

(A) X = $COOCH_3$, Y = H_2/Ni/heat (B) X = $CONH_2$, Y = H_2/Ni/heat

(C) X = $CONH_2$, Y = Br_2/NaOH (D) X = CN, Y = H_2/Ni/heat

(I.I.T. 2011)

Hint: Condensation polymers are formed by condensation of a diol or diamine with dicarboxylic acids.

Hence X may be $-\overset{O}{\overset{\|}{C}}-OR$ or $-\overset{O}{\overset{\|}{C}}-NH_2$ or $-C\equiv N$

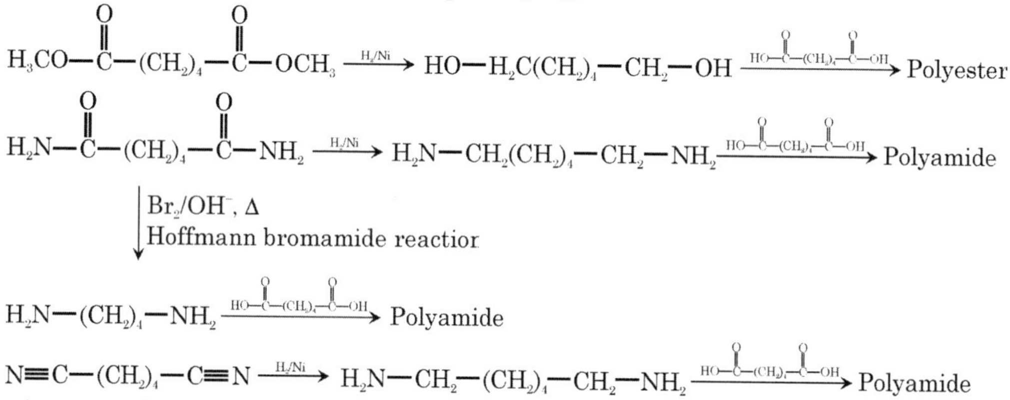

111. Amongst the compounds given, the one that would form a brilliant colored dye on treatment with $NaNO_2$ in dil. HCl followed by addition to an alkaline solution of β-naphthol is

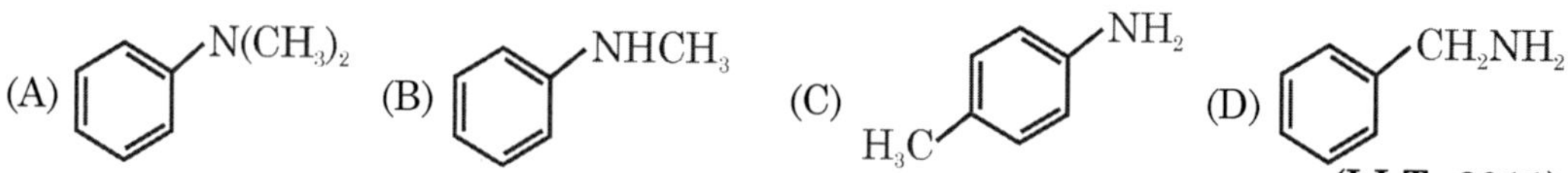

(I.I.T. 2011)

Hint:

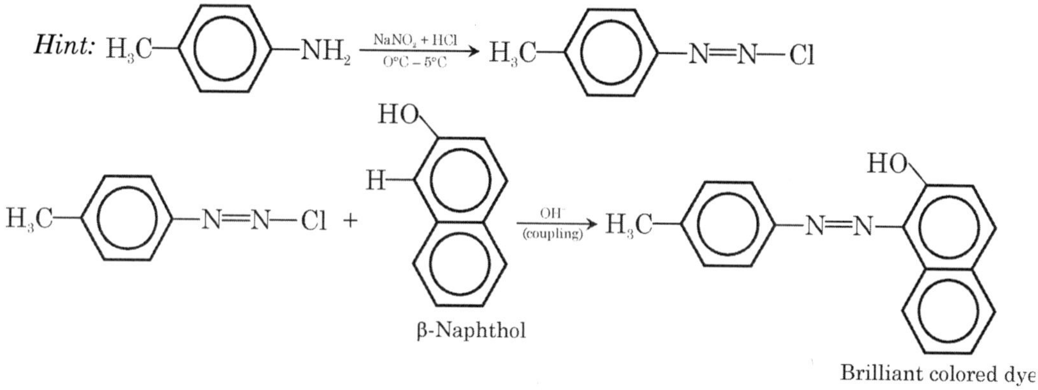

ANSWERS

1. A	2. B	3. D	4. A	5. B	6. D	7. B, C
8. A	9. D	10. A	11. D	12. D	13. B	14. C
15. A	16. B, D	17. A, C	18. D	19. A	20. A	21. A
22. A	23. A	24. C	25. A	26. B	27. B	28. C
29. D	30. C	31. B	32. A	33. A, B	34. A, B	35. A
36. D	37. B	38. B	39. D	40. B	41. A	42. A
43. C	44. A	45. B	46. A	47. B	48. D	49. A
50. C	51. C	52. A	53. D	54. C	55. B	56. C
57. B	58. A	59. D	60. C	61. D	62. D	63. A
64. A	65. A	66. D	67. D	68. C	69. B	70. B
71. A	72. A	73. C	74. A	75. B	76. A	77. B
78. D	79. D	80. C	81. A, B	82. B	83. A	84. A
85. B	86. D	87. D	88. D	89. B	90. B	91. B, D
92. A	93. B	94. A	95. C	96. B	97. D	98. C
99. D	100. A	101. B	102. B	103. B	104. D	105. C
106.B	107. C	108. C	109. A	110. A, B, C, D		111. C

42

Biomolecules (Carbohydrates, Aminoacids, Peptides, Proteins)

OBJECTIVE TYPE MULTIPLE CHOICE QUESTIONS

1. Cellulose is formed from
   (A) D-galactose units
   (B) 50% of D-glucose and 50% of L-glucose units
   (C) L-glucose units
   (D) D-glucose units.
2. Cane sugar on hydrolysis gives
   (A) glucose and glucose
   (B) glucose and lactose
   (C) fructose and fructose
   (D) glucose and fructose
3. Carbohydrates which cannot be hydrolysed by human digestive system is
   (A) cellulose (B) glycogen (C) starch (D) all of these.
4. α-D(+)-glucose and β-D(+)-glucose are
   (A) anomers
   (B) epimers
   (C) geometrical isomers
   (D) enantiomers.
5. In polysaccharides, the linkage connecting monosaccharide units is
   (A) peptide linkage
   (B) glycogen linkage
   (C) nucleoside linkage
   (D) glycolic linkage.
6. In sucrose there is
   (A) tail to tail linkage
   (B) head to head linkage
   (C) head to tail linkage
   (D) 50% of head to head and 50% of head to tail linkages.
7. Mutarotation does not occur in
   (A) L-glucose (B) D-glucose (C) sucrose (D) none of these.
8. Glucose molecule reacts with x number of molecules of phenylhydrazine to form osazone. The value of x is
   (A) 4 (B) 6 (C) 2 (D) 1

9. Identify the linear polymer out of the following
   (A) amylose (B) starch (C) glycogen (D) amylopectin
10. Which of the following is sweetest sugar?
   (A) Maltose (B) Fructose (C) Glucose (D) Sucrose
11. Glucose reacts with ethanoic anhydride to form
   (A) monoacetate (B) tetraacetate (C) pentaacetate (D) hexaacetate
12. Which of the following sugars is least sweet?
   (A) Sucrose (B) Maltose (C) Fructose (D) Glucose
13. Which of the following has branched chain structure?
   (A) Cellulose (B) Amylose (C) Amylopectin (D) Nylon
14. The amino acid $H_2N\overset{COOH}{\overset{|}{C}}HCH_2CH_2COOH$ exists at high pH as
   (A) $H_3\overset{+}{N}\overset{COO^-}{\overset{|}{C}}HCH_2CH_2COO^-$ (B) $H_3\overset{+}{N}\overset{COO^-}{\overset{|}{C}}HCH_2CH_2COOH$
   (C) $H_2N\overset{COO^-}{\overset{|}{C}}HCH_2CH_2COO^-$ (D) $H_3\overset{+}{N}\overset{COOH}{\overset{|}{C}}HCH_2CH_2COOH$
15. The amino acid $H_2N\overset{COOH}{\overset{|}{C}}HCH_2CH_2COOH$ exists at low pH as
   (A) $H_2N\overset{COO^-}{\overset{|}{C}}HCH_2CH_2COO^-$ (B) $H_3\overset{+}{N}\overset{COOH}{\overset{|}{C}}HCH_2CH_2COOH$
   (C) $H_3\overset{+}{N}\overset{COO^-}{\overset{|}{C}}HCH_2CH_2COO^-$ (D) $H_3\overset{+}{N}\overset{COO^-}{\overset{|}{C}}HCH_2CH_2COOH$
16. If K_{a_1} and K_{a_2} are the ionisation constants of $H_3\overset{+}{N}\underset{CH_3}{\underset{|}{C}}HCOOH$ and $H_3\overset{+}{N}\underset{CH_3}{\underset{|}{C}}HCOO^-$ respectively, the pH of the solution at the isoelectric point is
   (A) $p\text{H} = (pK_{a_1} + pK_{a_2})/2$ (B) $p\text{H} = (pK_{a_1} - pK_{a_2})^{1/2}$
   (C) $p\text{H} = (pK_{a_1} + pK_{a_2})^{1/2}$ (D) $p\text{H} = pK_{a_1} + pK_{a_2}$
17. C-terminus (or carbonyl terminus) of the peptide structure is on the
   (A) both sides (B) right hand side
   (C) left hand side (D) random way (non-specific)
18. Invert sugar is
   (A) lactose (B) sucrose (C) maltose (D) cellobiose
19. Cellobiose is a
   (A) trisaccharide (B) disaccharide (C) monosaccaride (D) polymer
20. Which of the following carbohydrates does not give positive Fehling's solution test?
   (A) lactose (B) maltose (C) sucrose (D) glucose

21. Which of the following is obtained by the hydrolysis of starch in the presence of enzyme diastase?
(A) Lactose (B) Maltose (C) Sucrose (D) All of these.

22. Cellobiose is a
(A) 1,2′-β-glycoside (B) 1,2′-α-glycoside
(C) 1,4′-β-glycoside (D) 1,4′-α-glycoside

23. Amylopectin and amylose on hydrolysis with dilute acids give
(A) fructose only (B) glucose only
(C) fructose and glucose respectively (D) glucose and fructose respectively.

24. Inversion of cane sugar is effected by
(A) hydrolysis in the presence of enzyme invertase
(B) hydrohysis in the presence of dilute acid
(C) both (A) and (B) (D) none of these.

25. Sucrose is a
(A) monosachharide (B) mixture of glucose and fructose
(C) non-reducing sugar (D) reducing sugar.

26. Sucrose is a
(A) 1,2′-glycoside (B) 1,2′-β-glycoside
(C) 1,4′-β-glycoside (D) 1,4′-α-glycoside.

27. Amylopectin and amylose are two fractions of starch. Which fraction of starch gives a violet colour with iodine?
(A) Only amylopectine (B) Only amylose
(C) Both (A) and (B) (D) None of these.

28. Maltose is a
(A) 1,2′-β-glycoside (B) 1,2′-α-glycoside
(C) 1,4′-β-glycoside (D) 1,4′-α-glycoside.

29. Lactose on hydrolysis forms
(A) two fructose molecules (B) two glucose molecules
(C) equal mixture of glucose and galactose
(D) a fructose and a galactose molecules

30. Which of the following gives only glucose on hydrolysis?
(A) Cane sugar (B) Lactose (C) Maltose (D) Sucrose

31. In alkaline medium, fructose behaves as
(A) a furanose (B) an aldose
(C) a non-reducing sugar (D) a reducing sugar.

32. Identify the reagent with which glucose, fructose and maltose react.
(A) $[Ag(NH_3)_2]OH$ (B) Phenyl hydrazine.
(C) NH_2OH/HCl (D) $CuSO_4$/sodium potassium tartrate

33. The reaction of glucose with hydroxylamine to form an oxime indicates that glucose contains a/an
(A) carboxylic acid group (B) carbonyl group
(C) ketone group (D) carbonyl group.

34. D-Ribose contains
(A) eight carbon atoms (B) six carbon atoms
(C) five carbon atoms (D) four carbon atoms.

35. The reagent which converts aldoses to gluconic acid is
(A) conc. HNO_3 (B) Br_2/H_2O
(C) Fehling's solution (D) Tollen's reagent

36. The reagent which converts aldoses to saccharic acid is
(A) conc. HNO_3 (B) Br_2/H_2O
(C) Fehling's solution (D) Tollen's reagent.

37. Identify the substance which is formed when cane sugar is heated above its melting point.
(A) Brown sugar (B) soot (C) carbon black (D) caramel

38. D-Galactose contains
(A) six carbon atoms (B) five carbon atoms
(C) four carbon atoms (D) three carbon atoms.

39. To become a carbohydrate a compound must contains at least
(A) 2 carbons (B) 3 carbons (C) 4 carbons (D) 6 carbons

40. Oxidation of glucose is one of the most important reactions in a living cell. What is the number of ATP molecules generated in cells from one molecule of glucose?
(A) 12 (B) 18 (C) 28 (D) 38

41. Invertase brings about the conversion of
(A) starch to glucose (B) glucose to C_2H_5OH and CO_2
(C) sucrose to glucose and fructose (D) maltose to glucose.

42. The intermediate compound formed in the conversion of starch to glucose?
(A) lactose (B) fructose (C) sucrose (D) maltose

43. Carbohydrates are sweet because they have
(A) hydrogen bonding (B) electrovalent bonding
(C) covalent bonding (D) coordinate bonding.

44. The change in optical rotation (with time) of freshly prepanol solution of sugar is called
(A) specific rotation (B) inversion
(C) mutarotation (D) rotatory motion.

45. Which of the following statements about ribose is correct?
(A) It is a polyhydroxy compound (B) It is an aldehyde sugar
(C) It has six carbon atoms (D) It exhibits optical activity.

46. Glycogen is a
    (A) monosaccharide (B) disacharide
    (C) trisaccharide (D) polysaccharide
47. Which carbohydrate is used in silvering of mirror?
    (A) Glucose (B) Sucrose (C) Fructose (D) Starch
48. Cellulose on complete hydrolysis gives
    (A) β-D-glucose (B) α-D-glucose (C) α-D-fructose (D) β-D-fructose
49. Starch on complete hydrolysis gives
    (A) α-D-glucose (B) β-D-glucose (C) β-D-fructose (D) α-D-fructose
50. α-D-glucose and β-D-glucose differ from each other due to the difference in one of the carbon with respect to its
    (A) size of hemiacetal ring (B) number of OH groups
    (C) configuration (D) conformation.
51. Identify the amino acid which is not optically active?
    (A) Gluctamic acid (B) Valine
    (C) Alanine (D) Glycine
52. An oligopeptide may contain
    (A) 2-12 amino acids (B) 2-10 amino acids
    (C) 2-8 amino acids (D) 2-6 amino acids
53. Peptides on hydrolysis gives
    (A) amines (B) amino acids (C) ammonia (D) alcohols
54. Proteins give a purple colour when treated with
    (A) Benedict solution (B) Iodine solution
    (C) ninhyrin solution (D) biuret
55. Destruction of activity of proteins by chemical agent or heat is called
    (A) denaturation (B) deamination (C) denitrification (D) dehydration
56. Naturally occurring amino acids have
    (A) L-configuration (B) D-configuration
    (C) both (A) and (B) (D) meso isomerism.
57. Which of the amino acids have a polar group?
    (A) Glutamine (B) Glycine (C) Proline (D) Methionine
58. Which of the amino acids have a non-polar group?
    (A) Cystein (B) Isobucine (C) Agrinine (D) Serine
59. The peptide bond is
    (A) $-\underset{\displaystyle O}{\underset{|}{C}}-NH-$ (B) $H_2N-\underset{\displaystyle O}{\underset{\|}{C}}-NH-R$
    (C) $R-\underset{\displaystyle O}{\underset{\|}{C}}-NH-R'$ (D) $-CONH_2$
60. Zwitter ion is formed by
    (A) lysine (B) benzoic acid (C) aniline (D) acetanilide

61. A zwitterion is
    (A) a negatively charged ion without a metal atom in it
    (B) a positively charged ion without a metal atom in it
    (C) an ion with positive and negative charges at different points in it
    (D) a heavy ion with a small charge in it.

62. Which of the following is correct regarding proteins?
    (A) Proteins contain exclusively L isomers of amino acids
    (B) Proteins contain exclusively D isomers of amino acids
    (C) Proteins contain 50% each of D- and L-amino acids
    (D) Proteins contain 67% of L-form and 33% of D-form of amino acids

63. Number of essential amino acids is
    (A) 24 (B) 20 (C) 15 (D) 10

64. The helical structure of proteins is stabilised by
    (A) hydrogen bond (B) van der Waals forces
    (C) peptide bond (D) dipeptide bond

65. High dipole moment value of glycine in water indicates the formation of
    (A) hydrogen bonding (B) zwitterion
    (C) cyclic dimer (D) polymer

66. In amino acids, the amino group is attached to
    (A) α-carbon atom (B) β-carbon atom
    (C) δ-carbon atom (D) γ-carbon atom.

67. What structural feature distinguishes proline from other natural amino acids?
    (A) It is optically inactive (B) It contains aromatic group
    (C) It is a dicarboxylic acid (D) It has a secondary

68. The number of tripeptides formed by three different amino acids are
    (A) 3 (B) 4 (C) 5 (D) 6

69. Which amino acid is chiral?
    (A) Alanine (B) Valine (C) Proline (D) Histidine

70. The main structural features of proteins are
    (A) ehter linkage (B) ester linkage
    (C) peptide linkage (D) all of these.

71. Which of the following is a protein?
    (A) Nucleic acid (B) Keratin (C) Glycogen (D) Lecithin

72. Proteins are polyamides of
    (A) α-amino acids (B) α-hydroxy acids
    (C) β-amino acids (D) β-hydroxy acids

73. Which of the following is α-amino acid?
    (A) 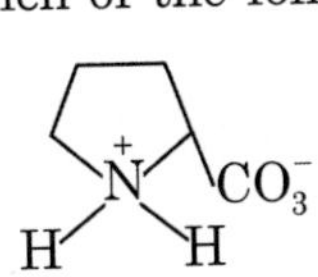

    (B) 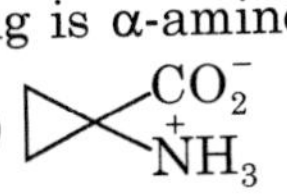

    (C) Both (D) None of these.

74. Two forms of D-glucopyranose an called
 (A) enantiomers (B) anomers
 (C) epimers (D) diastereoisomers. **(I.I.T. 2005)**
 Hint: D-glucopyranose is the cyclic form of glucose. Two isomers are observed around C_1 containing the chiral centre. These isomers are called α-anomer and β-anomer.
75. The protein responsible for the maintenance of osmotic presence in the blood is
 (A) aluminium (B) α-globulin (C) β-globulin (D) γ-globulin
76. Dencentration of protein causes the breakdown of
 (A) primary structure (B) secondary structure
 (C) tertiary structure (D) none of these.
77. Denaturation of protein is caused by
 (A) treatment with acid (B) treatment with alkali
 (C) treatment with alcohol (D) all of these.
78. Secondary structure of protein refers to
 (A) mainly denatured proteins and structure of prosthetic groups
 (B) three dimensional structure specially the bonds between amino acid residues that are distinct from each other in the peptide chain.
 (C) linear sequence of amino acid residues in the peptide chain
 (D) regular following pattern of contiguous portions of the peptide chain.
79. Secondary structure of proteins is produced through the formation of
 (A) —S—S— linkages (B) —O—O— linkages
 (C) —C—C— linkages (D) —N—N— linkages.
80. Peptides are examples of
 (A) simple protein (B) conjugated protein
 (C) derived protein (D) none of these.
81. The relation of isoelectric point for the amino acid to solubility is
 (A) the two are not related
 (B) an amino acid is least soluble at its isoelectric point
 (C) an amino acid is least soluble at its isoelectric point
 (D) solubility of only some amino acids depend on it.
82. Which of the following statements about protein is not true?
 (A) Eggs are rich in protein
 (B) Pulses are good sources of protein
 (C) Amino acid residues join together to make a protein molecule
 (D) Proteins are polymers with formula $(C_6H_{10}O_5)_n$
83. Which of the following proteins transport O_2 in the blood stream?
 (A) Insulin (B) Oxytocin (C) Allumin (D) Haemoglobin
84. Molecular mass of protein is
 (A) > 1000 (B) > 10000 (C) < 1000 (D) > 1000 and < 1000
85. Which of the following is not a classification of protein?
 (A) Enzymes (B) Antigens (C) Antibodies (D) Hormones

86. Which of the following statements is true of proteins?
    (A) They are hormones (B) They act as antibodies
    (C) They catalyse biochemical reactions
    (D) All of these.
87. Starch contains
    (A) 1,4′-α-glycoside bond (B) 1,4′-β-glycoside land
    (C) 1,6′-α-glycoside bond (D) 1,6′-β-glycoside bond
88. Cellulose contains
    (A) 1,4′-α-glycoside bond (B) 1,4′-β-glycoside bond
    (C) 1,6′-α-glycoside bond (D) 1,6′-β-glycoside bond
89. Which of the following gives positive Tollen's test?
    (A) Glucose, sucrose (B) Glucose, fructose
    (C) Hexanal, acetophenone (D) Fructose, sucrose
    Hint: Aldehydes and α-hydroxyketones give positive Tollen's test. Glucose has an aldehydic group and fructose in an α-hydroxyketone.
90. Which of the following is a basic amino acid?
    (A) Leucine (B) Valine (C) Histidne (D) Aspartic acid
91. Which of the following is an acidic amino acid?
    (A) Isoleucine (B) Proline (C) Histidine (D) Aspartic acid
92. Which of the following is a neutral amino acid?
    (A) Histidine (B) Lysine (C) Glutamic acid (D) Serine
93. Statement–1: Glucose gives a reddish brown precipitate with Fehling's solution.
    Statement–2: Reaction of glucose with Fehling's solution gives CuO and gluconic acid.
    (A) Statement–1 is true, Statement–2 is true; Statement–2 is a correct explanation for Statement–1
    (B) Statement–1 is true, Statement–2 is true; Statement–2 is not a correct explanation for Statement–1
    (C) Statement–1 is true, Statement–2 is false
    (D) Statement–1 is false, Statement–2 is true. **(I.I.T. 2007)**
    Hint: (C) $C_6H_{12}O_6$ + Fehling's solution $\longrightarrow (C_6H_{11}O_7)^- + Cu_2O \downarrow$ (Red)
94. The correct statement(s) about the following sugars X and Y is(are)

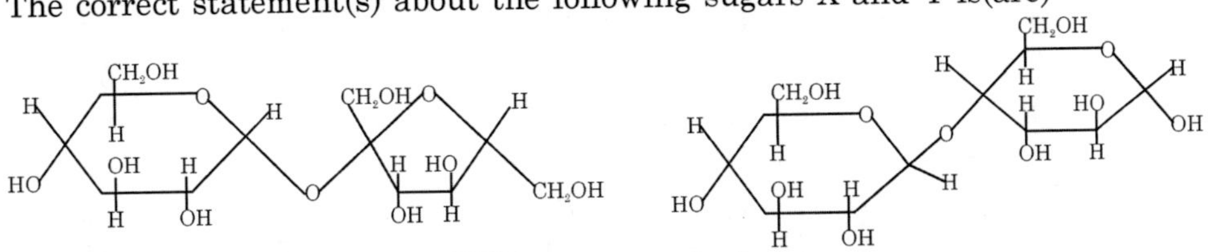

    (A) X is a reducing sugar and Y is a non-reducing sugar.
    (B) X is a non-reducing sugar and Y is a reducing sugar.
    (C) The glucosidic linkages in X and Y are α and β, respectively.
    (D) The glucosidic linkages in X and Y are β and α, respectively. **(I.I.T. 2009)**

95. The correct statement about the following disaccharide is

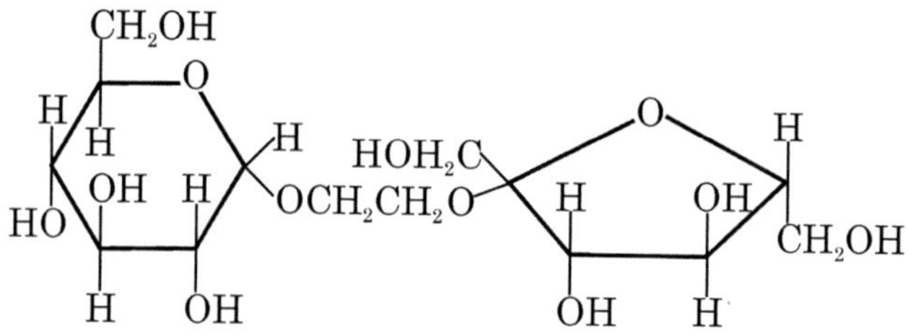

(A) Ring (a) is pyranonse with α-glycosidic link
(B) Ring (a) is furanonse with α-glycosidic link
(C) Ring (b) is furanonse with α-glycosidic link
(D) Ring (b) is pyranonse with β-glycosidic link **(I.I.T. 2010)**

Hint: Ring (a) is pyranose with α-glycosidic link.

96. The following carbohydrate is

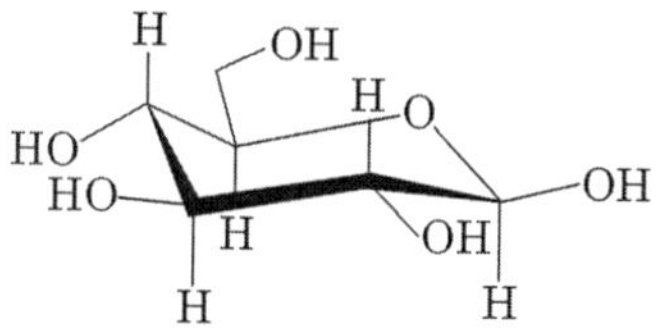

(A) A ketohexose (B) an aldohexose (C) an α-furanose (D) an α-pyranose.

(I.I.T. 2011)

Hint: The carbohydrate is an aldohexose.

ANSWERS

1. D	2. D	3. A	4. A	5. D	6. B
7. C	8. B	9. A	10. B	11. C	12. B
3. C	14. C	15. B	16. A	17. A	18. B
19. B	20. C	21. B	22. C	23. B	24. C
25. C	26. A	27. C	28. D	29. C	30. C
31. D	32. C	33. B	34. C	35. B	36. A
37. D	38. A	39. B	40. D	41. C	42. D
43. A	44. C	45. C	46. D	47. A	48. A
49. A	50. C	51. D	52. B	53. B	54. D
55. A	56. A	57. A	58. B	59. A	60. A
61. C	62. A	63. D	64. A	65. B	66. A
67. D	68. D	69. B	70. C	71. B	72. A
73. C	74. A	75. A	76. B, D	77. D	78. D
79. A	80. C	81. C	82. D	83. D	84. B
85. B	86. D	87. A	88. B	89. B	90. C
91. D	92. D	93. D	94. B, C	95. A	96. B

43

Polymers

OBJECTIVE TYPE MULTIPLE CHOICE QUESTIONS

1. Which of the following is a branched chain polymer?
   (A) PVC (B) Nylon
   (C) Polyester (D) Low density polythene
2. Which of the following is a chain-growth polymer?
   (A) Polypropylene (B) Glyptal (C) Dacron (D) Nylon-6
3. Which of the following is a step-growth polymer?
   (A) Polythene (B) Nylon (C) Polyisoprene (D) Polyacrylonitrile
4. Which of the following is a natural polymer?
   (A) Dacron (B) Starch (C) Orlon (D) Terylene
5. Which of the following is a thermoplantic polymer?
   (A) Natural rubber (B) Nylon-6
   (C) Polythene (D) Bakelite
6. Which of the following is thermosetting polymer?
   (A) Polythene (B) Bakelite (C) Terylene (D) Nylon-6
7. Which of the following is the monomer of natural rubber?
   (A) 2-Methylbuta-1,2-diene (B) 2-Methylbuta-1,3-diene
   (C) Chloroprene (D) Buta-1,3-diene
8. Polymerisation of two or more different monomers is called
   (A) copolymerisation (B) addition polymerisation
   (C) chain polymerisation (D) homo polymerisation.
9. Identify the incorrect statement.
   (A) Thermosetting polymers are crosslinked polymers
   (B) Thermosetting polymers soften on heating and harden on cooling irreversibly
   (C) Thermosetting polymers are insoluble in organic solvents
   (D) Thermosetting polymers can be reclaimed.

10. Identify the incorrect statement.
    (A) Thermoplastics soften on healing and harden on cooling reversibly
    (B) Thermoplastics possess linear structure without cross linking
    (C) Thermoplastics cannot be reclaimed
    (D) Thermoplastics are weak and soft.
11. Chlorophyll is a
    (A) coordination compound (B) elastomer
    (C) polymer (D) macromolecules.
12. The aid used in the synthesis of nylon-66 is
    (A) formic acid (B) stearic acid (C) palmitic acid (D) adipic acid
13. Which of the following intermolecular forces is present in nylon-66?
    (A) Hydrogen bonding (B) van der Waals
    (C) Dipole-dipole interaction (D) None of these
14. Nylon-6 is prepared by heating
    (A) ethylene glycol and adipic acid
    (B) hexamethylemediamine acid adipic acid
    (C) caprolactum and adipic acid (D) caprolactum
15. Nylon-66 is prepared by heating
    (A) caprolactum (B) ethylene glycol and adipic acid
    (C) hexamethylene diamine and adipic acid
    (D) caprolactum and adipic acid
16. Which is not true about polymers?
    (A) They have low M.W. (B) They do not carry any charge
    (C) They have high viscosity (D) They scatter light.
17. Bakelite is prepared by heating
    (A) phenol and propanone (B) phenol and ethanol
    (C) phenol and formaldehyde (D) phenol and H_2SO_4
18. Crosslinking is present in
    (A) nylon-6 (B) polythene (C) bakelite (D) all of these.
19. Dacron is a polymer of
    (A) ethylene glycol + vinyl acetate (B) glycol + pathalic anhydride
    (C) ethylene glycol + malic anhydride
    (D) ethylene glycol + trephthalic acid
20. Zeigler–Natta catalyst is
    (A) $TiCl_4$ and $(CH_3)_3Al$ (B) $TiCl_4$ and $(C_6H_5)_3Al$
    (C) titanium isopropoxide (D) TiO_2
21. Orlon is a polymer of
    (A) Vinyl chloride (B) acrylomitrile
    (C) tetrafluoroethylene (D) neoprene

22. The rubber used in cycle are manufactured from
    (A) styrene-butadiene rubber (B) natural rubber
    (C) neoprene rubber (D) butyl rubber
23. Type rubber is prepared by vulcanising natural rubber with
    (A) 1.5% S (B) 2% S (C) 3% S (D) 3.5% S
24. Which one of the following is a plasticiser?
    (A) Phthalic acid (B) Phthalic anhydride
    (C) Di-*n*-butyle phthalate (D) Cresyl phosphate
25. Buna-S is a polymer of
    (A) butadiene (B) butadiene and chloroprene
    (C) styrene (D) butadiene and styrene
26. Natural rubber is a polymer of
    (A) *cis*-isoprene (B) trans-isoprene
    (C) co-*cis*-trans-isoprene (D) none of these.
27. Glyptal is obtained from glycerol by reacting with
    (A) malonic acid (B) phthalic acid (C) malic acid (D) ethanoic acid
28. Which of the following is an example of condensation polymer?
    (A) Polythene (B) Terylene (C) Orlon (D) PVC
29. If N_1, N_2, N_3, ... are the number of molecules with molecular masses M_1, M_2, M_3, ... respectively, the number average molecular mass of a polymer is given by
    (A) $\frac{\Sigma N_i M_i}{\Sigma N_i}$ (B) $\frac{\Sigma N_i M_i^2}{\Sigma N_i M_i}$ (C) $\frac{\Sigma N_i M_i}{\Sigma M_i}$ (D) $\frac{\Sigma N_i M_i^2}{\Sigma N_i M_i}$
30. If N_1, N_2, N_3, ... are the number of molecules with molecular masses M_1, M_2, M_3, ... respectively, the weight average molecular mass of a polymer is given by
    (A) $\frac{\Sigma N_i M_i}{\Sigma N_i}$ (B) $\frac{\Sigma N_i M_i}{\Sigma M_i}$ (C) $\frac{\Sigma M_i^2}{\Sigma N_i}$ (D) $\frac{\Sigma N_i M_i^2}{\Sigma N_i M_i}$
31. Which of the following has amide links?
    (A) Nylon-66 (B) Peptide (C) Protein (D) All of these.
32. Identify the incorrect statement.
    (A) The order of intermolecular forces in the polymers nylon-6, neoprene and *PVC* is: neoprene < *PVC* < nylon-6
    (B) Plastic has plasticity but no elasticity
    (C) Rubber has elasticity and no plasticity
    (D) Natural rubber is a step-growth polymer.
33. Identify the incorrect statement.
    (A) Scrap thermosetting polymers can be reused
    (B) Silk is a polysaccharide polymer
    (C) Catalyst used in the manufacture of orlon is benzoyl peroxide
    (D) Nylon is dyed with acid dyes.

35. Identify the incorrect statement.
   (A) Polyester is a thermosetting polymer
   (B) Teflon is a thermosetting polymer
   (C) In the manufacture of PVC, the catalyst used is H_2O_2
   (D) The temperature at which vulcanisation rubber by sulphur occurs is 110–140°C.
36. Identify the incorrect statement.
   (A) Plywood sheets are stuck together by unpolymerised urea-formaldehyde resin
   (B) Polymer used for making non-breakable plastic crockery is melamine-formaldehyde
   (C) The name of commercially important polymer perspex is polymethylmethacrylate
   (D) Polyvinyl chloride is a thermosetting polymer.
37. $CF_2—CF_2$ is a unit of
   (A) Buna-S (B) bakelite (C) polythene (D) teflon
38. The monomer of tiflon is
   (A) $F_2C{=}CF_2$ (B) $FClC{=}CF_2$ (C) $Cl_2C{=}CCl_2$ (D) $FClC{=}CClF$
39. The polymer having amide linkage is
   (A) PVC (B) terylene (C) lakelite (D) nylon-66
40. The polymer having ester linkage is
   (A) PVC (B) terylene (C) bakelite (D) nylon-66
41. The hardest polymer among the following is
   (A) thermoplastic (B) branched chain polymer
   (C) crosslinked polymer (D) linear polymer.
42. Which of the polymers has ester linkage $\left(-\overset{\overset{\displaystyle O}{\|}}{C}-O-\right)$?
   (A) Urea-formaldehyde (B) Alkyl resin
   (C) Terylene (D) Bakelite
43. The catalyst used in the manufacturer of stereospecific polypropylene is
   (A) benzoyl peroxide (B) H_2/Ni
   (C) $LiAlH_4$ (D) $TiCl_4$, $(CH_3)_3Al$
44. Which of the following polymers is not an example of interpolymerisation?
   (A) PVC (B) Nylon-66 (C) SBR (D) Buna-N
45. Which of the following is not an example of addition polymer?
   (A) Neoprene (B) Glyptal (C) PVC (D) polytetrafluorethylene
46. Identify the thermosetting polymer(s).
   (A) Melamine-formaldehyde (B) Acrylite
   (C) Alitrocellulose (D) Phenol-formaldehyde
47. Identify the thermoplastic polymer(s).
   (A) Silicone (B) Cellulose acetate
   (C) Perspex (D) Polythene
48. Identify the polymer(s) which is not a polyester.
   (A) Terylene (B) Glyptal (C) Bakelite (D) melamine-formaldeyde

49. The polymer having strongest interparticle forces is
    (A) thermosetting plastics (B) thermoplastics
    (C) elastomers (D) fibers.
50. Cellulose is a polymer of
    (A) β-L-glucose (B) β-D-glucose (C) α-D-glucose (D) both (B) and (C)
51. Among cellulose, poly(vinyl chloride), nylon and natural rubber, the polymer in which the intermolecular force of attraction is weakest is
    (A) nylon (B) poly(vinyl chloride)
    (C) cellulose (D) natural rubber. **(I.I.T. 2009)**

    Hint: Both nylon and callulose have intramolecular H-bonding, while poly(vinyl chloride) has dipole-dipole interaction. Natural rubber has the London forces which are the weakest.

ANSWERS

1. D	2. A	3. B	4. B	5. C	6. B
7. B	8. A	9. D	10. C	11. A	12. D
13. A	14. D	15. C	16. A	17. A	18. C
19. D	20. A	21. B	22. D	23. D	24. C, D
25. D	26. A	27. B	28. B	29. A	30. D
31. D	32. D	33. A	34. A	35. A	36. A
37. D	38. A	39. D	40. B	41. C	42. B, C
43. D	44. A	45. B	46. A	47. A, B, C	48. C
49. A	50. B	51. D			

44

Matching Type Questions

	X	Y
1.	(A) Heisenberg	(a) $\lambda = h/p$
	(B) de Broglie equation	(b) spectrum of H_2
	(C) Schrodinger	(c) stationary orbits
	(D) Rydberg	(d) uncertainty principle
	(E) Bohr	(e) wave theory
2.	(A) Soddy	(a) cathode rays
	(B) Thomson	(b) transuranium elements
	(C) Moseley	(c) molecular orbital theory
	(D) Seaborg	(d) atomic number
	(E) Mulliken	(e) isotopes
3.	(A) valence bond theory	(a) Gillespe and Nyholm
	(B) group displacement law	(b) Soddy and Rutherford
	(C) radioactive disintegration	(c) Aston and Soddy
	(D) isotopes	(d) Soddy, Fajans and Russel
	(E) VSEPR theory	(e) Heiteler and London
4.	(A) concept of hybridisation	(a) Guldberg and Waage
	(B) equation of state	(b) Thomson
	(C) law of mass action	(c) Sidgwick and Powel
	(D) electron	(d) Priestley
	(E) laughing gas	(e) van der Waals
5.	(A) kinetic theory of gas	(a) Sommerfield
	(B) idea of elliptical orbit	(b) Urey
	(C) nuclear fission	(c) Thomson
	(D) cathode rays	(d) Bernoulli
	(E) heavy water	(e) Otto Hahn and Strassman

6. (A) isotopes — (a) $^{30}_{14}Si$, $^{31}_{15}P$
   (B) isobars — (b) CO, CN^-
   (C) isotones — (c) Fe_2O_3, Al_2O_3
   (D) isoelectronic — (d) $^{76}_{34}Sc$, $^{76}_{32}Ge$
   (E) isomorphs — (e) $^{35}_{17}Cl$, $^{37}_{17}Cl$
7. (A) isomers — (a) graphite, diamond
   (B) isotopes — (b) CH_3CH_2OH, CH_3OCH_3
   (C) isoelectronic — (c) $^{40}_{20}Ca$, $^{40}_{19}K$
   (D) isomorphs — (d) $^{16}_{8}O$, $^{18}_{8}O$
   (E) isobars — (e) $FeSO_4 \cdot 7H_2O$, $MgSO_4 \cdot 7H_2O$
   (F) allotropes — (f) Na^+, Mg^{2+} **(I.I.T. 1973)**
8. (A) three electron bond — (a) AgI
   (B) covalent bond — (b) KI
   (C) ionic bond — (c) BDE = 2–10 kcal/mol
   (D) ionic, covalent and coordinate bonds — (d) NO
   (E) hydrogen bond — (e) NH_4I
9. (A) superconductor — (a) antiferromagnetic
   (B) $Cu(CH_3COO)_2 \cdot 5H_2O$ — (b) antiferroelectric
   (C) AgI — (c) paramagnetic
   (D) $PbZrO_3$ — (d) diamagnetic
   (E) F-centre — (e) artificial rain
10. (A) $YBa_2Cu_3O_7$ — (a) antiferromagnet
    (B) MnO_2 — (b) ferrimagnet
    (C) Fe_3O_4 — (c) ferromagnet
    (D) CrO_2 — (d) superconductor
    (E) CsCl — (e) Schottky defect
11. (A) shape selective catalyst — (a) $k = A\exp(-E_a/RT)$
    (B) solar cell — (b) zeolite
    (C) Wilkinson catalyst — (c) Si
    (D) Arrhenius equation — (d) $(Ph_3P)_3RhCl$
    (E) hydrolysis of peptide — (e) pepsin
12. (A) non-zero entropy at absolute zero — (a) a process in which volume remains constant
    (B) adiabatic process — (b) a process in which pressure remains constant
    (C) isochoric process — (c) a process in which temperature remains constant
    (D) isobaric process — (d) NO

(E) isothermal process	(e) a process in which heat does not enter or leave the system
13. (A) endothermic reaction	(a) a process which is carried out infinitesimally slowly
(B) exothermic reaction	(b) $\Delta E = q - W$
(C) reversible process	(c) $(\Delta H_2 - \Delta H_1)/(T_2 - T_1) = \Delta C_p$
(D) first law of thermodynamics	(d) ΔH is positive
(E) Kirchhoff's equation	(e) ΔH is negative
14. (A) piezoelectricity	(a) ZnS
(B) pyroelectricity	(b) CsCl
(C) ferroelectricity	(c) application of heat
(D) Frenkel defect	(d) mechanical stress
(E) the 8 : 8 type of packing	(e) $BaTiO_3$
15. (A) Aufbau	(a) Raoult's law
(B) de Broglie	(b) azeotropic mixture
(C) isotonic solutions	(c) electron has dual character
(D) ideal solutions	(d) 4*s* orbital is filled before 3*d* orbital
(E) non-ideal solutions	(e) osmotic pressure
16. (A) number of mols of solute per kg of solvent	(a) conductance of ions
(B) number of mols of solute perlitre of solution	(b) *p*H scale
(C) Kohlrausch	(c) anode rays
(D) Sorensen	(d) molality
(E) Goldstein	(e) molarity
17. (A) gold number	(a) cold cream
(B) water in oil emulsion	(b) vanishing cream
(C) oil in water emulsion	(c) negative sol
(D) $Fe(OH)_3$ sol	(d) protective colloid
(E) sulphur sol	(e) positive sol

18. (A) Ziegler–Natta catalyst	(a) positive electron	(i) monoclinic S, orthorhombic S
(B) NaCl structure	(b) negative electron	(ii) -1.6×10^{-19} coulomb
(C) allotropes	(c) high pressure	(iii) 1.6×10^{-19} coulomb
(D) positron	(d) $(C_2H_5)_3Al + TiCl_4$	(iv) CsCl structure
(E) β-rays	(e) two or more forms of same element differing in physical properties	(v) hydrogenation of ethylene

19.

(A) Planck	(a) non-ideal solution	(i) $\Delta H_{mixing} = 0$
(B) plot of P_{total} *vs* x_A is non-linear	(b) ΔH unaffected	(ii) $\Delta V_{mixing} = 0$
(C) catalyst	(c) $E = h\upsilon$	(iii) 1.66×10^{-27} g
(D) 1 u	(d) ideal solution	(iv) ΔG unaffected
(E) Raoult's law	(e) 931 MeV	(v) 6.6×10^{-34} J s

20.

(A) The property dependent upon the amount of substance present in the system	(a) intensive property	(i) ice in a closed beaker
(B) The property independent of the amount of substance present in the system	(b) isolated system	(ii) ice in an open beaker
(C) a system which does not exchange mass and energy with the surroundings	(c) open system	(iii) ice in a thermos
(D) a system which exchanges energy with the surroundings and not mass	(d) extensive property	(iv) surface tension
(E) a system which exchanges energy and mass with the surroundings	(e) closed system	(v) entropy

21.

(A) a reaction which is spontaneous at all temperatures	(a) ABCABC ... type of packing	(i) $\Delta G = +$ ve
(B) a reaction which is non-spontaneous at all temperatures	(b) AB AB ... type of packing	(ii) Cu
(C) Cubic close packed arrangement	(c) $\Delta H = +$ ve	(iii) Zn
(D) hexagonal close packed arrangement	(d) anion vacancies	(iv) $\Delta G = -$ ve
(E) LiCl	(e) $\Delta H = -$ ve	(v) Schottky defect

22.

(A) Bohr	(a) vapour pressure
(B) Roentgen	(b) colloids
(C) Tyndall	(c) chemical bond
(D) Raoult	(d) frequency condition
(E) Pauling	(e) X-rays

(M.L.N.R. 1997)

23.

(A) carbonium ions	(a) *sp*, sp^2, sp^3
(B) free radicals	(b) π bonds
(C) Baver and Nathan effect	(c) diamagnetic

(D) hybridisation of C atom	(d) paramagnetic	
(E) *sp*, sp^2 hybridisations	(e) hyperconjugation	
24. (A) relative energy of orbitals	(a) $sp > sp^2 > sp^3$	
(B) bond angle	(b) $s > sp > sp^2 > sp^3 > p$	
(C) bond length	(c) $sp^3 > sp^2 > sp$	
(D) *s*-character	(d) $sp > sp^2 > sp^3$	
(E) relative size of orbitals	(e) $sp^3 > sp^2 > sp$	
25. (A) allotropes	(a) NO_3^-, CO_3^{2-}	
(B) isoelectronic	(b) $^{76}_{32}Ge$, $^{77}_{32}Ge$	
(C) isotopes	(c) $^{40}_{20}Ca$, $^{40}_{19}K$	
(D) isotones	(d) $^{76}_{32}Ge$, $^{77}_{33}As$, $^{78}_{34}Sc$	
(E) isobars	(e) O_2, O_3	
26. (A) neutron	(a) Kohlraush	
(B) molecular speed	(b) van der Waals	
(C) intermolecular forces	(c) Maxwell	
(D) conductance of ions	(d) Chadwick	**(I.IT. 1982)**
27. (A) ice	(a) insulator	
(B) graphite	(b) ionic compound	
(C) solid CO_2	(c) conducting covalent solid	
(D) magnesiam oxide	(d) hydrogen bonded structure	
(E) silica	(e) molecular crystal	
(F) SiO_4^{4-}	(f) tetrahedra	
28. (A) Bleaching agent	(i) Aluminium	
(B) Smelling salt	(ii) Carbon	
(C) Cryolite	(iii) Zinc	
(D) Bell Metal	(iv) Ammonium carbonate	
(E) Fluorspar	(v) Ammonium phosphate	
(F) Fertiliser	(vi) Calcium	
(G) Anthracite	(vii) Chlorine	**(I.I.T. 1980)**
29. (A) Liquid air	(i) Deacon's process	
(B) Na_2CO_3	(ii) Parken's process	
(C) Nitric oxide	(iii) Claude process	
(D) Silver	(iv) Ostwald process	
(E) Chlorine	(v) Solvay process	**(I.I.T. 1985)**
30. (A) Hg_2Cl_2	(i) Cassiterite	
(B) $(NaPO_3)_6$	(ii) Lunar caustic	
(C) NO_3^-	(iii) Producer gas	
(D) SnO_2	(iv) Water softener	
(E) $KCl \cdot MgCl_2 \cdot 6H_2O$	(v) Calomel	**(I.I.T. 1982)**

31. (A) Al — (i) Calamine
(B) Cu — (ii) Cryolite
(C) Mg — (iii) Malachite
(D) Zn — (iv) Carnallite **(I.I.T. 1983)**

32. (A) Haber — (i) Activation energy
(B) Graham — (ii) Diffusion of gases
(C) Arrhenius — (iii) Octet reule
(D) Lewis — (iv) Ammonia synthesis **(I.I.T. 1983)**

33. (A) Spinel — (i) $MgAl_2O_4$
(B) Feldspar — (ii) $PbCO_3$
(C) Cerrusite — (iii) $KAlSi_3O_8$
(D) Malachile — (iv) $MgSO_4 \cdot H_2O$
(E) Epsom salt — (v) $Cu(OH)_2 \cdot Cu_2CO_3$ **(I.I.T. 1983)**

34. (A) Self-reduction — (i) Lead
(B) Carbon-reduction — (ii) Silver
(C) Complex formation and displacement by metal — (iii) Copper
(D) Decomposition of iodide — (iv) Boron **(I.I.T. 2006)**

35. (A) $Bi_2O_3 \longrightarrow (BiO)^+$ — (i) Heat
(B) $[AlO_2]^- \longrightarrow Al(OH)_3$ — (ii) $PbCO_3$
(C) $SiO_4^{4-} \longrightarrow Si_2O_7^-$ — (iii) $KAlSi_3O_8$
(D) $B_4O_7^{2-} \longrightarrow [B(OH)_3]$ — (iv) $MgSO_4 \cdot H_2O$ **(I.I.T. 2006)**

36. According to Bohr's theory of H-atom, E_n = total energy, K_n = kinetic energy of nth orbit, V_n = potential energy of nth orbit, γ_n = radius of nth orbit.
(A) $V_n/K_n = ?$ — (i) 0
(B) If radius of nth orbit $\propto E_n^x, x = ?$ — (ii) -1
(C) Orbital angular momentum — (iii) -2
(D) $\frac{1}{V^x} \propto Z^y, y = ?$ — (iv) 1 **(I.I.T. 2006)**

37. (A) $[CuCl_4]^{2-}$ — (i) sp^3d^2 hybridisation
(B) $[Fe(H_2O)_5(NO)^+]SO_4$ — (ii) dsp^3 hybridisation
(C) $[Ni(CN)_4]^{2-}$ — (iii) sp hybridisation
(D) $[Fe(CO)_5]$ — (iv) dsp^2 hybridisation
(E) $[Ag(NH_3)_2]^+$ — (v) sp^3 hybridisation

38. (A) $K_3[CoF_6]$ — (i) $\mu = 0$
(B) $K_2[Ni(CN)_4]$ — (ii) $\mu = 1.73$ B.M.
(C) $[VO(H_2O)_5]SO_4$ — (iii) $\mu = 2.83$ B.M.
(D) $K_3[Mn(CN)_6]$ — (iv) $\mu = 3.87$ B.M.

(E) $Na_3[FeF_6]$	(v) μ = 4.90 B.M.
(F) $[Cr(NH_3)_6]Br_3$	(vi) μ = 5.92 B.M.
39. (A) $[Zn(NH_3)_4]Cl_2$	(i) Yellow
(B) $CuSO_4 \cdot 5H_2O$	(ii) Green
(C) $[Ni(H_2O)_6]SO_4$	(iii) Pink
(D) $[Co(H_2O)_6]SO_4$	(iv) Colourless
(E) $[Fe(H_2O)_6]Cl_3$	(v) Blue
40. (A) C_2H_5Br + alc. KOH	(i) cracking
(B) preparation of alkanes	(ii) Friedel–Crafts reaction
(C) phenol + $CHCl_3$ + NaOH	(iii) Reimer–Tiemann reaction
(D) $CH_3COOC_2H_5$ + NaOH	(iv) Wurtz reaction
(E) $C_6H_6 + CH_3Cl$ + anhy. $AlCl_3$	(v) saponification
(F) pyrolysis of alkanes	(vi) elimination reaction **(I.I.T. 1982)**
41. (A) dehydrohalogenation	(i) Kolbe reaction
(B) electrolysis of sodium salt	(ii) alcoholic KOH
(C) decarboxylation	(iii) addition reaction of ethylene
(D) ozonolysis	(iv) sodalime
(E) Williamson synthesis	(v) structure of alkene
(F) dichloroethylene	(vi) ether **(I.I.T. 1983)**
42. (A) Lucas test	(i) phenol
(B) neutral $FeCl_3$ test	(ii) glucose
(C) Dye test	(iii) tertiary alcohol
(D) Tollen's test	(iv) aniline **(I.I.T. 1983)**
43. (A) ethanol-methanol	(i) condensation reaction
(B) phenol-methanol	(ii) homologous series
(C) propanone sodium bisulphite	(iii) nucleophilic addition
(D) carbon tetrachloride	(iv) fire extinguishing **(I.I.T. 1981)**
44. (A) Cyclopropane, Cl_2, light	(i) homolytic addition
(B) propene, HBr, peroxide	(ii) eletrophilic substitution
(C) benzene, HNO_3, H_2SO_4	(iii) free radical substitution
(D) polymerisation	(iv) alcoholic KOH
(E) dehydrohalogenation	(iv) conversion of C_2H_2 to C_6H_6 **(I.I.T. 1981)**
45. (A) phenol	(i) coloured glass
(B) $Na_2S_2O_3$	(ii) antichlor
(C) salicylic acid	(iii) refractory material
(D) quicklime	(iv) antiseptic
(E) CuO	(v) analgesic **(I.I.T. 1985)**

46.

(A) freon	(i) antiseptic
(B) dacron	(ii) polymer of ethylene glycol and terephthalic acid
(C) Grignard reagent	(iii) refrigeration
(D) iodoform	(iv) CH_3MgX
(E) DDT	(v) insecticide

47.

(A) phenetole	(i) anaesthetic
(B) Williamson synthesis	(ii) estimation of alkoxy group
(C) oxonium salt	(iii) reaction of either with inorganic acid
(D) diethyl ether	(iv) reaction of alkyl halide with C_2H_5ONa
(E) Zeisel method	(v) reaction of phenol with diethyl sulphate and NaOH

48.

(A) detergent	(i) Ni catalyst
(B) hydrogenation	(ii) lubricant
(C) bee wax	(iii) hydrocarbons
(D) chromium soap	(iv) myricyl palmitate
(E) paraffin wax	(v) sodium lauryl sulphonate

49.

(A) Friedel–Crafts reaction	(a) oil	(i) alkenes
(B) fermentation	(b) Lewis acid	(ii) soap
(C) dehydrohalogenation	(c) copper(I) chloride	(iii) anhydrous $AlCl_3$
(D) Sandmeyer	(d) yeast	(iv) chlorobenzene
(E) saponification	(e) alcoholic alkali	(v) ethanol **(I.I.T. 1988)**

50.

(A) Fehling solution	(a) mesityle oxide	(i) reddish brown precipitate of Cu_2O
(B) Tollen's reagent	(b) reducing agent	(ii) phorone
(C) acetone	(c) Rochelle salt	(iii) ammoniacal $AgNO_3$
(D) benzaldehyde	(d) 50% NaOH	(iv) urinary antiseptic
(E) formaldehyde + NH_3	(e) urotropine	(v) $C_6H_5CH_2OH$ + C_6H_5COONa

51.

(A) Hell–Volhard–Zelinsky reaction	(a) aqueous CH_3COONa	(i) C_2H_5OH
(B) Kolbe reaction	(b) $LiAlH_4$	(ii) α-chloroacid
(C) CH_3COOH	(c) Cl_2 + P	(iii) carboxylic acid
(D) Grignard reagent	(d) solid CO_2	(iv) ethane
(E) decarboxylation	(e) sodalime	(v) removal of CO_2

52.

(A) acetamide	(a) H_2, $Pd/BaSO_4$ + S	(i) methylamine
(B) ethyl acetate	(b) $LiAlH_2$	(ii) acetophenone
(C) acetic anhydride	(c) Br_2 + NaOH	(iii) C_2H_5OH
(D) benzene + acetic anhydride	(d) anhy. $AlCl_3$	(iv) CH_3CHO
(E) acetyl chloride	(e) C_2H_5ONa	(v) acetoacetic ester

53.	(A) carbylamine	(a) primary amine	(i) diazonium salt
	(B) aromatic amine	(b) boiling aqueous HCl	(ii) $CHCl_3$ + alc. KOH
	(C) CH_3CN	(c) HN_2	(iii) $NH_3 + CH_3COONa$
	(D) CH_3CH_2CN	(d) boiling aqueous	(iv) butanone
	(E) CH_3NC	(e) CH_3MgBr, hydrolysis	(v) $HCOOH + CH_3NH_3$
54.	(A) *m*-dinitrobenzene	(a) HNO_2	(i) less basic than aniline
	(B) *o*-nitroaniline	(b) + I group at ortho position	(ii) *m*-nitroaniline
	(C) *o*-toluidine	(c) $(NH_4)_2S$	(iii) more basic than aniline
	(D) *N*-phenylhydroxylamine	(d) ammoniacal $AgNO_3$	(iv) nitrolic acid
	(E) primary nitroalkane	(e) – I group at ortho position	(v) silver mirror
55.	(A) Hinsberg's reagent	(a) $K_2Cr_2O_7$, H^+	(i) *N*-nitrosoamine
	(B) secondary amine	(b) phenol	(ii) distinction between amines
	(C) Hofmann bromamide reaction	(c) HNO_2	(iii) primary amine
	(D) coupling reaction	(d) benzenesulphonyl chloride	(iv) aniline black
	(E) aniline	(e) Br_2 + KOH	(v) azo dye
56.	(A) Sabatier–Senderens reaction	(a) iodobenzene	(i) benzanilide
	(B) Schotten–Baumann reaction	(b) benzenediazonium chloride	(ii) CuBr/HBr
	(C) Sandmeyer reaction	(c) acetophenone	(iii) Cu powder, heat
	(D) Wolff–Kishner reduction	(d) aniline + benzoyl chloride	(iv) ethyne to ethane
	(E) Ullmann reaction	(e) H_2/Ni, 250–300°C	(v) NH_2NH_2, KOH, ethylene glycol, 200°C
57.	(A) Rosenmund reduction	(a) aromatic aldeyhde, aliphatic acid anhydride	(i) Na/C_2H_5OH
	(B) Reimer–Tiemann reaction	(b) $C_6H_5ONa + CO_2$	(ii) H_2, $Pd/BaSO_4$ + S
	(C) Perkin reaction	(c) acetonitrile	(iii) salicylic acid
	(D) Mendius reaction	(d) phenol, $CHCl_3$, NaOH	(iv) α, β-unsaturated acid
	(E) Kolbe–Schmidt reaction	(e) acetyl chloride	(v) salicylaldehyde
58.	(A) Claisen condensation	(a) Zn/Hg, HCl	(i) benzaldehyde
	(B) Clemmensen reduction	(b) aldehyde	(ii) Cu power, HCl
	(C) Cannizzaro reaction	(c) KCN	(iii) β-ketoester
	(D) Benzoin condensation	(d) benzenediazonium chloride	(iv) ketone to alkane
	(E) Gattermann reaction	(e) ethyl acetate, C_2H_5ONa	(v) 50% NaOH

59. (A) Bergius process (a) diene, dienophile (i) ethylene
(B) Fries reaction (b) alkaline $KMnO_4$ (ii) distinction between alcohols
(C) Diels–Alder reaction (c) hyrogenation of coal (iii) adduct
(D) Baeyer's reagent (d) conc. HCl + anhy. $ZnCl_2$ (iv migration of acyl group
(E) Lucas reagent (e) phenyl acetate (v) petrol

60. (A) TNT (i) phenol-formaldehyde plastic
(B) teflon (ii) antiknock compound
(C) tetraethyl lead (iii) polyamide polymer
(D) nylon-66 (iv) corrosion resistant
(E) bakelite (v) explosive

61. (A) Liebig method (i) estimation of halogens
(B) Ziesel method (ii) estimation of N
(C) Carius method (iii) estimation of C, H
(D) Kjeldahl method (iv) detection of N, S, halogens
(E) Lassaigne's test (v) estimation of OCH_3 group

62. (A) *n*-heptane (i) cetane No. 0
(B) isooctane (ii) cetane No. 100
(C) hexadecane (iii) octane No. 0
(D) α-methylnaphthalene (iv) octane No. 100
(E) triplane (v) octane No. 124

63. (A) metamers (i) $CH_3COCH_2COCH_3$, $CH_3COCHC(OH)CH_3$
(B) functional isomers (ii) isopentane, neopentane
(C) position isomers (iii) CH_3COOCH_3, CH_3CH_2COOH
(D) chain isomers (iv) *n*-propyl chloride, isopropyl chloride
(E) tautomers (v) $C_2H_5OC_2H_5$, $CH_3OC_3H_7$

64. (A) $CHCl_3$ (i) addition polymer
(B) aspirin (ii) vitamin C
(C) PVC (iii) polymer of glucose
(D) starch (iv) antipyretic
(E) ascorbic acid (v) anaesthetic

65. (A) sublimation (i) H_2O_2
(B) steam distillation (ii) benzoic acid
(C) fractional crystallisation (iii) benzene and toluene
(D) fractional distillation (iv) Na_2SO_4, $Na_2Cr_2O_7$
(E) distillation under reduced pressure (v) turpentine oil

66. (A) caprolactum (i) antimalerial
(B) zwitterion (ii) protective action of colloid
(C) CFC (iii) fire extinguisher
(D) nitrous oxide (iv) amino acid
(E) ampicillin (v) green house effect

(F) CCl_4

(vi) nylon-66

(vii) general anaesthetic

(viii) broad spectrum anaesthetic

(ix) antipyretic **(Assam J.A.T. 1999)**

67. (A) Isobars

(B) Isomorphs

(C) Alkynes

(D) Isomers

(E) Cycloalkanes

(i) C_nH_{2n-2} where $n = 1, 2, 3$

(ii) $^{40}_{20}Ca$ and $^{20}_{19}K$

(iii) $FeSO_4 \cdot 7H_2O$ and $MgSO_4 \cdot 7H_2O$

(iv) C_nH_{2n-2} where $n = 3, 4$

(v) C_2H_5OH and CH_3OCH_3

(Bihar C.E.E. 1999)

68. (A) R—CH=CH_2

(B) 

(C) 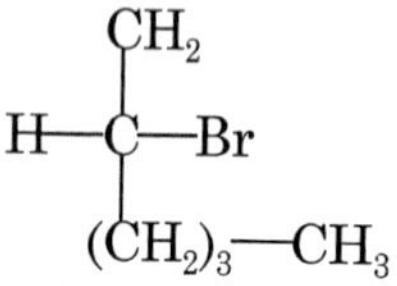

(D) 

(E) 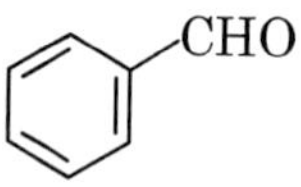

(F) 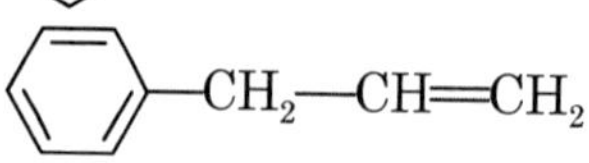

(i) 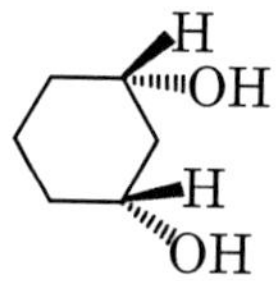

(ii) 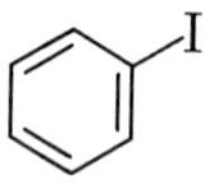

(iii) 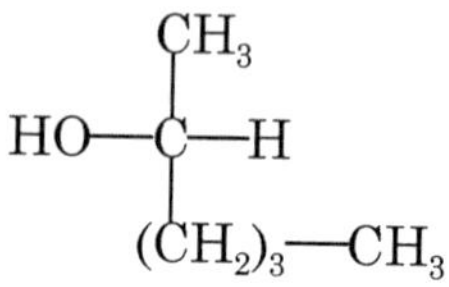

(iv) 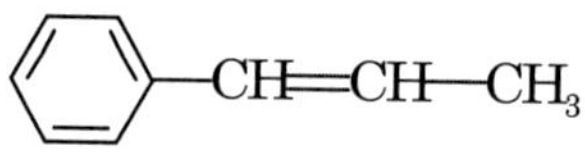

(v) 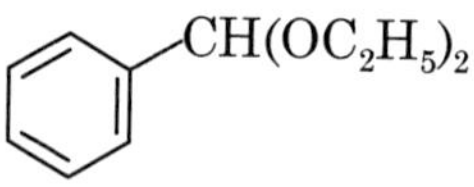

(vi) R—CHCl—CH_2—CCl_3

Suggest a suitable reagent for conversion of each pair. **(Roorkee 2001)**

69. (A) Benzoic acid

(B) 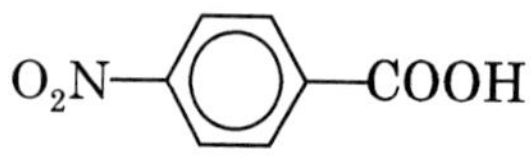

(C) 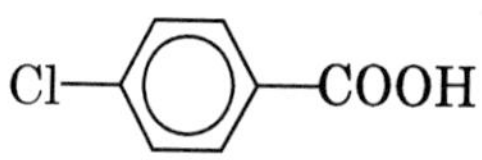

(D) 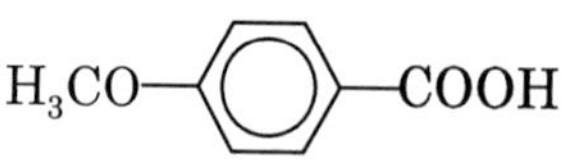

(E) 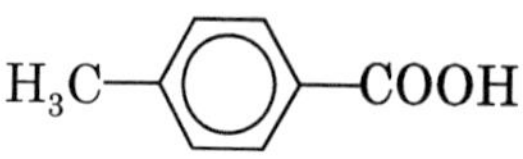

(i) $K_a = 3.3 \times 10^{-5}$

(ii) $K_a = 6.3 \times 10^{-5}$

(iii) $K_a = 30.6 \times 10^{-5}$

(iv) $K_a = 6.4 \times 10^{-5}$

(v) $K_a = 4.2 \times 10^{-5}$ **(I.I.T. 2006)**

70. (A) Compound X, B.P. = 100°C

(B) Compound Y, B.P. = 27°C

(i) $K_b = 0.53$

(i) $K_b = 0.68$

(C) Compound Z, B.P. = 253°C (i) $K_b = 0.98$

(Assume the molecular masses of X, Y and Z same.) **(I.I.T. 2003)**

Hint: $K_b = RT_b^2/1000 l_v = RT_b^2 M/\Delta H_v$ (since $l_v = \Delta H_v/M$)

71. (A) $PhCH_2CD_2Br$ on reaction with C_2H_5O gives $PH—CH{=}CD_2$ (i) *E*1 reaction
(B) $Ph—CH_2CH_2Br$ reacts faster than $Ph—CD_2CH_2Br$ on reaction with $C_2H_5O^-$ in ethanol (ii) *E*2 reaction
(C) $Ph—CHD—CH_2—Br$ on treatment with $C_2H_5D/C_2H_5O^-$ gives $Ph—CD{=}CH_2$ as the major product. (iii) *E*1*CB* reaction
(D) $PhCHBrCH_3$ and $PhCHBrCD_3$ react with the same rate. (iv) First order reaction

(I.I.T. 2006)

72. (A) Aufbau principle (i) Atomic orbital with one radial node
(B) Pauli exclusion principle (ii) 3*d* orbital is of higher energy than 4*s* ortital
(C) Hund's rule (iii) Atomic orbital defined by $n = 7,\ \ell = 1$
(D) 3*p* orbitals (iv) No two electrons in an atom can have all four quantum numbers identical
(E) *d*-orbitals (v) Distribution of elections is symmetrical
(F) 7*p* orbitals (vi) electron pairing is not possible until each orbital of the same subshell is occupied by a single election.

73. (A) XeF_2 (i) 180°
(B) XeF_4 (ii) 106°45′
(C) XeO_3 (iii) 109.5°
(D) XeO_4 (iv) 90°
(E) NH_3 (v) 104°31′
(F) H_2O (vi) 103°

74. (A) NH_4^+ (i) Polar covalent bond
(B) S_8 (ii) Coordinate bond
(C) HBr (iii) Covalent bond
(D) $Al_2(CH_3)_6$ (iv) Intermolecular H-bond
(E) HF (v) Electron deficient

75. (A) Ni (i) Covalent bond
(B) AgCl (ii) Ionic bond
(C) SiC (iii) van der Waals forces
(D) $NH_3(g)$ (iv) coordinate bond
(E) $H_3N—BF_3$ (v) metallic bond

76.

Column I	Column II
(A) The charge carried by	(i) Strongest reducing agent
(B) For a galvanic cell $E_{cell} > 0$	(ii) Liberation of O_2 at anode
(C) Li has the lowest standard electrode potential.	(iii) The cell is at equilibrium.
(D) Electrolysis of aqueous $CuCl_2$ using platinum electrodes	(iv) 1 Faraday
(E) Electrolysis of aqueous Na_2SO_4 using platinum electrodes	(v) Liberation of Cl_2 gas at anode

77.

Column I	Column II
(A) $[Co(NH_3)_4(H_2O)_2]Cl_2$	(i) Geometrical isomers
(B) $[Pt(NH_3)_2Cl_2]$	(ii) Paramagnetic
(C) $[Co(H_2O)_5Cl]Cl$	(iii) Diamagnetic
(D) $[Ni(H_2O)_6]Cl_2$	(iv) Metal ion with + 2 oxidation state

(I.I.T. 2007)

78.

Column I	Column II
(A) Hydrogen gas (P = 200 atm, T = 273 K)	(i) Compressibility factor = 1
(B) Hydrogen gas (P = 0, T = 273 K)	(ii) Attractive forces are dominant
(C) CO_2 (P = 1 atm, T = 273 K)	(iii) $PV = nRT$
(D) Real gas with very large molar volume	(iv) $P(V - ab) = nRT$

(I.I.T. 2007)

79.

Column I	Column II
(A) Cellulose	(i) Natural polymer
(B) Nylon-66	(ii) Synthetic polymer
(C) Protein	(iii) Amide linkage
(D) Sucrose	(iv) Glucoside linkage

(I.I.T. 2007)

80.

Column I	Column II
(A) $CH_3CH_2CH_2CN$	(p) Reduction with Pd–C/H_2
(B) $CH_3CH_2OCOCH_3$	(q) Reduction with $SnCl_2$/HCl
(C) $CH_3—CH{=}CH—CH_2OH$	(r) Development of foul smell on treatment with chloroform and alcohalic KOH
(D) $CH_3CH_2CH_2CH_2NH_2$	(s) Reduction with diisobutylaluminium hydride (DIBAL-H)
	(t) Alkaline hydrolysis

(I.I.T. 2009)

81.

Column I	Column II
(A) B_2	(p) Paramagnetic
(B) N_2	(q) Undergoes oxidation
(C) O_2^-	(r) Undergoes reduction
(D) O_2	(s) Bond order $\geq$ 2
	(t) Mixing of s and p orbitals

(I.I.T. 2009)

82. (A) 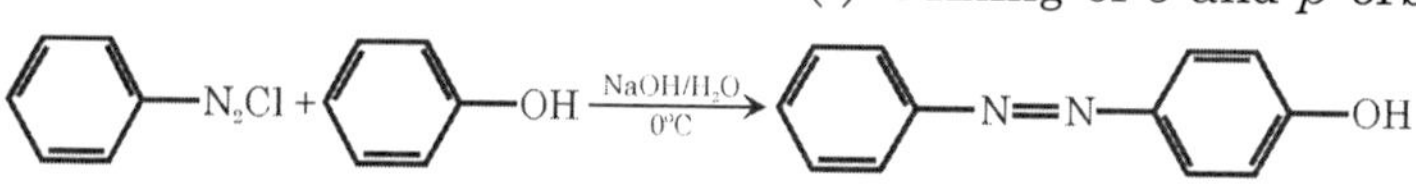

(p) Racemic mixture

(B) 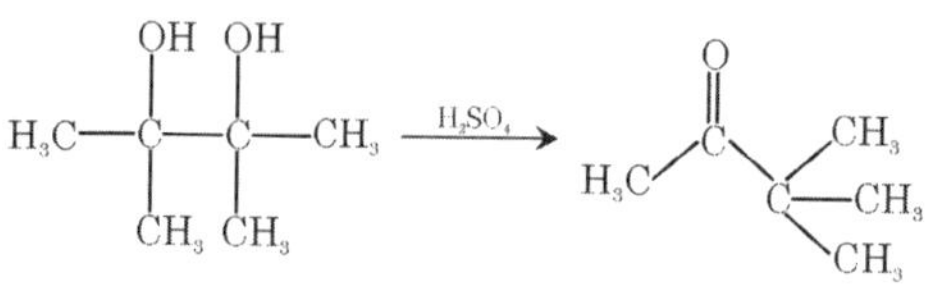

(q) Addition reaction

(C) 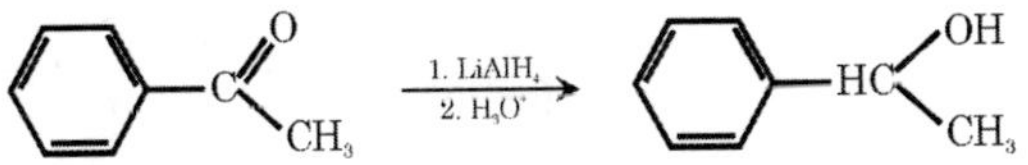

(D) 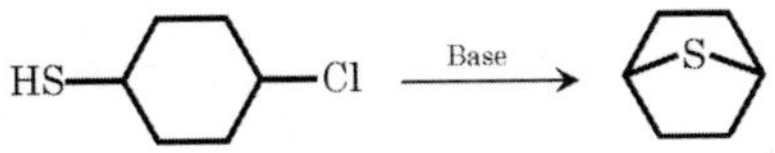

(r) Substitution reaction

(s) Coupling reaction

(t) Carbocation intermediate **(I.I.T. 2010)**

83. (A) $(CH_3)_2SiCl_2$

(B) XeF_4

(C) Cl_2

(D) VCl_5

(p) Hydrogen halide formation

(q) Redox reaction

(r) Reacts with glass

(s) Polymerization

(t) O_2 formation **(I.I.T. 2010)**

84. (A) $CO_2(s) \longrightarrow CO_2(g)$

(B) $CaCO_3(s) \longrightarrow CaO(s) + CO_2(g)$

(C) $2H\cdot \longrightarrow H_2(g)$

(D) $P_{(white,\ solid)} \longrightarrow P_{(red,\ solid)}$

(p) phase transition

(q) allotropic change

(r) ΔH is positive

(s) ΔS is positive

(t) ΔS is negative **(I.I.T. 2011)**

85. (A) 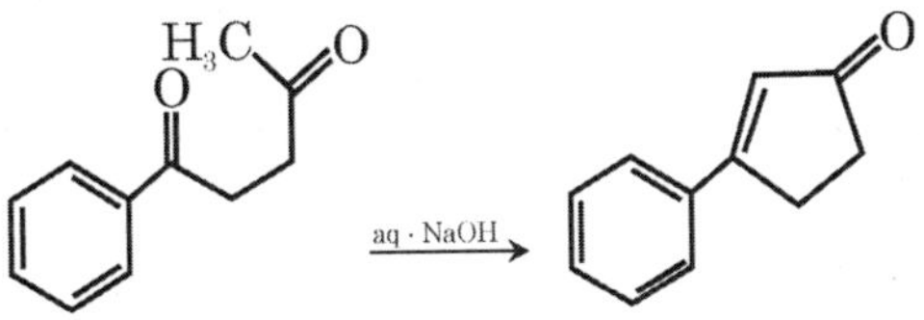

(B) 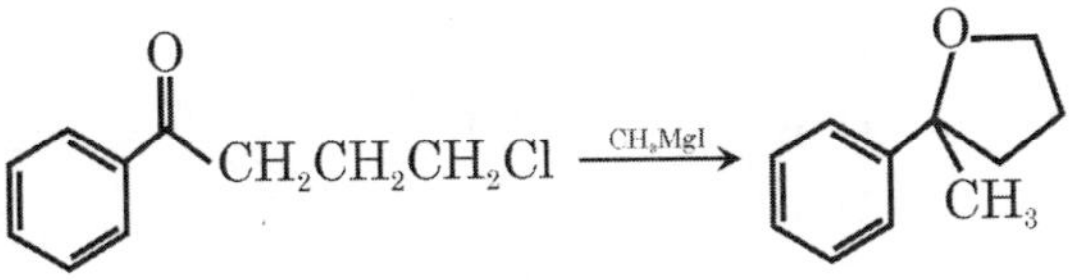

(C) 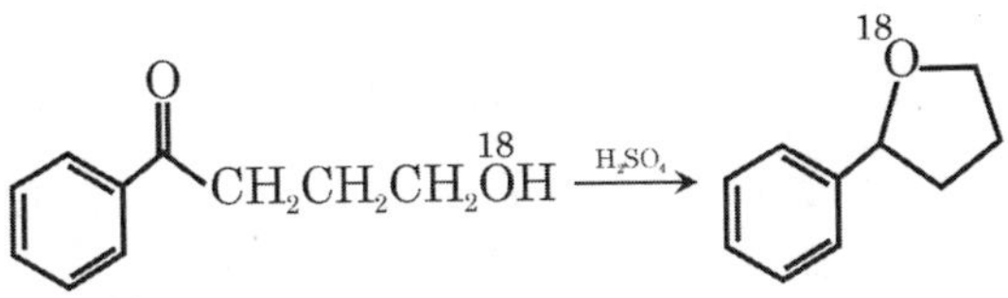

(D) 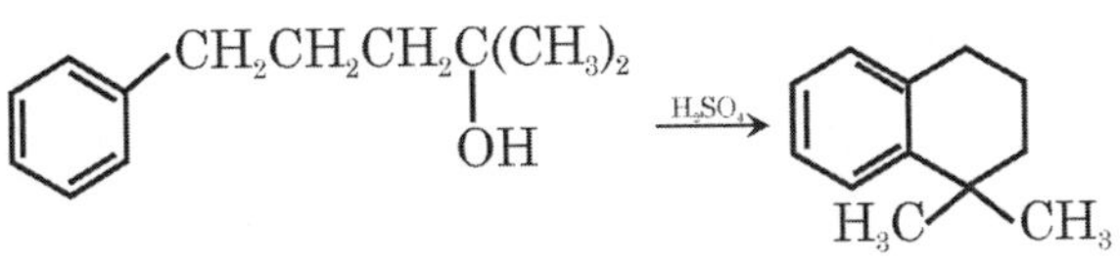

(p) Nucleophilic substitution

(q) Electrophilic substitution

(r) Dehydration

(s) Nucleophilic addition

(t) Carbanion **(I.I.T. 2011)**

86. (A) Cu + dil. HNO_3
    (B) Cu + conc. HNO_3
    (C) Zn + dil. HNO_3
    (D) Zn + conc. HNO_3

(p) NO
(q) NO_2
(r) N_2O
(s) $Cu(NO_3)_2$
(t) $Zn(NO_3)_2$ **(I.I.T. 2009)**

87. (A) 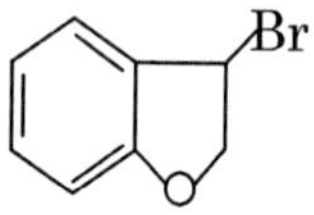

(B) 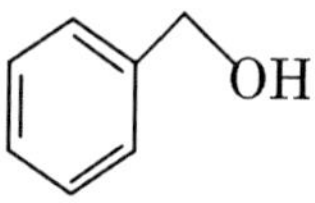

(C) 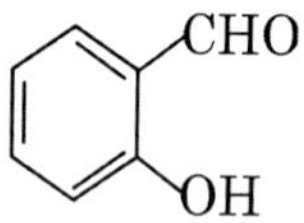

(D) 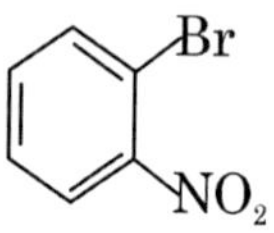

(p) Nucleophilic substitution
(q) Elimination
(r) Nucleophilic addition
(s) Esterification with acetic anhydride
(t) Dehydrogenation **(I.I.T. 2009)**

ANSWERS

1.	A, d	B, a	C, e	D, b	E, c	
2.	A, e	B, a	C, d	D, b	E, c	
3.	A, e	B, b	C, d	D, c	E, a	
4.	A, c	B, e	C, a	D, b	E, d	
5.	A, d	B, a	C, e	D, c	E, b	
6.	A, e	B, d	C, a	D, b	E, c	
7.	A, b	B, d	C, f	D, e	E, c	F, a
8.	A, d	B, a	C, b	D, e	E, c	
9.	A, d	B, a	C, e	D, b	E, c	
10.	A, d	B, a	C, b	D, c	E, e	
11.	A, b	B, c	C, d	D, a	E, e	
12.	A, d	B, e	C, a	D, b	E, c	
13.	A, d	B, e	C, a	D, b	E, c	
14.	A, d	B, c	C, e	D, a	E, b	
15.	A, d	B, c	C, e	D, a	E, b	
16.	A, d	B, e	C, a	D, b	E, c	
17.	A, d	B, a	C, b	D, e	E, c	

18. A, d, v	B, c, iv	C, e, i	D, a, iii	E, b, ii		
19. A, c, v	B, a, ii	C, b, iv	D, e, iii	E, d, i		
20. A. d, v	B, a, iv	C, b, iii	D, e, i	E, c, ii		
21. A, c, iv	B, c, i	C, a, ii	D, b, iii	E, d, v		
22. A, d	B, e	C, b	D, a	E, c		
23. A, c	B, d	C, e	D, a	E, b		
24. A, a	B, d	C, c	D, b	E, e		
25. A, e	B, a	C, b	D, d	E, c		
26. A. d	B, c	C, b	D, a			
27. A, d	B, c	C, e	D, b	E, a	F, f	
28. A, vii	B, iv	C, i	D, iii	E, vi	F, v	G, ii
29. A, iii	B, v	C, iv	D, ii	E, i		
30. A, vii	B, iv	C, v	D, i	E, vi	F, ii	G, iii
31. A, ii	B, iii	C, iv	D, i			
32. A, iv	B, ii	C, i	D, iii			
33. A, i	B, iii	C, ii	D, v	E, iv		
34. A, i, iii	B, i, iii	C, ii	D, iv			
35. A, ii	B, iv	C, i	D, ii, iii			
36. A, iii	B, ii	C, i	D, iv			
37. A, v	B, i	C, iv	D, ii	E, iii		
38. A, v	B, i	C, ii	D, iii	E, vi	F, iv	
39. A, iv	B, v	C, ii	D, iii	E, i		
40. A, vi	B, iv	C, iii	D, v	E, ii	F, i	
41. A, ii	B, i	C, iv	D, v	E, vi	F, iii	
42. A, iii	B, i	C, iv	D, ii			
43. A, ii	B, iii	C, i	D, iv			
44. A, iii	B, i	C, ii	D, v	E, iv		
45. A, iv	B, ii	C, v	D, iii	E, i		
46. A, iii	B, ii	C, iv	D, i	E, v		
47. A, v	B, iv	C, iii	D, i	E, ii		
48. A, v	B, i	C, iv	D, ii	E, iii		
49. A, b, iii	B, d, v	C, e, i	D, c, iv	E, a, ii		
50. A, c, i	B, b, iii	C, a, ii	D, d, v	E, e, iv		
51. A, c, ii	B, a, iv	C, b, i	D, d, iii	E, e, v		
52. A, c, i	B, e, v	C, b, iii	D, d, ii	E, a, iv		
53. A, a, ii	B, c, i	C, d, iii	D, e, iv	E, b, v		
54. A, c, ii	B, e, i	C, b, iii	D, d, v	E, a, iv		
55. A, d, ii	B, c, i	C, e, iii	D, b, v	E, a, iv		
56. A, e, iv	B, d, i	C, b, ii	D, c, v	E, a, iii		

57. A, e, ii	B, d, v	C, a, iv	D, c, i	E, b, iii	
58. A, e, iii	B, a, iv	C, b, v	D, c, i	E, d, ii	
59. A, c, v	B, e, iv	C, a, iii	D, b, i	E, d, ii	
60. A, v	B, iv	C, ii	D, iii	E, i	
61. A, iii	B, v	C, i	D, ii	E, iv	
62. A, iii	B, iv	C, ii	D, i	E, v	
63. A, v	B, iii	C, iv	D, ii	E, i	
64. A, v	B, iv	C, i	D, iii	E, ii	
65. A, ii	B, v	C, iv	D, iii	E, i	
66. A, vi	B, iv	C, v	D, v, vii	E, viii	F, iii
67. A, ii	B, iii	C, v	D, i	E, iv	
68. A, vi	B, i	C, iii,	D, ii	E, v	F, iv
69. A, ii	B, iii	C, iv	D, i	E, v	
70. A, ii	B, i	C, iii			
71. A, ii	B, ii	C, ii	D, i, iv		
72. A, ii	B, iv	C, vi	D, i	E, v	F, iii
73. A, i	B, iv	C, vi	D, iii	E, v	F, ii
74. A, ii	B, iii	C, i	D, v	E, iv	
75. A, v	B, ii	C, i	D, iii	E, iv	
76. A, iv	B, iii	C, i	D, v	E, ii	
77. A, i, ii, iv	B, i, iii, iv	C, ii, iv	D, ii, iv		
78. A, i, iv	B, iii	C, i, ii	D, i, iv		
79. A, i, iv	B, ii, iii	C, i, iii	D, iv		
80. A, p, q, s, t	B, s, t	C, p	D, r		
81. A, p, r, t	B, s, t	C, p, q	D, p, q, s		
82. A, r, s, t	B, t	C, p, q	D, r		
83. A, p, s	B, p, q, r, t	C, p, q, t	D, p		
84. A, p, r, s	B, r, s	C, t	D, p, q, t		
85. A, r, s, t	B, p, s	C, r, s	D, q, r		
86. A, p, s	B, q, s	C, r, t	D, q, t		
87. A, p, q, t	B, p, s, t	C, r, s	D, p		

Part—IV

Practice Tests and Question Papers

Practice Tests

PRACTICE TEST 1

1. When an electron comes from $2E$ energy state to E energy state, the emitted wavelength is λ. If the transition occurs from $4E/3$ to E energy state, the emitted wavelength is
   (A) 3λ (B) $3\lambda/4$ (C) $\lambda/3$ (D) $4\lambda/3$
2. Naturally occurring boron consists of two isotopes whose atomic weights are 10.01 and 11.01. If the atomic weight of natural boron is 10.81, the percentage of each isotope in natural boron is
   (A) 50, 55 (B) 40, 60 (C) 20, 80 (D) 90, 10
3. The most ionic oxide among the following is
   (A) Mn_2O_7 (B) CrO_3 (C) P_4O_{10} (D) MnO
4. Identify the strongest reducing agent?
   (A) Na (B) K (C) Mg (D) I_2
5. The number of moles of any gas is given by
   (A) mnN_A (B) nN_A (C) n/V (D) N/N_A
6. A face centred cube is formed by atoms A and B. If atom A is present at the corner of a cube and the atom B is present at the faces of the cube, the resultant stoichiometry of the solid is
   (A) AB_4 (B) AB_3 (C) A_2B_3 (D) A_3B
7. Hydrochloric acid solutions in X and Y have concentrations $5N$ and $2N$ respectively. The volume of the solutions X and Y required to make 1 L of a $4N$ HCl in X and Y together is
   (A) (1/5) L of X and (1/2) L of Y (B) (1/3) L of X and (2/3) L of Y
   (C) (2/3) L of X and(1/3) L of Y (D) (1/4) L of X and (3/4) L of Y
8. The freezing point of a solution of glucose in water which has a boiling point of 100.36°C is ($K_f = 1.86$ K kg^{-1}, $K_b = 0.512$ K m^{-1})
   (A) – 1.30°C (B) – 13.0°C (C) – 0.13°C (D) – 1.13°C
9. If a solute undergoes dimerisation and trimerisation, the minimum values of the van't Hoff factor will be respectively
   (A) 0.25, 0.67 (B) 0.50, 0.33 (C) 1.50, 1.33 (D) 0.50, 1.50
10. K_p for the eqmuilibrium, $HgO(s) \rightleftharpoons Hg(g) + (1/2)O_2(g)$ is
    (A) $K_p = [1/3^{2/3}]p$ (B) $K_p = [1/3^{2/3}]p^{3/2}$ (C) $K_p = [2/3^{1/2}]p^{1/2}$ (D) $K_p = [2/3^{3/2}]p^{3/2}$

11. Which of the following buffers has a *p*H greater than 7?
    (A) $(NH_4)_2SO_4 + NH_3$ (B) $NaH_2PO_4 + Na_3PO_4$
    (C) $CH_3COOH + CH_3COONa$ (D) $Na_2CO_3 + NaHCO_3$
12. When the activation energies of the forward and backward reactions are equal, we have
    (A) $\Delta E = 0$ (B) $\Delta G = 0$, $\Delta E = 0$
    (C) $\Delta E = \infty$, $\Delta S = 0$ (D) $\Delta E = 0$, $\Delta S = 0$
13. Which of the following aqueous solutions will remain alkaline after electrolysis?
    (A) LiF (B) NaCl (C) KNO_3 (D) CH_3COONa
14. The equivalent conductance of 0.1*N* CH_3COOH is 50 $cm^2\ ohm^{-1}$ at 298 K while Λ_m° is 390 $cm^2\ ohm^{-1}$. The degree of dissociation of 0.1*N* CH_3COOH will be
    (A) 0.13 (B) 0.013 (C) 0.0013 (D) 0.05
15. Which of the following compounds is not easily hydrolysed?
    (A) $AsCl_3$ (B) NF_3 (C) $SnCl_4$ (D) PCl_3
16. Alumina is insoluble in water because
    (A) Al^{3+} and O^{2-} ions are not extensively hydrated
    (B) Al_2O_3 has low lattice energy and high enthalpy of hydration
    (C) Al_2O_3 has high lattice energy and low enthalpy of hydration
    (D) Al_2O_3 is a covalent compound.
17. The formula of inorganic benzene is
    (A) $B_3N_3H_3$ (B) $B_3N_6H_6$ (C) $B_3N_3H_6$ (D) $B_6N_6H_3$
18. The formula of the complex formed by the reaction of S^{2-} ion on sodium nitroprusside is
    (A) $Na_2[Fe(CN)_4SCN]$ (B) $Na_3[Fe(CN)_4NOS]$
    (C) $Na_2[Fe(CN)_5(NOS)_2]$ (D) $Na_4[Fe(CN)_5NOS]$
19. Which of the following oxides is basic in nature?
    (A) TiO_2 (B) V_2O_5 (C) NiO (D) FeO
20. Out of the following reactions, identify the reactions which are not electrophilic addition reactions.
    (A) $CH_3CH_3 + Cl_2 \longrightarrow C_2H_5Cl + HCl$
    (B) $CH_3CHO + HCN \longrightarrow CH_3CH(OH)CN$
    (C) $CH_3COCH_3 + HCN \longrightarrow (CH_3)_2C(OH)CN$
    (D) $CH_2{=}CH_2 + Br_2 \longrightarrow BrCH_2CH_2Br$
21. Anisole reacts with HI to form
    (A) iodobezene (B) methyl iodide and phenol
    (C) iodobenzene and methanol (D) phenol and methanol
22. D.D.T. is prepared by the action of
    (A) toluene and chlorine (B) benzene and chlorine
    (C) chloral and chlorobenzene (D) chloral and bromobenzene

23. The ester RCOOR on catalytic hydrogenation in presence of cupric oxide and copper chromite gives
    (A) RCH_2OR (B) RCOOH, R′OH
    (C) ROH, R′COOH (D) RCH_2OH, R′OH
24. The reduction of benzoyl chloride with $Pd/BaSO_4$ gives
    (A) benzoic acid (B) benzaldehyde (C) benzoyl chloride (D) none of these.
25. The electrolytic reduction of nitrobenzene gives
    (A) hydrazobenzene (B) *p*-aminophenol
    (C) phenylhydroxylamine (D) aniline
26. Which of the following will liberate CO_2 from a solution of $NaHCO_3$?
    (A) $C_6H_5NH_3^+Cl^-$ (B) $C_6H_5NH_2$ (C) C_6H_5COOH (D) CH_3CONH_2
27. Which of the following is used as a catalyst in Friedel–Crafts reaction?
    (A) HNO_3 (B) H_2SO_4 (C) BF_3 (D) $AlCl_3$
28. Heating of calcium formate and calcium benzoate gives
    (A) formaldehyde (B) benzophenone (C) acetophenone (D) benzaldehyde
29. Ethylamine on oxidation with acidified $KMnO_4$ produces a/an
    (A) ester (B) aldehyde (C) alcohol (D) nitro compound
30. The product obtained when nitrobenzene is reduced with $LiAlH_4$ is
    (A) Azobenzene (B) Aniline (C) hydrazobenzene (D) Azoxybenzene

The questions below (31 to 35) consist of an *assertion* in column 1 and reason in *column 2*. Use the following key to choose the appropriate answer.

(A) If both *assertion* and *reason* are correct, and the *reason* is the correct explanation of the assertion.
(B) If both *assertion* and *reason* are correct, but the *reason* is not the correct explanation of the *assertion*.
(C) If the *assertion* is correct, but the *reason* is incorrect.
(D) If the *assertion* is incorrect, but the *reason* is correct.

Assertion (column 1)	*Reason* (column 2)
31. $Li^+(aq)$ has larger ionic radius than that of $Na^+(aq)$.	$Li^+(aq)$ is relatively more extensively hydrated as compared to $Na^+(aq)$.
32. According to Le Chatelier's principle, increase in temperature shifts the equilibrium in the direction of endothermic reaction.	The enthalpy change for the endothermic reactions is always negative.
33. Na_2SO_4 is soluble in water but $BaSO_4$ is insoluble.	Lattice energy of $BaSO_4$ exceeds its hydration energy.
34. CH_3MgI should be prepared under perfectly anhydrous conditions.	Grignard reagents react very rapidly with water.
35. Benzaldehyde is more reactive than ethanol towards nucleophilic attack.	All the carbon atoms of benzaldehyde molecule are sp^3 hybridised.

36. Match the items in Column I with the items in Column II.

Column I	*Column II*
(A) Ice	(i) Ionic compound
(B) Solid CO_2	(ii) Insulator
(C) Graphite	(iii) Conducting covalent solid
(D) Silica	(iv) Molecular crystal
(E) Calcium Oxide	(v) Hydrogen bonded structure

37. Match the items in Column I with the items in Column II.

Column I (Electrical unit)	*Column II* (Single unit)
(A) volt-ampere	(i) volt
(B) ampere-second	(ii) watt
(C) joule/ampere : second	(iii) coulomb
(D) watt-ampere : ohm	(iv) ohm
(E) volt/ampere	(v) ampere

38. Match the items in Column I with the items in Column II.

Column I	*Column II*
(A) Free energy change of a reaction in the standard state (ΔG^0)	(i) $q_{rev}/T = (\Delta H/T)$
(B) Entropy change of a reaction (ΔS)	(ii) $-\left(\frac{d\Delta G}{dT}\right)_P$
(C) Enthalpy change of a reaction in the standard state (ΔH^0)	(iii) $-nFE$
(D) Free energy change of a reaction (ΔG)	(iv) $RT^2\left(\frac{d\ln K}{dT}\right)_P$
(E) Entropy change for an isoberic process (ΔS_i)	(v) $-RT\ln K$

ANSWERS

1. A	2. C	3. D	4. B	5. D	6. B
7. B	8. A	9. B	10. D	11. B, D	12. A
13. B, D	14. A	15. B	16. A	17. C	18. D
19. C, D	20. B, C, D	21. B	22. C	23. D	24. B
25. B, D	26. A,C	27. C, D	28. A, B, D	29. B	30. A
31. A	32. C	33. A	34. A	35. C	

36. A, v; B, iv; C, iii; D, ii; E, i

37. A, ii; B, iii; C, i; D, v; E, iv

38. A, v; B, ii; C, iv; D, iii; E, i

PRACTICE TEST 2

1. The energy levels of X, Y, Z of an atom correspond to the increasing energy values $E_X < E_Y < E_Z$. If λ_1, λ_2 and λ_3 are the wavelength of radiation corresponding to the transitions Z ⟶ Y, Y ⟶ X and Z ⟶ X respectively, identify the correct relation.
   (A) $\lambda_3^2 = \lambda_1^2\lambda_2^2$ (B) $\lambda_1 + \lambda_2 + \lambda_3 = 0$ (C) $\lambda_3 = \lambda_1 + \lambda_2$ (D) $\lambda_3 = \lambda_1\lambda_2(\lambda_1 + \lambda_2)$
2. In what ratio ${}^{35}_{17}Cl$ and ${}^{37}_{17}Cl$ be present so as to obtain ${}^{35.5}_{17}Cl$?
   (A) 3 : 1 (B) 1 : 3 (C) 1 : 2 (D) 1 : 1/3
3. Naturally occurring carbon consists of two isotopes ^{12}C and ^{13}C. Calculate the percentage abundance of ^{12}C and ^{13}C in a sample of carbon whose atomic weight is 12.01112. Assume the masses of ^{12}C and ^{13}C as 12 and 13.00334 u respectively.
   (A) 65%, 35% (B) 60%, 30% (C) 20%, 80% (D) 98.89%, 1.11%
4. The ∠N—O—N bond angle is maximum in
   (A) NO_2 (B) N_2O_3 (C) NO_2 (D) NO_2^+
5. Identify the strongest oxidising agent among the following
   (A) $K_2Cr_2O_7$ (B) $KMnO_4$ (C) H_2O_2 (D) O_3
6. At low pressure van der Waals equation is reduced to
   (A) $PV_m = RT$ (B) $Z = \frac{PV_m}{RT} = 1 - \frac{a}{RT}$
   (C) $Z = \frac{PV_m}{RT} = 1 + \frac{bP}{RT}$ (D) $Z = \frac{PV_m}{RT} = 1 - \frac{aP}{RT}$
7. The ratio of close packed atoms to tetrahedral holes in cubic close packing is
   (A) 2 : 1 (B) 1 : 3 (C) 1 : 2 (D) 1 : 1
8. A sample of $Na_2CO_3 \cdot xH_2O$ weighing 1.43 g is added to 100 mL of 0.1N H_2SO_4 solution. The value of x is
   (A) 6 (B) 10 (C) 5 (D) 7
9. For the reaction, $2A(g) + B(g) \rightleftharpoons 3C(g) + D(g)$, two moles of A and two moles of B were taken in 1 L flask. When the system attains equilibrium, the following must always be true.
   (A) [A] < [B] (B) [A] = [B] (C) [B] = [C] (D) [A] + [B] < [C] + [D]
10. Which of the following solutions will exhibit the highest osmotic pressure?
    (A) 0.05M NaCl (B) 0.05M $FeCl_3$ (C) 0.10M $BaCl_2$ (D) 0.05M Na_2SO_4
11. K_p for the equilibrium, $X(g) \rightleftharpoons Y(g) + C(g)$ is
    (A) $K_p = \alpha^2[(K_p + p)/p]$ (B) $K_p = \alpha^2(K_p + p)$
    (C) $K_p = \alpha^3 p$ (D) $K_p = \alpha^2(K_p + p + 1)$
12. Which of the following solutions can be titrated with sodium hydroxide as well as hydrochloric acid?
    (A) Aniline in water (B) Triethylamine in water
    (C) Pyruvic acid in water (D) Glycine in water.
13. The half-life period of a zero order reaction is
    (A) $t_{1/2} = a/2k$ (B) $t_{1/2} = 2k/a$ (C) $t_{1/2} = a^2/2k$ (D) $t_{1/2} = (1/2)[k/a^2]$

14. In an isothermal process, when heat is supplied to an ideal gas
    (A) the gas obeys the law of conservation of energy
    (B) the kinetic energy of the gas increases
    (C) the gas does negative work (D) the gas does positive work.
15. The standard EMF for the cell reaction, $Zn + Cu^{2+} \longrightarrow Zn^{2+} + Cu$ is 1.1 V at 25°C. The EMF of the cell reaction when 0.1*M* Cu^{2+} and 0.1*M* Zn^{2+} solutions are used at 25°C, is
    (A) 1.10 V (B) 0.10 V (C) – 1.10 V (D) – 0.11 V
16. The ionic size decreases as
    (A) $K^+ > S^{2-} > Sc^{3+} > V^{5+} > Mn^{7+}$ (B) $S^{2-} > K^+ > Sc^{3+} > V^{s+} > Mn^{7+}$
    (C) $Mn^{7+} > V^{5+} > Sc^{3+} > K^+ > S^{2-}$ (D) $Mn^{7+} > V^{5+} > Sc^{3+} > S^{2-} > K^+$
17. Strong heating of $MgCl_2 \cdot 6H_2O$ gives
    (A) Mg(OH)Cl (B) Mg(OH)·HCl
    (C) anhydrous $MgCl_2$ (D) MgO
18. Which of the following complex ions has square planar structure?
    (A) $[CoCl_4]^{2-}$ (B) $[Ni(CO)_4]$ (C) $[Cu(NH_3)_4]^{2+}$ (D) $[Ni(CN)_4]^{2-}$
19. The addition of 6*N* HC1 to a solution containing Hg_2^{2+}, Hg^{2+}, Pb^{2+} and Cd^{2+} leads to the precipitation of
    (A) $PbCl_2$ (B) Hg_2Cl_2 (C) Hg_2Cl_2, $PbCl_2$ (D) $HgCl_2$, $PbCl_2$
20. The compound formed by passing electric discharge through graphite in the presence of H_2 is
    (A) CH_4 (B) C_2H_2 (C) C_2H_6 (D) none of these.
21. Ethyl alcohol is obtained when ethyl chloride is boiled with
    (A) water (B) aqueous KOH (C) H_2O_2 (D) alcoholic KOH
22. When *n*-butyl cyclopentyl ether is heated with excess HBr at 373 K, the products formed are
    (A) *n*-butanol + bromocyclopentane (B) 1-bromobutane + bromocyclopentane
    (C) 1-bromobutane + cyclopentanol (D) *n*-butanol + cydopentanol
23. The major product formed when 2-chlorobutane is treated with alcoholic KOH, is
    (A) but-2-yne (B) but-1-yne (C) but-2-ene (D) but-1-ene
24. Which of the following reagents reduces carboxylic acid to primary alcohol?
    (A) H_2 (B) $NaBH_4$ (C) $Na + C_2H_5OH$ (D) $LiAlH_4$
25. The product formed when benzaldehyde reacts with cone. NaOH is
    (A) phenol (B) benzyl alcohol
    (C) sodium benzoate (D) none of these.
26. Which of the following will react with ethanolic KCN?
    (A) benzaldehyde (B) C_6H_SCl (C) CH_3COCl (D) CH_3Cl
27. Methyl iodide on treatment with ammonia gives
    (A) methylamine (B) dimethylamine
    (C) trimethylamine (D) none of these.

28. Which of the following reagents is used to convert *m*-dinitrobenzene to *m*-nitroaniline?
    (A) Fe/HCl (B) Sn/HCl (C) H_2/Pt (D) $(NH_4)_2S$
29. Which of the following reagents is used for converting $C_6H_5NO_2$ to $C_6H_5\underset{\underset{O}{\downarrow}}{N}{=}NC_6H_5$?
    (A) $LiAlH_4$/ether (B) Zn/NaOH
    (C) Glucose/NaOH (D) Na_3AsO_3/NaOH
30. Benzaldehyde reacts with ammonia to give
    (A) $C_6H_5CH{=}NH$ (B) $(C_6H_5CH{=}N)_2CHC_6H_5$
    (C) $C_6H_5CH(OH)NH_2$ (D) none of these.

The questions below (31 to 35) consist of an *assertion* in column 1 and *reason* in column 2. Use the following key to choose the appropriate answer.

(A) If both *assertion* and *reason* are correct, and the *reason* is the correct explanation of the *assertion*.
(B) If both *assertion* and *reason* are correct, but the *reason* is not the correct explanation of the *assertion*.
(C) If the *assertion* is incorrect, but the *reason* is incorrect.
(D) If the *assertion* is incorrect, but the *reason* is correct.

Assertion (column 1)	*Reason* (column 2)
31. Lithium resembles Mg.	Li^+ has almost same charge density as Mg^{2+}.
32. Heating finely divided silica with Mg produces crystalline variety of silicon.	Silicon crystallises in *fcc* lattice.
33. Electron affinity of oxygen is less than that of fluorine but greater than that of nitrogen.	Ionisation energy of nitrogen, oxygen and fluorine is in the order: N > O < F.
34. Cyclobutane is less stable than cyclopentane.	Presence of bent bonds causes loss of orbital overlap.
35. Ethyl bromide and alcoholic silver nitrate react to produce nitromethane as the major product.	NO_2^- ion is an ambident nucleophile.

36. Match the items in Column I with the items in Column II.

Column I (Reaction)	*Column II* (Method for determination of rate constant)
(A) $H_2O_2 \longrightarrow H_2O + \frac{1}{2}O_2$	(i) Radioactive decay method
(B) 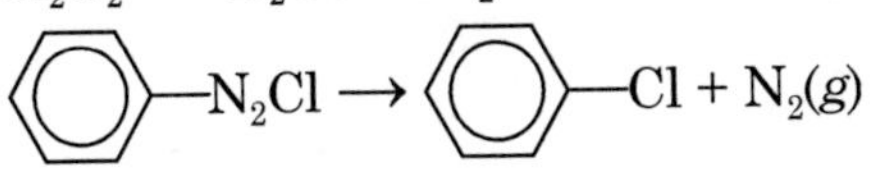	(ii) Volume measurement
(C) $CH_3COOC_2H_5 + H_2O \xrightarrow{H^+} CH_3COOH + C_2H_5OH$	(iii) Pressure measurement
(D) ${}^7_3Li + {}^1_1H \longrightarrow 2\,{}^4_2He$	(iv) Titration method
(E) $CH_3COOH(g) \longrightarrow CO_2(g) + CH_3CH_3(g)$	

37. Match the items in Column I with the items in Column II.

Column I	*Column II*
(A) Poling	(i) ZnS

(B) Cupellation — (ii) Refining of silver

(C) Froth floatation — (iii) Manufacture of Na_2CO_3

(D) Solvay process — (iv) Purification of copper

(E) van Arkel method — (v) Manufacture of NaOH

(vi) Purification of titanium

38. Match the items in Column I with the items in Column II.

Column I (Catalyst)	*Column II* (Use)
(A) $(C_2H_5)_3Al + TiCl_4$	(i) Wake process for conversion of C_2H_4 to CH_3CHO
(B) $[(C_6H_5)_3P]_2RhCl$	(ii) In the preparation of $(CH_3)_2SiCl_2$
(C) $PdCl_2$	(iii) Manufacture of synthetic petrol
(D) Cu	(iv) Hydrogenation of ethene
(E) 100 parts Co + 8 parts MgO + 5 parts ThO_2 + 200 parts keiselguhr	(v) Polymerisation of ethene

ANSWERS

1. D	2. A	3. D	4. D	5. D	6. D
7. C	8. B	9. A	10. C	11. B	12. D
13. A	14. D	15. A	16. B	17. D	18. C, D
19. C	20. A, B, C	21. A, B	22. B	23. C	24. B, C, D
25. B, C	26. A, D	27. C	28. D	29. C, D	30. B
31. A	32. D	33. B	34. C	35. B	

36. A, iv; B, ii; C, iv; D, i; E, iii

37. A, iv; B, ii; C, i; D, iii; E, vi

38. A, v; B, iv; C, i; D, ii; E, iii

PRACTICE TEST 3

1. If an electromagnetic radiation of $\lambda = 242$ nm is just sufficient to ionise sodium, the ionisation energy of sodium atom in kJ mol^{-1} is ($h = 6.6256 \times 10^{-34}$ J s)
(A) 0.494 kJ mol^{-1} (B) 4.94 kJ mol^{-1} (C) 494 kJ mol^{-1} (D) 49.4 kJ mol^{-1}
2. If N_0 is the number of initial nuclei present, the number of nuclei remaining undecayed at the end of nth half-life is
(A) n^2N_0 (B) $n^{-2}N_0$ (C) $2^{-n}N_0$ (D) 2^nN_0
3. Which of the following oxoacids of sulphur contains S—S bonds?
(A) $H_2S_2O_7$ (B) $H_2S_2O_3$ (C) $H_2S_2O_5$ (D) $H_2S_2O_4$
4. Consider the overlaps: *2s-2s, 2p-2p* and *2p-2s*. The bond strength decreases in the order
(A) *s-s* > *p-s* > *p-p* (B) *s-s* > *p-p* > *p-s* (C) *p-p* > *p-s* > *s-s* (D) *p-p* > *s-s* > *p-s*
5. Iron exhibits the + 1 oxidation state in
(A) $Fe(CO)_5$ (B) $[FeCl_4]^-$ (C) $Fe_4[Fe(CN)_6]_3$ (D) $[Fe(H_2O)_5NO]SO_4$
6. Two separate bulbs contain ideal gases X and Y. The molecular mass of X is half that of Y. If two gases are at the same temperature, the ratio of the pressure of X to that of Y is (Assume the density of X is twice that of Y.)
(A) ¼ (B) 4 (C) ½ (D) 2
7. In a face centred cubic crystal containing A and B atoms, A atoms are present at the corners of the unit cell and B atoms are present at the face centres. One of the A atoms is missing from one comer in the unit cell. The stoichiometry of the compound is
(A) AB_6 (B) $A_{24}B_7$ (C) A_7B_{24} (D) A_7B_3
8. Excess hydrogen sulphide is passed through an aqueous solution containing 1.245 g sample of $CuSO_4 \cdot xH_2O$. The filtrate contains the liberated H_2SO_4 which required 20 mL 0.5*N* NaOH solution for neutralisation. The value of x is
(A) 2 (B) 3 (C) 4 (D) 5
9. An aqueous solution contains 15% urea and 20% glucose. The freezing point of the solution is ($K_f = 1.86$ K kg^{-1})
(A) – 0.671°C (B) 6.71°C (C) 67.1°C (D) – 67.1°C
10. The partial pressure of hydrogen (p_{H_2}) in the equilibrium, $C(s) + H_2O(g) \rightleftharpoons CO(g) + H_2(g)$, is
(A) $p_{H_2} \propto p^2_{H_2O}/p_{CO}$ (B) $p_{H_2} \propto p^2_{H_2O}$
(C) $p_{H_2} \propto (p^2_{H_2O})^{1/2}$ (D) $p_{H_2} \propto p_{H_2O}$
11. Which of the following salts is expected to produce a neutral'solution in water?
(A) $HCOONH_4$ (B) $(NH_4)_2CO_3$ (C) CH_3COONH_4 (D) CH_3COONa
12. For the reaction, $N_2O_5 \rightleftharpoons 2NO_2 + (½)O_2$, $d[N_2O_5]/dt = k_1[N_2O_5]$, $d[N_2]/dt = k_2[N_2O_5]$ and $d[O_2]/dt = k_3[N_2O_5]$. The relation between k_1, k_2 and k_3 is
(A) $3k_1 = 2k_2 = 4k_3$ (B) $2k_1 = k_2 = 4k_3$
(C) $2k_1 = 4k_2 = k_3$ (D) $k_1 = k_2 = k_3$

13. A catalyst in a reaction cannot change
    (A) the enthalpy of the reaction
    (B) the equilibrium constant of the reaction
    (C) the activation energy of the reaction
    (D) the rate of reaction.
14. When a gas is ideal and the process is isothermal, we have
    (A) $w = 0$ (B) $H_1 \neq H_2$ (C) $E_1 = E_2$ (D) $P_1V_1 = P_2V_2$
15. If half-cell reaction $A + e^- \longrightarrow A^-$ has a large negative reduction potential, it follows that
    (A) A is readily reduced (B) A is readily oxidised
    (C) A^- is readily reduced (D) A^- is readily oxidised.
16. Which of the following has the highest negative electron affinity?
    (A) Zn (B) $O \longrightarrow O^{2-}$ (C) $S \longrightarrow S^{2-}$ (D) $Se \longrightarrow Se^{2-}$
17. Which of the complexes is a high-spin complex?
    (A) $[Co(NH_3)_6]Cl_3$ (B) $K_3[CoF_6]$ (C) $K_3[Fe(CN)_6]$ (D) $K_2[Ni(CN)_4]$
18. Fe^{2+} and Fe^{3+} ions are distinguished by using the reagent
    (A) $Fe_4[Fe(CN)_6]$ (B) $K_3[Fe(CN)_6]$ (C) KSCN (D) dimethylglyoxime
19. Which of the following is/are covalent?
    (A) $SiCl_4$ (B) $SnCl_4$ (C) $SnCl_2$ (D) NaCl
20. The compound which gives most stable carbonium ion is
    (A) $CH_3CH_2CH_2CH_2OH$ (B) $(CH_3)_2CHCH_2OH$
    (C) $(CH_3)_3COH$ (D) $CH_3CH(OH)CH_2CH_3$
21. The impurity of ethyne in ethene is removed by
    (A) conc. H_2SO_4 (B) copper sulphate
    (C) ammoniacal $AgNO_3$ (D) ammoniacal Cu_2Cl_2
22. Identify the reducing agent for the transformation, $C_6H_5CH{=\!=}CHNO_2 \longrightarrow$ $C_6H_5CH_2CH_2NO_2$
    (A) $LiAlH_4$ (B) $NaBH_4$ (C) Pd/H_2 (D) $[(C_6H_5)_3P]_3RhCl/H_2$
23. Which of the following methods is used to prepare chlorobenzene?
    (A) Direct chlorination of toluene
    (B) Action of benzenediazonium chloride with CuCl/HCl
    (C) Action of HCl on phenol in presence of $ZnCl_2$
    (D) Action of PCl_5 on phenol.
24. Dichloromethane on heating with aqueous NaOH gives
    (A) HCHO (B) HCOONa (C) CH_3OH (D) CH_3COONa
25. Acetic acid forms acetyl chloride on reaction with
    (A) $COCl_2$ (B) $SOCl_2$ (C) PCl_3 (D) PCl_5
26. Which of the following reagents is used for the oxidation of toluene to benzaldehyde?
    (A) CrO_3 (B) CrO_2Cl_2
    (C) $KMnO_4 + H_2SO_4$ (D) $K_2Cr_2C_7 + NaCl + H_2SO_4$

27. Benzene on heating in the presence of oxygen using V_2O_5 catalyst produces
    (A) CO (B) $CO_2 + H_2O$
    (C) maleic anhydride (D) none of these.
28. Ethylamine reacts with nitrosyl chloride to give
    (A) $C_2H_5NO_2$ (B) C_2H_5ONO (C) C_2H_5OH (D) C_2H_5Cl
29. The order of acidity of (I) formic acid, (II) benzoic acid, (III) acetic acid and (IV) phenylacetic acid is
    (A) III > IV > II > I (B) I > II > IV > III
    (C) IV > III > II > I (D) I > II > III > IV
30. Methylethyl ketone is reduced to n-butane by
    (A) HI/red P at 423 K (B) Mg/Hg, H_2O
    (C) Wolff–Kishner reduction (D) Meerwein–Pondorf reduction.

The question below (31 to 35) consist of an *assertion* in column 1 and *reason* in column 2. Use the following key to choose the appropriate answer.
(A) If both *assertion* and *reason* are correct, and the *reason* is the correct explanation of the *assertion.*
(B) If both *assertion* and *reason* are correct, but the *reason* is not the correct explanation of the *assertion.*
(C) If the *assertion* is correct, but the *reason* is incorrect.
(D) If the *assertion* is incorrect, but the *reason* is correct.

Assertion (column 1)	*Reason* (column 2)
31. Among alkali metals the tendency to form peroxides increases with the increase in atomic number.	Larger cations form more stable crystal lattice with larger anions.
32. H_3PO_3 is a dibasic acid.	Two hydrogen atoms are linked directly to phosphorus atoms in H_3PO_3.
33. Ionisation energy of Be is less than that of boron.	The first electron released from Be is of *s*-orbital but that from B is of *p*-orbital.
34. Aniline hydrogen sulphate on heating forms a mixture of *o*- and *p*-amino-benzene sulphonic acids.	The sulphonic acid group is an electron withdrawing group.
35. Benzonitrile is prepared by the reaction of chlorobenzene with KCN.	Cyanide ion is a strong nucleophile.

36. Match the items in Column I with the items in Column II.

Column I	*Column II*
(A) XeF_2	(i) Trigonal pyramidal
(B) XeF_4	(ii) Linear
(C) XeF_6	(iii) Square planar
(D) XeO_3	(iv) Tetrahedral
(E) XeO_4	(v) Distorted octahedral

37. Match the items in Column I with the items in Column II.

Column I (Catalyst)	*Column II* (Use)
(A) V_2O_5	(i) Manufacture of H_2
(B) H_3PO_4 + keiselguhr	(ii) Manufacture of CH_3OH from CO and H_2
(C) Cr_2O_3 + ZnO	(iii) Manufacture of Cl_2
(D) $CuCl_2$	(iv) Polymerisation of alkenes
(E) Cr_2O_3 + Fe_2O_3	(v) Manufacture of CH_3COOH from CH_3CHO and O_2

38. Match the items in Column I with the items in Column II.

Column I (Compound)	*Column II* (Metal ion)
(A) Nitrogenase	(i) No metal ion is present
(B) Chlorophyll	(ii) Co(III)
(C) Vitamin B_{12}	(iii) Fe(III)
(D) Haemoglobin	(iv) Mg(II)
(E) Vitamin B_6	(v) Fe(II) and Mo(V)

ANSWERS

1. B	2. C	3. A	4. C	5. D	6. C
7. C	8. D	9. A	10. C	11. C	12. B
13. A	14. C, D	15. D	16. B	17. B	18. C
19. A, B	20. C	21. C, D	22. D	23. B	24. B, C
25. B, C, D	26. A, B, D	27. B	28. D	29. B	30. A, C
31. A	32. C	33. D	34. B	35. D	

36. A, ii; B, iii; C, v; D, i; E, iv
37. A, v; B, iv; C, ii; D, iii; E, i
38. A, v; B, iv; C, ii; D, iii; E, i

I.I.T. OBJECTIVE QUESTION PAPER—1999

1. The electrons, identified by quantum numbers n and l, (i) $n = 4, l = 1$ (ii) $n = 4, l = 0$ (iii) $n = 3, l = 2$ (iv) $n = 3, l = 1$ can be placed in order of increasing energy, from the lowest to highest, as
   (A) (iv) < (ii) < (iii) < (i) (B) (ii) < (iv) < (i) < (iii)
   (C) (i) < (iii) < (ii) < (iv) (D) (iii) < (i) < (iv) < (ii)
2. The number of neutrons accompanying the formation of $^{139}_{54}Xe$ and $^{94}_{38}Sr$ from the absorption of a slow neutron by $^{235}_{92}U$, followed by nuclear fission is
   (A) 0 (B) 2 (C) 1 (D) 3
3. The correct order of increasing C—O bond length of CO, CO_3^{2-}, CO_2 is
   (A) $CO_3^{2-} < CO_2 < CO$ (B) $CO_2 < CO_3^{2-} < CO$
   (C) $CO < CO_3^{2-} < CO_2$ (D) $CO < CO_2 < CO_3^{2-}$
4. A gas will approach ideal behaviour at
   (A) low temperature and low pressure
   (B) low temperature and high pressure
   (C) high temperature and low pressure
   (D) high temperature and high pressure.
5. The normality of $0.3M$ phosphorus acid (H_3PO_3) is
   (A) 0.1 (B) 0.9 (C) 0.3 (D) 0.6
6. The coordination number of a metal crystallising in a hexagonal close-packed structure is
   (A) 12 (B) 4 (C) 8 (D) 6
7. A gas X at 1 atm is bubbled through a solution containing a mixture of 1 MY^- and 1 MZ^- at 25°C. If the reduction potential of Z > Y > X, then
   (A) Y will oxidise X and not Z (B) Y will oxidise Z and not X
   (C) Y will oxidise both X and Z (D) Y will reduce both X and Z.
8. The pH of $0.1M$ solution of the following salts increases in the order
   (A) $NaCl < NH_4Cl < NaCN < HCl$ (B) $HCl < NH_4Cl < NaCl < NaCN$
   (C) $NaCN < NH_4Cl < NaCl < HCl$ (D) $HCl < NaCl < NaCN < NH_4Cl$
9. For the chemical reaction $3X(g) + Y(g) \rightleftharpoons X_3Y(g)$, the amount of X_3Y at equilibrium is affected by
   (A) temperature and pressure (B) temperature only
   (C) pressure only (D) temperature, pressure and catalyst.

10. In the dichromate dianion,
    (A) 4 Cr—O bonds are equivalent (B) 6 Cr—O bonds are equivalent
    (C) all Cr—O bonds are equivalent (D) all Cr—O bonds are non-equivalent.
11. One mole of calcium phosphide on reaction with excess water gives
    (A) one mole of phosphine (B) two moles of phosphoric acid
    (C) two moles of phosphine (D) one mole of phosphorous pentaoxide.
12. The oxidation number of sulphur in S_8, S_2F_2, H_2S respectively, is
    (A) 0, + 1 and – 2 (B) + 2, + 1 and – 2
    (C) 0, + 1 and + 2 (D) – 2, + 1 and – 2
13. On heating ammonium dichromate, the gas evolved is
    (A) oxygen (B) ammonia (C) nitrous oxide (D) nitrogen
14. In the commercial electrochemical process for aluminium extraction the electrolyte used is
    (A) $Al(OH)_3$ in NaOH solution
    (B) an aqueous solution of $Al_2(SO_4)_3$
    (C) a molten mixture of Al_2O_3 and Na_3AlF_6
    (D) a molten mixture of AlO(OH) and $Al(OH)_3$
15. The geometry of H_2S and its dipole moment are
    (A) angular and non-zero (B) angular and zero
    (C) linear and non-zero (D) linear and zero.
16. The geometry of $Ni(CO)_4$ and $Ni(PPh_3)_2Cl_2$ is
    (A) both square planar (B) tetrahedral and square planar respectively
    (C) both tetrahedral (D) square planar and tetrahedral respectively.
17. In compounds of type ECl_3, where E = B, P, As or Bi, the angles Cl—E—Cl for different E are in the order
    (A) B > P = As = Bi (B) B > P > As > Bi
    (C) B < P = As = Bi (D) B < P < As < Bi
18. In the compound $CH_2{=}CH{-}CH_2{-}CH_2{-}C{\equiv}CH$, the C_2—C_3 bond is of the type,
    (A) sp-sp^2 (B) sp^3-sp^3 (C) sp-sp^3 (D) sp^2-sp^3
19. When propanoic acid is treated with aqueous sodium bicarbonate, CO_2 is liberated. The C of CO_2 comes from
    (A) methyl group (B) carboxylic acid group
    (C) methylene group (D) bicarbonate.
20. The enol form of acetone, after treatment with D_2O, gives:
    (A) 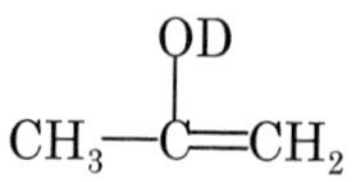 (B) 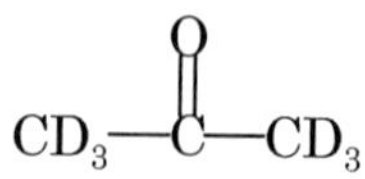
    (C) 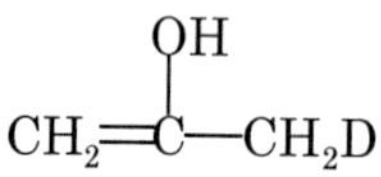 (D) 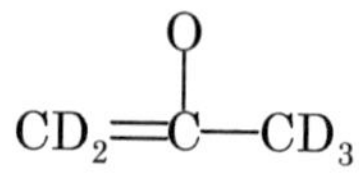

21. A positive carbylamine test is given by
    (A) N,N-dimethylaniline (B) 2,4-dimethylaniline
    (C) N-methyl-*o*-methylaniline (D) *p*-methylbenzylamine
22. The optically active tartaric acid is named as D-(+)-tartaric acid because it has a positive
    (A) optical rotation and is derived from D-glucose
    (B) *p*H in organic solvent
    (C) optical rotation and is derived from D-(+)-glyceraldehye
    (D) optical rotation only when substituted by deuterium.
23. A solution if (+)-2-chloro-2-phenylethene in toluene racemises slowly in the presence of small amount of $SbCl_5$ due to the formation of
    (A) carbanion (B) carbene (C) free radical (D) carbocation.
24. The product(s) obtained via oxymercuration ($HgSO_4 + H_2SO_4$) of but-l-yne would be
    (A) $CH_3—CH_2—\overset{\overset{O}{\|}}{C}—CH_3$ (B) $CH_3—CH_2—CH_2—CHO$
    (C) $CH_3—CH_2—CHO + HCHO$ (D) $CH_3CH_2COOH + HCOOH$
25. The most unlikely representation of resonance structures of *p*-nitrophenoxide ion is:
    (A) 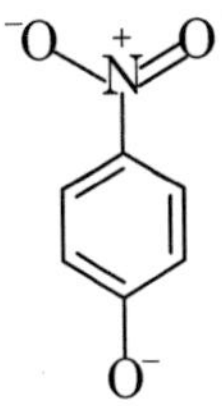 (B) 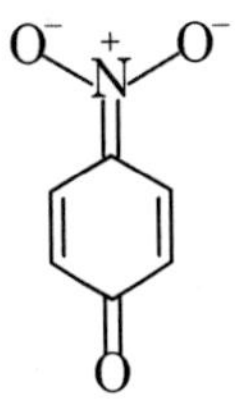 (C) 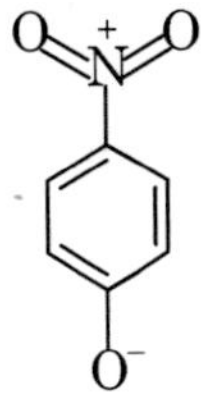 (D) 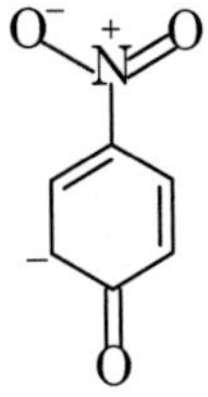

Directions: Question numbers 26–35 carry 3 marks each and may have more than one correct answer. All correct answers must be marked to get any credit in these questions.

26. The ether 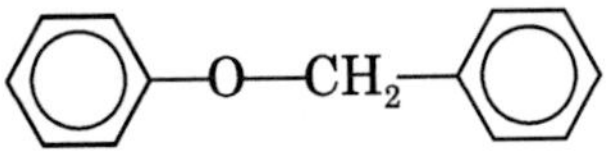 when treated with HI produces
    (A) 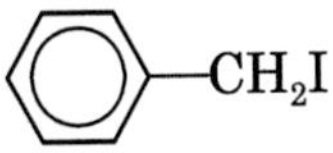 (B) 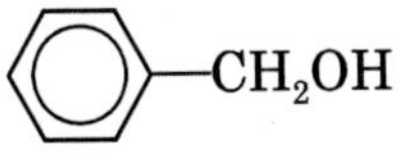
    (C) 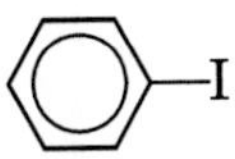 (D) 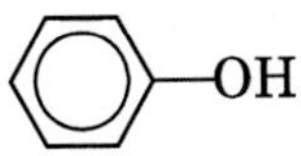
27. Toluene, when treated with Br_2/Fe, gives *p*-bromotoluene as the major product because the CH_3 group
    (A) is *para* directing (B) is *meta* directing
    (C) deactivates the ring (D) activates the ring by hyperconjugation.
28. The following statements(s) is(are) correct
    (A) A plot of log K_p versus $1/T$ is linear
    (B) A plot of log [X] *versus* time is linear for a first order reaction, $X \longrightarrow P$

(C) A plot of P *versus* $1/T$ is linear at constant volume

(D) A plot of P *versus* $1/V$ is linear at constant temperature

29. The following is(are) endothermic reaction(s):

(A) Combustion of methane (B) Decomposition of water

(C) Dehydrogenation of ethane to ethylene

(D) Conversion of graphite to diamond.

30. Ground state electronic configuration of nitrogen atom can be represented by

(A) [↓↑] [↓↑] [↑][↑][↑] (B) [↑↓] [↑↓] [↑][↓][↑] (C) [↓↑] [↓↑] [↑][↓][↓] (D) [↑↓] [↑↓] [↓][↓][↓]

31. In the depression of freezing point experiment, it is found that the

(A) vapour pressure of the solution is less than that of pure solvent

(B) vapour pressure of the solution is more than that of pure solvent

(C) only solute molecules solidify at the freezing point

(D) only solvent molecules solidify at the freezing point.

32. Ionic radii of

(A) $Ti^{4+} < Mn^{7+}$ (B) $^{35}Cl^- < {}^{37}Cl^-$ (C) $K^+ > Cl^-$ (D) $P^{3+} > P^{5+}$

33. Ammonia, on reaction with hypochlorite anion, can form

(A) NO (B) NH_4Cl (C) N_2H_4 (D) HNO_2

34. A buffer solution can be prepared from a mixture of

(A) sodium acetate and acetic acid in water

(B) sodium acetate and hydrochloric acid in water

(C) ammonia and ammonium chloride in water

(D) ammonia and sodium hydroxide in water.

35. An aromatic molecule will

(A) have $4n$ π electrons (B) have $(4n + 2)$ π electrons

(C) be planar (D) be cyclic.

ANSWERS

1. A	2. D	3. D	4. C	5. D	6. A
7. A	8. B	9. A	10. B	11. C	12. A
13. D	14. C	15. A	16. C	17. B	18. D
19. D	20. B	21. B, D	22. C	23. D	24. A
25. C	26. A, D	27. A, D	28. A, B, D	29. B, C, D	30. A, D
31. A, D	32. D	33. C	34. A, C	35. B, C, D	

MODEL SOLUTIONS—I.I.T. 1999

1. A

(i) value of $(n + l) = 4 + 1 = 5$ (ii) value of $(n + l) = 4 + 0 = 4$

(iii) value of $(n + l) = 3 + 2 = 5$ (iv) value of $(n + l) = 3 + 1 = 4$

Thus $(n + l)$ value is lower for (ii) and (iv) but is equal in both. In (iv) the n value is lower than that in (ii) and hence (iv) is lower energy than (ii). Similarly, (iii) is lower energy than (i). Hence the order of increasing energy is : (iv) < (ii) < (iii) < (i).

2. D

$${}^{235}_{92}U + {}^{1}_{0}n \longrightarrow {}^{139}_{54}Xe + {}^{94}_{38}Sr + 3{}^{1}_{0}n$$

3. D

Bond order in CO_2 = 3 (from molecular orbital theory)

$$\text{Bond order in } CO_2 = \frac{\text{No. of bonds in all possible sides}}{\text{No. of resonating structures}} \text{ (from resonance)}$$

$$= \frac{4}{2} = 2$$

$$\text{Bond order in } CO_3^{2-} = \frac{4}{3} = 1.33 \text{ (from resonance)}$$

$$\text{Bond length} \propto \frac{1}{\text{Bond order}}$$

Hence the order of increasing bond length: $CO < CO_2 < CO_3^{2-}$

4. C

A gas will approach ideal behaviour at high temperature and low pressure.

5. D

Phosphorus acid is a dibasic acid and hence its molecular weight = 2 × Eq. wt. Thus for H_3PO_3, $1M = 2N$. Hence $0.3M = 0.6N$.

6. A

The coordination number of a metal crystallising in a hexagonal close-packed structure is 12.

7. A

The reduction potential of X, Y and Z is in the order: Z > Y > X.

On the basis of reduction potential, the following spontaneous reactions can be predicted:

(a) Z is reduced and X is oxidised. (b) Y is reduced and X is oxidised.

(c) Z is reduced and Y is oxidised.

Hence Y will oxidise X and not Z.

8. B

Among the given substances, HCl is the strongest acid. An aqueous solution of NH_4Cl is slightly acidic due to hydrolysis. NH_4Cl is a salt of a strong acid and a weak base.

$$NH_4Cl + H_2O \rightleftharpoons NH_4OH + HCl \qquad ([OH^+] > [OH^-])$$

An aqueous solution of NaCN is basic due to hydrolysis. NaCN is salt of a weak acid and a strong base.

$$NaCN + H_2O \rightleftharpoons NaOH + HCN \qquad ([OH^-] > [H^+])$$

An aqueous solution of NaCl is neutral since it is a salt of a strong acid and a strong base. Hence the increasing order of pH is:

$$HCl < NH_4Cl < NaCl < NaCN$$

9. A

The equilibrium is affected by temperature and pressure due to change in heat and change in volume of substance, respectively.

10. B

The structure of dichromate ion $\left(Cr_2O_7^{2-}\right)$ is as follows:

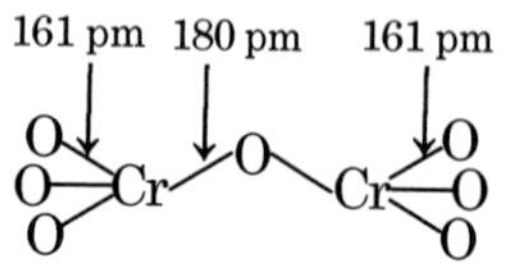

In dichromate ion, all the six non-bridging Cr—O bonds are equivalent and two bridged Cr—O bonds are equivalent.

11. C

$$Ca_3P_2 + 6H_2O \longrightarrow 3Ca(OH)_2 + \underset{\text{two moles}}{2PH_3}$$

12. A

The oxidation number of S in $S_8 = 0$

The oxidation number of S in $S_2F_2 = +1$

The oxidation number of S in $H_2S = -2$

13. D

$$(NH_4)_2Cr_2O_7 \xrightarrow{\Delta} Cr_2O_3 + 4H_2O + N_2$$

14. C

Na_3AlF_6 is used to decrease the melting point of alumina and to increase the conductivity of molten mass.

15. A

The S atom in H_2S is sp^3 hybridised and H_2S molecule is angular (V-shaped) giving rise to non-zero value of dipole moment.

16. C

Both $Ni(CO)_4$ and $Ni(PPh_3)_2Cl_2$ have sp^3 hybridisation and hence are tetrahedral.

${}_{28}Ni: 1s^2 2s^2 2p^6 3s^2 3p^6 3d^8 4s^2 \qquad Ni^{2+}: 1s^2 2s^2 2p^6 3s^2 3p^6 3d^8$

Ni atom in the ground state

$Ni(CO)_4$

Ni^{2+} ion

$Ni(PPh_3)_2Cl_2$

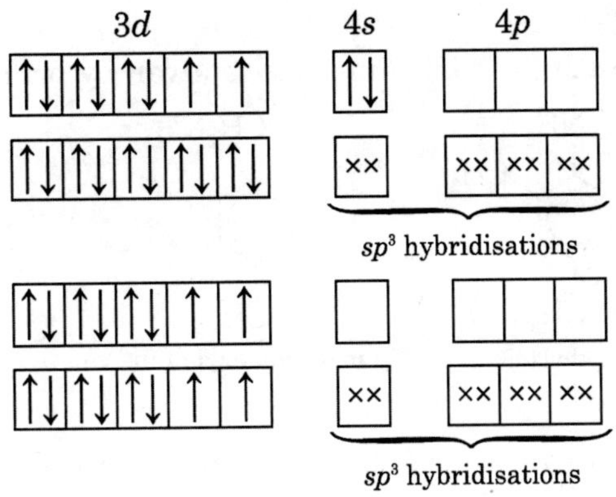

17. B

The structure, hybridisation and bond angles of BCl_3, PCl_3, $AsCl_3$ $BiCl_3$ are as follows:

Molecule	Structure	Hybridyisation	Bond angle
BCl_3	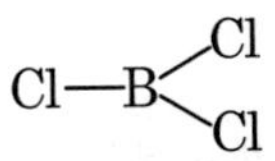 Triangular planar	sp^2	120°
PCl_3	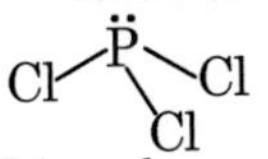 Trigonal pyramid	sp^3	100°
$AsCl_3$	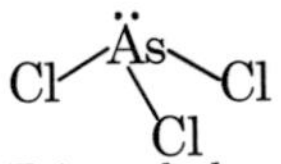 Trigonal planar	sp^3	98°
$BiCl_3$	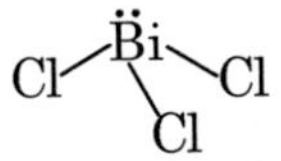	sp^3	96°

The electronegativity value of P, As and Bi is in the order Bi < As < P. As the electronegativity of the central atom increases, there occurs more repulsion between two bonding pairs (*bp-bp*) of electrons and this leads to greater bond angle in PCl_3 in comparison to $AsCl_3$ and $BiCl_3$. Hence the order of bond angle ($\angle$Cl—E—Cl) is $BCl_3 > PCl_3 > AsCl_3 > BiCl_3$.

18. D

$$\underset{sp^2}{\overset{1}{CH_2}}=\underset{sp^2}{\overset{2}{CH}}-\underset{sp^3}{\overset{3}{CH_2}}-\underset{sp^3}{\overset{4}{CH_2}}-\underset{sp}{\overset{5}{C}}\equiv\underset{sp}{\overset{6}{CH}}$$

Hence the C_2—C_3 bond is of the type sp^2-sp^3.

19. D

$$C_2H_5COOH + NaHCO_3 \longrightarrow C_2H_5COONa + H_2O + CO_2$$

20. B

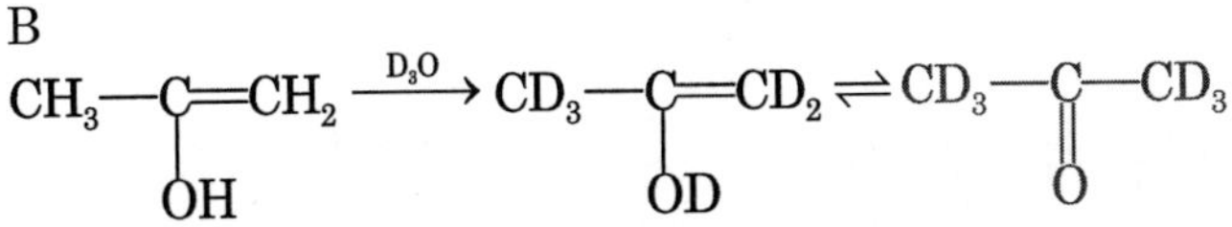

21. B, D

A positive carbylamine test is given by primary amines.

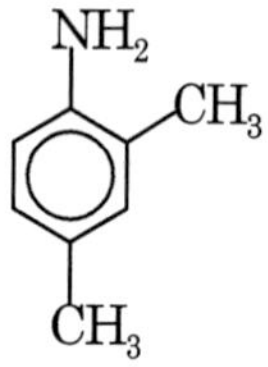

2,4-dimethylaniline

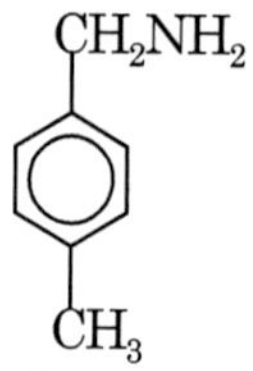

p-methylbenzylamine

22. C

D-(+)-tartaric acid has a positive optical rotation and its name is derived from D-(+)-glyceraldehye. The symbol D is assigned to the arrangement of —OH group on the right and the hydrogen atom on the left.

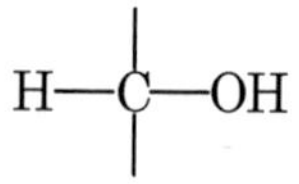

The + sign is used to represent the isomer which rotates the plane of polarised light towards right. D-(+)-tartaric acid is derived from D-(+)-glyceraldehye as shown below.

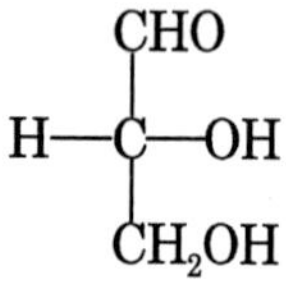

D-(+)-glyceraldehye

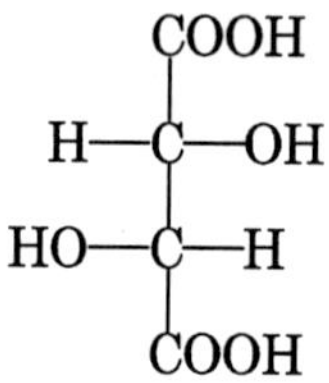

D-(+)-tartaric acid

23. D

$SbCl_5$ is used for the formation of intermediate carbocation.

24. A

$$CH_3—CH_2—C\equiv CH + H_2O \xrightarrow[\text{Markownikoff addition}]{HgSO_4 + H_2SO_4} \left[CH_3—CH_2—\underset{OH}{\underset{|}{C}}=CH_2\right] \xrightarrow{\text{tautomerisation}}$$

$$CH_3—CH_2—\underset{O}{\underset{\|}{C}}—CH_3$$

The enol form tautomerises because the keto form is more stable than the enol form.

25. C

The structure C is most unlikely because N atom has five bonds and a positive charge.

26. A, D

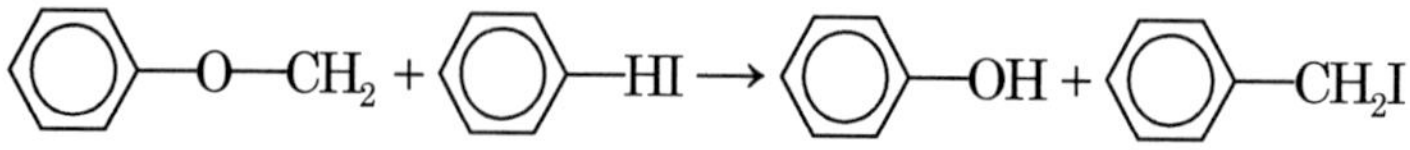

D A

27. A, D

The —CH_3 group being an electron-releasing group activates the benzene ring by hyperconjugation. The electron-releasing or activating group —CH_3 is *ortho* and *para* directing.

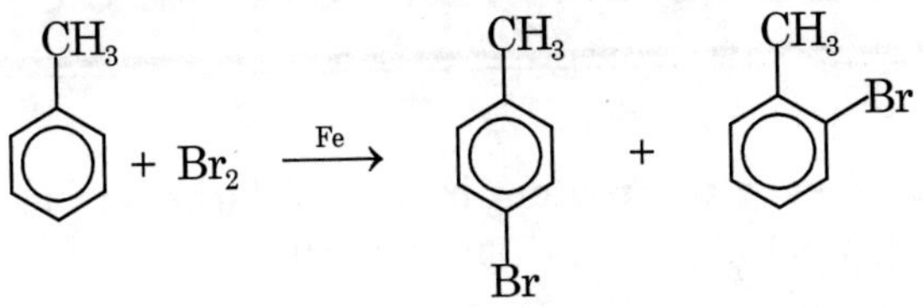

p-bromotoluene major product *o*-bromotoluene minor product

28. A, B, D

The relevant expressions are given below:

(A) $\log K_p = -\frac{\Delta H}{R} \cdot \frac{1}{T} + 1$

(B) $\log [X] = \log [X_0] + kt$

(C) $\frac{P}{T}$ = constant (at constant V)

(D) PV = constant (at constant T)

29. B, C, D

All combustion reactions are exothermic. The other three reactions are endothermic.

30. A, D

A and D follow Hund's rules. B and C do not follow Hund's rule.

31. A, D

32. D

Larger the effective nuclear charge, smaller is the radius.

33. C

$2NH_3 + OCl^- \longrightarrow NH_2—NH_2 + H_2O + Cl^-$

34. A, C

The mixture given in A and C can constitute buffer solution because a buffer solution is prepared by mixing a weak acid/weak base with salt of its conjugate base/acid.

35. B, C, D

An aromatic molecule will have:

(B) ($4n$ π) electrons (by Huckel's rule).

(C) planar structure (due to resonance).

(D) cyclic structure (due to the presence of sp^2 hybridised carbon atoms).

I.I.T. OBJECTIVE QUESTION PAPER—2000

1. Amongst the following identify the species with an atom in + 6 oxidation state.
   (A) MnO_4^- (B) $Cr(CN)_6^{3-}$ (C) NiF_6^{2-} (D) CrO_2Cl_2
2. The correct order of acidic strength is
   (A) $Cl_2O_7 > SO_2 > P_4O_{10}$ (B) $CO_2 > N_2O_5 > SO_3$
   (C) $Na_2O > MgO > Al_2O_3$ (D) $K_2O > CaO > MgO$
3. The correct order of radii is
   (A) N < Be < B (B) $F^- < O^{2-} < N^{3-}$ (C) Na < Li < K (D) $Fe^{3+} < Fe^{2+} < Fe^{4+}$
4. The number of P—O—P bonds in cyclic metaphosphoric acid is
   (A) zero (B) two (C) three (D) four
5. Ammonia can be dried by
   (A) conc. H_2SO_4 (B) P_4O_{10} (C) CaO (D) anhydrous $CaCl_2$
6. Molecular shapes of SF_4, CF_4, and XeF_4 are
   (A) the same, with 2, 0 and 1 lone pairs of electrons respectively
   (B) the same, with 1, 1 and 1 lone pairs of electrons respectively
   (C) different, with 0, 1, and 2 lone pairs of electrons respectively
   (D) different, with 1, 0 and 2 lone pairs of electrons respectively.
7. The chemical processes in the production of steel from haematite ore involve
   (A) reduction (B) oxidation
   (C) reduction followed by oxidation
   (D) oxidation followed by reduction.
8. The hybridisations of atomic orbitals of nitrogen in NO_2^+, NO_2^- and NH_4^+ are
   (A) *sp*, sp^3 and sp^2 respectively (B) *sp*, sp^2 and sp^3 respectively
   (C) sp^2, *sp* and sp^3 respectively (D) sp^2, sp^3 and *sp* respectively.
9. Amongst H_2O, H_2S, H_2Se and H_2Te, the one with the highest boiling point is
   (A) H_2O because of hydrogen bonding
   (B) H_2Te because of higher molecular weight
   (C) H_2S because of hydrogen bonding
   (D) H_2Se because of lower molecular weight.
10. Electrolytic reduction of alumina to aluminium by Hall–Heroult process is carried out
    (A) in the presence of NaCl (B) in the presence of fluorite
    (C) in the presence of cryolite which forms a melt with lower melting temperature
    (D) in the presence of cryolite which forms a melt with higher melting temperature.

11. The ΔH_f^0 for $CO_2(g)$, $CO(g)$ and $H_2O(g)$ is – 393.5, – 110.5 and – 241.8 kJ mol^{-1} respectively. The standard enthalpy change (in kJ) for the reaction $CO_2(g) + H_2(g) \longrightarrow CO(g) + H_2O(g)$ is

(A) 524.1 (B) 41.2 (C) – 262.5 (D) – 41.2

12. The rate constant for the reaction $2N_2O_5 \longrightarrow 4NO_2 + O_2$, is 3.0×10^{-5} s^{-1}. If the rate is 2.4×10^{-5} mol L^{-1} s^{-1}, then the concentration of N_2O_5 (in mol L^{-1}) is

(A) 1.4 (B) 1.2 (C) 0.04 (D) 0.8

13. For the electrohemical cell, $M|M^+||X^-|X$, E^0 (M^+/M) = 0.44 V and E^0 (X/X^-) = 0.33 V. From this data one can deduce that

(A) $M + X \longrightarrow M^+ + X^-$ is the spontaneous reaction

(B) $M^+ + X^- \longrightarrow M + X$ is the spontaneous reaction

(C) E_{cell} = 0.77 V (D) E_{cell} = – 0.77 V

14. For the reversible reaction, $N_2(g) + 3H_2(g) \rightleftharpoons 2NH_3(g)$ at 500°C, the value of K_p is 1.44×10^{-5} when partial pressure is measured in atmospheres. The corresponding value of K_c with concentration in mole $litre^{-1}$ is

(A) $1.44 \times 10^{-5}/(0.082 \times 500)^{-2}$ (B) $1.44 \times 10^{-5}/(8.314 \times 773)^{-2}$

(C) $1.44 \times 10^{-5}/(0.082 \times 773)^{2}$ (D) $1.44 \times 10^{-5}/(0.082 \times 773)^{-2}$

15. The rms velocity of hydrogen is $\sqrt{7}$ times the rms velocity of nitrogen. If T is the temperature of the gas,

(A) $T(H_2) = T(N_2)$ (B) $T(H_2) > T(N_2)$ (C) $T(H_2) < T(N_2)$ (D) $T(H_2) = \sqrt{7}\,T(N_2)$

16. The compressibility factor of a gas is less than unity at STP. Therefore.

(A) $V_m > 22.4$ litres (B) $V_m < 22.4$ litres

(C) $V_m = 22.4$ litres (D) $V_m = 44.8$ litres.

17. At 100°C and 1 atm, if the density of liquid water is 1.0 g cm^{-3} and that of water vapours is 0.0006 g cm^{-3}, then the volume occupied by water molecules in 1 L of steam at that temperature is

(A) 6 cm^3 (B) 60 cm^3 (C) 0.6 cm^3 (D) 0.06 cm^3

18. The electronic configuration of an element is $1s^2 2s^2 2p^6 3s^2 3p^6 3d^5 4s^1$. This represents its

(A) excited state (B) ground state (C) cationic form (D) anionic form.

19. The number of nodal planes in a p_x orbital is

(A) one (B) two (C) three (D) zero

20. When two reactants, A and B are mixed to give products C and D, the reaction quotient Q, at the initial stages of the reaction

(A) is zero (B) decreases with time

(C) is independent of time (D) increases with time.

21. Which of the following compounds will exhibit geometrical isomerism?

(A) 1-phenylbut-2-ene (B) 3-phenylbut-1-ene

(C) 2-phenylbut-1-ene (D) 1,1-diphenylprop-1-ene

22. Which of the following has the most acidic hydrogen?
(A) hexan-3-one (B) hexane-2,4-dione
(C) hexane-2,5-dione (D) hexane-2,3-dione

23. Among the following, the strongest base is
(A) $C_6H_5NH_2$ (B) p-NO_2—$C_6H_4NH_2$
(C) m-NO_2—$C_6H_4NH_2$ (D) $C_6H_5CH_2NH_2$

24. Which one of the following will most readily be dehydrated in acidic condition?
(A) 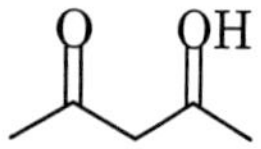 (B) 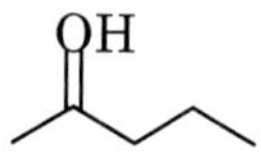 (C) 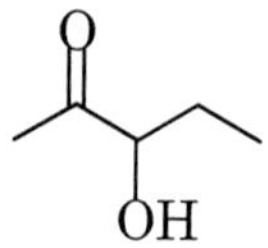 (D) 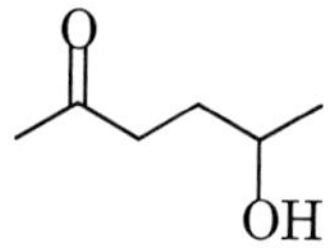

25. Which of the following has the highest nucleophilicity?
(A) F^- (B) OH^- (C) CH_3^- (D) NH_2^-

26. Propyne and propene can be distinguished by
(A) conc. H_2SO_4 (B) Br_2 in CCl_4 (C) dil. $KMnO_4$ (D) $AgNO_3$ in ammonia.

27. The order of reactivities of the following alkyl halides for an S_{N^2} reaction is
(A) RF > RCl > RBr > RI (B) RF > RBr > RCl > RI
(C) RCl > RBr > RF > RI (D) RI > RBr > RCl > RF

28. The appropriate reagent for the following transformation is

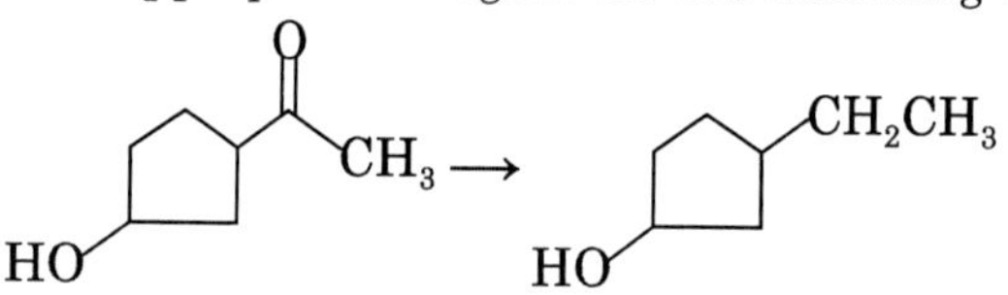

(A) Zn(Hg), HCl (B) NH_2NH_2, OH^- (C) H_2/Ni (D) $NaBH_4$

29. Benzoyl chloride is prepared from benzoic acid by
(A) Cl_2, $h\nu$ (B) SO_2Cl_2 (C) $SOCl_2$ (D) Cl_2, H_2O

30. Which one of the following alkenes will react fastest with H_2 under catalytic hydrogenation condition?
(A) 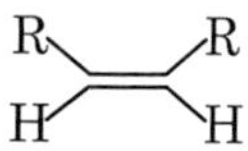 (B) 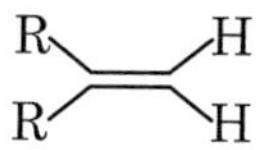 (C) 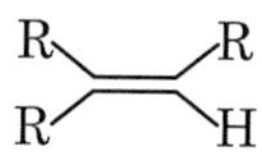 (D) 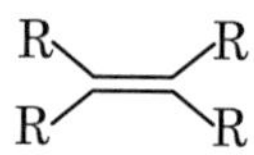

The questions below (31 to 35) consist of an *Assertion* and the *Reason*. Use the following key to choose the appropriate answer.

(A) If both *assertion* and *reason* are CORRECT, and *reason* is the CORRECT explanation of the *assertion*.
(B) If both *assertion* and *reason* are CORRECT, but reason is NOT the CORRECT explanation of the *assertion*.
(C) If *assertion* is CORRECT, but *reason* is INCORRECT.
(D) If *assertion* is INCORRECT, but *reason* is CORRECT.

31. *Assertion*: The pressure of a fixed amount of an ideal gas is proportional to its temperature.

Reason: Frequency of collisions and their impact both increase in proportion to the square root of temperature.

32. *Assertion*: The heat absorbed during the isothermal expansion of an ideal gas against vacuum is zero.

    Reason: The volume occupied by the molecules of an ideal gas is zero.

33. *Assertion*: But-1-ene on reaction with HBr in the presence of a peroxide produces 1-bromobutane.

    Reason: It involves the formation of a primary radical.

34. *Assertion*: Phenol is more reactive than benzene towards electrophilic substitution reaction.

    Reason: In the case of phenol, the intermediate carbocation is more resonance stabilised.

35. *Assertion*: The first ionization energy of Be is greater than that of B.

    Reason: 2*p* orbital is lower in energy than 2*s*.

ANSWERS

1. D	2. A	3. B	4. C	5. C	6. D
7. D	8. B	9. A	10. C	11. B	12. D
13. B	14. D	15. C	16. B	17. C	18. B
19. A	20. D	21. A	22. B	23. D	24. A
25. C	26. D	27. D	28. B	29. C	30. A
31. B	32. B	33. C	34. C	35. C	

MODEL SOLUTIONS—I.I.T. 2000

1. D

   The oxidation states of the central atom in the given species are as follows:

   $\overset{+7}{Mn}O_4^-$, $\overset{+3}{Cr}(CN)_6^{3-}$, $\overset{+4}{Ni}F_6^{2-}$, $\overset{+6}{Cr}O_2Cl_2$

   In CrO_2Cl_2, Cr has + 6 oxidation state.

2. A

   The oxides Cl_2O_7, SO_2 and P_4O_{10} are the anhydrides of $HClO_4$, H_2SO_3 and H_3PO_3 respectively. The acidic character of acids increases along a period from left to right:

   $HClO_4 > H_2SO_3 > H_3PO_3$

   The corresponding anhydrides also follow the same order: $Cl_2O_7 > SO_2 > P_4O_{10}$.

3. B

   F^-, O^{2-} and N^{3-} are isoelectronic species and in these ions the radius decreases with increase in atomic number due to increase in effective nuclear charge (Z_{eff}).

   $Z_{eff} = Z - \sigma$

   where Z = atomic number and σ = screening constant. Due to th presence of equal number of electrons, σ is constant in F^-, O^{2-} and N^{3-}. Hence the decreasing order of Z_{eff} is: $F^- > O^{2-} > N^{3-}$. Since radius $\propto \frac{1}{Z_{eff}}$, the increasing order of radius is: $F^- < O^{2-} < N^{3-}$.

4. C

   The structure of metaphosphoric acid is

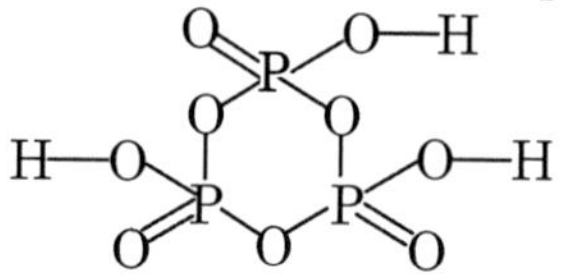

   Hence the number of P—O—P bonds is three.

5. C

6. D

   The hybridisation of the central atom in SF_4, CF_4 and XeF_4 is sp^3d, sp^3 and sp^3d^2 respectively. Hence the structure of SF_4, CF_4 XeF_4 is trigonal bipyramidal, tetrahedral and octahedral respectively. The structures of these molecules are as follows:

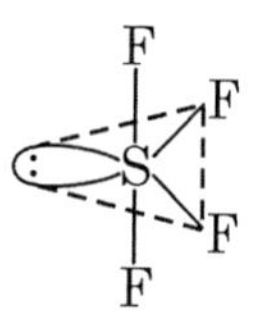

sp^3d, trigonal pyramidal, one lone pair

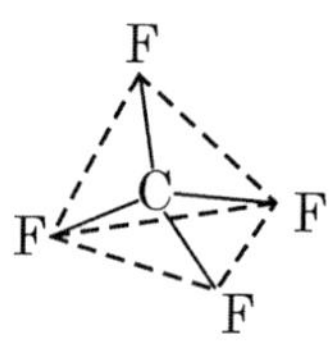

sp^3d, tetrahedral no lone pair

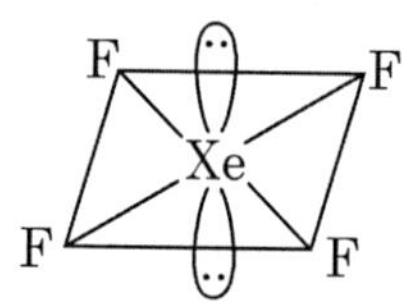

sp^3d^2, octahedral two lone pairs

7. D

Carbon which is added to haematite is first oxidised to CO and CO_2. The CO formed acts as a reducing agent for the reduction of haematite to Fe.

8. B

The structures of NO_2^+, NO_3^- and NH_4^+ ions are linear, triangular planar and tetrahedral respectively. Hence the hybridisation of atomic orbitals of nitrogen in these ions is sp, sp^2 and sp^3 respectively,

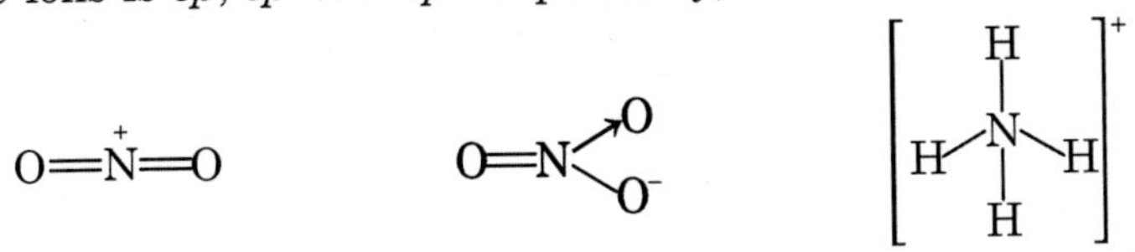

linear, sp triangular planar, sp^2 tetrahedral, sp^3

9. A

In spite of lower molecular mass, H_2O exhibits the highest boiling point due to the presence of strong intermolecular hydrogen bonding. H_2S, H_2Se and H_2Te do not form hydrogen bond.

10. C

Cryolite (Na_3AlF_6) is used in Hall–Heroult process to lower the melting point of alumina as well as to make the molten mass a good conductor of electric current.

11. B

$CO_2(g) + H_2(g) \longrightarrow CO(g) + H_2O(g),\ \Delta H^0 = ?$

$\Delta H^0 = \Delta H_f^0(CO, g) + \Delta H_f^0(H_2O, g) - \Delta H_f^0(CO_2, g)$

$= [-110.5 - 241.8 - (-393.5)]$ kJ mol^{-1} = 41.2 kJ mol^{-1}

12. D

$2N_2O_5(g) \longrightarrow 4NO_2(g) + O_2(g)$

The rate expression is given by

$r = K[N_2O_5]$

2.40×10^{-5} mol L^{-1} s^{-1} = $(3.0 \times 10^{-5}\ s^{-1})\ [N_2O_5]$

or $[N_2O_5] = \dfrac{2.40 \times 10^{-5}}{3.0 \times 10^{-5}}$ mol L^{-1} = 0.8 mol L^{-1}

13. B

For the electrochemical cell $M|M^+||X^-|X$, the cell reaction is derived as follows:

	$X^{-1} \longrightarrow X + e^-$	$E^0 = -0.33$ V
	$M^+ + e^- \longrightarrow M$	$E^0 = 0.44$ V
Add	$M^+ + X^- \longrightarrow M + X$	$E_{cell} = 0.11$ V

Since E^0_{cell} is positive, the cell reaction is spontaneous.

14. D

$N_2(g) + 3H_2(g) \rightleftharpoons 2NH_3(g)$

K_p and K_c are related by the equation,

$K_p = K_c(RT)^{\Delta n}$

where Δn is the change in stoichiometric number of gaseous speices in the equation and it is 2 – 4 = – 2. Hence

$$K_p = K_c(RT)^{-2}$$

$$\text{or} \quad K_c = \frac{K_p}{(RT)^{-2}} = \frac{1.44 \times 10^{-5}}{(0.082 \times 773)^{-2}}$$

15. C

Given that

$$u_{rms}(H_2) = \sqrt{7}\, u_{rms}(N_2)$$

$$\sqrt{\frac{3RT_{H_2}}{M_{H_2}}} = \sqrt{7}\sqrt{\frac{3RT_{N_2}}{M_{N_2}}}$$

$$\text{or} \quad \frac{T_{H_2}}{M_{H_2}} = 7 \times \frac{T_{N_2}}{M_{N_2}}$$

$$\text{or} \quad T_{H_2} = 7 \times \frac{M_{H_2}}{M_{N_2}} \times T_{N_2} = 7 \times \frac{2}{28} \times T_{N_2} = \frac{1}{2} T_{N_2}$$

Hence $T_{H_2} < T_{N_2}$

16. B

The compressibility factor of a gas,

$$Z = \frac{V_{m,\ real}}{V_{m,\ real}}$$

It is given that $Z < 1$ at STP. Hence

$V_{m,\ real} < V_{m,\ ideal}$ as in

But for an ideal gas at STP, $V_{m,\ ideal} = 22.4$ L

Hence $V_{m,\ real} < 22.4$ L

17. C

$$\text{Density } (\rho) = \frac{\text{Mass}}{\text{Volume}} = \frac{m}{V}$$

The ratio of denisty of liquid and vapour having same mass is

$$\frac{\rho_l}{\rho_v} = \frac{V_v}{V_l}$$

$$\text{or} \quad V_l = \frac{\rho_v \times V_v}{\rho_l} = \frac{0.0006 \times 1000 \text{ cm}^3}{1} = 0.6 \text{ m}^3$$

18. B

The electronic configuration represents the ground state of chromium. The half-filled orbital (d^5) is more stable.

19. A

Number of nodal planes in p_x orbital is one in the *yz* plane.

20. D

Reaction quotient, $Q = \frac{[C][D]}{[A][B]}$

In the initial stages of the reaction, the concentrations of C and D increase and the concentrations of A and B decrease. Hence the value of Q increases with time.

21. A

Only 1-phenylbut-2-ene will exhibit geometrical isomerism.

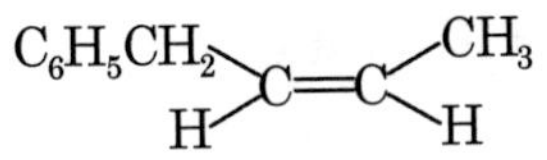

cis-1-phenylbut-2-ene

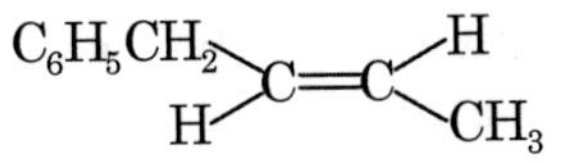

trans-1-phenylbut-2-ene

22. B

Only in hexane-2,4-dione, a methylene group is surrounded by two electron-withdrawing carbonyl groups. Hence its both hydrogen atoms are acidic in character.

23. D

Aliphatic amine is a stronger base than aromatic amine which is stabilised due to resonance. The electron-withdrawing benzene ring in aniline reduces the availability of lone pair of electron on nitrogen atom for protonation.

24. A

The dehydration of the alcohol (A) takes places readily due to the acidity of the α-hydrogen atom and also the product is stabilised due to the presence of conjugate double bonds.

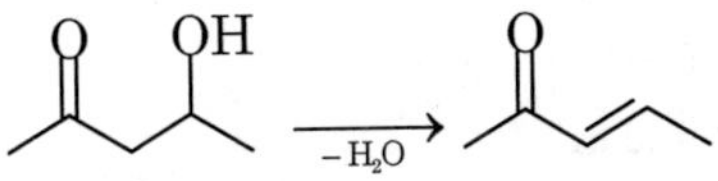

25. C

The increasing order of basic strength (nucleophilicity) is:

$F^- < OH^- < NH_3^- < CH_3^-$

Hence CH_3^- has the highest nucleophilicity.

26. D

Propyne forms white pecipitate of silver methyl acetylide on reaction with ammoniacal silver nitrate. Propene does not form acetylide due to the absence of acidic hydrogen atom.

$CH_3—C{\equiv}CH + NH_3 + AgNO_3 \rightarrow CH_3—C{\equiv}CAg\downarrow + NH_4NO_3$

$CH_3—CH{=}CH_2 + NH_3 + AgNO_3 \rightarrow$ No reaction

27. D

Of the halide ions, the iodide ion (I^-) is the best leaving group and the fluoride ion (F^-) is the poorest.

28. B

Wolff–Kishner reduction occurs in strongly basic medium and can be used for reduction of compounds which are sensitive to acid.

29. C

Thionyl chloride readily replaces —OH group of an acid by —Cl group producing acid chloride.

$$\underset{\text{Benzoic acid}}{C_6H_5COOH} + SOCl_2 \longrightarrow \underset{\text{Benzoyl chloride}}{C_6H_5COCl} + HCl + SO_2$$

30. A

A *cis*-alkene is more reactive than a *trans*-alkene due to greater strain at the double bond and least steric hindrance. The smaller the number of R substituents, lesser is the steric hindrance and faster is the rate of hydrogenation.

31. B

Both assertion and reason are correct but reason is not the correct explanation of assertion.

32. B

Both assertion and reason are correct but reason is not the correct explanation of assertion.

For an isothermal expansion, the work done by the gas is zero (w = 0) and internal energy of an ideal gas depends only on temperature.

33. C

Assertion is correct but reason is incorrect. In this reaction a secondary free radical is formed as an intermediate.

$$(C_6H_5CO)_2O_2 \longrightarrow 2C_6H_5COO\cdot$$

$$C_6H_5COO\cdot + HBr \longrightarrow C_6H_5COOH + Br\cdot$$

$$CH_3—CH_2—CH{=}CH_2 + Br\cdot \longrightarrow \underset{\text{Secondary free radical}}{CH_3—CH_2—\dot{C}H—CH_2Br}$$

$$CH_3—CH_2—\dot{C}H—CH_2Br + HBr \longrightarrow CH_3—CH_2—CH_2—CH_2Br + Br\cdot$$

34. C

Assertion is correct but reason is incorrect. In the electrophilic substitution reaction of phenol, the intermediates formed are oxonium ions, instead of carbocations.

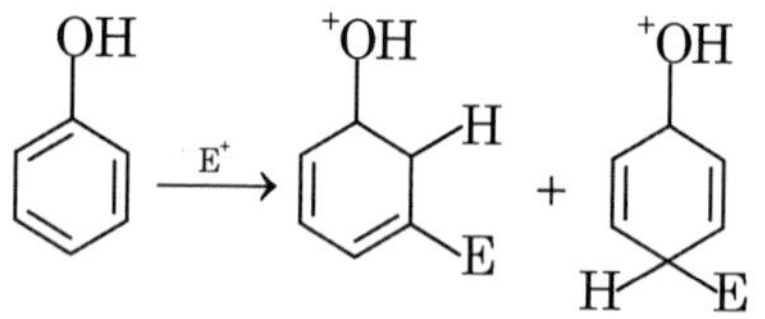

35. C

Assertion is correct but reason is incorrect. Boron ($1s^22s^22p^1$) contains one extra electron in $2p$ orbital in comparison to beryllium ($1s^22s^2$). The removal of $2p$ electron requires less energy in comparison to the removal of an electron from the more stable completely filled $2s$ orbital.

I.I.T. OBJECTIVE QUESTION PAPER—2001

Useful Data

Planck's constant, $h = 9.2626 \times 10^{-34}$ J s

Atomic numbers: Cr = 24, Mn = 25, Fe = 26, Co = 27, Pt = 78

1. In the presence of peroxide, hydrogen chloride and hydrogen iodide do not give anti-Markownikoff addition to alkenes because
   (A) both are highly ionic
   (B) one is oxidising and the other is reducing
   (C) one of the steps is endothermic in both the cases
   (D) all the steps are exothermic in both the cases.
2. The correct order of basicities of the following compounds

   1. $CH_3-C(=NH)NH_2$ 2. $CH_3-CH_2-NH_2$ 3. $(CH_3)_2NH$ 4. CH_3CONH_2 is

   (A) 2 > 1 > 3 > 4 (B) 1 > 3 > 2 > 4 (C) 3 > 1 > 2 > 4 (D) 1 > 2 > 3 > 4
3. The compound that will react most readily with NaOH to form methanol is
   (A) $(CH_3)_4N^+I^-$ (B) CH_3OCH_3 (C) $(CH_3)_3S^+I^-$ (D) $(CH_3)_3CCl$
4. The number of isomers for the compound with molecular formula $C_2BrClFI$ is
   (A) 3 (B) 4 (C) 5 (D) 6
5. In thermodynamics, a process is called reversible when
   (A) surroundings and system change into each other
   (B) there is no boundary between system and surroundings
   (C) the surroundings are always in equilibrium with the system
   (D) the system changes into the surroundings spontaneously.
6. In a solid AB having the NaCl structure, A atoms occupy the corners of the cubic unit cell. If all the face-centred atoms along one of the axes are removed, then the resultant stoichiometry of the solid is
   (A) AB_2 (B) A_2B (C) A_4B_3 (D) A_3B_4
7. The wavelength associated with a golf ball weighing 200 g and moving at a speed of 5 m/h is of the order
   (A) 10^{-10} m (B) 10^{-20} m (C) 10^{-30} m (D) 10^{-40} m
8. The quantum numbers + 1/2 and – 1/2 for the electron spin represent
   (A) rotation of the electron in clockwise and anticlockwise direction respectively
   (B) rotation of the electron in anticlockwise and clockwise direction respectively
   (C) magnetic moment of the electron pointing up and down respectively
   (D) two quantum mechanical spin states which have no classical analogue.

9. Which one of the following statements is false?
   (A) Work is a state function (B) Temperature is a state function
   (C) Change in the state is completely defined when the initial and final states are specified
   (D) Work appears at the boundary of the system.
10. The RMS velocity of an ideal gas at constant pressure varies with density (d) as
   (A) d^2 (B) d (C) $\sqrt{d}$ (D) $1\sqrt{d}$
11. The reaction, $3ClO^-(aq) \longrightarrow ClO_3^-(aq) + 2Cl^-(aq)$, is an example of
   (A) oxidation reaction (B) reduction reaction
   (C) disproportionation reaction (D) decomposition reaction.
12. The set with correct order of acidity is
   (A) $HClO < HClO_2 < HClO_3 < HClO_4$
   (B) $HClO_4 < HClO_3 < HClO_2 < HClO$
   (C) $HClO < HClO_4 < HClO_3 < HClO_2$
   (D) $HClO_4 < HClO_2 < HClO_3 < HClO$
13. An aqueous solution of 6.3 g oxalic acid dihydrate is made up to 250 mL. The volume of 0.1N NaOH required to completely neutralise 10 mL of this solution is
   (A) 40 mL (B) 20 mL (C) 10 mL (D) 4 mL
14. The number of S—S bonds in sulphur trioxide trimer (S_3O_9) is
   (A) three (B) two (C) one (D) zero
15. The chemical composition of slag formed during the smelting process in the extraction of copper is
   (A) $Cu_2O + FeS$ (B) $FeSiO_3$ (C) $CuFeS_2$ (D) $Cu_2S + FeO$
16. The common features among the species CN^-, CO and NO^+ are
   (A) bond order three and isoelectronic
   (B) bond order two and π-acceptors
   (C) isoelectronic and weak field ligands
   (D) bond order three and weak field ligands.

The questions below (17–21) consist of an *Assertion* in column 1 and the *Reason* in column 2. Use the following key to choose the appropriate answer.

(A) If both *assertion* and *reason* are correct, and *reason* is the correct explanation of the *assertion*
(B) If both *assertion* and *reason* are correct, but *reason* is not the correct explanation of the *assertion*
(C) If *assertion* is correct, but *reason* is incorrect
(D) If *assertion* is incorrect, but *reason* is correct.

Assertion (column 1)	*Reason* (column 2)
17. Addition of bromine to *trans* but-2-ene yields *meso*-2,3-dibromobutane.	Bromine addition to an alkene is an electrophilic addition.
18. Dimethylsulphide is commonly used for the reduction of an ozonide of an alkene to get the carbonyl compounds.	It reduces the ozonide giving water soluble dimethyl sulphoxide and excess of it evaporates.

19. In any ionic solid [MX] with Schottky defects, the number of positive and negative ions are same. Equal number of cation and anion vacancies are present.

20. In strongly acidic solutions, aniline becomes more reactive towards electrophilic reagents. The amino group being completely protonated in strongly acidic solution, the lone pair of electrons on the nitrogen is no longer available for resonance.

21. Between $SiCl_4$ and CCl_4, only $SiCl_4$ reacts with water. $SiCl_4$ is ionic and CCl_4 is covalent.

22. In the standardisation of $Na_2S_2O_3$ using $K_2Cr_2O_7$ by iodometry, the equivalent weight of $K_2Cr_2O_7$ is
(A) (molecular weight)/2 (B) (molecular weight)/6
(C) (molecular weight)/3 (D) same as molecular weight.

23. The complex ion which has no *d* electron in the central metal atom is
(A) $[MnO_4]^-$ (B) $[Co(NH_3)_6]^{3+}$ (C) $[Fe(CN)_6]^{3-}$ (D) $[Cr(H_2O)_6]^{3+}$

24. The set representing the correct order of first ionisation energy is
(A) K > Na > Li (B) Be > Mg > Ca (C) B > C > N (D) Ge > Si > C

25. The order of hybridisation of the central atom in the species NH_3, $[PtCl_4]^{2-}$, PCl_5 and BCl_3 is
(A) dsp^2, dsp^3, sp^2, sp^3 (B) sp^3, dsp^2, dsp^3, sp^2
(C) dsp^2, sp^2, sp^3, dsp^3 (D) dsp^2, sp^3, sp^2, dsp^3

26. For a sparingly soluble salt A_pB_q, the relationship of its solubility product (L_s) with its solubility (S) is
(A) $L_s = s^{p+q} \cdot p^q \cdot q^q$ (B) $L_s = s^{p+q} \cdot p^q \cdot q^p$
(C) $L_s = s^{pq} \cdot p^p \cdot q^p$ (D) $L_s = s^{pq} \cdot (pq)^{p+q}$

27. Saturated solution of KNO_3 is used to make salt-bridge because
(A) velocity of K^+ is greater than that of NO_3^-
(B) velocity of NO_3^- is greater than that of K^+
(C) velocity of K^+ and NO_3^- is nearly the same
(D) KNO_3 is highly soluble in water.

28. The order of equivalent conductance at infinite dilution of LiCl, NaCl and KCl is
(A) LiCl > NaCl > KCl (B) KCl > NaCl > LiCl
(C) NaCl > KCl > LiCl (D) LiCl > KCl > NaCl

29. If I is the intensity of absorbed light and C is the concentration of AB for the photochemical process $AB + h\upsilon \longrightarrow AB\cdot$, the rate of formation of AB· is directly proportional to
(A) C (B) I (C) I^2 (D) $C.I$

30. At constant temperature, the equilibrium constant (K_p) for the decomposition reaction $N_2O_4 \rightleftharpoons 2NO_2$ is expressed by $K_p = (4x^2P)/(1 - x^2)$, where P = pressure, x = extent of decomposition. Which of the following statements is true?
(A) K_p increases with increase of P (B) K_p increases with increase of x
(C) K_p increases with decrease of x (D) K_p remains constant with change in P and x.

31. A mixture of benzaldehyde and formaldehyde on heating with aqueous NaOH solution gives
    (A) benzyl alcohol and sodium formate
    (B) sodium benzoate and methyl alcohol
    (C) sodium benzoate and sodium formate
    (D) benzyl alcohol and methyl alcohol.
32. An S_{N^2} reaction at an asymmetric carbon of a compound always gives
    (A) an enantiomer of the substrate (B) a product with opposite optical rotation
    (C) a mixture of diastereomers (D) a single stereoisomer.
33. 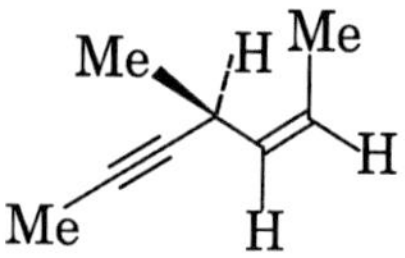

    Hydrogenation of the above compound in the presence of poisoned palladium catalyst gives
    (A) an optically active compound (B) an optically inactive compound
    (C) a racemic mixture (D) a diastereomeric mixture.
34. The reaction of propene with HOCl proceeds via the addition of
    (A) H^+ in the first step (B) Cl^+ in the first step
    (C) OH^- in the first step (D) Cl^- and OH^- in a single step.
35. 1-Propanol and 2-propanol can be best distinguished by
    (A) oxidation with alkaline $KMnO_4$ followed by reaction with Fehling solution
    (B) oxidation with acidic dichromate followed by reaction with Fehling solution
    (C) oxidation by heating with copper followed by reaction with Fehling solution
    (D) oxidation with concentrated H_2SO_4 followed by reaction with Fehling solution.

ANSWERS

1. C	2. B	3. A	4. D	5. C	6. D
7. C	8. D	9. A	10. D	11. C	12. A
13. A	14. D	15. B	16. A	17. B	18. A
19. A	20. D	21. C	22. B	23. A	24. B
25. B	26. A	27. C	28. B	29. B	30. D
31. A	32. D	33. B	34. B	35. C	

MODEL SOLUTIONS—I.I.T. 2001

1. C

   One of the steps in the reaction is endothermic and hence sufficient energy is not available for anti-Markownikoff addition of HCl or HI to alkenes in the presence of peroxide.

2. B

   The amine 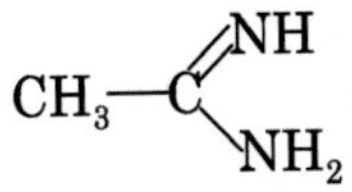 is most basic as its protonation leads to resonating cation having a large delocalisation energy.

   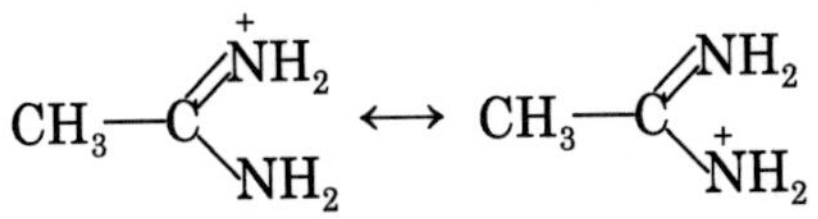

   Secondary amine is more basic than primary amine. Amides are less basic than primary amines. Hence the order of decreasing basicity is: 1 > 3 > 2 > 4.

3. A

   The positive change on nitrogen makes $(CH_3)_4N^+I^-$ most reactive towards nucleophilic attack.

   $(CH_3)_4N^+I^- + NaOH \longrightarrow (CH_3)_4\overset{+}{N}\overset{-}{O}H + NaI$

   $(CH_3)_4\overset{+}{N}\overset{-}{O}H \xrightarrow{\Delta} (CH_3)_3N + CH_3OH$

4. D

   The six isomers are:

   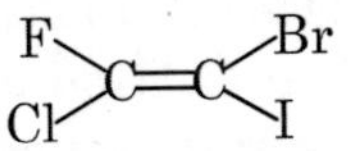
   1. Z isomer

   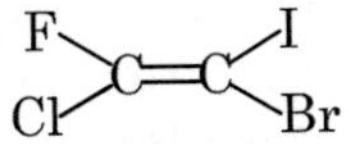
   2. E isomer

   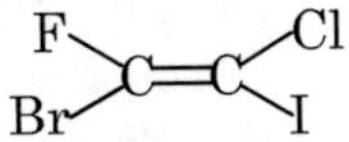
   3. Z isomer

   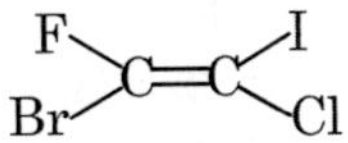
   4. E isomer

   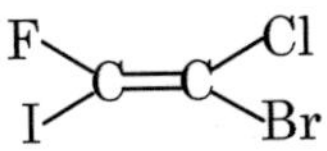
   5. Z isomer

   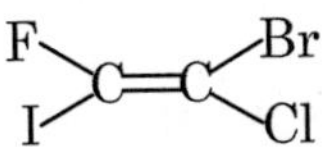
   6. E isomer

5. C

   In thermodynamics, a process is called reversible when the surroundings are always in equilibrium with the system.

6. D

   The structure of AB after the removal of all the face-central atoms along one of the axis is:

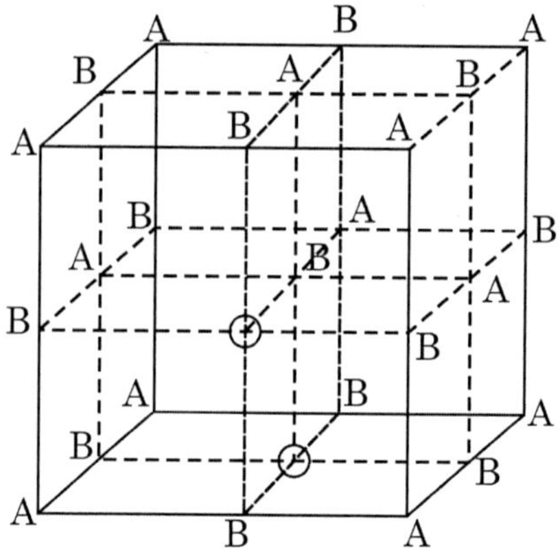

Number of A atoms per unit cell $= \underset{\text{from corners}}{8 \times \frac{1}{8}} + \underset{\text{from face-centre}}{4 \times \frac{1}{2}} = 1 + 2 = 3$

Number of B atoms per unit cell $= \underset{\text{from edges}}{12 \times \frac{1}{4}} + \underset{\text{from body-centre}}{1} = 3 + 1 = 4$

Hence the stoichiometry of the solid is A_3B_4.

7. C

de Broglie wavelength,

$$\lambda = \frac{h}{mv} = \frac{6.626 \times 10^{-34}\,\text{J s}}{(0.200\,\text{kg})[5\,\text{m}/(60 \times 60.5)]} = 2.4 \times 10^{-30}\ \text{m}$$

8. D

The magnetic spin quantum numbers $m_s = +\frac{1}{2}$ and $m_s = -\frac{1}{2}$ for the electron spin represent two quantum mechanical spin states which have no classical analogues. Classically, the two spin states are considered as the rotation of an electron on its own axis either in clockwise or anticlockwise direction. But spin is a purely quantum mechanical property and differs from its classical namesake.

9. A

Work is not a state function. Work has to be performed.

10. D

$$u_{rms} = \sqrt{\frac{3RT}{M}} = \sqrt{\frac{3PV}{M}} = \sqrt{\frac{3P}{M/V}} = \sqrt{\frac{3P}{d}}$$

Hence at constant pressure,

$$u_{rms} \propto \frac{1}{\sqrt{d}}$$

11. C

The oxidation number of Cl in ClO^-, ClO_3^-, and Cl^- is + 1, + 5 and – 1 respectively. In the reaction $3ClO^- \longrightarrow ClO_3^-, + 2Cl^-$, Cl in ClO^- undergoes oxidation as well as reduction reaction as follows:

oxidation: $ClO^- + 4OH^- \longrightarrow ClO_3^-, + 2H_2O + 4e^-$
reduction: $ClO^- + H_2O + 2e^- \longrightarrow Cl^- + 2OH^-$
Hence it is a disproportionation reaction.

12. A
The larger the oxidation number of Cl atom, the larger is the acidity. Hence the order of acidity is:
$HClO < HClO_2 < HClO_3 < HClO_4$

13. A

$$\text{Normality of oxalic acid} = \frac{6.3\ \text{g}/63\ \text{g equiv.}^{-1}}{0.25\ \text{L}} = 0.4N$$

$$N_1V_1 = N_2V_2$$
$$0.1 \times V_1 = 0.4 \times 10$$
or $$V_1 = 40\ \text{mL}$$

14. D
Number of S—S bonds = 0
The structure of sulphur trioxide trimer (S_3O_9) is:

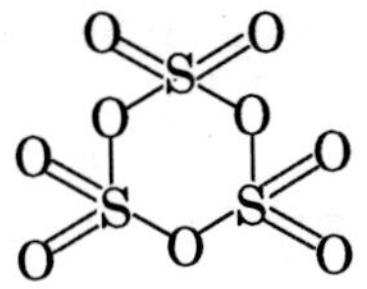

15. B
$FeO + SiO_2 \longrightarrow FeSiO_3$ (slag)

16. A
All the three species contain 14 electrons. Their molecular electronic configuration are:

$$(\sigma^b_{1s})^2 (\sigma^*_{1s})^2 (\sigma^b_{2s})^2 (\sigma^*_{2s})^2 (\pi^b_{p_x})^2 (\pi^b_{p_y})^2 (\pi^b_{p_z})^2$$

$$\text{Bond order} = \frac{\text{No. of bonding electrons} - \text{No. of antibonding electrons}}{2} = \frac{10-4}{2} = 3$$

17. B
Both assertion and reason are correct and reason is not the correct explanation of assertion.

18. A
Both assertion and reason are correct and reason is the correct explanation of assertion.

19. A
Both assertion and reason are correct and reason is the correct explanation of assertion.

20. D
Assertion is incorrect but reason is correct. Due to protonation of the $—NH_2$ group, the resultant $—NH_3^+$ becomes *meta* director. The reactivity of aniline decreases in concentrated acidic solution due to the protonation of the lone pair of electrons by H^+ ions and the lone pair of electrons is no longer available for resonance. Resonance is responsible for activity of a compound.

21. C

Assertion is correct but reason is incorrect. Hydrolysis of $SiCl_4$ is due to the presence of *d*-orbitals.

22. B

The reaction undergone by $Cr_2O_7^{2-}$ ion is:

$$Cr_2O_7^{2-} + 14H^+ + 6e^- \longrightarrow 2Cr^{3+} + 7H_2O$$

Equivalent weight of $K_2Cr_2O_7 = \dfrac{\text{Mol. wt.}}{\text{Decrease in O. N.} \times n}$

where n = number of atoms of Cr.

$$= \frac{\text{Mol. wt.}}{3 \times 2} = \frac{\text{Mol. wt.}}{6}$$

23. A

Contral atom	*Oxidation state*	*Number of electrons*	*Electronic Configuration*	*No. of d-electrons*
Mn	+ 7	25 – 7 = 18	$1s^2 2s^2 2p^6 3s^2 3p^6$	0
Co	+ 3	27 – 3 = 24	$1s^2 2s^2 2p^6 3s^2 3p^6 3d^6$	6
Fe	+ 3	26 – 3 = 23	$1s^2 2s^2 2p^6 3s^2 3p^6 3d^5$	5
Cr	+ 3	24 – 3 = 21	$1s^2 2s^2 2p^6 3s^2 3p^6 3d^3$	3

Hence MnO_4^- has no *d*-electron.

24. B

Ionisation energy decreases on moving down a group from top to bottom. Hence the order of ionisation energy is: Be > Mg > Ca.

25. B

The structures of NH_3, $[PtCl_4]^{2-}$, PCl_5 and BCl_3 are tetrahedral, square planar, trigonal bipyramid and triangular planar respectivily. Hence the hybridisation of the central atoms in the above compounds is sp^3, dsp^2, dsp^3, sp^2 respectively.

26. A

$A_pB_q \rightleftharpoons pA^{q+} + qB^{p-}$

If s is the solubility of A_pB_q, then $[A^{q+}] = ps$ and $[B^{p-}] = qs$

$K_{sp} = [A^{q+}]^p [B^{p-}]^q = (ps)^p (qs)^q = s^{p+q} \cdot p^p \cdot q^q$

27. C

Saturated solution of KNO_3 is used to make salt bridge because the velocities of K^+ and NO_3^- ions are nearly the same. Hence the conductivity in the cathodic and anodic compartments is not affected.

28. B

The size of hydrated ions in solution is as follows:

$Li^+ > Na^+ > K^+$

Hence the conductivity of these ions follow the reverse order:

KCl > NaCl > LiCl

29. B

Rate $\propto I$

30. D

K_p is not dependent on pressure (P) and concentration (x) of substances.

31. A

Formaldehyde is always converted into formic acid.

$C_6H_5CHO + HCHO + NaOH \rightarrow C_6H_5CH_2OH + HCOONa$

It is a crossed Cannizzaro reaction.

32. D

In an S_{N^2} reaction, optical inversion takes place. Hence an S_{N^2} reaction always gives a product with opposite optical rotation and the product is a single stereoisomer of the reactant optically active compound.

33. B

An optically inactive alkene is formed.

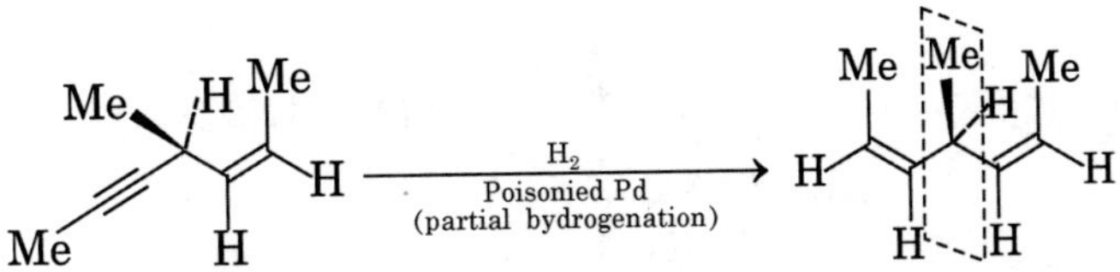

Optically active Optically inactive

Due to the *cis*-addition of hydrogen to the triple bond, the reduction product has a plane of symmetry and hence is optically inactive.

34. B

It is an electrophilic addition reaction (Markownikoff addition).

$$CH_3—CH{=}CH_2 + \overset{+}{Cl} \xrightarrow{\text{slow}} CH_3—\overset{+}{C}H—\underset{\substack{|\\Cl}}{C}H_2 \xrightarrow[\text{fast}]{HO^-} CH_3—\underset{\substack{|\\OH}}{C}H—\underset{\substack{|\\Cl}}{C}H_2$$

35. C

Heating of an alcohol with copper causes dehydration.

1-Propanol on dehydration gives propanol while 2-propanol on dehydration gives propanone. Only propanol reacts with Fehling solution.

$$\underset{\text{1-Propanol}}{CH_3—CH_2—CH_2—OH} \xrightarrow{\text{Cu, 300°C}} CH_3—CH_2—CHO \xrightarrow[\text{Fehling solution}]{2CuO} \underset{\text{Red}}{Cu_2O\downarrow}$$

$$\underset{\text{2-Propanol}}{CH_3—\underset{\substack{|\\OH}}{C}H—CH_3} \xrightarrow{\text{Cu, 300°C}} \underset{\text{Propanone}}{CH_3—CO—CH_3} \xrightarrow[\text{Fehling solution}]{2CuO} \text{No reaction.}$$

I.I.T. OBJECTIVE QUESTION PAPER—2002

1. How many moles of electron weigh one kilogram?
   (A) 6.023×10^{23} (B) $(1/9.108) \times 10^{31}$
   (C) $(6.023/9.108) \times 10^{54}$ (D) $(1/9.108 \times 6.023) \times 10^{8}$
2. Which of the following volume (V)-temperature (T) plots represents the behaviour of one mole of an ideal gas at one atmospheric pressure?

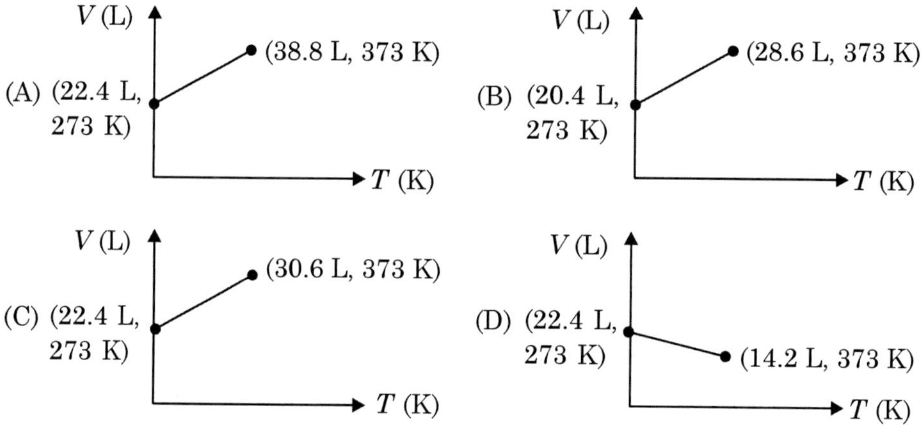

3. A substance A_xB_y crystallises in a face centred cubic (*fcc*) lattice in which atoms A occupy each corner of the cube and atoms B occupy the centres of each face of the cube. Identify the correct composition of the substance A_xB_y.
   (A) AB_3 (B) A_4B_3
   (C) A_3B (D) Composition cannot be specified.
4. When the temperature is increased, surface tension of water
   (A) increases (B) decreases
   (C) remains constant (D) shows irregular behaviour.
5. Rutherford's experiment, which established the nuclear model of the atom, used a beam of
   (A) β-particles, which impinged on a metal foil and got absorbed
   (B) γ-rays, which impinged on a metal foil and ejected electrons
   (C) helium atoms, which impinged in a metal foil and got scattered
   (D) helium nuclei, which impinged on a metal foil and got scattered.
6. If the nitrogen atom had electronic configuration $1s^7$, it would have energy lower than that of the normal ground state configuration $1s^22s^22p^3$, because the electrons would be closer to the nucleus. Yet $1s^7$ is not observed because it violates

(A) Heisenberg uncertainty principle (B) Hund's rule
(C) Pauli exclusion principle (D) Bohr postulate of stationary orbits.

7. Specify the coordination geometry around and hybridisation of N and B atoms in a 1 : 1 complex of BH_3 and NH_3.
(A) N: tetrahedral, sp^3, B: tetrahedral, sp^3
(B) N: pyramidal, sp^3, B: pyramidal, sp^3
(C) N: pyramidal, sp^3, B: planar, sp^2
(D) N: pyramidal, sp^3, B: tetrahedral, sp^3

8. One mole of a non-ideal gas undergoes a change of state (2.0 atm, 3.0 L, 95 K) $\longrightarrow$ (4.0 atm, 5.0 L, 245 K) with a change in internal energy, ΔU = 30.0 L atm. The change in enthalpy (ΔH) of the process in L atm is
(A) 40.0 (B) 42.3 (C) 44.0
(D) not defined, because pressure is not constant.

9. Consider the following equilibrium in a closed container: $N_2O_4(g) \rightleftharpoons 2NO_2(g)$. At a fixed temperature, the volume of the reaction container is halved. For the change, which of the following statements holds true regarding the equilibrium constant (K_p) and degree of dissociation (α)?
(A) neither K_p nor α changes (B) both K_p and α change
(C) K_p changes, but α does not change
(D) K_p does not change, but α changes.

10. Standard electrode potential data are useful for understanding the suitability of an oxidant in a redox titration. Some half-cell reactions and their standard potentials are given below:

$MnO_4^-(aq) + 8H^+(aq) + 5e^- \longrightarrow Mn^{2+}(aq) + 4H_2O(l)$ $E^0 = 1.51$ V
$Cr_2O_7^{2-}(aq) + 14H^+(aq) + 6e^- \longrightarrow 2Cr^{3+}(aq) + 7H_2O(l)$ $E^0 = 1.38$ V
$Fe^{3+}(aq) + e^- \longrightarrow Fe^{2+}(aq)$ $E^0 = 0.77$ V
$Cl_2(g) + 2e^- \longrightarrow 2Cl^-(aq)$ $E^0 = 1.40$ V

Identify the only incorrect statement regarding the quantitative estimation of aqueous $Fe(NO_3)_2$.
(A) MnO_4^- can be used in aqueous HCl
(B) $Cr_2O_7^{2-}$ can be used in aqueous HCl
(C) MnO_4^- can be used in aqueous H_2SO_4
(D) $Cr_2O_7^{2-}$ can be used in aqueous H_2SO_4

11. Consider the chemical reaction, $N_2(g) + 3H_2(g) \longrightarrow 2NH_3(g)$. The rate of this reaction can be expressed in terms of time derivative of concentration of $N_2(g)$, $H_2(g)$ or $NH_3(g)$. Identify the correct relationship amongst the rate expressions.
(A) Rate $= -d[N_2]/dt = -1/3d[H_2]/dt = 1/2d[NH_3]/dt$
(B) Rate $= -d[N_2]/dt = -3d[H_2]/dt = 2d[NH_3]/dt$
(C) Rate $= d[N_2]/dt = 1/3d[H_2]/dt = 1/2d[NH_3]/dt$
(D) Rate $= -d[N_2]/dt = -d[H_2]/dt = d[NH_3]/dt$

12. Anhydrous ferric chloride is prepared by
(A) heating hydrated ferric chloride at a high temperature in a stream of air

(B) heating metallic iron in a stream of dry chlorine gas
(C) reaction of metallic iron with hydrochloric acid(*aq*)
(D) reaction of metallic iron with hydrochloric acid(*g*).

13. Identify the correct order of solubility of Na_2S, CuS and ZnS in aqueous medium
(A) CuS > ZnS > Na_2S (B) ZnS > Na_2S > CuS
(C) Na_2S > CuS > ZnS (D) Na_2S > ZnS > CuS

14. Which of the following processes is used in the extractive metallurgy of magnesium?
(A) fused salt electrolysis (B) self-reduction
(C) aqueous solution electrolysis (D) thermite reduction.

15. An aqueous solution of a substance gives a white precipitate on treatment with dilute hydrochloric acid, which dissolves on heating. When hydrogen sulfide is passed through the hot acidic solution, a black precipitate is obtained. The substance is a
(A) Hg_2^{2+} salt (B) Cu^{2+} salt (C) Ag^+ salt (D) Pb^{2+} salt

16. A gas X is passed through water to form a saturated solution. The aqueous solution on treatment with silver nitrate gives a white precipitate. The saturated aqueous solution also dissolves magnesium ribbon with evolution of a colourless gas Y. Identify X and Y.
(A) X = CO_2, Y = O_2 (B) X = Cl_2, Y = CO_2
(C) X = Cl_2, Y = H_2 (D) X = H_2, Y = Cl_2

17. Polyphosphates are used as water softening agents because they
(A) form soluble complexes with anionic species
(B) precipitate anionic species
(C) form soluble complexes with cationic species
(D) precipitate cationic species.

18. Identify the correct order of acidic strengths of CO_2, CuO, CaO, H_2O.
(A) CaO < CuO < H_2O < CO_2 (B) H_2O < CuO < CaO < CO_2
(C) CaO < H_2O < CuO < CO_2 (D) H_2O < CO_2 < CaO < CuO

19. Identify the least stable ion amongst the following.
(A) Li^- (B) Be^- (C) B^- (D) C^-

20. Which of the following molecular species has unpaired electron(s)?
(A) N_2 (B) F_2 (C) O_2^- (D) O_2^{2-}

21. The nodal plane in the π-bond of ethene is located in
(A) the molecular plane
(B) a plane parallel to the molecular plane
(C) a plane perpendicular to the molecular plane which bisects the carbon-carbon σ-bond at right angle
(D) a plane perpendicular to the molecular plane which contains the carbon-carbon σ-bond.

22. Which of the following compounds exhibits stereoisomerism?
(A) 2-methylbut-1-ene (B) 3-methylbut-1-yne
(C) 3-methylbutanoic acid (D) 2-methylbutanoic acid

23. Which of the following acids has the smallest dissociation constant?
(A) $CH_3CHFCOOH$ (B) FCH_2CH_2COOH
(C) $BrCH_2CH_2COOH$ (D) $CH_3CHBrCOOH$

24. Consider the reaction: $H_3C-\underset{D}{CH}-\underset{CH_3}{CH}-CH_3 + Br\cdot \rightarrow X + HBr$. Identify the structure of the major product X.

(A) $H_3C-\underset{D}{CH}-\underset{CH_3}{CH}-\dot{C}H_2$

(B) $H_3C-\underset{D}{CH}-\underset{CH_3}{\dot{C}}-CH_3$

(C) $H_3C-\underset{D}{\dot{C}}-\underset{CH_3}{CH}-CH_3$

(D) $H_3C-\dot{C}H-\underset{CH_3}{CH}-CH_3$

25. Identify the reagent from the following list which can easily distinguish between but-1-yne and but-2-yne.

(A) bromine, CCl_4

(B) H_2, Lindlar catalyst

(C) dilute H_2SO_4, $HgSO_4$

(D) ammoniacal Cu_2Cl_2 solution.

26. Identify the correct order of boiling points of the following compounds:

$\underset{1}{CH_3CH_2CH_2OH}$ $\quad \underset{2}{CH_3CH_2CH_2CHO}$ $\quad \underset{3}{CH_3CH_2CH_2COOH}$

(A) 1 > 2 > 3 (B) 3 > 1 > 2 (C) 1 > 3 > 2 (D) 3 > 2 > 1

27. Identify the set of reagent/reaction conditions X and Y in the following set of transformations:

$$CH_3-CH_2-CH_2Br \xrightarrow{X} \text{Product} \xrightarrow{Y} CH_3-\underset{Br}{CH}-CH_3$$

(A) X = dilute aqueous NaOH, 20°C; Y = HBr/acetic acid, 20°C

(B) X = concentrated alcoholic NaOH, 80°C; Y = HBr/acetic acid, 20°C

(C) X = dilute aqueous NaOH, 20°C; Y = $Br_2/CHCl_3$, 0°C

(D) X = concentrated alcoholic NaOH, 80°C; Y = $Br_2/CHCl_3$, 0°C

28. Identify the correct order of reactivity in electrophilic substitution reactions of the following compounds.

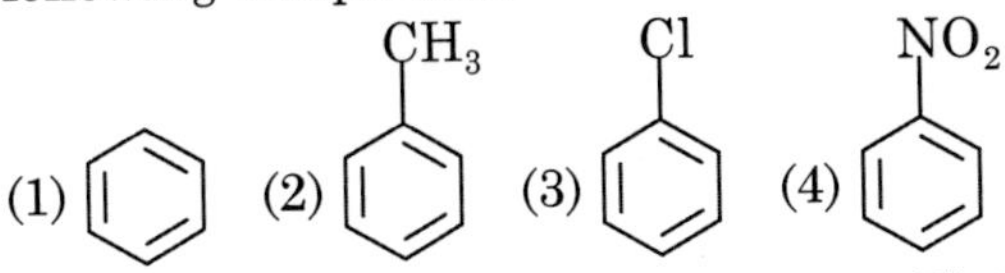

(A) 1 > 2 > 3 > 4 (B) 4 > 3 > 2 > 1 (C) 2 > 1 > 3 > 4 (D) 2 > 3 > 1 > 4

29. Compound *A* (molecular formula C_3H_8O) is treated with acidified potassium dichromate to form a product *B* (molecular formula C_3H_6O). *B* forms a shining silver mirror on warming with ammoniacal silver nitrate. *B* when treated with an aqueous solution of $H_2NCONHNH_2\cdot HCl$ and sodium acetate gives a product *C*. Identify the structure of *C*.

(A) $CH_3CH_2CH{=}NNHCONH_2$

(B) $CH_3-\underset{CH_3}{C}{=}NNHCONH_2$

(C) $CH_3-\underset{CH_3}{C}{=}NCONHNH_2$

(D) $CH_3CH_2CH{=}NCONHNH_2$

30. Which of the following hydrocarbons has the lowest dipole moment?

(A) 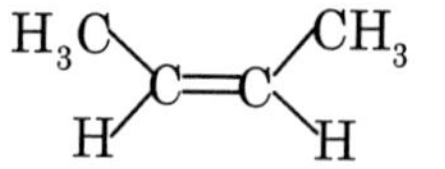

(B) $CH_3C\equiv CCH_3$

(C) $CH_3CH_2C\equiv CH$

(D) $CH_2=CH-C\equiv CH$

ANSWERS

1. D	2. C	3. A	4. B	5. D	6. C
7. A	8. D	9. D	10. A	11. A	12. B
13. D	14. A	15. D	16. C	17. C	18. A
19. B	20. C	21. A	22. D	23. C	24. B
25. D	26. B	27. B	28. C	29. A	30. B

MODEL SOLUTIONS—I.I.T. 2002

1. D

Mass of electron = 9.108×10^{31} kg

9.108×10^{31} kg of electron contain $= \dfrac{1}{6.023 \times 10^{23}}$ moles

1 kg of electron contains $= \dfrac{1}{6.023 \times 10^{23} \times 9.108 \times 10^{-31}}$ moles

$= \dfrac{1}{6.023 \times 9.108} \times 10^{8}$ moles

2. C

At constant pressure

$$\frac{V_1}{T_1} = \frac{V_2}{T_2}$$

$$\frac{224}{273} = \frac{V_2}{373}$$

or $$V_2 = \frac{224 \times 373}{273} = 20.6 \text{ L}$$

3. A

A atoms occupy eight corners of the face centred cube. Hence the number of A atoms in the unit cell $= 8 \times \dfrac{1}{8} = 1$. B atoms occupy the centres of six faces of the cube. Hence the number of B atoms sharing the unit cell $= 6 \times \dfrac{1}{2} = 3$. Then the composition of the substance A_xB_y is AB_3.

4. B

When the temperature is increased the ionic mobility of ions increases. Hence the surface tension increases.

5. D

Rutherford's experiment used a beam of α-particles and an alpha particle is represented by doubly charged helium nucleus containing two protons and two neutrons. The scattering of α-particles established the nuclear model of atom according to which an atom consists of positively charged nucleus.

6. C

According to the Pauli exclusion principle not more than two electrons can be placed in one and the same ortital and the two electrons in an ortital must have opposite spins. Hence placing of seven electrons in an ortital ($1s^7$) violates the Pauli exclusion principle.

7. A

The 1 : 1 complex of BF_3 and NH_3 is represented as

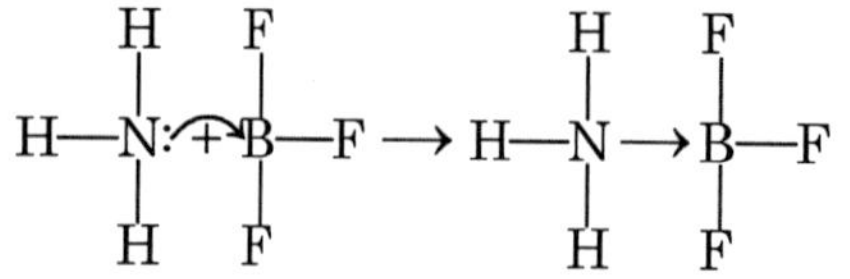

In this coordination compound, both B and N attain tetrahedral geometry and sp^3 hybridisation.

8. D

To calculate ΔH, pressure should remain constant.

9. D

$$N_2O_4(g) \rightleftharpoons 2NO_2(g)$$

Initial moles	1	0
Moles at equation	$1-\alpha$	2α

Total number of moles at equilibrium $= 1-\alpha+2\alpha = 1+\alpha$

If P is the total pressure at equilibrium,

$$p_{N_2O_4} = \frac{1-\alpha}{1-\alpha}\times P \text{ and } p_{NO_2} = \frac{2\alpha}{1+\alpha}\times P$$

$$K_p = \frac{(p_{NO_2})^2}{p_{N_2O_4}} = \frac{4\alpha^2 \times p}{(1-\alpha)(1+\alpha)}$$

If volume is halved, the pressure will be doubled. Since K_p is constant, α will be lowered in order to maintain the constant value of K_p. Hence K_p does not change but α changes.

10. A

The ionic equiation for the reaction of MnO_4^- and HCl is as follows:

$2\,MnO_4^- + 16H^+ + 10Cl^- \longrightarrow 2Mn^{2+} + 8H_2O + Cl_2$

The electrochemical cell is represented as

$Pt|Cl_2(g)$ (1 atm) $Pt|Cl^-(aq)||MnO_4^-(aq)|Mn^{2+}(aq)$

$$E^0_{cell} = E^0_{cathode} - E^0_{anode}$$
$$= 1.51\text{ V} - 1.4\text{ V} = 0.11\text{ V}$$

Since E^0_{cell} is positive, ΔH^0 is negative. Hence the above cell reation is feasible. But MnO_4^- is an oxidising agent and it can ozidise Fe^{2+} to Fe^{3+} and Cl^- to Cl_2. Hence MnO_4^-/HCl is not a suitable reagent for quentitative estimation of $Fe(NO_3)_2$.

11. A

For the reaction,

$N_2(g) + 3H_2(g) \longrightarrow 2NH_3(g)$

the rate of reaction is expressed as

$$\text{rate} = -\frac{d[N_2]}{dt} = -\frac{1}{3}\frac{d[H_2]}{dt} = \frac{1}{2}\frac{d[NH_3]}{dt}$$

The negative sign is due to the decrease in concentration of N_2 and H_2 with time and the positive rign is due to the increase in the concentration with time.

12. B

Anhydrous iron (III) chloride is prepared by beating metallic iron in a stream of dry chlorine gas.

$2Fe + 3Cl_2 \xrightarrow{\text{heat}} 2FeCl_3$

13. D

Sodium sulphide is the strongest electrolyte among Na_2S, ZnS and CuS, and sodium being an alkali metal, Na_2S is highly soluble in water. The solubility product of ZnS and CuS is in the order: CuS < ZnS. Hence CuS is less soluble than ZnS. Thus the order of solubility is: Na_2S > ZnS > CuS.

14. A

Magnesium being a reactive metal and its position being higher in the electrochemical series, it is extracted by the electrolysis of fused oxide or fused anhydrous chloride.

$MgCl_2 \longrightarrow Mg^{2+} + 2Cl^-$

At anode: $2Cl^- \longrightarrow Cl_2 + 2e^-$

At cathode: $M^{2+} + 2e^- \longrightarrow Mg$

15. D

$Pb^{2+} + HCl\ (dil.) \longrightarrow \underset{\text{White}}{PbCl_2\downarrow} + 2H^+$

The white precipitate of $PbCl_2$ dissolves on heating. When H_2S is passed through the hot acidic solution of $PbCl_2$, a black precipitate of PbS is formed.

$PbCl_2 + H_2S \xrightarrow[\text{medium}]{\text{Acidic}} \underset{\text{Black}}{PbS\downarrow} + 2HCl$

Hence the substance is a Pb^{2+} salt.

16. C

$Cl_2 + H_2O \longrightarrow HCl + HOCl$

X

$HCl + AgNO_3 \longrightarrow \underset{\text{White}}{AgCl\downarrow} + HNO_3$

$2HCl + Mg \longrightarrow MgCl_2\downarrow + H_2$

Hence X = Cl_2, Y = H_2

17. C

Polyphosphates are used as water softeners because they form soluble complexes with cationic species (Ca^{2+}, Mg^{2+}) present in hard water.

$\underset{\text{Sodium polyhexametasphosphate}}{Na_2[Na_4(PO_3)_6]} + CaCl_2 \longrightarrow \underset{\text{Soluble complex}}{Na_2[CaNa_2)PO_3]_6} + 2NaCl$

18. A

The acidic/basic properties of CO_2, CuO, CaO, H_2O are as follows: CO_2—acidic, CuO—weakly basic, CaO—strongly basic, H_2O—neutral, CO_2 is an anhydride of carbonic acid (H_2CO_3) and is an acidic oxide. Calcium is a group 2 metal and its oxide (CaO) is strongly basic. Copper is a transition metal (group 11) and its oxide is weakly basic. Hence the order of acidic strength is: CaO < CuO < H_2O < CO_2.

19. B

The electronic configuration of the ions are as follows:

Li^- : $1s^2 2s^2$ Stable due to presence of completely-filled $2s$ subshell

Be^- : $1s^2 2s^2 2p^1$ Least stable due to the presence of only one electron

B^- : $1s^2 2s^2 2p^2$ Less stable due to the presence of two electrons in $2p$ subshell

C^- : $1s^2 2s^2 2p^3$ Stable due to the presence of half-filled $2p$ subshell

20. C

The species O_2^- has one unpaired electron.

Molecular orbital configuration of O_2^-:

$$(\sigma_{1s}^b)^2 (\sigma_{1s}^*)^2 (\sigma_{2s}^b)^2 (\sigma_{2s}^*)^2 (\sigma_{p_z}^b)^2 (\pi_{2p_x}^b)^2 (\pi_{2p_y}^b)^2 (\pi_{2p_x}^*)^2 (\pi_{2p_y}^*)^1$$

21. A

The carbon atoms in ethane is sp^2 bybridised. The unhybridised $2p_x$ orbital of each carbon atom houses one unpaired eletron. A π-bond is formed by the $2p_x$(C)–$2p_x$(C) overlap as shown below:

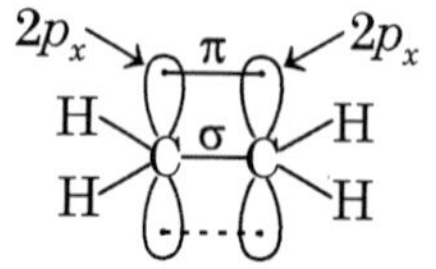

The $2p_x$ orbitals are pendicular to the plane containing ethane molecule. Hence the molecular plane does not contain any π-electron density and the nodal plane lies along the molecular plane.

22. D

The structures of the conpounds are as follows:

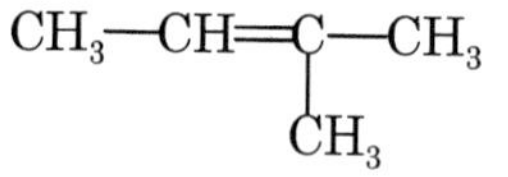

A

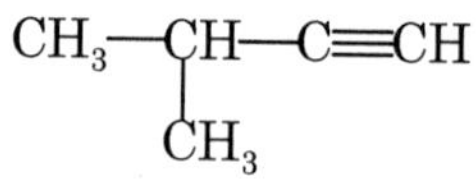

B

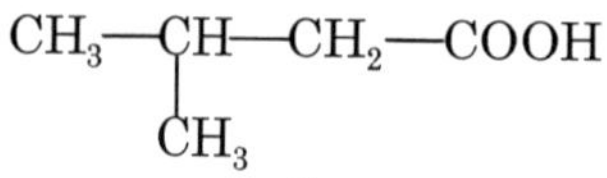

C

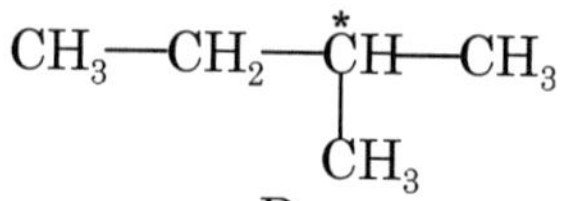

D

In D, there is one asymmetric carbon atom. Hence it exhibits the property of stereoisomerism. The compound A is of the type abC=Ca$_2$ and it does not exhibit the property of geometrical isomerism since 180° rotation about C=C produces the same compound.

23. C

Acidic strength is proportional to dissociation constant (K_a) of acid. In the given acids, the order of increasing acidic strength is as follows:

$BrCH_2CH_2COOH < FCH_2CH_2COOH < CH_3CHBrCOOH < CH_3CHFCOOH$

Hence the acid $BrCH_2CH_2COOH$ has the smallest dissociation constant.

24. B

The tertiary free radical $CH_3—CH(D)—\dot{C}(CH_3)—CH_3$ is more stable among all other free radicals because the order of stability of free radicals is: primary < secondary < tertiary.

25. D

Acidic hydrogen is present in but-1-yne and is absent in but-2-yne. Hence but-1-yne gives a red precipitate with ammoniacal Cu_2Cl_2 solution while but-2-yne does not form such a precipitate.

$$\underset{\text{But-1-yne}}{2CH_3—CH_2—C{\equiv}CH} + Cu_2Cl_2 + 2NH_4OH \rightarrow \underset{\text{Red}}{2CH_3—CH_2—C{\equiv}CCu\downarrow} + 2NH_4Cl + 2H_2O$$

$$CH_3—C{\equiv}C—CH_3 + Cu_2Cl_2 + 2NH_4OH \rightarrow \text{No reaction}$$

26. B

The compound 3 forms intermolecular hydrogen bonding and exists as a dimer. The compound 1 forms intermolecular hydrogen bonds. The compound 2 is not involved in hydrogen bonding. Hence the decreasing order of molecular mass of these compounds is: 3 > 2 > 1. So they follow the same order in their boiling point: 3 > 2 > 1.

27. B

$$CH_3—CH_2—CH_2Br \xrightarrow[80^\circ]{\text{alcoholic NaOH}} CH_3—CH{=}CH_2 \xrightarrow[20^\circ]{HBr/CH_3COOH} CH_3—\underset{\displaystyle Br}{CH}—CH_3$$

Hence X = alcoholic NaOH, 80°C and Y = HBr/CH_3COOH, 20°C.

28. C

The $—NO_2$ group is an electron-withdrawing group and hence it deactivates the benzene ring: The —Cl group also deactivates the benzene ring but its deactivation of the benzene ring is less than that of the $—NO_2$ group. The $—CH_3$ group is an electron-releasing group and hence it activates the benzene ring. Thus on the basis of activation and deactivation of the benzne ring, the order of reactivity in electrophilic substitution reaction is: 2 > 1 > 3 > 4.

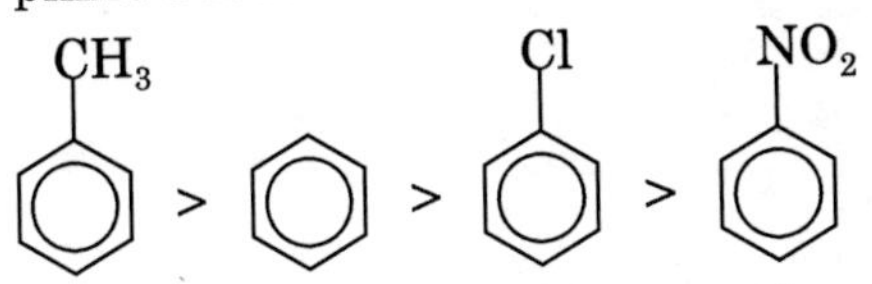

2 1 3 4

29. A

Since compound B forms a shining mirror on warming with ammoniacal silver nitrate, B is an aldehyde. Since the compound A on oxidation gives an aldehyde B, the compound A is a primary alcohol. Hence B (C_3H_6O) is CH_3CH_2CHO and A (C_3H_8O) is $CH_3CH_2CH_2OH$. The compound B reacts with an aqueous solution of $H_2NCONHNH_2$

and sodium acctate to form C. The compound C is $CH_3CH_2CH{=}NNHCONH_2$. The involved reactions are as follows:

$$\underset{A}{CH_3CH_2CH_2OH} \xrightarrow{K_2Cr_2O_7/H^+} \underset{B}{CH_3CH_2CHO} \xrightarrow[AgNO_3]{\text{Ammoniacal}} Ag\downarrow + CH_3CH_2COOH$$

$$\downarrow H_2NCONHNH_2$$

$$\underset{C}{CH_3CH_2CH{=}NNHCONH_2}$$

30. B

The dipole moment of $CH_3{-}C{\equiv}C{-}CH_3$ is zero because the molecule has a symmetrical planar structure.

I.I.T. OBJECTIVE QUESTION PAPER—2003

1. Among the following, the molecule with the highest dipole moment is
   (A) CH_3Cl (B) CH_2Cl_2 (C) $CHCl_3$ (D) CCl_4
2. Which of the following are isoelectronic and isostructural?
   NO_3^-, CO_3^{2-}, ClO_3^-, SO_3
   (A) NO_3^-, CO_3^{2-} (B) SO_3, NO_3^- (C) ClO_3^-, CO_3^{2-} (D) CO_3^{2-}, SO_3
3. In the reaction, 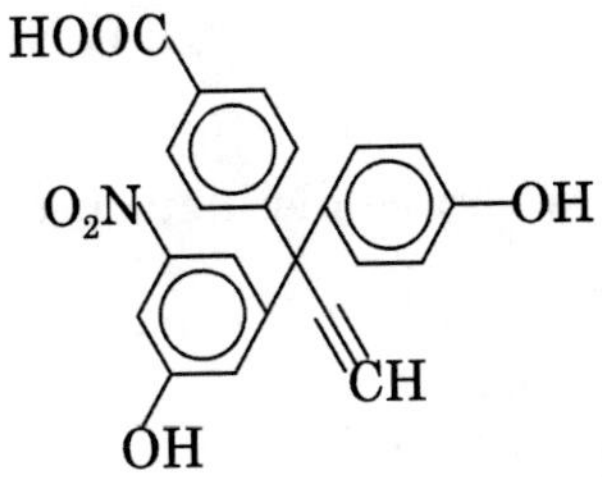 $\xrightarrow{\text{2 moles NaNH}_2}$ A, the product is

(A) 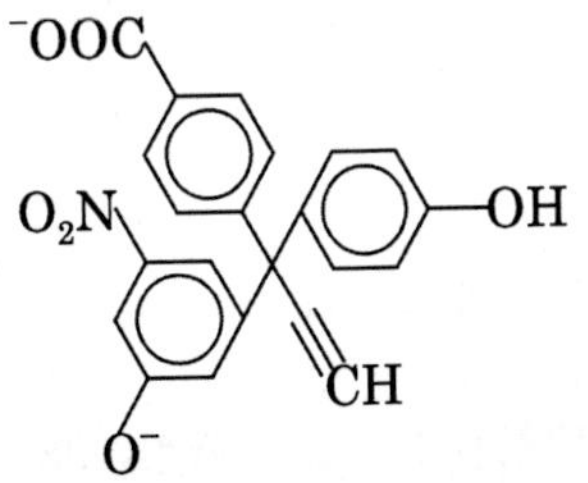

(B) 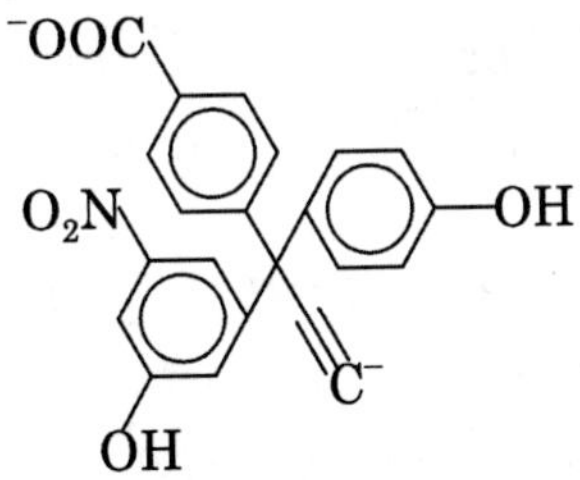

(C) 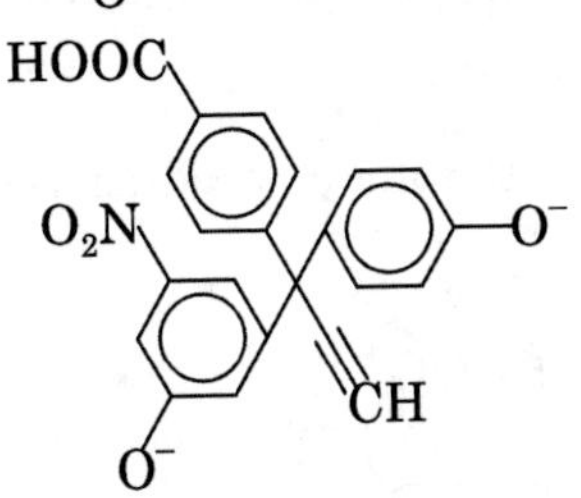

(D) 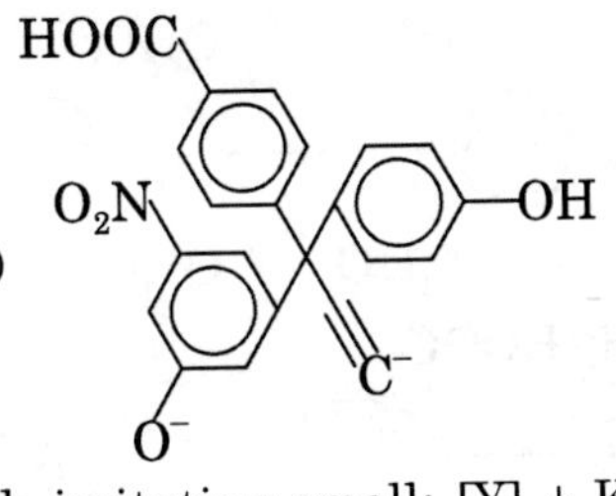

4. [X] + H_2SO_4 ⟶ [Y] a colourless gas with irritating smell; [Y] + $K_2Cr_2O_7$ + H_2SO_4 ⟶ green solution. [X] and [Y] are
   (A) SO_3^{2-}, SO_2 (B) Cl^-, HCl (C) S^{2-}, H_2S (D) CO_3^{2-}, CO_2
5. For H_3PO_3 and H_3PO_4 the correct choice is
   (A) H_3PO_3 is dibasic and reducing (B) H_3PO_3 is dibasic and non-reducing
   (C) H_3PO_4 is tribasic and reducing (D) H_3PO_3 is tribasic and non-reducing.
6. When MnO_2 is fused with KOH, a coloured compound is formed. The product and colour is
   (A) Mn_3O_4, black (B) $KMnO_4$, purple (C) Mn_2O_3, brown (D) K_2MnO_4, purple green

7. Rate of physisorption increases with
   (A) decrease in temperature (B) increase in temperature
   (C) decrease in pressure (D) decrease in surface area.

8. Which of the following represents the given mode of hybridisation sp^2-sp^2-sp-sp from left to right?
   (A) $H_2C{=}CH{-}C{\equiv}N$ (B) $HC{\equiv}C{-}C{\equiv}CH$
   (C) $H_2C{=}C{=}C{=}CH_2$ (D) $H_2C{=}CH{-}CH{=}CH_2$

9. In the sequence of reactions, 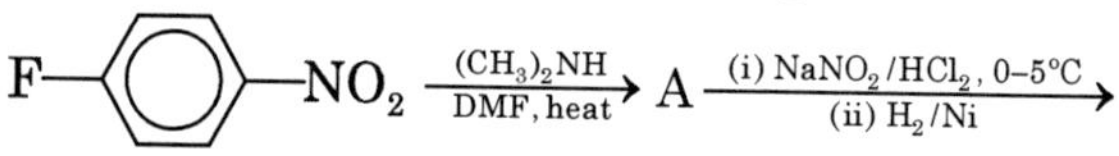 B, B is

   (A) 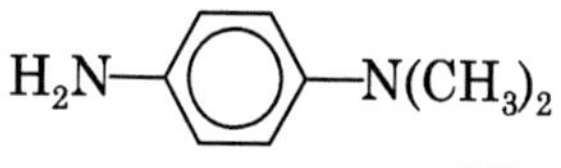 (B) 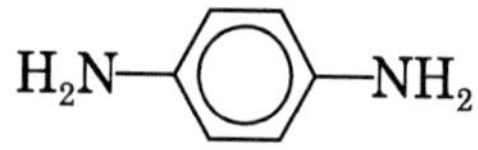

   (C) 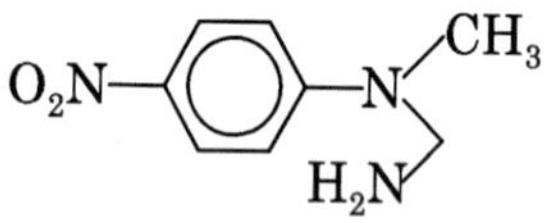 (D) 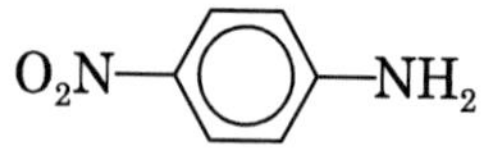

10. The product of acid hydrolysis of P and Q can be distinguished by

    P = 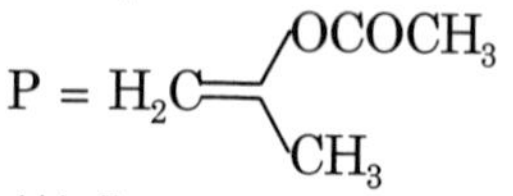 Q = 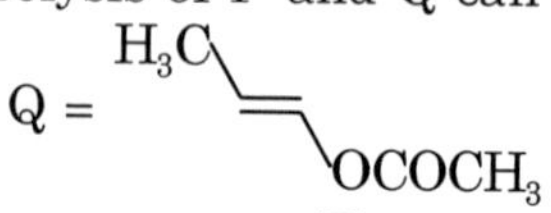

    (A) Lucas reagent (B) 2,4-DNP
    (C) Fehling's solution (D) $NaHSO_3$

11. 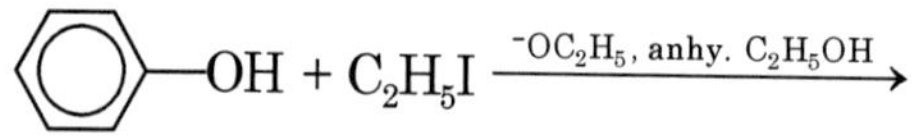

    (A) $C_6H_5OC_2H_5$ (B) $C_2H_5OC_2H_5$ (C) $C_6H_5OC_6H_5$ (D) C_6H_5I

12. Which has the maximum number of atoms?
    (A) 24 g of C (12) (B) 56 g of Fe (56) (C) 27 g of Al (27) (D) 108 g of Ag (108)

13. In the reaction, 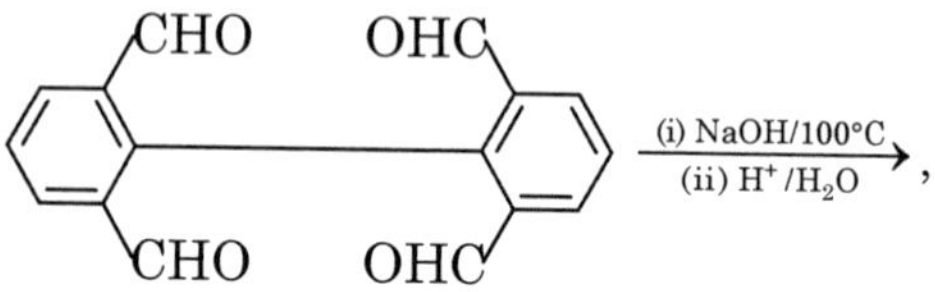, the major product is

    (A) 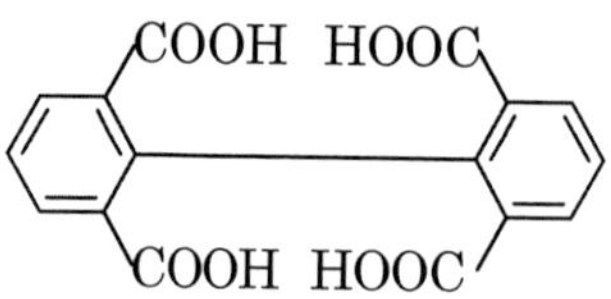 (B) 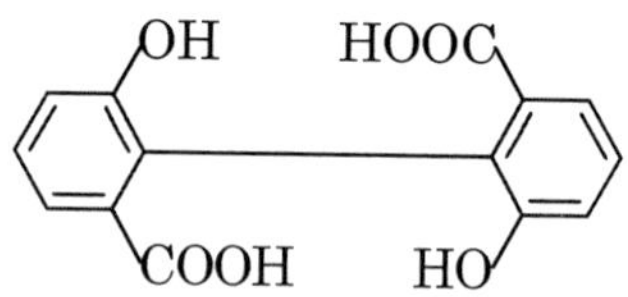

    (C) 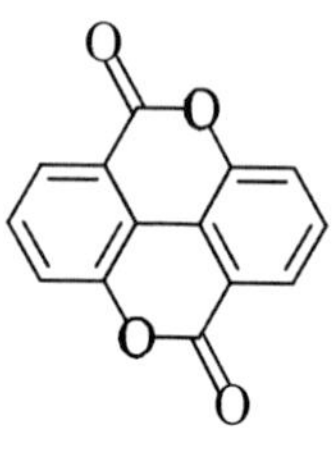 (D) 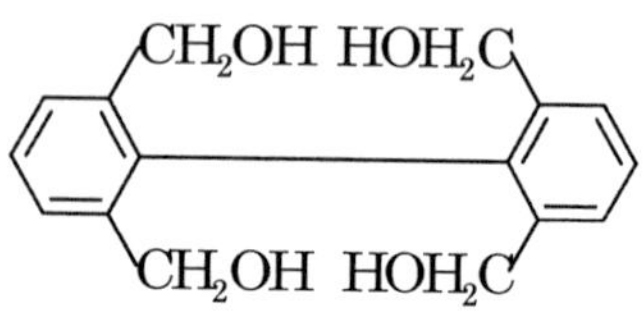

14. In the electrolytic cell, the flow of electrons is from
    (A) cathode to anode in solution
    (B) cathode to anode through external supply
    (C) cathode to anode through internal supply
    (D) anode to cathode through internal supply.

15. In a first order reaction, the concentration of reactant decreases from 800 mol/dm^3 to 50 mol/dm^3 is 2×10^4 sec. The rate constant of reaction in sec^{-1} is
    (A) 2×10^4 (B) 3.45×10^{-5} (C) 1.386×10^{-4} (D) 2×10^{-4}

16. During depression of freezing point in a solution, the following are in equilibrium
    (A) liquid solvent, solid solvent (B) liquid solvent, solid state
    (C) liquid solute, solid solute (D) liquid solute, solid solvent

17. H_3BO_3 is
    (A) monobasic and weak Lewis acid
    (B) monobasic and weak Bronsted acid
    (C) monobasic and strong Lewis acid
    (D) tribasic and weak Bronsted acid.

18. In the reaction, $Ph—C{\equiv}C—CH_3 \xrightarrow{Hg^{2+}/H^+}$ A, A is

(A) 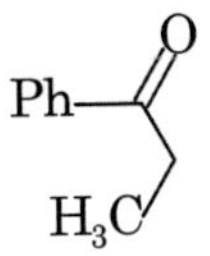 (B) 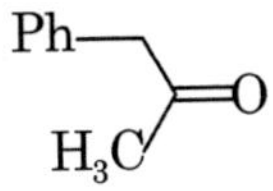 (C) 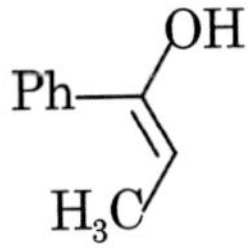 (D) 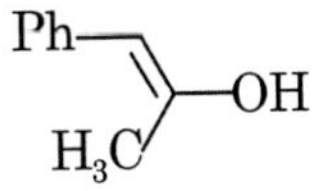

19. In the reaction, ethyl ester $\xrightarrow{\text{excess } CH_3MgBr}$ P. The product P is

(A) 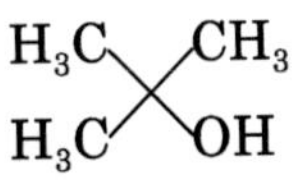 (B) 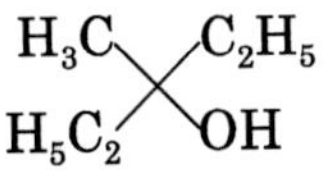

(C) 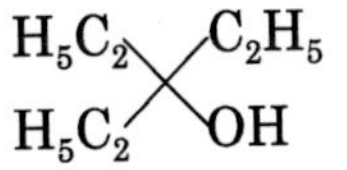 (D) 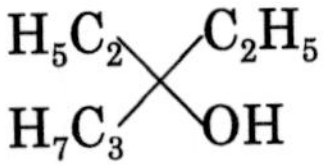

20. Mixture X = 0.02 mol of $[Co(NH_3)_5SO_4]Br$ and 0.02 mol of $[Co(NH_3)_5Br]SO_4$ was prepared in 2 L of solution. 1 L of mixture X + excess $AgNO_3 \longrightarrow$ Y. 1 L of mixture X + excess of $BaCl_2 \longrightarrow$ Z. Number of moles of Y and Z are
    (A) 0.01, 0.01 (B) 0.02, 0.01 (C) 0.01, 0.02 (D) 0.02, 0.02

21. Which of the reaction defines ΔH_f^0?
    (A) $C_{diamond} + O_2(g) \longrightarrow CO_2(g)$ (B) $\frac{1}{2}H_2(g) + \frac{1}{2}F_2(g) \longrightarrow HF(g)$
    (C) $N_2(g) + 3H_2(g) \longrightarrow 2NH_3(g)$ (D) $CO(g) + 1/2O_2(g) \longrightarrow CO_2(g)$.

22. ^{23}Na is more stable isotope of Na. Find out the process by which $^{24}_{11}$Na can undergo radioactive decay.
    (A) β^- emission (B) α emission (C) β^+ emission (D) K electron capture

23. $(Me)_2SiCl_2$ on hydrolysis will produce
    (A) $(Me)_2Si(OH)_2$ (B) $(Me)_2Si{=}O$
    (C) $\left[O—(Me)_2Si—O \right]_n$ (D) $Me_2SiCl(OH)$

24. A solution which is $10^{-3}\,M$ in Mn^{2+}, Fe^{2+}, Zn^{2+} and Hg^{2+} is treated with $10^{-16}\,M$ sulphide solution. If K_{sp} of MnS, FeS, ZnS and HgS is 10^{-15}, 10^{-23}, 10^{-20} and 10^{-54} respectively, which one will precipitate first?
(A) FeS (B) MnS (C) HgS (D) ZnS

25. In the process of extraction of gold, roasted gold ore + CN^- + $H_2O \xrightarrow{O_2}$ [X] + OH^-. [X] + Zn $\longrightarrow$ [Y] + Au, identify the complexes [X] and [Y].
(A) [X] = $[Au(CN)_2]^-$, Y = $[Zn(CN)_4]^{2-}$ (B) X = $[Au(CN)_4]^{2-}$, Y = $[Zn(CN)_4]^{2-}$
(C) X = $[Au(CN)_2]^-$, Y = $[Zn(CN)_6]^{4-}$ (D) X = $[Au(CN)_4]^-$, Y = $[Zn(CN)_4]^{2-}$

26. Positive deviation from ideal behaviour takes place because of
(A) molecular interaction between atoms and $PV/nRT > 1$
(B) molecular interaction between atoms and $PV/nRT < 1$
(C) Finite size of atoms and $PV/nRT > 1$
(D) Finite size of atoms and $PV/nRT < 1$

27. 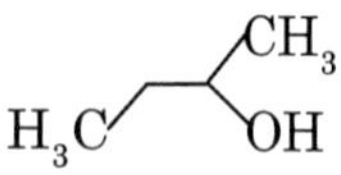

$\xrightarrow{H^+/H_2O}$ [F] $\xrightarrow{Br_2/CCl_4}$ C_4H_8Br (5 such products are possible). How many structures of F are possible?
(A) 2 (B) 5 (C) 6 (D) 3

28. An enantiomerically pure acid is treaterd with racemic mixture of an alcohol having one chiral carbon. The ester formed will be
(A) optically active mixture (B) pure enantiomer
(C) meso compound (D) racemic mixture.

ANSWERS

1. A	2. A	3. A	4. A	5. A	6. D
7. A	8. A	9. A	10. C	11. B	12. A
13. B	14. C	15. C	16. A	17. A	18. A
19. A	20. A	21. B	22. A	23. C	24. C
25. A	26. A	27. D	28. A		

MODEL SOLUTIONS—I.I.T. 2003

1. A

Methane is a symmetrical molecule and hence its dipole moment is zero. The replacement of one of the hydrogen atoms by a chlorine atom increases the dipole moment in CH_3Cl. This is because the bond dipole moment of C—H bond and that of C—Cl bond reinforce one another and the resultant of two C—H dipoles adds to the resultant of C—H and C—Cl dipoles.

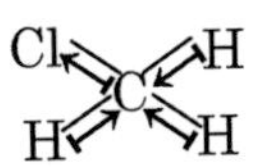

CH_3Cl (μ = 1.86 D)

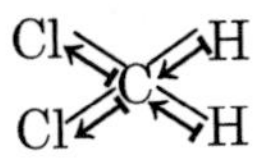

CH_2Cl_2 (μ = 1.62 D)

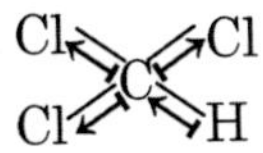

$CHCl_3$ (μ = 1.03 D)

The replacement of another hydrogen atom by a chlorine atom causes the resultant of two C—H dipoles to add to the resultant of two C—Cl dipoles. But due to bond pair-bond pair repulsion between C—Cl bonds, the ∠Cl—C—Cl increases leading to reduction of dipole moment of the molecule. In $CHCl_3$, the resultant of C—H and C—Cl dipoles opposes the resultant of two C—Cl dipoles and increase in bond angles is there due to the introduction of the third chlorine atom. On introduction of the fourth chlorine atom, the molecule (CCl_4) again becomes symmetric and dipole moment reduces to zero. Hence CH_3Cl has the maximum dipole moment.

2. A

Both NO_3^- and CO_3^{2-} ions have the same number of electrons (32) and the central atom in each is sp^2 hybridised. Hence NO_3^- and CO_3^{2-} ions are isoelectronic and isostructural.

3. A

The order of acidic strength of the various groups present in the compound is:

—COOH > —OH (phenolic, deactivated by NO_2 group) > —OH (phenolic) > alkynic H.

Two moles of NH_2^- ions abstract two moles of most acidic hydrogen out of the four moles of hydrogen present in one mole of the compound. This would lead to the formation of A.

4. A

Since both SO_2 and H_2S being reducing agents, can turn acidified dischromate solution green, [Y] may be SO_2 or H_2S. SO_2 is produced by the action of acid on sulphites while H_2S is produced by the action of acid in sulphides. But SO_2 has an odour of burning sulphur, which is irritating. H_2S has the odour of rotten egg. Hence [Y] is SO_2.

$$\underset{[X]}{SO_3^{2-}} + H_2S \longrightarrow SO_4^{2-} + H_2O + \underset{[Y]}{SO_2}$$

$$SO_3^{2-} + K_2Cr_2O_7 + H_2SO_4 \longrightarrow K_2SO_4 + \underset{\text{Green solution}}{Cr_2(SO_4)_3} + H_2O$$

5. A

The structure of H_3PO_3 indicates that there are two —OH groups and hence it is a dibasic acid. The oxidisation number of P in H_3PO_3 is + 3. We know that phosphorus exhibits the highest oxidation state of + 5. Hence H_3PO_3 can be oxidised and this means that H_3PO_3 is a reducing agent.

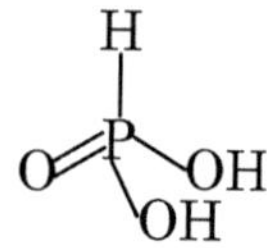

6. A

In alkaline medium, the stable oxidation state of manganese is + 6. Hence MnO_2 is oxidised to K_2MnO_4 by atmospheric oxygen in KOH medium.

$2MnO_2 + 4KOH + O_2 \longrightarrow 2K_2MnO_4 + 2H_2O$

7. A

Physisorption is an exothermic process. Hence according to the Le Chatelier's principle, a decrease in temperature will favour a direction in which heat is evolved. So the rate of physisorption increases with decrease of temperature.

8. A

$H_2C{=}CH{-}C{\equiv}N$

sp^2 *sp*^{2} *sp* *sp*

9. A

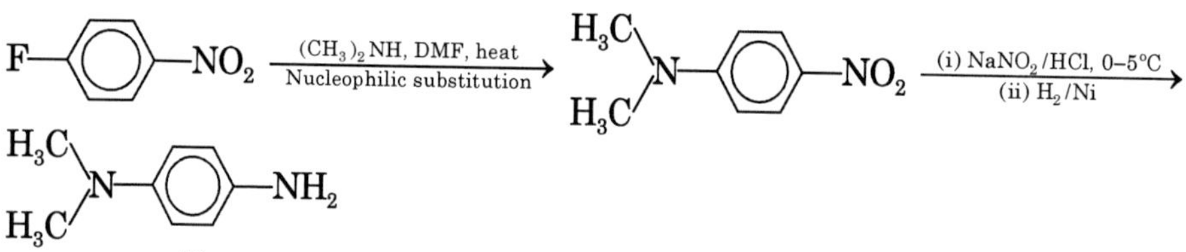

B

10. C

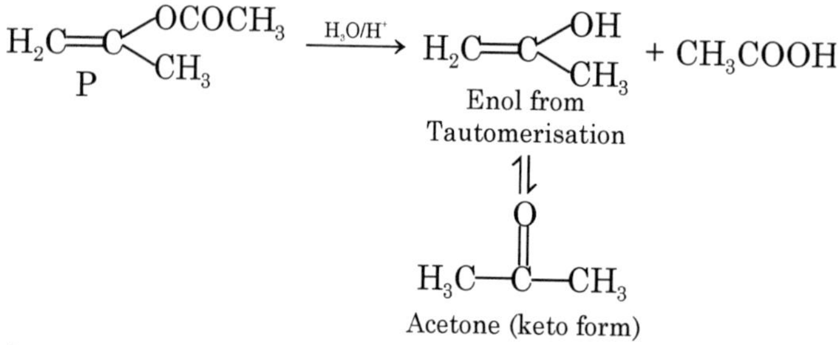

Acetone does not reduce Fehling solution.

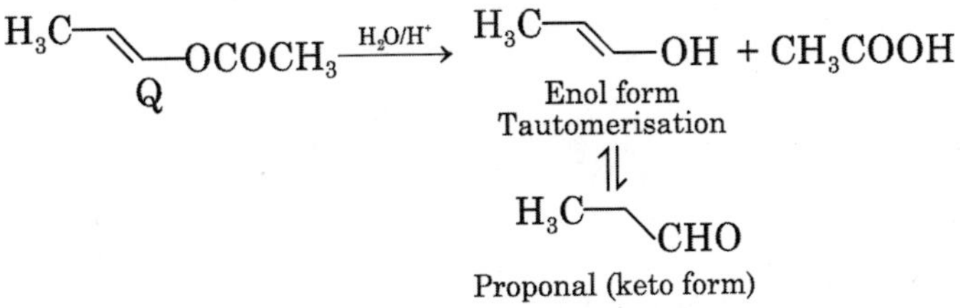

Propanal reduces Fehling solution. Hence ketone and aldehyde can be distinguished by Fehling solution.

11. B

$C_2H_5O^-$ abstracts the proton from phenol and converts phenol to phenoxide ion. The phenoxide ion then makes nucleophilic attack on the methylene carbon of iodoethane. But $C_2H_5O^-$ ion is in excess. It is a better nucleophile than $C_6C_5O^-$ ion since in $C_2H_5O^-$ the negative charge is localised over oxygen and in $C_6H_5O^-$ the negative charge is delocalised over the whole molecular framework. Hence the $C_2H_5O^-$ ion makes the nucleophilic attack at iodoethane to give ethoxyethane (Williamson's synthesis).

$$H_3C—CH_2—\ddot{O}:^- + H_3C—CH_2—I \rightarrow CH_3—CH_2—O—CH_2—CH_3 + I^-$$

12. A

$$\text{Number of atoms in 24 g of C} = \frac{24}{12} \times 6.023 \times 10^{23} = 2 \times 6.023 \times 10^{23}$$

$$\text{Number of atoms in 56 g of Fe} = \frac{56}{56} \times 6.023 \times 10^{23} = 1 \times 6.023 \times 10^{23}$$

$$\text{Number of atoms in 27 g of Al} = \frac{27}{27} \times 6.023 \times 10^{23} = 1 \times 1.6023 \times 10^{23}$$

$$\text{Number of atoms in 108 g of Ag} = \frac{108}{108} \times 6.023 \times 10^{23} = 1 \times 6.023 \times 10^{23}$$

Hence the maximum number of atoms is present in 24 g of carbon.

13. B

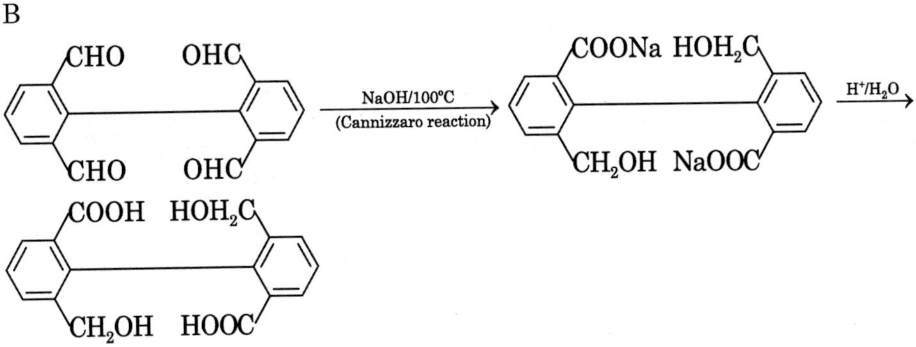

14. C

In an electrochemical cell, the flow of electron occurs from cathode to anode through internal supply of cations.

15. C

For a first order reaction,

$$k = \frac{2.303}{t}\log\frac{[A]_0}{[A]_t}$$

$$= \frac{2.303}{2\times10^4}\log\frac{800}{50} = 1.386 \times 10^4\ s^{-1}$$

16. A

At the freezing point liquid and solid are in equilibrium. When a solution of a non-volatile solute is cooled to a temperature below the freezing point of a solution, some of liquid solvent separates as a solid solvent. Hence the concentration of solution increases.

17. A

The boron atom in H_3BO_3 is electron-deficient with one vacant *p*-orbital. Hence boric acid behaves as Lewis acid. Boron does not have a *d*-orbital to accept electrons from a Lewis base. Hence it can accommodate only one additional pair of electrons in its outermost shell. Thus it behaves as a monobasic weak Lewis acid.

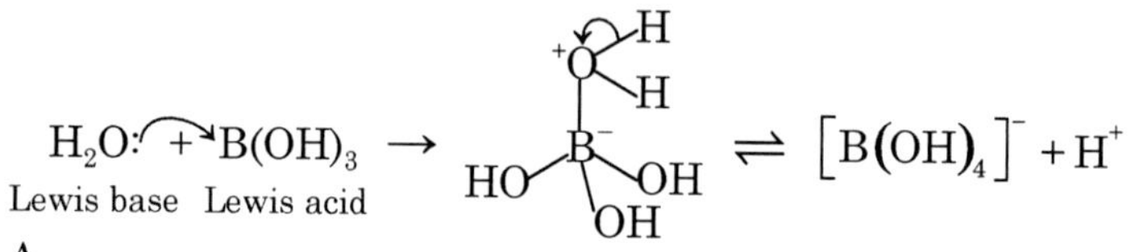

18. A

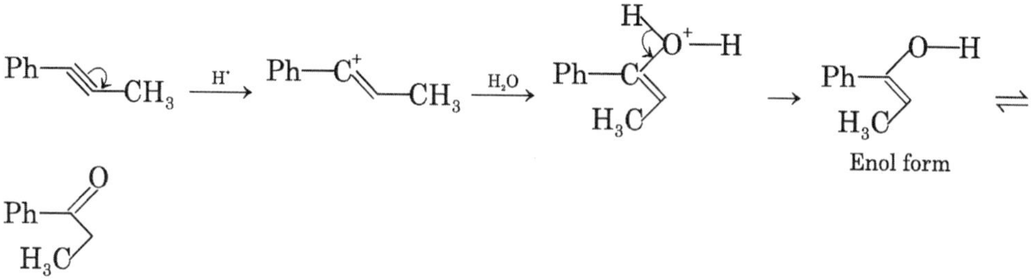

19. A

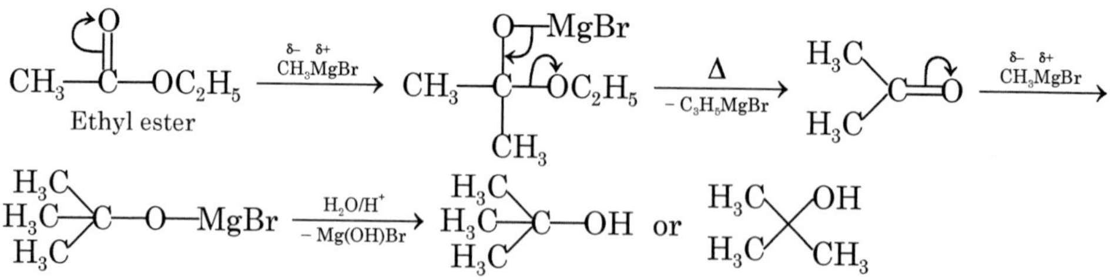

20. A

A mixture containing 0.02 mol of $[Co(NH_3)_5SO_4]Br$ and 0.02 mol of $[Co(NH_3)_5Br]SO_4$ was prepared in 2 L of solution. Hence the concentrations of $[Co(NH_3)_5SO_4]Br$ and

$[Co(NH_3)_5Br]SO_4$ in solution are 0.01 mol L^{-1} and 0.01 mol L^{-1} respectively. The following reaction occurs when excess of $AgNO_3$ is added to the mixture X:

$$\underset{0.01\ \text{mol L}^{-1}}{[Co(NH_3)_5SO_4]Br} + \underset{\text{Excess}}{AgNO_3} \longrightarrow \underset{\text{Y (soluble)}}{[Co(NH_3)_5SO_4]NO_3} + \underset{0.01\ \text{mol L}^{-1}}{AgBr\downarrow}$$

Hence the number of moles of Y = 0.01.

The following reaction occurs when excess of $BaCl_2$ is added to the mixture X.

$$\underset{0.01\ \text{mol L}^{-1}}{[Co(NH_3)_5Br]SO_4} + \underset{\text{Excess}}{BaCl_2} \longrightarrow \underset{\text{Z (soluble)}}{[Co(NH_3)_5Br]Cl_2} + \underset{0.01\ \text{mol L}^{-1}}{BaSO_4\downarrow}$$

Hence the number of moles of Z = 0.01.

21. B

The standard enthalpy of formation (ΔH_f^0) is the amount of heat absorbed or evolved at constant pressure when 1 mole of a compound is formed from its elements in their standard state at 1 atom and 298 K. The correct equation which defines (ΔH_f^0) is (B). At 298 K and 1 atom, the standard state of carbon is graphite and hence (A) is incorrect. In (C), two moles of $NH_3(g)$ are formed and hence (C) is incorrect. In (D), $CO_2(g)$ gas is formed from a compound $CO(g)$ and hence (D) is incorrect.

22. A

The n/p ratio in ^{24}Na is 13/11 which is greater than 1 and hence ^{24}Na is radioactive. In order to achieve stability, ^{24}Na adjusts its n/p ratio to the proper value of unity. It is done by breaking a neutron to a proton and electron $^{1}_{0}n \longrightarrow {}^{1}_{1}p + {}^{0}_{-1}e$ or β^-. Hence ^{24}Na is changed into ^{23}Na by means of β^- emission and proton emission.

23. C

Me_2SiCl_2 on hydrolysis gives an unstable compound $Me_2Si(OH)_2$. Due to very large size of silicon in comparison to oxygen, silicon cannot form π-bond with oxygen atom $(\rangle Si{=}O)$. Hence $Me_2Si(OH)_2$ attains stabilisation through loss of water and polymerisation.

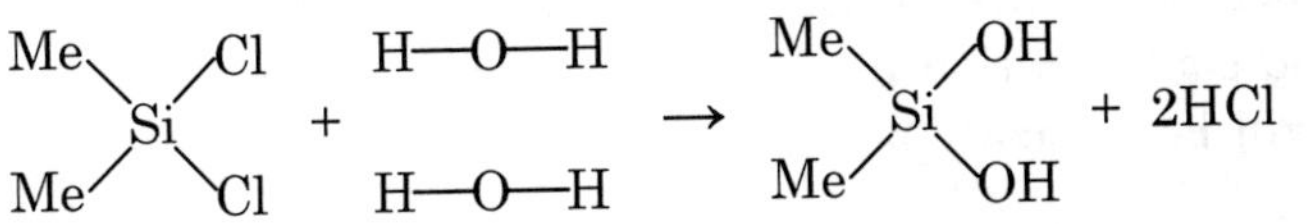

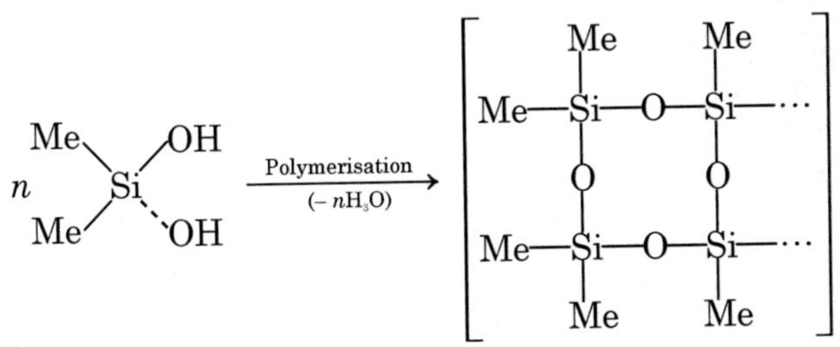

24. C

The precipitation of a salt occurs when the ionic product exceeds the solubility product. Hence HgS having the lowest K_{sp} among the given salts will precipitate first.

25. A

$2Au + 4CN^- + H_2O + \frac{1}{2}O_2 \longrightarrow 2[Au(CN)_2]^- + 2OH^-$

$2[Au(CN)_2]^- + Zn \longrightarrow [Zn(CN)_4]^{2-} + 2Au$

Hence $X = [Au(CN)_2]^-$ and $Y = [Zn(CN)_4]^{2-}$

26. A

For an ideal gas the compressibility factor, $Z = \dfrac{PV}{nRT} = 1$. If a gas deviates from ideal behaviour, $Z > 1$. Hence for positive deviation from ideal behaviour, $Z > 1$. For positive deviation of gas from ideal behaviours,

$PV = nRT + nPb$

where b is the effective volume of the molecule. The factor nPb is responsible for increasing the value of PV above the ideal value. Hence it is the finite size of the molecules that is responsible for origin of b and positive deviation at high pressure.

27. D

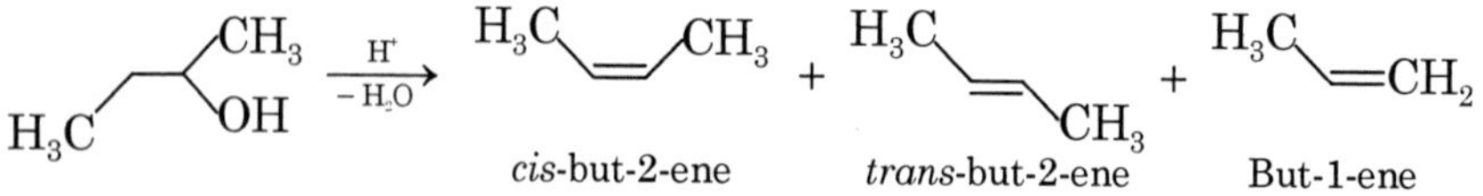

These there alkenes on reaction with Br_2/CCl_4 give addition products having the formula $C_4H_8Br_2$. The five dibromo derivatives are:

$CH_3-CH(Br)-CH(Br)-CH_3$ $\quad CH_3-CBr_2-CH(Br)-CH_3$ $\quad BrH_2C-CH(Br)-CH_2-CH_3$

$BrH_2C-CH_2-CH(Br)-CH_3$ $\quad BrH_2C-CH_2-CH_2-CH_2Br$

28. A

The optically active acid reacts with racemic mixture of alcohol to form two types of isomeric esters. In each the configuration of the chiral centre of acid remains the same. Hence the mixture will be optically active.

I.I.T. OBJECTIVE QUESTION PAPER—2004

1. 2-Phenylpropene on acidic hydration gives
   (A) 2-phenylpropan-2-ol (B) 2-phenylpropan-1-ol
   (C) 3-phenylpropan-1-ol (D) 1-phenylpropan-2-ol

2. 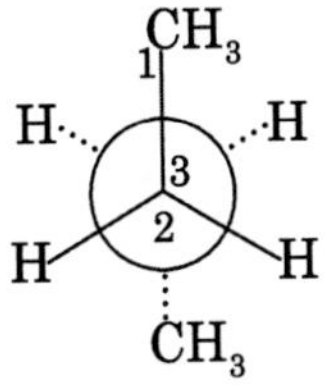

   C_2 is rotated anticlockwise 120° about C_2—C_3 bond. The resulting conformer is
   (A) staggered (B) eclipsed (C) gauche (D) partially eclipsed

3. Benzamide on treatment with $POCl_3$ gives
   (A) aniline (B) benzonitrile (C) chlorobenzene (D) benzylamine

4. The methods chiefly used for the extraction of lead and tin from their ores are respectively
   (A) self-reduction and carbon reduction
   (B) self-reduction and electrolytic reduction
   (C) carbon reduction and self-reduction
   (D) cyanide process and carbon reduction.

5. 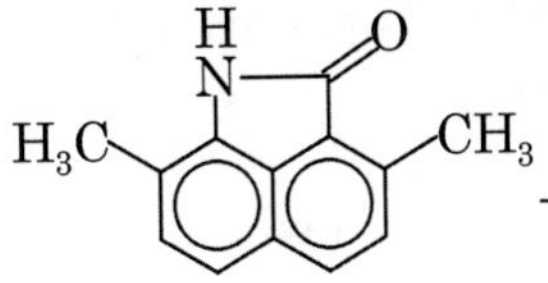

$\xrightarrow{Fe/Br_2}$. The product of monobromination of this compound is

(A) 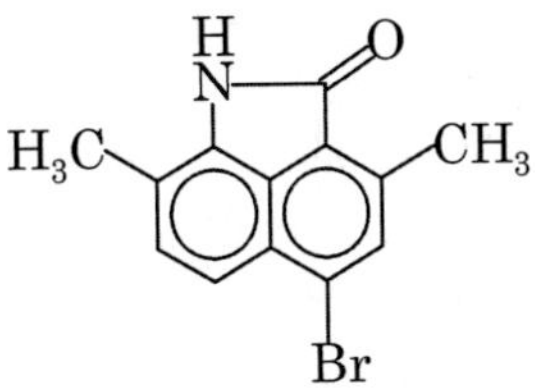

(B) 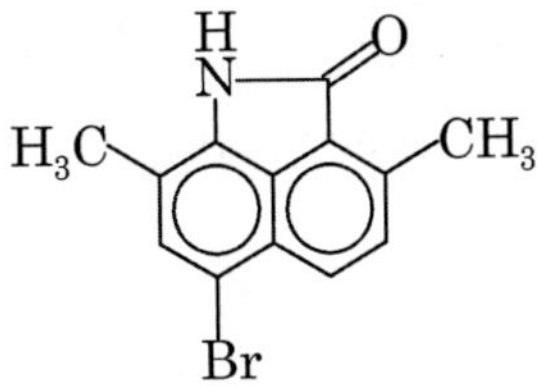

(C) 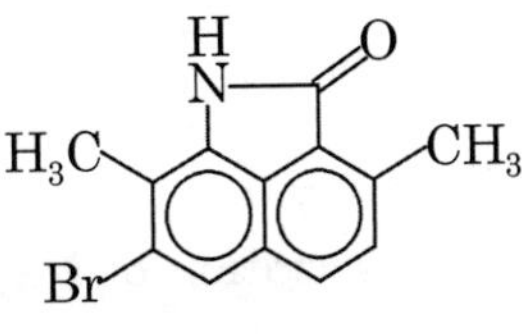

(D) 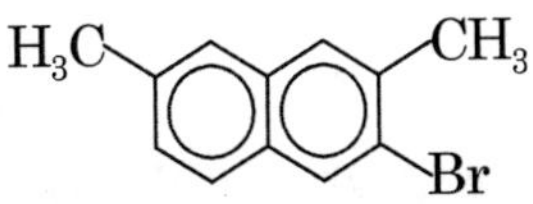

6. The order of reactivity of phenyl magnesium bromide with the following compounds is

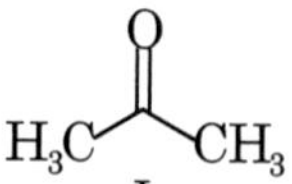
I

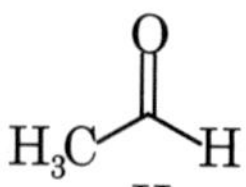
II

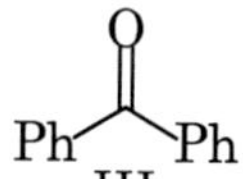
III

(A) II > III > I (B) I > III > II (C) II > I > III (D) All react with same rate.

7. Ammonium dichromate on heating gives a gas which is also given by
(A) heating NH_4NO_2 (B) heating NH_4NO_3
(C) $Mg_3N_2 + H_2O$ (D) Na (comp.) + H_2O_2

8. The emf of the cell $Zn|Zn^{2+}$ ($a = 0.1M$) || Fe^{2+} ($a = 0.01M$)|Fe is 0.2905 V. Equilibrium constant for the cell reaction is
(A) $10^{0.32/0.591}$ (B) $10^{0.32/0.0295}$ (C) $10^{0.26/0.0295}$ (D) $e^{0.32/0.295}$

9. HX is a weak acid ($K_a = 10^{-5}$). It forms a salt NaX (0.1*M*) on reacting with sodium hydroxide. The degree of hydrolysis of NaX is
(A) 0.01% (B) 0.001% (C) 0.1% (D) 0.5%

10. Spontaneous adsorption of a gas in solid surface is an exothermic process because
(A) ΔH increases for system (B) ΔS increases for gas
(C) ΔS decreases for gas (D) ΔG increases for gas

11. For a monoatomic gas, kinetic energy = E. The relation of E with rms velocity is
(A) $u = \left(\frac{2E}{m}\right)^{1/2}$ (B) $u = \left(\frac{3E}{2m}\right)^{1/2}$ (C) $u = \left(\frac{E}{2m}\right)^{1/2}$ (D) $u = \left(\frac{E}{3m}\right)^{1/2}$

12. The pair of compounds having metals in their highest oxidation state is
(A) MnO_2, $FeCl_3$ (B) MnO_4^-, CrO_2Cl_2
(C) $[Fe(CN)_6]^{3-}$, $[Co(CN)_3]$ (D) $[NiCl_4]^{2-}$, $[CoCl_4]^-$

13. The compound with tetrahedral geometry is
(A) $[Ni(CN)_4]^{2-}$ (B) $[Pd(CN)_4]^{2-}$ (C) $[PdCl_4]^{2-}$ (D) $[NiCl_4]^{2-}$

14. Spin-only magnetic moment of the compound $Hg[Co(NCS)_4]$ is
(A) $\sqrt{3}$ (B) $\sqrt{15}$ (C) $\sqrt{24}$ (D) $\sqrt{8}$

15. A sodium salt of an unknown anion when heated with $MgCl_2$ gives a white precipitate only on boiling. The anion is
(A) SO_4^{2-} (B) HCO_3^- (C) CO_3^{2-} (D) NO_3^-

16. Which hydrogen-like species will have the same radius as that of Bohr orbit of hydrogen atom?
(A) $n = 2$, Li^{2+} (B) $n = 2$, Be^{3+} (C) $n = 2$, He^+ (D) $n = 3$, Li^{2+}

17. In 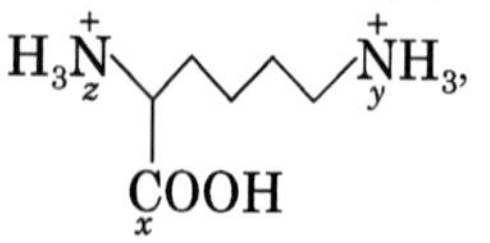 arrange x, y, z in the order of increasing acidic strength.
(A) $x > z > y$ (B) $z < x > y$ (C) $x > y > z$ (D) $z > x > y$

18. 0.004*M* Na_2SO_4 is isotonic with 0.01*M* glucose. Degree of dissociation of Na_2SO_4 is
(A) 75% (B) 50% (C) 25% (D) 85%

19. ΔH_{vap} = 30 kJ mol^{-1} and ΔS_{vap} = 75 J mol^{-1} K^{-1}. Find the temperature of vapour at 1 atmosphere.
(A) 400 K (B) 350 K (C) 298 K (D) 250 K
20. 2 mol of an ideal gas expanded isothermally and reversibly from 1 L to 10 L at 300 K. What is the enthalpy change?
(A) 4.98 kJ (B) 11.47 kJ (C) – 11.47 kJ (D) 0 kJ
21. (A) follows the first order reaction, (A) ⟶ Product. The concentration of A changes from 0.1*M* to 0.025*M* in 40 min. Find the rate of reaction of A when [A] = 0.01*M*.
(A) 3.47×10^{-4} *M* min^{-1} (B) 3.47×10^{-5} *M* min^{-1}
(C) 1.73×10^{-4} *M* min^{-1} (D) 1.73×10^{-5} *M* min^{-1}
22. Hex-2-yne gives *trans*-hex-2-ene with
(A) Li/NH_3 (B) $Pd/BaSO_4$ (C) $LiAlH_4$ (D) Pt/H_2
23. How many chiral compounds are possible on mono chlorination of 2-methylbutane?
(A) 2 (B) 4 (C) 6 (D) 8
24. Which of the following pairs give positive Tollen's test?
(A) Glucose, sucrose (B) glucose, fructose
(C) Hexanal, acctophenone (D) Fructose, sucrose
25. Which of the following has —O—O— linkage?
(A) $H_2S_2O_6$ (B) $H_2S_2O_8$ (C) $H_2S_2O_3$ (D) $H_2S_4O_6$
26. When I^- ion is oxidised by MnO_4^- ion in alkaline medium, I^- ion converts to
(A) IO_3^- (B) I_2 (C) IO_4^- (D) IO^-
27. Number of lone pair(s) in $XeOF_4$ is
(A) 0 (B) 1 (C) 2 (D) 3
28. According to MO theory,
(A) O_2^+ is paramagnetic and bond order is greater than O_2
(B) O_2^+ is paramagnetic and bond order is less than O_2
(C) O_2^+ is diamagnetic and bond order is less than O_2
(D) O_2^+ is diamagnetic and bond order is more than O_2

ANSWERS

1. A	2. C	3. B	4. A	5. B	6. C
7. A	8. B	9. A	10. C	11. A	12. B
13. D	14. B	15. B	16. B	17. A	18. A
19. A	20. D	21. A	22. A	23. B	24. B
25. B	26. A	27. B	28. A		

MODEL SOLUTIONS—I.I.T. 2004

1. A

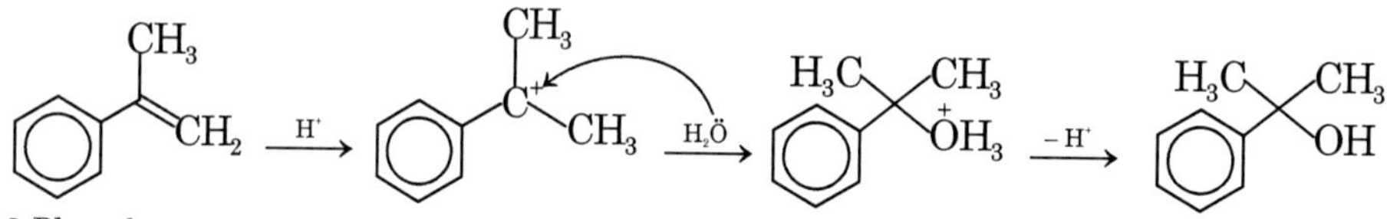

2-Phenylpropene 2-Phenhylpropan-2-ol

2. C

When C_2 is rotated anticlockwise 120° about $C_2—C_3$ bond, the resulting conformer is a gauche conformer.

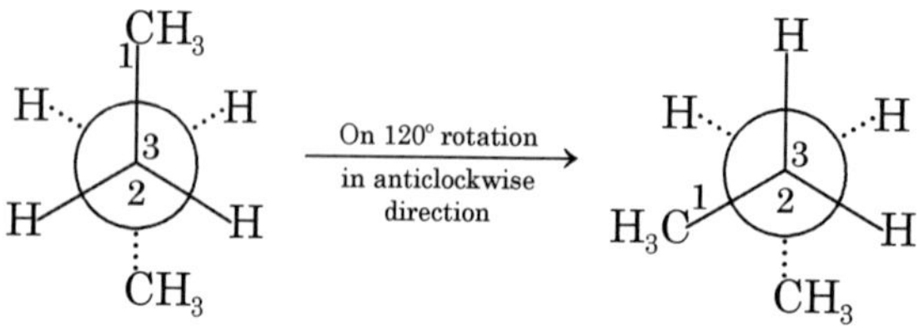

Gauche conformer

3. B

Benzamide on treatment with $POCl_3$ gives benzonitrile because dehydration occurs since $POCl_3$ is a dehydrating agent.

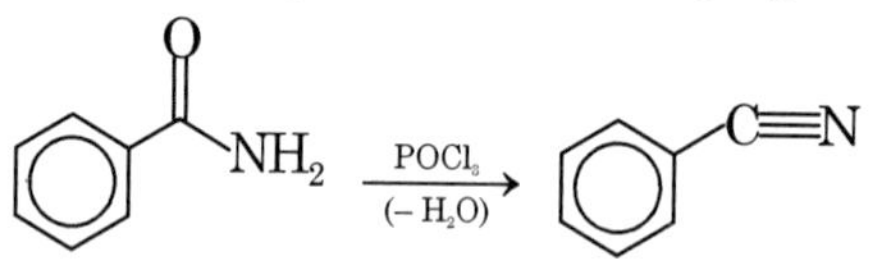

4. A

The methods chiefly used for the extraction of lead and tin from their ores are respectively self-reduction and carbon reduction.

(a) $2PbS + 3O_2 \longrightarrow 2PbO + 2SO_2$

$PbS + 2O_2 \longrightarrow PbSO_4$

$2PbO + PbS \longrightarrow 3Pb + SO_2$

$PbS + PbSO_4 \longrightarrow 2Pb + 2SO_2$

(b) $SnO_2 + 2C \longrightarrow Sn + 2CO$

5. B

The ring with maximum electron density will be substituted by an electrophile. The ring attached to —NH— will have rich electron density due to resonance. Since the *ortho* position is blocked, the electrophile attacks the *para* position.

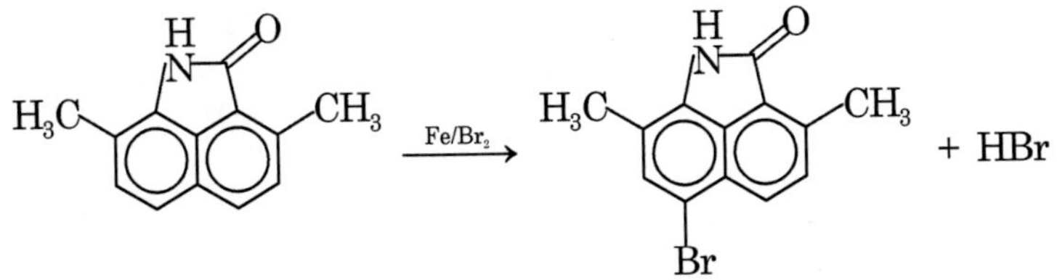

6. C

A nucleophile attacks the most electrophilic site first. Among aldehyde and ketone, the aldehyde is more electrophilic since in the ketone the δ+ charge on carbonyl carbon is decreased by the + I-effect of both the alkyl groups.

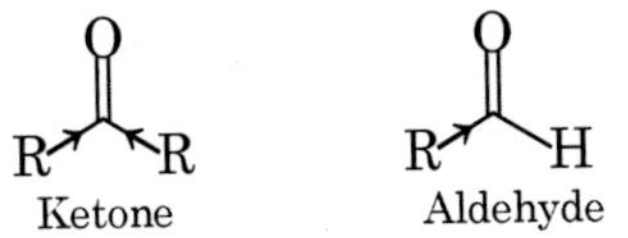

Moreover, in the tetrahedral intermediate, the aldehyde has less steric repulsion than that in ketone. This facilitates the formation of intermediate in the aldehyde. Hence the order of reactivity is: II > I > III.

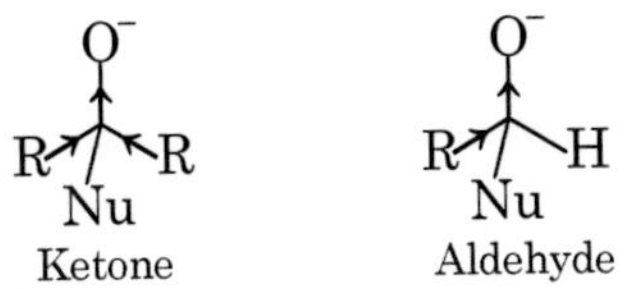

7. A

Ammonium dichromate on heating gives a gas which is also given by heating ammonium nitrite.

$$(NH_4)_2Cr_2O_7 \xrightarrow{\Delta} Cr_2O_3 + 4H_2O + N_2\uparrow$$

$$NH_4NO_2 \xrightarrow{\Delta} 2H_2O + N_2\uparrow$$

8. B

For the cell $Zn|Zn^{2+}$ ($a = 0.1M$) || Fe^{2+} ($a = 0.01M$)|Fe the cell reaction is

$$\begin{array}{l} Zn(s) \longrightarrow Zn^{2+}(aq) + 2e^- \\ \underline{Fe^{2+}(aq) + 2e^- \longrightarrow Fe(s)} \\ Zn(s) + Fe^{2+}(aq) \longrightarrow Zn^{2+}(aq) + Fe(s) \end{array}$$

According to the Nernst equation,

$$E_{cell} = -\frac{0.0591}{n}\log\frac{[Zn^{2+}]}{Fe^{2+}}$$

$$0.2905 = E^0_{cell} - \frac{0.0591}{2}\log\frac{0.1}{0.01}$$

$$E^0_{cell} = 0.2905 + 0.0295 \log 10 = 0.2905 + 0.0295 = 0.32 \text{ V}$$

At equilibrium, $E_{cell} = 0$

$$E_{cell} = E^0_{cell} - \frac{0.0591}{n}\log K$$

$$0 = E^0_{\text{cell}} - \frac{0.0591}{n}\log K$$

$$\text{or} \quad E^0_{\text{cell}} = \frac{0.0591}{2}\log K$$

$$0.32 = \frac{0.0591}{2}\log K$$

$$\text{or} \quad K = 10^{0.32/0.0295}$$

9. A

$NaX + H_2O \longrightarrow NaOH + HX$

Since HX is a weak acid, NaX is a salt of a weak acid and a strong base. The hydrolysis constant (K_h) for such a salt is given by

$$K_h = \frac{K_w}{K_a} = \frac{1\times10^{-14}}{10^{-5}} = 1\times10^{-9}$$

For this type of salt,

$$K_h = Ch^2$$

where h is the degree of hydrolysis. Hence

$1 \times 10^{-9} = 0.1 \times h^2$

$$\text{or} \quad h^2 = \frac{1\times10^{-9}}{0.1} = 1\times10^{-8}$$

$$\text{or} \quad h = 1\times10^{-4}$$

Hence % degree of hydrolysis = $1 \times 10^{-4} \times 100 = 0.01\%$

10. C

For a spontaneous adsorption, ΔG is negative. Since the degree of randomness of gas molecules on the surface of solid decreases, ΔS is negative. Hence $T\Delta S$ is negative. Since adsorption is an exothermic process, ΔH is negative.

$\Delta H = \Delta G + T\,\Delta S$

Hence during adsorption, ΔS decreases.

11. A

Root main square velocity, $u_{rms} = \sqrt{\frac{3PV}{m}}$

For 1 molecule of gas, $PV = kT$

Hence $u_{rms} = \sqrt{\frac{3kT}{m}}$

But K.E. $= \frac{3}{2}kT$

or 2 K.E. $= 3kT$

Hence $u_{rms} = \sqrt{\frac{2\text{ K.E.}}{m}}$

12. B

In MnO_2 and $FeCl_3$, the oxidation states of Mn and Fe are + 4 and + 3 respectively. In MnO_4^- and CrO_2Cl_2, the oxidation states of Mn and Cr are + 7 and + 6 respectively. In $[Fe(CN)_6]^{3-}$ and $[Co(CN)_3]$, the oxidation states of Fe and Co are + 3 and + 3 respectively.
In $[NiCl_4]^{2-}$ and $[CoCl_4]^-$, the oxidation states of Ni and Co are + 2 and + 3 respectively. Hence the pair of compounds having metals in their highest oxidation state is (B) MnO_4^-, CrO_2Cl_2.

13. D

Electronic configuration of Ni^{2+}: $1s^2 2s^2 2p^6 3s^2 3p^6 3d^8$

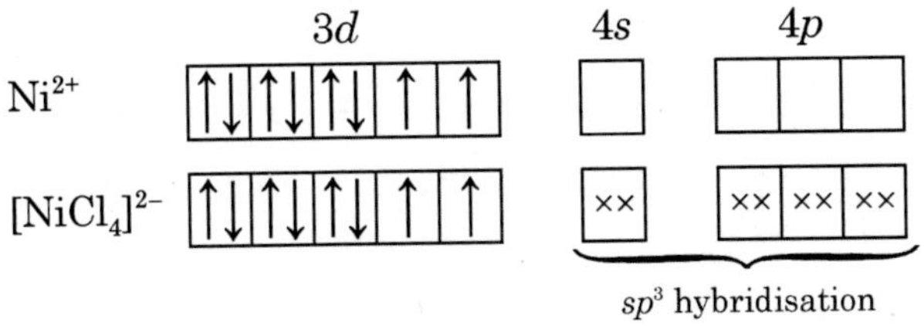

Since the hybridisation of $[NiCl_4]^{2-}$ is sp^3, the shape of $[NiCl_4]^{2-}$ is tetrahedral.

14. A

The oxidation state of Co in $Hg[Co(NCS)_4]$ is + 2, Co^{2+} ion belongs to $3d^7$ system.

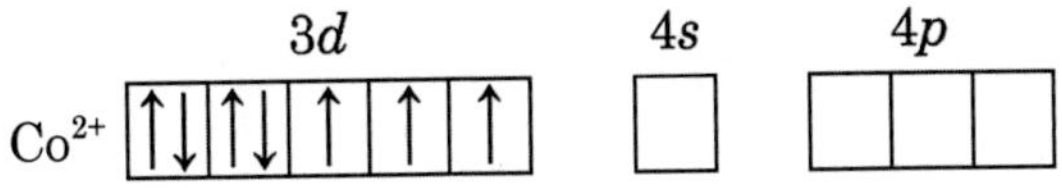

Hence $n = 3$.

Spin-only magnetic moment, $\mu_s = \sqrt{n(n+2)} = \sqrt{3(3+2)}$ B.M. $= \sqrt{15}$ B.M.

15. B

The anion is HCO_3^-.

$$MgCl_2 + 2NaHCO_3 \longrightarrow \underset{\text{Soluble}}{Mg(HCO_3)_2} + 2NaCl$$

$$Mg(HCO_3)_2 \xrightarrow{\Delta} \underset{\text{White}}{MgCO_3\downarrow} + H_2O + CO_2$$

16. B

We know $r = \dfrac{n^2 r_{(H)}}{Z}$

Given that $r = r_{(H)}$

Hence $n^2 = Z$

For Be^{3+}, $Z = 4$

Hence $n = 2$

Thus $r_{2\,(Be^{3+})} = r_{1(H)}$

17. A

The pK_a value of carboxylic group is less than the pK_a value of $—\overset{+}{N}H_3(y)$ in amino acid and $—\overset{+}{N}H_3(z)$ will have comparatively less pK_a value than $—\overset{+}{N}H_3(y)$ due to the

– I-effect of the carboxylate group. Since the acid strength is inversely proportional to pK_a, the order of acid strength is

$—COOH\ (x) > —\overset{+}{N}H_3\ (z) > —\overset{+}{N}H_3\ (y)$

18. A

Since $0.004M$ Na_2SO_4 solution is isotonic with $0.01M$ solution of glucose,

$\pi_{Na_2SO_4} = \pi_{glucose} = 0.01 \times RT$

$$Na_2SO_4 \rightleftharpoons 2Na^+ + SO_4^{2-}$$

At $t = 0$ 1 0 0

At equilibrium $1 - \alpha$ 2α α (where α is the degree of dissociation of Na_2SO_4)

$(\pi_{cal})_{Na_2SO_4} = CRT = 0.004 \times R \times T$

By van't Hoff factor,

$$\frac{(\pi_{obs})_{Na_2SO_4}}{(\pi_{calc})_{Na_2SO_4}} = \frac{\text{No. of particles after dissociation}}{\text{No. of particles before dissociation}} = \frac{1 - \alpha + 2\alpha + \alpha}{1}$$

But $(\pi_{obs})_{Na_2SO_4} = \pi_{glucose} = 0.01 \times RT$

Hence $\dfrac{0.01 \times R \times T}{0.004 \times R \times T} = \dfrac{1 + 2\alpha}{1}$

or $\dfrac{10}{4} = \dfrac{1 + 2\alpha}{1}$

or $10 = 4 + 8\alpha$

or $\alpha = \dfrac{10 - 4}{8} = 0.75$

$\% \alpha = 0.75 \times 100 = 75$

19. A

$$\Delta G = \Delta H - T\Delta S$$

At equilibrium, $\Delta G = 0$. Hence

$$T\Delta S = \Delta H$$

or $T = \dfrac{\Delta H}{\Delta S} = \dfrac{30 \times 10^3}{75} = 400\ K$

or $T = 400$ K

20. D

For an isothermal process, $\Delta T = 0$ and $\Delta E = 0$.

$$H = E + PV$$

$$\Delta H = \Delta E + \Delta(PV)$$

or $\Delta H = \Delta E + nR\Delta T = 0 + nR \times 0$

or $\Delta H = 0$

21. A

For first order reaction,

$$k = \frac{2.303}{t} \log \frac{[A]_0}{[A]_t}$$

At $t = 40$ minute,

$$k = \frac{2.303}{40}\log\frac{0.1}{0.025} = \frac{2.303}{40}\log 4 = \frac{2.303}{40}\times 2\log 2$$

$$= \frac{2.303}{4}\times 2\times 0.3010 = 0.0347\ \text{min}^{-1}$$

$$r = k[\text{A}]$$

$$= 0.0347\times 0.01 = 3.47\times 10^{-4}\ \text{mol L}^{-1}\ \text{min}^{-1}$$

22. A

Li/NH_3 brings about *trans* addition of H_2 to an alkyne.

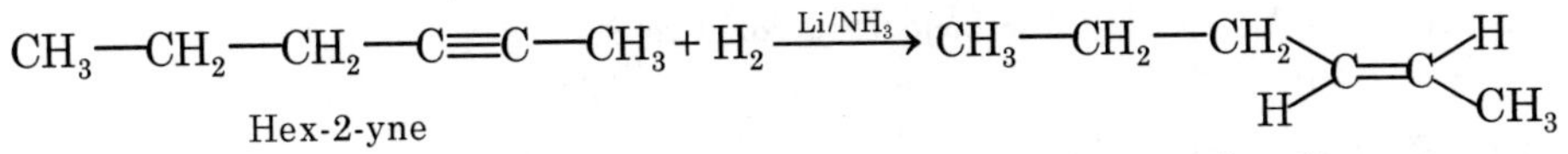

Hex-2-yne — *trans*-hex-2-ene

23. B

On chlorination of 2-methylbutane, four chiral compounds are formed.

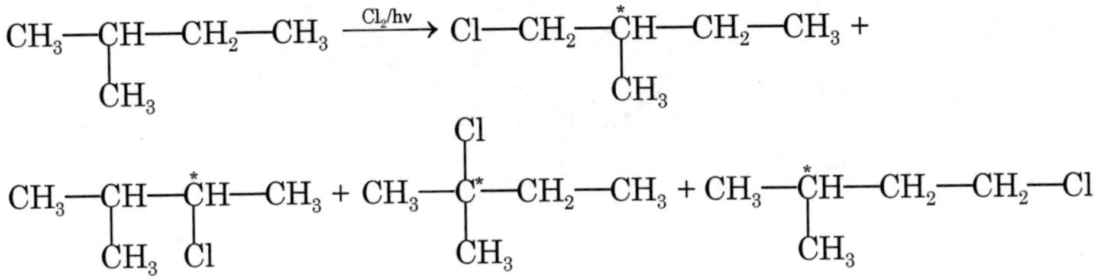

24. B

Aldehydes and α-hydroxyketones exhibit positive Tollen's test. Glucose has an aldehyde group and fructose is an α-hydroxyketone. Hence both glucose and fructose give positive Tollen's test.

25. B

$H_2S_2O_8$ has —O—O—linkage.

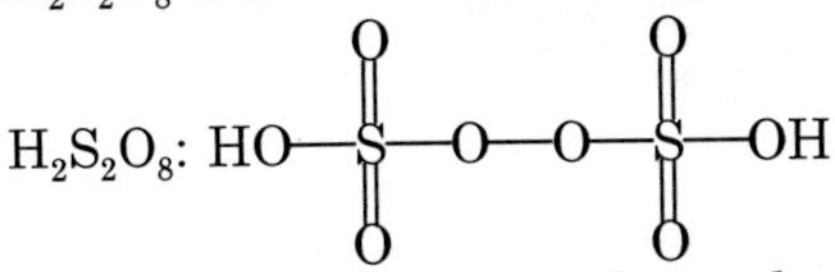

Other three compounds do not have —O—O—llinkage.

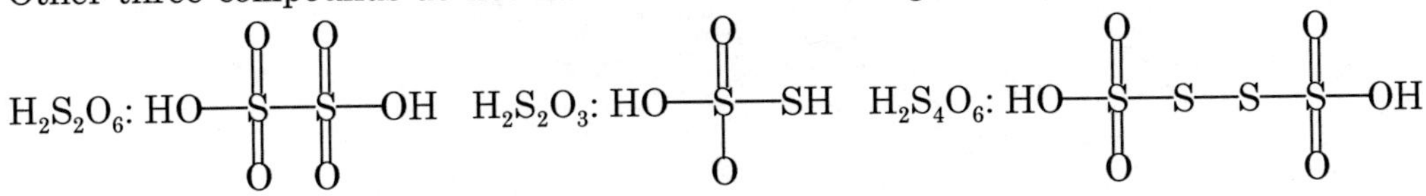

26. A

When I^- ion is oxidised by MnO_4^- ion in alkaline medium, I^- ion is converted into IO_3^- ion.

$$KI + 2KMnO_4 + H_2O \rightarrow 2MnO_2 + KIO_3 + 2KOH$$

27. B

Number of lone pair on Xe in $XeOF_4$ is 1. The structure of $XeOF_4$ is as follows:

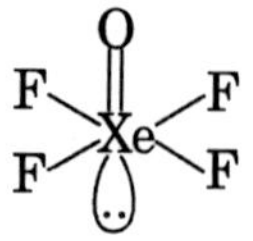

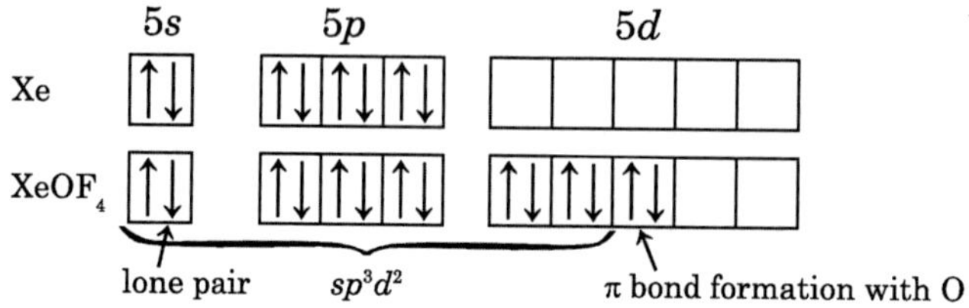

28. A

$O_2^+: (\sigma_{1s}^b)^2 (\sigma_{1s}^*)^2 (\sigma_{2s}^b)^2 (\sigma_{2s}^*)^2 (\sigma_{2p_z}^b)^2 (\pi_{2p_x}^b)^2 (\pi_{2p_y}^b)^2 (\pi_{2p_x}^*)^1$

B. O. $= \frac{1}{2}$ (Number of bonding electrons – Number of antibonding electrons)

$= \frac{1}{2}(10 - 5) = 2.5$

$O_2: (\sigma_{1s}^b)^2 (\sigma_{1s}^*)^2 (\sigma_{2s}^b)^2 (\sigma_{2s}^*)^2 (\sigma_{2p_z}^b)^2 (\pi_{2p_x}^b)^2 (\pi_{2p_y}^b)^2 (\pi_{2p_x}^*)^1 (\pi_{2p_y}^*)^1$

B. O. $= \frac{1}{2}(10 - 6) = 2$

I.I.T. OBJECTIVE QUESTION PAPER—2005

1. When phenyl magnesium bromide reacts with tertiary butanol, which of the following is formed?
(A) Tertiary butyl methyl ether (B) Benzene
(C) Tertiary butyl benzene (D) Phenol

2. 1-Bromo-3-chlorocyclobutane when treated with two equivalents of Na in the presence of ether, which of the following will be formed?
(A) 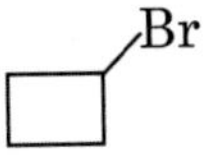 (B) 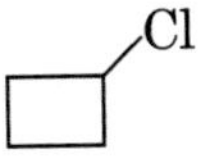 (C) 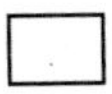 (D) 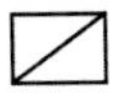

3. In the reaction, 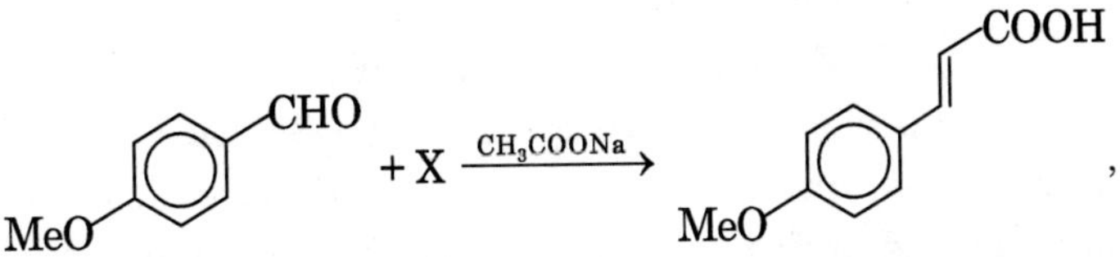 , what is X?
(A) CH_3COOH (B) $BrCH_2COOH$ (C) $(CH_3CO)_2O$ (D) OHC—COOH

4. Cyclohexene is best prepared from cyclohexanol by which of the following?
(A) Conc. H_3PO_4 (B) Conc. $HCl/ZnCl_2$ (C) Conc. HCl (D) Conc. HBr

5. 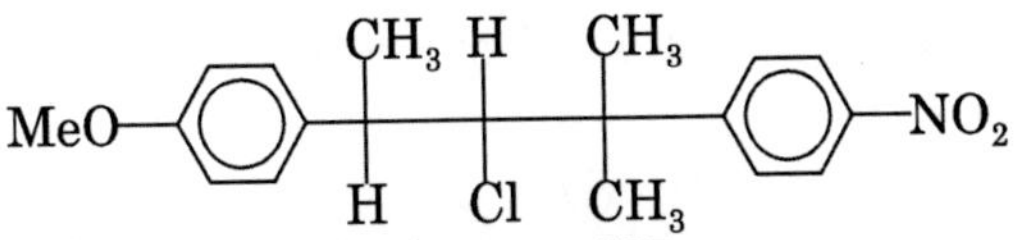 on hydrolysis in aqueous acetone gives

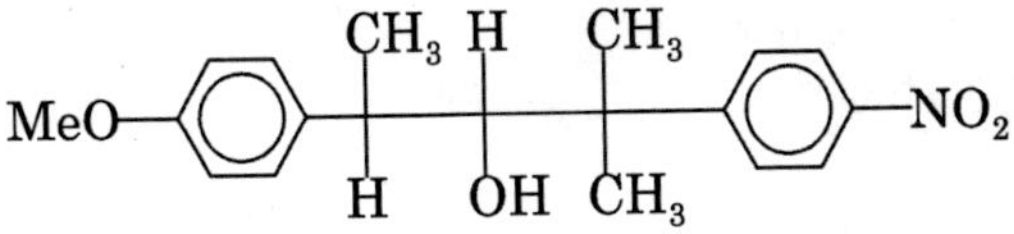 (K)

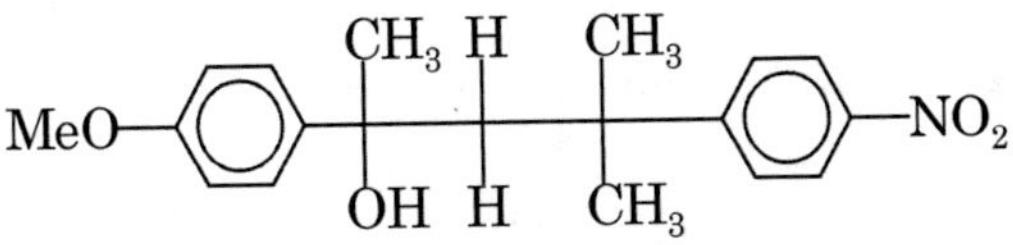 (L)

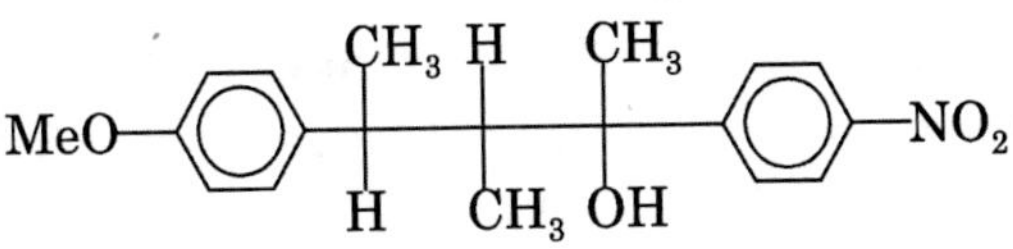 (M)

It mainly gives
(A) K and L (B) only K (C) L and M (D) only M

6. For 1-methoxy-1,3-butadiene, which of the following resonating structures is the least stable?

(A) $H_2\bar{C}-\overset{+}{C}H-CH=CH-O-CH_3$ (B) $H_2\bar{C}-CH=CH-CH=\overset{+}{O}-CH_3$

(C) $H_2C=CH-\overset{+}{C}H-\bar{C}H-O-CH_3$ (D) $H_2C=CH-\bar{C}H-CH=\overset{+}{O}-CH_3$

7. 2-Butanone can be converted into propanoic acid by which of the following?
   (A) NaOH, NaI/H^+ (B) Fehling solution
   (C) NaOH, I_2/H^+ (D) Tollen's reagent.
8. Two forms of D-glucopyranose are called
   (A) enantiomers (B) anomers (C) epimers (D) diastereoisomers.
9. Which of the following pairs is expected to exhibit same colour in solution?
   (A) $VOCl_2$; $FeCl_2$ (B) $CuCl_2$; $VOCl_2$ (C) $MnCl_2$; $FeCl_2$ (D) $FeCl_2$; $CuCl_2$
10. Which of the following allotropes of phosphorus is thermodynamically most stable?
    (A) Red (B) White (C) Black (D) Yellow
11. A metal nitrate gives black precipitate with KI and on adding excess of KI gives orange colour. It is
    (A) Hg^{2+} (B) Bi^{3+} (C) Sn^{2+} (D) Pb^{2+}
12. Which of the following will not be oxidised by O_3?
    (A) KI (B) $FeSO_4$ (C) $KMnO_4$ (D) K_2MnO_4
13. Which type of isomerism is shown by $[Co(NH_3)_4Br_2]Cl$?
    (A) Geometrical and ionisation (B) Optical and ionisation
    (C) Geometrical and optical (D) Geometrical only.
14. Which of the following *fcc* structures contains cations in alternate tetrahedral voids?
    (A) NaCl (B) ZnS (C) Na_2O (D) CaF_2
15. The elevation in boiling point, when 13.44 g of freshly prepared $CuCl_2$ are added to one kilogram of water, is (some useful data: $K_b = 0.52$ kg K mol^{-1}, molecular mass of $CuCl_2 = 134$ g]
    (A) 0.05 (B) 0.1 (C) 0.16 (D) 0.21
16. The half-cell reactions for resulting of iron ore are:
    $2H^+ + \frac{1}{2}O_2 + 2e^- \longrightarrow H_2O$; $E^0 = +1.23$ V, $Fe^{2+} + 2e^- \longrightarrow Fe(s)$; $E^0 = -0.44$ V ΔG^0 (in kJ) for the reaction is
    (A) – 76 (B) – 322 (C) – 122 (D) – 176
17. The number of radial nodes in 3*s* and 2*p* orbitals respectively are
    (A) 2 and 0 (B) 1 and 2 (C) 0 and 2 (D) 2 and 1
18. Which of the following ones contains both copper and iron?
    (A) Cuprite (B) Chalcocite (C) Chalcopyrite (D) Malachite.
19. A pale blue liquid which is obtained by equimolar mixture of two gases at – 30° is
    (A) N_2O (B) N_2O_3 (C) N_2O_4 (D) N_2O_5
20. Which of the following is obtained when 4-methylbenzenesulphonic acid is reacted with excess of sodium acetate solution?

(A) 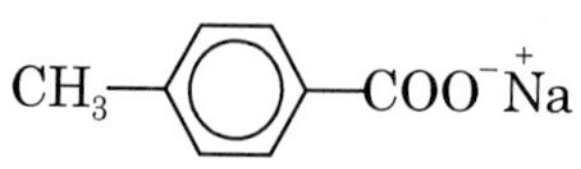

(B) 

(C) 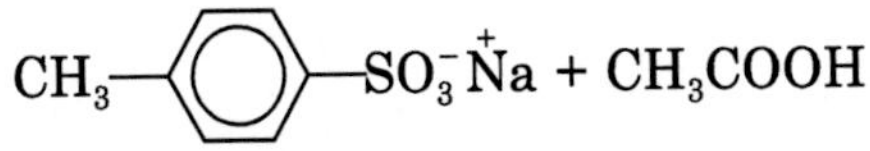

(D) 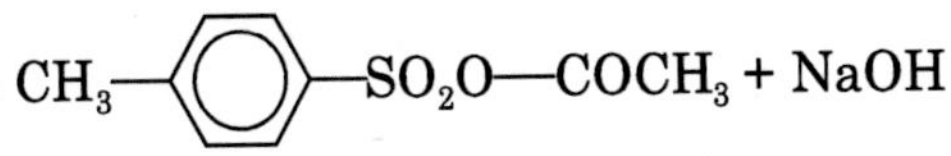

21. CH_3NH_2 (0.1 mole, $K_b = 5 \times 10^{-4}$) is added to 0.08 mole of HCl and the solution is diluted to one litre. The resulting hydrogen ion concentration is

(A) 1.6×10^{-11} (B) 8×10^{-11} (C) 5×10^{-5} (D) 8×10^{-2}

22. Which silicates are formed from $[SiO_4]^{4-}$ tetrahedral units by sharing three oxygen atoms?

(A) Sheet silicates (B) Pyro silicates

(C) Linear chain silicates (D) 3-Dimensional silicates.

23. Which gas is evolved when PbO_2 is treated with conc. HNO_3?

(A) NO_2 (B) O_2 (C) N_2 (D) N_2O

24. If helium and methane are allowed to diffuse out of the container under the similar conditions of temperature and pressure, then the ratio of rate of diffusion of helium to methane is

(A) 2.0 (B) 1.0 (C) 0.5 (D) 4.0

25. Which of the following contains maximum number of lone pairs on the central atom?

(A) ClO_3^- (B) XeF_4 (C) SF_4 (D) I_3^-

26. Which of the following is correct for lyophilic sols?

(A) They are irreversible

(B) They are formed by inorganic substances

(C) They are readily coagulated by addition of electrolytes

(D) They are self stabilised.

27. Which of the following statements is incorrect about order of reaction?

(A) Order of reaction is determined experimentally

(B) It is the sum of power of concentration terms in the rate law expression

(C) It does not necessarily depend on stoichiometric coefficients

(D) Order of reaction cannot have fractional value.

28. One mole of a monoatomic ideal gas expands adiabatically at initial temperature T against a constant external pressure of 1 atm from one litre to two litres. Find out the final temperature. (R = 0.0821 L atm K^{-1} mol^{-1})

(A) T (B) $\dfrac{T}{2^{[(5/3)\,-\,1]}}$

(C) $T - \dfrac{2}{3 \times 0.0821}$ (D) $T + \dfrac{2}{3 \times 0.0821}$

ANSWERS

1. B	2. D	3. C	4. A	5. A	6. C
7. C	8. B	9. B	10. C	11. B	12. C
13. A	14. B	15. C	16. B	17. A	18. C
19. B	20. C	21. B	22. A	23. B	24. A
25. D	26. D	27. D	28. C		

MODEL SOLUTIONS—I.I.T. 2005

1. B

$PhMgBr + Me_2COH \rightarrow Ph—H + Me_3COMgBr$

2. D

It is an intermolecular Wurtz reaction.

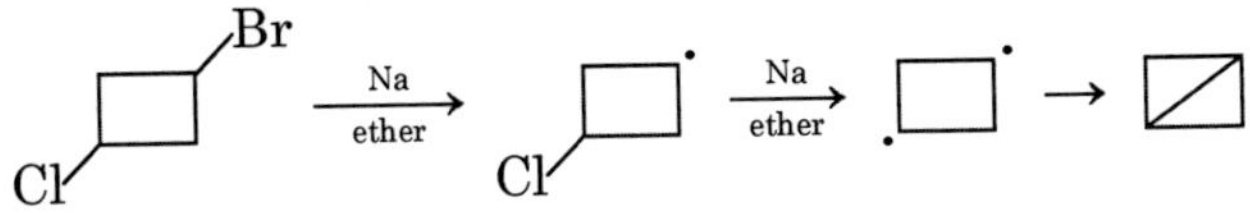

3. C

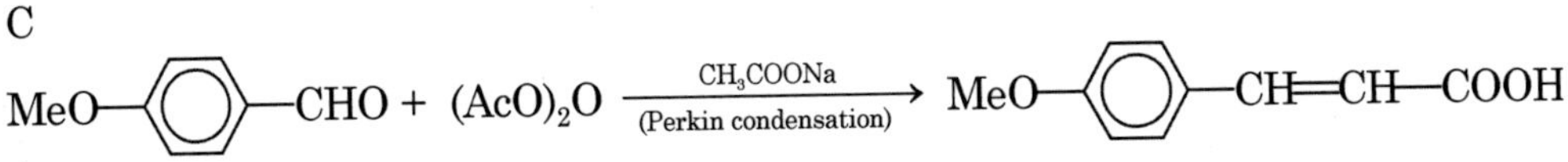

4. A

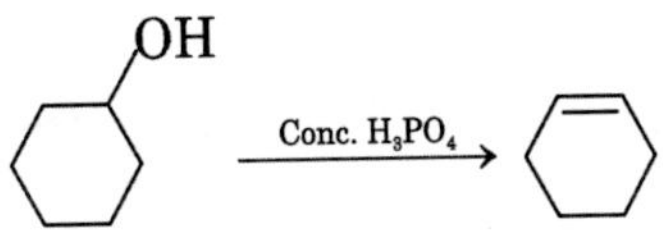

H_3PO_4 acts as a dehydrating agent.

5. A

Both S_{N^1} and S_{N^2} reactions are possible due to aqueous acetone solution.

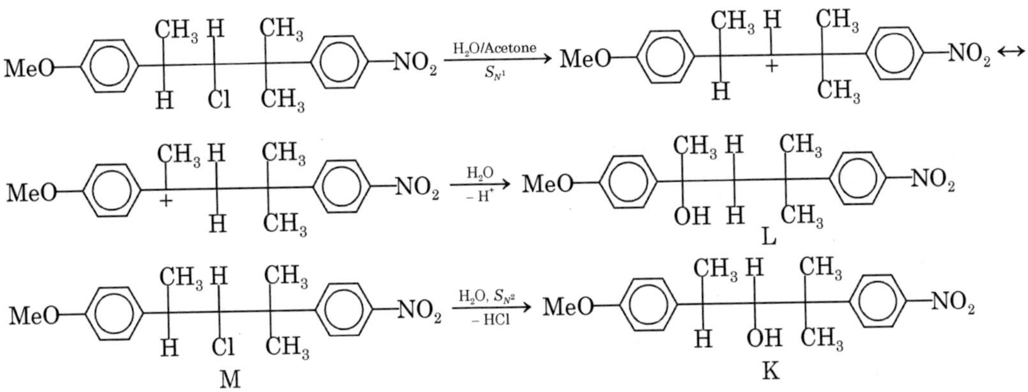

6. C

The point of difference is nature of anion. 2° carbanions are less stable than 1° carbanions.

7. A

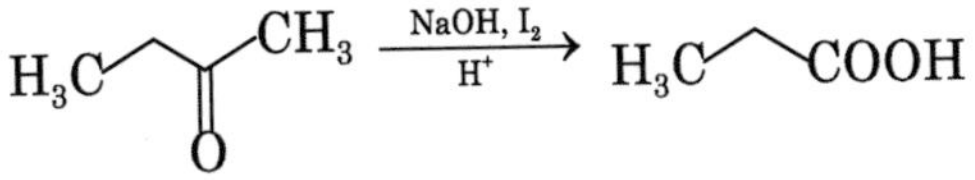

8. B

D-Glucopyranose is the cyclic form of glucose. Two isomers are observed around C_1 containing the chiral centre. These isomers are calledα-anomer and β-anomer. Anomers have different configurations at the C_1 atom.

9. B

Both Cu^{2+} and V^{4+} ions have one unpaired electron and hence are coloured. Both $CuCl_2$ and $VOCl_2$ are green coloured compounds.

10. C

Black phosphorus is thermodynamically most stable due to the presence of layered structure. It is inert.

11. B

$$Bi^{3+} + KI \longrightarrow \underset{\text{Black}}{BiI_3}$$

$$BiI_3 + KI \longrightarrow \underset{\text{Orange solution}}{K[BiI_4]}$$

12. C

Since manganese can exhibit the maximum possible oxidation state of + 7, $KMnO_4$ containing Mn(VII) cannot be oxidised by any oxidising agent.

13. A

$[Co(NH_3)_4Br_2]Cl$ can exhibit both geometrical isomerism (*cis*-$[Co(NH_3)_4Br_2]Cl$, *trans*-$[Co(NH_3)_4Br_2]Cl$) ionisation isomerism ($[Co(NH_3)_4Br_2]Cl$, $[Co(NH_3)_4BrCl]Br$).

14. B

In ZnS, the anions (S^{2-}) are placed in *fcc* manner and the cations (Zn^{2+}) are placed in the alternate tetrahedral voids.

15. C

$$\Delta T_b = i \times K_b \times m = 3 \times 0.52 \times \left(\frac{13.44}{134.4} \times \frac{1000}{1000}\right) = 0.16$$

16. B

$$2H^+ + \frac{1}{2}O_2 + 2e^- \longrightarrow H_2O;\ E^0 = +\ 1.23\ V$$

$$Fe^{2+} + 2e^- \longrightarrow Fe(s);\ E^0 = -\ 0.44\ V$$

Subtract

$$Fe(s) + 2H^+ + \frac{1}{2}O_2 \longrightarrow Fe^{2+} + H_2O;\ E^0_{cell} = 1.67\ V$$

$$\Delta G^0 = -\ nFE^0_{cell} = -\ 2 \times 96500 + 1.67 = -\ 322\ kJ$$

17. A

Number of radial nodes = $n - l - 1$

For 3*s* orbital, the number of radial nodes = 3 – 0 – 1 = 2

For 2*p* orbital, the number of radial nodes = 2 – 1 – 1 = 0

18. C

Chalcopyrite ($CuFeS_2$) contain both copper and iron.

19. B

$$NO + NO_2 \xrightarrow{-30^\circ} \underset{\text{Pale blue liquid}}{N_2O_3}$$

20. C

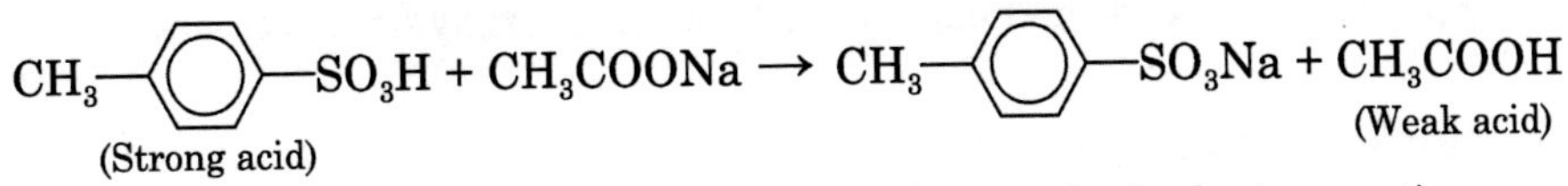

This reaction is an acid-base reaction and not a hydrolysis reaction.

21. B

	CH_3NH_2 +	HCl →	$CH_3\overset{+}{N}H_3$ + Cl^-
Initial mole	0.1	0.08	
Moles at equi.	0.02	–	0.08

$$[OH^-] = K_b \frac{[CH_3NH_2]}{[CH_3\overset{+}{N}H_3]} = \frac{5 \times 10^{-4} \times 0.02}{0.08} = \frac{5}{4} \times 10^{-4}$$

$$[H^+] = \frac{K_w}{[OH^-]} = \frac{10^{-14} \times 4}{5 \times 10^{-4}} = 8 \times 10^{-11}$$

22. A

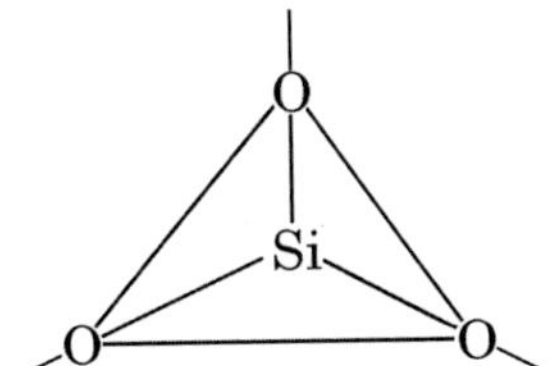

Sheet structure

23. B

$$2PbO_2 + 4HNO_3 \longrightarrow 2Pb(NO_3)_2 + 2H_2O + O_2$$

24. A

$$\frac{r_{He}}{r_{CH_4}} = \sqrt{\frac{16}{4}} = \sqrt{4} = 2$$

25. D

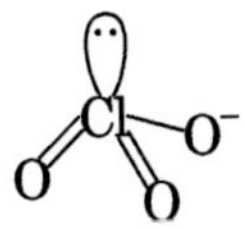

(One lone pair)

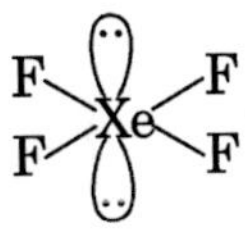

(Two lone pairs)

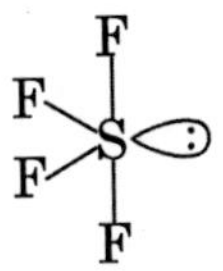

(One lone pair)

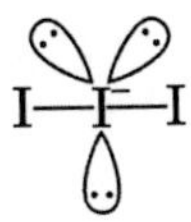

(Three lone pairs)

26. D

Lyophilic sols are solvent loving in nature and they are extensively solvated. Due to this property, the lyophilic sols are self-stabilised and they cannot be readily coagulated.

27. D

The order of reaction is determined experimentally and it may be fractional.

28. C

Work done against constant pressure $= -P_{\text{ext}}(V_2 - V_1)$
Under adiabatic condition, $\Delta q = 0$. Hence $w = \Delta U$.

$$\Delta U = nC_{v,m}(T_2 - T_1) = 1 \times \frac{3}{2}R(T_2 - T_1)$$

$$-P(V_2 - V_1) = \frac{3}{2}R(T_2 - T_1) \text{ (Expansion work is negative.)}$$

or $$-1(2-1) = \frac{3}{2}R(T_2 - T_1)$$

or $$T_2 - T_1 = -\frac{2}{3R}$$

or $$T_2 = T_1 - \frac{2}{3R} = T_1 - \frac{2}{3 \times 0.0821}$$

I.I.T. OBJECTIVE QUESTION PAPER—2006

Section-I (MCQ with only one choice correct)

1. $B(OH)_3 + NaOH \rightleftharpoons Na[B(OH)_4]$

   How can this reaction be made to proceed in forward direction?

   (A) Addition of *cis*-1,2-diol (B) Addition of borax
   (C) Addition of *trans*-1,2-diol (D) Addition of Na_2HPO_4

   Hint: $Na[B(OH)_4]$ is removed from the equilibrium due to the formation of chelate compound with *cis*-1,2-diol and the reaction moves in the forward direction.

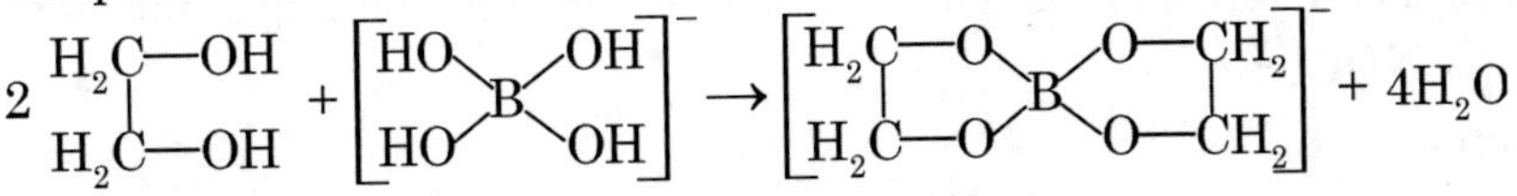

2. A solution when diluted with H_2O and boiled, it gives a white precipitate. On addition of excess NH_4Cl/NH_4OH, the volume of precipitate decreases leaving behind a white gelatinous precipitate. Identify the precipitate which dissolves in NH_4OH/NH_4Cl.

   (A) $Zn(OH)_2$ (B) $Al(OH)_3$ (C) $Mg(OH)_2$ (D) $Ca(OH)_2$

   Hint: White precipitate of $Zn(OH)_2$ dissolves in NH_4OH/NH_4Cl to form tetraamminezinc (II) hydroxide $[Zn(NH_3)_4](OH)_2$: $Zn(OH)_2 + 4NH_4OH \longrightarrow [Zn(NH_3)_4](OH)_2 + 4H_2O$

3. When benzenesulphonic acid and *p*-nitrophenol are treated with $NaHCO_3$, the gases released respectively are

   (A) SO_2, NO_2 (B) SO_2, NO (C) SO_2, CO_2 (D) CO_2, CO_2

   Hint:

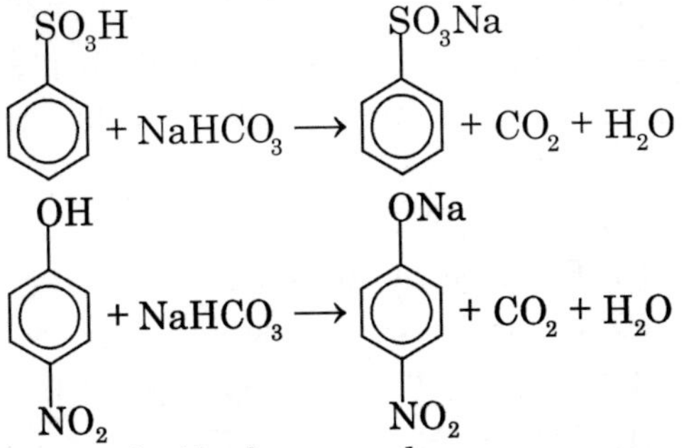

4. A monoatomic ideal gas undergoes a process in which the ratio of P to V at any instant is constant and equals to 1. What is the molar heat capacity of the gas?

   (A) $4R/2$ (B) $3R/2$ (C) $5R/2$ (D) 0

   Hint: Equation for the process: PV^{-1} = constant

Molar heat capacity for the process $= \dfrac{R}{\gamma - 1} + \dfrac{R}{1-(-1)} = \dfrac{R}{\frac{5}{3}-1} + \dfrac{R}{2} = \dfrac{3R}{2} + \dfrac{R}{2} = \dfrac{4R}{2}$

5. Arrange the following compounds according to decreasing boiling point.

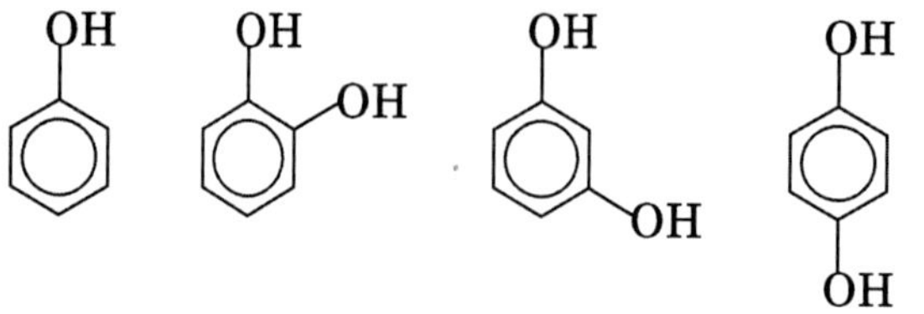

I II III IV

(A) IV > III > II > I (B) III > IV > II > I

(C) I > II > III > IV (D) II > III > I > IV

Hint: Increase in the number of hydroxyl groups increases the boiling point. Among the isomeric dihydroxybenzenes, as the distance between the hydroxyl groups increases, the repulsion between them decreases and hence the boiling point increases.

6. $CH_3-CH=CH_2 + NOCl \rightarrow P$

Identify the adduct.

(A) $CH_3-CH(Cl)-CH_2(NO)$ (B) $CH_3-CH(NO)-CH_2(Cl)$

(C) $CH_3-CH_2-CH(NO)(NO)$ (D) $CH_2(NO)-CH_2-CH_2(Cl)$

Hint: The reaction occurs through Markownikoff addition.

$$CH_3-CH=CH_2 + NO^+ \rightarrow CH_3-\overset{+}{C}H-CH_2(NO) \xrightarrow{Cl^-} CH_3-CH(Cl)-CH_2(NO)$$

The 2° carbocation ($CH_3-\overset{+}{C}H-CH_2-NO$) formed is more stable than 1° carbocation ($CH_3-CH_2-\overset{+}{C}H_2$). The chloride ion attacks the more stable 2° carbocation to give the major product (A).

7. The IUPAC name of C_6H_5COCl is

(A) Benzoyl chloride (B) Benzenechloroketone

(C) Benzenecarbonyl chloride (D) Chlorophenylketone

8. $CH_3NH_2 + CHCl_3 + 3KOH \rightarrow (P) + 3KCl + H_2O$. The missing product is

(A) $CH_3-C\equiv N$ (B) $CH_3-NH-Cl$

(C) $CH_3-\bar{N}\equiv\overset{+}{C}$ (D) $CH_3-\overset{+}{N}\equiv\overset{-}{C}$

Hint: This is carlylamine reaction and alkyl isocyanide is formed.

9. The blue solution of $CuSO_4$ is decolourised on addition of excess KCN. The product is

(A) $[Cu(CN)_4]^{2-}$ (B) Cu^{2+} (C) $Cu(OH)_2$

(D) CuCN is reduced by CN^- to Cu^+ which forms the complex $[Cu(CN)_4]^{3-}$.

Hint: $2CuSO_4 + 4KCN \longrightarrow 2CuCN + (CN)_2 + 2K_2SO_4$

$CuCN + 4KCN \longrightarrow [Cu(CN)_4]^{3-} + 4K^+ + CN^-$

10. The species present in solution when CO_2 is dissolved in water are

(A) CO_2, H_2CO_3, HCO_3, CO_3^{2-} (B) H_2CO_3, CO_3^{2-}

(C) CO_3^{2-}, HCO_3^- (D) CO_2, H_2CO_3

Hint: $CO_2 + H_2 \rightleftharpoons H_2CO_3 \rightleftharpoons H^+ + HCO_3^-; HCO_3^- \rightleftharpoons H^+ + CO_3^{2-}$

11. The direct converion of A to B is difficult. Hence it is carried out by the following shown path:

$A \longrightarrow C \longrightarrow D \longrightarrow B$

Given: $\Delta S_{(A \to C)} = 50\,\text{eu}$, $\Delta S_{(C \to D)} = 30\,\text{eu}$, $\Delta S_{(B \to D)} = 20\,\text{eu}$, where eu is entropy unit. The value of $\Delta S_{(A \to B)}$ is

(A) + 100 eu (B) + 60 eu (C) – 100 eu (D) – 60 eu

Hint: $\Delta S_{(A \to B)} = \Delta S_{(A \to C)} + \Delta S_{(C \to D)} - \Delta S_{(B \to D)} = 50 + 30 - 20 = 60\,\text{eu}$

12. Haber–Bosch process for the formation of NH_3 at 298 K is $N_2(g) + 3H_2(g) \rightleftharpoons 2NH_3(g)$; $\Delta H = -\ 46.0$ kJ; $K_p = 14$

The correct statement is

(A) On adding N_2, the equilibrium will spontaneously shift to the right and ΔS will be positive

(B) Even though at 298 K, the yield is more, still industrial manufacture of NH_3 is carried out at 400 K ($K_p = 41$). This is because at 298 K, the catalyst increases the rate by a factor of two for both forward and backward reaction. But at 400 K, the iron catalyst increases the rate of forward reaction by a factor of two and that of backward reactio by a factor of 1.7

(C) At equilibrium, $2G_{NH_3} = G_{N_2} + 3G_{H_2}$ where G is the Gibbs free energy

(D) Addition of catalyst does not alter K_p but changes ΔH.

Section-II (MCQ with one or more than one choice correct)

13. $Ag^+ + NH_3 \rightleftharpoons [Ag(NH_3)]^+ ; k_1 = 6.8 \times 10^3$

$[Ag(NH_3)]^+ + NH_3 \rightleftharpoons [Ag(NH_3)_2]^+ ; k_2 = 1.6 \times 10^3$

The overall formation constant of $[Ag(NH_3)_2]^+$ is

(A) 1.08×10^7 (B) 1.08×10^3 (C) 6.8×10^6 (D) 6.8×10^7

Hint: $\beta_k = k_1 \times k_2 = 6.8 \times 10^3 \times 1.6 \times 10^3 = 1.08 \times 10^7$

14. If the bond length of CO bond in carbon monoxide is 1.128 Å, then what is the value of CO bond length in $Fe(CO)_5$?

(A) 1.13 Å (B) 1.128 Å (C) 1.10 Å (D) 1.115 Å

Hint: Due to the back donation of electrons from M to CO, the M—C bond becomes stronger and the C≡O bond becomes weaker leading to a shorter M—C bond and longer C—O bond as compared to M—C single bond and C≡O triple bond respectively.

15. Which of the following compounds on reaction with aqueous alkali followed by acidification gives the following lactone?

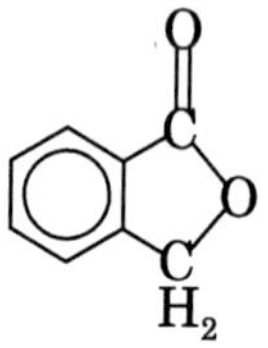

(A) 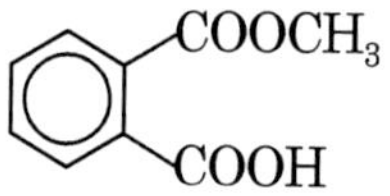

(B) 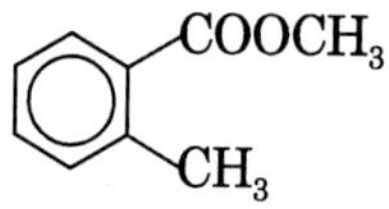

(C) 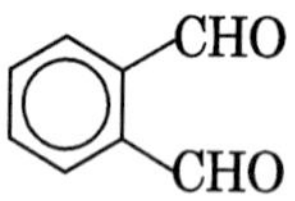

(D) 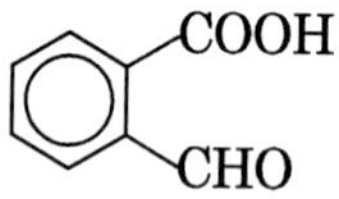

Hint: 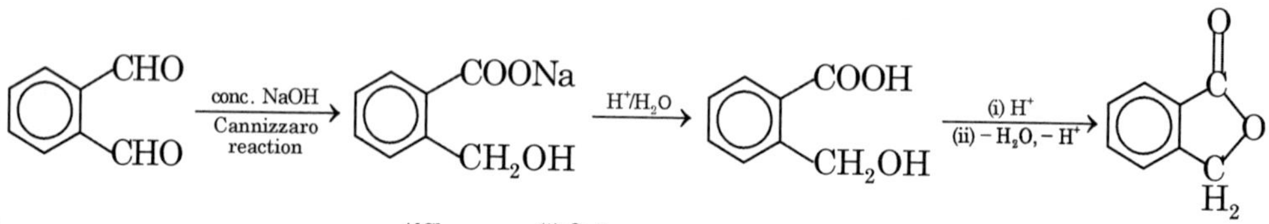

16.  $+ \text{Cl—CH}_2\text{—CH}_2\text{—CH}_3 \xrightarrow{\text{AlCl}_3} \text{P} \xrightarrow[\text{(ii) H}_3\text{O}^+]{\text{(i) O}_2/\Delta} \text{Q} + \text{Phenol}$

The major products P and Q are

(A) 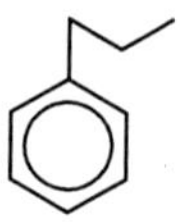 and CH_3CH_2CHO

(B) 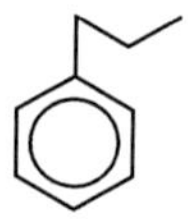 and CH_3COCH_3

(C)  and CH_3COCH_3

(B)  and CH_3CH_2CHO

Hint: 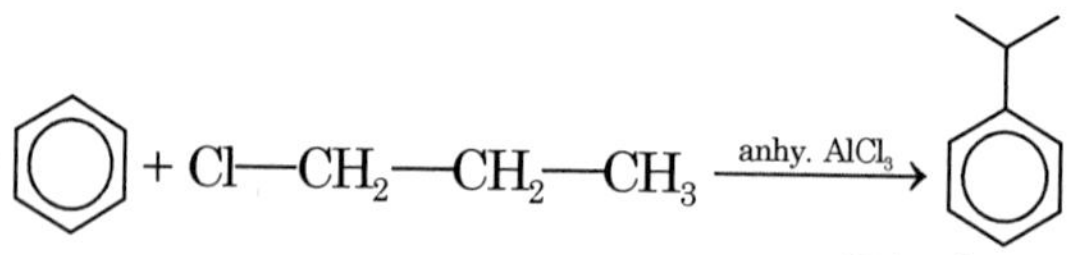

P (major product)

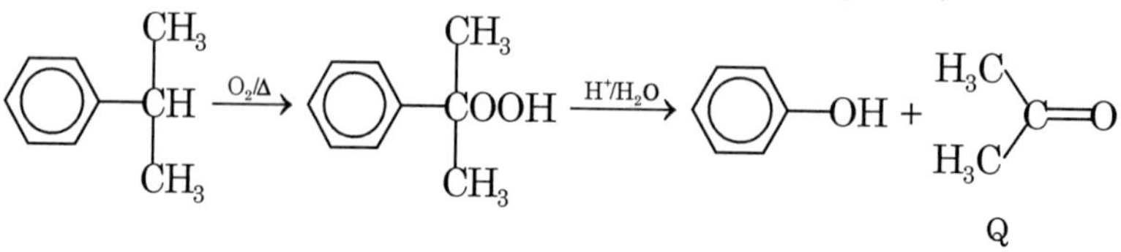

Q

17. The given graph represents the variation of Z (compressibility factor) *versus* P for three real gases A, B and C. Identify the incorrect statement.

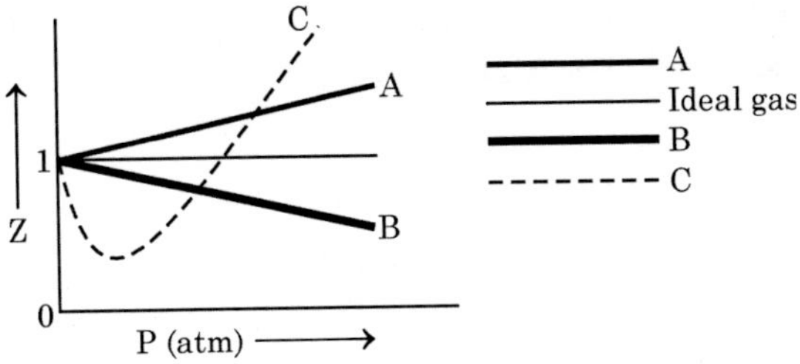

(A) For the gas A, $a = 0$ and the dependence of Z on P is linear
(B) For the gas B, $b = 0$ and the dependence of Z on P is linear
(C) Gas C is a real gas and we can find a and b if intersection data is given
(D) All van der Waals gases will behave like gas C and give positive slope at high pressure.

Hint: For gas B, $b = 0$, $Z = 1 - \dfrac{a}{VRT}$. Hence Z does not vary linearly with pressure.

For gas A, $a = 0$, $Z = 1 + \dfrac{Pb}{RT}$ implies that Z varies linearly with pressure. Given the intersection data for gas C we can find the values of a and b. All van der Waals gases like gas C give positive slope at high pressure.

18. The lowest molecular mass ketone and its next homologue are reacted with NH_2OH to form oximes. Which of the following is/are correct?

(A) Two oximes are formed (B) Three oximes are formed
(C) One oxime is optically active (D) Two oximes are optically active.

Hint:

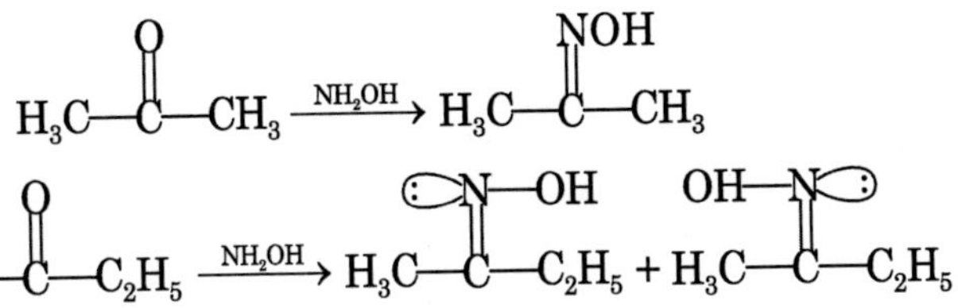

19. 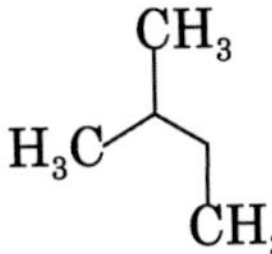 $\xrightarrow{Cl_2,\ h\nu}$ N isomers of $C_5H_{11}Cl$ $\xrightarrow{\text{fractional distillation}}$ M distilled products.

The values of N and M are:

(A) 6, 6 (B) 6, 4 (C) 4, 4 (D) 3, 3

Hint:

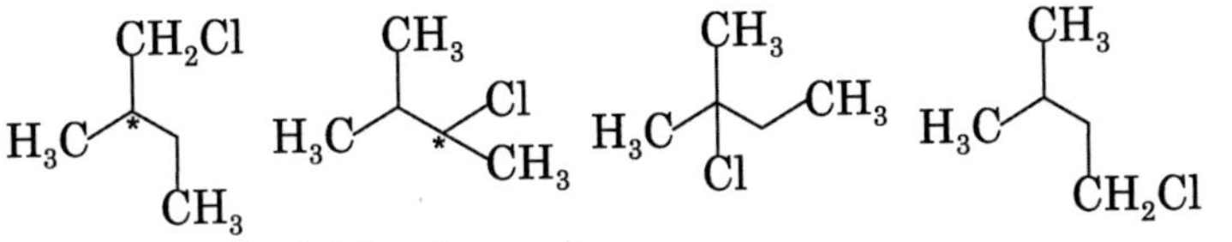

d, l (two isomers) *d, l* (two isomers)

M: Two *d, l* compounds (four isomers) cannot be separated by fractional distillation.

20. Ammoniacal solution of $MgSO_4$ in the presence of NH_4Cl is heated with Na_2HPO_4. A white crystalline precipitate is formed of

(A) $Mg(NH_4)PO_4$ (B) $Mg_3(PO_4)_2$ (C) $MgCl_2 \cdot MgSO_4$ (D) $MgSO_4 \cdot MgPO_4$

Hint: $Mg^{2+} + NH_3 + HPO_4^{2-} \longrightarrow Mg(NH_4)PO_4$

Section-III (Passage)

Passage I

$RCONH_2$ is converted into RNH_2 by means of Hofmann bromamide degradation.

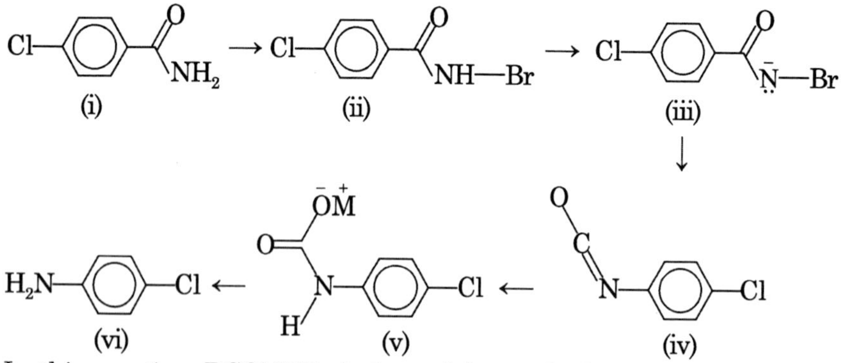

In this reaction, RCONHBr is formed from which this reaction has derived its name. Electron donating group at phenyl activates the reaction. Hofmann bromamide reaction is an intramolecular reaction.

21. How can the conversion of (i) to (ii) be brought about?

(A) NBS (B) $KBr + CH_3ONa$

(C) KBr + NaOH (D) Br_2 + NaOH

Hint: Br_2 + NaOH is the required reagent.

22. Which is the rate determining step in Hofmann bromamide raction?

(A) Formation of (v) (B) Formation of (ii)

(C) Formation of (iii) (D) Formation of (iv).

Hint: Formation of compound (iv) will occur in the rate determining step.

23. What are the constituent amines when the mixture of vii and viii undergoes Hofmann bromamide degradation?

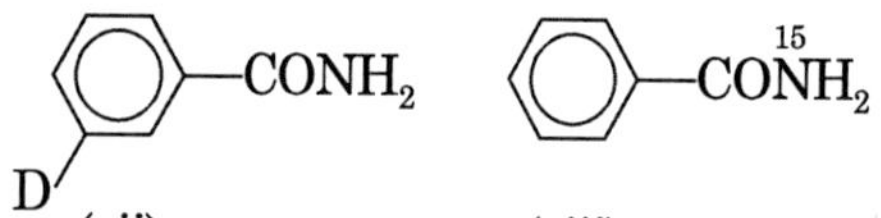

(vii) (viii)

(A) 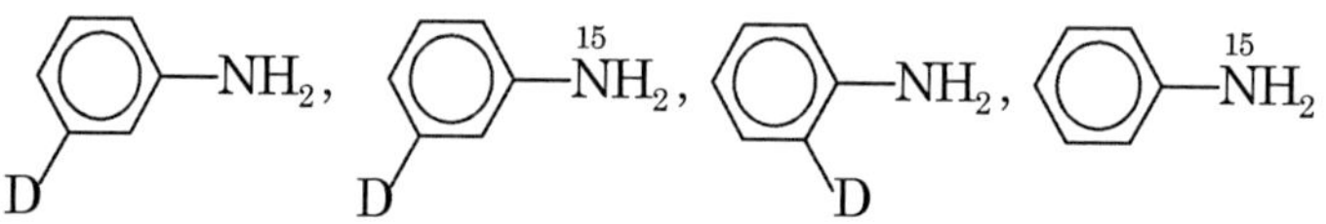

(B) 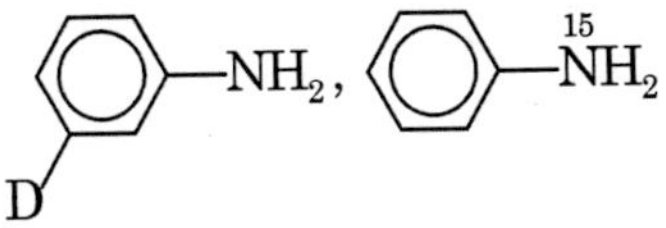

(C) 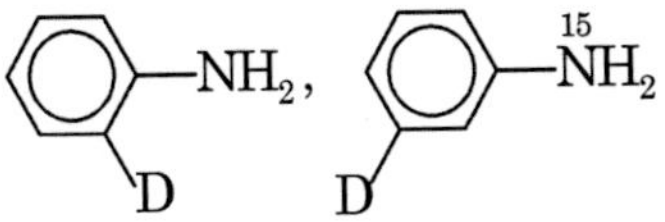

(D) 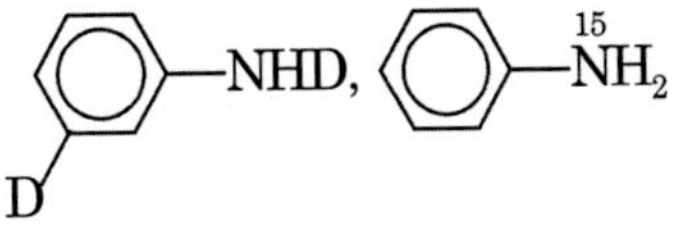

Hint:

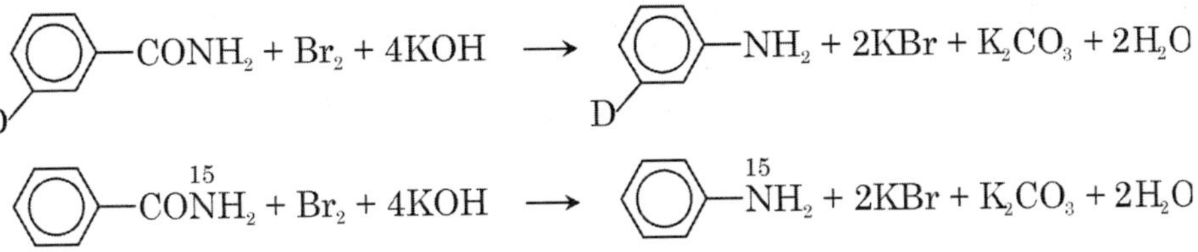

Passage II

Tollen's reagent is used for the detection of aldehyde. When a solution of $AgNO_3$ is added to glucose with NH_4OH, gluconic acid is formed.

$Ag^+ - e^- \longrightarrow Ag, E^0_{cell} = 0.8$ V

$C_6H_{12}O_6 + H_2O \longrightarrow$ Gluconic acid $(C_6H_{12}O_7) + 2H^+ + 2e^-$, $E^0_{ox} = -0.05$ V

$\left[Ag(NH_3)_2\right]^+ + e^- \longrightarrow Ag(s) + 2NH_3$, $E^0_{red} = 0.373$ V

$\left[\text{Use } 2.303 \times \frac{RT}{F} = 0.0592 \text{ and } \frac{F}{RT} = 38.9 \text{ V}^{-1} \text{ at } 298\text{ K}\right]$

24. $2Ag^+ + C_6H_{12}O_6 + H_2O \longrightarrow 2Ag(s) + C_6H_{12}O_7 + 2H^+$

Find $\ln K$ of this reaction.

(A) 66.00 (B) 58.35 (C) 29.60 (D) 55.60

Hint: $E^0_{cell} = \frac{RT}{nF} \ln K$

$$(0.8 - 0.05) = \frac{1}{2} \times \frac{0.0592}{2.303} \ln K$$

$$\text{or } \ln K = \frac{0.75 \times 2 \times 2.303}{0.0592} = 58.35$$

25. When ammonia is added to the solution, the pH increases to 11. Which half-cell reaction is affected by pH and by how much?

(A) E_{ox} will increase by a factor of 0.65 from E^0_{ox}

(B) E_{ox} will decrease by a factor of 0.65 from E^0_{ox}

(C) E_{red} will increase by a factor of 0.65 from E^0_{red}

(D) E_{red} will decrease by a factor of 0.65 from E^0_{red}

Hint: On increasing the concentration of NH_3, the concentration of H^+ ion decreases. Hence E_{red} increases.

$$C_6H_{12}O_6 + H_2O \longrightarrow C_6H_{12}O_7 + 2H^+ + 2e^-$$

$$E = E^0 - \frac{RT}{2F}\ln\left[H^+\right]^2 = E^0 - \frac{2}{2\times 38.9}\ln\left[H^+\right] = E^0 + \frac{2.303}{38.9}pH$$

$$= E^0 + \frac{2.303\times 11}{38.9} = E^0 + \frac{1.302}{2} = E^0 + 0.65$$

$$E - E^0 = 0.65 \text{ V}$$

26. Ammonia is used in this reaction than any other base. What is the correct reason for this?
    (A) $[Ag(NH_3)_2]^+$ is a weaker oxidising agent than Ag^+
    (B) NH_3 prevents the decomposition of gluconic acid
    (C) Ag^+ precipitates gluconic acid as its silver salt
    (D) NH_3 changes the standard reduction potential of $[Ag(NH_3)_2]^+$.
    Hint: Since $E^0_{Ag^+/Ag} > E^0_{[Ag(NH_3)_2]^+/Ag}$, the correct answer is (A).

Passage III

The coordination number of Ni^{2+} is 4 in A and B.

$NiCl_2$ + KCN (excess) $\longrightarrow$ A (cyano complex)

$NiCl_2$ + conc. HCl (excess) $\longrightarrow$ B (chloro complex)

27. The IUPAC names of A and B are respectively
    (A) potassium tetracyanonickelate(II), potassium tetrachloronickelate(II)
    (B) potassium tetracyanonickel(II), potassium tetrachloronickel(II)
    (C) potassium tetracyanonickel, potassium tetrachloronickel
    (D) potassium tetracyanonickelate(II), potassium tetrachloronickel(II)
    Hint: A = $K_2[Ni(CN)_4]$, B = $K_2[NiCl_4]$
28. Predict the magnetic nature of A and B.
    (A) Both are diamagnetic (B) Both are paramagnetic
    (C) A is paramagnetic and B is diamagnetic
    (D) A is diamagnetic and B is paramagnetic.
    Hint: CN^- ion being a strong field ligand, it forms low-spin square planar (dsp^2 hybridrisation) $K_2[Ni(CN)_4]$ which is diamagnetic. Cl^- ion being a weak field ligand, it forms high-spin tetrahedral (sp^3 hybridisation) $K_2[NiCl_4]$ which is paramagnetic with two unpaired electrons.
29. The hybridisations of A and B are respectively
    (A) dsp^2, sp^3 (B) sp^3, sp^3 (C) dsp^2, dsp^2 (D) sp^3, dsp^2
    Hint: $K_2[Ni(CN)_4]$: | ↑↓ | ↑↓ | ↑↓ | ↑↓ | ×× | | ×× | | ×× | ×× | | (dsp^2 hybridisation)
    $K_2[NiCl_4]$: | ↑↓ | ↑↓ | ↑↓ | ↑ | ↑ | | ×× | | ×× | ×× | ×× | (sp^3 hybridisation)

Passage IV

Due to the neutrons present in cosmic rays, the following reaction takes place by the bombardment of nitrogen with neutrons.

$^{14}_{7}N + ^{1}_{0}n \rightarrow ^{14}_{6}C + ^{1}_{1}H;\ ^{14}_{6}C \rightarrow ^{14}_{7}N + ^{0}_{-1}e$

Cosmic rays form $^{14}_{6}C$ which gets circulated in the atmosphere as well as in living species. Whenever there is a nuclear explosion, the concentration of ^{14}C increases both in the atmosphere as will as in living species. When a species dies the ^{14}C concentration decreases and hence this decrease can be measured and the time estimated as to when the organism died. For this we require (1) half-life of ^{14}C (5760 years) (2) activity of ^{14}C in living species (3) activity of ^{14}C in fossil. Beyond 30000 years the age of fossil cannot be accurately determined as the activity then would be too low.

30. In radiocarbon dating for finding the age of fossils, the most appropriate statement is

(A) During the life-time ^{14}C assimilated by the human beings is in equilibrium with the ^{14}C that decomposes by β^- emission. At any time in a human life $^{14}C/^{12}C$ is always constant at a particular instant

(B) ^{14}C dating method is inappropriate for finding the life of a given sample because ^{14}C undergoes β^- emission and hence the ratio of $^{14}C/^{12}C$ in human beings in not constant

(C) Radiocarbon dating is inappropriate to determine the age because ^{14}C concentration varies from place to place

(D) None of these.

Hint: The correct statement is (C). (A) is incorrect because after nuclear explosion $^{14}C/^{12}C$ will not remain constant. (B) is incorrect because ^{14}C will always remain in equilibrium with the surroundings.

31. A fossil whose age can be determined using carbon dating must be

(A) 6 years old (B) 6000 years old (C) 60000 years old

(D) can be accurately calculated for any age of the sample.

Hint: For effective carbon dating, the fossil must be minimum 1000 years old.

32. Two organisms died on the same day. One died at a place where nuclear explosion had taken place while the other died at a place where no such explosion had occurred. The ratio of ^{14}C during life to that present in the fossil at an instant is C_1 for the former and C_2 for the latter. The age of the former was calculated as t_1 and for the latter as t_2. The correct answer is

(A) There is an increae in t_1 w.r.t. t_2 and $t_1 - t_2 = \frac{1}{\lambda}\ln\frac{C_1}{C_2}$

(B) There is a decreae in t_1 w.r.t. t_2 and $t_1 - t_2 = \frac{1}{\lambda}\ln\frac{C_1}{C_2}$

(C) $\frac{t_1}{t_2} = \frac{C_1}{C_2}$ (D) Cannot be determined.

Hint: $t_1 = \frac{1}{\lambda}\ln C_1$; $t_2 = \frac{1}{\lambda}\ln C_2$. Hence $t_1 - t_2 = \frac{1}{\lambda}\ln\frac{C_1}{C_2}$. Since due to nuclear explosion the initial activity A_0 would be greater, it would take more time to reach the same value of activity A.

33. Match the extraction processes listed in Column I with metals listed in Column II.

Column I	*Column II*
(A) Self-reduction	(i) Lead
(B) Carbon reduction	(ii) Silver
(C) Complex formation and displacement by metal	(iii) Copper
(D) Decomposition of iodide	(iv) Boron

Hint: A, i , iii; B, i, iii; C, ii; D, iv

34. Match the following:

Column I	*Column II*
(A) $Bi_2O_3 \longrightarrow (BiO)^+$	(i) Heating
(B) $[AlO_2]^- \longrightarrow Al(OH)_3$	(ii) Hydrolysis
(C) $SiO_4^{4-} \longrightarrow Si_2O_7^-$	(iii) Acidification
(D) $B_4O_7^{2-} \longrightarrow [B(OH)_3]$	(iv) Dilution with water

Hint: A, ii; B, iv; C, i; D, ii, iii

35. According to Bohr's theory of H-atom, E_n = total energy, K_n = kinetic energy of nth orbit, V_n = potential energy of nth orbit. r_n = radius of nth orbit. Match the following:

Column I	*Column II*
(A) $V_n/K_n = ?$	(i) 0
(B) If radius of nth orbit $\propto E_n^x$, x = ?	(ii) – 1
(C) Orbital angular momentum in lowest energy orbit	(iii) – 2
(D) $\frac{1}{V^n} \propto Z^y$, y = ?	(iv) 1

Hint: A, iii; B, ii; C, i; D, iv

36. Match the following:

Column I	*Column II*
(A) $PhCH_2CD_2Br$ on reaction with $C_2H_5O^-$ gives $Ph—CH=CD_2$	(i) *E*1
(B) $PhCHBrCH_3$ and $PhCHBrCD_3$ both react with the same rate	(ii) *E*2

(C) $PhCHDCH_2Br$ on treatement with $C_2H_5O^-$ and C_2H_5OD gives $Ph—CD{=}CD_2$	(iii) *E*1CB
(D) $PhCH_2CH_2Br$ reacts faster than $PhCD_2CH_2Br$ on reaction with $C_2H_5O^-$ in ethanol	(iv) First order reaction

Hint: A, ii; B, i, iv; C, ii; D, ii

ANSWERS

1. A	2. A	3. D	4. A	5. A	6. A
7. C	8. D	9. D	10. A	11. B	12. C
13. A	14. A	15. C	16. C	17. B	18. B
19. B	20. A	21. B	22. D	23. B	24. B
25. C	26. A	27. A	28. D	29. A	30. C
31. B	32. A				

Matching Type Qustions

33. A, i , iii; B, i, iii; C, ii; D, iv
34. A, ii; B, iv; C, i; D, ii, iii
35. A, iii; B, ii; C, i; D, iv
36. A, ii; B, i, iv; C, ii; D, ii

I.I.T. JEE Objective Question Paper—2007

PAPER—I

Note:

(i) The question paper has 4 sections.

(ii) Section I contains 9 multiple choice questions which have only one correct answer. Each question carries + 3 marks for each correct answer and – 1 mark for each wrong answer.

(iii) Section II contains 4 questions. Each question contains Statement–1 (Assertion) and Statement–2 (Reason).

Bubble (A) if both the Statements are true and Statement–2 is the correct explanation of Statement–1.

Bubble (B) if both the Statements are true but Statement–2 is not the correct explanation of Statement–1.

Bubble (C) if the Statement–1 is true and Statement–2 is false.

Bubble (D) if the Statement–1 is false and Statement–2 is true.

Carries + 3 marks for each correct answer and – 1 mark for each wrong answer.

(iv) Section III contains 2 paragraphs. Based upon each paragraph, 3 multiple choice questions have to be answered. Each question has only one correct answer and carries + 4 marks for correct answer and – 1 mark for wrong answer.

(v) Section IV contains 3 questions. Each question contains statements given in 2 columns. Statements in the first column have to be matched with statements in the second column and each question carries + 6 marks and marks will be awarded if all the four parts are correctly matched. No marks will be given for any wrong match in any question. There is no negative marking.

Section–I: Straight Objective Type

This section contains 9 multiple choice questions numbered 1 to 9. Each question has 4 choices (A), (B), (C) and (D), out of which ONLY ONE is correct.

1. Among the following, the least stable resonance structure is

(A) 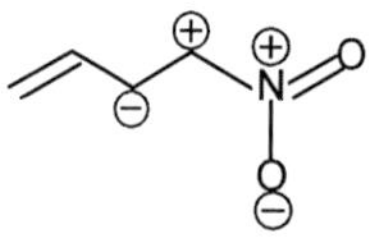

(B) 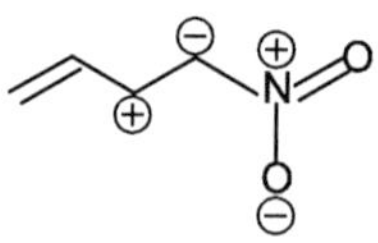

(C) 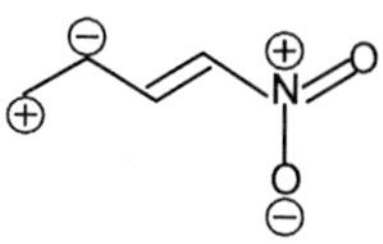

(D) 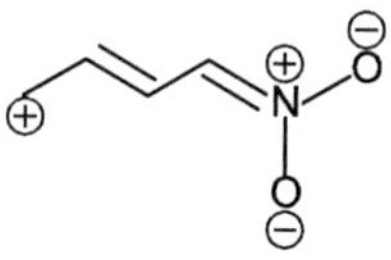

2. For the process $H_2O(l)$ (1 bar, 373 K) $\longrightarrow$ $H_2O(g)$ (1 bar, 373 K), the correct set of thermodynamic parameters is

(A) $\Delta G = 0$, ΔS = +ve (B) $\Delta G = 0$, ΔS = –ve

(C) ΔG = +ve, $\Delta S = 0$ (D) ΔG = –ve, ΔS = +ve

3. Cyclohexene on ozonolysis followed by reaction with zinc dust and water gives compound E. Compound E on further treatment with aqueous KOH yields compound F. Compound F is

(A) 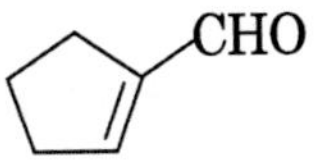 (B) 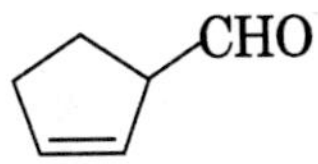 (C) 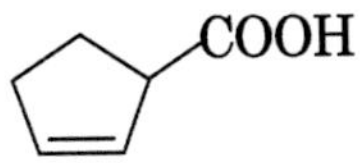 (D) 

4. Consider a reaction aG + bH $\longrightarrow$ Products. When concentration of both the reactants G and H is doubled, the rate increases by eight times. However, when concentration of G is doubled keeping the concentration of H fixed, the rate is doubled. The overall order of the reaction is

(A) 0 (B) 1 (C) 2 (D) 3

5. Among the following metal carbonyls, the C—O bond order is lowest in

(A) $[Mn(CO)_6]^+$ (B) $Fe(CO)_5$ (C) $[Cr(CO)_6]$ (D) $[V(CO)_6]^-$

6. A positron is emitted from ${}^{23}_{11}Na$. The ratio of the atomic mass and atomic number of the resulting nuclide is

(A) 22/10 (B) 22/11 (C) 23/10 (D) 23/12

7. Consider a titration of potassium dichromate solution with acidified Mohr's salt solution using diphenylamine as indicator. The number of moles of Mohr's salt required per mole of dichromate is

(A) 3 (B) 4 (C) 5 (D) 6

8. The number of stereoisomers obtained by bromination of *trans*-but-2-ene is

(A) 1 (B) 2 (C) 3 (D) 4

9. A solution of metal ion when treated with KI gives a red precipitate which dissolves in excess KI to give a colourless solution. Moreover, the solution of metal ion on treatment with a solution of cobalt(II) thiocyanate gives rise to a deep blue crystalline precipitate. The metal ion is

(A) Pb^{2+} (B) Hg^{2+} (C) Cu^{2+} (D) Co^{2+}

Section–II: Assertion-Reason Type

This section contains 4 questions numbered 10 to 13. Each question contains Statement–1 (Assertion) and Statement–2 (Reason). Each question has four choices (A), (B), (C) and (D), out of which ONLY ONE is correct.

10. Statement–1: Molecules that are not superimposable on their mirror images are chiral.

because

Statement–2: All chiral molecules have chiral centres.

(A) Statement–1 is true, Statement–2 is true; Statement–2 is a correct explanation for Statement–1
(B) Statement–1 is true, Statement–2 is true; Statement–2 is not a correct explanation for Statement–1
(C) Statement–1 is true, Statement–2 is false
(D) Statement–1 is false, Statement–2 is true.

11. Statement–1: Alkali metals dissolve in liquid ammonia to give blue solution.
because
Statement–2: Alkali metals in liquid ammonia give solvated species of the type $[M(NH_3)_n]^+$ (M = alkali metals).
(A) Statement–1 is true, Statement–2 is true; Statement–2 is a correct explanation for Statement–1
(B) Statement–1 is true, Statement–2 is true; Statement–2 is not a correct explanation for Statement–1
(C) Statement–1 is true, Statement–2 is false
(D) Statement–1 is false, Statement–2 is true.

12. Statement–1: Band gap in germanium is small.
because
Statement–2: The energy spread of each germanium atomic energy level is infinitesimally small.
(A) Statement–1 is true, Statement–2 is true; Statement–2 is a correct explanation for Statement–1
(B) Statement–1 is true, Statement–2 is true; Statement–2 is not a correct explanation for Statement–1
(C) Statement–1 is true, Statement–2 is false
(D) Statement–1 is false, Statement–2 is true.

13. Statement–1: Glucose gives a reddish-brown precipitate with Fehling's solution.
because
Statement–2: Reaction of glucose with Fehling's solution gives CuO and gluconic acid.
(A) Statement–1 is true, Statement–2 is true; Statement–2 is a correct explanation for Statement–1
(B) Statement–1 is true, Statement–2 is true; Statement–2 is not a correct explanation for Statement–1
(C) Statement–1 is true, Statement–2 is false
(D) Statement–1 is false, Statement–2 is true.

Section–III: Linked Comprehensive Type

This section contains 2 paragraphs C_{14-16} and C_{17-19}. Based upon each paragraph, 3 multiple choice questions have to be answered. Each question has 4 choices (A), (B), (C) and (D), out of which ONLY ONE is correct.

C_{14-16}: Paragraph for Question Nos. 14 to 16

Reimer–Tiemann reaction introduces an aldehyde group, on to the aromatic ring of phenol, *ortho* to the hydroxyl group. This reaction involves electrophilic aromatic substitution. This is a general method for the synthesis of salicylaldehydes as depicted below:

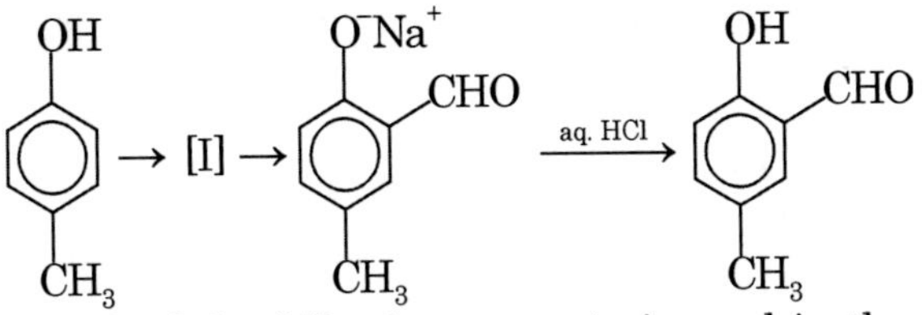

14. Which of the following reagents is used in the above reaction?
(A) aq. NaOH + CH_3Cl (B) aq. NaOH + CH_2Cl_2
(C) aq. NaOH + $CHCl_3$ (D) aq. NaOH + CCl_4

15. The electrophile in this reaction is
(A) :CHCl (B) $\overset{+}{C}HCl_2$ (C) $:CCl_2$ (D) $\cdot CCl_3$

16. The structure of the intermediate I is
(A) 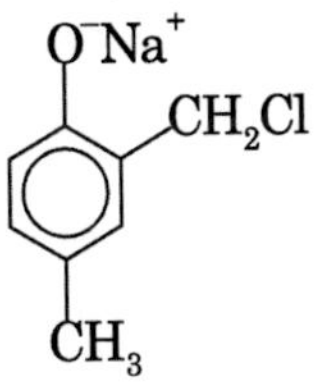 (B) 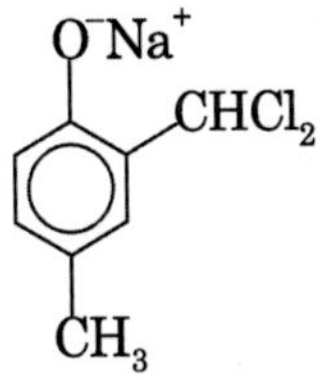 (C) 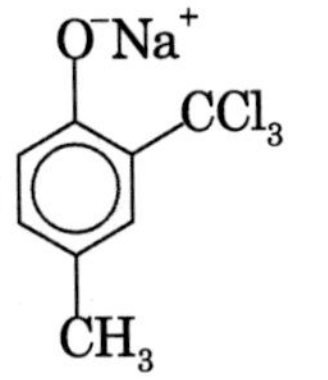 (D) 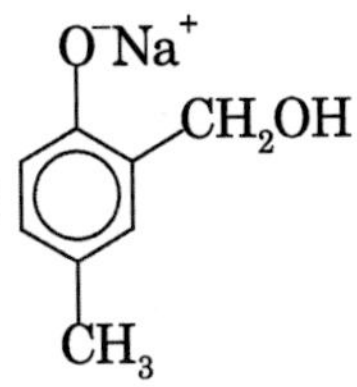

C_{17-19}: Paragraph for Question Nos. 17 to 19

Redox reactions play a pivotal role in chemistry and biology. The values of standard redox potential (E^0) of two half-cell reactions decide which way the reaction is expected to proceed. A simple example is a Daniel cell in which zinc goes into solution and copper gets deposited. Given below are a set of half-cell reactions (acidic medium) along with their E^0 (V with respect to normal hydrogen electrode) values. Using this data obtain correct explanations to Questions 17 to 19. $I_2 + 2e^- \longrightarrow 2I^-$, $E^0 = 0.54$ V; $Cl_2 + 2e^- \longrightarrow 2Cl^-$, $E^0 = 1.36$ V; $Mn^{3+} + e^- \longrightarrow Mn^{2+}$, $E^0 = 1.50$ V; $Fe^{3+} + e^- \longrightarrow Fe^{2+}$, $E^0 = 0.77$ V; $O_2 + 4H^+ + 4e^- \longrightarrow 2H_2O$, $E^0 = 1.23$ V

17. Among the following, identify the correct statement.
(A) Chloride ion is oxidised by O_2 (B) Fe^{2+} is oxidised by iodine
(C) Iodide ion is oxidised by chlorine (D) Mn^{2+} is oxidised by chlorine.

18. While Fe^{3+} is stable, Mn^{3+} is not stable in acid solution because
(A) O_2 oxidises Mn^{2+} to Mn^{3+} (B) O_2 oxidises both Mn^{2+} and Fe^{2+} to Fe^{3+}
(C) Fe^{3+} oxidises H_2O to O_2 (D) Mn^{3+} oxidises H_2O and O_2

19. Sodium fusion extract obtained from aniline, on treatment with iron(II) sulphate and H_2SO_4 in the presence of air gives a prussian blue precipitate. The blue colour is due to the formation of
(A) $Fe_4[Fe(CN)_6]_3$ (B) $Fe_3[Fe(CN)_6]_2$ (C) $Fe_4[Fe(CN)_6]_2$ (D) $Fe_3[Fe(CN)_6]_3$

Section–IV: Matrix-Match Type

This section contains 3 questions. Each question contains statements given in two columns which have to be matched. Statements (A, B, C, D) in Column I have to be matched with statements (p, q, r, s) in Column II. The answers to these questions have to be appropriately bubbled as illustrated in the following example.

If the correct matches are A-p, A-s, B-q, B-r, C-p, C-q and D-s, then the correctly bubbled 4×4 matrix should be as follows:

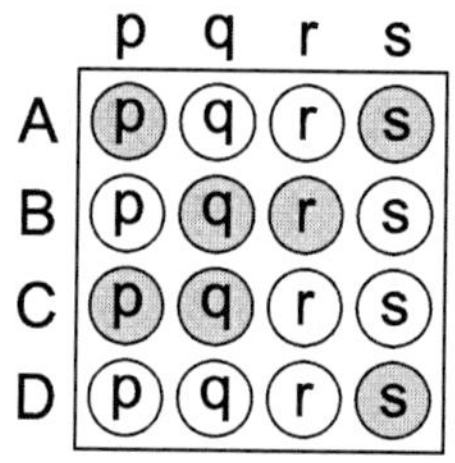

20. Match the reactions in Column I with nature of the reactions/type of the products in Column II. Indicate your answer by darkening the appropriate bubbles of the 4×4 matrix given in the ORS.

Column I	*Column II*
(A) $O_2^- \longrightarrow O_2 + O_2^{2-}$	(p) redox reaction
(B) $CrO_4^{2-} + H^+ \longrightarrow$	(q) one of the products has trigonal planar structure
(C) $MnO_4^- + NO_2^- + H^+ \longrightarrow$	(r) dimeric bridged tetrahedral metal ion
(D) $NO_3^- + H_2SO_4 + Fe^{2+} \longrightarrow$	(s) disproportionation

21. Match the compounds/ions in Column I with their properties/reactions in Column II. Indicate your answer by darkening the appropriate bubbles of the 4×4 matrix given in the ORS.

Column I	*Column II*
(A) C_6H_5CHO	(p) gives precipitate with 2,4-dinitrophenylhydrazine
(B) $CH_3C{\equiv}CH$	(q) gives precipitate with $AgNO_3$
(C) CN^-	(r) is a nucleophile
(D) I^-	(s) is involved in cyanohydrin formation

22. Match the crystal system/unit cells mentioned in Column I with their characteristic features mentioned in Column II. Indicate your answer by darkening appropriate bubbles of the 4×4 matrix given in the ORS.

Column I	*Column II*
(A) Simple cubic and face-centred cubic	(p) have these cell parameters: $a = b = c$ and $\alpha = \beta = \gamma$
(B) Cubic and rhombohedral	(q) are two crystal systems
(C) Cubic and tetragonal	(r) have only two crystallography angles of 90°
(D) Hexagonal and monoclinic	(s) belong to same crystal system

ANSWERS

1. A	2. A	3. A	4. D	5. B	6. C
7. D	8. A	9. B	10. C	11. B	12. C
13. C	14. C	15. C	16. B	17. C	18. D

19. A
20. A-p, s; B-r; C-p, q; D-p
21. A-p, q, s; B-q; C-q, r, s; D-q, r
22. A-p, s; B-p, q; C-q; D-q, r

PAPER—II

Section–I: Straight Objective Type

This section contains 9 multiple choice questions numbered 1 to 9. Each question has 4 choices (A), (B), (C) and (D), out of which ONLY ONE is correct.

1. The number of structural isomers for C_6H_{14} is

(A) 3 (B) 4 (C) 5 (D) 6

2. In the reaction, 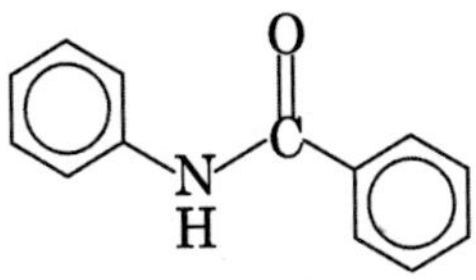 $\xrightarrow[\text{conc. } H_2SO_4]{\text{conc. } HNO_3}$ X, the structure of the major product X is

(A) 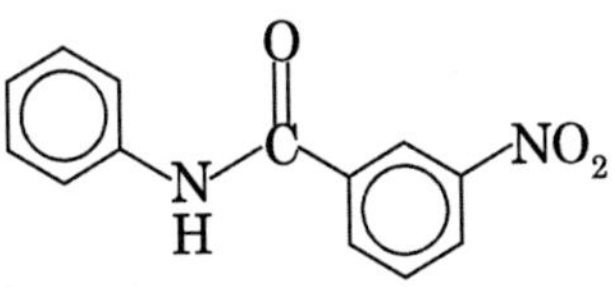

(B) 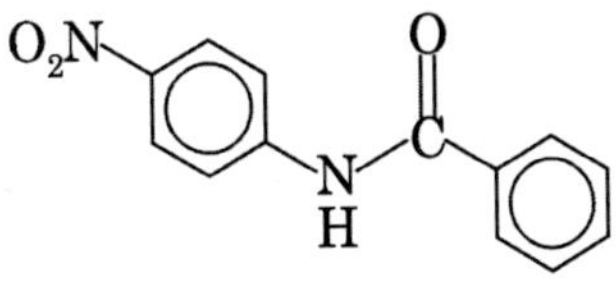

(C) 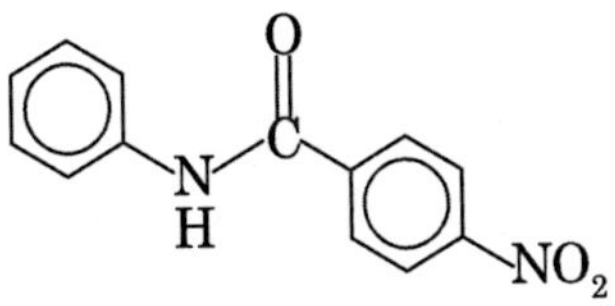

(D) 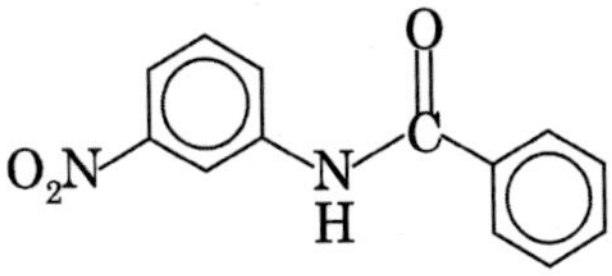

3. When 20 g of naphthoic acid ($C_{11}H_8O_2$) is dissolved in 50 g of benzene (K_f = 1.72 K kg mol^{-1}), a freezing point depression of 2 K is observed. The van't Hoff factor (i) is

(A) 0.5 (B) 1 (C) 2 (D) 3

4. Among the following the paramagnetic compound is

(A) Na_2O_2 (B) O_3 (C) N_2O (D) KO_2

5. The value of $\log_{10} K$ for a reaction $A \rightleftharpoons B$ is (Given: $\Delta H^0_{298\text{ K}} = -54.0$ kJ mol^{-1}, $\Delta S^0_{298\text{ K}} = 10$ J K^{-1} mol^{-1} and $R = 8.314$ J K^{-1} mol^{-1}; $2.303 \times 8.314 \times 298 = 5705$)

(A) 5 (B) 10 (C) 95 (D) 100

6. The species having bond order different from that in CO is

(A) NO^- (B) NO^+ (C) CN^- (D) N_2

7. The percentage of *p*-character in the orbitals forming P—P bonds in P_4 is
   (A) 25 (B) 33 (C) 50 (D) 75
8. Extraction of zinc from zinc blende is achieved by
   (A) electrolytic reduction
   (B) roasting followed by reduction with carbon
   (C) roasting followed by reduction with another metal
   (D) roasting followed by self-reduction.
9. The reagent(s) for the conversion, 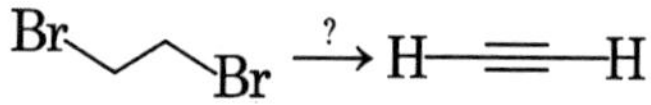 is/are
   (A) alcoholic KOH
   (B) alcoholic KOH followed by $NaNH_2$
   (C) aqueous KOH followed by $NaNH_2$
   (D) Zn/CH_3OH

Section–II: Assertion-Reason Type

This section contains 4 questions numbered 10 to 13. Each question contains Statement–1 (Assertion) and Statement–2 (Reason). Each question has 4 choices (A), (B), (C) and (D), out of which ONLY ONE is correct.

10. Statement–1: *p*-Hydroxybenzoic acid has a lower boiling point then *o*-hydroxybenzoic acid.
    because
    Statement–2: *o*-Hydroxybenzoic acid has intramolecular hydrogen bonding.
    (A) Statement–1 is true, Statement–2 is true; Statement–2 is a correct explanation for Statement–1
    (B) Statement–1 is true, Statement–2 is true; Statement–2 is not a correct explanation for Statement–1
    (C) Statement–1 is true, Statement–2 is false
    (D) Statement–1 is false, Statement–2 is true.
11. Statement–1: Micelles are formed by surfactant molecules above the critical micellar concentration (CMC).
    because
    Statement–2: The conductivity of a solution having surfactant molecules decreases sharply at the CMC.
    (A) Statement–1 is true, Statement–2 is true; Statement–2 is a correct explanation for Statement–1
    (B) Statement–1 is true, Statement–2 is true; Statement–2 is not a correct explanation for Statement–1
    (C) Statement–1 is true, Statement–2 is false
    (D) Statement–1 is false, Statement–2 is true.
12. Statement–1: Boron always forms covalent bond.
    because
    Statement–2: The small size of B^{3+} favours the formation of covalent bond.
    (A) Statement–1 is true, Statement–2 is true; Statement–2 is a correct explanation for Statement–1

(B) Statement–1 is true, Statement–2 is true; Statement–2 is not a correct explanation for Statement–1
(C) Statement–1 is true, Statement–2 is false
(D) Statement–1 is false, Statement–2 is true.

13. Statement–1: In water, orthoboric acid behaves as a weak monobasic acid.
because
Statement–2: In water, orthoboric acid acts as a proton donor.
(A) Statement–1 is true, Statement–2 is true; Statement–2 is a correct explanation for Statement–1
(B) Statement–1 is true, Statement–2 is true; Statement–2 is not a correct explanation for Statement–1
(C) Statement–1 is true, Statement–2 is false
(D) Statement–1 is false, Statement–2 is true.

Section–III: Linked Comprehensive Type

This section contains 2 paragraphs C_{14-16} and C_{17-19}. Based upon each paragraph, 3 multiple choice questions have to be answered. Each question has 4 choices (A), (B), (C) and (D), out of which ONLY ONE is correct.

C_{14-16}: Paragraph for Question Nos. 14 to 16

Chemical reactions involve interaction of atoms and molecules. A large number of atoms/molecules (approximately 6.023×10^{23}) are present in a few grams of any chemical compound varying with their atomic/molecular masses. To handle such large numbers conveniently, the mole concept was introduced. This concept has implications in diverse areas such as analytical chemistry, biochemistry, electrochemistry and radiochemistry. The following examples illustrate a typical case, involving chemical/electrochemical reaction, which requires a clear understanding of the mole concept.

A 4.0 molar aqueous solution of NaCl is prepared and 500 ml of this solution is electrolysed. This leads to the evolution of chlorine gas at one of the electrodes (atomic mass: Na = 23, Hg = 200; 1 Faraday = 96500 Coulombs).

14. The total number of moles of chlorine gas evolved is
(A) 0.5 (B) 1.0 (C) 2.0 (D) 3.0

15. If the cathode is a Hg electrode, the maximum weight (g) of amalgam formed from the solution is
(A) 200 (B) 225 (C) 400 (D) 446

16. The total charge (coubombs) required for complete electrolysis is
(A) 24125 (B) 48250 (C) 96500 (D) 193000

C_{17-19}: Paragraph for Question Nos. 17 to 19

The noble gases have closed-shell electronic configuration and are monoatomic gases under normal conditions. The low boiling points of the lighter noble gases are due to weak dispersion forces between the atoms and the absence of other interatomic interactions.

The direct reactions of xenon with fluorine lead to a series of compounds with oxidation numbers + 2, + 4 and + 6. XeF_4 reacts violently with water to give XeO_3. The compounds of xenon exhibit rich stereochemistry and their geometries can be deduced considering the total number of electron pairs in the valence shell.

17. Argon is used in arc welding because of its
    (A) low reactivity with metal
    (B) ability to lower the melting point of metal
    (C) flammability
    (D) high calorific value.

18. The structure of XeO_3 is
    (A) linear (B) planar (C) pyramidal (D) T-shaped.

19. XeF_4 and XeF_6 are expected to be
    (A) oxidising (B) reducing (C) unreactive (D) strongly basic.

Section–IV: Matrix-Match Type

This section contains 3 questions. Each question contains statements given in two columns which have to be matched. Statements (A, B, C, D) in Column I have to be matched with statements (p, q, r, s) in Column II. The answers to these questions have to be appropriately bubbled as illustrated in the following example.
If the correct matches are A-p, A-s, B-q, B-r, C-p, C-q and D-s, then the correctly bubbled 4×4 matrix should be as follows:

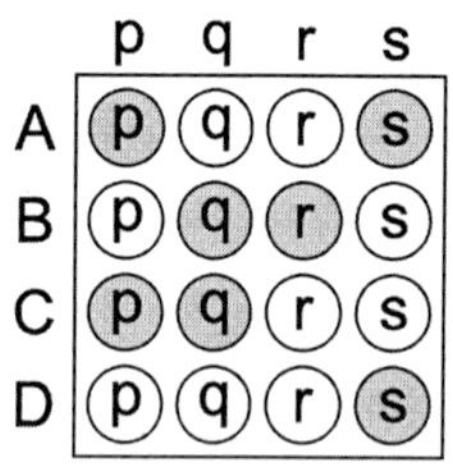

20. Match the complexes in Column I with their properties listed in Column II. Indicate your answer by darkening the appropriate bubbles of the 4×4 matrix given in the ORS.

Column I	*Column II*
(A) $[Co(NH_3)_4(H_2O)_2]Cl_2$	(p) geometrical isomers
(B) $[Pt(NH_3)_2Cl_2]$	(q) paramagnetic
(C) $[Co(H_2O)_5Cl]Cl$	(r) diamagnetic
(D) $[Ni(H_2O)_6]Cl_2$	(s) metal ion with + 2 oxidation state

21. Match gases under specified conditions listed in Column I with their properties/laws in Column II. Indicate your answer by darkening the appropriate bubbles of the 4×4 matrix given in the ORS.

Column I	*Column II*
(A) hydrogen gas (P = 200 atm, T = 273 K)	(p) compressibility factor = 1
(B) hydrogen gas (P = 0, T = 273 K)	(q) attractive forces are dominant

(C) CO_2 (P = 1 atm, T = 273 K) (r) $PV = nRT$

(D) real gas with very large molar volume (s) $P(V - ab) = nRT$

22. Match the chemical substances in Column I with type of polymers/type of bonds in Column II. Indicate your answer by darkening appropriate bubbles of the 4 × 4 matrix given in the ORS.

Column I	*Column II*
(A) Cellulose	(p) natural polymer
(B) nylon-66	(q) synthetic polymer
(C) protein	(r) amide linkage
(D) sucrose	(s) glycoside linkage

ANSWERS

1. C	2. B	3. A	4. D	5. B	6. A
7. D	8. B	9. B	10. D	11. B	12. A
13. C	14. B	15. D	16. D	17. A	18. C

19. A

20. A-p, q, s; B-p, r, s; C-q, s; D-q, s

21. A-p, s; B-r; C-p, q; D-p, s

22. A-p, s; B-q, r; C-p, r; D-s

MODEL SOLUTIONS—I.I.T. 2007

Paper-1
Section–I: Straight Objective Type

1. (A) In 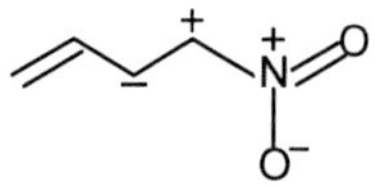 same type of charge (+) is present at the nearest position.

   Hence it is less stable.

2. $H_2O(l) \rightleftharpoons H_2O(g)$
   (1 bar, 373 K) (1 bar, 373 K)

   At 100°C, $H_2O(l)$ is in equilibrium with $H_2O(g)$. Hence $\Delta G = 0$. Since the liquid molecules get converted into gaseous molecules, ΔG = +ve. Hence (A) is correct.

3. 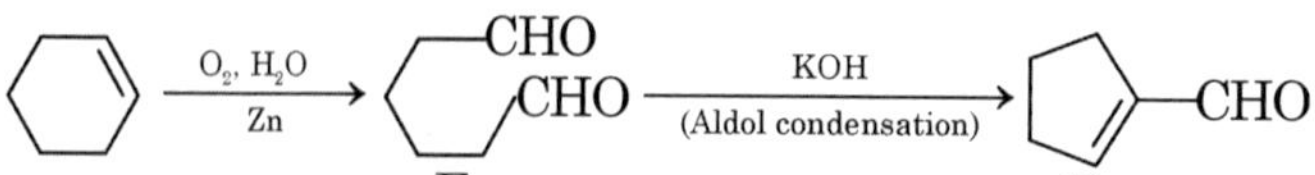

   E F

4. $aG + bH \longrightarrow$ Product

   Rate of reaction $\propto [G]^a[H]^b$

   $a = 1, \ b = 2$

   Overall order of reaction $= a + b = 1 + 2 = 3$

5. (A) Mn^+: $3d^54s^1$

   In the presence of strong field ligand CO, the effective configuration of Mn^+ is $3d^64s^0$. There are three lone pairs of electrons (d^6) for back bonding with the vacant orbital of carbon in CO.

   (B) Fe: $3d^64s^2$

   In the presence of CO, the effective configuration of Fe(0) is $3d^8$. There are four lone pairs of electrons (d^8) for back bonding with the vacant orbital of carbon in CO.

   (C) Cr: $3d^54s^1$

   In the presence of CO, the effective configuration of Cr(0) is $3d^6$. There are three lone pairs of electrons (d^6) for back bonding with the vacant orbital of carbon in CO.

   (D) V: $3d^44s^2$

   In the presence of CO, the effective configuration of V(0) is $3d^6$. There are three lone pairs of electrons for back bonding with the vacant orbital of carbon of CO.

   Hence the C—O back bonding is the lowest in $Fe(CO)_5$.

6. When a positron is emitted from a nucleus, a proton gets converted into a neutron. Hence the atomic number decreases by one but the atomic mass remains constant.

   Thus, $\dfrac{\text{Atomic mass}}{\text{Atomic number}} = \dfrac{23}{10}$

7. $Cr_2O_7^{2-} + Fe^{2+} \longrightarrow 2Cr^{3+} + Fe^{3+}$
   n factor of $Cr_2O_7^{2-} = 6$; n factor of $Fe^{2+} = 1$
   Hence to reduce one mole of potassium dichromate six moles of Fe^{2+} are required.
   Hence (D) is correct.

8. 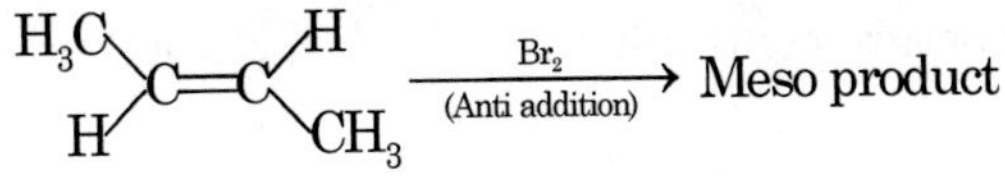

   trans-but-2-ene

   Hence the bromination of *trans*-but-2-ene gives only one stereoisomer.

9. $Hg^{2+} + 2KI \longrightarrow \underset{\text{Red}}{HgI_2\downarrow} + 2K^+$

   $HgI_2 + \underset{\text{(excess)}}{2KI} \longrightarrow \underset{\text{Soluble}}{K_2HgI_4}$

   $HgI_4^{2+} + Co(SCN)_2 \longrightarrow \underset{\text{Blue crystalline precipitate}}{Hg[Co(NCS)_4]}$

Section–II: Assertion-Reason Type

10. Statement–1 is true and Statement–2 is false.
11. Statement–1 in true, Statement–2 is true and Statement–2 is not the correct explanation for Statement–1.
    Blue colour of the solution is due to the solvated electrons.
12. Statement–1 is true and Statement–2 is false.
13. Statement–1 is true and Statement–2 is false.
    $C_6H_{12}O_6 + \text{Fehling's solution} \longrightarrow (C_6H_{11}O_7)^- + \underset{\text{Red}}{Cu_2O\downarrow}$

Section–III: Linked Comprehensive Type

14. In the given reaction, the reagent used is aqueous NaOH and $CHCl_3$.
15. The electrophile in the given reaction is $:CCl_2$.
    $OH^-(aq) + CHCl_3 \rightleftharpoons \bar{C}Cl_3 + H_2O$
    $\bar{C}Cl_3 \longrightarrow Cl^- + \underset{\text{Dichlorocarbene intermediate}}{:CCl_2}$

16. 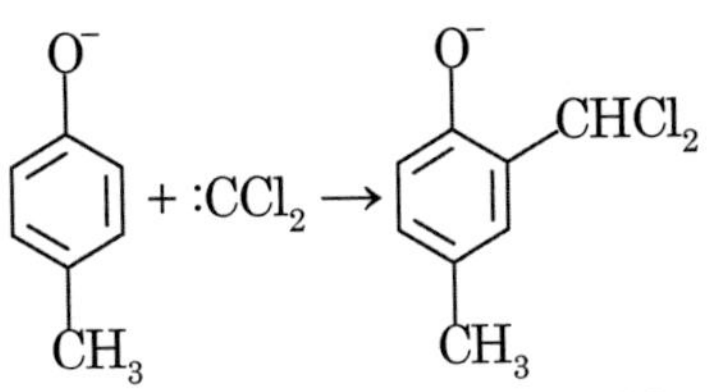

17. Iodide ion is oxidised by chlorine since the electrode potential of I_2 is less than that of Cl_2.
18. Mn^{3+} ion is not stable in acid medium since the reaction of Mn^{3+} ion with H_2O is spontaneous.
19. The blue colour is due to the formation of $Fe_4[Fe(CN)_6]_3$.

Section–IV: Matrix-Match Type

20. (A) The reaction, $2O_2^- \rightarrow O_2 + O_2^{2-}$, is a redox reaction as well as a disproportionation reaction.

(B) $2CrO_4^{2-} + 2H^+ \rightarrow Cr_2O_7^{2-} + H_2O$

$Cr_2O_7^{2-}$ ion has a dimeric bridged structure containing tetrahedral chromium ion.

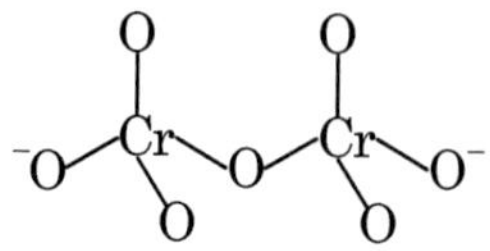

(C) $MnO_4^- + NO_2^- + H^+ \rightarrow MnO_2 + NO_3^- + OH^-$

It is a redox reaction and one of the products (NO_3^-) has trigonal planar structure:

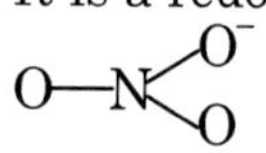

(D) $2HNO_3^- + 3H_2SO_4 + 6FeSO_4 \rightarrow 2NO + 3Fe_2(SO_4)_3 + 4H_2O$

It is a redox reaction.

21. (A) C_6H_5CHO gives precipitates with 2,4-dinitrophenylhydrazine or ammoniacal $AgNO_3$, and is involved in cyanohydrin formation.

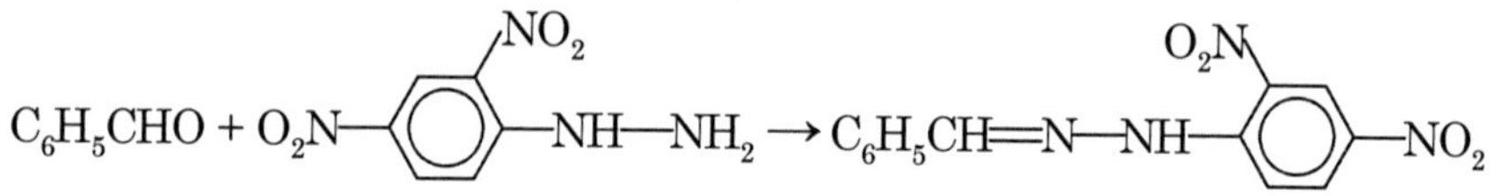

Orange-red precipitate

$$C_6H_5CHO + Ag_2O \xrightarrow{NH_4OH} C_6H_5COO^- + Ag\downarrow$$

Grey-white precipitate

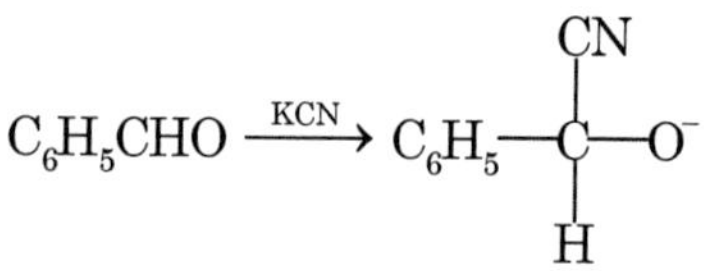

(B) $CH_3C{\equiv}CH$ gives precipitate with ammoniacal $AgNO_3$ solution.

$$CH_3C{\equiv}CH \xrightarrow{\text{ammoniacal } AgNO_3} CH_3C{\equiv}C^-Ag^+\downarrow$$

White precipitate

(C) CN^- ion gives precipitate with ammoniacal $AgNO_3$ solution, is a nucleophile and is involved in cyanohydrin formation

$$AgNO_3 + CN^- \rightarrow AgCN\downarrow$$

White precipitate

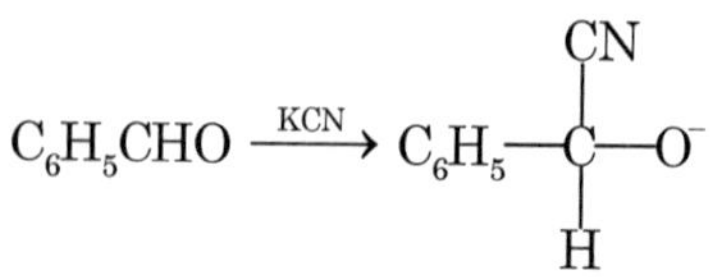

(D) I^- ion gives precipitate with ammoniacal $AgNO_3$ solution and is a nucleophile.

$$AgNO_3 + I^- \rightarrow AgI\downarrow$$

Yellow precipitate

22. (A) Simple cubic and face centred cubic unit cells have the cell parameters: $a = b = c$ and $\alpha = \beta = \gamma$, and they belong to the same crystal systems.

(B) Cubic and rhombohedral unit cells have the cell parameters: $a = b = c$ and $\alpha = \beta = \gamma$, and they are two different crystal systems.
(C) Cubic and tetragonal unit cells are two different crystal systems.
(D) Hexagonal and monoclinic unit cells are two different crystal systems and have only two crystallography angles of 90° ($\alpha = \beta = 90°$).

PAPER—II

Section–I: Straight Objective Type

1. The structural isomers for C_6H_{14} are

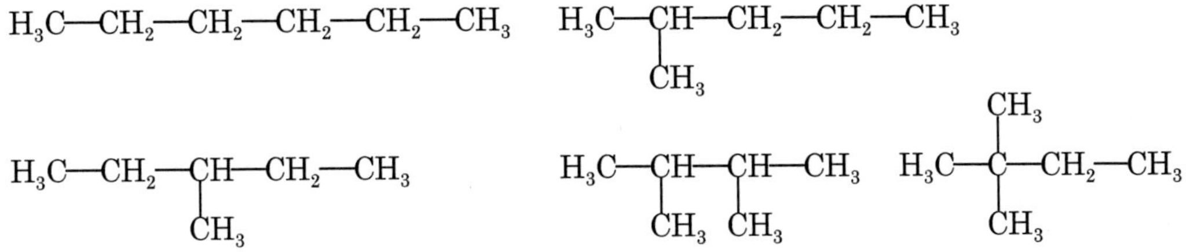

2. 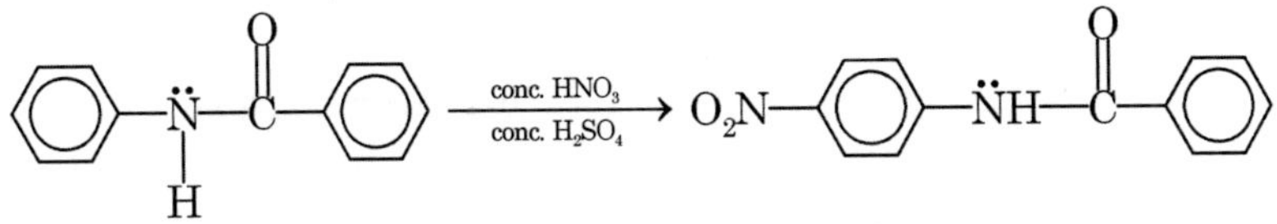

Due to the presence of a lone pair of electrons on the nitrogen atom, it will activate the ring and it will stabilise the intermediate cation at the *o*- and *p*-positions.

3. $\Delta T_f = i \times K_f \times m$

or $2 = i \times 1.72 \times \dfrac{20}{172} \times \dfrac{1000}{50}$

or $i = \dfrac{2 \times 172 \times 50}{1.72 \times 20 \times 1000} = 0.5$

4. $O_2^{2-} : (\sigma_{1s})^2 (\sigma_{1s}^*)^2 (\sigma_{2s})^2 (\sigma_{2s}^*)^2 (\sigma_{2p_z})^2 (\pi_{2p_x})^2 (\pi_{2p_y})^2 (\pi_{2p_x}^*)^2 (\pi_{2p_y}^*)^2$; No. of unpaired electrons in $O_2^{2-} = 0$.

$N{=}N{\rightarrow}O$; No. of unpaired electrons in $N_2O = 0$

$O{=}\overset{+}{O}{-}\overset{-}{O} \leftrightarrow O{=}O{\rightarrow}O$; No. of unpaired electron in $O_3 = 0$

$O_2^- : (\sigma_{1s})^2 (\sigma_{1s}^*)^2 (\sigma_{2s})^2 (\sigma_{2s}^*)^2 (\sigma_{2p_z})^2 (\pi_{2p_x})^2 (\pi_{2p_y})^2 (\pi_{2p_x}^*)^2 (\pi_{2p_y}^*)^1$; No. of unpaired electrons in $O_2^- = 0$

Hence O_2^- is paramagnetic.

5. $\Delta G = \Delta H - T\Delta S = -54.07 \times 1000 - 298 \times 10 = -57050 \text{ J mol}^{-1}$

$\Delta G = -2.303\, RT \log K$

$= -2.303 \times 8.314 \times 298 \log K = -5705 \log K$

or $-57050 = -5705 \log K$

or $\log K = 10$

6. NO^+, CN^- and N_2 are isoelectronic with CO (fourteen electron system) and hence they all have the same bond order (= 3).

N═O$^-$ is sixteen electron system and its bond order is 2. Hence NO^- has bond order different from that in CO.

7. The hybridisation of P in P_4 is sp^3.

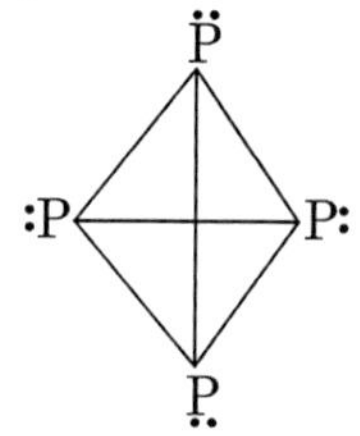

Hybridisation: sp^3; structure: tetrahedral

8. Extraction of zinc from zinc blende is achieved by roasting followed by reduction with carbon.
9. $Br—CH_2—CH_2—Br \xrightarrow{\text{alc. KOH}} CH_2═CH—Br \xrightarrow{NaNH_2} HC≡CH$

 The compound $CH_2═CH—Br$ has partial C—Br double bond and it requires a stronger base than alcoholic KOH to remove HBr. Hence the reagent for the conversion is alcoholic KOH followed by $NaNH_2$.
10. Statement–1 is false and Statement–2 is true.

 o-Hydroxybenzoic acid has a lower boiling point than *p*-hydroxybenzoic acid. This is because *o*-hydroxybenzoic acid is involved in intramolecular hydrogen bonding and is a monomer, while *p*-hydroxybenzoic acid is involved in intramolecular hydrogen bonding and is a dimer.

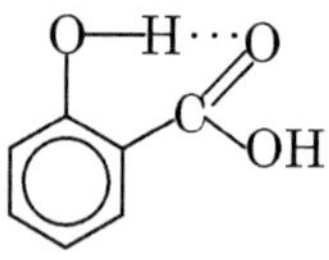

o-Hydroxybenzoic acid (monomer)

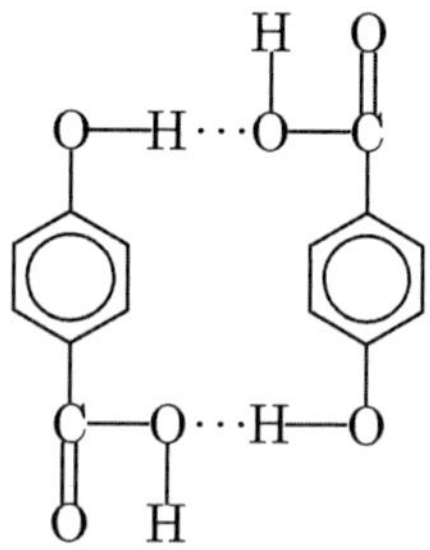

p-Hydroxybenzoic acid (dimer)
More stronger intermolecular forces increase the boiling point.

Section–II: Assertion-Reason Type

11. Statement–1 is true, Statement–2 is true and Statement–2 is not a correct explanation for Statement–1. The formation of micelles occurs only above a particular temperature called Kraft temperature (T_k) and above a particular concentration called critical micelle concentration. Each micelle contains at least one hundred molecules. Hence the conductivity of the solution decreases sharply at the critical micelle concentration.
12. Statement–1 is true, Statement–2 is true and Statement–2 is a correct explanation for Statement–1.

 According to the Fajan's rule, the small cations having high charge density have tendency to form always covalent bonds.

13. Statement–1 is true and Statement–2 is false. Orthobric acid (H_3BO_3) is a weak Lewis acid: $H_3BO_3 + H_2O \rightleftharpoons B(OH)_4 + H^+$
Orthoboric acid accepts OH^- ion from water and it does not donate proton.

Section–III: Linked Comprehensive Type

14. $NaCl \longrightarrow Na^+ + Cl^-$
At anode: $2Cl^- \longrightarrow Cl_2$
Moles of Cl^- ion = 2 in 500 mL
Hence 1 mole of Cl_2 gas is evolved.
15. Na/Hg (amalgam) formed at the cathode = 2 moles
Weight of 2 moles of Na/Hg = 2 × 223 = 446 g
16. $2Cl^- - 2e^- \longrightarrow Cl_2$
2 moles of electrons (2 Faraday) are required.
1 F = 96500 C
2 F = 193000 C
Hence total charge required for complete electrolysis of NaCl is 193000 C.
17. In high temperature metallergical extraction by arc welding, argon is mainly used to provide an inert atmosphere so that oxygen cannot react with metals to form metal oxide. The use of argon in arc welding in due to its low reactivity with metals even at high temperature.
18. The hybridisation of Xe in XeO_3 is sp^3 and it has a pyramidal structure.

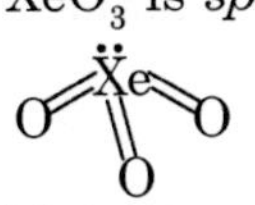

Pyramidal structure of XeO_3

19. XeF_4 and XeF_6 are expected to undergo oxidation.
$6XeF_4 + 12H_2O \longrightarrow 2XeO_3 + 4Xe + 24HF + 3O_2$
$XeF_6 + 3H_2O \longrightarrow XeO_3 + 6HF$

Section–IV: Matrix-Match Type

20. (A) $[Co(NH_3)_4(H_2O)_2]Cl_2$ exhibits geometrical isomerism, is paramagnetic and has metal ion with + 2 oxidation state, Co^{2+} is a $3d^7$ system having three unpaired electrons.

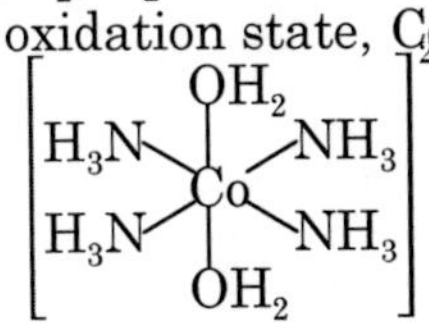

trans-$[Co(NH_3)_4(H_2O_2)]^{2+}$

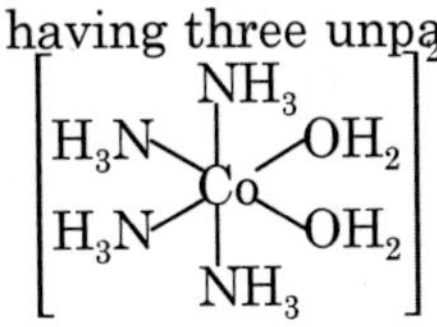

cis-$[Co(NH_3)_4(H_2O)_2]^{2+}$

(B) $[Pt(NH_3)_2Cl_2]$ exhibits geometrical isomerism, is diamagnetic and has metal ion with + 2 oxidation state. Pt^{2+} is a $5d^8$ system having no unpaire electron (dsp^2 hybridisation).

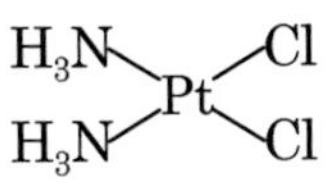

cis-$[Pt(NH_3)_2Cl_2]$

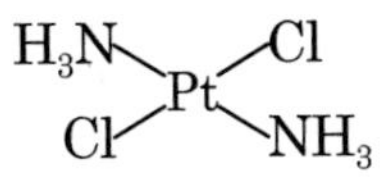

trans-$[Pt(NH_3)_2Cl_2]$

(C) $[Co(H_2O)_5Cl]Cl$ has metal ion with + 2 oxidation state and is paramagnetic. Co^{2+} is a $3d^7$ system having three unpaired electrons.

(D) $[Ni(H_2O)_6]Cl_2$ has metal ion with + 2 oxidation state and is paramagnetic. Ni^{2+} is a $3d^8$ system having two unpaired electrons. H_2O being a weak field ligand, the hybridisation of Ni is $[Ni(H_2O)_6]Cl_2$ is sp^3d^2.

21. (A) The compressibility factor of H_2 gas ($P = 200$ atm, $T = 273$ K) $\neq 1$. $Z = PV_m/RT$ at high pressure and low temperature. It follows the real gas equation $P(V - nb) = RT$. van der Waals equation $\left(P + \frac{an^2}{V^2}\right)(V - nb) = RT$ reduces to $P(V - nb) = RT$ at high pressure and low temperature since $a = 0$ at high pressure and low temperature.

(B) At $P = 0$, Z for H_2 gas is 1 and it increases continuously with increasing pressure.

(C) The molecules of CO_2 have larger intermolecular attractive forces under normal conditions of temperature and pressure (1 atm, 273 K).

(D) At very larger molecular volume, $Z \neq 1$. $Z = PV_m$/RT.

22. (A) Cellulose is a natural polymer and has glycoside linkage.

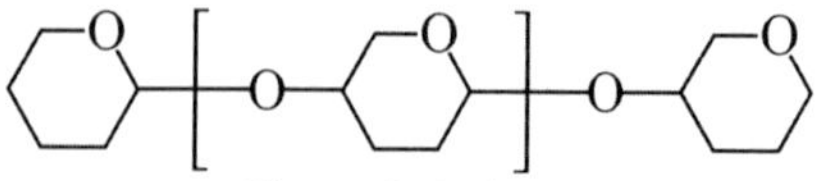

Glycoside linkage

(B) Nylon-6,6 is a synthetic polymer and has amide linkage.

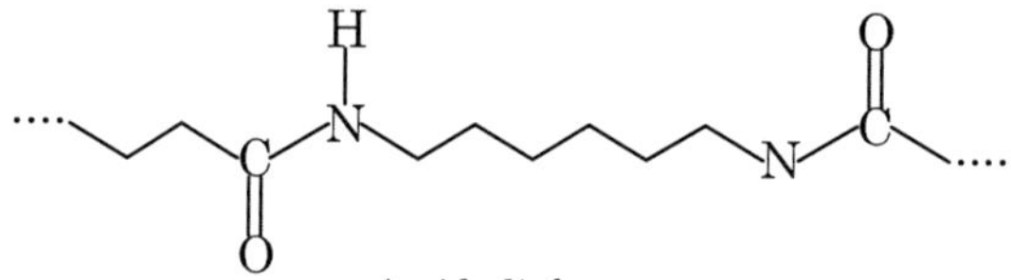

Amide linkage

(C) Protein is a natural polymer and has amide linkage.

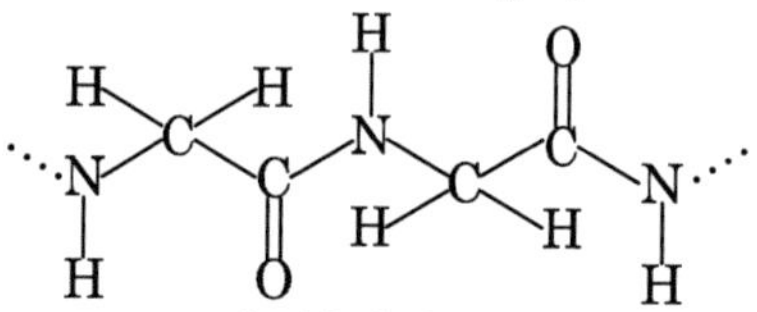

Amide linkage

(D) Sucrose has glycoside linkage.

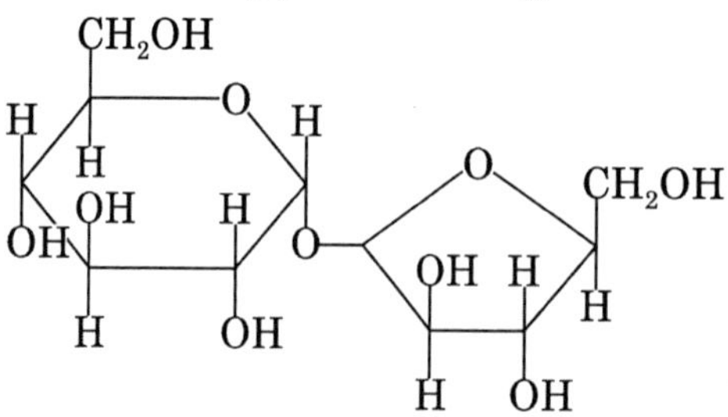

Glycoside linkage

I.I.T. JEE Objective Question Paper—2008

Question paper format:

The question paper consists of 4 sections.

1. Section I contains 6 multiple choice questions. Each question has 4 choices (A), (B), (C) and (D), out of which only one is correct.
2. Section II contains 4 multiple choice correct answer type questions. Each question has 4 choices (A), (B), (C) and (D), out of which only one or more answers are correct.
3. Section III contains 4 questions. Each question contains Statement–1 (Assertion) and Statement–2 (Reason).
   Bubble (A) if both the Statements are true and Statement–2 is the correct explanation of Statement–1.
   Bubble (B) if both the Statements are true but Statement–2 is not the correct explanation of Statement–1.
   Bubble (C) if Statement–1 is true and Statement–2 is false.
   Bubble (D) if Statement–1 is false and Statement–2 is true.
4. Section IV contains 3 sets of Linked-Comprehension type questions. Each set consists of a paragraph followed by three questions. Each question has 4 choices (A), (B), (C) and (D), out of which only one is correct.

Marking scheme:

1. For each question in Section I, you will be awarded 3 marks if you have darkened only the bubble corresponding to the correct answer and zero mark if no bubble is darkened. In all other cases, minus one (– 1) mark will be awarded.
2. For each question in Section II, you will be awarded 4 marks if you have darkened all the bubble(s) corresponding to the correct answer and zero mark for all other cases. It may be noted that there is no negative marking for wrong answer.
3. For each question in Section III, you will be awarded 3 marks if you darken only the bubble corresponding to the correct answer and zero mark if no bubble is darkened. In all other cases, minus one (– 1) mark will be awarded.
4. For each question in Section IV, you will be awarded 4 marks if you darken only the bubble corresponding to the correct answer and zero mark if no bubble is darkened. In all other cases, minus one (– 1) mark will be awarded.

Section–I: Straight Objective Type

This section contains 6 multiple choice questions. Each question has 4 choices (A), (B), (C) and (D), out of which ONLY ONE is correct.

1. Hyperconjugation involves overlap of the following orbitals

   (A) $\sigma-\sigma$ (B) $\sigma-p$ (C) $p-p$ (D) $\pi-\pi$

2. The major product of the following reaction is

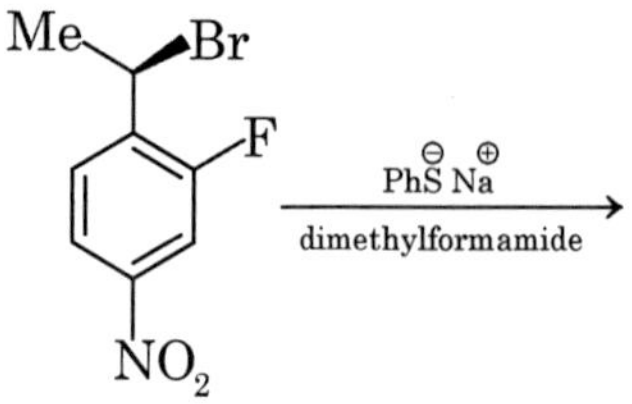

(A) 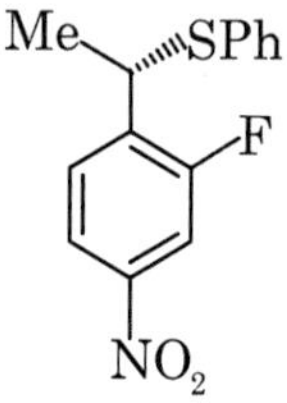 (B) 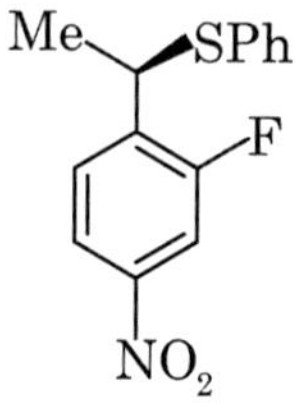 (C) 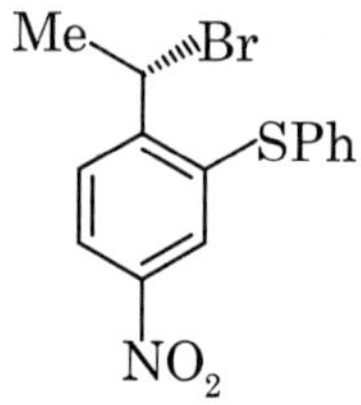 (D) 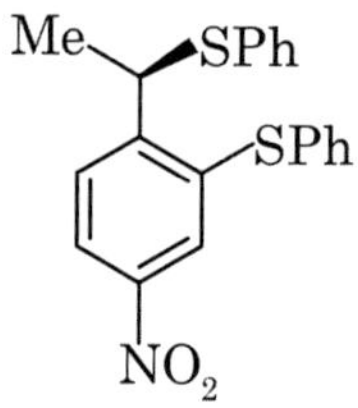

3. Aqueous solution of $Na_2S_2O_3$ on reaction with Cl_2 gives

   (A) $Na_2S_4O_6$ (B) $NaHSO_4$ (C) NaCl (D) NaOH

4. Native silver metal forms a water soluble complex with a dilute aqueous solution of NaCN in the presence of

   (A) nitrogen (B) oxygen (C) carbon dioxide (D) argon

5. Under the same reaction conditions, initial concentration of 1.386 mol dm^{-1} of a substance becomes half in 40 seconds and 20 seconds through first order and zero order kinetics, respectively. Ratio (k_1/k_0) of the rate constants for first order (k_1) and zero order (k_0) of the reactions is

   (A) 0.5 mol^{-1} dm^3 (B) 1.0 mol dm^{-3} (C) 1.5 mol dm^{-3} (D) 2.0 mol^{-1} dm^3

6. 2.5 mL of (2/5) M weak monoacidic base ($K_b = 1 \times 10^{-12}$ at 25°C) is titrated with (2/15) M HCl in water at 25°C. The concentration of H^+ at equivalence point is ($K_w = 1 \times 10^{-14}$ at 25°C)

   (A) 3.7×10^{-13} M (B) 3.2×10^{-7} M (C) 3.2×10^{-2} M (D) 2.7×10^{-2} M

Section–II: Multiple Correct Answers Type

This section contains 4 multiple correct answer(s) type questions. Each question has 4 choices (A), (B), (C) and (D), out of which ONE or MORE is/are correct.

7. The correct statement(s) about the compound given below is(are)

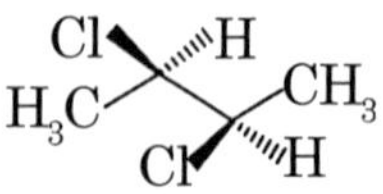

   (A) The compound is optically active

   (B) The compound possesses centre of symmetry

   (C) The compound possesses plane of symmetry

   (D) The compound possesses axis of symmetry.

8. The correct statement(s) concerning the structures E, F and G is(are)

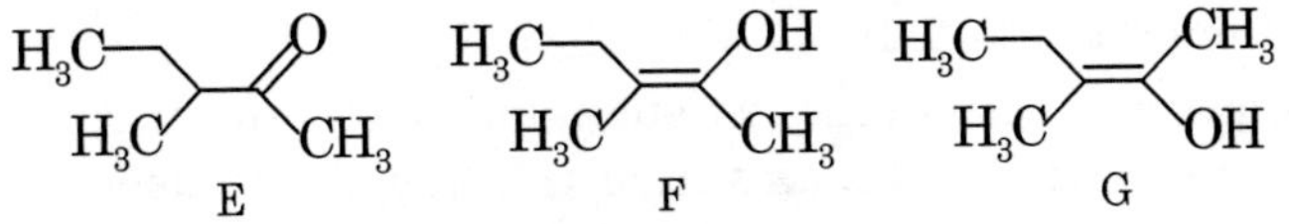

(A) E, F and G are resonance structures
(B) E, F and G are tautomers
(C) F and G are geometrical isomers
(D) F and G are diastereomers.

9. A solution of colourless salt H on boiling with excess NaOH produces a non-flammable gas. The gas evolution ceases after some time. Upon addition of Zn dust to the same solution, the gas evolution restarts. The colourless salt(s) H is(are)

(A) NH_4NO_3 (B) NH_4NO_2 (C) NH_4Cl (D) $(NH_4)_2SO_4$

10. A gas described by van der Waals equation
(A) behaves similar to an ideal gas in the limit of large molar volumes
(B) behaves similar to an ideal gas in the limit of large pressures
(C) is characterized by van der Waals coefficients that are dependent on the identity of the gas but are independent of the temperature
(D) has the pressure that is lower than the pressure exerted by the same gas behaving ideally.

Section–III: Reasoning Type

This section contains 4 reasoning type questions. Each question has 4 choice (A), (B), (C) and (D), out of which ONLY ONE is correct.

11. Statement–1: Bromobenzene upon reaction with Br_2/Fe gives 1,4-dibromobenzene as the major product.
Statement–2: In bromobenzene, the inductive effect of the bromo group is more dominant than the mesomeric effect in directing the incoming electrophile.
(A) Statement–1 is true, Statement–2 is true; Statement–2 is a correct explanation for Statement–1
(B) Statement–1 is true, Statement–2 is true; Statement–2 is not a correct explanation for Statement–1
(C) Statement–1 is true, Statement–2 is false
(D) Statement–1 is false, Statement–2 is true.

12. Statement–1: Pb^{4+} compounds are stronger oxidizing agents than Sn^{4+} compounds.
Statement–2: The higher oxidation states for the group 14 elements are more stable for the heavier members of the group due to 'inert pair effect'.
(A) Statement–1 is true, Statement–2 is true; Statement–2 is a correct explanation for Statement–1
(B) Statement–1 is true, Statement–2 is true; Statement–2 is not a correct explanation for Statement–1

(C) Statement–1 is true, Statement–2 is false

(D) Statement–1 is false, Statement–2 is true.

13. Statement–1: The plot of atomic number (y-axis) versus number of neutrons (x-axis) for stable nuclei shows a curvature towards x-axis from the line of 45° slope as the atomic number is increased.

Statement–2: Proton-proton electrostatic repulsions begin to overcome attractive forces involving protons and neutrons in heavier nuclides.

(A) Statement–1 is true, Statement–2 is true; Statement–2 is a correct explanation for Statement–1

(B) Statement–1 is true, Statement–2 is true; Statement–2 is not a correct explanation for Statement–1

(C) Statement–1 is true, Statement–2 is false

(D) Statement–1 is false, Statement–2 is true.

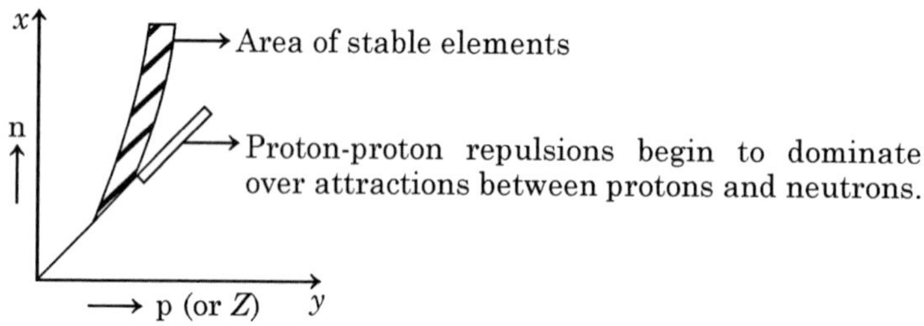

14. Statement–1: For every chemical reaction at equilibrium, standard Gibbs energy of reaction is zero.

Statement–2: At constant temperature and pressure, chemical reactions are spontaneous in the direction of decreasing Gibbs energy.

(A) Statement–1 is true, Statement–2 is true; Statement–2 is a correct explanation for Statement–1

(B) Statement–1 is true, Statement–2 is true; Statement–2 is not a correct explanation for Statement–1

(C) Statement–1 is true, Statement–2 is false

(D) Statement–1 is false, Statement–2 is true.

Section–IV: Linked Comprehension Type

This section contains 3 paragraphs. Based upon each paragraph, 3 multiple choice questions have to be answered. Each question has 4 choices (A), (B), (C) and (D), out of which ONLY ONE is correct.

Paragraph for Question Nos. 15 to 17

In the following reaction sequence, products I, J and L are formed. K represents a reagent.

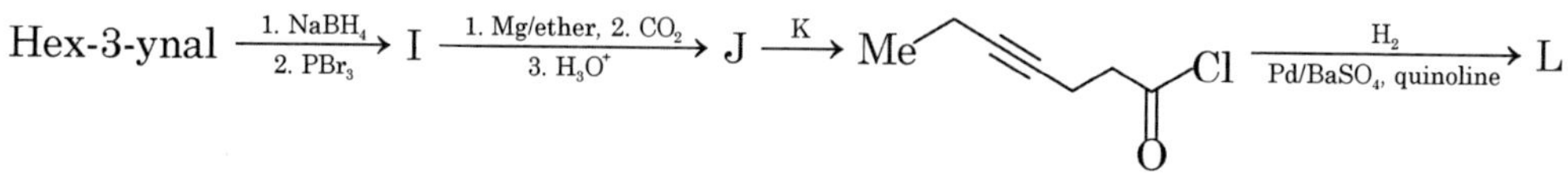

15. The structure of the product I is

(A) 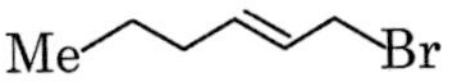

(B) 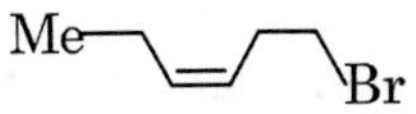

(C) 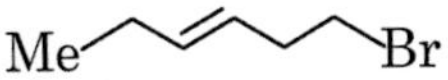

(D) 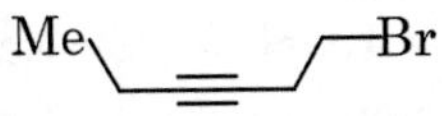

16. The structures of compounds J and K, respectively, are

(A) 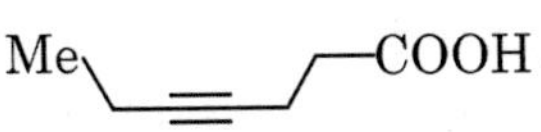 and $SOCl_2$

(B) 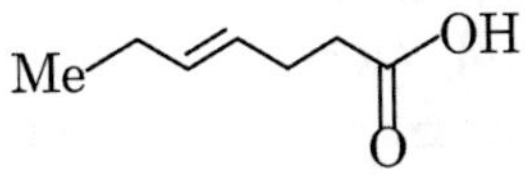 and SO_2Cl_2

(C) 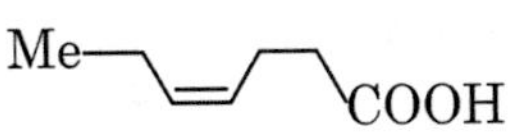 and $SOCl_2$

(D) 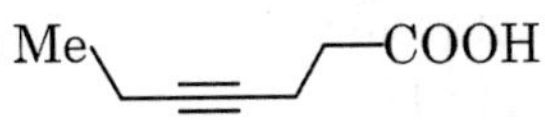 and CH_3SO_2Cl

17. The structure of product L is

(A) 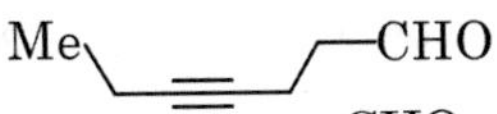

(B) 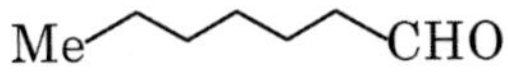

(C) 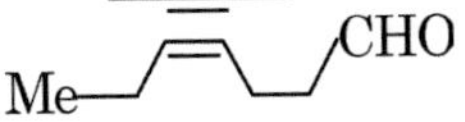

(D) 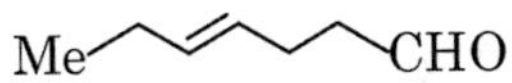

Paragraph for Question Nos. 18 to 20

There are some deposits of nitrates and phosphates in earth's crust. Nitrates are more soluble in water. Nitrates are difficult to reduce under the laboratory conditions but microbes do it easily. Ammonia forms large number of complexes with transition metal ions. Hybridization easily explains the ease of sigma donation capability of NH_3 and PH_3. Phosphine is a flammable gas and is prepared from white phosphorus.

18. Among the following, the correct statement is
   (A) Phosphates have no biological significance in humans
   (B) Between nitrates and phosphates, phosphates are less abundant in earth's crust
   (C) Between nitrates and phosphates, nitrates are less abundant in earth's crust
   (D) Oxidation of nitrates is possible in soil.

19. Among the following, the correct statement is
   (A) Between NH_3 and PH_3, NH_3 is a better electron donor because the lone pair of electrons occupies spherical *s* orbital and is less directional
   (B) Between NH_3 and PH_3, PH_3 is a better electron donor because the lone pair of electrons occupies sp^3 orbital and is more directional
   (C) Between NH_3 and PH_3, NH_3 is a better electron donor because the lone pair of electrons occupies sp^3 orbital and is more directional
   (D) Between NH_3 and PH_3, PH_3 is a better electron donor because the lone pair of electrons occupies spherical *s* orbital and is less directional.

20. White phosphorus on reaction with NaOH gives PH_3 as one of the products. This is a
   (A) dimerization reaction
   (B) disproportionation reaction
   (C) condensation reaction
   (D) precipitation reaction.

Paragraph for Question Nos. 21 to 23

Properties such as boiling point, freezing point and vapour pressure of a pure solvent change when solute molecules are added to get homogeneous solution. These are called colligative properties. Applications of colligative properties are very useful in day-to-day life. One of its examples is the use of ethylene glycol and water mixture as anti-freezing liquid in the radiator of automobiles. A solution M is prepared by mixing ethanol and water. The mole fraction of ethanol in the mixture is 0.9.

Given: Freezing point depression constant of water (K_f^{water}) = 1.86 K kg mol^{-1}; Freezing point depression constant of ethanol (K_f^{ethanol}) = 2.0 K kg mol^{-1}; Boiling point elevation constant of water (K_b^{water}) = 0.52 K kg mol^{-1}; Boiling point elevation constant of ethanol (K_b^{ethanol}) = 1.2 K kg mol^{-1}; Standard freezing point of water = 273 K; Standard freezing point of ethanol = 155.7 K; Standard boiling point of water = 373 K; Standard boiling point of ethanol = 351.5 K; Vapour pressure of pure water = 32.8 mmHg; Vapour pressure of pure ethanol = 40 mmHg; Molecular weight of water = 18 g mol^{-1}; Molecular weight of ethanol = 46 g mol^{-1}.

In answering the following questions, consider the solutions to be ideal dilute solutions and solutes to be non-volatile and non-dissociative.

21. The freezing point of the solution M is
    (A) 268.7 K (B) 268.5 K (C) 234.2 K (D) 150.9 K
22. The vapour pressure of the solution M is
    (A) 39.3 mmHg (B) 36.0 mmHg (C) 29.5 mmHg (D) 28.8 mmHg
23. Water is added to the solution M such that the mole fraction of water in the solution becomes 0.9. The boiling point of this solution is
    (A) 380.4 K (B) 376.2 K (C) 375.5 K (D) 354.7 K

ANSWERS

1. A	2. B	3. B	4. B	5. A	6. D
7. A, D	8. B, C, D	9. A, B	10. A, C, D	11. B	12. C
13. A	14. D	15. D	16. A	17. C	18. C
19. C	20. B	21. D	22. B	23. B	

MODEL SOLUTIONS—I.I.T. JEE 2008

Section–I: Straight Objective Type

1. Hyperconjugation involves the overlap of σC—H bonding electrons with the *p*-orbital of unsaturated system, carbocation or carbon free radical.
2. The given organic compound is a haloalkane as well as haloarene. Since PhS^- is a weak base, it cannot cause nucleophilic substitution by benzyne mechanism. Since the NO_2 group is present at the *meta* position with respect to the fluorine atom, S_NAr mechanism is also ruled out. Hence, the given compound reacts with PhS^- by S_{N^2} mechanism.

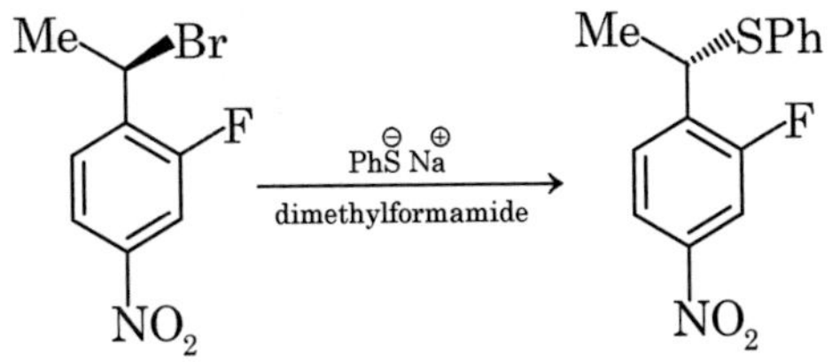

3. Chlorine is a stronger oxidising agent than iodine. Hence $Na_2S_2O_3$ on oxidation forms $NaHSO_4$ rather than sodium tetrathionate.
$Na_2SO_2O_3 + 4Cl_2 + 5H_2O \longrightarrow 2NaHSO_4 + 8HCl$
4. Silver is oxidised to Ag^+ by oxygen in the presence of cyanide ion. The reaction of Ag^+ with CN^- gives a complex ion, $[Ag(CN)_2]^-$.
5. For first order reaction, $k_1 = \frac{0.693}{t_{1/2}} = \frac{0.693}{40}$ sec^{-1}. For zero order reaction,

$$k_0 = \frac{[A_0]}{2t_{1/2}} = \frac{1.386}{2 \times 20} \text{ mol dm}^{-3} \text{ sec}^{-1}$$

$$\frac{k_1}{k_0} = \frac{0.693 \times 2 \times 20}{40 \times 1.386} = \frac{1}{2} = 0.5 \text{ mol}^{-1} \text{ dm}^3$$

6. $BOH + HCl \longrightarrow BCl + H_2O$

$$2.5 \times 10^{-3} \times \frac{2}{5} = x \times 10^{-3} \times \frac{2}{15}$$

or $$x = 7.5 \text{ mL}$$

$$[BCl] = \frac{2.5 \times \frac{2}{5}}{10} = 0.1M$$

$$\underset{0.1(1-h)}{B^+} + H_2O \rightleftharpoons \underset{0.1h}{BOH} + \underset{0.1h}{H^+}$$

$$K_h = \frac{K_w}{K_b} = \frac{[BOH][H^+]}{[B^+]} = \frac{0.1\,h^2}{(1-h)}$$

$$\frac{1\times10^{-14}}{1\times10^{-12}} = 1\times10^{-2} = \frac{0.1\,h^2}{(1-h)}$$

or h = 0.27 (We cannot ignore the value of h with respect to 1 because it is more than 0.1)

Hence $[H^+] = 0.1 \times 0.27 = 2.7 \times 10^{-2} M$

7. The given compound is represented in saw-horse projection as

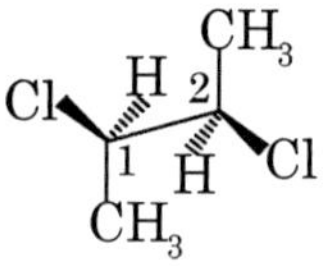

It does not have a plane of symmetry or a centre of symmetry. Hence it is optically active. The compound on rotation about C_1–C_2, keeping C_2 fixed and C_1 rotated by 180°, gives the following compound:

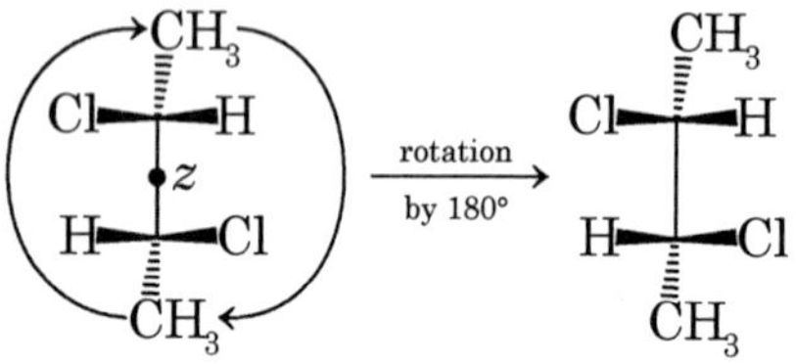

The compound possesses 180° axis of symmetry (two fold axis of symmetry) passing through the point z perpendicular to the plane of paper.

8. E is a keto form and F and G are its enol forms. F and G are also geometrical isomers and diastereoisomers.

9. Since the colourless salt produces NH_3 gas (non-inflammable) on boiling with excess of NaOH, it is an ammonium salt. On addition of zinc dust (a reducing agent) to this solution, only the nitrites and nitrates will liberate NH_3 gas again.

$$3Zn + NO_2^- + 5OH^- \longrightarrow 3ZnO_2^{2-} + H_2O + NH_3$$

$$4Zn + NO_3^- + 7OH^- \longrightarrow 4ZnO_2^{2-} + 2H_2O + NH_3$$

10. A real gas following the van der Waals equation behaves similar to an ideal gas at low pressures (or large molar volumes). The van der Waals coefficients a and b are independent of temperature and characteristic of a gas. The pressure exerted by a real gas is lower than the pressure exerted by the same gas under ideal conditions since some pressure is lost by the real gas due to the presence of intermolecular force of attraction.

11. The Statement–2 is correct because the inductive effect is a permanent effect and the inductive effect of halogen atom always dominates its mesomeric effect. But the Statement–2 cannot explain the formation of 1,4-dibromobenzene as the major product. The reason for the formation of 1,4-dibromobenzene is that Br is an *ortho* and *para* directing group due to the resonance stabilization of arenium ion. The prominent attack of Br^+ occurs at the *para* position due to the steric effect.

12. Statement–1 is true and Statement–2 is false. Since Pb^{4+} ion is less stable than Sn^{4+} ion, Pb^{4+} ion is a stronger oxidising agent than Sn^{4+} ion. Due to the inert pair effect, the stability of higher oxidation state of group 14 elements decreases on moving down the group.
13. Proton-proton repulsions begin to dominate over proton-neutron attractions with the increase of value of Z. Hence the Statement–2 is correct. In order to stabilize the heavier nuclides, more number of neutrons are to be added so that the proton-neutron attractions overcome the proton-proton repulsions. Hence the Statement–2 is the correct explanation of Statement–1.
14. For a chemical reaction at equilibrium, Gibbs free energy change of reaction (ΔG) is zero but the standard Gibbs free energy change of reaction (ΔG^0) may not be zero. Hence the Statement–1 is false but the Statement–2 is true.

15. 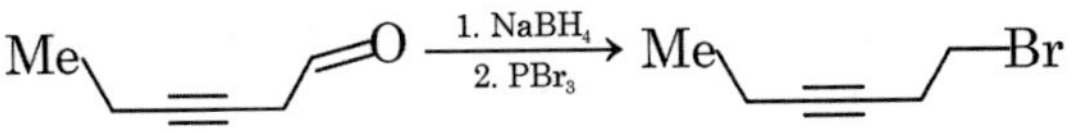

I

16. 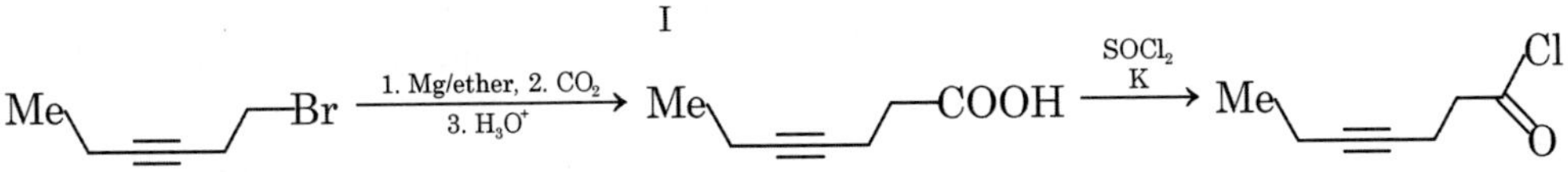

J

17. 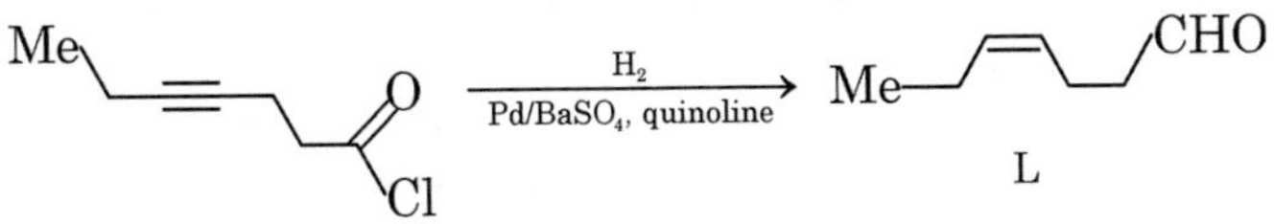

L

18. Nitrates are less abundant than phosphates in the earth's crust because they are soluble in water. Moreover, they can also be reduced by the numerous microbes present in the earth's crust.
19. NH_3 is a better electron donor than PH_3 because the lone pair of electrons on the nitrogen atom occupies an sp^3 hybridised orbital and is more directional due to the smaller size of nitrogen atom.
20. $\overset{0}{P}_4 + 3NaOH + 3H_2O \longrightarrow 3NaH_2\overset{+1}{P}O_2 + \overset{-1}{P}H_3$

    This is a disproportionation reaction.
21. Since the mole fraction of ethanol in the mixture is 0.9, ethanol is the solvent and water is the solute. Assuming water to be a non-volatile solute, the freezing point of the solution should be less than the freezing point of the pure solvent (ethanol). Thus, the freezing point of the solution is less than 155.7 K.
22. It is given that the solute is to be treated as non-volatile and non-dissociative. Thus, vapour pressure of solution = $P^0_{\text{solvent}}\, x_{\text{solvent}} = 40 \times 0.9 = 36.0$ mmHg.
23. Since the mole fraction of water in solution is 0.9, water is the solvent and ethanol is the solute and has to be considered as non-volatile and non-dissociative. Mole fraction of ethanol in the solution = 0.1. Molality of ethanol in solution = 0.1/(0.9 × 18) = 6.17. $\Delta T_b = K_b m = 0.52 \times 6.17 = 3.2$. Hence the boiling point of solution = 373 + 3.2 = 376.2 K.

I.I.T. JEE Objective Question Paper—2009

PAPER—I

Question paper format:

1. The question paper consists of 4 sections.
2. Section I contains 8 multiple choice questions. Each question has 4 choices (A), (B), (C), (D) for its answer, out of which only one is correct.
3. Section II contains 4 multiple choice questions. Each question has 4 choices (A), (B), (C), (D) for its answer, out of which one or more is/are correct.
4. Section III contains 2 group of questions. Each group has 3 questions based on a paragraph. Each question has 4 choices (A), (B), (C) and (D) for its answer, out of which only one is correct.
5. Section IV contains 2 questions. Each question has four statements (A, B, C and D) given in column I and five statements (p, q, r, s and t) in Column II. Any given statement in column I can have correct matching with one or more statement(s) given in column II. For example, if for a given question, statement B matches with the statements given in q and r, then for that particular question, against statement B, darken the bubbles corrresponding to q and r in the ORS.

Marking scheme:

6. For each question in Section I you will be awarded 3 marks if you darken the bubble corresponding to the correct answer and zero mark if no bubble is darkened. In case of bubbling of incorrect answer, minus one (– 1) mark will be awarded.
7. For each question in Section II, you will be awarded 4 marks if you darken the bubble(s) corresponding to the correct choice(s) for the answer, and zero mark if no bubble is darkened. In all other cases, minus one (– 1) mark will be awarded.
8. For each question in Section III, you will be awarded 4 marks if you darken the bubble(s) corresponding to the correct answer and zero mark if no bubble is darkened. In all other cases, minus one (– 1) mark will be awarded.
9. For each question in Section IV, you will be awarded 2 marks for each row in which you have darkened the bubble(s) corresponding to the correct answer. Thus, each question in this section carries a maximum of 8 marks. There is no negative marking for incorrect answer(s) for this section.

Section–I: Single Correct Choice Type

This section contains 8 multiple choice questions. Each question has 4 choices (A), (B), (C) and (D), for its answere out of which ONLY ONE is correct.

1. Among the electrolytes Na_2SO_4, $CaCl_2$, $Al_2(SO_4)_3$ and NH_4Cl, the most effective coagulating agent for Sb_2S_3 sol is

   (A) Na_2SO_4 (B) $CaCl_2$ (C) $Al_2(SO_4)_3$ (D) NH_4Cl

2. Given that the abundances of isotopes ^{54}Fe, ^{56}Fe and ^{57}Fe are 5%, 90% and 5%, respectively, the atomic mass of Fe is

(A) 55.85 (B) 55.95 (C) 55.75 (D) 56.06

3. The correct acidity order of the following is

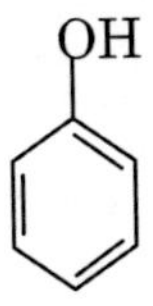
(I)

(II)
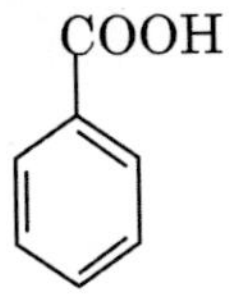
(III)
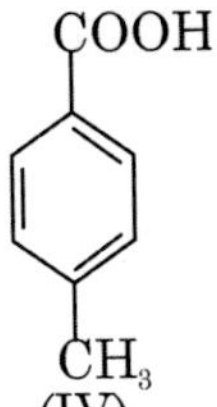
(IV)

(A) (III) > (IV) > (II) > (I) (B) (IV) > (III) > (I) > (II)

(C) (III) > (II) > (I) > (IV) (D) (II) > (III) > (IV) > (I)

4. The IUPAC name of the following compound is

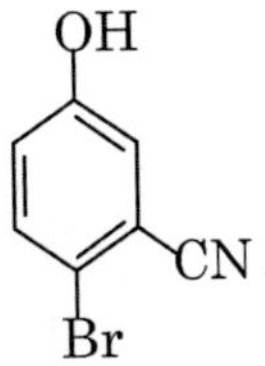

(A) 4-Bromo-3-cyanophenol

(B) 2-Bromo-5-hydroxybenzonitrile

(C) 2-Cyano-4-hydroxybromobenzene

(D) 6-Bromo-3-hydroxybenzonitrile

5. The term that corrects for the attractive forces present in a real gas in the van der Waals equation is

(A) nb (B) $\frac{an^2}{V^2}$ (C) $-\frac{an^2}{V^2}$ (D) $-nb$

6. Among cellulose, poly(vinyl chloride), nylon and natural rubber, the polymer in which the intermolecular force of attraction is weakest is

(A) nylon (B) poly(vinyl chloride)

(C) cellulose (D) natural rubber.

7. The reaction of P_4 with X leads selectively to P_4O_6. The X is

(A) dry O_2 (B) a mixture of O_2 and N_2

(C) moist O_2 (D) O_2 in the presence of aqueous NaOH

8. The Henry's law constant for the solubility of N_2 gas in water at 298 K is 1.0×10^5 atm. The mole fraction of N_2 in air is 0.8. The number of moles of N_2 from air dissolved in 10 moles of water at 298 K and 5 atm pressure is

(A) 4.0×10^{-4} (B) 4.0×10^{-5} (C) 5.0×10^{-4} (D) 4.0×10^{-6}

Section–II: Multiple Correct Choice Type

This section contains 4 multiple choice questions. Each question has 4 choices (A), (B), (C) and (D) for its answer, out of which ONE OR MORE is/are correct.

9. The compound(s) that exhibit(s) geometrical isomerism is(are)
   (A) $[Pt(en)Cl_2]$ (B) $[Pt(en)_2]Cl_2$
   (C) $[Pt(en)_2Cl_2]Cl_2$ (D) $[Pt(NH_3)_2Cl_2]$
10. The compound(s) formed upon combustion of sedium metal in excess air is(are)
    (A) Na_2O_2 (B) Na_2O (C) NaO_2 (D) NaOH
11. The correct statement(s) regarding defects in solids is(are)
    (A) Frenkel defect is usually favoured by a very small difference in the sizes of cation and anion.
    (B) Frenkel defect is a dislocation defect.
    (C) Trapping of an electron in the lattice leads to the formation of F-center.
    (D) Schottky defects have no effect on the physical properties of solids.
12. The correct statement(s) about the compound $H_3C(HO)HC—CH{=}CH—CH(OH)CH_3$ (X) is(are)
    (A) The total number of stereoisomers possible for X is 6.
    (B) The total number of diastereomers possible for X is 3.
    (C) If the stereochemistry about the double bond in X is *trans*, the number of enantiomers possible for X is 4.
    (D) If the stereochemistry about the double bond in X is *cis*, the number of enantiomers possible for X is 2.

Section–III: Comprehension Type

This section contains 2 groups of questions. Each group has 3 multiple choice questions based on a paragraph. Each question has 4 choices (A), (B), (C) and (D) for its answer, out of which ONLY ONE is correct.

Paragraph for Question Nos. 13 to 15

p-Amino-*N*,*N*-dimethylaniline is added to a strongly acidic solution of X. The resulting solution is treated with a few drops of aqueous solution of Y to yield blue coloration due to the formation of methylene blue. Treatement of the aqueous solution of Y with the reagent potassium hexacyanoferrate(II) leads to the formation of an intense blue precipitate. The precipitate dissolves on excess addition of the reagant. Similarly, treatment of the solution of Y with the solution of potassium hexacyanoferrate(III) leads to a brown coloration due to the formation of Z.

13. The compound X is
    (A) $NaNO_3$ (B) NaCl (C) Na_2SO_4 (D) Na_2S
14. The compound Y is
    (A) $MgCl_2$ (B) $FeCl_2$ (C) $FeCl_3$ (D) $ZnCl_2$
15. The compound Z is
    (A) $Mg_2[Fe(CN)_6]$ (B) $Fe[Fe(CN)_6]$ (C) $Fe_4[Fe(CN)_6]_3$ (D) $K_2Zn_3[Fe(CN)_6]_2$

Paragraph for Question Nos. 16 to 18

A carbonyl compound P, which gives positive iodoform test, undergoes reaction with MeMgBr followed by dehydration to give an olefin Q. Ozonolysis of Q leads to a dicarbonyl compound R, which undergoes intramolecular aldol reaction to give predominantly S.

$$P \xrightarrow[\substack{2.\ H^+,\ H_2O \\ 3.\ H_2SO_4,\ \Delta}]{1.\ MeMgBr} Q \xrightarrow[2.\ Zn,\ H_2O]{1.\ O_3} R \xrightarrow[2.\ \Delta]{1.\ OH^-} S$$

16. The structure of the carbonyl compound P is

(A) 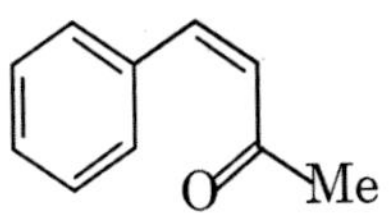

(B) 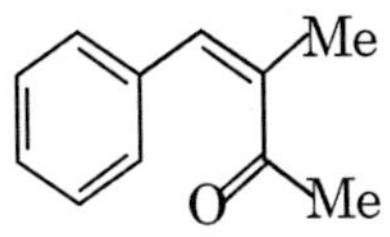

(C) 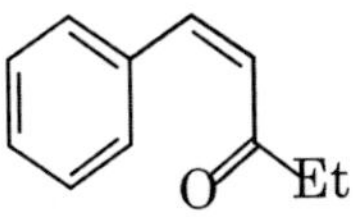

(D) 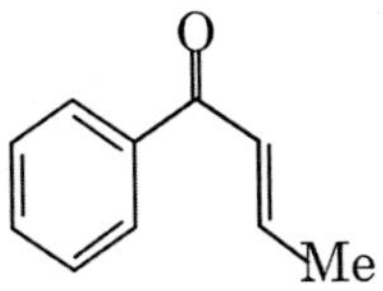

17. The structures of the products Q and R, respectively, are

(A) 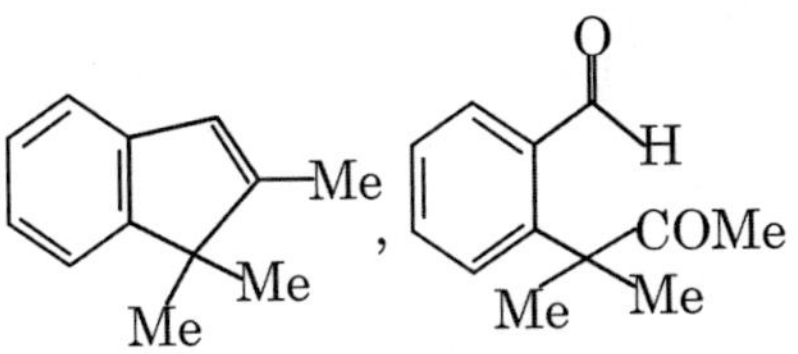

(B) 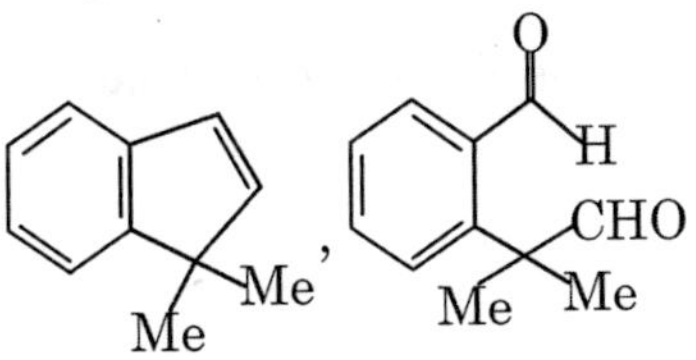

(C) 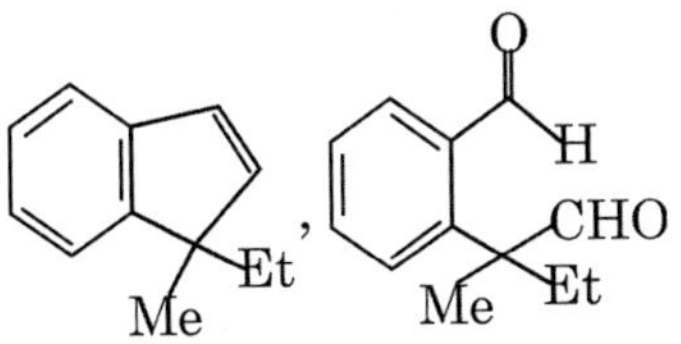

(D) 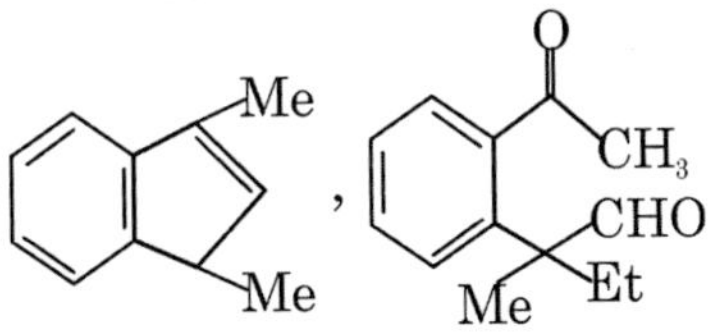

18. The structure of the product S is

(A) 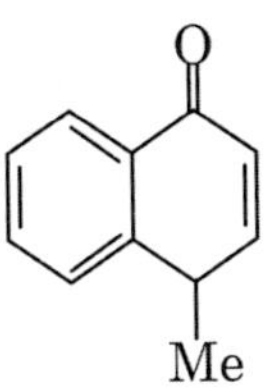

(B) 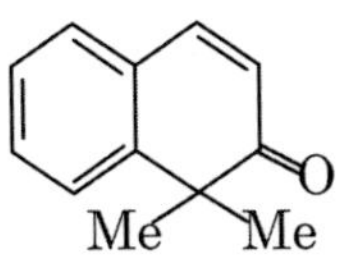

(C) 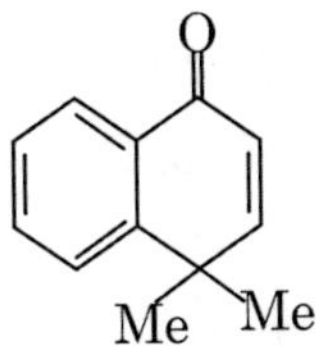

(D) 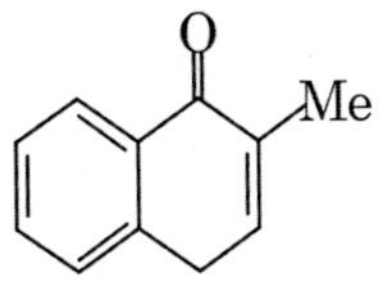

Section–IV: Matrix-Match Type

This section contains 2 questions. Each question contains statements given in two columns, which have to be matched. The statements in Column I are labelled A, B, C and D, while the statements in Column II are labelled p, q, r, s and t. Any given statement in Column I can have correct matching with ONE OR MORE statement(s) in Column II. The appropriate bubbles corresponding to the answers to these questions have to be darkened as illustrated in the following example:

If the correct matches are A – p, s and t; B – q and r; C – p and q; and D – s and t; then the correct darkening of bubbles will look like the following:

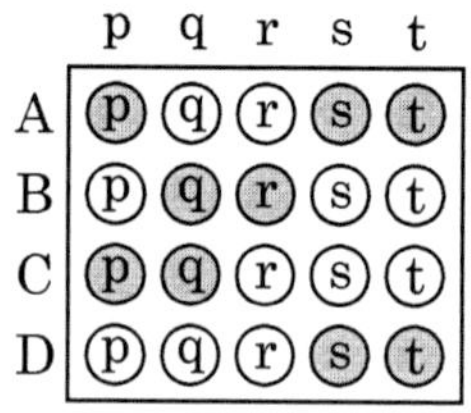

19. Match each of the compounds in Column I with its characteristic reaction(s) in Column II.

Column I	Column II
(A) $CH_3CH_2CH_2CN$	(p) Reduction with Pd–C/H_2
(B) $CH_3CH_2OCOCH_3$	(q) Reduction with $SnCl_2$/HCl
(C) $CH_3—CH═CH—CH_2OH$	(r) Development of foul smell on treatment with chloroform and alcohalic KOH
(D) $CH_3CH_2CH_2CH_2NH_2$	(s) Reduction with diisobutylaluminium hydride (DIBAL-H)
	(t) Alkaline hydrolysis

20. Match each of the diatomic molecules in Column I with its property/properties in Column II.

Column I	Column II
(A) B_2	(p) Paramagnetic
(B) N_2	(q) Undergoes oxidation
(C) O_2^-	(r) Undergoes reduction
(D) O_2	(s) Bond order $\geq$ 2
	(t) Mixing of *s* and *p* orbitals

ANSWERS

1. C	2. B	3. A	4. B	5. B	6. D
7. B	8. A	9. C, D	10. A, B	11. B, C	12. A, D
13. D	14. C	15. B	16. B	17. A	18. B

19. A – p, q, s, t; B – s, t; C – p; D – r

20. A – p, r, t; B – s, t; C – p, q; D – p, q, s

MODEL SOLUTIONS—I.I.T. JEE 2009

Section–I: Straight Objectiv Type

1. Since Sb_2S_3 is a negative sol, $Al_2(SO_4)_3$ will be the most effective coagulant due to higher charge density on Al^{3+} in accordance with Hardy–Schulze rule. Order of effectiveness of cations: $Al^{3+} > Ca^{++} > Na^+ > NH_4^+$.

2. $\overline{A} = \dfrac{\sum A_i X_i}{\sum X_i}$

   $\overline{A} = 54 \times 0.05 + 56 \times 0.90 + 57 \times 0.05$ (where $\overline{A}$ is atomic mass of Fe); $\sum x_i = 1$

   $\overline{A} = 55.95$

3. Carboxylic acid is a stronger acid than phenol. The presence of electron donating methyl group decreases acidic strength while presence of electron withdrawing halogen increases acidic strength. Decreasing order of acid strength: III > IV > II.

4. The priority of CN is the highest.

5. The measure of force of attraction for n moles of real gas $= \dfrac{an^2}{V^2}$

   $$\left(P + \frac{an^2}{V^2}\right)(V - nb) = nRT$$

6. Both nylon and cellulose have intermolecular hydrogen bonding whereas polyvinyl chloride has dipole-dipole interaction. Natural rubber has London forces which are the weakest.

7. $P_4 + 3O_2 \xrightarrow{N_2} P_4O_6$ (exclusively)

   (N_2 is used to retard the further oxidation)

8. According to Henry's law

   $P_{N_2} = K_{N_2} X_{N_2}$ where K is the Henry's constant (in atm) and X_{N_2} is mole fraction of N_2.

   $P_{N_2} = X_{N_2} P = 0.8 \times 5 \text{ atm} = 4.0 \text{ atm}$

   Hence 4 atm = 1.0×10^5 atm $\times X_{N_2}$

   $$4 \times 10^{-5} = \frac{n_{N_2}}{n_{N_2} + n_{H_2O}} = \frac{n_{N_2}}{n_{N_2} + 10}$$

   $n_{N_2} = 4 \times 10^{-4}$ moles

Section–II: Multiple Choice Correct Type

9. 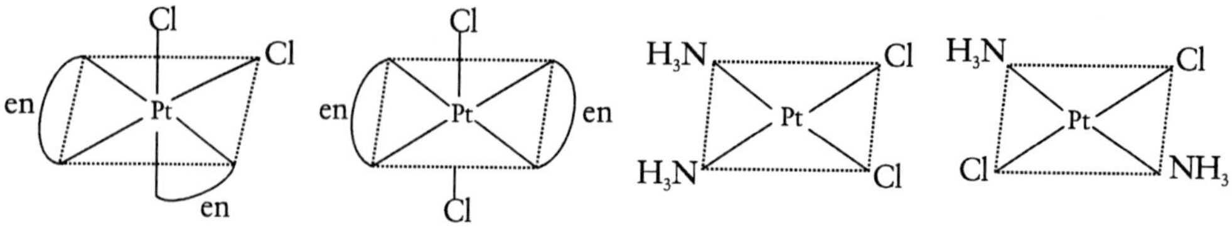

cis *trans* *cis* *trans*

10. Combustion of sodium metal in excess of air yields Na_2O and Na_2O_2.

$$4Na + O_2 \xrightarrow{\Delta} 2Na_2O$$

$$2Na + O_2 \xrightarrow{\Delta} Na_2O_2$$

11. Frenkel defect is a dislocation defect, observed when the size of cation and anion differ significantly. F-center is formed when an anion is lost from the lattice and vacancy is filled by trapping an electron. Schottky defect changes the density of a crytalline solid.

12. 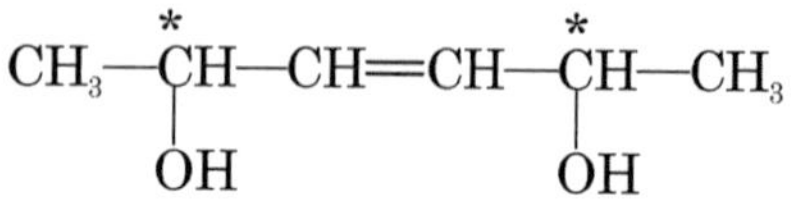

(I)	*d*	*cis*	*d*
(II)	*d*	*trans*	*d*
(III)	*l*	*cis*	*l*
(IV)	*l*	*trans*	*l*
(V)	*d*	*cis*	*l*
(VI)	*d*	*trans*	*l*

I and II are enantiomers, II and IV are enatiomers, V is meso due to the presence of plane of symmetry while VI is meso due to the presence of center of symmetry. A total of six stereoisomers are possible while number of diastereomers are four (I or III, II or IV, V and VI). If double bond in X is *trans*, number of enantiomers is two (II and IV). If double bond in X is *cis*, number of enantiomers is two (I and III).

Section–III: Comprehensive Type

13. $Na_2S + 2H^+ \longrightarrow H_2S + 2Na^+$

$FeCl_3 + H_2S \longrightarrow FeCl_2 + 2HCl + S$

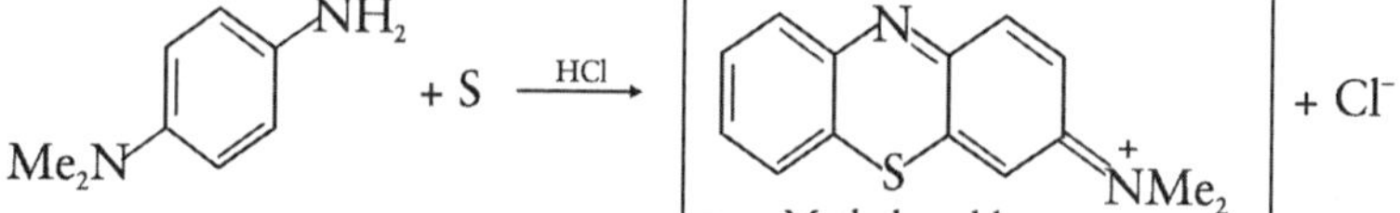

Methylene blue

14. The compound Y is $FeCl_3$ beacuse when it is treated with $K_4[Fe(CN)_6]$ it gives intense blue precipitate of $Fe_4[Fe(CN)_6]_3$, which dissolves in excess $K_4[Fe(CN)_6]$.

$$4Fe^{3+} + 3\left[Fe(CN)_6\right]^{4-} \longrightarrow Fe_4\left[Fe(CN)_6\right]_3$$

15. The compound Z is $Fe[Fe(CN)_6]$ (brown colour) is formed due to reaction of $K_3[Fe(CN)_6]$ and $FeCl_3$ (Y).

$$Fe^{3+} + \left[Fe(CN)_6\right]^{3-} \rightarrow Fe\left[Fe(CN)_6\right]$$

Solution for the question Nos. 16 to 18

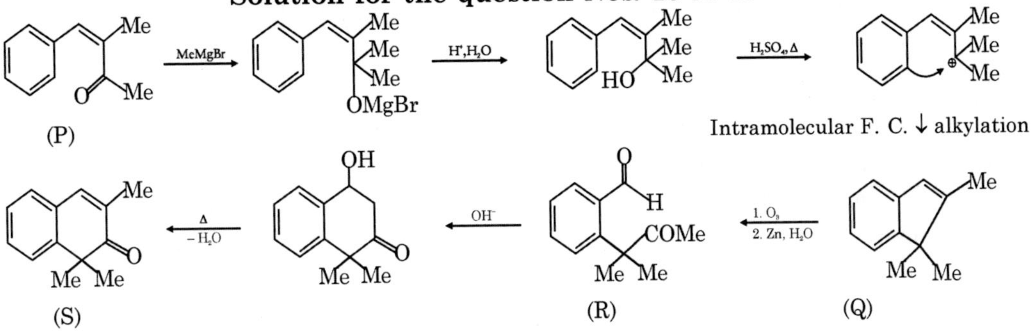

Section–IV: Matrix-Match Type

19. A – p, q, s, t; B – s, t; C – p; D – r

20. A – p, r, t; B – s, t; C – p, q; D – p, q, s (According to MOT)

PAPER—II

Section–I: Single Correct Choice Type

This section contains 4 multiple choice questions. Each question has 4 choices (A), (B), (C) and (D) for its answer, out of which ONLY ONE is correct.

1. In the following carbocation, H/CH_3 that is most likely to migrate to the positively charged carbon is

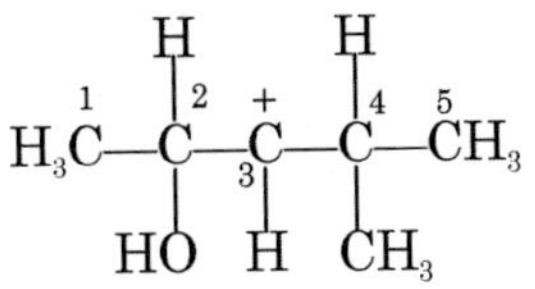

(A) CH_3 at C–4 (B) H at C–4 (C) CH_3 at C–3 (D) H at C–2

2. The spin-only magnetic moment value (in Bohr Megneton units) of $Cr(CO)_6$ is

(A) 0 (B) 2.84 (C) 4.90 (D) 5.92

3. The correct stability order of the following resonance structures is

$H_2C=\overset{+}{N}=\overset{-}{N}$ (I) $H_2\overset{+}{C}-N=\overset{-}{N}$ (II) $H_2\overset{-}{C}-\overset{+}{N}\equiv N$ (III) $H_2\overset{-}{C}-N=\overset{+}{N}$ (IV)

(A) (I) > (II) > (IV) > (III) (B) (I) > (III) > (II) > (IV)

(C) (II) > (I) > (III) > (IV) (D) (III) > (I) > (IV) > (II)

4. For a first order reaction A $\longrightarrow$ P, the temperature (T) dependent rate constant (k) was found to follow the equation $\log k = -(2000)\dfrac{1}{T} + 6.0$. The pre-exponential factor A and the activation energy E_a, respectively, are

   (A) $1.0 \times 10^6\ s^{-1}$ and 9.2 kJ mol^{-1} (B) 6.0 s^{-1} and 16.6 kJ mol^{-1}
   (C) $1.0 \times 10^6\ s^{-1}$ and 16.6 kJ mol^{-1} (D) $1.0 \times 10^6\ s^{-1}$ and 38.3 kJ mol^{-1}

Section–II: Multiple Correct Choice Type

This section contains 5 multiple choice questions. Each question has 4 choices (A), (B), (C) and (D) for its answer, out of which ONE OR MORE is/are correct.

5. The nitrogen oxide(s) that contain(s) N—N bond(s) is(are)

   (A) N_2O (B) N_2O_3 (C) N_2O_4 (D) N_2O_5

6. The correct statement(s) about the following sugars X and Y is(are)

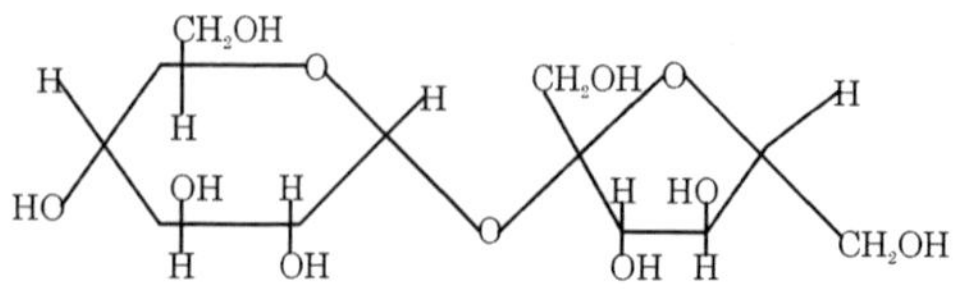

X

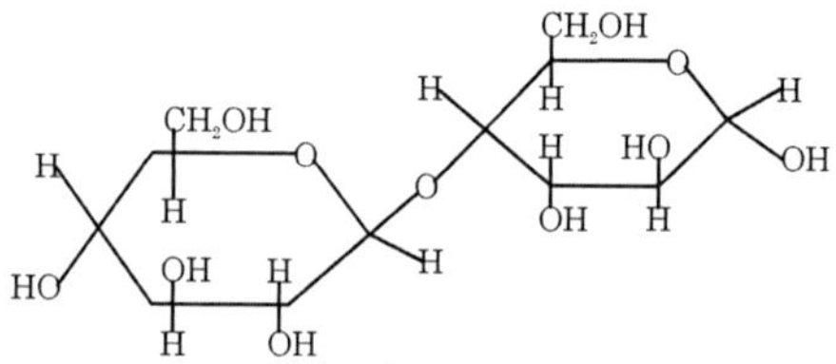

Y

   (A) X is a reducing sugar and Y is non-reducing sugar.
   (B) X is a non-reducing sugar and Y is a reducing sugar.
   (C) The glucosidic linkages in X and Y are α and β, respectively.
   (D) The glucosidic linkages in X and Y are β and α, respectively.

7. In the reaction,

   $2X + B_2H_6 \longrightarrow [BH_2(X)_2]^+[BH_4]^-$

   the amine(s) X is(are)

   (A) NH_3 (B) CH_3NH_2 (C) $(CH_3)_2NH$ (D) $(CH_3)_3N$

8. Among the following, the state function(s) is(are)

   (A) Internal energy (B) Irreversible expansion work
   (C) Reversible expansion work (D) Mole enthalpy.

9. For the reduction of NO_3^- ion in an aqueous solution, E^0 is + 0.96 V. Values of E^0 for some metal ions are given below.

   $V^{2+}(aq) + 2e^- \longrightarrow V$ $E^0 = -1.19$ V
   $Fe^{3+}(aq) + 3e^- \longrightarrow Fe$ $E^0 = -0.04$ V
   $Au^{3+}(aq) + 3e^- \longrightarrow Au$ $E^0 = +1.40$ V
   $Hg^{2+}(aq) + 2e^- \longrightarrow Hg$ $E^0 = +0.86$ V

   The pair(s) of metals that is(are) oxidized by NO_3^- in aqueous solution is(are)

   (A) V and Hg (B) Hg and Fe (C) Fe and Au (D) Fe and V

Section–III: Matrix–Match Type

This section contains 2 questions. Each question contains statements given in two columns, which have to be matched. The statements in Column I are labelled A, B, C and D, while the statements in Column II are labelled p, q, r, s and t. Any given statement in Column I can have correct matching with ONE OR MORE statement(s) in Column II. The appropriate bubbles corresponding to the answers to these questions have to be darkened as illustrated in the following example:

If the correct matches are A—p, s and t; B—q and r; C—p and q; and D—s and t; then the correct darkening of bubbles will look like the following:

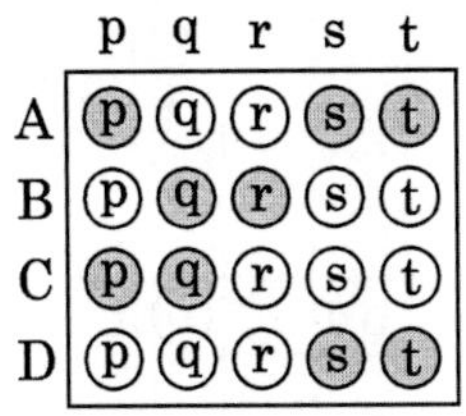

10. Match each of the reactions given in Column I with the corresponding product(s) in Column II.

Column I	Column II
(A) Cu + dil. HNO_3	(p) NO
(B) Cu + conc. HNO_3	(q) NO_2
(C) Zn + dil. HNO_3	(r) N_2O
(D) Zn + conc. HNO_3	(s) $Cu(NO_3)_2$
	(t) $Zn(NO_3)_2$

11. Match each of the compound in Column I with its characteristic reaction(s) in Column II.

Column I	Column II
(A) 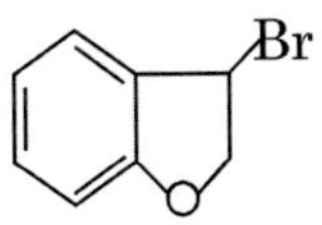	(p) Nucleophilic substitution
(B) 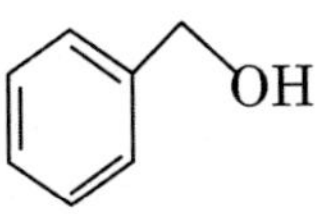	(q) Elimination
(C) 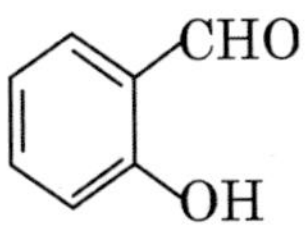	(r) Nucleophilic addition
(D) 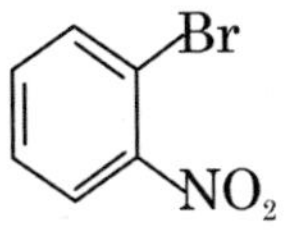	(s) Esterification with acetic anhydride
	(t) Dehydrogenation

Section–IV: Integer Answer Type

This section contains 8 questions. The answer to each of the questions is a single digit integer, ranging from 0 to 9. The appropriate bubbles below the respective question numbers in the ORS have to be darkened. For example, if the correct answers to question numbers X, Y, Z and W (say) are 6, 0, 9 and 2, respectively, then the correct darkening of bubbles will look like the following:

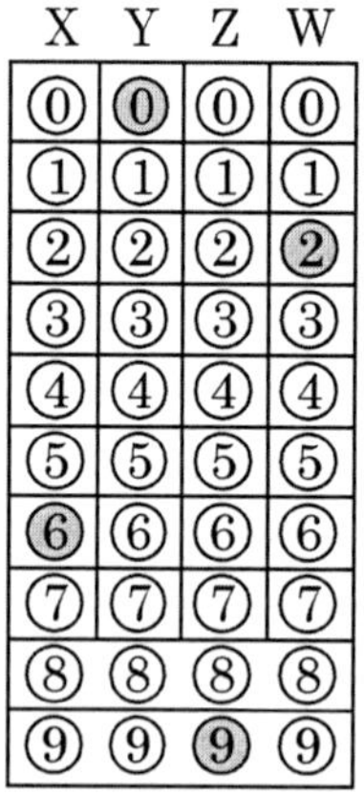

12. The coordination number of Al in the crystalline state of $AlCl_3$ is
13. The oxidation number of Mn in the product of alkaline oxidative fusion of MnO_2 is
14. The total number of α and β particles emitted in the nuclear reaction ${}^{238}_{92}U \rightarrow {}^{214}_{82}Pb$ is
15. The dissociation constant of a substituted benzoic acid at 25°C is 1.0×10^{-4}. The pH of a 0.01M solution of its sodium salt is
16. The number of water molecule(s) directly bonded to the metal centre in $CuSO_4 \cdot 5H_2O$ is
17. At 400 K, the root mean square (rms) speed of a gas X (molecular weight = 40) is equal to the most probable speed of gas Y is
18. The total number of cyclic structural as well as stereoisomers possible for a compound with the molecular formula C_5H_{10} is
19. In a constant volume calorimeter, 3.5 g of a gas with molecular weight 28 was burnt in excess oxygen at 298.0 K. The temperature of the calorimeter was found to increase from 298.0 K to 298.45 K due to the combustion process. Given that the heat capacity of the calorimeter is 2.5 kJ K^{-1}, the numerical value for the enthalpy of combustion of the gas in kJ mol^{-1} is

ANSWERS

1. D 2. A 3. B 4. D 5. A, B, C 6. B, C
7. A, B, C 8. A, D 9. A, B, D
10. A–p, s; B–q, s; C–r, t; D–q, t
11. A–p, q, t; B–p, s, t; C–r, s; D–p
12. 6 13. 4 14. 6 15. 9 16. 8 17. 7
18. 8 19. 4

PAPER—II

Section–I: Single Correct Choice Type

1. 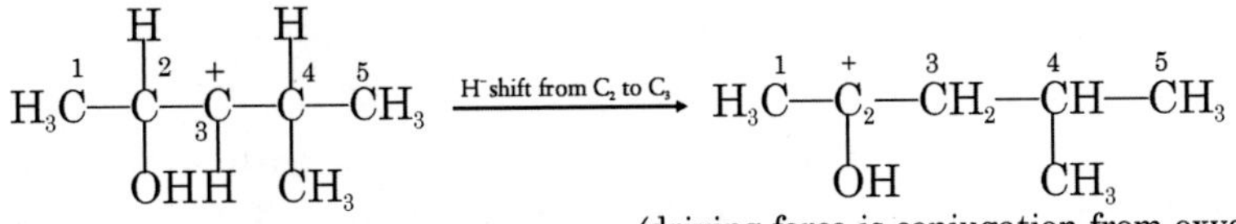

(driving force is conjugation from oxygen)

Migrating tendency of hydride is greater than that of alkyl group. Migration of hydride from second carbon gives more stable carbocation (stabilized by + R effect of OH group and + I effect of methyl group).

2. $Cr(CO)_6$ 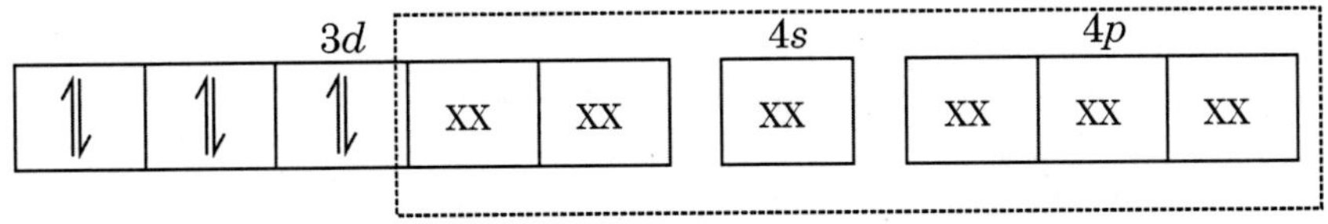

d^2sp^3 hybridisation

CO is a strong field ligand and it forces the unpaired electrons of Cr to pair up. As the coordination compound does not have any unpaired electron, its magnetic moment is zero.

3. The structure having complete octet of all the atoms and having greater number of cavalent bonds are more stable than the others. The structure with the negative charge on more electronegative atom is more stable than the one in which the negative charge is present on less electronegative atom. Thus, I is more stable. On the basis stability of resonance structures, the stability order is : I > III > II > IV.

4. $\log k = \log A - \dfrac{E_a}{2.303\ RT}$; $\log k = 6.0 - (2000)\dfrac{1}{T}$; $\log A = 6.0$; $A = 10^6 s^{-1}$

$\dfrac{E_a}{2.303R} = 2000$ or $E_a = 2000 \times 2.303 \times 8.314 = 38.3$ kJ mol^{-1}.

5. N_2O, N_2O_3 and N_2O_4 contain N—N bonds. Their structures are as follows:

N_2O $N\equiv N\rightarrow O$ N_2O_3 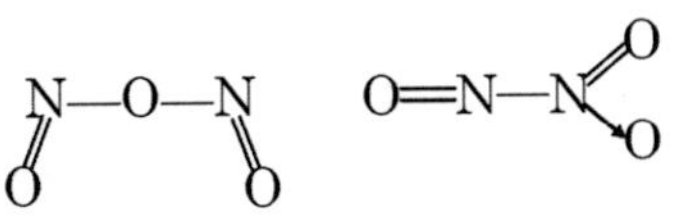

N_2O_4 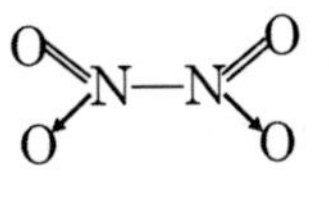 N_2O_5 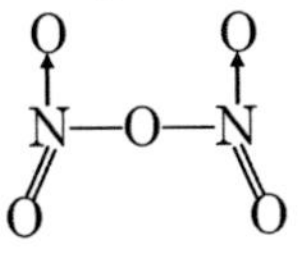

6. The choice is B and C.
7. Due to bulkyness of $(CH_3)_3N$, it does not react.
8. Internal energy and molar enthalpy are state functions.

9. $E^0_{cell} = E^0_{cathode} - E^0_{anode}$

For the reduction to occur $E^0_{cathode}$ should be greater than E^0_{anode}. Hence, V, Fe and Hg can be oxidized by NO_3^- ion in aqueous solution.

10. A–p, s; B–q, s; C–r, t; D–q, t
11. A–p, q, t; B–p, s, t; C–r, s; D–p
12. In the crystalline state of $AlCl_3$, the Cl^- ions from space lattice with Al^{3+} occupying octahedral voids. Hence, the coordination number of Al in the crystalline state of $AlCl_3$ is 6.
13. $4KOH + 2MnO_2 + O_2 \longrightarrow 2K_2MnO_4 + 2H_2O$

The oxidation number of Mn in the product formed in the above reaction is 6.

14. ${}^{238}_{92}U \longrightarrow {}^{214}_{82}Pb + 6{}^{4}_{2}He + 2{}^{0}_{-1}\beta$

Hence, total number of particles emitted is 8.

15. pH of sodium salt of weak acid $= \frac{1}{2}(pK_w + pK_a + \log C) = \frac{1}{2}(14 + 4 - 2) = 8$
16. The number of water molecules directly bonded to the metal centre in $CuSO_4 \cdot 5H_2O$ is 4.

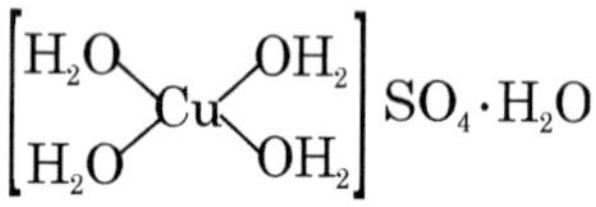

17. U_{rms} of X $= \sqrt{\dfrac{3RT_X}{M_X}}$

U_{rms} of Y $= \sqrt{\dfrac{3RT_Y}{M_Y}}$

$$\sqrt{\frac{3RT_X}{M_X}} = \sqrt{\frac{3RT_Y}{M_Y}}$$

$$M_Y = \frac{2T_Y M_X}{3T_X} = \frac{2 \times 60 \times 40}{3 \times 400} = 4$$

18. 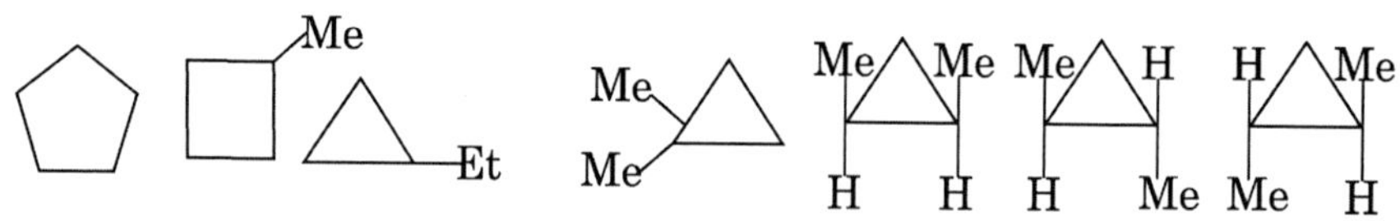

Seven cyclic isomers are possible for the compound having molecular formula C_5H_{10}.

19. Heat absorbed by the calorimeter = 2.5 × (298.45 – 298) kJ

Heat evolved by the combustion of 3.5 g gas = 2.5 × 0.45 kJ

Heat evolved by the combustion of 1 mol gas $= \dfrac{2.5 \times 0.45}{\dfrac{3.5}{28}} = 9 \text{ kJ mol}^{-1}$

Heat of combustion at constant volume = 9 kJ mol^{-1}.

I.I.T. JEE Objective Question Paper—2010

PAPER—1

A. General:

1. This Question Paper contains 32 pages having 84 questions.
2. The question paper CODE is printed on the right hand top corner of this sheet and also on the back page (page no. 32) of this booklet.
3. No additional sheets will be provided for rough work.
4. Blank papers, clipboards, log tables, slide rules, calculators, cellular phones, pagers and electronic gadgets in any form are not allowed.
5. The answer sheet, a machine-gradable Objective Response Sheet (ORS), is provided separately.
6. Do not tamper/mutilate the ORS or this booklet.
7. Do not break the seals of the question-paper booklet before instructed to do so by the invigilators.

B. Filling the bottom-half of the ORS:

8. The ORS has CODE printed on its lower and upper parts.
9. Make sure the CODE on the ORS is the same as that on this booklet. If the Codes do not match, ask for a change of the booklet.
10. Write your Registration No., Name and Name of centre and sign with pen in appropriate boxes. Do not write these any where else.
11. Darken the appropriate bubbles below your registration number with HB pencil.

C. Question paper format and Marking scheme:

12. The question paper consists of four Sections.
13. For each question in Section I, you will be awarded 3 marks if you darken only the bubble corresponding to the correct answer and zero mark if no bubbles are darkened. In all other cases, minus one (– 1) mark will be awarded.
14. For each question in Section II, you will be awarded 3 marks if you darken only the bubble corresponding to the correct answer and zero mark if no bubbles are darkened. Partial marks will be awarded for partially correct answers. No negative marks will be awarded in this Section.
15. For each question in Section III, you will be awarded 3 marks if you darken only the bubble corresponding to the correct answer and zero mark if no bubbles are darkened. In all other cases, minus one (– 1) mark will be awarded.

16. For each question in Section IV, you will be awarded 3 marks if you darken the bubble corresponding to the correct answer and zero mark if no bubble is darkened. No negative marks will be awarded for this Section.

Useful Data

Atomic Numbers: Be 4; C 6; N 7; O 8; Al 13; Si 14; Cr 24; Fe 26; Zn 30; Br 35.

1 amu	=	1.66×10^{-27} kg	R	=	0.082 L-atm K^{-1} mol^{-1}
h	=	6.626×10^{-34} J s	N_A	=	6.022×10^{23}
m_e	=	9.1×10^{-31} kg	e	=	1.6×10^{-19} C
c	=	3.0×10^{8} m s^{-1}	F	=	96500 C mol^{-1}
R_H	=	2.18×10^{-18} J	$4\pi\epsilon_0$	=	1.11×10^{-1} J^{-1} C^2 m^{-1}

SECTION – I (Single Correct Choice Type)

This Section contains 8 multiple choice questions. Each question has four choices (A), (B), (C) and (D) out of which only one is correct.

1. The synthesis of 3-octyne is achieved by adding a bromoalkane into a mixture of sodium amide and alkyne. The bromoalkane and alkyne respectively are
   (A) $BrCH_2CH_2CH_2CH_2CH_3$ and $CH_3CH_2C{\equiv}CH$
   (B) $BrCH_2CH_2CH_3$ and $CH_3CH_2CH_2C{\equiv}CH$
   (C) $BrCH_2CH_2CH_2CH_2CH_3$ and $CH_3C{\equiv}CH$
   (D) $BrCH_2CH_2CH_2CH_3$ and $CH_3CH_2C{\equiv}CH$

2. The correct statement about the following disaccharide is

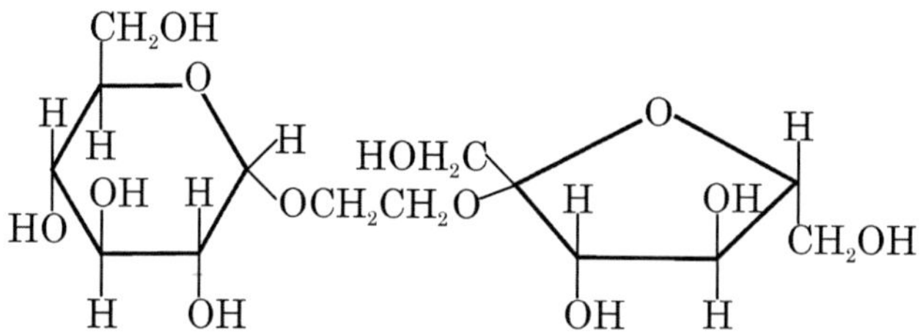

   (A) Ring (a) is pyranonse with α-glycosidic link
   (B) Ring (a) is furanonse with α-glycosidic link
   (C) Ring (b) is furanonse with α-glycosidic link
   (D) Ring (b) is pyranonse with β-glycosidic link

3. In the reaction 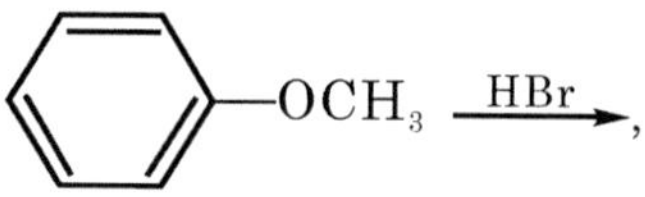 , the products are

(A) 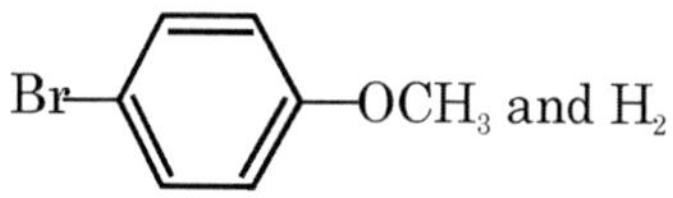

(B) 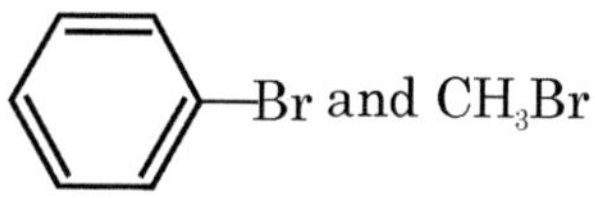

(C) 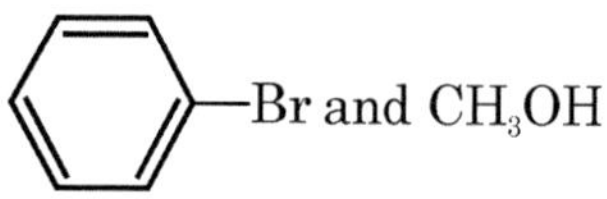

(D) 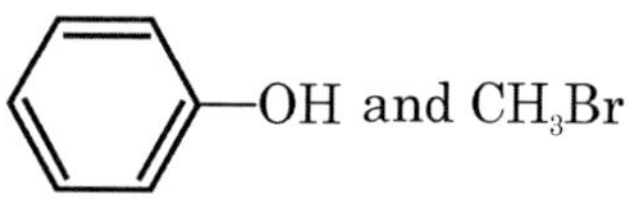

4. Plots showing the variation of the rate constant (k) with temperature (T) are given below. The plot that follows Arrhenius equation is

(A)

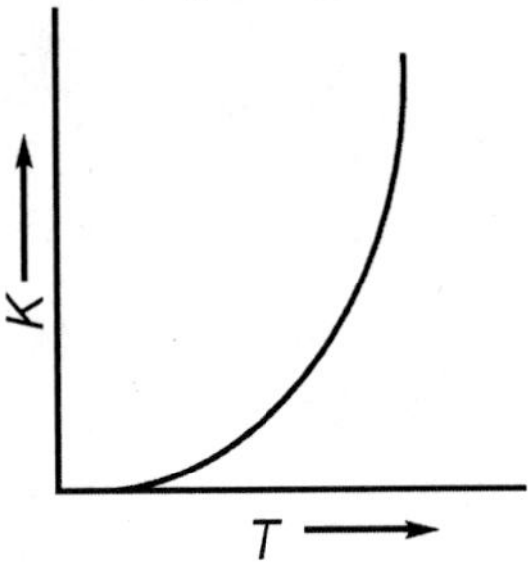

(B)

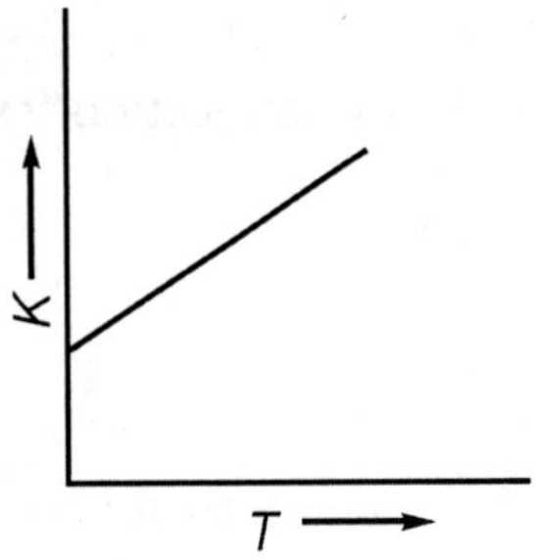

(C)

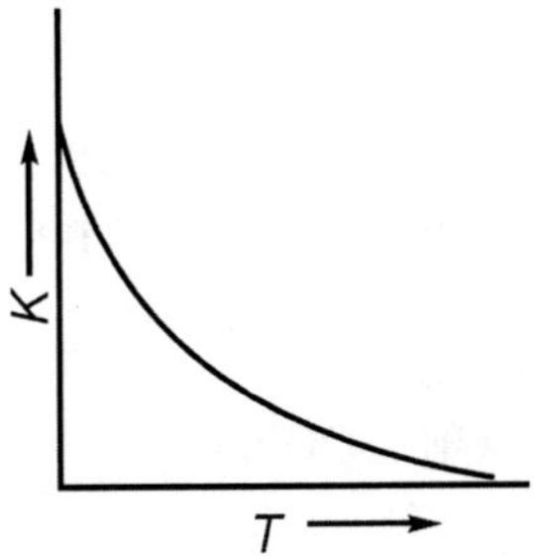

(D)

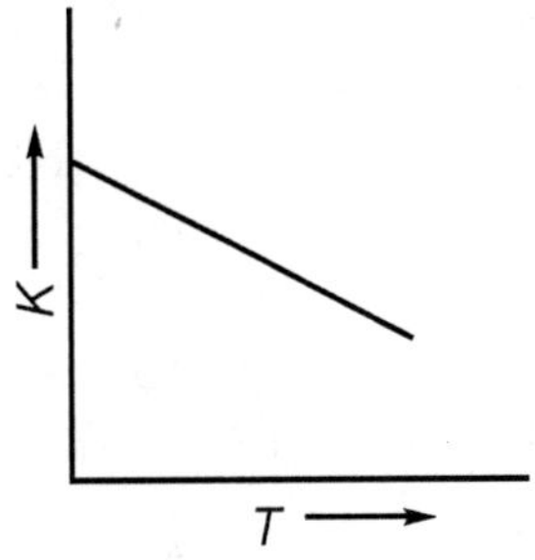

5. The species which by definition has ZERO standard molar enthalpy of formation at 298 K is

(A) $Br_2(g)$ (B) $Cl_2(g)$ (C) $H_2O(g)$ (D) $CH_4(g)$

6. The bond energy (in kcal mol^{-1}) of a C—C single bond is approximately

(A) 1 (B) 10 (C) 100 (D) 1000

7. The correct structure of ethylenediaminetetraacetic acid (EDTA) is

(A) 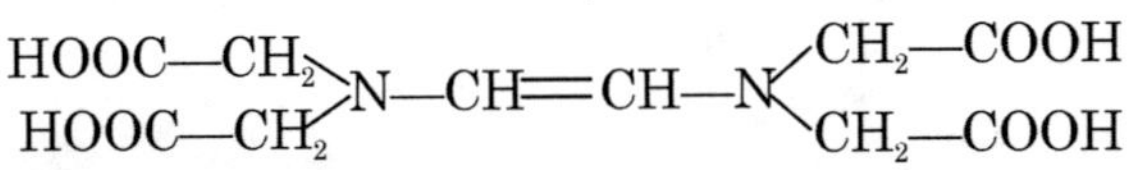

(B) 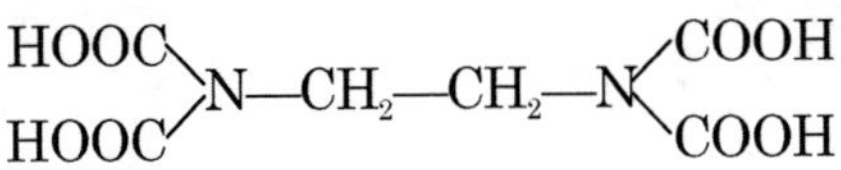

(C) 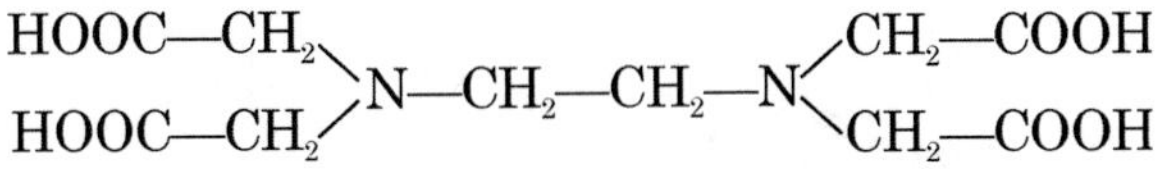

(D) 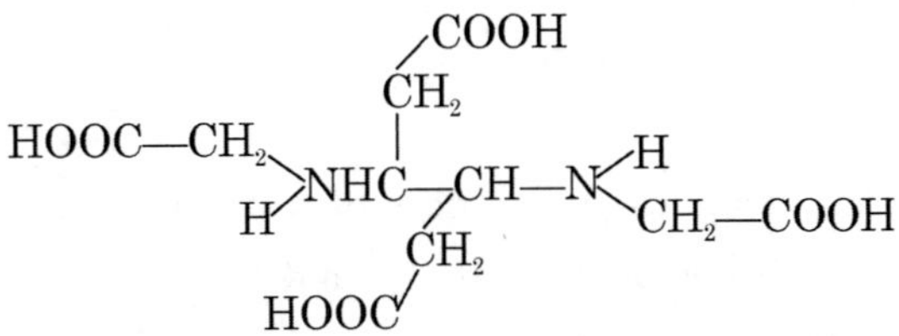

8. The ionization isomer of $[Cr(H_2O)_4Cl(NO_2)]Cl$ is

(A) $[Cr(H_2O)_4(O_2N)]Cl_2$ (B) $[Cr(H_2O)_4Cl_2](NO_2)$

(C) $[Cr(H_2O)_4Cl(ONO)]Cl$ (D) $[Cr(H_2O)_4Cl_2(NO_2)]H_2O$

SECTION – II (Multiple Correct Choice Type)

This Section contains 5 multiple choice questions. Each question has four choices (A), (B), (C) and (D) out of which one or more may be correct.

9. In the Newman projection for 2,2-dimethylbutane,

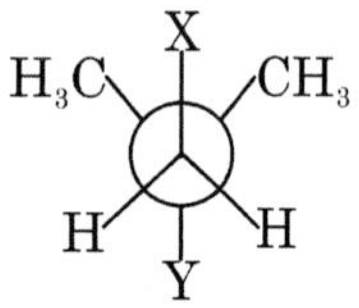

X and Y can respectively be

(A) H and H (B) H and C_2H_5

(C) C_2H_5 and H (D) CH_3 and CH_3

10. Aqueous solutions of HNO_3, KOH, CH_3COOH and CH_3COONa of identical concentrations are provided. The pair(s) of solutions which form a buffer upon mixing is (are)

(A) HNO_3 and CH_3COOH (B) KOH and CH_3COONa

(C) HNO_3 and CH_3COONa (D) CH_3COOH and CH_3COONa

11. In the reaction 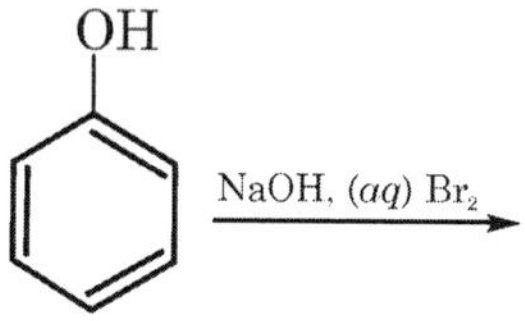 the intermediate(s) is(are)

(A) 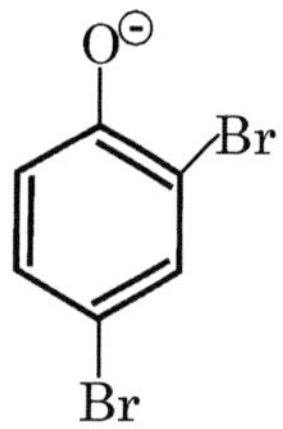

(B) 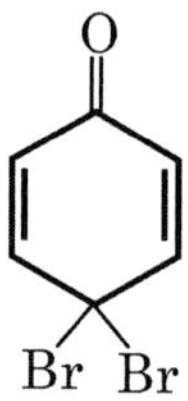

(C) 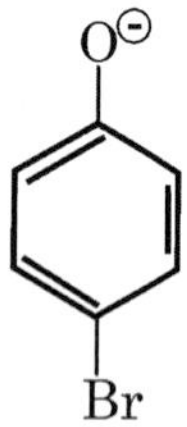

(D) 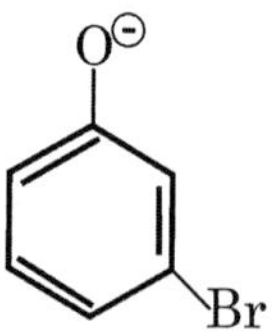

12. The reagent(s) used for softening the temporary hardness of water is(are)

(A) $Ca_3(PO_4)_2$ (B) $Ca(OH)_2$ (C) Na_2CO_3 (D) NaOCl

13. Among the following, the intensive property is (properties are)

(A) molar conductivity (B) electromotive force

(C) resistance (D) heat capacity

SECTION – III (Paragraph Type)

This section contains 2 paragraphs. Based upon the first paragraph 2 multiple choice questions and based upon the second paragraph 3 multiple choice questions have to be answered. Each of these questions has four choices (A), (B), (C) and (D) out of which only one is correct.

Paragraph for Question Nos. 14 to 15

The concentration of potassium ions inside a biological cell is at least twenty times higher than the outside. The resulting potential difference across the cell is important in several processes such as transmission of nerve impulses and maintaining the ion balance. A simple model for such a concentration cell involving a metal M is:

$M(s) | M^+(aq; 0.05 \text{ molar}) || M^+(aq), 1 \text{ molar}) | M(s)$

For the above electrolytic cell, the magnitude of the cell potential $|E_{cell}| = 70$ mV.

14. For the above cell
    (A) $E_{cell} < 0; \Delta G > 0$
    (B) $E_{cell} > 0; \Delta G < 0$
    (C) $E_{cell} < 0; \Delta G^0 > 0$
    (D) $E_{cell} > 0; \Delta G^0 > 0$

15. If the 0.05 molar solution of M^+ is replaced by 0.0025 molar M^+ solution, then the magnitude of the cell potential would be
    (A) 35 mV (B) 70 mV (C) 140 mV (D) 700 mV

Paragraph for Question Nos. 16 to 18

Copper is the most noble of the first row transition metals and occurs in small deposits in several countries. Ores of copper include chalcanthite ($CuSO_4 \cdot 5H_2O$), atacamite ($Cu_2Cl(OH)_3$), cuprite (Cu_2O), copper glance (Cu_2S) and malachite ($Cu_2(OH)_2CO_3$). However, 80% of the world copper production comes from the ore of chalcopyrite ($CuFeS_2$). The extraction of copper from chalcopyrite involves partial roasting, removal of iron and self-reduction.

16. Partial roasting of chalcopyrite produces
    (A) Cu_2S and FeO
    (B) Cu_2O and FeO
    (C) CuS and Fe_2O_3
    (D) Cu_2O and Fe_2O_3

17. Iron is removed from chalcopyrite as
    (A) FeO (B) FeS (C) Fe_2O_3 (D) $FeSiO_3$

18. In self-reduction, the reducing species is
    (A) S (B) O^{2-} (C) S^{2-} (D) SO_2

SECTION - IV (Integer Type)

This section contains TEN questions. The answer to each question is a single digit integer ranging from 0 to 9. The correct digit below the question number in the ORS is to be bubbled.

19. A student performs a titration with different burettes and finds titre values of 25.2 mL, 25.25 mL and 25.0 mL. The number of significant figures in the average titre value is

20. The concentration of R in the reaction R $\longrightarrow$ P was measured as a function of time and the following data is obtained:

[R] (molar)	1.0	0.75	0.40	0.10
t (min.)	0.0	0.05	0.12	0.18

The order of the reaction is

21. The number of neutrons emitted when ${}^{235}_{92}U$ undergoes controlled nuclear fission to ${}^{142}_{54}Xe$ and ${}^{90}_{38}Sr$ is

22. The total number of basic groups in the following form of lysine is

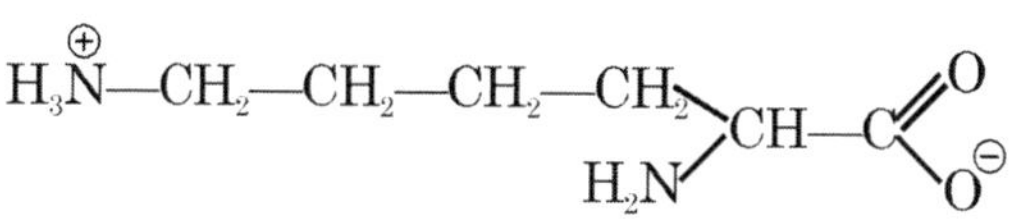

23. The total number of cyclic isomers possible for a hydrocarbon with the molecular formula C_4H_6 is

24. In the scheme given below, the total number of intramolecular aldol condensation products formed from Y is

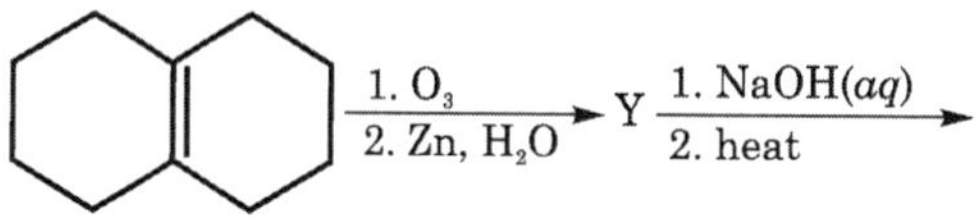

25. Amongst the following, the total number of compounds soluble in aqueous NaOH is

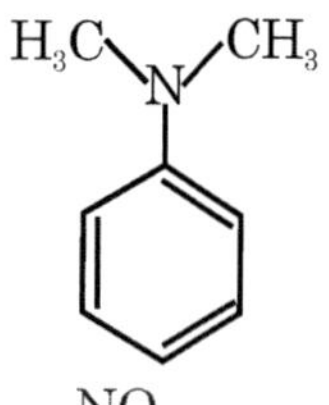
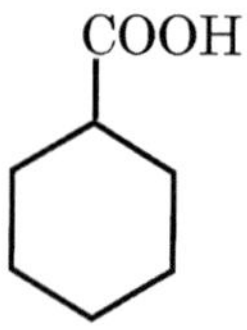
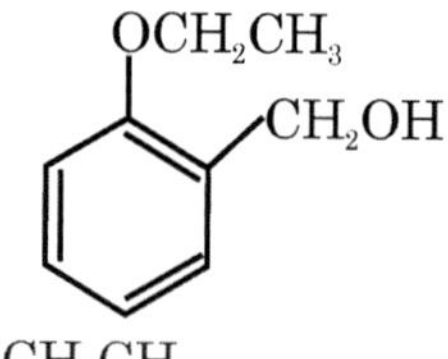
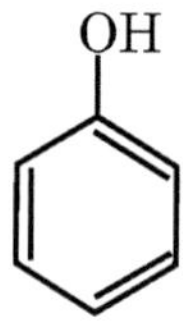
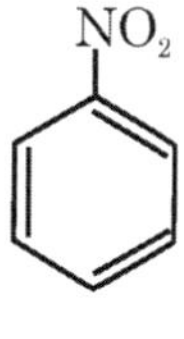
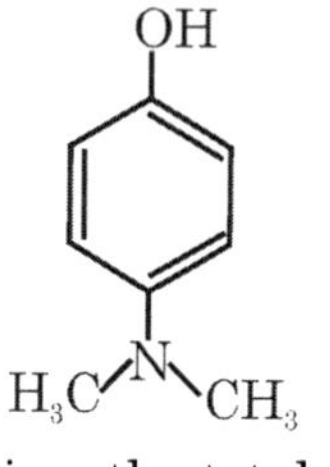
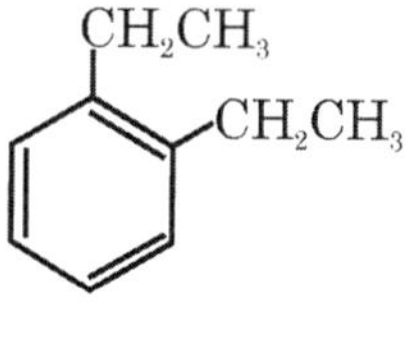
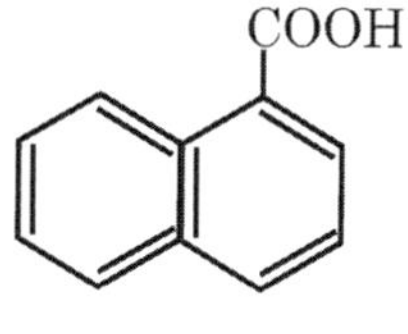

26. Amongst the following, the total number of compounds whose aqueous solution turns red litmus paper blue is

KCN K_2SO_4 $(NH_4)_2C_2O_4$ NaCl $Zn(NO_3)_2$ $FeCl_3$ K_2CO_3 NH_4NO_3 LiCN

27. Based on VSEPR theory, the number of 90 degree F—Br—F angles in BrF_5 is

28. The value of n in the molecular formula $Be_nAl_2Si_6O_{18}$ is

ANSWERS

1. D	2. A	3. D	4. A	5. B	6. C
7. C	8. B	9. B, D	10. C, D	11. A, B, C	12. B, C, D
13. A, B	14. B	15. C	16. B	17. D	18. C
19. 3	20. 0	21. 3	22. 2	23. 5	24. 1
25. 4	26. 3	27. 0	28. 3		

Model Solutions—I.I.T. JEE 2010

Paper–1

1. $CH_3—CH_2—C{\equiv}CH \xrightarrow{NaNH_2} CH_3—CH_2—C{\equiv}C^-Na^+$

$CH_3—CH_2—CH_2—CH_2\boxed{Br + Na^+}{}^-C{\equiv}C—CH_2—CH_3 \longrightarrow$

$CH_3—CH_2—CH_2—CH_2—C{\equiv}C—CH_2—CH_3$

3-octyne

2. Ring (a) is pyranose with α-glycosidic link.

3. 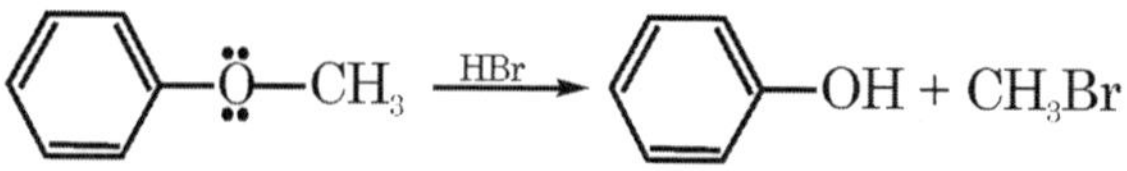

4. $k = Ae^{-E_a/RT}$

As $T\uparrow$; $k\uparrow$ exponentially

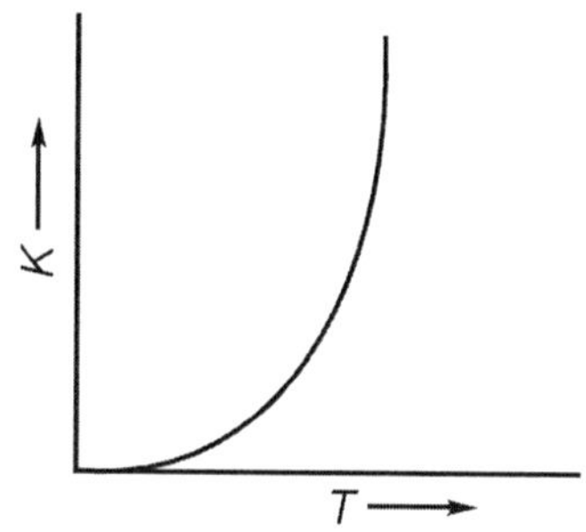

5. Chlorine is a gas at 298 K and bromine is a liquid at 298 K.
6. The bond energy of a C—C single bond is approximately 100 kcal mol^{-1}.
7. The correct structure of ethylenediaminetetraacetic acid (EDTA) is (C).
8. Cl^- ion is replaced by NO_2^- ion in the coordination sphere.
9. 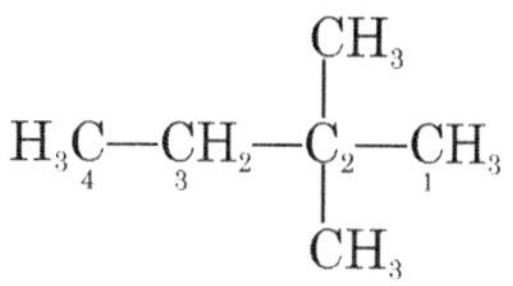

On $C_2—C_3$ bond axis, X = CH_3, Y = CH_3 and on $C_1—C_2$ bond axis, X = H, Y = C_2H_5.

10. In the option (C), if HNO_3 is present in limiting amount, then the mixture will behave as a buffer. In the option (D), the mixture contains a weak acid (CH_3COOH) and its salt CH_3COONa and this mixture behaves as a buffer.

11. The phenoxide ion is para* and ortho directing (*preferably),

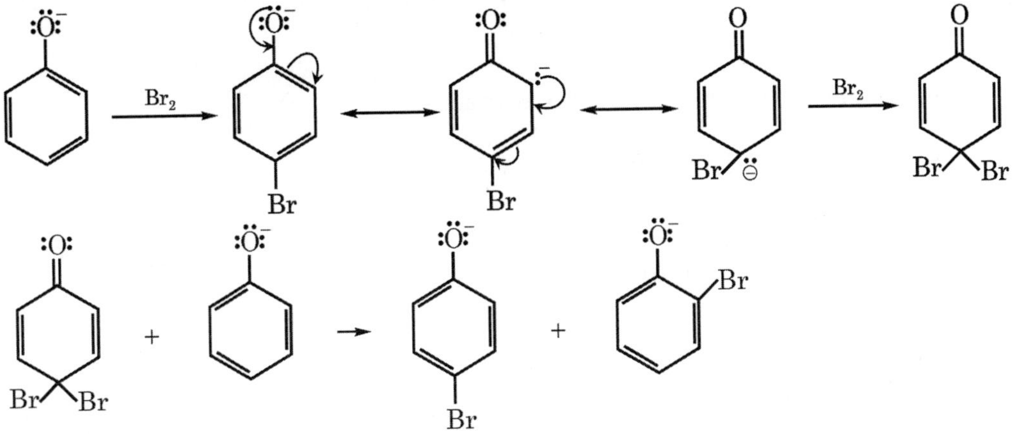

12. $Ca(HCO_3)_2 + Ca(OH)_2 \rightarrow 2CaCO_3 \downarrow + 2H_2O$
(Clarke's method)

$NaOCl + H_2O \rightleftharpoons HOCl + NaOH$

$OH^- + HCO_3^- \rightarrow CO_3^{2-} + H_2O$

$Ca(HCO_3)_2 + Na_2CO_3 \rightarrow CaCO_3 \downarrow + 2NaHCO_3$

13. Resistance and heat capacity are mass dependent properties and hence are extensive property.

14. $M(s) + M^+_{(aq)1M} \rightarrow M^+_{(aq).05M} + M(s)$

According to Nernst equation, $E_{cell} = 0 - \dfrac{2.303RT}{F}\log\dfrac{M^+_{.05M}}{M^+_{1M}}$

$$= 0 - \frac{2.303RT}{F}\log(5 \times 10^{-2}) = +\text{ve}$$

Hence $|E_{cell}| = E_{cell} = 0.70$ V and $\Delta G < 0$ for the feasibility of the reaction.

15. From the above equation $\dfrac{2.303RT}{F} = 0.0538$. Hence $E_{cell} = E^0_{cell} - \dfrac{0.0538}{1}\log 0.0025$

$$= 0 - \frac{0.0538}{1}\log 0.0025 \approx 0.13988 \text{ V} \approx 140 \text{ mV}$$

16. $2CuFeS_2 + O_2 \rightarrow Cu_2S + 2FeS + SO_2 \uparrow$

$2Cu_2S + 3O_2 \rightarrow 2Cu_2O + 2SO_2 \uparrow$

$2FeS + 3O_2 \rightarrow 2FeO + 2SO_2 \uparrow$

17. $FeO + SiO_2 \rightarrow FeSiO_3$ (slag)

18. $Cu_2S + 2Cu_2O \rightarrow \underset{\text{(Blister copper)}}{6Cu} + SO_2 \uparrow$

$S^{2-} \rightarrow S^{4+}$ is oxidation, hence S^{2-} is the reducing agent.

19. 3

20. 0

From the given data, (for zero order reaction)

$$k_I = \frac{x}{t} = \frac{0.25}{0.05} = 5 \quad \text{and} \quad k_{II} = \frac{x}{t} = \frac{0.60}{0.12} = 5$$

21. ${}^{235}_{92}U \longrightarrow {}^{142}_{54}Xe + {}^{90}_{38}Sr + 3\,{}^{1}_{0}n$

Hence 3 neutrons are emitted.

22. 2

$-\overset{\overset{\displaystyle O}{\|}}{C}-O^{\ominus}$ and $-NH_2$ are the basic groups in lysine.

23. 5

In C_4H_6, the possible cyclic isomers are

24. 1

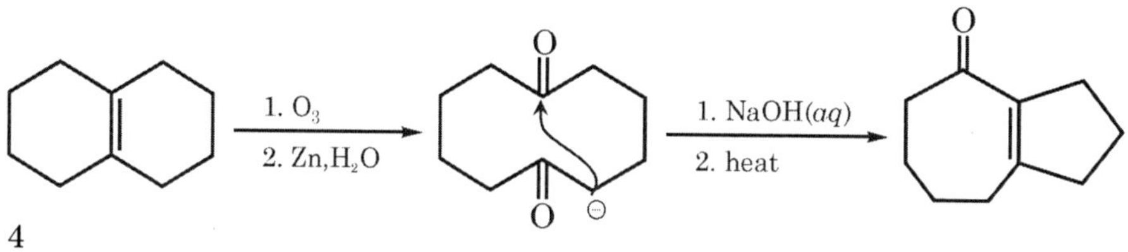

25. 4

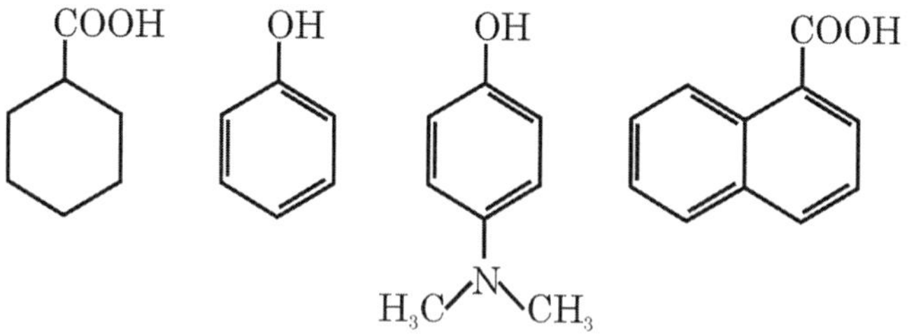

These four compounds are soluble in aqueous NaOH.

26. 3

KCN, K_2CO_3 and LiCN are basic in nature and their aqueous solution turns red litmus paper blue.

27. 0

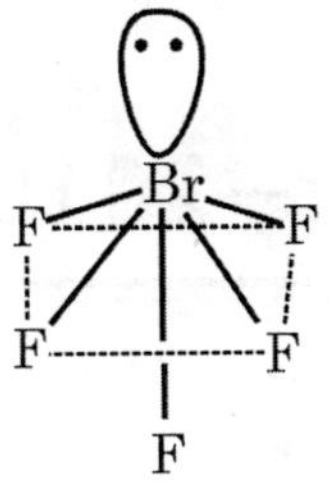

All the four planar bonds (F—Br—F) will reduce from 90° to 84.8° after lp–bp repulsion.

28. 3

$Be_3Al_2Si_6O_{18}$ (beryl)

According to charge balance in a molecule, $2n + 6 + 24 - 36 = 0$ or $n = 3$.

I.I.T. JEE Objective Question Paper—2010

PAPER—2

A. General:

1. This Question Paper contains 32 pages having 57 questions.
2. The question paper CODE is printed on the right hand top corner of this sheet and also on the back page (page no. 32) of this booklet.
3. No additional sheets will be provided for rough work.
4. Blank papers, clipboards, log tables, slide rules, calculators, cellular phones, pagers and electronic gadgets in any form are not allowed.
5. Log and Antilog tables are given in page numbers 30 and 31 respectively.
6. The answer sheet, a machine-gradable Objective Response Sheet (ORS), is provided separately.
7. Do not tamper/mutilate the ORS or this booklet.
8. Do not break the seals of the question-paper booklet before instructed to do so by the invigilators.

B. Filling the bottom-half of the ORS:

9. The ORS has CODE printed on its lower and upper Parts.
10. Make sure the CODE on the ORS is the same as that on this booklet. If the Codes do not match, ask for a change of the booklet.
11. Write your Registration No., Name and Name of centre and sign with pen in appropriate boxes. Do not write these anywhere else.
12. Darken the appropriate bubbles under each digit of your Registration No. with HB pencil.

C. Question paper format and marking scheme:

13. The question paper consists of four Sections.
14. For each question in Section I: you will be awarded 5 marks if you darken only the bubble corresponding to the correct answer and zero mark if no bubbles are darkened. In all other cases, minus two (– 2) mark will be awarded.
15. For each question in Section II: you will be awarded 3 marks if you darken the bubble corresponding to the correct answer and zero mark if no bubbles are darkened. No negative marks will be awarded for incorrect answers in this Section.
16. For each question in Section III: you will be awarded 3 marks if you darken only the bubble corresponding to the correct answer and zero mark if no bubbles are darkened. In all other cases, minus one (– 1) mark will be awarded.

17. For each question in Section IV: you will be awarded 2 marks for each row in which you darkened the bubbles(s) corresponding to the correct answer. Thus each question in this section carries a maximum of 8 marks. There is no negative marks awarded for incorrect answer(s) in this Section.

Useful Data

Atomic Numbers: B 5; C 6; N 7; O 8; F 9; Na 11; Si 14; P 15; S 16; Cl 17; Ti 22; V 23; Cr 24; Ni 28; Cu 29; Br 35; Rh 45; Sn 50; Xe 54; Tl 81.

1 amu	=	1.66×10^{-27} kg	e	=	1.6×10^{-19} C
R	=	0.082 L-atm K^{-1} mol^{-1}	c	=	3.0×10^{8} m s^{-1}
h	=	6.626×10^{-34} J s	F	=	96500 C mol^{-1}
N_A	=	6.022×10^{23}	R_H	=	2.18×10^{-18} J
m_e	=	9.1×10^{-31} kg	$4\pi\epsilon_0$	=	1.11×10^{-10} J^{-1} C^2 m^{-1}

SECTION – I (Single Correct Choice Type)

This section contains 6 multiple choice questions. Each question has four choices (A), (B), (C) and (D) out of which only one is correct.

1. The compounds P, Q and S were separately subjected to nitration using HNO_3/H_2SO_4 mixture. The major product formed in each case respectively, is

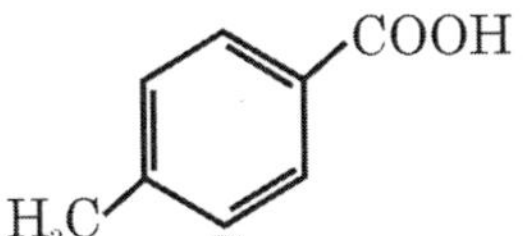
P

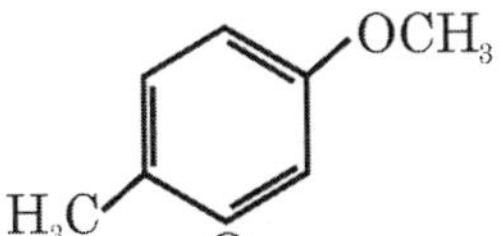
Q

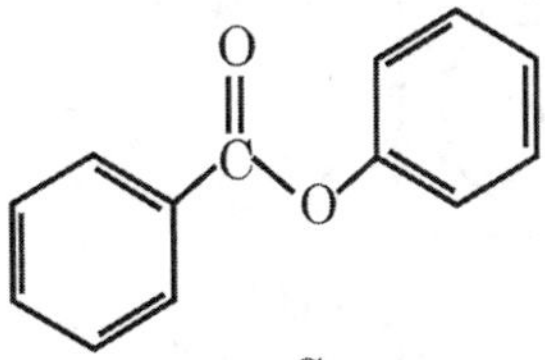
S

(A) 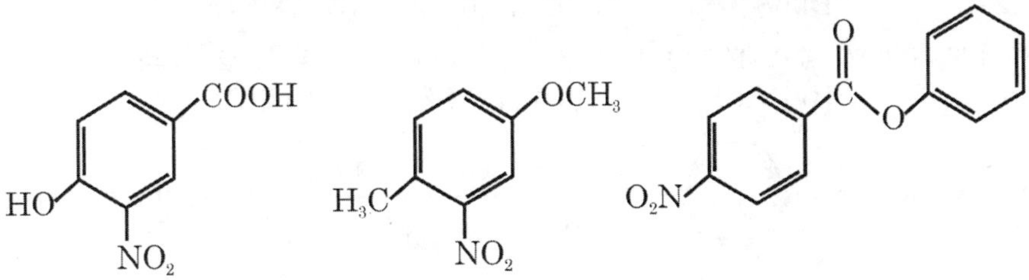

(B) 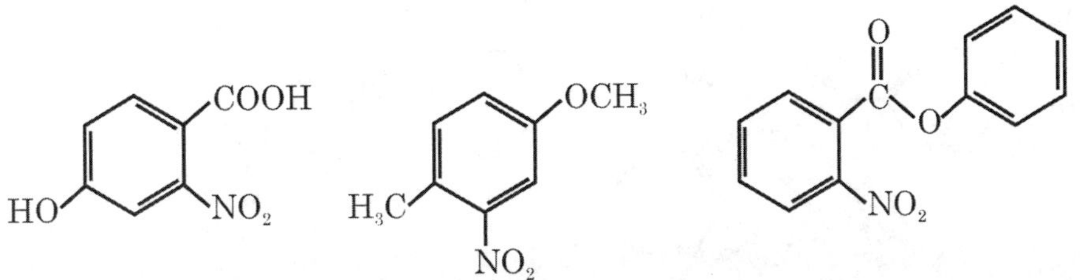

(C) 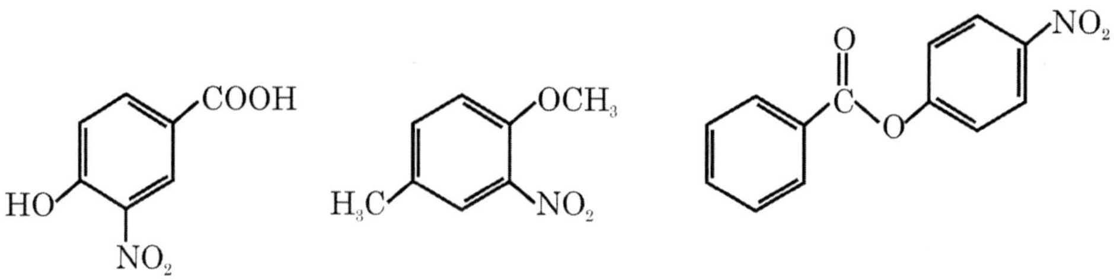

(D) 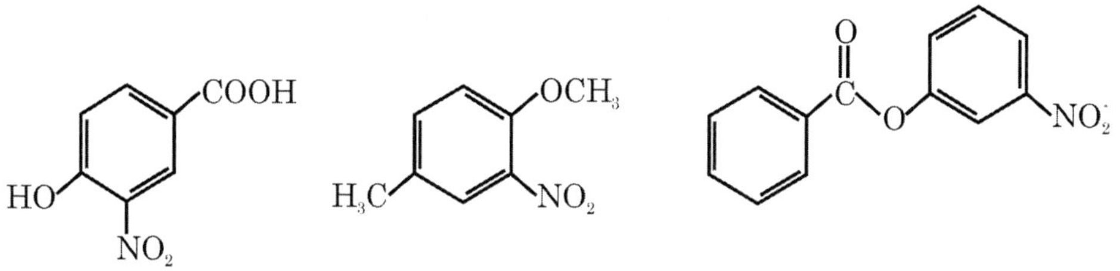

2. Assuming that Hund's rule is violated, the bond order and magnetic nature of the diatomic molecule B_2 is
   (A) 1 and diamagnetic (B) 0 and diamagnetic
   (C) 1 and paramagnetic (D) 0 and paramagnetic
3. The packing efficiency of the two-dimensional square unit cell shown below is

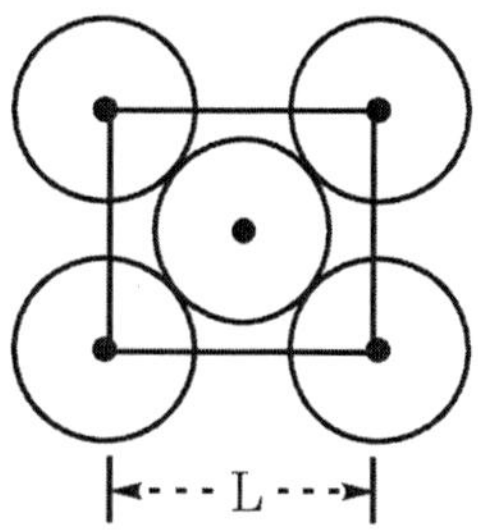

   (A) 39.27% (B) 68.02% (C) 74.05% (D) 78.54%
4. The complex showing a spin-only magnetic moment of 2.82 B.M. is
   (A) $Ni(CO)_4$ (B) $[NiCl_4]^{2-}$ (C) $Ni(PPh_3)_4$ (D) $[Ni(CN)_4]^{2-}$
5. In the reaction 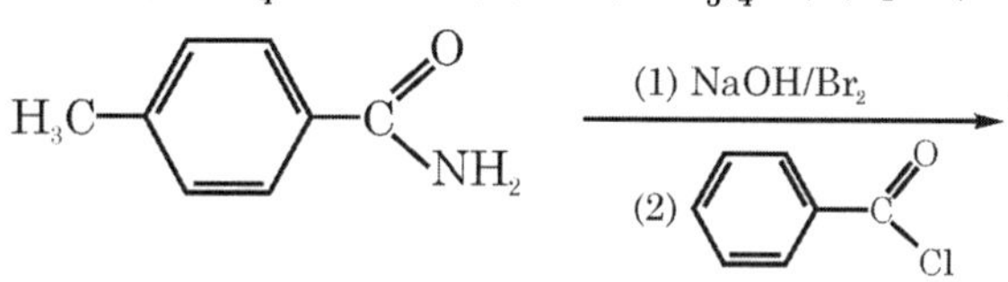 T, the structure of the product T is

   (A) 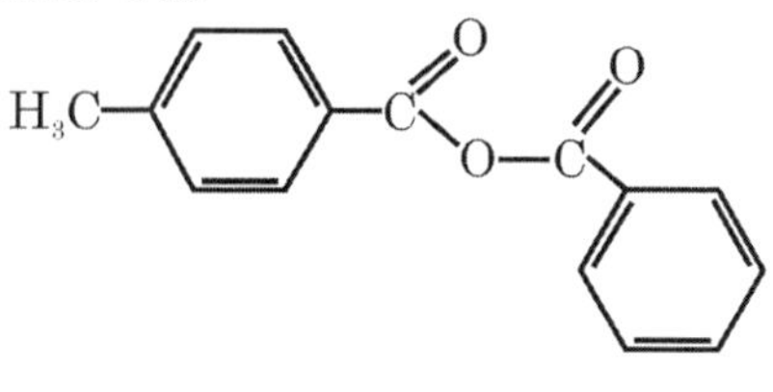 (B) 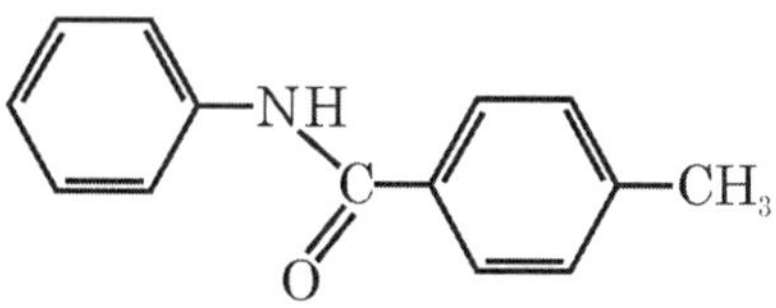

(C) 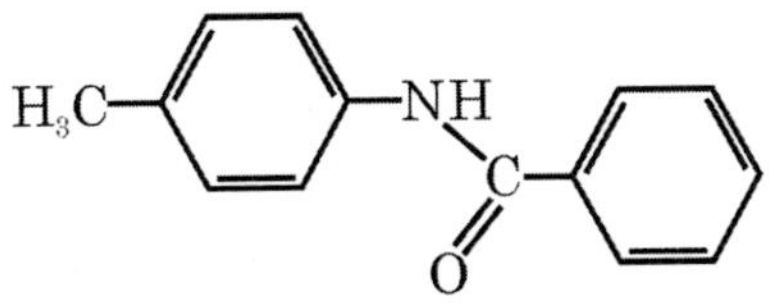 (D) 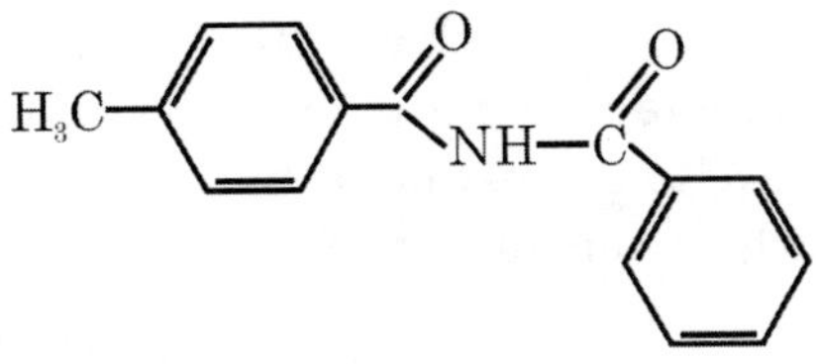

6. The species having pyramidal shape is

(A) SO_3 (B) BrF_3 (C) SiO_3^{2-} (D) OSF_2

SECTION – II (Integer Type)

This section contains 5 questions. The answer to each question is a single-digit integer, ranging from 0 to 9. The correct digit below the question no. in the ORS is to be bubbled.

7. Silver (atomic weight = 108 g mol^{-1}) has a density of 10.5 g cm^{-3}. The number of silver atoms on a surface of area 10^{-12} m^2 can be expressed in scientific notation as $y \times 10^x$. The value of x is

8. Among the following, the number of elements showing only one non-zero oxidation state is
O, Cl, F, N, P, Sn, Tl, Na, Ti

9. One mole of an ideal gas is taken from a to b along two paths denoted by the solid and the dashed lines as shown in the graph below. If the work done along the solid line path is w_s and that along the dotted line path is w_d, then the integer closest to the ratio w_d/w_s is

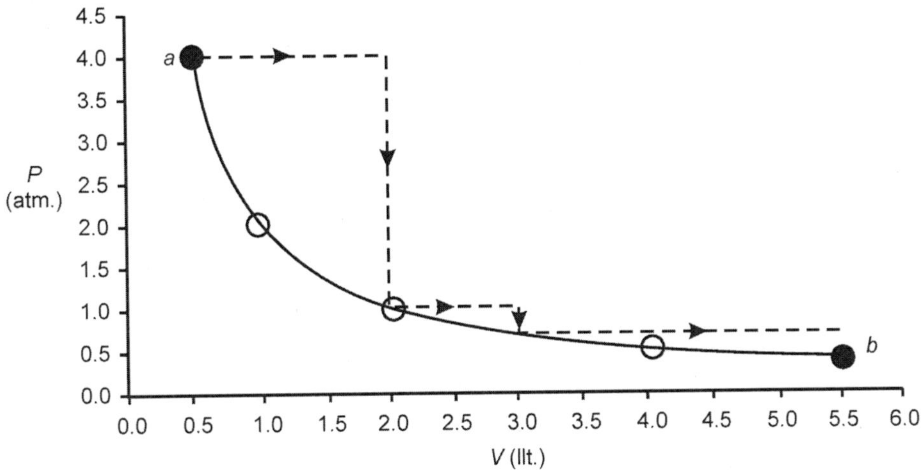

10. The total number of diprotic acids among the following is

H_3PO_4 H_2SO_4 H_3PO_3 H_2CO_3 $H_2S_2O_7$

H_3BO_3 H_3PO_2 H_2CrO_4 H_2SO_3

11. Total number of geometrical isomers for the complex $[RhCl(CO)(PPh_3)(NH_3)]$ is

SECTION – III (Paragraph Type)

This section contains 2 paragraphs. Based upon each of the paragraphs 3 multiple choice questions have to be answered. Each of these questions has four choices (A), (B), (C) and (D) out of which only one is correct.

Paragraph for questions 12 to 14

Two aliphatic aldehydes P and Q react in the presence of aqueous K_2CO_3 to give compound R, which upon treatment with HCN provides compound S. On acidification and heating, S gives the product shown below:

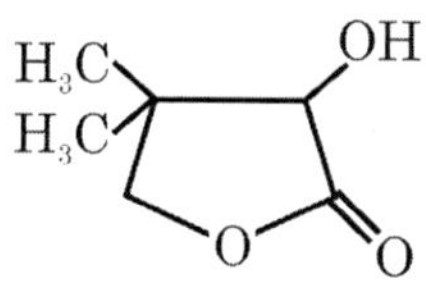

12. The compounds P and Q respectively are

(A) 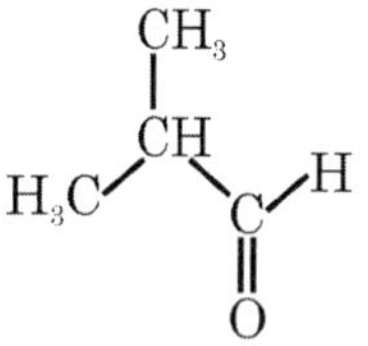 and 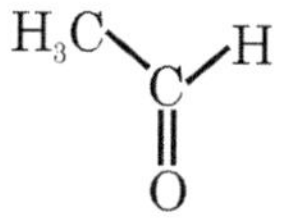

(B) 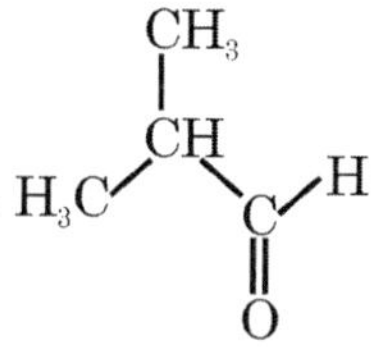 and 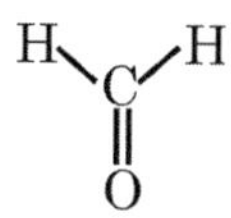

(C) 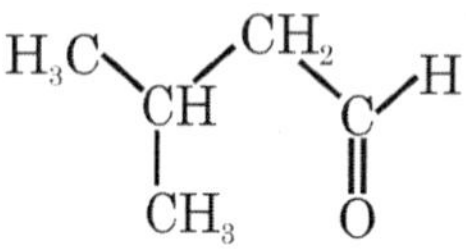 and 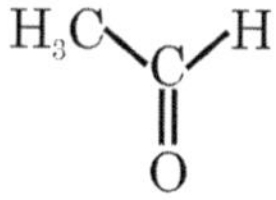

(D) 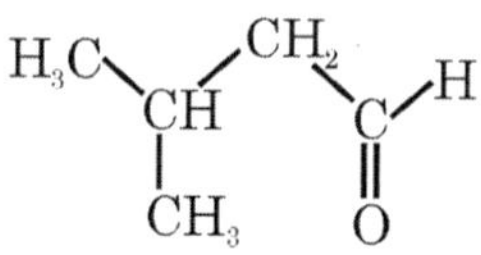 and 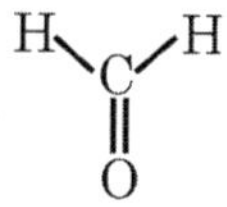

13. The compound R is

(A) 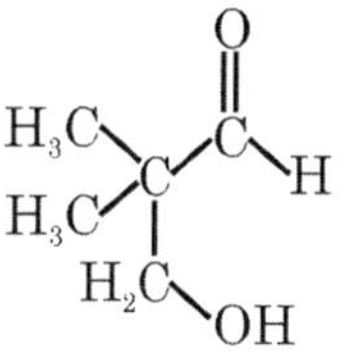

(B) 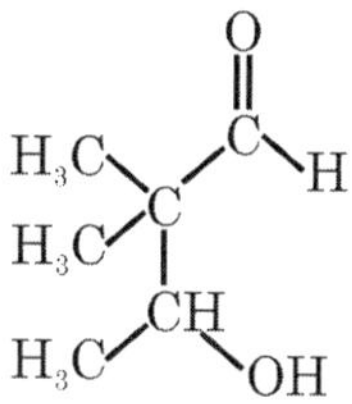

(C) 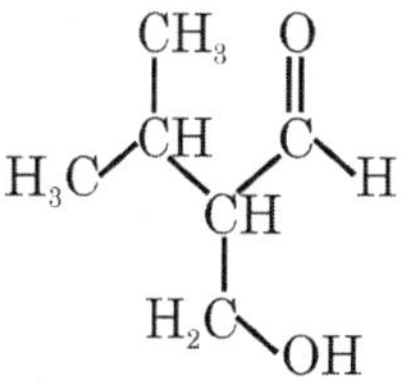

(D) 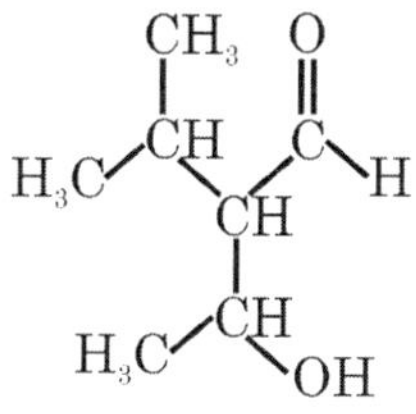

14. The compound S is

(A) 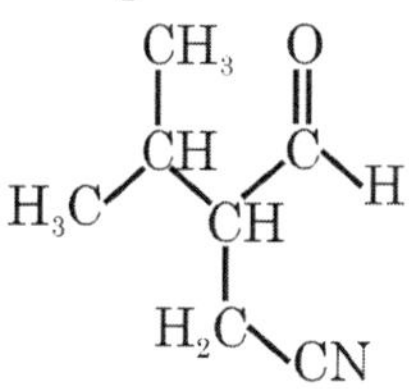

(B) 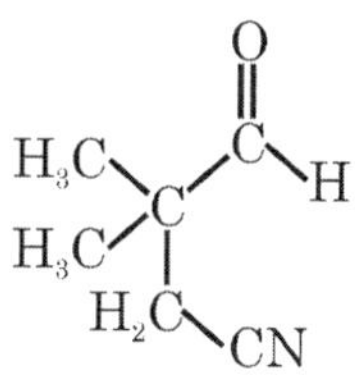

(C) 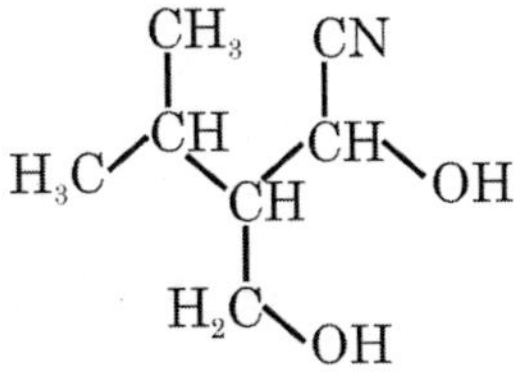

(D) 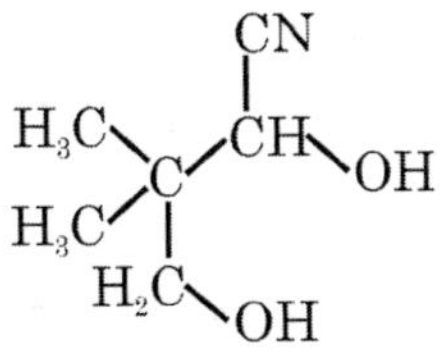

Paragraph for questions 15 to 17

The hydrogen-like species Li^{2+} is in a spherically symmetric state S_1 with one radial node. Upon absorbing light the ion undergoes transition to a state S_2. The state S_2 has one radial node and its energy is equal to the ground state energy of the hydrogen atom.

15. The state S_1 is

(A) 1*s* (B) 2*s* (C) 2*p* (D) 3*s*

16. Energy of the state S_1 in units of the hydrogen atom ground state energy is

(A) 0.75 (B) 1.50 (C) 2.25 (D) 4.50

17. The orbital angular momentum quantum number of the state S_2 is

(A) 0 (B) 1 (C) 2 (D) 3

SECTION – IV (Matrix Type)

This section contains 2 questions. Each question has four statements (A, B, C and D) given in Column I and five statements (p, q, r, s and t) in Column II. Any given statement in Column I can have correct matching with one or more statement(s) given in Column II. For example, if for a given question, statement B matches with the statements given in q and r, then for that particular question, against statement B, darken the bubbles corresponding to q and r in the ORS.

18. Match the reactions in Column I with appropriate options in Column II.

Column I	Column II
(A) 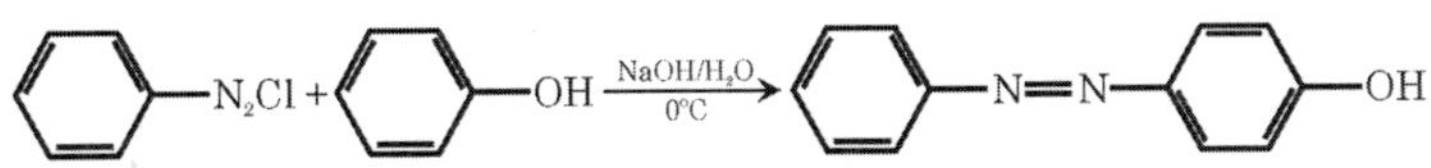	(p) Racemic mixture
(B) 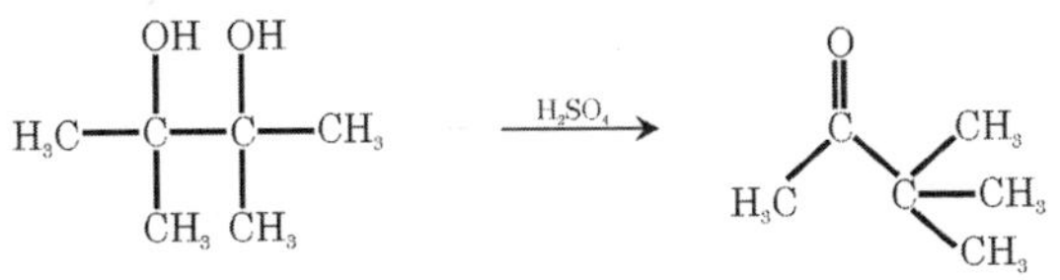	(q) Addition reaction
(C) 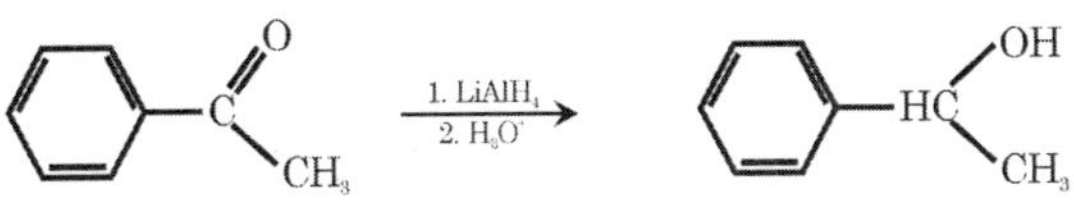	(r) Substitution reaction
(D) 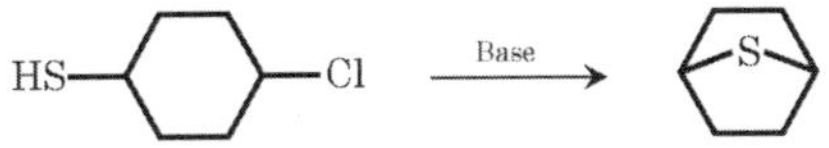	(s) Coupling reaction
	(t) Carbocation intermediate

19. All the compounds listed in Column I react with water. Match the result of the respective reactions with the appropriate options listed in Column II.

Column I	Column II
(A) $(CH_3)_2SiCl_2$	(p) Hydrogen halide formation
(B) XeF_4	(q) Redox reaction
(C) Cl_2	(r) Reacts with glass
(D) VCl_5	(s) Polymerization
	(t) O_2 formation

ANSWERS

1. C	2. A	3. D	4. B	5. C	6. D
7. 7	8. 2	9. 2	10. 6	11. 3	12. B
13. A	14. D	15. B	16. C	17. B	

18. A–r, s, t; B–t; C–p, q; D–r

19. A–p, s; B–p, q, r, t; C–p, q, t; D–p

Model Solutions—I.I.T. JEE 2010

Paper–2

1. 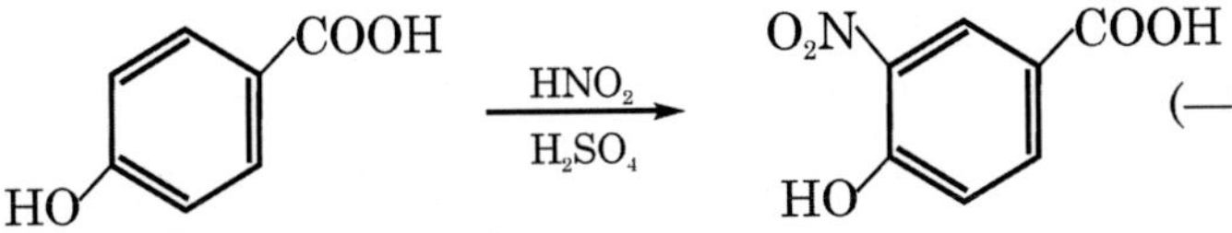

P

(—OH group is more activating)

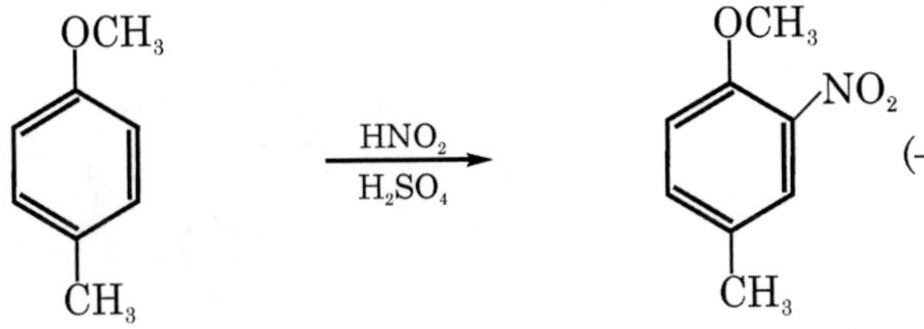

Q

(—OCH_3 group is more activating)

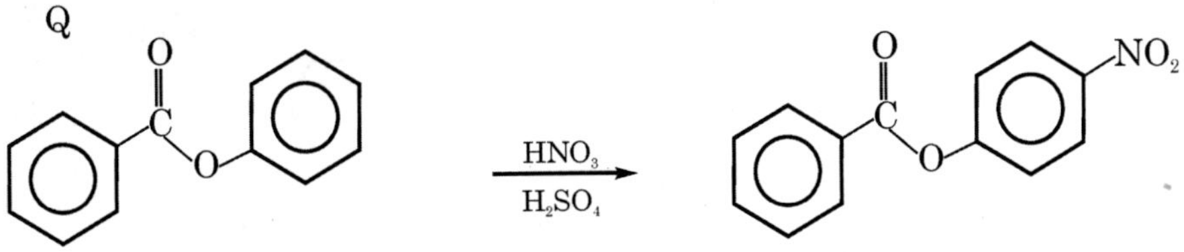

2. $B_2\ (10) = \sigma_{1s^2}\sigma^*_{1s^2}\sigma_{2s^2}\sigma^*_{2s^2}\pi_{2p_x^2}$

Bond order $= \dfrac{6-4}{2} = 1$

Nature of B_2 molecule is diamagnetic since it does not contain any unpaired electron.

3. 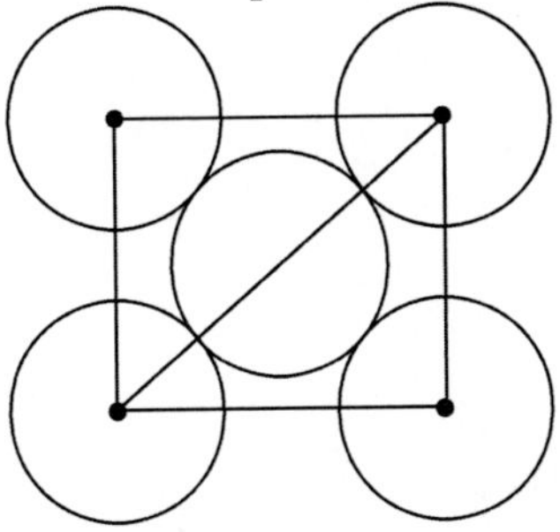

$a = (2\sqrt{2r})$

$$P.F = \frac{2 \times \pi r^2}{\left(2\sqrt{2r}\right)^2} = \frac{2\pi r^2}{8r^2} = \frac{\pi}{4} = \frac{3.14159}{4} = 0.7854$$. Hence the packing efficiency is 78.54%.

4. $[NiCl_4]^{2-}$, oxidation state of Ni = + 2

Ni (28) = $3d^8 4s^2$

$Ni^{2+} = 3d^8$

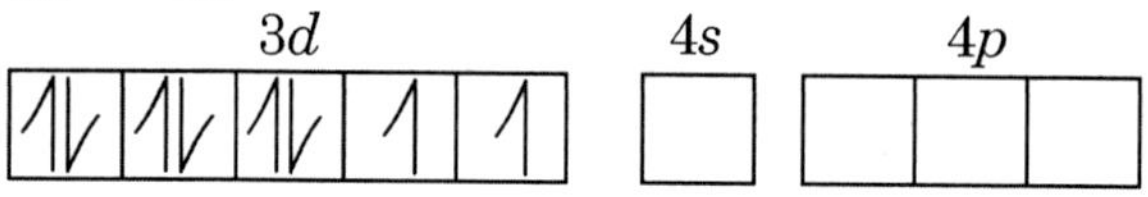

Cl^- being a weak ligand, outer orbitals are involved in the hybridisation.

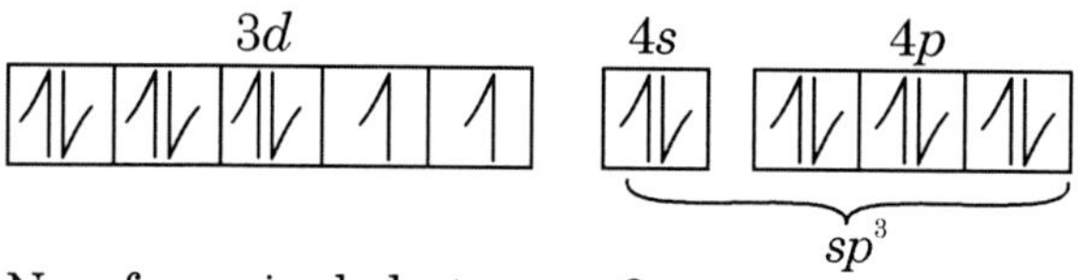

No. of unpaired electrons = 2

Magnetic moment μ = 2.82 B. M.

5. 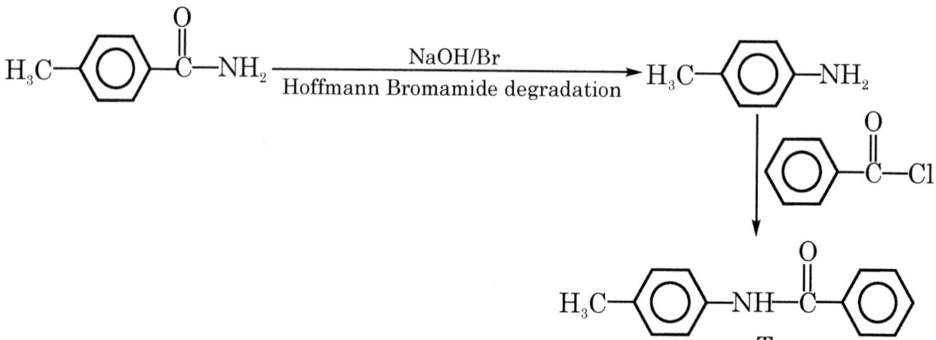

T

6. 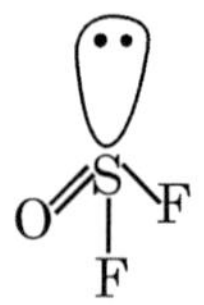 Trignonal pyramid

7. $d = \frac{\text{mass}}{V} \Rightarrow 10.5$ g/cc means in 1 cc $\Rightarrow$ 10.5g of Ag present

Number of atoms of Ag in 1 cc $\Rightarrow \frac{10.5}{108} \times N_A$

In 1 cm, number of atoms of Ag $= \sqrt[3]{\frac{10.5}{108} N_A}$

In 1 cm², number of atoms of Ag $= \left(\frac{10.5}{108} N_A\right)^{2/3}$

In 10^{-12} m² or 10^{-8} cm², number of atoms of Ag $= \left(\frac{10.5}{108} N_A\right)^{2/3} \times 10^{-8}$

$$= \left(\frac{1.05 \times 6.022 \times 10^{24}}{108}\right)^{2/3} \times 10^{-8} = 1.5 \times 10^7$$

Hence, $x = 7$

8. Na and F exhibit only one non-zero oxidation state.

9. $w_d = 4 \times 1.5 + 1 \times 1 + 2.5 \times 2/3 = 8.65$

Thus the process is isothermal.

$$w_s = 2 \times 2.303 \log \frac{5.5}{0.5} = 2 \times 2.303 \times \log 11 = 2 \times 2.303 \times 1.0414 = 4.79$$

$$\frac{w_d}{w_s} = \frac{8.65}{4.79} = 1.80 \simeq 2$$

10. 6 : H_2SO_4, H_2CO_3, $H_2S_2O_7$, H_2CrO_4, H_3PO_3, H_2SO_3

11. 3

12. B

13. A

14. D

Solution to Questions 12-14:

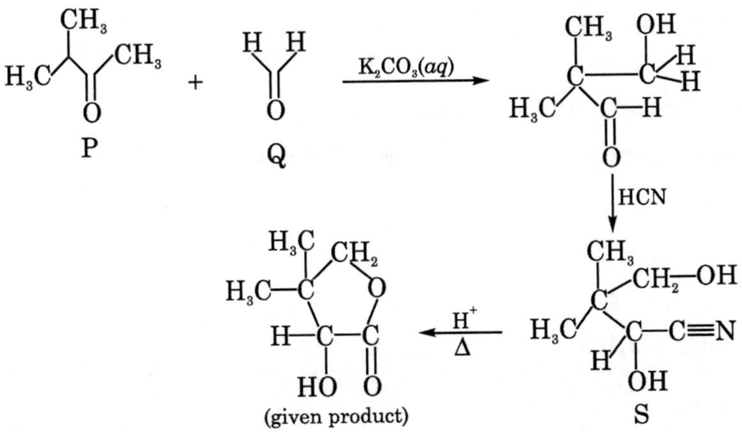

15. For S_1 (spherically symmetrical), radial node = 1

Hence $n - 1 = 1$

Or $n = 2$

For S_2, radial node = 1

$$E_{S_2} = \frac{-13.6 \times 2^2}{n^2} = E_H \text{ in ground state} = -13.6$$

$$E_{S_2} = \frac{-13.6 \times 9}{n^2} = -13.6 \text{ Or } n = 3$$

Hence state S_1 is 2s and state S_2 is 3p.

16. $\dfrac{E_{S_1}}{E_{H(ground)}} = \dfrac{-13.6 \times 9}{4 \times (-13.6)} = 2.25$

17. The azimuthal quantum number for $S_2 = l = 1$

18. A–r, s, t; B–t; C–p, q; D–r

(A) This is an electrophilic substitution reaction and a coupled product is formed. Hence this is also a coupling reaction.

(B) This is pinacole-pinacolone rearrangement and in this reaction the intermediate is a carbocation.

(C) This is an addition of the carbonyl compound and both the enanciomers are formed. Hence mixture is formed.

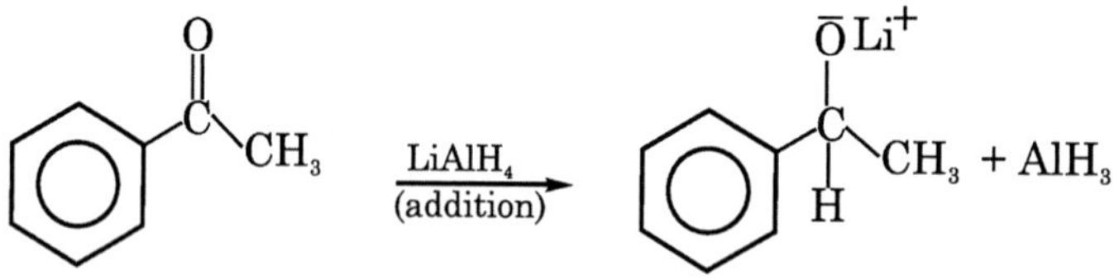

(D) This is a nucleophilic substitution reaction.

19. A–p, s; B–p, q, r, t; C–p, q, t; D–p

(A) $(CH_3)_2SiCl_2 + 2H_2O \longrightarrow (CH_3)_2Si(OH)_2 + 2HCl$

$(CH_3)_2Si(OH)_2$ undergoes polymerization to form silicones.

(B) $3XeF_4 + 6H_2O \longrightarrow XeO_3 + 2Xe + 12HF + 1\frac{1}{2}O_2$

(C) $Cl_2 + H_2O \longrightarrow 2HCl + \frac{1}{2}O_2$

(D) $VCl_5 + H_2O \longrightarrow VOCl_3 + 2HCl$

I.I.T. JEE Objective Question Paper—2011

PAPER—1

A. General:

1. The question paper code is printed on the right hand top corner of this sheet and on the back page (page No. 36) of this booklet.
2. No additional sheets will be provided for rough work.
3. Blank papers, clipboards, log tables, slide rules, calculators, cellular phones, pagers and electronic gadgets are not allowed.
4. Write your name and registration number in the space provided on the back page of this booklet.
5. The answer sheet, a machine-gradable Optical Response Sheet (ORS), is provided separately.
6. Do not tamper with/multilate the ORS or the booklet.
7. Do not break the seals of the question-paper booklet before being instructed to do so by the invigilators.
8. This question paper contains 36 pages having 69 questions.
9. On breaking the seals, please check that all the questions are legible.

B. Filling the bottom-half of the ORS:

10. The ORS also has a code printed on its Left and Right parts.
11. Make sure the code on the ORS is the same as that on this booklet. If the codes do not match ask for a change of the booklet.
12. Write your Name, Registration No. and the name of centre and sign with pen in the boxes provided. Do not write them anywhere else. Darken the appropriate bubble under each digit of your Registration No. with a good quality HB pencil.

C. Question paper format and marking scheme:

13. The question paper consists of four sections.
14. In Section I (Total marks: 21), for each question you will be awarded 3 marks if you darken only the bubble corresponding to the correct answer and zero marks if no bubble is darkened. In all other cases, minus one (– 1) mark will be awarded.
15. In Section II (Total marks: 16), for each question you will be awarded 4 marks if you darken all the bubble(s) corresponding to the correct answer(s) only and zero marks otherwise. There are no negative marks in this section.

16. In Section III (Total marks: 15), for each question you will be awarded 3 marks if you darken only the bubbles corresponding to the correct answer and zero marks if no bubble is darkened. In all other cases, minus one (– 1) marks will be awarded.
17. In Section IV (Total marks: 28), for each question you will be awarded 4 marks if you darken only the bubbles corresponding to the correct answer and zero marks otherwise. There are no negative marks in this section.

SECTION – I (Total marks: 21)
(Single Correct Answer Type)

This section contains 7 multiple choice questions. Each question has four choices (A), (B), (C) and (D) out of which only one is correct.

1. Extra pure N_2 can be obtained by heating
   (A) NH_3 with CuO (B) NH_4NO_3 (C) $(NH_4)_2CrO_7$ (D) $Ba(N_3)_4$
2. Dissolving 120 g of urea (mol. wt. 60) in 1000 g of water gave a solution of density 1.15 g/mL. The molarity of the solution is
   (A) 1.78M (B) 2.00M (C) 2.05M (D) 2.22M
3. Bombardment of aluminium by α-particle leads to its artificial disintegration in two ways, (i) and (ii) as shown. Products X, Y and Z respectively are,

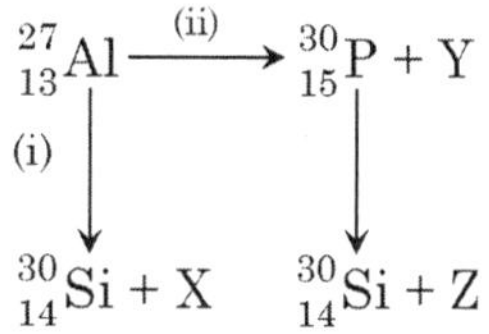

   (A) proton, neutron, positron (B) neutron, positron, proton
   (C) proton, positron, neutron (D) positron, proton, neutron
4. Geometrical shapes of the complexes formed by the reaction of Ni^{2+} with Cl^-, CN^- and H_2O, respectively, are
   (A) octahedral, tetrahedral and square planar
   (B) tetrahedral, square planar and octahedral
   (C) square planar, tetrahedral and octahedral
   (D) octahedral, square planar and tetrahedral
5. The major product of the following reaction is

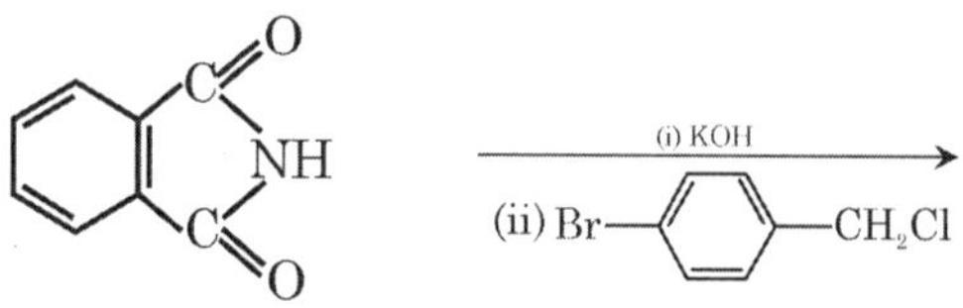

(A) 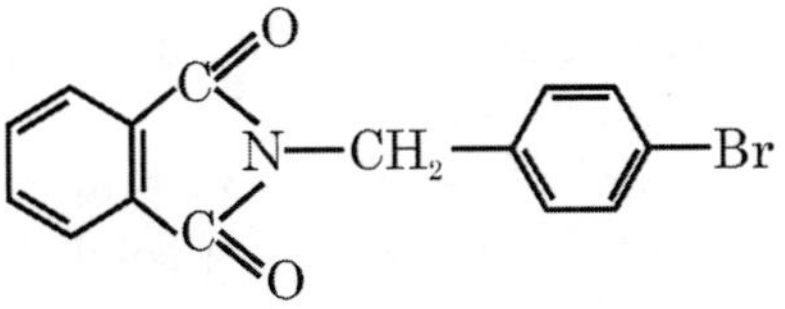
(B) 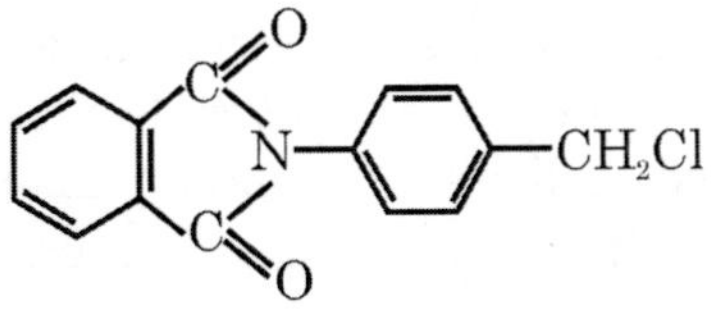

(C) 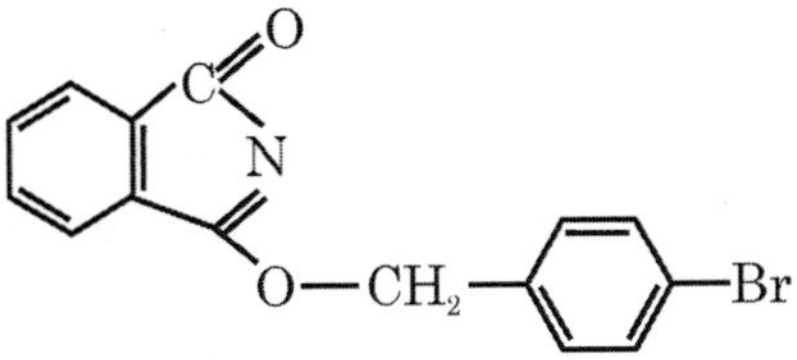
(D) 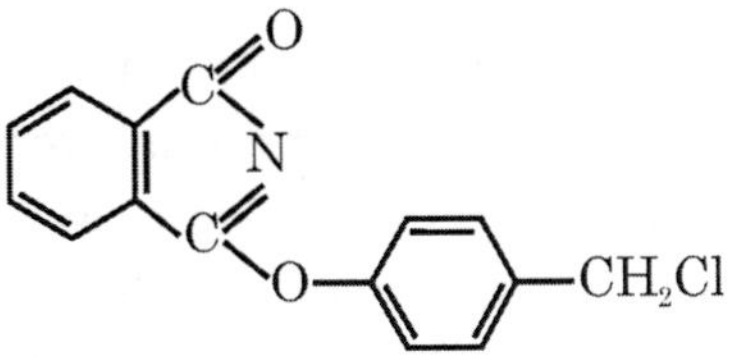

6. Among the following compounds, the most acidic is

(A) *p*-nitrophenol (B) *p*-hydroxybenzoic acid

(C) *o*-hydroxybenzoic acid (D) *p*-toluic acid.

7. $AgNO_3(aq)$ was added to an aqueous KCl solution gradually and the conductivity of the solution was measured. The plot of conductance (Λ) versus the volume of $AgNO_3$ is

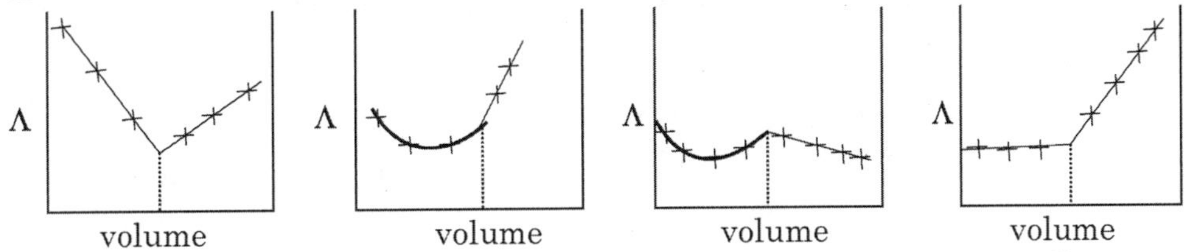

(P) (Q) (R) (S)

(A) (P) (B) (Q) (C) (R) (D) (S)

SECTION – II (Total marks: 16)
(Multiple Correct Answer Type)

This section contains 4 multiple choice questions. Each question has four choices (A), (B), (C) and (D) out of which one or more may be correct.

8. Amongst the given options, the compound(s) in which all the atoms are in one plane in all the possible conformations (if any), is (are)

(A) 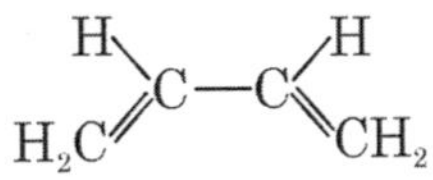
(B) 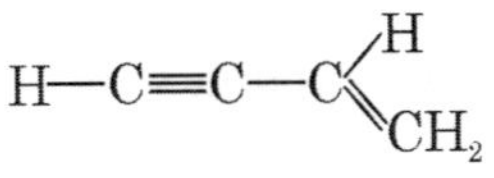

(C) $H_2C{=}C{=}O$ (D) $H_2C{=}C{=}CH_2$

9. Extraction of metal from the ore cassiterite involves

(A) carbon reduction of an oxide ore (B) self-reduction of a sulphide ore

(C) removal of copper impurity (D) removal of iron impurity.

10. According to kinetic theory of gases

(A) collisions are always elastic

(B) heavier molecules transfer more momentum to the wall of the container
(C) only a small number of molecules have very high velocity
(D) between collisions, the molecules move in straight lines with constant velocities.

11. The correct statement(s) pertaining to the adsorption of a gas on a solid surface is (are)
(A) Adsorption is always exothermic
(B) Physisorption may transform into chemisorption at high temperature
(C) Physiosorption increases with increasing temperature but chemisorption decreases with increasing temperature
(D) Chemisorption is more exothermic than physisorption, however it is very slow due to higher energy of activation.

SECTION – III (Total marks: 15)
(Paragraph Type)

This section contains 2 paragraphs. Based upon one of paragraphs 2 multiple choice questions and based on the other paragraph 3 multiple choice questions have to be answered. Each of these questions has four choice (A), (B), (C) and (D) out of which only one is correct.

Paragraph for Question Nos. 12 and 13

An acyclic hydrocarbon P, having molecular formula C_6H_{10}, gave acetone as the only organic product through the following sequence of reactions, in which Q is an intermediate organic compound.

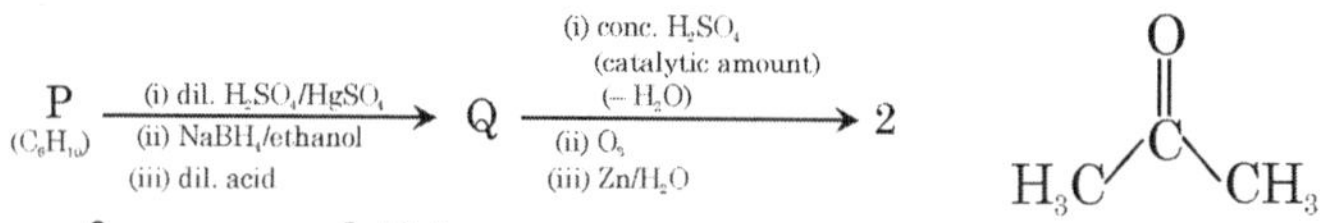

12. The structure of compound P is
(A) $CH_3CH_2CH_2CH_2—C\equiv C—H$ (B) $H_3CH_2C—C\equiv C—CH_2CH_3$
(C) 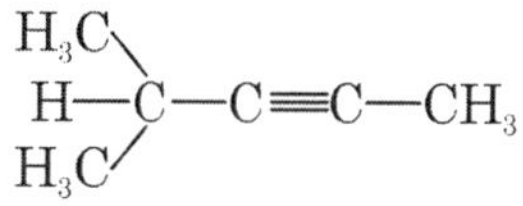
(D) 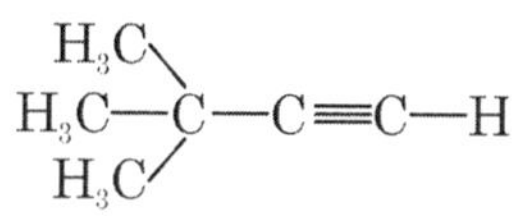

13. The structure of the compound Q is
(A) 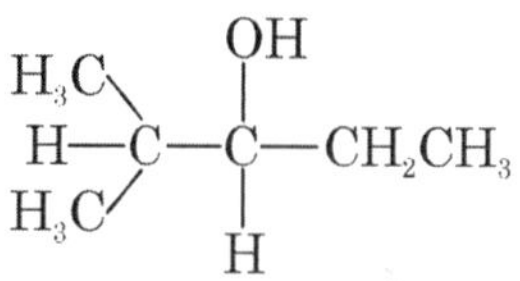
(B) 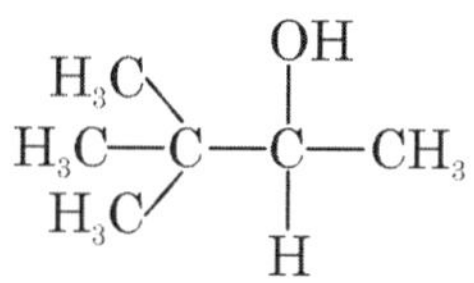
(C) 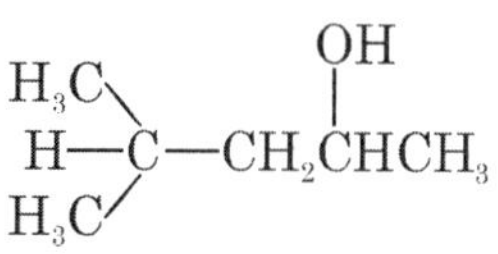
(D) 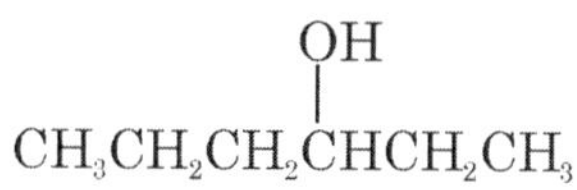

Paragraph for Question Nos. 14 to 16

When a metal rod M is dipped into an aqueous colourless concentrated solution of compound N, the solution turns light blue. Addition of aqueous NaCl to the blue solution gives a white precipitate O. Addition of aqueous NH_3 dissolves O and gives an intense blue solution.

14. The metal rod M is

   (A) Fe (B) Cu (C) Ni (D) Co

15. The compound N is

   (A) $AgNO_3$ (B) $Zn(NO_3)_2$ (C) $Al(NO_3)_3$ (D) $Pb(NO_3)_2$

16. The final solution contains

   (A) $[Pb(NH_3)_4]^{2+}$ and $[CoCl_4]^{2-}$ (B) $[Al(NH_3)_4]^{3+}$ and $[Cu(NH_3)_4]^{2+}$

   (C) $[Ag(NH_3)_2]^{+}$ and $[Cu(NH_3)_4]^{2+}$ (D) $[Ag(NH_3)_2]^{+}$ and $[Ni(NH_3)_6]^{2+}$

SECTION – IV (Total marks: 28)
(Integer Answer Type)

This section contains 7 questions. The answer to each of the question is a single digit integer, ranging from 0 to 9. The bubble corresponding to the correct is to darkened in the ORS.

17. The difference in the oxidation numbers of the types of sulphur atoms in $Na_2S_4O_6$ is

18. A decapeptide (Mol. Wt. 796) on complete hydrolysis gives glycine (Mol. Wt. 75), alanine and phenylalanine. Glycine contributes 47.0% to the total weight of the hydrolysed products. The number of glycine units present in the decapeptide is

19. The work function (ϕ) of some metals is listed below. The number of metals which will show photoelectric effect when light of 300 nm wavelenght falls on the metal is

Metal	Li	Na	K	Mg	Cu	Ag	Fe	Pt	W
ϕ (eV)	2.4	2.3	2.2	3.7	4.8	4.3	4.7	6.3	4.75

20. The maximum number of electrons that can have principal quantum number, $n = 3$, and spin quantum number, $m_s = -\frac{1}{2}$, is

21. Reaction of Br_2 with Na_2CO_3 in aqueous solution gives sodium bromide and sodium bromate with evolution CO_2 gas. The number of sodium bromide molecules involved in the balanced chemical equation is

22. To an evacuated vessel with movable piston under external pressure of 1 atm, 0.1 mol of He and 1.0 mol of an unknown compound (vapour pressure 0.68 atm. at 0°C) are introduced. Considering the ideal gas behaviour, the total volume (in L) of the gases at 0°C is close to

23. The total number of alkenes possible by dehydrobromination of 3-bromo-3-cyclopentylhexane using alcoholic KOH is

ANSWERS

1. D	2. C	3. A	4. B	5. A	6. C
7. D	8. B, C	9. A, C, D	10. A, B, C, D	11. A, B, D	12. D
13. B	14. B	15. A	16. C	17. 5	18. 6
19. 4	20. 9	21. 5	22. 7	23. 5	

Model Solutions—I.I.T. JEE 2011

Paper–1

1. (D)

$$Ba(N_3)_2 \xrightarrow{\Delta} Ba + 3N_2$$

2. (C)

Total mass of solution = 1000 + 120 = 1120 g

$$\text{Total volume of solution in L} = \frac{1120}{1.15} \times 10^3$$

$$M = \frac{W}{M.W.} \times \frac{1}{V(\text{in L})} = \frac{120}{60} \times \frac{1.15 \times 10^3}{1120} = 2.05M$$

3. (A)

$${}^{27}_{13}Al + {}^{4}_{2}\alpha \longrightarrow {}^{30}_{14}Si + {}^{1}_{1}p\,(X)$$

$${}^{27}_{13}Al + {}^{4}_{2}\alpha \longrightarrow {}^{30}_{15}P + {}^{1}_{0}n\,(Y)$$

$${}^{30}_{15}P \longrightarrow {}^{30}_{14}Si + {}^{0}_{+1}\beta\,(Z)$$

4. (B)

$[NiCl_4]^{2-} \longrightarrow$ Tetrahedral

$[Ni(CN)_4]^{2-} \longrightarrow$ Square planar

$[Ni(H_2O)_6]^{2+} \longrightarrow$ Octahedral

5. (A)

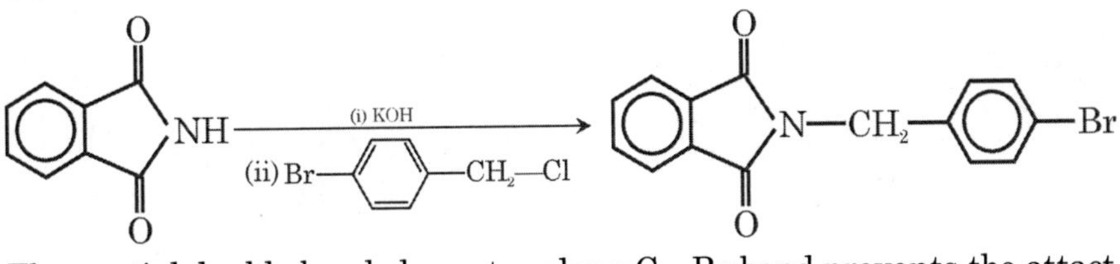

The partial double bond character along C—Br bond prevents the attact of nucleophile at phenylic position.

6. (C)

Due to the ortho effect *o*-hydroxybenzoic acid is the strongest acid and the correct order of decreasing K_a is

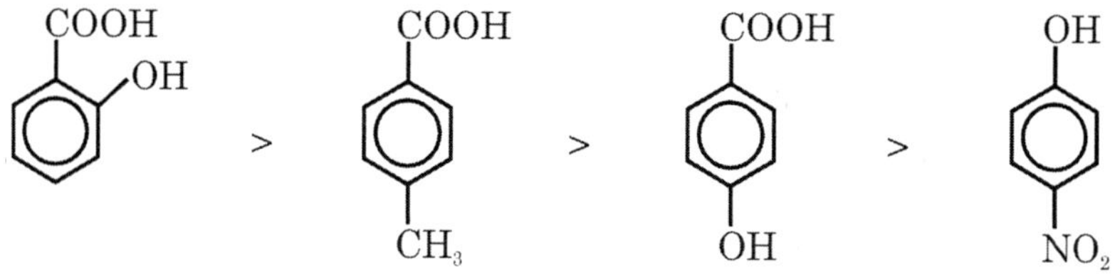

7. (D)

$AgNO_3(aq) + KCl(aq) \longrightarrow AgCl(s) + KNO_3(aq)$

Initially there is aqueous KCl solution. As solution of $AgNO_3$ is added, $AgCl(s)$ is formed. Hence the conductivity of solution is almost compensated (or slightly increased) by the formation of KNO_3. After the end point the conductivity increases more repidly because of addition of excess $AgNO_3$ solution.

8. (B, C)

Along the C—C single bond conformations are possible in butadiene in which all the atoms may not lie in the same plane. In $H_2C{=}C{=}CH_2$, the hydrogen atoms are not in the same plane.

9. (A, C, D)

$SnO_2 + 2C \longrightarrow 2CO + Sn$

The ore cassiterite contains the impurity of Fe, Mn, W and traces of Cu.

10. (A, B, C, D)
11. (A, B, D)
12. (D)
13. (B)

Solution for the Q. Nos. 12 to 13:

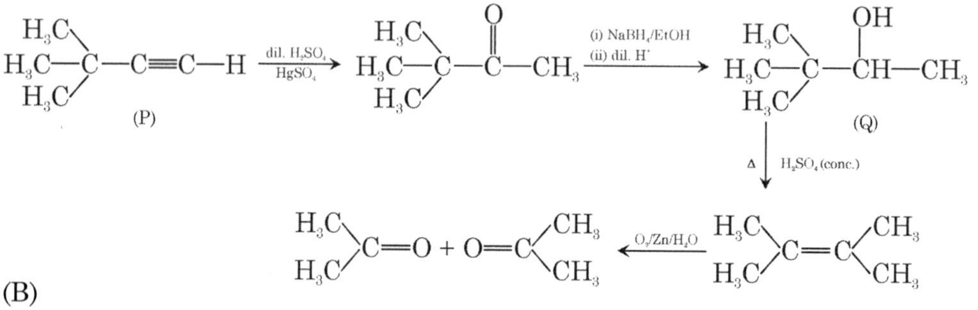

14. (B)

$\underset{M}{Cu} + \underset{N}{2AgNO_3} \longrightarrow \underset{Blue}{Cu(NO_3)_2} + 2Ag$

While Cu is partially oxidized to $Cu(NO_3)_2$, the remaining $AgNO_3$ reacts with NaCl.

15. (A)
16. (C)

$\underset{N}{AgNO_3} + NaCl \longrightarrow \underset{O}{AgCl} \downarrow + NaNO_3$

$AgCl + 2NH_3 \longrightarrow [Ag(NH_3)_2]^+Cl^-$

$Cu(NO_3)_2 + 4NH_4OH \longrightarrow [Cu(NH_3)_4]^{2+}$

17. (5)

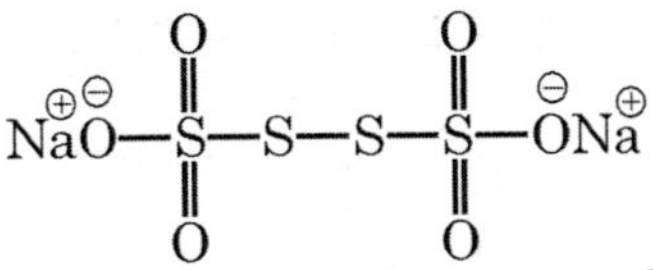

S will have oxidation number = + 5, 0

Difference in oxidation number = 5

18. (6)

For n-units of glycine, $\dfrac{n \times 75}{(796 + 9 \times 18)} \times 100 = 47$

or $n = 6$

19. (4)

The energy associated with incident photon = $\dfrac{hc}{\lambda}$

or $\quad E = \dfrac{6.6 \times 10^{-34} \times 3 \times 10^{8}}{300 \times 10^{-9}}$ J

E in eV = $\dfrac{6.6 \times 10^{-34} \times 3 \times 10^{8}}{300 \times 10^{-9} \times 1.6 \times 10^{-19}} = 4.16$ eV

Hence the number of metals showing photoelectric effects will be 4, i.e. Li, Na, K, Mg.

20. (9)

For principal quantum number (n = 3), number of orbitals = n^2 = 9.

Hence the number of electrons with $m_s = -\dfrac{1}{2}$ will be 9.

21. (5)

$3Br_2 + 3Na_2CO_3 \longrightarrow 5NaBr + NaBrO_3 + 3CO_2$

Hence the number of NaBr molecules is 5.

22. (7)

For any ideal gas, $PV = nRT$

$0.32 \times V = 0.1 \times 0.0821 \times 273$

$V = 7$ L

(unknown compound X will not follow ideal gas equation)

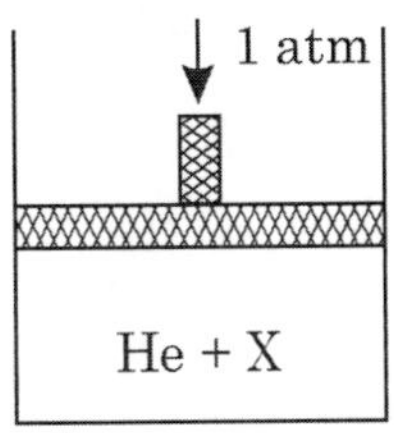

For He, $n = 0.1$, $P = 0.32$ atm.
$V = ?$, $T = 273$ K

23. (5)

Total no. of alkenes will be 5.

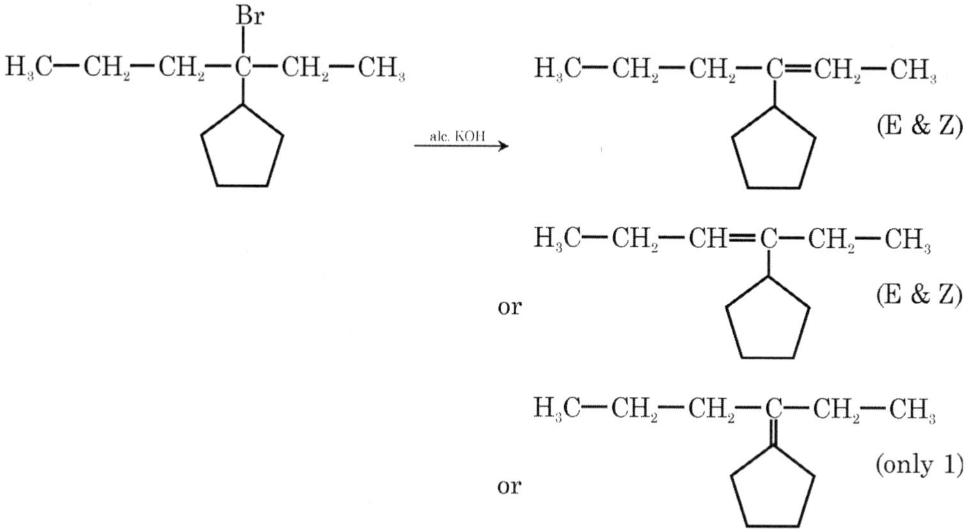

I.I.T. JEE Objective Question Paper—2011

PAPER—2

A. General:

1. The question paper code is printed on the right hand top corner of this sheet and on the back page (page No. 36) of this booklet.
2. No additional sheets will be provided for rough work.
3. Blank papers, clipboards, log tables, slide rules, calculators, cellular phones, pagers and electronic gadgets are not allowed.
4. Write your name and registration number in the space provided on the back page of this booklet.
5. The answer sheet, a machine-gradable Optical Response Sheet (ORS), is provided separately.
6. Do not tamper with/multilate the ORS or the booklet.
7. Do not break the seals of the question-paper booklet before being instructed to do so by the invigilators.
8. This question paper contains 36 pages having 69 questions.
9. On breaking the seals, please check that all the questions are legible.

B. Filling the bottom-half of the ORS:

10. The ORS also has a code printed on its Left and Right parts.
11. Make sure the code on the ORS is the same as that on this booklet. If the codes do not match ask for a change of the booklet.
12. Write your Name, Registration No. and the name of centre and sign with pen in the boxes provided. Do not write them anywhere else. Darken the appropriate bubble under each digit of your Registration No. with a good quality HB pencil.

C. Question paper format and marking scheme:

13. The question paper consists of four sections.
14. In Section I (Total marks: 21), for each question you will be awarded 3 marks if you darken only the bubble corresponding to the correct answer and zero marks if no bubble is darkened. In all other cases, minus one (– 1) mark will be awarded.
15. In Section II (Total marks: 16), for each question you will be awarded 4 marks if you darken all the bubble(s) corresponding to the correct answer(s) only and zero marks otherwise. There are no negative marks in this section.

16. In Section III (Total marks: 24), for each question you will be awarded 4 marks if you darken only the bubbles corresponding to the correct answer and zero marks otherwise. There are no negative marks in this sections.
17. In Section IV (Total marks: 16), for each question you will be awarded 2 marks for each row in which you have darken all the bubble(s) corresponding to the correct answer(s) only and zero marks otherwise. Thus each question in this section carries a maximum of 8 marks. There are no negative marks in this section.

SECTION – I (Total marks: 24)
(Single Correct Answer Type)

This Section contains 8 multiple choice questions. Each question has four choices (A), (B), (C) and (D) out of which only one is correct.

1. The freezing point (in °C) of a solution containing 0.1 g of $K_3[Fe(CN)_6]$ (Mol. Wt. 329) in 100 g of water (K_f = 1.86 K kg mol^{-1}) is

   (A) -2.3×10^{-2} (B) -5.7×10^{-2} (C) -5.7×10^{-3} (D) -1.2×10^{-2}

2. Amongst the compounds given, the one that would form a brilliant colored dye on treatment with $NaNO_2$ in dil. HCl followed by addition to an alkaline solution of β-naphthol is

   (A) 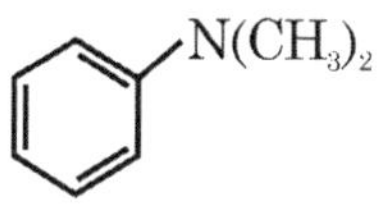

 (B) 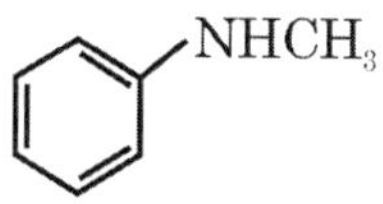

 (C) 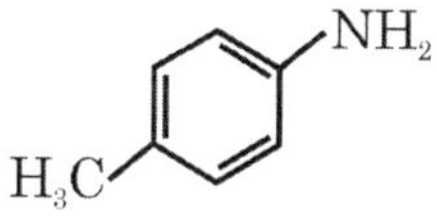

 (D) 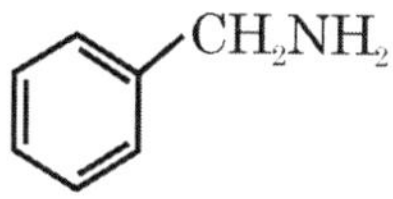

3. The major product of the following reaction is

   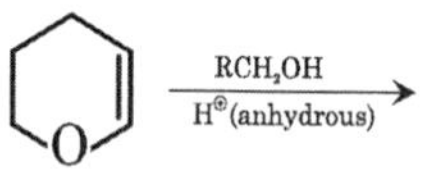

   (A) a hemiacetal (B) an acetal (C) an ether (D) an ester.

4. The following carbohydrate is

   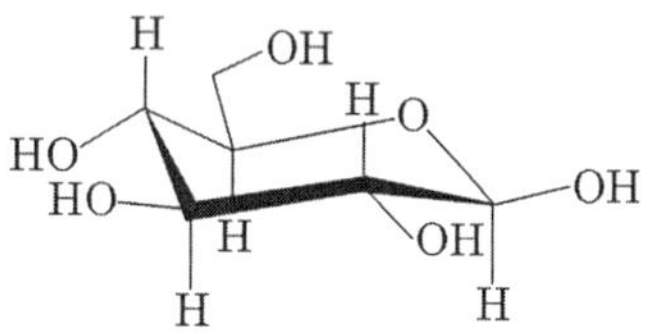

   (A) a ketohexose (B) an aldohexose (C) an α-furanose (D) an α-pyranose.

5. Oxidation states of the metal in the minerals haematite and magnetite, respectively, are

   (A) II, III in haematite and III in magnetite (B) II, III in haematite and II in magnetite

   (C) II in haematite and II, III in magnetite (D) III in haematite and II, III in magnetite

6. Among the following complexes (K–P)

   $K_3[Fe(CN)_6]$ (K), $[Co(NH_3)_6]Cl_3$ (L), $Na_3[Co(oxalate)_3]$ (M), $[Ni(H_2O)_6]Cl_2$ (N), $K_2[Pt(CN)_4]$ (O) and $[Zn(H_2O)_6](NO_3)_2$ (P), which are diamagnetic?

(A) K, L, M, N (B) K, M, O, P (C) L, M, O, P (D) L, M, N, O

7. Passing H_2S gas into a mixture of Mn^{2+}, Ni^{2+}, Cu^{2+} and Hg^{2+} ions in an acidified aqueous solution precipitates

(A) CuS and HgS (B) MnS and CuS (C) MnS and NiS (D) NiS and HgS

8. Consider the following cell reaction : $2Fe(s) + O_2(g) + 4H^+(aq) \rightarrow 2Fe^{2+}(aq) + 2H_2O(l)$ $E^0 = 1.67$ V

At $[Fe^{2+}] = 10^{-3}M$, $P(O_2) = 0.1$ atm and $pH = 3$, the cell potential at 25°C is

(A) 1.47 V (B) 1.77 V (C) 1.87 V (D) 1.57 V

SECTION – II (Total marks: 16)
(Multiple Correct Answer(s) Type)

This section contains 4 multiple choice questions. Each question has four choices (A), (B), (C) and (D) out of which one or more may be correct.

9. Reduction of the metal centre in aqueous permanganate ion involves

(A) 3 eletrons in neutral medium (C) 5 electrons in neutral medium

(D) 3 eletrons in alkaline medium (D) 5 electrons in acidic medium.

10. The correct functional group X and the reagent/reaction conditions Y in the following scheme are

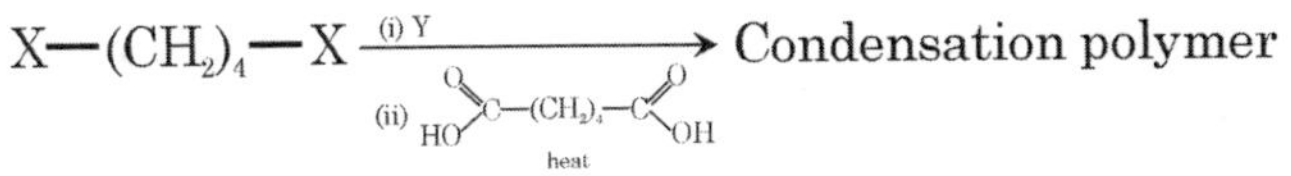

(A) X = $COOCH_3$, Y = H_2/Ni/heat (B) X = $CONH_2$, Y = H_2/Ni/heat

(C) X = $CONH_2$, Y = Br_2/NaOH (D) X = CN, Y = H_2/Ni/heat.

11. For the first order reaction, $2N_2O_5(g) \longrightarrow 4NO_2(g) + O_2(g)$

(A) the concentration of the reactant decreases exponentially with time

(B) the half-life of the reaction decreases with increasing temperature

(C) the half-life of the reaction depends on the initial concentration of the reactant

(D) the reaction proceeds to 99.6% completion in eight half-life duration.

12. The equilibrium $2Cu^I \rightleftharpoons Cu^0 + Cu^{II}$, in aqueous medium at 25°C shifts towards the left in the presence of

(A) NO_3^- (B) Cl^- (C) SCN^- (D) CN^-

SECTION – III (Total marks: 24)
(Integer Answer Type)

This section contains 6 questions. The answer to each of the questions is a single-digit integer, ranging from 0 to 9. The bubble corresponding to the correct answer is to be darkened in the ORS.

13. The maximum number of isomers (including stereoisomers) that are possible on monochlorination of the following compound is

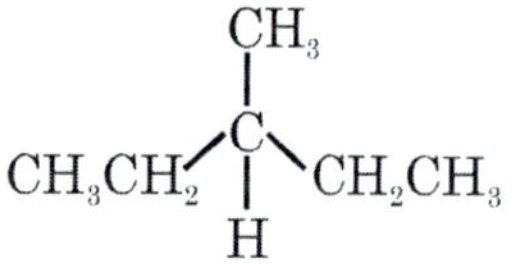

14. The total number of contributing structure showing hyperconjugation (involving C—H bonds) for the following carbocation is

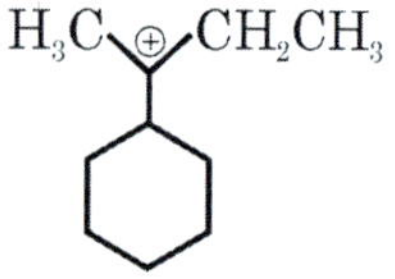

15. Among the following, the number of compounds than can react with PCl_5 to give $POCl_3$ is

    O_2, CO_2, SO_2, H_2O, H_2SO_4, P_4O_{10}

16. The volume (in mL) of $0.1M$ $AgNO_3$ required for complete precipitation of chloride ions present in 30 mL of $0.01M$ solution of $[Cr(H_2O)_5Cl]Cl_2$, as silver chloride is close to

17. In 1 L saturated solution of AgCl [K_{sp}(AgCl) = 1.6×10^{-10}], 0.1 mol of CuCl [K_{sp}(CuCl) = 1.0×10^{-6}] is added. The resultant concentration of Ag^+ in the solution is 1.6×10^{-x}. The value of x is

18. The number of hexagonal faces that are present in a truncated octahedron is

SECTION – IV (Total marks: 16)
(Matrix-Match Type)

This section contains 2 questions. Each question has four statements (A, B, C and D) given in Column I and five statements (p, q, r, s and t) in Column II. Any given statement in Column I can have correct matching with one or more statement(s) given in Column II. For example, if for a given question, statement B matches with the statement given q and r, then for the particular question, against statement B, darken the bubbles corresponding to q and r in the ORS.

19. Match the transformations in column I with appropriate options in column II.

Column I	Column II
(A) $CO_2(s) \longrightarrow CO_2(g)$	(p) phase transition
(B) $CaCO_3(s) \longrightarrow CaO(s) + CO_2(g)$	(q) allotropic change
(C) $2H \longrightarrow H_2(g)$	(r) ΔH is positive
(D) $P_{(\text{white, solid})} \longrightarrow P_{(\text{red, solid})}$	(s) ΔS is positive
	(t) ΔS is negative

20. Match the reactions in column I with appropriate types of steps/reactive intermediate involved in these reactions as given in column II.

Column I	Column II
(A) 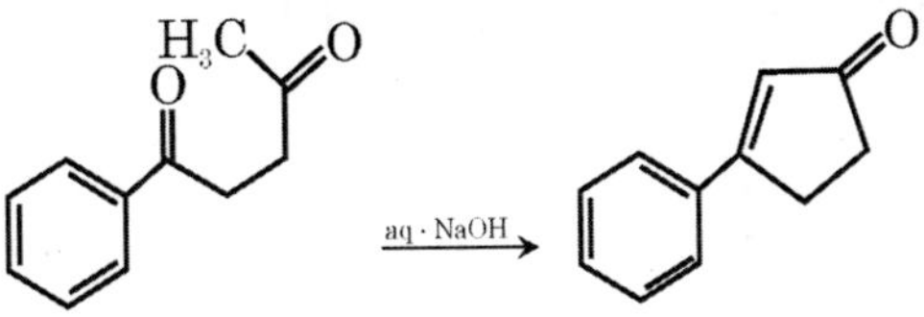	(p) Nucleophilic substitution
(B) 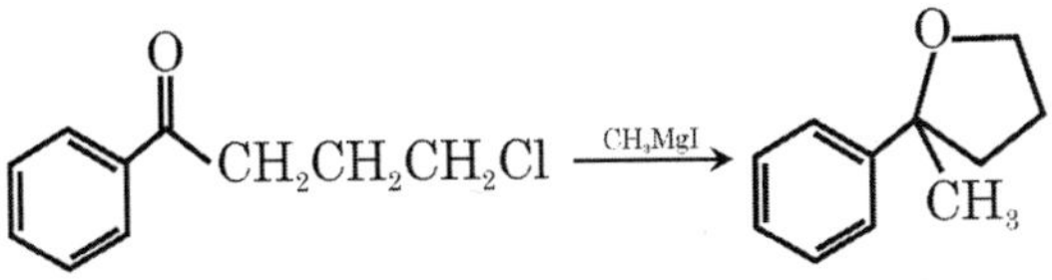	(q) Electrophilic substitution
(C) 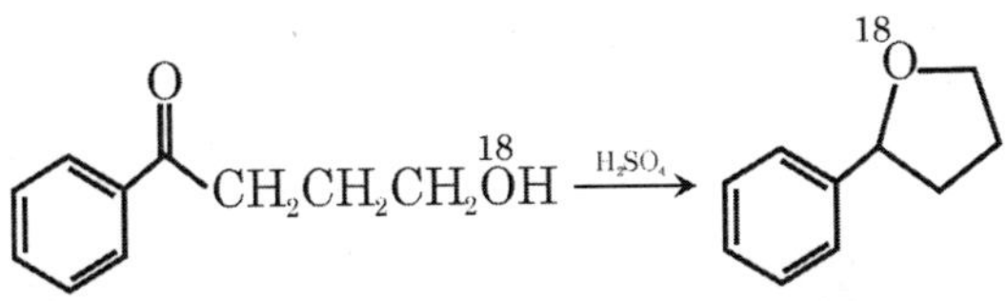	(r) Dehydration
(D) 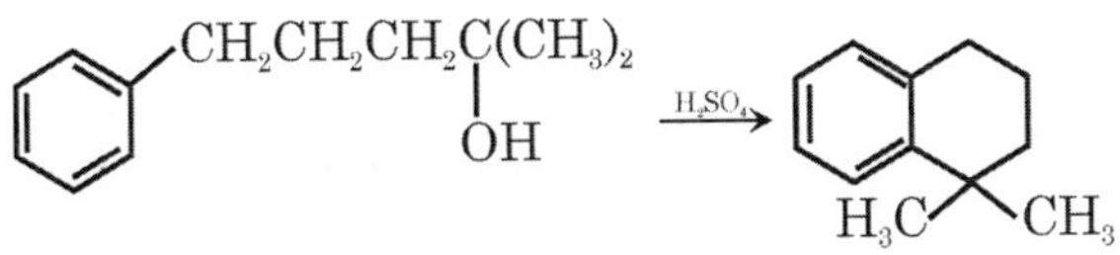	(s) Nucleophilic addition
	(t) Carbanion

ANSWERS

1. A	2. C	3. B	4. B	5. D	6. C
7. A	8. D	9. A, D	10. A, B, C, D	11. A, B, D	12. B, C, D
13. 8	14. 6	15. 5	16. 6	17. 7	18. 8

19. A – p, r, s; B – r, s; C – t; D – p, q, t

20. A – r, s, t; B – p, s; C – r, s; D – q, r

Model Solutions—I.I.T. JEE 2011

Paper–2

1. (A)

$K_3[Fe(CN)_6] \longrightarrow 3K^+ + [Fe(CN)_6]^{3-}$

$i = 4$

$$\Delta T_f = K_f \times i \times \frac{m}{M} \times \frac{1000}{W} = 1.86 \times 4 \times \frac{0.1}{329} \times \frac{1000}{100} = 2.3 \times 10^{-2}$$

$T_f = -\,2.3 \times 10^{-2}$

2. (C)

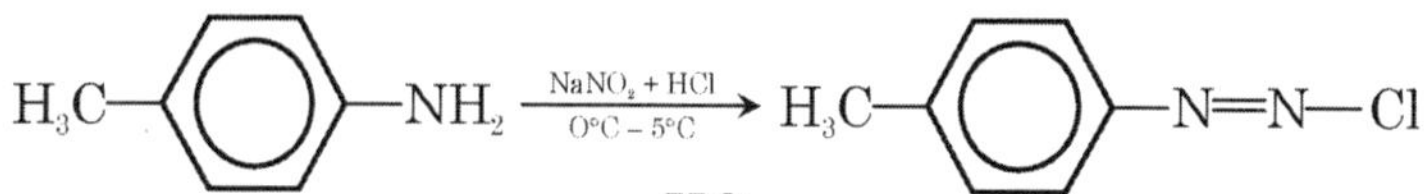

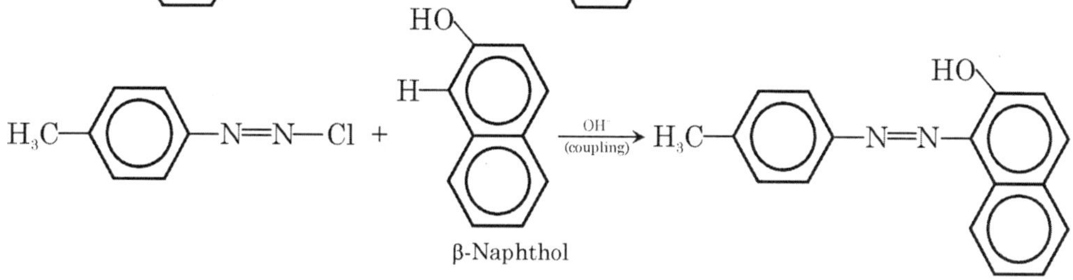

β-Naphthol

Brilliant colored red dye

3. (B)

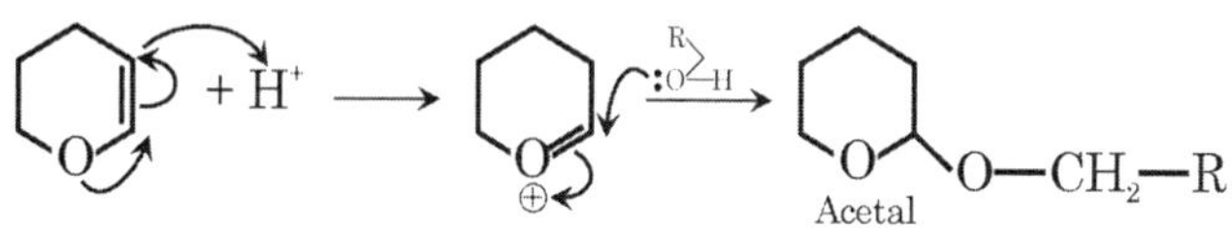

Acetal

4. (B)

The carbohydrate is an aldohexose.

5. (D)

Haematite: Fe_2O_3: $2x + 3 \times (-2) = 0$

$x = 3$

Magnetite: Fe_3O_4 [an equimolar mixture of FeO and Fe_2O_3]

FeO: $x - 2 = 0$ or $x = 2$

Fe_2O_3: $x = 3$

6. (A)

Following compounds are diamagnetic: L: $[Co(NH_3)_6]Cl_3$; M: $Na_3[Co(ox)_3]$; O: $K_2[Pt(CN)_4]$; P: $[Zn(H_2O)_6](NO_3)_2$.

7. (A)

H_2S in presence of aqueous acidified solution precipitates as sulphide of Cu and Hg other than Pb^{2+}, Bi^{3+}, Cd^{2+}, As^{3+}, Sb^{3+} and Sn^{2+}.

8. (D)

$2Fe(s) + O_2(g) + 4H^+(aq) \longrightarrow 2Fe^{2+}(aq) + 2H_2O(l)$

$n = 4$ (no. of moles of electron involved)

$$E_{cell} = E^0_{cell} - \frac{0.0591}{n}\log Q = 1.67 - \frac{0.0591}{4}\log\frac{(10^{-3})^2}{0.1\times(10^{-3})^4} \quad \{\because [H^+] = 10^{-pH}\}$$

$$= 1.67 - 0.10 = 1.57 \text{ V}$$

9. (A, D)

In acidic medium,

$$MnO_4^- + 8H^+ + 5e^- \longrightarrow Mn^{2+} + 4H_2O$$

In neutral medium,

$$MnO_4^- + 2H_2O + 3e^- \longrightarrow MnO_2 + 4OH^-$$

Hence the number of electrons loose in acidic and neutral medium is 5 and 3 electrons respectively.

10. (A, B, C, D)

Condensation polymers are formed by condensation of a diol or diamine with dicarboxylic acids.

Hence X may be $-\overset{O}{\overset{\|}{C}}-OR$ or $-\overset{O}{\overset{\|}{C}}-NH_2$ or $-C\equiv N$

$$H_3CO-\overset{O}{\overset{\|}{C}}-(CH_2)_4-\overset{O}{\overset{\|}{C}}-OCH_3 \xrightarrow{H_2/Ni} HO-H_2C(CH_2)_4-CH_2-OH \xrightarrow{HO-\overset{O}{\overset{\|}{C}}-(CH_2)_4-\overset{O}{\overset{\|}{C}}-OH} \text{Polyester}$$

$$H_2N-\overset{O}{\overset{\|}{C}}-(CH_2)_4-\overset{O}{\overset{\|}{C}}-NH_2 \xrightarrow{H_2/Ni} H_2N-CH_2(CH_2)_4-CH_2-NH_2 \xrightarrow{HO-\overset{O}{\overset{\|}{C}}-(CH_2)_4-\overset{O}{\overset{\|}{C}}-OH} \text{Polyamide}$$

$\downarrow$ Br_2/OH^-, Δ Hoffmann bromamide reaction

$$H_2N-(CH_2)_4-NH_2 \xrightarrow{HO-\overset{O}{\overset{\|}{C}}-(CH_2)_4-\overset{O}{\overset{\|}{C}}-OH} \text{Polyamide}$$

$$N\equiv C-(CH_2)_4-C\equiv N \xrightarrow{H_2/Ni} H_2N-CH_2-(CH_2)_4-CH_2-NH_2 \xrightarrow{HO-\overset{O}{\overset{\|}{C}}-(CH_2)_4-\overset{O}{\overset{\|}{C}}-OH} \text{Polyamide}$$

11. (A, B, D)

For first order reaction, $[A] = [A]_0 e^{-kt}$

Hence the concentration of $[NO_2]$ decreases exponentially.

Also, $t_{1/2} = \dfrac{0.693}{K}$, which is independent of concentration and $t_{1/2}$ decreases with the increase of temperature.

$$t_{99.6} = \frac{2.303}{K}\log\left(\frac{100}{0.4}\right)$$

$$t_{99.6} = \frac{2.303}{K}(2.4) = 8\times\frac{0.693}{K} = 8t_{1/2}$$

12. (B, C, D)

Cu^{2+} ions will react with CN^- and SCN^- forming $[Cu(CN)_4]^{3-}$ and $[Cu(SCN)_4]^{3-}$ leading the reaction in the backward direction.

$Cu^{2+} + 2CN^- \longrightarrow Cu(CN)_2$

$2Cu(CN)_2 \longrightarrow 2CuCN + (CN)_2$

$CuCN + 3CN^- \longrightarrow [Cu(CN)_4]^{3-}$

$Cu^{2+} + 4SCN^- \longrightarrow [Cu(SCN)_4]^{3-}$

Cu^{2+} also combines with Cl^- to give $CuCl_2$ which reacts with Cu to produce CuCl pushing the reaction in the backward direction.

$CuCl_2 + Cu \longrightarrow 2CuCl \downarrow$

13. (8)

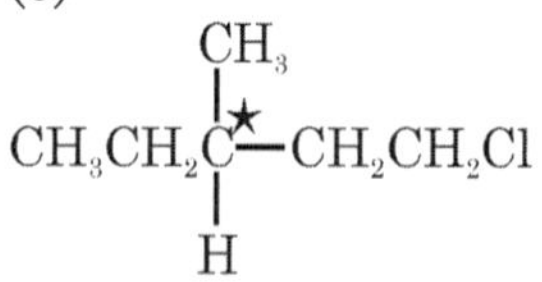

Enantiomeric pair = 2

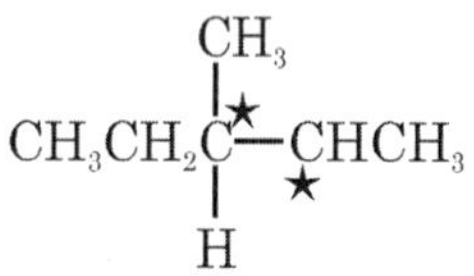

Two enantiomeric pairs = 4

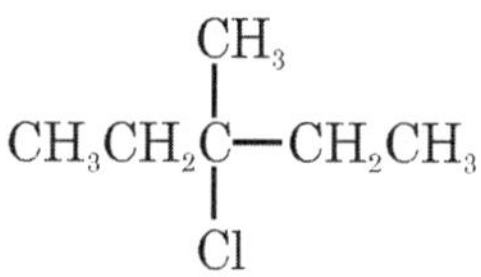

1

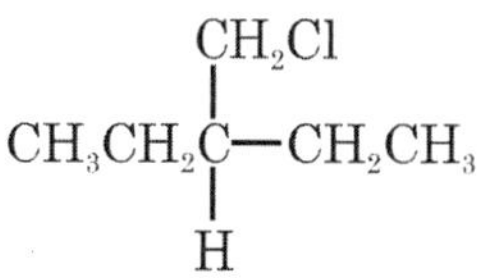

1

Total number of isomers = 2 + 4 + 1 + 1 = 8

14. (6)

Six H atoms are there.

15. (5)

H_2O, O_2, SO_2, CO_2, H_2SO_4

$$PCl_5 + H_2O \xrightarrow{\text{Insufficient water}} POCl_3 + 2HCl$$

$$2PCl_5 + O_2 \longrightarrow 2POCl_3 + 2Cl_2$$

$$PCl_5 + SO_2 \longrightarrow POCl_3 + SOCl_2$$

$$PCl_5 + CO_2 \longrightarrow POCl_3 + COCl_2$$

$$2PCl_5 + H_2SO_4 \longrightarrow 2POCl_3 + SO_2Cl_2 + 2HCl$$

16. (6)

Number of ionisable Cl^- in $[Cr(H_2O)_5Cl]Cl_2$ is 2.

Hence millimoles of $Cl^- = 30 \times 0.01 \times 2 = 0.6$.

Hence millimoles of Ag^+ required $= 0.6$.

Thus $0.6 = 0.1\ V$

$V = 6$ mL.

17. (7)

Let the solubility of AgCl x be mol litre^{-1} and that of CuCl be y mol litre^{-1}.

$$AgCl \rightleftharpoons \underset{x}{Ag^+} + \underset{x}{Cl^-}$$

$$CuCl \rightleftharpoons \underset{y}{Cu^+} + \underset{y}{Cl^-}$$

Hence K_{sp} of AgCl $= [Ag^+][Cl^-]$

$1.6 \times 10^{-10} = x(x + y)$ ...(i)

Similarly K_{sp} of CuCl $= [Cu^+][Cl^-]$

$1.6 \times 10^{-6} = y(x + y)$ ...(ii)

On solving (i) and (ii), we get $[Ag^+] = 1.6 \times 10^{-7}$

Hence $x = 7$

18. (8)

19. (A)–(p, r, s) (B)–(r, s) (C)–(t) (D)–(p, q, t)

20. (A)–(r, s, t) (B)–(p, s) (C)–(r, s) (D)–(q, r)

I.I.T. JEE Objective Question Paper—2012

PAPER—1

INSTRUCTIONS

A. General:

1. This booklet is your Question paper. Do not break the seals of this booklet before being instructed to do so by the invigilators.
2. The question paper code is printed on the right hand top corner of this page and on the back page of this booklet.
3. Blank spaces and blank pages are provided in this booklet for your rough work. No additional sheets will be provided for rough work.
4. Blank papers, clipboards, log tables, slide rules, calculators, cameras, cellular phones, pagers, and electronic gadgets are not allowed inside the examination hall.
5. Answers to the questions and personal details are to be filled on a two-part carbon-less paper, which is provided separately. You should not separate these parts. The invigilator will separate them at the end of examination. The upper sheet is machine-gradable Objective Reponse Sheet (ORS) which will be taken back by the invigilator.
6. Using a black ball point pen, darken the bubbles on the upper original sheet. Apply sufficient pressure so that the impression is created on the bottom sheet.
7. Do not tamper with/mutilate the ORS or the booklet.
8. On breaking the seals of the booklet check that it contains 28 pages and all 60 questions and corresponding answer choices are legible. Read carefully the instructions printed at the beginning of each section.

B. Filling the Right Part of the ORS:

9. The ORS also has a code printed on its left and right parts.
10. Check that the same code is printed on the ORS and on this booklet. If it is not then ask for a change of the booklet. Sign at the place provided on the ORS affirming that you have verified that all the codes are same.
11. Write your Name, Registration Number and the name of examination centre and sign with pen in the boxes provided on the right part of the ORS . Do not write any of this information anywhere else. Darken the appropriate bubble under each digit of your Registration Number in such a way that the impression is created on the bottom sheet. Also darken the paper code given on the right side of ORS.

C. Question paper format and Marking scheme:

The question paper consists of 3 parts (Physics, Chemistry and Mathematics). Each part consists of three sections.

12. Section I contains 10 multiple choice questions. Each question has four choices (A), (B), (C) and (D) out of which only one is correct.
13. Section II contains 5 multiple choice questions. Each question has four choices (A), (B), (C) and (D) out of which one or more are correct.
14. Section III contains 5 questions. The answer to each question is a single digit integer, ranging from 0 to 9 (both inclusive).

D. Marking Scheme:

15. For each question in Section I, you will be awarded 3 marks if you darken the bubble corresponding to the correct answer only and zero marks if no bubbles are darkened. In all other cases, minus one (– 1) mark will be awarded in this section.
16. For each question in Section II, you will be awarded 4 marks if you darken all the bubble(s) corresponding to the correct answer(s) only. In all other cases zero (0) marks will be awarded. No negative marks will be awarded for incorrect answer in this section.
17. For each question in Section III, you will be awarded 4 marks if you darken the bubble corresponding to the correct answer only. In all other cases zero (0) marks will be awarded. No negative marks will be awarded for incorrect answer in this section.

SECTION – I
(Single Correct Answer Type)

This section contains 10 multiple choice questions. Each question has four choices (A), (B), (C) and (D) out of which only one is correct.

1. A compound M_pX_q has cubic close packing (ccp) arrangement of X. Its unit cell structure is shown below. The empirical formula of the compound is

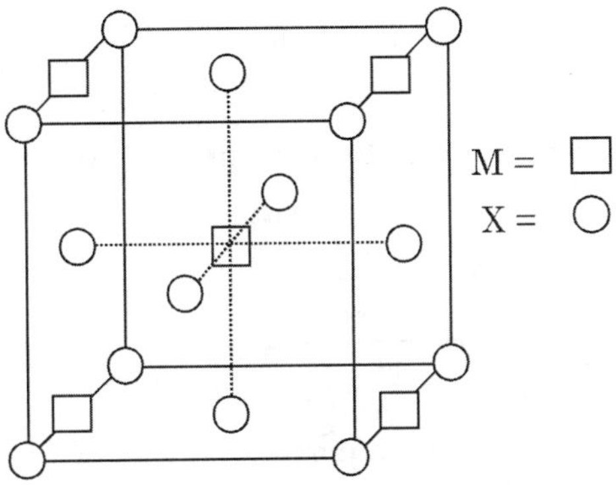

(A) MX (B) MX_2 (C) M_2X (D) M_5X_{14}

2. The carboxyl functional group (—COOH) is present in

(A) picric acid (B) barbituric acid (C) ascorbic acid (D) aspirin.

3. As per IUPAC nomenclature, the name of the complex $[Co(H_2O)_4(NH_3)_2]Cl_3$ is

(A) Tetraaquadiaminecobalt(III) chloride (B) Tetraaquadiamminecobalt(III) chloride
(C) Diaminetetraaquacobalt(III) chloride (D) Diamminetetraaquacobalt(III) chloride.

4. In allene (C_3H_4), the type(s) of hybridization of the carbon atoms is(are)
(A) *sp* and sp^3 (B) *sp* and sp^2 (C) only sp^2 (D) sp^2 and sp^3

5. The kinetic energy of an electron in the second Bohr orbit of a hydrogen atom is [a_0 is Bohr radius]
(A) $\frac{h^2}{4\pi^2 m a_0^2}$ (B) $\frac{h^2}{16\pi^2 m a_0^2}$ (C) $\frac{h^2}{32\pi^2 m a_0^2}$ (D) $\frac{h^2}{64\pi^2 m a_0^2}$

6. Which ordering of compounds is according to the decreasing order of the oxidation state of nitrogen?
(A) HNO_3, NO, NH_4Cl, N_2 (B) HNO_3, NO, N_2, NH_4Cl
(C) HNO_3, NH_4Cl, NO, N_2 (D) NO, HNO_3, NH_4Cl, N_2

7. For one mole of a van der Waals gas when $b = 0$ and $T = 300$ K, the *PV* *vs* 1/*V* plot is shown below. The value of the van der Waals constant *a* (atm litre^2 mol^{-2}) is

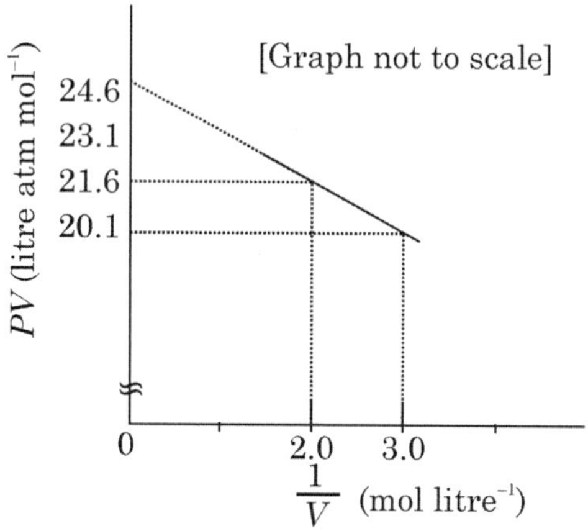

(A) 1.0 (B) 4.5 (C) 1.5 (D) 3.0

8. The number of aldol reaction(s) that occurs in the given transformation is

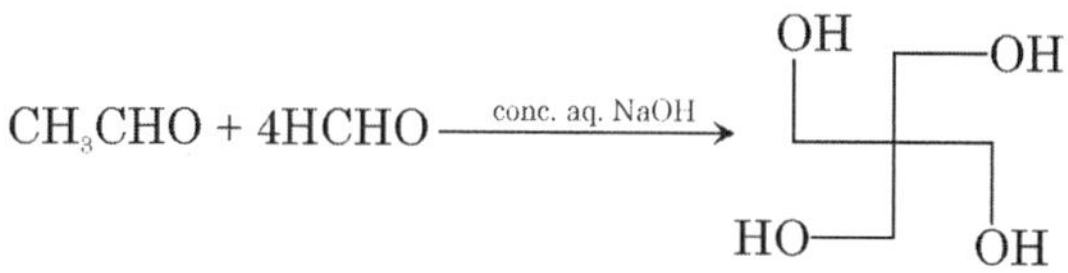

(A) 1 (B) 2 (C) 3 (D) 4

9. The colour of light absorbed by an aqueous solution of $CuSO_4$ is
(A) orange-red (B) blue-green (C) yellow (D) violet.

10. The number of optically active products obtained from the complete ozonolysis of the given compound is

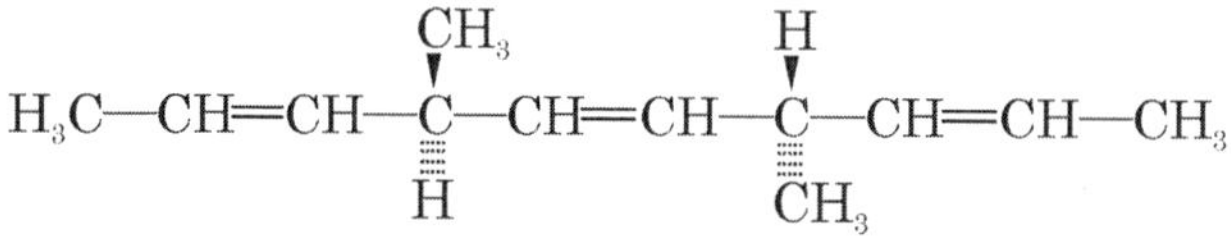

(A) 0 (B) 1 (C) 2 (D) 4

SECTION – II
(Multiple Correct Answer(s) Type)

This section contains 5 multiple choice questions. Each question has four choices (A), (B), (C) and (D) out of which one or more are correct.

11. Which of the following hydrogen halides react(s) with $AgNO_3(aq)$ to give a precipitate that dissolves in $Na_2S_2O_3(aq)$?

(A) HCl (B) HF (C) HBr (D) HI

12. Identify the binary mixture(s) that can be separated into individual compounds, by differential extraction, as shown in the given scheme.

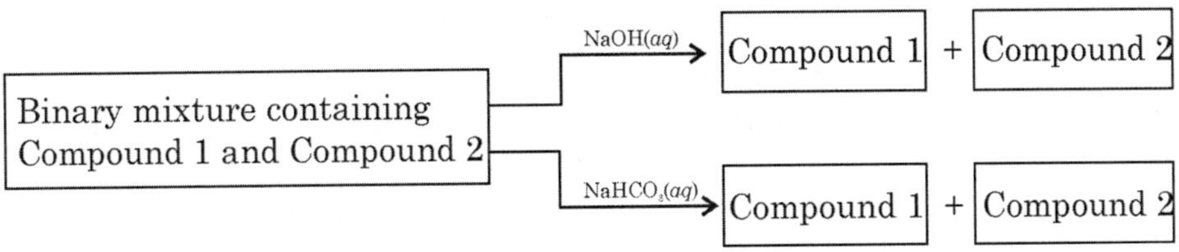

(A) C_6H_5OH and C_6H_5COOH (B) C_6H_5COOH and $C_6H_5CH_2OH$

(C) $C_6H_5CH_2OH$ and C_6H_5OH (D) $C_6H_5CH_2OH$ and $C_6H_5CH_2COOH$

13. For an ideal gas, consider only *P-V* work in going from an initial state *X* to the final state *Z*. The final state *Z* can be reached by either of the two paths shown in the figure. Which of the following choice(s) is(are) correct? [Take ΔS as change in entropy and w as work done]

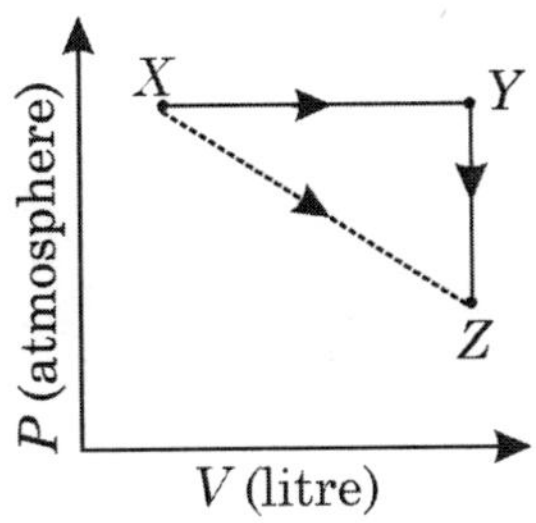

(A) $\Delta S_{x \to z} = \Delta S_{x \to y} + \Delta S_{y \to z}$ (B) $w_{x \to z} = w_{x \to y} + w_{y \to z}$

(C) $w_{x \to y \to z} = w_{x \to y}$ (D) $\Delta S_{x \to y \to z} = \Delta S_{x \to y}$

14. Which of the following molecules, in pure form, is(are) unstable at room temperature?

(A)  (B)  (C) 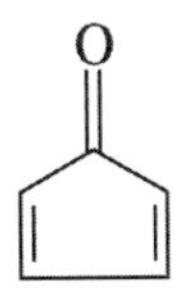 (D) 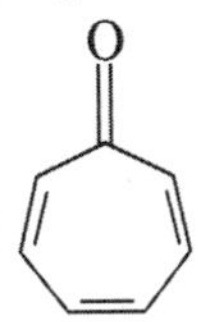

15. Choose the correct reason(s) for the stability of the lyophobic colloidal particles.

(A) Preferential adsorption of ions on their surface from the solution

(B) Preferential adsorption of solvent on their surface from the solution

(C) Attraction between different particles having opposite charges on their surface

(D) Potential difference between the fixed layer and the diffused layer of opposite charges around the colloidal particles.

SECTION – III
(Integer Answer Type)

This section contains 5 questions. The answer to each question is a single digit integer, ranging from 0 to 9 (both inclusive).

16. 29.2% (*w/w*) HCl stock solution has density of 1.25 g mL^{-1}. The molecular mass of HCl is 36.5 g mol^{-1}. The volume (mL) of stock solution required to prepare a 200 mL solution of 0.4*M* HCl is

17. The substituents R_1 and R_2 for nine peptides are listed in the table given below. How many of these peptides are positively charged at *p*H = 7.0?

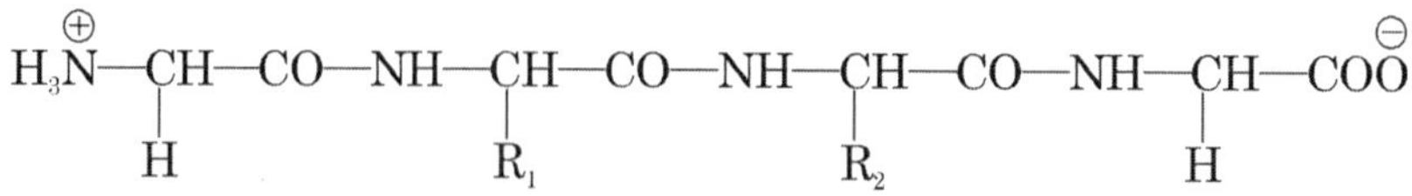

Peptide	R_1	R_2
I	H	H
II	H	CH_3
III	CH_2COOH	H
IV	CH_2CONH_2	$(CH_2)_4NH_2$
V	CH_2CONH_2	CH_2CONH_2
VI	$(CH_2)_4NH_2$	$(CH_2)_4NH_2$
VII	CH_2COOH	CH_2CONH_2
VIII	CH_2OH	$(CH_2)_4NH_2$
IX	$(CH_2)_4NH_2$	CH_3

18. An organic compound undergoes first-order decomposition. The time taken for its decomposition to 1/8 and 1/10 of its initial concentration are $t_{1/8}$ and $t_{1/10}$ respectively. What is the value of $([t_{1/8}]/[t_{1/10}] \times 10)$? (take $\log_{10} 2 = 0.3$)

19. When the following aldohexose exists in its D-configuration, the total number of stereoisomers in its pyranose form is

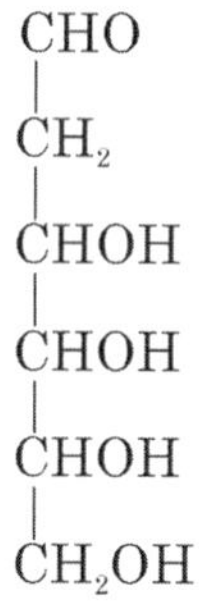

20. The periodic table consists of 18 groups. An isotope of copper, on bombardment with protons, undergoes a nuclear reaction yielding element X as shown below. To which group, element X belongs in the periodic table?

$^{63}_{29}Cu + ^{1}_{1}H \longrightarrow 6^{1}_{0}n + \alpha + 2^{1}_{1}H + X$

ANSWERS

1. B	2. D	3. D	4. B	5. C	6. B
7. C	8. C	9. A	10. A	11. A, C, D	12. B, D
13. A, C	14. B, C	15. A, D	16. 8	17. 4	18. 9
19. 8	20. 8				

Model Solutions—I.I.T. JEE 2012

Paper—1

1. (B)

$$X = 8 \times \frac{1}{8} + 6 \times \frac{1}{2} = 4$$

$$M = 4 \times \frac{1}{4} + 1 = 2$$

Hence, unit cell formula of the compound is M_2X_4 and the empirical formula of the compound is MX_2.

2. (D)

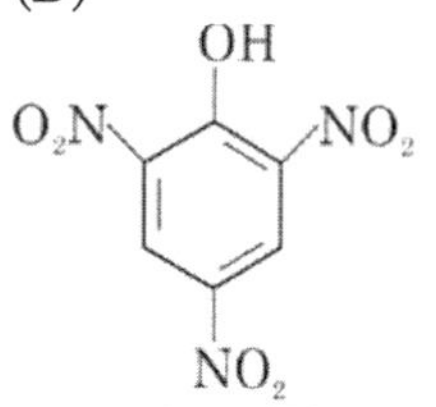

(Picric Acid)

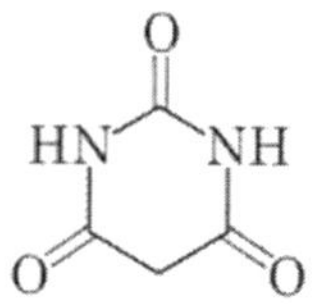

(Barbituric Acid)

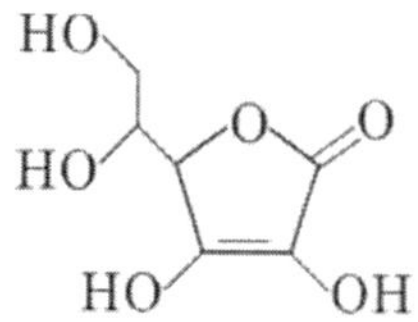

(Ascorbic Acid)

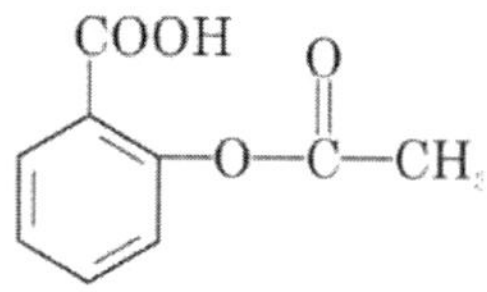

(Aspirin)

3. (D)

$[Co(H_2O)_4(NH_3)_2]Cl_3$

Diamminetetraaquacobalt(III) chloride

4. (B)

$$\overset{sp^2}{H_2C}=\overset{sp}{C}=\overset{sp^2}{CH_2}$$

5. (C)

According to Bohr's postulate,

$$mvr = \frac{nh}{2\pi}$$

Hence, $$v = \frac{nh}{2\pi mr}$$

$$KE = \frac{1}{2}mv^2 = \frac{1}{2}m\left(\frac{nh}{2\pi mr}\right)^2$$

But $$r = \frac{a_0 \times n^2}{Z}$$

Hence, for second Bohr orbit

$$r = \frac{a_0 \times 2^2}{1} = 4a_0$$

$$KE = \frac{1}{2}m\left(\frac{2^2h^2}{4\pi^2m^2 \times (4a_0)^2}\right) = \frac{h^2}{32\pi^2ma_0^2}$$

6. (B)

$\overset{+5}{HNO_3}, \overset{+2}{NO}, \overset{0}{N_2}, \overset{-3}{NH_4Cl}$

7. (C)

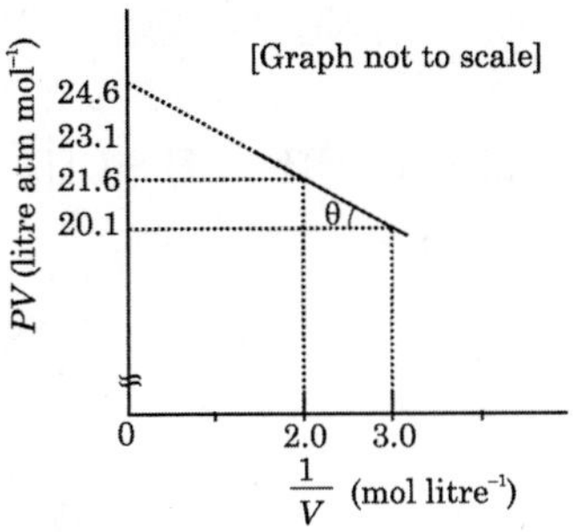

van der Waals equation for 1 mole of real gas is

$$\left(P+\frac{a}{V^2}\right)(V-b) = RT$$

But $b = 0$ (given)

or $\left(P+\frac{a}{V^2}\right)(V) = RT$

or $PV = -a \times \frac{1}{V} + RT$

$y = mx + c$

Slope = tan (π – θ) = – a

$$\tan\theta = a = \frac{24.6-20.1}{3-2} = 1.5$$

or $\tan\theta = \frac{24.6-20.1}{3-2} = 1.5$

8. (C)

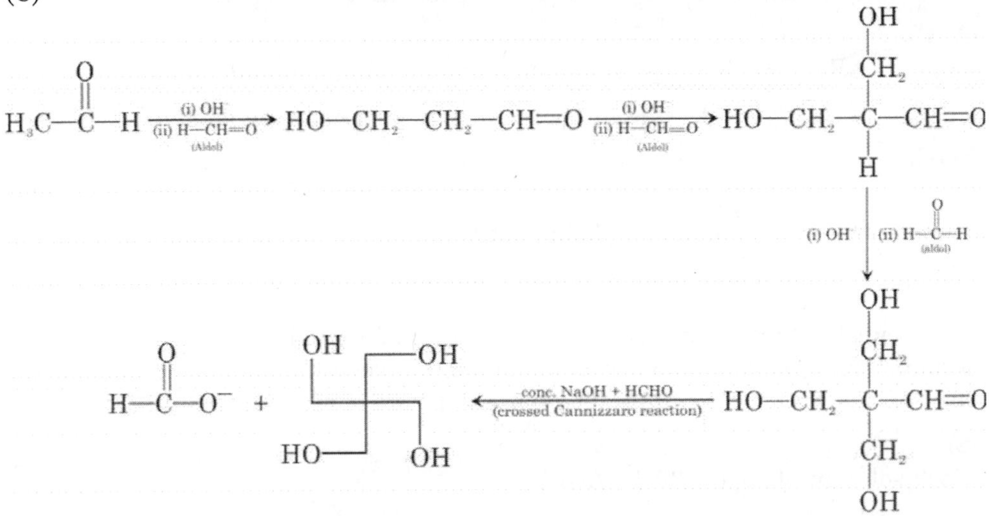

9. (A)

Aqueous solution of copper(II) sulphate absorbs orange red light and appears blue (complementary colour).

10. (A)

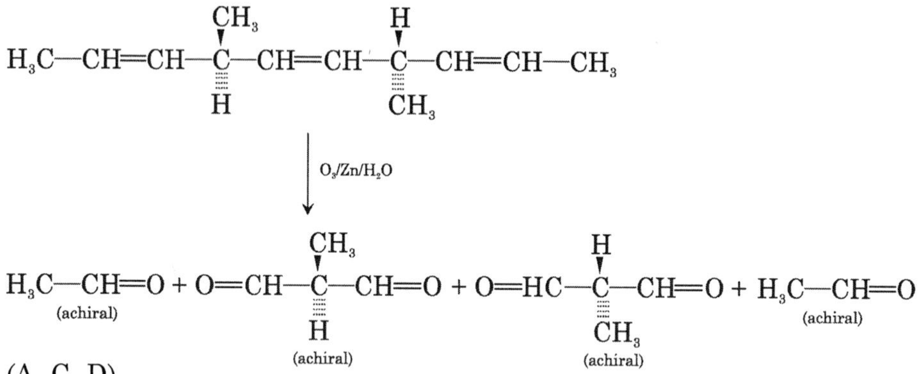

11. (A, C, D)
$HX + AgNO_3 \longrightarrow AgX \downarrow + HNO_3$ (X = Cl, Br, I)
$AgX + 2Na_2S_2O_3 \longrightarrow Na_3[Ag(S_2O_3)_2] + NaX$

12. (B, D)
(A) Both are soluble in NaOH and hence they are inseparable.
(B) Only benzoic acid (C_6H_5COOH) is soluble in NaOH and $NaHCO_3$, while benzyl alcohol ($C_6H_5CH_2OH$) is not. Hence, they are separable.
(C) Although NaOH can enable separation between benzyl alcohol ($C_6H_5CH_2OH$) and phenol (C_6H_5OH) since only the latter is soluble in NaOH. However, in $NaHCO_3$, both are insoluble. Hence, they are inseparable.
(D) α-Phenyl acetic acid ($C_6H_5CH_2COOH$) is soluble in NaOH and $NaHCO_3$, while benzyl alcohol ($C_6H_5CH_2OH$) is not. Hence, they are separable.

13. (A, C)
$\Delta S_{X \to Z} = \Delta S_{X \to Y} + \Delta S_{Y \to Z}$ [entropy (S) is a state function and hence it is additive]
$w_{X \to Y \to Z} = w_{X \to Y}$ (work done in $Y \to Z$ is zero since it is an isochoric process)

14. (B, C)
The compounds 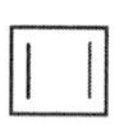 and 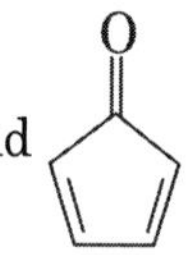 being antiaromatic are unstable at room temperature.

15. (A, D)
The lyophobic colloids are stable due to the preferential adsorption of ions on their surface from solution and potential difference between the fixed layer and the diffused layer of opposite charges around the colloidal particles which makes lyophobic sol stable.

16. (8)
Stock solution of HCl = 29.2% (w/w)

Molarity of stock solution of HCl = $\frac{29.2 \times 1000 \times 1.25}{36.5 \times 100}$

If volume of stock solution required = V mL,

$$V \times \frac{29.2}{36.5} \times \frac{1000}{80} = 200 \times 0.4$$

or $V = 8$ mL

17. (4)

Peptides with isoelectric point > 7, would exist as cation in neutral solution (pH = 7). Positively charged peptides = 4 (IV, VI, VIII and IX)

18. (9)

$$t_{1/8} = \frac{2.303 \log 8}{k} = \frac{2.303 \times 3 \log 2}{k}$$

$$t_{1/10} = \frac{2.303}{k} \log 10 = \frac{2.303}{k}$$

$$\left[\frac{t_{1/8}}{t_{1/10}}\right] \times 10 = \frac{\left(\frac{2.303 \times 3 \log 2}{k}\right)}{\left(\frac{2.303}{k}\right)} \times 10 = 9$$

19. (8)

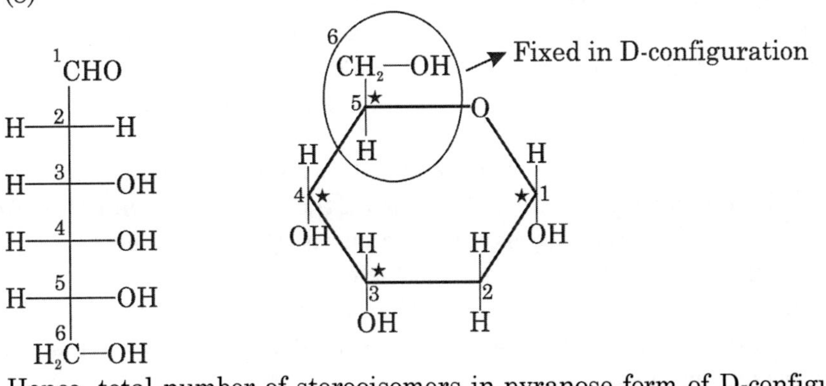

Hence, total number of stereoisomers in pyranose form of D-configuration = 2^3 = 8.

20. (8)

$${}^{63}_{29}\text{Cu} + {}^{1}_{1}\text{H} \longrightarrow 6\,{}^{1}_{0}\text{n} + {}^{4}_{2}\text{He} + 2\,{}^{1}_{1}\text{H} + {}^{A}_{Z}\text{X}$$

Mass number: 63 + 1 = 1 × 6 + 4 + 1 × 2 + A

or A = 64 – 12 = 52

Atomic number: 29 + 1 = 6 × 0 + 2 + 2 × 1 + Z

or Z = 30 – 4 = 26

Hence, ${}^{A}_{Z}\text{X} = {}^{52}_{26}\text{Fe}$

Hence, X is in the group 8 in the periodic table.

I.I.T. JEE Objective Question Paper—2012

PAPER—2

INSTRUCTIONS

A. General:

1. This booklet is your Question paper. Do not break the seals of this booklet before being instructed to do so by the invigilators.
2. The question paper code is printed on the right hand top corner of this page and on the back page of this booklet.
3. Blank spaces and blank pages are provided in this booklet for your rough work. No additional sheets will be provided for rough work.
4. Blank papers, clipboards, log tables, slide rules, calculators, cameras, cellular phones, pagers, and electronic gadgets are not allowed inside the examination hall.
5. Answers to the questions and personal details are to be filled on a two-part carbon-less paper, which is provided separately. You should not separate these parts. The invigilator will separate them at the end of examination. The upper sheet is machine-gradable Objective Reponse Sheet (ORS) which will be taken back by the invigilator. You will be allowed to take away the bottom sheet at the end of the examination.
6. Using a black ball point pen, darken the bubbles on the upper original sheet. Apply sufficient pressure so that the impression is created on the bottom sheet.
7. Do not tamper with/mutilate the ORS or the booklet.
8. On breaking the seals of the booklet check that it contains 28 pages and all 60 questions and corresponding answer choices are legible. Read carefully the instructions printed at the beginning of each section.

B. Filling the Right Part of the ORS:

9. The ORS has code printed on its left and right parts.
10. Check that the same code is printed on the ORS and on this booklet. If it is not then ask for a change of the booklet. Sign at the place provided on the ORS affirming that you have verified that all the codes are same.
11. Write your Name, Registration Number and the name of examination centre and sign with pen in the boxes provided on the right part of the ORS. Do not write any of this information anywhere else. Darken the appropriate bubble under each digit of your Registration Number in such a way that the impression is created on the bottom sheet. Also darken the paper code given on the right side of ORS.

C. Question Paper Format:

The question paper consists of 3 parts (Physics, Chemistry and Mathematics). Each part consists of three sections.

12. Section I contains 8 multiple choice questions. Each question has four choices (A), (B), (C) and (D) out of which only one is correct.
13. Section II contains 3 paragraphs each describing theory, experiment, data, etc. There are 6 multiple choice questions relating to three paragraphs with 2 questions on each paragraph. Each question of a particular paragraphs has four choices (A), (B), (C) and (D) out of which only one is correct.
14. Section III contains 6 multiple choice questions. Each question has four choices (A), (B), (C) and (D) out of which one or more are correct.

D. Marking Scheme:

15. For each question in Section I and Section II, you will be awarded 3 marks if you darken the bubble corresponding to the correct answer only and zero (0) marks if no bubbles are darkened. In all other cases, minus one (– 1) mark will be awarded in these sections.
16. For each question in Section III, you will be awarded 4 marks if you darken all the bubble(s) corresponding to the correct answer(s) only. In all other cases zero (0) marks will be awarded. No negative marks will be awarded for incorrect answer(s) in this section.

SECTION – I
(Single Correct Answer Type)

This section contains 8 multiple choice questions. Each question has four choices (A), (B), (C) and (D) out of which only one is correct.

1. The major product H of the given reaction sequence is

$$CH_3—CH_2—CO—CH_3 \xrightarrow{CN^-} G \xrightarrow[\text{Heat}]{95\%\ H_2SO_4} H$$

(A) $CH_3—CH{=}\underset{\displaystyle CH_3}{\underset{|}{C}}—COOH$

(B) $CH_3—CH{=}\underset{\displaystyle CH_3}{\underset{|}{C}}—CN$

(C) $CH_3—CH_2—\overset{\displaystyle OH}{\overset{|}{\underset{\displaystyle CH_3}{\underset{|}{C}}}}—COOH$

(D) $CH_3—CH{=}\underset{\displaystyle CH_3}{\underset{|}{C}}—CO—NH_2$

2. $NiCl_2\{P(C_2H_5)_2(C_6H_5)\}_2$ exhibits temperature dependent magnetic behaviour (paramagnetic/diamagnetic). The coordination geometries of Ni^{2+} in the paramagnetic and diamagnetic states are respectively

(A) tetrahedral and tetrahedral (B) square planer and square planar

(C) tetrahedral and square planar (D) square planer and tetrahedral.

3. In the cyanide extraction process of silver from argentite ore, the oxidising and reducing agents used are

(A) O_2 and CO respectively (B) O_2 and Zn dust respectively
(C) HNO_3 and Zn dust respectively (D) HNO_3 and CO respectively.

4. The reaction of white phosphorous with aqueous NaOH gives phosphine along with another phosphorous containing compound. The reaction type, the oxidation states of phosphorus in phosphine and the other product are respectively
   (A) redox reaction, – 3, – 5 (B) redox reaction, + 3, + 5
   (C) disproportionation reaction, – 3, + 5 (D) disproportionation reaction, – 3, + 3

5. The shape of XeO_2F_2 molecule is
   (A) trigonal bipyramidal (B) square planar (C) tetrahedral (D) see-saw.

6. For a dilute solution containing 2.5 g of a non-volatile non-electrolyte solute in 100 g of water, the elevation in boiling point at 1 atm pressure is 2°C. Assuming concentration of solute is much lower than the concentration of solvent, the vapour pressure (mm of Hg) of the solution is (K_b = 0.76 K kg mol^{-1})
   (A) 724 (B) 740 (C) 736 (D) 718

7. The compound that undergoes decarboxylation most readily under mild condition is

(A) 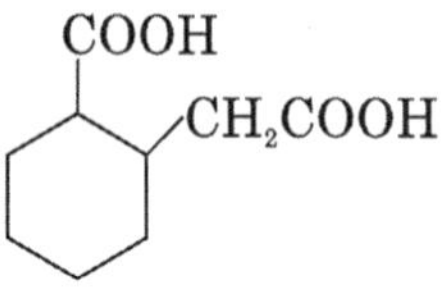

(B) 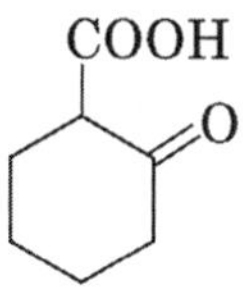

(C) 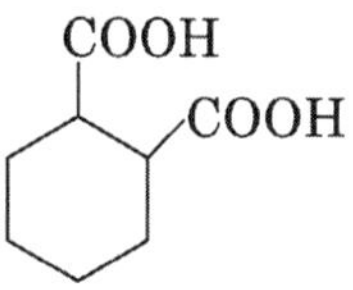

(D) 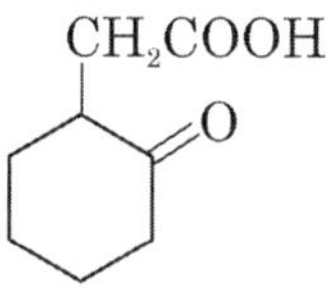

8. Using the data provided, calculate the multiple bond energy (kJ mol^{-1}) of a C═C bond in C_2H_2. That energy is (bond energy of C—H bond 350 kJ mol^{-1})

$2C(s) \longrightarrow 2C(g) \quad \Delta H = 1410 \text{ kJ mol}^{-1}$

$2C(s) \longrightarrow 2C(g) \quad \Delta H = 1410 \text{ kJ mol}^{-1}$

$H_2(g) \longrightarrow 2H(g) \quad \Delta H = 330 \text{ kJ mol}^{-1}$

   (A) 1165 (B) 837 (C) 865 (D) 815

SECTION – II
(Paragraph Type)

This section contains 6 multiple choice questions relating to three paragraphs with two questions on each paragraph. Each question has four choices (A), (B), (C) and (D) out of which only one is correct.

Paragraph for Questions 9 and 10

In the following reaction sequence, the compound J is an intermediate.

$$I \xrightarrow[CH_3COONa]{(CH_3CO)_2O} J \xrightarrow[\text{(ii) } SOCl_2 \text{ (iii) anlyd. } AlCl_3]{\text{(i) } H_2,\ Pd/C} K$$

J ($C_9H_8O_2$) gives effervescence on treatment with $NaHCO_3$ and a positive Baeyer's test.

9. The compound I is

(A) 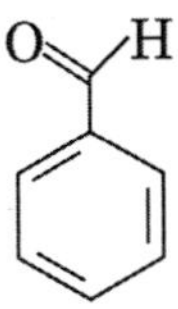 (B) 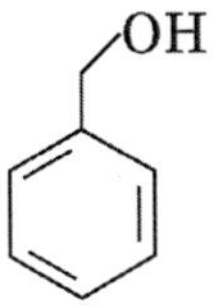 (C) 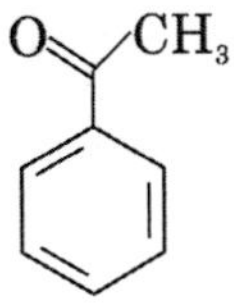 (D) 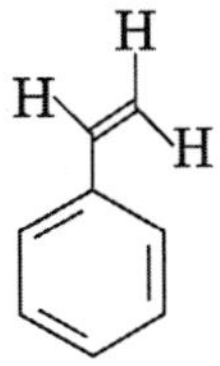

10. The compound K is

(A) 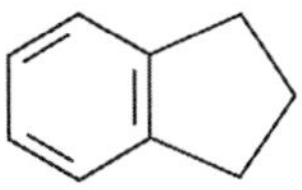 (B) 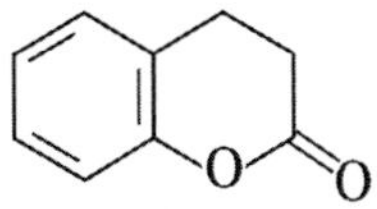

(C) 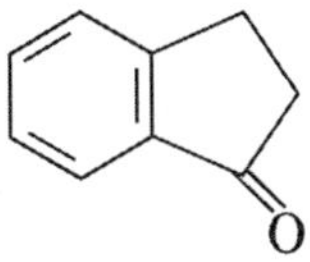 (D) 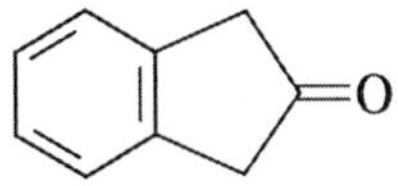

Paragraph for Questions 11 and 12

The electrochemical cell shown below is a concentration cell.
$M|M^{2+}$ (saturated solution of a sparingly soluble salt, MX_2)$||M^{2+}$ (0.001 mol dm^{-3})$|M$
The emf of the cell depends on the difference in concentrations of M^{2+} ions at the two electrodes. The emf of the cell at 298 K is 0.059 V.

11. The value of ΔG (kJ mol^{-1}) for the given cell is (1 F = 96500 C mol^{-1})
(A) – 5.7 (B) 5.7 (C) 11.4 (D) – 11.4

12. The solubility product (K_{sp}; $mol^3\ dm^{-9}$) of MX_2 at 298 K based on the information available for the given concentration cell is ($2.303 \times R \times 2.98/F = 0.059\ V$)
(A) 1×10^{-15} (B) 4×10^{-15} (C) 1×10^{-12} (D) 4×10^{-12}

Paragraph for Questions 13 and 14

Bleaching powder and bleach solution are produced on a large scale and used in several household products. The effectiveness of bleach solution is often measured by iodometry.

13. Bleaching powder contains is salt of an oxoacid as one of its components. The anhydride of that oxoacid is
(A) Cl_2O (B) Cl_2O_7 (C) ClO_2 (D) Cl_2O_6

14. 25 mL of household solution was mixed with 30 mL of 0.50M KI and 10 mL of 4N acetic acid. In the titration of the liberated iodine, 48 mL of 0.25N $Na_2S_2O_3$ was used to reach the end point. The molarity of the household bleach solution is
(A) 0.48M (B) 0.96M (C) 0.24M (D) 0.024M

SECTION – III
(Multiple Correct Answer(s) Type)

The section contains 6 multiple choice questions. Each question has four choices (A), (B), (C) and (D) out of which one or more are correct.

15. The reversible expansion of an ideal gas under adiabatic and isothermal conditions is shown in the figure. Which of the following statement(s) is (are) correct?

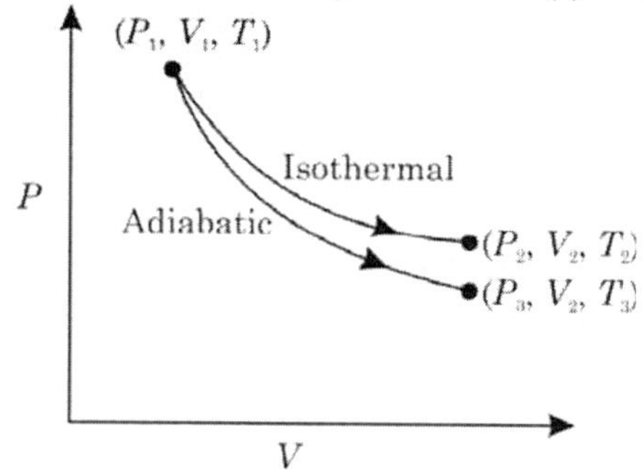

(A) $T_1 = T_2$ (B) $T_3 > T_1$

(C) $w_{isothermal} > w_{adiabatic}$ (D) $\Delta U_{isothermal} > \Delta U_{adiabatic}$

16. For the given aqueous reactions, which of the statement(s) is (are) true?

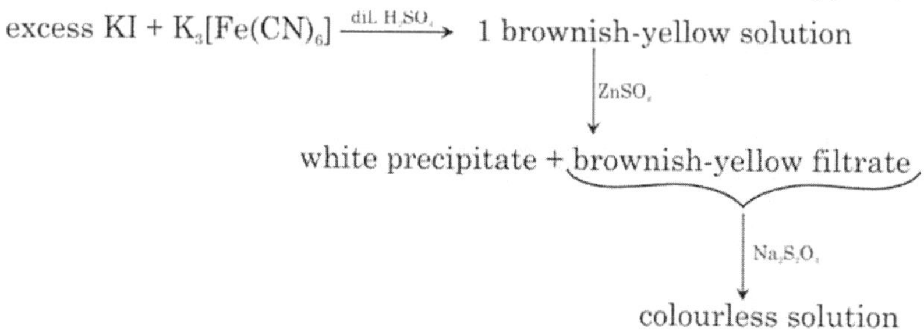

(A) The first reaction is a redox reaction (B) White precipitate is $Zn_3[Fe(CN)_6]_2$

(C) Addition of filtrate to starch solution gives blue colour

(D) White precipitate is soluble in NaOH solution.

17. With reference to the scheme given, which of the given statement(s) about T, U, V and W is (are) correct?

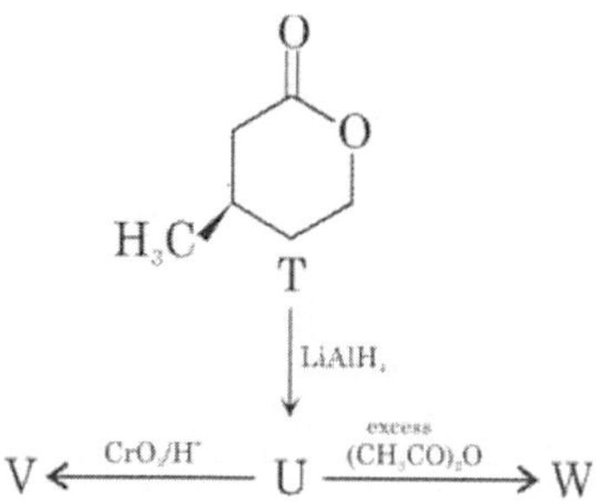

(A) T is soluble in hot aqueous NaOH (B) U is optically active

(C) Molecular formula of W is $C_{10}H_{18}O_4$

(D) V gives effervescence on treatment with aqueous $NaHCO_3$.

18. Which of the given statement(s) about N, O, P and Q with respect to M is (are) correct?

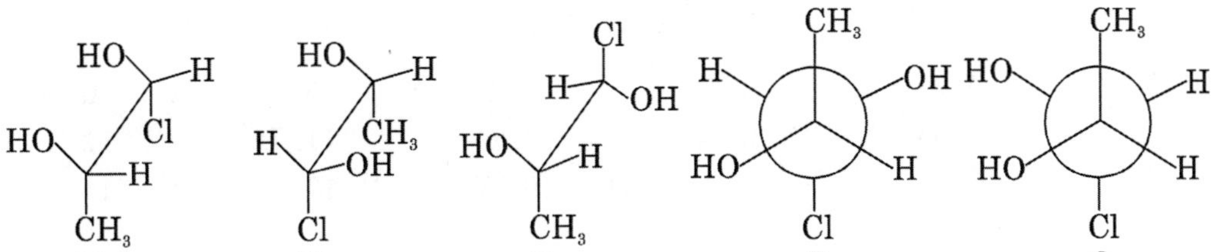

M N O P Q

(A) M and N are non-mirror image stereoisomers

(B) M and O are identical

(C) M and P are enantiomers

(D) M and Q are identical.

19. With respect to graphite and diamond, which of the statement(s) given below is (are) correct?

(A) Graphite is harder than diamond.

(B) Graphite has higher electrical conductivity than diamond.

(C) Graphite has higher thermal conductivity than diamond.

(D) Graphite has higher C—C bond order than diamond.

20. The given graphs/data I, II, III and IV represent general trends observed for different physisorption and chemisorption processes under mild conditions of temperature and pressure. Which of the following choice(s) about I, II, III and IV is (are) correct?

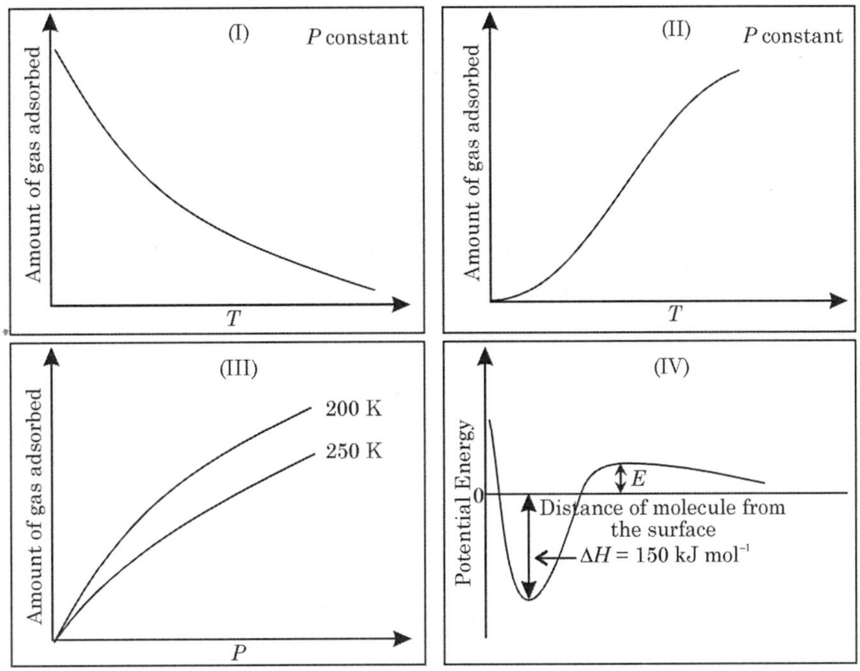

(A) I is physisorption and II is chemisorption

(B) I is physisorption and III is chemisorption

(C) IV is chemisorption and II is chemisorption

(D) IV is chemisorption and III is chemisorption.

ANSWERS

1. A	2. C	3. B	4. C	5. D	6. A
7. B	8. D	9. A	10. C	11. D	12. B
13. A	14. C	15. A, C, D	16. A, C, D	17. A, C, D	18. A, B, C
19. B, D	20. A, C				

Model Solutions—I.I.T. JEE 2012

Paper—2

1. (A)

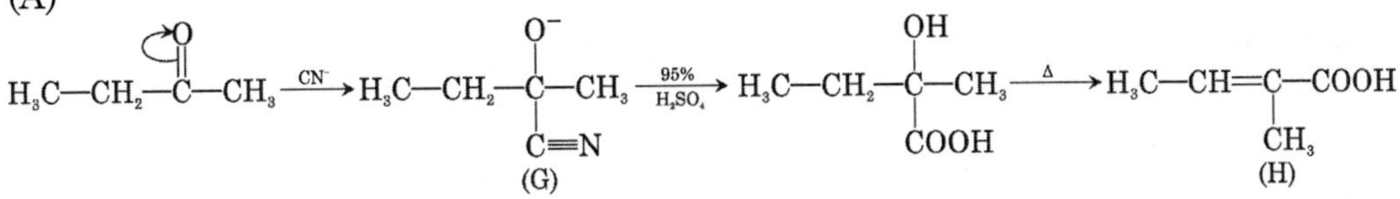

2. (C)

In both the complexes, Ni exists as Ni^{2+}

In sp^3 (tetrahedral geometry)

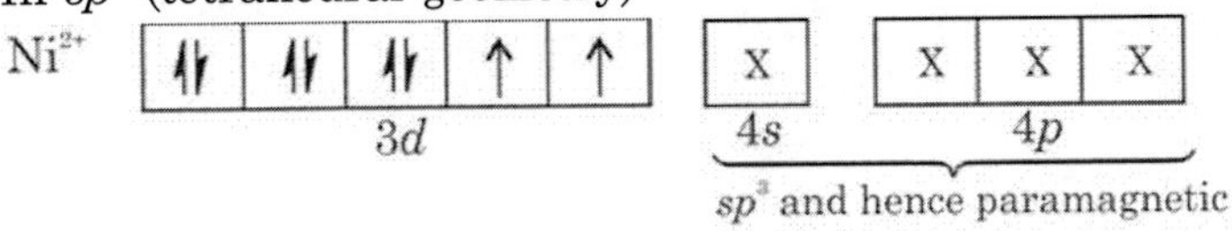

(x represents electron pair donated by the ligand)

In dsp^2 (square planar geometry)

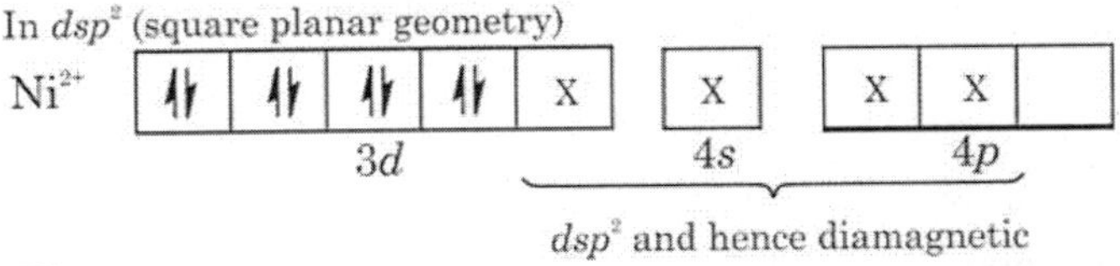

3. (B)

The reactions involved in cyanide extraction process are:

$$\underset{\text{(argentite ore)}}{Ag_2S} + 4NaCN \longrightarrow 2Na[Ag(CN)_2] + Na_2S$$

$$4Na_2S + \underset{\text{(oxidising agent)}}{5[O_2]} + 2H_2O \longrightarrow 2Na_2SO_4 + 4NaOH + 2S$$

$$2Na[Ag(CN)_2] + \underset{\text{(reducing agent)}}{Zn} \longrightarrow Na_2[Zn(CN)_4] + 2Ag\downarrow$$

4. (C)

The balanced disproportionation reaction involving white phosphorus with aqueous NaOH is

$$\overset{0}{P}_4 + 3NaOH + 3H_2O \longrightarrow \overset{-3}{P}H_3 + 3NaH_2\overset{+1}{P}O_2$$

Since the option involving + 1 oxidation state is missing, it may be considered that NaH_2PO_2 formed has undergone thermal decomposition as follows:

$$2NaH_2PO_2 \xrightarrow{\Delta} Na_2H\overset{+5}{P}O_4 + PH_3$$

Although heating is not mentioned in the question, the other product as per available options seems to be Na_2HPO_4 (oxidation state of P = + 5).

5. (D)

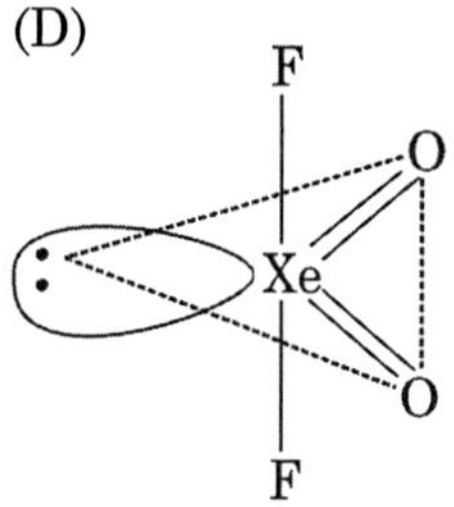

Hybridization: sp^3d
Shape: see-saw

6. (A)

B $\longrightarrow$ Solute; A $\longrightarrow$ Solvent
$W_B = 2.5$ g, $W_A = 100$ g
$\Delta T_b = 2^0$

$$\frac{p^0 - p_s}{p^0} = x_B = \frac{n_B}{n_B + n_A}$$

$$\frac{p^0 - p_s}{p^0} = \frac{n_B}{n_A} \quad \text{Hence, } n_B \ll n_A$$

$$\frac{p^0 - p_s}{p^0} = \frac{n_B}{n_A}$$

$$\frac{760 - P_{soln}}{760} = \frac{2.5/M}{\frac{100}{18} \times \frac{1000}{1000}} = \frac{m \times 18}{1000} \quad \text{...(i)}$$

From boiling point elevation, we have

$$2 = 0.76 \times m$$

or $$m = \frac{2}{0.76} \quad \text{...(ii)}$$

On equating Equns. (i) and (ii), we get
$P_{soln} = 724$ mm

7. (B)

β-Keto acids undergo decarboxylation easily.

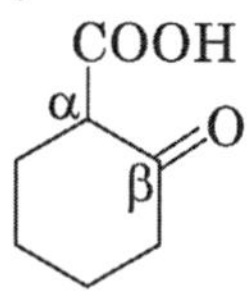

(β-keto acid)

8. (D)

(i) $2C(s) + H_2(g) \longrightarrow H—C \equiv C—H(g)$ $\Delta H = 225$ kJ mol^{-1}

(ii) $2C(s) \longrightarrow 2C(g)$ $\Delta H = 1410$ kJ mol^{-1}

(iii) $H_2(g) \longrightarrow 2H(g)$ $\Delta H = 330$ kJ mol^{-1}

From Equn. (i), we have

$$225 = [2 \times \Delta H_{C(s) \to C(g)} + 1 \times BE_{H-H}] - [2 \times BE_{C-H} + 1 \times BE_{C\equiv C}]$$

or $$225 = [1410 + 1 \times 330] - [2 \times 350 + 1 \times BE_{C\equiv C}]$$

or $$225 = [1410 + 330] - [700 + BE_{C\equiv C}]$$

or $$225 = 1740 - 700 - BE_{C\equiv C}$$

or $$225 = 1040 - BE_{C\equiv C}$$

or $$BE_{C\equiv C} = 1040 - 225 = 815 \text{ kJ mol}^{-1}$$

9. (A)

10. (C)

Sol. 9–10

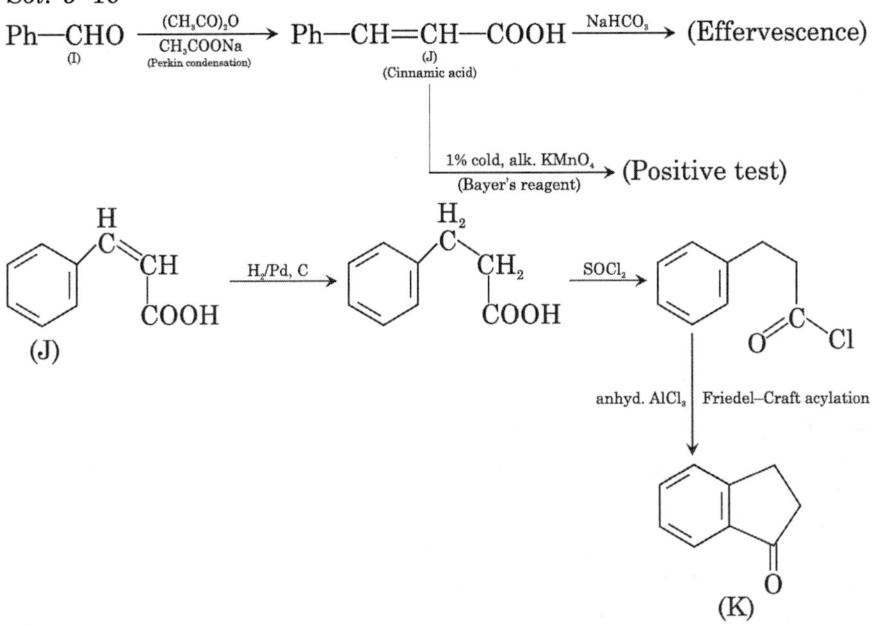

11. (D)

At anode: $M(s) + 2X^-(aq) \longrightarrow MX_2(aq) + 2e^-$

At cathode: $M^{+2}(aq) + 2e^- \longrightarrow M(s)$

n factor of the cell reaction is 2.

$\Delta G = -nFE_{cell} = -2 \times 96500 \times 0.059 = -11387$ J/mole $= -11.387$ KJ/mole $= -11.4$ KJ/mole

12. (B)

$M|M^+(\text{sat.})\|M^{2+}(0.001M)$

$$E_{cell} = \frac{-0.059}{n}\log\frac{[M^{2+}]_a}{[M^{2+}]_c}$$

$$0.059 = \frac{0.059}{2}\log\frac{[0.001]}{[M^{2+}]_a}$$

$[M^{2+}]_a = 10^{-5} = S$ (solubility of salt in saturated solution)

$$\underset{(S)}{MX_2} = \underset{(S)}{M^{2+}} + \underset{(2S)}{2X^-(aq)}$$

$K_{sp} = 4S^3 = 4 \times (10^{-5})^3 = 4 \times 10^{-15}$

13. (A)

$$\underset{\text{(Bleaching powder)}}{Ca(OCl)Cl} \longrightarrow Ca^{2+} + OCl^- + Cl^-$$

$$\underset{\text{(oxoacid)}}{HOCl} \longrightarrow H^+ + OCl^-$$

$$2HOCl \xrightarrow{\Delta} H_2O + Cl_2O$$

14. (C)

$$N_1V_1 = N_2V_2$$

$N_1 \times 25 = 0.25 \times 48$

Or $N_1 = \frac{0.25 \times 48}{25}$

Or $N_1 = 0.48M$

Hence, molarity = 0.48/2 = 0.24

15. (A, C, D)

$T_1 = T_2$ since the process is isothermal.

The work done in adiabatic process is less than in isothermal process because area covered by isothermal curve is more than the area covered by the adiabatic curve. In an adiabatic process expansion occurs by using internal energy and hence it decreases while in an isothermal process temperature remains constant and as a result there is no change in internal energy.

16. (A, C, D)

$$K_3\left[\overset{+3}{Fe}(CN)_6\right] + KI(\text{excess}) \longrightarrow K_4\left[\overset{+2}{Fe}(CN)_6\right] + \underset{\text{Brownish yellow solution}}{KI_3} \quad \text{(redox reaction)}$$

$$\underset{\text{(Brownish yellow filtrate)}}{I_3^-} + 2Na_2S_2O_3 \longrightarrow \underset{\text{Clear solution}}{Na_2S_4O_6} + 2NaI + I^-$$

$$K_4[Fe(CN)_6] + ZnSO_4 \longrightarrow \underset{\text{(White ppt.)}}{K_2Zn_3[Fe(CN)_6]_2} \xrightarrow{NaOH} \underset{\text{Soluble}}{Na_2[Zn(OH)_4]}$$

17. (A, C, D)

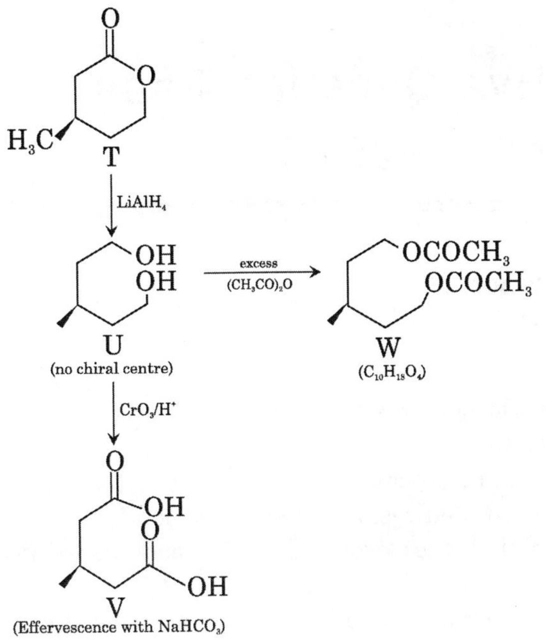

18. (A, B, C)

Converting all the structures in the Fischer projection, we have

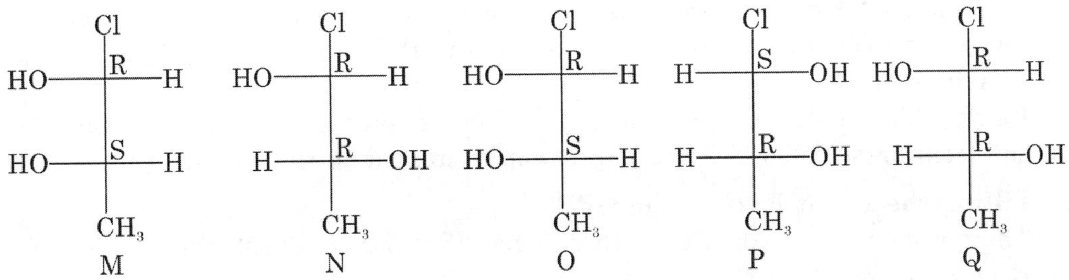

M and N are diastereoisomers. M and O are identical. M and P are enantiomers. M and Q are diastereoisomers.

19. (B, D)

Diamond is harder than graphite. Graphite is good conductor of electricity than diamond. Diamond is better thermal conductor than graphite. In graphite, the C—C bond acquires double bond character and hence higher bond order than in diamond.

20. (A, C)

Under mild conditions of temperature, the extent of physisorption decreases with increase in temperature and hence the graph (I) is correct. Since the initial temperature provided meets the necessary activation energy needed for chemical bond formation for chemisorption, the graph (II) is correct. The graph (IV) is correct for chemisorption since the plot is now of potential energy *vs* distance of molecule from surface.

I.I.T. JEE Objective Question Paper (Advanced)—2013

Paper–1

INSTRUCTIONS

A. General

1. This booklet is your Question Paper. Do not break the seals of this booklet before being instructed to do so by the invigilators.
2. Blank papers, clipboards, log tables, slide rules, calculators, cameras, cellular phones, pagers, and electronic gadgets are not allowed inside the examination hall.
3. Write your name and roll number in the space provided on the back cover of this booklet.
4. Answers to the questions and personal details are to be filled on a two-part carbon-less paper, which is provided separately. These parts should only be separated at the end of the examination when instructed by the invigilator. The upper sheet is a machine-gradable Objective Response Sheet (ORS) which will be ratained by the invigilator. You will be allowed to take away the bottom sheet at the end of the examination.
5. Using a black ball point pen, darken the bubbles on the upper original sheet. Apply sufficient pressure so that the impression is created on the bottom sheet.

B. Filling the Right Part of the ORS

The question paper consists of three parts (Physics, Chemistry and Mathematics). Each part consists of three sections.

Section 1 constrains 10 multiple choice question. Each question has four choices (A), (B), (C) and (D) out of which only one is correct.

Section 2 contains 5 multiple choice questions. Each question has four choices (A), (B), (C) and (D) out of which one or more are correct.

Section 3 contains 5 question. The answer to each question is a single-digit integer, ranging from 0 to 9 (both inclusive).

C. Marking Scheme

For each question in **Section 1**, you will be awarded 2 marks if you darken the bubble corresponding to the correct answer and zero mark if no bubbles are darkened. No negative marks will be awarded for incorrect answers in this section.

For each question in **Section 2**, you will be awarded 4 marks if you darken all the bubble(s) corresponding to only the correct answer(s) and zero mark if no bubbles are darkened. In all other cases, minus one (– 1) mark will be awarded.

For each question in **Section 3**, you will be awarded 4 marks if you darken the bubble corresponding to only the correct answer and zero mark if no bubbles are darkened. In all other cases, minus one (– 1) mark will be awarded.

SECTION – I
(Only One Option Correct Type)

This section contains 10 multiple choice questions. Each question has four choices (A), (B), (C) and (D) out of which only one is correct.

1. In the reaction,

   $P + Q \longrightarrow R + S$

   the time taken for 75% reaction of P is twice the time taken for 50% reaction of P. The concentration of Q varies with reaction time as shown in the figure. The overall order of the reaction is

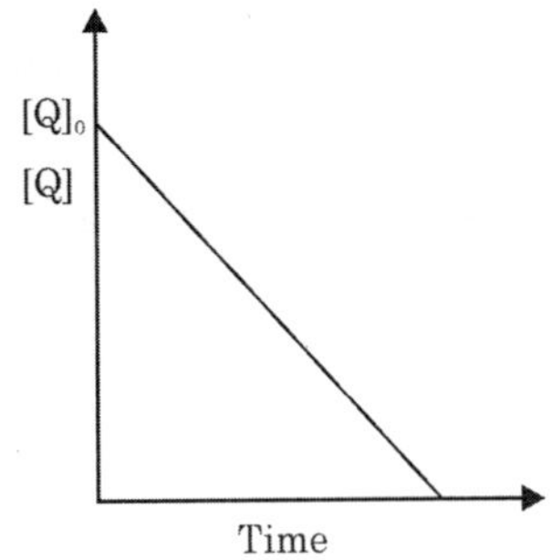

   (A) 2 (B) 3 (C) 0 (D) 1

2. Consider the following complex ions, P, Q and R

   $P = [FeF_6]^{3-}$, $Q = [V(H_2O)_6]^{2+}$ and $R = [Fe(H_2O)_6]^{2+}$

   The correct order of the complex ions, according to their spin-only magnetic moment values (in B.M.) is

   (A) $R < Q < P$ (B) $Q < R < P$ (C) $R < P < Q$ (D) $Q < P < R$

3. The arrangement of X^- ions around A^+ ion in solid AX is given in the figure (not drawn to scale). If the radius of X^- is 250 pm, the radius of A^+ is

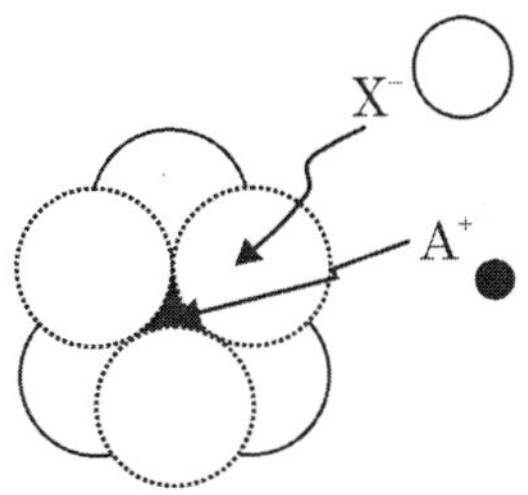

   (A) 104 pm (B) 125 pm (C) 183 pm (D) 57 pm

4. Concentrated nitric acid, upon long standing, turns yellow-brown due to the formation of
   (A) NO (B) NO_2 (C) N_2O (D) N_2O_4
5. The compound that does not liberate CO_2, on treatment with aqueous sodium bicarbonate solution, is
   (A) Benzoic acid (B) Benzenesulphonic acid
   (C) Salicylic acid (D) Carbolic acid (phenol)
6. Sulfide ores are common for the metals
   (A) Ag, Cu and Pb (B) Ag, Cu and Sn (C) Ag, Mg and Pb (D) Al, Cu and Pb
7. Methylene blue, from its aqueous solution, is adsorbed on activated charcoal at 25°C. For this process, the correct statement is
   (A) The adsorption requires activation at 25°C.
   (B) The adsorption is accompanied by a decrease in enthalpy.
   (C) The adsorption increases with increase of temperature.
   (D) The adsorption is irreversible.
8. KI in acetone, undergoes S_{N^2} reaction with each of P, Q, R and S. The rates of the reaction vary as

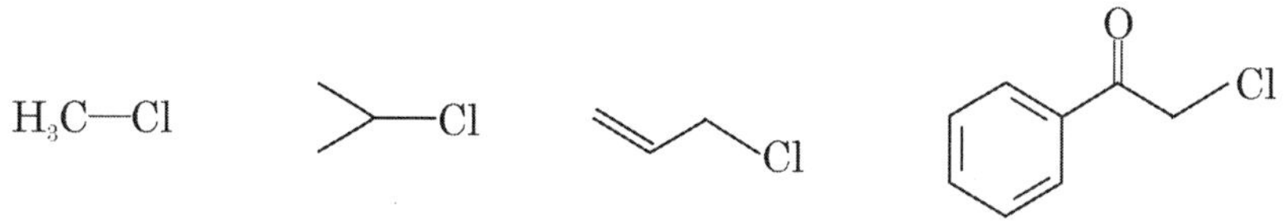

P Q R S

   (A) P > Q > R > S (B) S > P > R > Q (C) P > R > Q > S (D) R > P > S > Q
9. The standard enthalpies of formation of $CO_2(g)$, $H_2O(l)$ and glucose(*s*) at 25°C are – 400 kJ/mol, – 300 kJ/mol, respectively. The standard enthalpy of combusion per gram of glucose at 25°C is
   (A) + 2900 kJ (B) – 2900 kJ (C) – 16.11 kJ (D) + 16.11 kJ
10. Upon treatment with ammoniacal H_2S, the metal ion that precipitates as a sulfide is
   (A) Fe(III) (B) Al(III) (C) Mg(II) (D) Zn(II)

SECTION – II
(One or More Options Correct Type)

This section contains 5 multiple choice questions. Each question has four choices (A), (B), (C) and (D) out of which one or more are correct.

11. The initial rate of hydrolysis of methyl acetate (1*M*) by a weak acid (HA, 1*M*) is 1/100th of that of a strong acid (HX, 1*M*), at 25°C. The K_a of HA is
   (A) 1×10^{-4} (B) 1×10^{-5} (C) 1×10^{-6} (D) 1×10^{-3}
12. The hyperconjugative stabilities of *tert*-butyl cation and 2-butene, respectively, are due to

(A) $\sigma \rightarrow p$ (empty) and $\sigma \rightarrow \pi^*$ electron delocalisations.

(B) $\sigma \rightarrow \sigma^*$ and $\sigma \rightarrow \pi$ electron delocalisations.

(C) $\sigma \rightarrow p$ (filled) and $\sigma \rightarrow \pi$ electron delocalisations.

(D) p (filled) $\rightarrow \sigma^*$ and $\sigma \rightarrow \pi^*$ electron delocalisations.

13. The pair(s) of coordination complexes/ions exhibiting the same kind of isomerism is(are)

(A) $[Cr(NH_3)_5Cl]Cl_2$ and $[Cr(NH_3)_4Cl_2]Cl$ (B) $[Co(NH_3)_4Cl_2]^+$ and $[Pt(NH_3)_2(H_2O)Cl]^+$

(C) $[CoBr_2Cl_2]^{2-}$ and $[PtBr_2Cl_2]^{2-}$ (D) $[Pt(NH_3)_3(NO_3)]Cl$ and $[Pt(NH_3)_3Cl]Br$

14. Among P, Q, R and S, the aromatic compound(s) is/are

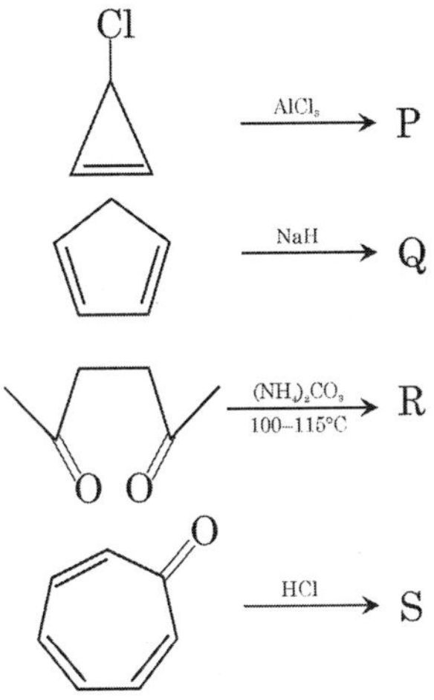

(A) P (B) Q (C) R (D) S

15. Benzene and naphthalene form an ideal solution at room temperature. For this process, the true statement(s) is(are)

(A) ΔG is positive (B) ΔS_{system} is positive (C) $\Delta S_{surroundings} = 0$ (D) $\Delta H = 0$

SECTION – III
(Integer Value Correct Type)

This section contains 5 questions. The answer to each of the questions is a single digit integer, ranging from 0 to 9 (both inclusive).

16. The atomic masses of He and Ne are 4 and 20 amu, respectively. The value of the de Broglie wavelength of He gas at – 73°C is "M" times that of the de Broglie wavelength of Ne at 727°C. M is

17. $EDTA^{4-}$ is ethylenediaminetetraacetate ion. The total number of N—Co—O bond angles in $[Co(EDTA)]^{1-}$ complex ion is

18. The total number of carboxylic acid groups in the product P is

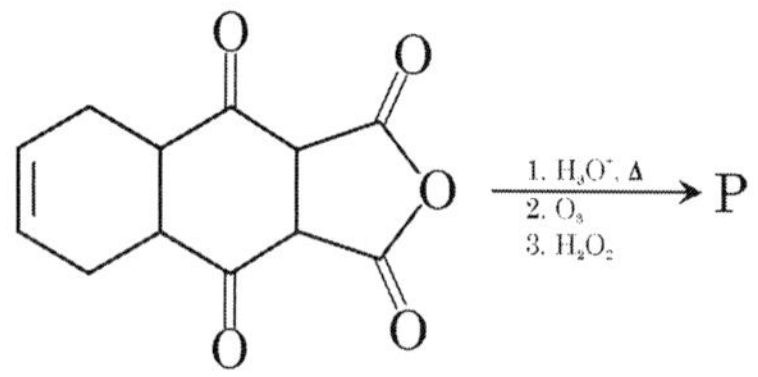

19. A tetrapeptide has —COOH group on alanine. This produces glycine (Gly), valine (Val), phenyl alanine (Phe) and alanine (Ala), on complete hydrolysis. For this tetrapeptide, the number of possible sequences (primary structures) with —NH_2 group attached to a chiral center is

20. The total number of lone-pairs of electrons in melamine is

ANSWERS

1. D	2. B	3. A	4. B	5. D	6. A
7. B	8. B	9. C	10. D	11. A	12. A
13. B, D	14. A, B, C, D	15. B, C, D	16. 5	17. 8	18. 2
19. 4	20. 6				

Model Solutions—I.I.T. JEE (Advanced) 2013

Paper–1

1. (D)
Overall order of reaction can be decided by the data given: $t_{75\%} = 2t_{50\%}$
Hence it is a first order reaction with respect to P.
From graph [Q] is linearly decreasing with time, i.e. the order of reaction with respect to Q is zero and the rate expression is: $r = k[\text{P}]^1[\text{Q}]^0$.

2. (B)
P = Fe^{3+} (no. of unpaired electrons = 5)
Q = V^{2+} (no. of unpaired electrons = 3)
R = Fe^{2+} (no. of unpaired electrons = 4)
Since all ligands are weak field, the number of unpaired electrons remains same in the complex ion. Magnetic moment is calculated using the equation: $\mu = \sqrt{n(n+2)}$ B.M.

3. (A)
According to the given figure, A^+ is present in the octahedral void of X^-. The limiting radius in octahedral void is related to the radius of sphere as
$r_{\text{void}} = 0.414 r_{\text{sphere}}$
$r_{A^+} = 0.414 r_{X^-} = 0.414 \times 250$ pm = 103.5 ≈ 104 pm

4. (B)
$4HNO_3 \longrightarrow 2H_2O + 4NO_2 + O_2$
NO_2 remains dissolved in nitric acid colouring it yellow or even red at higher temperature.

5. (D)
pK_a of PhOH (carbolic acid) is 9.98 and that of carbonic acid (H_2CO_3) is 6.63. Hence, phenol does not produce effervescence with HCO_3^- ion.

6. (A)
Sulfide ore of Ag → Argentite (Ag_2S), Pb → Galena (PbS), Cu → Copper glance (Cu_2S)

7. (B)
Adsorption of methylene blue on activated charcoal is physical adsorption. Hence, it is characterised by decrease in enthalpy.

8.(B)

Relative reactivity for S_N2 reaction in the given structures is

Substrate	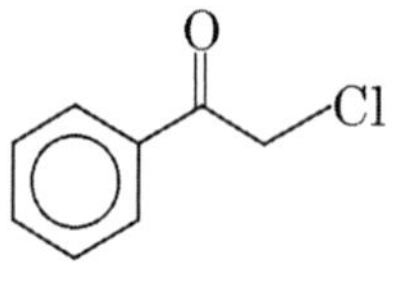 (S)	CH_3Cl (P)	$H_2C{=}CH{-}CH_2Cl$ (R)	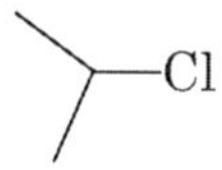 (Q)
Relative rates towards S_N2	100000	200	79	0.02

9. (C)

Combustion of glucose

$C_6H_{12}O_6 + 6O_2 \longrightarrow 6CO_2 + 6H_2O$

$\Delta H_{combustion} = (6 \times \Delta H_f CO_2 + 6 \times \Delta H_f H_2O) - \Delta H_f C_6H_{12}O_6$

$= (6 \times -400 + 6 \times -300) - (-1300) = -2900$ kJ/mol $= -2900/180$ kJ/g $= -16.11$ kJ/g

10. (D)

Among Fe^{3+}, Al^{3+}, Mg^{2+}, Zn^{2+} ions only Zn^{2+} ion is precipitated with ammonical H_2S as ZnS.

11. (A)

Rate in weak acid $= \frac{1}{100}$ (rate in strong acid)

Hence $[H^+]_{weak\ acid} = \frac{1}{100}[H^+]_{strong\ acid}$

or $[H^+]_{weak\ acid} = \frac{1}{100}M = 10^{-2}M$

Hence, $C\alpha = 10^{-2}$ and $K_a = 10^{-4}$

12. (A)

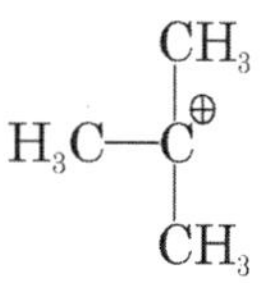

$\sigma - p$(empty)

$H_3C{-}CH{=}CH{-}CH_3$

$\sigma - \pi^*$(antibonding)

13. (B, D)

$[Co(NH_3)_4Cl_2]^+$ (an octahedral complex) and $[Pt(NH_3)_2(H_2O)Cl]^+$ (a square planar complex) will exhibit geometrical isomerism.

$[Pt(NH_3)_3(NO)_3]Cl$ and $[Pt(NH_3)_3Cl]Br$ will show ionization isomerism.

14. (A, B, C, D)

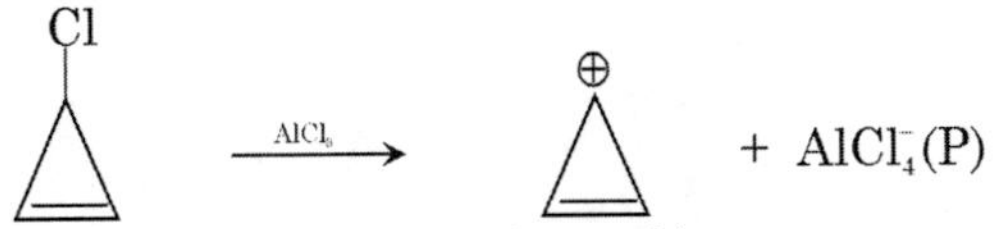

(aromatic)

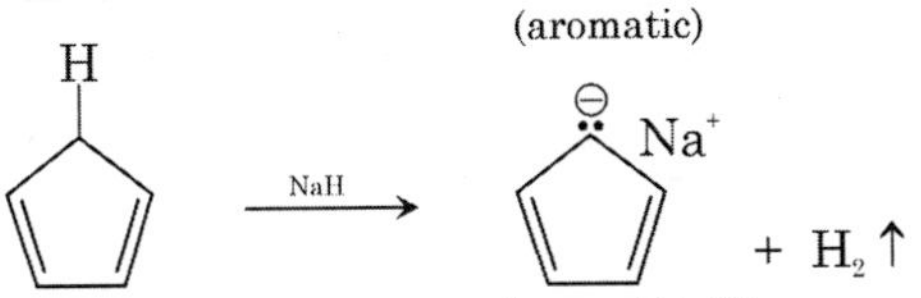

(aromatic) (Q)

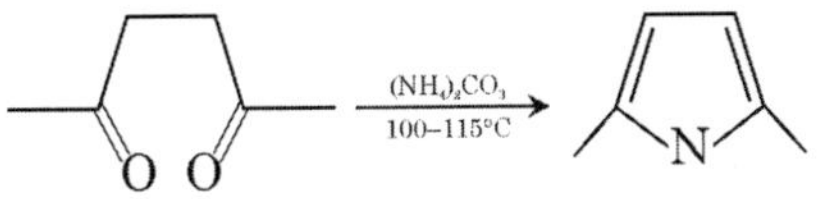

(R) aromatic

Mechanism

$$(NH_4)_2CO_3 \xrightarrow{\Delta} 2NH_3 + CO_2 + H_2O$$

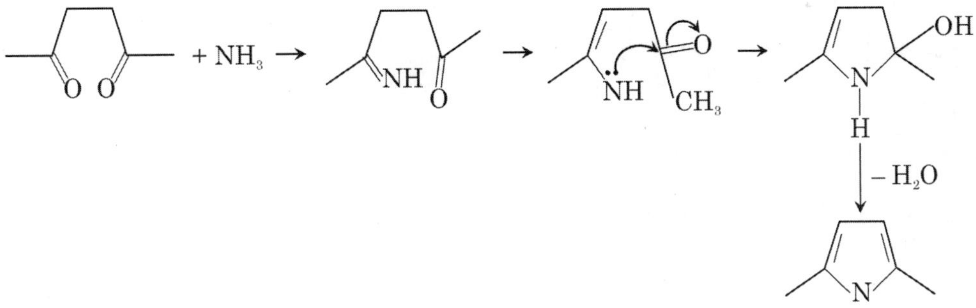

(aromatic)

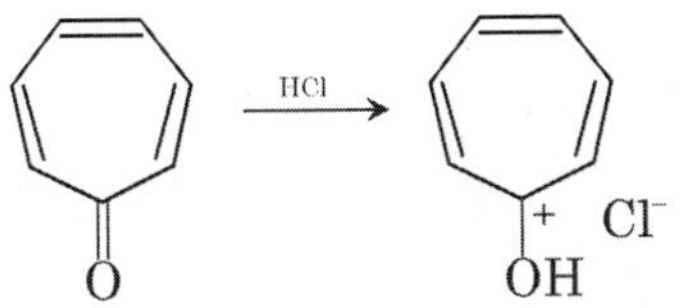

(S) aromatic

15. (B, C, D)

For ideal solution, $\Delta S_{system} > 0$, $\Delta S_{surrounding} = 0$, $\Delta H_{mixing} = 0$

16. (5)

$$\lambda = \frac{h}{mV} = \frac{h}{\sqrt{2M\,K.E.}} \quad \text{(since K.E. } \propto T)$$

or $\lambda \propto \dfrac{1}{\sqrt{MT}}$

For two geses,

$$\frac{\lambda_{He}}{\lambda_{Ne}} = \sqrt{\frac{M_{Ne}T_{Ne}}{M_{He}T_{He}}} = \sqrt{\frac{20}{4} \times \frac{1000}{200}} = 5$$

17. (8)

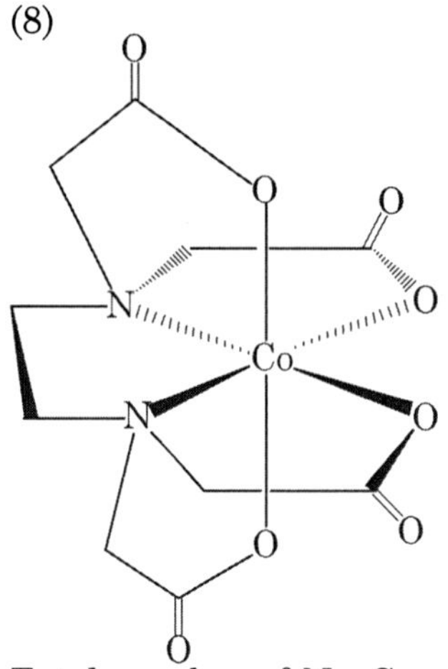

Total number of N—Co—O bond angles is 8.

18. (2)

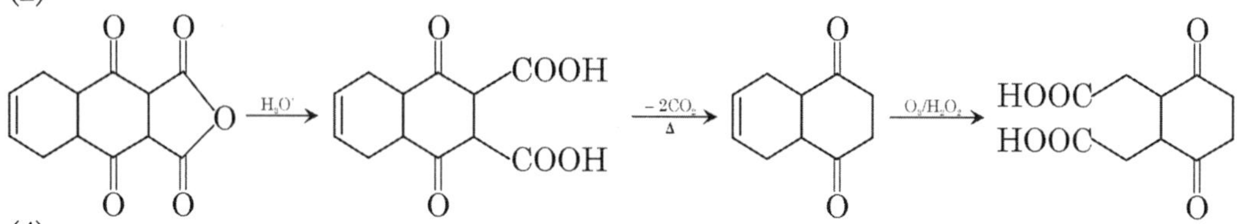

19. (4)

Because —COOH group of tetrapeptide is intact on alanine, its NH_2 must be participating in condensation.

Hence alanine is at one terminus, – – – A.

To fill the 3 blanks, the possible options are

(i) When NH_2 group attached to non-chiral carbon

G	V	P
G	P	V

(ii) When NH_2 group attached to chiral carbon

V	G	P	P	V	G
V	P	G	P	G	V

where, Glycine (G)
Valine (V)
Phenyl alanine (P)
Alanine (A)

Hence the number of possible sequence is 4.

20. (6 lone pairs)

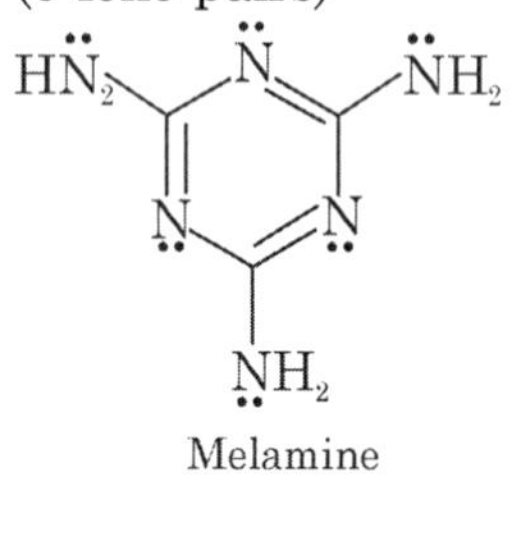

Melamine

I.I.T. JEE Objective Question Paper (Advanced)—2013

Paper–2

INSTRUCTIONS

A. General

1. This booklet is your Question Paper. Do not break the seals of this booklet before being instructed to do so by the invigilators.
2. The question paper CODE is printed on the right hand top corner of this sheet and on the back page of this booklet.
3. Blank spaces and blank pages are provided in the question paper for your rough work. No additional sheets will be provided for rough work.
4. Blank papers, clipboards, log tables, slide rules, calculators, cameras, cellular phones, pagers and electronic gadgets are not allowed inside the examination hall.
5. Write your name and roll number in the space provided on the back cover of this booklet.
6. Answers to the questions and personal details are to be filled on a two-part carbon-less paper, which is provided separately. These parts should only be separated at the end of the examination when instructed by the invigilator. The upper sheet is a machine-gradable Objective Response Sheet (ORS) which will be ratained by the invigilator. You will be allowed to take away the bottom sheet at the end of the examination.
7. Using a black ball point pen, darken the bubbles on the upper original sheet. Apply sufficient pressure so that the impression is created on the bottom sheet.
8. Do not tamper with/mutilate the ORS or the booklet.
9. On breaking the seals of the booklet check that it contains 44 pages and all the 60 questions and corresponding answer choices are legible. Read carefully the instructions printed at the beginning of each section.

B. Filling the Right Part of the ORS

10. The ORS has code printed on its left and right parts.
11. Check that the same code is printed on the ORS (on both sheets) is the same as that on this booklet and put your signature affirming that you have verified this.
12. If the codes do not match, ask for a change of the booklet.

SECTION – I (One or More Options Correct Type)

This section contains 8 multiple choice questions. Each question has four choices (A), (B), (C) and (D) out of which one or more are correct.

1. The K_{sp} of Ag_2CrO_4 is 1.1×10^{-12} at 298 K. The solubility (in mol/L) of Ag_2CrO_4 in a $0.1M$ $AgNO_3$ solution is

   (A) 1.1×10^{-11} (B) 1.1×10^{-10} (C) 1.1×10^{-12} (D) 1.1×10^{-9}

2. In the following reaction, the product(s) formed is(are)

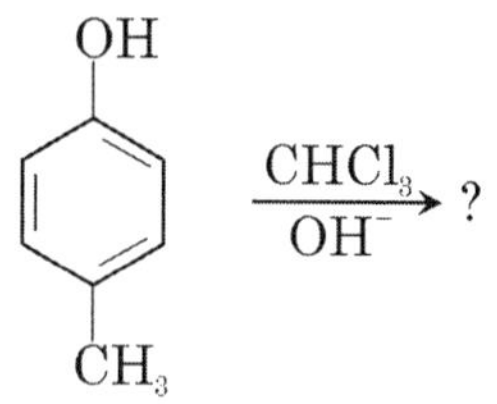

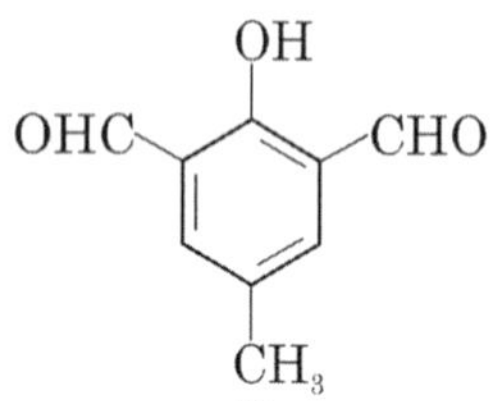

P

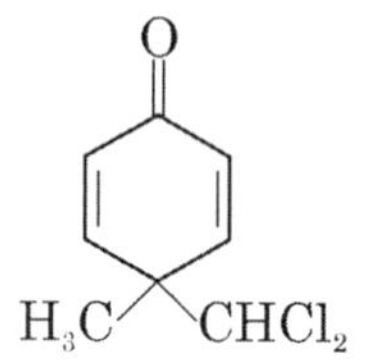

Q

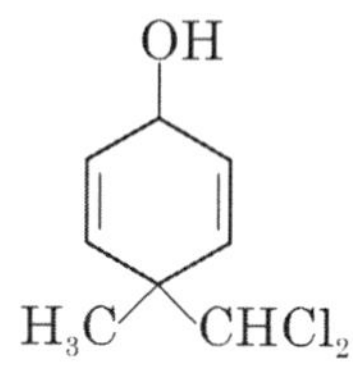

R

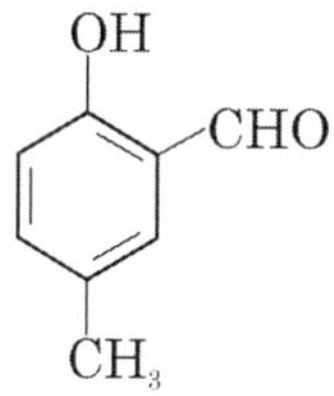

S

   (A) P (major) (B) Q (minor) (C) R (minor) (D) S (major)

3. The major product(s) of the following reaction is (are)

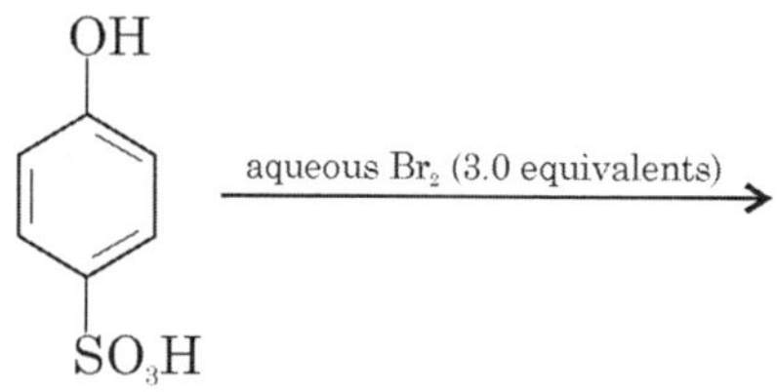

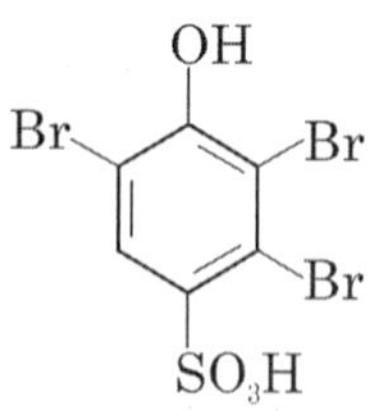

P

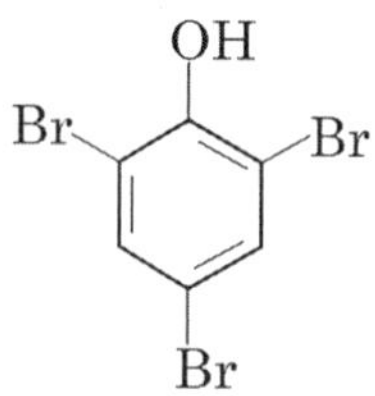

Q

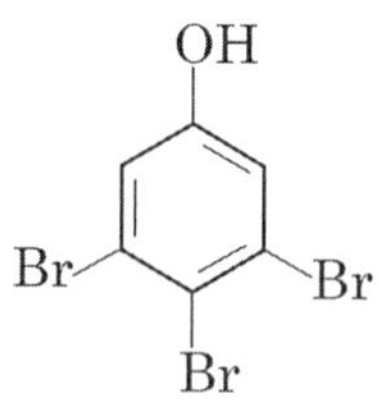

R

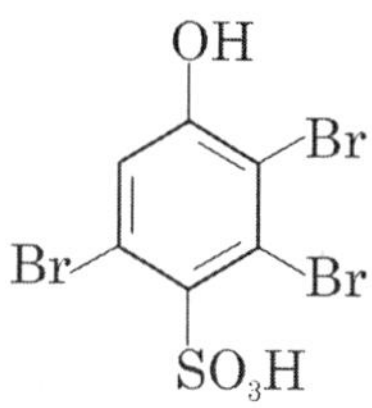

S

   (A) P (B) Q (C) R (D) S

4. After completion of the reactions (I and II), the organic compound(s) in the reaction mixtures is(are)

Reaction I : 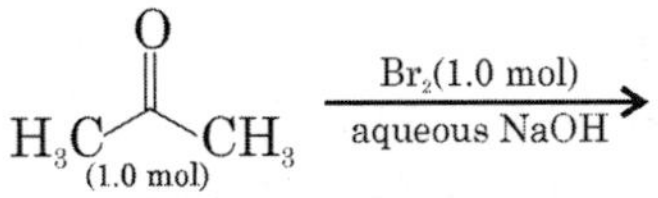

Reaction I : 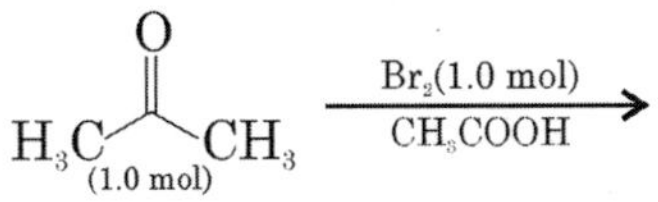

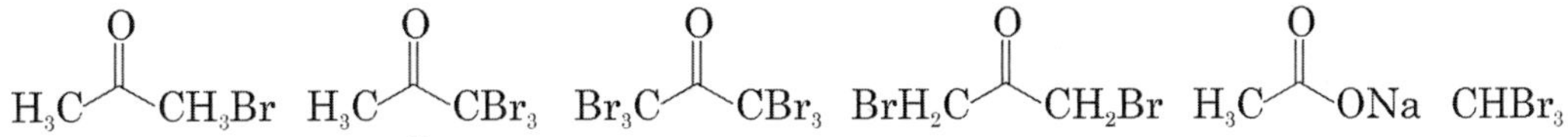

P Q R S T U

(A) Reaction I: P and Reaction II: P
(B) Reaction I: U, acetone and Reaction II: Q, acetone
(C) Reaction I: T, U, acetone and Reaction II: P
(D) Reaction I: R, acetone and Reaction II: S

5. The correct statement(s) about O_3 is(are)
   (A) O—O bond lengths are equal.
   (B) Thermal decomposition of O_3 is endothermic.
   (C) O_3 is diamagnetic in nature.
   (D) O_3 has a bent structure.

6. In the nuclear transmutation,
   $^{9}_{4}Be + X \longrightarrow {}^{8}_{4}Be + Y$
   (X, Y) is(are)
   (A) (γ, n) (B) (p, D) (C) (n, D) (D) (γ, p)

7. The carbon-based reduction method is not used for the extraction of
   (A) tin from SnO_2 (B) iron from Fe_2O_3
   (C) aluminium from Al_2O_3(D)magnesium from $MgCO_3 \cdot CaCO_3$

8. The themal dissociation equilibrium of $CaCO_3(s)$ is studied under different conditions.
   $CaCO_3(s) \rightleftharpoons CaO(s) + CO_2(g)$
   For this equilibrium, the correct statement(s) is(are)
   (A) ΔH is dependent on T
   (B) K is independent of the initial amount of $CaCO_3$
   (C) K is dependent on the pressure of CO_2 at a given T
   (D) ΔH is independent of the catelyst, if any.

SECTION – II (Paragrap Type)

This section contains 4 paragraphs each describing theory, experiment, data, etc. Eight questions relate to four paragraphs with two questions on each paragraph. Each question of a paragraph has only one correct answer among the four choices (A), (B), (C) and (D).

Paragraph for Questions 9 and 10

An aqueous solution of a mixture of two inorganic salts, when treated with dilute HCl, gave a precipitate (P) and a filtrate (Q). The precipitate P was found to dissolve in hot water. The filtrate (Q) remained unchanged, when treated with H_2S in a dilute mineral acid medium. However, it gave a precipitate (R) with H_2S in an ammoniacal medium. The precipitate R gave a coloured solution (S), when treated with H_2O_2 in an aqueous NaOH medium.

9. The precipitate P contains

(A) Pb^{2+} (B) Hg_2^{2+} (C) Ag^+ (D) Hg^{2+}

10. The coloured solution S contains

(A) $Fe_2(SO_4)_3$ (B) $CuSO_4$ (C) $ZnSO_4$ (D) Na_2CrO_4

Paragraph for Questions 11 and 12

P and Q are isomers of dicarboxylic acid $C_4H_4O_4$. Both decolorize Br_2/H_2O. On heating, P forms the cyclic anhydride.
Upon treatment with dilute alkaline $KMnO_4$, P as well as Q could produce one or more than one from S, T and U.

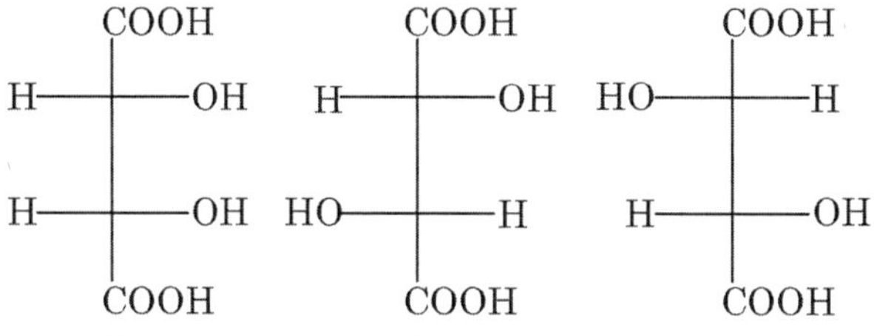

S T U

11. Compounds formed form P and Q are, respectively

(A) Optically active S and optically active pair (T, U)
(B) Optically inactive S and optically inactive pair (T, U)
(C) Optically active pair (T, U) and optically active S
(D) Optically inactive pair (T, U) and optically inactive S.

12. In the following reaction sequences V and W are respectively:

$$Q \xrightarrow[\Delta]{H_2/Ni} V$$

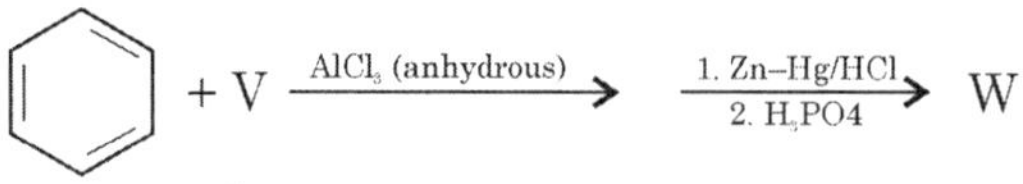

(A) 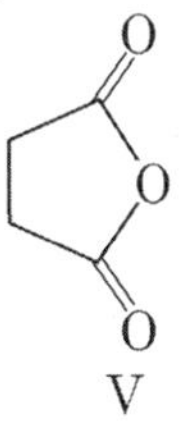

and 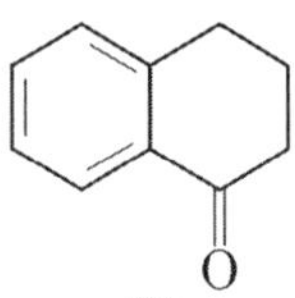

V W

(B) 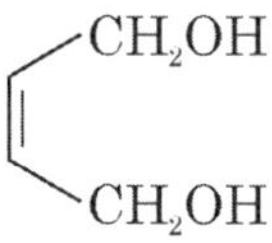

and 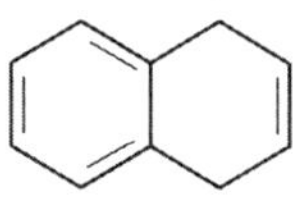

V W

(C) 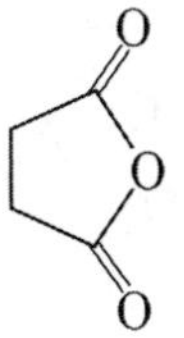 and 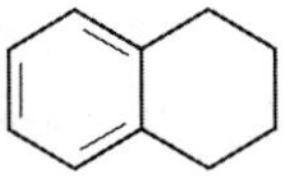

V W

(D) 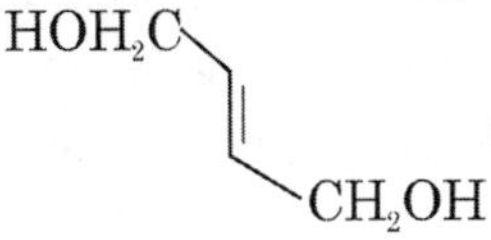 and 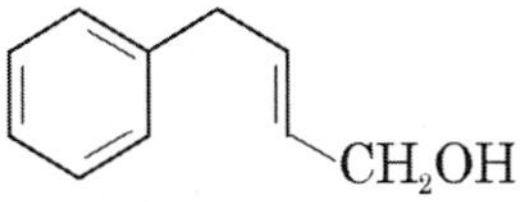

V W

Paragraph for Questions 13 and 14

A fixed mass *m* of a gas is subjected to transformation of states from K to L to M to N and back to K as shown

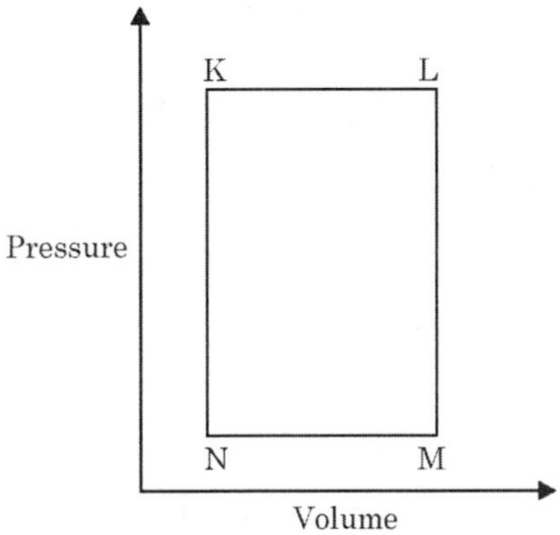

13. The succeeding operations that enable this transformation of states are
    (A) Heating, cooling, heating, cooling
    (B) Cooling, heating, cooling, heating
    (C) Heating, cooling, cooling, heating
    (D) Cooling, heating, heating, cooling.

14. The pair of isochoric processes among the transormation of states is
    (A) K to L and L to M
    (B) L to M and N to K
    (C) L to M and M to N
    (D) M to N and N to K

Paragraph for Questions 15 and 16

The reactions of Cl_2 gas with cold-dilute and hot-concentrated NaOH in water give sodium salts to two (different) oxoacids of chlorine, P and Q, respectively. The Cl_2 gas reacts with SO_2 gas, in presence of charcoal, to give a product R. R reacts with white phosphorus to give a compound S. On hydrolysis, S gives an oxoacid of phosphorus T.

15. P and Q, respectively, are the sodium salts of
    (A) hypochlorous and chloric acids
    (B) hypochlorous and chlorous acids
    (C) chloric and perchloric acids
    (D) chloric and hypochlorous acids

16. R, S and T, respectively, are

(A) SO_2Cl_2, PCl_5 and H_3PO_4 (B) SO_2Cl_2, PCl_3 and H_3PO_3

(C) $SOCl_2$, PCl_3 and H_3PO_2 (D) $SOCl_2$, PCl_5 and H_3PO_4

SECTION – III (Matching List Type)

This section contains 4 multiple choice questions. Each question has matching lists. The codes for the lists have choces (A), (B), (C) and (D) out of which only one is correct.

17. The unbalanced chemical reactions given in List I show missing reagent or condition (?) which are provided in List II. Match List I with List II and select the correct answer using the code given below the lists:

List I	List II
P. $PbO_2 + H_2SO_4 \xrightarrow{?} PbSO_4 + O_2$ + other product	1. NO
Q. $Na_2S_2O_3 + H_2O \xrightarrow{?} NaHSO_4$ + other product	2. I_2
R. $N_2H_4 \xrightarrow{?} N_2$ + other product	3. Warm
S. $XeF_2 \xrightarrow{?} Xe$ + other product	4. Cl_2

Codes:	P	Q	R	S
(A)	4	2	3	1
(B)	3	2	1	4
(C)	1	4	2	3
(D)	3	4	2	1

18. Match the chemical conversions in List I with the appropriate reagent in List II and select the correct answer using the code given below the lists:

List I	List II
P. 	1. (i) $Hg(OAc)_2$; (ii) $NaBH_4$
Q. 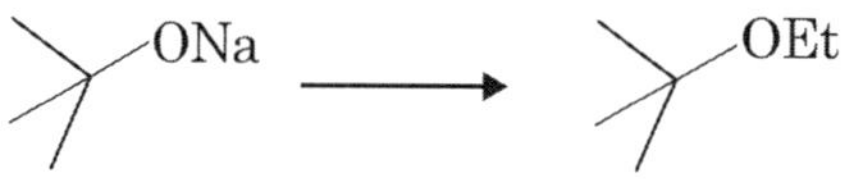	2. NaOEt
R. 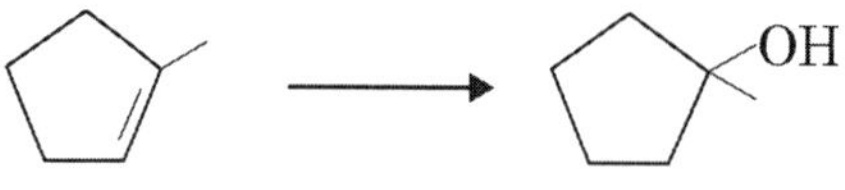	3. EtBr
S. 	4. (i) BH_3; (ii) H_2O_2/NaOH

Codes:	**P**	**Q**	**R**	**S**
(A)	2	3	1	4
(B)	3	2	1	4
(C)	2	3	4	1
(D)	3	2	4	1

19. An aqueous solution of X is added slowly to an aqueous solution of Y as shown in List I. The variation in conductivity of these reactions is given in List II. Match List I with List II and select the correct answer using the code given below the lists:

List I	**List II**
P. $(C_2H_5)_3N$ + CH_3COOH (X, Y)	1. Conductivity decreases and then increases
Q. KI (0.1*M*) + $AgNO_3$(0.01*M*) (X, Y)	2. Conductivity decreases and then does not change much
R. CH_3COOH + KOH (X, Y)	3. Conductivity increases and then does not change much
S. NaOH + HI (X, Y)	4. Conductivity does not change much and then increases

Codes:	**P**	**Q**	**R**	**S**
(A)	3	4	2	1
(B)	4	3	2	1
(C)	2	3	4	1
(D)	1	4	3	2

20. The standard reduction potential data at 25°C is given below.

E^0 (Fe^{3+}, Fe^{2+}) = + 0.77 V;

E^0 (Fe^{2+}, Fe) = – 0.44 V;

E^0 (Cu^{2+}, Cu) = + 0.34 V;

E^0 (Cu^+, Cu) = + 0.52 V;

E^0 ($O_2(g) + 4H^+ + 4e^- \longrightarrow 2H_2O$) = + 1.23 V;

E^0 ($O_2(g) + 2H_2O + 4e^- \longrightarrow 4OH^-$) = + 0.40 V;

E^0 (Cr^{3+}, Cr) = – 0.74 V;

E^0 (Cr^{2+}, Cr) = – 0.91 V

Match E^0 of the rebox pair in List I with the values given in List II and select the correct answer using the code given below the lists:

List I	**List II**
P. E^0 (Fe^{3+}·Fe)	1. – 0.18 V
Q. E^0 ($4H_2O \rightleftharpoons 4H^+ + 4OH^-$)	2. – 0.4 V
R. E^0 ($Cu^{2+} + Cu \longrightarrow 2Cu^+$)	3. – 0.04 V
S. E^0 (Cr^{3+}, Cr^{2+})	4. – 0.83 V

Codes:	**P**	**Q**	**R**	**S**
(A)	4	1	2	3
(B)	2	3	4	1
(C)	1	2	3	4
(D)	3	4	1	2

ANSWERS

1. B	2. B, D	3. B	4. C	5. A, C, D	6. A, B
7. C, D	8. A, B, D	9. A	10. D	11. B	12. A
13. C	14. B	15. A	16. A	17. D	18. A
19. A	20. D				

Model Solutions—I.I.T. JEE (Advanced) 2013

Paper–2

1. (B)

$$Ag_2CrO_4 \rightleftharpoons 2Ag^+ + CrO_4^{2}$$
$$s \qquad 0.1 + 2s \qquad s$$
$$\approx 0.1$$
$$1.1 \times 10^{-12} = (0.1)^2 s$$
$$s = 1.1 \times 10^{-10}$$

2. (B, D)

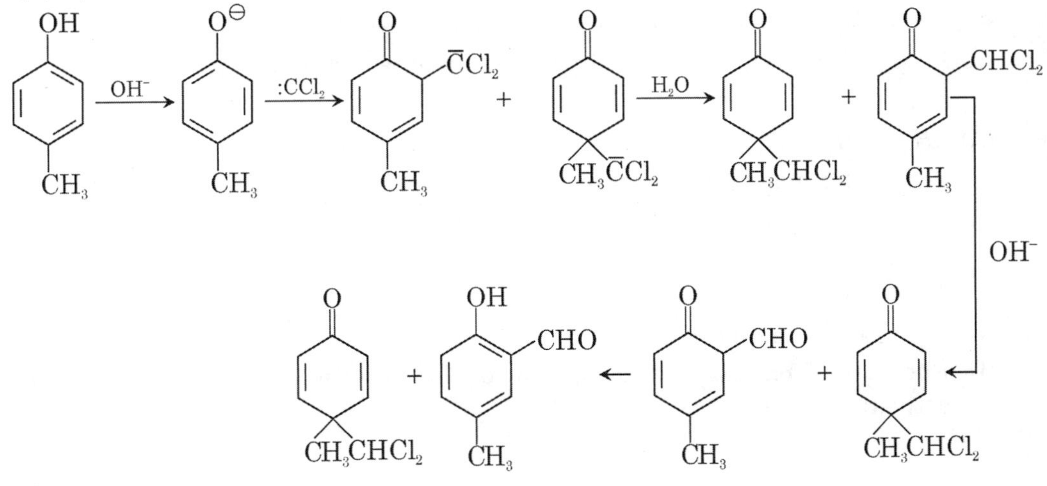

3. (B)

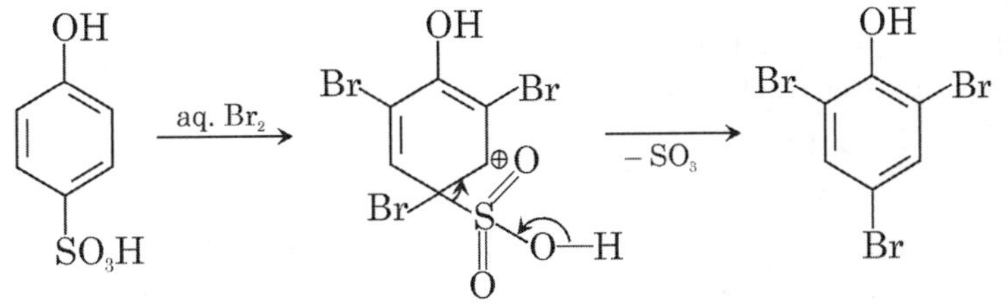

4. (C)

Reaction I: 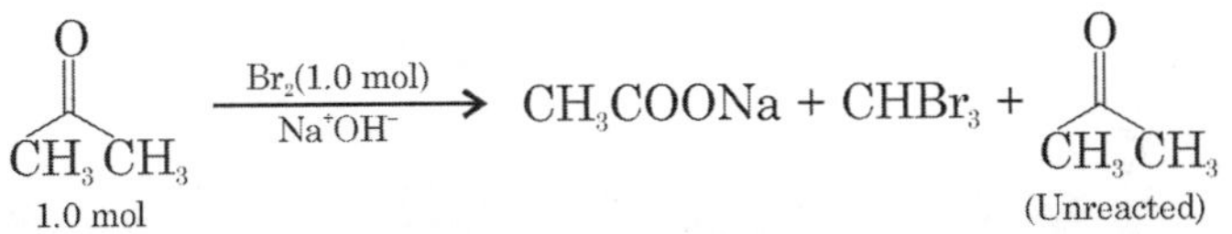

In basic medium complete haloform reaction takes place since the rate of reaction increases with each α-halogenation.

Reaction II: 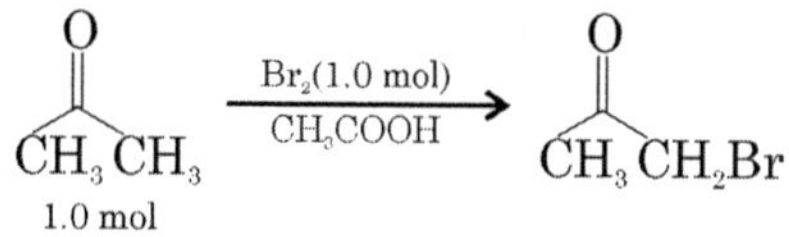

In acidic medium, monohalogenation takes place with 1 mol of halogen.

5. (A, C, D)

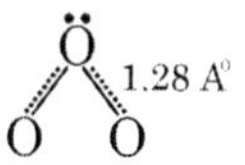

All electrons are paired and hence diamagnetic.

$3O_2 \longrightarrow 2O_3$ $\Delta H = +\ 142$ kJ/mol.

6. (A, B)

${}^{9}_{4}Be + X \longrightarrow {}^{8}_{4}Be + Y$

If X is ${}^{0}_{0}\gamma$ then Y is ${}^{1}_{0}n$

If X is ${}^{1}_{1}p$ then Y is ${}^{2}D$

7. (C, D)

SnO_2 and Fe_2O_3 are reduced by carbon reduction method.

Al_2O_3, $MgCO_3{\cdot}CaCO_3$ are reduced by electrolytic reduction method.

8. (A, B, D)

(A) $\Delta H_2 - \Delta H_1 = C_{P(\gamma \times n)}\,(T_2 - T_1)$

and C_p depends on temperature. Hence enthalpy also depends on temperature.

(B) $CaCO_3(s) \rightleftharpoons CaO(s) + CO_2(g)$ $k_p = (p_{CO_2})_{\text{at equlibrium}}$

For a given reaction.

K_{eq} depends only on temperature.

(C) K_{eq} depends only on temperature.

(D) Enthalpy of reaction is independent of the catalyst. Catalyst generaly changes activation energy.

9. (A)

10. (D)

(9 and 10)

$Pb^{2+} + 2Cl^{-} \longrightarrow PbCl_2$ (white ppt.)
soluble in hot water

$Pb^{2+}/Cr^{3+} \xrightarrow{HCl(aq)} PbCl_2 + CrCl_3$
(P) (Q) White ppt. Filtrate

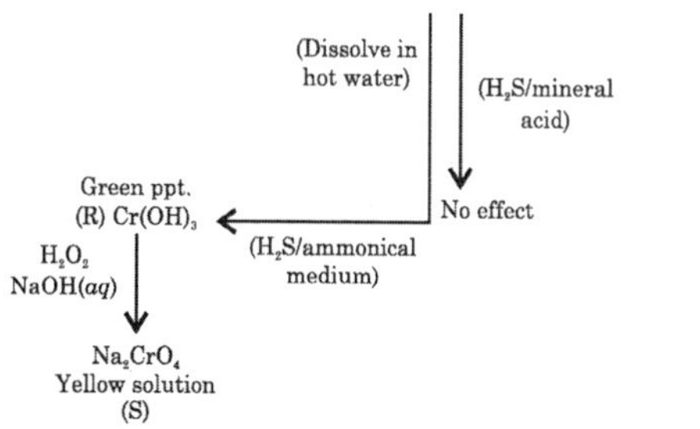

11. (B)

12. (A) (11 and 12)

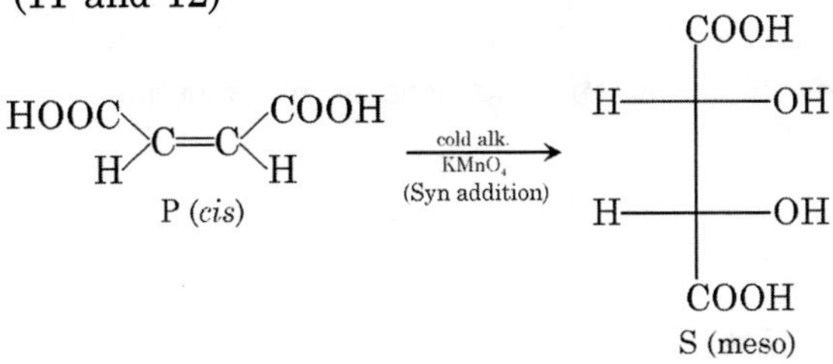

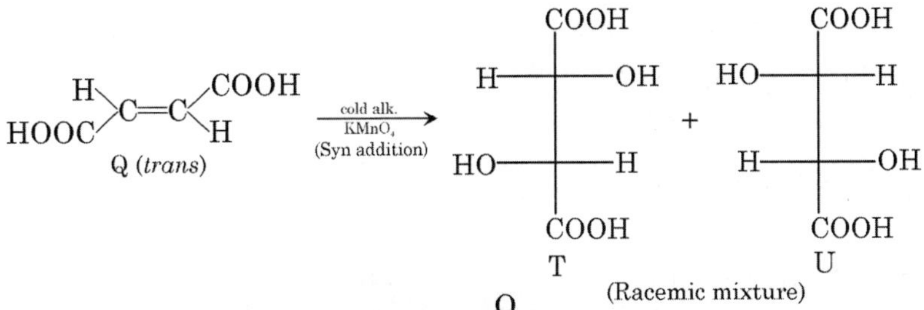

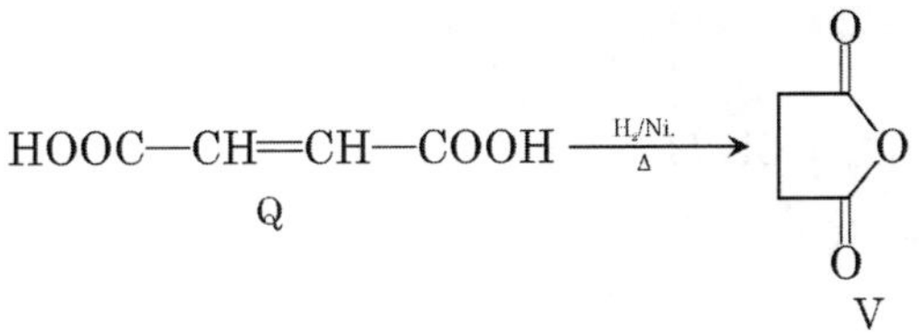

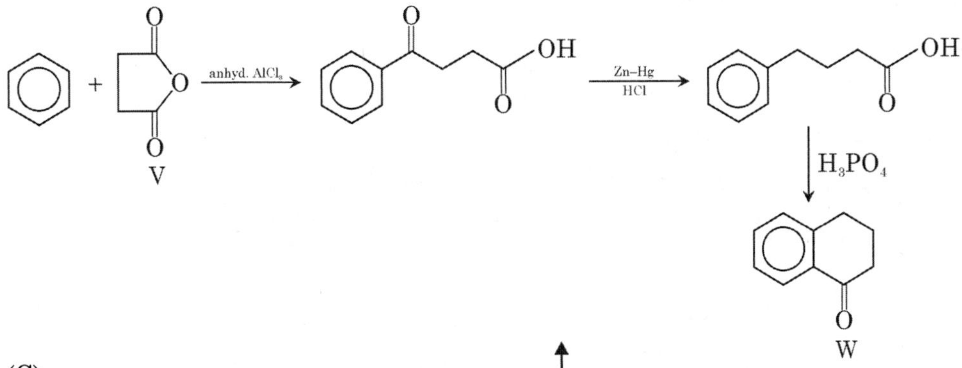

13. (C)

K → L ; $V \uparrow$ at constant P

Hence $T \uparrow$ (Heating)

L → M ; $P \downarrow$ at constant V

Hence $T \downarrow$ (Cooling)

M → N ; $V \downarrow$ at constant P

Hence $T \downarrow$ (Cooling)

N → K ; $P \uparrow$ at constant V

Hence $T \uparrow$ (Heating)

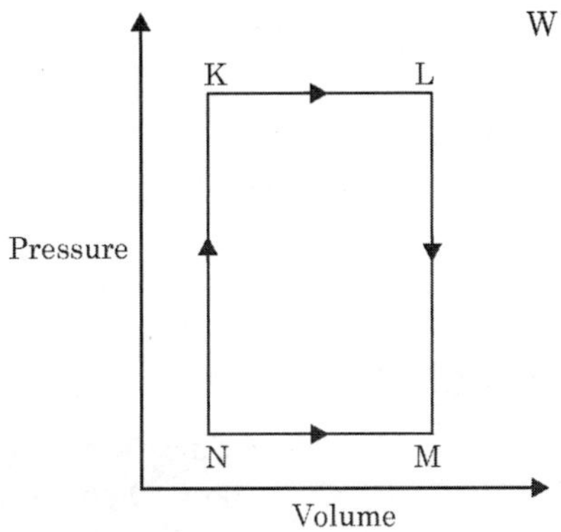

14. (B)

L → M

M → K

Both are having constant volume, therefore these processes are isochoric.

15. (A)

16. (A)

(15 and 16)

Cl_2 + cold dil. NaOH ⟶ NaOCl + NaCl

Cl_2 + hot conc. NaOH ⟶ $NaClO_3$ + NaCl

NaOCl is salt of hypochlorous acid = P.

$NaClO_3$ is salt of chloric acid = Q.

$Cl_2 + SO_2 \xrightarrow{\text{charcoal}} SO_2Cl_2$ (R)

$SO_2Cl_2 + P_4 \longrightarrow PCl_5$ (S) $+ SO_2$

$PCl_5 + H_2O \longrightarrow H_3PO_4$ (T) + HCl

17. (D)

(P) $2PbO_2 + H_2SO_4 \xrightarrow{\text{warm}} 2PbSO_4 + O_2 + 3H_2O$

(Q) $Na_2S_2O_3 + H_2O \xrightarrow{+Cl_2} NaHSO_4 + HCl$

(R) $N_2H_4 \xrightarrow{+I_2} NaHSO_4 + HCl$

(S) $XeF_2 \xrightarrow{+NO} Xe + NOF$

18. (A)

P. 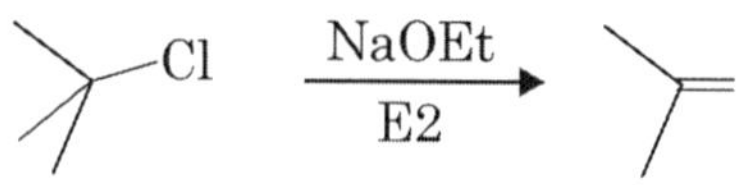

Q. 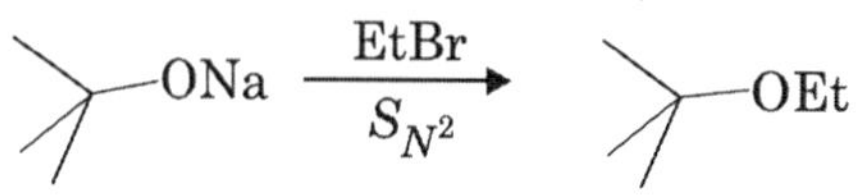

R. 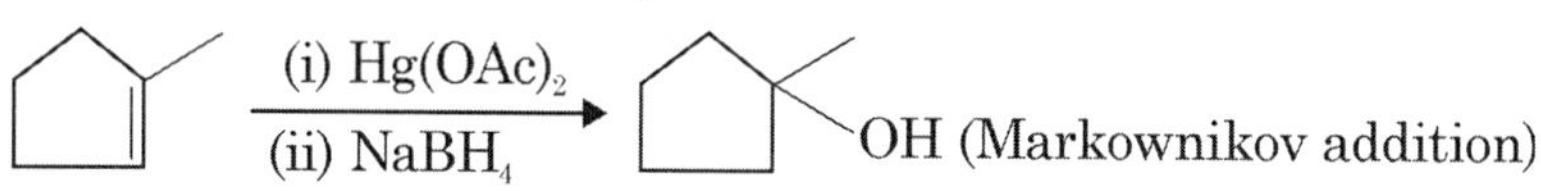

S. 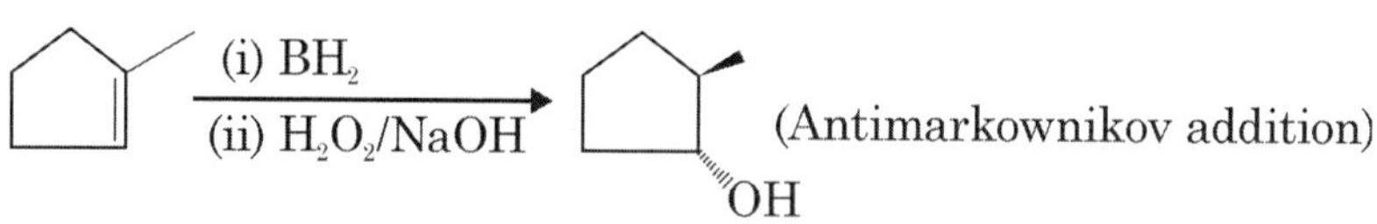

19. (A)

X Y

(P) $(C_2H_5)_3N + CH_3COOH \longrightarrow CH_3COO^-(aq) + (C_2H_5)_3NH^+(aq)$

Since CH_3COOH is a weak acid, its conductivity is already less. On addition of weak base, acid-base reaction takes place and new ions are created. Hence, conductivity increases.

(Q) KI (0.1*M*) + $AgNO_3$ (0.01*M*) $\longrightarrow$ AgI $\downarrow$ + $KNO_3(aq)$

Since the only reaction taking place is precipitation of AgI and in place of Ag^+, K^+ is coming in the solution, conductivity remains nearly constant and then increases.

(R) $CH_3COOH + KOH \longrightarrow CH_3COOH(aq) + H_2O$

$OH^-(aq)$ is getting replaced by CH_3COO^-, which has poorer conductivity. Hence, conductivity dereases and then after the end point, due to common ion effect, no further creation of ions takes place. Hence, conductivity remain nearly same.

(S) NaOH + HI $\longrightarrow$ NaI(aq) + H_2O

Since H^+ is getting replaced by Na^+ conductivity dereases and after end point due to OH^-, conductivity increases.

Hence, answer is: (P) – (3) ; (Q) – (4) ; (R) – (2) ; (S) – (1)

20. (D)

(P) $E^0_{Fe^{3+},Fe}$ $Fe^{3+} \xrightarrow[n=1]{+0.77\text{ V}} Fe^{2+} \xrightarrow[n=2]{-0.44\text{ V}} Fe$; overall $Fe^{3+} \to Fe$: x V, $n = 3$

Hence $1 \times 0.77 + 2 \times (-0.44) = 3 \times x$

or $x = -\dfrac{0.11}{3}\text{ V} \simeq -0.04\text{ V}$

(Q) $4H_2O \rightleftharpoons 4H^+ + 4OH^-$

$$\begin{array}{lr} 2H_2O \longrightarrow O_2 + 4H^+ + 4e^- & -1.23\text{ V} \\ + & \\ O_2 + 2H_2O + 4e^- \longrightarrow 4OH^- & +0.4\text{ V} \\ \hline 4H_2O \rightleftharpoons 4H^+ + 4OH^- & -0.83\text{ V} \end{array}$$

(R) $E^0_{(Cu^{2+}+Cu\to 2Cu^+)}$ $Cu^{2+} \xrightarrow[n=1]{x\text{ V}} Cu^+ \xrightarrow[n=1]{-0.52\text{ V}} Cu$; overall $Cu^{2+} \to Cu$: $+0.34$ V, $n = 2$

$x \times 1 + 0.52 \times 1 = 0.34 \times 2$

$x = 0.16$ V

$$\begin{array}{lr} Cu^{2+} + e^- \longrightarrow Cu^+ & +0.16\text{ V} \\ Cu \longrightarrow Cu^+ + e^- & -0.52\text{ V} \\ \hline Cu^{2+} + Cu \longrightarrow 2Cu^+ & -0.36\text{ V} \end{array}$$

However, in the given option, – 0.18 V is printed.

(S) $E^0_{(Cr^{3+}, Cu^{2+})}$

$$Cr^{3+} \xrightarrow[n=1]{x} Cr^{2+} \xrightarrow[n=2]{-0.91\ V} Cr$$

$$Cr^{3+} \longrightarrow Cr \qquad -0.74\ V, \quad n = 3$$

$$x \times 1 + 2 \times (-0.91) = 3 \times (-0.74)$$

or $x - 1.82 = -2.22$

or $x = -0.4$ V

Hence, (P) – (3) ; (Q) – (4) ; (R) – (1) ; (S) – (2)

I.I.T. JEE Objective Question Paper (Main)—2013

INSTRUCTIONS

1. Immediately fill the particulars on this page of the Test Booklet with blue/black ball point pen. Use of pencil is strictly prohibited.
2. The Answer Sheet is kept inside this Test Booklet. When you are directed to open the Test Booklet, take out the Anser Sheet and fill in the particulars carefully.
3. The test is of 3 hours duration.
4. The Test Booklet consists of 90 questions. The maximum marks are 360.
5. There are three parts in the question paper A, B, C consisting of Physics, Chemistry and Mathematics having 30 questions in each part of equal weightage. Each question is allotted 4 (four) marks for correct response.
6. Candidates will be awarded marks as stated above in instruction no. 5 for correct response of each question. ¼ (one fourth) marks will be duducted for indicating incorrect response of each question. No deduction from the total score will be made if no response is indicated for an item in the answer sheet.
7. There is only one correct response for each question. Filling up more than one response in any question will be treated as wrong response and marks for wrong response will be deducted accordingly as per instruction 6 above.
8. Use blue/black ball point pen only for writing particulars/marking responses on Side-1 and Side-2 of the Answer Sheet. Use of pencil is strictly prohibited.
9. No candidate is allowed to carry any textual material, printed or written, bits of paper, paper, mobile phone, any electronic device, etc., except the Admit card inside the examination hall/room.
10. Rough work is to be done on the space provided for this purpose in the Test Booklet only. This space is given at the bottom of each page and in 3 pages at the end of the booklet.
11. On completion of the test, the candidate must hand over the Answer Sheet to the invigilator on duty in the room/hall. However, the candidates are allowed to take away this Test Booklet with them.
12. The CODE for this Booklet is P. Make sure that the CODE printed on Side-2 of the Answer Sheet is the same as that on this booklet. In case of discrepancy, the candidate should immediately report the matter to the invigilator for replacement of both the Test Booklet and the Answer Sheet.
13. Do not fold or make any stray marks on the Answer Sheet.

1. Which of the following complex species is not expected to exhibit optical isomerism?

(A) $[Co(en)_3]^{3+}$ (B) $[Co(en)_2Cl_2]^+$
(C) $[Co(NH_3)_3Cl_3]$ (D) $[Co(en)(NH_3)_2Cl_2]^+$

2. Which one of the following molecules is expected to exhibit diamagnetic behaviour?

(A) C_2 (B) N_2 (C) O_2 (D) S_2

3. A solution of (–)-1-chloro-1-phenylethane in toluene recemises slowly in the presence of a small amount of $SbCl_5$, due to the formation of

(A) carbanion (B) carbene (C) carbocation (D) free radical.

4. Given: $E^0_{Cr^{3+}/Cr} = -0.74$ V ; $E^0_{MnO_4^-/Mn^{2+}} = 1.51$ V ; $E^0_{Cr_2O_7^{2-}/Cr^{3+}} = 1.33$ V ; $E^0_{Cl/Cl^-} = 1.36$ V

Based on the data given above, strongest oxidising agent will be

(A) Cl (B) Cr^{3+} (C) Mn^{2+} (D) MnO_4^-

5. A piston filled with 0.04 mol of an ideal gas expands reversibly from 50.0 mL to 375 mL at a constant temperature of 37.0°C. As it does so, it absorbs 208 J of heat. The values of q and w for the process will be (R = 8.314 J/mol K) (ln 7.5 = 2.01)

(A) $q = +208$ J, $w = -208$ J (B) $q = -208$ J, $w = -208$ J
(C) $q = -208$ J, $w = +208$ J (D) $q = +208$ J, $w = +208$ J

6. The molarity of a solution obtained by mixing 750 mL of 0.5M HCl with 250 mL of 2M HCl will be

(A) 0.875M (B) 1.00M (C) 1.75M (D) 0.975M

7. Arrange the following compounds in order of decreasing acidity.

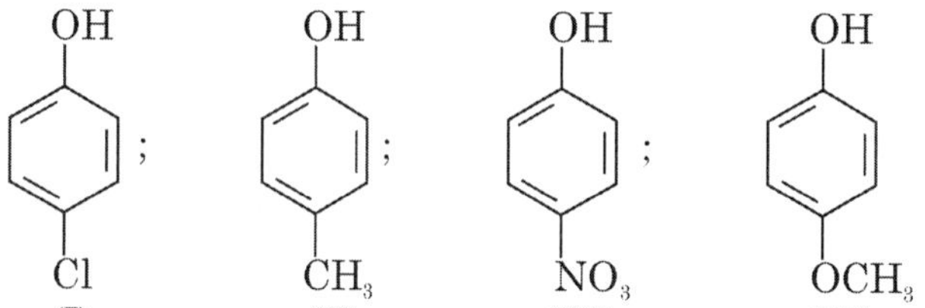

(I) (II) (III) (IV)

(A) II > IV > I > III (B) I > II > III > IV (C) III > I > II > IV (D) IV > III > I > II

8. For gaseous state, if most probable speed is denoted by C*, average speed by $\bar{C}$ and root mean square speed by C, then for a large number of molecules the ratios of these speed are

(A) $C^* : \bar{C} : C = 1.225 : 1.128 : 1$ (B) $C^* : \bar{C} : C = 1.128 : 1.225 : 1$
(C) $C^* : \bar{C} : C = 1 : 1.128 : 1.225$ (D) $C^* : \bar{C} : C = 1 : 1.225 : 1.228$

9. The rate of a reaction doubles when its temperature changes from 300 K to 310 K. Activation energy of such a reaction will be (R = 8.314 J K^{-1} mol^{-1} and log 2 = 0.301)

(A) 53.6 kJ mol^{-1} (B) 48.6 kJ mol^{-1} (C) 58.5 kJ mol^{-1} (D) 60.5 kJ mol^{-1}

10. A compound with molecular mass 180 is acylated with CH_3COCl to get a compound with molecular mass 390. The number of amino groups present per molecule of the former compound is

(A) 2 (B) 5 (C) 4 (D) 6

11. Which of the following arrangements does not represent the correct order of the property stated against it?

(A) $V^{2+} < Cr^{2+} < Mn^{2+} < Fe^{2+}$: paramagnetic behaviour

(B) $Ni^{2+} < Co^{2+} < Fe^{2+} < Mn^{2+}$: ionic size

(C) $Co^{3+} < Fe^{3+} < Cr^{3+} < Sc^{3+}$: stability in aqueous solution

(D) Sc < Ti < Cr < Mn : number of oxidation states.

12. The order of stability of the following carbocations is

$H_2C{=}CH{-}\overset{\oplus}{C}H_2$; $H_3C{-}CH_2{-}\overset{\oplus}{C}H_2$; 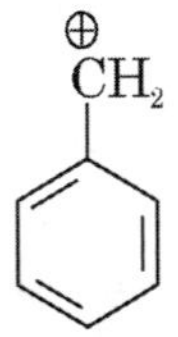

I II III

(A) II > III > I (B) I > II > III (C) III > I > II (D) III > II > I

13. Consider the following reaction

$$xMnO_4^- + yC_2O_4^{2-} + zH^+ \longrightarrow xMn^{2+} + 2yCO_2 + \frac{z}{2}H_2O$$

The values of x, y and z in the reaction are, respectively

(A) 5, 2 and 16 (B) 2, 5 and 8 (C) 2, 5 and 16 (D) 5, 2 and 8

14. Which of the following is the wrong statement?

(A) ONCl and ONO^- are not isoelectronic (B) O_3 molecule is bent

(C) Ozone is violet-black in solid state (D) Ozone is a diamagnetic gas

15. A gaseous hydrocarbon gives upon combustion 0.72 g of water and 3.08 g. of CO_2. The empirical formula of the hydrocarbon is

(A) C_2H_4 (B) C_3H_4 (C) C_6H_5 (D) C_7H_8

16. In which of the following pairs of molecules/ions, both the species are not likely to exist?

(A) H_2^+, He_2^{2-} (B) H_2^-, He_2^{2-} (C) H_2^{2+}, He_2 (D) H_2^-, He_2^{2+}

17. Which of the following exists as covalent crystals in the solid state?

(A) Iodine (B) Silicon (C) Sulphur (D) Phosphorus

18. Synthesis of each molecule of glucose in photosynthesis involves

(A) 18 molecules of ATP (B) 10 molecules of ATP

(C) 8 molecules of ATP (D) 6 molecules of ATP

19. The coagulating power of electrolytes having ions Na^+, Al^{3+} and Ba^{2+} for arsenic sulphide sol increases in the order

(A) $Al^{3+} < Ba^{2+} < Na^+$ (B) $Na^+ < Ba^{2+} < Al^{3+}$ (C) $Ba^{2+} < Na^+ < Al^{3+}$ (D) $Al^{3+} < Na^+ < Ba^{2+}$

20. Which of the following represents the correct order of increasing first ionization enthalpy for Ca, Ba, S, Se and Ar?

(A) Ca < S < Ba < Se < Ar (B) S < Se < Ca < Ba < Ar
(C) Ba < Ca < Se < S < Ar (D) Ca < Ba < S < Se < Ar

21. Energy of an electron is given by $E = -2.178 \times 10^{-18}\,\text{J}\left(\frac{Z^2}{n^2}\right)$. Wavelength of light required to excite an electron in a hydrogen atom from level $n = 1$ to $n = 2$ will be ($h = 6.62 \times 10^{-34}$ J s and $c = 3.0 \times 10^8$ m s^{-1})

(A) 1.214×10^{-7} m (B) 2.816×10^{-7} m (C) 6.500×10^{-7} m (D) 8.500×10^{-7} m

22. Compound (A), C_8H_9Br, gives a white precipitate when warmed with alcoholic $AgNO_3$. Oxidation of (A) gives an acid (B), $C_8H_6O_4$. (B) easily forms anhydride on heating. Identify the compound (A).

(A) 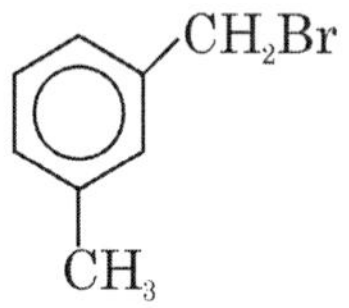 (B) 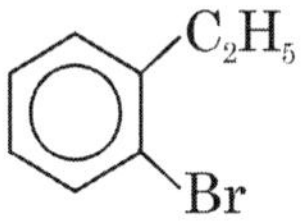 (C) 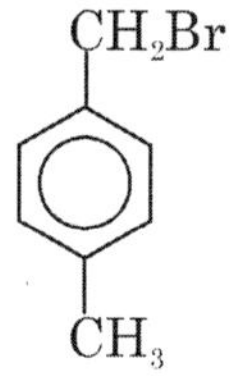 (D) 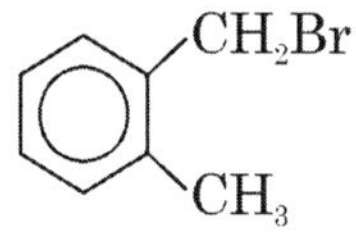

23. Four successive members of the first row transition elements are listed below with atomic numbers. Which one of them is expected to have the highest $E^0_{M^{3+}/M^{2+}}$ value?

(A) Cr(Z = 24) (B) Mn(Z = 25) (C) Fe(Z = 26) (D) Co(Z = 27)

24. How many litres of water must be added to 1 litre of an aqueous solution of HCl with a *p*H of 1 to create an aqueous solution with *p*H of 2?

(A) 0.1 L (B) 0.9 L (C) 2.0 L (D) 9.0 L

25. The first ionisation enthalpy of Na is 5.1 eV. The value of electron gain enthalpy of Na^+ will be

(A) – 2.55 eV (B) – 5.1 eV (C) – 10.2 eV (D) + 2.55 eV

26. An organic compound A upon reacting with NH_3 gives B. On heating B gives C. C in presence of KOH reacts with Br_2 to give $CH_3CH_2NH_2$. A is

(A) CH_3COOH (B) $CH_3CH_2CH_2COOH$
(C) $CH_3—CH(CH_3)—COOH$ (D) CH_3CH_2COOH

27. Stability of the species Li_2, Li_2^-, and Li_2^+ increases in the order of

(A) $Li_2 < Li_2^+ < Li_2^-$ (B) $Li_2^- < Li_2^+ < Li_2$ (C) $Li_2 < Li_2^- < Li_2^+$ (D) $Li_2^- < Li_2 < Li_2^+$

28. An unknown alochol is treated with the "Lucas reagent" to determine whether the alcohol is primary, secondary or tertiary. Which alcohol reacts fastest and by what mechanism?

(A) secondary alcohol by S_{N^1} (B) tertiary alcohol by S_{N^1}

(C) secondary alcohol by S_{N^2} (D) tertiary alcohol by S_{N^2}

29. The gas leaked from a storage tank of the Union Carbide plant in Bhopal gas tragedy was

(A) Methylisocyanate (B) Methylamine (C) Ammonia (D) Phosgene

30. Experimentally it was found that a metal oxide has formula $M_{0.98}O$. Metal M is present as M^{2+} and M^{3+} in its oxide. Fraction of the metal which exists as M^{3+} would be

(A) 7.01% (B) 4.08% (C) 6.05% (D) 5.08%

ANSWERS

1. B	2. A, B	3. C	4. D	5. A	6. A
7. C	8. C	9. A	10. B	11. A	12. D
13. C	14. All statements are correct and there is no answer.				15. D
16. C	17. B	18. A	19. B	20. C	21. A
22. D	23. D	24. D	25. B	26. D	27. B
28. B	29. A	30. B			

Model Solutions—I.I.T. JEE (Main) 2013

1. (B)

   $[Co(NH_3)_3Cl_3]$ exists in two forms (facial and meridomial).

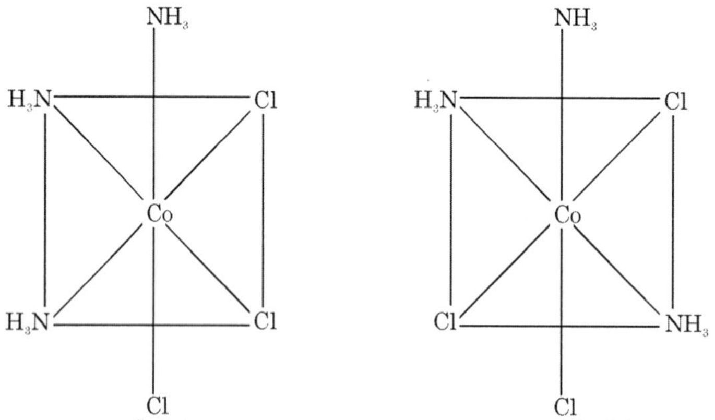

facial meridonial

   Both of these forms contain plane of symmetry and are achiral. Hence, $[Co(NH_3)_3Cl_3]$ does not show optical isomerism.

2. (A) and (B)

   C_2—Diamagnetic; N_2—Diamagnetic; O_2—Paramagnetic; S_2—Paramagnetic

3. (C)

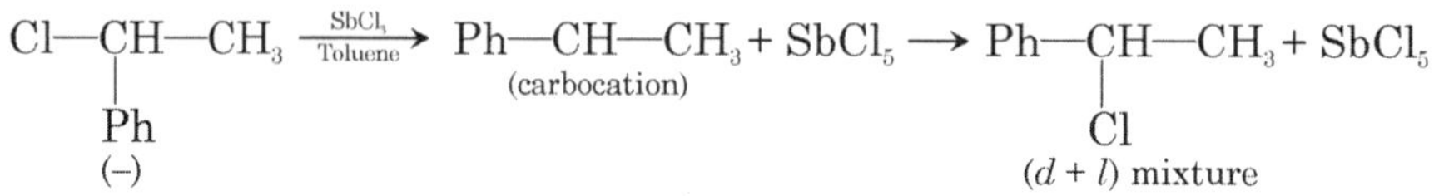

4. (D)

   Higher the E^0 value better is the oxidising agent. Hence, MnO_4^- is the strongest oxidising agent.

5. (A)

   The process is isothermal expansion. Hence, $q = -w$, $\Delta u = 0$.
   $q = +208$ J, $w = -208$ J (expansion work)

6. (A)

$$M_f = \frac{M_1V_1 + M_2V_2}{V_1 + V_2} = \frac{0.5 \times \frac{3}{4} + 2 \times \frac{1}{4}}{1} = 0.875M$$

7. (C)

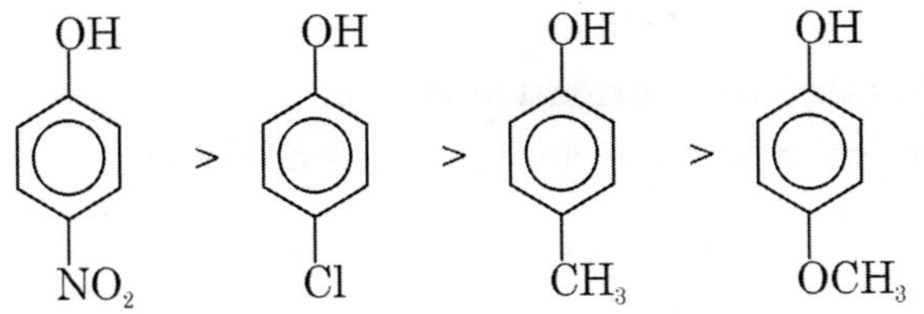

Electron releasing group decreases and electron withdrawing group increases acidic strength.

8. (C)

$$C^* = \text{most probable speed} = \sqrt{\frac{2RT}{M}}$$

$$\bar{C} = \text{root mean square speed} = \sqrt{\frac{8RT}{\pi M}}$$

$$C^* < \bar{C} < C$$

$$C^* : \bar{C} : C = 1 : \sqrt{\frac{4}{\pi}} : \sqrt{\frac{3}{2}} = 1 : 1.128 : 1.225$$

9. (A)

$$\log \frac{K_2}{K_1} = \frac{-E_a}{2.030R}\left(\frac{1}{T_2} - \frac{1}{T_1}\right)$$

$$\frac{K_2}{K_1} = 2; \quad T_2 = 310 \text{ K}, \quad T_1 = 300 \text{ K}$$

$$\log 2 = \frac{-E_a}{2.303 \times 8.134}\left(\frac{1}{310} - \frac{1}{300}\right)$$

$$E_a = 53598.6 \text{ J/mol} = 53.6 \text{ kJ/mol}$$

10. (B)

$$R-NH_2 + CH_3-\overset{\overset{O}{\|}}{C}-Cl \longrightarrow R-NH-\overset{\overset{O}{\|}}{C}-CH_3 + HCl$$

Each $CH_3-\overset{\overset{O}{\|}}{C}$ addition increases the molecular weight by 42.

Total increase in molecular weight = 390 – 180 = 210

Hence, the number of NH_2 groups $= \frac{210}{42} = 5$

11. (A)

(A) V^{2+} = 3 unpaired electrons

Cr^{2+} = 4 unpaired electrons

Mn^{2+} = 5 unpaired electrons

Fe^{2+} = 4 unpaired electrons

Hence, the order of paramagnetic behaviour is: $V^{2+} < Cr^{2+} = Fe^{2+} < Mn^{2+}$

(B) Ionic size decreases from left to right in a period.

(C) $Co^{3+}/Co^{2+} = 1.97$ V; $Fe^{3+}/Fe^{2+} = 0.77$ V; $Cr^{3+}/Cr^{2+} = -0.41$ V

Sc^{3+} is highly stable (It does not exhibit + 2 oxidation state).

(D) The oxidation state increases as we move from group 3 to group 7 in the same period.

12. (D)

The order of stability of carbocations is

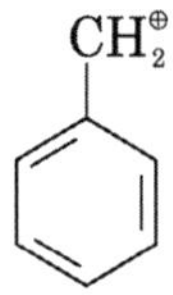

$> CH_2{=}CH{-}\overset{\oplus}{C}H_2 > CH_3{-}CH_2{-}\overset{\oplus}{C}H_2$

Benzyl Allyl Propyl

13. (C)

$$2MnO_4^- + 5C_2O_4^{2-} + 16H^+ \longrightarrow 2Mn^{2+} + 10CO_2 + 8H_2O$$

$x = 2$, $y = 5$ and $z = 16$

14. All statements are correct and there is no answer.

(A) ONCl = 8 + 7 + 17 = 32 electrons

ONO^- = 8 + 7 + 8 + 1 = 24 electrons

(B) 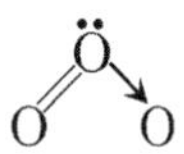 Central atom O is sp^2 hybridised with 1 lone pair, hence bent shape.

(C) Ozone is violet-black in solid state.

(D) O_3 has no unpaired electrons, hence is diamagnetic.

15. (D)

18 g H_2O contains 2 g H

$\therefore$ 0.72 g H_2O contains 0.08 g H

44 g CO_2 contains 12 g C

$\therefore$ 3.08 g CO_2 contains 0.84 g C

$$\therefore \quad C : H = \frac{0.84}{12} : \frac{0.08}{1} = 0.07 : 0.08 = 7 : 8$$

$\therefore$ Empirical formula = C_7H_8

16. (C)

H_2^{2+} : Bond order = 0

$$He_2 : \text{Bond order} = \frac{2-2}{2} = 0$$

Hence, both H_2^{2+} and He_2 do not exist.

17. (B)

Silicon exists as covalent crystal in solid state (Network like structure, like diamond).

18. (A)

$$6CO_2 + 12NADPH + 18ATP \longrightarrow C_6H_{12}O_6 + 12NADP + 18\ ADP$$

19. (B)

According to Hardy Schulze rule, greater the charge on cation, greater is its coagulating power for negetively charged solution. Hence, order of coagulating power : $Na^+ < Ba^{2+} < Al^{3+}$.

20. (C)

Order of increasing ΔH_{IE_1} : Ba < Ca < Se < S < Ar

Ba < Ca ; Se < S. On moving top to bottom in a group, size increases. So ionisation enthalpy decreases.

Ar: Maximum value of ionisation enthalpy since it is a noble gas.

21. (A)

$$\Delta E = 2.178 \times 10^{-18}\left(\frac{1}{1^2} - \frac{1}{2^2}\right) = \frac{hc}{\lambda}$$

$$2.178 \times 10^{-18}\left(\frac{1}{1^2} - \frac{1}{2^2}\right) = \frac{6.62 \times 10^{-34} \times 3.0 \times 10^8}{\lambda}$$

Hence, $\lambda \approx 1.214 \times 10^{-7}$ m

22. (D)

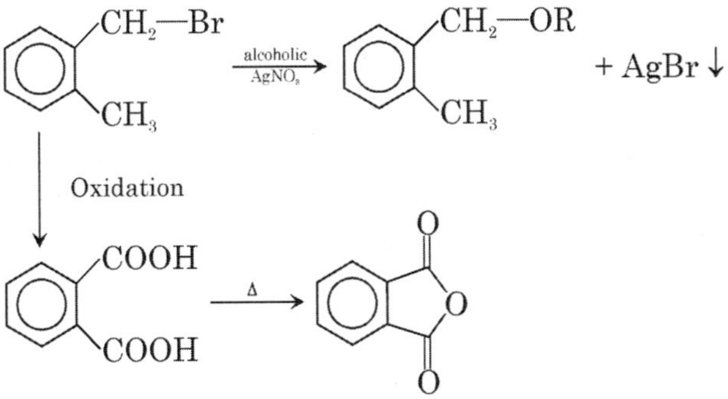

Phthalic anhydride

23. (D)

$$E^0_{Cr^{3+}/Cr^{2+}} = -0.41\ V;\ E^0_{Mn^{3+}/Mn^{2+}} = +1.57\ V;\ E^0_{Fe^{3+}/Fe^{2+}} = +0.77\ V;\ E^0_{Co^{3+}/Co^{2+}} = +1.97\ V$$

E^0 value normaly increases from left to right in the period of *d*-block elements. Some E^0 values are exceptionally higher due to stability of product ion. For example,

$$E^0_{Mn^{3+}/Mn^{2+}} = +1.57\ V; E^0_{Co^{3+}/Co^{2+}} = +1.97\ V.$$

24. (D)

pH = 1, $[H^+] = 10^{-1} = 0.1M$

pH = 2, $[H^+] = 10^{-2} = 0.01M$

$$M_1V_1 = M_2V_2$$
$$0.1 \times 1 = 0.01 \times V_2$$
$$V_2 = 10 \text{ L}$$

Volume of water added = 10 – 1 = 9 L

25. (B)

$Na \longrightarrow Na^+ + e^-$ 1^{st}I.E

$Na^+ + e^- \longrightarrow Na$ Electron gain enthalpy of Na^+, ΔH_{eg} = – 5.1 eV (because the reaction is reversed)

26. (D)

$$\underset{(A)}{CH_3CH_2{-}\overset{\overset{O}{\|}}{C}{-}OH} \xrightarrow{NH_3} \underset{(B)}{CH_3CH_2COONH_4} \xrightarrow{\Delta} \underset{(C)}{CH_3{-}CH_2{-}\overset{\overset{O}{\|}}{C}{-}NH_2} \xrightarrow[\text{Hofmann bromamide reaction}]{Br_2, KOH} CH_3{-}CH_2{-}NH_2$$

27. (B)

Li_2 (6) = $\sigma 1s^2 \sigma^* 1s^2 \sigma 2s^2$; B.O. = $\dfrac{4-2}{2} = 1$

Li_2^+ (5) = $\sigma 1s^2 \sigma^* 1s^2 \sigma 2s^1$; B.O. = $\dfrac{3-2}{2} = 0.5$

Li_2^- (7) = $\sigma 1s^2 \sigma^* 1s^2 \sigma 2s^2\, \sigma 2s^1$; B.O. = $\dfrac{4-3}{2} = 0.5$

Li_2^+ is more stable than Li_2^- because Li_2^- has more number of antibonding electrons.

28. (B)

The reaction of alcohol with Lucas reagent is mostly an S_N1 reaction and the rate of reaction is directly proportional to the carbocation stability formed in the reaction. Since 3° R—OH forms 3° carbocation, it will react fastest.

29. (A)

Methyl isocyanate, $CH_3{-}N{=}C{=}O$ (gas)

30. (B)

Consider one mole of the oxide.

Moles of M = 0.98, Moles of O^{2-} = 1

Let moles of $M^{3+} = x$

Moles of $M^{2+} = 0.98 - x$

Doing charge balance, we have

$(0.98 - x) \times 2 + 3x - 2 = 0$

or $1.96 - 2x + 3x - 2 = 0$

or $x = 0.04$

Hence, percentage of $M^{3+} = \dfrac{0.04}{0.98} \times 100 = 4.08\%$

I.I.T. JEE Objective Question Paper (Main)—2014

IMPORTANT INSTRUCTIONS

1. The test is of 3 hours duration.
2. The Test Booklet consists of 90 questions. The maximum marks are 360.
3. There are three parts in the question paper A, B, C consisting of Physics, Mathematics and Chemistry having 30 questions in each part of equal weightage. Each question is allotted 4 (four) marks for correct response.
4. Candidates will be awarded marks as stated above in instruction No. 3 for correct response of each question. (1/4) (one-fourth) marks will be deducted for indicating incorrect response of each question. No deduction from the total score will be made if no response is indicated for an item in the answer sheet.
5. There is only one correct response for each question. Filling up more than one response in each question will be treated as wrong response and marks for wrong response will be deducted accordingly as per instruction 4 above.

1. Which one of the following properties is not shown by NO?
(A) It combines with oxygen to form nitrogen dioxide
(B) Its bond order is 2.5
(C) It is diamagnetic in gaseous state
(D) It is a neutral oxide.

2. If Z is a compressibility factor, van der Waals equation at low pressure can be written as
(A) $Z = 1 - \frac{Pb}{RT}$ (B) $Z = 1 + \frac{Pb}{RT}$ (C) $Z = 1 + \frac{RT}{Pb}$ (D) $Z = 1 - \frac{a}{VRT}$

3. The metal that cannot be obtained by electrolysis of an aqueous solution of its salts is
(A) Cu (B) Cr (C) Ag (D) Ca

4. Resistance of $0.2M$ solution of an electrolyte is 50 Ω. The specific conductance of the solution is 1.4 S m^{-1}. The resistance of $0.5M$ solution of the same electrolyte is 280 Ω. The molar conductivity of $0.5M$ solution of the electrolyte in S m^2 mol^{-1} is
(A) 5×10^3 (B) 5×10^2 (C) 5×10^{-4} (D) 5×10^{-3}

5. CsCl crystallises in body centred cubic lattice. If a is its edge length then which of the following expressions is correct?

(A) $r_{Cs^+} + r_{Cl^-} = \frac{\sqrt{3}}{2}a$ (B) $r_{Cs^+} + r_{Cl^-} = \sqrt{3}a$ (C) $r_{Cs^+} + r_{Cl^-} = 3a$ (D) $r_{Cs^+} + r_{Cl^-} = \frac{3a}{2}$

6. Consider separate solutions of $0.500M$ $C_2H_5OH(aq)$, $0.100M$ $Mg_3(PO_4)_2(aq)$, $0.250M$ $KBr(aq)$ and $0.125M$ $Na_3OP_4(aq)$ at 25°C. Which statement is true about these solutions, assuming all salts to be strong electrolytes?
   (A) $0.125M$ $Na_3PO_4(aq)$ has the highest osmotic pressure.
   (B) $0.500M$ $C_2H_5OH(aq)$ has the highest osmotic pressure.
   (C) They all have the same osmotic pressure.
   (D) $0.100M$ $Mg_3(PO_4)_2(aq)$ has the highest osmotic pressure.

7. In which of the following reactions H_2O_2 acts as a reducing agent?
   (a) $H_2O_2 + 2H^+ + 2e^- \longrightarrow 2H_2O$
   (b) $H_2O_2 - 2e^- \longrightarrow O_2 + 2H^+$
   (c) $H_2O_2 + 2e^- \longrightarrow 2OH^-$
   (d) $H_2O_2 + 2OH^- - 2e^- \longrightarrow O_2 + 2H_2O$
   (A) (a), (c) (B) (b), (d) (C) (a), (b) (D) (c), (d)

8. In S_N2 reactions, the correct order of reactivity for the following compounds CH_3Cl, CH_3CH_2Cl, $(CH_3)_2CHCl$ and $(CH_3)_3CCl$ is
   (A) $CH_3CH_2Cl > CH_3Cl > (CH_3)_2CHCl > (CH_3)_3CCl$
   (B) $(CH_3)_2CHCl > CH_3CH_2Cl > CH_3Cl > (CH_3)_3CCl$
   (C) $CH_3Cl > (CH_3)_2CHCl > CH_3CH_2Cl > (CH_3)_3CCl$
   (D) $CH_3Cl > CH_3CH_2Cl > (CH_3)_2CHCl > (CH_3)_3CCl$

9. The Octahedral complex of a metal ion M^{3+} with four monodentate ligands L_1, L_2, L_3 and L_4 absorb wavelenghts in the region of red, green, yellow and blue, respectively. The increasing order of ligand strength of the four ligands is
   (A) $L_3 < L_2 < L_4 < L_1$ (B) $L_1 < L_2 < L_4 < L_3$
   (C) $L_4 < L_3 < L_2 < L_1$ (D) $L_1 < L_3 < L_2 < L_4$

10. For the estimation of nitrogen, 1.4 g of organic compound was digested by Kjeldahl method and the evolved ammonia was absorbed in 60 mL of $\frac{M}{10}$ sulphuric acid. The unreacted acid required 20 mL of $\frac{M}{10}$ sodium hydroxide for complete neutralization. The percentage of nitrogen in the compound is
   (A) 3% (B) 5% (C) 6% (D) 10%

11. The equivalent conductances of NaCl at concentration C and at infinite dilution are λ_C and λ_∞, respectively. The correct relationship between λ_C and λ_∞ is given as (where the constant B is positive)
   (A) $\lambda_C = \lambda_\infty - (B)\sqrt{C}$ (B) $\lambda_C = \lambda_\infty + (B)\sqrt{C}$
   (C) $\lambda_C = \lambda_\infty + (B)C$ (D) $\lambda_C = \lambda_\infty - (B)C$

12. For the reaction, $SO_2(g) + \frac{1}{2} O_2(g) \rightleftharpoons SO_3(g)$, if $K_P = K_C(RT)^x$ where the symbols have usual meaning then the value of x is (assuming ideality)

(A) $\frac{1}{2}$ (B) 1 (C) -1 (D) $-\frac{1}{2}$

13. In the reaction, $CH_3COOH \xrightarrow{LiAlH_4} A \xrightarrow{PCl_5} B \xrightarrow{Alc.\ KOH} C$, the product C is

(A) Ethylene (B) Acetyl chloride (C) Acetaldehyde (D) Acetylene

14. Sodium phenoxide when heated with CO_2 under pressure at 125°C yields a product which on acetylation produces C.

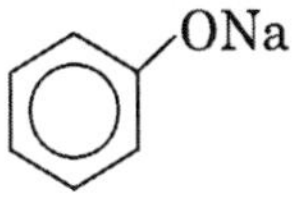

$+ CO_2 \xrightarrow[5\ atm]{125°C} B \xrightarrow[AC_2O]{H^+} C$

The major product C would be

(A) 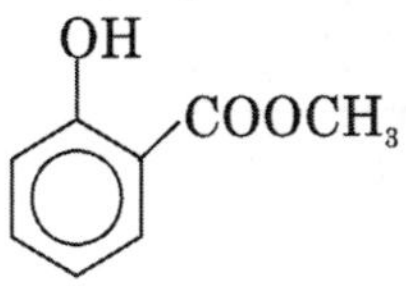

(B) 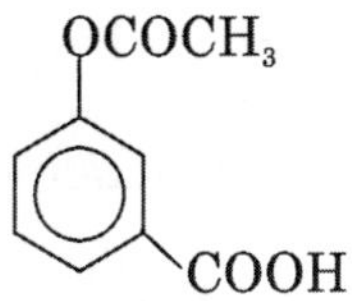

(C) 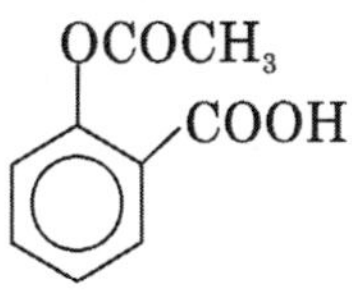

(D) 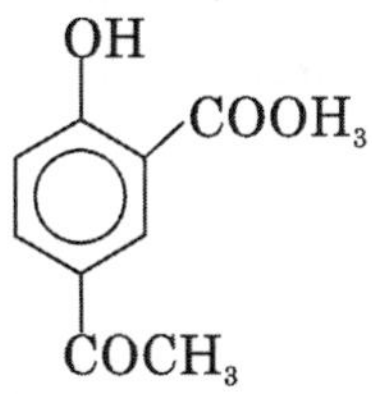

15. On heating an aliphatic primary amine with chloroform and ethanolic potassium hydroxide, the organic compound formed is

(A) an alkyl cyanide (B) an alkyl isocyanide
(C) an alkanol (D) an alkanediol

16. The correct statement for the molecule, CsI_3 is

(A) it contains Cs^{3+} and I^- ions.

(B) it contains Cs^{3+}, I^- and lattice I_2 molecule.

(C) it is a covalent molecule.

(D) it contains Cs^+ and I_3^- ions.

17. The equation which is balanced and represents the correct product(s) is

(A) $[Mg(H_2O)_6]^{2+} + (EDTA)^{4-} \xrightarrow{excess\ NaOH} [Mg(EDTA)]^{2+} + 6H_2O$

(B) $CuSO_4 + 4KCN \longrightarrow K_2[Cu(CN)_4] + K_2SO_4$

(C) $Li_2O + 2KCl \longrightarrow 2LiCl + K_2O$

(D) $[CoCl(NH_3)_5]^+ + 5H^+ \longrightarrow Co^{2+} + 5NH_4^+ + Cl^-$

18. For which of the following molecule significant $\mu \neq 0$?

(a)  (b)  (c)  (d) 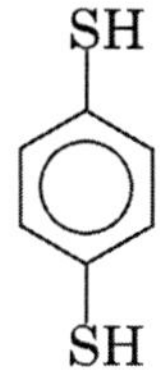

(A) Only (c) (B) (c) and (d) (C) Only (a) (D) (a) and (b)

19. For the non-stoichiometric reaction, $2A + B \longrightarrow C + D$, the following kinetic data were obtained in three separate experiments, all at 298 K.

Initial concentration (A)	Initial concentration (B)	Initial rate of formation of C(mol L^{-} S^{-})
0.1*M*	0.1*M*	1.2×10^{-3}
0.1*M*	0.2*M*	1.2×10^{-3}
0.2*M*	0.1*M*	2.4×10^{-3}

The rate law for the formation of C is

(A) $\frac{dc}{dt} = k[A][B]^2$ (B) $\frac{dc}{dt} = k[A]$ (C) $\frac{dc}{dt} = k[A][B]$ (D) $\frac{dc}{dt} = k[A]^2[B]$

20. Which series of reactions correctly represents chemical reactions related to iron and its compound?

(A) $Fe \xrightarrow{Cl_2,\text{heat}} FeCl_3 \xrightarrow{\text{heat, air}} FeCl_2 \xrightarrow{Zn} Fe$

(B) $Fe \xrightarrow{O_2,\text{heat}} Fe_3O_4 \xrightarrow{CO,\ 600°C} FeO \xrightarrow{CO,\ 700°C} Fe$

(C) $Fe \xrightarrow{\text{dil. } H_2SO_4} FeSO_4 \xrightarrow{H_2SO_4,\ O_2} Fe_2(SO_4)_3 \xrightarrow{\text{heat}} Fe$

(D) $Fe \xrightarrow{O_2,\text{heat}} FeO \xrightarrow{\text{dil. } H_2SO_4} FeSO_4 \xrightarrow{\text{heat}} Fe$

21. Considering the basic strength of amines in aqueous solution, which one has the smallest pK_b value?

(A) $(CH_3)_3N$ (B) $C_6H_5NH_2$ (C) $(CH_3)_2NH$ (D) CH_3NH_2

22. Which one of the following bases is not present in DNA?

(A) Cytosine (B) Thymine (C) Quinoline (D) Adenine

23. The correct set of four quantum numbers for the valence elections of rubidium atom ($Z = 37$) is

(A) $5,1,1,+\frac{1}{2}$ (B) $5,0,1,+\frac{1}{2}$ (C) $5,0,0,+\frac{1}{2}$ (D) $5,1,0,+\frac{1}{2}$

24. The major organic compound formed by the reaction of 1,1,1-trichloroethane with silver powder is

(A) 2-Butyne (B) 2-Butene (C) Acetylene (D) Ethene

25. Given below are the half-cell reactions:

$Mn^{2+} + 2e^- \longrightarrow Mn$; $E^0 = -1.18$ V

$2(Mn^{3+} + e^- \longrightarrow Mn^{2+})$; $E^0 = +1.51$ V

The E^0 for $3Mn^{2+} \longrightarrow Mn + 2Mn^{3+}$ will be

(A) – 0.33 V; the reaction will not occur (B) – 0.33 V; the reaction will occur

(C) – 2.69 V; the reaction will not occur (D) – 2.69 V; the reaction will occur.

26. The ratio of masses of oxygen and nitrogen in a particular gaseous mixture is 1 : 4. The ratio of number of their molecule is

(A) 1 : 8 (B) 3 : 16 (C) 1 : 4 (D) 7 : 32

27. Which one is classified as a condensation polymer?

(A) Teflon (B) Acrylonitrile (C) Dacron (D) Neoprene

28. Among the following oxoacids, the correct decreasing order of acid strength is

(A) $HClO_4 > HClO_3 > HClO_2 > HOCl$ (B) $HClO_2 > HClO_4 > HClO_3 > HOCl$

(C) $HOCl > HClO_2 > HClO_3 > HClO_4$ (D) $HClO_4 > HOCl > HClO_2 > HClO_3$

29. For complete combustion of ethanol, $C_2H_5OH(\ell) + 3O_2(g) \longrightarrow 2CO_2(g) + 3H_2O(\ell)$, the amount of heat produced as measured in bomb calorimeter is 1364.47 kJ mol^{-1} at 25°C. Assuming ideality the enthalpy of combustion, $\Delta_C H$, for the reaction will be ($R = 8.314$ kJ mol^{-1})

(A) – 1460.50 kJ mol^{-1} (B) – 1350.50 kJ mol^{-1}

(C) – 1366.95 kJ mol^{-1} (D) – 1361.95 kJ mol^{-1}

30. The most suitable reagent for the conversion of R—CH_2—OH ⟶ R—CHO is

(A) CrO_3 (B) PCC (Pyridinium chlorochromate)

(C) $KMnO_4$ (D) $K_2Cr_2O_7$

ANSWERS

1. C	2. D	3. D	4. C	5. A	6. C
7. B	8. D	9. D	10. D	11. A	12. D
13. A	14. C	15. B	16. D	17. D	18. B
19. B	20. B	21. C	22. C	23. C	24. A
25. C	26. D	27. C	28. A	29. C	30. B

Model Solutions—I.I.T. JEE (Main) 2014

Paper–1

1. (C)
NO is paramagnetic in gaseous state due to the presence of unpaired electron in its structure.

2. (D)
$$\left(P + \frac{n^2a}{V^2}\right)(V - nb) = nRT$$
For 1 mole, $\left(P + \frac{a}{V^2}\right)(V - b) = RT$
$$PV = RT + Pb - \frac{a}{V} + \frac{ab}{V^2}$$
At low pressure, terms Pb and $\frac{ab}{V^2}$ will be negligible as compared to RT.
Hence, $PV = RT - \frac{a}{V}$
$$Z = 1 - \frac{a}{RTV}$$

3. (D)
During the electrolysis of aqueous solution of s-block elements, H_2 gas is obtained at cathode.

4. (C)
$$50 = \frac{1}{K} \times \frac{\ell}{A}$$
$$50 = \frac{1}{1.4} \times \frac{\ell}{A}$$
$$\frac{\ell}{A} = 70 \text{ m}^{-1}$$
$$280 = \frac{1}{K} \times 70$$
$$K = \frac{1}{4} \text{ S m}^{-1}$$
$$\Lambda_m = \frac{1}{4} \times \left(\frac{1000}{M}\right) (10^{-2} \text{ m})^3$$
$$= \frac{1}{4} \times \frac{1000}{0.5} \times 10^{-6} = 500 \times 10^{-6} = 5 \times 10^{-4} \text{ S m}^2 \text{ mol}^{-1}$$

5. (A)

In CsCl structure, Cs^+ ion is in contact with Cl^- ion at the nearest distance which is equal to $\sqrt{3}\frac{a}{2}$.

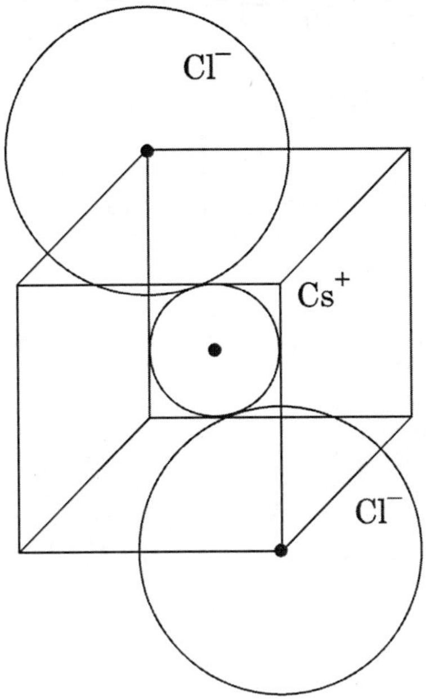

6. (C)

For C_2H_5OH, $\pi = 1 \times 0.5 \times RT$
For KBr, $\pi = 2 \times 0.25 \times RT$
For $Mg_3(PO_4)_2$, $\pi = 5 \times 0.1 \times RT$
For Na_3PO_4, $\pi = 4 \times 0.125 \times RT$
Hence, all are isotonic solutions.

7. (B)

A reducing agent loses electrons during a redox reaction.
Hence, (b, d) is correct.

8. (D)

$$\text{Rate of } S_{N^2} \text{ reaction} \propto \frac{1}{\text{steric over crowding in transition state}}$$

9. (D)

Strong field ligands cause higher magnitude of crystal field splitting which is accompanied by the absorption of higher energy radiation.

$\xrightarrow{\text{VIBGYOR}}$ decreasing energy

10. (D)

$$\% \text{ of N} = \frac{1.4 \times \text{milliequivalents of acid consumed}}{\text{mass of organic compound}}$$

$$\text{Meq of acid consumed} = \left(60 \times \frac{1}{10} \times 2\right) - \left(20 \times \frac{1}{10} \times 1\right) = 10$$

$$\text{Hence, } \% \text{ of N} = \frac{1.4 \times 10}{1.4} = 10\%$$

11. (A)

According to Debye Huckel's theory for a strong electrolyte, $\lambda_C = \lambda_\infty - B\sqrt{C}$

12. (D)

For the reaction,

$$SO_2(g) + \frac{1}{2} O_2(g) \rightleftharpoons SO_3(g)$$

$$\Delta n_g = -\frac{1}{2} = x$$

13. (A)

$$CH_3COOH \xrightarrow{LiAlH_4} CH_3CH_2OH \xrightarrow{PCl_5} CH_3CH_2Cl \xrightarrow{\text{alc. KOH}} \underset{\text{(Ethylene)}}{H_2C{=}CH_2}$$

14. (C)

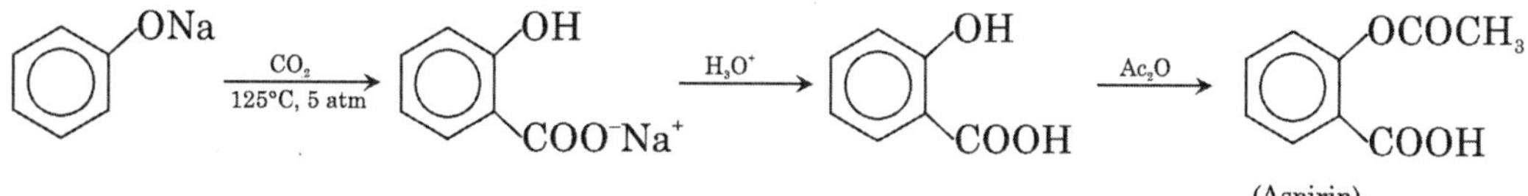

(Aspirin)

15. (B)

$$RNH_2 + CHCl_3 + \underset{\text{(alc.)}}{3KOH} \longrightarrow \underset{\text{(Alkyl isocyanide)}}{RNC\uparrow} + 3KCl + 3H_2O$$

16. (D)

$$CsI_3 \longrightarrow Cs^+ + I_3^-$$

Cs cannot show +3 oxidation state.

I_2 molecules are too large to be accommodated in lattice.

17. (D)

Equation–1 is not balanced w.r.t. charge.

Equation–2 gives $K_3[Cu(CN)_4]$ as product.

Equation–3 reaction is unfavourable in the forward direction (K_2O is unstable, while Li_2O is stable).

Equation–4 is correct and balanced.

18. (B)

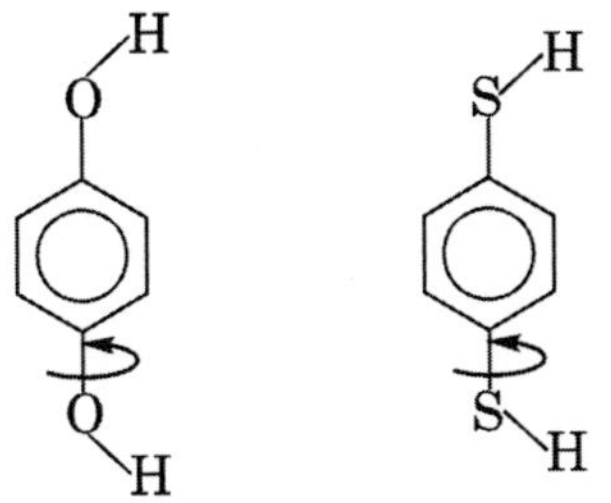

Due to infinite possible conformations in the above cases (of which only one has zero μ); a weighted μ will finally exist.

19. (B)

$R = k[A]^x [B]^y$

$1.2 \times 10^{-3} = k[0.1]^x [0.1]^y$

$1.2 \times 10^{-3} = k[0.1]^x [0.2]^y$

$2.4 \times 10^{-3} = k[0.2]^x [0.1]^y$

Solving gives $x = 1$, $y = 0$

$R = k[A]$

20. (B)

In Eq. (A) $FeCl_3$ cannot be reduced when heated in air.

In Eq. (C) $Fe_2(SO_4)_3$ cannot convert to Fe on heating; instead oxide(s) will be formed.

In Eq. (D) $FeSO_4$ cannot be converted to Fe on heating; instead oxide(s) will be formed.

Hence, Eq. (B) is correct.

21. (C)

Aliphatic amines are more basic than aromatic amines.
$(CH_3)_2NH > CH_3NH_2 > (CH_3)_3N$ (among aliphatic amines in water).

22. (C)

Adenine, thymine, cytosine, guanine are bases present in DNA.
Quinoline being an aromatic compound is NOT present in DNA.

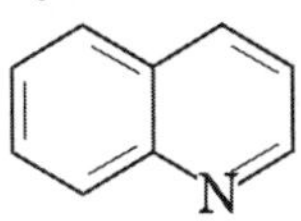

(Quinoline)

23. (C)

$_{37}Rb = [Kr]\ 5s^1$

$n = 5,\ l = 0,\ m = 0,\ s = +\frac{1}{2}$

24. (A)

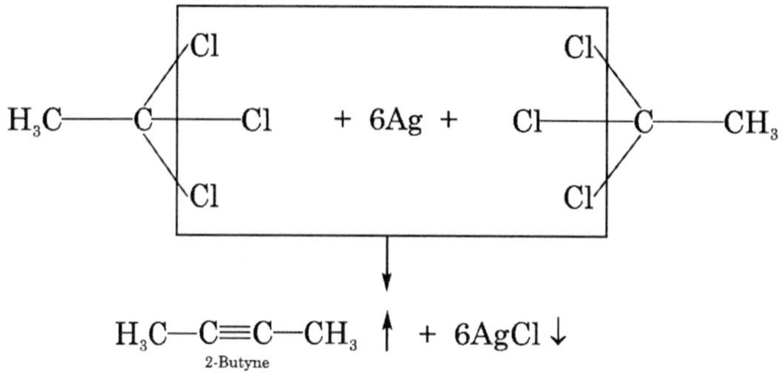

25. (C)

$$Mn^{2+} + 2e^- \longrightarrow Mn \quad E^0 = -1.18\ V$$
$$2Mn^{2+} \longrightarrow 2Mn^{3+} + 2e^- \quad E^0 = -1.51\ V$$
$$3Mn^{2+} \longrightarrow Mn + 2Mn^{3+} \quad E^0 = SOP + SRP$$
$$= -1.18 + (-1.51) = -2.69\ V$$

Negative EMF reflects non-spontaneous cell reaction.

26. (D)

Moles of $O_2 = \frac{w}{32}$ and moles of $N_2 = \frac{4w}{28}$

$$\frac{n_{O_2}}{n_{N_2}} = \frac{w}{32} \times \frac{28}{4w} = \frac{7}{32}$$

27. (C)

Teflon, acrylonitrile and neoprene are addition polymers while dacron is a condensation polymer.

28. (A)

$HClO < HClO_2 < HClO_3 < HClO_4$

Increasing acid strength is due to increase in oxidation state of central atom.

29. (C)

$$C_2H_5OH(\ell) + 3O_2(g) \longrightarrow 2CO_2(g) + 3H_2O(\ell)$$

$\Delta E = -1364.47$ kJ mol^{-1}

$\Delta H = ?$

$T = 298$ K

$\Delta n_g = -1$

$\Delta H = \Delta E + \Delta n_g RT$

Hence, $\Delta H = -1366.95$ kJ mol^{-1}

30. (B)

$$RCH_2OH \xrightarrow{PCC} RCHO$$

I.I.T. JEE Objective Question Paper (Advanced)—2014

Paper–1

INSTRUCTIONS

A. General

1. This booklet is your Question Paper. Do not break the seal of this booklet before being instructed to do so by the invigilators.
2. The question paper code is printed on the left hand top corner of this sheet and on the back cover page of this booklet.
3. Blank space and blank pages are provided in the question paper for your rough work. No additional sheets will be provided for rough work.
4. Blank papers, clipboards, log tables, slide rules, calculators, cameras, cellular phones, pagers, and electronic gadget of any kind are not allowed inside the examination hall.
5. Write your name and roll number in the space provided on the back cover of this booklet.
6. Answers to the questions and personal details are to be filled on an Optical Response Sheet, which is provided separately. The ORS is a doublet of two sheets—upper and lower, having identical layout. The upper sheet is a machine-gradable Objective Response Sheet (ORS) which will be collected by the invigilator at the end of the exmination. The upper sheet is designed in such a way that darkening the bubble with a ball point pen will leave an identical impression at the corresponding place on the lower sheet. You will be allowed to take away the lower sheet at the end of the examination (see Figure-1 on the back cover page for the correct way of darkening the bubbles for valid answers).
7. Use a black ball point pen only to darken the bubbles on the upper original sheet. Apply sufficient pressure so that the impression is created on the lower sheet. See Figure-1 on the back cover page for appropriate way to darkening the bubbles for valid answers.
8. Do not tamper with/mutilate the ORS so this booklet.
9. On breaking the seal of the booklet check that it contains 28 pages and all the 60 questions and corresponding answer choices are legible. Read carefully the instruction printed at the beginning of each section.

B. Filling the Right Part of the ORS

10. The ORS also has a code printed on its left and right parts.
11. Verify that the code printed on the ORS (on both the left and right parts) is the same as that on this booklet and put your signature in the Box designated as R4.

12. If the codes do not match, ask for a change of the Booklet/ORS as applicable.
13. Write your Name, Roll No. and the name of centre and sign with pen in the boxes provided on the upper sheet of ORS. Do not write any of this anywhere else. Darken the appropriate bubble under each digit of your Roll No. in such way that the impresssion is created on the bottom sheet. (see example in Figure-2 on the back cover).

C. Question Paper Format

The question paper consists of three parts (Physics, Chemistry and Mathematics). Each part consists of two sections.

14. **Section 1** contains 10 multiple choice questions. Each question has four choices (A), (B), (C) and (D) out of which one or more than one are correct.
15. **Section 2** contains 10 questions. The answer to each questions is a single-digit integer, ranging from 0 to 9 (both inclusive).

SECTION–I
(One or More Than One Options Correct Type)

This section contains 10 multiple choice questions. Each question has four choices (A), (B), (C) and (D) out of which one or more than one are correct.

1. The correct combination of names for isomeric alcohols with molecular formula $C_4H_{10}O$ is/are

    (A) *tert*-butanol and 2-methylpropan-2-ol
    (B) *tert*-butanol and 1,1-dimethylethan-1-ol
    (C) *n*-butanol and butan-1-ol
    (D) isobutyl alcohol and 2-methylpropan-1-ol

2. An ideal gas in a thermally insulated vessel at internal pressure = P_1, volume = V_1 and absolute temperature = T_1 expands irreversibly against zero external pressure, as shown in the diagram. The final internal pressure, volume and absolute temperature of the gas are P_2, V_2 and T_2, respectively. For this expansion,

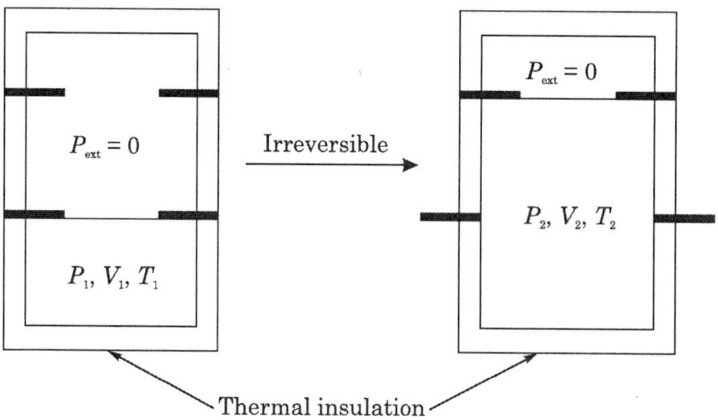

    (A) $q = 0$ (B) $T_2 = T_1$ (C) $P_2V_2 = P_1V_1$ (D) $P_2V_2^{\gamma} = P_1V_1^{\gamma}$

3. Hydrogen bonding plays a central role in the following phenomena
   (A) Ice floats in water.
   (B) Higher Lewis basicity of primary amines than tertiary amines in aqueous solutions.
   (C) Formic acid is more acidic than acetic acid.
   (D) Dimerisation of acetic acid in benzene.

4. In a galvanic cell, the salt bridge
   (A) does not participate chemically in the cell reaction.
   (B) stops the diffusion of ions from one electrode to another.
   (C) is necessary for the occurrence of the cell reaction.
   (D) ensures mixing of the two electrolytic solutions.

5. For the reaction, $I^- + ClO_3^- + H_2SO_4 \longrightarrow Cl^- + HSO_4^- + I_2$

   The correct statement(s) in the balanced equation is/are

   (A) Stoichiometric coefficient of HSO_4^- is 6.
   (B) Iodide is oxidized.
   (C) Sulphur is reduced.
   (D) H_2O is one of the products.

6. The reactivity of compound Z with different halogens under appropriate conditions is given below:

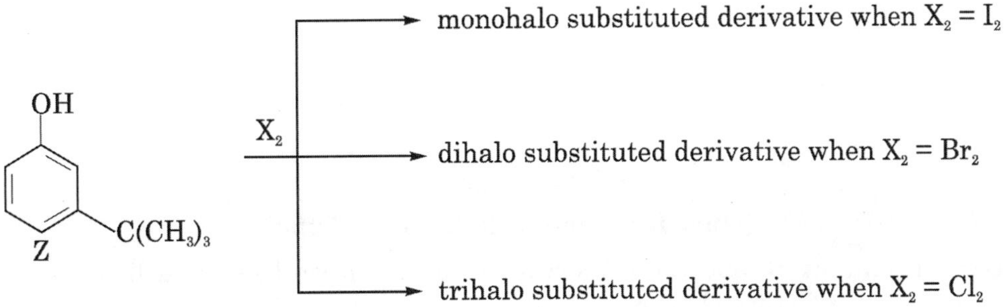

   The observed pattern of electrophilic substitution can be explained by
   (A) the steric effect of the halogen
   (B) the steric effect of the *tert*-butyl group
   (C) the electronic effect of the phenolic group
   (D) the electronic effect of the *tert*-butyl group

7. The correct statement(s) for orthoboric acid is/are
   (A) It behaves as a weak acid in water due to self-ionization.
   (B) Acidity of its aqueous solution increases upon addition of ethylene glycol.
   (C) It has a three dimensional structure due to hydrogen bonding.
   (D) It is a weak electrolyte in water.

8. Upon heating with Cu_2S, the reagent(s) that give copper metal is/are
   (A) $CuFeS_2$ (B) CuO (C) Cu_2O (D) $CuSO_4$

9. The pair(s) of reagents that yield paramagnetic species is/are
   (A) Na and excess of NH_3 (B) K and excess of O_2
   (C) Cu and dilute HNO_3 (D) O_2 and 2-ethylanthraquinol
10. In the reaction shown below, the major product(s) formed is/are

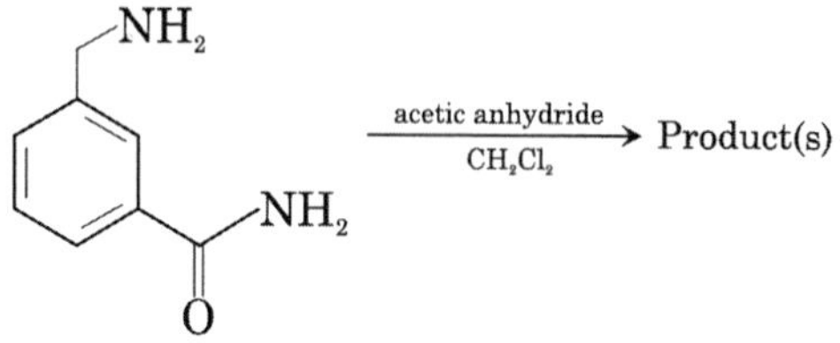

(A)

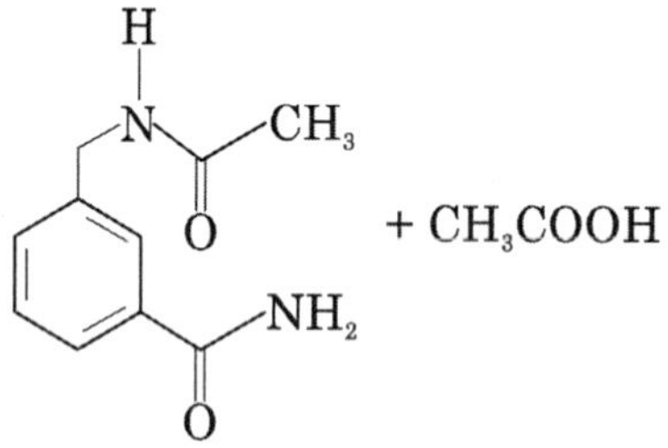

(B)

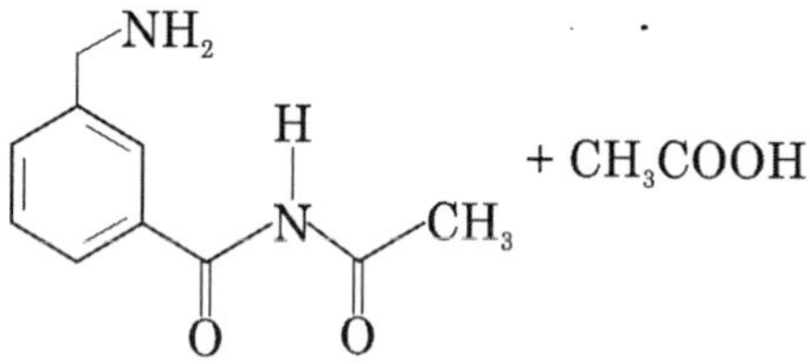

(C)

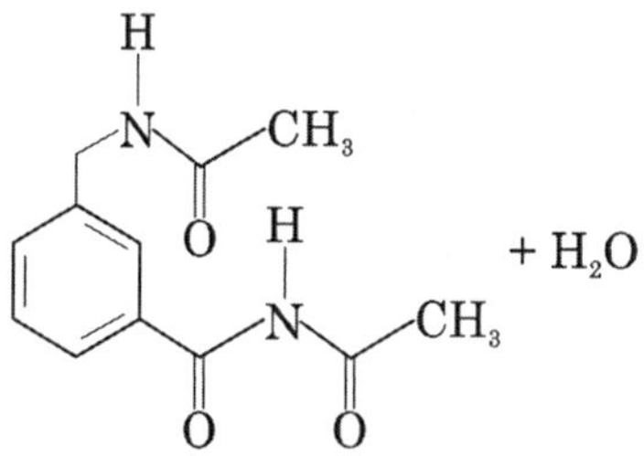

(D)

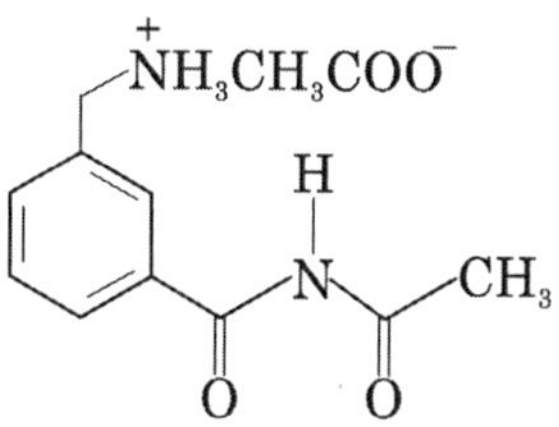

SECTION–II
(Only Integer Value Correct Type)

This section contains 10 questions. Each question, when worked out will result in one integer from 0 to 9 (both inclusive).

11. Among PbS, CuS, HgS, MnS, Ag_2S, NiS, CoS, Bi_2S_3 and SnS_2, the total number of black coloured sulphides is
12. The total number(s) of stable conformers with non-zero dipole moment for the following compound is(are)

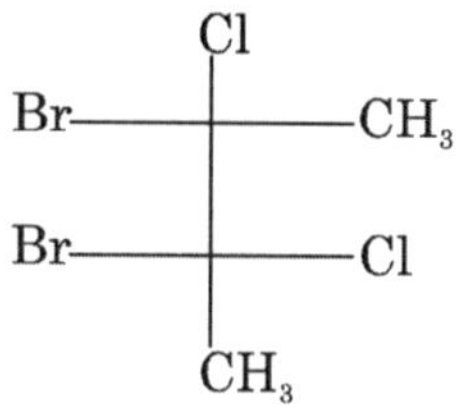

13. Consider the following list of reagents:
    Acidified $K_2Cr_2O_7$, alkaline $KMnO_4$, $CuSO_4$, H_2O_2, Cl_2, O_3, $FeCl_3$, HNO_3 and $Na_2S_2O_3$. The total number of reagents that can oxidise aqueous iodide to iodine is

14. A list of species having the formula XZ_4 is given below.

XeF_4, SF_4, SiF_4, BF_4^-, BrF_4^-, $[Cu(NH_3)_4]^{2+}$, $[FeCl_4]^{2-}$, $[CoCl_4]^{2-}$ and $[PtCl_4]^{2-}$.

Defining shape on the basis of the location of X and Z atoms, the total number of species having a square planar shape is

15. Consider all possible isomeric ketones, including stereoisomers of MW = 100. All these isomers are independently reacted with $NaBH_4$ (Note: stereoisomers are also reacted separately). The total number of ketones that give a racemic product(s) is/are

16. In an atom, the total number of electrons having quantum numbers $n = 4$, $|m_l| = 1$ and $m_s = -1/2$ is

17. If the value of Avogadro number is 6.023×10^{23} mol^{-1} and the value of Boltzmann constant is 1.380×10^{-23} J K^{-1}, then the number of significant digits in the calculated value of the universal gas constant is

18. MX_2 dissociates to M^{2+} and X^- ions in an aqueous solution, with a degree of dissociation (α) of 0.5. The ratio of the observed depression of freezing point of the acqueous solution to the value of the depression of freezing point in the absence of ionic dissociation is

19. The total number of *distinct naturally occurring amino acids* obtained by complete acidic hydrolysis of the peptide shown below is

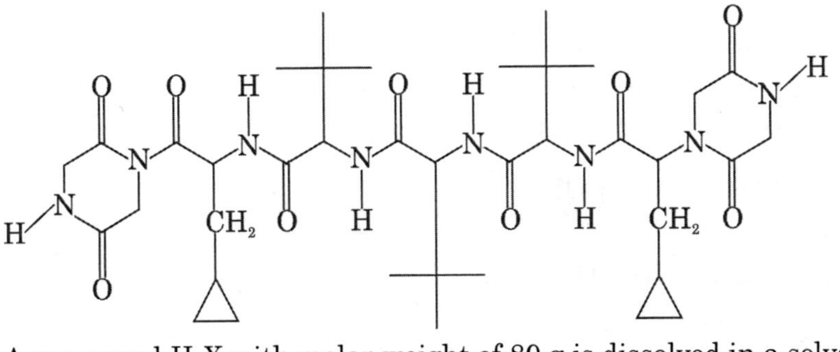

20. A compound H_2X with molar weight of 80 g is dissolved in a solvent having density of 0.4 g mL^{-1}. Assuming no change in volume upon dissolution, the molality of a 3.2 molar solution is

ANSWERS

1. A, C, D	2. A, B, C	3. A, B, D	4. A	5. A, B, D	6. A, B, C
7. B, D	8. B, C, D	9. A, B, C	10. A	11. 6	12. 3
13. 7	14. 4	15. 5	16. 6	17. 4	18. 2
19. 1	20. 8				

Model Solutions—I.I.T. JEE (Advanced) 2014

Paper–1

1. (A, C, D)

   Isomeric alcohols of $C_4H_{10}O$ are

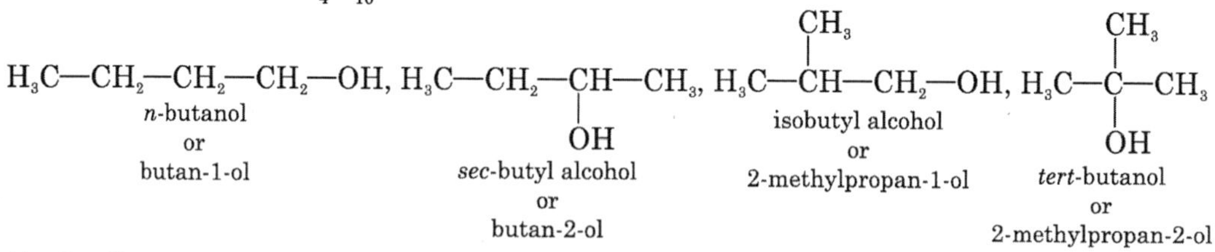

2. (A, B, C)

   Since container is thermally insulated. So, $q = 0$, and it is a case of free expansion. Therefore $W = 0$ and $\Delta E = 0$. Hence, $T_1 = T_2$. Also, $P_1V_1 = P_2V_2$

3. (A, B, D)

   (A) Ice has cage-like structure in which each water molecule is surrounded by four other water molecules tetrahedrally through hydrogen bonding. Due to this density of ice is less than water and it floats on water.

   (B) $R—NH_2 + H—OH \rightleftharpoons \underset{\text{(I)}}{R—\overset{+}{N}H_3} + OH^-$

   $(R)_3N + H—OH \rightleftharpoons \underset{\text{(II)}}{(R)_3—\overset{+}{N}H} + OH^-$

   The cation (I) is more stabilized through hydrogen bonding than cation (II). Hence, $R—NH_2$ is a better base than $(R)_3N$ in aqueous solution.

   (C) HCOOH is stronger acid than CH_3COOH due to the inductive effect and not due to hydrogen bonding.

   (D) Acetic acid dimerizes in benzene through intermolecular hydrogen bonding.

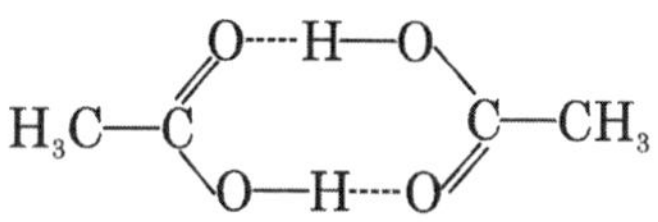

4. (A)

5. (A, B, D)

   The balanced equation is: $ClO_3^- + 6I^- + 6H_2SO_4 \longrightarrow 3I_2 + Cl^- + 6HSO_4^- + 3H_2O$

6. (A, B, C)

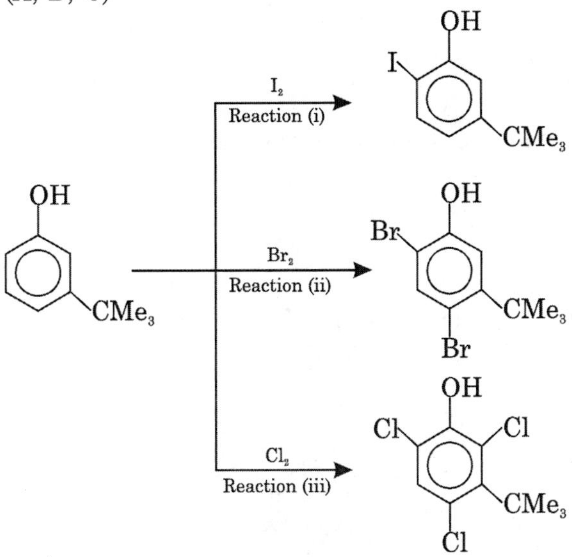

7. (B, D)

(A) H_3BO_3 is a weak monobasic Lewis acid.

$H_3BO_3 + H—OH \rightleftharpoons B(OH)_4^- + H^+$ ...(i)

(B) Equilibrium (i) shifts in the forward direction by the addition of syn-diols like ethylene glycol which forms a stable complex with $B(OH)_4^-$.

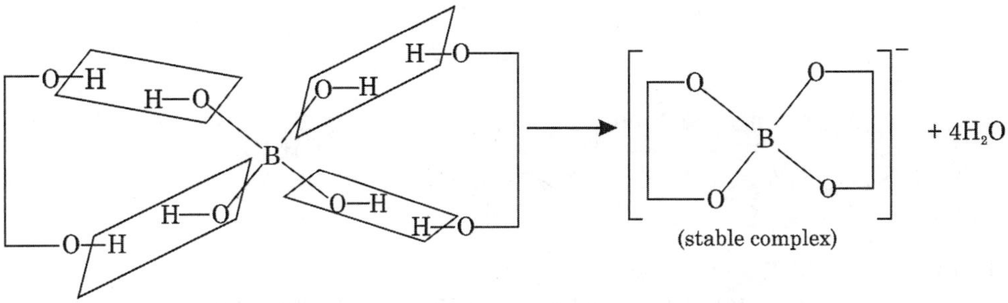

(stable complex)

(C) It has a planar sheet like structure due to hydrogen bonding.

(D) H_3BO_3 is a weak electrolyte in water.

8. (B, C, D)

(A) $2CuFeS_2 + O_2 \xrightarrow{\Delta} Cu_2S + 2FeS + SO_2$

(B) $4CuO \xrightarrow{1100°C} 2Cu_2O + O_2$

$2Cu_2O + Cu_2S \xrightarrow{\Delta} 6Cu + SO_2$

(C) $Cu_2S + 2Cu_2O \xrightarrow{\Delta} 6Cu + SO_2$

(D) $CuSO_4 \xrightarrow{720°C} CuO + SO_2 + \frac{1}{2}O_2$

$$4CuO \xrightarrow{1100°C} 2Cu_2O + O_2$$

$$2Cu_2O + Cu_2S \xrightarrow{\Delta} 6Cu + SO_2$$

9. (A, B, C)

(A) Sodium (Na) when dissolved in excess liquid ammonia, forms a blue coloured paramagnetic solution.

(B) $K + O_2 \longrightarrow \underset{\text{(Potassium superoxide)}}{KO_2}$

KO_2 is paramagnetic.

(C) $3Cu + \underset{\text{(dilute)}}{8HNO_3} \longrightarrow 3Cu(NO_3)_2 + 2NO + 4H_2O$

NO is paramagnetic.

(D) 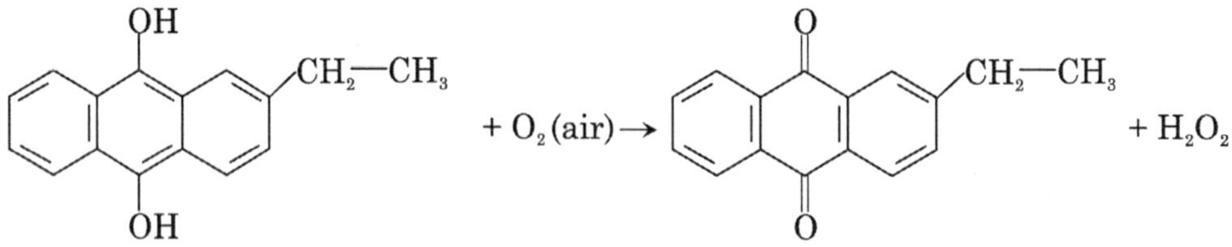

(2-Ethylanthraquinol)

H_2O_2 is diamagnetic

10. (A)

Only amines undergo acetylation and not acid amides.

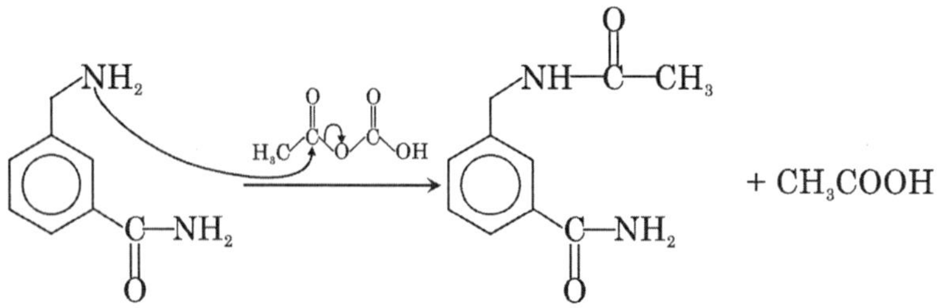

11. (6)

Black coloured sulphides are PbS, CuS, HgS, Ag_2S, NiS, CoS.

Bi_2S_3 in its crystalline form is dark brown but Bi_2S_3 precipitate obtained is black in colour.

12. (3)

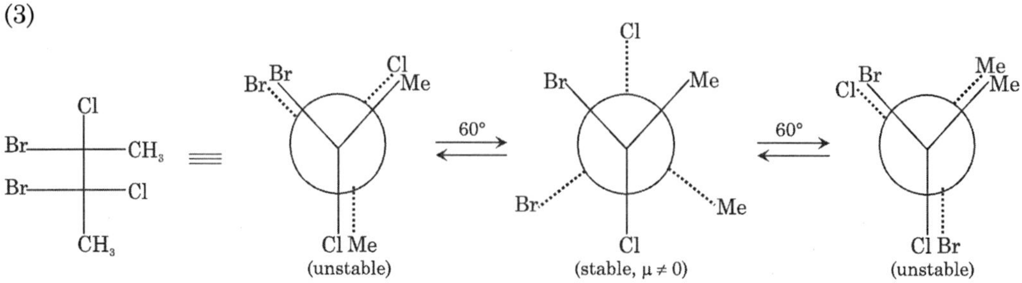

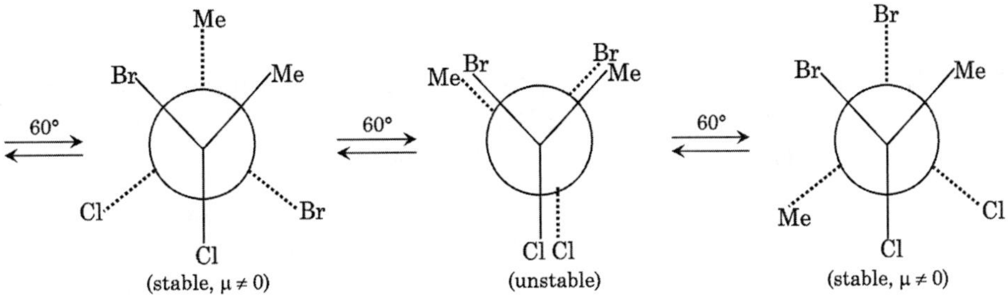

13. (7)

$K_2Cr_2O_7$, $CuSO_4$, H_2O_2, Cl_2, O_3, $FeCl_3$, HNO_3

$K_2Cr_2O_7, 7H_2SO_4 + 6KI \longrightarrow 4K_2SO_4 + Cr_2(SO_4)_3 + 3I_2 + 7H_2O$

$2CuSO_4 + 4KI \longrightarrow Cu_2I_2 + I_2 + 2K_2SO_4$

$H_2O_2 + 2KI \longrightarrow I_2 + 2KOH$

$Cl_2 + 2KI \longrightarrow 2KCl + I_2$

$O_3 + 2KI + H_2O \longrightarrow 2KOH + I_2 + O_2$

$2FeCl_3 + 2KI \longrightarrow 2FeCl_2 + I_2 + 2KCl$

$8HNO_3 + 6KI \longrightarrow 6KNO_3 + 2NO + 4H_2O + 3I_2$

$2KMnO_4 + KI + H_2O \longrightarrow KIO_3 + 2MnO_2 + 2KOH$

14. (4)

$XeF_4 \longrightarrow$ Square planar

$BrF_4^- \longrightarrow$ Square planar

$[Cu(NH_3)_4]^{2+} \longrightarrow$ Square planar

$[PtCl_4]^{2-} \longrightarrow$ Square planar

$SF_4 \longrightarrow$ See-saw

$SiF_4 \longrightarrow$ Tetrahedral

$BF_4^- \longrightarrow$ Tetrahedral

$[FeCl_4]^{2-} \longrightarrow$ Tetrahedral

$[CoCl_4]^{2-} \longrightarrow$ Tetrahedral

15. (5)

(1) $CH_3-CH_2-\underset{\substack{|\\CH_3}}{CH}-\overset{\substack{O\\||}}{C}-CH_3$ Will not give a racemic mixture on reduction with $NaBH_4$.

(2) 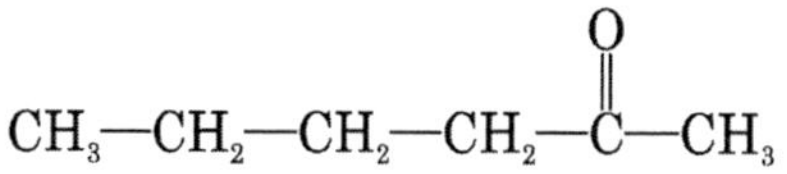 Will give a racemic mixture on reduction with $NaBH_4$.

(3) 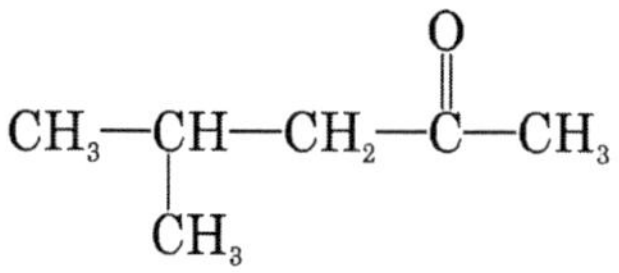 Will give a racemic mixture on reduction with $NaBH_4$.

(4) 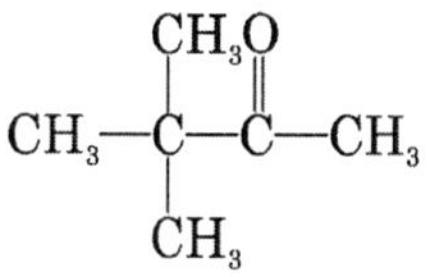 Will give a racemic mixture on reduction with $NaBH_4$.

(5) 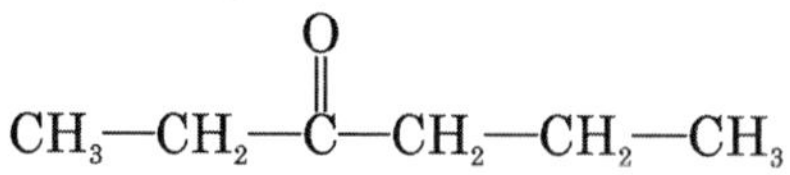 Will give a racemic mixture on reduction with $NaBH_4$.

(6) 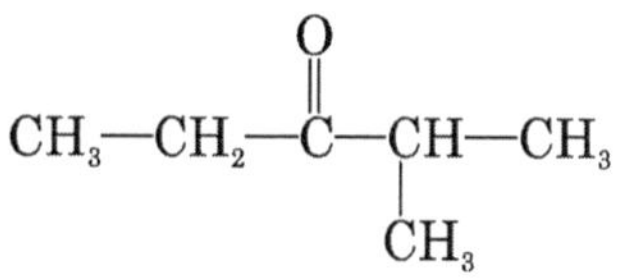 Will give a racemic mixture on reduction with $NaBH_4$.

16. (6)

$n = 4$

$\ell = 0, 1, 2, 3$

$|m_\ell| = 1 \Rightarrow \pm 1$

$m_s = -\frac{1}{2}$

For $\ell = 0,\ m_\ell = 0$

$\ell = 1,\ m_\ell = -1, 0, +1$

$\ell = 2,\ m_\ell = -2, -1, 0, +1, +2$

$\ell = 3,\ m_\ell = -3, -2, -1, 0, +1, +2, +3$

Hence, six electrons can have $|m_\ell| = 1$ and $m_s = -\frac{1}{2}$.

17. (4)

$k = \frac{R}{N_A}$

$R = kN_A$

$= 1.380 \times 10^{-23} \times 6.023 \times 10^{23} = 8.31174 \approx 8.312$

18. (2)

$MX_2 \rightleftharpoons M^{2+} + 2X^-$

$1 - \alpha \qquad \alpha \qquad 2\alpha$

$i = 1 + 2\alpha \qquad (\alpha = 0.5)$

$i = 2$

19. (1)

This peptide on complete hydrolysis produced 4 distinct amino acids which are given below:

(1) 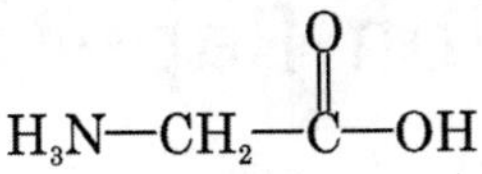

Glycine
(natural)

(2) 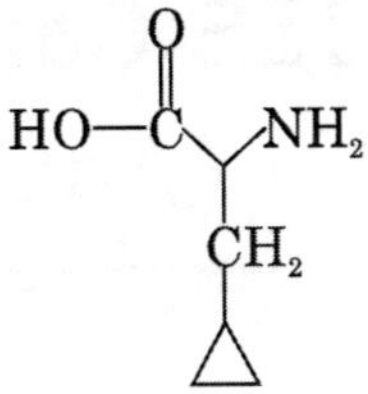

(3) 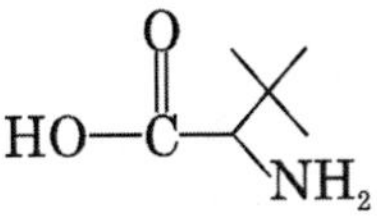

(4) 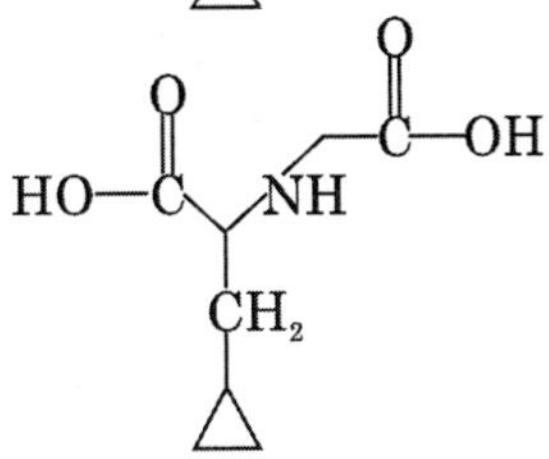

Only glycine is naturally occurring amino acid.

20. (8)

$V_{solution} \approx V_{solvent}$

Since in 1 L solution, 3.2 moles of solute are present, 1 L solution ≈ solvent (d = 0.4 g/mL) ≈ 0.4 kg

$$\text{Hence, molality } (m) = \frac{\text{moles of solute}}{\text{mass of solvent (kg)}} = \frac{3.2}{0.4} = 8$$

I.I.T. JEE Objective Question Paper (Advanced)—2014

Paper–2

INSTRUCTIONS

A. General

1. This booklet is your Question Paper. Do not break the seals of this booklet before being instructed to do so by the invigilators.
2. The question paper code is printed on the left hand top corner of this sheet and on the back page of this booklet.
3. Blank space and blank pages are provided in the question paper for your rough work. No additional sheets will be provided for rough work.
4. Blank papers, clipboards, log tables, slide rules, calculators, cameras, cellular phones, pagers and electronic gadgets are not allowed inside the examination hall.
5. Write your name and roll number in the space provided on the back cover of this booklet.
6. Answers to the questions and personal details are to be filled on an Optical Response Sheet, which is provided separately. The ORS is doublet of two sheets—upper and lower, having identical layout. The upper sheet is a machine-gradable Objective Response Sheet (ORS) which will be collected by the invigilator at the end of the examination. The upper sheet is designed in such a way that darkening the bubble with a ball point pen will leave an identical impression at the corresponding place on the lower sheet. You will be allowed to take away the lower sheet at the end of the examination (see Figure-1 on the back cover page for the correct way of darkening the bubbles for valid answers).
7. Use a black ball point pen only to darken the bubbles on the upper original sheet. Apply sufficient pressure so that the impression is created on the lower sheet. See Figure-1 on the back cover page for appropriate way of darkening the bubbles for valid answers.
8. Do not tamper with/mutilate the ORS so this booklet.
9. On breaking the seals of the booklet check that it contains 28 pages and all the 60 questions and corresponding answer choices are legible. Read carefully the instructions printed at the beginning of each section.

B. Filling the Right Part of the ORS

10. The ORS also has a code printed on its left and right parts.
11. Verify that the code printed on the ORS (on both the left and right parts) is the same as that on this booklet and put your signature in the Box designated as R4.
12. If the codes do not match, ask for a change of the Booklet/ORS as applicable.
13. Write your Name, Roll No. and the name of centre and sign with pen in the boxes provided on the upper sheet of ORS. Do not write any of this anywhere else. Darken the appropriate bubble under each digit of your Roll No. in such way that the impression is created on the bottom sheet. (see example in Figure-2 on the back cover).

C. Question Paper Format

The question paper consists of three parts (Physics, Chemistry and Mathematics). Each part consists of two sections.

14. **Section 1** contains 10 multiple choice questions. Each question has four choices (A), (B), (C) and (D) out of which one is correct.
15. **Section 2** contains 3 paragraphs each describing theory, experiment and data, etc. Six questions relate to three paragraphs with two questions on each paragraph. Each question pertaining to a particular passage should have only one correct answer among the four given choices (A), (B), (C) and (D).
16. **Section 3** contains 4 multiple choice questions. Each questions has two lists (List-1: P, Q, R and S; List-2: 1, 2, 3 and 4). The options for the correct match are provided as (A), (B), (C) and (D) out of which only one is correct.

SECTION – I (Only One Option Correct Type)

This section contains 10 multiple choice questions. Each question has four choices (A), (B), (C) and (D) out of which only one is correct.

1. Assuming 2*s*-2*p* mixing is not operative, the paramagnetic species among the following is

(A) Be_2 (B) B_2 (C) C_2 (D) N_2

2. For the process, $H_2O(\ell) \longrightarrow H_2O(g)$

at $T = 100°C$ and 1 atmosphere pressure, the correct choice is

(A) $\Delta S_{system} > 0$ and $\Delta S_{surrounding} > 0$ (B) $\Delta S_{system} > 0$ and $\Delta S_{surrounding} < 0$

(C) $\Delta S_{system} < 0$ and $\Delta S_{surrounding} > 0$ (D) $\Delta S_{system} < 0$ and $\Delta S_{surrounding} < 0$

3. For the elementary reaction $M \longrightarrow N$, the rate of disappearance of M increases by a factor of 8 upon doubling the concentration of M. The order of the reaction with respect of M is

(A) 4 (B) 3 (C) 2 (D) 1

4. For the identification of β-naphthol using dye test, it is necessary to use
   (A) dichloromethane solution of β-naphthol. (B) acidic solution of β-naphthol.
   (C) neutral solution of β-naphthol. (D) alkaline solution of β-naphthol.

5. Isomers of hexane, based on their branching, can be devided into three distinct classes as shown in the figure.

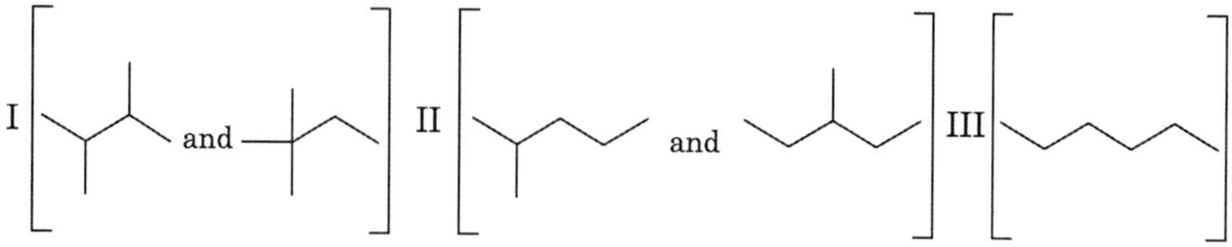

The correct order of their boiling point is
(A) I > II > III (B) III > II > I (C) II > III > I (D) III > I > II

6. The major product in the following reaction is

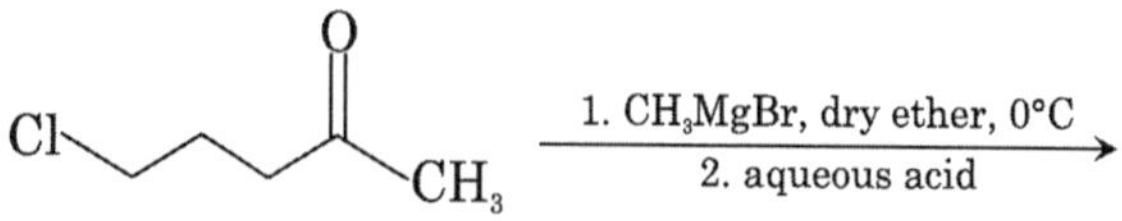

(A) 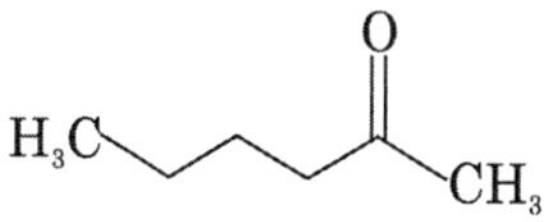

(B) 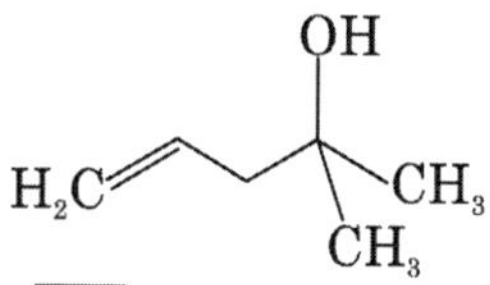

(C) 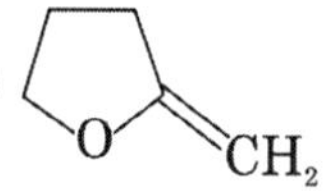

(D) 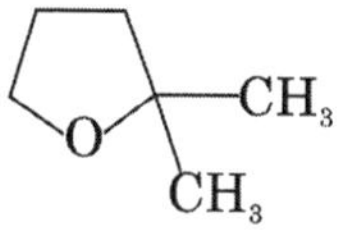

7. Under ambient conditions, the total number of gases released as products in the final step of the reaction scheme shown below is

XeF_6 —(Complete hydrolysis)→ P + other product
P —(OH^-/H_2O)→ Q
Q —(slow disproportionation in OH^-/H_2O)→ Products

(A) 0 (B) 1 (C) 2 (D) 3

8. The product formed in the reaction of $SOCl_2$ with white phosphorous is
   (A) PCl_3 (B) SO_2Cl_2 (C) SCl_2 (D) $POCl_3$

9. Hydrogen peroxide in its reaction with KIO_4 and NH_2OH, respectively, is acting as a
   (A) reducing agent, oxidising agent (B) reducing agent, reducing agent
   (C) oxidising agent, oxidising agent (D) oxidising agent, reducing agent.

10. The acidic hydrolysis of ether (X) shown below is fastest when

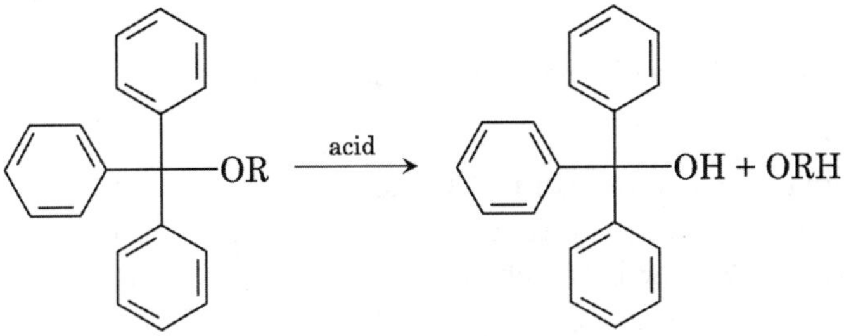

   (A) one phenyl group is replaced by a methyl group.
   (B) one phenyl group is replaced by a para-methoxyphenyl group.
   (C) two phenyl group are replaced by two para-methoxyphenyl groups.
   (D) no structural change is made to X.

SECTION – II Comprehension Type (Only One Option Correct)

This section contains three paragraphs, each describing theory, experiments, data, etc. Six questions relate to the three paragraphs with two questions on each paragraph. Each question has only one correct answer among the four given options (A), (B), (C) and (D).

Paragraph for Questions 11 and 12

X and Y are two volatile liquids with molar weights of 10 g mol^{-1} and 40 g mol^{-1}, respectively. Two cotton plugs, one soaked in X and the other soaked in Y, are simultaneously placed at the ends of a tube of length L = 24 cm, as shown in the figure. The tube is filled with an inert gas at 1 atmosphere pressure and a temperature of 300 K. Vapours to X and Y react to form a product which is first observed at a distance d cm from the plug soaked in X. Take X and Y to have equal molecular diameters and assume ideal behaviour for the inert gas and the two vapours.

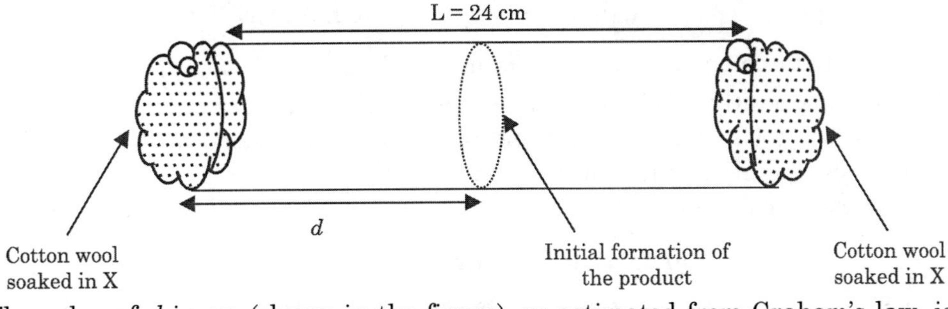

11. The value of *d* in cm (shown in the figure), as estimated from Graham's law, is
   (A) 8 (B) 12 (C) 16 (D) 20

12. The experimental value of *d* is found to be smaller than the estimate obtained using Graham's law. This is due to
    (A) larger mean free path for X as compared to that of Y.
    (B) larger mean free path for Y as compared to that of X.
    (C) increased collision frequency of Y with the inert gas as compared to that of X with the inert gas.
    (D) increased collision frequency of X with the inert gas as compared to that of Y with the inert gas.

Paragraph for Questions 13 and 14

Schemes 1 and 2 describe sequential transformation of alkynes M and N. Consider only the major products formed in each step for both schemes.

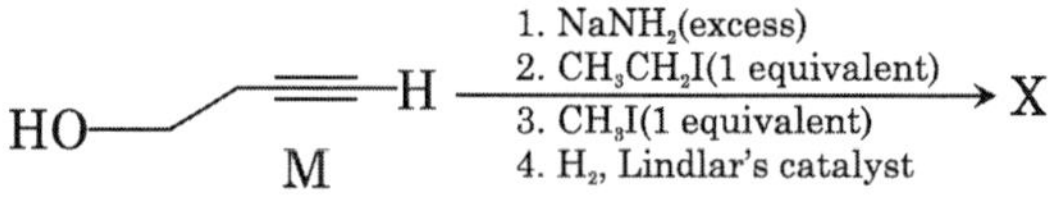

Scheme–1

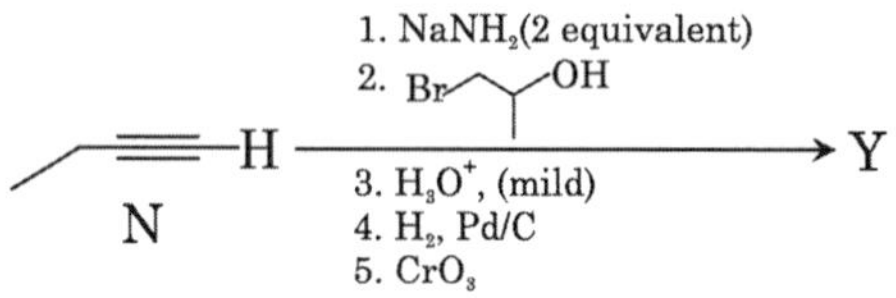

Scheme–2

13. The product X is

(A) 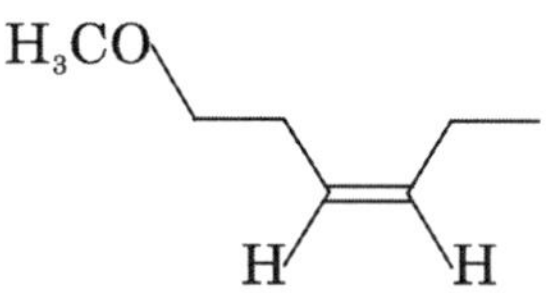

(B) 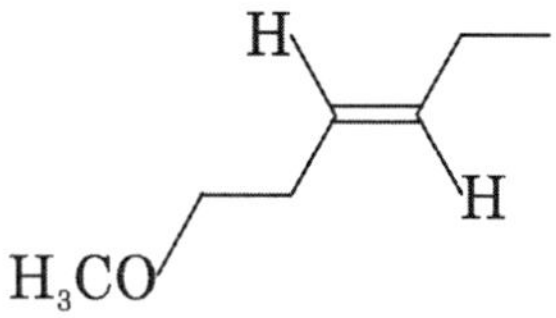

(C) 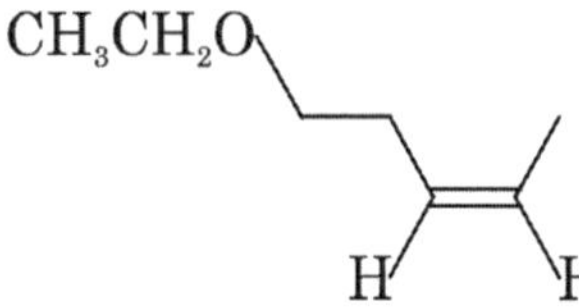

(D) 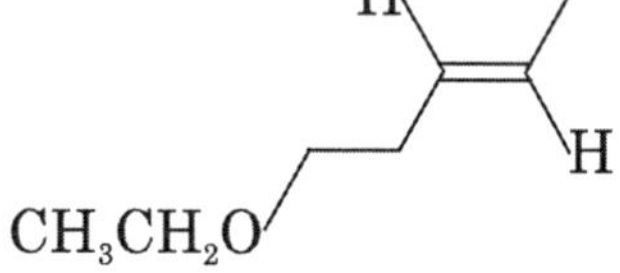

14. The correct statement with respect to product Y is
    (A) It gives a positive Tollens test and is a functional isomer of X.
    (B) It gives a positive Tollens test and is a geometrical isomer of X.
    (C) It gives a positive iodoform test and is a functional isomer of X.
    (D) It gives a positive iodoform test and is a geometrical isomer of X.

Paragraph for Questions 15 and 16

An aqueous solution of metal ion M1 reacts separately with reagents Q and R in excess to give tetrahedral and square planar complexes, respectively. An aqueous solution of another metal ion M2 always forms tetrahedral complexes with these reagents. Aqueous solution of M2 on reaction with reagent S gives white precipitate which dissolves in exess of S. The reactions are summarized in the scheme given below.

SCHEME:

$$\text{Tetrahedral} \xleftarrow[\text{excess}]{Q} \text{M1} \xrightarrow[\text{excess}]{R} \text{Square planar}$$

$$\text{Tetrahedral} \xleftarrow[\text{excess}]{Q} \text{M2} \xrightarrow[\text{excess}]{R} \text{Tetrahedral}$$

$$\text{M2} \xrightarrow{\text{S, stoichiometric amount}} \text{White precipitate} \xrightarrow[\text{excess}]{S} \text{precipitate dissolves}$$

15. M1, Q and R, respectively are
(A) Zn^{2+}, KCN and HCl
(B) Ni^{2+}, HCl and KCN
(C) Cd^{2+}, KCN and HCl
(D) Co^{2+}, HCl and KCN

16. Reagent S is
(A) $K_4[Fe(CN)_6]$
(B) Na_2HPO_4
(C) K_2CrO_4
(D) KOH

SECTION – III (Matching List Type) (Only One Option Correct)

This section contains four questions, each having two matching lists. Choices for the correct combination of elements from List-I and List-II are given as options (A), (B), (C) and (D) out of which only one is correct.

17. Match each coordination compound in List-I with an appropriate pair of characteristics from List-II and select the correct answer using the code given below the lists (en = $H_2NCH_2CH_2NH_2$; atomic numbers: Ti = 22, Cr = 24; Co = 27; Pt = 78)

List-I	List-II
P. $[Cr(NH_3)_4Cl_2]Cl$	1. Paramagnetic and exhibits ionisation isomerism
Q. $[Ti(H_2O)_5Cl](NO_3)_2$	2. Diamagnetic and exhibits *cis-trans* isomerism
R. $[Pt(en)(NH_3)Cl]NO_3$	3. Paramagnetic and exhibits *cis-trans* isomerism
S. $[Co(NH_3)_4(NO_3)_2]NO_3$	4. Diamagnetic and exhibits ionisation isomerism

Codes:	P	Q	R	S
(A)	4	2	3	1
(B)	3	1	4	2
(C)	2	1	3	4
(D)	1	3	4	2

18. Match the orbital overlap figures shown in List-I with the description given in List-II and select the correct answer using the code given below the lists.

List-I	List-II
P. 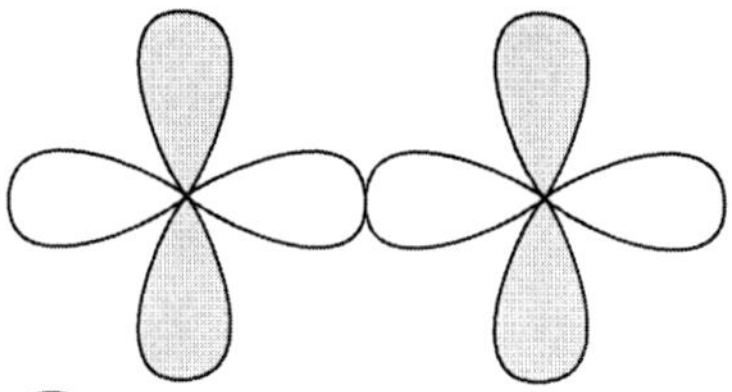	1. *p-d* π antibonding
Q. 	2. *d-d* σ bonding
R. 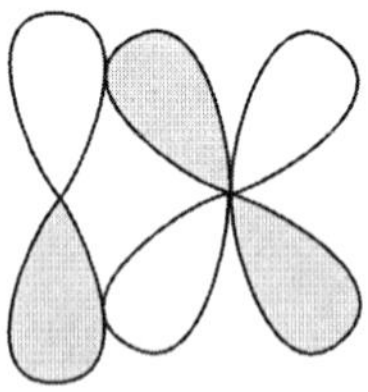	3. *p-d* π bonding
S. 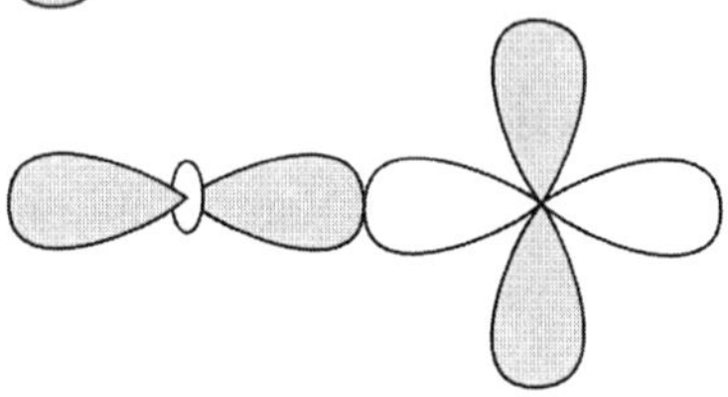	4. *d-d* σ antibonding

Codes:	**P**	**Q**	**R**	**S**
(A)	2	1	3	4
(B)	4	3	1	2
(C)	2	3	1	4
(D)	4	1	3	2

19. Different possible thermal decomposition pathways for peroxyesters are shown below. Match each pathway from List-I with an appropriate structure from List-II and select the correct answer using the code given below the lists.

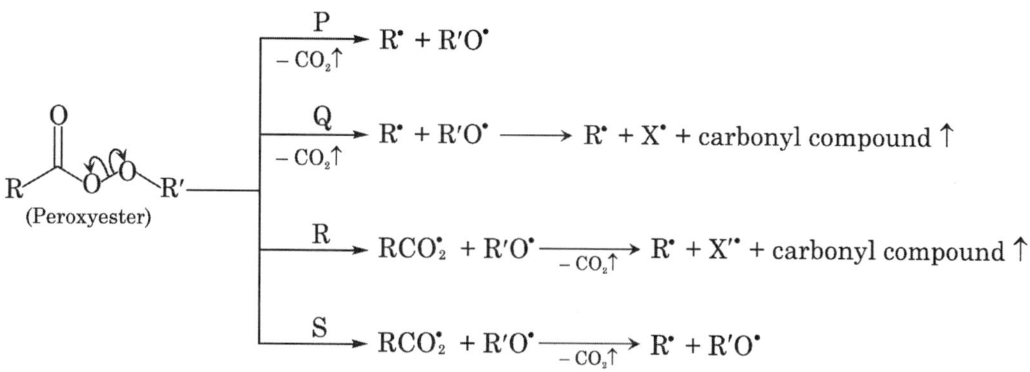

List-I	List-II
P. Pathway P	1. 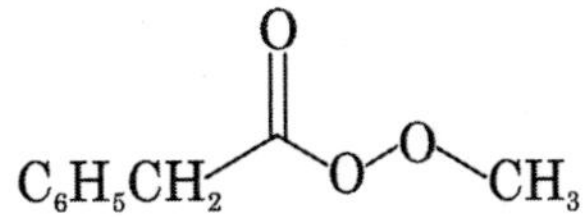
Q. Pathway Q	2. 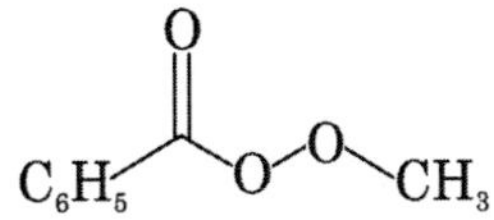
R. Pathway R	3. 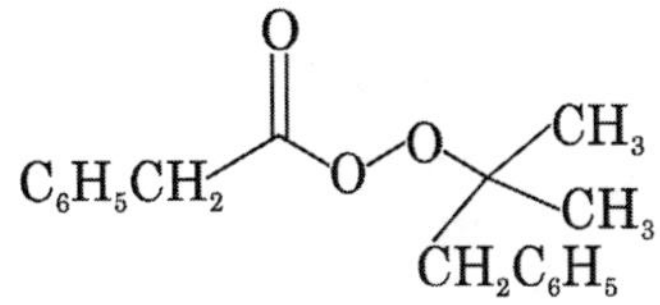
S. Pathway S	4. 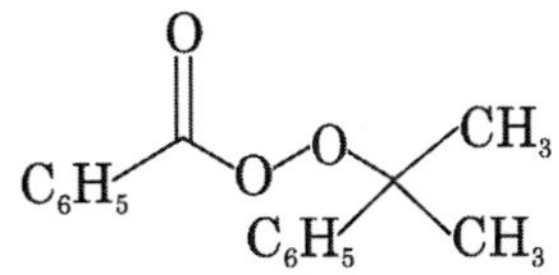

Codes:	**P**	**Q**	**R**	**S**
(A)	1	3	4	2
(B)	2	4	3	1
(C)	4	1	2	3
(D)	3	2	1	4

20. Match the four starting materials (P, Q, R, S) given in List-I with the corresponding reaction schemes (I, II, III, IV) provided in List-II and select the correct answer using the code given below the lists.

List-I

P. $H-C\equiv C-H$

Q. 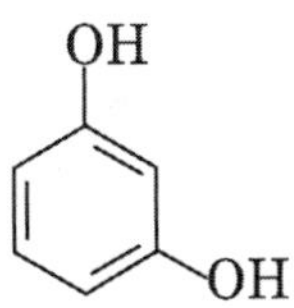

R. 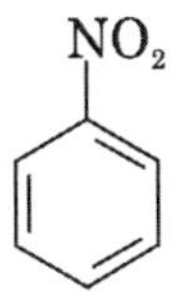

S. 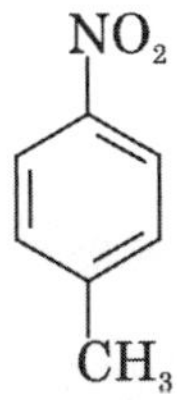

List-II

1. Scheme I

$$? \xrightarrow[\text{(ii) SOCl}_2\text{ (iv) NH}_3]{\text{(i) KMnO}_4\text{, HO}^-\text{, heat (ii) H}^+\text{, H}_2\text{O}} C_7H_6N_2O_3$$

2. Scheme II

$$? \xrightarrow[\text{(ii) HNO}_3\text{ (v) dil. H}_2\text{SO}_4\text{, heat (iv) HO}^-]{\text{(i) Sn/HCl (ii) CH}_3\text{COCl (iii) conc. H}_2\text{SO}_4} C_6H_6N_2O_2$$

3. Scheme III

$$? \xrightarrow[\text{(iii) H}_2\text{S, NH}_3\text{ (iv) NaNO}_2\text{, H}_2\text{SO}_4\text{ (v) hydrolysis}]{\text{(i) red hot iron, 873 K (ii) fuming HNO}_3\text{, H}_2\text{SO}_4\text{, heat}} C_6H_5NO_3$$

4. Scheme IV

$$? \xrightarrow[\text{(ii) conc. HNO}_3\text{, conc. H}_2\text{SO}_4\text{ (iii) dil. H}_2\text{SO}_4\text{, heat}]{\text{(i) conc. H}_2\text{SO}_4\text{, 60°C}} C_6H_5NO_4$$

Codes:	**P**	**Q**	**R**	**S**
(A)	1	4	2	3
(B)	3	1	4	2
(C)	3	4	2	1
(D)	4	1	3	2

ANSWERS

1. C	2. B	3. B	4. D	5. B	6. D
7. C	8. A	9. A	10. C	11. C	12. D
13. A	14. C	15. B	16. D	17. B	18. C
19. A	20. C				

Model Solutions—I.I.T. JEE (Advanced) 2014

Paper–2

1. (C)

   Assuming that no 2*s*-2*p* mixing takes place

   (A) $Be_2 \longrightarrow \sigma 1s^2, \sigma^* 1s^2, \sigma 2s^2, \sigma^* 2s^2$ (diamagnetic)

   (B) $B_2 \longrightarrow \sigma 1s^2, \sigma^* 1s^2, \sigma 2s^2, \sigma^* 2s^2, \sigma 2p_z^2, {}^{\pi 2p_x^0}_{\pi 2p_y^0}$ (diamagnetic)

   (C) $C_2 \longrightarrow \sigma 1s^2, \sigma^* 1s^2, \sigma 2s^2, \sigma^* 2s^2, \sigma 2p_z^2, {}^{\pi 2p_x^1}_{\pi 2p_y^1}, {}^{\pi^* 2p_x^0}_{\pi^* 2p_y^0}, \sigma^* 2p_z^0$ (paramagnetic)

   (D) $N_2 \longrightarrow \sigma 1s^2, \sigma^* 1s^2, \sigma 2s^2, \sigma^* 2s^2, \sigma 2p_z^2, {}^{\pi 2p_x^2}_{\pi 2p_y^2}, {}^{\pi^* 2p_x^0}_{\pi^* 2p_y^0}, \sigma^* 2p_z^0$ (diamagnetic)

2. (B)

   At 100°C and 1 atmosphere pressure $H_2O(\ell) \rightleftharpoons H_2O(g)$ is at equilibrium. For equilibrium $\Delta S_{total} = 0$ and $\Delta S_{system} + \Delta S_{surrounding} = 0$

   Hence, $\Delta S_{system} > 0$ and $\Delta S_{surrounding} < 0$

3. (B)

   $$\frac{r_1}{r_2} = \frac{1}{8} = \frac{[M]^n}{[2M]^n} \text{ or } n = 3$$

4. (D)

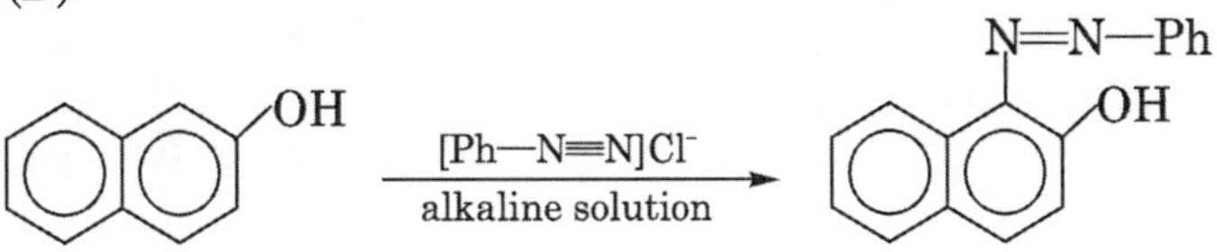

5. (B)

   III > II > I

   More the branching in an alkane, lesser will be the surface area and lesser will be the boiling point.

6. (D)

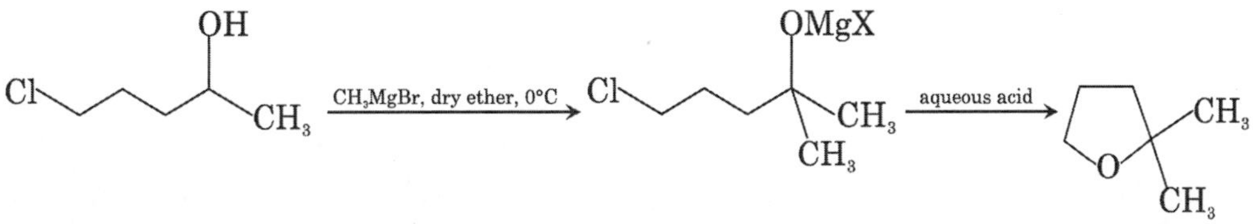

7. (C)

$$XeF_6 + 3H_2O \longrightarrow XeO_3 + 3H_2F_2$$

$$XeO_3 \xrightarrow{OH^-} HXeO_4^-$$

$$HXeO_4^- \xrightarrow{OH^-/H_2O \ \text{(disproportionation)}} XeO_6^{4-}(g) + Xe(g) + H_2O(\ell) + O_2(g)$$

8. (A)

$$P_4(s) + 8SOCl(\ell) \longrightarrow 4PCl_3(\ell) + 4SO_2(g) + 2S_2Cl(g)$$

9. (A)

$$KIO_4 + H_2O_2 \longrightarrow KIO_3 + H_2O + O_2$$

$$NH_2OH + 3H_2O_2 \longrightarrow HNO_3 + 4H_2O$$

10. (C)

When two phenyl groups are replaced by two para methoxy group, carbocation formed will be more stable.

11. (C)

$$\frac{r_X}{r_Y} = \frac{d}{24 - d} = \sqrt{\frac{40}{10}} = 2$$

or $\quad d = 48 - 2d$

or $\quad 3d = 48$

or $\quad d = 16$ cm

12. (D)

As the collision frequency increases then molecular speed decreases than the expected.

13. (A)

14. (C)

Solution for Q. No. 13 to 14

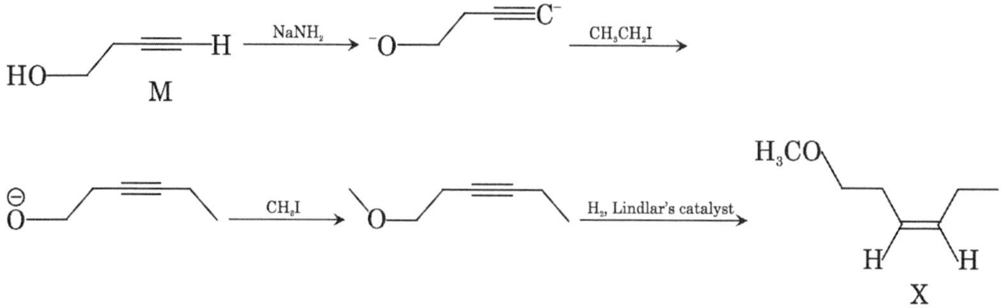

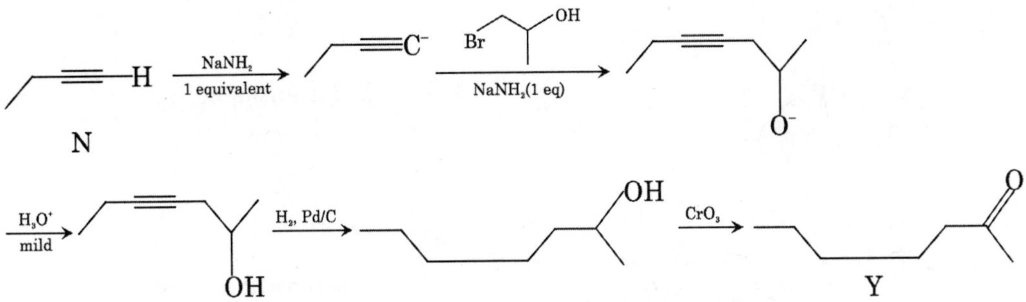

X and Y are functional isomers of each other and Y gives iodoform test.

15. (B)
16. (D)

Solution for the Q. No. 15 to 16.

$$\underset{\text{Tetrahedral}}{[NiCl_4]^{2-}} \xleftarrow[\text{excess}]{HCl(Q)} \underset{(M_1)}{Ni^{2+}} \xrightarrow[\text{excess}]{KCN(R)} \underset{\text{Square planar}}{[Ni(CN)_4]^{2-}}$$

$$\underset{\text{Tetrahedral}}{[ZnCl_4]^{2-}} \xleftarrow[\text{excess}]{HCl(Q)} \underset{(M_2)}{Zn^{2+}} \xrightarrow[\text{excess}]{KCN(R)} \underset{\text{Tetrahedral}}{[Zn(CN)_4]^{2-}}$$

$$Zn^{2+} \xrightarrow{\text{KOS (S)}} \underset{\text{White ppt.}}{Zn(OH)_2} \xrightarrow[\text{excess}]{KOH} \underset{\text{Soluble}}{[Zn(OH)_4]^{2-}}$$

17. (B)

(P) $[Cr(NH_3)_4Cl_2]Cl \rightarrow$ Paramagnetic and exhibits cis-trans isomerism

(Q) $[Ti(H_2O)_5Cl](NO_3)_2 \rightarrow$ Paramagnetic and exhibits ionization isomerism

(R) $[Pt(en)(NO_3)Cl]NO_3 \rightarrow$ Diamagnetic and exhibits ionization isomerism

(S) $[Co(NH_3)_4(NO_3)_2]NO_3 \rightarrow$ Diamagnetic and exhibits cis-trans isomerism

18. (C)

P. 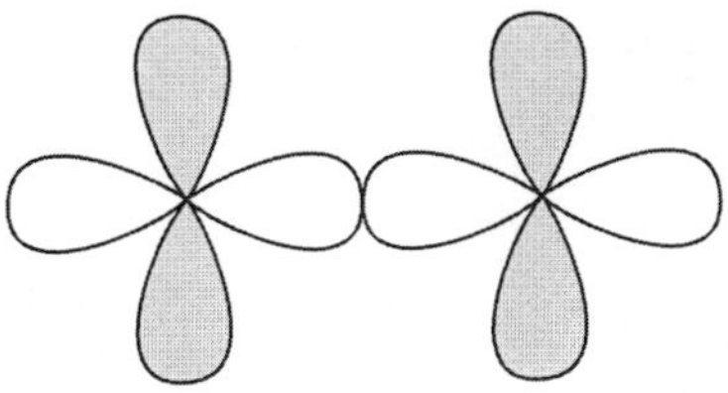 → *d*-*d* σ bonding

Q. 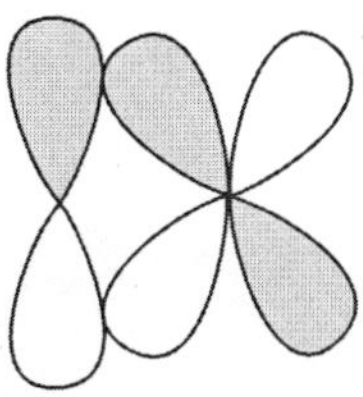 → *p*-*d* π bonding

R. 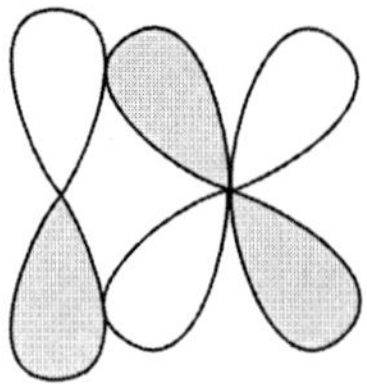 → p-d π antibonding

S. 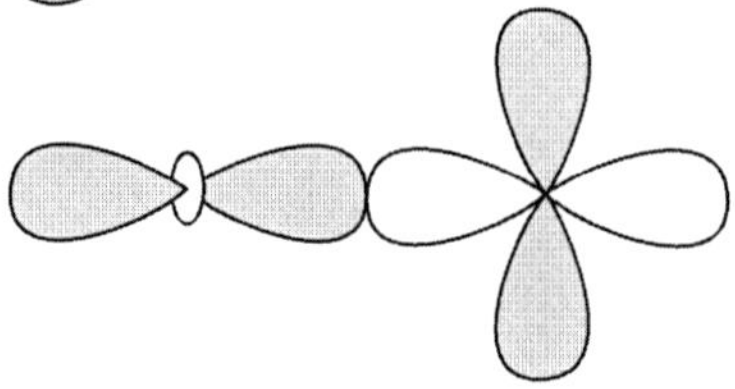 → d-d σ antibonding

19. (A)

(P) – 1; (Q) – 3; (R) – 4; (S) – 2

(P) 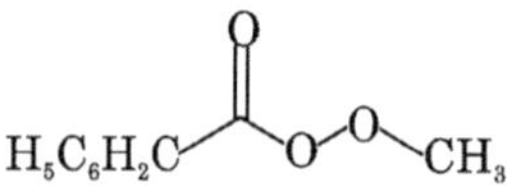 $C_6H_5—\dot{C}H_2 + CO_2 + CH_3O^\bullet$

(Q) 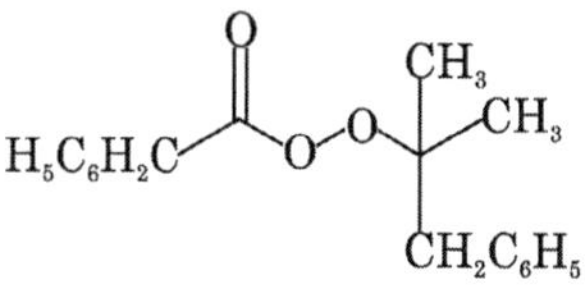 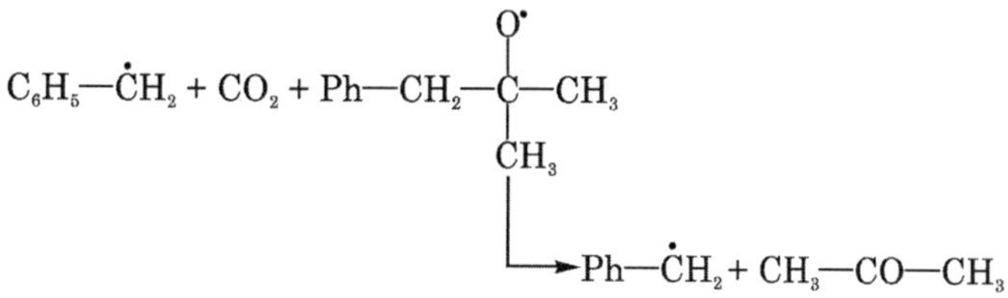

(R) 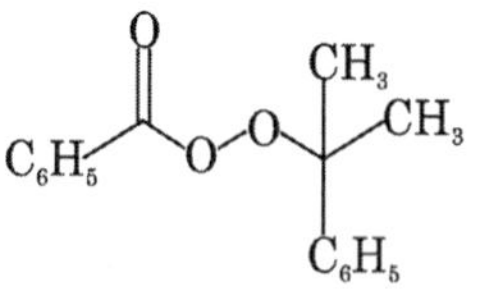 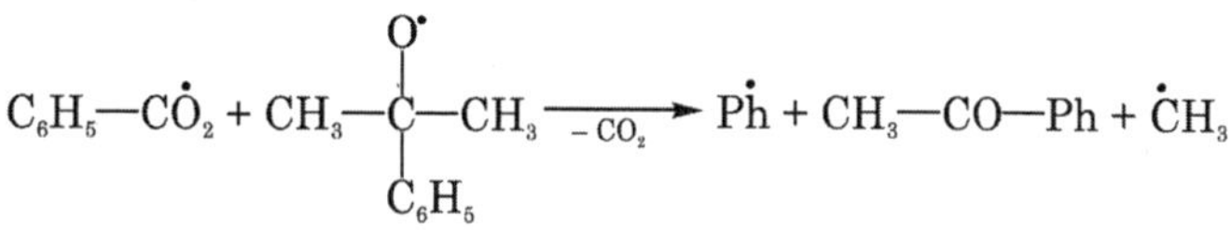

(S) 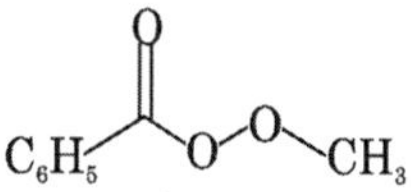 $C_6H_5—C\dot{O}_2 + CH_3O^\bullet$ → $C_6H_5^\bullet + CO_2$

20. (C)

(P) 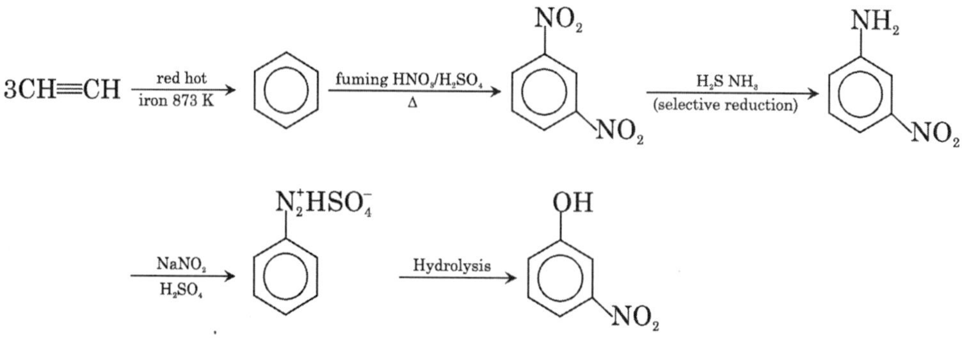

(Q) 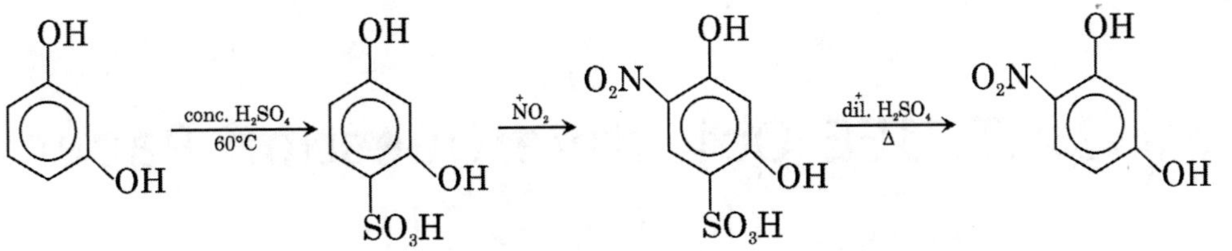

(R) 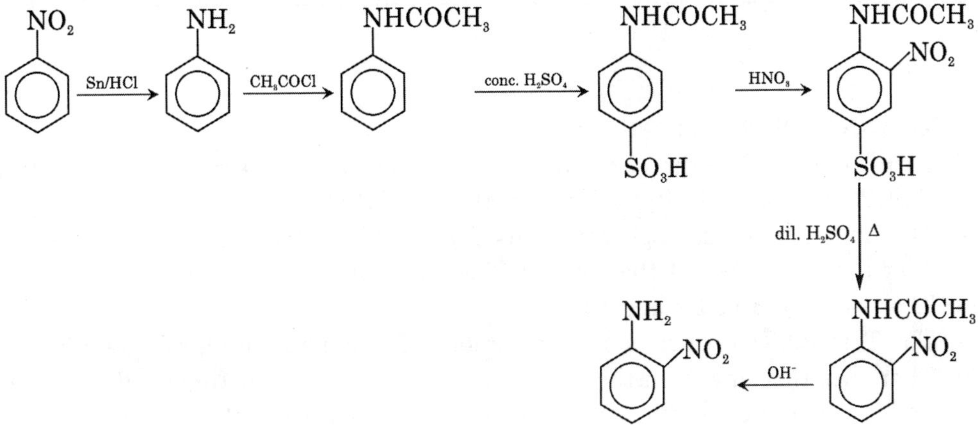

(S) 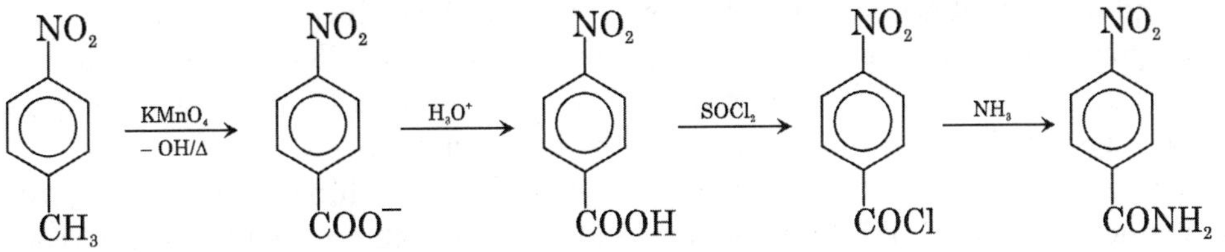

I.I.T. JEE Objective Question Paper (Main)—2015

IMPORTANT INSTRUCTIONS

1. Immediately fill in the particulars on this page of the Test Booklet with Blue/Black Ball Point Pen. Use of pencil is strictly prohibited.
2. The Answer Sheet is kept inside this Test Booklet. When you are directed to open the Test Booklet, take out the Answer Sheet and fill in the particulars carefully.
3. The test is of 3 hours duration.
4. The Test Booklet consists of 90 questions. The maximum marks are 360.
5. There are three parts in the question paper A, B, C consisting of Physics, Chemistry and Mathematics having 30 questions in each part of equal weightage. Each question is allotted 4 (four) marks for each correct response.
6. Condidates will be awarded marks as stated above in instruction No. 5 for correct response of each question. 1/4 (one fourth) marks will be deducted for indicating incorrect response of each question. No deduction from the total score will be made if no response is indicated for an item in the answer sheet.
7. There is only one correct response for each question. Filling up more than one response in each question will be treated as wrong response and marks for wrong response will be deducted accordingly as per instruction 6 above.

1. Which of the following is the energy of a possible excited state of hydrogen?

   (A) – 6.8 eV (B) – 3.4 eV (C) + 6.8 eV (D) + 13.6 eV

2. In the following sequence of reactions:

   $$\text{Toluene} \xrightarrow{KMnO_4} A \xrightarrow{SOCl_2} B \xrightarrow[BaSO_4]{H_2/Pd} C$$

   The product C is

   (A) $C_6H_5CH_3$ (B) $C_6H_5CH_2OH$ (C) C_6H_5CHO (D) C_6H_5COOH

3. Which compound would give 5-keto-2-methylhexanal upon ozonolysis?

   (A) 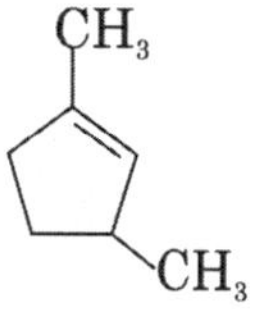

   (B) 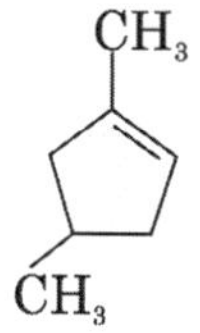

   (C) 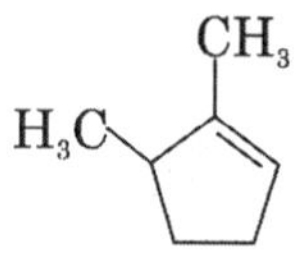

   (D) 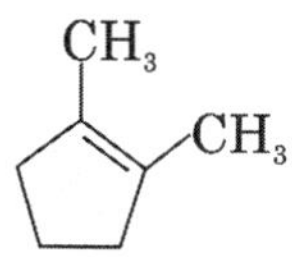

4. The ionic radii (in Å) of N^{3-}, O^{2-} and F^- are respectively
   (A) 1.36, 1.71 and 1.40
   (B) 1.71, 1.40 and 1.36
   (C) 1.71, 1.36 and 1.40
   (D) 1.36, 1.40 and 1.71

5. The colour of $KMnO_4$ is due to
   (A) *d–d* transition
   (B) L → M charge transfer transition
   (C) $\sigma-\sigma^*$ transition
   (D) M → L charge transfer transition.

6. Assertion: Nitrogen and oxygen are the main components in the atmosphere but these do not react to form oxides of nitrogen.
   Reason: The reaction between nitrogen and oxygen requires high temperature.
   (A) Both assertion and reason are correct, but the reason is not the correct explanation for the assertion
   (B) The assertion is incorrect, but the reason is correct
   (C) Both the assertion and reason are incorrect
   (D) Both assertion and reason are correct, and the reason is the correct explanation for the assertion.

7. Which of the following compounds is not an antacid?
   (A) Cimetidine
   (B) Phenelzine
   (C) Ranitidine
   (D) Aluminium hydroxide

8. In the context of the Hall–Heroult process for the extraction of Al, which of the following statements is false?
   (A) Al_2O_3 is mixed with CaF_2 which lowers the melting point of the mixture and brings conductivity
   (B) Al^{3+} is reduced at the cathode to form Al
   (C) Na_3AlF_6 serves as the electrolyte
   (D) CO and CO_2 are produced in this process.

9. Match the catalysts to the correct processes:

Catalyst	Process
(A) $TiCl_3$	(i) Wacker process
(B) $PdCl_2$	(ii) Ziegler–Natta polymerization
(C) $CuCl_2$	(iii) Contact process
(D) V_2O_5	(iv) Deacon's process

   (A) (A)–(ii), (B)–(i), (C)–(iv), (D)–(iii)
   (B) (A)–(ii), (B)–(iii), (C)–(iv), (D)–(i)
   (C) (A)–(iii), (B)–(i), (C)–(ii), (D)–(iv)
   (D) (A)–(iii), (B)–(ii), (C)–(iv), (D)–(i)

10. In the reaction,

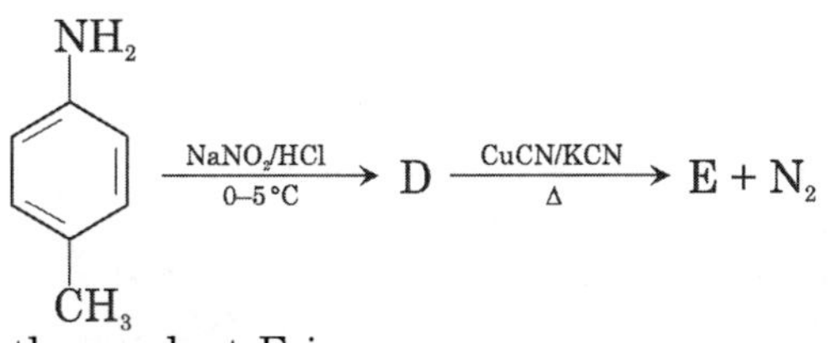

the product E is

(A) 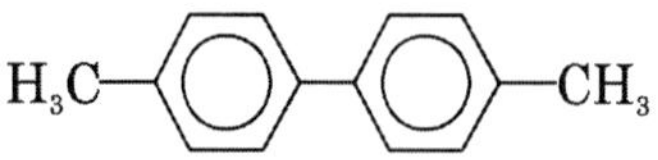

(B) 

(C) 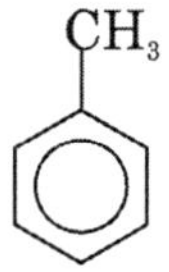

(D) 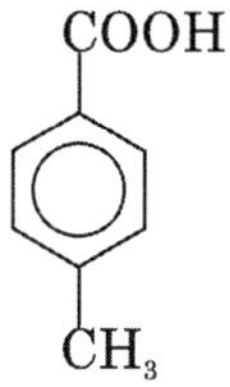

11. Which polymer is used in the manufacture of paints and lacquers?
    (A) Glyptal (B) Polypropene
    (C) Poly vinyl chloride (D) Bakelite
12. The number of geometric isomers that can exist for square planar $[Pt(Cl)(py)(NH_3)(NH_2OH)]^+$ is (py = pyridine):
    (A) 3 (B) 4 (C) 6 (D) 2
13. Higher order (> 3) reactions are rare due to
    (A) increase in entropy and activation energy as more molecules are involved
    (B) shifting of equilibrium towards reactants due to elastic collisions
    (C) loss of active species on collision
    (D) low probability of simultaneous collision of all the reacting species.
14. Which among the following is the most reactive?
    (A) Br_2 (B) I_2 (C) ICl (D) Cl_2
15. Two Faraday of electricity is passed through a solution of $CuSO_4$. The mass of copper deposited at the cathode is (At. Mass of Cu = 63.5 amu)
    (A) 63.5 g (B) 2 g (C) 127 g (D) 0 g
16. 3 g of activated charcoal was added to 50 mL of acetic acid solution (0.06*N*) in a flask. After an hour it was filtered and the strength of the filtrate was found to be 0.042*N*. The amount of acetic acid adsorbed (per gram of charcoal) is
    (A) 36 mg (B) 42 mg (C) 54 mg (D) 18 mg
17. The synthesis of alkyl fluorides is best accomplished by
    (A) Sandmeyer's reaction (B) Finkelstein reaction
    (C) Swarts reaction (D) Free radical fluorination.
18. The molecular formula of a commercial resin used for exchanging ions in water softening is $C_8H_7SO_3Na$ (Mol. Wt. 206). What would be the maximum uptake of Ca^{2+} ions by the resin when expressed in mole per gram resin?
    (A) $\frac{1}{206}$ (B) $\frac{2}{309}$ (C) $\frac{1}{412}$ (D) $\frac{1}{103}$

19. Which of the vitamins given below is water soluble?

(A) Vitamin D (B) Vitamin E (C) Vitamin K (D) Vitamin C

20. The intermolecular interaction that is dependent on the inverse cube of distance between the molecules is

(A) ion-dipole interaction (B) London force
(C) hydrogen bond (D) ion-ion interaction.

21. The following reaction is performed at 298 K.

$2NO(g) + O_2(g) \rightleftharpoons 2NO_2(g)$

The standard free energy of formation of NO(*g*) is 86.6 kJ/mol at 298 K. What is the standard free energy of formation of $NO_2(g)$ at 298 K? ($K_P = 1.6 \times 10^{12}$)

(A) $86600 + R(298) \ln (1.6 \times 10^{12})$ (B) $86600 \dfrac{\ln (1.6 \times 10^{12})}{R(298)}$
(C) $0.5[2 \times 86{,}600 - R(298) \ln (1.6 \times 10^{12})]$ (D) $R(298) \ln (1.6 \times 10^{12}) - 86600$

22. Which of the following compounds is not colored yellow?

(A) $K_3[Co(NO_2)_6]$ (B) $(NH_4)_3[As(Mo_3O_{10})_4]$
(C) $BaCrO_4$ (D) $Zn_2[Fe(CN)_6]$

23. In Carius method of estimation of halogens, 250 mg of an organic compound gave 141 mg of AgBr. The percentage of bromine in the compound is (At. Mass Ag = 108; Br = 80)

(A) 36 (B) 48 (C) 60 (D) 24

24. Sodium metal crystallizes in a body centred cubic lattice with a unit cell edge of 4.29 Å. The radius of sodium atom is approximately

(A) 3.22 Å (B) 5.72 Å (C) 0.93 Å (D) 1.86 Å

25. Which of the following compounds will exhibit geometrical isomerism?

(A) 3-Phenyl-1-butene (B) 2-Phenyl-1-butene
(C) 1,1-Diphenyl-1-propane (D) 1-Phenyl-2-butene

26. The vapour pressure of acetone at 20 °C is 185 torr. When 1.2 g of a non-volatile substance was dissolved in 100 g of acetone at 20 °C, its vapour pressure was 183 torr. The molar mass (g mol^{-1}) of the substance is

(A) 64 (B) 128 (C) 488 (D) 32

27. From the following statements regarding H_2O_2, choose the incorrect statement

(A) It decomposes on exposure to light
(B) It has to be stored in plastic or wax lined glass bottles in dark
(C) It has to be kept away from dust
(D) It can act only as an oxidizing agent.

28. Which one of the following alkaline earth metal sulphates has its hydration enthalpy greater than its lattice enthalpy?

(A) $BeSO_4$ (B) $BaSO_4$ (C) $SrSO_4$ (D) $CaSO_4$

29. The standard Gibbs energy change at 300 K for the reaction $2A \rightleftharpoons B + C$ is 2494.2 J. At a given time, the composition of the reaction mixture is $[A] = \frac{1}{2}$, $[B] = 2$ and $[C] = \frac{1}{2}$. The reaction proceeds in the [R = 8.314 J/K/mol, e = 2.718]
   (A) reverse direction because $Q > K_c$
   (B) forward direction because $Q < K_c$
   (C) reverse direction because $Q < K_c$
   (D) forward direction because $Q > K_c$

30. Which one has the highest boiling point?
   (A) Ne (B) Kr (C) Xe (D) He

ANSWERS

1. B	2. C	3. A	4. B	5. B	6. D
7. B	8. C	9. A	10. B	11. A	12. A
13. D	14. C	15. A	16. D	17. C	18. C
19. D	20. A	21. C	22. D	23. D	24. D
25. D	26. A	27. D	28. A	29. A	30. C

Model Solutions—I.I.T. JEE (Main) 2015

1. (B)

$$E_n = -13.6 \times \frac{Z^2}{n^2} \text{ eV}$$

For hydrogen $Z = 1$

$$E_n = -13.6 \times \frac{1}{n^2} \text{ eV}$$

$$E_2 = \frac{-13.6}{4} = -3.4 \text{ eV}$$

2. (C)

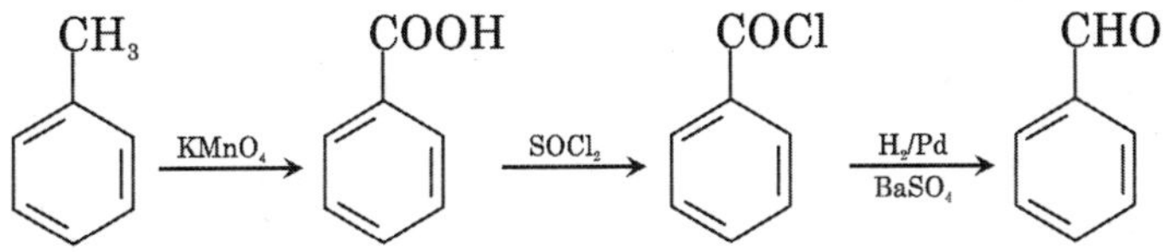

(A) (B) (C)

3. (A)

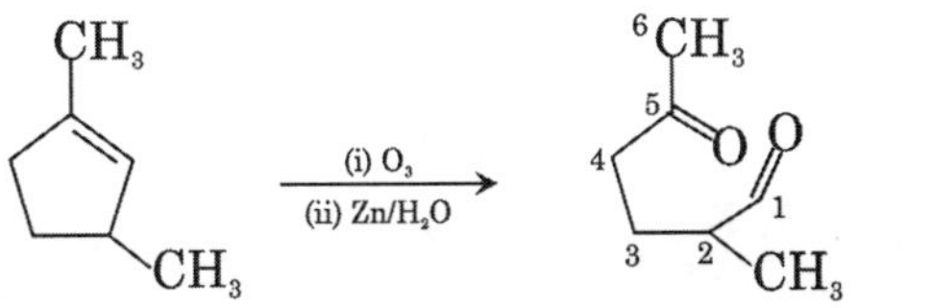

5-keto-2-methylhexanal

4. (B)
   Ionic radii
   $N^{3-} > O^{2-} > F^-$
5. (B)
   L $\longrightarrow$ M charge transfer transition
6. (D)
7. (B)
   Phenelzine is anti-depressant.
8. (C)
   $2Al_2O_3 + 3C \longrightarrow 4Al + 3CO_2$
   Na_3AlF_6 or CaF_2 is mixed with purified Al_2O_3 to lower the melting point and to bring conductivity. Oxygen liberated at anode reacts with carbon anode to liberate CO and CO_2.

9. (A)

$TiCl_4 + (C_2H_5)_3Al$: Ziegler–Natta catalyst, used for coordination polymerization.

$$CH_2{=}CH_2 \xrightarrow[H_2O + PdCl_2]{\text{Wacker process}} CH_3{-}CHO + 2HCl$$

$CuCl_2$ is used as catalyst in Deacon's process of production of Cl_2.

V_2O_5 is used as catalyst in Contact process of manufacturing of H_2SO_4.

10. (B)

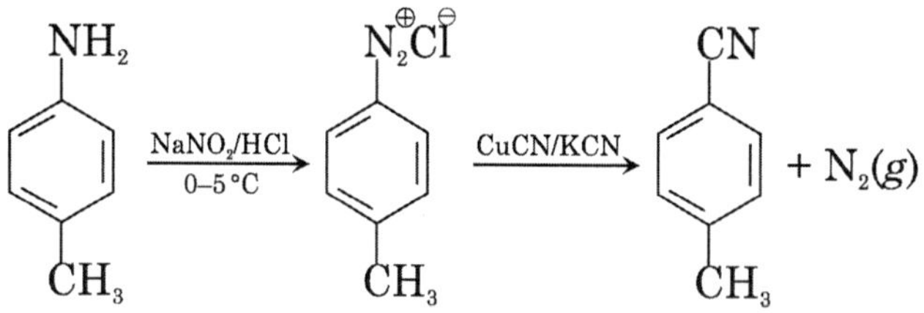

(D) (E)

11. (A)

Glyptal: Manufacture of paints and lacquers

Polypropene: Manufacture of ropes, toys

Poly vinyl chloride: Manufacture of raincoat, handbags

Bakelite: Making combs and electrical switch

12. (A)

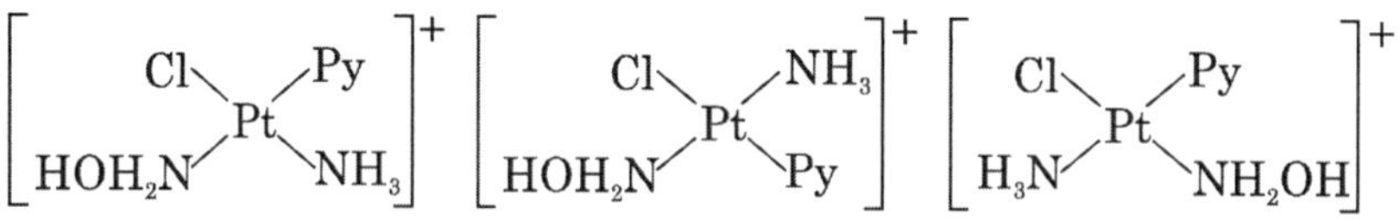

13. (D)

Higher order (> 3) reactions are less probable due to low probability of simultaneous collision of all the reaction species.

14. (C)

Interhalogen compounds are more reactive than the corresponding halogen molecules due to polarity of bond.

15. (A)

$Cu^{2+} + 2e^- \longrightarrow Cu$

2F of electricity will give 1 mole of Cu.

16. (D)

Amount adsorbed

$= (0.060 - 0.042) \times 50 \times 10^{-3} \times 60$

$= 0.018 \times 50 \times 60 \times 10^{-3} = 0.018 \times 3 = 0.054$ gm $= 54$ mg

Amount adsorbed per gram of activated charcoal $= \frac{54}{3} = 18$ mg

17. (C)

$2R{-}Cl + Hg_2F_2 \xrightarrow{\Delta} 2F{-}F + Hg_2Cl_2$

18. (C)

$2C_8H_7SO_3Na + Ca^{2+} \rightarrow (C_8H_7SO_3)_2Ca$

2 mole 1 mole

412 gm 1 mole

Maximum uptake of Ca^{2+} ions by the resin = 1/412 (mole per gm resin)

19. (D)

Vitamin C is water soluble.

20. (A)

For ion-dipole interaction,

$F \propto \mu \frac{dE}{dr}$ (where μ is dipole moment of dipole and r is distance between ion and dipole)

$\propto \mu \frac{d}{dr}\left(\frac{1}{r^2}\right) \propto \frac{\mu}{r^3}$

21. (C)

$\Delta G^0_{(\text{Re x}^0)} = 2\Delta G^0_f(NO_2) - 2\Delta G^0_f(NO) - \Delta G^0_f(O_2)$

$2\Delta G^0_f(NO_2) = \Delta G^0_{(\text{Re x}^0)} + 2\Delta G^0_f(NO) + \Delta G^0_f(O_2)$

$\Delta G = \Delta G^0 + RT \ln k_p$

At equilibrium,

$\Delta G = 0,\ Q = k_p,\ \Delta G^0 = -RT \ln k_p,\ \Delta G^0(O_2) = 0$

$\Delta G^0_f(NO_2) = 0.5[2 \times 86{,}600 - R(298) \ln (1.6 \times 10^{12})]$

22. (D)

$Zn_2[Fe(CN)_6]$ is bluish-white and rest are yellow coloured compounds.

23. (D)

Let the organic compound is R—Br.

$R{-}Br + AgNO_3 \rightarrow AgBr(\downarrow)$

Hence, number of moles of AgBr $\equiv$ Number of moles of R—Br $\equiv$ Number of moles of Br

$\frac{141 \times 10^{-3}}{188}$ = number of moles of Br $= 0.75 \times 10^{-3}$

Mass of bromine = $0.75 \times 80 \times 10^{-3}$ g = 60 mg

Percentage of bromine = $\frac{60}{250} \times 100 = 24\%$

24. (D)

For body center unit cell, $4r = a\sqrt{3}$

Hence, radius of Na = $\frac{\sqrt{3} \times 4.29}{4} = 1.857 \approx 1.86$ Å

25. (D)

Only 1-phenyl-2-butene can show geometrical isomerism.

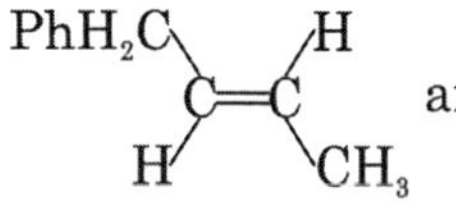

and

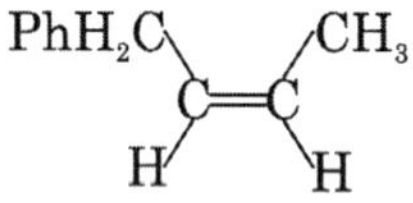

trans *cis*

26. (A)

Vapour pressure of pure acetone ($p°$) = 185 torr

Vapour pressure of solution (p) = 183 torr

Then from Raoult's law,

$$\frac{p° - p}{p} = \frac{n_{solute}}{n_{solvent}}$$

$$\frac{185 - 183}{183} = \frac{1.2/MW}{100/58}$$

$$\frac{2}{183} = \frac{1.2 \times 58}{100 \times MW}$$

$$MW = \frac{1.2 \times 58 \times 183}{200} = 63.68 \approx 64$$

27. (D)

H_2O_2 can act both as oxidizing agent and reducing agent and H_2O_2 decomposes on exposures to light and dust. Hence, it has to be stored in plastic or wax lined glass bottles in dark.

28. (A)

Because of smallest size of Be^{2+} ion, its hydration energy is maximum and is greater than the lattice energy of $BeSO_4$.

29. (A)

$\Delta G^0 = - RT \ln K$

$2494.2 = - 8.314 \times 300 \ln K$

$\ln K = - 1$

$\log_{2.718} K = -1$

$$K = (2.718)^{-1} = \frac{1}{2.718} = 0.3679$$

$$Q_c = \frac{[B][C]}{[A]^2} = \frac{2\times\frac{1}{2}}{\left(\frac{1}{2}\right)^2} = 4$$

The reaction proceeds in the reverse direction because $Q_c > K_c$.

30. (C)

The boiling point of noble gases in increasing order: He < Ne < Ar < Kr < Xe < Rn

Boiling point (K): 4.2 < 27.1 < 87.2 < 119.7 < 165 < 211

I.I.T. JEE Objective Question Paper (Advanced)—2015

Paper–1

INSTRUCTIONS

Question Paper Format and Marking Scheme

1. The question paper has three parts: Physics, Chemistry and Mathematics. Each part has three sections.
2. Section 1 contains 8 questions. The answer to each question is a single digit integer ranging from 0 to 9 (both inclusive).

   Marking Scheme: + 4 for correct answer and 0 in all other cases.
3. Section 2 contains 10 multiple choice questions with one or more than one correct option.

   Marking Scheme: + 4 for correct answer, 0 if not attempted and – 2 in all other cases.
4. Section 3 contains two "match the following" type questions and you will have to match entries in Column I with the entries in Column II.

   Marking Scheme: for each entry in Column I, + 2 for correct answer, 0 if not attempted and – 1 in all other cases.

SECTION–1
(Maximum Marks: 32)

This section contains eight questions.

The answer to each question is a single digit integer ranging from 0 to 9, both inclusive.

For each question, darken the bubble corresponding to the correct integer in the ORS

Marking scheme:

+ 4 if the bubble corresponding to the answer is darkened; 0 in all other cases

1. If the freezing point of a 0.01 molal aqueous solution of a cobalt(III) chloride-ammonia complex (which behaves as a strong electrolyte) is – 0.0558 °C, the number of chloride(s) in the coordination sphere of the complex is (K_f of water = 1.86 K kg mol^{-1})

2. The total number of stereoisomers that can exist for M is

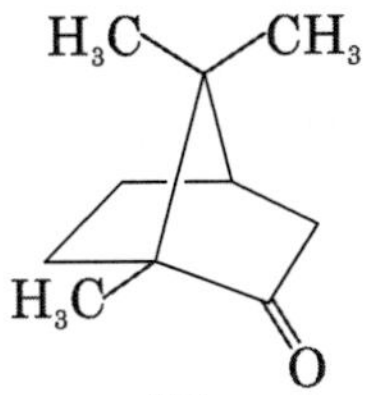

M

3. The number of resonance structures for N is

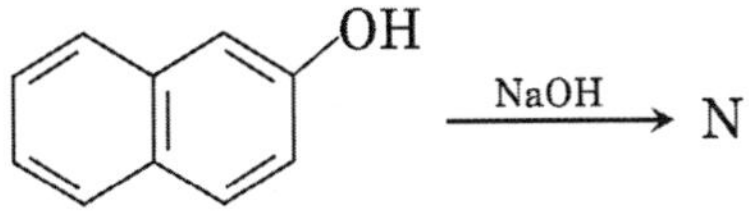

4. The total number of lone pairs of electrons in N_2O_3 is
5. For the octahedral complexes of Fe^{3+} in SCN^- (thiocyanato-S) and in CN^- ligand environments, the difference between the spin-only magnetic moments in Bohr magnetons (when approximated to the nearest integer) is (Atomic number of Fe = 26)
6. Among the triatomic molecules/ions, $BeCl_2$, N_3^-, N_2O, NO_2^+, O_3, SCl_2, ICl_2^-, I_3^- and XeF_2, the total number of linear molecule(s)/ion(s) where the hybridization of the central atom does not have contribution from the *d*-orbital(s) is (Atomic number: S = 16, Cl = 17, I = 53 and Xe = 54)
7. Not considering the electronic spin, the degeneracy of the second excited state ($n = 3$) of H atom is 9, while the degeneracy of the second excited state of H^- ion is
8. All the energy released from the reaction $X \longrightarrow Y$, $\Delta_r G^0 = -193$ kJ mol^{-1} is used for oxidizing M^+ as $M^+ \longrightarrow M^{3+} + 2e^-$, $E^0 = -0.25$ V.
Under standard conditions, the number of moles of M^+ oxidized when one mole of X is converted to Y is (F = 96500 C mol^{-1})

SECTION–2
(Maximum Marks: 40)

This section contains ten questions.
Each question has four options (A), (B), (C) and (D). One or more than one of these four option(s) is(are) correct.
For each question, darken the bubble(s) corresponding to the correct option(s) in the ORS
Marking scheme:
+ 4 if only the bubble(s) corresponding to all the correct option(s) is (are) darkened.
0 if none of the bubbles is darkened; – 2 in all other cases.

9. If the unit cell of a mineral has cubic close packed (*ccp*) array of oxygen atoms with m fraction of octahedral holes occupied by aluminium ions and n fraction of tetrahedral holes occupied by magnesium ions, m and n, respectively, are

(A) $\frac{1}{2}, \frac{1}{8}$ (B) $1, \frac{1}{4}$ (C) $\frac{1}{2}, \frac{1}{2}$ (D) $\frac{1}{4}, \frac{1}{8}$

10. Compound(s) that on hydrogenation produce(s) optically inactive compound(s) is(are)

(A) 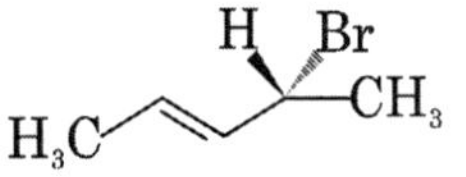

(B) 

(C) 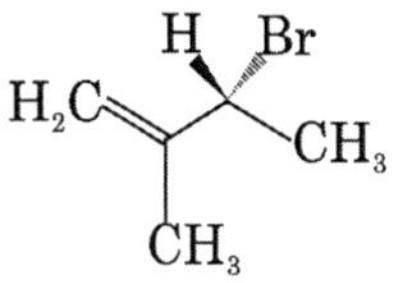

(D) 

11. Tha major product of the following reaction is

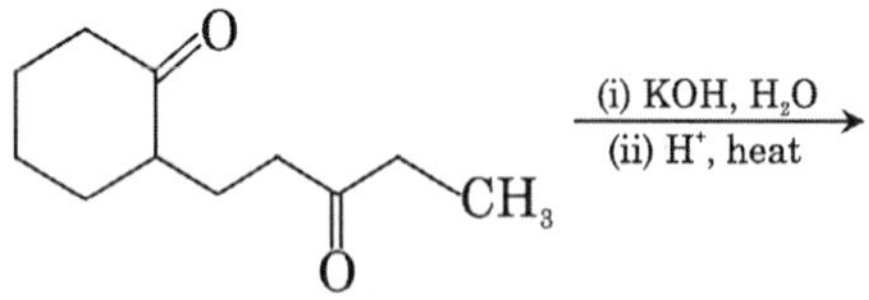

(A) 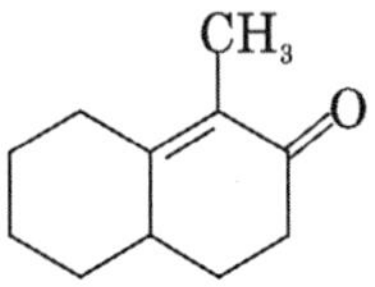

(B) 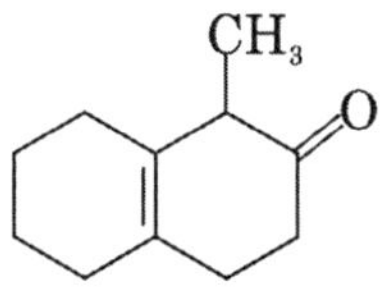

(C) 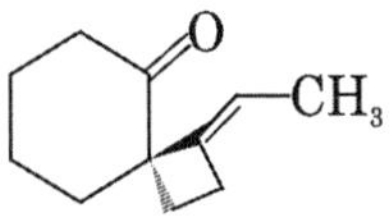

(D) 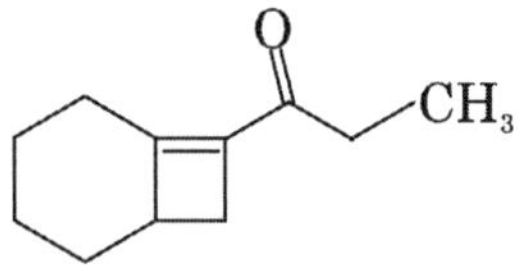

12. In the following reaction, the major product is

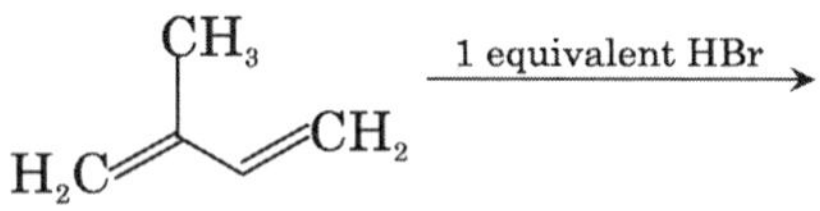

(A) 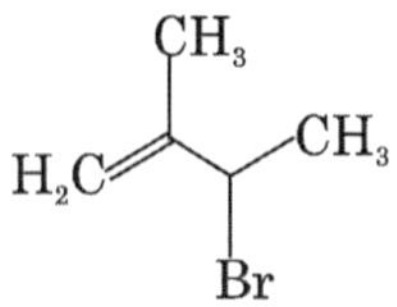

(B) 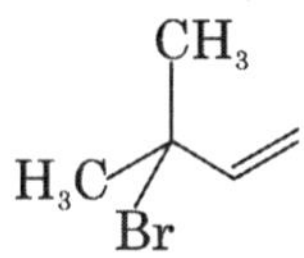

(C) 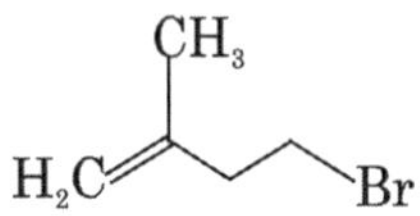

(D) 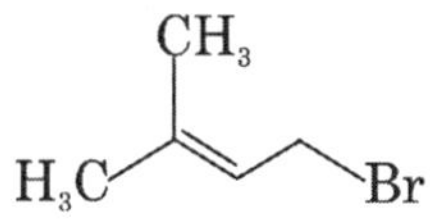

13. The structure of D-(+)-glucose is

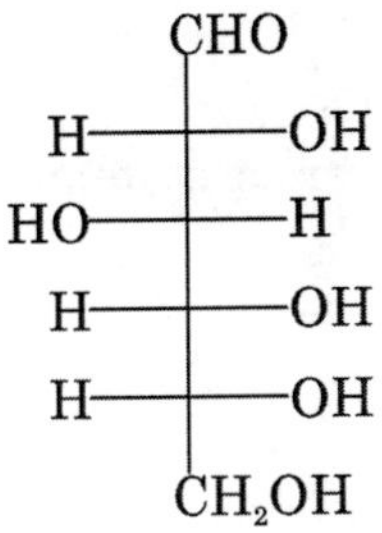

The structure of L-(–)-glucose is

(A) 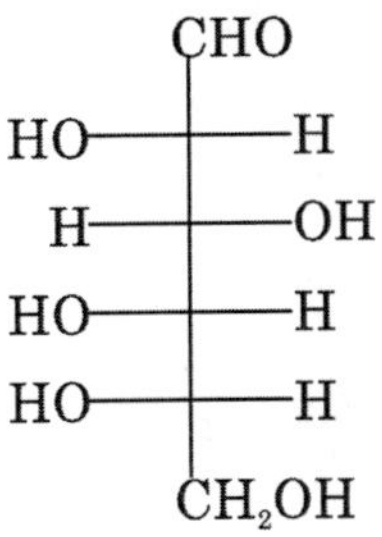

(B) 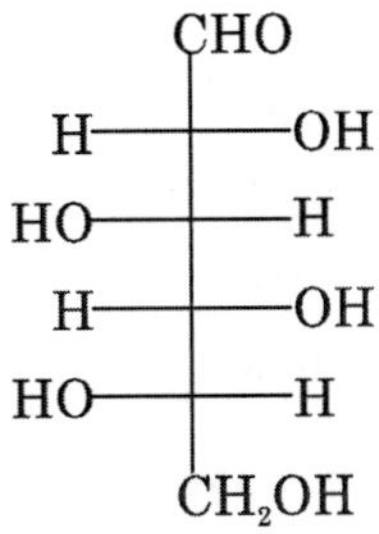

(C) 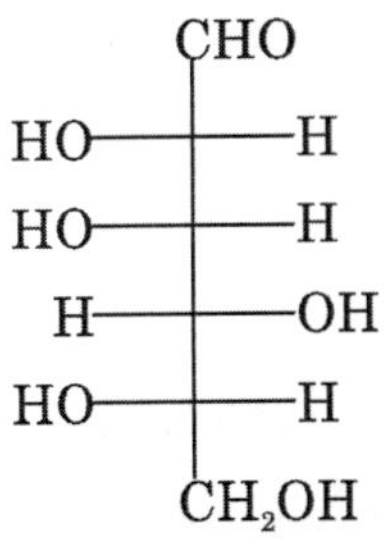

(D) 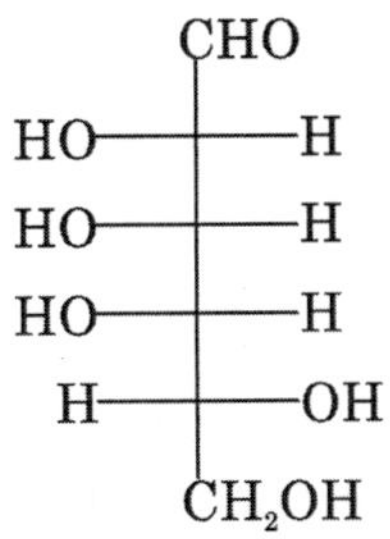

14. The major product of the following reaction is

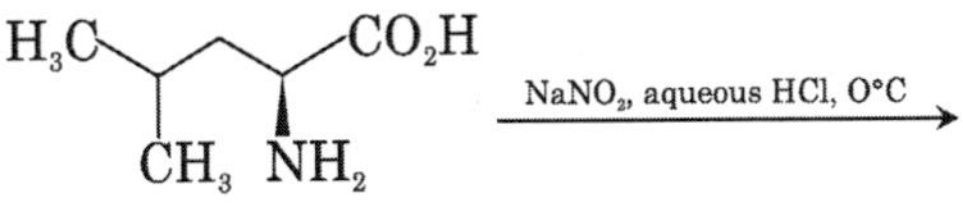

(A) 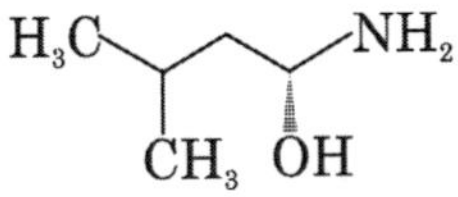

(B) 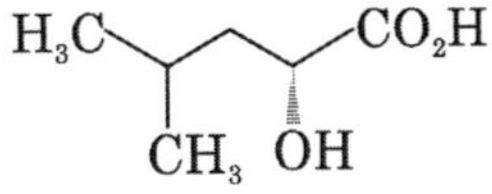

(C) 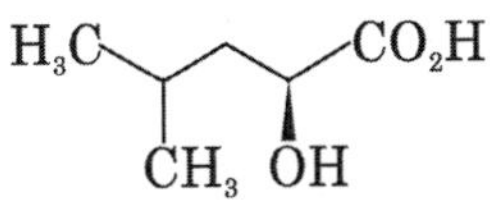

(D) 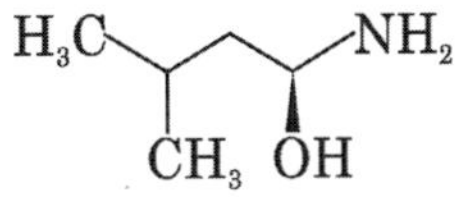

15. The correct statement(s) about Cr^{2+} and Mn^{3+} is(are) (Atomic number of Cr = 24 and Mn = 25)

(A) Cr^{2+} is a reducing agent

(B) Mn^{3+} is an oxidizing agent

(C) Both Cr^{2+} and Mn^{3+} exhibit d^4 electronic configuration
(D) When Cr^{2+} is used as a reducing agent, the chromium ion attains d^5 electronic configuration.

16. Copper is purified by electrolytic refining of blister copper. The correct statement(s) about this process is(are)
(A) Impure Cu strip is used as cathode
(B) Acidified aqueous $CuSO_4$ is used as electrolyte
(C) Pure Cu deposits at cathode
(D) Impurities settle as anode mud.

17. Fe^{3+} is reduced to Fe^{2+} by using
(A) H_2O_2 in presence of NaOH
(B) Na_2O_2 in water
(C) H_2O_2 in presence of H_2SO_4
(D) Na_2O_2 in presence of H_2SO_4

18. The % yield of ammonia as a function of time in the reaction
$N_2(g) + 3H_2(g) \rightleftharpoons 2NH_3(g)$, $\Delta H < 0$
at (P, T_1) is given below:

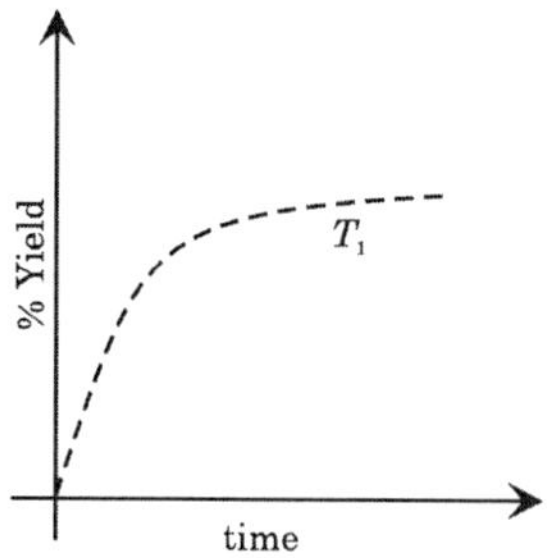

If this reaction is conducted at (P, T_2), with $T_2 > T_1$, the % yield of ammonia as a function of time is represented by

(A)
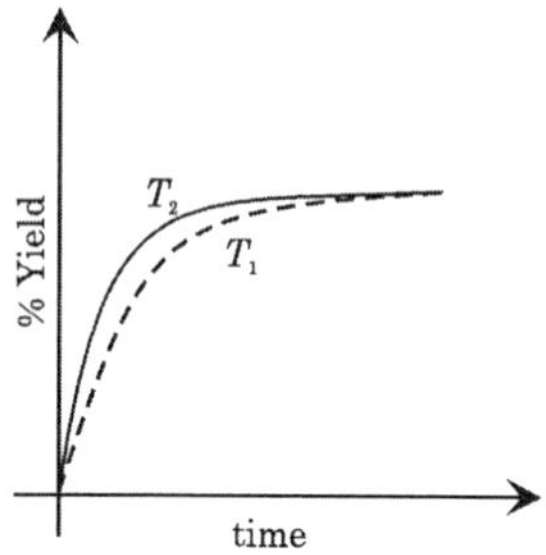

(B)
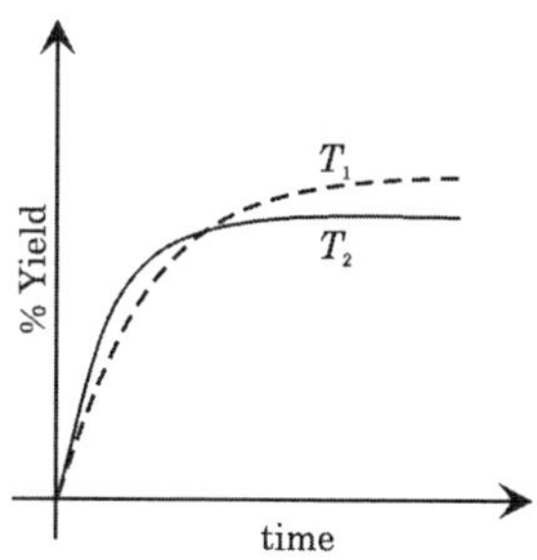

(C)
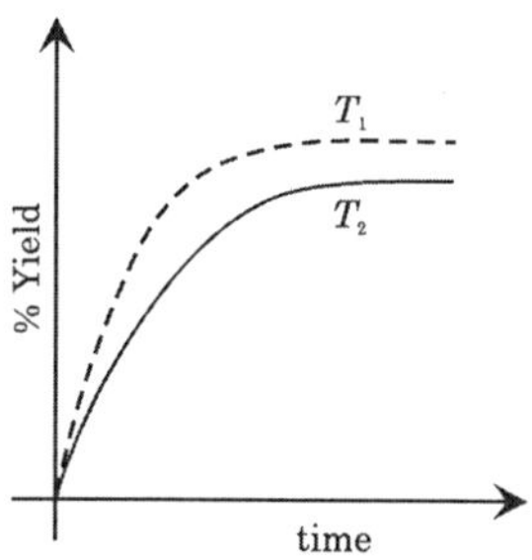

(D)
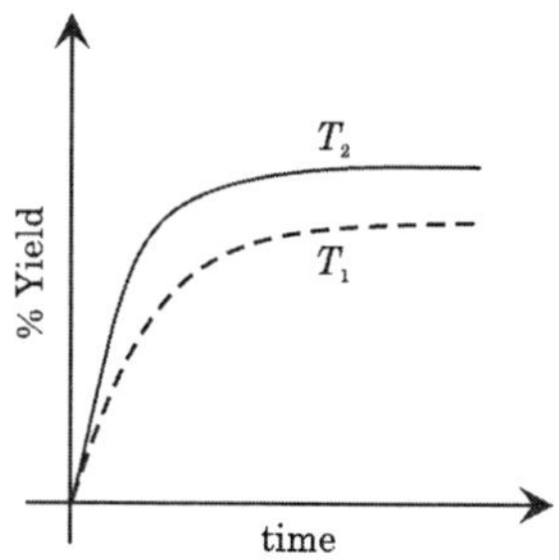

SECTION–3
(Maximum Marks: 16)

This section contain two questions.
Each question contains two columns, **Column I** and **Column II**
Column I has four entries (A), (B), (C) and (D).
Column II has five entries (P), (Q), (R), (S) and (T).
Match the entries in **Column I** with the entries in **Column II.**
One or more entries in **Column I** may match with one or more entries in **Column II.**
The ORS contains a 4 × 5 matrix whose layout will be similar to the one shown below:

(A)	(P)	(Q)	(R)	(S)	(T)
(B)	(P)	(Q)	(R)	(S)	(T)
(C)	(P)	(Q)	(R)	(S)	(T)
(D)	(P)	(Q)	(R)	(S)	(T)

For each entry in **Column I**, darken the bubbles of all the matching entries. For example, if entry (A) in **Column I** matches with entries (Q), (R) and (T), then darken these three bubbles in the ORS. Similarly, for entries (B), (C) and (D).
Marking scheme:
For each entry in **Column I**,
+ 2 if only the bubble(s) corresponding to all the correct match(es) is(are) darkened.
0 if none of the bubbles is darkened; – 1 in all other cases.

19. Match the anionic species gives in Column I that are present in the ore(s) gives in Column II.

Column I	**Column II**
(A) Carbonate	(P) Siderite
(B) Sulphide	(Q) Malachite
(C) Hydroxide	(R) Bauxite
(D) Oxide	(S) Calamine
	(T) Argentite

20. Match the thermodynamic processes given under Column I with the expression given under Column II.

Column I	**Column II**
(A) Freezing of water at 273 K and 1 atm	(P) $q = 0$
(B) Expansion of 1 mol of an ideal gas into a vacuum under isolated conditions	(Q) $w = 0$
(C) Mixing of equal volumes of two ideal gases at constant temperature and pressure in an isolated container	(R) $\Delta S_{sys} < 0$

(D) Reversible heating of $H_2(g)$ at 1 atm from 300 K to 600 K, followed by reversible cooling to 300 K at 1 atm	(S) $\Delta U = 0$
	(T) $\Delta G = 0$

ANSWERS

1. 1	2. 2	3. 9	4. 8	5. 4	6. 4
7. 3	8. 4	9. A	10. B, D	11. A	12. D
13. A	14. C	15. A, B, C	16. B, C, D	17. A, B	18. B

19. (A) $\longrightarrow$ (P, Q, S); (B) $\longrightarrow$ (T); (C) $\longrightarrow$ (Q, R); (D) $\longrightarrow$ (R)

20. (A) $\longrightarrow$ (R, T); (B) $\longrightarrow$ (P, Q, S); (C) $\longrightarrow$ (P, Q, S); (D) $\longrightarrow$ (P, Q, S, T)

Model Solutions—I.I.T. JEE (Advanced) 2015

Paper–1

1. (1)

$\Delta T_f = iK_f m$

$0.0558 = i \times 1.86 \times 0.01$

$i = 3$

Hence, the complex is $[Co(NH_3)_5Cl]Cl_2$

2. (2)

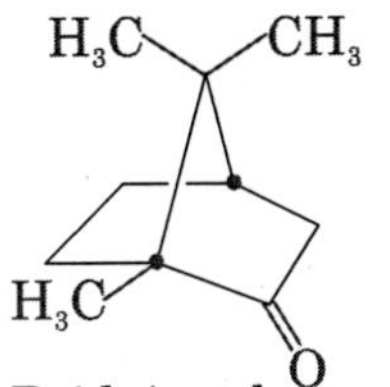

Bridging does not allow the other 2 variants to exist.
Total number of stereoisomers of M = 2

3. (9) 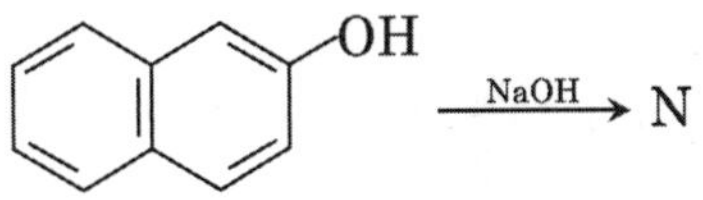

N is 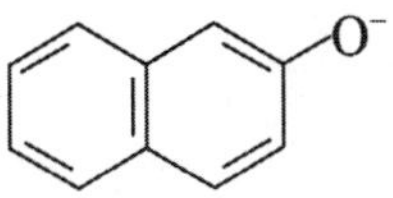

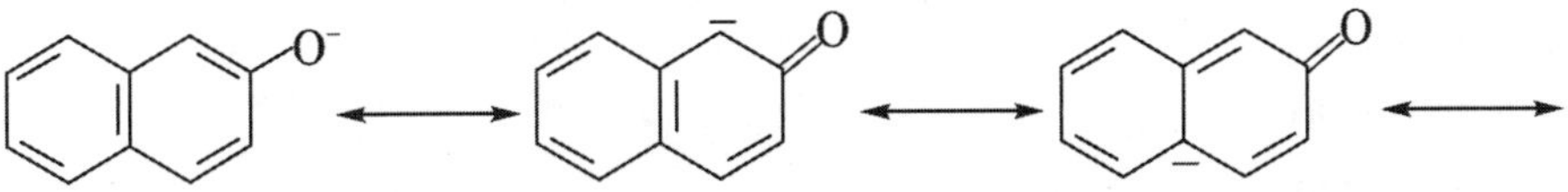

1 2 3

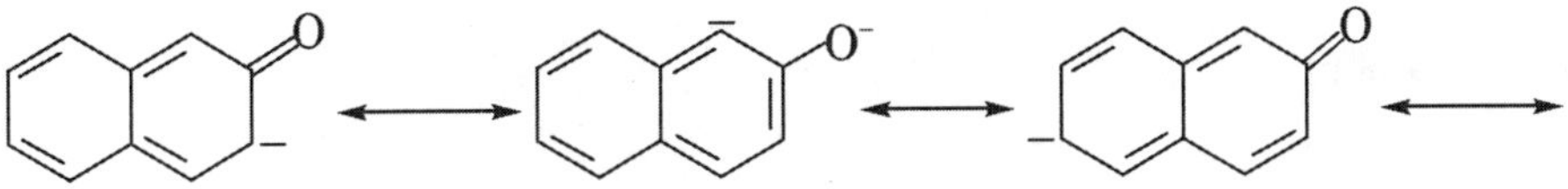

4 5 6

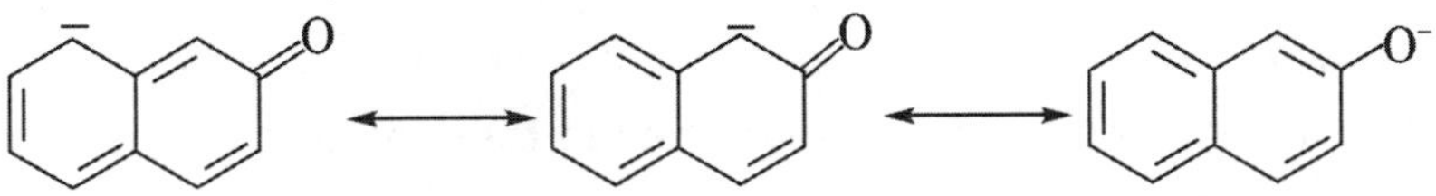

7 8 9

4. (8)

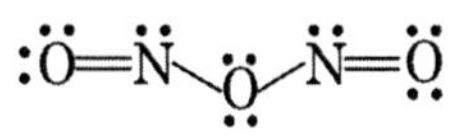
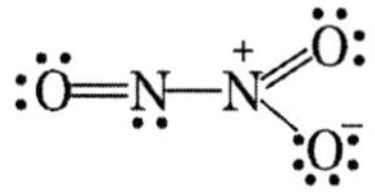

Total number of lone pairs = 8

5. (4)

$[Fe(SCN)_6]^{3-}$ and $[Fe(CN)_6]^{3-}$

In both the cases the electronic configuration of Fe^{3+} will be $1s^2 2s^2 2p^6 3s^2 3p^6 3p^5$.

Since SCN is a weak field ligand and $\bar{C}N$ is a strong field ligand, the pairing will occur only in case of $[Fe(CN)_6]^{3-}$.

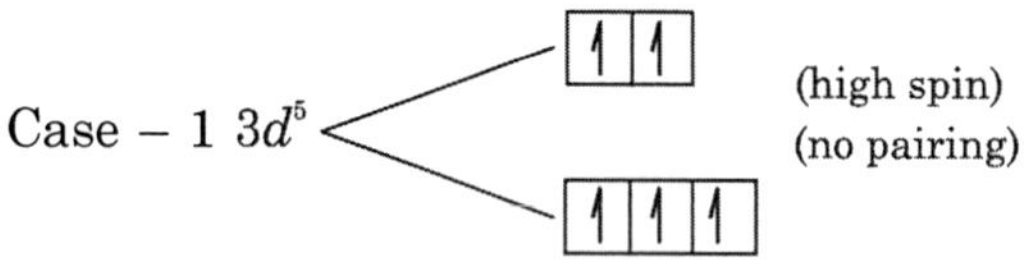

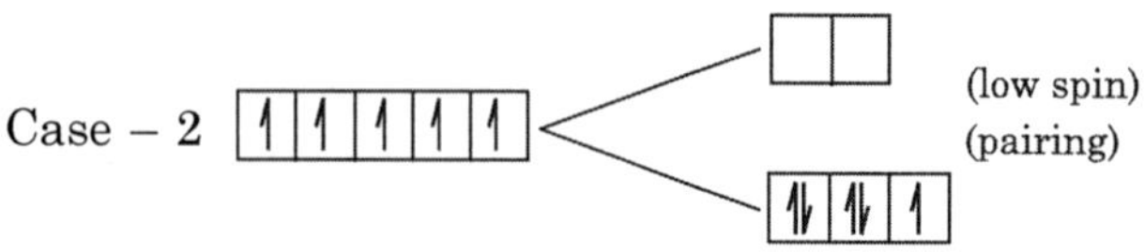

Case – 1 $\mu = \sqrt{n(n+2)} = \sqrt{5(5+2)} = \sqrt{35} = 5.91$ BM

Case – 2 $\mu = \sqrt{n(n+2)} = \sqrt{1(1+2)} = \sqrt{3} = 1.73$ BM

Difference in spin-only magnetic moment = 5.91 – 1.73 = 4.18 ≈ 4 BM

6. (4)

$BeCl_2$, N_3^-, N_2O, NO_2^+, O_3, SCl_2, ICl_2^-, I_3^-, XeF_2

$BeCl_2 \longrightarrow sp \longrightarrow$ linear

$N_3^- \longrightarrow sp \longrightarrow$ linear

$N_2O \longrightarrow sp \longrightarrow$ linear

$NO_2^+ \longrightarrow sp \longrightarrow$ linear

$O_3 \longrightarrow sp^2 \longrightarrow$ bent

$SCl_2 \longrightarrow sp^3 \longrightarrow$ bent

$I_3^- \longrightarrow sp^3d \longrightarrow$ linear

$ICl_2^- \longrightarrow sp^3d \longrightarrow$ linear

$XeF_2 \longrightarrow sp^3d \longrightarrow$ linear

Hence, only four (4) molecules have linear shape and in these no d-orbital is involved in hybridization.

7. (3)

Single electron species do not follow the $(n + \ell)$ rule but multi electron species do.

Ground state of H^- is $1s^2$.

First excited state of H^- is $1s^1$, $2s^1$.

Second excited state of H^- is $1s^1$, $2s^0$, $2p^1$.

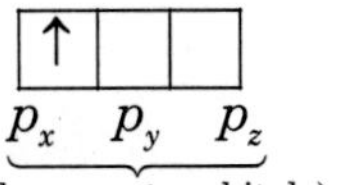

(3 degenerate orbitals)

8. (4)

$X \longrightarrow Y \qquad \Delta_r G^0 = -193$ kJ/mol

$M^+ \longrightarrow M^{3+} + 2e^- \qquad E^0 = -0.25$ V

ΔG^0 for the this reaction is: $\Delta G^0 = -nFE^0 = -2 \times (-0.25) \times 96500 = 48250$ J/mol $= 48.25$ kJ/mole

Hence, the number of moles of M^+ ion oxidized using $X \longrightarrow Y$ will be

$= \dfrac{193}{48.25} = 4$ moles

9. (A)

In *ccp* lattice:

Number of O atoms: 4

Number of octahedral voids: 4

Number of tatrahedral voids: 8

Number of $Al^{3+} = 4 \times m$

Number of $Mg^{2+} = 8 \times n$

Due to charge neutrality,

$4(-2) + 4m(+3) + 8n(+2) = 0$

Hence, $m = \dfrac{1}{2}$ and $n = \dfrac{1}{8}$

10. (B, D)

(1) 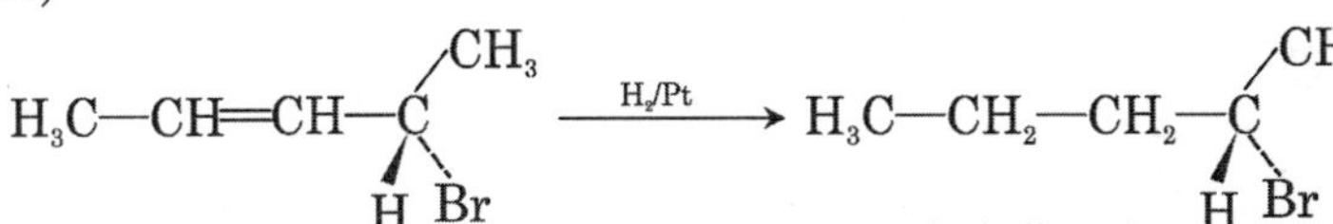

Optically active

(2) 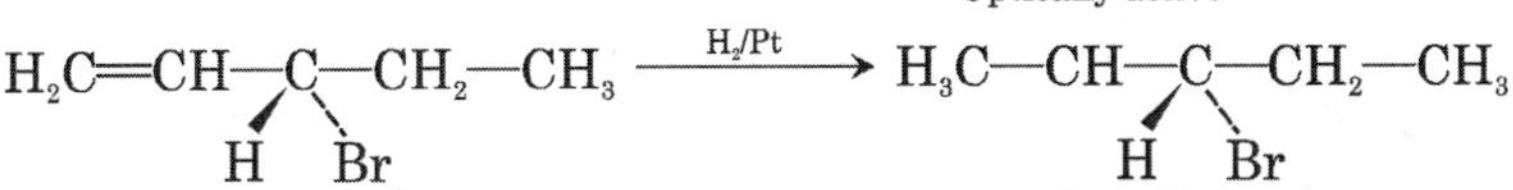

Optically inactive

(3) 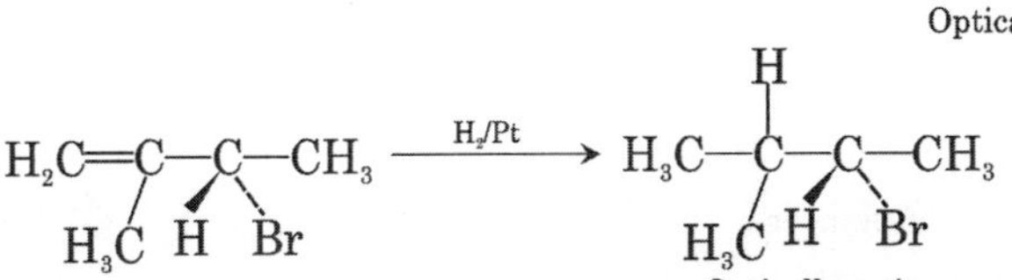

Optically active

(4) 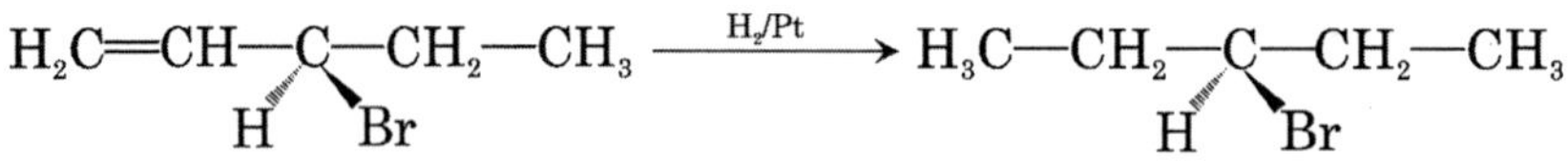

Optically inactive

11. (A)

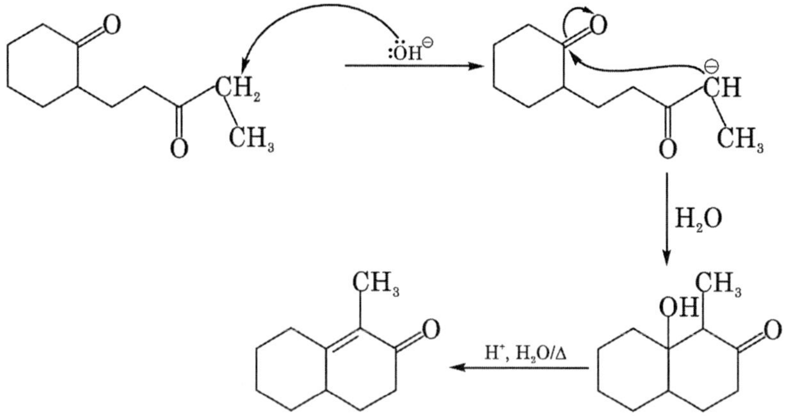

12. (D)

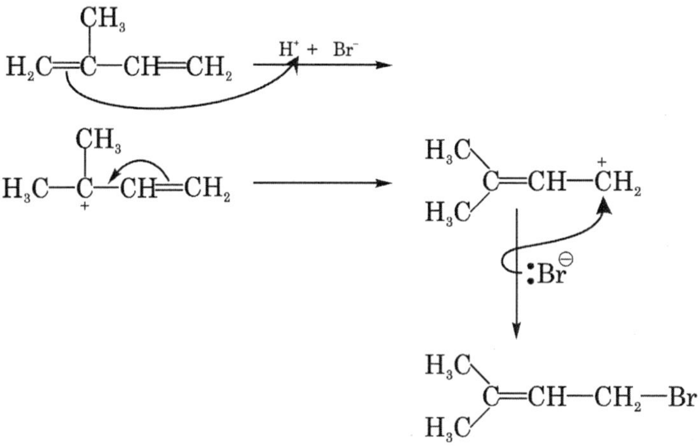

13. (A)

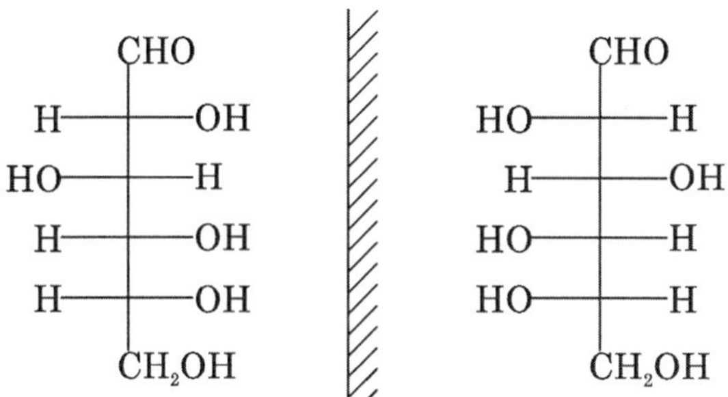

D(+)glucose L(–)glucose

14. (C)

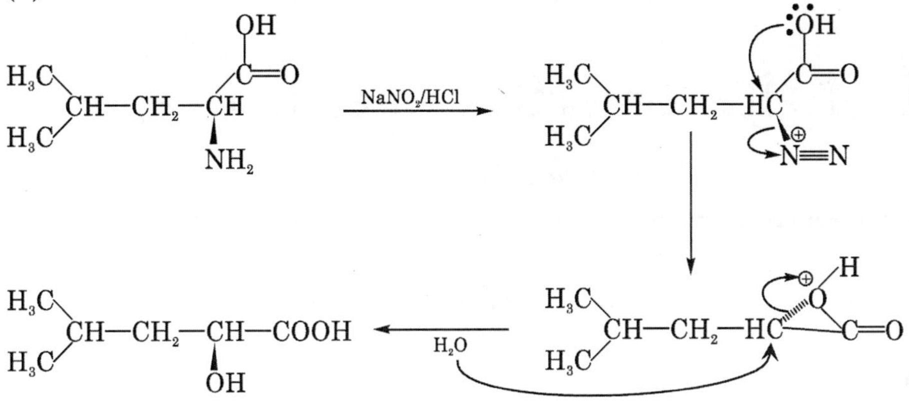

15. (A, B, C)

(1) Cr^{2+} is a reducing agent because Cr^{3+} is more stable.

(2) Mn^{3+} is an oxidizing agent because Mn^{2+} is more stable.

(3) Cr^{2+} and Mn^{3+} have d^4 eletronic configuration.

16. (B, C, D)

(1) Impure Cu strip is used as anode and impurities settle as anode mud.

(2) Pure Cu deposits at the cathode.

(3) Acidified aqueous $CuSO_4$ is used as an electrolyte.

17. (A, B)

$2Fe^{3+} + H_2O_2 + 2OH^- \longrightarrow 2Fe^{2+} + 2H_2O + O_2$

$Na_2O_2 + H_2O \longrightarrow H_2O_2 + NaOH$

18. (B)

$N_2(g) + 3H_2(g) \rightleftharpoons 2NH_3(g); \Delta H < 0$

Increasing the temperature lowers equilibrium yield of ammonia. However, at higher temperature the initial rate of forward reaction would be greater than at lower temperature. That is why the percentage yield of NH_3 would be more initially.

19. (A) ⟶ (P, Q, S); (B) ⟶ (T); (C) ⟶ (Q, R); (D) ⟶ (R)

Siderite — $FeCO_3$

Malachite — $CuCO_3 \cdot Cu(OH)_2$

Bauxite — $AlO_x(OH)_{3-2x}$; $0 < x < 1$

Calamine — $ZnCO_3$

Argentite — Ag_2S

20. (A) ⟶ (R, T); (B) ⟶ (P, Q, S); (C) ⟶ (P, Q, S); (D) ⟶ (P, Q, S, T)

I.I.T. JEE Objective Question Paper (Advanced)—2015

Paper–2

INSTRUCTIONS

Question Paper Format and Marking Scheme

1. The question paper has three parts: Physics, Chemistry and Mathematics. Each part has three sections.
2. Section 1 contains 8 questions. The answer to each question is a single digit integer ranging from 0 to 9 (both inclusive).
   Marking Scheme: + 4 for correct answer and 0 in all other cases.
3. Section 2 contains 8 multiple choice questions with one or more than one correct option.
   Marking Scheme: + 4 for correct answer, 0 if not attempted and – 2 in all other cases.
4. Section 3 contains 2 "paragraph" type questions. Each paragraph describes an experiment, a situation or a problem. Two multiple choice questions will be asked based on this paragraph. One or more than one option can be correct.
   Marking Scheme: + 4 for correct answer, 0 if not attempted and – 2 in all other cases.

SECTION–1
(Maximum Marks: 32)

This section contains eight questions.
The answer to each question is a single digit integer ranging from 0 to 9, both inclusive.
For each question, darken the bubble corresponding to the correct integer in the ORS.
Marking Scheme: + 4 if the bubble corresponding to the answer is darkened and 0 in all other cases.

1. In dilute aqueous H_2SO_4, the complex diaquodioxalatoferrate(II) is oxidized by MnO_4^-. For this reaction, the ratio of the rate of change of $[H^+]$ to the rate of change of $[MnO_4^-]$ is

2. The number of hydroxyl group(s) in Q is

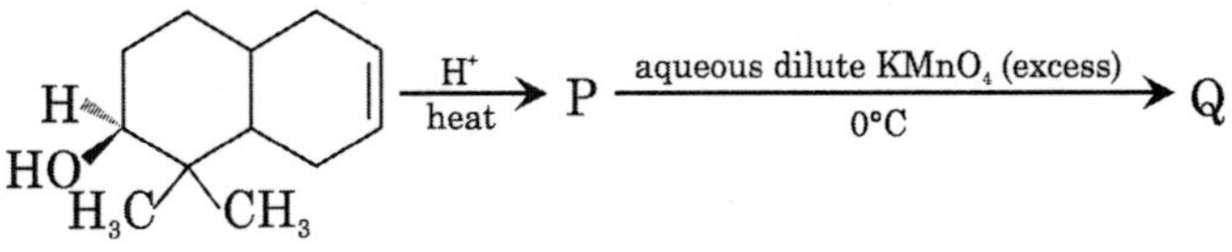

3. Among the following, the number of reaction(s) that produce(s) benzaldehyde is

I. 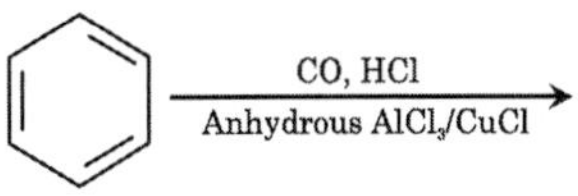

II. 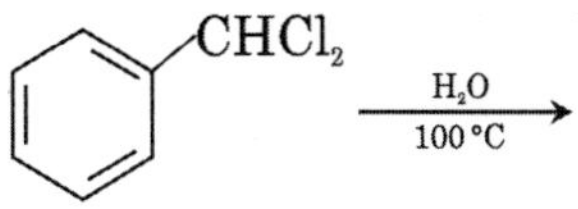

III. 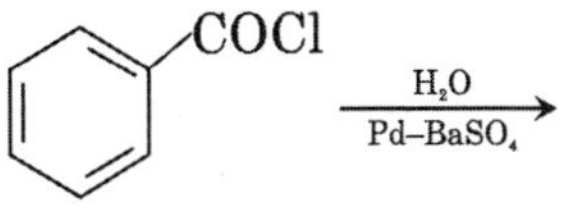

IV. 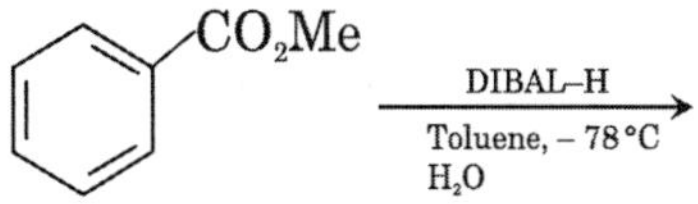

4. In the complex acetylbromidodicarbonylbis(triethylphosphine)iron(II), the number of Fe—C bond(s) is

5. Among the complex ions, $[Co(NH_2—CH_2—CH_2—NH_2)_2Cl_2]^+$, $[CrCl_2(C_2O_4)_2]^{3-}$, $[Fe(H_2O)_4(OH)_2]^+$, $[Fe(NH_3)_2(CN)_4]^-$, $[Co(NH_2—CH_2—CH_2—NH_2)_2(NH_3)Cl]^{2+}$ and $[Co(NH_3)_4(H_2O)Cl]^{2+}$, the number of complex ion(s) that show(s) *cis-trans* isomerism is

6. Three moles of B_2H_6 are completely reacted with methanol. The number of moles of boron containing product formed is

7. The molar conductivity of a solution of a weak acid HX (0.01*M*) is 10 times smaller than the molar conductivity of a solution of a weak acid HY (0.10*M*). If $\lambda^0_{X^-} \approx \lambda^0_{Y^-}$, the difference in their pK_a values, $pK_a(HX) - pK_a(HY)$, is (consider degree of ionisation of both acids to be << 1)

8. A closed vessel with rigid walls contains 1 mol of $^{238}_{92}U$ and 1 mol of air at 298 K. Considering complete decay of $^{238}_{92}U$ to $^{206}_{82}Pb$, the ratio of the final pressure to the initial pressure of the system at 298 K is

SECTION–2
(Maximum Marks: 32)

This section contains eight questions.

Each question has four options (A), (B), (C) and (D). One or more than one of these four option(s) is (are) correct.

For each question, darken the bubble(s) corresponding to the correct option(s) in the ORS.

Marking scheme:

+ 4 if only the bubble(s) corresponding to all the correct option(s) is(are) darkened

0 if none of the bubbles is darkened and – 2 in all other cases.

9. One mole of a monoatomic real gas satisfies the equation $P(V - b) = RT$ where b is a constant. The relationship of interatomic potential $V(r)$ and interatomic distance r for the gas is given by

(A) 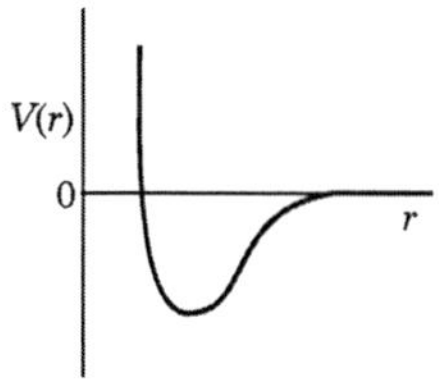

(B) 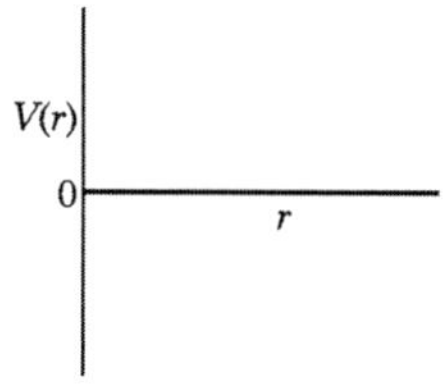

(C) 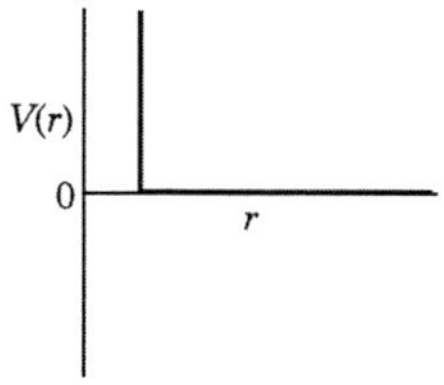

(D) 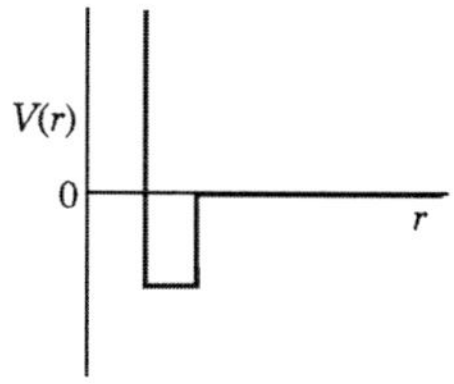

10. In the following reactions, the product S is

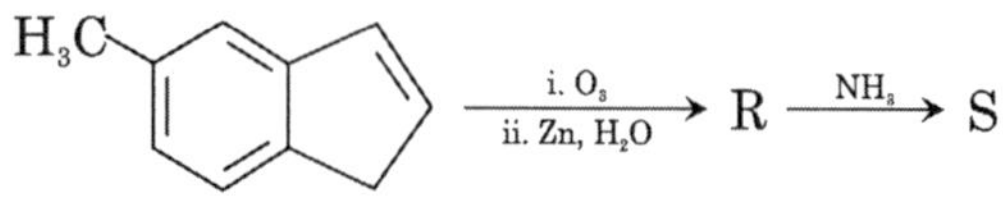

(A) 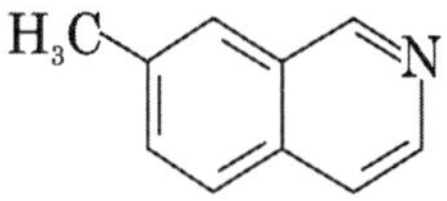

(B) 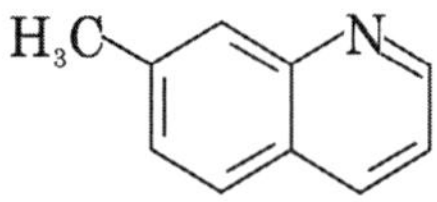

(C) 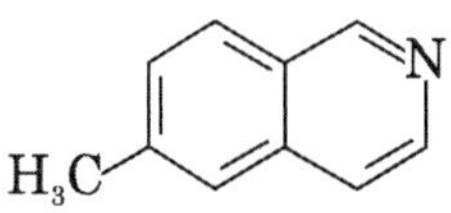

(D) 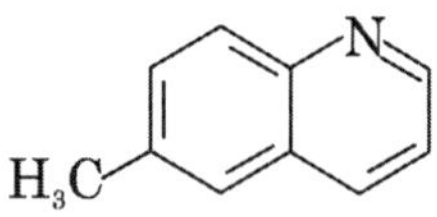

11. The major product U in the following reactions is

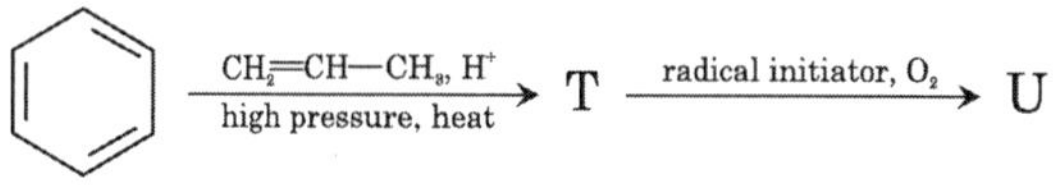

(A) 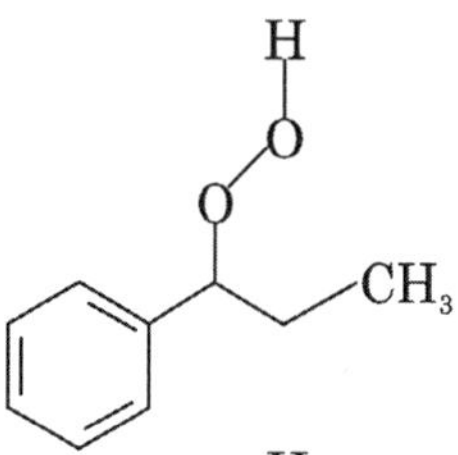

(B) 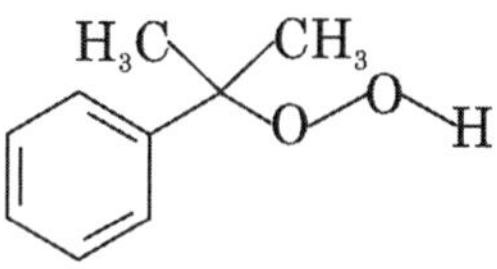

(C) 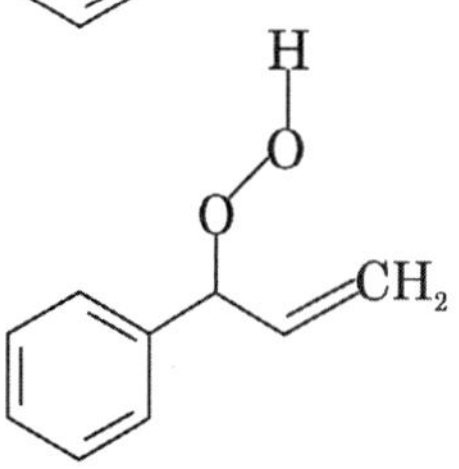

(D) 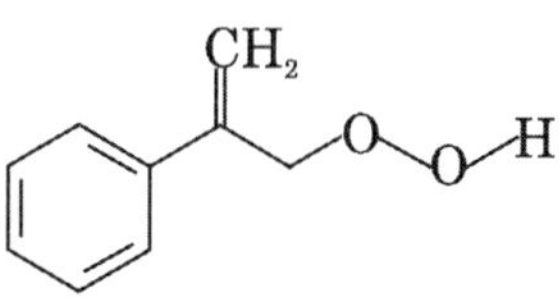

12. In the following reactions, the major product W is

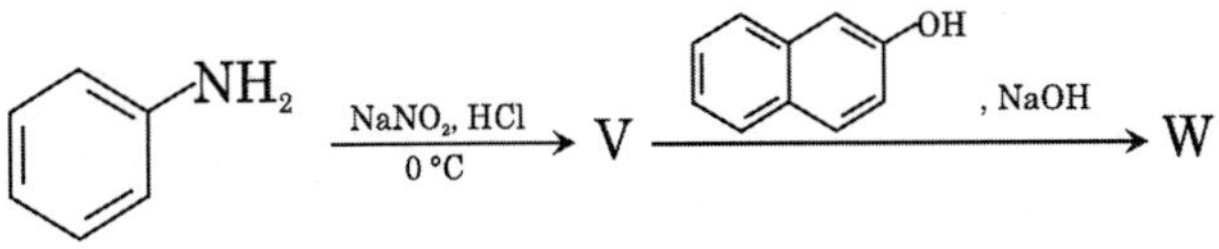

(A) 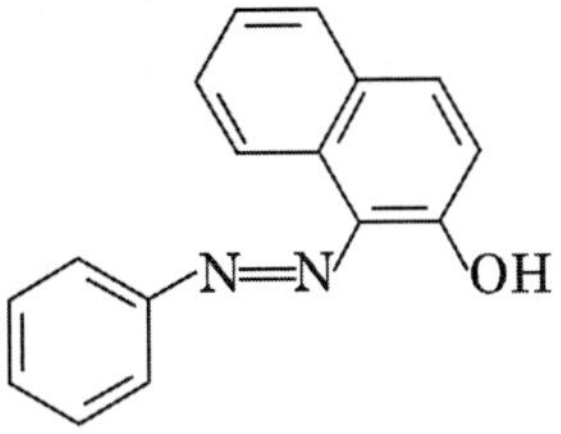

(B) 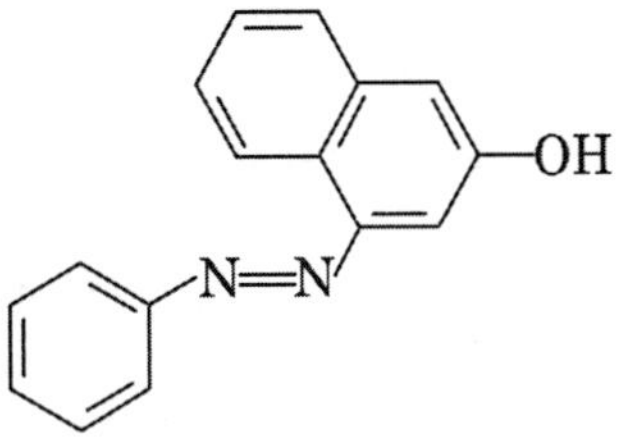

(C) 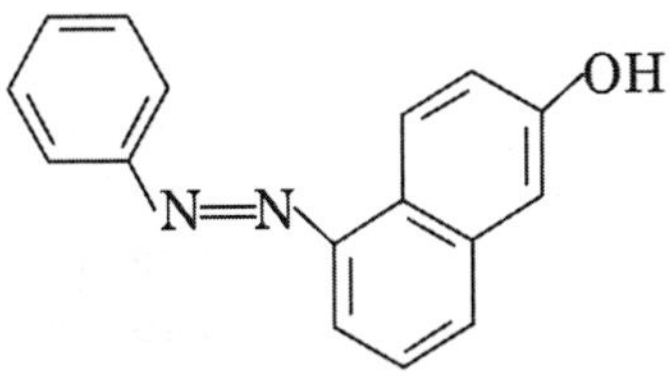

(D) 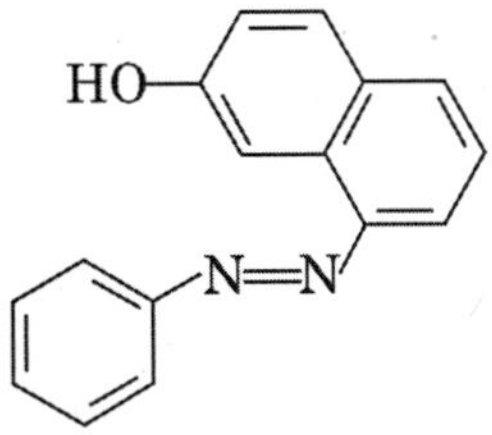

13. The correct statement(s) regarding, (i) HClO, (ii) $HClO_2$, (iii) $HClO_3$, (iv) $HClO_4$, is (are)

(A) The number of Cl═O bonds in (ii) and (iii) together is two.

(B) The number of lone pairs of electrons on Cl in (ii) and (iii) together is three.

(C) The hybridization of Cl in (iv) is sp^3.

(D) Amongst (i) to (iv), the strongest acid is (i).

14. The pair(s) of ions where both the ions are precipitated upon passing H_2S gas in presence of dil. HCl, is(are)

(A) Ba^{2+}, Zn^{2+} (B) Bi^{3+}, Fe^{3+} (C) Cu^{2+}, Pb^{2+} (D) Hg^{2+}, Bi^{3+}

15. Under hydrolytic conditions, the compounds used for preparation of liner polymer and for chain termination, respectively, are

(A) CH_3SiCl_3 and $Si(CH_3)_4$ (B) $(CH_3)_2SiCl_2$ and $(CH_3)_3SiCl$

(C) $(CH_3)_2SiCl_2$ and CH_3SiCl_3 (D) $SiCl_4$ and $(CH_3)_3SiCl$

16. When O_2 is adsorbed on a metallic surface, electron transfer occurs from the metal to O_2. The true statement(s) regarding this adsorption is(are)

(A) O_2 is physisorbed (B) heat is released

(C) occupancy of π^*_{2p} of O_2 is increased (D) bond length of O_2 is increased.

SECTION–3
(Maximum Marks: 16)

This section contain two questions.
Based on each paragraph, there will be two questions.
Each question has four options (A), (B), (C) and (D). One or more than one of these four option(s) is(are) correct. For each question, darken the bubble(s) corresponding to all the correct option(s) in the ORS.
Marking scheme:
+ 4 if only the bubble(s) corresponding to all the correct option(s) is (are) darkened.
0 if none of the bubbles is darkened and – 2 in all other cases.

Paragraph 1

When 100 mL of 1.0*M* HCl was mixed with 100 mL of 1.0*M* NaOH in an insulated beaker at constant pressure, a temperature increase of 5.7 °C was measured for the beaker and its contents (Expt. 1). Because the enthalpy of neutralization of a strong acid with a strong base is a constant (– 57.0 kJ mol^{-1}), this experiment could be used to measure the calorimeter constant. In a second experiment (Expt. 2), 100 mL of 2.0*M* acetic acid (K_a = 2.0 × 10^{-5}) was mixed with 100 mL of 1.0*M* NaOH (under identical conditions to Expt. 1) where a temperature rise of 5.6 °C was measured. (Consider heat capacity of all solutions as 4.2 J g^{-1} K^{-1} and density of all solutions as 1.0 g mL^{-1})

17. Enthalpy of dissociation (in kJ mol^{-1}) of acetic acid obt ained from the Expt. 2 is
(A) 1.0 (B) 10.0 (C) 24.5 (D) 51.4

18. The *p*H of the solution after Expt. 2 is
(A) 2.8 (B) 4.7 (C) 5.0 (D) 7.0

Paragraph 2

In the following reactions

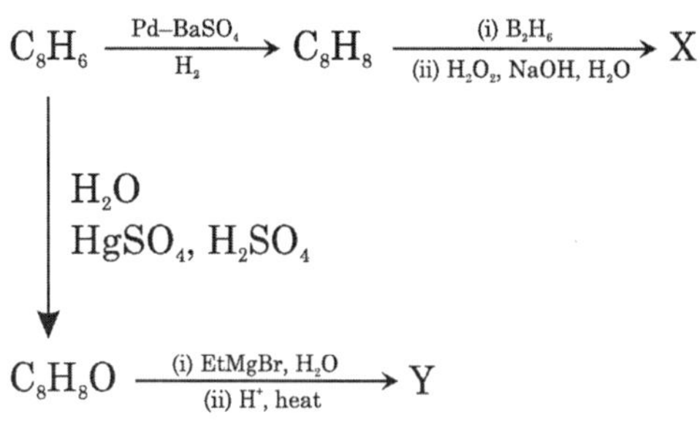

19. Compound X is

(A) 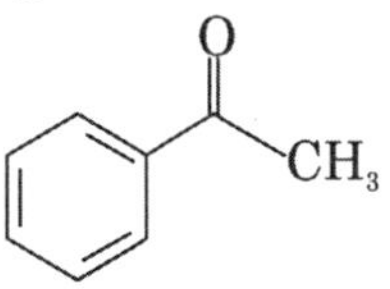

(B) 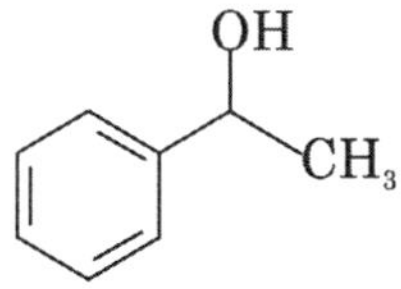

(C) 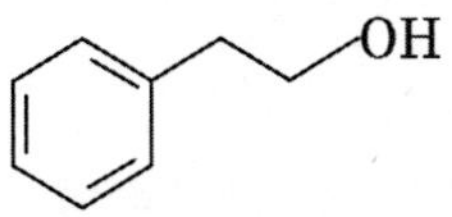

(D) 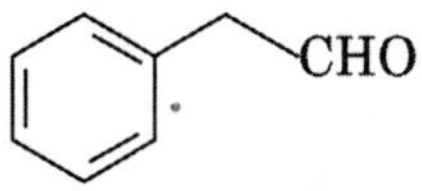

20. The major compound Y is

(A) 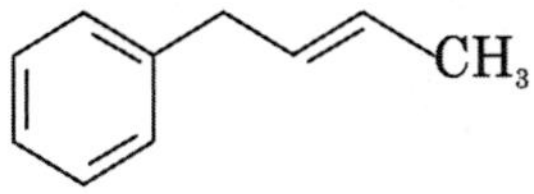

(B) 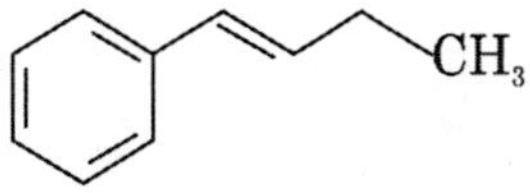

(C) 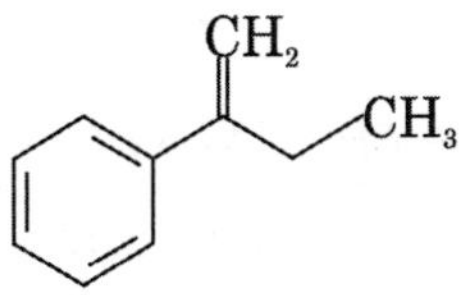

(D) 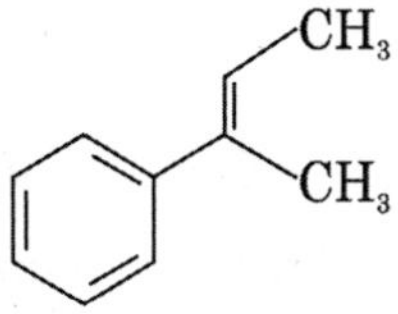

ANSWERS

1. 8	2. 4	3. 4	4. 3	5. 5	6. 6
7. 3	8. 9	9. C	10. A	11. B	12. A
13. B, C	14. C, D	15. B	16. B, C, D	17. A	18. B
19. C	20. D				

Model Solutions—I.I.T. JEE (Advanced) 2015

Paper–2

1. (8)

$[Fe(C_2O_4)_2(H_2O)_2]^{2-} + MnO_4^- + 8H^+ \longrightarrow Mn^{2+} + Fe^{3+} + 4CO_2 + 6H_2O\,[MnO_4^-]$

Hence, the ratio of rate of change of $[H^+]$ to that of rate of change of $[MnO_4^-]$ is 8.

2. (4)

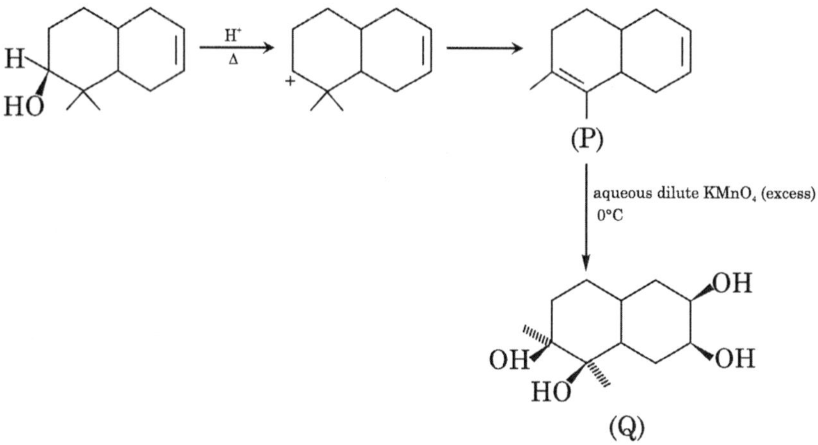

3. (4)

I. 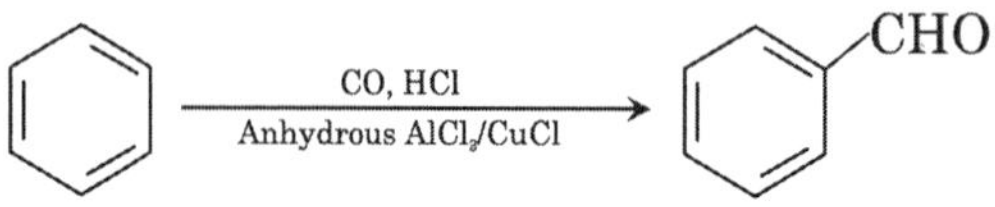

II. 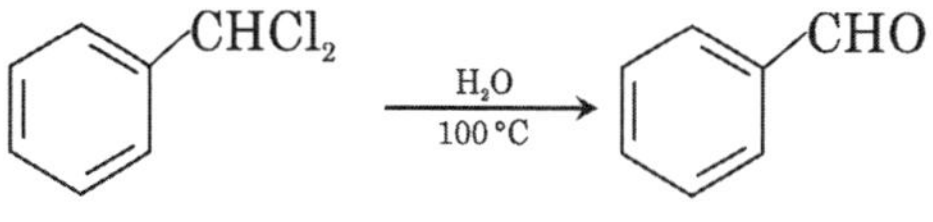

III. 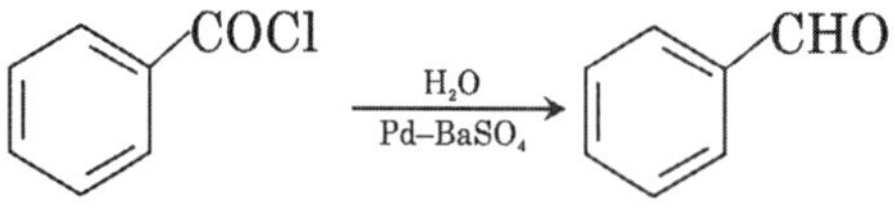

IV. 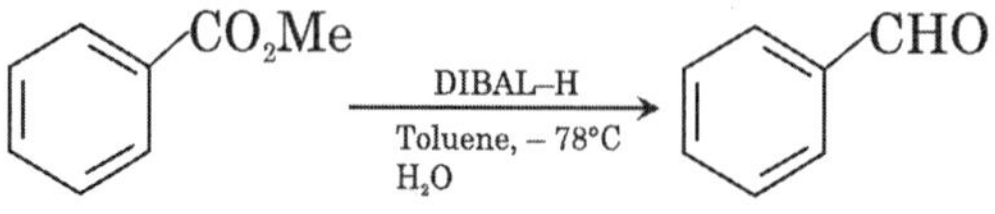

4. (3)

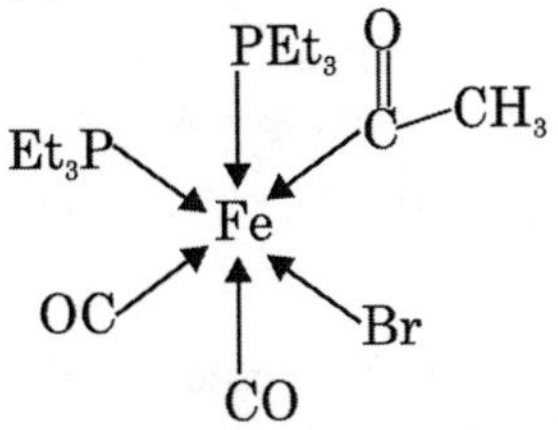

The number of Fe—C bonds is 3.

5. (5)

$[Co(en)_2Cl_2]^+ \longrightarrow$ will show *cis-trans* isomerism

$[CrCl_2(C_2O_4)_2]^{3-} \longrightarrow$ will show *cis-trans* isomerism

$[Fe(H_2O)_4(OH)_2]^+ \longrightarrow$ will show *cis-trans* isomerism

$[Fe(CN)_4(NH_3)_2]^- \longrightarrow$ will show *cis-trans* isomerism

$[Co(en)_2(NH_3)Cl]^{2+} \longrightarrow$ will show *cis-trans* isomerism

$[Co(NH_3)_4(H_2O)Cl]^{2+} \longrightarrow$ will not show *cis-trans* isomerism (Although it will show geometrical isomerism)

6. (6)

$B_2H_6 + 6MeOH \longrightarrow 2B(OMe)_3 + 6H_2$

1 mole of B_2H_6 reacts with 6 mole of MeOH to give 2 moles of $B(OMe)_3$.

3 mole of B_2H_6 will react with 18 mole of MeOH to give 6 moles of $B(OMe)_3$.

7. (3)

$HX \rightleftharpoons H^+ + X^-$ $\qquad$ $HY \rightleftharpoons H^+ + Y^-$

$$K_a = \frac{[H^+][X^-]}{[HX]} \qquad K_a = \frac{[H^+][Y^-]}{[HY]}$$

Λ_m for HX $= \Lambda_{m_1}$

Λ_m for HY $= \Lambda_{m_2}$

$$\Lambda_{m_1} = \frac{1}{10}\Lambda_{m_2}$$

$$K_a = C\alpha^2$$

$$K_{a_1} = C_1 \times \left(\frac{\Lambda_{m_1}}{\Lambda^0_{m_1}}\right)^2$$

$$K_{a_2} = C_2 \times \left(\frac{\Lambda_{m_2}}{\Lambda^0_{m_2}}\right)^2$$

$$\frac{K_{a_1}}{K_{a_2}} = \frac{C_1}{C_2} \times \left(\frac{\Lambda_{m_1}}{\Lambda_{m_2}}\right)^2 = \frac{0.01}{0.1} \times \left(\frac{1}{10}\right)^2 = 0.001$$

$$pK_{a_1} - pK_{a_2} = 3$$

8. (9)

In conversion of ${}^{238}_{92}\text{U}$ to ${}^{206}_{82}\text{Pb}$, 8 α-particles and 6 β-particles are ejected.

The number of gaseous moles initially = 1 mol

The number of gaseous moles finally = 1 + 8 mol; (1 mol from air and 8 mol of ${}^{4}_{2}\text{He}$)

Hence, the ratio = 9/1 = 9

9. (C)

At large interionic distances (because $a \longrightarrow 0$) the P.E. would remain constant. However, when $r \longrightarrow 0$, repulsion would suddenly increase.

10. (A)

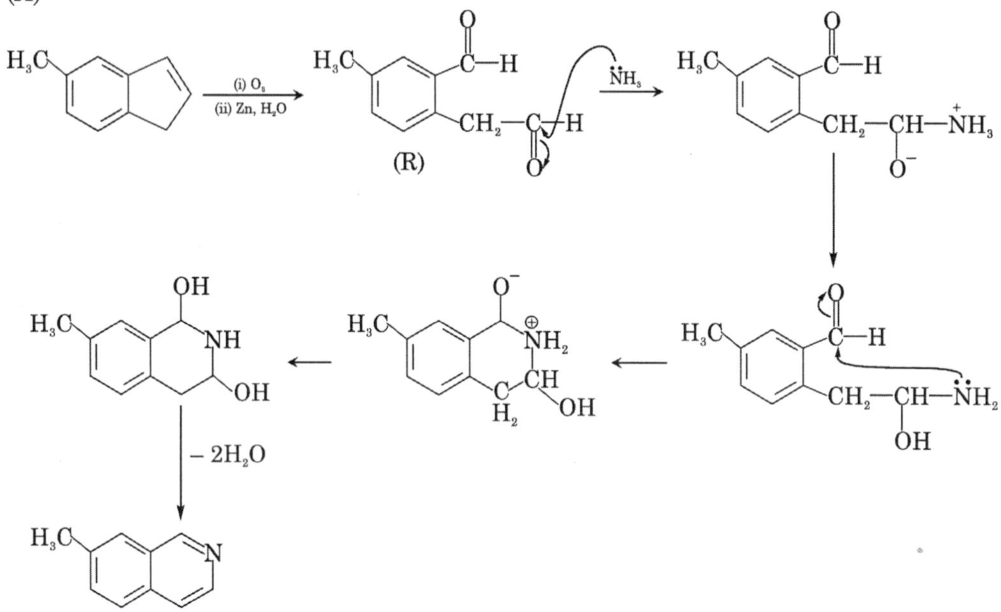

(S)

11. (B)

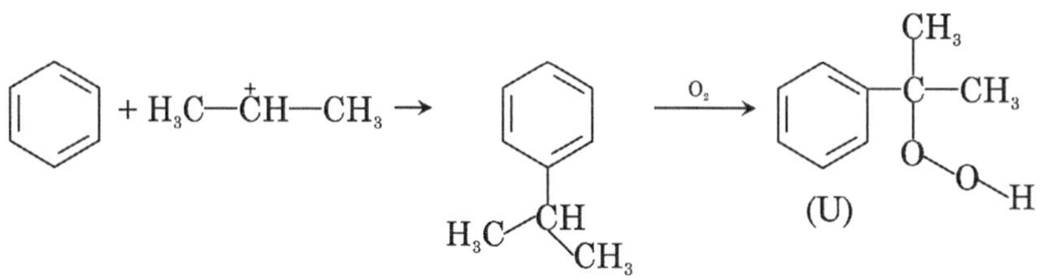

(T)

12. (A)

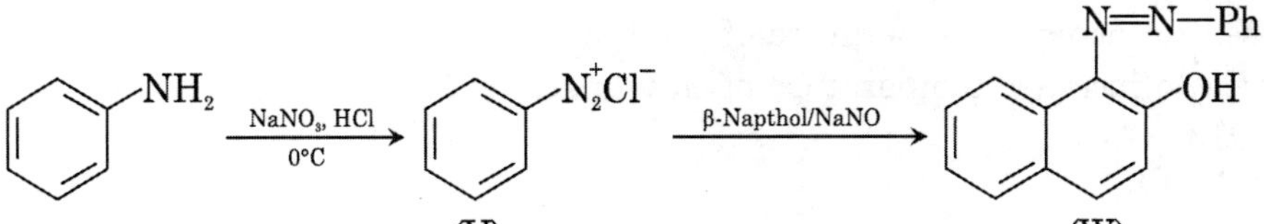

(V) (W)

13. (B, C)

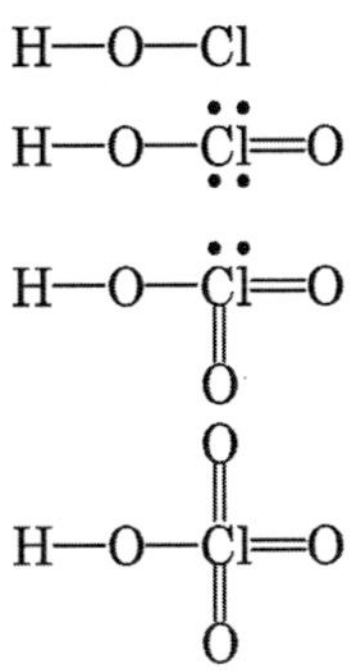

(I) (II) (III) (IV)

14. (C, D)

Cu^{2+}, Pb^{2+}, Hg^{2+}, Bi^{3+} give precipitate with H_2S in presence to dil. HCl.

15. (B)

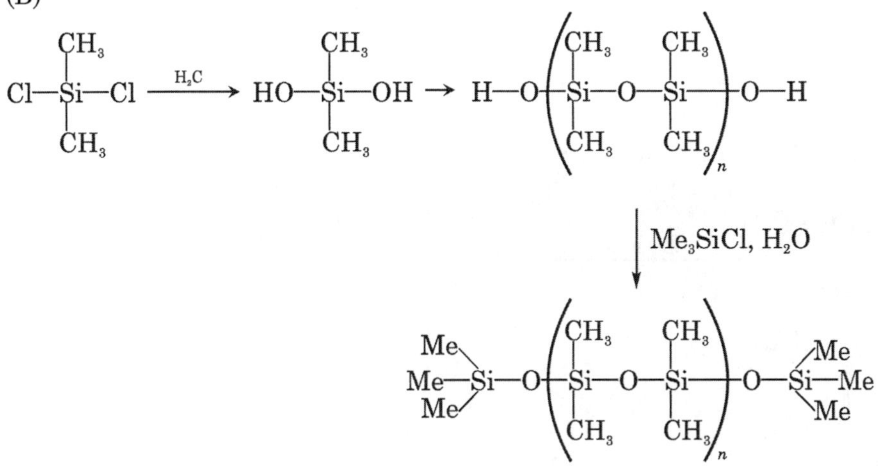

16. (B, C, D)

Adsorption of O_2 on metal surface is exothermic.

During the electron transfer metal to O_2 electron occupies π^*_{2p} orbital of O_2.

Due to the electron transfer to O_2 the bond order of O_2 decreases and hence bond length increases.

17. (A)

$HCl + NaOH \longrightarrow NaCl + H_2O$

$n = 100 \times 1 = 100$ m mole $= 0.1$ mole

Energy evolved due to neutralization of HCl and NaOH = $0.1 \times 57 = 5.7$ kJ = 5700 J
Energy used to increase temperature of solution = $200 \times 4.2 \times 5.7 = 4788$ J
Energy used to increase temperature of calorimeter = $5700 - 4788 = 912$ J
m.s. $\Delta t = 912$
m.s. $\times 5.7 = 912$
ms = 160 J/°C [Calorimeter constant]
Energy evolved by neutralization of CH_3COOH and NaOH
$= 200 \times 4.2 \times 5.6 + 160 \times 5.6 = 5600$ J

So energy used in dissociation of 0.1 mole $CH_3COOH = 5700 - 5600 = 100$ J
Enthalpy of dissociation = 1 kJ/mole

18. (B)

$$CH_3COOH = \frac{1\times100}{200} = \frac{1}{2}$$

$$CH_3CONa = \frac{1\times100}{200} = \frac{1}{2}$$

$$pH = pK_a + \log\frac{[\text{salt}]}{[\text{acid}]}$$

$$pH = 5 - \log 2 + \log\frac{1/2}{1/2}$$

$$pH = 4.7$$

19. (C)

C_8H_6: double bond equivalent $= 8 + 1 - \frac{6}{2} = 6$

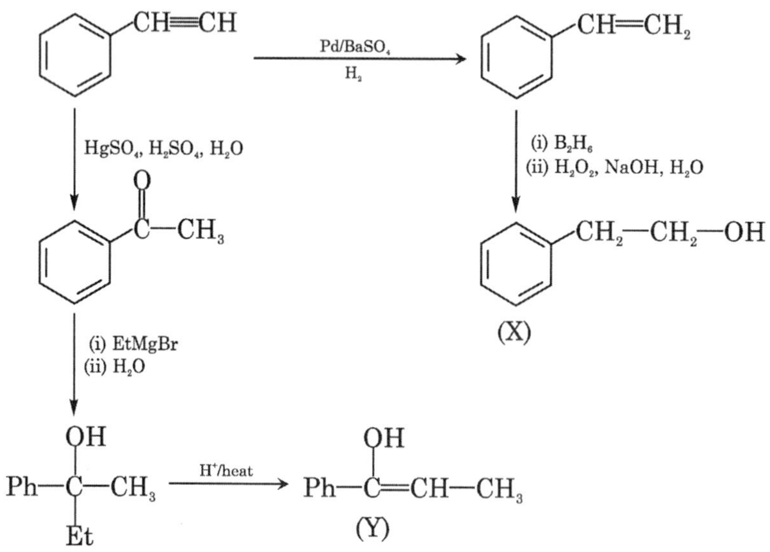

20. (D)

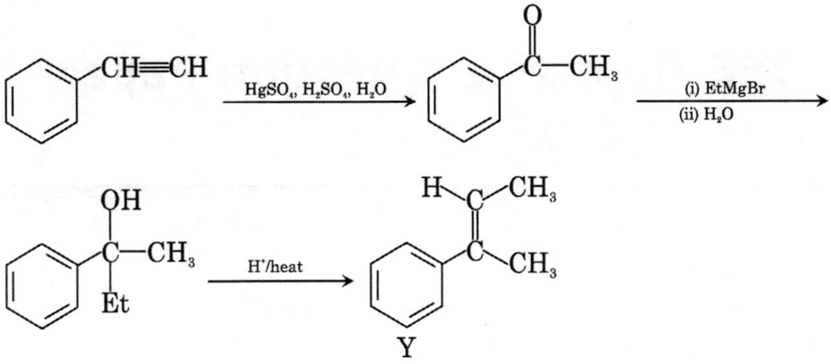
CH≡CH
HgSO4, H2SO4, H2O
O
C—CH3
(i) EtMgBr
(ii) H2O
OH
C—CH3
Et
H+/heat
H
C
CH3
C
CH3
Y

I.I.T. JEE Objective Question Paper (Main)—2016

IMPORTANT INSTRUCTIONS

1. Immediately fill in the particular on this page of the Test Booklet with only Blue/ Black Ball Point Pen provided by the Board.
2. The Answer Sheet is kept inside this Test Booklet. When you are directed to open the Test Booklet, take out the Answer Sheet and fill in the particulars carefully.
3. The test is of 3 hours duration.
4. The Test Booklet consists of 90 questions. The maximum marks are 360.
5. There are three parts in the question paper A, B, C consisting of Chemistry, Mathematics and Physics having 30 questions in each part of equal weightage. Each question is allotted 4 (four) marks for each correct response.
6. Candidates will be awarded marks as started above in instruction No. 5 for correct response of each question. 1/4 (one-fourth) marks will be deducted for indicating incorrect response of each question. No deduction from that total score will be made if no response is indicated for an item in the Answer Sheet.
7. There is only one correct response for each question. Filling up more than one response in each question will be treated as wrong response and marks for wrong response will be deducted accordingly as per instruction 6 above.
8. For writing particulars/marking responses on Side-1 and Side-2 of the Answer Sheet use only Blue/Black Ball Point Pen provided by the Board.
9. No candidate is allowed to carry any textual material, printed or written, bits of papers, pager, mobile phone, any electronic device, etc. except the Admit Card inside the examination room/hall.
10. Rough work is to be done on the space provided for this purpose in the Test Booklet only. This space is given at the bottom of each page.
11. On completion of the test, the candidate must hand over the Answer Sheet to the Invigilator on duty in the room/hall. However, the candidates are allowed to take away this Test Booklet with them.
12. The CODE for this Booklet is F. Make sure that the CODE printed on Side-2 of the Answer Sheet and also tally the serial number of the Test Booklet and Answer Sheet are the same as that on this booklet. In case of discrepancy, the candidate should immediately report the matter to the Invigilator for replacement of both the Test Booklet and the Answer Sheet.
13. Do not fold or make any stray mark on the Answer Sheet.

1. A stream of electrons from a heated filament was passed between two charged plates kept at a potential difference V esu. If e and m are charge and mass of an electron, respectively, then the value of h/λ (where λ is wavelength associated with electron wave) is given by
   (A) $2meV$ (B) $\sqrt{meV}$ (C) $\sqrt{2meV}$ (D) meV
2. 2-Chloro-2-methylpentane on reaction with sodium methoxide in methanol yields

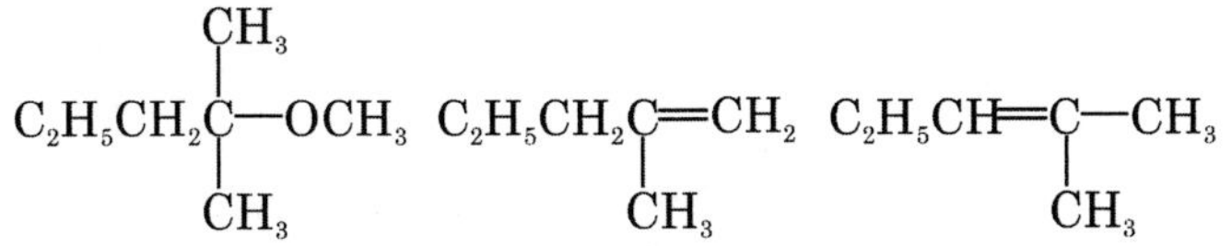

(a) (b) (c)

   (A) (a) and (c) (B) (c) only (C) (a) and (b) (D) All of these.
3. Which of the following compounds is metallic and ferromagnetic?
   (A) CrO_2 (B) VO_2 (C) MnO_2 (D) TiO_2
4. Which of the following statements about low density polythene is false?
   (A) It is a poor conductor of electricity.
   (B) Its synthesis requires dioxygen or a peroxide initiator as a catalyst.
   (C) It is used in the manufacture of buckets, dustbins, etc.
   (D) Its synthesis requires high pressure.
5. For a linear plot of log (x/m) versus log p in a Freundlich adsorption isotherm, which of the following statements is correct? (k and n are constants)
   (A) $1/n$ appears as the intercept.
   (B) Only $1/n$ appears as the slope.
   (C) log $(1/n)$ appears as the intercept.
   (D) Both k and $1/n$ appear in the slope term.
6. The heats of combustion of carbon and carbon monoxide are – 393.5 and – 285.3 kJ mol^{-1}, respectively. The heat of formation (in kJ) of carbon monoxide per mole is
   (A) 676.5 (B) – 676.5 (C) – 110.5 (D) 110.5
7. The hottest region of Bunsen flame shown in the figure below is

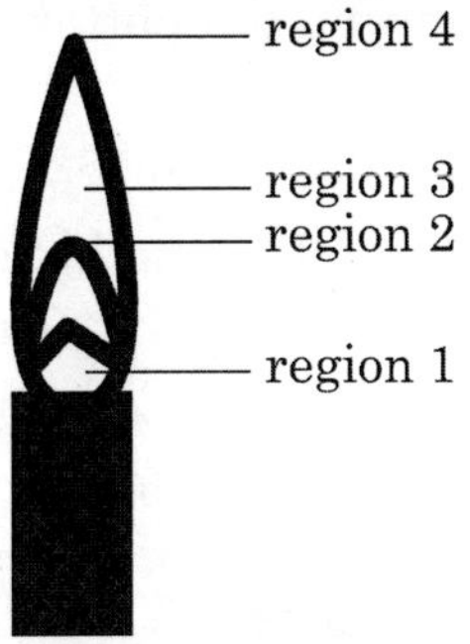

   (A) region 2 (B) region 3 (C) region 4 (D) region 1

8. Which of the following is an anionic detergent?
   (A) Sodium lauryl sulphate (B) Cetyltrimethyl ammonium bromide
   (C) Glyceryl oleate (D) Sodium stearate

9. 18 g glucose ($C_6H_{12}O_6$) is added to 178.2 g water. The vapor pressure of water (in torr) for this aqueous solution is
   (A) 76.0 (B) 752.4 (C) 759.0 (D) 7.6

10. The distillation technique most suited for separating glycerol from spent-lye in the soap industry is
   (A) Fractional distillation (B) Steam distillation
   (C) Distillation under reduced pressure (D) Simple distillation.

11. The species in which the N atom is in a state of *sp* hybridization is
   (A) NO_2^- (B) NO_3^- (C) NO_2 (D) NO_2^+

12. Decomposition of H_2O_2 follows a first order reaction. In fifty minutes the concentration of H_2O_2 decreases from 0.5 to 0.125*M* in one such decomposition. When the concentration of H_2O_2 reaches 0.05*M*, the rate of formation of O_2 will be
   (A) 6.93×10^{-4} mol min^{-1} (B) 2.66 L min^{-1} at STP
   (C) 1.34×10^{-2} mol min^{-1} (D) 6.93×10^{-2} mol min^{-1}

13. The pair having the same magnetic moment is [At. No.: Cr = 24, Mn = 25, Fe = 26, Co = 27]
   (A) $[Cr(H_2O)_6]^{2+}$ and $[Fe(H_2O)_6]^{2+}$ (B) $[Mn(H_2O)_6]^{2+}$ and $[Cr(H_2O)_6]^{2+}$
   (C) $[CoCl_4]^{2-}$ and $[Fe(H_2O)_6]^{2+}$ (D) $[Cr(H_2O)_6]^{2+}$ and $[CoCl_4]^{2-}$

14. The absolute configuration of

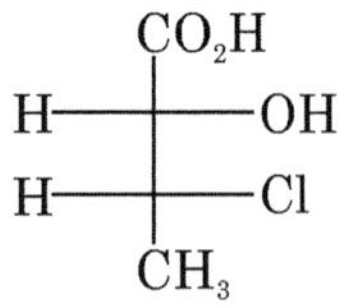

is

   (A) (2S, 3R) (B) (2S, 3S) (C) (2R, 3R) (D) (2R, 3S)

15. The equilibrium constant at 298 K for a reaction A + B $\rightleftharpoons$ C + D is 100. If the initial concentration of all the four species were 1*M* each, then equilibrium concentration of D (in mol L^{-1}) will be
   (A) 0.818 (B) 1.818 (C) 1.182 (D) 0.182

16. Which one of the following ores is best concentrated by froth floatation method?
   (A) Siderite (B) Galena (C) Malachite (D) Magnetite

17. At 300 K and 1 atm, 15 mL of a gaseous hydrocarbon requires 375 mL air containing 20% O_2 by volume for complete combustion. After combustion the gases occupy 330 mL. Assuming that the water formed is in liquid form and the volumes were measured at the same temperature and pressure, the formula of the hydrocarbon is
   (A) C_3H_6 (B) C_3H_8 (C) C_4H_8 (D) C_4H_{10}

18. The pair in which phosphorus atoms have a formal oxidation state of + 3 is
    (A) Pyrophosphorous and hypophosphoric acids
    (B) Orthophosphorous and hypophosphoric acids
    (C) Pyrophosphorous and pyrophosphoric acids
    (D) Orthophosphorous and pyrophosphorous acids
19. Which one of the following complexes shows optical isomerism?
    (A) *cis*-$[Co(en)_2Cl_2]Cl$ (B) *trans*-$[Co(en)_2Cl_2]Cl$
    (C) $[Co(NH_3)_4Cl_2]Cl$ (D) $[Co(NH_3)_3Cl_3]$
    (en = ethylenediamine)
20. The reaction of zinc with dilute and concentrated nitric acid, respectively, produces
    (A) NO_2 and NO (B) NO and N_2O (C) NO_2 and N_2O (D) N_2O and NO_2
21. Which one of the following statements about water is false?
    (A) Water can act both as an acid and as a base.
    (B) There is extensive intramolecular hydrogen bonding in the condensed phase.
    (C) Ice formed by heavy water sinks in normal water.
    (D) Water is oxidised to oxygen during photosynthesis.
22. The concentration of fluoride, lead, nitrate and iron in a water sample from an underground lake was found to be 1000 ppb, 40 ppb, 100 ppm and 0.2 ppm, respectively. This water is unsuitable for drinking due to high concentration of
    (A) Lead (B) Nitrate (C) Iron (D) Fluoride
23. The main oxides formed on combustion of Li, Na and K in excess of air are, respectively
    (A) LiO_2, Na_2O_2 and K_2O (B) Li_2O_2, Na_2O_2 and KO_2
    (C) Li_2O, Na_2O_2 and KO_2 (D) Li_2O, Na_2O and KO_2
24. Thiol group is present in
    (A) Cystine (B) Cysteine (C) Methionine (D) Cytosine
25. Galvanisation is applying a coating of
    (A) Cr (B) Cu (C) Zn (D) Pb
26. Which of the following atoms has the highest first ionization energy?
    (A) Na (B) K (C) Sc (D) Rb
27. In the Hofmann bromamide degradation reaction, the number of moles of NaOH and Br_2 used per mole of amine produced are
    (A) Four moles of NaOH and two moles of Br_2
    (B) Two moles of NaOH and two moles of Br_2
    (C) Four moles of NaOH and one mole of Br_2
    (D) One mole of NaOH and one mole of Br_2
28. Two closed bulbs of equal volume (V) containing an ideal gas initially at pressure p_i and temperature T_1 are connected through a narrow tube of negligible volume as

shown in the figure below. The temperature of one of the bulbs is then raised to T_2. The final pressure p_f is

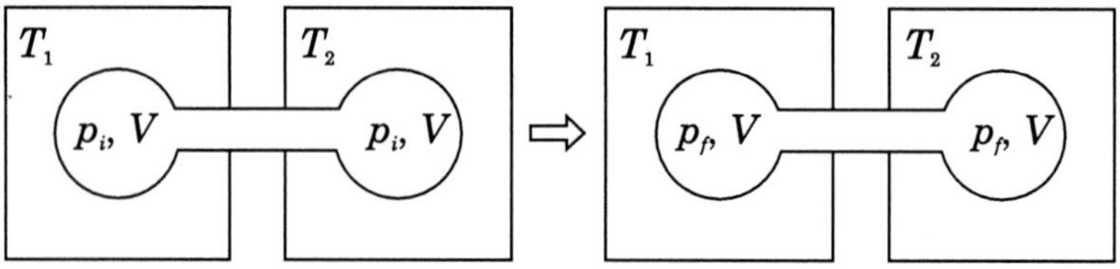

(A) $2p_i\left(\frac{T_1}{T_1+T_2}\right)$ (B) $2p_i\left(\frac{T_2}{T_1+T_2}\right)$ (C) $2p_i\left(\frac{T_1T_2}{T_1+T_2}\right)$ (D) $p_i\left(\frac{T_1T_2}{T_1+T_2}\right)$

29. The reaction of propene with HOCl ($Cl_2 + H_2O$) proceeds through the intermediate

(A) $CH_3—CH^+—CH_2—Cl$ (B) $CH_3—CH(OH)—CH_2^+$

(C) $CH_3—CHCl—CH_2^+$ (D) $CH_3—CH^+—CH_2—OH$

30. The product of the reaction given below is

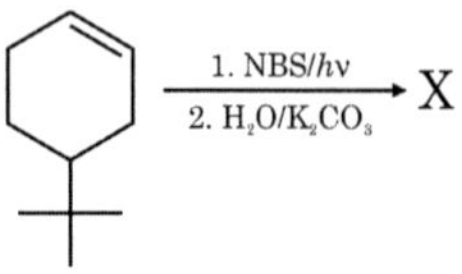

(A) 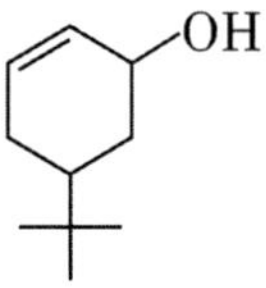

(B) 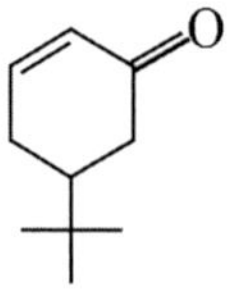

(C) 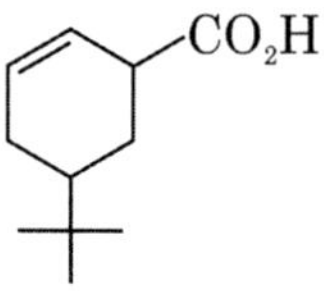

(D) 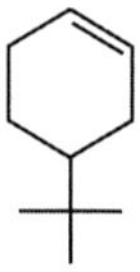

ANSWERS

1. C	2. D	3. A	4. C	5. B	6. C
7. A	8. A	9. B	10. C	11. D	12. A
13. A	14. A	15. B	16. B	17. B	18. D
19. A	20. D	21. B	22. B	23. C	24. B
25. C	26. C	27. C	28. B	29. A	30. A

Model Solutions—I.I.T. JEE (Main) 2016

1. (C)

$$\lambda = \frac{h}{mv} = \frac{h}{\sqrt{2m(KE)}}$$

Hence, $\frac{h}{\lambda} = \sqrt{2meV}$

2. (D)

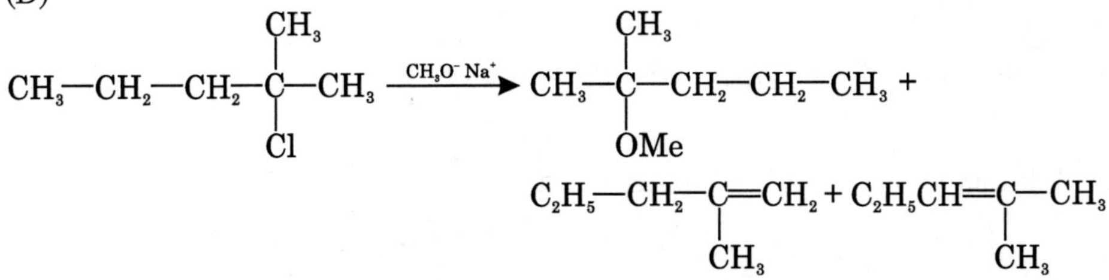

3. (A)

CrO_2 is metallic and ferromagnetic. CrO_2 like metals conducts electricity.

4. (C)

HDPE is used in making buckets and dustbins.

5. (B)

$$\frac{x}{m} = kp^{\frac{1}{n}}$$

$$\log\frac{x}{m} = \log k + \frac{1}{n}\log p$$

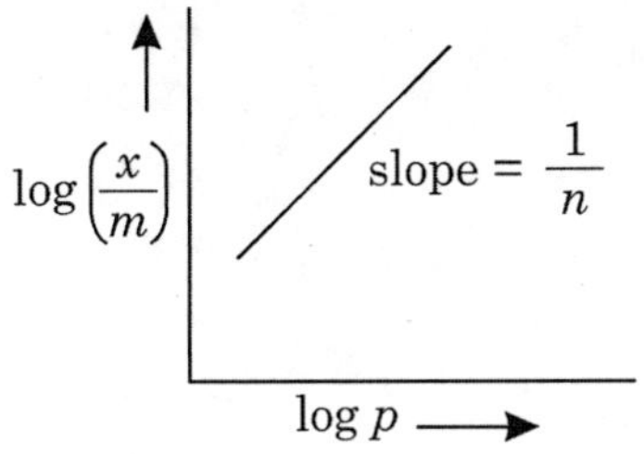

6. (C)

$C(g) + O_2(g) \longrightarrow CO_2(g)$ ...(1)

$CO(g) +– O_2(g) \longrightarrow CO_2(g)$ ...(2)

$\Delta H_1 = -\ 393.5$ kJ/mol; $\Delta H_2 = -\ 283.5$ kJ/mol

$C(g) + \frac{1}{2}\ O_2(g) \longrightarrow CO_2(g)$

$\Delta_f H_{(CO,\ g)} = \Delta H_1 - \Delta H_2$

$= -\ 393.5 - (-\ 283.5) = -\ 393.5 + 283.5 = -\ 110$ kJ/mol

7. (A)
Region 2 (blue flame)

8. (A)
Sodium lauryl sulphate ($CH_3(CH_2)_{11}OSO_3Na$) is an anionic detergent.

9. (B)

$$n_{C_6H_{12}O_6} = 0.1; \quad n_{H_2O} = \frac{178.2}{18} = 9.9$$

$$\frac{p^0 - p}{p^0} = \frac{0.1}{10}$$

$$p^0 - p = \frac{0.1}{10} \times 760 = 7.6$$

$$p = p^0 - 7.6 = 760 - 7.6 = 752.4 \text{ torr}$$

10. (C)
Glycerol decomposes before its boiling point.

11. (D)

$$O{=}\overset{\oplus}{N}{=}O$$

sp hybridisation

12. (A)

$$H_2O_2 \longrightarrow H_2O + \frac{1}{2}O_2$$

$$2t_{1/2} = 50 \text{ mins}; \quad t_{1/2} = 25 \text{ mins}$$

$$\text{Hence, } k = \frac{0.693}{25}$$

$$-\frac{d[H_2O_2]}{dt} = \frac{d[O_2]}{dt} \times 2$$

$$\text{Hence, } \frac{d[O_2]}{dt} = \frac{1}{2} \times \frac{0.693}{25} \times 0.05$$

$$= 6.93 \times 10^{-4} \text{ mol min}^{-1}$$

13. (A)
$Cr^{2+} \longrightarrow d^4 \longrightarrow$ Number of unpaired electrons 4
$Fe^{2+} \longrightarrow d^6 \longrightarrow$ Number of unpaired electrons 4
Both configurations have four unpaired electrons due to weak field ligand H_2O.

14. (A)

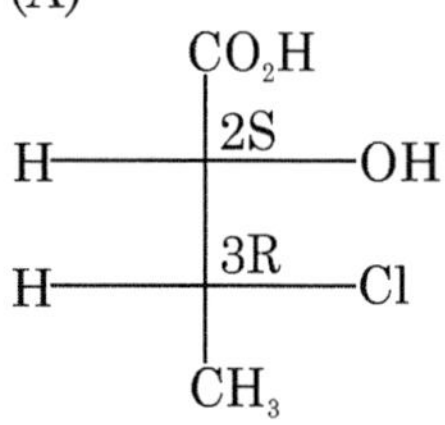

(2S, 3R)

15. (B)

$$A + B \rightleftharpoons C + D$$

A	B	C	D
1	1	1	1
$1-x$	$1-x$	$1+x$	$1+x$

Given $K_{eq} = 100$

$$\frac{(1+x)^2}{(1-x)^2} = 100$$

or $\dfrac{1+x}{1-x} = 10$

or $1 + x = 10 - 10x$

or $11x = 9$

or $x = \dfrac{9}{11}$

Hence, $[D] = 1 + \dfrac{9}{11}$

$= 1 + 0.818 = 1.818M$

16. (B)

Galena (PbS) is a sulphide ore and it is best concentrated by froth floatation method.

17. (B)

$$C_xH_y + \left(x + \frac{y}{4}\right)O_2 \longrightarrow xCO_2 + \frac{y}{2}H_2O$$

15 mL $\quad 15\left(x + \frac{y}{4}\right)$ mL $\quad 15x$ mL

$$V_{O_2} = \frac{20}{100} \times 375 = 75 \text{ mL} = 15\left(x + \frac{y}{4}\right)$$

or $x + \dfrac{y}{4} = 5$

If we put $x = 3$ and $y = 8$, then the above equation is satisfied and the answer will be C_3H_8.

18. (D)

$H_4P_2O_5$ (pyrophosphorous acid) (+ 3 state)

H_3PO_3 (orthophosphorous acid) (+ 3 state)

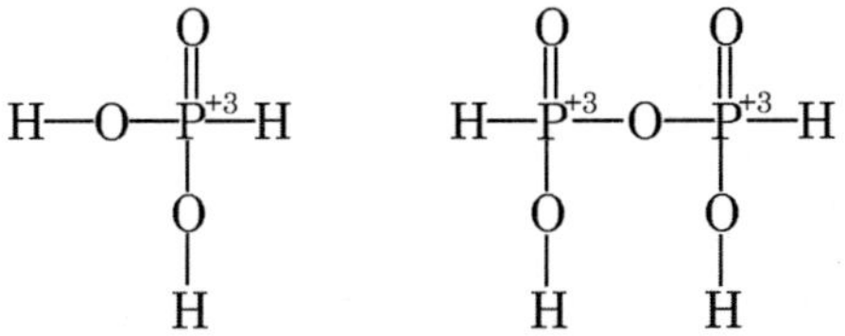

Orthophosphorous acid Pyrophosphorous acid

19. (A)

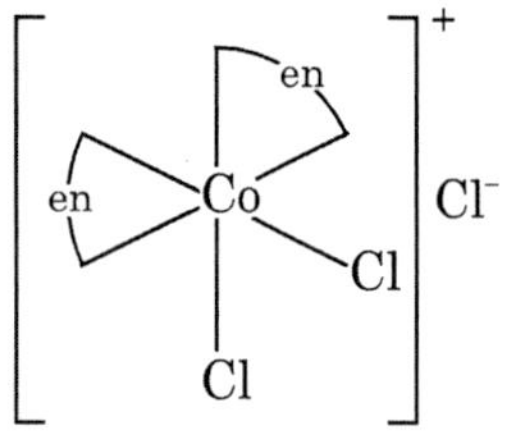

cis-[CO(en)$_2$Cl$_2$]Cl
(optically active)

20. (D)

$Zn + 4HNO_3$ (conc.) $\longrightarrow Zn(NO_3)_2 + 2H_2O + 2NO_2(\uparrow)$

$4Zn + 10HNO_3$ (dil.) $\longrightarrow 4Zn(NO_3)_2 + N_2O(\uparrow) + 5H_2O$

21. (B)

Ice shows intermolecular hydrogen bonding.

22. (B)

Maximum limit of nitrate in potable water is 50 ppm.

23. (C)

24. (B)

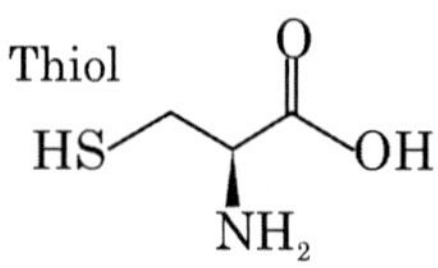

(Cysteine)

25. (C)

26. (C)

Ionisation energy (kJ/mol): Na, 496; K, 419; Rb, 403; Sc, 631.

27. (C)

$$R{-}\overset{\overset{\displaystyle O}{\|}}{C}{-}NH_2 + 4NaOH + Br_2 \longrightarrow RNH_2 + 2NaBr + Na_2CO_3 + 2H_2O$$

28. (B)

Initial total moles:

$$n = \frac{p_iV}{RT_1} + \frac{p_iV}{RT_1} = \frac{2p_iV}{RT_1} \quad ...(1)$$

Final total moles:

$$n = \frac{p_fV}{RT_1} + \frac{p_fV}{RT_2} \quad ...(2)$$

Equating the two equations, we get

$$\frac{2p_i}{T_1} = \frac{p_f}{T_1} + \frac{p_f}{T_2}$$

or $$p_f = \frac{2p_i}{T_1} \times \frac{T_1 T_2}{(T_1 + T_2)}$$

or $$p_f = 2p_i \times \frac{T_2}{(T_1 + T_2)}$$

29. (A)

$$CH_3—CH{=}CH_2 + Cl^{\oplus} \longrightarrow CH_3—\overset{\oplus}{C}H—CH_2Cl$$ (Markownikoff's addition)

30. (A)

Allylic bromination followed by hydrolysis.

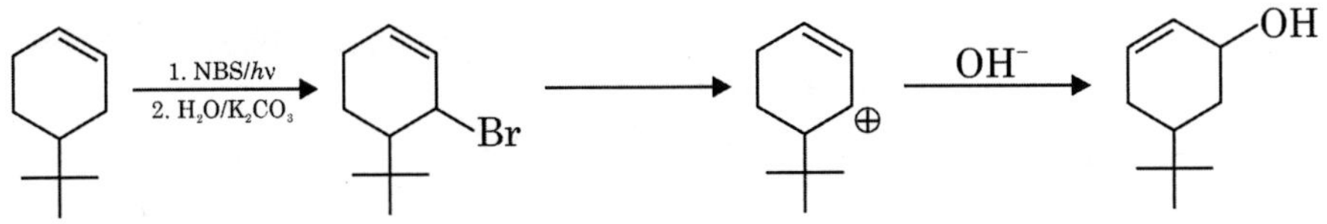

I.I.T. JEE Objective Question Paper (Advanced)—2016

Paper–1

INSTRUCTIONS

General

1. This sealed booklet is your Question Paper. Do not break the seal till you are told to do so.
2. The paper CODE is printed on the right hand top corner of this sheet and the right hand top corner of the back cover of this booklet.
3. Use the Optical Response Sheet (ORS) provided separately for answering the questions.
4. The paper CODE is printed on the left part as well as the right part of the ORS. Ensure that both these codes are identical and same as that on the question paper booklet. If not, contact the invigilator for change of ORS.
5. Blank spaces are provided within this booklet for rough work.
6. Write your name, roll number and sign in the space provided on the back cover of this booklet.
7. After breaking the seal of the booklet at 9:00 am, verify that the booklet contains 36 pages and that all the 54 questions along with the options are legible. If not, contact the invigilator for replacement of the booklet.
8. You are allowed to take away the Question Paper at the end of the examination.

Optical Response Sheet

9. The ORS (top sheet) will be provided with an attached Candidate's Sheet (bottom sheet). The Candidate's Sheet is a carbonless copy of the ORS.
10. Darken the appropriate bubbles on the ORS by applying sufficient pressure. This will leave an impression at the corresponding place on the Candidate's Sheet.
11. The ORS will be collected by the invigilator at the end of the examination.
12. You will be allowed to take away the Candidate's Sheet at the end of the examination.
13. Do not tamper with or mutilate the ORS. Do not use the ORS for rough work.
14. Write your name, roll number and code of the examination center, and sign with pen in the space provided for this purpose on the ORS. Do not write any of these details anywhere else on the ORS. Darken the appropriate bubble under each digit of your roll number.

Darkening the Bubbles on the ORS

15. Use a Black Ball Point Pen to darken the bubbles on the ORS.
16. Darken the bubble ○ completely.
17. The correct way of darkening a bubble is as: ●
18. The ORS is machine-gradable. Ensure that the bubbles are darkened in the correct way.
19. Darken the bubbles only if you are sure of the answer. There is no way to erase or "un-darken" a darkened bubble.

SECTION–1
(Maximum Marks: 15)

This section contains five questions.

Each question has four options (A), (B), (C) and (D). Only one of these four options is correct.

For each question, darken the bubbles corresponding to the correct option in the ORS.

For each question, marks will be awarded in one of the following categories:

Full Marks : + 3 If only the bubbles corresponding to the correct option is darkened.

Zero Marks : 0 If none of the bubbles is darkened.

Negative Marks : – 1 In all other cases.

1. P is the probability of finding the 1s electron of hydrogen atom in a spherical shell or infinitesimal thickness, dr, at a distance r from the nucleus. The volume of this shell is $4\pi r^2 dr$. The qualitative sketch of the dependence of P on r is

(A) 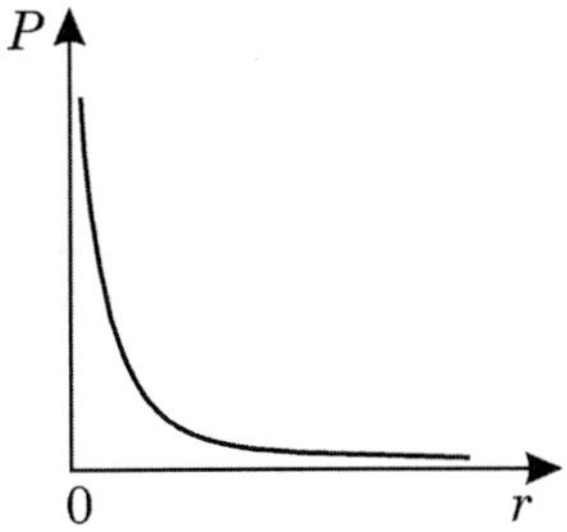

(B) 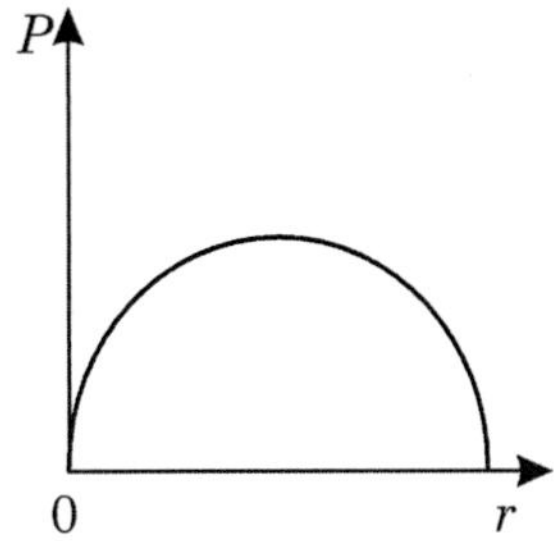

(C) 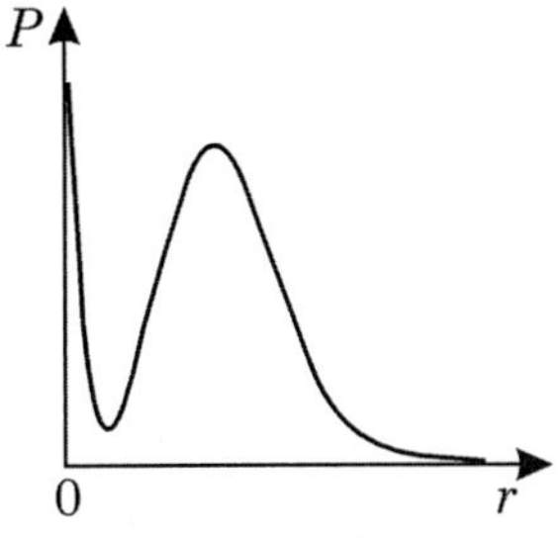

(D) 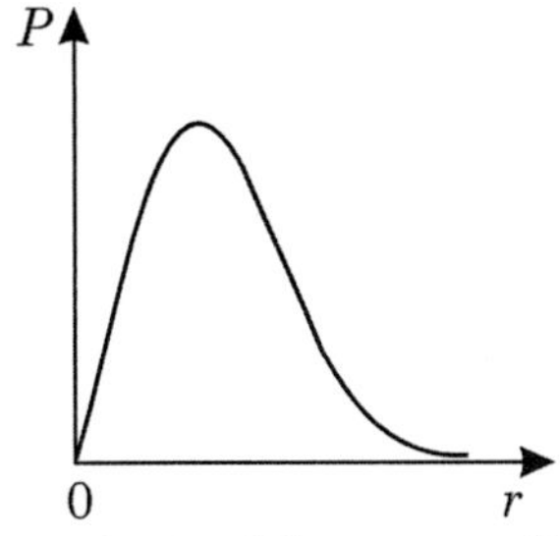

2. One mole of an ideal gas at 300 K in thermal contact with surroundings expands isothermally from 1.0 L to 2.0 L against a constant pressure of 3.0 atm. In this process, the change in entropy of surrounding (ΔS_{surr}) in JK^{-1} is (1 L atm = 101.3 J)

(A) 5.763 (B) 1.013 (C) – 1.013 (D) – 5.763

3. The increasing order of atomic radii of the following Group 13 elements is

   (A) Al < Ga < In < T1
   (B) Ga < Al < In < T1
   (C) Al < In < Ga < T1
   (D) Al < Ga < T1 < In

4. Among $[Ni(CO)_4]$, $[NiCl_4]^{2-}$, $[Co(NH_3)_4Cl_2]Cl$, $Na_3[CoF_6]$, Na_2O_2 and CrO_2, the total number of paramagnetic compounds is

   (A) 2
   (B) 3
   (C) 4
   (D) 5

5. On complete hydrogenation, natural rubber produces

   (A) ethylene-propylene copolymer
   (B) vulcanised rubber
   (C) polypropylene
   (D) polybutylene

SECTION–2
(Maximum Marks: 32)

This section contains EIGHT questions.

Each question has FOUR options (A), (B), (C) and (D). ONE OR MORE THAN ONE of these four option(s) is(are) correct.

For each question, darken the bubble(s) corresponding to all the correct option(s) in the ORS.

For each question, marks will be awarded in one of the following categories:

Full Marks : + 4 If only the bubble(s) corresponding to all the correct option(s) is (are) darkened.

Partial Marks : + 1 For darkening a bubble corresponding to each correct option, provided NO incorrect option is darkened.

Zero Marks : 0 If none of the bubbles is darkened.

Negative Marks : – 2 In all other cases.

For example, if (A), (C) and (D) are all the correct options for a question, darkening all these three will result in + 4 marks; darkening only (A) and (D) will result in + 2 marks; and darkening (A) and (B) result in – 2 marks, as a wrong option is also darkened.

6. According to the Arrhenius equation,

   (A) a high activation energy usually implies a fast reaction.
   (B) rate constant increases with increase in temperature. This is due to a greater number of collisions whose energy exceeds the activation energy.
   (C) higher the magnitude of activation energy, stronger is the temperature dependence of the rate constant.
   (D) the pre-exponential factor is a measure of the rate at which collisions occur, irrespective of their energy.

7. A plot of the number of neutrons (N) against the number of protons (P) of stable nuclei exhibits upward deviation from linearity for atomic number, $Z > 20$. For an unstable nucleus having N/P ratio less than 1, the possible mode(s) of decay is(are)

   (A) β^--decay (β emission)
   (B) orbital or K-electron capture
   (C) neutron emission
   (D) β^+-decay (positron emission).

8. The crystalline form of borax has
   (A) tetranuclear $[B_4O_5(OH)_4]^{2-}$ unit
   (B) all boron atoms in the same plane
   (C) equal number of sp^2 and sp^3 hybridised boron atoms
   (D) one terminal hydroxide per boron atom.

9. The compound(s) with TWO lone pairs of electrons on the central atom is(are)
   (A) BrF_5 (B) ClF_3 (C) XeF_4 (D) SF_4

10. The reagent(s) that can selectively precipitate S^{2-} from a mixture of S^{2-} and SO_4^{2-} in aqueous solution is(are)
    (A) $CuCl_2$ (B) $BaCl_2$
    (C) $Pb(OOCCH_3)_2$ (D) $Na_2[Fe(CN)_5NO]$

11. Positive Tollen's test is observed for

    (A) 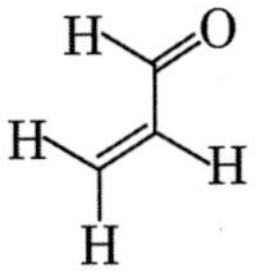

    (B) 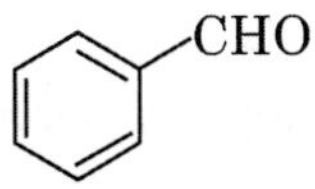

    (C) 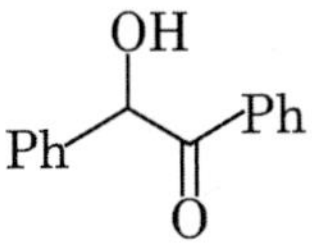

    (D) 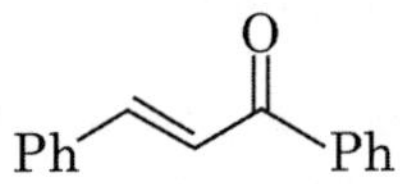

12. The product(s) of the following reaction sequence is(are)

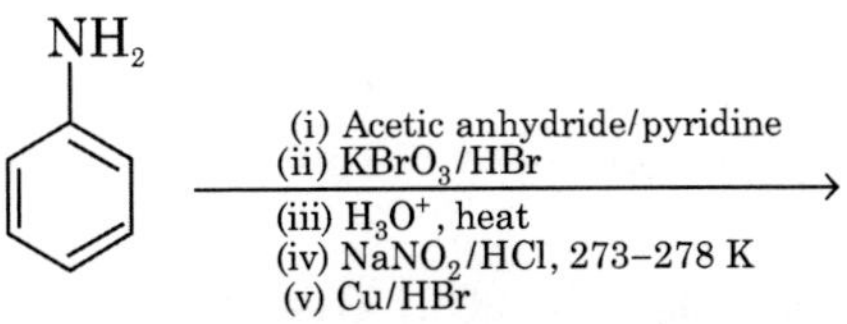

(A) 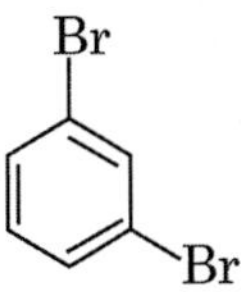

(B) 

(C) 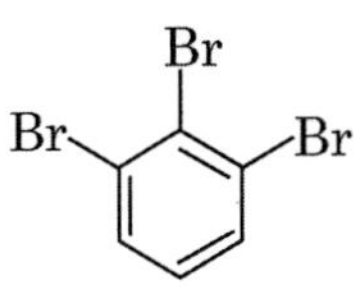

(D) 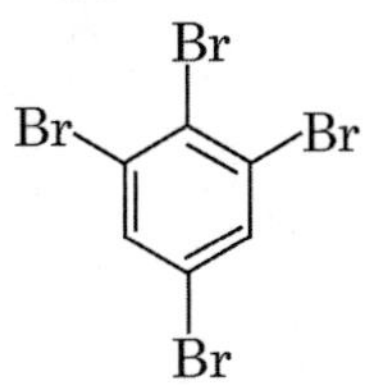

13. The correct statement(s) about the following reaction sequence is(are)

$$\text{Cumene } (C_9H_{12}) \xrightarrow[\text{(ii) } H_3O^+]{\text{(i) } O_2} P \xrightarrow{CHCl_3/NaOH} Q \text{ (major)} + R \text{ (minor)}$$

$$Q \xrightarrow[PhCH_2Br]{NaOH} S$$

(A) R is steam volatile
(B) Q gives dark violet coloration with 1% aqueous $FeCl_3$ solution
(C) S gives yellow precipitate with 2,4-dinitrophenylhydrazine
(D) S gives dark violet coloration with 1% aqueous $FeCl_3$ solution.

SECTION–3
(Maximum Marks: 15)

This section contains five questions.

The answer to each question is a single digit integer ranging from 0 to 9, both inclusive.

For each question, darken the bubble corresponding to the correct integer in the ORS.

For each question, marks will be awarded in one of the following categories:

Full Marks : + 3 If only the bubble corresponding to the correct answer is darkened.

Zero Marks : 0 In all other cases.

14. The mole fraction of a solute in a solution is 0.1. At 298 K, molarity of this solution is the same as its molality. Density of this solution at 298 K is 2.0 g cm^{-3}. The ratio of the molecular weights of the solute and solvent, $\left(\frac{MW_{solute}}{MW_{solvent}}\right)$, is

15. The diffusion coefficient of an ideal gas is proportional to its mean free path and mean speed. The absolute temperature of an ideal gas is increased 4 times and its pressure is increased 2 times. As a result, the diffusion coefficient of this gas increases x times. The value of x is

16. In neutral or faintly alkaline solution, 8 moles of permanganate anion quantitatively oxidise thiosulphate anions to produce X moles of a sulphur containing product. The magnitude of X is

17. The number of geometric isomers possible of the complex $[CoL_2Cl_2]^-$ ($L = H_2NCH_2CH_2O^-$) is

18. In the following monobromination reaction, the number of possible chiral products is

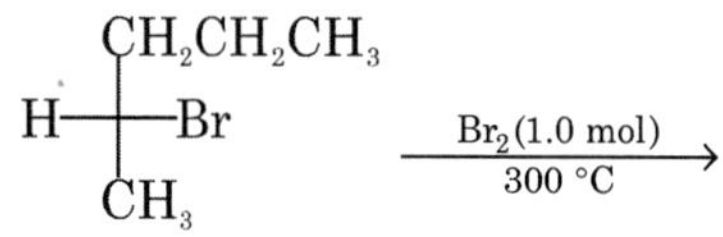

(0.1 mol)
(enantiomerically pure)

ANSWERS

1. D	2. C	3. B	4. B	5. A	6. B, C, D
7. B, D	8. A, C, D	9. B, C	10. A	11. A, B	12. B
13. B, C	14. 9	15. 4	16. 6	17. 5	18. 5

Model Solutions—I.I.T. JEE (Advanced) 2016

Paper–1

1. (D)

The probability distribution curve for 1*s* electron of hydrogen atom.

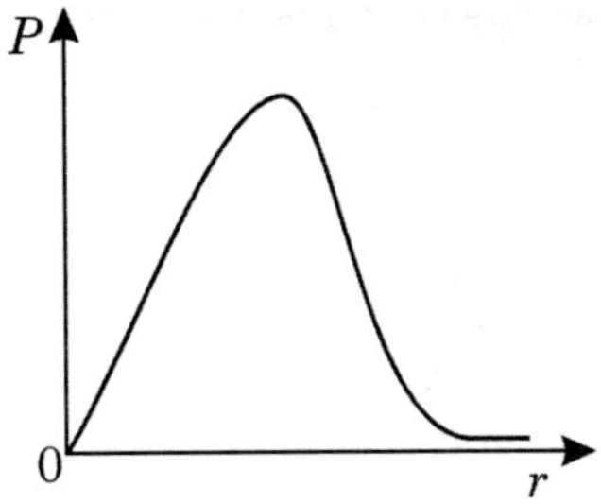

2. (C)

Isothermal process, $\Delta U = 0$

$dq = -\,dW = P_{ext}\,(V_2 - V_1) = 3\text{ L atm} = 3 \times 101.3\text{ J}$

$\Delta S_{surrounding} = -\dfrac{3 \times 101.3}{300}\text{JK}^{-1} = -\,1.013\text{ JK}^{-1}$

Hence, $\Delta S_{surr} = -\,1.013\text{ JK}^{-1}$

3. (B)

Increasing order of atomic radius of group 13 elements: Ga < Al < In < Tl.
Due to poor shielding of *d*-electrons in Ga, its radius decreases below Al.

4. (B)

Number of paramagnetic compounds are 3. The paramagnetic compounds are:
$[NiCl_4]^{2-}$, $Na_3[CoF_6]$, CrO_2

5. (A)

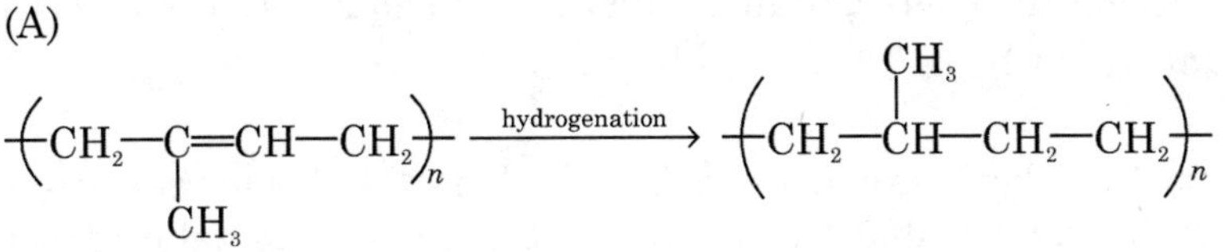

Natural rubber

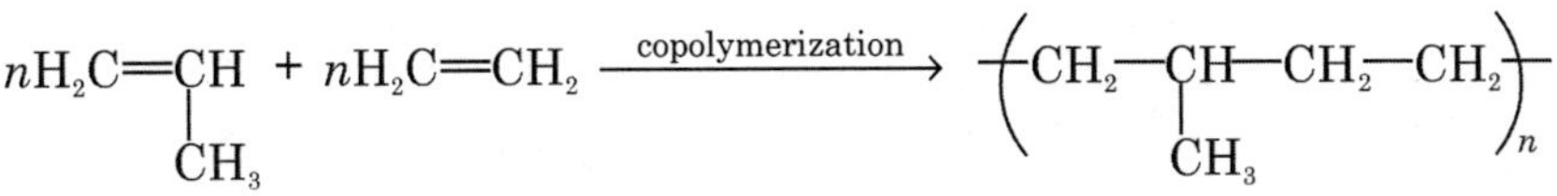

propylene ethylene propylene-ethylene copolymer

6. (B, C, D)
7. (B, D)
$\frac{n}{p} < 1$
Positron, emission
$_aX^b \longrightarrow {}_{+1}^{0}e + {}_{a-1}Y^b$
K-electron capture
$_aX^b + {}_{-1}e^0 \longrightarrow {}_{a-1}Y^b$
Both the processes cause an increase in n/p ratio towards 1, thus stabilising the nucleus.
8. (A, C, D)
The structure of anion of borax is

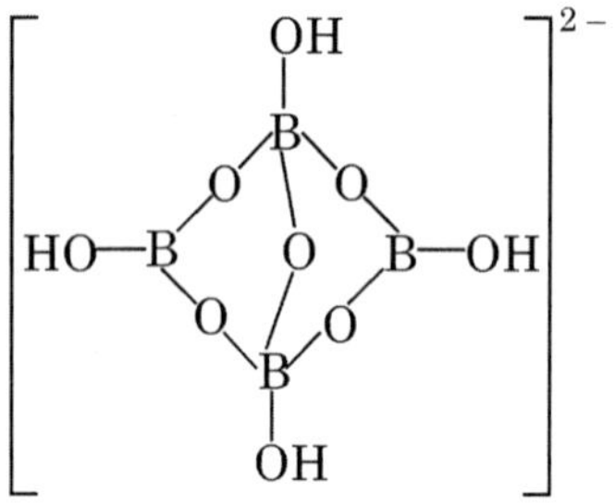

9. (B, C)
ClF_3 and XeF_4 contain two lone pair of electrons on the central atom.

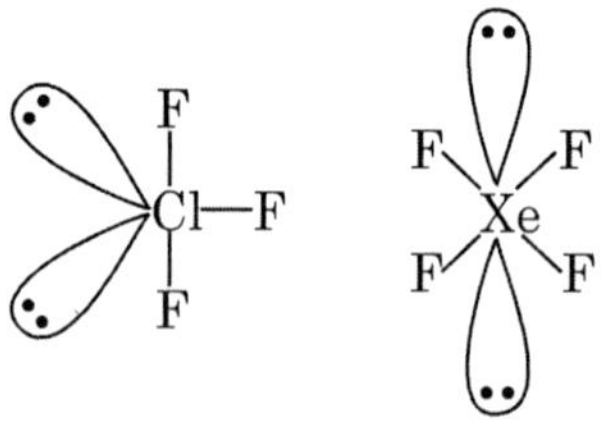

10. (A)
The reagent that can selectively precipitate S^{2-} and SO_4^{2-} in aqueous solution is $CuCl_2$.
$S^{2-} + CuCl_2 \longrightarrow CuS(\downarrow) + 2Cl^-$
11. (A, B)
Beside aldehyde α-hydroxy ketones can also exhibit Tollen's test due to rearrangement in aldehyde via ene diol intermediate. (However, this needs a terminal α-carbon.)
12. (B)

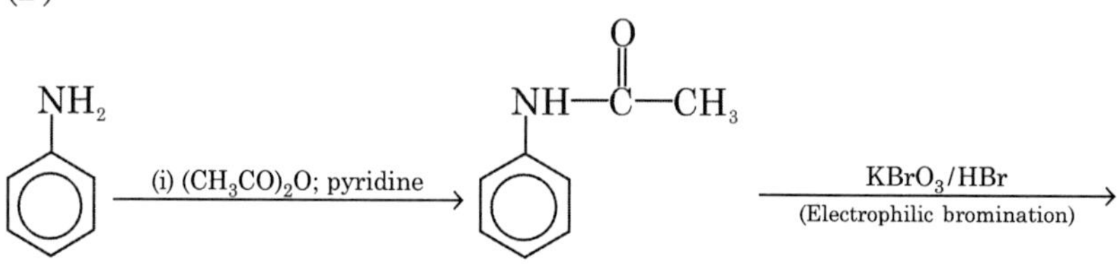

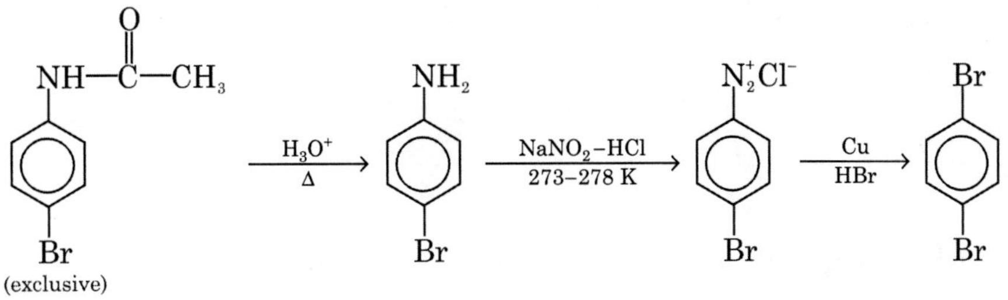

13. (B, C)

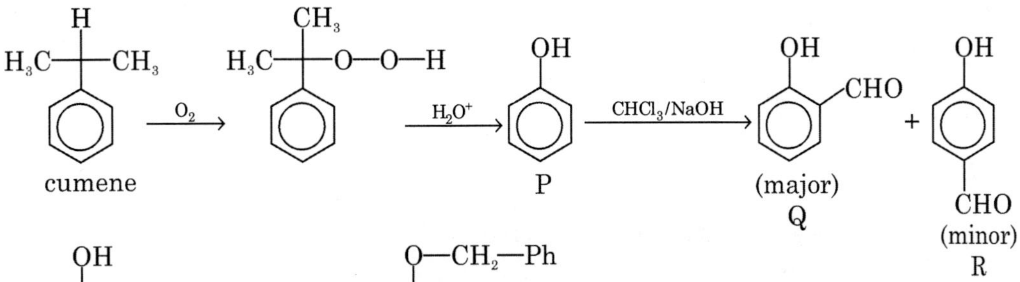

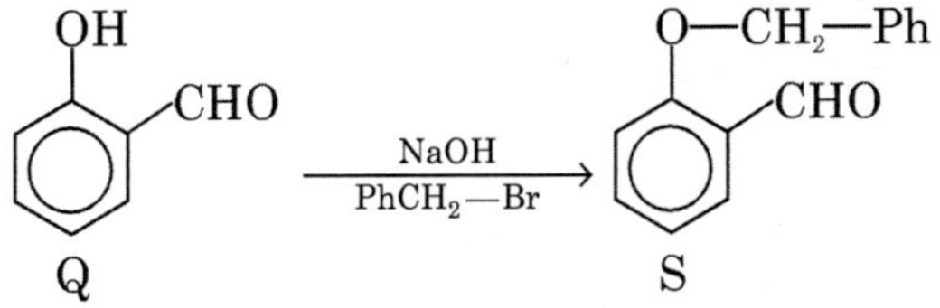

Q (not R) is steam volatile due to intramolecular hydrogen bonding.
Q gives violet colouration due to phenolic functional group.
S gives yellow precipitate due to aldehydic group with 2,4-DNP.
S does not have free phenolic group to respond to $FeCl_3$ test.

14. (9)

$$m = \frac{X_A \times 1000}{X_B \times M_A}$$

$$m = \frac{1000}{9M_A} \quad \text{...(i)}$$

$$M = \frac{n_B \times 1000 \times d}{n_A \times M_A + n_B \times M_B} = \frac{X_B \times 1000 \times d}{X_A \times M_A + X_B \times M_B}$$

$$= \frac{200}{0.9M_A + 0.1M_B}$$

$$= \frac{2000}{9M_A + M_B} \quad \text{...(ii)}$$

Since $m = M$, we have

$$\frac{1000}{9M_A} = \frac{2000}{9M_A + M_B}$$

$9M_A + M_B = 18M_A$

Hence, $9M_A = M_B$

or $\dfrac{M_B}{M_A} = 9$

15. (4)

Mean free path $(\ell) \propto \dfrac{T}{P}$; Mean speed $(C_{av}) \propto \sqrt{T}$

Diffusion coefficient (D) is proportional to both mean free path (ℓ) and mean speed (C_{av}).

Hence, $D \propto \dfrac{T^{3/2}}{P}$

At temperature T_1 and P_1,

$$D_1 = \frac{KT_1^{3/2}}{P_1} \qquad ...(1)$$

If T_1 is increased 4 times and P_1 is increased 2 times,

$$D_2 = \frac{K(4T_1)^{3/2}}{2P_1} \qquad ...(2)$$

Hence, $\dfrac{D_2}{D_1} = \dfrac{(4)^{3/2}}{2} = \dfrac{2^3}{2} = 4$

16. (6)

$$8MnO_4^- + 3S_2O_3^{2-} + H_2O \xrightarrow{\text{faintly alkaline}} 8MnO_2 + 6SO_4^{2-} + 2OH^-$$

Hence, 8 moles MnO_4^- produce 6 moles SO_4^{2-}.

17. (5)

Number of geometrical isomers = 5

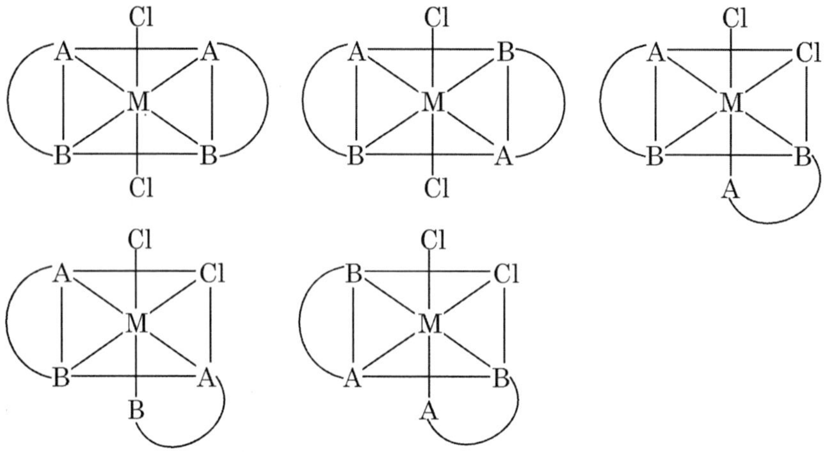

AB = $H_2NCH_2CH_2O^-$; A site is N and B site is O.

18. (5)

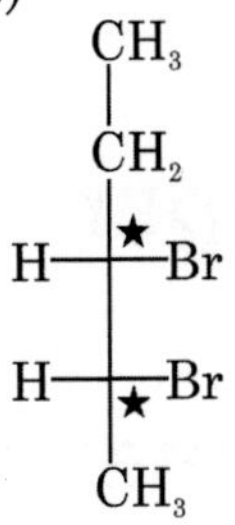

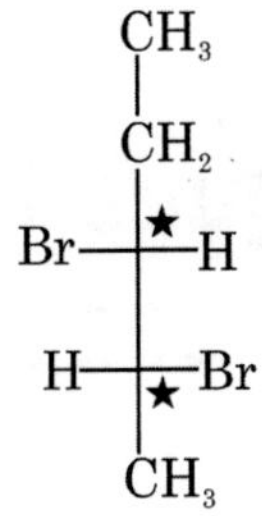

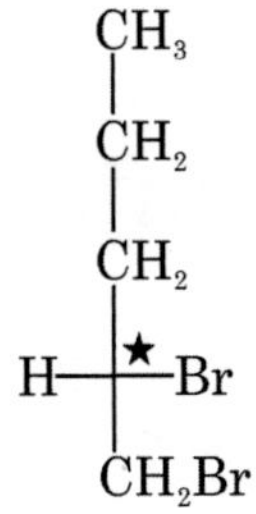

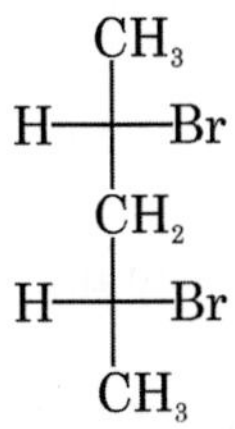

achiral (meso)
(not to be counted)

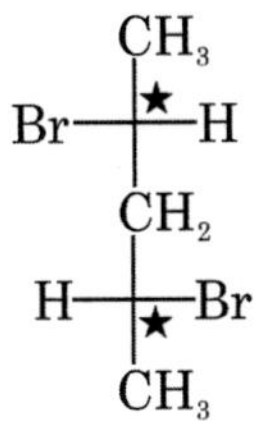

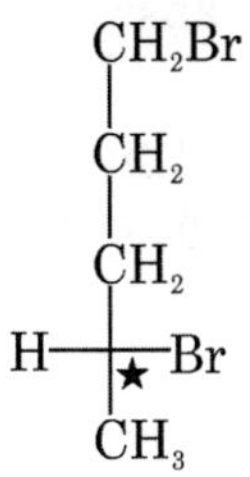

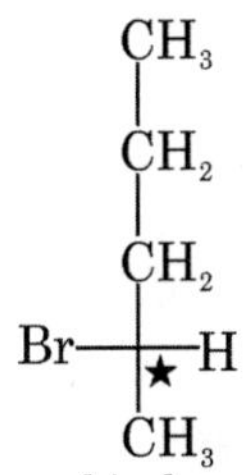

no chiral centre
(not to be counted)

I.I.T. JEE Objective Question Paper (Advanced)—2016

Paper–2

READ THE INSTRUCTIONS CAREFULLY

General

1. This sealed booklet is your Question Paper. Do not break the seal till you are told to do so.
2. The paper CODE is printed on the right hand top corner of this sheet and the right hand top corner of the back cover of this booklet.
3. Use the Optical Response Sheet (ORS) provided separately for answering the questions.
4. The paper CODE is printed on the left part as well as the right part of the ORS. Ensure that both these codes are identical and same as that on the question paper booklet. If not, contact the invigilator for change of ORS.
5. Blank spaces are provided within this booklet for rough work.
6. Write your name, roll number and sign in the space provided on the back cover of this booklet.
7. After breaking the seal of the booklet at 2:00 pm, verify that the booklet contains 36 pages and that all the 54 questions along with the options are legible. If not, contact the invigilator for replacement of the booklet.
8. You are allowed to take away the Question Paper at the end of the examination.

Optical Response Sheet

9. The ORS (top sheet) will be provided with an attached Candidate's Sheet (bottom sheet). The Candidate's Sheet is a carbonless copy of the ORS.
10. Darken the appropriate bubbles on the ORS by applying sufficient pressure. This will leave an impression at the corresponding place on the Candidate's Sheet.
11. The ORS will be collected by the invigilator at the end of the examination.
12. You will be allowed to take away the Candidate's Sheet at the end of the examination.
13. Do not tamper with or mutilate the ORS. Do not use the ORS for rough work.
14. Write your name, roll number and code of the examination center, and sign with pen in the space provided for this purpose on the ORS. Do not write any of these details anywhere else on the ORS. Darken the appropriate bubble under each digit of your roll number.

Darkening the Bubbles on the ORS

15. Use a black ball point pen to darken the bubbles on the ORS.
16. Darken the bubble ○ completely.
17. The correct way of darkening a bubble is as: ●
18. The ORS is machine-gradable. Ensure that the bubbles are darkened in the correct way.
19. Darken the bubbles only if you are sure of the answer. There is no way to erase of "un-darken" a darkened bubble.

SECTION–1
(Maximum Marks: 18)

This section contains six questions.
Each question has FOUR options (A), (B), (C) and (D). Only one of these four options is correct.
For each question, darken the bubble corresponding to the correct option in the ORS.
For each question, marks will be awarded in one of the following categories:
Full Marks : + 3 If only the bubble corresponding to the correct option is darkened.
Zero Marks : 0 If none of the bubbles is darkened.
Negative Marks : – 1 In all other cases.

1. The correct order of acidity for the following compounds is

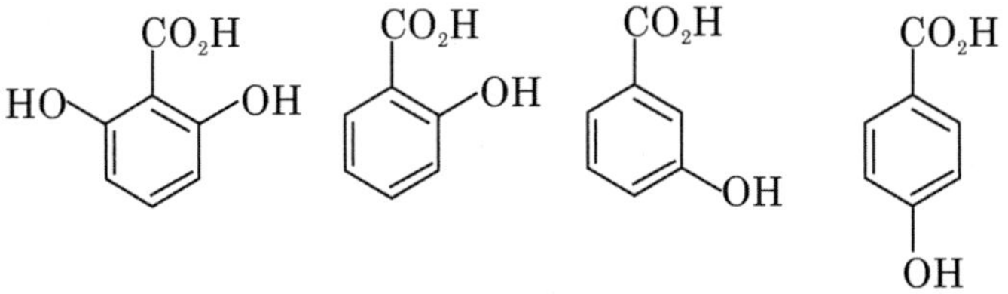

I II III IV

(A) I > II > III > IV (B) III > I > II > IV
(C) III > IV > II > I (D) I > III > IV > II

2. The geometries of the ammonia complexes of Ni^{2+}, Pt^{2+} and Zn^{2+}, respectively, are
(A) octahedral, square planar and tetrahedral
(B) square planar, octahedral and tetrahedral
(C) tetrahedral, square planar and octahedral
(D) octahedral, tetrahedral and square planar.

3. For the following electrochemical cell at 298 K,

$Pt(s) | H_2(g, 1 \text{ bar}) | H^+(aq, 1M) || M^{4+}(aq), M^{2+}(aq) | Pt(s)$

$$E_{cell} = 0.092 \text{ V when } \frac{[M^{2+}(aq)]}{[M^{4+}(aq)]} = 10^x$$

$$\text{Given: } E^0_{M^{4+}/M^{2+}} = 0.151 \text{ V}; 2.303\frac{RT}{F} = 0.059 \text{ V}$$

The value of x is

(A) – 2 (B) – 1 (C) 1 (D) 2

4. The major product of the following reaction sequence is

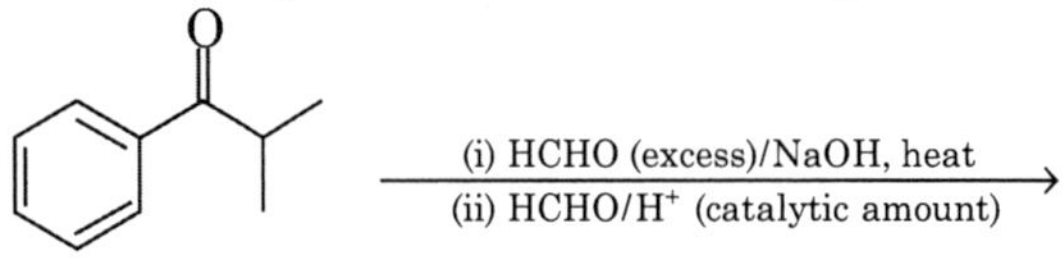

(A) 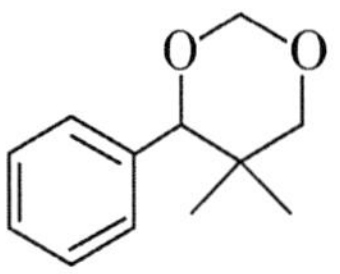

(B) 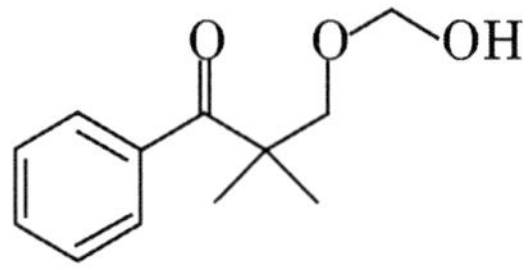

(C) 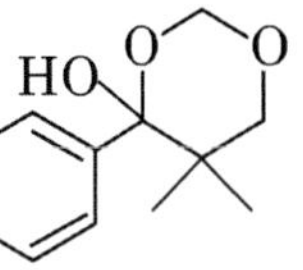

(D) 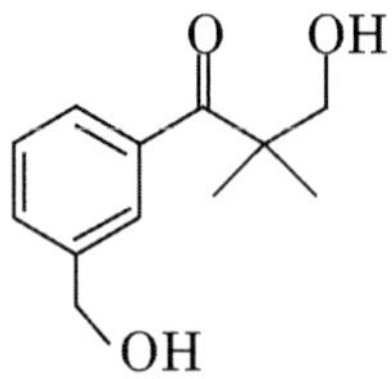

5. In the following reaction sequence in aqueous solution, the species X, Y and Z, respectively, are

$$S_2O_3^{2-} \xrightarrow{Ag^+} \underset{\text{clear solution}}{X} \xrightarrow{Ag^+} \underset{\text{white precipitate}}{Y} \xrightarrow{\text{with time}} \underset{\text{black precipitate}}{Z}$$

(A) $[Ag(S_2O_3)_2]^{3-}$, $Ag_2S_2O_3$, Ag_2S (B) $[Ag(S_2O_3)_3]^{5-}$, Ag_2SO_3, Ag_2S

(C) $[Ag(SO_3)_2]^{3-}$, $Ag_2S_2O_3$, Ag (D) $[Ag(SO_3)_3]^{3-}$, Ag_2SO_4, Ag

6. The qualitative sketches I, II and III given below show the variation of surface tension with molar concentration of three different aqueous solutions of KCl, CH_3OH and $CH_3(CH_2)_{11}OSO_3^-Na^+$ at room temperature. The correct assignment of the sketches is

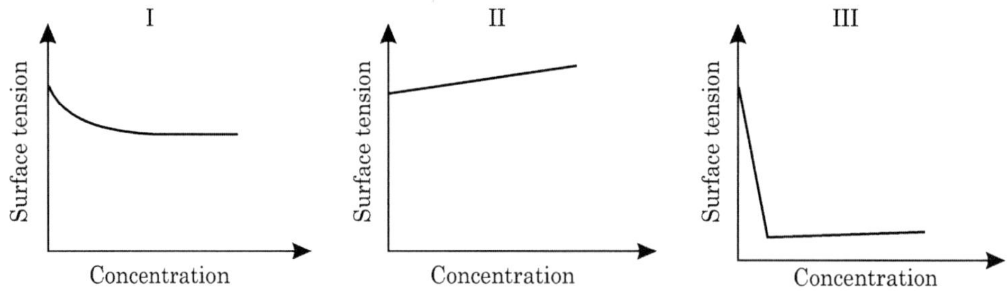

(A) I : KCl II : CH_3OH III : $CH_3(CH_2)_{11}OSO_3^-Na^+$

(B) I : $CH_3(CH_2)_{11}OSO_3^-Na^+$ II : CH_3OH III : KCl

(C) I : KCl II : $CH_3(CH_2)_{11}OSO_3^-Na^+$ III : CH_3OH

(D) I : CH_3OH II : KCl III : $CH_3(CH_2)_{11}OSO_3^-Na^+$

SECTION–2
(Maximum Marks: 32)

This section contains eight questions.

Each question has four options (A), (B), (C) and (D). One or more than one of these four option(s) is(are) correct.

For each question, darken the bubble(s) corresponding to all the correct option(s) in the ORS.

For each question, marks will be awarded in one of the following categories:

Full Marks : + 4 If only the bubble(s) corresponding to all the correct option(s) is(are) darkened.

Partial Marks : + 1 For darkening a bubble corresponding to each correct option, provided NO incorrect option is darkened.

Zero Marks : 0 If none of the bubbles is darkened.

Negative Marks : – 2 In all other cases.

For example, if (A), (C) and (D) are all the correct options for a question, darkening all these three will result in + 4 marks; darkening only (A) and (D) will result in + 2 marks; and darkening (A) and (B) will result in – 2 marks, as a wrong option is also darkened.

7. For 'invert sugar', the correct statement(s) is(are)

(Given: specific rotations of (+)-sucrose, (+)-maltose, L-(–)-glucose and L-(+)-fructose in aqueous solution are + 66°, + 140°, – 52° and + 92°, respectively)

(A) 'invert sugar' is prepared by acid catalyzed hydrolysis of maltose

(B) 'invert sugar' is an equimolar mixture of D-(+)-glucose and D-(–)-fructose

(C) specific rotation of 'invert sugar' is – 20°

(D) on reaction with Br_2 water, 'invert sugar' forms saccharic acid as one of the products.

8. Among the following reaction(s) which gives(give) *tert*-butylbenzene as the major product is(are)

(A) 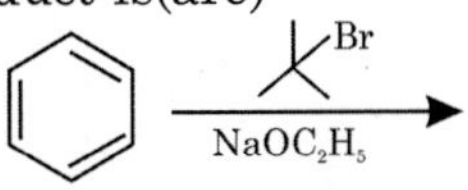

(B) 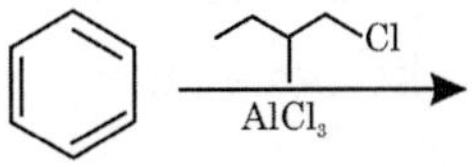

(C) 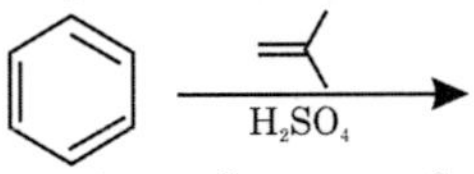

(D) 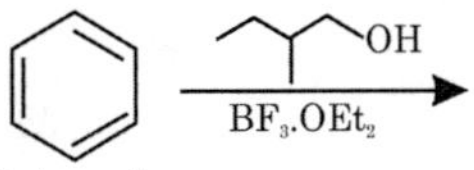

9. Extraction of copper from copper pyrite ($CuFeS_2$) involves

(A) crushing followed by concentration of the ore by froth floatation

(B) removal of iron as slag

(C) self-reduction step to produce 'blister copper' following evolution of SO_2

(D) refining of 'blister copper' by carbon reduction.

10. The correct statement(s) for cubic close packed (*ccp*) three dimensional structure is(are)

(A) The number of the nearest neighbours of an atom present in the topmost layer is 12

(B) The efficiency of atom packing is 74%

(C) The number of octahedral and tetrahedral voids per atom are 1 and 2, respectively

(D) The unit cell edge length is $2\sqrt{2}$ times the radius of the atom.

11. Reagent(s) which can be used to bring about the following transformation is(are)

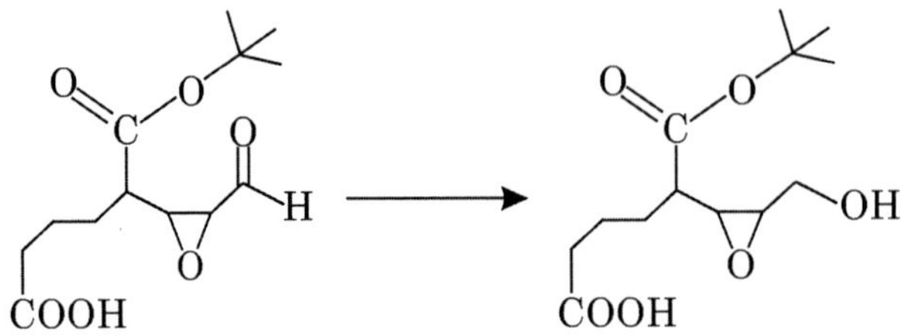

(A) $LiAlH_4$ in $(C_2H_5)_2O$ (B) BH_3 in THF

(C) $NaBH_4$ in C_2H_5OH (D) Raney Ni/H_2 in THF

12. Mixture(s) showing positive deviation from Raoult's law at 35°C is(are)

(A) carbon tetrachloride + methanol (B) carbon disulphide + acetone

(C) benzene + toluene (D) phenol + aniline

13. The nitrogen containing compound produced in the reaction of HNO_3 with P_4O_{10}

(A) can also be prepared by reaction of P_4 and HNO_3

(B) is diamagnetic

(C) contains one N—N bond

(D) reacts with Na metal producing a brown gas.

14. According to Molecular Orbital Theory

(A) C_2^{2-} is expected to be diamagnetic

(B) O_2^{2+} is expected to have a longer bond length than O_2

(C) and N_2^- have the same bond order

(D) He_2^+ has the same energy as two isolated He atoms.

SECTION–3
(Maximum Marks: 12)

This section contains two questions.
Based on each paragraph, there will be two questions.
Each question has four options (A), (B), (C) and (D). Only one of these four options is correct.
For each question, darken the bubble(s) corresponding to all the correct option(s) in the ORS.

For each question, marks will be awarded in one of the following categories:
Full Marks : + 3 If only the bubble corresponding to all the correct option is darkened.
Zero Marks : 0 In all other cases.

Paragraph 1

Thermal decomposition of gaseous X_2 to gaseous X at 298 K takes place according to the following equation:

$$X_2(g) \rightleftharpoons 2X(g)$$

The standard reaction Gibbs energy, $\Delta_r G^0$, of this reaction is positive. At the start of the reaction, there is one mole of X_2 and no X. As the reaction proceeds, the number of moles of X formed is given by β. Thus, $\beta_{\text{equilibrium}}$ is the number of moles of X formed at equilibrium. The reaction is carried out at a constant total pressure of 2 bar. Consider the gases to behave ideally. (Given: $R = 0.083$ L bar K^{-1} mol^{-1})

15. The equilibrium constant K_p for this reaction at 298 K, in terms of $\beta_{\text{equilibrium}}$, is

(A) $\dfrac{8\beta^2_{\text{equilibrium}}}{2-\beta_{\text{equilibrium}}}$ (B) $\dfrac{8\beta^2_{\text{equilibrium}}}{4-\beta^2_{\text{equilibrium}}}$

(C) $\dfrac{4\beta^2_{\text{equilibrium}}}{2-\beta_{\text{equilibrium}}}$ (D) $\dfrac{4\beta^2_{\text{equilibrium}}}{4-\beta^2_{\text{equilibrium}}}$

16. The incorrect statement among the following, for this reaction is

(A) decrease in the total pressure will result in formation of more moles of gaseous X

(B) at the start of the reaction, dissociation of gaseous X_2 takes place spontaneously

(C) $\beta_{\text{equilibrium}} = 0.7$

(D) $K_c < 1$

Paragraph 2

Treatment of compound O with $KMnO_4/H^+$ gave P, which on heating with ammonia gave Q. The compound Q on treatment with Br_2/NaOH produced R. On strong heating, Q gave S, which on further treatment with ethyl-2-bromopropanoate in the presence of KOH followed by acidification, gave a compound T.

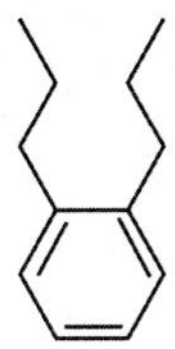

[O]

17. The compound R is

(A) 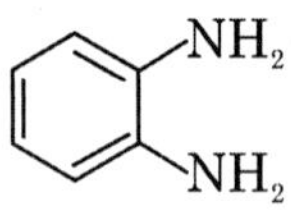

(B) 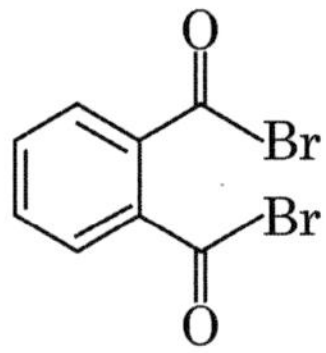

(C) 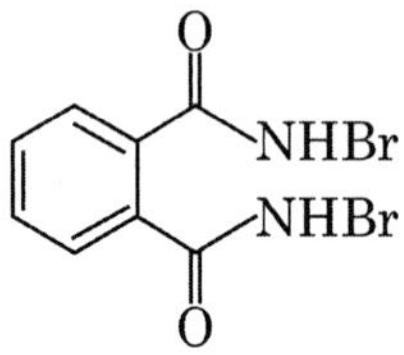

(D) 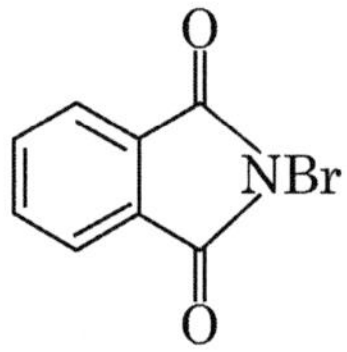

18. The compound T is

(A) glycine (B) alanine (C) valine (D) serine

ANSWERS

1. A	2. A	3. D	4. A	5. A	6. D
7. B, C	8. B, C, D	9. A, B, C	10. B, C, D	11. C, D	12. A, B
13. B, D	14. A, C	15. B	16. C	17. A	18. B

Model Solutions—I.I.T. JEE (Advanced) 2016

Paper–2

1. (A)

Stabler the conjugate base, stronger the acid.

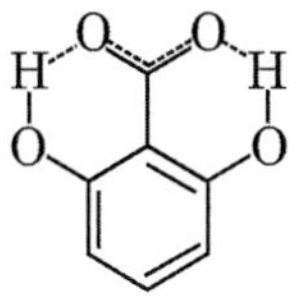

Conjugate base is stabilized by intramolecular H-bond from both the sides.

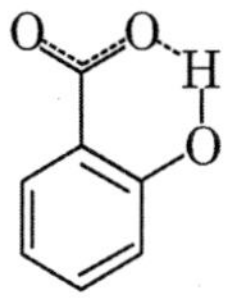

Conjugate base is stabilized by intramolecular H-bond from one side.

2. (A)

$[Ni(NH_3)_6]^{2+}$; $[Pt(NH_3)_4]^{2+}$; $[Zn(NH_3)_4]^{2+}$

octahedral square planer tetrahedral

3. (D)

Anode: $H_2 - 2e \rightleftharpoons 2H^+$

Cathode: $M^{4+} + 2e \rightleftharpoons M^{2+}$

Net cell reaction : $H_2 + M^{4+} \rightleftharpoons 2H^+ + M^{2+}$

$$E_{cell} = E^0_{cell} - \frac{0.059}{2}\log\frac{[H^+]^2[M^{2+}]}{[M^{4+}] \times P_{H_2}}$$

$$0.092 = 0.151 - \frac{0.059}{2}\log 10^x$$

or $$0.092 = 0.151 - \frac{0.059}{2}x$$

or $$\frac{0.059x}{2} = 0.151 - 0.092$$

or $0.059x = 0.059 \times 2$

or $x = 2$

4. (A)

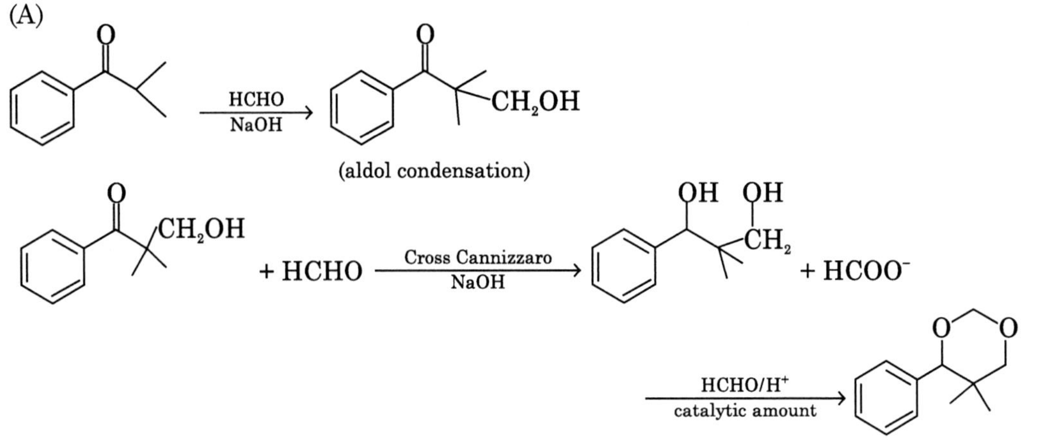

(acetal formation)

5. (A)

$$Ag^+ + 2S_2O_3^{2-} \longrightarrow \underset{\substack{X \\ \text{(clear solution)}}}{[Ag(S_2O_3)_2]^{3-}} \xrightarrow{Ag^+} \underset{\substack{Y \\ \text{(white)}}}{Ag_2S_2O_3} \xrightarrow{\text{with time}} \underset{\substack{Z \\ \text{(black)}}}{Ag_2S}$$

$$Ag_2S_2O_3 + H_2O \longrightarrow Ag_2S + H_2SO_4$$

6. (D)

Strong electrolytes such as KCl increase the surface tension slightly. Low molar mass organic compounds usually decrease the surface tension. Surface active organic compounds such as detergents sharply decrease surface tension.

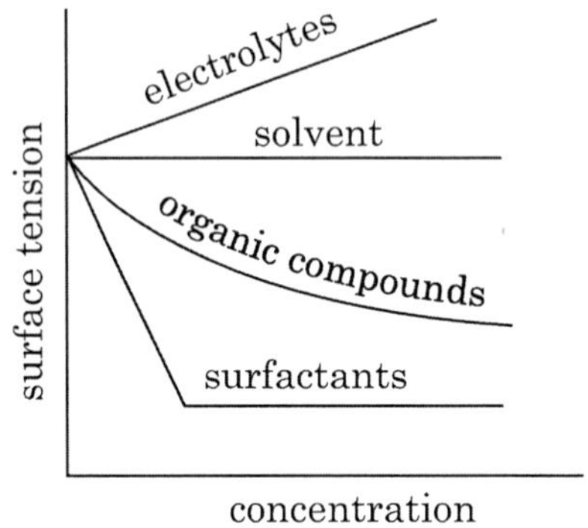

7. (B, C)

$$C_{12}H_{22}O_{11} + H_2O \xrightarrow{H^+} \underset{\substack{\text{D-(+)-glucose} \\ +52^\circ}}{C_6H_{12}O_6} + \underset{\substack{\text{D-(−)-fructose} \\ -92^\circ}}{C_6H_{12}O_6}$$

$$\alpha_{\text{invert sugar}} = \frac{+52^\circ - 92^\circ}{2} = -20^\circ$$ (Average is taken as both monomers are one mole each.)

8. (B, C, D)

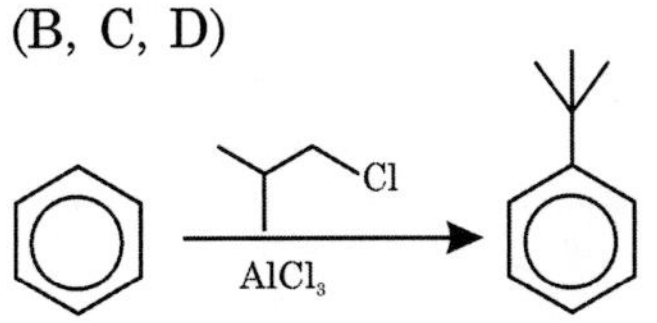

via rearrangement
of carbocation

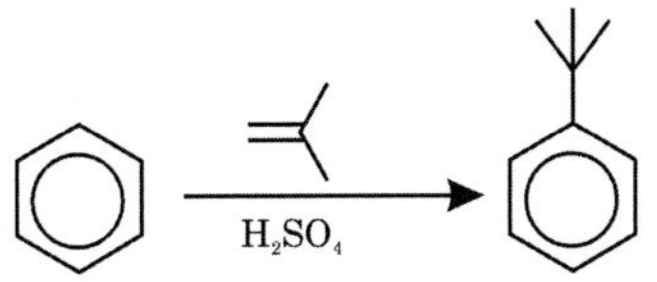

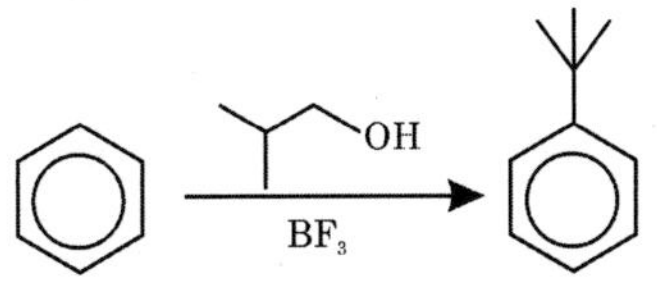

via rearrangement
of carbocation

9. (A, B, C)
Refining of blister copper is carried out by poling technique.

10. (B, C, D)
The middle layers will have twelve nearest neighbours. The top-most layer will have nine nearest neighbours. $4r = a\sqrt{2}$, where a = edge length of unit cell and r = radius of atom.

11. (C, D)
$NaBH_4$ and raney Ni/H_2 do not react with acid, ester or epoxide entities of an organic compound.

12. (A, B)
Benzene and toluene will form ideal solution, and phenol and aniline will exhibit negative deviation.

13. (B, D)

$$2HNO_3 \xrightarrow[\text{(dehydration, } - H_2O)]{P_4O_{10}} N_2O_5$$

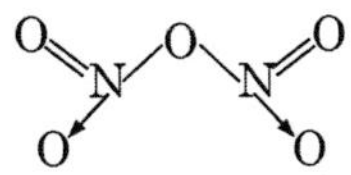

$$Na + N_2O_5 \rightarrow NaNO_3 + \underset{\text{(brown)}}{NO_2(\uparrow)}$$

14. (A, C)

15. (B)

$$X_2(g) \rightleftharpoons 2X(g)$$

$$1-\frac{\beta_e}{2} \qquad \beta_e$$

Total number of moles at equilibrium

$$= 1 - \frac{\beta_e}{2} + \beta_e = 1 + \frac{\beta_e}{2}$$

$$K_p = \frac{(p_x)^2}{p_{x_2}} = \frac{\left(\dfrac{\beta_e \times 2}{1+\dfrac{\beta_e}{2}}\right)^2}{\dfrac{\left(1-\dfrac{\beta_e}{2}\right)\times 2}{1+\dfrac{\beta_e}{2}}} = \frac{2\beta_e^2}{1-\dfrac{\beta_e^2}{4}}$$

$$K_p = \frac{8\beta_e^2}{4-\beta_e^2}$$

16. (C)

No data is given to find the exact value of $\beta_{equilibrium}$

$\Delta G_c^\circ = -2.303RT \log K_c$

$\log K_c = -1$

or $K_c < 1$

17. (A)

18. (B)

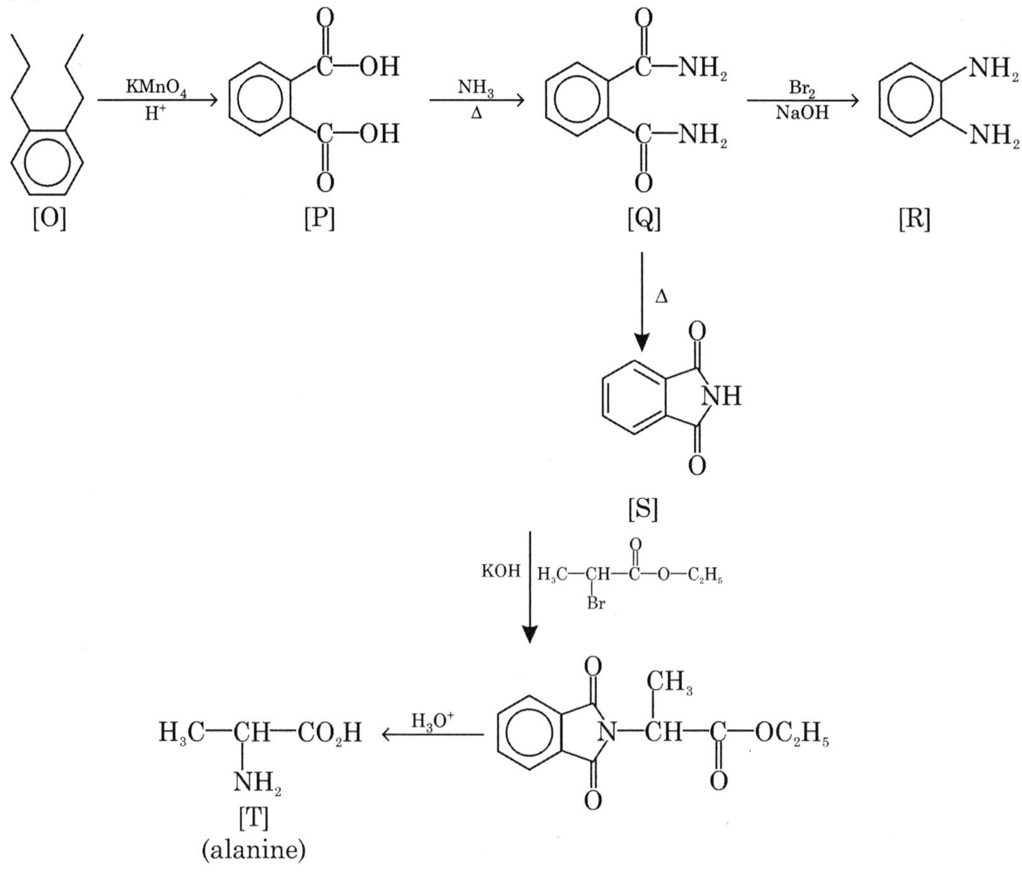

[O] [P] [Q] [R] [S] [T] (alanine)

I.I.T. JEE Objective Question Paper (Main)—2017

IMPORTANT INSTRUCTIONS

1. Immediately fill in the particulars on this page of the Test Booklet with only Blue/ Black Ball Point Pen provided in the examination hall.
2. The Answer Sheet is kept inside this Test Booklet. When you are directed to open the Test Booklet, take out the Answer Sheet and fill in the particulars carefully.
3. The test is of 3 hours duration.
4. The Test Booklet consists of 90 questions. The maximum marks are 360.
5. There are three parts in the question paper A, B, C consisting of Chemistry, Mathematics and Physics having 30 questions in each part of equal weightage. Each question is allotted 4 (four) marks for each correct response.
6. Candidates will be awarded, marks as stated above in instruction No. 5 for correct response of each question. 1/4 (one-fourth) marks of the total marks allotted to the question (i.e. 1 mark) will be deducted for indicating incorrect response for each question. No deduction from that total score will be made if no response is indicated for an item in the Answer Sheet.
7. There is only one correct response for each question. Filling up more than one response in each question will be treated as wrong response and marks for wrong response will be deducted accordingly as per instruction 6 above.
8. For writing particulars/marking responses on Side-1 and Side-2 of the Answer Sheet use only Black Ball Point Pen provided in the examination hall.
9. No candidate is allowed to carry any textual material, printed or written, bits of papers, pager, mobile phone, any electronic device, etc. except the Admit Card inside the examination room/hall.
10. Rough work is to be done on the space provided for this purpose in the Test Booklet only. This space is given at the bottom of each page and in four pages (Pages 20-23) at the end of the booklet.
11. On completion of the test, the candidate must hand over the Answer Sheet to the Invigilator on duty in the room/hall. However, the candidates are allowed to take away this Test Booklet with them.
12. The CODE for this booklet is C. Make sure that the CODE printed on Side-2 of the Answer Sheet also tally the serial number of the Test Booklet and Answer Sheet is the same as that on this booklet. In case of discrepancy, the candidate should immediately report the matter to the Invigilator for replacement of both the Test Booklet and the Answer Sheet.
13. Do not fold or make any stray mark on the Answer Sheet.

1. The freezing point of benzene decreases by 0.45 °C when 0.2 g of acetic acid is added to 20 g of benzene. If acetic acid associates to form a dimer in benzene, percentage association of acetic acid in benzene will be (K_f for benzene = 5.12 K kg mol^{-1})
   (A) 94.6% (B) 64.6% (C) 80.4% (D) 74.6%
2. On treatment of 100 mL of 0.1M solution of $CoCl_3 \cdot 6H_2O$ with excess $AgNO_3$ 1.2×10^{22} ions are precipitated. The complex is:
   (A) $[Co(H_2O)_5Cl]Cl_2 \cdot H_2O$ (B) $[Co(H_2O)_4Cl_2]Cl \cdot 2H_2O$
   (C) $[Co(H_2O)_3Cl_3] \cdot 3H_2O$ (D) $[Co(H_2O)_6]Cl_3$
3. Which of the following compounds will form significant amount of meta product during mono-nitration reaction?
   (A) 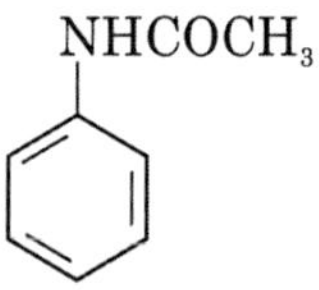

   (B) 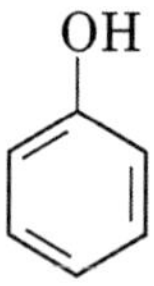

   (C) 

   (D) 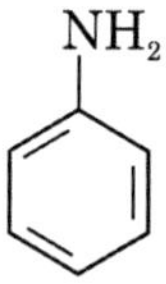

4. The products obtained when chlorine gas reacts with cold and dilute aqueous NaOH are:
   (A) Cl^- and ClO_2^- (B) ClO^- and ClO_3^- (C) ClO_2^- and ClO_3^- (D) Cl^- and ClO^-
5. Both lithium and magnesium display several similar properties due to the diagonal relationship. However, the one which is incorrect, is
   (A) nitrates of both Li and Mg yield NO_2 and O_2 on heating
   (B) both form basic carbonates
   (C) both form soluble hydrogencarbonates
   (D) both form nitrides.
6. A water sample has ppm level concentration of following anions
   $F^- = 10;\ SO_4^{2-} = 100;\ NO_3^- = 50$
   The anion/anions that make/makes the water sample unsuitable for drinking is/are
   (A) Only SO_4^{2-} (B) Only NO_3^-
   (C) Both SO_4^{2-} and NO_3^- (D) Only F^-
7. The formation of which of the following polymers involves hydrolysis reaction?
   (A) Terylene (B) Nylon 6 (C) Bakelite (D) Nylon 6, 6
8. The Tyndall effect is observed only when the following conditions are satisfied.
   (a) The diameter of the dispersed particles is much smaller than the wavelength of the light used.
   (b) The diameter of the dispersed particle is not much smaller than the wavelength of the light used.
   (c) The refractive indices of the dispersed phase and dispersion medium are almost similar in magnitude.
   (d) The refractive indices of the dispersed phase and dispersion medium differ greatly in magnitude.
   (A) (b) and (c) (B) (a) and (d) (C) (b) and (d) (D) (a) and (c)

9. pK_a of pH_a weak acid (HA) and pK_b of a weak base (BOH) are 3.2 and 3.4, respectively. The pH of their salt (AB) solution is

   (A) 1.0 (B) 7.2 (C) 6.9 (D) 7.0

10. The major product obtained in the following reaction is

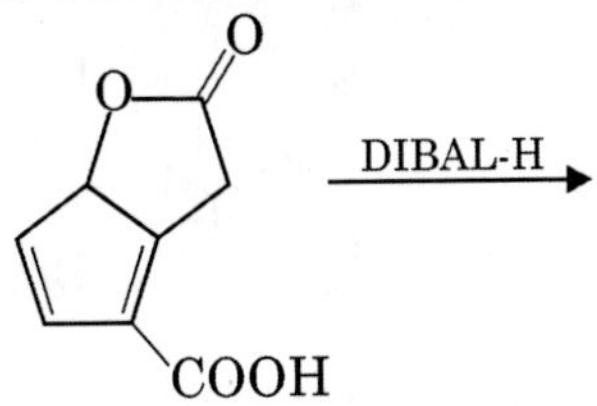

(A) 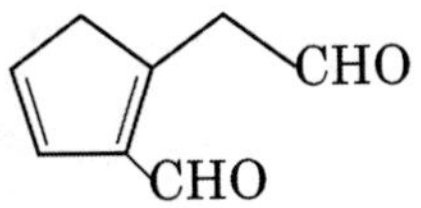

(B) 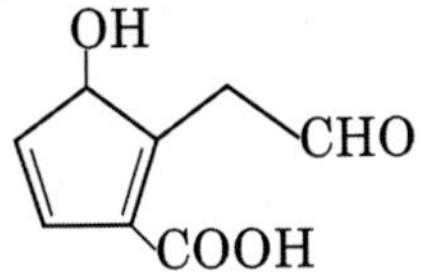

(C) 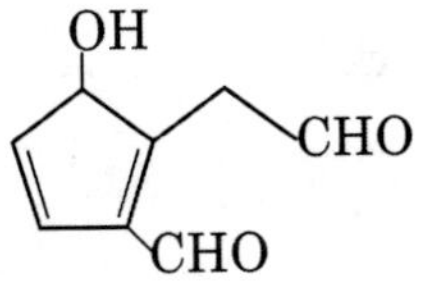

(D) 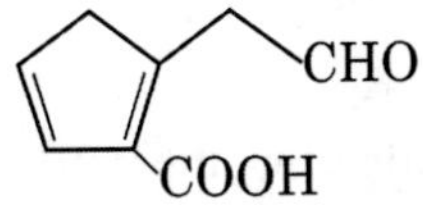

11. Which of the following compounds will behave as a reducing sugar in an aqueous KOH solution?

(A) 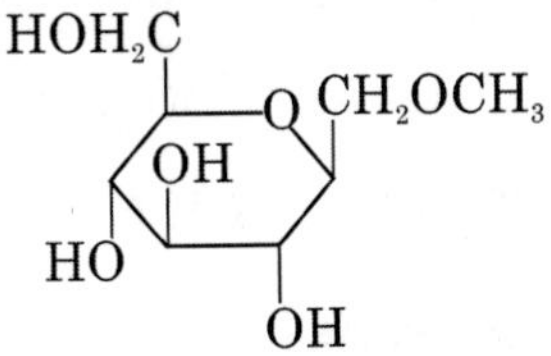

(B) 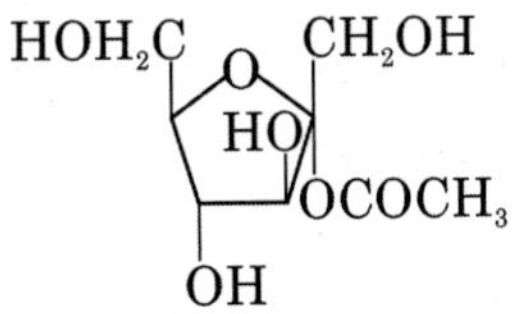

(C) 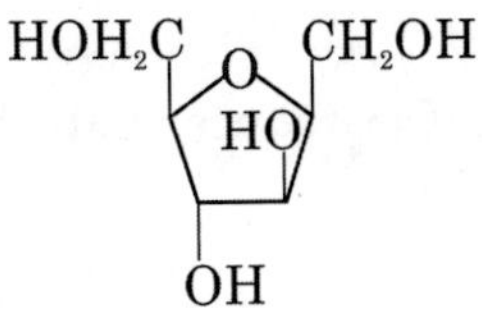

(D) 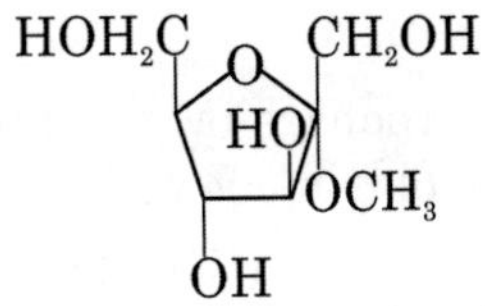

12. The correct sequence of reagents for the following conversion will be

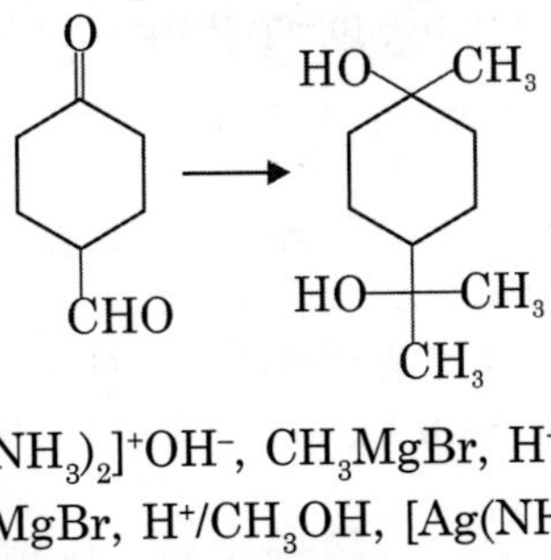

   (A) $[Ag(NH_3)_2]^+OH^-$, CH_3MgBr, H^+/CH_3OH (B) $[Ag(NH_3)_2]^+OH^-$, H^+/CH_3OH, CH_3MgBr

   (C) CH_3MgBr, H^+/CH_3OH, $[Ag(NH_3)_2]^+OH^-$ (D) CH_3MgBr, $[Ag(NH_3)_2]^+OH^-$, H^+/CH_3OH

13. Which of the following species is not paramagnetic?

   (A) B_2 (B) NO (C) CO (D) O_2

14. Which of the following, upon treatment with *tert*-BuONa followed by addition of bromine water, fails to decolourize the colour of bromine?

(A) 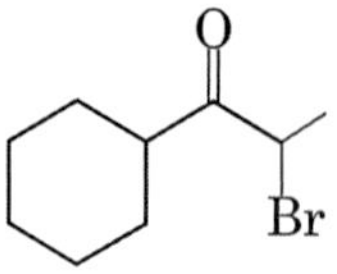

(B) 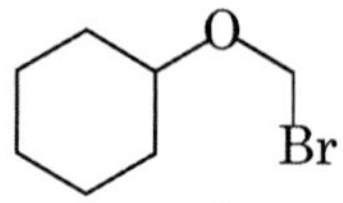

(C) 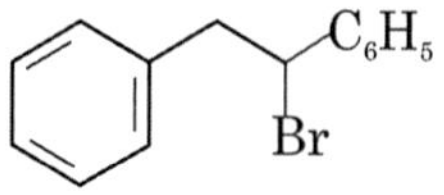

(D) 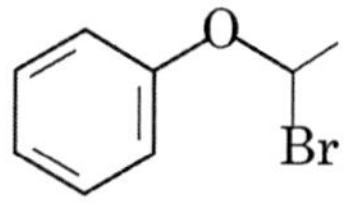

15. Which of the following reactions is an example of a redox reaction?

(A) $XeF_6 + 2H_2O \longrightarrow XeO_2F_2 + 4HF$ (B) $XeF_4 + O_2F_2 \longrightarrow XeF_6 + O_2$

(C) $XeF_2 + PF_5 \longrightarrow [XeF]^+PF_6^-$ (D) $XeF_6 + H_2O \longrightarrow XeOF_4 + 2HF$

16. ΔU is equal to

(A) Isothermal work (B) Isochoric work

(C) Isobaric work (D) Adiabatic work.

17. Which of the following molecules is least resonance stabilized?

(A) 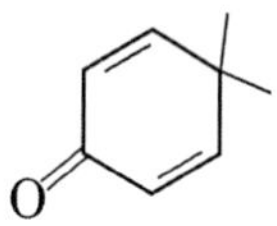 (B)  (C) 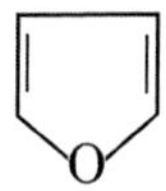 (D) 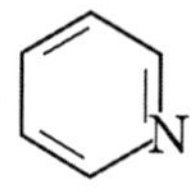

18. The increasing order of the reactivity of the following halides for the S_N1 reaction is

I. $CH_3\underset{\displaystyle Cl}{\underset{|}{C}}HCH_2CH_3$ II. $CH_3CH_2CH_2Cl$ III. p-$H_3CO—C_6H_4—CH_2Cl$

(A) (II) < (III) < (I) (B) (III) < (II) < (I)

(C) (II) < (I) < (III) (D) (I) < (III) < (II)

19. 1 gram of a carbonate (M_2CO_3) on treatment with excess HCl produces 0.01186 mole of CO_2. The molar mass of M_2CO_3 in g mol^{-1} is

(A) 11.86 (B) 1186 (C) 84.3 (D) 118.6

20. Sodium salt of an organic acid X produces effervescence with conc. H_2SO_4. X reacts with the acidified aqueous $CaCl_2$ solution to give a white precipitate which decolourises acidic solution of $KMnO_4$. X is

(A) $Na_2C_2O_4$ (B) C_6H_5COONa (C) HCOONa (D) CH_3COONa

21. The most abundant elements by mass in the body of a healthy human adult are: oxygen (61.4%); carbon (22.9%), hydrogen (10.0%) and nitrogen (2.6%). The weight which a 75 kg person would gain if all ^1H atoms are replaced by ^2H atoms is

(A) 10 kg (B) 15 kg (C) 37.5 kg (D) 7.5 kg

22. The major product obtained in the following reaction is

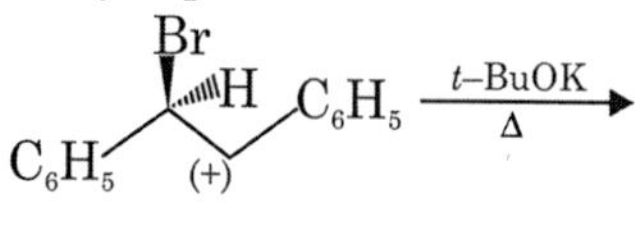

(A) (–)$C_6H_5CH(O^tBu)CH_2C_6H_5$ (B) (±)$C_6H_5CH(O^tBu)CH_2C_6H_5$
(C) $C_6H_5CH{=}CHC_6H_5$ (D) (+)$C_6H_5CH(O^tBu)CH_2C_6H_5$

23. Given: $C_{(graphite)} + O_2(g) \longrightarrow CO_2(g);\ \Delta_r H^0 = -\ 393.5\ kJ\ mol^{-1}$

$$H_2(g) + \frac{1}{2}O_2(g) \longrightarrow H_2O(l);\ \Delta_r H^0 = -\ 285.8\ kJ\ mol^{-1}$$

$$CO_2(g) + 2H_2O(l) \longrightarrow CH_4(g) + 2O_2(g);\ \Delta_r H^0 = +\ 890.3\ kJ\ mol^{-1}$$

Based on the above thermochemical equations, the value of $\Delta_r H^0$ at 298 K for the reaction $C_{(graphite)} + 2H_2(g) \longrightarrow CH_4(g)$ will be

(A) – 144.0 kJ mol^{-1} (B) + 74.8 kJ mol^{-1}
(C) + 144.0 kJ mol^{-1} (D) – 74.8 kJ mol^{-1}

24. In the following reactions, ZnO is respectively acting as a/an

(a) $ZnO + Na_2O \longrightarrow Na_2ZnO_2$ (b) $ZnO + CO_2 \longrightarrow ZnCO_3$

(A) acid and base (B) base and acid (C) base and base (D) acid and acid

25. The radius of the second Bohr orbit for hydrogen atom is (Planck's const. $h = 6.6262 \times 10^{-34}$ Js, mass of electron = 9.1091×10^{-31} kg, charge of electron, $e = 1.60210 \times 10^{-19}$ C, permittivity of vacuum $\epsilon_0 = 8.854185 \times 10^{-12}\ kg^{-1}\ m^{-3}\ A^2$)

(A) 2.12 Å (B) 1.65 Å (C) 4.76 Å (D) 0.529 Å

26. Two reactions R_1 and R_2 have identical pre-exponential factors. Activation energy of R_1 exceeds that of R_2 by 10 kJ mol^{-1}. If k_1 and k_2 are rate constants for reactions R_1 and R_2 respectively at 300 K, then $\ln(k_2/k_1)$ is equal to (R = 8.314 J mol^{-1} K^{-1})

(A) 4 (B) 8 (C) 12 (D) 6

27. A metal crystallizes in a face-centred cubic structure. If the edge length of its unit cell is a, the closest approach between two atoms in metallic crystal will be

(A) $\frac{a}{\sqrt{2}}$ (B) $2a$ (C) $2\sqrt{2}a$ (D) $\sqrt{2}a$

28. The group having isoelectronic species is

(A) O^-, F^-, Na^+, Mg^{2+} (B) O^{2-}, F^-, Na^+, Mg^{2+}
(C) O^-, F^-, Na, Mg^+ (D) O^{2-}, F^-, Na, Mg^{2+}

29. Given $E^0_{Cl_2/Cl^-} = 1.36$ V, $E^0_{Cr^{3+}/Cr} = -\ 0.74$ V

$E^0_{Cr_2O_7^{2-}/Cr^{3+}} = 1.33$ V, $E^0_{MnO_4^-/Mn^{2+}} = 1.51$ V

Among the following, the strongest reducing agent is

(A) Cl^- (B) Cr (C) Mn^{2+} (D) Cr^{3+}

30. 3-Metylpent-2-ene on reaction with HBr in presence of peroxide forms an addition product. The number of possible stereoisomers for the product is

(A) Four (B) Six (C) Zero (D) Two

ANSWERS

1. A	2. A	3. D	4. D	5. B	6. D
7. B	8. C	9. C	10. B	11. B	12. B
13. C	14. B	15. B	16. D	17. A	18. C
19. C	20. A	21. D	22. C	23. D	24. A
25. A	26. A	27. A	28. B	29. B	30. A

Model Solutions—I.I.T. JEE (Main) 2017

1. (A)

$\Delta T_f = i \times K_f \times m$

or $0.45 = i \times 5.12 \times \frac{0.2}{60} \times \frac{1000}{20}$

$i = 0.527$

$\underset{1-\alpha}{2\,\text{AcOH}} \rightleftharpoons \underset{\alpha/2}{(\text{AcOH})_2}$

$i = 1 - \frac{\alpha}{2}$ (where α is degree of association)

$0.527 = 1 - \frac{\alpha}{2}$

$\alpha = 0.946$ or 94.6%

2. (A)

$\underset{0.01M}{CoCl_3 \cdot 6H_2O} + AgNO_3 \longrightarrow nAgCl \downarrow$

$\frac{1.2 \times 10^{22}}{6.0 \times 10^{23}} = 2 \times 10^{-2}$ moles $= 0.02$ moles

$0.01 \times n = 0.02$

$n = 2$

Hence, complex is $[Co(H_2O)_5Cl]Cl_2 \cdot H_2O$.

3. (D)

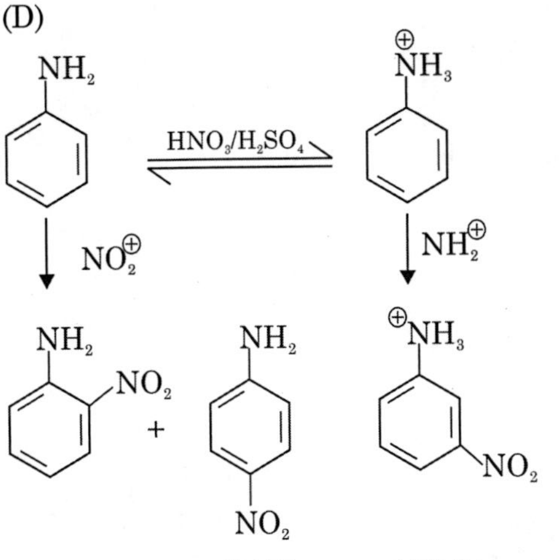

(2%) (51%) (47%)

Anilinium ion is strongly deactivating group and is a meta directing in nature. Hence, it gives meta nitration product in significant amount (47%).

4. (D)

$$Cl_2 + \underset{\text{(cold \& dil.)}}{NaOH} \longrightarrow NaCl + NaClO + H_2O$$

5. (B)

A, C and D are correct statements.

Li forms only carbonate (Li_2CO_3) and not basic carbonate. But Mg forms basic carbonate.

$$5Mg^{2+} + 6CO_3^{2-} + 7H_2O \longrightarrow 4MgCO_3 \cdot Mg(OH)_2 \cdot 5H_2O \downarrow + 2HCO_3^-$$

6. (D)

Permissible limit of F^- ion is 1 ppm.

7. (B)

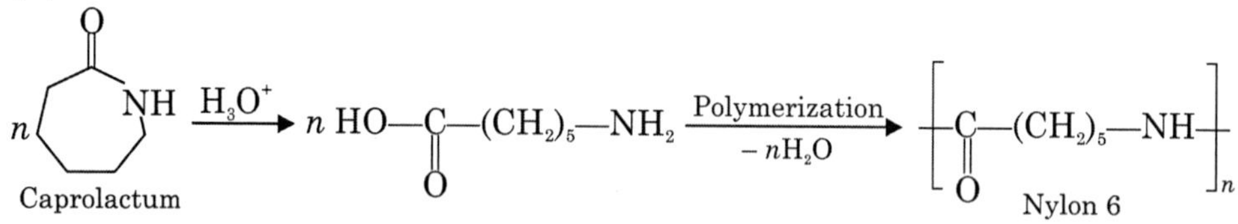

8. (C)

Tyndall effect is observed only when the following two conditions are satisfied.

(i) The diameter of the dispersed particle is not much smaller than the wavelength of the light used.

(ii) The refractive indices of the dispersed phase and dispersion medium differ greatly in magnitude.

9. (C)

pH of a salt made up of weak acid and weak base is calculated by using the expression

$$pH = 7 + \frac{1}{2}pK_a - \frac{1}{2}pK_b = 7 + \frac{3.2}{2} - \frac{3.4}{2} = 6.9$$

10. (B)

DIBAL-H reduces only ester group to an aldehyde and alcohol.

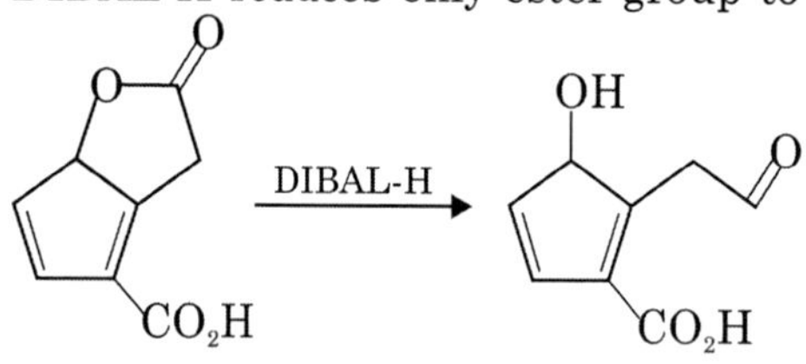

11. (B)

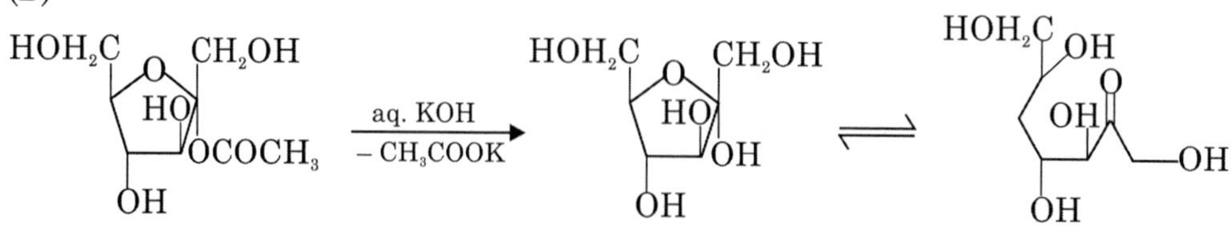

(α-Hydroxyketone)
(a reducing sugar)

12. (B)

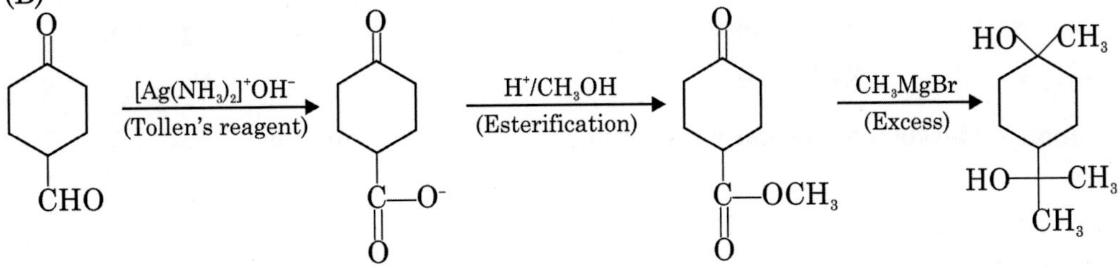

13. (C)

(1) $B_2 : \sigma_{1s}^2, \sigma_{1s}^{*2}, \sigma_{2s}^2, \sigma_{2s}^{*2}, \pi_{2p_x}^1, \pi_{2p_y}^1$

B_2 contains two unpaired electrons. Hence, it is paramagnetic.

(2) $\dot{N}O$ — odd electron species. Hence, it is paramagnetic.

(3) $CO : \sigma_{1s}^2, \sigma_{1s}^{*2}, \sigma_{2s}^2, \sigma_{2s}^{*2}, \pi_{2p_x}^2, \pi_{2p_y}^2, \sigma_{2p_z}^2$

CO contains no unpaired electron. Hence, it is diamagnetic.

(4) $O_2 : \sigma_{1s}^2, \sigma_{1s}^{*2}, \sigma_{2s}^2, \sigma_{2s}^{*2}, \sigma_{2p_z}^2, \pi_{2p_x}^2 \pi_{2p_y}^2, \pi_{2p_x}^{*1} \pi_{2p_y}^{*1}$

O_2 contains two unpaired electrons. Hence, it is paramagnetic.

14. (B)

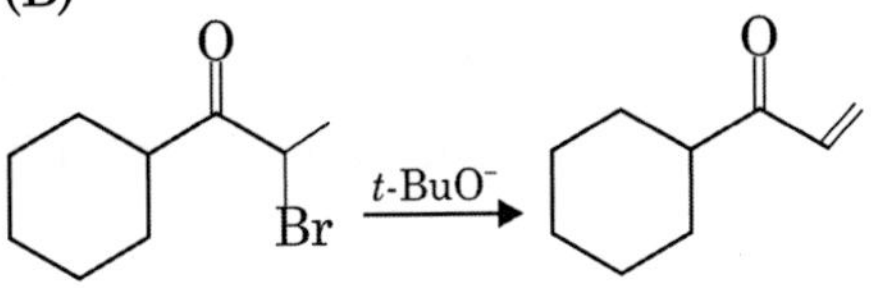

(decolourizes Br_2 water)

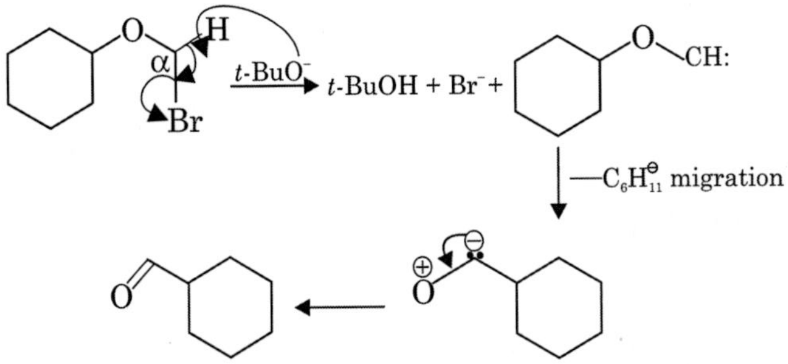

(cannot decolourize Br_2 water)

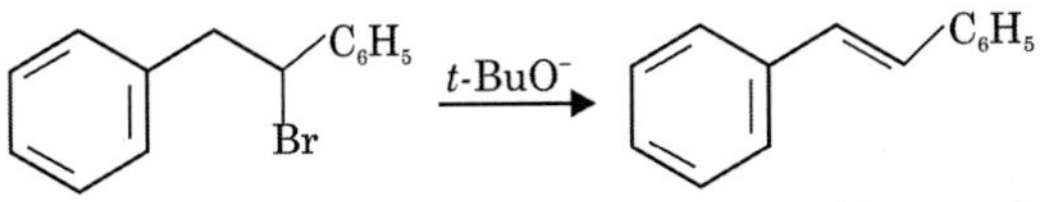

(decolourizes Br_2 water)

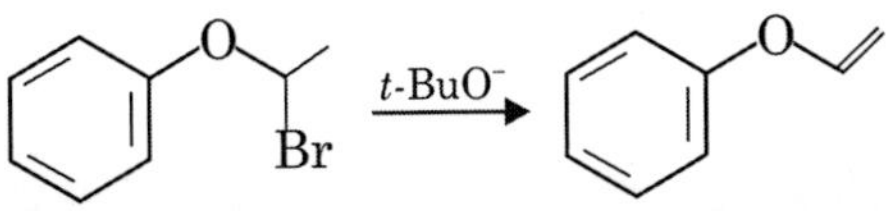

(decolourizes Br_2 water)

15. (B)

Any reaction having a substance in its elemental form is a redox reaction.

$$\overset{+4}{Xe}F_4 + \overset{+1}{O_2}F_2 \longrightarrow \overset{+6}{Xe}F_6 + \overset{0}{O_2}$$

Xenon undergoes oxidation while oxygen undergoes reduction.

16. (D)

For adiabatic process: $q = 0$

From 1st law $\Delta U = q + w$, we can write $\Delta U = w$

Work involved in adiabatic process is at the expense of change in interval energy of the system.

17. (A)

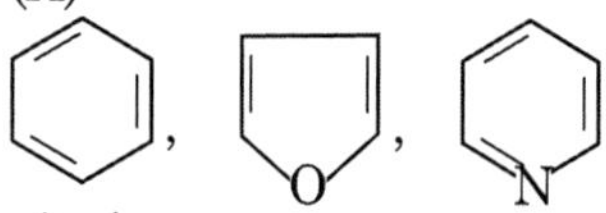 are aromatic and stabilized according to Huckel's rule.

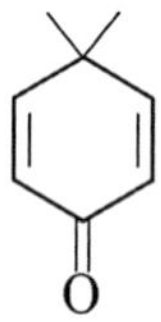 is non-aromatic. Hence, it is least stabilized by resonance.

18. (C)

Order of reactivity for S_N1 mechanism is in accordance with order of stability of carbocation involved.

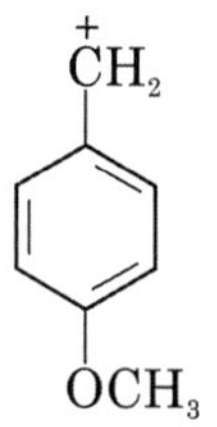 $> CH_3—\overset{+}{C}H—CH_2—CH_3 > CH_3CH_2\overset{+}{C}H_2$

Hence, the order of reactivity is as follows:

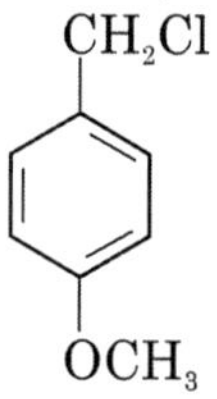 $> CH_3—\underset{}{\overset{Cl}{\overset{|}{C}}}H—CH_2—CH_3 > CH_3CH_2CH_2Cl$

19. (C)

$$M_2CO_3 + 2HCl \longrightarrow 2MCl + H_2O + CO_2$$

No. of moles of M_2CO_3 = No. of moles of CO_2 evolved

$\frac{1}{M} = 0.01186$ (M = molar mass of M_2CO_3)

Or $M = \frac{1}{0.01186} = \frac{10^5}{1186} = 84.3$ g/mol

20. (A)

$$\underset{(X)}{Na_2C_2O_4} + H_2SO_4 \text{ (conc.)} \longrightarrow Na_2SO_4 + H_2O + CO_2\uparrow + CO\uparrow$$

$$CaCl_2 + Na_2C_2O_4 \longrightarrow \underset{\text{white precipitate}}{CaC_2O_4} + 2NaCl$$

$$C_2O_4^{2-} + MnO_4^- \xrightarrow{H^+} CO_2 + Mn^{2+}$$

21. (D)

Mass of ^1H in body $= 75 \times 10^3 \times \dfrac{10}{100}$ g $= 7.5 \times 10^3$ g

No. of moles ^1H replaced by $^2H = 7.5 \times 10^3$

Hence, mass increased $= 7.5 \times 10^3$ g $= 7.5$ kg

22. (C)

Elimination reaction is highly favoured if (i) bulkier base is used, (ii) higher temperature is used. Hence, 1,2-elimination provides the major product.

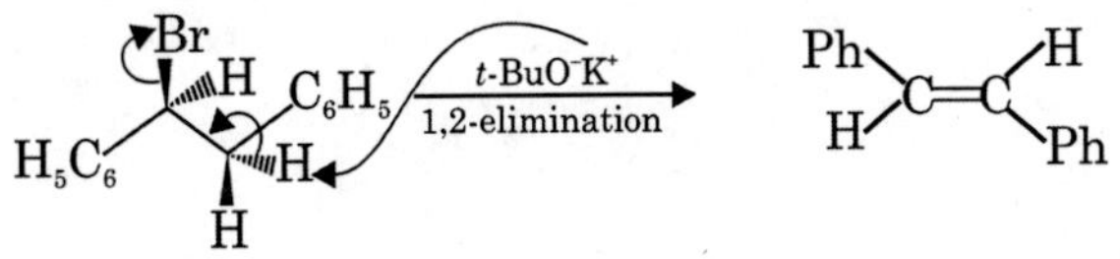

$+ t\text{-BuOH} + Br^-$

23. (D)

$$C_{(graphite)} + O_2(g) \longrightarrow CO_2(g);\ \Delta_r H^0 = -393.5 \text{ kJ mol}^{-1} \quad ...(i)$$

$$H_2(g) + \frac{1}{2}O_2(g) \longrightarrow H_2O(l);\ \Delta_r H^0 = -285.8 \text{ kJ mol}^{-1} \quad ...(ii)$$

$$CO_2(g) + 2H_2O(l) \longrightarrow CH_4(g) + 2O_2(g);\ \Delta_r H^0 = +890.3 \text{ kJ mol}^{-1} \quad ...(iii)$$

$$CO_2(g) + 2H_2O(l) \longrightarrow CH_4(g) + 2O_2(g) \quad ...(iv)$$

$\Delta_r H_4 = \Delta_r H_1 + 2\Delta_r H_2 + \Delta_r H_3$

$= -393.5 + (-285.8 \times 2) + 890.3 = -74.8$ kJ/mol

24. (A)

$$\underset{\text{Acid}}{ZnO} + \underset{\text{Base}}{Na_2O} \rightarrow \underset{\text{Salt}}{Na_2ZnO_2}$$

$$\underset{\text{Base}}{ZnO} + \underset{\text{Acid}}{CO_2} \rightarrow \underset{\text{Salt}}{ZnCO_3}$$

Non-metal oxides are generally acidic while alkali metal oxides are basic. ZnO is an amphoteric oxide.

25. (A)

$$r_n = 0.529 \times \frac{n^2}{Z}$$

$r_2 = 0.529 \times 4 = 2.116$ Å

26. (A)

Arrhenius equation, $k = Ae^{-E_a/RT}$

$$\ln\left(\frac{k_2}{k_1}\right) = \frac{1}{RT}[(E_a)_1 - (E_a)_2] = \frac{1 \times 10 \times 10^3}{8.314 \times 300} = 4$$

27. (A)

In FCC, distance of closest approach between two atoms = $2r$

In FCC, atoms are in close contact along face diagonal of FCC unit cell.

$$4r = \sqrt{2}a$$

$$\text{Or} \quad 2r = \frac{\sqrt{2}a}{2} = \left(\frac{a}{\sqrt{2}}\right)$$

28. (B)

Isoelectronic species have same number of electrons. All these species contains ten electrons.

29. (B)

$$E^0_{\text{Red}} \propto \frac{1}{\text{Reducing strength}}$$

$$E^0_{Cl_2/Cl^-} = 1.36\ \text{V};\ E^0_{Cr^{3+}/Cr} = -\,0.74\ \text{V};$$

$$E^0_{MnO_4^-/Mn^{2+}} = 1.51\ \text{V};\ E^0_{Cr_2O_7^{2-}/Cr^{3+}} = 1.33\ \text{V}$$

30. (A)

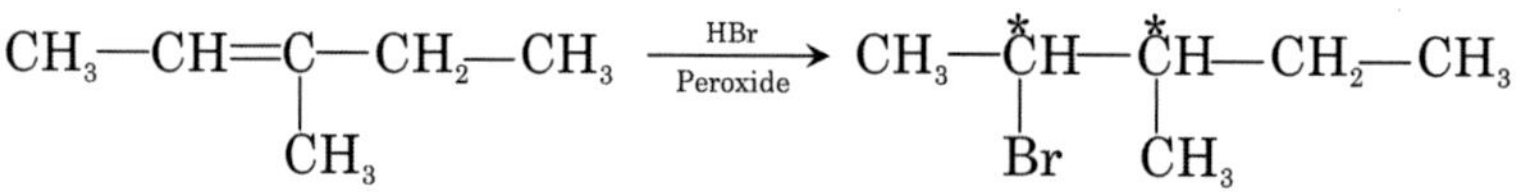

No. of stereoisomers = $2^n = 2^2 = 4$ [n = No. of chiral carbon atoms]

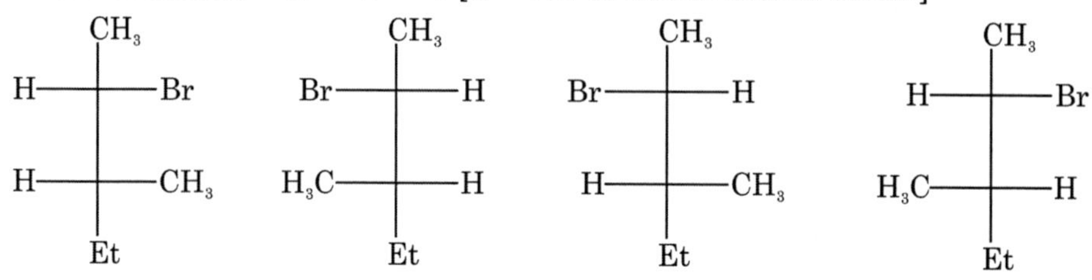

(I) (II) (III) (IV)

I.I.T. JEE Objective Question Paper (Advanced)—2017

Paper–1

INSTRUCTIONS

General

1. This sealed booklet is your Question Paper. Do not break the seal till you are told to do so.
2. The paper CODE is printed on the right hand top corner of this sheet and the right hand top corner of the back cover of this booklet.
3. Use the Optical Response Sheet (ORS) provided separately for answering the questions.
4. The paper CODE is printed on the left part as well as the right part of the ORS. Ensure that both these codes are identical and same as that on the question paper booklet. If not, contact the invigilator for change of ORS.
5. Blank spaces are provided within this booklet for rough work.
6. Write your name, roll number and sign in the space provided on the back cover of this booklet.
7. After breaking the seal of the booklet at 9:00 am, verify that the booklet contains 36 pages and that all the 54 questions along with the options are legible. If not, contact the invigilator for replacement of the booklet.
8. You are allowed to take away the Question Paper at the end of the examination.

Optical Response Sheet

9. The ORS (top sheet) will be provided with an attached Candidate's Sheet (bottom sheet). The Candidate's Sheet is a carbonless copy of the ORS.
10. Darken the appropriate bubbles on the ORS by applying sufficient pressure. This will leave an impression at the corresponding place on the Candidate's Sheet.
11. The ORS will be collected by the invigilator at the end of the examination.
12. You will be allowed to take away the Candidate's Sheet at the end of the examination.
13. Do not tamper with or mutilate the ORS. Do not use the ORS for rough work.
14. Write your name, roll number and code of the examination center, and sign with pen in the space provided for this purpose on the ORS. Do not write any of these details anywhere else on the ORS. Darken the appropriate bubble under each digit of your roll number.

Darkening the Bubbles on the ORS

15. Use a Black Ball Point Pen to darken the bubbles on the ORS.
16. Darken the bubble ○ completely.
17. The correct way of darkening a bubble is as: ●
18. The ORS is machine-gradable. Ensure that the bubbles are darkened in the correct way.
19. Darken the bubbles only if you are sure of the answer. There is no way to erase or "un-darken" a darkened bubble.

SECTION–1
(Maximum Marks: 28)

- This section contains SEVEN questions.
- Each question has FOUR options (A), (B), (C) and (D). ONE OR MORE THAN ONE of these four options is (are) correct.
- For each question, darken the bubble(s) corresponding to the correct option(s) in the ORS.
- For each question, marks will be awarded in one of the following categories:

  Full Marks : + 4 If only the bubbles corresponding to the correct option(s) is (are) darkened.

  Partial Marks : + 1 For darkening a bubble corresponding to each correct option, provided NO incorrect option is darkened.

  Zero Marks : 0 If none of the bubbles is darkened.

  Negative Marks : – 2 In all other cases.
- For example, if (A), (C) and (D) are all the correct options for a question, darkening all these three will get + 4 marks; darkening only (A) and (D) will get + 2 marks; and darkening (A) and (B) will get – 2 marks, as a wrong option is also darkened.

1. The IUPAC name(s) of the following compound is (are)

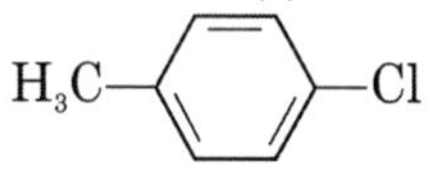

(A) 4-methylchlorobenzene (B) 4-chlorotoluene

(C) 1-chloro-4-methylbenzene (D) 1-methyl-4-chlorobenzene

2. The correct statement(s) for the following addition reactions is (are)

(i) 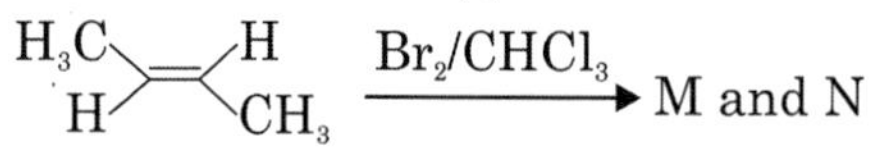

(ii) 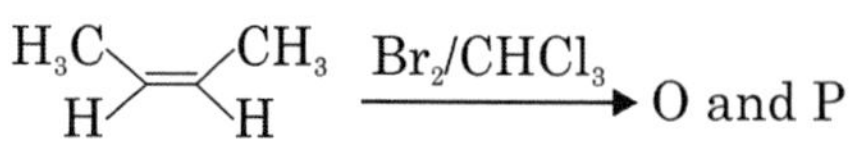

(A) (M and O) and (N and P) are two pairs of diastereomers

(B) Bromination proceeds through *trans*-addition in both the reactions

(C) O and P are identical molecules

(D) (M and O) and (N and P) are two pairs of enantiomers.

3. Addition of excess aqueous ammonia to a pink coloured aqueous solution of $MCl_2 \cdot 6H_2O$ (X) and NH_4Cl gives an octahedral complex Y in the presence of air. In aqueous solution, complex Y behaves as 1 : 3 electrolyte. The reaction of X with excess HCl at room temperature results in the formation of a blue coloured complex Z. The calculated spin-only magnetic moment of X and Z is 3.87 B.M., whereas it is zero for the complex Y.
Among the following options, which statements is (are) correct?
(A) The hybridization of the central metal ion in Y is d^2sp^3
(B) Z is a tetrahedral complex
(C) Addition of silver nitrate to Y gives only two equivalents of silver chloride
(D) When X and Z are in equilibrium at 0°C, the colour of the solution is pink.

4. For a solution formed by mixing liquids L and M, the vapour pressure of L plotted against the mole fraction of M in solution is shown in the following figure. Here P_L and x_M represent mole fractions of L and M, respectively, in the solution. The correct statement(s) applicable to this system is (are)

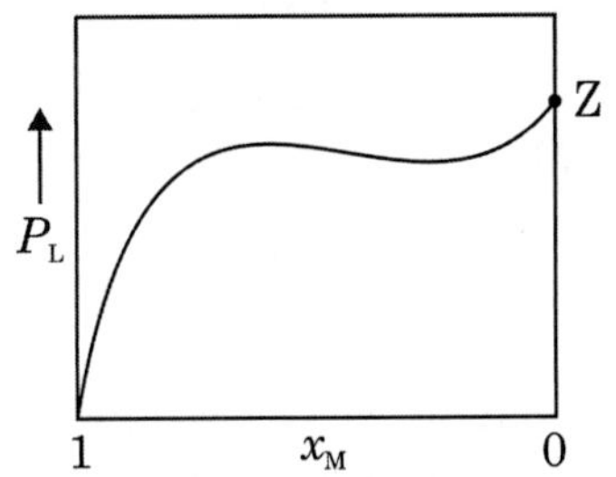

(A) Attractive intramolecular interactions between L–L in pure liquid L and M–M in pure liquid M are stronger than those between L–M when mixed in solution
(B) The point Z represents vapour pressure of pure liquid M and Raoult's law is obeyed when $P_L \longrightarrow 0$
(C) The point Z represents vapour pressure of pure liquid L and Raoult's law is obeyed when $P_L \longrightarrow 1$
(D) The point Z represents vapour pressure of pure liquid M and Raoult's law is obeyed from $P_L = 0$ to $P_L = 1$.

5. An ideal gas is expanded from (P_1, V_1, T_1) to (P_2, V_2, T_2) under different conditions. The correct statement(s) among the following is (are)
(A) The work done on the gas is maximum when it is compressed irreversibly from (P_2, V_2) to (P_1, V_1) against constant pressure P_1
(B) The work done on the gas is less when it is expanded reversibly from V_1 to V_2 under adiabatic conditions as compared to that when expanded reversibly from V_1 to V_2 under isothermal conditions

(C) The change in internal energy of the gas (i) is zero, if it is expanded reversibly with $T_1 = T_2$, and (ii) is positive, if it is expanded reversibly under adiabatic conditions with $T_1 \neq T_2$

(D) If the expansion is carried out freely, it is simultaneously both isothermal as well as adiabatic.

6. The correct statement(s) about the oxoacids, $HClO_4$ and HClO, is (are)
   (A) $HClO_4$ is more acidic than HClO because of the resonance stabilization of its anion
   (B) $HClO_4$ is formed in the reaction between Cl_2 and H_2O
   (C) The central atom in both $HClO_4$ and HClO is sp^3 hybridized
   (D) The conjugate base of $HClO_4$ is a weaker base than H_2O.

7. The colour of the X_2 molecules of group 17 elements changes gradually from yellow to violet down the group. This is due to
   (A) the physical state of X_2 at room temperature changes from gas to solid down the group
   (B) decrease in HOMO-LUMO gap down the group
   (C) decrease in π^*-σ^* down the group
   (D) decrease in ionization energy down the group.

SECTION–2
(Maximum Marks: 15)

- This section contains FIVE questions.
- The answer to each question is a SINGLE DIGIT INTEGER ranging from 0 to 9, both inclusive.
- For each question, darken the bubble corresponding to the correct integer in the ORS.
- For each question, marks will be awarded in one of the following categories:

Full Marks : + 3 If only the bubble corresponding to the correct answer is darkened.
Zero Marks : 0 In all other cases.

8. Among H_2, He_2^+, Li_2, Be_2, B_2, C_2, N_2, O_2^- and F_2, the number of diamagnetic species is (Atomic number): H = 1, He = 2, Li = 3, Be = 4, B = 5, C = 6, N = 7, O = 8, F = 9)

9. Among the following, the number of aromatic compound(s) is

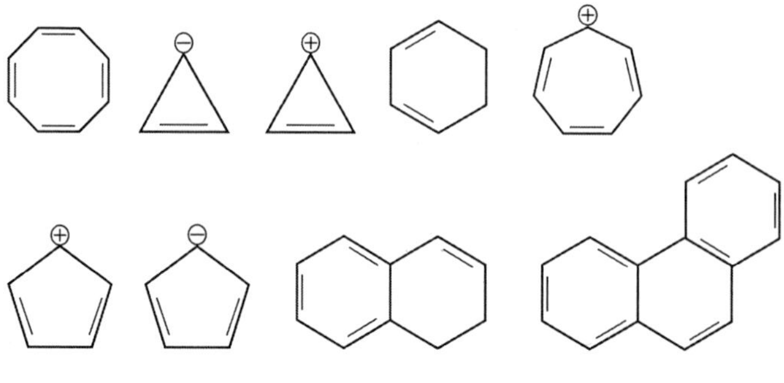

10. The conductance of a $0.0015M$ aqueous solution of a weak monobasic acid was determined by using a conductivity cell consisting of platinized Pt electrodes. The distance between the electrodes is 120 cm with an area of cross section of 1 cm^2. The conductance of this solution was found to be 5×10^{-7} S. The pH of the solution is 4. The value of limiting molar conductivity $\left(\Lambda_m^0\right)$ of this weak monobasic acid in aqueous solution is $Z \times 10^2$ S cm^{-1} mol^{-1}. The value of Z is

11. The sum of the number of lone pairs of electrons on each central atom in the following species is $[TeBr_6]^{2-}$, $[BrF_2]^+$, SNF_3 and $[XeF_3]^-$
[Atomic number: N = 7, F = 9, S = 16, Br = 35, Te = 52, Xe = 54]

12. A crystalline solid of a pure substance has a face-centred cubic structure with a cell edge of 400 pm. If the density of the substance in the crystal is 8 g cm^{-3}, then the number of atoms present in 256 g of the crystal is $N \times 10^{24}$. The value of N is

SECTION–3
(Maximum Marks: 18)

- This section contains SIX questions of matching type.
- This section contains TWO tables (each having 3 columns and 4 rows).
- Based on each table, there are THREE questions.
- Each question has FOUR options (A), (B), (C) and (D). ONLY ONE of these four options is correct.
- For each question, darken the bubble corresponding to the correct option in the ORS.
- For each question, marks will be awarded in one of the following categories:

Full Marks : + 3 If only the bubble corresponding to the correct option is darkened.

Zero Marks : 0 If none of the bubbles is darkened.

Negative Marks : – 1 In all other cases.

Answer Q. 13, Q. 14 and Q. 15 by appropriately matching the information given in the three columns of the following table.

The wave function Ψ_{n, l, m_l} is a mathematical function whose value depends upon spherical polar coordinates (r, θ, ϕ) of the electron and characterized by the quantum numbers n, l and m_1. Here r is distance from nucleus, θ is colatitude and ϕ is azimuth. In the mathematical functions given in the table, Z is atomic number and a_0 is Bohr radius.

Column–1	Column–2	Column–3
(I) 1s orbital	(i) $\Psi_{n, l, m_l} \propto \left(\frac{Z}{a_0}\right)^{\frac{3}{2}} e^{-\left(\frac{Zr}{a_e}\right)}$	(P) 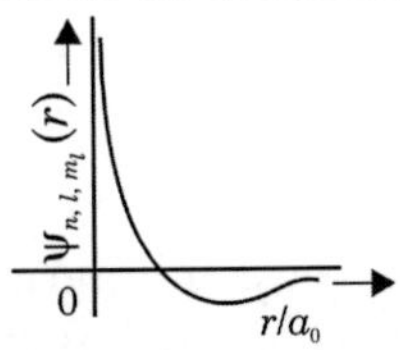

(II) 2*s* orbital	(ii) One radial node	(Q) Probability density at nucleus $\propto \frac{1}{a_0^3}$
(III) $2p_z$ orbital	(iii) $\psi_{n,l,m_l} \propto \left(\frac{Z}{a_0}\right)^{\frac{5}{2}} re^{-\left(\frac{Zr}{2a_0}\right)} \cos\theta$	(R) Probability density is maximum at nucleus.
(IV) $3d_z^2$ orbital	(iv) *xy*-plane is a nodal plane.	(S) Energy needed to excite an electron from $n = 2$ state to $n = 4$ state is 27/32 times the energy needed to excite an electron from $n = 2$ state to $n = 6$ state.

13. For the given orbital in column 1, the only correct combination for any hydrogen-like species is
   (A) (IV) (iv) (R) (B) (II) (ii) (P) (C) (III) (iii) (P) (D) (I) (ii) (S)
14. For He^+ ion, the only incorrect combination is
   (A) (II) (ii) (Q) (B) (I) (i) (S) (C) (I) (i) (R) (D) (I) (iii) (R)
15. For hydrogen atom, the only correct combination is
   (A) (I) (iv) (R) (B) (I) (i) (P) (C) (II) (i) (Q) (D) (I) (i) (S)

Answer Q. 16, Q. 17 and Q. 18 by appropriately matching the information given in the three columns of the following table.

Columns 1, 2 and 3 contain starting materials, reaction conditions and type of reactions, respectively		
Column–1	**Column–2**	**Column–3**
(I) Toluene	(i) $NaOH/Br_2$	(P) Condensation
(II) Acetophenone	(ii) $Br_2/h\varpi\nu$	(Q) Carboxylation
(III) Banzaldehyde	(iii) $(CH_3CO)_2O/CH_3COOK$	(R) Substitution
(IV) Phenol	(iv) $NaOH/CO_2$	(S) Haloform

16. For the synthesis of benzoic acid, the only correct combination is
   (A) (III) (iv) (R) (B) (IV) (ii) (P) (C) (I) (iv) (Q) (D) (II) (i) (S)
17. The only correct combination in which the reaction proceeds through radical mechanism is
   (A) (I) (ii) (R) (B) (II) (iii) (R) (C) (III) (ii) (P) (D) (IV) (i) (Q)
18. The only correct combination that gives two different carboxylic acids is
   (A) (IV) (iii) (Q) (B) (III) (iii) (P) (C) (II) (iv) (R) (D) (I) (i) (S)

ANSWERS

1. B, C	2. A, B	3. A, B, D	4. A, C	5. A, B, D	6. A, C, D
7. B, C	8. 6	9. 5	10. 6	11. 6	12. 2
13. B	14. D	15. D	16. D	17. A	18. B

Model Solutions—I.I.T. JEE (Advanced) 2017

Paper–1

1. (B, C)

   Toluene is accepted by IUPAC as a name of parent carbon atom. Hence, the IUPAC name of the compound is 4-chlorotoluene and 1-chloro-4-methyl benzene.

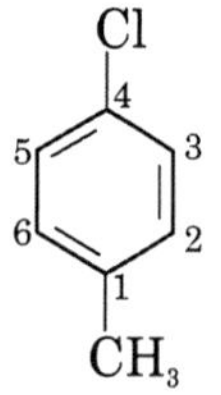

2. (A, B)

(i)

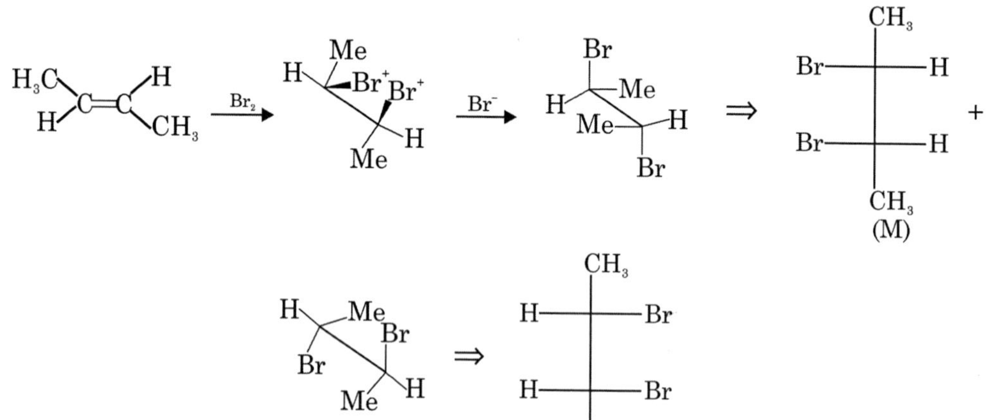

(M) and (N) are identical meso compounds.

(ii)

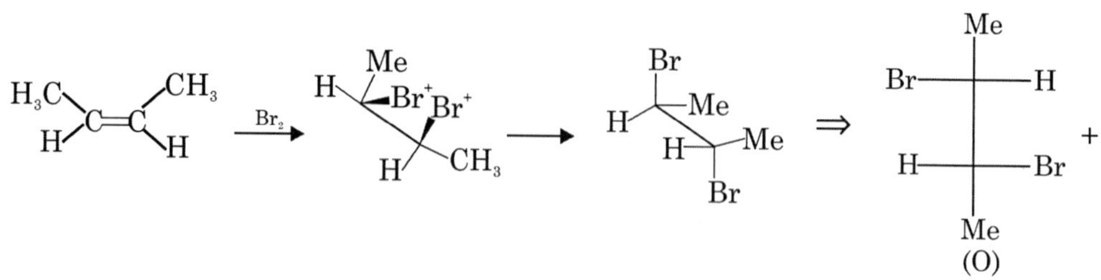

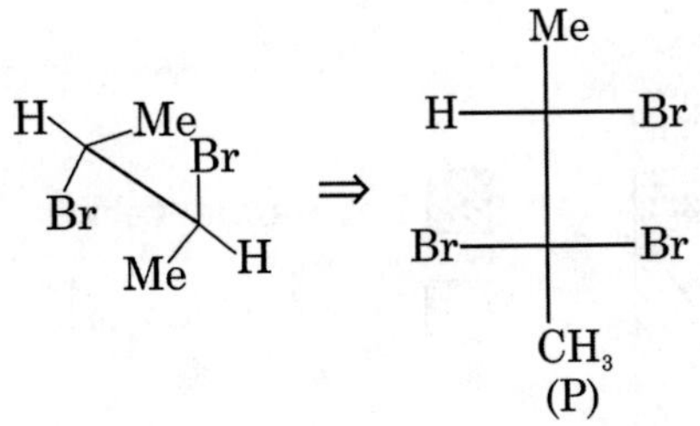

3. (A, B, D)

$$[Co^{II}(H_2O)_6]Cl_2 \xrightarrow[\text{Air}/O_2]{\text{Excess } NH_4OH/NH_4Cl} [Co^{III}(NH_3)_6]Cl_3$$

Pink (X) (Y)

$$[Co\,(H_2O)_6]^{2+} + 4Cl^- \longrightarrow [CoCl_4]^{2-}$$

(X) (excess) (Z) blue

(A) Hybridization of Y is d^2sp^3 as NH_3 is a strong field ligand.

(B) $[CoCl_4]^{2-}$ has sp^3 hybridization as Cl^- is a weak field ligand.

(C) $[Co\,(NH_3)_6]Cl_3 + 3AgNO_3(aq) \rightarrow 3AgCl$
(Y)

(D) $[CoCl_4]^{2-} + 6H_2O \rightleftharpoons [Co(H_2O)_6]^{2+} + 4Cl^-$ $\Delta H = (-)$ve (exothermic)
(Blue) (Pink)

When ice is added to the solution, the equilibrium shifts to the right. Hence pink colour will remain predominant. Hence, the correct answer is (A, B, D).

4. (A, C)

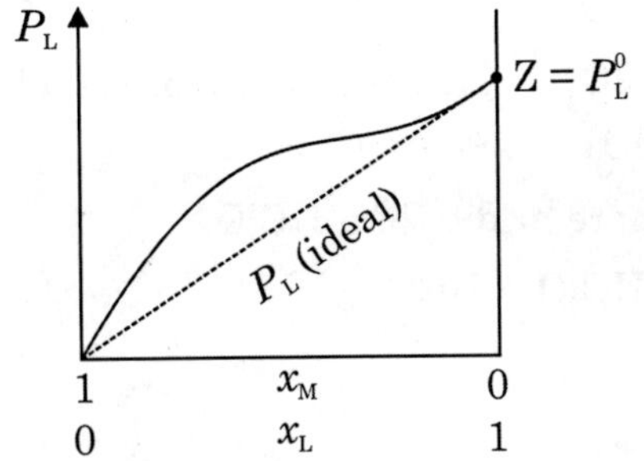

(A) This is a case of positive deviation. Hence, F_{L-L}, $F_{M-M} > F_{L-M}$

(C) $P_L \geq P_L^0 x_L$

But when $x_L \rightarrow 1$, the mixture contains almost pure liquid L. Hence, $P_L \rightarrow P_L^0$

5. (A, B, D)

(A) 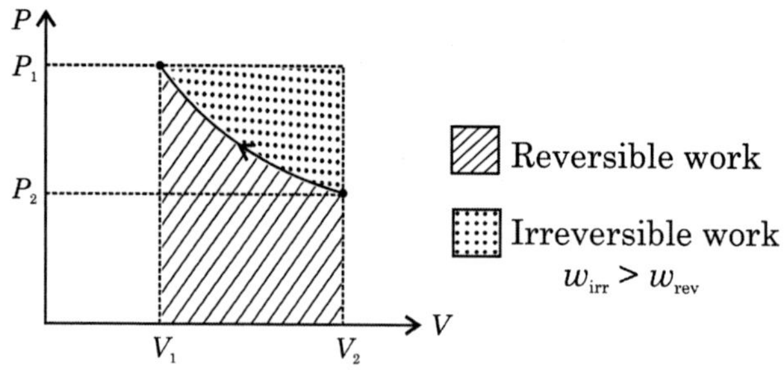

(B) 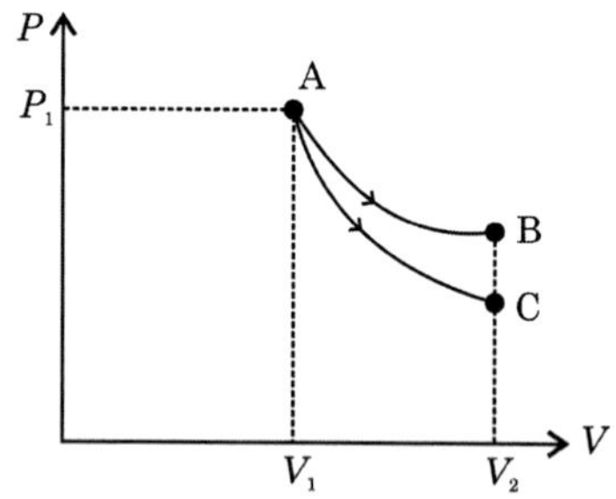

AB: Reversible isothermal
AC: Reversible adiabatic
$|w_{AB}| > |w_{AC}|$

(C) (i) $\Delta U = nC_V\Delta T = 0$ (isothermal, hence $\Delta T = 0$)

(ii) $\Delta U = q + w = -ve$ ($q = 0$, $w < 0$)

$\Delta U = nC_V\Delta T \Rightarrow \Delta T < 0$

(D) $q = 0$ (adiabatic), $w = 0$ (free expansion)

$\Delta U = 0 \Rightarrow \Delta T = 0$ (isothermal)

6. (A, C, D)

(A) $HClO_4 \rightleftharpoons H^+ + ClO_4^-$ ClO_4^- is a resonance stablized anion

$HClO \rightleftharpoons H^+ + ClO^-$

Hence, $HClO_4$ is more acidic than HClO.

(B) $Cl_2 + H_2O \longrightarrow HClO + HCl$

(C) $HClO_4$

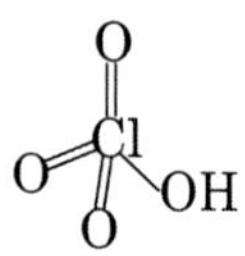

Central atom is sp^3 hybridised.

HClO 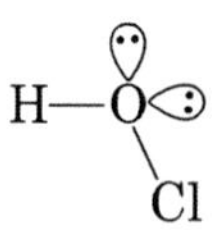

Central atom is sp^3 hybridised.

(D) $HClO_4 + H_2O \rightarrow H_3O^{\oplus} + ClO_4^-$. Since H_2O accepts H^+ from $HClO_4$, H_2O is stronger base compared to ClO_4^-.

7. (B, C)

Halogens are coloured due to HOMO-LUMO transition of electrons.

M.O. (configuration) : $F_2 = \sigma 1s^2, \sigma^* 1s^2, \sigma 2s^2, \sigma^* 2s^2, \sigma 2p_z^2, \pi 2p_x^2 = \pi 2p_y^2, \pi^* 2p_x^2 = \pi^* 2p_y^2, \sigma^* 2p_z$

filled (HOMO) vacant (LUMO)

On moving down the group HOMO-LUMO energy gap decreases. Hence, the transition of electrons becomes easier from π^*2p to σ^*2p. Therefore, colour intensifies.

8. (6)

H_2	$\Rightarrow$	$\sigma 1s^2$	(Diamagnetic)
He_2^+	$\Rightarrow$	$\sigma 1s^2 \sigma * 1s^1$	(Paramagnetic)
Li_2	$\Rightarrow$	$\sigma 1s^2 \sigma * 1s^2 \sigma 2s^2$	(Diamagnetic)
Be_2	$\Rightarrow$	$\sigma 1s^2 \sigma * 1s^2 \sigma 2s^2 \sigma * 2s^2$	(Diamagnetic)
B_2	$\Rightarrow$	$\sigma 1s^2 \sigma * 1s^2 \sigma 2s^2 \sigma * 2s^2 \pi 2p_x^1 = \pi 2p_y^1$	(Paramagnetic)
C_2	$\Rightarrow$	$\sigma 1s^2 \sigma * 1s^2 \sigma 2s^2 \sigma * 2s^2 \pi 2p_x^2 = \pi 2p_y^2$	(Diamagnetic)
N_2	$\Rightarrow$	$\sigma 1s^2 \sigma * 1s^2 \sigma 2s^2 \sigma * 2s^2 \pi 2p_x^2 = \pi 2p_y^2 \sigma 2p_z^2$	(Diamagnetic)
O_2^-	$\Rightarrow$	$\sigma 1s^2 \sigma * 1s^2 \sigma 2s^2 \sigma * 2s^2 \pi 2p_z^2 \pi 2p_x^2 = \pi 2p_y^2 \pi * 2p_x^2 = \pi * 2p_y^1$	(Paramagnetic)
F_2	$\Rightarrow$	$\sigma 1s^2 \sigma * 1s^2 \sigma 2s^2 \sigma * 2s^2 \sigma 2p_z^2 \pi 2p_x^2 = \pi 2p_y^2 \pi * 2p_x^2 = \pi * 2p_y^2$	(Diamagnetic)

9. (5)

Cyclooctatetraene, non-aromatic

Due to the non-planarity of ring, the π-electrons are not delocalized.

Cyclopropenyl anion, non-aromatic

4π-electrons are delocalized.

Cyclopropenyl cation; aromatic

2π-electrons are delocalized.

Cyclohexadiene, non-aromatic

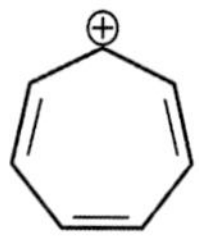

Tropylium ion: aromatic
6π-electrons are delocalized.

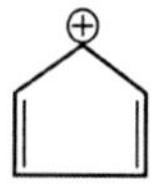

Cyclopentadienyl cation; non-aromatic
4π-electrons are delocalized.

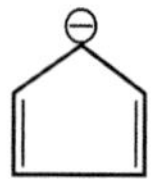

Cyclopentadienyl anion; aromatic
6π-electrons are delocalized.

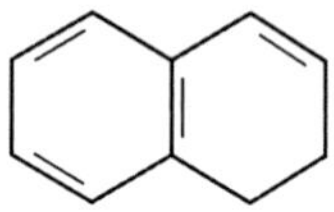

aromatic

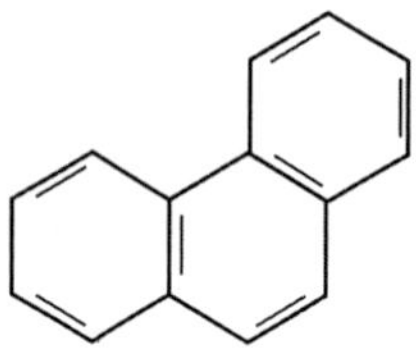

aromatic
14 πe^-s
(delocalized)

10. (6)

For weak acid $[H^+] = c\alpha = c\dfrac{\Lambda_m^C}{\Lambda_m^\infty} = c \times \dfrac{\kappa \times \dfrac{1000}{c}}{\Lambda_m^\infty} = \dfrac{\kappa \times 1000}{\Lambda_m^\infty} = \dfrac{G \times \left(\dfrac{l}{a}\right) \times 1000}{\Lambda_m^\infty}$

$$10^{-4} = \frac{5 \times 10^{-7} \times \left(\dfrac{120}{1}\right) \times 1000}{Z \times 10^2} \Rightarrow Z = 6$$

11. (6)

	Number of σ-bonds formed by central atom	Number of lone pairs on central atom
(i) In $[TeBr_6]^{2-}$	6	1
(ii) In $[BrF_2]^+$	2	2
(iii) In SNF_3	4	0
(iv) In $[XeF_3]^-$	3	3

Hence, total number of lone pairs of electrons = 1 + 2 + 0 + 3 = 6

12. (2)

$$d = \frac{Z \times M}{N_A \times a^3}$$

For FCC unit cell $Z = 4$. Edge length $a = 4 \times 10^{-8}$ cm

Hence, $M = \dfrac{d \times N_A \times a^3}{Z} = \dfrac{8 \times 6 \times 10^{23} \times 64 \times 10^{-24}}{4}$ g/mol

No. of atoms $= \dfrac{\text{wt (g)}}{\text{molar mass}} \times N_A = \dfrac{256 \times 10 \times 6 \times 10^{23}}{8 \times 6 \times 16} = 2 \times 10^{24}$ (value of $N = 2$)

13. (B)

(A) (IV) (iv) (R) ⇒ incorrect, because, d_{z^2} has no nodal plane.

(B) (II) (ii) (P) ⇒ correct, because 2s orbital has 1 radial node.

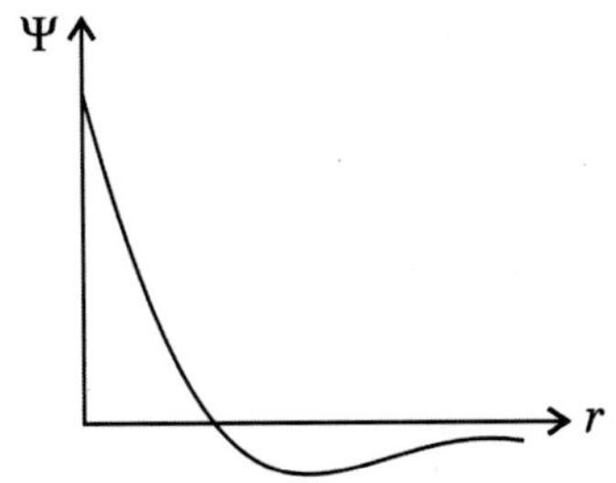

(C) (III) (iii) (P) ⇒ incorrect, because probability density for 2p at nucleus is zero.

(D) (I) (ii) (S) ⇒ incorrect, because 1s orbital has no radial node.

14. (D)

The option (D) is incorrect because in the wave function of 1s orbital, no angular function is present.

15. (D)

For 1s orbital: $\Psi_{n, l, m_l} \, \alpha \left(\dfrac{Z}{a_0}\right)^{3/2 - zr} e^{a_0}$

Energy needed to excite an electron from $n = 2$ to $n = 4$,

$\Delta E_{2-4} = 13.6Z^2 \times \dfrac{3}{16}$ eV

Energy needed to excite an electron from $n = 2$ to $n = 6$,

$\Delta E_{2-6} = 13.6Z^2 \times \dfrac{8}{36}$ eV

$\Delta E_{2-4} = \dfrac{27}{32} E_{2-6}$ (hence, true)

16. (D)

(II) (i) (S)

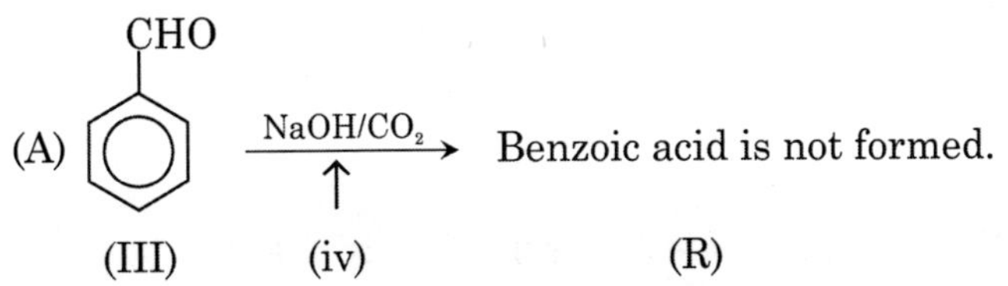

(B) 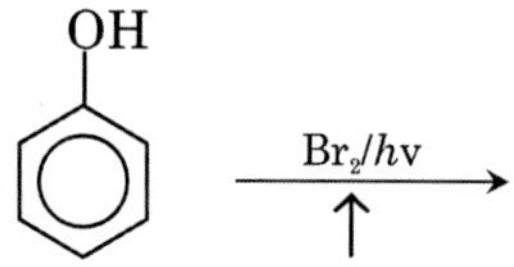 Benzoic acid is not formed.

(I) (ii) (P)

(C) 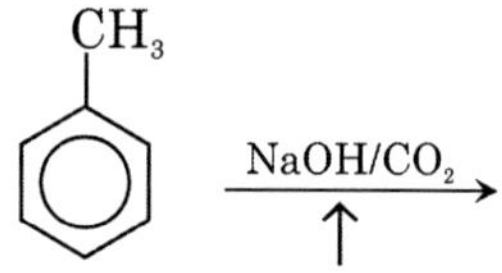 Benzoic acid is not formed.

(IV) (iv) (Q)

(D) 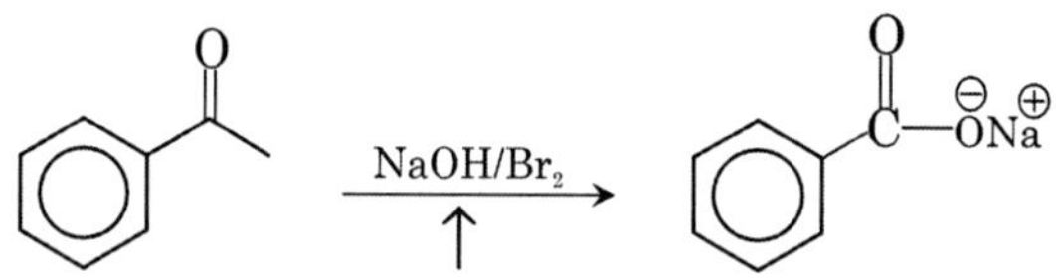 + $CHBr_3$ (haloform reaction)

(II) (i) (S)

17. [A] (I) (ii) (R)

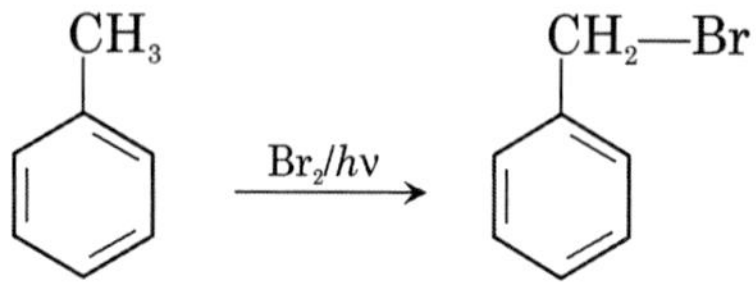

The mechanism involved is free radical substitution.

(B) 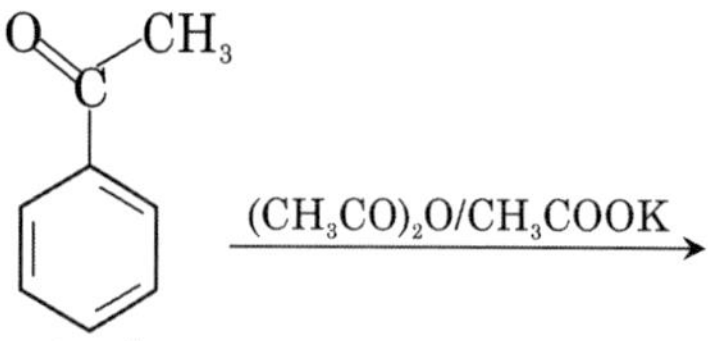 Condensation reaction.

Acetophenone

(No radical mechanism is involved.)

(C) 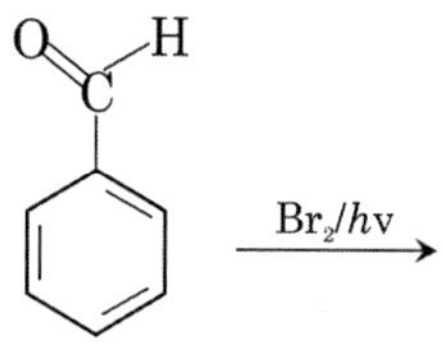 Condensation does not occur.

(D) 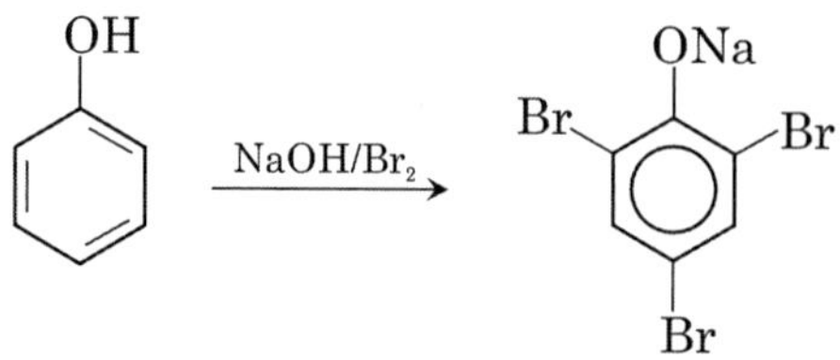

(No free radical mechanism is involved.)

18. (B)

$$C_6H_5CHO + CH_3-\underset{\underset{O}{\|}}{C}-O-\underset{\underset{O}{\|}}{C}-CH_3 \xrightarrow{CH_3COOK} C_6H_5CH{=}CH-COOH$$

Cinnamic acid
(*cis* and *trans*)

It is a Perkin condensation reaction.
Mechanism

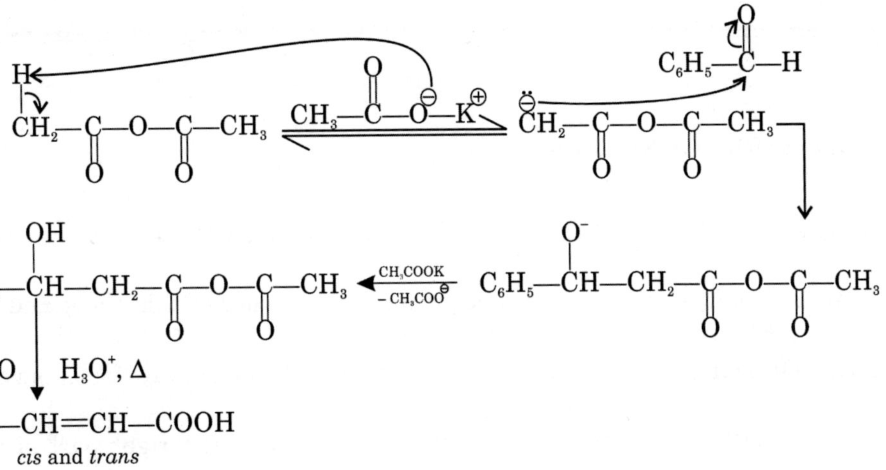

I.I.T. JEE Objective Question Paper (Advanced)—2017

Paper–2

READ THE INSTRUCTIONS CAREFULLY

General

1. This sealed booklet is your Question Paper. Do not break the seal till you are told to do so.
2. The paper CODE is printed on the right hand top corner of this sheet and the right hand top corner of the back cover of this booklet.
3. Use the Optical Response Sheet (ORS) provided separately for answering the questions.
4. The paper CODE is printed on the left part as well as the right part of the ORS. Ensure that both these codes are identical and same as that on the question paper booklet. If not, contact the invigilator for change of ORS.
5. Blank spaces are provided within this booklet for rough work.
6. Write your name, roll number and sign in the space provided on the back cover of this booklet.
7. After breaking the seal of the booklet at 2:00 pm, verify that the booklet contains 36 pages and that all the 54 questions along with the options are legible. If not, contact the invigilator for replacement of the booklet.
8. You are allowed to take away the Question Paper at the end of the examination.

Optical Response Sheet

9. The ORS (top sheet) will be provided with an attached Candidate's Sheet (bottom sheet). The Candidate's Sheet is a carbonless copy of the ORS.
10. Darken the appropriate bubbles on the ORS by applying sufficient pressure. This will leave an impression at the corresponding place on the Candidate's Sheet.
11. The ORS will be collected by the invigilator at the end of the examination.
12. You will be allowed to take away the Candidate's Sheet at the end of the examination.
13. Do not tamper with or mutilate the ORS. Do not use the ORS for rough work.
14. Write your name, roll number and code of the examination center, and sign with pen in the space provided for this purpose on the ORS. Do not write any of these details anywhere else on the ORS. Darken the appropriate bubble under each digit of your roll number.

Darkening the Bubbles on the ORS

15. Use a black ball point pen to darken the bubbles on the ORS.
16. Darken the bubble ○ completely.
17. The correct way of darkening a bubble is as: ●
18. The ORS is machine-gradable. Ensure that the bubbles are darkened in the correct way.
19. Darken the bubbles only if you are sure of the answer. There is no way to erase of "un-darken" a darkened bubble.

SECTION–1
(Maximum Marks: 21)

- This section contains SEVEN questions.
- Each question has FOUR options (A), (B), (C) and (D). ONLY ONE of these four options is correct.
- For each question, darken the bubble corresponding to the correct option in the ORS.
- For each question, marks will be awarded in one of the following categories:
  Full Marks : + 3 If only the bubble corresponding to the correct option is darkened.
  Zero Marks : 0 If none of the bubbles is darkened.
  Negative Marks : – 1 In all other cases.

1. Which of the following combinations will produce H_2 gas?
   (A) Zn metal and NaOH(*aq*)
   (B) Au metal and NaCN(*aq*) in the presence of air
   (C) Cu metal and conc. HNO_3
   (D) Fe metal and conc. HNO_3

2. Pure water freezes at 273 K and 1 bar. The addition of 34.5 g of ethanol to 500 g of water changes the freezing point of the solution. Use the freezing point depression constant of water as 2 K kg mol^{-1}. The figures shown below represents plots of vapour pressure (V. P.) versus temperature (*T*). [Molecular weight of ethanol is 46 g mol^{-1}.] Among the following, the option representing change in the freezing point is

(A) 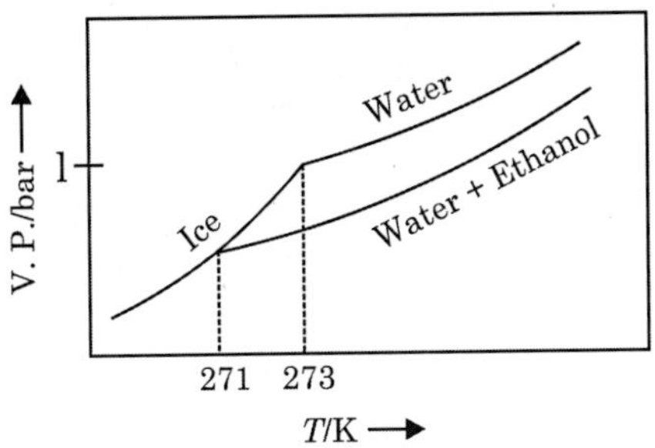

(B) 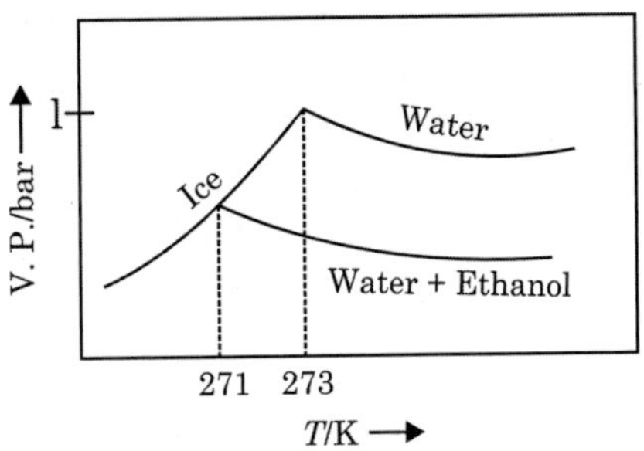

(C) 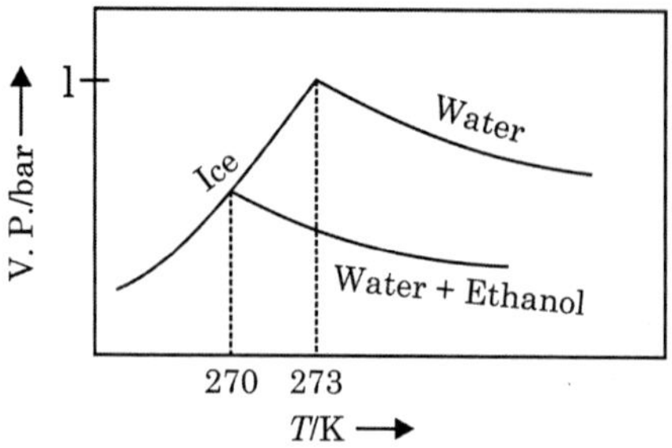

(D) 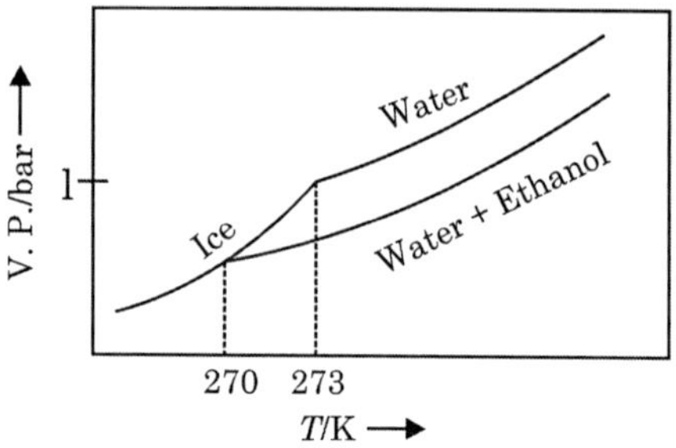

3. The order of the oxidation state of the phosphorus atom in H_3PO_2, H_3PO_4, H_3PO_3 and $H_4P_2O_6$ is

(A) $H_3PO_4 > H_4P_2O_6 > H_3PO_3 > H_3PO_2$ (B) $H_3PO_3 > H_3PO_2 > H_3PO_4 > H_4P_2O_6$

(C) $H_3PO_2 > H_3PO_3 > H_4P_2O_6 > H_3PO_4$ (D) $H_3PO_4 > H_3PO_2 > H_3PO_3 > H_4P_2O_6$

4. The standard state Gibbs free energies of formation of C (graphite) and C (diamond) at $T = 298$ K are

$\Delta_f G^0$ [C (graphite)] = 0 kJ mol^{-1}; $\Delta_f G^0$ [C (diamond)] = 2.9 kJ mol^{-1}

The standard state means that the pressure should be 1 bar and the substance should be pure at a given temperature. The conversion of graphite [C (graphite)] to diamond [C (diamond)] reduces its volume by 2×10^{-6} m^3 mol^{-1}. If C (graphite) is converted to C (diamond) isothermally at $T = 298$ K, the pressure at which C (graphite) is in equilibrium with C (diamond), is

[Useful information: 1 J = 1 kg m^2 s^{-2}; 1 Pa = 1 kg m^{-1} s^{-2}; 1 bar = 10^5 Pa]

(A) 14501 bar (B) 29001 bar (C) 58001 bar (D) 1405 bar

5. The major product of the following reaction is

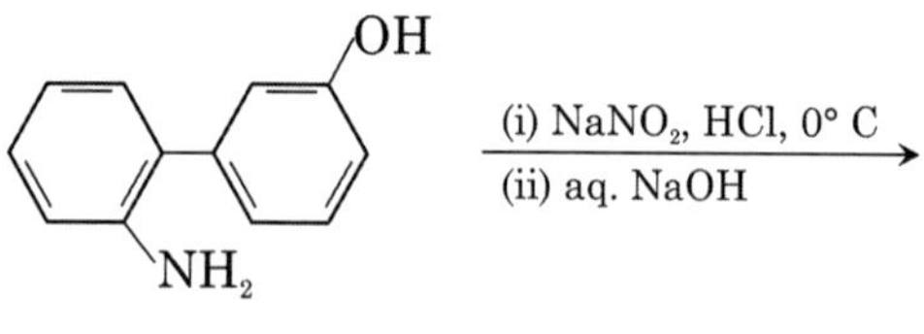

(A) 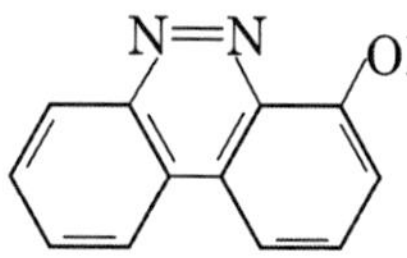

(B) 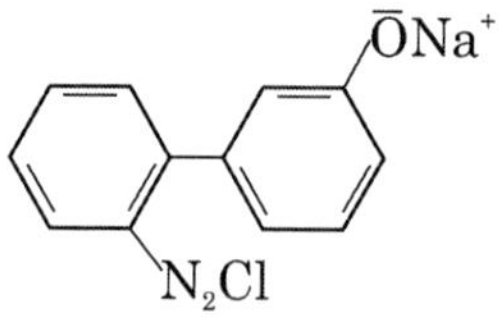

(C) 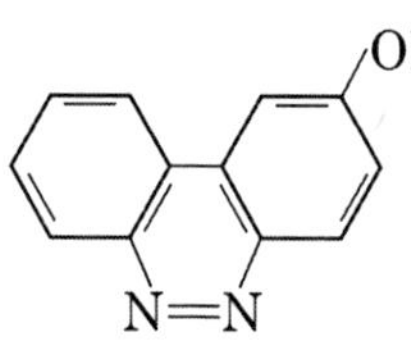

(D) 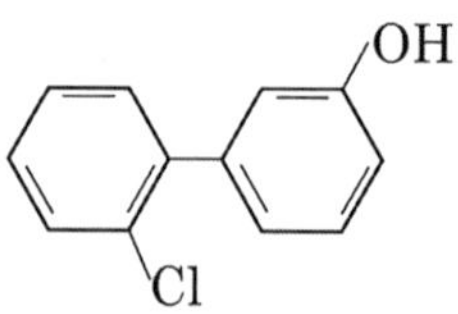

6. The order of basicity among the following compounds is

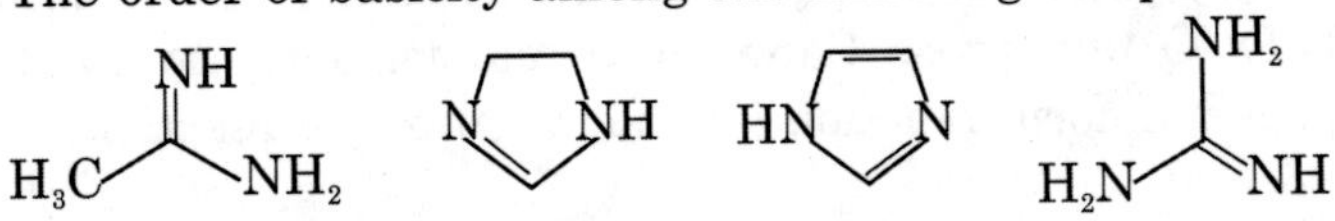

(I) (II) (III) (IV)

(A) II > I > IV > III
(B) IV > II > III > I
(C) I > IV > III > II
(D) IV > I > II > III

7. For the following cell:
$Zn(s)\ |\ ZnSO_4(aq)\ ||\ CuSO_4(aq)\ |\ Cu(s)$
When the concentration of Zn^{2+} is 10 times the concentration of Cu^{2+}, the expression of ΔG (in J mol^{-1}) is
[F is Faraday constant, R is gas constant, T is temperature, E^0(cell) = 1.1 V]
(A) $2.303RT + 1.1$ F (B) $2.303RT - 2.2$ F (C) 1.1 F (D) – 2.2 F

SECTION–2
(Maximum Marks: 28)

- This section contains SEVEN questions.
- Each question has FOUR options (A), (B), (C) and (D). ONE OR MORE THAN ONE of these four options is (are) correct.
- For each question, darken the bubble(s) corresponding to all the correct option(s) in the ORS.
- For each question, marks will be awarded in one of the following categories:
  Full Marks : + 4 If only the bubble(s) corresponding to all the correct option(s) is (are) darkened.
  Partial Marks : + 1 For darkening a bubble corresponding to each correct option, provided NO incorrect option is darkened.
  Zero Marks : 0 If none of the bubbles is darkened.
  Negative Marks : – 2 In all other cases.
- For example, if (A), (C) and (D) are all the correct options for a question, darkening all these three will result in + 4 marks; darkening only (A) and (D) will get in + 2 marks; and darkening (A) and (B) will get in – 2 marks, as a wrong option is also darkened.

8. In a bimolecular reaction, the steric factor P was experimentally determined to be 4.5. The correct option(s) among the following is (are)
(A) The value of frequency factor predicted by Arrhenius equation is higher than that determined experimentally
(B) The activation energy of the reaction is unaffected by the value of the steric factor
(C) Since P = 4.5, the reaction will not proceed unless an effective catalyst is used
(D) Experimentally determined value of frequency factor is higher than that predicted by Arrhenius equation.

9. Compounds P and R upon ozonolysis produce Q and S, respectively. The molecular formula of Q and S is C_8H_8O. Q undergoes Cannizzaro reaction but not haloform reaction, whereas S undergoes haloform reaction but not Cannizzaro reaction.

(i) $P \xrightarrow[\text{(ii) Zn/H}_2\text{O}]{\text{(i) O}_3\text{/CH}_2\text{Cl}_2} \underset{(C_8H_8O)}{Q}$ (ii) $R \xrightarrow[\text{(ii) Zn/H}_2\text{O}]{\text{(i) O}_3\text{/CH}_2\text{Cl}_2} \underset{(C_8H_8O)}{S}$

The option(s) with suitable combination of P and R, respectively, is (are)

(A) 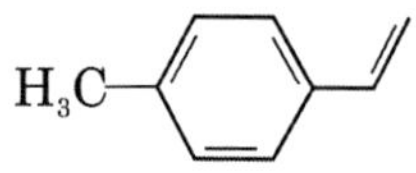 and 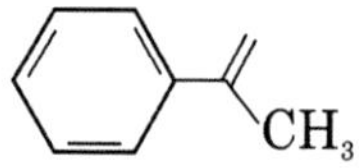

(B) 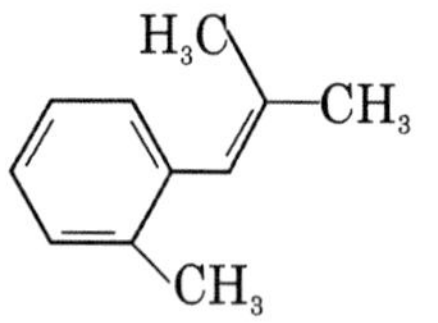 and 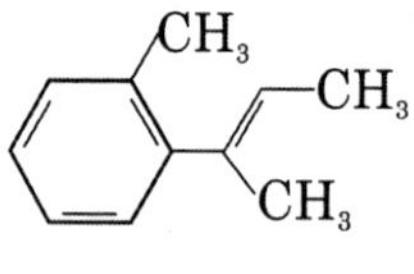

(C) 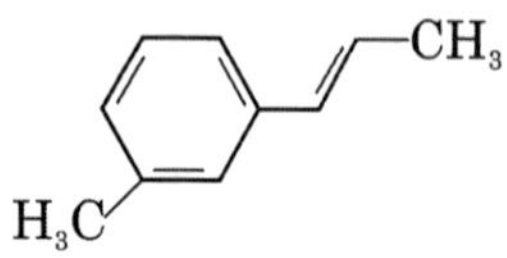 and 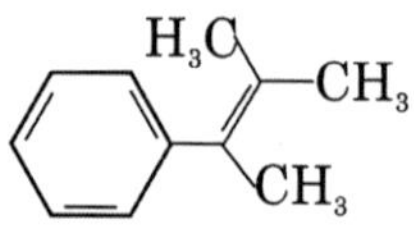

(D) 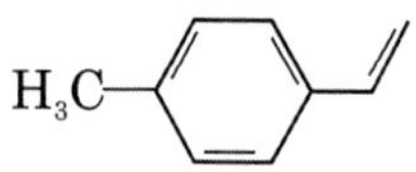 and 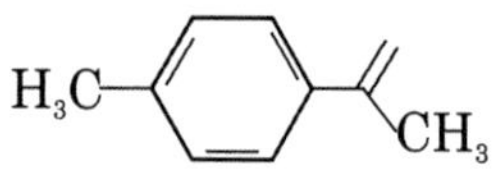

10. For a reaction taking place in a container in equilibrium with its surroundings, the effect of temperature on its equilibrium constant K in terms of change in entropy is described by
    (A) With increase in temperature, the value of K for exothermic reaction decreases because the entropy change of the system is positive
    (B) With increase in temperature, the value of K for endothermic reaction increases because unfavourable change in entropy of the surroundings decreases
    (C) With increase in temperature, the value of K for exothermic reaction decreases because favourable change in entropy of the surroundings decreases
    (D) With increase in temperature, the value of K for endothermic reaction increases because the entropy change of the system is negative.

11. The option(s) with only amphoteric oxides is (are)
    (A) Cr_2O_3, CrO, SnO, PbO   (B) NO, B_2O_3, PbO, SnO_2
    (C) Cr_2O_3, BeO, SnO, SnO_2 (D) ZnO, Al_2O_3, PbO, PbO_2

12. Among the following, the correct statement(s) is (are)
    (A) $Al(CH_3)_3$ has the three-centre two-electron bonds in its dimeric structure
    (B) $AlCl_3$ has the three-centre two-electron bonds in its dimeric structure

(C) BH_3 has the three-centre two-electron bonds in its dimeric structure

(D) The Lewis acidity of BCl_3 is greater than that of $AlCl_3$.

13. The correct statement(s) about surface properties is (are)

(A) Cloud is an emulsion type of colloid in which liquid is dispersed phase and gas is dispersion medium

(B) Adsorption is accompanied by decrease in enthalpy and decrease in entropy of the system

(C) Brownian motion of colloidal particles does not depend on the size of the particles but depends on viscosity of the solution

(D) The critical temperatures of ethane and nitrogen are 563 K and 126 K, respectively. The adsorption of ethane will be more than that of nitrogen on same amount of activated charcoal at a given temperature.

14. For the following compounds, the correct statement(s) with respect of nucleophilic substitution reactions is (are)

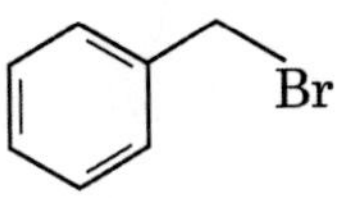

(I)

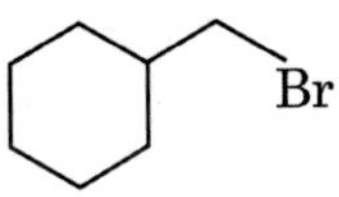

(II)

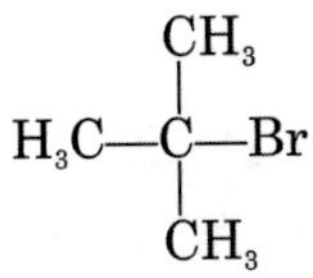

(III)

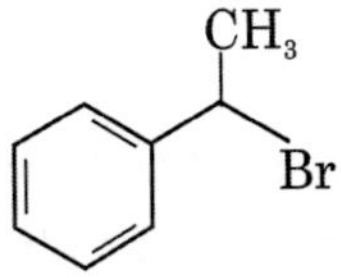

(IV)

(A) I and II follow S_N2 mechanism

(B) The order of reactivity for I, III and IV is: IV > I > III

(C) I and III follow S_N1 mechanism

(D) Compound IV undergoes inversion of configuration.

SECTION–3
(Maximum Marks: 12)

- This section contains TWO paragraphs.
- Based on each paragraph, there are TWO questions.
- Each question has FOUR options (A), (B), (C) and (D). ONLY ONE of these four options is correct.
- For each question, darken the bubble corresponding to the correct option in the ORS.
- For each question, marks will be awarded in one of the following categories:
  Full Marks : + 3 If only the bubble corresponding to all the correct option is darkened.
  Zero Marks : 0 In all other cases.

Paragraph for Questions 15 & 16

Upon heating $KClO_3$ in the presence of catalytic amount of MnO_2, a gas W is formed. Excess amount of W reacts with white phosphorus to give X. The reaction of X with pure HNO_3 gives Y and Z.

15. W and X are, respectively

(A) O_3 and P_4O_6 (B) O_2 and P_4O_{10} (C) O_3 and P_4O_{10} (D) O_2 and P_4O_6

16. Y and Z are, respectively

(A) N_2O_4 and H_3PO_3 (B) N_2O_4 and HPO_3

(C) N_2O_5 and HPO_3 (D) N_2O_3 and H_3PO_4

Paragraph for Questions 17 & 18

The reaction of compound P with CH_3MgBr (excess) in $(C_2H_5)_2O$ followed by addition of H_2O gives Q. The compound Q on treatment with H_2SO_4 at 0 °C gives R. The reaction of R with CH_3COCl in the presence of anhydrous $AlCl_3$ in CH_2Cl_2 followed by treatment with H_2O produces compound S. [Et in compound P is ethyl group]

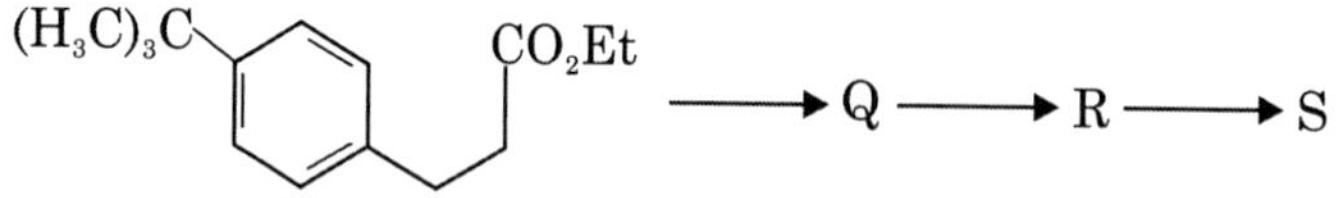

P

17. The reactions, Q to R and S to S, are

(A) Dehydration and Friedel–Crafts acylation

(B) Friedel–Crafts alkylation, dehydration and Friedel–Crafts acylation

(C) Aromatic sulfonation and Friedel–Crafts acylation

(D) Friedel–Crafts alkylation and Fridel–Crafts acylation.

18. The product S is

(A) 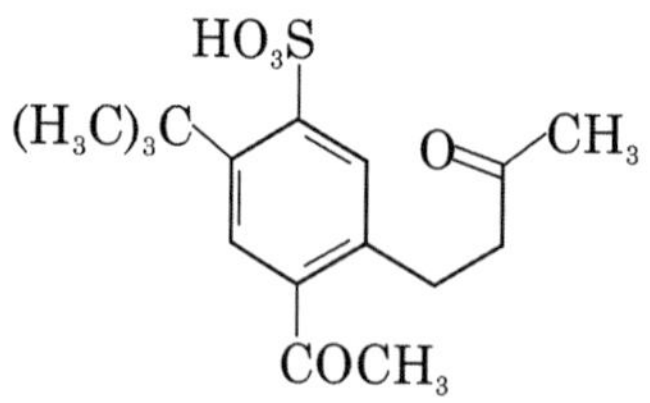

(B) 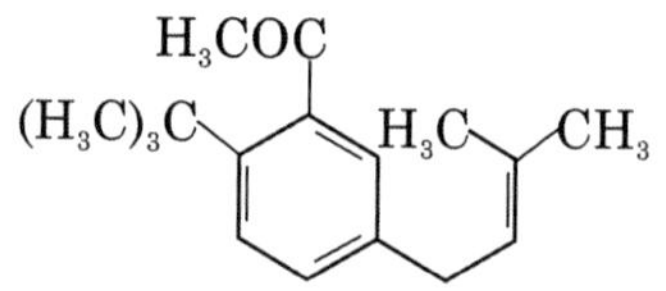

(C) 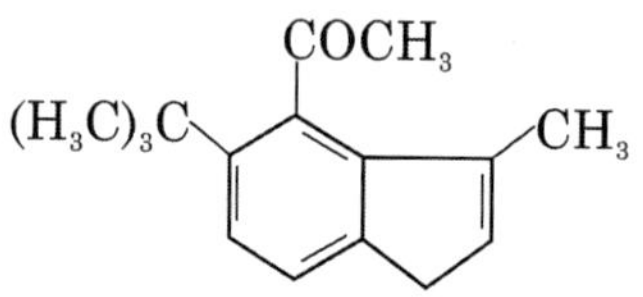

(D) 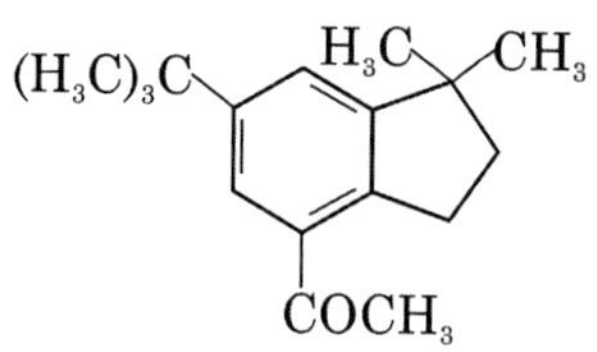

ANSWERS

1. A	2. D	3. A	4. A	5. C	6. D
7. B	8. B, D	9. A, C	10. B, C	11. C, D	12. A, C, D
13. B, D	14. A, B, C, D	15. B	16. C	17. B	18. D

Model Solutions—I.I.T. JEE (Advanced) 2017

Paper–2

1. (A)

(A) $Zn + 2NaOH \longrightarrow Na_2ZnO_2 + H_2$

(B) $4Au + 8NaCN + O_2 + 2H_2O \longrightarrow 4Na[Au(CN)_2] + 4NaOH$

(C) $\underset{\text{(conc.)}}{Cu + 4HNO_3} \longrightarrow Cu(NO_3)_2 + 2NO_2 + 2H_2O$

(D) Formation of passive layer of Fe_2O_3 on the surface of Fe and NO_2 gas is evolved.

2. (D)

Ethanol should be considered non-volatile as per given option.

$\Delta T_f = K_f \times m$

$$\Delta T_f = 2 \times \frac{34.5}{46 \times 0.5} = 3 \text{ K}$$

$T_f^0 = 273$ K

$T_f = 270$ K

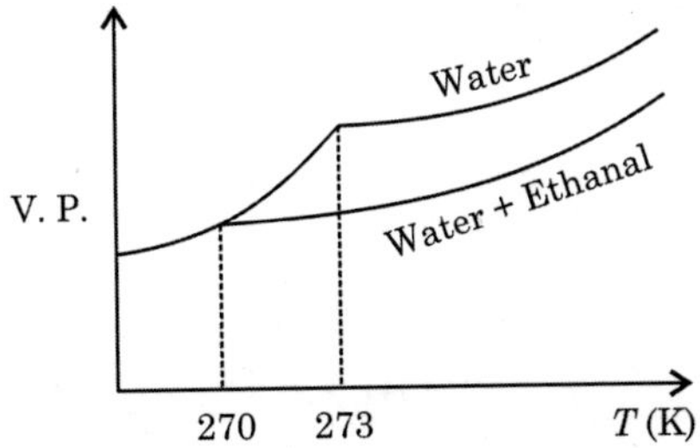

3. (A)

$H_3PO_4 \rightarrow$ 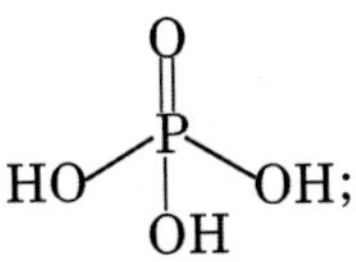 ; oxidation state of P = + 5

$H_4P_2O_6 \rightarrow$ 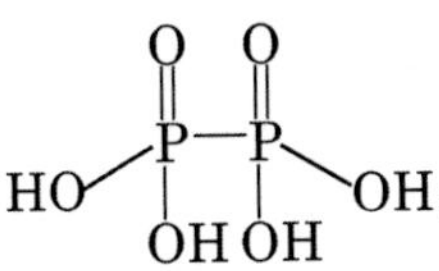 ; oxidation state of P = + 4

$H_3PO_4 \rightarrow$ 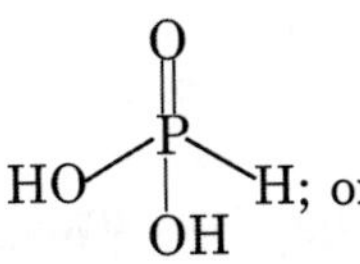 ; oxidation state of P = + 3

$H_3PO_2 \rightarrow$ 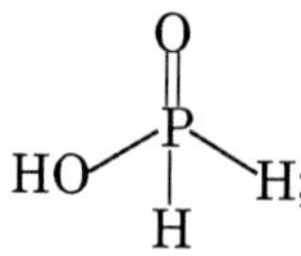 H; oxidation state of P = + 1

Hence, Ans. (A).

4. (A)

C(graphite) $\longrightarrow$ C(diamond); $\Delta G^0 = \Delta_f G^0_{\text{diamond}} - \Delta_f G^0_{\text{graphite}} = 2.9$ kJ/mol at 1 bar.

As $dG_T = V.dP$,

$$\int_{\Delta G_1}^{\Delta G_2} d(\Delta G_T) = \int_{P_1}^{P_2} \Delta V.dP$$

$$\Delta G_2 - \Delta G_1 = \Delta V.(P_2 - P_1)$$

$$(2.9 \times 10^3 - 0) = (-2 \times 10^{-6})\,(1 - P_2)$$

$$P_2 - 1 = \frac{2.9 \times 10^3}{2 \times 10^{-6}}\,\text{Pa} = 1.45 \times 10^4 \text{ bar}$$

$$P_2 = 14501 \text{ bar}$$

5. (C)

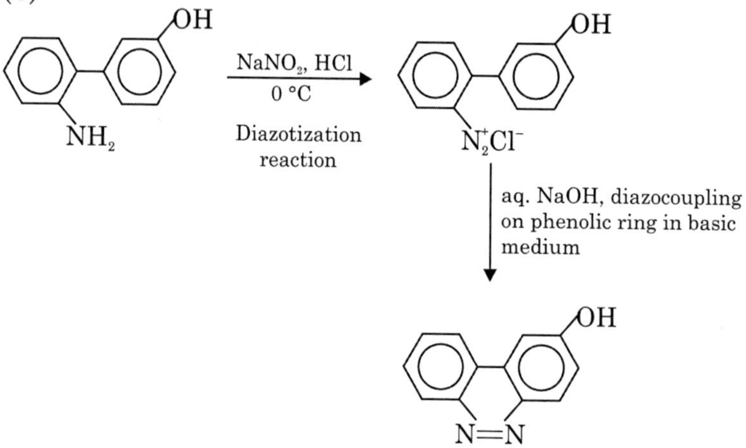

Para is less hindered.

6. (D) (IV > I > II > III)

Basic strength ∝ stability of conjugated acid

∝ + M / + H / + I

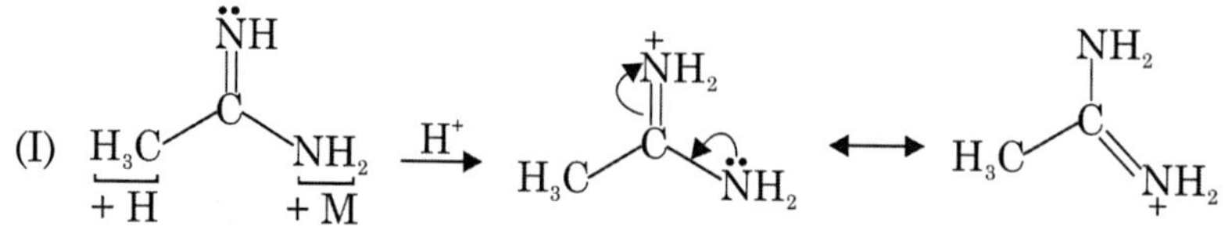

Conjugated acid stabilized by 2 equivalent resonating structures.

(II) 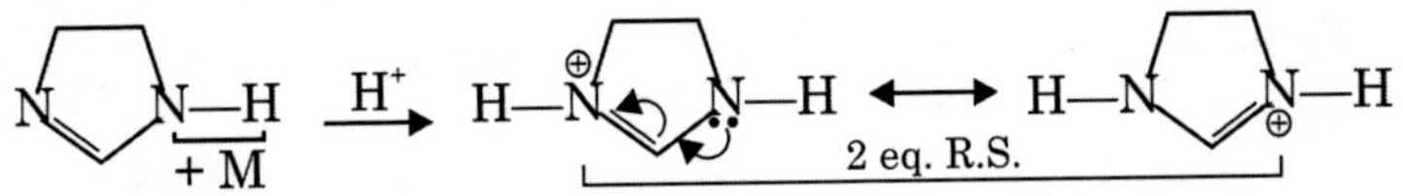

(III) 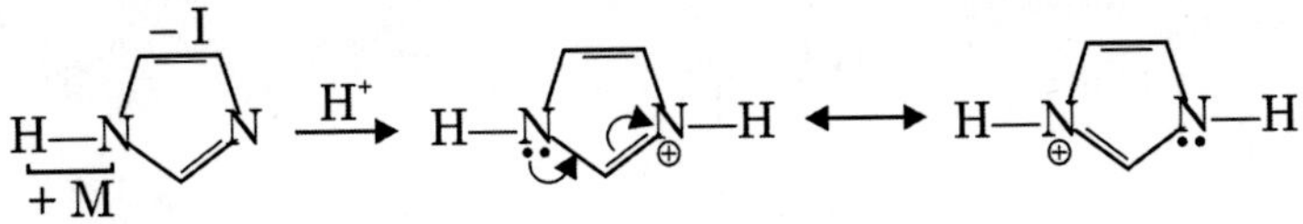

(IV) 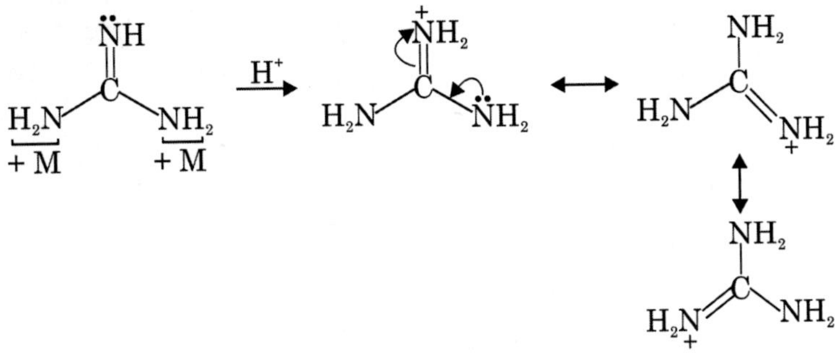

Conjugated acid stabilized by 3 equivalent resonating structures.

7. (B)

$\Delta G = \Delta G^0 + 2.303RT \log Q$

$\Delta G = -nFE^0 + 2.303RT \log Q$

Given: $E^0 = 1.1$ V and $n = 2$

$$\Delta G = (-2 \times 1.1 \times F) + 2.303RT \log \frac{[Zn^{2+}]}{[Cu^{2+}]}$$

$\Delta G = -2.2\ F + 2.303RT$

8. (B, D)

$K = P.\ A.\ e^{-E_a/RT}$

(A) If $P < 1$, $A_{arr} > A_{expt}$

(D) If $P > 1$, $A_{arr} < A_{expt}$

(C) If P is very small, then catalyst is required to carry out the reaction at measurable rate.

9. (A, C)

(A) 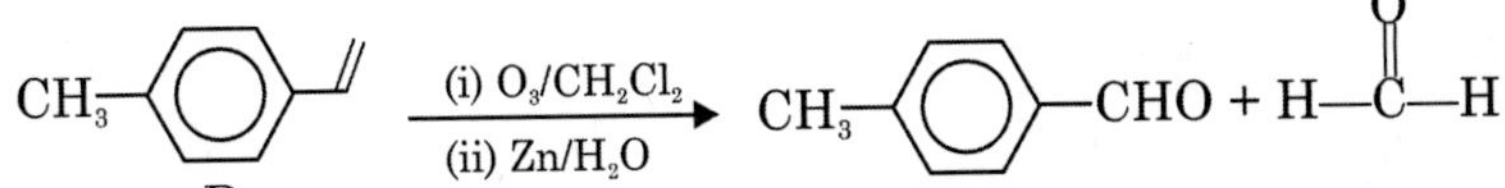

P Q

give Cannizzaro reaction
but no haloform reaction

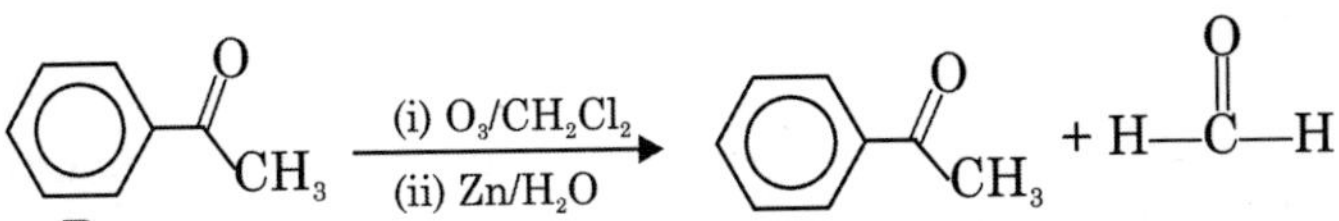

R S

no Cannizzaro reaction
but give haloform reaction

(B) Products of ozonolysis of R are having 9 carbon atoms.

(C) 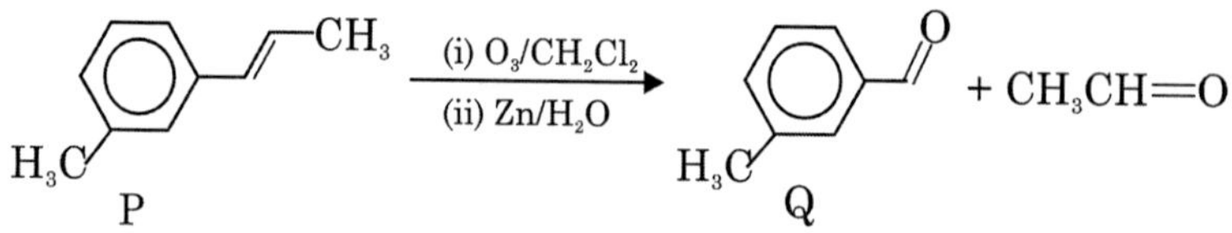

gives Cannizzaro reaction
but no haloform reaction

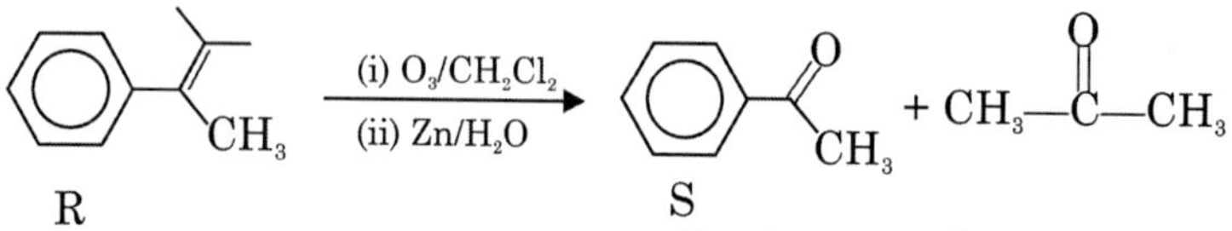

no Cannizzaro reaction
but give haloform reaction

(D) Products of ozonolysis of R are having 9 carbon atoms.

10. (B, C)

$$\Delta S_{\text{surr.}} = \frac{-q_{\text{process}}}{T_{\text{surr.}}}$$

If $\Delta H > 0$ on $T \uparrow K_{eq} \uparrow$, $\Delta S_{\text{surr.}} < 0$ (Surrounding is unfavourable.)
If $\Delta H < 0$ on $T \uparrow K_{eq} \downarrow$, $\Delta S_{\text{surr.}} > 0$ (Surrounding is favourable.)

11. (C, D)

(C) Cr_2O_3, BeO, SnO, SnO_2
All are amphoteric oxides.

(D) ZnO, Al_2O_3, PbO, PbO_2
All are amphoteric oxides.

12. (A, C, D)

(A) 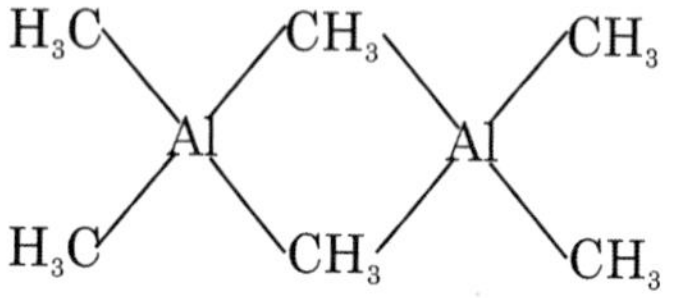

Two 3c – $2e^-$ bond
(three-centre-two-electron bond)

(B) 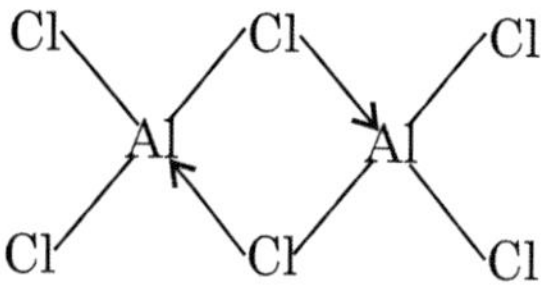

Two 3c – $4e^-$ bond
(three-centre-four-electron bond)

(C) 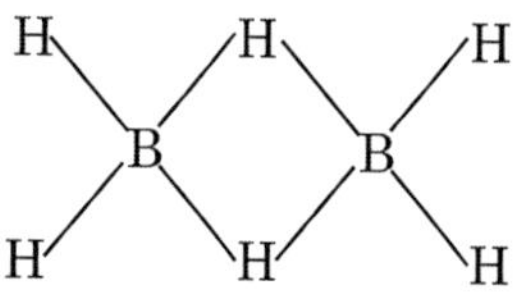

Two 3c – $2e^-$ bond
(three-centre-two-electron bond)

(D) Lewis acidic strength decreases down the group. The decrease in acid strength occurs because as size increases, the attraction between the incoming electron pair and the nucleus weakens. Hence, Lewis acidic strength of BCl_3 is more than that of $AlCl_3$.

13. (B, D)
   (A) Emulsion is liquid in liquid type colloid.
   (B) For adsorption, $\Delta H < 0$ and $\Delta S < 0$.
   (C) Smaller the size and less viscous the dispersion medium, more will be the Brownian motion.
   (D) Higher the T_C, greater will be the extent of adsorption.

14. (A, B, C, D)

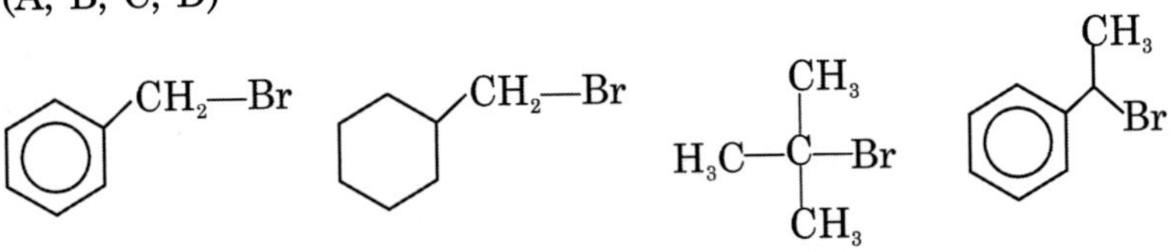

   (A) I and II follow S_N2 mechanism as they are primary
   (B) Reactivity order: IV > I > III
   (C) I and III follow S_N1 mechanism as they form stable carbocation.
   (D) Compound IV undergoes inversion of configuration.

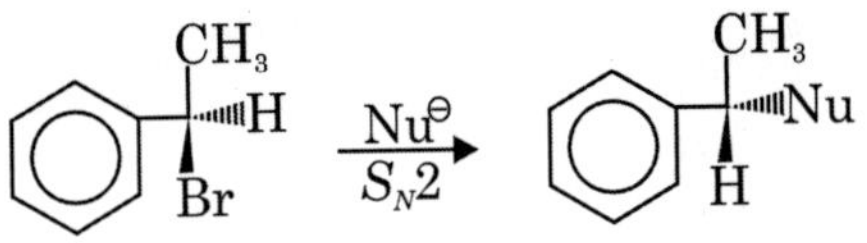

(inverted product)

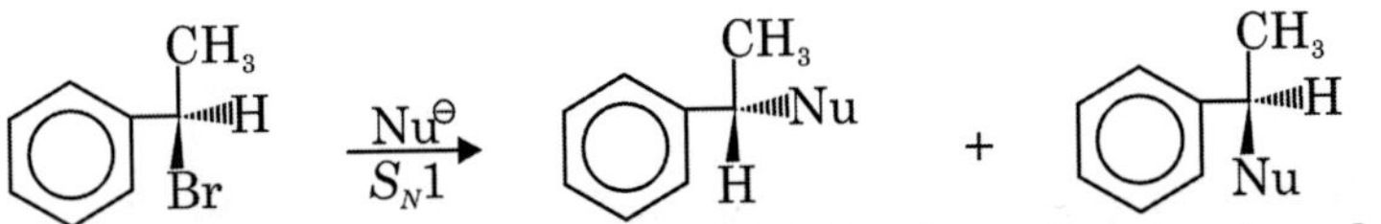

(inverted product) (retained product)

Solutions for paragraph Q. 15 & 16

15. (B)
   W and X are, respectively, W = O_2 and X = P_4O_{10}

16. (C)
   Y and Z are, respectively, Y = N_2O_5, Z = HPO_3

$$2KClO_3 \xrightarrow[\Delta]{MnO_2} 2KCl + \underset{(W)}{3O_2}$$

$$\underset{\text{white phosphorus}}{P_4} + \underset{\substack{\text{(excess)} \\ (W)}}{5O_2} \rightarrow \underset{(X)}{P_4O_{10}}$$

$$\underset{(X)}{P_4O_{10}} + 4HNO_3 \rightarrow \underset{(Y)}{2N_2O_5} + \underset{(Z)}{(HPO_3)_4}$$

Solutions for paragraph Q. 17 & 18

17. (B)

The reaction Q to R is Friedel–Crafts alkylation and hydration.

The reaction R to S is Fiedel–Crafts alkylation.

18. (D)

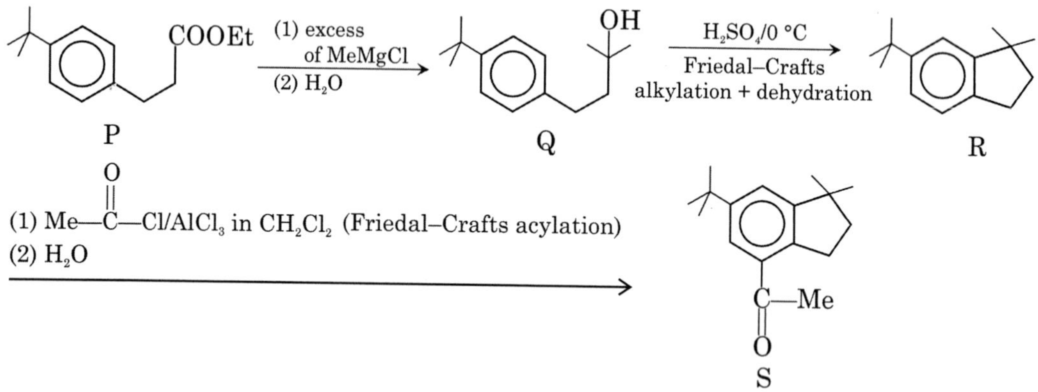

I.I.T. JEE Objective Question Paper (Main)—2018

IMPORTANT INSTRUCTIONS

1. Immediately fill in the particulars on this page of the Test Booklet with only Black Ball Point Pen provided in the examination hall.
2. The Answer Sheet is kept inside this Test Booklet. When you are directed to open the Test Booklet, take out the Answer Sheet and fill in the particulars carefully.
3. The test is of 3 hours duration.
4. The Test Booklet consists of 90 questions. The maximum marks are 360.
5. There are three parts in the question paper A, B, C consisting of Chemistry, Physics and Mathematics having 30 questions in each part of equal weightage. Each question is allotted 4 (four) marks for each correct response.
6. Candidates will be awarded marks as stated above in instruction No. 5 for correct response of each question. 1/4 (one-fourth) marks of the total marks allotted to the question (i.e. 1 mark) will be deducted for indicating incorrect response of each question. No deduction from that total score will be made if no response is indicated for an item in the Answer Sheet.
7. There is only one correct response for each question. Filling up more than one response in any question will be treated as wrong response and marks for wrong response will be deducted accordingly as per instruction No. 6 above.
8. For writing particulars/marking responses on Side-1 and Side-2 of the Answer Sheet use only Black Ball Point Pen provided in the examination hall.
9. No candidate is allowed to carry any textual material, printed or written, bits of papers, pager, mobile phone, any electronic device, etc. except the Admit Card inside the examination room/hall.
10. Rough work is to be done on the space provided for this purpose in the Test Booklet only. This space is given at the bottom of each page and in four pages at the end of the booklet.
11. On completion of the test, the candidate must hand over the Answer Sheet to the Invigilator on duty in the room/hall. However, the candidates are allowed to take away this Test Booklet with them.
12. The CODE for this Booklet is B. Make sure that the CODE printed on Side-2 of the Answer Sheet also tally the serial number of the Test Booklet and Answer Sheet is the same as that on this Booklet. In case of discrepancy, the candidate should immediately report the matter to the Invigilator for replacement of both the Test Booklet and the Answer Sheet.
13. Do not fold or make any stray marks on the Answer Sheet.

1. Total number of lone pair of electron in I_3^- ion is
   (A) 9 (B) 12 (C) 3 (D) 6
2. Which of the following salts is the most basic in aqueous solution?
   (A) $FeCl_3$ (B) $Pb(CH_3COO)_2$ (C) $Al(CN)_3$ (D) CH_3COOK
3. Phenol reacts with methyl chloroformate in the presence of NaOH to form product A. A reacts with Br_2 to form product B. A and B are respectively

   (A) 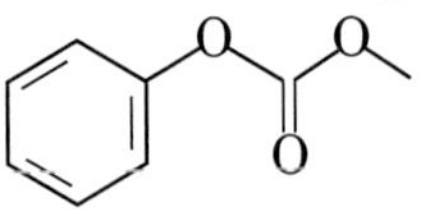 and 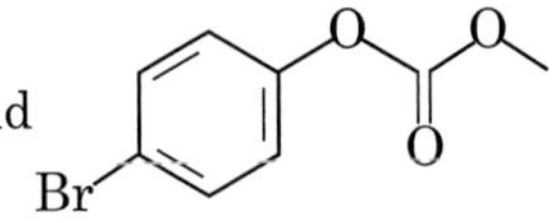

   (B) 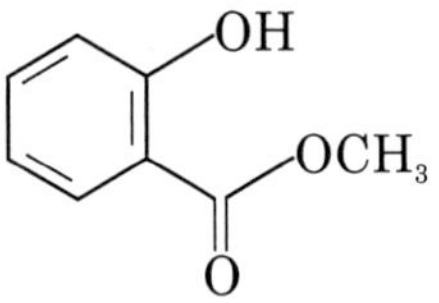 and 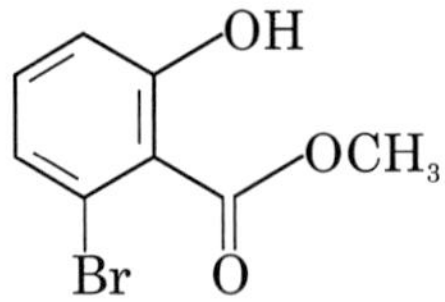

   (C) 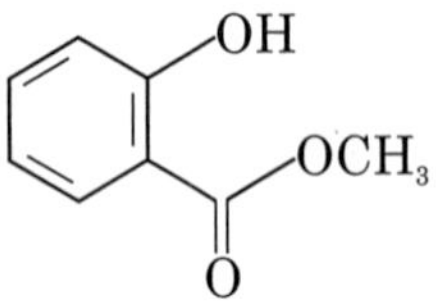 and 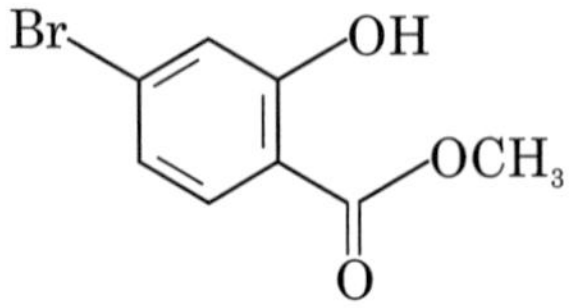

   (D) 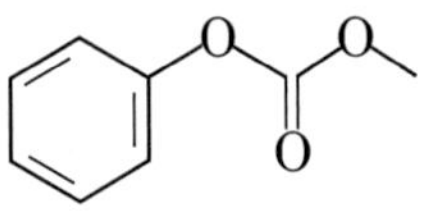 and 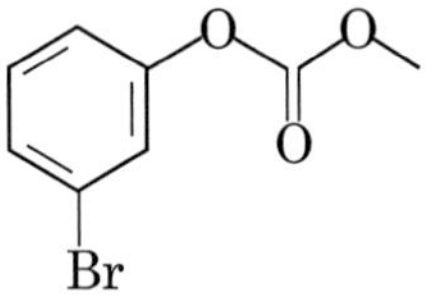

4. The increasing order of basicity of the following compounds is

   (a) 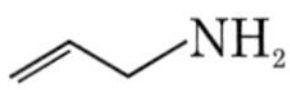 (b) 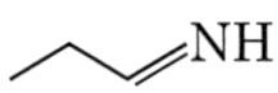 (c) 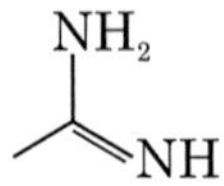 (d) 

   (A) (b) < (a) < (d) < (c) (B) (d) < (b) < (a) < (c)
   (C) (a) < (b) < (c) < (d) (D) (b) < (a) < (c) < (d)
5. An alkali is titrated against an acid with methyl orange as indicator. Which of the following is a correct combination?

	Base	Acid	End Point
(A)	Weak	Strong	Yellow to pinkish red
(B)	Strong	Strong	Pink to colourless
(C)	Weak	Strong	Colourless to pink
(D)	Strong	Strong	Pinkish red to yellow

6. The trans-alkenes are formed by the reduction of alkynes with
   (A) Na/liq.NH_3 (B) Sn–HCl (C) H_2–Pd/C, $BaSO_4$ (D) $NaBH_4$

7. The ratio of mass percent of C and H of an organic compound ($C_XH_YO_Z$) is 6 : 1. If one molecule of the above compound ($C_XH_YO_Z$) contains half as much oxygen as required to burn one molecule of compound C_XH_Y completely to CO_2 and H_2O. The empirical formula of compound $C_XH_YO_Z$ is

(A) $C_3H_4O_2$ (B) $C_2H_4O_3$ (C) $C_3H_6O_3$ (D) C_2H_4O

8. Hydrogen peroxide oxidises $[Fe(CN)_6]^{4-}$ to $[Fe(CN)_6]^{3-}$ in acidic medium but reduces $[Fe(CN)_6]^{3-}$ to $[Fe(CN)_6]^{4-}$ in alkaline medium. The other products formed are, respectively

(A) H_2O and ($H_2O + O_2$) (B) H_2O and ($H_2O + OH^-$)
(C) ($H_2O + O_2$) and H_2O (D) ($H_2O + O_2$) and ($H_2O + OH^-$)

9. The major product formed in the following reaction is

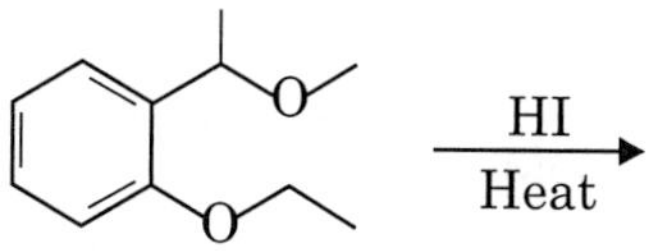

(A) 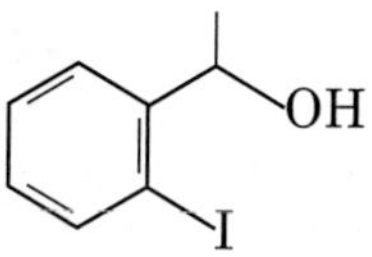

(B) 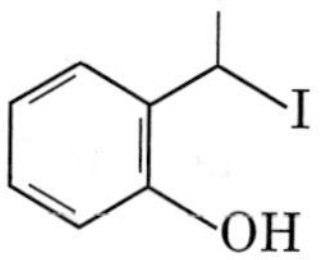

(C) 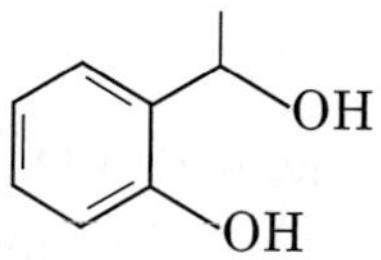

(D) 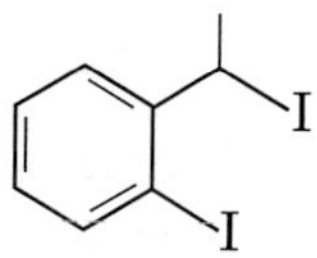

10. How long (approximate) should water be electrolysed by passing through 100 amperes current so that the oxygen released can completely burn 27.66 g of diborane? (Atomic weight of B = 10.8 u)

(A) 3.2 hours (B) 1.6 hours (C) 6.4 hours (D) 0.8 hours

11. Which of the following lines correctly show the temperature dependence of equilibrium constant *K*, for an exothermic reaction?

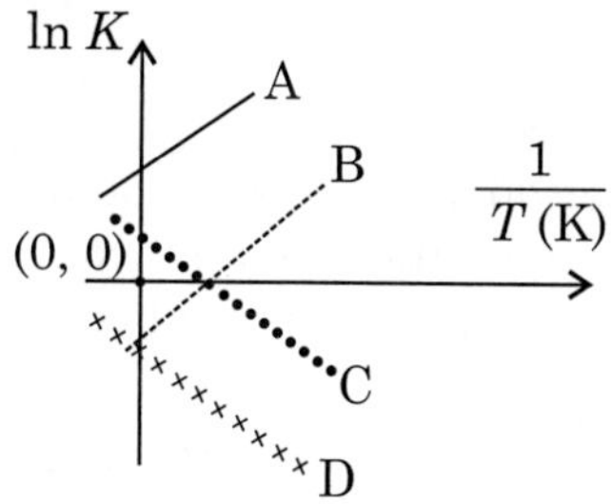

(A) C and D (B) A and D (C) A and B (D) B and C

12. At 518°C, the rate of decomposition of a sample of gaseous acetaldehyde, initially at a pressure of 363 Torr, was 1.00 Torr s^{-1} when 5% had reacted and 0.5 Torr s^{-1} when 33% had reacted. The order of the reaction is

(A) 1 (B) 0 (C) 2 (D) 3

13. Glucose on prolonged heating with HI gives

(A) Hexanoic acid (B) 6-iodohexanal (C) *n*-Hexane (D) 1-Hexene

14. Consider the following reaction and statements:

$[Co(NH_3)_4Br_2]^+ + Br^- \rightarrow [Co(NH_3)_3Br_3] + NH_3$

(I) Two isomers are produced if the reactant complex ion is a *cis*-isomer

(II) Two isomers are produced if the reactant complex ion is a *trans*-isomer

(III) Only one isomer is produced if the reactant complex ion is a *trans*-isomer

(IV) Only one isomer is produced if the reactant complex ion is a *cis*-isomer.

The correct statements are:

(A) (III) and (IV) (B) (II) and (IV) (C) (I) and (II) (D) (I) and (III)

15. The major product of the following reaction is

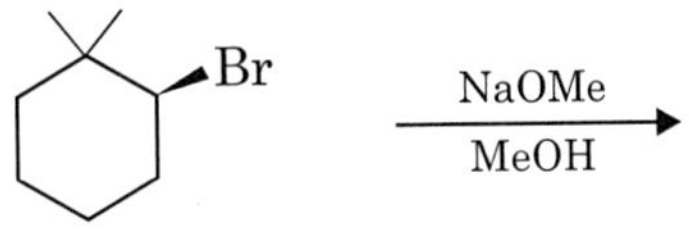

(A) 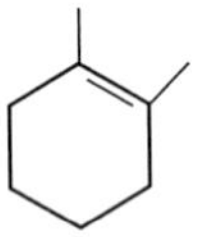 (B) 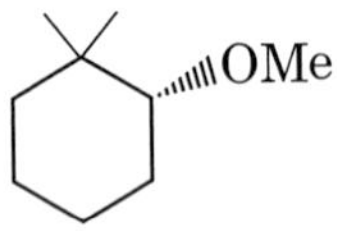

(C) 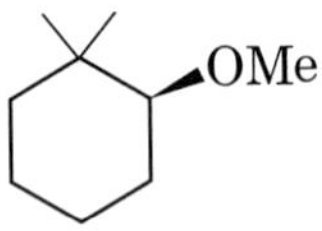

(D) 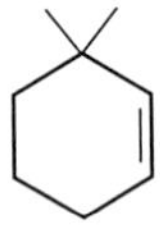

16. Phenol on treatment with CO_2 in the presence of NaOH followed by acidification produces compound X as the major product. X on treatment with $(CH_3CO)_2O$ in the presence of catalytic amount of H_2SO_4 produces

(A) 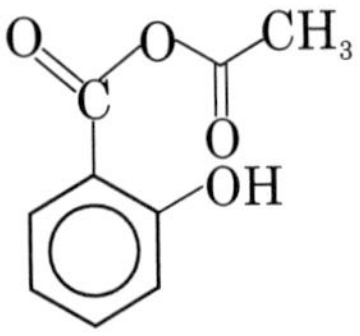

(B) 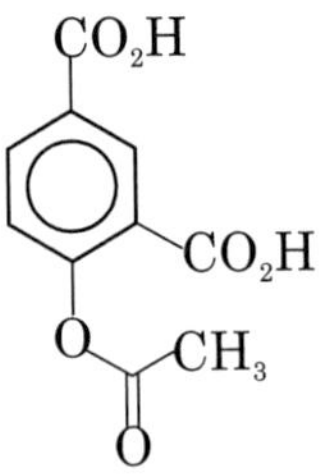

(C) 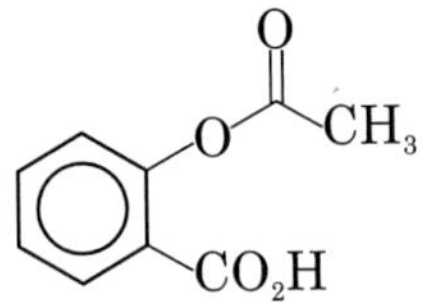

(D) 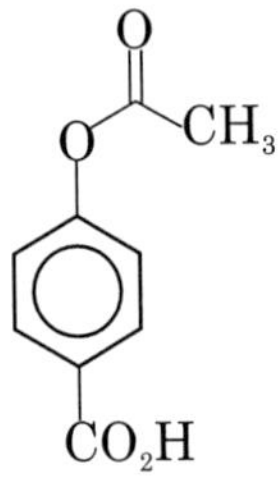

17. An aqueous solution contains an unknown concentration of Ba^{2+}. When 50 mL of a 1*M* solution of Na_2SO_4 is added, $BaSO_4$ just begins to precipitate. The final volume is 500 mL. The solubility product of $BaSO_4$ is 1×10^{-10}. What is the original concentration of Ba^{2+}?

(A) $1.1 \times 10^{-9}M$ (B) $1.0 \times 10^{-10}M$ (C) $5 \times 10^{-9}M$ (D) $2 \times 10^{-9}M$

18. Which of the following compounds will be suitable for Kjeldahl's method for nitrogen estimation?

(A) 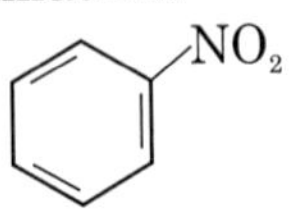

(B) 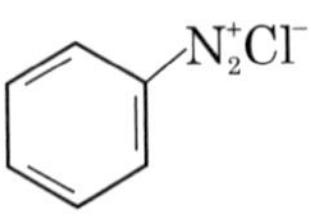

(C) 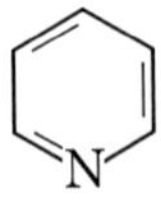

(D) 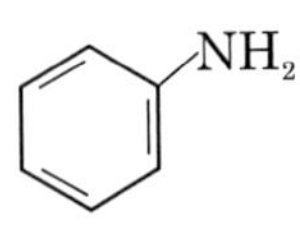

19. When metal M is treated with NaOH, a white gelatinous precipitate X is obtained, which is soluble in excess of NaOH. Compound X when heated strongly gives an oxide which is used in chromatography as an adsorbent. The metal M is

(A) Al (B) Fe (C) Zn (D) Ca

20. An aqueous solution contains 0.10*M* H_2S and 0.20*M* HCl. If the equilibrium constants for the formation of HS^- from H_2S is 1.0×10^{-7} and that of S^{2-} from HS^- ions is 1.2×10^{-13}, then the concentration of S^{2-} ions in aqueous solution is

    (A) $6 \times 10^{-21}M$ (B) $5 \times 10^{-19}M$ (C) $5 \times 10^{-8}M$ (D) $3 \times 10^{-20}M$

21. The recommended concentration of fluoride ion in drinking water is up to 1 ppm as fluoride ion is required to make teeth enamel harder by converting $[3Ca_3(PO_4)_2 \cdot Ca(OH)_2]$ to

    (A) $[3Ca_3(PO_4)_2 \cdot CaF_2]$ (B) $[3\{Ca(OH)_2\} \cdot CaF_2]$
    (C) $[CaF_2]$ (D) $[3(CaF_2) \cdot Ca(OH)_2]$

22. The compound that does not produce nitrogen gas by the thermal decomposition is

    (A) NH_4NO_2 (B) $(NH_4)_2SO_4$ (C) $Ba(N_3)_2$ (D) $(NH_4)_2Cr_2O_7$

23. The predominant form of histamine present in human blood is (pK_a, histidine = 6.0)

    (A) 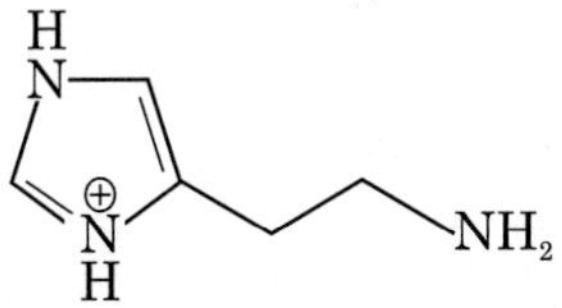
    (B) 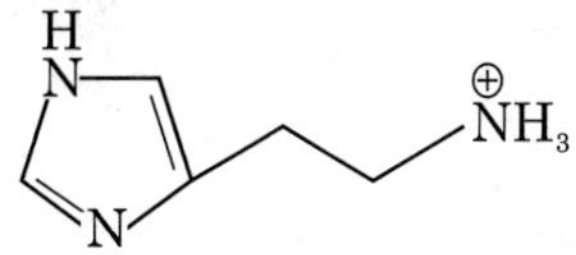
    (C) 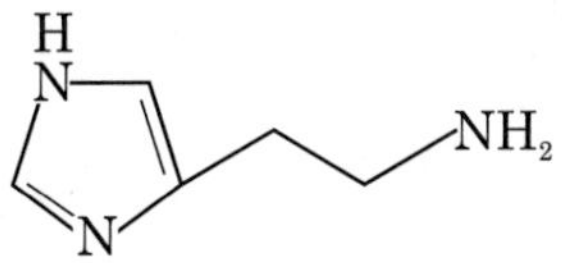
    (D) 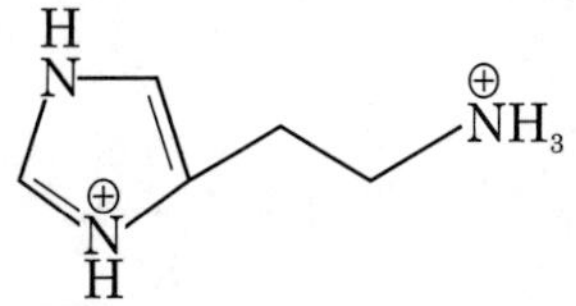

24. The oxidation states of Cr in $[Cr(H_2O)_6]Cl_3$, $[Cr(C_6H_6)_2]$, and $K_2[Cr(CN)_2(O)_2(O_2)(NH_3)]$ respectively are

    (A) + 3, 0, and + 6 (B) + 3, 0, and + 4
    (C) + 3, + 4, and + 6 (D) + 3, + 2 and + 4

25. Which type of 'defect' has the presence of cations in the interstitial sites?

    (A) Frenkel defect (B) Metal deficiency defect
    (C) Schottky defect (D) Vacancy defect.

26. The combustion of benzene (ℓ) gives $CO_2(g)$ and $H_2O(\ell)$. Given that heat of combustion of benzene at constant volume is – 3263.9 kJ mol^{-1} at 25°C. Heat of combustion (in kJ mol^{-1}) of benzene at constant pressure will be (R = 8.314 J K^{-1} mol^{-1})

    (A) 3260 (B) – 3267.6 (C) 4152.6 (D) – 452.46

27. Which of the following are Lewis acids?

    (A) PH_3 and $SiCl_4$ (B) BCl_3 and $AlCl_3$ (C) PH_3 and BCl_3 (D) $AlCl_3$ and $SiCl_4$

28. Which of the following compounds contain(s) no covalent bond(s)?
    KCl, PH_3, O_2, B_2H_6, H_2SO_4

    (A) KCl (B) KCl, B_2H_6 (C) KCl, B_2H_6, PH_3 (D) KCl, H_2SO_4

29. For 1 molal aqueous solution of the following compounds, which one will show the highest freezing point?

(A) $[Co(H_2O)_4Cl_2]Cl\cdot 2H_2O$ (B) $[Co(H_2O)_3Cl_3]\cdot 3H_2O$
(C) $[Co(H_2O)_6]Cl_3$ (D) $[Co(H_2O)_5Cl]Cl_2\cdot H_2O$

30. According to molecular orbital theory, which of the following will not be a viable molecule?
(A) H_2^- (B) H_2^{2-} (C) He_2^{2+} (D) He_2^+

Answers

1. A	2. D	3. A	4. A	5. A	6. A
7. B	8. A	9. B	10. A	11. C	12. C
13. C	14. D	15. D	16. C	17. A	18. D
19. A	20. D	21. A	22. B	23. B	24. A
25. A	26. B	27. B	28. A	29. B	30. B

Model Solutions—I.I.T. JEE (Main) 2018

1. (A)

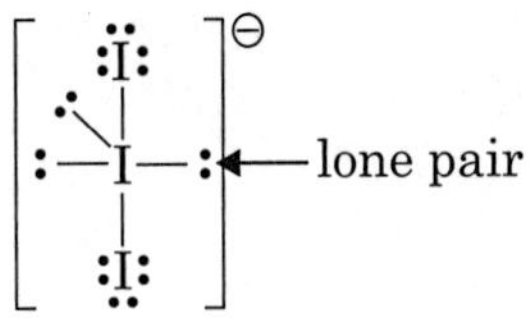

2. (D)

CH_3COOK is most basic among the given options.

3. (A)

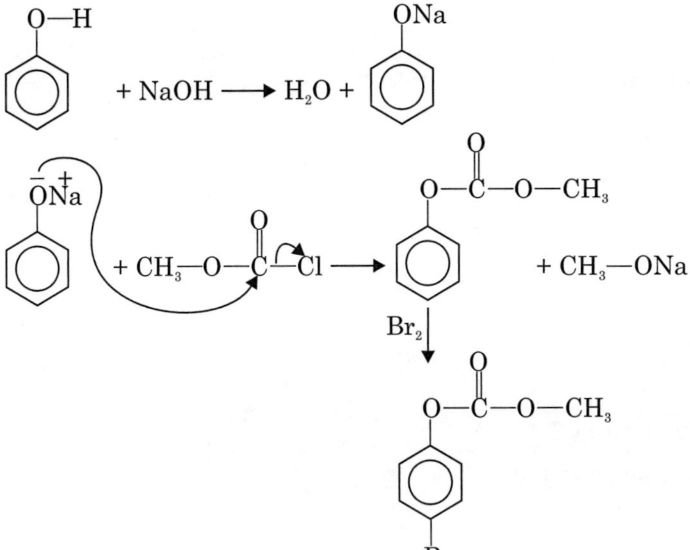

4. (A)

Basic Nature: (b) < (a) < (d) < (c)

5. (A)

When a strong acid is titrated against a weak base using methyl orange as an indicator, the end point is yellow to pinkish red.

6. (A)

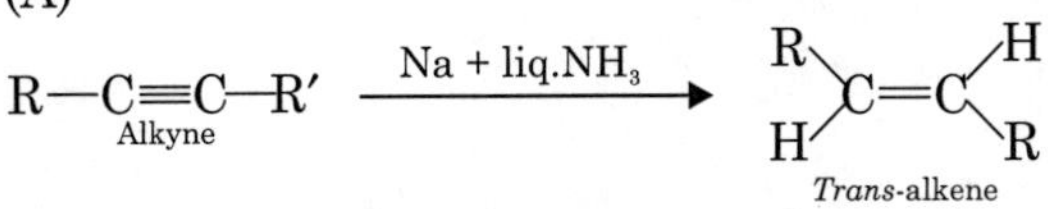

7. (B)

$C_XH_YO_Z$

$$\frac{w_C}{w_H} = \frac{12X}{Y} = \frac{6}{1} \quad \text{or} \quad \frac{X}{Y} = \frac{1}{2}$$

$$C_XH_Y + \left(X + \frac{Y}{4}\right)O_2 \rightarrow XCO_2 + \frac{Y}{2}H_2O$$

X : Y : Z = 2 : 4 : 3

8. (A)

$$\underset{(OA)}{H_2\overset{-1}{O_2}} + [Fe(CN)_6]^{4-} \xrightarrow{H^+} [Fe(CN)_6]^{3-} + H_2\overset{-2}{O}$$

$$\underset{(RA)}{H_2\overset{-1}{O_2}} + [Fe(CN)_6]^{3-} \longrightarrow [Fe(CN)_6]^{4-} + \overset{0}{O_2}$$

9. (B)

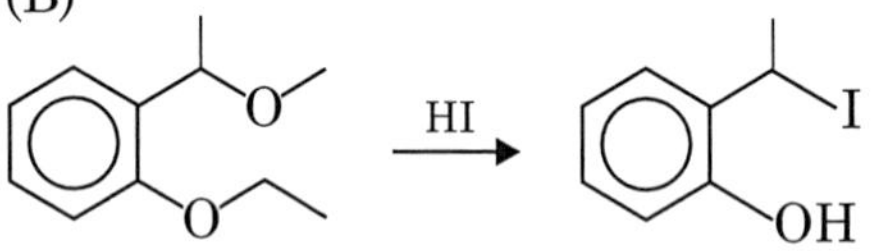

10. (A)

$$B_2H_6 + 3O_2 \longrightarrow B_2O_3 + 3H_2O$$

$$\frac{27.6}{27.6} = 3 \text{ mol}$$

1 mol

$$\frac{It}{F} = \frac{\theta}{F} = \frac{w}{E} = n_{O_2} \times 4$$

$t = 965 \times 12$ sec = 3.2 hr

11. (C)

$$\log_{10} K_{eq} = \text{constant} - \left(\frac{\Delta H}{2.303R}\right)\left(\frac{1}{T}\right)$$

Given: = ΔH = –ve i.e. slope = + ve

12. (C)

Rate of a reaction = $k[P]^n$

$$R_1 = 1 \text{ torr/sec} = k\left[365 \times \frac{95}{100}\right]^n$$

$$R_2 = 0.5 \text{ torr/sec} = k\left[365 \times \frac{67}{100}\right]^n$$

$$\frac{1}{0.5} = \left(\frac{95}{67}\right)^n$$

$$2 = \left(\frac{96}{67}\right)^n = (1.43)^n \quad \text{or} \quad n = 2$$

13. (C)

$$\text{Glucose} + \text{HI} \xrightarrow{\Delta} n\text{-Hexane}$$

14. (D)

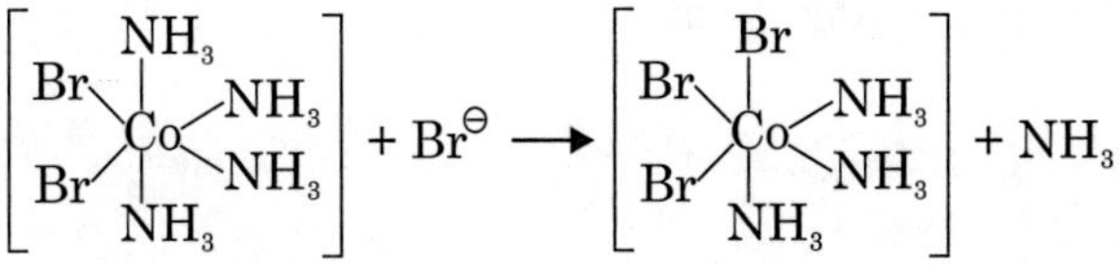

cis

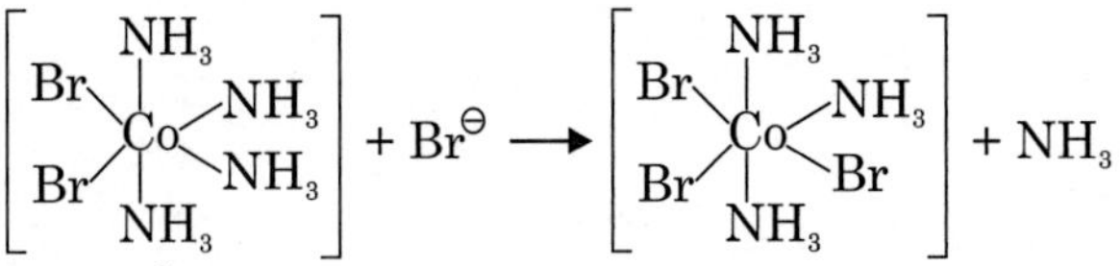

cis

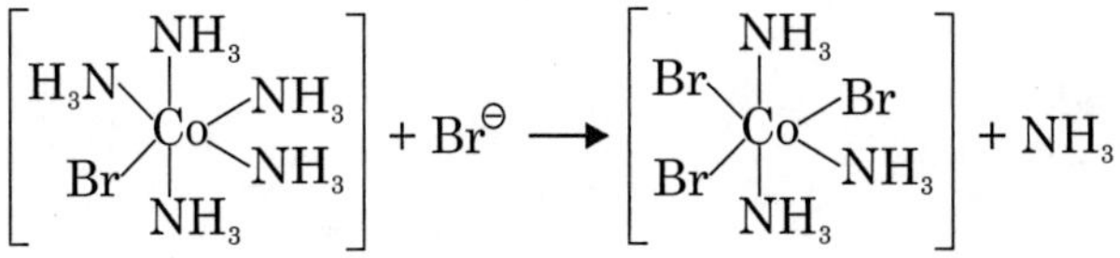

trans one product

15. (D)

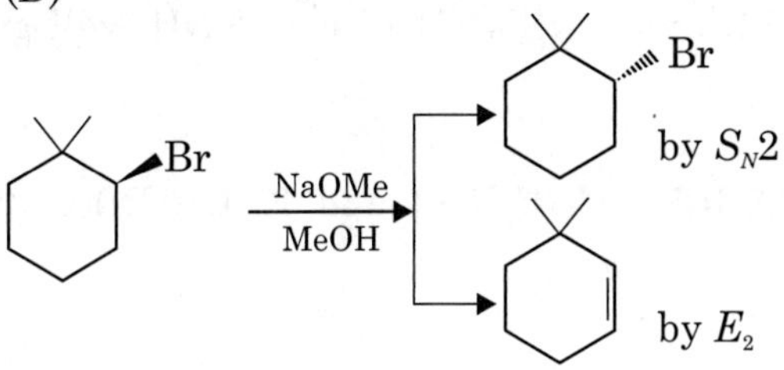

16. (C)

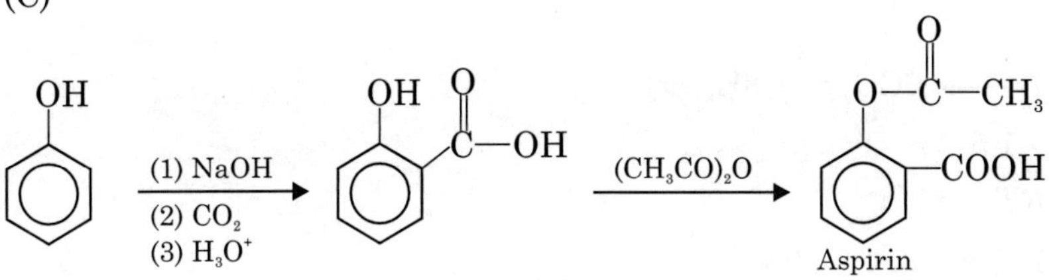

17. (A)

$$\left(\frac{MV}{V+50}\right)\left(\frac{1\times 50}{V+50}\right) = 10^{-10} \quad \text{...(1)}$$

$V = 450$ mL as total volume = 500 mL

$$\left(\frac{\text{M}\times 450}{500}\right)\left(\frac{50}{100}\right) = 10^{-10}$$

$$\text{M} = \frac{50}{45}\times 10\times 10^{-10} = \frac{50}{45}\times 10^{-9} = 1.1\times 10^{-9}\,M$$

18. (D)

A compound containing primary amine N atom is suitable for estimation of nitrogen by Kjeldahl's method.

19. (A)
Alumina is used in chromatography as an adsorbent.

20. (D)
$K_1 = 10^{-7}$, $K_2 = 1.2 \times 10^{-13}$

$$K_1 K_2 = \frac{[H^+]^2\,[S^{2-}]}{[H_2S]}$$

$$10^{-7} \times 1.2 \times 10^{-13} = \frac{(0.2)^2\,(S^{2-})}{(0.1)}$$

$$[S^{2-}] = \frac{1.2 \times 10^{-13} \times 10^{-7} \times 0.1}{4 \times 10^{-2}} = \frac{12}{4} \times 10^{-20} = 3 \times 10^{-20} M$$

21. (A)
Solid $[3Ca_3(PO_4)_2 \cdot Ca(F_2)]$ is formed.

22. (B)
$(NH_4)_2SO_4 \longrightarrow NH_3\uparrow$

23. (B)
Blood is slightly basic in nature (pH 7.35) and at this pH terminal NH_2 will get + charge due to more basic nature.

24. (A)
The oxidation states of Cr in $[Cr(H_2O)_6]Cl_3$, $[Cr(C_6H_6)_2]$ and $K_2[Cr(CN)(O_2)(O_2)]$ respectively are + 3, 0 and + 6.

25. (A)
In Frenkel defect, cations are present in interstitial sites.

26. (B)
$C_6H_6(\ell) + 7.5O_2(g) \longrightarrow 6CO_2(g) + 3H_2O(\ell)$

$\Delta n_g = 6 - 7.5 = -1.5$

$\Delta H = \Delta U + (\Delta_{ng})RT$

$$= -3263.9 + \frac{(-1.5) \times 8.314 \times 298}{1000} = -3267.6 \text{ kJ / mol}^{-1}$$

27. (B)
In both BCl_3 and $AlCl_3$, the octet is not fulfilled and hence they behave as Lewis acid.

28. (A)
KCl contains ionic bond.

29. (B)
1 molal aqueous solution of the non-electrolytic molecule will exhibit the highest freezing point.

30. (B)
Bond order of $H_2^{2-} = 0$
Hence, according to the molecular orbital theory, H_2^{2-} is not a viable molecule.

I.I.T. JEE Objective Question Paper (Advanced)—2018

Paper–1

INSTRUCTIONS

- This question paper has three (3) parts: PART-I: Physics, PART-II: Chemistry and PART-III: Mathematics.
- Each part has total of eighteen (18) questions divided into three (3) sections (Section-1, Section-2 and Section-3).
- Total number of questions in Paper-1: Fifty-four (54).
- Paper-1 Maximum Marks: One Hundred Eighty (180).

Instructions for Section-1: Questions and Marking Scheme

SECTION–1 (Maximum Marks: 24)

- This section contains **SIX (6)** questions.
- Each question has **FOUR options** for correct answer(s). **ONE OR MORE THAN ONE** of these four option(s) is (are) correct option(s).
- For each question, choose the correct option(s) to answer the question.
- Answer to each question will be evaluated according to the following marking scheme:

Full Marks	: + 4	If only (all) the correct option(s) is (are) chosen.
Partial Marks	: + 3	If all the four options are correct but ONLY three options are chosen.
Partial Marks	: + 2	If three or more options are correct but ONLY two options are chosen, both of which are correct options.
Partial Marks	: + 1	If two or more options are correct but ONLY one option is chosen and it is a correct option.
Zero Marks	: 0	If none of the options is chosen (i.e. the question is unanswered).
Negative Marks	: – 2	In all other cases.

For Example: If first, third and fourth are the ONLY three correct options for a question with second option being an incorrect option; selecting only two of the three correct options (e.g., the first and fourth options), without selecting any incorrect option (second option in this case), will result in + 2 marks. Selecting only one of the three correct options (either first or third or fourth option), without selecting any incorrect option (second option in this case), will result in + 1 marks. Selecting any incorrect option(s) (second option in this case), with or without selection of any correct option(s) will result in – 2 marks.

Answering Section-1 Questions:

- To select the option(s), **using the mouse click** on the corresponding button(s) of the option(s).
- To deselect chosen option(s), click on the button(s) of the chosen options(s) again or click on the **Clear Response** button to clear all the chosen options.
- To change the option(s) of a previously answered question, if required, first click on the **Clear Response** button to clear all the chosen options and then select the new option(s).
- To mark a question ONLY for review (i.e. without answering it), click on the **Mark for Review & Next button**.
- To mark a question for review (after answering it), click on **Mark for Review & Next button** – answered question which is also marked for review will be evaluated.
- To save the answer, click on the **Save & Next button** – the answered question will be evaluated.

SECTION–2: (Maximum Marks: 44)

- This section contains **ELEVEN (11)** questions.
- Each question has **FOUR options** for correct answer(s). **ONE OR MORE THAN ONE** of these four option(s) is (are) correct option(s).
- For each question, choose the correct option(s) to answer the question.
- Answer to each question will be evaluated according to the following marking scheme.

Full Marks	: + 4	If only (all) the correct option(s) is (are) chosen.
Partial Marks	: + 3	If all the four options are correct but ONLY three options are chosen.
Partial Marks	: + 2	If three or more options are correct but ONLY two options are chosen, both of which are correct options.
Partial Marks	: + 1	If two or more options are correct but ONLY one option is chosen and it is a correct option.
Zero Marks	: 0	If none of the options is chosen (i.e. the question is unanswered).
Negative Marks	: – 2	In all other cases.

For Example: If first, third and fourth are the ONLY three correct options for a question with second option being an incorrect option; selecting only two of the three correct options (e.g., the first and fourth options), without selecting any incorrect option (second option in this case), will result in + 2 marks. Selecting only one of the three correct options (either first or third or fourth option), without selecting any incorrect option (second option in this case), will result in + 1 marks. Selecting any incorrect option(s) (second option in this case), with or without selection of any correct option(s) will result in – 2 marks.

1. The compound(s) which generate(s) N_2 gas upon thermal decomposition below 300°C is (are)

   (A) NH_4NO_3 (B) $(NH_4)_2Cr_2O_7$ (C) $Ba(N_3)_2$ (D) Mg_3N_2

2. The correct statement(s) regarding the binary transition metal carbonyl compounds is (are)

   (Atomic numbers: Fe = 26, Ni = 28)

   (A) Total number of valence shell electrons at metal centre in $Fe(CO)_5$ or $Ni(CO)_4$ is 16

   (B) These are predominantly low spin in nature

   (C) Metal—carbon bond strengthens when the oxidation state of the metal is lowered

   (D) The carbonyl C—O bond weakens when the oxidation state of the metal is increased

3. Based on the compounds of group 15 elements, the correct statement(s) is (are)

   (A) Bi_2O_5 is more basic than N_2O_5

   (B) NF_3 is more covalent than BiF_3

   (C) PH_3 boils at lower temperature than NH_3

   (D) The N—N single bond is stronger than the P—P single bond

4. In the following reaction sequence, the correct structure(s) of X is (are)

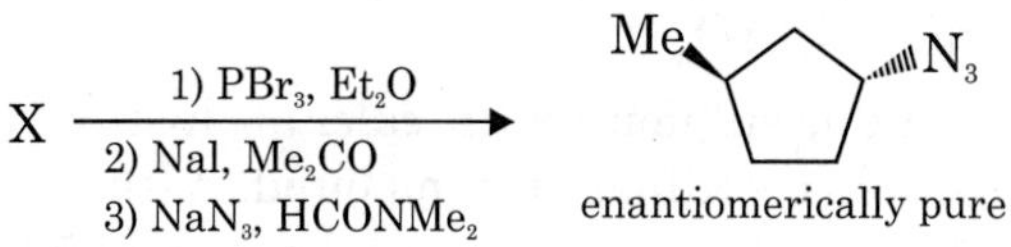

(A) 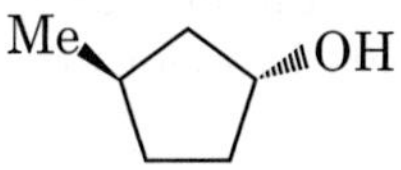

(B) 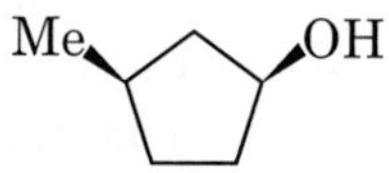

(C) 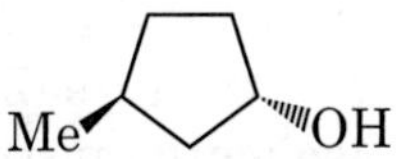

(D) 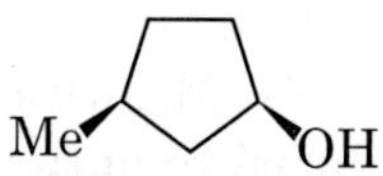

5. The reaction(s) leading to the formation of 1,3,5-trimethylbenzene is (are)

(A) 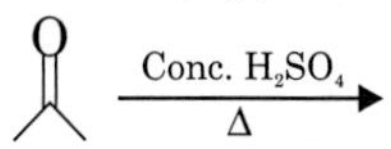

(B) Me—≡—H $\xrightarrow[873\ K]{\text{heated iron tube}}$

(C) 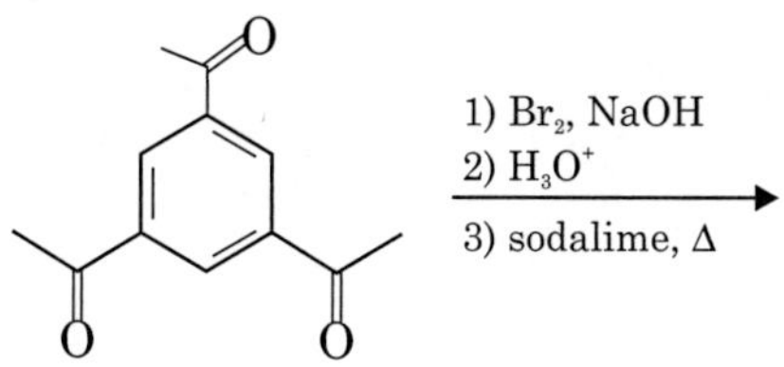

(D) 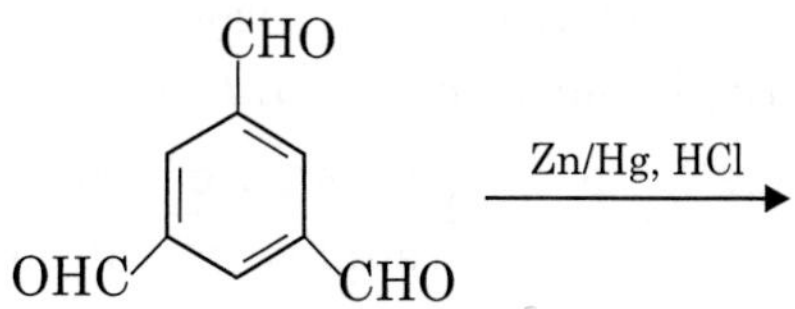

6. A reversible cyclic process for an ideal gas is shown below. Here, *P*, *V*, and *T* are pressure, volume and temperature, respectively. The thermodynamic parameters *q*, *w*, *H* and *U* are heat, work, enthalpy and internal energy, respectively.

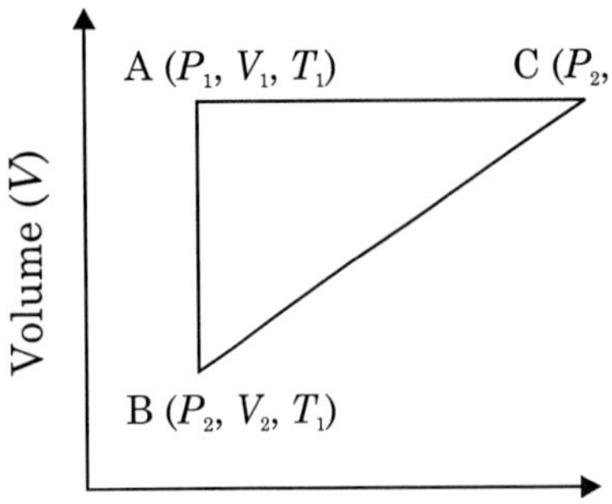

The correct option(s) is (are)

(A) $q_{AC} = \Delta U_{BC}$ and $w_{AB} = P_2(V_2 - V_1)$

(B) $w_{BC} = P_2(V_2 - V_1)$ and $q_{BC} = \Delta H_{AC}$

(C) $\Delta H_{CA} < \Delta U_{CA}$ and $q_{AC} = \Delta U_{BC}$

(D) $q_{BC} = \Delta H_{AC}$ and $\Delta H_{CA} > \Delta U_{CA}$

7. Among the species given below, the total number of diamagnetic species is _____.
H atom, NO_2 monomer, O_2^- (superoxide), dimeric sulphur in vapour phase, Mn_3O_4, $(NH_4)_2[FeCl_4]$, $(NH_4)_2[NiCl_4]$, K_2MnO_4, K_2CrO_4

8. The ammonia prepared by treating ammonium sulphate with calcium hydroxide is completely used by $NiCl_2 \cdot 6H_2O$ to form a stable coordination compound. Assume that both the reactions are 100% complete. If 1584 g of ammonium sulphate and 952 g of $NiCl_2 \cdot 6H_2O$ are used in the preparation, the combined weight (in gram) of gypsum and the nickel-ammonia coordination compound thus produced is ______.
(Atomic weights in g mol^{-1}: H = 1, N = 14, O = 16, S = 32, Cl = 35.5, Ca = 40, Ni = 59)

9. Consider an ionic solid MX with NaCl structure. Construct a new structure (Z) whose unit cell is constructed from the unit cell of MX following the sequential instructions given below. Neglect the charge balance.
   (i) Remove all the anions (X) except the central one
   (ii) Replace all the face centered cations (M) by anions (X)
   (iii) Remove all the corner cations (M)
   (iv) Replace the central anion (X) with cation (M).

   The value of $\left(\dfrac{\text{number of anions}}{\text{number of cations}}\right)$ in Z is ______.

10. For the electrochemical cell,

$$Mg(s) \,|\, Mg^{2+}\,(aq,\ 1M) \,\|\, Cu^{2+}\,(aq,\ 1M) \,|\, Cu(s)$$

the standard emf of the cell is 2.70 V at 300 K. When the concentration of Mg^{2+} is changed to xM, the cell potential changes to 2.67 V at 300 K. The value of x is ____.
[Given, $\dfrac{F}{R} = 11500\ KV^{-1}$, where F is the Faraday constant and R is the gas constant, ln(10) = 2.30.]

11. A closed tank has two compartments A and B, both filled with oxygen (assumed to be ideal gas). The partition separating the two compartments is fixed and is a perfect heat insulator (Figure 1). If the old partition is replaced by a new partition which can slide and conduct heat but does NOT allow the gas to leak across (Figure 2), the volume (in m^3) of the compartment A after the system attains equilibrium is ____.

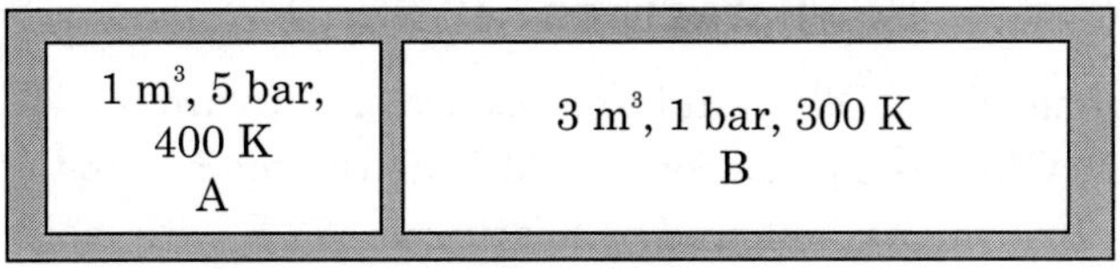

Figure 1

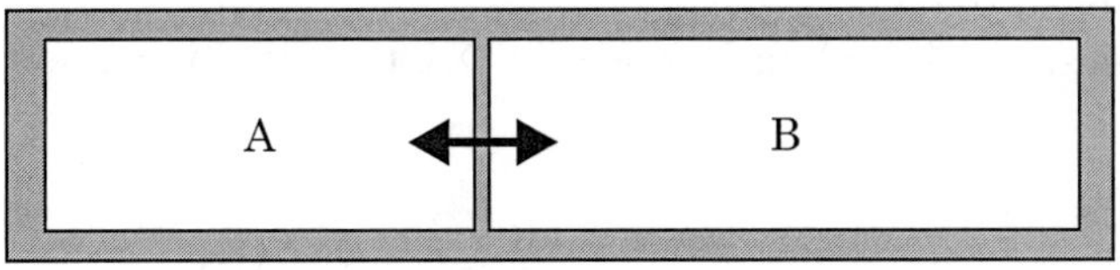

Figure 2

12. Liquids A and B form ideal solution over the entire range of composition. At temperature T, equimolar binary solution of liquids A and B has vapour pressure 45 Torr. At the same temperature, a new solution of A and B having mole fractions x_A and x_B, respectively, has vapour pressure of 22.5 Torr. The value of x_A/x_B in the new solution is _____.
(Given that the vapour pressure of pure liquid A is 20 Torr at temperature T)

13. The solubility of a salt of weak acid (AB) at pH 3 is Y $\times 10^{-3}$ mol L^{-1}. The value of Y is _____.
(Given that the value of solubility product of AB (K_{sp}) $= 2 \times 10^{-10}$ and the value of ionization constant of HB (K_a) $= 1 \times 10^{-8}$)

14. The plot given below shows $P - T$ curves (where P is the pressure and T is the temperature) for two solvents X and Y and isomolal solutions of NaCl in these solvents. NaCl completely dissociates in both the solvents.

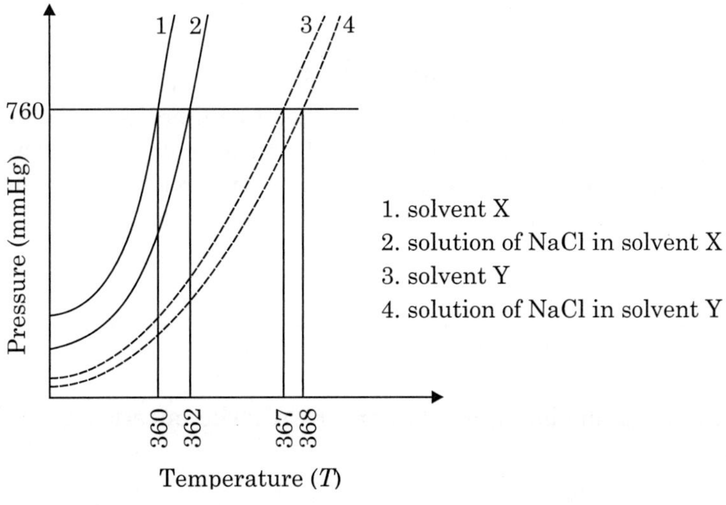

On addition of equal number of moles of a non-volatile solute S in equal amount (in kg) of these solvents, the elevation of boiling point of solvent X is three times that of solvent Y. Solute S is known to undergo dimerization in these solvents. If the degree of dimerization is 0.7 in solvent Y, the degree of dimerization in solvent X is _____.

Paragraph "X"

Treatment of benzene with CO/HCl in the presence of anhydrous $AlCl_3$/CuCl followed by reaction with Ac_2O/NaOAc gives compound **X** as the major product. Compound **X** upon reaction with Br_2/Na_2CO_3, followed by heating at 473 K with moist KOH furnishes **Y** as the major product. Reaction of **X** with H_2/Pd-C, followed by H_3PO_4 treatment gives **Z** as the major product.

(There are two questions based on Paragraph "X", the question given below is one of them.)

15. The compound **Y** is

(A) 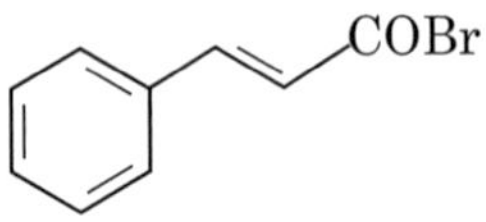

(B) 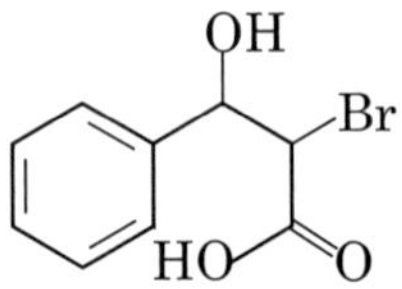

(C) 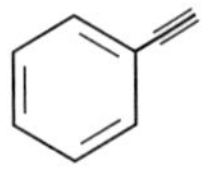

(D) 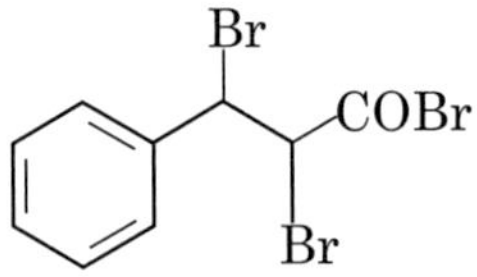

16. The compound **Z** is

(A) 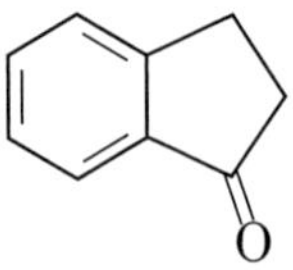

(B) 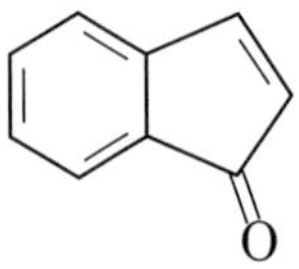

(C) 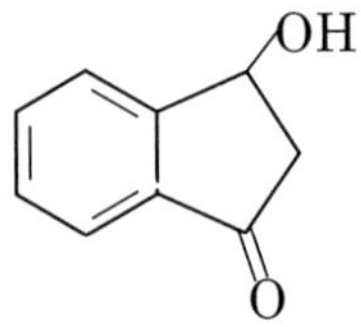

(D) 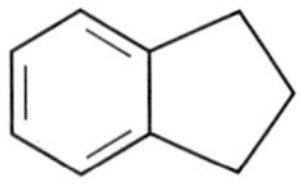

Paragraph "A"

An organic acid $P(C_{11}H_{12}O_2)$ can easily be oxidized to a dibasic acid which reacts with ethyleneglycol to produce a polymer dacron. Upon ozonolysis, **P** gives an aliphatic ketone as one of the products. **P** undergoes the following reaction sequences to furnish **R** via **Q**. The compound **P** also undergoes another set of reactions to produce **S**.

S ← [(1) H_2/Pd-C; (2) NH_3/Δ; (3) Br_2/NaOH; (4) $CHCl_3$, KOH, Δ; (5) H_2/Pd-C] — P — [(1) H_2/Pd-C; (2) $SOCl_2$; (3) MeMgBr, $CdCl_2$; (4) $NaBH_4$] → Q — [(1) HCl; (2) Mg/Et_2O; (3) CO_2 (dry ice); (4) H_3O^+] → R

(There are two questions based on Paragraph "A". The question given on next page is one of them.)

17. The compound **R** is

(A) 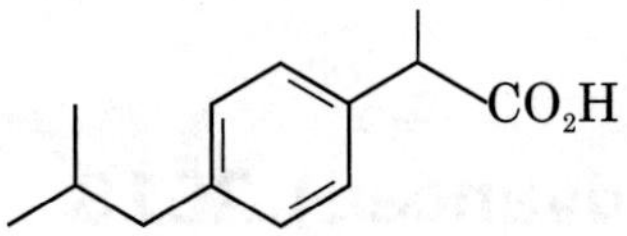

(B) 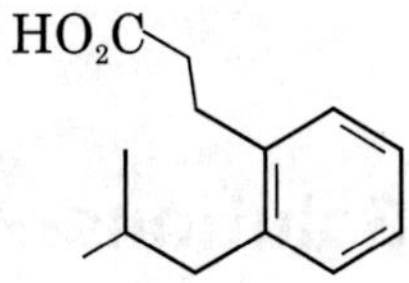

(C) 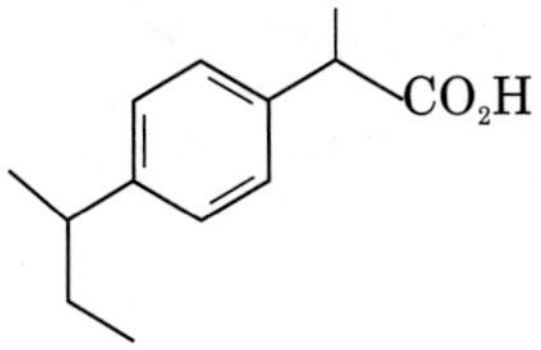

(D) 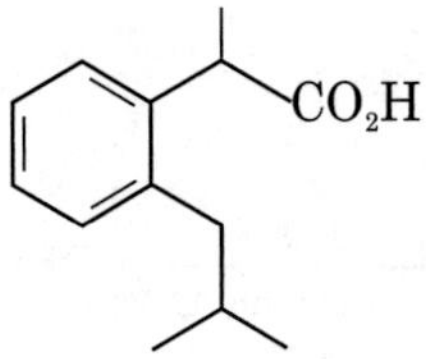

18. The compound **S** is

(A) 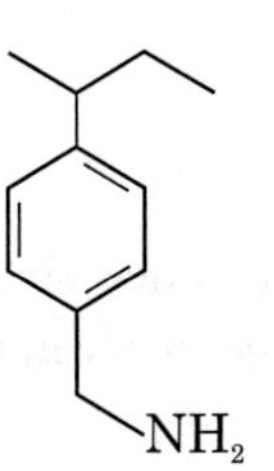

(B) 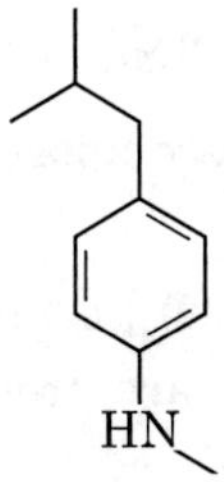

(C) 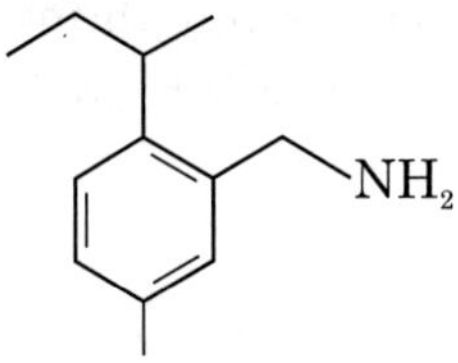

(D) 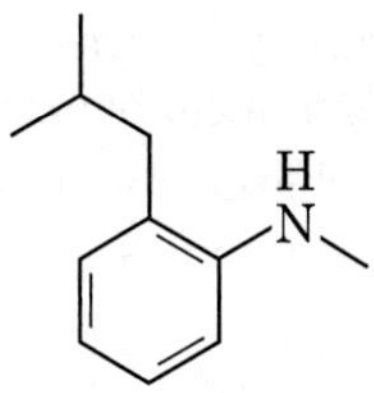

Answers

1. B, C	2. B, C	3. A, B, C	4. B	5. A, B, D	6. B, C
7. 1	8. 2992	9. 3	10. 10	11. 2.22	12. 19
13. 4.47	14. 0.05	15. C	16. A	17. A	18. B

Model Solutions—I.I.T. JEE (Advanced) 2018

Paper–1

1. (B, C)

   (A) $NH_4NO_3 \xrightarrow[\text{below } 300°C]{\Delta} N_2O + 2H_2O$

   (B) $(NH_4)_2Cr_2O_7 \xrightarrow{\Delta} N_2 + Cr_2O_3 + 4H_2O$

   (C) $Ba(N_3)_2 \xrightarrow{\Delta} Ba + 3N_2$

   (D) Mg_3N_2 (it does not decompose into N_2)

2. (B, C)

   (A) $[Fe(CO)_5]$ and $[Ni(CO)_4]$ complexes have 18-electrons in their valence shell.

   (B) Carbonyl complexes are predominantly low spin complexes due to strong ligand field.

   (C) As electron density increases on metals (with lowering oxidation state on metals), the extent of synergic bonding increases. Hence M—C bond strength increases.

   (D) While positive charge on metals increases and the extent of synergic bond decreases, the C—O bond becomes stronger.

3. (A, B, C)

   (A) Bi_2O_5 is metallic oxide but N_2O_5 is non-metallic oxide. Therefore, Bi_2O_5 is basic but N_2O_5 is acidic.

   (B) In NF_3, N and F are non-metals but in BiF_3, Bi is metal but F is non-metal. Therefore, NF_3 is more covalent than BiF_3.

   (C) In PH_3, hydrogen bonding is absent but in NH_3 hydrogen bonding is present. Therefore, PH_3 boils at lower temperature than NH_3.

   (D) Due to small size in N—N single bond *lp-lp* repulsion is more than P—P single bond. Therefore, N—N single bond is weaker than the P—P single bond.

4. (B)

   X $\xrightarrow[\text{(2) NaI, } Me_2C{=}O \text{ (3) } NaN_3\text{, } HCONMe_2]{(1)\ PBr_3,\ Et_2O}$ 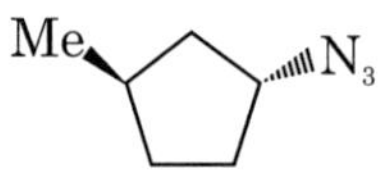

   All the three reactions are S_N2. Hence, X is 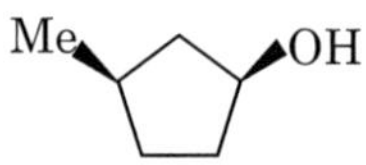

5. (A, B, D)

   (A) 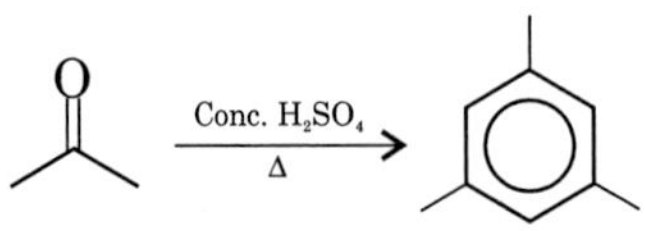

   (B) 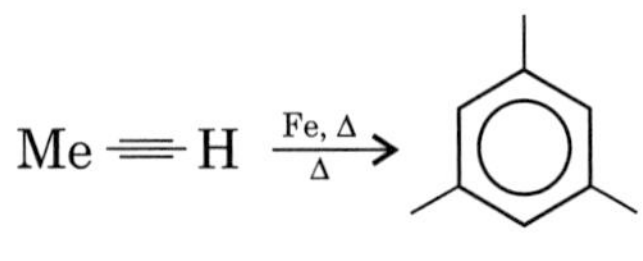

(C) 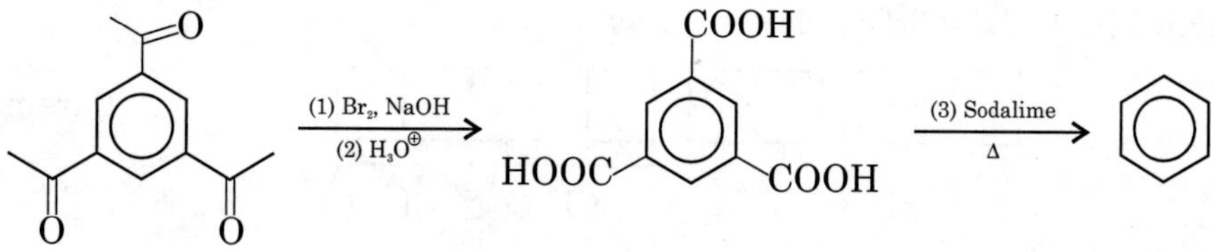

(D) 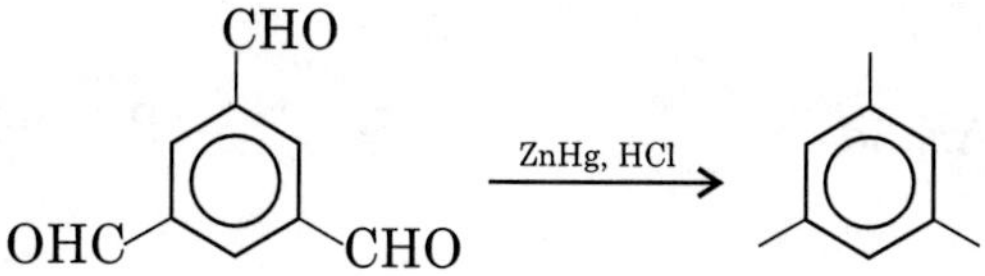

6. (B, C)

AC → Isochoric

AB → Isothermal

BC → Isobaric

$q_{AC} = \Delta U_{BC} = nC_V (T_2 - T_1)$

$W_{AB} = nRT_1 \ln\left(\frac{V_2}{V_1}\right)$ A (wrong)

$q_{BC} = \Delta H_{AC} = nC_P (T_2 - T_1)$

$W_{BC} = - P_2 (V_1 - V_2)$ B (correct)

$nC_P (T_1 - T_2) < nC_V (T_1 - T_2)$ C (correct)

$\Delta H_{CA} < \Delta U_{CA}$

D (wrong)

7. (1)

H-atom = $\boxed{1}$ $1s^1$ Paramagnetic

NO_2 = 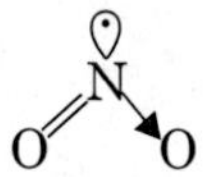 odd electron species Paramagnetic

O_2^- (superoxide) = One unpaired electrons in π^* M.O. Paramagnetic

S_2 (in vapour phase) = same as O_2, two unpaired electrons are present in π^* M.O. Paramagnetic

$Mn_3O_4 = 2\ \overset{2+}{Mn}O.\overset{4+}{Mn}O_2$ Paramagnetic

$(NH_4)_2[FeCl_4]$ = $Fe^{2+} = 3d^6 4s^0$

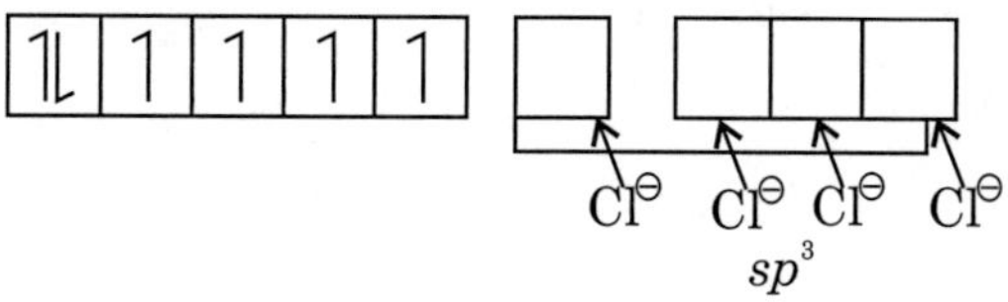

Paramagnetic

$(NH_4)_2[NiCl_4]$: $Ni = 3d^8 4s^2$; $Ni^{2+} = 3d^8 4s^0$

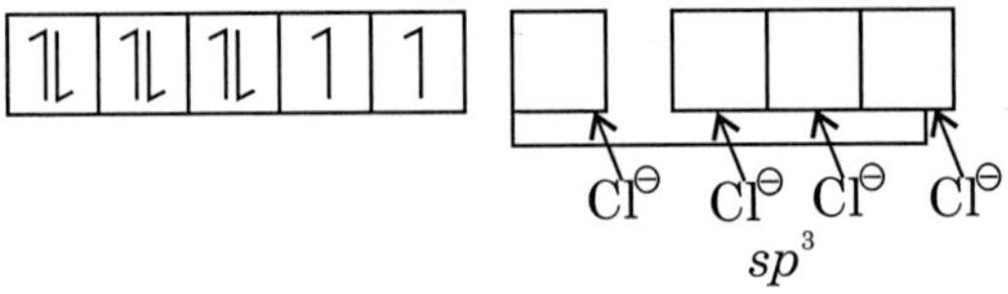

Paramagnetic

$K_2MnO_4 = 2K^+$ 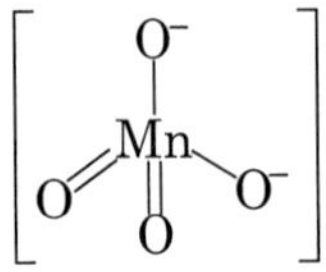, $Mn^{6+} = [Ar]\ 3d^1$ Paramagnetic

$K_2CrO_4 = 2K^+$ 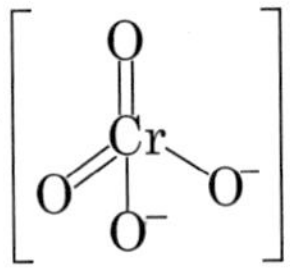, $Cr^{6+} = [Ar]\ 3d^0$ Diamagnetic

8. (2992)

$$\underset{\substack{1584\,g\\=12\,mol}}{(NH_4)_2SO_4} + Ca(OH)_2 \rightarrow \underset{\substack{\text{gypsum } (M=172)\\12\,mol}}{CaSO_4\cdot 2H_2O} + \underset{24\,mol}{2NH_3}$$

$$\underset{952\,g=4\,mol}{NiCl_2\cdot 6H_2O} + \underset{24\,mol}{6NH_3} \rightarrow \underset{(M=232);\,4\,mol}{Ni(NH_3)_6]Cl_2} + 6H_2O$$

Total mass = 12 × 172 + 4 × 232 = 2992 g

9. (3)

$X^{\ominus}$: O.V.

M^+: FCC

M^+	X^-
(i) 4	1
(ii) 4 – 3	3 + 1
(iii) 4 – 3 – 1	3 + 1
(iv) 1	3

$$Z = \frac{3}{1} = 3$$

10. (10)

$Mg(s) + Cu^{2+}(aq) \rightarrow Mg^{2+}(aq) + Cu(s)$

$E^0_{Cell} = 2.70$ $\qquad Mg^{2+} = xM$

$E_{Cell} = 2.67$ $\qquad Cu^{2+} = 1M$

$$E_{Cell} = E^0_{Cell} - \frac{RT}{nF}\ln x$$

$$2.67 = 2.70 - \frac{RT}{2F}\ln x$$

$$-\ 0.03 = -\frac{R \times 300}{2F} \times \ln x$$

$$\ln x = \frac{0.03 \times 2}{300} \times \frac{F}{R} = \frac{0.03 \times 2 \times 11500}{300 \times 1}$$

$\ln x = 2.30 = \ln(10)$

$x = 10$

11. (2.22)

$P_1 = 5$ $\quad$ $P_2 = 1$

$v_1 = 1$ $\quad$ $v_2 = 3$

$T_1 = 400$ $\quad$ $T_2 = 300$

$n_1 = \frac{5}{400R}$ $\quad$ $n_2 = \frac{3}{300R}$

Let volume be $(v + x)$ $\quad$ $v = (3 - x)$ $\quad$ $15 - 5x = 4 + 4x$

$$\frac{P_A}{T_A} = \frac{P_B}{T_B}$$

$$\frac{n_{b_1} \times R}{v_{b_1}} = \frac{n_{b_2} \times R}{v_{b_2}}$$

$$\frac{5}{400(4 + x)} = \frac{3}{300R(3 - x)}$$

or $5(3 - x) = 4 + 4x$

or $x = \frac{11}{9}$

$$v = 1 + x = 1 + \frac{11}{9} = \left(\frac{20}{9}\right) = 2.22$$

12. (19)

$$45 = P_A^0 \times \frac{1}{2} + P_B^0 \times \frac{1}{2}$$

$$P_A^0 + P_B^0 = 90 \qquad ...(1)$$

Given $P_A^0 = 20$ torr; $P_B^0 = 70$ torr

22.5 torr $= 20\ x_A + 70\ (1 - x_A) = 70 - 50\ x_A$

or $x_A = \left(\frac{70 - 22.5}{50}\right) = 0.95$

$x_B = (1 - 0.95) = 0.05$

Hence, $\frac{x_A}{x_B} = \frac{0.95}{0.05} = 19$

13. (4.47)

$$S = \sqrt{K_{sp}\left(\frac{[H^+]}{K_a} + 1\right)} = \sqrt{2 \times 10^{-10}\left(\frac{10^{-3}}{10^{-8}} + 1\right)} \simeq \sqrt{2 \times 10^{-5}} = 4.47 \times 10^{-3} M$$

14. (0.05)

From graph

For solvent X $\Delta T_{bx} = 2$

$\Delta T_{bx} = m_{NaCl} \times K_{b(x)}$ ...(1)

For solvent Y $\Delta T_{by} = 1$

$\Delta T_{b(y)} = m_{NaCl} \times K_{b(y)}$ ...(2)

Dividing equation (1) by equation (2), we have

$$\frac{K_{b(x)}}{K_{b(y)}} = 2$$

For solute S

$$\underset{1-\alpha}{2(S)} \rightarrow \underset{\alpha/2}{S_2}$$

$i = (1 - \alpha/2)$

$$\Delta T_{b(x)(s)} = \left(1 - \frac{\alpha_1}{2}\right) K_{b(x)}$$

$$\Delta T_{b(y)(s)} = \left(1 - \frac{\alpha_2}{2}\right) K_{b(y)}$$

Given $\Delta T_{b(x)(s)} = 3\Delta T_{b(y)(s)}$

$$\left(1 - \frac{\alpha_1}{2}\right) K_{b(x)} = 3 \times \left(1 - \frac{\alpha_2}{2}\right) \times K_{b(y)}$$

or $$2\left(1 - \frac{\alpha_1}{2}\right) = 3\left(1 - \frac{\alpha_2}{2}\right)$$

$\alpha_2 = 0.7$

Hence, $\alpha_1 = 0.05$

15. (C)

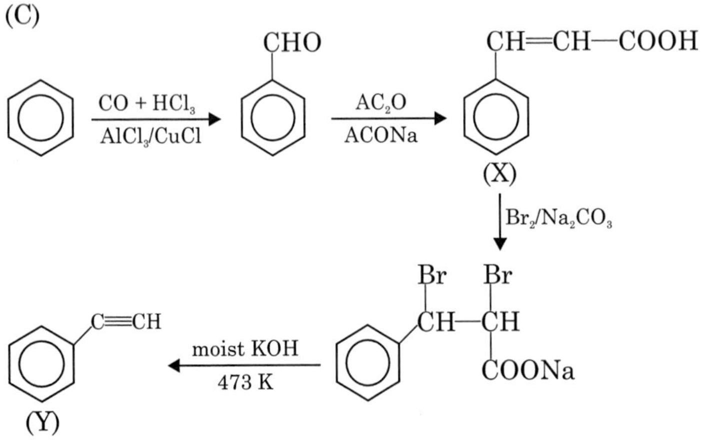

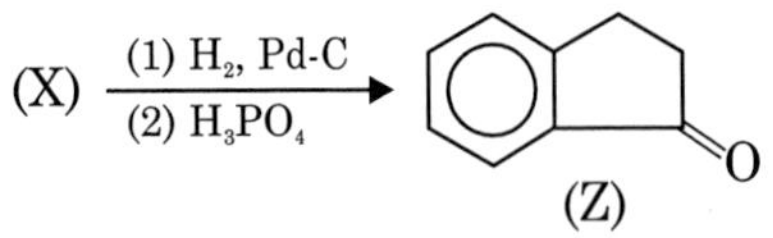

16. (A)

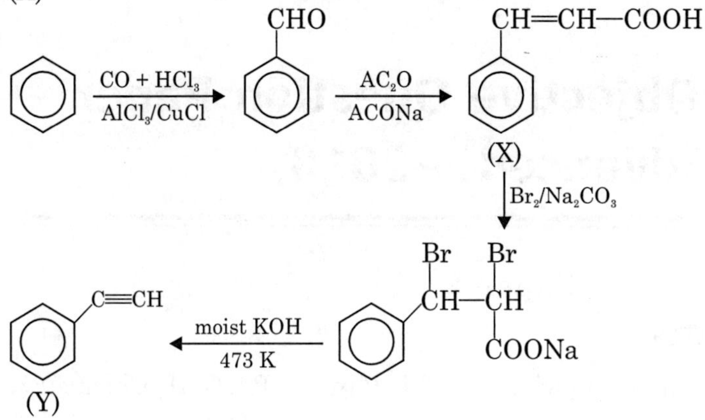

17. (A)

18. (B) **Solution 17 and 18:**

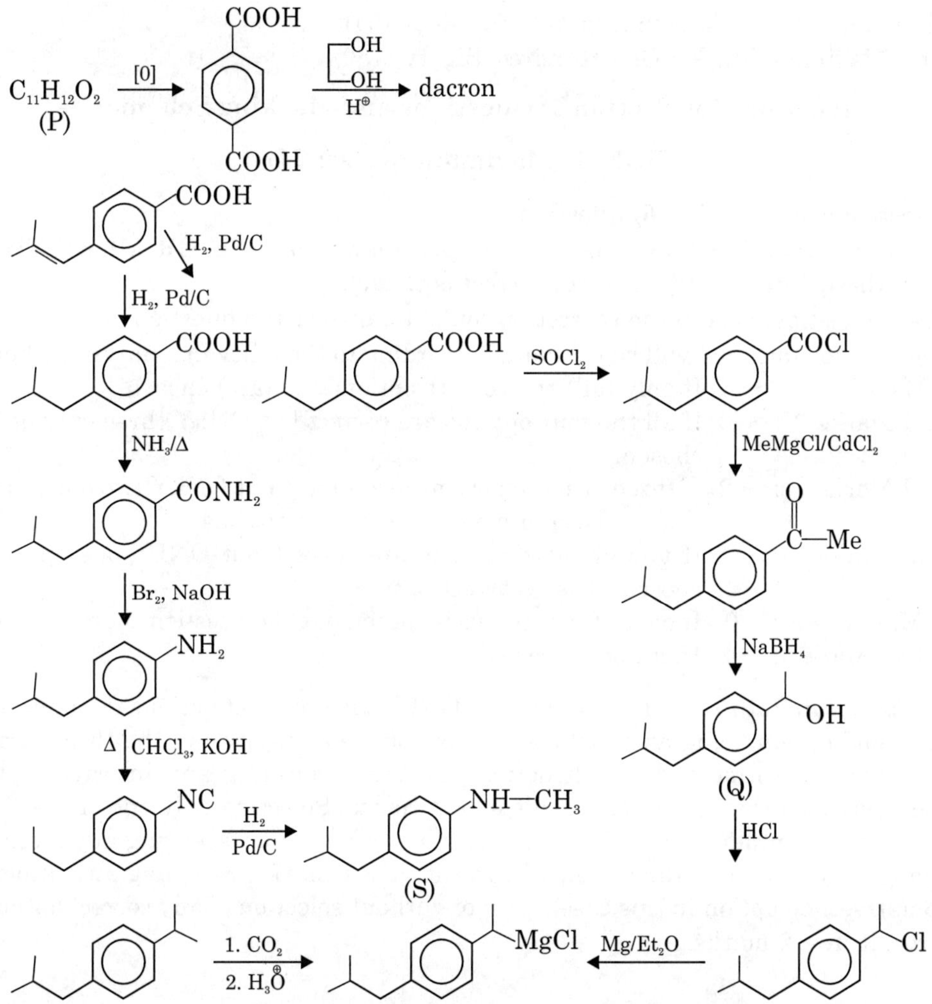

I.I.T. JEE Objective Question Paper (Advanced)—2018

Paper–2

INSTRUCTIONS TO CANDIDATES

- This question paper has three (3) parts: PART-I: Physics, PART-II: Chemistry and PART-III: Mathematics.
- Each part has total of eighteen (18) questions divided into three (3) sections (Section-1, Section-2 and Section-3).
- Total number of questions in Paper-2: Fifty-four (54).
- Paper-2 Maximum Marks: One Hundred Eighty (180).

Instructions for Section-1: Questions and Marking Scheme

SECTION–1 (Maximum Marks: 24)

- This section contains **SIX (6)** questions.
- Each question has **FOUR options** for correct answer(s). **ONE OR MORE THAN ONE** of these four option(s) is (are) correct option(s).
- For each question, choose the correct option(s) to answer the question.
- Answer to each question will be evaluated according to the following marking scheme:

Full Marks	:	+ 4	If only (all) the correct option(s) is (are) chosen.
Partial Marks	:	+ 3	If all the four options are correct but ONLY three options are chosen.
Partial Marks	:	+ 2	If three or more options are correct but ONLY two options are chosen, both of which are correct options.
Partial Marks	:	+ 1	If two or more options are correct but ONLY one option is chosen, it is a correct option.
Zero Marks	:	0	If none of the options is chosen (i.e. the question is unanswered).
Negative Marks	:	– 2	In all other cases.

For Example: If first, third and fourth are the ONLY three correct options for a question with second option being an incorrect option; selecting only two of the three correct options (e.g., the first and fourth options), without selecting any incorrect option (second option in this case), will result in + 2 marks. Selecting only one of the three correct options (either first or third or fourth option), without selecting any incorrect option (second option in this case), will result in + 1 marks. Selecting any incorrect option(s) (second option in this case), with or without selection of any correct option(s) will result in – 2 marks.

Answering Section-1 Questions

- To select the option(s), **using the mouse click** on the corresponding button(s) of the option(s).
- To deselect chosen option(s), click on the button(s) of the chosen option(s) again or click on the **Clear Response** button to clear all the chosen options.
- To change the option(s) of a previously answered question, if required, first click on the **Clear Response** button to clear all the chosen options and then select the new option(s).
- To mark a question ONLY for review (i.e. without answering it), click on the **Mark for Review & Next button**.
- To mark a question for review (after answering it), click on **Mark for Review & Next button** – answered question which is also marked for review will be evaluated.
- To save the answer, click on the **Save & Next button** – the answered question will be evaluated.

Instructions for Section-2: Questions and Marking Scheme

SECTION–2 (Maximum Marks: 24)

- This section contains **EIGHT (8)** questions. The answer to each question is **NUMERICAL VALUE.**
- For each question, enter the correct numerical value (in decimal notation, truncated/ rounded-off to the second decimal place; e.g., 6.25, 7.00, – 0.33, – 0.30, 30.27, – 127.30) using the mouse and the on-screen virtual numeric keypad in the place designated to enter the answer.
- Answer to each question will be evaluated according to the following marking scheme.
  Full Marks : + 3 If ONLY the correct numerical value is entered as answer.
  Zero Marks : 0 In all other cases.

Answering Section-2 Questions

- Using the attached computer mouse, click on numbers (and/or symbols) on the on-screen virtual numeric keypad to enter the numerical value as answer in the space provided for answer.
- To change the answer, if required, first click on the **Clear Response** button to clear the entered answer and then enter the new numerical value.
- To mark a question ONLY for review (i.e. answering it), click on the **Mark for Review & Next button** – the answered question which is also marked for review will be evaluated.
- To mark a question for review (after answering it), click **Mark for Review & Next button** – the answered question which is also marked for review will be evaluated.
- To save the answer, click on the **Save & Next button** – the answered question will be evaluated.

Instructions for Section-3: Questions and Marking Scheme

SECTION–3 (Maximum Marks: 12)

- This section contains **TWO (2)** paragraphs. Based on each paragraph, there are **TWO (2)** questions.
- Each question has **FOUR** options. **ONLY ONE** of these four options corresponds to the correct answer.
- For each question, choose the option corresponding to the correct answer.
- Answer to each question will be evaluated according to the following marking scheme:

  Full Marks : + 3 If ONLY the correct option is chosen.

  Zero Marks : 0 If none of the options is chosen (i.e. the question is unanswered).

  Negative Marks : – 1 In all other cases.

Answering Section-3 Questions

- To select an option, using the mouse click on the corresponding button of the option.
- To deselect the chosen answer, click on the button of the chosen option again or click on the **Clear Response button**.
- To change the chosen answer, click on the button of another option.
- To mark a question ONLY for review (i.e. without answering it), click on **Mark for Review & Next button**.
- To mark a question for review (after answering it), click on **Mark for Review & Next button** – the answered question which is also marked for review will be evaluated.
- To save the answer, click on the **Save & Next button** – the answered question will be evaluated.

1. The correct option(s) regarding the complex $[Co(en)(NH_3)_3(H_2O)]^{3+}$ (en = $H_2NCH_2CH_2NH_2$) is (are)

   (A) It has two geometrical isomers

   (B) It will have three geometrical isomers if bidentate 'en' is replaced by two cyanide ligands

   (C) It is paramagnetic

   (D) It absorbs light at longer wavelength as compared to $[Co(en)(NH_3)_4]^{3+}$

2. The correct option(s) to distinguish nitrate salts of Mn^{2+} and Cu^{2+} taken separately is (are):

   (A) Mn^{2+} shows the characteristic green colour in the flame test

   (B) Only Cu^{2+} shows the formation of precipitate by passing H_2S in acidic medium

   (C) Only Mn^{2+} shows the formation of precipitate by passing H_2S in faintly basic medium

   (D) Cu^{2+}/Cu has higher reduction potential than Mn^{2+}/Mn (measured under similar conditions).

3. Aniline reacts with mixed acid (conc. HNO_3 and conc. H_2SO_4) at 288 K to give P (51%), Q (47%) and R (2%). The major product(s) of the following reaction sequence is (are)

$$R \xrightarrow[\text{(3) } H_3O^+ \text{ (4) } NaNO_2,\ HCl/273\text{-}278\ K \text{ (5) EtOH, } \Delta]{\text{(1) } Ac_2O, \text{ pyridine (2) } Br_2,\ CH_3CO_2H} S \xrightarrow[\text{(3) } NaNO_2,\ HCl/273\text{-}278\ K \text{ (4) } H_3PO_2]{\text{(1) Sn/HCl (2) } Br_2/H_2O \text{ (excess)}} \text{major product(s)}$$

(A) 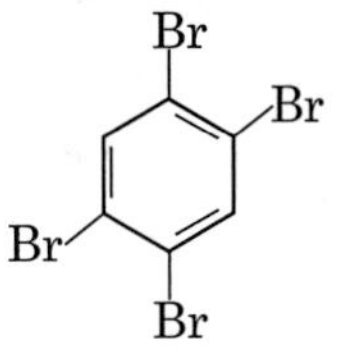 (B) 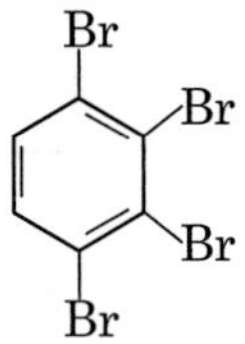 (C) 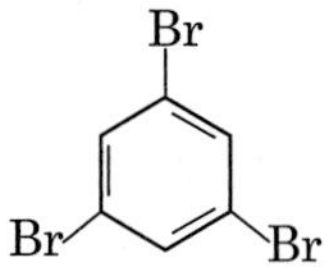 (D) 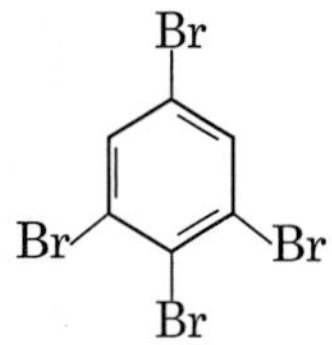

4. The Fischer presentation of D-glucose is given below.

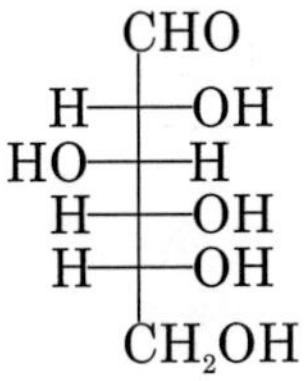

D-glucose

The correct structure(s) of β-glucopyranose is (are)

(A) 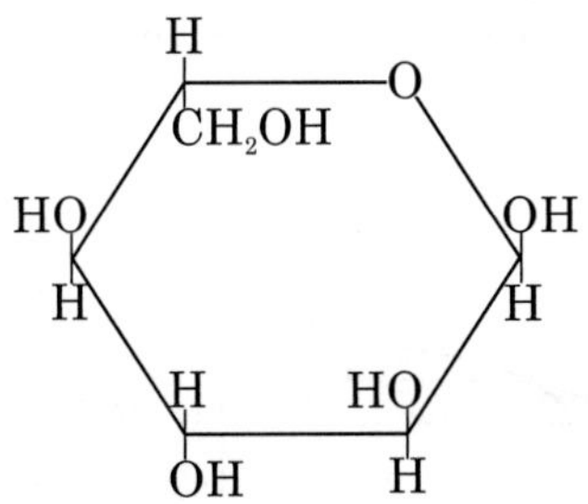

(B) 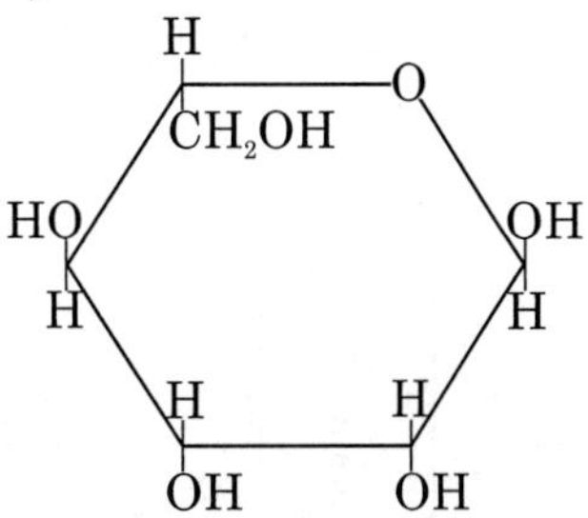

(C) 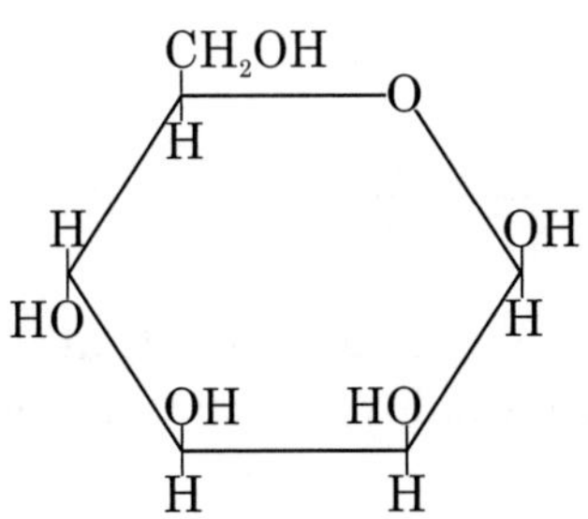

(D) 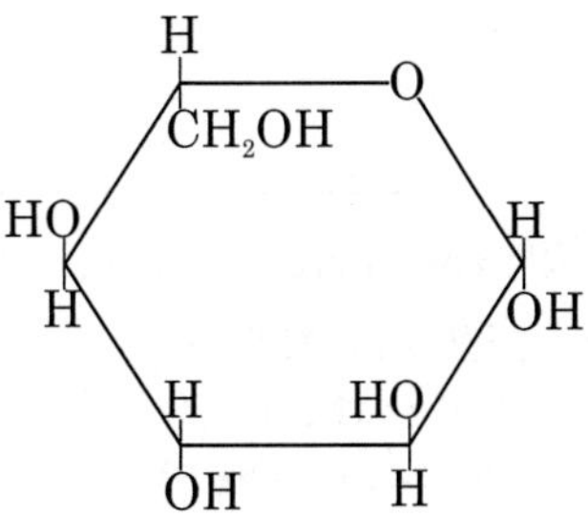

5. For a first order reaction $A(g) \rightarrow 2B(g) + C(g)$ at constant volume and 300 K, the total pressure at the beginning ($t = 0$) and at time t are P_0 and P_t, respectively. Initially, only A is present with concentration $[A]_0$, and $t_{1/3}$ is the time required for the partial pressure of A to reach 1/3rd of its initial value. The correct option(s) is (are) (Assume that all these gases behave as ideal gases.)

(A) 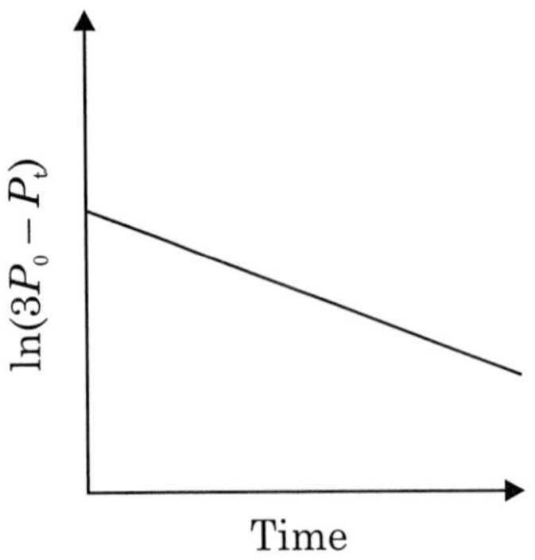

(B) 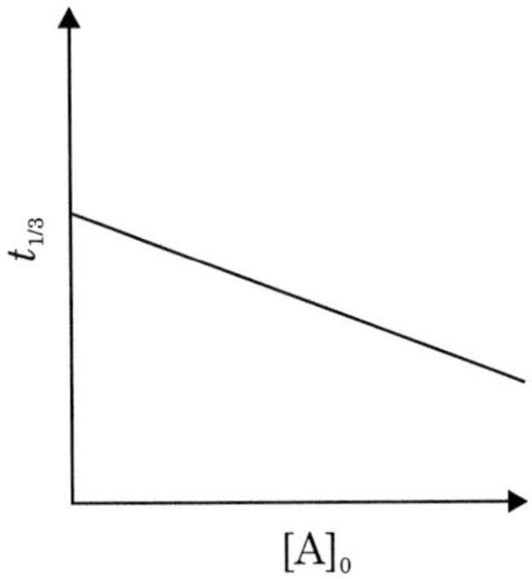

(C) 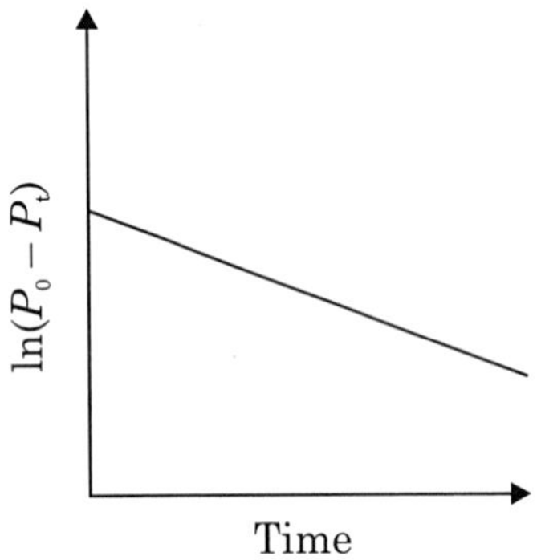

(D) 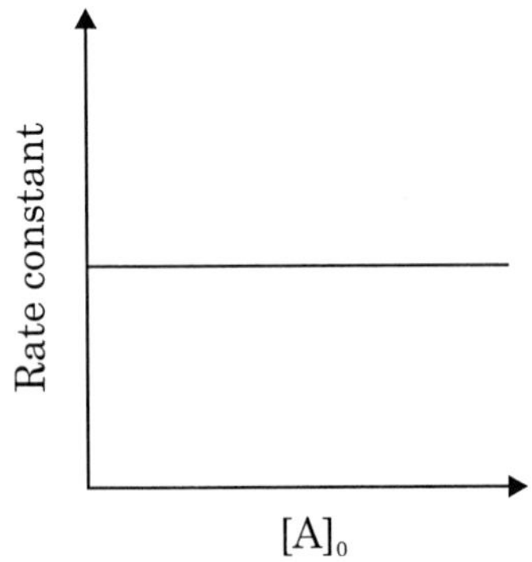

6. For a reaction, $A \rightleftharpoons P$, the plots of [A] and [P] with time at temperatures T_1 and T_2 are given below.

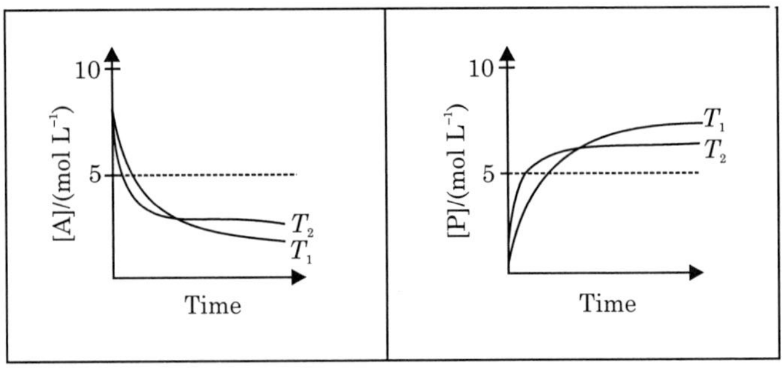

If $T_2 > T_1$, the correct statement(s) is (are)

Assume ΔH^θ and ΔS^θ are independent of temperature and ratio of $\ln K$ at T_1 to $\ln K$ at T_2 is greater than T_2/T_1. Here H, S, G and K are enthalpy, entropy, Gibbs energy and equilibrium constant, respectively.

(A) $\Delta H^\theta < 0$, $\Delta S^\theta < 0$ (B) $\Delta G^\theta < 0$, $\Delta H^\theta > 0$

(C) $\Delta G^\theta < 0$, $\Delta S^\theta < 0$ (D) $\Delta G^\theta < 0$, $\Delta S^\theta > 0$

7. The total number of compounds having at least one bridging oxo group among the molecules given below is ______.

N_2O_3, N_2O_5, P_4O_6, P_4O_7, $H_4P_2O_5$, $H_5P_3O_{10}$, $H_2S_2O_3$, $H_2S_2O_5$

8. Galena (an ore) is partially oxidized by passing air through it at high temperature. After some time, the passage of air is stopped, but the heating is continued in a closed furnace such that the contents undergo self-reduction. The weight (in kg) of Pb produced per kg of O_2 consumed is ______.

   (Atomic weights in g mol^{-1}: O = 16, S = 32, Pb = 207)

9. To measure the quantity of $MnCl_2$ dissolved in an aqueous solution, it was completely converted to $KMnO_4$ using the reaction,

   $MnCl_2 + K_2S_2O_8 + H_2O \rightarrow KMnO_4 + H_2SO_4 + HCl$ (equation not balanced).

   Few drops of concentrated HCl were added to this solution and gently warmed. Further, oxalic acid (225 g) was added in portions till the colour of the permanganate ion disappeared. The quantity of $MnCl_2$ (in mg) present in the initial solution is _____.

   (Atomic weights in g mol^{-1}: Mn = 55, Cl = 35.5)

10. For the given compound X, the total number of optically active stereoisomers is ____.

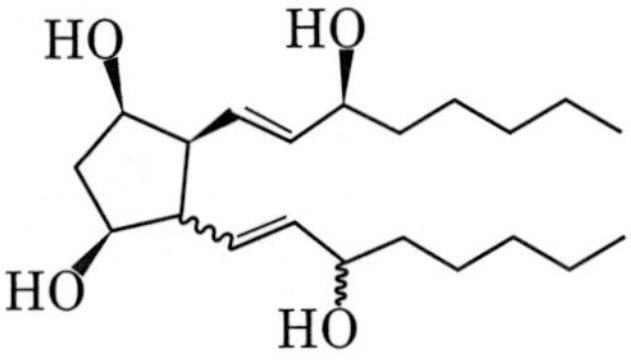

X

(wedge bond) This type of bond indicates that the configuration at the specific carbon and the geometry of the double bond is fixed.

(wavy bond) This type of bond indicates that the configuration at the specific carbon and the geometry of the double bond is NOT fixed.

11. In the following reaction sequence, the amount of D (in g) formed from 10 moles of acetophenone is _____.

    (Atomic weight in g mol^{-1}: H = 1, C = 12, N = 14, O = 16, Br = 80. The yield (%) corresponding to the product in each step is given in the parenthesis.)

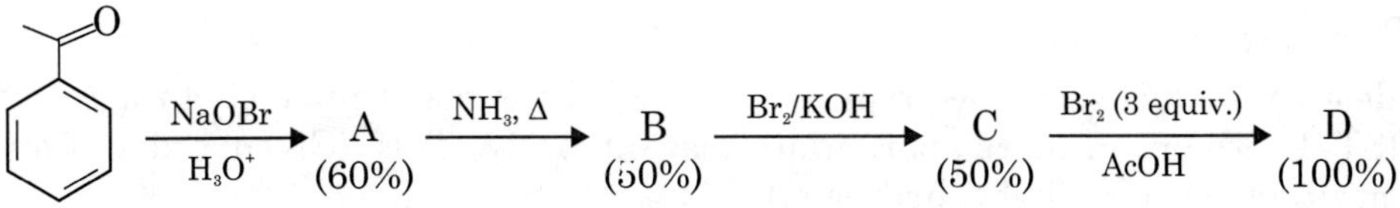

12. The surface of copper gets tarnished by the formation of copper oxide. N_2 gas was passed to prevent the oxide formation during heating of copper at 1250 K. However, the N_2 gas contains 1 mole % of water vapour as impurity. The water vapour oxidises copper as per the reaction given below:

    $2Cu(s) + H_2O(g) \rightarrow Cu_2O(s) + H_2(g)$

    p_{H_2} is the minimum partial pressure of H_2 (in bar) needed to prevent the oxidation at 1250 K. The value of ln (p_{H_2}) is ____.

    (Given: total pressure = 1 bar, R (universal gas constant) = 8 J K^{-1} mol^{-1}, ln(10) = 2.3. Cu(*s*) and $Cu_2O(s)$ are mutually immiscible.)

    At 1250 K: $2Cu(s) + 1/2O_2(g) \rightarrow Cu_2O(s)$; $\Delta G^{\theta} = -$ 78,000 J mol^{-1}

    $H_2(g) + 1/2O_2(g) \rightarrow H_2O(g)$; $\Delta G^{\theta} = -$ 1,78,000 J mol^{-1}; G is the Gibbs energy.

13. Consider the following reversible reaction,

$A(g) + B(g) \rightleftharpoons AB(g)$.

The activation energy of the backward reaction exceeds that of the forward reaction by $2RT$ (in J mol^{-1}). If the pre-exponential factor of the forward reaction is 4 times that of the reverse reaction, the absolute value of ΔG^θ (in J mol^{-1}) for the reaction at 300 K is _____.

(Given: ln(2) = 0.7, $RT = 2500$ J mol^{-1} at 300 K and G is the Gibbs energy.)

14. Consider an electrochemical cell: $A(s) \mid A^{n+}$ $(aq, 2M) \parallel B^{2n+}$ $(aq, 1M) \mid B(s)$. The value of ΔH^θ for the cell reaction is twice that of ΔG^θ at 300 K. If the emf of the cell is zero, the ΔS^θ (in J K^{-1} mol^{-1}) of the cell reaction per mole of B formed at 300 K is ____.

(Given: ln(2) = 0.7, R (universal gas constant) = 8.3 J K^{-1} mol^{-1}. H, S and G are enthalpy, entropy and Gibbs energy, respectively.)

15. Match each set of hybrid orbitals from LIST–I with complex(es) given in LIST–II.

LIST–I	LIST–II
P. dsp^2	1. $[FeF_6]^{4-}$
Q. sp^3	2. $[Ti(H_2O)_3Cl_3]$
R. sp^3d^2	3. $[Cr(NH_3)_6]^{3+}$
S. d^2sp^3	4. $[FeCl_4]^{2-}$
	5. $Ni(CO)_4$
	6. $[Ni(CN)_4]^{2-}$

The correct option is

(A) P → 5; Q → 4, 6; R → 2, 3; S → 1
(B) P → 5, 6; Q → 4; R → 3; S → 1, 2
(C) P → 6; Q → 4, 5; R → 1; S → 2, 3
(D) P → 4, 6; Q → 5, 6; R → 1, 2; S → 3

16. The desired product X can be prepared by reacting the major product of the reactions in LIST–I with one or more appropriate reagents in LIST–II. (Given: order of migratory aptitude: aryl > alkyl > hydrogen)

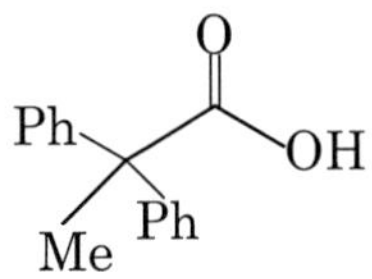

LIST–I	LIST–II
P. 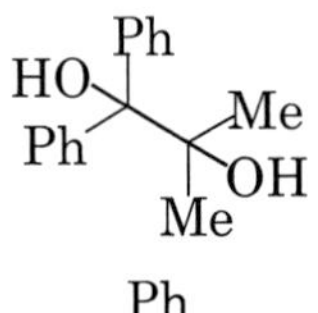 $+ H_2SO_4$	1. I_2, NaOH
Q. 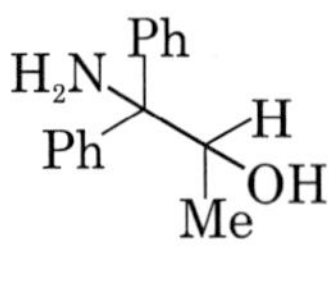 $+ HNO_2$	2. $[Ag(NH_3)_2]OH$

R. 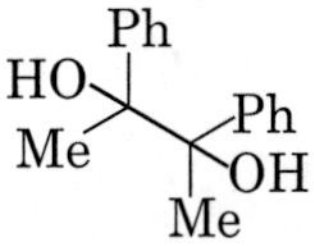 + H_2SO_4 3. Fehling solution

S. 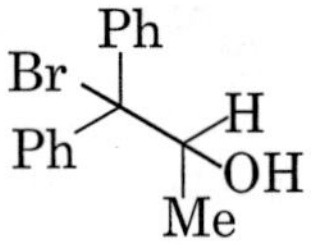 + $AgNO_3$ 4. HCHO, NaOH

5. NaOBr

The correct option is

(A) P → 1; Q → 2, 3; R → 1, 4; S → 2, 4 (B) P → 1, 5; Q → 3, 4; R → 4, 5; S → 3

(C) P → 1, 5; Q → 3, 4; R → 5; S → 2, 4 (D) P → 1, 5; Q → 2, 3; R → 1, 5; S → 2, 3

17. LIST–I contains reactions and LIST–II contains major products.

LIST–I	LIST–II
P. 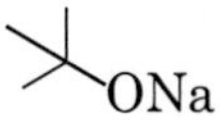 + 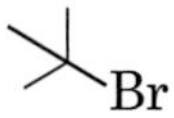 ⟶	1. 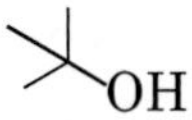
Q. 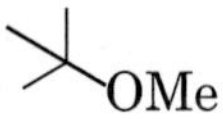 + HBr ⟶	2. 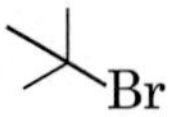
R. 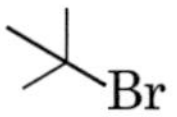 + NaOMe ⟶	3. 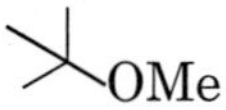
S. 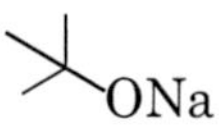 + MeBr ⟶	4. 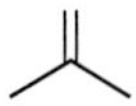
	5. 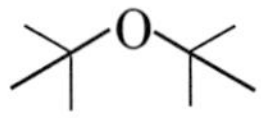

Match each reaction in LIST–I with one or more products in LIST–II and choose the correct option.

(A) P → 1, 5; Q → 2; R → 3; S → 4 (B) P → 1, 4; Q → 2; R → 4; S → 3

(C) P → 1, 4; Q → 1, 2; R → 3, 4; S → 4 (D) P → 4, 5; Q → 4; R → 4; S → 3, 4

18. Dilution process of different aqueous solutions, with water, are given in LIST–I. The effects of dilution of the solutions on $[H^+]$ are given in LIST–II.

(Note: Degree of dissociation (α) of weak acid and weak base is << 1; degree of hydrolysis of salt << 1; $[H^+]$ represents the concentration of H^+ ions)

LIST–I	LIST–II
P. (10 mL of 0.1*M* NaOH + 20 mL of 0.1*M* acetic acid) diluted to 60 mL	1. the value of $[H^+]$ does not change on dilution
Q. (20 mL of 0.1*M* NaOH + 20 mL of 0.1*M* acetic acid) diluted to 80 mL	2. the value of $[H^+]$ changes to half of its initial value on dilution
R. (20 mL of 0.1*M* HCl + 20 mL of 0.1*M* ammonia solution) diluted to 80 mL	3. the value of $[H^+]$ changes to two times of its initial value on dilution

S. 10 mL saturated solution of $Ni(OH)_2$ in equilibrium with excess solid $Ni(OH)_2$ is diluted to 20 mL (solid $Ni(OH)_2$ is still present after dilution).

4. the value of $[H^+]$ changes to $\frac{1}{\sqrt{2}}$ times of its initial value on dilution

5. the value of $[H^+]$ changes to $\sqrt{2}$ times of its initial value on dilution

Match each process given in LIST–I with one or more effect(s) in LIST–II. The correct option is

(A) P → 4; Q → 2; R → 3; S → 1
(B) P → 4; Q → 3; R → 2; S → 3
(C) P → 1; Q → 4; R → 5; S → 3
(D) P → 1; Q → 5; R → 4; S → 1

Answers

1. A, B, D	2. B, D	3. D	4. D	5. A, D	6. A, C
7. 5 or 6	8. 6.47	9. 126	10. 7	11. 495	12. – 14.6
13. 8500	14. – 11.62	15. C	16. D	17. B	18. D

Model Solutions—I.I.T. JEE (Advanced) 2018

Paper–2

1. (A, B, D)

(A) The complex $[Co(en)(NH_3)_3(H_2O)]^{+3}$ is of the type $[M(AA)b_3c]$ and has two geometrical isomers:

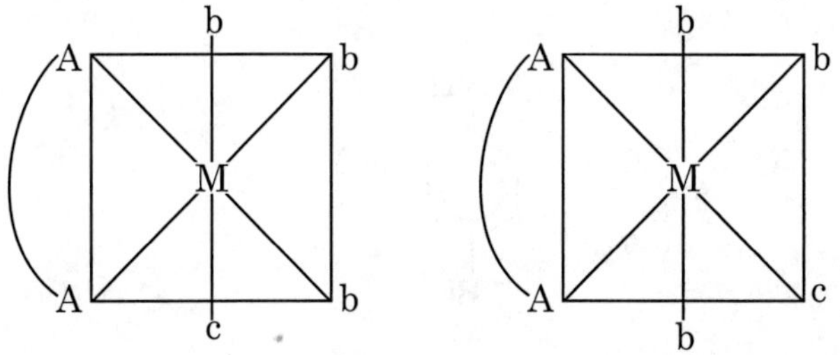

(B) If (en) is replaced by two cyanide ligand, complex will be of the type $[Ma_3b_2c]$ and has three geometrical isomers:

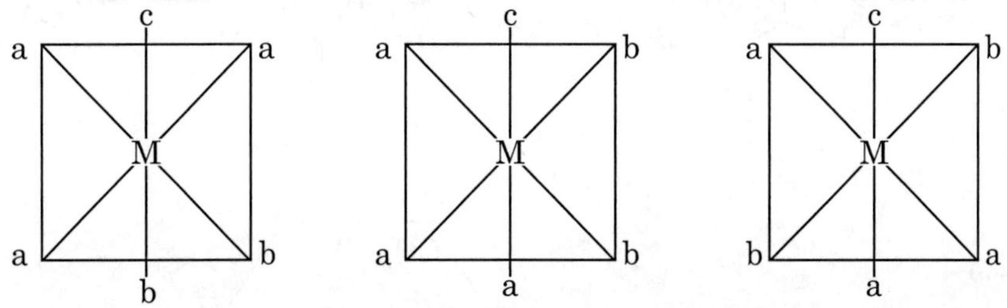

(C) The complex $[Co(en)(NH_3)_3(H_2O)]^{3+}$ has d^6 configuration (t_{2g}^6) on central metal with three strong field ligands. Hence, it is dimagnetic in nature.

(D) The complex $[Co(en)(NH_3)_3(H_2O)]^{3+}$ has lesser CFSE (Δ_o) value than that of $[Co(en)(NH_3)_4]^{3+}$. Hence, the complex $[Co(en)(NH_3)_3(H_2O)]^{+}$ absorbs longer wavelength for *d-d* transition.

2. (B, D)

(A) Both Cu^{2+} and Mn^{2+} give green colour flame in flame test and hence they cannot be distinguished.

(B) Cu^{2+} belongs to group-II of cationic radical and will give precipitate of CuS in acidic medium.

(C) Cu^{2+} and Mn^{2+} both form precipitate in basic medium.

(D) $Cu^{2+}/Cu = +\ 0.34$ V; $Mn^{2+}/Mn = -\ 1.18$ V.

3. (D)

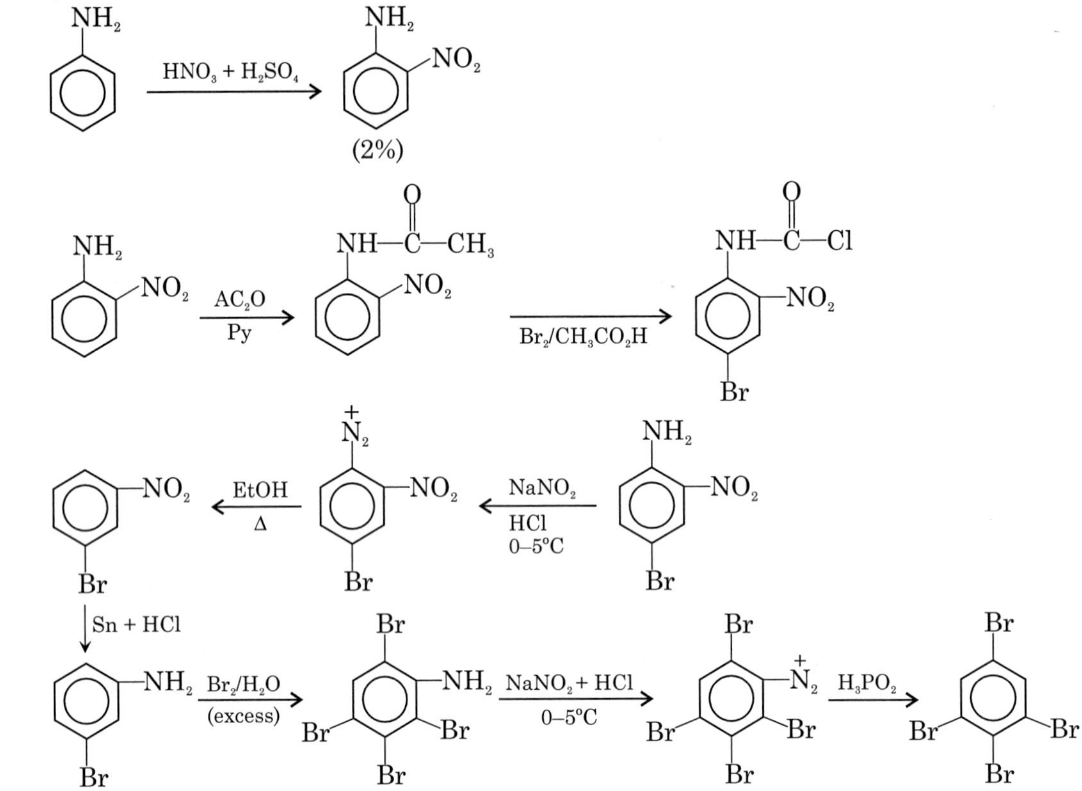

4. (D)

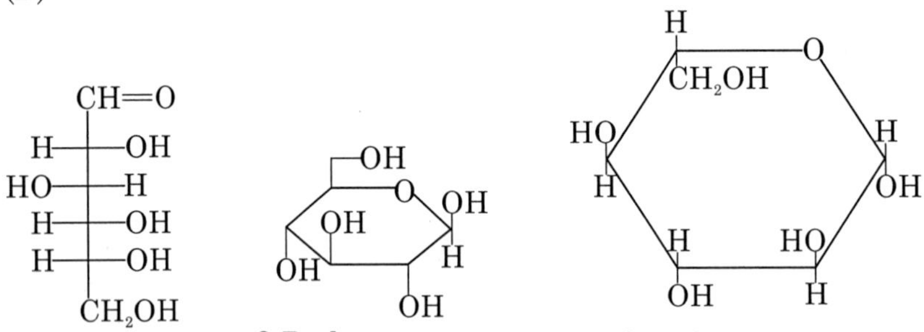

D-glucose β-D-glucopyranose β-L-glucopyranose

5. (A, D)

	A →	2B	+ C
$t = 0$	P_0	–	–
$t = t$	$P_0 - P$	$2P$	P

$P_0 + 2P = P_t$

$$K = \frac{1}{t}\ln\frac{P_0}{P_0 - P} = \frac{1}{t}\ln\frac{P_0}{P_0 - \frac{(P_t - P_0)}{2}}$$

$$K = \frac{1}{t}\ln\frac{2P_0}{3P_0 - P_t} \quad \text{or} \quad Kt + \ln 2P_0 = \ln(3P_0 - P_t)$$

$$t_{1/3} = \frac{1}{K}\ln\frac{P_0}{P_0/3} = \frac{1}{K}\ln 3 = \text{constant}$$

Rate constant does not depend on concentration.

6. (A, C)

$$A \rightleftharpoons P$$

Given: $T_2 > T_1$

$$\frac{\ln K_1}{\ln K_2} > \frac{T_2}{T_1}$$

$T_1 \ln K_1 > T_2 \ln K_2$

$-\Delta G_1^0 > -\Delta G_2^0$

$(-\Delta H^0 + T_1\Delta S^0) > (-\Delta H^0 + T_2\Delta S^0)$

$T_1\Delta S^0 > T_2\Delta S^0$

$\Delta S^0 < 0$

7. (5 or 6)

N_2O_3 = 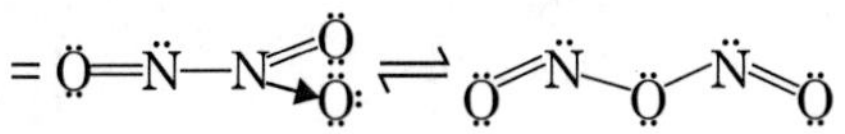

N_2O_5 = 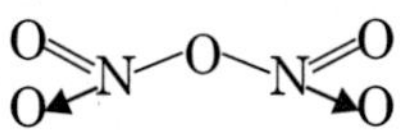

P_4O_6 = 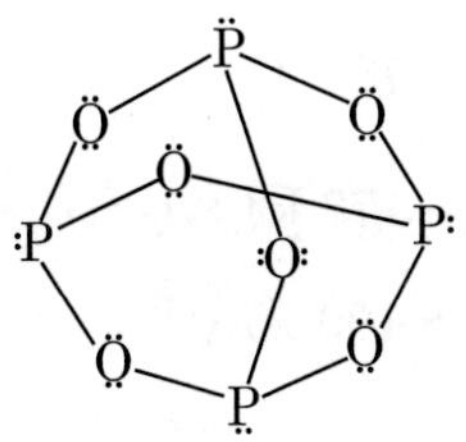

P_4O_7 = 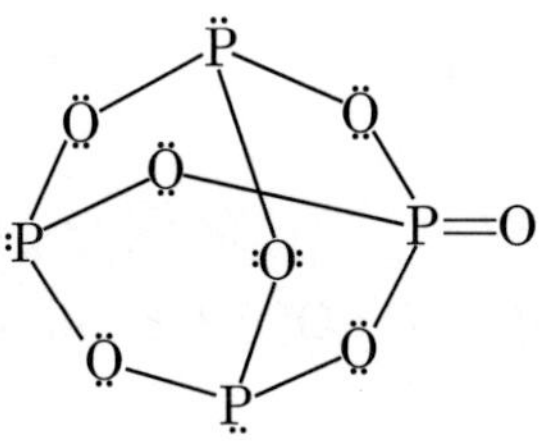

$H_4P_2O_5$ = 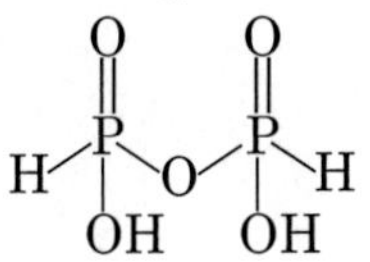

$H_2S_2O_3$ = 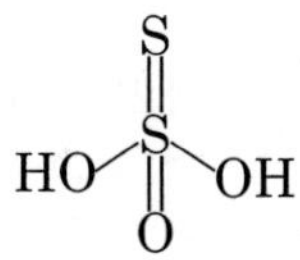

$H_5P_3O_{10}$ = 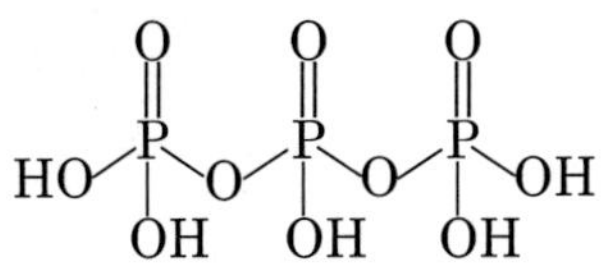

$H_2S_2O_5$ = 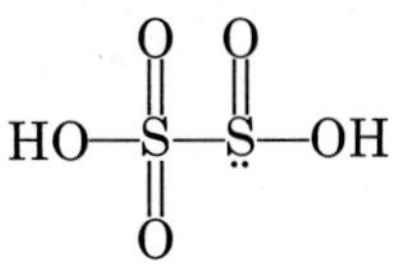

8. (6.47)

$PbS + O_2 \rightarrow Pb + SO_2$

$\frac{1000}{32}$ mol $\quad \frac{1000}{32} \times 207$ g

mol of Pb = mol of $O_2 = \frac{1000}{32}$ mol

Hence, mass of Pb $= \frac{1000}{32} \times 207$ g $= \frac{207}{32}$ kg $= 6.47$ kg

9. (126)

$$\underset{a\text{ mole}}{MnCl_2} + K_2S_2O_8 + H_2O \rightarrow \underset{a\text{ mole}}{KMnO_4} + H_2SO_4 + HCl$$

$$C_2O_4^{2-} + MnO_4^- \xrightarrow{H^+} CO_2$$

m_{eq} of $C_2O_4^{2-} = m_{eq}$ of MnO_4^-

$2 \times 0.225/90 = a \times 5$

$a = 1 \times [55 + 71] = 126$ mg

10. (7)

11. (495)

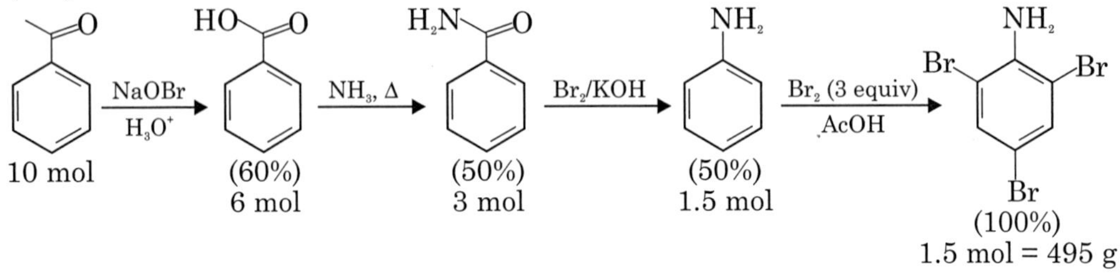

12. (– 14.6)

$$2Cu(s) + \frac{1}{4}O_2(g) \rightarrow Cu_2O(s) \qquad \Delta G^0 = -78 \text{ kJ}$$

$$H_2(g) + \frac{1}{2}O_2 \rightarrow H_2O(g) \qquad \Delta G^0 = -178 \text{ kJ} \times (-1)$$

Hence, $2Cu(s) + H_2O(g) \rightarrow Cu_2O(s) + H_2(g) \qquad \Delta G^0 = +100$ kJ

$\Delta G = \Delta G^0 + RT \ln Q$

$$0 = 100 + \frac{8}{1000} \times 1250 \ln \frac{p_{H_2}}{p_{H_2O}}$$

$$-\frac{100 \times 1000}{8} = 1250 \ln \frac{p_{H_2}}{\left(\frac{1}{100} \times 1\right)}$$

$\ln p_{H_2} = -14.6$

13. (8500)

$A(g) + B(g) \rightleftharpoons AB(g)$

$E_{ab} - E_{af} = 2RT$ or $\Delta H = -2RT$

$$\frac{A_f}{A_b} = 4$$

$$K_{eq} = \left(\frac{K_f}{K_b}\right) = \frac{A_f e^{-E_{af}/RT}}{A_b e^{-E_{ab}/RT}} = 4(e^2)$$

$\Delta G^0 = -RT \ln K = -2500 \times \ln(4 \times e^2) = -8500$ J/mol

Hence, absolute value of $\Delta G^0 = 8500$ J/mol

14. (– 11.62)

$$A(s)\,|\,A^{n+}(aq, 2M)\,\|\,B^{2n+}(aq, \underset{E_{cell}=0}{1M})\,|\,B(s)$$

$\Delta H^\circ = 2\Delta G_0^0$

Cell reaction: $A \rightarrow A^{\Delta+} + ne^- \times 2$

$B^{2n+} + 2ne^- \rightarrow B(s)$

$$2A(s) + \underset{1M}{B}^{2n+}(aq) \rightarrow \underset{2M}{2A}^{n+}(aq) + B(s)$$

$$\Delta G = \Delta G^0 + RT \ln \frac{[A^{n+}]^2}{[B^{2n+}]}$$

$$\Delta G^0 = -RT \ln \frac{[A^{n+}]^2}{[B^{2n+}]} = -RT \ln \frac{2^2}{1} = -RT \ln 4$$

$\Delta G^0 = \Delta H^0 - T\Delta S^0$

$\Delta G^0 = 2\Delta G^0 - T\Delta S^0$

$$\Delta S^0 = \frac{\Delta G^0}{T} = -\frac{RT \ln 4}{T} = -8.3 \times 2 \times 0.7 = -11.62 \text{ J K}^{-1} \text{ mol}^{-1}$$

15. (C)

(1) $[FeF_6]^{4-}$

$Fe^{2+} = [Ar]\ 3d^6$, Fe^{2+}

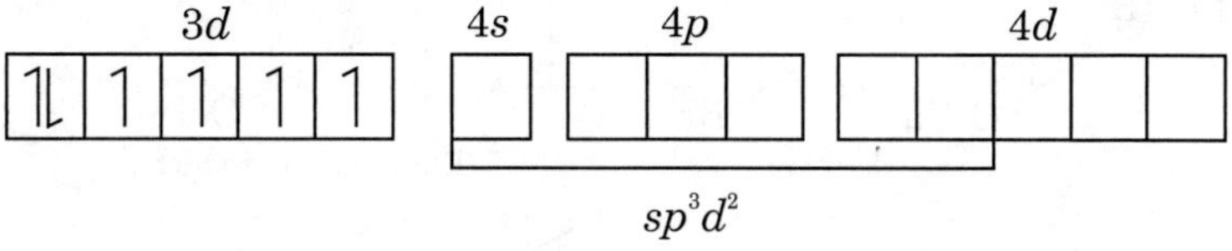

(2) $[Ti(H_2O)_3Cl_3]$

$Ti^{3+} = [Ar]\ 3d^1$, Ti^{3+}

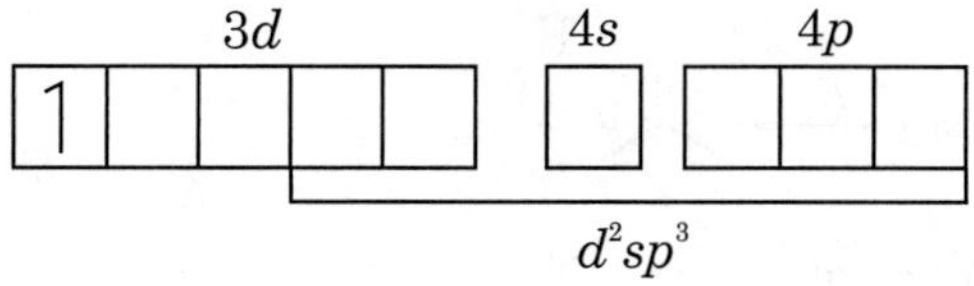

(3) $Cr(NH_3)_6]^{3+}$

$Cr^{3+} = [Ar]\ 3d^3$, Cr^{3+}

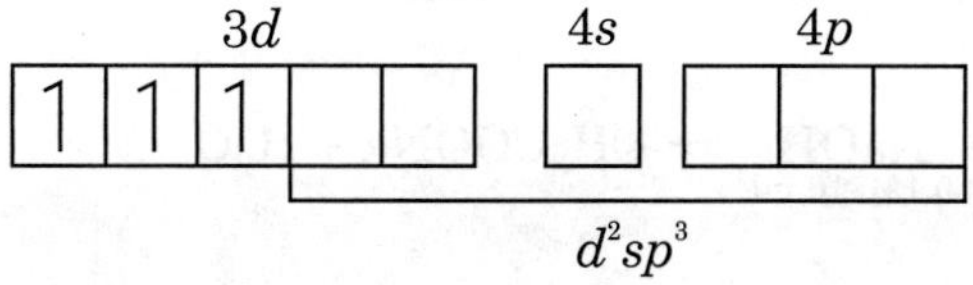

(4) $[FeCl_4]^{2-}$

Fe^{2+} = [Ar] $3d^6$, Fe^{2+} 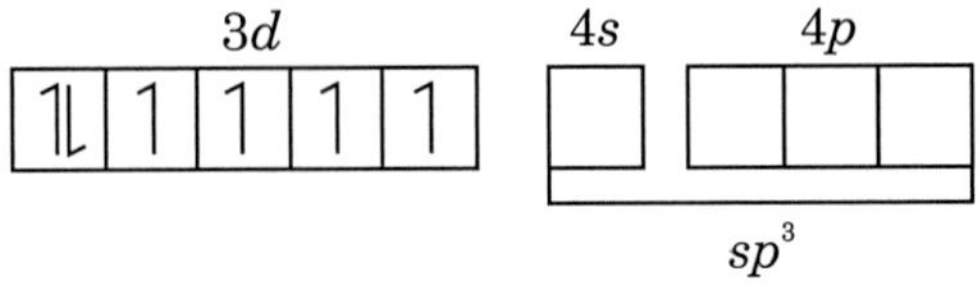

(5) $[Ni(CO)_4]$

Ni: $3d^8 4s^2$

Ni 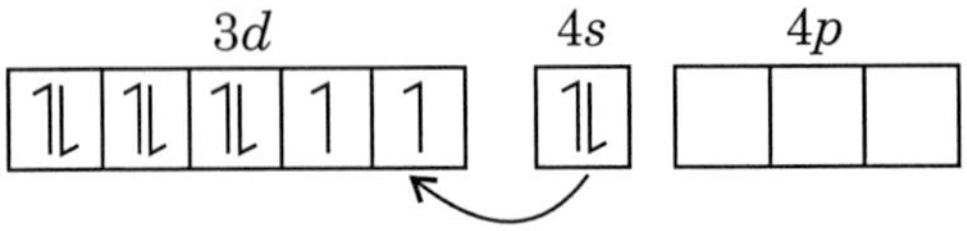

Back pairing of electrons due to the presence of strong field ligand CO

Ni 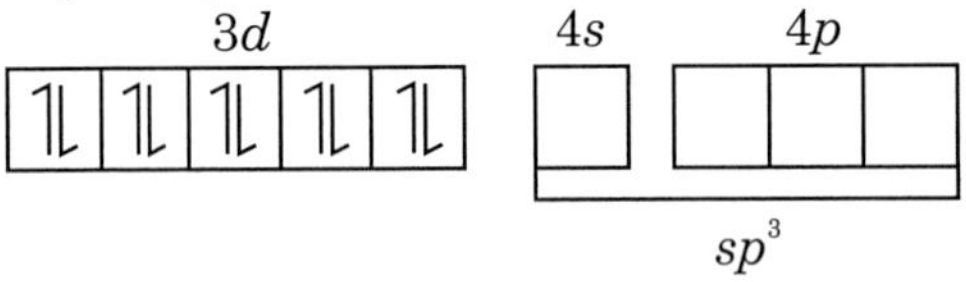

(6) $[Ni(CN_4)]^{2-}$

Ni^{2+}: $3d^8$

Ni^{2+} 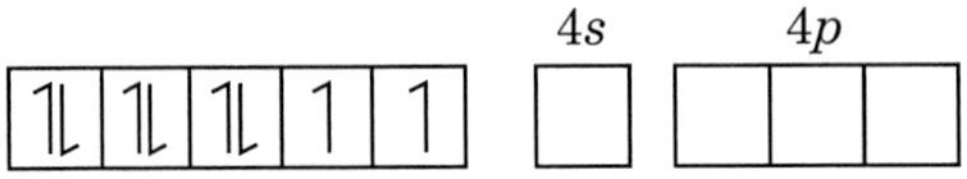

Electron pairing takes place due to the presence of strong field ligand CO.

Ni^{2+} 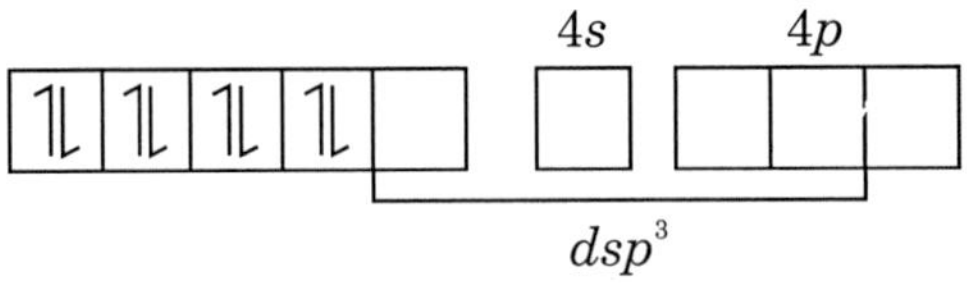

16. (D)

17. (B)

P. 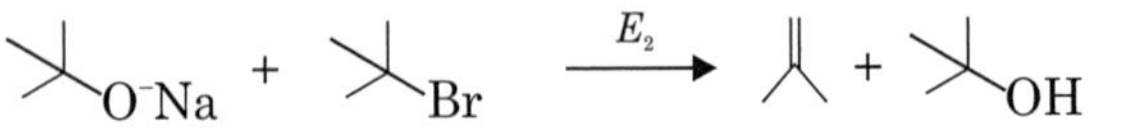

(Elimination products)

Q. 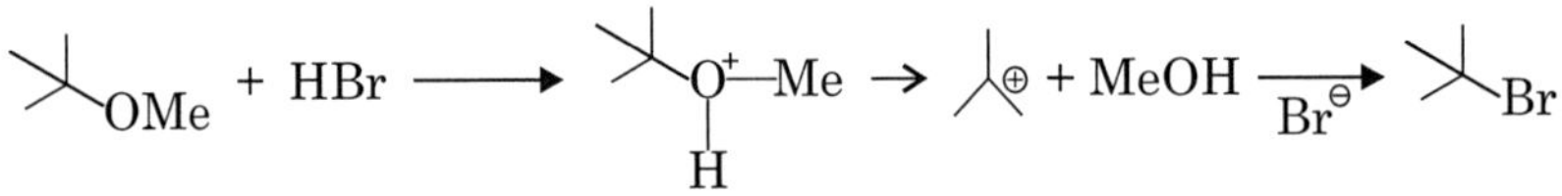

R. 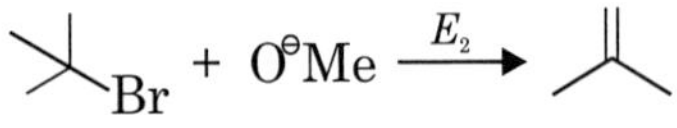

S. 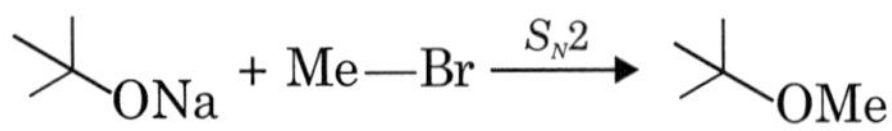

18. (D)

P. $\underset{0.1M,\ 20\ ml}{CH_3COOH} + \underset{0.1M,\ 10\ ml}{NaOH} \rightarrow CH_3COONa + H_2O$

$pH = pK_0$

$[H^+]$ will not change on dilution.

Correct match: P-1.

Q. $\underset{0.1M,\ 20\,ml}{CH_3COOH} + \underset{0.1M,\ 20\,ml}{NaOH} \rightarrow \underset{0.05M}{CH_3COONa} + H_2O$

$$[OH^-] = \sqrt{K_H C} = \sqrt{\left(\frac{k_w}{k_a} C\right)}$$

$$[H^+]_1 = \sqrt{\frac{k_w k_a}{C}}$$

$$\frac{[H^+]_2}{[H^+]_1} = \sqrt{\frac{C_1}{C_2}} = \sqrt{\frac{0.05}{0.025}} = \sqrt{2}$$

Correct match: Q-5

R. $\underset{0.1M,\ 20\ ml}{NH_4OH} + \underset{0.1M,\ 20\ ml}{HCl} \rightarrow \underset{0.05M}{NH_4Cl}$

$$[H^+] = \sqrt{K_H C}$$

$$\frac{[H^+]_2}{[H^+]_1} = \sqrt{\frac{C_2}{C_1}} = \frac{1}{\sqrt{2}}$$

Correct match: R-4

S. Because of dilution solubility does not change and hence $[H^+]$ = constant.

APPENDICES

Appendix—I: Long Form of the Periodic Table

Atomic number → 6
Symbol → C
12.011 ← Atomic mass

GROUP NUMBER

IUPAC (1984)	1	2	3	4	5	6	7	8	9	10	11	12	13	14	15	16	17	18
Period number 1	1 H 1.0079																	2 He 4.0026
2	3 Li 6.941	4 Be 9.01218											5 B 10.811	6 C 12.011	7 N 14.0067	8 O 15.9994	9 F 18.9984	10 Ne 20.179
3	11 Na 22.98977	12 Mg 24.305											13 Al 26.98154	14 Si 28.0855	15 P 30.9734	16 S 32.066	17 Cl 35.453	18 Ar 39.948
4	19 K 39.0983	20 Ca 40.08	21 Sc 44.9559	22 Ti 47.88	23 V 50.9415	24 Cr 51.996	25 Mn 54.9380	26 Fe 55.847	27 Co 58.9332	28 Ni 58.69	29 Cu 63.546	30 Zn 65.38	31 Ga 69.72	32 Ge 72.59	33 As 74.9216	34 Se 78.96	35 Br 79.909	36 K 83.80
5	37 Rb 85.4678	38 Sr 87.62	39 Y 88.9059	40 Zr 91.22	41 Nb 92.9064	42 Mo 95.94	43 Tc (99)	44 Ru 101.07	45 Rh 102.905	46 Pd 106.42	47 Ag 107.87	48 Cd 112.40	49 In 114.82	50 Sn 118.69	51 Sb 121.75	52 Te 127.60	53 I 126.904	54 Xe 131.29
6	55 Cs 132.9054	56 Ba 137.34	57 La* 138.91	72 Hf 178.49	73 Ta 180.95	74 W 183.85	75 Re 186.206	76 Os 190.2	77 Ir 192.22	78 Pt 195.09	79 Au 196.967	80 Hg 200.59	81 Tl 204.383	82 Pb 207.19	83 Bi 208.9804	84 Po (209)	85 At (210)	86 Rn (222)
7	87 Fr (223)	88 Ra 226.0254	89 Ac** 227.028	104 Rf 261.11	105 Db 262.1	106 Sg (263)	107 Bh 262.12	108 Hs (265)	109 Mt (266)	110 Ds (269)	111 Rg (272.2)	112 Cn (285)	113 Uut (284)	114 Fl (289)	115 Uup (289)	116 Lv (291)	117 Uus	118 Uuo (294)

*Lanthanoids	58 Ce 140.12	59 Pr 140.907	60 Nd 144.24	61 Pm (145)	62 Sm 150.36	63 Eu 151.96	64 Gd 157.25	65 Tb 158.92	66 Dy 162.50	67 Ho 164.9304	68 Er 167.26	69 Tm 168.93	70 Yb 173.04	71 Lu 174.967
**Actinoids	90 Th 232.0381	91 Pa 231.0359	92 U 238.0289	93 Np 237.0482	94 Pu (242)	95 Am (243)	96 Cm (247)	97 Bk (247)	98 Cf (251)	99 Es (254)	100 Fm (257)	101 Md (258)	102 No (259)	103 Lr (260)

The element 117 is not known but is included to show its expected position.

Appendix—II
Electronic Configuration of Elements

Atomic number	Element	1	2		3			4				5				6				7			
		s	s	p	s	p	d	s	p	d	f	s	p	d	f	s	p	d	f	s	p	d	f
1	H	1																					
2	He	2																					
3	Li	2	1																				
4	Be	2	2																				
5	B	2	2	1																			
6	C	2	2	2																			
7	N	2	2	3																			
8	O	2	2	4																			
9	F	2	2	5																			
10	Ne	2	2	6																			
11	Na	2	2	6	1																		
12	Mg	2	2	6	2																		
13	Al	2	2	6	2	1																	
14	Si	2	2	6	2	2																	
15	P	2	2	6	2	3																	
16	S	2	2	6	2	4																	
17	Cl	2	2	6	2	5																	
18	Ar	2	2	6	2	6																	
19	K	2	2	6	2	6		1															
20	Ca	2	2	6	2	6		2															
21	Sc	2	2	6	2	6	1	2															
22	Ti	2	2	6	2	6	2	2															
23	V	2	2	6	2	6	3	2															
24	Cr	2	2	6	2	6	5	1															
25	Mn	2	2	6	2	6	5	2															
26	Fe	2	2	6	2	6	6	2															
27	Co	2	2	6	2	6	7	2															
28	Ni	2	2	6	2	6	8	2															
29	Cu	2	2	6	2	6	10	1															
30	Zn	2	2	6	2	6	10	2															
31	Ga	2	2	6	2	6	10	2	1														

Atomic number	Element	1	2		3			4				5				6				7			
		s	*s*	*p*	*s*	*p*	*d*	*s*	*p*	*d*	*f*	*s*	*p*	*d*	*f*	*s*	*p*	*d*	*f*	*s*	*p*	*d*	*f*
32	Ge	2	2	6	2	6	10	2	2														
33	As	2	2	6	2	6	10	2	3														
34	Se	2	2	6	2	6	10	2	4														
35	Br	2	2	6	2	6	10	2	5														
36	Kr	2	2	6	2	6	10	2	6														
37	Rb	2	2	6	2	6	10	2	6			1											
38	Sr	2	2	6	2	6	10	2	6			2											
39	Y	2	2	6	2	6	10	2	6	1		2											
40	Zr	2	2	6	2	6	10	2	6	2		2											
41	Nb	2	2	6	2	6	10	2	6	4		1											
42	Mo	2	2	6	2	6	10	2	6	5		1											
43	Tc	2	2	6	2	6	10	2	6	5		2											
44	Ru	2	2	6	2	6	10	2	6	7		1											
45	Rh	2	2	6	2	6	10	2	6	8		1											
46	Pd	2	2	6	2	6	10	2	6	10													
47	Ag	2	2	6	2	6	10	2	6	10		1											
48	Cd	2	2	6	2	6	10	2	6	10		2											
49	In	2	2	6	2	6	10	2	6	10		2	1										
50	Sn	2	2	6	2	6	10	2	6	10		2	2										
51	Sb	2	2	6	2	6	10	2	6	10		2	3										
52	Te	2	2	6	2	6	10	2	6	10		2	4										
53	I	2	2	6	2	6	10	2	6	10		2	5										
54	Xe	2	2	6	2	6	10	2	6	10		2	6										
55	Cs	2	2	6	2	6	10	2	6	10		2	6			1							
56	Ba	2	2	6	2	6	10	2	6	10		2	6			2							
57	La	2	2	6	2	6	10	2	6	10		2	6	1		2							
58	Ce	2	2	6	2	6	10	2	6	10	1	2	6	1		2							
59	Pr	2	2	6	2	6	10	2	6	10	3	2	6			2							
60	Nd	2	2	6	2	6	10	2	6	10	4	2	6			2							
61	Pm	2	2	6	2	6	10	2	6	10	5	2	6			2							
62	Sm	2	2	6	2	6	10	2	6	10	6	2	6			2							
63	Eu	2	2	6	2	6	10	2	6	10	7	2	6			2							
64	Gd	2	2	6	2	6	10	2	6	10	7	2	6	1		2							

Atomic number	Element	1	2		3			4				5				6				7			
		s	*s*	*p*	*s*	*p*	*d*	*s*	*p*	*d*	*f*	*s*	*p*	*d*	*f*	*s*	*p*	*d*	*f*	*s*	*p*	*d*	*f*
65	Tb	2	2	6	2	6	10	2	6	10	9	2	6			2							
66	Dy	2	2	6	2	6	10	2	6	10	10	2	6			2							
67	Ho	2	2	6	2	6	10	2	6	10	11	2	6			2							
68	Er	2	2	6	2	6	10	2	6	10	12	2	6			2							
69	Tm	2	2	6	2	6	10	2	6	10	13	2	6			2							
70	Yb	2	2	6	2	6	10	2	6	10	14	2	6			2							
71	Lu	2	2	6	2	6	10	2	6	10	14	2	6	1		2							
72	Hf	2	2	6	2	6	10	2	6	10	14	2	6	2		2							
73	Ta	2	2	6	2	6	10	2	6	10	14	2	6	3		2							
74	W	2	2	6	2	6	10	2	6	10	14	2	6	4		2							
75	Re	2	2	6	2	6	10	2	6	10	14	2	6	5		2							
76	Os	2	2	6	2	6	10	2	6	10	14	2	6	6		2							
77	Ir	2	2	6	2	6	10	2	6	10	14	2	6	7		2							
78	Pt	2	2	6	2	6	10	2	6	10	14	2	6	9		1							
79	Au	2	2	6	2	6	10	2	6	10	14	2	6	10		1							
80	Hg	2	2	6	2	6	10	2	6	10	14	2	6	10		2							
81	Tl	2	2	6	2	6	10	2	6	10	14	2	6	10		2	1						
82	Pb	2	2	6	2	6	10	2	6	10	14	2	6	10		2	2						
83	Bi	2	2	6	2	6	10	2	6	10	14	2	6	10		2	3						
84	Po	2	2	6	2	6	10	2	6	10	14	2	6	10		2	4						
85	At	2	2	6	2	6	10	2	6	10	14	2	6	10		2	5						
86	Rn	2	2	6	2	6	10	2	6	10	14	2	6	10		2	6						
87	Fr	2	2	6	2	6	10	2	6	10	14	2	6	10		2	6			1			
88	Ra	2	2	6	2	6	10	2	6	10	14	2	6	10		2	6			2			
89	Ac	2	2	6	2	6	10	2	6	10	14	2	6	10		2	6	1		2			
90	Th	2	2	6	2	6	10	2	6	10	14	2	6	10		2	6	2		2			
91	Pa	2	2	6	2	6	10	2	6	10	14	2	6	10	2	2	6	1		2			
92	U	2	2	6	2	6	10	2	6	10	14	2	6	10	3	2	6	1		2			
93	Np	2	2	6	2	6	10	2	6	10	14	2	6	10	4	2	6	1		2			
94	Pu	2	2	6	2	6	10	2	6	10	14	2	6	10	6	2	6			2			
95	Am	2	2	6	2	6	10	2	6	10	14	2	6	10	7	2	6			2			
96	Cm	2	2	6	2	6	10	2	6	10	14	2	6	10	7	2	6	1		2			
97	Bk	2	2	6	2	6	10	2	6	10	14	2	6	10	9	2	6			2			
98	Cf	2	2	6	2	6	10	2	6	10	14	2	6	10	10	2	6			2			
99	Es	2	2	6	2	6	10	2	6	10	14	2	6	10	11	2	6			2			

Atomic number	*Element*	1	2		3			4				5				6				7			
		s	*s*	*p*	*s*	*p*	*d*	*s*	*p*	*d*	*f*	*s*	*p*	*d*	*f*	*s*	*p*	*d*	*f*	*s*	*p*	*d*	*f*
100	Fm	2	2	6	2	6	10	2	6	10	14	2	6	10	12	2	6			2			
101	Md	2	2	6	2	6	10	2	6	10	14	2	6	10	13	2	6			2			
102	No	2	2	6	2	6	10	2	6	10	14	2	6	10	14	2	6			2			
103	Lr	2	2	6	2	6	10	2	6	10	14	2	6	10	14	2	6	1		2			
104	Rf	2	2	6	2	6	10	2	6	10	14	2	6	10	14	2	6	2		2			
105	Db	2	2	6	2	6	10	2	6	10	14	2	6	10	14	2	6	3		2			
106	Sg	2	2	6	2	6	10	2	6	10	14	2	6	10	14	2	6	4		2			
107	Bh	2	2	6	2	6	10	2	6	10	14	2	6	10	14	2	6	5		2			
108	Hs	2	2	6	2	6	10	2	6	10	14	2	6	10	14	2	6	6		2			
109	Mt	2	2	6	2	6	10	2	6	10	14	2	6	10	14	2	6	7		2			
110	Ds	2	2	6	2	6	10	2	6	10	14	2	6	10	14	2	6	8		2			
111	Rg	2	2	6	2	6	10	2	6	10	14	2	6	10	14	2	6	10		1			
112	Cn	2	2	6	2	6	10	2	6	10	14	2	6	10	14	2	6	10		2			
113	Uut †	2	2	6	2	6	10	2	6	10	14	2	6	10	14	2	6	10		2	1		
114	Fl	2	2	6	2	6	10	2	6	10	14	2	6	10	14	2	6	10		2	2		
115	Uup †	2	2	6	2	6	10	2	6	10	14	2	6	10	14	2	6	10		2	3		
116	Lv	2	2	6	2	6	10	2	6	10	14	2	6	10	14	2	6	10		2	4		
117*	Uus	2	2	6	2	6	10	2	6	10	14	2	6	10	14	2	6	10		2	5		
118	Duo †	2	2	6	2	6	10	2	6	10	14	2	6	10	14	2	6	10		2	6		

* Element yet to be discovered. † Official IUPAC name yet to be announced.

Appendix—III
Atomic Masses of Elements

Name	*Symbol*	*Atomic number*	*Atomic mass*
Actinium	Ac	89	(227)
Aluminium	Al	13	26.9815
Americium	Am	95	(243)
Antimony	Sb	51	121.75
Argon	Ar	18	39.948
Arsenic	As	33	74.9216
Astatine	At	85	(210)
Barium	Ba	56	137.34
Berkelium	Bk	97	(247)
Beryllium	Be	4	9.0122
Bismuth	Bi	83	208.980
Bohrium	Bh	107	262.12
Boron	B	5	10.811
Bromine	Br	35	79.909
Cadmium	Cd	48	112.40
Calcium	Ca	20	40.08
Californium	Cf	98	(251)
Carbon	C	6	12.01115

Name	*Symbol*	*Atomic number*	*Atomic mass*
Cerium	Ce	58	140.12
Caesium	Cs	55	132.905
Chlorine	Cl	17	35.453
Chromium	Cr	24	51.996
Cobalt	Co	27	58.9332
Copernicium	Cn	112	(285)
Copper	Cu	29	63.54
Curium	Cm	96	(247)
Darmstadtium	Ds	110	(269)
Dubnium	Db	105	262.1
Dysprosium	Dy	66	162.50
Einstenium	Es	99	(254)
Erbium	Er	68	167.26
Europium	Eu	63	151.96
Flerovium	Fl	114	(289)
Fermium	Fm	100	(253)
Fluorine	F	9	18.9984
Francium	Fr	87	(233)
Gadolinium	Gd	64	157.25
Gallium	Ga	31	69.72
Germanium	Ge	32	72.59
Gold	Au	79	196.967
Hafnium	Hf	72	178.49
Hassnium	Hs	108	(265)
Helium	He	2	4.0026
Holmium	Ho	67	164.930
Hydrogen	H	1	1.00797
Indium	In	49	114.82
Iodine	I	53	126.904
Iridium	Ir	77	192.2
Iron	Fe	26	55.847
Krypton	Kr	36	83.80
Lanthanum	La	57	138.91
Lawrencium	Lr	103	(257)
Lead	Pb	82	207.19
Lithium	Li	3	6.939
Livermorium	Lv	116	(291)
Lutetium	Lu	71	174.97
Magnesium	Mg	12	24.312
Manganese	Mn	25	54.9380

Name	*Symbol*	*Atomic number*	*Atomic mass*
Meitnerium	Mt	109	(266)
Mendeleevium	Md	101	(256)
Mercury	Hg	80	200.59
Molybdenum	Mo	42	95.94
Neodymium	Nd	60	141.24
Neon	Ne	10	20.183
Neptunium	Np	93	(237)
Nickel	Ni	28	58.71
Niobium	Nb	41	92.906
Nitrogen	N	7	14.0467
Nobelium	No	102	(254)
Osmium	Os	76	190.2
Oxygen	O	8	15.9994
Palladium	Pd	46	106.4
Phosphorus	P	15	30.9734
Platinum	Pt	78	195.09
Plutonium	Pu	94	(242)
Polonium	Po	84	(210)
Potassium	K	19	39.102
Praseodymium	Pr	59	104.907
Promethium	Pm	61	(145)
Protactinium	Pa	91	(231)
Radium	Ra	88	(226)
Radon	Rn	86	(222)
Rhenium	Re	75	186.2
Rhodium	Rh	45	102.905
Roentgenium	Rg	111	(272.2)
Rubidium	Rb	37	85.47
Ruthenium	Ru	44	101.07
Rutherfordium	Rf	104	261.11
Samarium	Sm	62	150.35
Scandium	Sc	21	44.946
Seaborgium	Sg	106	263.118
Selenium	Se	34	78.96
Silicon	Si	14	28.086
Silver	Ag	47	107.870
Sodium	Na	11	22.9898
Strontium	Sr	38	87.62
Sulphur	S	16	32.064
Tantalum	Ta	73	180.94

Name	Symbol	Atomic number	Atomic mass
Technetium	Tc	43	(99)
Tellurium	Te	51	127.604
Terbium	Tb	65	158.92
Thallium	Tl	81	204.37
Thorium	Th	90	232.038
Thulium	Tm	69	168.93
Tin	Sn	50	118.69
Titanium	Ti	22	47.90
Tungsten	W	74	183.85
Ununpentium	Uup	115	(289)
Ununoctium	Uuo	118	(294)
Ununtrium	Uut	113	(284)
Uranium	U	92	238.03
Vanadium	V	23	50.942
Xenon	Xe	54	131.30
Yetterbium	Yb	70	173.04
Yttrium	Y	39	88.905
Zinc	Zn	30	65.37
Zirconium	Zr	40	91.22

Number in parentheses gives the mass number of the most stable isotope.

Appendix—IV
Some Useful Constants and Conversion Factors

Constant	Symbol	Value
Acceleration due to gravity	g	$980.665\ \text{cm s}^{-2}$
Atmospheric pressure	P	1.01325×10^{6} dyne cm^{-2}
Avogadro number	N	$6.0225 \times 10^{23}\ \text{mol}^{-1}$
Bohr radius	a_0	0.529×10^{-8} cm
Boltzmann constant	k	1.3805×10^{-6} erg deg^{-1} = $1.3805 \times 10^{-23}\ \text{J K}^{-1}$
Charge on electron	e	-4.803×10^{-10} esu = -1.602×10^{-19} C
Faraday	F	$9.6487 \times 10^{4}\ \text{C mol}^{-1}$ = 82 mL atom/degree/mol
Gas constant	R	1.9872 cal $\text{deg}^{-1}\ \text{mol}^{-1}$ = 0.082057 L atm $\text{deg}^{-1}\ \text{mol}^{-1}$ = $8.314\ \text{J K}^{-1}\ \text{mol}^{-1}$
Mass of electron	m	9.1096×10^{-28} g = 5.4859×10^{-1} u
Mass of proton	–	1.6726×10^{-24} g = 1.007277 u
Mass of neutron	–	1.6749×10^{-24} g = 1.008665 u
Molar volume, ideal gas at STP	–	22.136 L
Planck constant	h	6.6256×10^{-27} erg s = 6.6256×10^{-34} J s
Rydberg constant	R_H	$1.097373 \times 10^{5}\ \text{cm}^{-1}$
Temperature	K	K = °C + 273.15°
Velocity of light in vacuum	c	$2.99795 \times 10^{10}\ \text{cm s}^{-1}$
Velocity of α-particle	–	$2.0 \times 10^{9}\ \text{cm s}^{-1}$
Velocity of β-ray	–	$2.36 \times 10^{10}\ \text{cm s}^{-1}$
Velocity of γ-ray	–	$3.0 \times 10^{10}\ \text{cm s}^{-1}$

Conversion Factors

Unit	*Abbreviation*	*Equivalent/Definition*
amu	u	1.66×10^{-24} g = 1.66×10^{-31} J = 931.48 MeV
angstrom	Å	10^{-8} cm = 10^{-10} m = 0.1 nm
atmosphere	atm	760 mmHg = 760 torr 1.01325×10^{5} N m^{-2} = 1.01325×10^{5} Pa
becquerel	bq	1 dps
calorie	cal	4.184 J
cm^{-1}	cm^{-1}	1.986×10^{-23} J
curie	c	3.7×10^{10} dps
dyne	dyne	10^{-5} N
electron volt	eV	1.6022×10^{-19} J = 1.6022×10^{-12} erg = 2.3061×10^{4} cal mol^{-1} = 8066 cm^{-1}
entropy	S	4184 J kg^{-1} K^{-1}
erg	erg	10^{-7} J = 2.390057×10^{-8} cal
Faraday	F	96487 coulomb
joule	J	10^{7} erg = 0.2390057 cal
mmHg (torr)	mmHg	1.31579×10^{-3} atm
nanometer	nm	10^{-7} cm
rutherford	rd	10^{6} dps
triple point of H_2O		273.16 K
volt	V	amp ohm
watt	W	V amp

Appendix—V
Eminent Scientists and Their Contributions

Name	*Contributions*
Alfred Rochow	Electronegativity scale
Arrhenius	Theory of electrolytic dissociation
Aston	Mass spectrograph
Avogadro	Avogadro hypothesis
Bartlett	Compounds of inert gases
Becquerel	Radioactivity
Bohr	Atomic structure
Boyle	Gas laws
Bronsted and Lowry	Acid-base concept
Brown	Brownian movement
Cavendish	Discovery of hydrogen

Name	*Contributions*
Chadwick	Discovery of neutron
Charles	Gas laws
Dalton	Atomic theory, law of partial pressures, law of multiple proportions
deBroglie	Wave nature of electron
Debye–Huckel	Theory of strong electrolytes
Dorn	Discovery of radon
Einstein	Theory of relativity, mass-energy relationship
Fajan	Polarisation of ions
Fajan, Russel and Soddy	Radiocative group displacement law
Faraday	Law of electrolysis
Gay-Lussac	Law of combining volumes
Goldberg and Waage	Law of mass action
Goldstein	Discovery of anode rays
Graham	Law of diffusion
Hann and Strassman	Nuclear fission
Helmholtz	First law of thermodynamics
Heisenberg	Uncertainty principle
Heitlor and London	Valence bond theory
Hund	Multiplicity rule
Hund and Mulliken	Molecular orbital theory
Irene Curie and F. Joliot	Artificial radioactivity
Kekule	Structure of benzene
Keyser	Adsorption
Kossel and Lewis	Electrovalency
Langmuir	Adsorption isotherm
Lavoisier	Laws of combination of mass
Lawrence	Cyclotron
Le Chatelier	Effect of *P*, *T* and C on equilibrium
Lewis	Convalency, acid-base concept
Libby	Radioactive dating
Lockyer and Frankland	Discovery of helium
Marie Curie	Discovery of radium and polonium
Maxwell	Kinetic theory of gases
Mendeleev	Periodic classification of elements
Milliken	Charge of electron
Moseley	Modern periodic law
Nernst	Electrode potential
Ostwald	Dilution law
Pauli	Exclusion principle
Pauling	Chemical bonding
Pauling and Slatter	Valence bond theory
Planck	Wave nature of light

Name	*Contributions*
Proust	Law of reciprocal proportions
Raman	Scattering of light by liquid molecules
Raoult	Relation between vapour pressure and mole fraction
Ramsay and Travers	Discovery of neon, krypton and xenon
Rayleigh and Ramsay	Discovery of argon
Ritcher	Law of definite proportions
Roentgen	Discovery of X-ray
Rutherford	Nuclear model of atom
Rutherford and Soddy	Radioactive disintegration
Schrodinger	Wave equation
Scheele	Discovery of nitrogen and chlorine
Schulze and Hardy	Coagulation of colloids by electrolytes
Seaborg	Transuranium elements
Sidgwick and Powell	Hybridisation
Smalley, Curt and Kroto	Buckminster fullerene
Soddy and Aston	Isotopes
Soddy and Fajan	Group displacement law
Sommerfield	Atomic model
Sorensen	*p*H
Sorel	Composition of ozone
Thenard	Discovery of hydrogen peroxide
Thomson	Discovery of electron, photoelectric effect
Tyndall	Scattering of light by colloidal particles
Urey	Discovery of D_2 and D_2O
van der Waals	Real gas equation
van't Hoff	Relation between rate constant, equilibrium constant and temperature
Werner	Theory of coordinate linkage
Wohler	Synthesis of first organic compound (urea)
Yukawa	Discovery of mesons

Appendix—VI
Some Important Ores/Minerals and Their Composition*

Name	*Composition*
1. Alumina	$\underline{Al}_2O_3$
2. Alunite	$K_2SO_4 \cdot \underline{Al}_2(SO_4)_3 \cdot 4\underline{Al}(OH)_3$
3. Azurite	$\underline{Cu}(OH)_2 \cdot 2\underline{Cu}CO_3$
4. Anhydrite	$\underline{Ca}SO_4$
5. Anglesite	$\underline{Pb}SO_4$

* The underlined metal may be manufactured from the ore/mineral.

	Name	*Composition*
6.	Argentite	$\underline{Ag}_2S$
7.	Asbestos	$Ca\underline{Mg}_3(SiO_3)_4$
8.	Bauxite	$\underline{Al}_2O_3\cdot 2H_2O$
9.	Bengal salt petre	$\underline{K}NO_3$
10.	Borax (tincal)	$Na_2\underline{B}_4O_7\cdot 10H_2O$
11.	Calcite	$\underline{Ca}CO_3$
12.	Calamine	$\underline{Zn}CO_3$
13.	Carnallite	$KCl\cdot\underline{Mg}Cl_2\cdot 6H_2O$
14.	Cassiterite	$\underline{Sn}O_2$
15.	Cerussite	$\underline{Pb}CO_3$
16.	Chalk (aragonite, marble, limestone)	$\underline{Ca}CO_3$
17.	Chile salt petre	$\underline{Na}NO_3$
18.	Chlorapatite	$\underline{Ca}_5(PO_4)_3Cl$
19.	Cinnabar	$\underline{Hg}S$
20.	Clay	$\underline{Al}_2O_3\cdot 2SiO_2\cdot 2H_2O$
21.	Copper glance	$\underline{Cu}_2S$
22.	Copper pyrites	$\underline{Cu}FeS_2$
23.	Corrundum	$\underline{Al}_2O_3$
24.	Cryolite	$Na_3\underline{Al}F_6$
25.	Cuprite (ruby copper)	$\underline{Cu}_2O$
26.	Diaspore	$\underline{Al}_2O_3\cdot H_2O$
27.	Dolomite	$\underline{Mg}CO_3\cdot\underline{Ca}CO_3$
28.	Epsomite (Epsom salt)	$\underline{Mg}SO_4\cdot 7H_2O$
29.	Fluorapatite	$\underline{Ca}_5(PO_4)_3F$
30.	Fluorspar	$\underline{Ca}F_2$
31.	Galena	$\underline{Pb}S$
32.	Greenockite	$\underline{Cd}S$
33.	Gypsum	$\underline{Ca}SO_4\cdot 2H_2O$
34.	Haematite (red)	$\underline{Fe}_2O_3$
35.	Heavy spar	$\underline{Ba}SO_4$
36.	Horn silver	$\underline{Ag}Cl$
37.	Iron pyrites	$\underline{Fe}S_2$
38.	Kieserite	$\underline{Mg}SO_4\cdot H_2O$
39.	Lanarkite	$\underline{Pb}O\cdot\underline{Pb}SO_4$
40.	Limonite (brown)	$\underline{Fe}_2O_3\cdot 3H_2O$
41.	Magnesite	$\underline{Mg}CO_3$
42.	Magnetite	$\underline{Fe}_3O_4$
43.	Malachite	$\underline{Cu}(OH)_2\cdot\underline{Cu}CO_3$
44.	Mica	$K_2O\cdot 3\underline{Al}_2O_3\cdot 6SiO_2\cdot 2H_2O$

	Name	*Composition*
45.	Phosphorite	$\underline{Ca}_3(PO_4)_2$
46.	Pitch blende	$\underline{U}_3O_8$
47.	Pollucite	$H_2Cs_4\underline{Al}_4(SiO_3)_9$
48.	Pyrargyrite (ruby silver)	$3\underline{Ag}_2S{\cdot}Sb_2O_3$
49.	Pyrolusite	$\underline{Mn}O_2$
50.	Rock salt	$\underline{Na}Cl$
51.	Schonite	$\underline{K}_2SO_4{\cdot}MgSO_4{\cdot}6H_2O$
52.	Siderite	$\underline{Fe}CO_3$
53.	Sylvanite	$(\underline{Ag}, Au)Te$
54.	Sylvine	$\underline{K}Cl$
55.	Talc	$\underline{Mg}_2(Si_2O_5)_2{\cdot}\underline{Mg}(OH)_2$
56.	Triphylite	$(\underline{Li}Na)_3(PO_4)_2{\cdot}(FeMn)_3(PO_4)_4$
57.	Trona (natron)	$\underline{Na}H_2CO_3{\cdot}NaHCO_3{\cdot}2H_2O$
58.	Whitherite	$\underline{Ba}CO_3$
59.	Willemite	$\underline{Zn}_2SiO_4$
60.	Zinc blende	$\underline{Zn}S$
61.	Zincite	$\underline{Zn}O$
62.	Zircon	$\underline{Zr}SiO_4$

Appendix—VII
Some Important Compounds and Their Formulae*

	Compound	*Formula*
1.	Absolute alcohol	C_2H_5OH
2.	Alum	$M^I_2SO_4 \cdot M^{III}_2(SO_4)_3 \cdot 24H_2O$ $(M^I = Na^+, K^+, NH_4^+;\ M^{III} = Al^{3+}, Cr^{3+}, Fe^{3+})$
3.	Aqua fortis	HNO_3
4.	Anhydrone	$Mg(ClO_4)_2$
5.	Aspirin	*o*-$CH_3OCO—C_6H_4—COOH$
6.	Baking soda	$NaHCO_3$
7.	Baryta	BaO
8.	Baryta water	A solution of $Ba(OH)_2$
9.	Black oxide	MnO_2
10.	Bleaching power	$CaOCl_2$
11.	Blue vitriol	$CuSO_4{\cdot}5H_2O$
12.	Borax (tincal)	$Na_2B_4O_7{\cdot}10H_2O$

* For more compounds and their formulae, see Appendix-VI.

	Compound	*Formula*
13.	Borazole (inorganic benzene)	$B_3N_3H_6$
14.	Brine	A solution of NaCl
15.	Brown ring compound	$[Fe(H_2O)_5NO]SO_4$
16.	Butter of tin (oxymuriate of tin)	$SnCl_4 \cdot 5H_2O$
17.	Calgon	$Na_2[Na_4(PO_3)_6]$
18.	Calomel	Hg_2Cl_2
19.	Carbolic acid	C_6H_5OH
20.	Carborundum	SiC
21.	Caro's acid	H_2SO_5
22.	Caustic potash	KOH
23.	Caustic soda	NaOH
24.	Chloropicrin	$CCl_3—NO_2$
25.	Cream of tarter	KOOCCH(OH)CH(OH)COOH
26.	Chrome alum	$K_2SO_4 \cdot Cr_2(SO_4)_3 \cdot 24H_2O$
27.	Chrome yellow	$PbCrO_4$
28.	Chromyl chloride	CrO_2Cl_2
29.	Corrosive sublimate	$HgCl_2$
30.	Crystal carbonate	$Na_2CO_3 \cdot H_2O$
31.	Dead burnt	$CaSO_4$ (anhydrous)
32.	DDT	Dichlorodiphenyltrichloroethane
33.	Dry ice	Solid CO_2
34.	Ferric alum	$K_2SO_4 \cdot Fe_2(SO_4)_3 \cdot 24H_2O$
35.	Fisher salt	$K_3[Co(NO_2)_6]$
36.	Fremy salt	KHF_2
37.	Freon	CCl_2F_2
38.	Fullerene	An allotrope (C_{60}) of carbon
39.	Gammexane (BHC)	$C_6H_6Cl_6$
40.	Glauber salt	$Na_2SO_4 \cdot 10H_2O$
41.	Graphite	An allotrope of carbon
42.	Grape sugar	$C_6H_{12}O_6$ (fructose)
43.	Gun power	KNO_3 (75%) + S (12%) + charcoal (13%)
44.	Gypsom salt (gypsum)	$CaSO_4 \cdot 2H_2O$
45.	Inorganic rubber	$(PNCl_2)_n$
46.	Hair salt	$Al_2(SO_4)_3 \cdot 18H_2O$
47.	Hydrolith	CaH_2

	Compound	Formula
48.	Hypo	$Na_2S_2O_3 \cdot 5H_2O$
49.	Jawelle water	A solution of KOCl
50.	Jeweller's rougue	Fe_2O_3
51.	Killed spirits	$ZnCl_2 + ZnO$
52.	King of chemicals (oil of vitriol)	H_2SO_4
53.	Laughing gas	N_2O
54.	Litharge	PbO
55.	Lithopone	$ZnS + BaSO_4$
56.	Lunar caustic	$AgNO_3$
57.	Marshal's acid	$H_2S_2O_8$ (perdisulphuric acid)
58.	Tear gas	$CCl_3—NO_2$
59.	Marsh gas	CH_4
60.	Microcosmic salt	$Na(NH_4)HPO_4 \cdot 4H_2O$
61.	Milk of lime (slaked lime or lime water)	$Ca(OH)_2$
62.	Milk of magnesia	$Mg(OH)_2$
63.	Minium	Pb_3O_4
64.	Mohr salt	$FeSO_4 \cdot (NH_4)_2SO_4 \cdot 6H_2O$
65.	Mosaic gold	SnS_2
66.	Muriatic acid	HCl
67.	Mustard gas	$ClCH_2CH_2SCH_2CH_2Cl$
68.	Nitre cake	$NaHSO_4$
69.	Oleum	$H_2S_2O_7$
70.	Oil of mirbane	$C_6H_5NO_2$
71.	Oil of vitriol	H_2SO_4
72.	Oil of winter green (methyl salicylate)	*o*-$HOC_6H_4COOCH_3$
73.	Oxone	Na_2O_2
74.	Pearl ash	K_2CO_3
75.	Perhydrol	30% H_2O_2
76.	Permutit (zeolite)	$Na_2O \cdot Al_2O_3 \cdot 2SiO_2 \cdot 6H_2O$
77.	Pharoh's serpent	$Hg(SCN)_2$
78.	Philosopher's wool	ZnO
79.	Phosgene	$COCl_2$
80.	Plaster of Paris	$CaSO_4 \cdot 0.5H_2O$
81.	Prussic acid	HCN
82.	Prussian blue	$Fe_4[Fe(CN)_6]_3$
83.	Quick lime (lime)	CaO

	Compound	Formula
84.	Realgar	As_2S_3
85.	Red oxide	Pb_3O_4
86.	Rinman's green	$[Co_2(ZnO_2)_3]$
87.	Rochelle salt	NaOOCCH(OH)CH(OH)COOK
88.	Rust	$Fe_2O_3 \cdot xH_2O$
89.	Salt cake	Na_2SO_4 (anhydrous)
90.	Scheele's green	$Cu_3(AsO_3)_2$
91.	Smelling salt	$(NH_4)_2CO_3$
92.	Soda ash	Na_2CO_3 (anhydrous)
93.	Solammoniac	NH_4Cl
94.	Spirit of wine (grain alcohol)	C_2H_5OH
95.	Sugar of lead	$Pb(CH_3COO)_2$
96.	Talc	$3\ MgO \cdot 4SiO_2 \cdot H_2O$
97.	Tarter emetic	KOOCCH(OH)CH(OH)COO(SbO)
98.	Teflon	$(C_2F_4)_n$
99.	TEL	$(C_2H_5)_4Pb$
100.	TNB	Trinitrobenzene (an explosive)
101.	TNT	Trinitrotoluene (an explosive)
102.	Turnbull's blue	$KFe[Fe(CN)_6]$
103.	Ultramarine	$Na_5Al_3Si_4S_3O_{12}$
104.	Vinegar	CH_3COOH (7-8%)
105.	Washing soda	$Na_2CO_3 \cdot 10H_2O$
106.	Water glass	Na_2SiO_3
107.	White vitriol	$ZnSO_4 \cdot 7H_2O$
108.	Wood spirit	CH_3OH
109.	Zeisse's salt	$K[PtCl_3(C_2H_4)]$
109.	Zinc white	ZnO

Appendix—VIII
Some Important Compounds and Their Uses

1. Acetaldehyde: as an antiseptic in nose drops.
2. Acetaldehyde ammonia: as a rubber accelerator.
3. Acetanilide: as an antiseptic under the name antifebrin.
4. Acetic anhydride: as an acetylating agent, in the manufacture of aspirin, rayon and dyes.
5. Acetone: for storing acetylene, nail polish.
6. Alum: as a mordant in dye industry, as a styptic to stop bleeding, for sizing of paper, in tanning of leather and purification of water.
7. Alumina: as an abrasive, refractory material, for preparing cement, artificial gems, in chromatography.

8. Aluminum chloride: as a catalyst in Friedel–Crafts reaction.
9. Ammonia: as a cleaner for removal of grease in dry cleaning, in the manufacture of explosive (NH_4NO_3), fertilizers [$(NH_4)_2SO_4$, $CaNH_4(NO_3)_3$, $Ca(NH_4)_2(SO_4)_2$], $NaHCO_3$, HNO_3, artificial silk.
10. Aniline: in the synthesis of dyes and dye intermediates, sulpha drugs, antioxidants and accelerators.
11. Bleaching powder: in disinfection of water, for making wool unshrinkable.
12. Butter of tin (oxymuriate of tin): as a mordant in dyeing.
13. Calcium ammonium nitrate (CAN): as a fertilizer.
14. Calcium carbonate: as a building material (marble), flux in metallurgy, in paints, distempers, tooth powder, tooth paste, medicine for indigestion, manufacture of CO_2.
15. Calcium sulphate: in the manufacture of black board chalks, plaster of Paris, H_2SO_4, $(NH_4)_2SO_4$ and cement.
16. Carbon black: as filler in rubber industry, in printer's ink and black paints.
17. Carbon dioxide: in the manufacture of Na_2CO_3 (Solvay process), white lead, aerated water, as refrigerant (dry ice), fire extinguisher and agent for artificial respiration (carbogen).
18. Carbon monoxide: as a reducing agent, in the metallurgy of Fe and Ni, in the manufacture of producer gas and water gas.
19. Carbon tetrachloride: as solvent, fire extinguisher (pyrene), fumigant and anthelmentic (wormicide).
20. Chloretone: as sedative and hypnotic.
21. Chlorine: as germicide, disinfectant, bleaching agent, in the preparation of bleaching powder, NaOCl (antiseptic), metal chlorates, tear gas, mustard gas, $COCl_2$, DDT and gammexane.
22. Chloroform: as anaesthetic, preservative for anatomical specimens, in the preparation of chloretone (hypnotic and sedative) and chloropicrin (insecticide and tear gas).
23. Chloropicrin: as insecticide and tear gas.
24. Chloroprene: in the manufacture of synthetic rubber neoprene.
25. Copper metal: in jewellery, ornaments, copper coin, calorimeter, electrical wire and cable, for electrotyping and electroplating.
26. Copper sulphate: as insecticide, fungicide (Bordeaux mixture), in dyeing, calico printing, electrotyping, electroplating, for detection of moisture and preparation of Fehling solution.
27. Ether: as anaesthetic, solvent, in perfumes, Wurtz reaction, in the preparation of Grignard reagent.
28. Ethyl acetate: in perfumes.
29. Ethylene: as an anaesthetic.
30. Ethylene dibromide: as a petrol additive.
31. Ferric oxide (Fe_2O_3): as catalyst for conversion of CO to CO_2, polishing powder in jewellery, brown pigment.
32. Formaldehyde: in silvering of mirror, in the preparation of bakelite, urotropine, as a preservative for biological specimens [formalin (40% HCHO + 8% CH_3OH + 52% H_2O)].
33. Freon-12 (CCl_2F_2): as propellant and refrigerant.
34. Gammexane (BHC): as an insecticide.
35. Green vitriol ($FeSO_4 \cdot 7H_2O$): as insecticide, mordant in dyeing, in ink and tanning industry.
36. Heavy water (D_2O): as a moderator in nuclear reactor.

37. Hydrogen peroxide: as rocket propellant, germicide, antiseptic, oxidising agent, bleaching agent, for restoring white colour of lead painting blackened by H_2S.
38. Hydrogen sulphide: as laboratory reagent and reducing agent.
39. Hypo: in the metallurgy of Ag and Au, as a fixing agent in photography, antichlor, in volumetric analysis.
40. Iodine: in medicine for treatment of goiter, for increasing the yield of egg in poultry.
41. Iodoform: as an antiseptic.
42. Lead peroxide: as oxidizing agent, in match industry and storage battery.
43. Litharge: in paints, varnishes, in the preparation of lead glass and glazed pottery.
44. Lunar caustic: as a caustic in surgery, in ink, hair dye, for silvering of mirror.
45. Magnesium metal: as fuse in aluminothermit process, in photography as source of light, in the synthesis of Grignard reagent.
46. Magnesia: in Sorel's cement ($MgCl_2 \cdot 5MgO \cdot xH_2O$) for filling of teeth, as substitute for tiles, filler in rubber industry, insulator, antacid, basic lining in steel industry, refractory lining.
47. Magnesia alba ($xMgCO_3 \cdot yMg(OH)_2 \cdot zH_2O$): in tooth paste, tooth powder, silver polish, as laxative and antacid.
48. Magnesium carbonate: as filler in paper and rubber industry, as refractory material.
49. Magnesium sulphate: as filler in paper industry, purgative, for fire proofing of fabrics.
50. Mssicot (PbO, yellow): along with glycerol used as a cement for stone and glass.
51. Metaldehyde [$(CH_2O)_3$]: as a solid fuel in lamps.
52. Nitric acid: as oxidising agent, solvent for etching names and designs on copper, brass and bronze articles, in the preparation of dyes, perfumes, drugs, artificial silk, fertilizers [$Ca(NO_3)_2$, NH_4NO_3, $CaNH_4(NO_3)_3$], explosives (dynamite, nitroglycerine, picric acid, TNB, TNT, RDX (cyclomite), gun cotton).
53. Nitric oxide: as catalyst in lead chamber process, in the manufacture of HNO_3, for detection of oxygen.
54. Nitrous acid: in the synthesis of azo dyes.
55. Nitrous oxide (laughing gas): as anaesthetic, rocket propellant.
56. Oil of vitriol (H_2SO_4): as dehydrating agent, pickling agent, in lead storage batteries, in the manufacture of fertilizers, explosives, detergents and dyes.
57. Oxygen: as rocket fuel (liquid O_2), explosive (liquid O_2 + C), in oxy-acetylene and oxy-hydrogen flames for cutting and welding of metals, for artificial respiration [O_2 + He or $O_2 + CO_2$ (carbogen)].
58. Ozone: as oxidizing agent, bleaching agent, disinfectant for water, in the manufacture of camphor, artificial silk, $KMnO_4$, for detecting the position of double bond.
59. Paraldehyde [$(CH_2O)_n$]: as sporofic and hypnotic.
60. Permutit (zeolite, $Na_2O \cdot Al_2O_3 \cdot 2SiO_2 \cdot 6H_2O$): for purification of water.
61. Phenol: as antiseptic, in the preparation of aspirin, picric acid, bakelite and phenolphthalein.
62. Phosphorus: in match industry (red P), for making insecticide (AlP), rat poison (Zn_3P_2).
63. Phosphorus pentaoxide: as dehydrating agent.
64. Phenacyl chloride ($C_6H_5COCH_2Cl$): as lachrymator for dispersing mobs.
65. Potassium metal: as reducing agent, in photoelectric cells.
66. Potassium bicarbonate: as antacid, in baking powder.
67. Potassium carbonate: for washing wool, in the preparation of fusion mixture, hard glass and soft soap.

68. Potassium sulphate: as purgative, fertilizer, in the preparation of potash alum and glass.
69. Quick lime: as basic lining in steel industry, for softening of water, purification of coal gas, sugar, in the manufacture of cement, CaC_2, slaked lime, lime colours and bleaching power.
70. Silver metal: in medicines, silver plating, ornaments, coins, for filling of teeth (silver alloy).
71. Slaked lime [$Ca(OH)_2$]: in softening of water, white washing, preparation of sodalime, NaOH and bleaching powder.
72. Silver bromide: in photography.
73. Sodium metal: as coolant in nuclear reactor, thermometric liquid (Na–K alloy), in sodium vapour lamp, in the manufacture of $(C_2H_5)_4Pb$ and Na/Hg.
74. Sodium bicarbonate (baking soda): as antacid, in fire extinguishers, for the preparation of soft drinks, fruit salts, baking powder and Siedlitz powder.
75. Sodium bromide: as sedative.
76. Sodium carbonate (washing soda): in laundry, in the manufacture of soap, fusion mixture and glass.
77. Sodium hydroxide: for mercerizing cotton, adsorbent for CO_2, X_2, SO_2 and NO_2.
78. Sodium peroxide: as a source of O_2 (oxone), oxidising agent, bleaching agent (Na_2O_2 + HCl), air purifier.
79. Sodium sulphate (Glauber salt): as laxative, in the preparation of window glass and craft paper.
80. Stannous chloride: as reducing agent, mordant in dyeing, for the manufacture of purple of cassius.
81. Sulphur: in vulcanisation of rubber, skin ointments, fireworks, match industry, in the preparation of dyes and sulpha drugs.
82. Sulphur dioxide: as solvent (liquid SO_2), antichlor, bleaching agent.
83. Sulphur trioxide: for drying of gases, manufacture of H_2SO_4 and oleum.
84. Tetraethyl lead (TEL): as an antiknocking agent in petrol.
85. Tin metal: for making tin foil for wrapping, squeezable tubes for toothpastes and ointments.
86. Trilead (red lead, minium): as oxidising agent, in the manufacture of matches, paints, flint glass and silver mirror.
87. 2,2,4-Trimethylpentane (isooctane): petrol additive for decrease of knocking.
88. Urotropine (hexamethylenetetramine): as urinary antiseptic.
89. Vinyl acetate ($CH_2{=}CHOCOCH_3$): plastic industry.
90. Vinyl chloride ($CH_2{=}CHCl$): in the manufacture of PVC.
91. Westron and westrosol: as solvents.
92. White lead (basic lead carbonate, $Pb(OH)_2 \cdot 2PbCO_3$): as white paint.
93. White metal (alloy of Sn, Sb, Cu): used to make artificial jewellery.
94. Zinc metal: as reducing agent, in galvanisation, metallurgy of Pb, Ag and Au, in the preparation of Zn–Cu couple.
95. Zinc oxide (zinc white): as a white paint.
96. Zinc sulphide: as luminous paint in dials of watches, in X-ray screens, in the manufacture of lithopone (ZnS + $BaSO_4$).

Appendix—IX
Some Important Reagents and Mixtures

1. Aqua regia	Conc. HNO_3 + conc. HCl (1:3) (used as a laboratory reagent)
2. Amatol	NH_4NO_3 (80%) + trinitrotoluene (20%) (used as an explosive)
3. Ammonal	Al powder + NH_4NO_3 (used as an explosive)
4. Baking powder	$NaHCO_3$ + sodium potassium tartrate
5. Baeyer reagent	Alkaline $KMnO_4$ solution (used for detecting ethylenic and acetylenic linkages)
6. Benedict solution	A solution of $CuSO_4 \cdot 5H_2O$ + NaOH + sodium citrate (used for detecting aldehydes)
7. Black ash	Na_2CO_3 + CaS (impure Na_2CO_3 obtained in Leblanc process)
8. Bordeaux mixture	A solution of $CuSO_4$ + lime (used to kill moulds and fungi on plants)
9. Carbogen	O_2 (90–95%) + CO_2 (5–10%) (used for artificial respiration)
10. Carbon oil	Vegetable oil + lime water (used for treatment of burns)
11. Coal gas	H_2 (47%) + CH_4 (32%) + CO (7%) + N_2 (4%) + C_2H_4 (3%) + C_2H_2 (2%) + CO_2 (1%) + other gases (4%) (used to produce reducing atmosphere in metallurgical operations)
12. Euchlorine	Cl_2 + ClO_2
13. Fenton reagent	H_2O_2 + few drops of $FeCl_3$
14. Fehling solution	$CuSO_4 \cdot 5H_2O$ + NaOH + sodium potassium tartrate (used for detecting aldehydes)
15. Freezing mixture	NaCl + ice (used for lowering temperature)
16. Fusion mixture	Na_2CO_3 + K_2CO_3 (used as a laboratory reagent)
17. Gobar gas	CH_4 + CO + H_2 (used as a domestic fuel)
18. Granite	Mica + rock + clay + sand
19. Gun powder	KNO_3 (75%)+S (12%) + charcoal (13%) (used as an explosive)
20. Ignition mixture	BaO_2 + Mg + Al (used in aluminothermit process)
21. Jones reagent	Chromic acid in aqueous acetone solution (used for oxidising 1° alcohol to aldihyde and 2° alcohol to ketone without affecting C$=$C bond)
22. Lindlar catalyst	Pd/$BaSO_4$, S (used for hydrogenation of alkyne to alkene)
23. Lithopone	ZnS + $BaSO_4$ (used as a white paint)
24. Lucas reagent	Conc. HCl + anhy. $ZnCl_2$ (used for distinguishing three types of alcohols)
25. Matte	Cu_2S + FeS
26. Methylated spirit	Rectified spirit (85–90%) + CH_3OH (10–15%) + pyridine + acetone (used as a solvent)
27. Milk of magnesia	An aqueous suspension of $Mg(OH)_2$ (used as an antacid)
28. Molish reagent	An ethanolic solution of α-naphthol (used for detecting carbohydrates)
29. Mortar	(slaked lime + sand) (1 : 3 in water)

30. Nessler reagent	A solution of K_2HgI_4 + KOH (used for detecting NH_4^+ ion)
31. Nitro chalk	NH_4NO_3 + $CaCO_3$ (used as a fertilizer)
32. Nitrolim	$CaCN_2$ + graphite (used as a fertilizer)
33. Nitrophos	$Ca(H_2PO_4)_2$ + $Ca(NO_3)_2$ (used as a fertilizer)
34. Oil gas	H_2 (50–55%) + CH_4 (25–30%) + CO (10–12%) + CO_2 (3%) (used in laboratory)
35. Power alcohol	(Petrol + C_2H_5OH) (4 : 1) + little benzene (used as a motor fuel)
36. Producer gas	N_2 (52–55%) + CO (22–30%) + H_2 (8–12%) + CO_2 (3%) (used as a fuel)
37. Purple of cassius	Colloidal sol of Au + $Sn(OH)_4$ (used for colouring of glass and pottery red)
38. Rectified spirit	C_2H_5OH (95.87%) + H_2O (4.13%) (used as a solvent)
39. Schiff reagent	An aqueous solution of rosaniline hydrochloride whose red colour has been discharged by SO_2 (used for detecting aldehydes)
40. Schweitzer reagent	$[Cu(NH_3)_4]SO_4$ (used in the manufacture of artificial silk)
41. Soda bleach	Na_2O_2 + HCl (used for bleaching of fabrics)
42. Sodalime	NaOH + $Ca(OH)_2$ (used in decarboxylation of carboxylic acids)
43. Sorel's cement	MgO + $MgCl_2$ (used as a substitute for tiles)
44. Sublimed white lead	PbO + $PbSO_4$ + ZnO (used as a white paint)
45. Superphosphate of lime	$Ca(H_2PO_4)_2$ + $CaSO_4$ (used as a fertilizer).
46. Thermite mixture	Al powder + metal oxide (used in metallurgy)
47. Tincture of iodine	I_2 + Nal + C_2H_5OH + water (used as an antiseptic)
48. Thomas slag	$Ca_3(PO_4)_2$, CaO (used as a fertilizer)
49. Water gas	H_2(51%) + CO(41%) + N_2(4%) + CO_2(4%) (used for manufacture of CH_3OH)

Appendix—X
Some Important Facts

Most electronegative element	F
Least electronegative element	Cs
Element having highest ionisation energy	He
Element having lowest ionisation energy	Cs
Element having highest electron affinity	Cl
Element having lowest electron affinity	Noble gases
Most reactive non-mental	F
Most reactive metal	Fr
Lightest metal	Li (density = 0.534 g/mL)
Heaviest metal	Os (density = 22.69 g/mL)

Most electropositive element	Fr
Least electropositive element	F
Metal having lowest M. P.	Hg (M. P. = – 38.9°C)
Metal having highest M. P.	W (M. P. = 3410°C)
Non-metal having lowest M. P. and B. P.	He (B. P. = – 268.9°C)
Most stable element	Te ($t_{½}$ = 2 × 10^{21} years)
Element having smallest atomic size	H
Element having largest atomic size	Cs
Smallest anion	H^-
Largest anion	At^-
Smallest cation	H^+
Largest cation	Cs^+
First man-made element	Tc
Most poisonous metal	Pu
Liquid non-metal	Bromine
Liquid metal	Hg, Ga, Cs, Fr
Best conductor among non-metals	Graphite
Least conductor among non-metals	S
Best conductor among metals	Ag
Least conductor among metals	Pb
Non-metal having highest M. P. and B. P.	Diamond
Hardest element	Diamond
Semi-metal elements	Si, Ge, As, Sb, Se, Te
Coinage metals	Cu, Ag, Au
Noble metals	Cu, Ag, Au, Rh, Ir, Ru, Os, Pd, Pt
Strongest oxidising agent	F_2
Strongest reducing agent	Li
Most malleable metal	Au
Most abundant element in the earth's crust	Oxygen
Rarest element in the earth's crust	At
Element exhibiting highest oxidation state	Ru (Ru^{8+}), Os (Os^{8+})
Polymorphic elements	O, S, P
Element having maximum number of naturally occurring isotopes	Sn (10)
Heaviest naturally occurring element	$^{228}_{92}U$
Element which sublimes on heating	Iodine
Element which is stored under water	Phosphorus
Element which is stored under kerosene	Na, K, Rb, Cs
Most abundant metal	Al
Most abundant element in the universe	Hydrogen

Element having maximum tendency for catenation	C
Element having highest tensile strength	B
Liquid radioactive element	Fr
Element which do not have neutron	${}_1^1H$
Metal exhibiting strongest covalency	Ru, Os
Amphoteric non-metal	Si
Amphoteric metal	Zn, Al, Sn, Pb, Cr, As
Elements showing diagonal relationship	Li–Mg; Be–Al; B–Si
Non-metals having metallic lustre	Iodine, graphite, diamond
Heaviest solid non-metal	At
Lightest solid non-metal	B
Heaviest gaseous non-metal	Rn
Lightest gaseous non-metal	Hydrogen
Heaviest semi-metal	Te
Lighest semi-metal	Si
Last discovered element in the periodic table	Uuo (118)
Latest discovered element	Uup (115)
Volatile *d*-block elements	Zn, Cd, Hg
Most commonly used metal	Fe
Element with electronegativity next to fluorine	Oxygen
Total number of liquid element in the periodic table	3 (Br, Fr, Hg)
Total number of gaseous elements in the periodic table	12 (H_2, He, N_2, O_2, F_2, Cl_2, Ne, Ar, Xe, Kr, Rn, Uuo)
Total number of solid elements in the periodic table	102 (= 117 – 3 – 2)
Number of metals in the periodic table	93 (= 117 – 18 – 6)
Number of non-metals in the periodic table	18
Group containing maximum number of gaseous elements	Group 18
Most abundant gas	N_2
Largest group in the periodic table	Group 13
Element which constitutes 90% of the mass of sun	H_2
Recipient of first Nobel prize in chemistry	van't Hoff
Lowest refractive index	He (1.000036)
Highest refractive index	Diamond (2.4)
Coolant in nuclear reactors	D_2O
Neutral oxides of non-metals	CO, NO, N_2O, H_2O
Amphoteric oxides	ZnO, PbO, BeO, Al_2O_3, SnO, V_2O_5
Oldest known organic acid	CH_3COOH

First organic compound synthesised from its elements	CH_3COOH
Natural explosive	NCl_3
Artificial explosive	Dynamite
Dry ice	$CO_2(s)$
Dry bleacher	O_3
Some isomorphous substances	$FeSO_4 \cdot 7H_2O$, $MgSO_4 \cdot 7H_2O$, $ZnSO_4 \cdot 7H_2O$
Some efflorescent substances	$Na_2CO_3 \cdot 10H_2O$, $Na_2SO \cdot 10H_2O$, $MgSO_4 \cdot 7H_2O$, $Na_2S_2O_3 \cdot 5H_2O$, $FeSO_4 \cdot 7H_2O$
Some commonly used oxidising agents	H_2O_2, SO_2, SO_3, Cl_2, O_3, $KMnO_4$, $K_2Cr_2O_7$, HNO_3
Some commonly used reductants	H_2S, SO_2, H_2SO_3, $SnCl_2$, H_2, CO, $LiAlH_4$, C
Some commonly used bleaching agents	H_2O_2, SO_2, Cl_2, bleaching powder
Compound used to disperse mobs (tear gas)	Phenacyl chloride ($C_6H_5COCH_2Cl$)
Compounds used as war gases	Phosgene ($COCl_2$), mustard gas, lewisite
Liquid silver	Hg
Mosaic gold	SnS_2
White gold	Pt
Fool's gold	FeS_2
Liquid gold	Petrol
Rolled gold	An alloy of Cu and Al
Fulminating gold	$H_2N{-}Au{=}NH$
Modern periodic table is based upon concept put forward by	Moseley
Long form of periodic table was put forward by	Bohr
Most stable carbonate	Cs_2CO_3
Strongest alkali	CsOH
Strongest basic oxide	Cs_2O
Strongest hydrogen bond is present in	HF_2^- (BDE = 113 kJ mol^{-1})
Compound with highest ionic character	CsF
Compound with highest covalent character	H_2, O_2, N_2, Cl_2, Br_2, I_2
Softest form of carbon	Lamp black
Strongest acid	HSO_3F + 90% SbF_5 called magic acid
Most reactive form of P	White
Least reactive form of P	Red
Most stable allotrope of phosphorus	Black (due to layer structure)
Purest form of silica	Quartz
Recently discovered allotrope of cabon	Buckminsterfullerene
Noble gases which form compounds	Xe, Kr
Most recent element named by IUPAC	Lv (116)

Appendix—XI
Important Industrial Processes

Baeyer process	Manufacture of Al
Bessemer process	Manufacture of steel
Birkeland–Eyde process	Manufacture of HNO_3
Bosch process	Manufacture of H_2
Carter process	Manufacture of white lead (basic lead carbonate)
Castner process	Manufacture of Na
Castner–Kellner process	Manufacture of NaOH
Contact process	Manufacture of H_2SO_4
Cupellation process	Purification of Ag
Deacon process	Manufacture of Cl_2
Dow process	Manufacture of phenol
Down process	Manufacture of Na
Goldschmidt process	Thermite welding
Haber process	Manufacture of NH_3
Hasenclever process	Manufacture of bleaching powder
Hoop process	Purification of Al
Lead chamber process	Manufacture of H_2SO_4
Leblanc process	Manufacture of K_2CO_3
L. D. process	Manufacture of steel
MacArthur–Forest process (cyanide process)	Manufacture of Ag
Merck process	Manufacture of H_2O_2
Nelson process	Manufacture of NaOH
Open hearth process	Manufacture of steel
Ostwald process	Manufacture of HNO_3
Parke process	Manufacture of Ag
Pattinson process	Manufacture of Ag
Serpeck process	Manufacture of Al
Solvay ammonia soda process	Manufacture of Na_2CO_3 and $NaHCO_3$

Appendix—XII
Catalysts Used in Some Industrial Processes

Process	*Catalyst*
Bergius process for manufacture of petrol from coal gas	Nickel oleate
Bosch process for manufacture of H_2	$Fe_2O_3 + Cr_2O_3$
Contact process for manufacture of H_2SO_4	V_2O_5 or Pt
Deacon process for manufacture of Cl_2	$CuCl_2$
Dehydrogenation of alcohol to aldehyde	Cu
Fischer–Tropsch process for manufacture of petrol	

from CO and H_2	Co + ThO_2 + MgO + keiselguhr
Friedel–Crafts process for alkylation of aromatic hydrocarbons	Anhydrous $AlCl_3$
Hydrogenation of alkenes	$(Ph_3P)_3RhCl$ (Wilkinson's catalyst)
Hydrogenation of alkynes to alkanes	Ni
Hydrogenation of alkynes to alkenes	Pd–$BaSO_4$, S
Hydrogenation of oils to fats	Ni
Lead chamber process for manufacture of H_2SO_4	NO
Manufacture of C_2H_5OH from starch	Diastase + maltase
Manufacture of C_2H_5OH from molasses	Invertase + zymase
Manufacture of CH_3OH from CO and H_2	ZnO + Cr_2O_3
Manufacture of CH_3COOH from CH_3CHO and O_2	V_2O_5
Manufacture of HCHO from CO and H_2	Cu
Manufacture of O_2 from $KClO_3$	MnO_2
Manufacture of phthalic acid from naphthalene	$HgSO_4$
Ostwald process for manufacture of HNO_3	Pt
Polymerisation of alkenes	H_3PO_4 + keiselguhr
Polymerisation of alkenes	$(C_2H_5)_3Al$ + $TiCl_4$ (Zeigler–Natta catalyst)

Appendix—XIII
Solubility of Some Common Inorganic Compounds in Water

Water Soluble

Acetate ($C_2H_3O_2^-$)	All acetates are soluble.
Bromide (Br^-)	All bromides are soluble except AgBr, $PbBr_2$, Hg_2Br_2 and $HgBr_2$.
Chloride (Cl^-)	All chlorides are soluble except AgCl, $PbCl_2$ and Hg_2Cl_2.
Nitrate (NO_3^-)	All nitrates are soluble.
Iodide (I^-)	All iodides are soluble except AgI, PbI_2, Hg_2I_2 and HgI_2.
Sulphate (SO_4^{2-})	All sulphates are soluble except Ag_2SO_4, $PbSO_4$, Hg_2SO_4, $CaSO_4$, $BaSO_4$ and $SrSO_4$.

Water Insoluble

Carbonate (CO_3^{2-})	All carbonates are insoluble except those of Group 1 elements and $(NH_4)_2CO_3$.
Hydroxide (OH^-)	All hydroxides are insoluble except those of Group 1 elements, $Ca(OH)_2$, $Ba(OH)_2$ and $Sr(OH)_2$.
Phosphate (PO_4^{3-})	All phosphates are insoluble except those of Group 1 elements and $(NH_4)_3PO_4$.
Sulphide (S^{2-})	All sulphides are insoluble except those of Group 1 and 2 elements and $(NH_4)_2S$.

Appendix—XIV
Some Common Adsorbent for Gases

Gas	*Adsorbent*
O_2	Alkaline pyrogallol
O_3	Turpentine oil
N_2	Heated magnesium
NO	A solution of $FeSO_4$
CO, C_2H_2	Ammoniacal solution of Cu_2Cl_2
CO_2, SO_2, F_2, Cl_2, Br, I_2, NO_2, acidic oxides	NaOH or KOH solution
NH_3, $H_2O(g)$	Conc. H_2SO_4
$H_2O(g)$	$CuSO_4$ (anhydrous), $CaCl_2$ (anhydrous). P_4O_{10}, silica gel

Appendix—XV
Some Common Gases and Their Characteristics

Gas	*Formula*	*Characteristics*
Ammonia	NH_3	Colourless, pungent odour, extremely soluble in water
Bromine	Br_2	Reddish brown, pungent and suffocating odour, toxic, sparingly soluble in water, fairly soluble in organic solvents
Carbon dioxide	CO_2	Colourless, odourless, toxic, fairly soluble in water
Carbon monoxide	CO	Colourless, odourless, toxic, sparingly soluble in water
Chlorine	Cl_2	Greenish yellow, choking and irritating smell, toxic, sparingly soluble in cold water and insoluble in hot water
Hydrogen chloride	HCl	Choking odour, corrosive, toxic, fairly soluble in water
Hydrogen cyanide	HCN	Slight odour of bitter almonds, very toxic
Hydrogen sulphide	H_2S	Odour of rotten egg, very toxic
Iodine	I_2	Violet, sparingly soluble in water, fairly soluble in organic solvents
Methane	CH_4	Colourless, odourless, inflammable
Nitric oxide	NO	Colourless, sparingly soluble in water, paramagnetic
Nitrogen dioxide	NO_2	Reddish brown, irritating odour
Nitrous oxide	N_2O	Colourless, sweet odour, laughing gas
Oxygen	O_2	Colourless, odourless, tasteless, sparingly soluble in water, paramagnetic, non-inflammable
Ozone	O_3	Pale blue, fishy smell, toxic, sparingly soluble in water, mercury loses its meniscus in contact with ozone
Sulphur dioxide	SO_2	Colourless, irritating odour

Appendix—XVI
Important Alloys and Their Composition

Alloy	*Composition*
Alclad	Al (95%) + Cu (4%) + Mg (0.5%) + Mn (0.5%)
Alnico	Fe (50%) + Al (20%) + Ni (20%) + Co (10%)
Aluminium bronge	Cu (90%) + Al (10%)
Babbitt metal	Sn (89%) + Sb (7.3%) + Cu (3.7%)
Bearing metal	Pb (98.66%) + Ca (0.7%) + Na (0.6%) + Li (0.04%)
Bell metal	Cu (78%) + Sn (22%)
Brass	Cu (60–80%) + Zn (20–40%)
Britania metal	Sn (86%) + Sb (12%) + Cu (2%)
Bronze	Cu (75-90%) + Sn (10–25%)
Constantan	Cu (60%) + Ni (40%)
Copper coin (brown)	Cu (95%) + Sn (4%) + Zn (1%)
Delta metal	Cu (60%) + Zn (38%) + Fe (2%)
Duralumin	Al(92.5–95.5%)+Cu (3.5–4.5%)+Mg (0.3–1.0%)+Mn (0.4–1.0%)+Si (0.3–1.0%)
Dutch metal	Cu (80%) + Zn (20%)
Elektron	Mg (95%) + Zn (4.5%) + Cu (0.5%)
Ferroaluminium	Fe (80%) + Al (20%)
Ferromanganese	Mn (70–80%) + Fe (20–30%)
Ferromdybdenum	Fe (35–40%) + Mo (55–65%)
Ferrovanadium	Fe (39.5–63.5%) + V (35–45%) + C (0.2–3.5%)
German silver	Cu (25–60%) + Zn (25–35%) + Ni (10–15%)
Gold 22 carat	Au (91.7%) + Ag (4.15%) + Cu (4.15%)
Gun metal	Cu (88%) + Sn (10%) + Zn (2%)
Invar	Fe (63.8%) + Ni (36%) + C (0.2%)
Magnalium	Al (70–90%) + Mg (10–30%)
Magnolia	Pb (85%) + Sb (15%)
Manganese bronze	Cu (55–65%) + Mn (16–17%) + Zn (20–30%) + Fe (0.5–2%)
Manganin	Cu (~ 84%) + Mn (~ 12%) + Ni (~ 4%) + Fe (trace)
Misch metal	Ce (50%) + La (25%) + Nd (10%) + Fe (4.5%) + (Ca, C, Si) (10.5%)
Monel metal	Ni (~ 67%) + Cu (~ 30%) + (Fe + Mn) (~ 3%)
Muntz metal	Cu (60%) + Zn (40%)
Nichrome	Ni (~ 60%) + Fe (~ 24%) + Cr (~ 16%) + C (0.1%)
Nickel coinage	Cu (80%) + Ni (20%)
Permalloy	Ni (78%) + Fe (21 %) + C (1 %)
Pewter	Sn (85%) + Cu (7%) + Bi (6%) + Sb (1.7%)
Phosphor bronze	Cu (79.7%) + Sn (10%) + Sb (9.5%) + P (0.8%)

Rhine metal	Sn (97%) + Cu (3%)
Rose metal	Bi (50%) + Pb (27.1 %) + Sn (22.9%)
Solder	Sn (~ 50%) + Pb (~ 50%)
Spiegeleisen	Fe (80–85%) + Mn (15–20%) + C (trace)
Stainless steel	Fe (71–75%) + Cr (17–19%) + Ni (7–9%)+ C (~0.2%)
Tin foil	Sn (88%) + Pb (8%) + Cu (3.5%) + Sb (0.5%)
Tin sel	Sn (60%) + Pb (40%)
Silicon bronze	Cu (97%) + Si (3%)
Type metal	Pb (60%) + Sb (30%) + Sn (10%)
White metal	Sn (82%) + Sb (12%) + Cu (6%)
Wood metal	Bi (50%) + Pb (25%) + Sn (12.5%) + Cd (12.5%)
γ alloy	Al (92%) + Cu (4%) + Ni (2%) + Mg (2%)

Appendix—XVII
Some Important Name Reactions

Name	*Starting material*	*Reaction condition*	*Synthesis of*
1. Aldol condensation	aldehyde having α-hydrogen atom	conc. NaOH	aldol and ketol
2. Beckmann rearrangement	aromatic ketoxime	180–200°C, PCl_5 or P_4O_{10} or SO_2Cl_2 or H_2SO_4	*N*-substituted acetamide
3. Benzoin condensation	aromatic aldehyde	KCN	benzoin
4. Cannizzaro reaction	aldehyde not having α-hydrogen atom	aq. KOH or alc. KOH	alcohol and acid
5. Carbylamine reaction	primary amine	$CHCl_3$ + NaOH	carbylamine (test for —NH_2 group)
6. Clemmensen reduction	aldehyde or ketone	Zn/Hg + conc. HCl	hydrocarbon
7. Claisen condensation	(aldehyde, ketone or ester containing α-hydrogen atom) + (aromatic aldehyde or ketone)	NaOH	unsaturated aromatic aldehyde or ketone
8. Coupling reaction	C_6H_5NNCl + phenol (or aniline)	NaOH (in case of phenol), HCl (in case of aniline)	azo dye (detection of—NH_2 and phenolic—OH group)
9. Diazotisation	aromatic primary amine	$NaNO_2$/conc. HCl, 0–5°C	C_6H_5NNCl
10. Diels–Alder reaction	α,β-unsaturated carbonyl compound + conjugated diene	100°C	cyclic addition product
11. Etard reaction	toluene	CrO_2Cl_2/CS_2	benzaldehyde
12. Fittig reaction	aryl halide	Na/dry ether	diphenyl

13. Friedel–Crafts reaction	C_6H_6 + RX (or RCOCl)	anhy. $AlCl_3$	alkyl and acyl benzene
14. Gattermann reaction	C_6H_5NNCl	Cu powder + HCl	halobenzene
15. Gattermann–Kotch reaction	C_6H_6 + CO + HCl	anhy. $AlCl_3$	aromatic aldehyde
16. Haloform reaction	CH_3CHO, methyl ketone, CH_3CH_2OH, $RCHOHCH_3$	$NaOH/X_2$ or NaOX	haloform (test for —$COCH_3$ group)
17. Hell–Volhard–Zelinsky reaction	carboxylic acid having α-hydrogen atom	Br_2/red P	α-halogenated carboxylic acid
18. Hofmann bromamide reaction	acid amide	$Br_2/NaOH$	primary amine
19. Hofmann mustard oil reaction	primary aliphatic amine + CS_2	$HgCl_2$	$CH_3CH_2—N{=}C{=}S$ (mustard oil)
20. Hunsdiecker reaction	Ag salt of carboxylic acid + Br_2	CCl_4, 80°C	alkyl or aryl bromide
21. Kolbe electrolytic reaction	alkali metal salt of carboxylic acid	electrolysis	alkane, alkene and alkyne
22. Kolbe–Schmidt reaction	C_6H_5OH	CO_2, 125°C, 4–7 atm, H^+	salicylic acid
23. Meerwein–Ponndorf–Verley reduction	ketone	$[(CH_3)_2CHO]_3Al$ + $(CH_3)_2CHOH$	secondary alcohol
24. Mendius reaction	alkyl or aryl cyanide	Na/C_2H_5OH	primary amine
25. Perkin condensation	aromatic aldehyde + $(CH_3CO)_2O$	CH_3COONa, H^+/H_2O	α,β-unsaturated aromatic acid
26. Reimer–Tiemann reaction	phenol	NaOH + $CHCl_3$ or CCl_4, dil. HCl	salicyaldehyde or salicylic acid
27. Rosenmund reduction	acid chloride	H_2, $Pd/BaSO_4$, S, boiling xylene	aldehyde
28. Sabatier–Senderens reaction	unsaturated hydrocarbon	raney Ni/H_2, 200–300°C	alkane
29. Sandmeyer reaction	C_6H_5NNCl	CuCl/HCl or CuBr/HBr or CuCN/KCN, heat	halo or cyanobenzene
30. Schieman's reaction	C_6H_5NNCl	HBF_4	Fluorobenzene
31. Schotten–Baumann reaction	(phenol or aniline or alcohol) + C_6H_5COCl	NaOH	benzoylated product
32. Stephen reaction	alkyl cyanide	$SnCl_2/HCl$	aldehyde
33. Ullmann reaction	iodobenzene + Cu	heat	diphenyl
34. Williamson synthesis	alkyl halide + sodium alkoxide or sodium phenoxie	heat	ether
35. Wurtz–Fittig reaction	alkyl halide + aryl halide	Na/dry ether	alkyl benzene

Appendix—XVIII
Wavelengths and Frequencies of Electromagnetic Radiations

Radiation	*Wavelength (Å)*	*Frequency (Hz or s^{-1})*
Radiowaves	3×10^9 to 3×10^{13}	1×10^5 to 1×10^9
Microwaves	6×10^6 to 3×10^9	1×10^9 to 5×10^{11}
Infrared rays	6×10^6 to 7.6×10^3	5×10^{11} to 3.95×10^{14}
Visible rays	3800 to 7600	3.95×10^{14} to 7.9×10^{14}
Ultraviolet rays	150 to 3800	7.9×10^{14} to 2×10^{16}
X-rays	0.1 to 150	2×10^{19} to 3×10^{19}
Gamma rays	0.01 to 0.1	3×10^{19} to 3×10^{20}
Cosmic rays	0.00 to 0.01	3×10^{20} to infinite

Appendix—XIX
Electrochemical Series

Half-reaction	*Standard electrode potential at 298 K (V)*	*Half-reaction*	*Standard electrode potential at 298 K (V)*
$Li^+ + e^- \longrightarrow Li$	– 3.05	$Sn^{2+} + 2e^- \longrightarrow Sn$	– 0.14
$K^+ + e^- \longrightarrow K$	– 2.93	$Pb^{2+} + 2e^- \longrightarrow Pb$	– 0.13
$Ba^{2+} + 2e^- \longrightarrow Ba$	– 2.91	$Fe^{3+} + 3e^- \longrightarrow Fe$	– 0.04
$Sr^{2+} + 2e^- \longrightarrow Sr$	– 2.89	$2H^+ + 2e^- \longrightarrow H_2$	0.00
$Ca^{2+} + 2e^- \longrightarrow Ca$	– 2.86	$Cu^{2+} + e^- \longrightarrow Cu^+$	+ 0.16
$Na^+ + e^- \longrightarrow Na$	– 2.71	$Sn^{4+} + 2e^- \longrightarrow Sn^{2+}$	+ 0.15
$Mg^{2+} + 2e^- \longrightarrow Mg$	– 2.36	$Bi^{3+} + 3e^- \longrightarrow Bi$	+ 0.20
$Al^{3+} + 3e^- \longrightarrow Al$	– 1.66	$Cu^{2+} + 2e^- \longrightarrow Cu$	+ 0.34
$Mn^{2+} + 2e^- \longrightarrow Mn$	– 1.18	$Cu^+ + e^- \longrightarrow Cu$	+ 0.52
$Zn^{2+} + 2e^- \longrightarrow Zn$	– 0.76	$I_2 + 2e^- \longrightarrow 2I^-$	+ 0.54
$Cr^{3+} + 3e^- \longrightarrow Cr$	– 0.74	$Fe^{3+} + e^- \longrightarrow Fe^{2+}$	+ 0.77
$U^{4+} + e^- \longrightarrow U^{3+}$	– 0.61	$Hg_2^{2+} + 2e^- \longrightarrow 2Hg$	+ 0.79
$S + 2e^- \longrightarrow S^{2-}$	– 0.48	$Ag^+ + e^- \longrightarrow Ag$	+ 0.80
$Fe^{2+} + 2e^- \longrightarrow Fe$	– 0.44	$Hg^{2+} + 2e^- \longrightarrow Hg$	+ 0.86
$In^{3+} + 2e^- \longrightarrow In^+$	– 0.44	$Br_2 + 2e^- \longrightarrow 2Br^-$	+ 1.09
$Cr^{3+} + e^- \longrightarrow Cr^{2+}$	– 0.41	$Pt^{2+} + 2e^- \longrightarrow Pt$	+ 1.20
$Cd^{2+} + 2e^- \longrightarrow Cd$	– 0.40	$Cl_2 + 2e^- \longrightarrow 2Cl^-$	+ 1.36
$Tl^{3+} + e^- \longrightarrow Tl^{2+}$	– 0.37	$Au^{3+} + 3e^- \longrightarrow Au$	+ 1.50
$Tl^+ + e^- \longrightarrow Tl$	– 0.34	$Ce^{4+} + e^- \longrightarrow Ce^{3+}$	+ 1.61
$Co^{2+} + 2e^- \longrightarrow Co$	– 0.27	$Co^{3+} + e^- \longrightarrow Co^{2+}$	+ 1.81
$V^{3+} + e^- \longrightarrow V^{2+}$	– 0.26	$Ag^{2+} + e^- \longrightarrow Ag^+$	+ 1.98
$Ni^{2+} + 2e^- \longrightarrow Ni$	– 0.23	$Au^+ + e^- \longrightarrow Au$	+ 1.69
$In^+ + e^- \longrightarrow In$	– 0.14	$F_2 + 2e^- \longrightarrow 2F^-$	+ 2.87

Appendix—XX
Effect of Strong Heat on Compounds

Compound	*Products*	*Compound*	*Products*
$AlCl_3 \cdot 6H_2O$	Al_2O_3, HCl, H_2O	$2Pb_3O_4$	PbO (yellow), O_2
NH_4HCO_3	NH_3, CO_2, H_2O	$Pb(CH_3COO)_2$	PbO, CO_2, CH_3COCH_3
NH_4Cl	NH_3, HCl	$Pb(NO_3)_2$	PbO (yellow), NO_2, O_2
$(NH_4)_2Cr_2O_7$	Cr_2O_3, N_2, H_2O	HgO	Hg, O_2
NH_4NO_3	N_2O, H_2O	Hg_2Cl_2	$HgCl_2$, Hg
NH_4NO_2	N_2, H_2O	$Hg(NO_3)_2$	Hg, NO_2, O_2
NH_3	N_2, H_2	Li_2CO_3	Li_2O, CO_2
NH_4VO_3	V_2O_5, NH_3, H_2O	$LiNO_3$	Li_2O, NO_2, O_2
BaO_2	BaO, O_2	$Mg(NO_3)_2$	MgO, NO_2, O_2
$Na_2B_4O_7 \cdot 10H_2O$	$NaBO_2$, B_2O_3	$MgCO_3$	MgO, CO_2
$CaCO_3$	CaO, CO_2	$MgCl_2 \cdot 6H_2O$	MgO, HCl
$Ca(NO_3)_2$	CaO, NO_2, O_2	$Na(NH_4)HPO_4$	$NaPO_3$, NH_3, H_2O
$CuCO_3$	CuO, CO_2	H_3PO_4	HPO_3, H_2O
$Cu(NO_3)_2$	CuO, NO_2	$H_2C_2O_4$	CO, CO_2, H_2O
$CuSO_4 \cdot 5H_2O$	CuO, SO_3	$P_4 + S_8$	$P_4S_3, P_4S_5, P_4S_7, P_4S_{10}$
$CoCO_3$	CoO, CO_2	Ag_2CO_3	Ag, CO_2, O_2
$Co(NO_3)_2$	Co_2O_3, NO_2, O_2	Ag_2O	Ag, O_2
$Ni(CO)_4$	Ni, CO	$AgNO_3$	Ag, NO_2, O_2
$Ni(NO_2)_3$	Ni_2O_3, NO, NO_2, O_2	NaN_3	Na, N_2
$ZnCl_2 \cdot 2H_2O$	ZnO, HCl	$NaHCO_3$	Na_2CO_3, CO_2, H_2O
$ZnSO_4 \cdot 2H_2O$	ZnO, SO_2, O_2	$Na_2HPO_4 \cdot 12H_2O$	$Na_4P_2O_7$, H_2O
$Fe_2(SO_4)_3$	Fe_2O_3, SO_3	$NaNO_3$	$NaNO_2$, O_2
$FeSO_4 \cdot 7H_2O$	Fe_2O_3, SO_2, SO_3	$Na_2CO_3 \cdot 10H_2O$	Na_2CO_3, H_2O
FeC_2O_4	FeO, CO, CO_2	$ZnCO_3$	ZnO, CO_2
$Fe(OH)_3$	Fe_2O_3, H_2O	$Zn(NO_3)_2$	ZnO, NO_2, O_2
$FeCl_3$	$FeCl_2$, Cl_2	$KMnO_4$	K_2MnO_4, MnO_2, O_2
$FeS_2 + O_2$	Fe_2O_3, SO_2	$Mn(NO_3)_2$	MnO_2, NO_2
$CaSO_4 \cdot 2H_2O$	CaO, SO_2, O_2	$MnCO_3$	MnO, CO_2
$Ca(HCOO)_2$	HCHO, CaO, CO_2	MnO_2	Mn_2O_3, O_2
$Ca(CH_3COO)_2$	CH_3COCH_3, CaO, CO_2	$K_2Cr_2O_7$	K_2CrO_4, Cr_2O_3, O_2
H_3PO_2	PH_3, H_3PO_4	$KClO_3$	KCl, O_2
PbO_2	PbO (black), NO_2, O_2	Urea	Cyanuric acid, NH_3

Appendix—XXI
Colour Changes and *p*H Ranges of Some Important Acid-Base Indicators

Indicator	*p*H *range*	*Colour in acid solution*	*Colour in alkaline solution*	*Used in the titration of*
Methyl orange	3.1–4.4	pink	yellow	strong acid and weak or strong base
Methyl red	4.2–6.3	red	yellow	strong acid and weak base
Phenolphthalein	8.2–10.0	colourless	pink	strong or weak acid and strong base
Litmus	6.0–6.8	colourless	pink	strong acid and strong base
Bromothymol blue	6.0–7.6	yellow	blue	strong acid and strong base

Appendix—XXII
Elements: Discoverers, Names and Symbols

Element	*Symbol*	*Atomic No.*	*Year of Discovery*	*Discoverer and Nationality*
Actinium	Ac	89	1899	A. Debierne (France)
Aluminium	Al	13	1827	F. Woehler (Germany)
Americium	Am	95	1944	A. Ghiorso (USA) R. A. James (USA) G. T. Seaborg (USA) S. G. Seaborg (USA)
Antimony	Sb	51	Ancient	
Argon	Ar	18	1894	Lord Raleigh (Great Britain) Sir William Ramsay (Great Britain)
Arsenic	As	33	1250	Albertus Magnus (Germany)
Astatine	At	85	1940	D. R. Corson (USA) K. R. Mackenzie (USA) E. Segre (USA)
Barium	Ba	56	1808	Sir Humphry Davy (Great Britain)
Berkelium	Bk	97	1950	G. T. Seaborg (USA) S. G. Thompson (USA) A. Ghiorso (USA)
Beryllium	Be	4	1828	F. Woehler (Germany) A. A. B. Bussy (France)

Element	*Symbol*	*Atomic No.*	*Year of Discovery*	*Discoverer and Nationality*
Bismuth	Bi	83	1753	Claude Geoffroy (France)
Bohrium	Bh	107	1976	USA
Boron	B	5	1808	Sir Humphry Davy (Great Britain) J. L. Gay-Lussac (France) L. J. Thenard (France)
Bromine	Br	35	1826	A. J. Balard (France)
Cadmium	Cd	48	1817	Fr. Stromeyer (Germany)
Calcium	Ca	20	1808	Sir Humphry Davy (Great Britain)
Californium	Cf	98	1950	G. T. Seaborg, S. G. Thompson, A. Ghiorso, K. Street, Jr. (USA)
Carbon	C	6	Ancient	
Cerium	Ce	58	1803	J. J. Berzelius (Sweden) William Hisinger (Sweden) M. H. Klaproth (Germany)
Caesium	Cs	55	1860	R. Bunsen, G. R. Kirchhoff (Germany)
Chlorine	Cl	17	1774	K. W. Scheele (Sweden)
Chromium	Cr	24	1797	L. N. Vauquelin (France)
Cobalt	Co	27	1735	G. Brandt (Germany)
Copernicium	Cn	112	1996	S. Hofmann *et al.* (Germany)
Copper	Cu	29	Ancient	
Curium	Cm	96	1944	G. T. Seaborg (USA) R. A. James (USA) A. Ghiorso (USA)
Darmstadtium	Ds	110	1994	P. Armbnuster, G. Munzenberg (Germany)
Dubnium	Db	105	1970	G. N. Flerov *et al.* (Russia)
Dysprosium	Dy	66	1886	Lecoq de Boisbaudran (France)
Einsteinium	Es	99	1952	A. Ghiorso (USA)
Erbium	Er	68	1843	C. G. Mosander (Sweden)
Europium	Eu	63	1896	E. Demarcay (France)
Fermium	Fm	100	1953	A. Ghiorso (USA)
Flerovium	Fl	114	2006	Kenton Moody *et al.* (USA)
Fluorine	F	9	1886	H. Moissan (France)
Francium	Fr	87	1939	Marguerite Perey (France)
Gadolinium	Gd	64	1880	J. C. Marignac (France)
Gallium	Ga	31	1875	Lecoq de Boisbaudran (France)

Element	*Symbol*	*Atomic No.*	*Year of Discovery*	*Discoverer and Nationality*
Germanium	Ge	32	1886	Clemens Winkler (Germany)
Gold	Au	79	Ancient	
Hafnium	Hf	72	1923	D. Coster (Dutch) G. von Hevesey (Hungary)
Helium	He	2	1868	P. Janseen (spectra) (France) Sir William Ramsay (isolation) (Great Britain)
Hassium	Hs	108	1984	P. Armbnuster, G. Munzenberg (Germany)
Holmium	Ho	67	1879	P. T. Cleve (Sweden)
Hydrogen	H	1	1766	Sir Henry Cavendish (Great Britain)
Indium	In	49	1863	E. Reich (Germany) T. Richter (Germany)
Iodine	I	53	1811	B. Courtois (France)
Iridium	Ir	77	1803	S. Tennant (Great Britain)
Iron	Fe	26	Ancient	
Krypton	Kr	36	1898	Sir William Ramsay (Great Britain) M. W. Travers (Great Britain)
Lanthanum	La	57	1839	C. G. Mosander (Sweden)
Lawrencium	Lr	103	1961	A. Ghiorso, T. Sikkeland, A. E. Larsh, R. M. Latimer (USA)
Lead	Pb	82	Ancient	
Lithium	Li	3	1817	A. Arfvedson (Sweden)
Livermorium	Lv	116	2006	Kenton Moody *et al.* (USA)
Lutetium	Lu	71	1907	G. Urbain (France) C. A. von Welsbach (Austria)
Magnesium	Mg	12	1808	Sir Humphry Davy (Great Britain)
Manganese	Mn	25	1774	J. G. Gahn (Sweden)
Mendeleevium	Md	101	1955	A. Ghiorso, G. R. Choppin, G. T. Seaborg, B. G. Harvey, S. G. Thompson (USA)
Meitnerium	Mt	109	1982	G. Munzenberg *et al.* (Germany)
Mercury	Hg	80	Ancient	
Molybdenum	Mo	42	1778	G. W. Scheele (Sweden)

Element	*Symbol*	*Atomic No.*	*Year of Discovery*	*Discoverer and Nationality*
Neodymium	Nd	60	1885	C. A. von Welsbach (Austria)
Neon	Ne	10	1898	Sir William Ramsay (Great Britain)
				M. W. Travers (Great Britain)
Neptunium	Np	93	1940	E. M. McMillan (USA)
				P. H. Abelson (USA)
Nickel	Ni	28	1751	A. E. Cronstedt (Sweden)
Niobium	Nb	41	1801	Charles Hatchett (Great Britain)
Nitrogen	N	7	1772	Daniel Rutherford (Great Britain)
Nobelium	No	102	1958	A. Ghiorso (USA)
				T. Sikkeland (USA)
				J. R. Walton (USA)
				G. T. Seaborg (USA)
Osmium	Os	76	1803	S. Tennant (Great Britain)
Oxygen	O	8	1774	Joseph Priestley (Great Britain)
				C. W. Scheele (Sweden)
Palladium	Pd	46	1803	W. H. Wollaston (Great Britain)
Phosphorus	P	15	1669	H. Brandt (Germany)
Platinum	Pt	78	1735	A. De Ullao (Spain)
			1741	Charles Wood (Great Britain)
Plutonium	Pu	94	1940	G. T. Seaborg (USA)
				E. M. McMillan (USA)
				J. W. Kennedy (USA)
				A. C. Wahl (USA)
Polonium	Po	84	1898	Marie Curie (France)
Potassium	K	19	1807	Sir Humphry Davy (Great Britain)
Praseodymium	Pr	59	1885	C. A. von Welsbach (Austria)
Promethium	Pm	61	1945	J. A. Marinsky (USA)
				L. E. Glendenin (USA)
				C. D. Coryell (USA)
Protactinium	Pa	91	1917	O. Hohn (Germany)
				L. Meitner (Austria)
Radium	Ra	88	1898	Pierre and Marie Curie (France)
Radon	Rn	86	1900	F. E. Dom (Germany)
Rhenium	Re	75	1925	W. Noddack (Germany)
				I. Tacke (Germany)
				Otto Berg (Germany)

Element	*Symbol*	*Atomic No.*	*Year of Discovery*	*Discoverer and Nationality*
Rhodium	Rh	45	1804	W. H. Wollaston (Great Britain)
Roentgenium	Rg	111	1995	S. Hofmann (Germany)
Rubidium	Rb	37	1861	R. W. Bunsen (Germany)
				G. Kirchhoff (Germany)
Ruthenium	Ru	44	1844	K. K. Klaus (Russia)
Rutherfordium	Rf	104	1965	USA
Samarium	Sm	62	1879	Lecoq de Boisbaurdran (France)
Scandium	Sc	21	1879	L. F. Nilson (Sweden)
Seaborgium	Sg	106	1976	A. Ghiorso *et al.* (USA)
Selenium	Se	34	1817	J. J. Berzelius (Sweden)
Silicon	Si	14	1824	J. J. Berzelius (Sweden)
Silver	Ag	47	Ancient	
Sodium	Na	11	1807	Sir Humphry Davy (Great Britain)
Strontium	Sr	38	1808	Sir Humphry Davy (Great Britain)
Sulphur	S	16	Ancient	
Tantalum	Ta	73	1802	A. G. Ekeberg (Sweden)
Technetium	Tc	43	1937	C. Perrier (Italy)
Tellurium	Te	52	1782	E. J. Muller (Austria)
Terbium	Tb	65	1843	C. G. Mosander (Sweden)
Thallium	Tl	81	1861	Sir William Crookes (Great Britain)
Thorium	Th	90	1828	J. J. Berzelius (Sweden)
Thulium	Tm	69	1879	P. T. Cleve (Sweden)
Tin	Sn	50	Ancient	
Titanium	Ti	22	1791	W. Gergor (Great Britain)
Tungsten	W	74	1783	J. J. de Elehujar and
				F. de Elehuyar (Spain)
Uranium	U	92	1789	M. H. Klaproth (Germany)
			1841	E. M. Peligot (France)
Vanadium	V	23	1801	A. M. Del Rio (Spain)
			1830	N. G. Sefstrom (Sweden)
Xenon	Xe	54	1898	Sir William Ramsay (Great Britain)
				M. W. Travers (Great Britain)
Ytterbium	Yb	70	1907	G. Urbain (France)
Yttrium	Y	39	1843	C. G. Mosander (Sweden)
Zinc	Zn	30	1746	A. S. Marggraf (Germany)
Zirconium	Zr	40	1789	M. H. Klaproth (Germany)

Appendix—XXIII

Some Important Rearrangement Reactions in Organic Chemistry

1. Benzil-benzilic acid Rearrangement

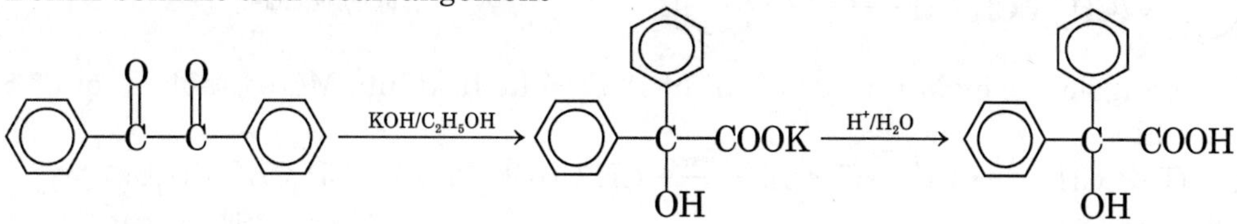

Benzil — Potassium benzilate — Benzilic acid

2. Benzidine Rearrangement

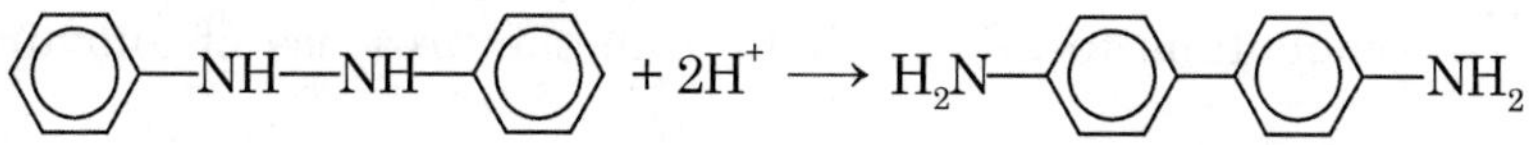

Hydrazobenzene — Benzidine

3. Pinacol-pinacolone Rearrangement

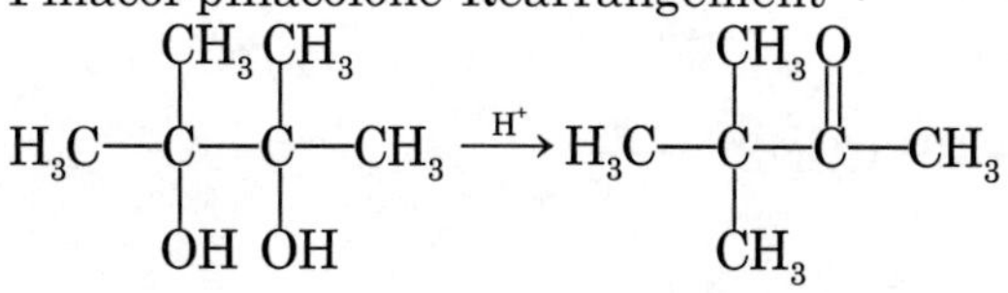

Pinacol — Pinacolone

4. Claisen Rearrangement

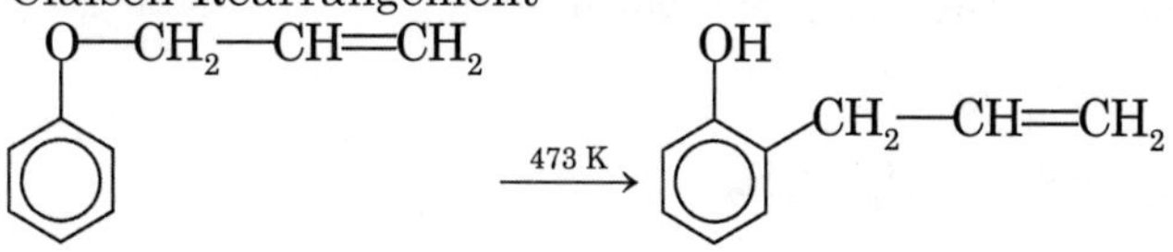

5. Beckmann Rearrangement

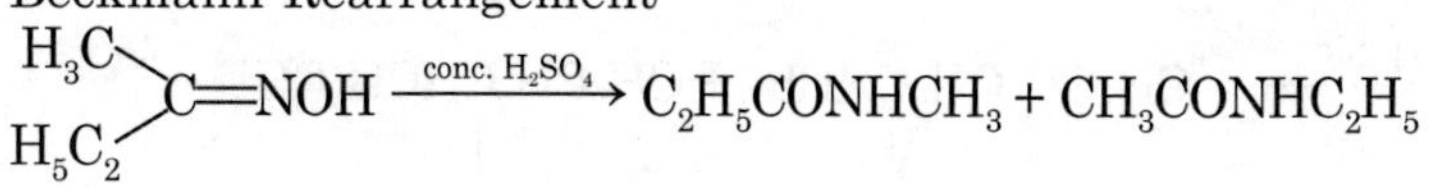

6. Fries Rearrangement

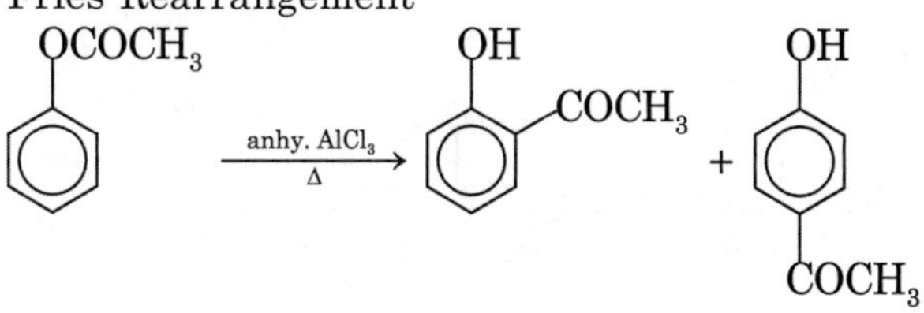

7. Friedel–Crafts Reaction

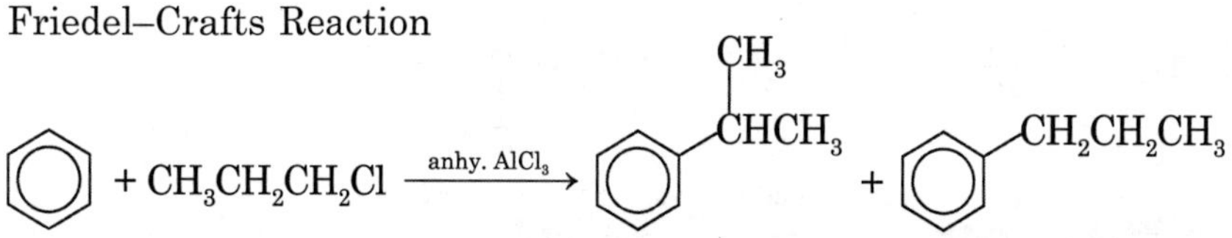

(major product) (minor product)

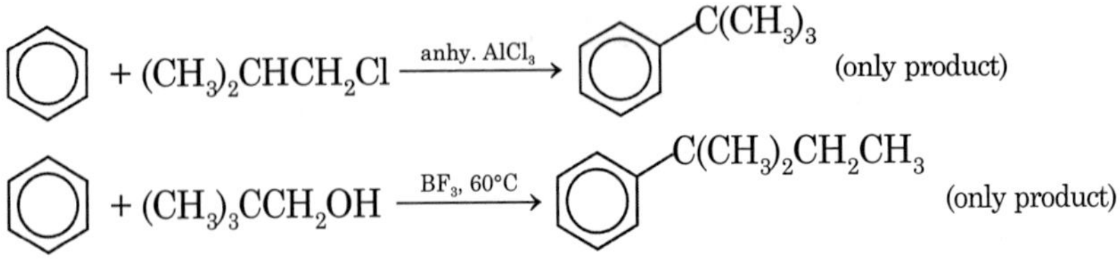

8. Rearrangement, by Migration of Br, of 1° Free Radical into More Stable 2° or 3° Free Radical

$$CH_3CHBrCH_3 \xrightarrow{Cl_2} CH_3CHBr\dot{C}H_2 \xrightarrow{rearr.} CH_3\dot{C}HCH_2Br \xrightarrow{Cl_2} CH_3CHClCH_2Br$$

2-Bromopropane → 1-Bromo-2-chloropropane

$$(CH_3)_3CBr \xrightarrow{Cl_2} (CH_3)_2CBr\dot{C}H_2 \xrightarrow{rearr.} (CH_3)_2\dot{C}CH_2Br \xrightarrow{Cl_2} (CH_3)_2CClCH_2Br$$

2-Bromo-2-methylpropane → 1-Bromo-2-chloro-2-methylpropane

9. Rearrangement of 1-Bromobutane to 2-Bromobutane in the presence of anhydrous $AlCl_3$ at 573 K

$$\underset{\text{1-Bromobutane}}{CH_3CH_2CH_2CH_2Br} \xrightarrow{\text{anhy. } AlCl_3,\ 573\ K} \underset{\text{2-Bromobutane}}{CH_3CH_2CHBrCH_3}$$

10. Hofmann Bromamide Reaction

$$\underset{\text{Acetamide}}{CH_3CONH_2} \xrightarrow{Br_2/KOH} [CH_3CONHBr \xrightarrow[-H_2O]{OH^-} CH_3CO\ddot{\bar{N}}Br \xrightarrow{-Br^-}$$

$$CH_3C(=O)-\ddot{N}: \xrightarrow{rearr.} [O=C=N-CH_3] \xrightarrow{KOH} \underset{\text{Methylamine}}{CH_3NH_2}$$

11. Nucleophilic Substitution Reactions Involving S_{N^1} Mechanism

$$H_3C-\overset{CH_3}{\underset{CH_3}{C}}-CH_2-Br \xrightarrow{C_2H_5OH} H_3C-\overset{CH_3}{\underset{OC_2H_5}{C}}-CH_2-CH_3$$

$$CH_3CH_2CH_2\underset{OH}{CH}CH_3 \xrightarrow{HCl,\ ZnCl_2,\ \Delta} \underset{\text{(major product)}}{CH_3CH_2CH_2\underset{Cl}{CH}CH_3} + \underset{\text{(minor product)}}{CH_3CH_2\underset{Cl}{CH}CH_2CH_3}$$

12. Elimination Reactions Involvnig $E1$ Mechanism

$$CH_3CH_2\overset{CH_3}{C}HCH_2OH \xrightarrow[-H_2O]{Al_2O_3,\ 623\ K} \underset{\text{(major product)}}{CH_3CH=\overset{CH_3}{C}HCH_3} + \underset{\text{(minor product)}}{CH_3CH_2\overset{CH_3}{C}=CH_2}$$

$$CH_3\overset{CH_3}{\underset{CH_3}{C}}CH(OH)CH_3 \xrightarrow[-H_2O]{Al_2O_3,\ 623\ K} \underset{\text{(major product)}}{CH_3\overset{CH_3}{C}=\underset{CH_3}{C}CH_3} + \underset{\text{(minor product)}}{CH_2=\overset{CH_3}{C}\underset{CH_3}{C}HCH_3}$$

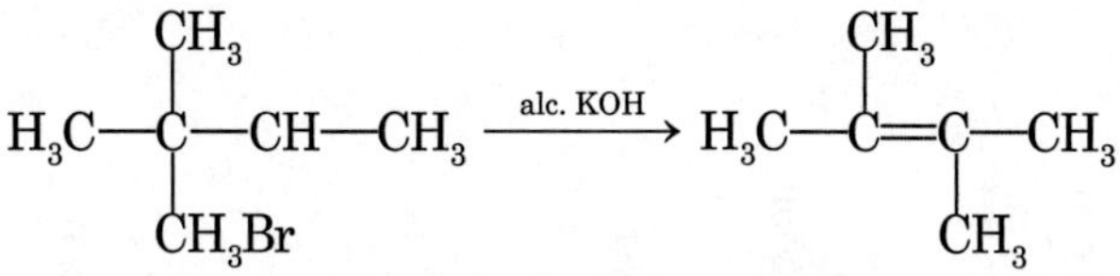

13. Electrophilic Addition Reaction

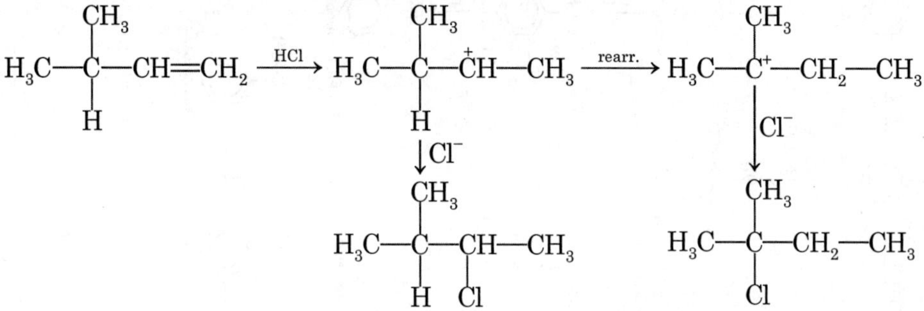

14. Hofmann–Martius Rearrangement

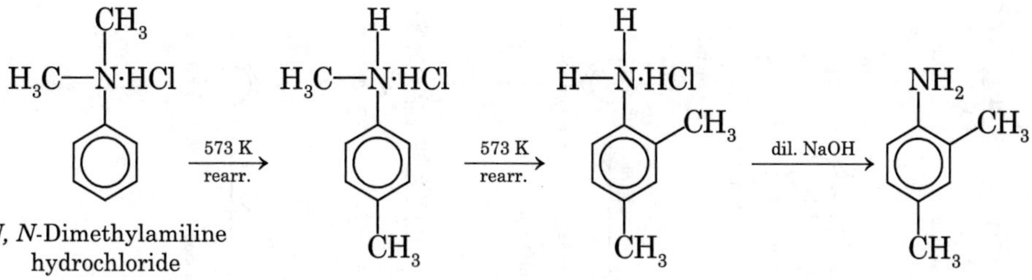

N, N-Dimethylamiline hydrochloride

2,4-Dimethylaniline

15. Rearrangement of Ammonium Cyanate to Urea on Heating

$$NH_4NCO \xrightarrow{\text{heat}} NH_2CONH_2$$

16. Rearrangement of Butane to 2-Methylpropane on Heating with Anhydrous $AlCl_3$

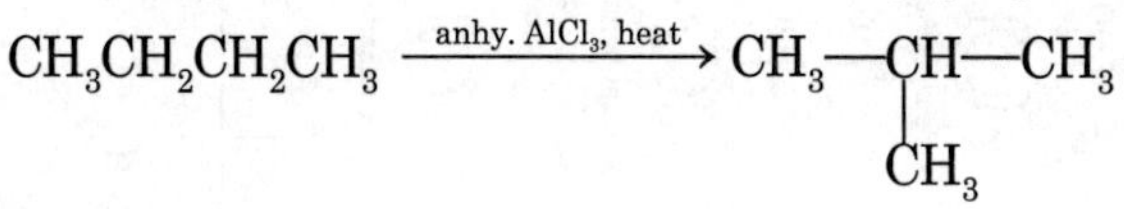

17. Rearrangement of But-1-ene to But-2-ene and 2-Methylpropene on Heating with Anhydrous $AlCl_3$

$$\underset{\text{But-1-ene}}{CH_3CH_2CH{=}CH_2} \xrightarrow{\text{anhy. } AlCl_3\text{, heat}} \underset{\text{But-2-ene}}{CH_3CH{=}CHCH_3} + \underset{\text{2-Methylpropene}}{(CH_3)_2C{=}CH_2}$$

18. Hydration of Alkenes

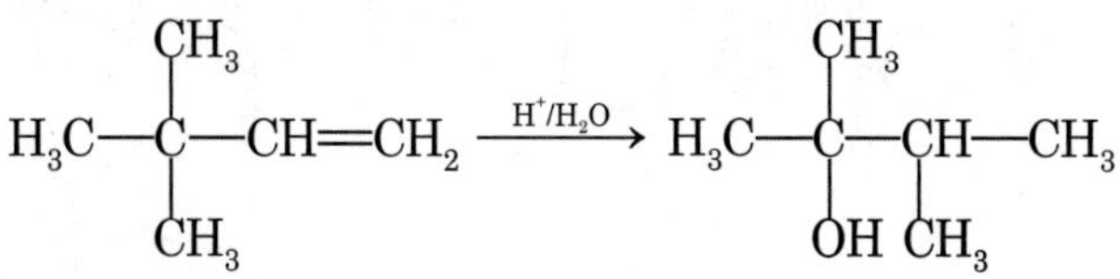

3,3-Dimethylbut-1-ene 2,3-Dimethylbutan-1-ol

Appendix—XXIV
Conversion Chart of Aromatic Compounds

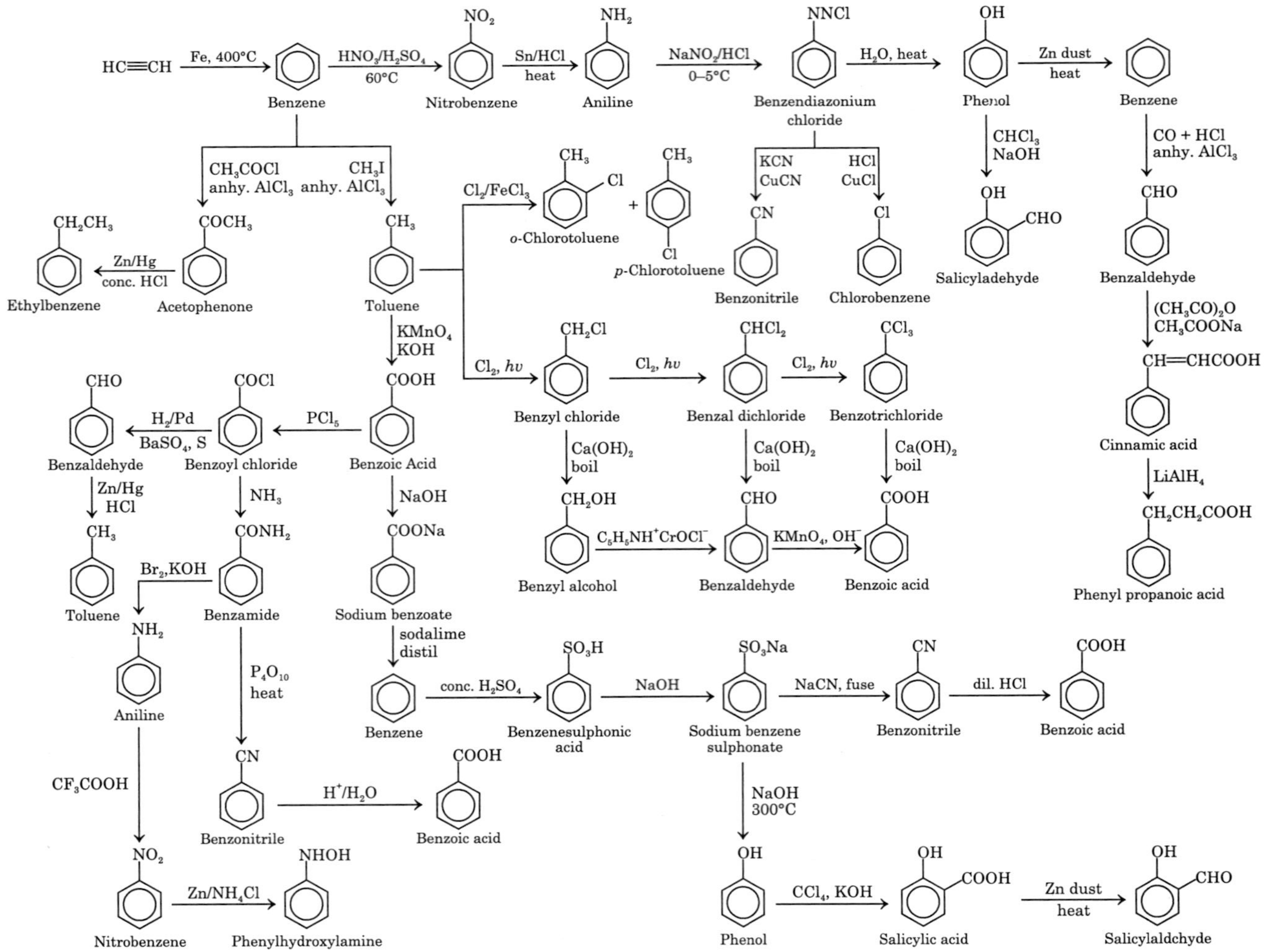

Appendix—XXV

Colour of Some Transition and Non-transition Metal Compounds

No.	Compound	Colour
1.	Chromium(II) acetate	Red
2.	Chromium(III) acetate	Greyish-green
3.	Chromium(III) chloride hexahydrate	Violet
4.	Chromium(III) nitrate nonahydrate	Brown
5.	Chromium(III) oxalate monohydrate	Yellow
6.	Chromium(III) oxide	Green
7.	Cobalt(II) chloride	Black
8.	Cobalt(III) chloride hexahydrate	Pink
9.	Cobalt(II) hydroxide	Rose-red
10.	Cobalt(III) hydroxide	Blackish-brown
11.	Cobalt(II) oxide	Greenish-brown
12.	Cobalt(III) oxide	Blackish-grey
13.	Cobalt(II) sulphate hexahydrate	Red
14.	Cobalt(III) sulphate nonahydrate	Blue
15.	Hexaamminecobalt(III) chloride	Orange
16.	Copper(II) acetate monohydrate	Dark green
17.	Copper(I) chloride	White
18.	Copper(II) chloride dihydrate	Green
19.	Copper(I) cyanide	White
20.	Copper(II) cyanide	Yellowish-green
21.	Copper(II) ferrocyanide	Yellowish-green
22.	Copper(I) hydroxide	Yellow
23.	Copper(II) hydroxide	Blue
24.	Copper(II) nitrate	Blue
25.	Copper(II) oxalate	Blue
26.	Copper(I) oxide	Red
27.	Copper(II) oxide	Black
28.	Bis(acetylacetonato)copper(II)	Blue
29.	Copper(I) sulphate	Grey
30.	Copper(II) sulphate (anhydrous)	Green

31. Copper(II) sulphate pentahydrate	Blue
32. Copper(II) sulphide	Black
33. Diamminedichlorocopper(II) chloride	Yellow
34. Iron(II) chloride (anhydrous)	Yellow
35. Iron(II) chloride tetrahydrate	Bluish-green
36. Iron(III) chloride (anhydrous)	Blackish-brown
37. Iron(III) chloride hexahydrate	Brownish-yellow
38. Iron(II) ferrocyanide	Blackish-white
39. Iron(III) ferrocyanide	Dark blue
40. Iron(II) hydroxide	Pale green
41. Iron(III) nitrate hexahydrate	Pale violet
42. Iron(III) oxalate pentahydrate	Yellow
43. Iron(II) oxide	Black
44. Iron(III) oxide	Reddish-brown
45. Tris(acetylacetonato)iron(III)	Deep red
46. Iron(II) sulphate	Light green
47. Iron(III) sulphate hexahydrate	Yellow
48. Manganese(II) acetate	Pale red
49. Manganese(II) carbonate	Rose pink
50. Manganese(II) chloride tetrahydrate	Rose
51. Manganese(III) chloride	Black
52. Manganese(II) hydroxide	Whitish-pink
53. Manganese(II) nitrate hexahydrate	Rose-white
54. Manganese(II) oxalate dihydrate	Reddish-white
55. Manganese(II) oxide	Green
56. Manganese(IV) oxide	Black
57. Manganese(II) sulphate	Pale red
58. Manganese(III) sulphate	Green
59. Molybdenum(V) chloride	Greyish-black
60. Molybdenum(V) oxide	Violet blue
70. Nickel(II) carbonate	Light green
71. Nickel(II) chloride hexahydrate	Green

72. Bis(dimethylglyoximato)nickel(II)	Scarlet red
73. Nickel(II) nitrate hexahydrate	Green
74. Nickel(II) oxalate dihydrate	Light green
75. Nickel(II) oxide	Greyish-black
76. Nickel(II) sulphate hexahydrate	Blue
77. Nickel(II) sulphate heptahydrate	Green
78. Hexaamminenickel(II) chloride	Blue
79. Silver(I) nitrate	White
80. Silver(I) cyanide	White
81. Silver(I) oxide	Black
82. Silver(II) sulphide	Black
83. Tin(II) oxide	Black
84. Tin(II) sulphide	Greyish-black
85. Vanadium(V) chloride	Reddish-brown
86. Vanadium(V) oxide	Yellowish-red
87. Vanadium(IV) oxochloride	Green
88. Vanadium(V) oxochloride	Yellow
89. Oxovanadium(IV) sulphate	Blue
90. Zinc(II) oxide	White
91. Zinc(II) sulphide	White
92. Zinc(II) chloride	White
93. Zinc(II) ferrocyanide	White
94. Bis(acetylacetonato)zinc(II)	White
95. Zirconium(IV) oxide	Yellow
96. Zirconyl chloride hexahydrate	White

Appendix—XXVI

Logarithms

	0	1	2	3	4	5	6	7	8	9	1	2	3	4	5	6	7	8	9
10	0000	0043	0086	0128	0170						5	9	13	17	21	26	30	34	38
						0212	0253	0294	0334	0374	4	8	12	16	20	24	28	32	36
11	0414	0453	0492	0531	0569						4	8	12	16	20	23	27	31	35
						0607	0645	0682	0719	0775	4	7	11	15	18	22	26	29	33
12	0792	0828	0864	0899	0934						3	7	11	14	18	21	25	28	32
						0969	1004	1038	1072	1106	3	7	10	14	17	20	24	27	31
13	1139	1173	1206	1239	1271						3	6	10	13	16	19	23	26	29
						1303	1335	1367	1399	1430	3	7	10	13	16	19	22	25	29
14	1461	1492	1523	1553	1584						3	6	9	12	15	19	22	25	28
						1614	1644	1673	1703	1732	3	6	9	12	14	17	20	23	26
15	1761	1790	1818	1847	1875						3	6	9	11	14	17	20	23	26
						1903	1931	1959	1987	2014	3	6	8	11	14	17	19	22	25
16	2041	2068	2095	2122	2148						3	6	8	11	14	16	19	22	24
						2175	2201	2227	2253	2279	3	5	8	10	13	16	18	21	23
17	2304	2330	2355	2380	2405						3	5	8	10	13	15	18	20	23
						2430	2455	2480	2504	2529	3	5	8	10	12	15	17	20	22
18	2553	2577	2601	2625	2648						2	5	7	9	12	14	17	19	21
						2672	2695	2718	2742	2765	2	4	7	9	11	14	16	18	21
19	2788	2810	2833	2856	2878						2	4	7	9	11	13	16	18	20
						2900	2923	2945	2967	2989	2	4	6	8	11	13	15	17	19
20	3010	3032	3054	3075	3096	3118	3139	3160	3181	3201	2	4	6	8	11	13	15	17	19
21	3222	3243	3363	3284	3304	3324	3345	3365	3385	3404	2	4	6	8	10	12	14	16	18
22	3424	3444	3464	3483	3502	3522	3541	3560	3579	3598	2	4	6	8	10	12	14	15	17
23	3617	3636	3655	3464	3692	3711	3729	3747	3766	3784	2	4	6	7	9	11	13	15	17
24	3802	3820	3838	3856	3874	3892	3909	3927	3945	3962	2	4	5	7	9	11	12	14	16
25	3979	3997	4014	4031	4048	4065	4082	4099	4116	4133	2	3	5	7	9	10	12	14	15
26	4150	4166	4183	4200	4216	4232	4249	4265	4281	4298	2	3	5	7	8	10	11	13	15
27	4314	4330	4346	4362	4378	4393	4409	4425	4440	4456	2	3	5	6	8	9	11	13	14
28	4472	4487	4502	4518	4533	4548	4564	4579	4594	4609	2	3	5	6	8	9	11	12	14
29	4624	4639	4654	4669	4683	4698	4713	4728	4742	4757	1	3	4	6	7	9	10	12	13
30	4771	4786	4800	4814	4829	4843	4857	4871	4886	4900	1	3	4	6	7	9	10	11	13
31	4914	4928	4942	4955	4969	4983	4997	5011	5024	5038	1	3	4	6	7	8	10	11	12
32	5051	5065	5079	5092	5105	5119	5132	5145	5159	5172	1	3	4	5	7	8	10	11	12
33	5185	5198	5211	5224	5237	5250	5263	5276	5289	5302	1	3	4	5	6	8	9	10	12
34	5315	5328	5340	5353	5366	5378	5391	5403	5416	5428	1	3	4	5	6	8	9	10	11
35	5441	5453	5465	5478	5490	5502	5514	5527	5539	5551	1	2	4	5	6	7	9	10	11
36	5563	5575	5587	5599	5611	5623	5635	5647	5658	5670	1	2	4	5	6	7	8	10	11
37	5682	5694	5705	5717	5729	5740	5752	5763	5775	5786	1	2	3	5	6	7	8	9	10
38	5798	5809	5821	5832	5843	5855	5866	5877	5888	5899	1	2	3	5	6	7	8	9	10
39	5911	5922	5933	5944	5955	5966	5977	5988	5999	6010	1	2	3	4	5	7	8	9	10
40	6021	6031	6042	6053	6064	6075	6085	6096	6107	6117	1	2	3	4	5	6	8	9	10
41	6128	6138	6149	6160	6170	6180	6191	6201	6212	6222	1	2	3	4	5	6	7	8	9
42	6232	6243	6253	6263	6274	6284	6294	6304	6314	6325	1	2	3	4	5	6	7	8	9
43	6335	6345	6355	6365	6375	6385	6395	6405	6415	6425	1	2	3	4	5	6	7	8	9
44	6435	6444	6454	6464	6474	6484	6493	6503	6513	6522	1	2	3	4	5	6	7	8	9
45	6532	6542	6551	6561	6571	6580	6590	6599	6609	6618	1	2	3	4	5	6	7	8	9
46	6628	6637	6646	6656	6665	6675	6684	6693	6702	6712	1	2	3	4	5	6	7	7	8
47	6721	6730	6739	6749	6758	6767	6776	6785	6794	6803	1	2	3	4	5	6	6	7	8
48	6812	6821	6830	6839	6848	6857	6866	6875	6884	6893	1	2	3	4	5	5	6	7	8
49	6902	6911	6920	6928	6937	6946	6955	6964	6972	6981	1	2	3	4	4	5	6	7	8

Logarithms

	0	1	2	3	4	5	6	7	8	9	1	2	3	4	5	6	7	8	9
50	6990	6998	7007	7016	7024	7033	7042	7050	7059	7067	1	2	3	4	5	6	6	7	8
51	7076	7084	7093	7101	7110	7118	7126	7135	7143	7152	1	2	3	3	4	5	6	7	8
52	7160	7168	7177	7185	7193	7202	7210	7218	7226	7235	1	2	2	3	4	5	6	7	7
53	7243	7251	7259	7267	7275	7284	7292	7300	7308	7316	1	2	2	3	4	5	6	6	7
54	7324	7332	7340	7348	7356	7364	7372	7380	7388	7396	1	2	2	3	4	5	6	6	7
55	7404	7412	7419	7427	7435	7443	7451	7459	7466	7474	1	2	2	3	4	5	5	6	7
56	7482	7490	7497	7505	7513	7520	7528	7536	7543	7551	1	2	2	3	4	5	5	6	7
57	7559	7566	7574	7582	7589	7597	7604	7612	7619	7627	1	1	2	3	4	5	5	6	7
58	7634	7642	7649	7657	7664	7672	7679	7686	7694	7701	1	1	2	3	4	4	5	6	7
59	7709	7716	7723	7731	7738	7745	7752	7760	7767	7774	1	1	2	3	4	4	5	6	7
60	7782	7789	7796	7803	7810	7818	7825	7832	7839	7846	1	1	2	3	4	4	5	6	6
61	7853	7860	7868	7875	7882	7889	7896	7903	7910	7917	1	1	2	3	4	4	5	6	6
62	7924	7931	7938	7945	7952	7959	7966	7973	7980	7987	1	1	2	3	3	4	5	6	6
63	7993	8000	8007	8014	8021	8028	8035	8041	8048	8055	1	1	2	3	3	4	5	5	6
64	8062	8069	8075	8082	8089	8096	8102	8109	8116	8122	1	1	2	3	3	4	5	5	6
65	8129	8136	8142	8149	8156	8162	8169	8176	8182	8189	1	1	2	3	3	4	5	5	6
66	8195	8202	8209	8215	8222	8228	8235	8241	8248	8254	1	1	2	3	3	4	5	5	6
67	8261	8267	8274	8280	8287	8293	8299	8306	8312	8319	1	1	2	3	3	4	5	5	6
68	8325	8331	8338	8344	8351	8357	8363	8370	8376	8382	1	1	2	3	3	4	4	5	6
69	8388	8395	8401	8407	8414	8420	8426	8432	8439	8445	1	1	2	2	3	4	4	5	6
70	8451	8457	8463	8470	8476	8482	8488	8494	8500	8506	1	1	2	2	3	4	4	5	6
71	8513	8519	8525	8531	8537	8543	8549	8555	8561	8567	1	1	2	2	3	4	4	5	6
72	8573	8579	8585	8591	8597	8603	8609	8615	8621	8627	1	1	2	2	3	4	4	5	5
73	8633	8639	8645	8651	8657	8663	8669	8675	8681	8686	1	1	2	2	3	4	4	5	5
74	8692	8698	8704	8710	8716	8722	8727	8733	8739	8745	1	1	2	2	3	4	4	5	5
75	8751	8756	8762	8768	8774	8779	8785	8791	8797	8802	1	1	2	2	3	3	4	5	5
76	8808	8814	8820	8825	8831	8837	8842	8848	8854	8859	1	1	2	2	3	3	4	5	5
77	8865	8871	8876	8882	8887	8893	8899	8904	8910	8915	1	1	2	2	3	3	4	4	5
78	8921	8927	8932	8938	8943	8949	8954	8960	8965	8971	1	1	2	2	3	3	4	4	5
79	8976	8982	8987	8993	8998	9004	9009	9015	9020	9025	1	1	2	2	3	4	4	4	5
80	9031	9036	9042	9047	9053	9058	9063	9069	9074	9079	1	1	2	2	3	3	4	4	5
81	9085	9090	9096	9101	9106	9112	9117	9122	9128	9133	1	1	2	2	3	3	4	4	5
82	9138	9143	9149	9154	9159	9165	9170	9175	9180	9186	1	1	2	2	3	3	4	4	5
83	9191	9196	9201	9206	9212	9217	9222	9227	9232	9238	1	1	2	2	3	3	4	4	5
84	9243	9248	9253	9258	9263	9269	9274	9279	9284	9289	1	1	2	2	3	3	4	4	5
85	9294	9299	9304	9309	9315	9320	9325	9330	9335	9340	1	1	2	2	3	3	4	4	5
86	9345	9350	9355	9360	9365	9370	9375	9380	9385	9390	1	1	2	2	3	3	4	4	5
87	9395	9400	9405	9410	9415	9420	9425	9430	9435	9440	0	1	1	2	2	3	3	4	4
88	9445	9450	9455	9460	9465	9469	9474	9479	9484	9489	0	1	1	2	2	3	3	4	4
89	9494	9499	9504	9509	9513	9518	9523	9528	9533	9538	0	1	1	2	2	3	3	4	4
90	9542	9547	9552	9557	9562	9566	9571	9576	9581	9586	0	1	1	2	2	3	3	4	4
91	9590	9595	9600	9605	9609	9614	9619	9624	9628	9633	0	1	1	2	2	3	3	4	4
92	9638	9643	9647	9652	9657	9661	9666	9671	9675	9680	0	1	1	2	2	3	3	4	4
93	9685	9689	9694	9699	9703	9708	9713	9717	9722	9727	0	1	1	2	2	3	3	4	4
94	9731	9736	9741	9745	9750	9754	9759	9763	9768	9773	0	1	1	2	2	3	3	4	4
95	9777	9782	9786	9791	9795	9800	9805	9809	9814	9818	0	1	1	2	2	3	3	4	4
96	9823	9827	9832	9836	9841	9845	9850	9854	9859	9863	0	1	1	2	2	3	3	4	4
97	9868	9872	9877	9881	9886	9890	9894	9899	9903	9908	0	1	1	2	2	3	3	4	4
98	9912	9917	9921	9926	9930	9934	9939	9943	9948	9952	0	1	1	2	2	3	3	4	4
99	9956	9961	9965	9969	9974	9978	9983	9987	9991	9996	0	1	1	2	2	3	3	3	4

Appendix—XXVII
Antilogarithms

	0	1	2	3	4	5	6	7	8	9	1	2	3	4	5	6	7	8	9
.00	1000	1002	1005	1007	1009	1012	1014	1016	1019	1021	0	0	1	1	1	1	2	2	2
.01	1023	1026	1028	1030	1033	1035	1038	1040	1042	1045	0	0	1	1	1	1	2	2	2
.02	1047	1050	1052	1054	1057	1059	1062	1064	1067	1069	0	0	1	1	1	1	2	2	2
.03	1072	1074	1076	1079	1081	1084	1086	1089	1091	1094	0	0	1	1	1	1	2	2	2
.04	1096	1099	1102	1104	1107	1109	1112	1114	1117	1119	0	1	1	1	1	2	2	2	2
.05	1122	1125	1127	1130	1132	1135	1138	1140	1143	1146	0	1	1	1	1	2	2	2	2
.06	1148	1151	1153	1156	1159	1161	1164	1167	1169	1172	0	1	1	1	1	2	2	2	2
.07	1175	1178	1180	1183	1186	1189	1191	1194	1197	1199	0	1	1	1	1	2	2	2	2
.08	1202	1205	1208	1211	1213	1216	1219	1222	1225	1227	0	1	1	1	2	2	2	2	3
.09	1230	1233	1236	1239	1242	1245	1247	1250	1253	1256	0	1	1	1	2	2	2	2	3
.10	1259	1262	1265	1268	1271	1274	1276	1279	1282	1285	0	1	1	1	2	2	2	2	3
.11	1288	1291	1294	1297	1300	1303	1306	1309	1312	1315	0	1	1	1	2	2	2	2	3
.12	1318	1321	1324	1327	1330	1334	1337	1340	1343	1346	0	1	1	1	2	2	2	2	3
.13	1349	1352	1355	1358	1361	1365	1368	1371	1374	1377	0	1	1	1	2	2	2	3	3
.14	1380	1384	1387	1390	1393	1396	1400	1403	1406	1409	0	1	1	1	2	2	2	3	3
.15	1413	1416	1419	1422	1426	1429	1432	1435	1439	1442	0	1	1	1	2	2	2	3	3
.16	1445	1449	1452	1455	1459	1462	1466	1469	1472	1476	0	1	1	1	2	2	2	3	3
.17	1479	1483	1486	1489	1493	1496	1500	1503	1507	1510	0	1	1	1	2	2	2	3	3
.18	1514	1517	1521	1524	1528	1531	1535	1538	1542	1545	0	1	1	1	2	2	2	3	3
.19	1549	1552	1556	1560	1563	1567	1570	1574	1578	1581	0	1	1	1	2	2	3	3	3
.20	1585	1589	1592	1596	1600	1603	1607	1611	1614	1618	0	1	1	1	2	2	3	3	3
.21	1622	1626	1629	1633	1637	1641	1644	1648	1652	1656	0	1	1	2	2	2	3	3	3
.22	1660	1663	1667	1671	1675	1679	1683	1687	1690	1694	0	1	1	2	2	2	3	3	3
.23	1698	1702	1706	1710	1714	1718	1722	1726	1730	1734	0	1	1	2	2	2	3	3	4
.24	1738	1742	1746	1750	1754	1758	1762	1766	1770	1774	0	1	1	2	2	2	3	3	4
.25	1778	1782	1786	1791	1795	1799	1803	1807	1811	1816	0	1	1	2	2	2	3	3	4
.26	1820	1824	1828	1832	1837	1841	1845	1849	1854	1858	0	1	1	2	2	3	3	3	4
.27	1862	1866	1871	1875	1879	1884	1888	1892	1897	1901	0	1	1	2	2	3	3	3	4
.28	1905	1910	1914	1919	1923	1928	1932	1936	1941	1945	0	1	1	2	2	3	3	4	4
.29	1950	1954	1959	1963	1968	1972	1977	1982	1986	1991	0	1	1	2	2	3	3	4	4
.30	1995	2000	2004	2009	2014	2018	2023	2028	2032	2037	0	1	1	2	2	3	3	4	4
.31	2042	2046	2051	2056	2061	2065	2070	2075	2080	2084	0	1	1	2	2	3	3	4	4
.32	2089	2094	2099	2104	2109	2113	2118	2123	2128	2133	0	1	1	2	2	3	3	4	4
.33	2138	2143	2148	2153	2158	2163	2168	2173	2178	2183	0	1	1	2	2	3	3	4	4
.34	2188	2193	2198	2203	2208	2213	2218	2223	2228	2234	1	1	2	2	3	3	4	4	5
.35	2239	2244	2249	2254	2259	2265	2270	2275	2280	2286	1	1	2	2	3	3	4	4	5
.36	2291	2296	2301	2307	2312	2317	2323	2328	2333	2339	1	1	2	2	3	3	4	4	5
.37	2344	2350	2355	2360	2366	2371	2377	2382	2388	2393	1	1	2	2	3	3	4	4	5
.38	2399	2404	2410	2415	2421	2427	2432	2438	2443	2449	1	1	2	2	3	3	4	4	5
.39	2455	2460	2466	2472	2477	2483	2489	2495	2500	2506	1	1	2	2	3	3	4	5	5
.40	2512	2518	2523	2529	2535	2541	2547	2553	2559	2564	1	1	2	2	3	4	4	5	5
.41	2570	2576	2582	2588	2594	2600	2606	2612	2618	2624	1	1	2	2	3	4	4	5	5
.42	2630	2636	2642	2649	2655	2661	2667	2673	2679	2685	1	1	2	2	3	4	4	5	6
.43	2692	2698	2704	2710	2716	2723	2729	2735	2742	2748	1	1	2	3	3	4	4	5	6
.44	2754	2761	2767	2773	2780	2786	2793	2799	2805	2812	1	1	2	3	3	4	4	5	6
.45	2818	2825	2831	2838	2844	2851	2858	2864	2871	2877	1	1	2	3	3	4	5	5	6
.46	2884	2891	2897	2904	2911	2917	2924	2931	2938	2944	1	1	2	3	3	4	5	5	6
.47	2951	2958	2965	2972	2979	2985	2992	2999	3006	3013	1	1	2	3	3	4	5	5	6
.48	3020	3027	3034	3041	3048	3055	3062	3069	3076	3083	1	1	2	3	4	4	5	6	6
.49	3090	3097	3105	3112	3119	3126	3133	3141	3148	3155	1	1	2	3	4	4	5	6	6

Antilogarithms

	0	1	2	3	4	5	6	7	8	9	1	2	3	4	5	6	7	8	9
.50	3162	3170	3177	3184	3192	3199	3206	3214	3221	3228	1	1	2	3	4	5	5	6	7
.51	3236	3243	3251	3258	3266	3273	3281	3289	3296	3304	1	2	2	3	4	5	5	6	7
.52	3311	3319	3327	3334	3342	3350	3357	3365	3373	3381	1	2	2	3	4	5	5	6	7
.53	3388	3396	3404	3412	3420	3428	3436	3443	3451	3459	1	2	2	3	4	5	6	6	7
.54	3467	3475	3483	3491	3499	3508	3516	3524	3532	3540	1	2	2	3	4	5	6	6	7
.55	3548	3556	3565	3573	3581	3589	3597	3606	3614	3622	1	2	2	3	4	5	6	7	7
.56	3631	3639	3648	3656	3664	3673	3681	3690	3698	3707	1	2	3	3	4	5	6	7	8
.57	3715	3724	3733	3741	3750	3758	3767	3776	3784	3793	1	2	3	3	4	5	6	7	8
.58	3802	3811	3819	3828	3837	3846	3855	3864	3873	3882	1	2	3	4	4	5	6	7	8
.59	3890	3899	3908	3917	3926	3936	3945	3954	3963	3972	1	2	3	4	5	5	6	7	8
.60	3981	3990	3999	4009	4018	4027	4036	4046	4055	4064	1	2	3	4	5	6	6	7	8
.61	4074	4083	4093	4102	4111	4121	4130	4140	4150	4159	1	2	3	4	5	6	7	8	9
.62	4169	4178	4188	4198	4207	4217	4227	4236	4246	4256	1	2	3	4	5	6	7	8	9
.63	4266	4276	4285	4295	4305	4315	4325	4335	4345	4355	1	2	3	4	5	6	7	8	9
.64	4365	4375	4385	4395	4406	4416	4426	4436	4446	4457	1	2	3	4	5	6	7	8	9
.65	4467	4477	4487	4498	4508	4519	4529	4539	4550	4560	1	2	3	4	5	6	7	8	9
.66	4571	4581	4592	4603	4613	4624	4634	4645	4656	4667	1	2	3	4	5	6	7	9	10
.67	4677	4688	4699	4710	4721	4732	4742	4753	4764	4775	1	2	3	4	5	7	8	9	10
.68	4786	4797	4808	4819	4831	4842	4853	4864	4875	4887	1	2	3	4	6	7	8	9	10
.69	4898	4909	4920	4932	4943	4955	4966	4977	4989	5000	1	2	3	5	6	7	8	9	10
.70	5012	5023	5035	5047	5058	5070	5082	5093	5105	5117	1	2	4	5	6	7	8	9	11
.71	5129	5140	5152	5164	5176	5188	5200	5212	5224	5236	1	2	4	5	6	7	8	10	11
.72	5248	5260	5272	5284	5297	5309	5321	5333	5346	5358	1	2	4	5	6	7	9	10	11
.73	5370	5383	5395	5408	5420	5433	5445	5458	5470	5483	1	3	4	5	6	8	9	10	11
.74	5495	5508	5521	5534	5546	5559	5572	5585	5598	5610	1	3	4	5	7	8	9	10	11
.75	5623	5636	5649	5662	5675	5689	5702	5715	5728	5741	1	3	4	5	7	8	9	10	12
.76	5754	5768	5781	5794	5808	5821	5834	5848	5861	5875	1	3	4	5	7	8	9	11	12
.77	5888	5902	5916	5929	5943	5957	5970	5984	5998	6012	1	3	4	5	7	8	10	11	12
.78	6026	6039	6053	6067	6081	6095	6109	6124	6138	6152	1	3	4	6	7	8	10	11	13
.79	6166	6180	6194	6209	6223	6237	6252	6266	6281	6295	1	3	4	6	7	9	10	12	13
.80	6310	6324	6339	6353	6368	6383	6397	6412	6427	6442	1	3	4	6	7	9	10	12	13
.81	6457	6471	6486	6501	6516	6531	6546	6561	6577	6592	2	3	5	6	8	9	11	12	14
.82	6607	6622	6637	6653	6668	6683	6699	6714	6730	6745	2	3	5	6	8	9	11	12	14
.83	6761	6776	6792	6808	6823	6839	6855	6871	6887	6902	2	3	5	6	8	9	11	13	14
.84	6918	6934	6950	6966	6982	6998	7015	7031	7047	7063	2	3	5	6	8	10	12	13	15
.85	7079	7096	7112	7129	7145	7161	7178	7194	7211	7228	2	3	5	7	8	10	12	-13	15
.86	7244	7261	7278	7295	7311	7328	7345	7362	7379	7396	2	3	5	7	8	10	12	13	15
.87	7413	7430	7447	7464	7482	7499	7516	7534	7551	7568	2	3	5	7	9	10	12	14	16
.88	7586	7603	7621	7638	7656	7674	7691	7709	7727	7745	2	4	5	7	9	11	12	14	16
.89	7762	7780	7798	7816	7834	7852	7870	7889	7907	7925	2	4	5	7	9	11	13	14	16
.90	7943	7962	7980	7998	8017	8035	8054	8072	8091	8110	2	4	6	7	9	11	13	15	17
.91	8128	8147	8166	8185	8204	8222	8241	8260	8279	8299	2	4	6	8	9	11	13	15	17
.92	8318	8337	8356	8375	8395	8414	8433	8453	8472	8492	2	4	6	8	10	12	14	15	17
.93	8511	8531	8551	8570	8590	8610	8630	8650	8670	8690	2	4	6	8	10	12	14	16	18
.94	8710	8730	8750	8770	8790	8810	8831	8851	8872	8892	2	4	6	8	10	12	14	16	18
.95	8913	8933	8954	8974	8995	9016	9036	9057	9078	9099	2	4	6	8	10	12	15	17	19
.96	9120	9141	9162	9183	9204	9226	9247	9628	9290	9311	2	4	6	8	11	13	15	17	19
.97	9333	9354	9376	9397	9419	9441	9462	9484	9506	9528	2	4	7	9	11	13	15	17	20
.98	9550	9572	9594	9616	9638	9661	9683	9705	9727	9750	2	4	7	9	11	13	16	18	20
.99	9772	9795	9817	9840	9863	9886	9908	9931	9954	9977	2	5	7	9	11	14	16	18	20